Advances in Volcanology

An Official Book Series of the International Association of Volcanology and Chemistry of the Earth's Interior

Advances in Volcanology is an official book series of the International Association of Volcanology and Chemistry of the Earth's Interior (IAVCEI). The aim of the book series is to publish scientific monographs on a varied array of topics or themes in volcanology. The series aims at building a varied library of reference works, by describing the current understanding of selected themes such as certain types of volcanism or regional aspects. The books in the series are prepared by leading experts actively working in the relevant field. The Advances in Volcanology series contains single and multi-authored books as well as edited volumes. The Series Editor, Dr. Karoly Nemeth, is currently accepting proposals and a proposal document can be obtained from the Publisher, Dr. Annett Buettner (annett.buettner@springer.com).

Dávid Karátson · Jean-Claude Thouret

Volcano Geomorphology

Landforms, Processes, and Hazards

Springer

Dávid Karátson
Department of Physical Geography
Eötvös Loránd University
Budapest, Hungary

Jean-Claude Thouret
Laboratoire Magmas et Volcans, CNRS,
OPGC and IRD
University of Clermont-Auvergne
Clermont-Ferrand, France

ISSN 2364-3277 ISSN 2364-3285 (electronic)
Advances in Volcanology
ISBN 978-3-031-77692-2 ISBN 978-3-031-77693-9 (eBook)
https://doi.org/10.1007/978-3-031-77693-9

This Springer imprint is published by the registered company Springer Nature Switzerland AG
The registered company address is: Gewerbestrasse 11, 6330 Cham, Switzerland

Foreword by Benjamin van Wyk de Vries

Volcano Geomorphology—Landforms, Processes and Hazards fills a gap in the scientific literature by providing the first really comprehensive modern overview in this field. From the time of Cotton's *Volcanoes as Landscape Forms* (1944, 1952), various books have been produced dealing in part with aspects of volcano geomorphology, but the field has evolved so much that a comprehensive and systematic work is now needed.

Dávid Karátson and Jean-Claude Thouret have taken a long time to produce this work, I know because I have seen them 'on the job' during many visits to our laboratory, and I have participated in a very small way in some parts. The detailed and painstaking work amazes me, and the dedication they have both shown to quality, accuracy, and in ensuring that the content is both up to date and covers all aspects of volcano geomorphology, is impressive.

The book has 11 chapters, starting at the plate tectonic scale and context, then working through the various basic or 'elementary' landforms, building in complexity to polygenetic and caldera landforms, then on to degradation, as volcanoes are built unlike other mountains and are destined to erode away. The work ends with the geomorphological hazards and methods and techniques of studying landforms. This progression allows the reader to build up a sound basis on the subject, using the observational, descriptive parts to gain experience before moving forward onto the methods and consequences.

The work takes examples from an immense swath of literature, is highly visual, with superb illustrations, and I recommend it for both a straight read through to gain a sound foundation in the subject, or as a reference source to turn to for specific questions.

I think this work will have a major impact and will be a reference for a long time, as it is not only up to date, but is so encompassing and complete.

July 2024

Benjamin van Wyk de Vries
Professor, Laboratoire Magmas et Volcans
Université Clermont-Auvergne
Clermont-Ferrand, France

Foreword by Cliff Ollier

Volcanoes make news. Volcanic eruptions are spectacular and can create beautiful landscapes. There are also many hazards associated with volcanic activity, and local people have to balance the dangers with advantages, such as good soil. Because of this there are many books about volcanoes, so why write another? This volume covers all aspects of volcano science. It is unusually detailed and quantitative. Definitions are dissected, detailed, and refined, and we are thoroughly informed on numbers, ages, height of volcanic constructs, length of lava flows, and more. It is very up to date with many modern references.

In 1944, the New Zealander Charles Cotton wrote a book *Volcanoes as Landscape Forms*. He wrote "An approach to volcanism may be made through either petrology or geomorphology, a volcano and its mechanism being regarded either as a geological crucible or as a builder of landforms".

That was a long time ago and the present book presents an up-to-date account of volcanic landforms and geomorphology. In brief, this is the most modern text on volcanoes and will probably be the leading reference for years to come.

September 2024

Cliff Ollier
Emeritus Professor
University of New England
Armidale, Australia

Series Editor's Preface

Volcano geomorphology sparks interest across the geoscience disciplines. Volcanoes are one of the most fascinating landforms as their formation operates over an unusual timescale in comparison to other geological processes. Volcanic eruptions can occur as a single explosion, while long lasting lava effusions may take years. These time frames are several magnitudes shorter in comparison to any other geological processes, leaving behind material built into the rock record of Earth. Volcanoes also appear as a very diverse spectrum of edifice structures ranging from very small volume monogenetic volcano types to massive volcanic edifices of lava shields that commonly occupy areas up to several thousands of km^2. Volcanoes also not only represented by constructional features in the landscape but can also form negative edifices cut into the syn-eruptive surface. Such volcanoes range from a few hundred metres to tens of kilometres across, with a hole-in-the-ground appearance, making them the perfect depocenters for sediments in continental terrains. The resultant maars or caldera volcanoes are thus excellent sedimentary record keepers of continental geological processes, which often represent the only evidence of past climate conditions and the physiography of large landmasses.

Volcanoes are also a source of fascination and have been part of human culture from the dawn of humanity. Volcanoes disperse material as fragmented tephra which fertilise the landscape to provide good soils with high agricultural potential. However, they are also geological phenomena that can impact large areas in an exceptionally rapid and destructive way. Eruptive products of volcanoes vary from small to vast volumes that can create new landscapes and remain for a long time as part of the evolving landscape, including extensive lava fields and continental-scale ignimbrite sheets. In addition, submarine volcanic landforms have recently drawn scientists' attention as there is an overwhelmingly large number of such volcanoes, and their eruptive mechanism is complex due to their extreme environments from deep to shallow subaqueous settings. The huge diversity of volcanic landforms has also recently been considered as part of Earth's geoheritage.

Volcanology as a dynamic science has evolved dramatically over the last few decades and the accumulated knowledge on how volcanoes work has generated numerous superb textbooks dealing with their geological context, eruption mechanisms, texture of the eruptive products, and their complex and unique geohazards.

Volcanoes, due to their special geomorphological aspects, have tended to appear in major geomorphology textbooks as standalone chapters outlining their various classifications based on purely morphological aspects. This trend has been altered by the expanding research based on Earth observations using various remote sensing and morphological analysis techniques. These have accelerated the development of our understanding of the morphological variations of the Earth's landscape. In parallel, there have been great advances in remote sensing techniques for mapping other planets and moons of the solar system, providing a wealth of new information that can be compared with and tested out on Earth. While these new advances in how we define various landforms has manifested itself in research papers, the volcano sections of geomorphological textbooks have not been updated, focusing only on very basic aspects of volcano geomorphology.

Hence the need to have a dedicated state-of-the-art summary of our current knowledge on volcano geomorphology. To fill these gaps, thematic or special issues of specialized journals such as geomorphology (Volcano Geomorphology: landforms, processes and hazards, Vol. 136, 1, pp. 1–176 (2012) *https://www.sciencedirect.com/journal/geomorphology/vol/136/issue/1*, Géomorphologie: relief, processus, environnement (Formes de relief volcaniques, processus et risques—Volcanic landforms, processes and hazards on volcanoes, Vol. 13, 3, pp. 211–282 (2007) (*https://doi.org/10.4000/geomorphologie.1443*), or Zeitschrift für Geomorphologie (Volcanic landforms, processes and hazards—International Association of Geomorphology, Working Group on "Volcanic Geomorphology" Suppl Bd XI, pp. 1–231 (2005) (*https://www.schweizerbart.de/publications/detail/isbn/9783443211400/Zeitschr_f_Geomorphologie_Suppl_Bd_1*) have provided excellent collection of highly specialized papers. Also, a state-of-the-art scientific review on volcanic geomorphology has been published (*Thouret, J.C., 1999. Volcanic geomorphology—an overview. Earth-Science Reviews 47(1):95–131. https://doi.org/10.1016/S0012-8252(99)00014-8*). A more recent analysis of the research trends within volcanic geomorphology also shows a very dynamically evolving research subject (*Carrión-Mero P. et al.: Volcanic Geomorphology: A Review of Worldwide Research. Geosciences. 2020, 10 (9): 347. https://doi.org/10.3390/geosciences10090347*).

Such volumes and research publications are always great milestones in the evolution of any scientific subject, but, by their nature, they can be a rather random collection of good results rather than a systematic summary of all known information. On the other hand, new editions of volcanology books (Sigurdsson, H. et al., eds., The Encyclopedia of Volcanoes, 2nd ed., Elsevier, 1456 p, ISBN: 978-0-12-385938-9,—*https://www.sciencedirect.com/book/9780123859389/the-encyclopedia-of-volcanoes;* Cas, R. A. F. et al.: Volcanology Processes, Deposits, Geology and Resources, Springer, 1833 p, ISBN: 978-3-319-66612-9, *https://link.springer.com/book/10.1007/978-3-319-66613-6);* series books of Advances in Volcanology (e.g. Rouwet, D. et al., eds., Volcanic Lakes, Springer, 533 p, ISBN: 978-3-642-36832-5; *https://link.springer.com/book/10.1007/978-3-642-36833-2),* and Special Issues within journals such as Journal of Volcanology and Geothermal Research (e.g. Martin, U. et al., eds., Maar diatreme volcanism and associated processes, 2007, vol 159, 1–3: 1–312) have provided some indications of the current development of volcano geomorphology, but those works tended also to be rather sporadic and incomplete. In addition, this type of work is nearly always specialized in some area within volcanology, approaching it from a geological angle.

Volcano Geomorphology: Landforms, Processes and Hazards is thus a completely new thematic book, which is dedicated entirely to the geomorphological aspects of volcanoes. The two authors, Dávid Karátson and Jean-Claude Thouret, are leading active experts on volcano geomorphology with a tremendous knowledge of the full domain of volcano geomorphology. They have worked and published dozens of highly cited specialized scientific papers on volcanoes from most of the known tectonic settings, covering ancient to modern volcanoes. Both authors also have decades of experiences in teaching geomorphology and acting as supervisors to numerous postgraduate students. Their excellent knowledge led to the creation of this new book which is included in the International Association of Volcanology and Chemistry of Earth's Interior *Advances in Volcanology* book series published by Springer. This series aims to create a platform for experts to summarise key research subjects within volcanology to provide textbooks that can be used at any level from undergraduate teaching to in-depth research projects in any situation.

The book focuses on volcano geomorphology, written from a volcanology perspective, while also treating volcanoes as significant, but unique geological features of Earth's terrestrial and marine sedimentary basins. The book combines a traditional geomorphological approach in describing volcanic landforms and related geomorphic processes in detail, while additionally providing a review of new research topics in geomorphology and physical volcanology. The book updates the current "wisdom" on volcano landform classification by means of an analytical approach, blending this with the most recent volcano classifications reflecting source to sink aspects of volcano systems. In this way, this book will form a strong basis for how we carry out research on volcanoes and how we link them with ancient volcanic rocks that can be used to shed light on old volcanic landforms and establish a volcanic history. Embracing the monogenetic to polygenetic volcanism concept through volcano geomorphology will provide an approach at the level of a volcano system to help understand the landscape evolution of regions impacted by volcanism.

This book also incorporates the latest advances in our understanding of mass wasting in volcanic terrains and specifically how a volcano is eroded and feeds detritus into all geological environments in various climate and hydrological systems, linking volcano geomorphology to basin evolution, uplift history, landscape inversion, and mass wasting associated with volcanic hazards, to mention just a few of the main subject areas. Such knowledge provides insight into the processes and rates of growth and destruction of volcanic landforms and landscapes. Lastly, the new remote sensing and terrain analysis techniques, including geographical information system (GIS) technologies, or landscape model calculation statistical methods are presented, with a clear list of potential paths that future generations can explore in volcanology. Overall, this book gives an overview of how to study volcanic landforms and processes by providing an updated review of a wide array of classical and modern methods.

Jeddah, Saudi Arabia
October 2024

Dr. Karoly Nemeth
nemeth.k@sgs.gov.sa
nemeth.karoly@epss.hun-ren.hu
Professor of Geology
Doctor of the Hungarian Academy of Sciences (D.Sc.)
National Program of Earthquakes and Volcanoes
Saudi Geological Survey
Jeddah, Saudi Arabia

MTA-FI FluidsByDepth Lendület Research Group
Institute of Earth Physics and Space Sciences
Hungarian Research Network (HUN-REN)
Sopron, Hungary

Acknowledgments

While this book has been written by two authors, we consider it as being the result of the collaboration, participation, support, discussions, suggestions, and comments of many colleagues and friends over the past few years.

First of all, we would like to mention the involvement of a number of authors who directly contributed to the book, including the whole of both chapters "Case Studies Illustrating the Role of Geomorphology in Assessing Volcanic Hazards" and "Methods and Techniques for Studying Volcanic Landforms". The contributors for the five case studies in Chapter "Case Studies Illustrating the Role of Geomorphology in Assessing Volcanic Hazards" are: Valentin Troll and Juan Carlos Carracedo, Gro B. M. Pedersen and Magnús T. Gudmundsson, Ralf Gertisser, Jonathan N. Procter, Sonia Calvari, Daniele Casalbore, Federico Di Traglia, and Antonio Costa. The contributors to Chapter "Methods and Techniques for Studying Volcanic Landforms" on methodology are: Pablo Grosse and Mathieu Kervyn, Tamás Telbisz, Pierre Lahitte, Daniel O'Hara, Simone Tarquini and Mattia de' Michieli Vitturi, Sylvain J. Charbonnier and Valentin Gueugneau, Benjamin van Wyk de Vries, Quentin Dumont, Valérie Cayol and Jean-Luc Froger, Pierre Lahitte and Xavier Quidelleur, Didier Miallier, Irene Schimmelpfennig and Gordon Bromley, Jean-Pascal Dumoulin and Jean-Claude Lefèvre.

Many colleagues and friends helped us review the successive versions of the chapters of this book, providing constructive criticism, ideas, and suggestions, and contributing important information that helped with specific parts. Among them, we would like to acknowledge Pierre Boivin, Jean-Louis Bourdier, Sylvain Charbonnier, Massimiliano Favalli, Pablo Grosse, Matthieu Kervyn, Pierre Lahitte, Cliff Ollier, Raphaël Paris, Vera Ponomareva, Benoît Smets, Balázs Székely, Tamás Telbisz, Benjamin van Wyk de Vries, and Gerhard Wörner. Several colleagues have provided interesting figures and nice photographs, in particular Pierre Boivin, Oryaëlle Chevrel, Larry Crumpler, Catherine Deniel, Tim Druitt, Massimiliano Favalli, Kaveh Ghahraman, Loraine Gourbet, Alain Gourgaud, Magnús Tumi Gudmundsson, Daniel O'Hara, Andrew Harris, Naomi Irapta, Szabolcs Kósik, Szabolcs Lövész, Balázs Nagy, Gábor Nagy, Raphaël Paris, Claus Siebe, and Hetu Sheth.

We owe the outstanding artwork to the talent of Mrs. Frédérique Van Celst, and we thank Ms. Fanni Vörös for her help with drafting many figures and plots. The English editorial work of Mrs. Frances Van Wyk de Vries has been critical in improving the English, though any remaining mistakes can be attributed to us.

This book has also benefitted from two grants, one from Centre Clermontois de Recherche sur le Volcanisme and Laboratory of excellence (CLERVOLC) from the LMV, University Clermont-Auvergne, and the other was provided by the Doctoral School of Earth Sciences of Eötvös Loránd University (ELTE), Budapest. We also thank Mrs. Sylvaine Jouhannel at LMV and Mr. Csongor Jávor at ELTE for their help with the references and literature search.

During the preparation of this book, we have appreciated the help and support of the Springer Team, in particular Annett Buettner, Julia Bauer, Hemavathy Manivannan, and A. Perumal Umamagesh.

Finally, on a more personal note, we would like to thank our families and those who shared our endeavour on a daily basis. Dávid Karátson cannot thank his wife Eszter enough for putting up with the extra demands of the book on top of his normal university workload, while Jean-Claude Thouret acknowledges the support of his wife, Liliane, who shared most of the burden imposed by the writing of this book.

Contents

Introduction 1

1 **Volcanism and Volcanic Landforms: Development of Volcano Geomorphology** 1

2 **Volcanoes and Volcanic Landforms: Toward a Definition** 2
- 2.1 What Is a Volcano? A Simple, Twofold Definition 2
- 2.2 Limitations to the Initial Definition 3
- 2.3 A Framework to Support the Definition of Volcanic Landforms 5
- 2.4 The 'Volcano-Magma System' 6

3 **Towards a Geomorphological Definition** 7
- 3.1 Construction and Destruction 7
- 3.2 A Volcanic Construct Involves Several Components 7

4 **Principles of Existing Volcano Classifications** 16

5 **Principles of Classification of Volcanic Landforms in Our Book** 18

6 **Objectives and Rationale of Volcano Geomorphology** 19

7 **Content of This Book** 19

References 20

Chapter 1 Spatial Distribution and Tectonic Setting of Volcanoes . . . 23

1 **Spatial Distribution of Volcanoes at the Earth's Surface** 23

2 **Volcanoes and Volcanic Fields in Five Tectonic Settings** 24
- 2.1 Divergent Plate Boundaries: Geomorphology and Volcanism at Spreading Mid-Ocean Ridges 27
- 2.2 Volcanoes Along Continental Rift Valleys 30
- 2.3 Volcanism at Subduction Zones Along Convergent Plates 35
- 2.4 Hybrid Settings of Volcanoes Near Subducting Plates 37
- 2.5 Intraplate Hot-Spot Volcanoes and Large Igneous Volcanic Fields 38

3 Tectonic Interactions with Magmatism at Volcanoes 50
3.1 Broad Scale Interactions 51
3.2 Structural Setting and Stress Field Control on Volcano Distribution 53
3.3 Volcanic Edifices Can Affect Their Tectonic/Magmatic System and Subsequent Growth 57
3.4 Magmatism Can Modify the Morphology, Structure and Development of Volcanoes. 57

4 Volcanoes and Magma Generation, Transport and Storage . . . 60

References 67

Chapter 2 Elementary Landforms and Processes I: Lava Flows, Lava Lakes, and Lava Domes. 73

1 Lava Flows 73
1.1 Definition of a Lava Flow 73
1.2 Four Types of Lava Flow 74
1.3 Characteristic Landforms of Lava Flows 84
1.4 Physical and Rheological Parameters Driving Lava Flows. 89
1.5 Thermal Structure and Inflation Mechanism of Lava Flows. 93
1.6 Columns and Columnar Jointing in Lava Flows. 95

2 Volcanic Lava Lakes 98
2.1 Active Lava Lakes. 98
2.2 Typical Features 99
2.3 Types of Lava Lakes 101
2.4 Dynamics of Persistent Lava Lakes 102

3 Lava Domes Created by Extrusive Activity 104
3.1 Definition 104
3.2 Geometric and Rheological Characteristics of Lava Dome Types 106
3.3 Principal Classes of Domes 109
3.4 Lava Dome Structures and Textures 111
3.5 Lava Dome Eruptions: Dynamics, Morphology, and Timescale 115

References 130

Chapter 3 Elementary Landforms and Processes II: From Simple, Terrestrial Landforms to Hydroclastic, Glaciovolcanic and Underwater Landforms 137

1 General Features of Simple Volcanoes in Terrestrial Environments. 137

2 Pyroclastic Cones. 140
2.1 Scoria Cones 140
2.2 Pumice Cones 148

3 **Hydroclastic Cones** 149
3.1 Maars. 149
3.2 Tuff Rings 154
3.3 Tuff Cones. 156
3.4 Factors Involved in the Formation and Evolution of Explosive Cones 160

4 **Basaltic Volcanic Fields**. 162

5 **Glaciovolcanic Landforms and Magma-Ice Interactions** 163
5.1 Main Glaciovolcanic Landforms 164
5.2 Volcano-Ice Interactions and Processes 168

6 **Elementary Volcanic Landforms in Underwater Environments**. 174
6.1 Constraints and Processes. 175
6.2 Abundant Submarine Landforms: Abyssal Hills, Seamounts, and Guyots. 175
6.3 Types of Submarine Eruptions and Products 177

7 **Shoaling and Emergent Volcanic Islands** 183

References 197

Chapter 4 Polygenetic Landforms I: Shield Volcanoes 207

1 **Introduction, Definition and General Features** 207

2 **Occurrence, Constraints on Morphology, and Main Types**. 208

3 **Hawaiian Shields**. 211
3.1 General Geomorphic Features 211
3.2 Constraints on Subsidence and Uplift. 215
3.3 Productivity. 216
3.4 Structure-Controlled Landforms: Rift Zones, Radial Vents, and Summit Calderas 216
3.5 Landslides and Fault Scarps 220
3.6 Evolutionary Scheme 221

4 **Galápagos Shields** 226
4.1 General Features of a Complex Tectonic Setting 226
4.2 Geomorphic and Structural Characteristics of the Western and Eastern Galápagos Shields. 228
4.3 Explaining the Contrasting Features of the Galápagos Shields. 229
4.4 Evolution of the Galapagos Shields 230

5 **Icelandic Shields** 230
5.1 General Features. 230
5.2 Morphometric Features as Constrained by Shield Formation 231

6 **Medium-Sized Continental Intraplate and Arc Shields**. 233
6.1 Intraplate Rift and Hot-Spot Shields (Turkana Subtype) 233
6.2 Arc Shields (Newberry Subtype) 235

7 **The Wide Diversity of Shield Volcanoes—Concluding Remarks** 238
7.1 Shape Variations of Shield Types and Subtypes 238
7.2 Coexistence of Shield- and Composite Volcanoes in Similar Settings 239

References 247

Chapter 5 Polygenetic Landforms II: Composite Cones: From Individual Constructs to Clusters 257

1 **Terminology: Strato- or Composite?** 257

2 **Classification of Landforms from Simple Cones to Compound Edifices** 260

3 **Eruptive Characteristics and Lithofacies Associations** 264
3.1 The Role of Eruptive Style on Cone Shape 265
3.2 Lithofacies Associations 266

4 **Overview of Main Occurrences by Geodynamic Setting** 268
4.1 Occurrences in Subduction and Collision Zones 268
4.2 Occurrences in Intraplate and Rift Zone Settings 280

5 **Characterization of Main Landform Types** 284
5.1 Simple Cones 286
5.2 Truncated Cones 290
5.3 Somma Volcanoes 292
5.4 Compound Cones 293

6 **Evolution, Lifetime, and Growth Rate of Composite Cones** . . . 293
6.1 Lifetime: Fast Growth of Composite Cones 294
6.2 Variability in Eruptive Activity 294
6.3 Periods or 'Cycles' of Eruptive Activity 296

7 **Large, Long-Lived Volcanic Clusters or Volcano Complexes** . . . 296
7.1 LVC Definition and Differences from Other Composite Cones 297
7.2 Case Studies of LVCs 297

References 311

Chapter 6 Calderas and Caldera Complexes 321

1 **Definition and Early Classification** 321

2 **Terminology** 323

3 **Geomorphic and Structural Boundaries of a Caldera** 326

4 **Geomorphic and Geological Features Within, Around, and Beneath Calderas** 328
4.1 Around the Caldera 328
4.2 Within the Caldera 329
4.3 Caldera Margins 329
4.4 Beneath the Caldera: Subcaldera Intrusions, and Cauldron Roots 330

5 Relationships with Tectonics and Regional Structures: Elliptical and Graben Calderas . . . 333

6 Caldera Development and Evolution . . . 333
6.1 Pre-caldera Volcanism . . . 335
6.2 Syn-caldera Stage . . . 335
6.3 Post-caldera Resurgence, Renewed Eruptions and Hydrothermal Activity . . . 335

7 A Simple Descriptive Distinction of Calderas Based on Magma Composition . . . 339
7.1 Mafic Calderas on Shield Volcanoes . . . 339
7.2 Felsic Calderas . . . 345

8 Two Categories of Explosive Calderas and Processes . . . 345
8.1 Small and Medium-Sized Calderas on Composite Cones . . . 346
8.2 Large, Ash-Flow or Ignimbrite Calderas with Rhyolitic Composition . . . 352

9 Resurgence Modes . . . 357
9.1 Resurgence Mechanism . . . 357
9.2 Resurgence Models . . . 359

10 Subsidence Mechanisms Inducing Distinct Caldera Geometries . . . 360
10.1 Interpretations of Mechanisms of Caldera Collapse . . . 360
10.2 Another Mechanism of Caldera Collapse . . . 363

11 Caldera Complexes and Volcano-tectonic Depressions . . . 365
11.1 Caldera Complexes in the Central Andes . . . 366
11.2 Volcano-tectonic Depressions . . . 370

12 Challenging Classification of Calderas and Caldera Complexes . . . 373

13 Concluding Remarks . . . 374

References . . . 384

Chapter 7 Short-Term Degradation of Active Volcanoes . . . 391

1 Overview of Degradation Processes . . . 391
1.1 Definition, Scope, and Facies Relationships . . . 391
1.2 Differences Between Terrestrial and Underwater Environments . . . 394
1.3 Short-Term Degradation Processes Specific to Volcanoes . . . 395

2 Volcanic Debris Avalanches . . . 397
2.1 General Features . . . 397
2.2 Causes of Debris Avalanches . . . 399
2.3 Frequency of Debris Avalanches: Cyclicity of Edifice Growth and Destruction . . . 402

2.4 Main Types of Lateral Flank Failures 404
2.5 Characteristics of Volcanic Debris-Avalanche Deposits 408
2.6 Movement of Volcanic Debris Avalanches: Transport Mechanisms 412

3 Lahars 413
3.1 Types of Lahars 415
3.2 Case Studies of Hazardous Lahars as Agents of Landscape Change 417
3.3 Lahar Emplacement and Associated Deposits 417
3.4 Effects and Impacts of Lahars 419
3.5 Characteristics of Lahars and Debris Flows 422
3.6 What Are the Most Significant Factors that Determine a Lahar's Ability to Cause Damage? 424
3.7 Lessons Gleaned from Surveys, Experiments and Case Studies 426

4 Subaqueous Mass Flows Related to Volcanic Activity 427
4.1 Terminology and Major Features 427
4.2 A Wide Variety of Subaqueous Processes 427
4.3 Submarine Mass-Transport Mechanisms: From Mass Flows to Turbidites 429

5 Concluding Remarks 431

References 439

Chapter 8 Long-Term Degradation of Volcanoes 453

1 Long-Term Erosion Processes Gradually Take Over in Volcanic Terrains 453

2 Weathering of Volcanic Rocks 454
2.1 Physical Weathering 457
2.2 Chemical Weathering 458
2.3 From Regolith to Volcanic Soils 459

3 Fluvial Erosion on Volcano Surfaces and Drainage Network Formation 460
3.1 Evolution of Drainage on Volcanoes 461
3.2 Fluvial Erosion of Volcanic Plateaus/Tablelands 465
3.3 Modification of the Drainage by Volcanic Activity 465

4 Degradation of Volcanic Cones and Surfaces, and Formation of Erosional Landforms 468
4.1 Erosion of Volcanic Cones Through Four 'Stages' 470
4.2 Erosion of Craters and Calderas 474
4.3 Erosion Rates in Volcanic Terrains 481

5 **Degradation of Volcanic Plateaus/Tablelands** . . . 485
5.1 Erosion of Lava Plains . . . 485
5.2 Erosion of Ignimbrite Sheets . . . 487

6 **Wind Erosion, Coastal Abrasion and the Effects of Glaciation on Volcanic Landscapes** . . . 488
6.1 Aeolian Landforms . . . 488
6.2 Coastal Erosional Landforms . . . 488
6.3 Landforms Created by Glacial Erosion . . . 488

7 **Concluding Remarks** . . . 489

References . . . 502

Chapter 9 The Role of Geomorphology in Assessing Volcanic Hazards . . . 509

1 **Introduction** . . . 509
1.1 Terminology . . . 510
1.2 Distinguishing Volcanic and Geomorphic Hazards . . . 511
1.3 Objectives and Outline . . . 511

2 **Interplay of Geomorphic Processes with Eruptive Activity** . . . 513
2.1 Effects of Slope, Topography, and Bedrock on Lava Flow Propagation . . . 513
2.2 Slope, Digital Topography and Lava Flow Hazard Monitoring . . . 514
2.3 Direct Interplay of Geomorphic Parameters with Hazards from PDCs . . . 515

3 **Geomorphic Impact of Indirect Volcanic Hazards** . . . 527
3.1 Volcanic Sediment-Laden Flows . . . 527
3.2 Assessment and Mitigation of Sediment-Laden Volcanic Flows . . . 532
3.3 Jökulhlaups . . . 535
3.4 Effects, Hazards and Source Mechanisms of Volcanogenic Tsunamis . . . 537

4 **Post-eruption Geomorphic Responses and Hazards** . . . 541
4.1 Hydro-morphological and Sedimentological Concepts . . . 542
4.2 Three Categories of Geomorphic Response to Volcanically-Disturbed Watersheds . . . 546

5 **Geomorphic Impacts of Very Large Eruptions (VEI>6)** . . . 556

6 **Delayed Geomorphic Hazards in Caldera and Ignimbrite Contexts** . . . 557

7 **Geomorphic Hazards on Dormant Volcanoes** . . . 558

8 **Concluding Remarks** . . . 560

References . . . 569

Chapter 10 Case Studies Illustrating the Role of Geomorphology in Assessing Volcanic Hazards 583

1 Case Study 1 583

2 Case Study 2 587

3 Case Study 3 589

4 Case Study 4 592

5 Case Study 5 595

References 598

Chapter 11 Methods and Techniques for Studying Volcanic Landforms 601

1 Introduction 601

2 DEM-Based Morphometry 602
2.1 Morphometric Analysis of Volcanic Landforms 602
2.2 Volcano Volumetry 606
2.3 Landform Reconstruction, Surface Fitting 610
2.4 Numerical Modelling of Volcanic Edifice Erosion 614

3 Numerical Modelling of Lava Flows 617

4 The Overspill Hazard of Small-volume PDCs 620
4.1 Introduction 620
4.2 Recent Findings and Perspectives 622

5 Approaches for Unravelling Volcano Structures 624
5.1 Volcano Structure—Methods of Analysis in Geomorphology 624
5.2 Monitoring Volcano Deformation and Internal Magmatic Processes Using InSAR 627

6 Geochronological Approaches for Studying Volcano Evolution 630
6.1 K–Ar and ^{40}Ar/^{39}Ar Dating of Volcanic Rocks 630
6.2 Dating Volcanoes and Volcanic Deposits by Means of Luminescence Techniques 634
6.3 Cosmogenic Nuclide Dating 638
6.4 Radiocarbon Dating 641

References 645

Appendix 657

Index 659

About the Authors

Dávid Karátson's career has been linked to Eötvös Loránd University (ELTE) in Budapest, Hungary. After graduating in geography, geomorphology, and taking an university teaching degree, he obtained his Ph.D. from ELTE in 1995 in comparative morphometric analysis of the Carpathian volcanic edifices.

He then became a lecturer at ELTE in the Department of Physical Geography, which he was made head of department of in 2007, becoming a full professor in 2012. Between 2016 and 2019, he was director of the Institute of Geography and Earth Sciences. Since the 2000s, he has been committed to the Erasmus programme, maintaining relations mainly with Italian institutions. He has supervised 65 Master theses and 10 Ph.Ds., two of which were co-supervised by professors from Université Clermont-Auvergne (France).

He has carried out extensive volcanological and volcano geomorphological studies worldwide, mostly in countries with active volcanoes. Highlights include a JSPS grant to complete comparative studies on Kyushu and Honshu, Japan (1997), a Fulbright Grant to the Northern Arizona University, USA (2004/5) to study the San Francisco Volcanic Field, and a Humboldt Fellowship to the Georg-August Universität Göttingen, Germany (2008/2010) to study volcano erosion in Chilean and Argentinian Andes. In the 2010s, he led an international team to reconstruct the palaeotopography of Late Bronze Age Santorini, Greece, and he was a co-leader of the CoV11 field trip to the island (2022).

He has been the convener of volcano geomorphology sessions at the EGU Congress, Vienna, Austria; he is international correspondent of Geology Today (Wiley-Blackwell). He is corresponding member of the HungarianAcademy of Sciences. He created the screenplay for the Kemenes Volcano Park Visitor Centre, Hungary (2013).

He has published 117 papers, 61 of which were in international journals. He has been the editor or author of 11 books, including *From Börzsöny to Harghita Mountains: Volcanology, Surface Evolution, Palaeogeography*, for which he obtained the highest doctoral degree (D.Sc.) issued by the Hungarian Academy of Sciences in 2008, and *Ciomadul: The Youngest Volcano in the Carpathians—Volcanism, Palaeoenvironment, Human Impact* (Springer, 2022).

Jean-Claude Thouret was for many years Professor at the Laboratoire Magmas et Volcans (UMR 6524 CNRS), Clermont-Auvergne University (UCA) Observatoire de Physique du Globe (OPGC), and Institut de Recherche pour le Développement (IRD), in Clermont-Ferrand, France, where he has been an Emeritus professor since 2017.

Following higher studies in Paris (Ecole Normale Supérieure), he obtained a university teacher's degree in 1974, before becoming a lecturer and assistant professor at the J. Fourier University of Grenoble in 1975. He then obtained a Ph.D. on the Nevado del Ruiz—Tolima volcanic massif (Colombia) in 1988, followed by a post-doc (NATO fellowship) at the USGS Cascades Volcano Observatory. From 1990, he became a professor at Blaise Pascal University in Clermont (since renamed UCA).

He taught geomorphology and volcanology at UCA, as well as supervising 11 Ph.D. students and 12 Master's students. His research has been devoted to active volcanoes, especially pyroclastic deposits and lahars, to volcanic hazards and geomorphic processes on volcanoes, in particular in the Andes (Peru, Colombia), Indonesia, and United States.

He led several national and international research projects funded by the French National Research Agency, INQUA and IAVCEI. He organized three international symposiums on natural hazards and risk in 1994, 2003, and 2011, the latter in Singapore. He was invited Professor at Massey University, Palmerston North, New Zealand, between 2005 and 2007. He spent four years in Lima, at the Geophysical Institute of Peru, on an IRD placement, where he supervised about twenty Peruvian M.Sc. and Ph.D. students and initiated research on composite volcanoes, hazards and ignimbrites.

He was Chief Editor (2004–2009) of the Journal *Géomorphologie: relief, processus et environnement*, and Guest Editor of five Special Issues on volcanoes in *GeoJournal*, *Géomorphologie*, *Zeitschrift für Geomorphologie*, and *Geomorphology* between 1999 and 2011.

He has authored 113 publications in international journals, 63 in national journals, 29 chapters in English, and 18 in French textbooks, and he has given 145 presentations at international conferences.

Introduction

1 Volcanism and Volcanic Landforms: Development of Volcano Geomorphology

Volcanism is the surface manifestation of internal heat-producing processes on a rocky planet or moon, with the release of solid, liquid or gaseous products (Francis 1993). On Earth, igneous processes including volcanic activity are of considerable interest to geologists because they generate new crust, form deposits of gold, silver and copper, contribute to fertile soil formation, and perturb the atmosphere and climate. At a more immediately practical level, an understanding of how volcanoes work and what triggers eruptions is essential in order to mitigate their negative effects on human society.

Volcanism creates *volcanic landforms*, of which the main elements are *volcanoes*. Volcanoes are complex, dynamic constructs, produced and evolving through the unique combination of Earth's internal (endogenous) and surface (exogenous) processes. The stunning morphological variety of volcanic landforms (Green and Short 1971) is due to a combination of magma composition and volume related to their tectonic setting, eruption style, plumbing system, as well as the age, and the lifetime over which a volcano has been active and has developed (Huff et al. 2022). While the apparent lack of plate tectonics on other rocky planets in the Solar System makes volcanism and volcano types relatively uniform (with basalt volcanoes prevailing), there is a considerable variety of volcanic landforms on Earth, reflecting the large variation in tectonic processes and magma generating parameters in time and space.

Our knowledge of volcanoes as landforms is a relatively young branch of Earth Sciences, as it has only been recognised as a separate domain for about 80 years: "strangely, volcanic landscapes have been little studied", as stated by Francis, who gave the title of his chapter on 'Volcanoes as landscape forms' in his book on volcanoes (Volcanoes–A Planetary Perspective 1993) in tribute to Charles A. Cotton's seminal work (1944, 1952) with the same title. It is indeed true that before Cotton there were no stand-alone books written on the geomorphology of volcanoes, only book chapters and short essays (e.g. those by Geikie 1903; Daly 1911; Tyrrell 1931; Escher 1933; Rittmann 1936). However, if we look at the structure of Cotton's book and in particular his volcano types, the terms are often quite different (for example 'lava volcano', 'pumice volcano', 'stratified volcano') from the terms we use today. Even in the next milestone in this field, Cliff Ollier's 'Volcanoes' (1969), the names of the volcano types, although overlapping with those used today, still differ.

Ollier significantly revised his Volcanoes textbook two decades later (1988), and Francis's

D. Karátson and J.-C. Thouret, *Volcano Geomorphology*, Advances in Volcanology,
https://doi.org/10.1007/978-3-031-77693-9_1

excellent book mentioned above was published in 1993 and reedited in 2004 together with C. Oppenheimer. The line of thinking of these scientists, even though many new findings have appeared in since publication, provides the framework for our book, and we follow the same basic approach to volcano geomorphology. Other important works have also been published and should be mentioned, such as Williams and McBirney's Volcanology (1979) with its improvements to volcano classification (e.g. Galápagos type, and caldera typology); Chester's Volcanoes and Society (1993) which includes volcano hazards among other topics; A. Scarth's Volcanoes—An Introduction (1993); Lockwood and Hazlett's Volcanoes–Global Perspectives (2010, 2022) with a part devoted to volcanic landforms and settings; and chapters on volcano types in The Encyclopedia of Volcanoes (Sigurdsson et al. 2000, 2015).

Volcanic activity exerts a significant influence on the surrounding landscape (Huff et al. 2022). A number of factors are involved, including the lifetime of volcano development, the nature and pattern of various eruptive fissures, vents and intrusions, and the composition and physical characteristics of erupted and intruded material (including emitted gases, which can affect the climate). In some cases, voluminous lava flows or a thick blanket of tephra can mantle or even bury the preexisting landscape, whereas, in others, a localised distribution of lavas and tephras may produce a distinctive assemblage of cones and lava flows, aerially congregated into volcanic fields and provinces over a few millions of years.

On the other hand, volcanic landforms are transient features on a geological timescale, partly because few of them exist for more than a few million years, and partly because their activity occurs nearly simultaneously with erosion, degrading the constructive landforms in millions or at most a few tens of millions of years. As a result, erosion processes, whose degree and extent is controlled mostly by the prevailing climate and by the rock composition, give way to a wide range of erosional volcanic landforms.

2 Volcanoes and Volcanic Landforms: Toward a Definition

A volcano is a complex geological and geomorphic object, and an active volcano is an evolving landform that can develop, become eroded and grow again in a relatively short time. Different definitions, based on the field of knowledge, have been provided by different disciplines, such as geology, geophysics, planetology, volcanology, and geomorphology.

2.1 What Is a Volcano? A Simple, Twofold Definition

The NASA website states the simplest definition: 'A volcano is an opening on the surface of a planet that allows material warmer than its surroundings to escape from its interior', or simply summarised as 'the site of emission of volcanic products' (de Silva and Lindsay 2015). This definition can be more precise: 'An opening or vent in the Earth's crust through which molten rock (lava), hot rock fragments (pyroclastic rocks from fragmented magma ejected explosively), and volatiles (hot gases) escape and are transported onto the Earth's surface' (e.g. Gudmundsson 2020, 2022).

In fact, a twofold definition of a volcano given by the Glossary of Geology (1997, p. 690) includes the initial one: 'a vent in the surface of the Earth through which magma, a high-temperature mixture of melt, crystals and gases erupt', but also the geomorphic result: 'the form or structure, usually conical that is produced by the ejected material'. The term volcano therefore refers to the landform created by the repeated accumulation of solidified lava and volcanic debris (or volcaniclastic material) near the vent (MacDonald 1972; Fisher et al. 1977; Francis 1993). The primary eruptive processes, either effusive, extrusive or explosive, create volcanic landforms made up of lava flows, lava domes, tephra and volatiles (e.g. pyroclastic or hydroclastic cones). Following the commonly held image shared by most people, a volcano is 'an edifice that develops when molten rock reaches, or closely approaches, Earth's surface and erupts' (Lockwood and Hazlett 2010). The second part of the definition is interesting, as it includes all types of shallow intrusions that do not reach the Earth's surface, a quite common situation in past and present volcanic areas.

However, such definitions do not encompass all volcanic landforms: for example submarine lava flows from mid-ocean ridges do not construct conspicuous near-vent edifices, while depressions such as maars ('holes in the ground') and calderas do not correspond to positive constructs, but are negative, excavational landforms. Some small volcanic-like landforms such as the mud volcanoes of sedimentary basins and geothermal areas are sometimes referred to as 'sedimentary volcanoes' (e.g. van Loon 2010; Mazzini et al. 2017), but they are not considered volcanoes by definition because they do not serve as vents for magma. Mud volcanoes, geyser and hot spring cones can fall outside of this volcano definition, because they are not, or at least not primarily, formed by eruptive activity and, similarly, have no real vent. These manifestations of volcanic activity result from the interaction of groundwater with magma or solidified but still-hot igneous rocks at shallow depths.

A simple geomorphological definition of a volcano is a surface landform associated with eruptive activity from a magmatic vent in subaerial, subglacial or subaqueous environments. Volcanoes come in many shapes and sizes, from small hills to the largest mountains on Earth and to the most extensive lava plateaus of Large Igneous Provinces (LIPs) (MacDonald 1972). Volcanic landforms evolve over time as a result of repeated eruptive activity, but also periods of erosion and collapse; hence they can become superimposed or nested. According to this definition, a volcano can be:

- a negative or excavational landform, i.e. a maar or caldera without a volcanic cone. Volcanoes do not only include constructional landforms, and they do not all resemble to the stereotypical towering conical, snow-capped giants like Fuji, Japan. They range from small scoria or cinder cones (the most common type of volcano) to huge piles of lava, reaching up to 10 km above the ocean floor. Calderas, caldera complexes (e.g. Yellowstone, USA) and volcano-tectonic depressions (e.g. La Primavera, Mexico or the Taupo volcanic zone, New Zealand), also called inverse volcanoes, do not rise above the surrounding landscape, instead forming negative relief (Siebert et al. 2015).
- of any size, from minor volcanic landforms on a lava flow, such as tumuli and pressure ridges, to broad oceanic shields almost 9 km in height on the ocean floor and up to 100,000 km^3 in volume and even to LIPs, with a total area of over one million km^2 and tens of millions of km^3.
- formed not only by newly erupted magma, but also by volatiles released as part of magmatic activity or by erupted debris (e.g. a maar).

2.2 Limitations to the Initial Definition

As early as 1993, Francis stated that discoveries made on other planets meant a broad enough term was necessary to cover all types of volcanism. Volcanism may be defined as the manifestation at the surface of a planet or satellite of internal thermal processes through the emission at the surface of solid, liquid or gaseous products. A volcano therefore is simply a site at which material reaches the surface of the planet from the interior. Two prerequisites must exist to feed volcanism: a source of heat and rock to melt. Volcanoes are the surface expression of the thermal process, that is, the partial melting of mantle material and the buoyant rise of magma to the surface through the crust.

Lopes et al. (2010) stated that 'Extraterrestrial volcanism, including active volcanoes, has changed our definition of a volcano' in their overview of past volcanism on the Moon, Mercury and Mars, active volcanism on Jupiter's moon Io, and cryovolcanism on

Saturn's moon in the outer solar system. These authors suggest that the definition of volcanoes should be extended to include all 'openings on a planet or moon's surface from which magma, as defined for that planetary body, and/or magmatic gas is erupted'. However, the present book does not deal with volcanic landforms on extra-terrestrial planets and moons; instead, we refer the readers to alternatives (e.g. Davies 2007; Ollivier et al. 2009; Wilson 2009; Gregg et al. 2021) and recognise the fast evolution of the field of volcanism in planetary geology (Head and Wilson 2022).

A number of shortcomings undermine the initial definition of a volcano: (1) a volcano cannot be reduced to the surface expression of magma/lava, as volcanic activity takes place below the surface, through the plumbing system down to the magma chamber (Fig. 1); (2) a large part of magma is intruded near the Earth's surface (depending on the tectonic setting and the local stress field), but does not necessarily lead to the eruption of material onto the surface, such as 'blind', shallow intrusions that can deform the upper crust; (3) there are single, individual volcanoes (such as monogenetic cones), but also eruptive fissures and crater rows, or volcanic clusters that comprise many vents and/or a collection of individual volcanoes. Structures considered as volcanoes but of different sizes can be spatially connected to each other, or even superimposed: e.g. scoria and spatter cones or adventitious landforms along fissures that open on a shield or composite volcano; (4) the location of the eruption centre, either point-like, e.g. a summit or flank vent, or elongated fissure also raise problems as to what should be considered a volcano.

Being aware of the shortcomings associated with the initial, simple definitions of a volcano, other authors have proposed alternative definitions. One of them is a holistic definition assuming that a volcano has a broad meaning as 'a geologic environment'.

In their essay 'What is a volcano?' Borgia et al. (2010) provide two versions of the same

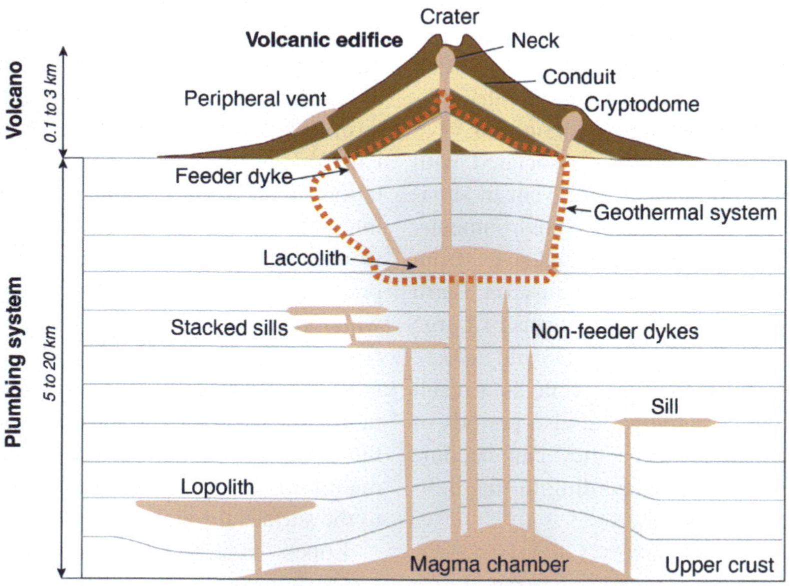

Fig. 1 Scheme of a volcano system, including the edifice or volcanic construct, the central conduit from the magma reservoir to the vent, and the relatively shallow plumbing system with a variety of intrusions. Adapted and modified after Acocella (2021)

definition of a volcano. The short one is as follows: 'A volcano is a geologic environment that, at any scale, is characterized by three linked elements: magma, eruptions, and edifice'. The long one reads as: 'Volcanoes are geologic environments where magma, generated at a source within the crust or mantle, flows upward and is subject to varying amounts of physicochemical evolution, intruding and reacting with the encasing rocks and other magma, and originating a geothermal system. Once near the lithosphere top, the magma erupts, piercing the interface. Volcanic deposits are accumulated by eruptions giving rise to a volcanic edifice. In turn, these deposits may become modified by magma intrusion, eruptions, geothermal fluids, tectonics, erosion, landslide, and all other kinds of geologic processes. The boundaries of this environment (volcano) are frequently time dependent, transitional, poorly defined, or unknown. However, working boundaries can be based on different arguments using such factors as geometry, morphology, and structure'.

In short, a volcano cannot be limited to just the volcanic edifice itself, but must include its basement, at least that part where volcanic and geologic processes occur, such as intrusions, geothermal and hydrothermal activity, and edifice-basement tectonic interactions. This definition has an unexpected consequence: all Earth Science disciplines can be applied to, and integrated in, the study of volcanoes.

However, such holistic definitions have at least two shortcomings: first, the indistinct term for environment may be confusing; second, no physical limits were assigned to a volcano environment, suggesting for example that the Earth itself may be one huge volcano! Additionally, it is not credible that volcanology encompasses all Earth Science disciplines, although volcanoes are multifaceted. In fact, volcanology is the study of volcanoes, magma and associated geological, geophysical, geomorphological and geochemical phenomena. As such, volcanology includes geology, petrology and geochemistry, to name a few disciplines, but this Earth Science field also covers other theoretical and applied fields such as monitoring and hazard studies associated with volcanism.

2.3 A Framework to Support the Definition of Volcanic Landforms

A useful initial framework consists of considering the main volcanic landforms as the result of magma composition and eruption style. Magma composition is important because it is the main control on the rheology (here deformation and flow of magma and lava) that ultimately control eruptive style (de Silva and Lindsay 2015). Thus, the general characteristics of the volcanic landforms reflect the fundamental control of magma composition and the properties that govern eruption styles. Processes of differentiation and magma movement from the reservoir to the surface and the nature of the central conduit are pivotal to the final volcanic landform that is created.

Volcanic landforms can be defined as distinct temporal and spatial identities at the scale being considered. On a local scale, a single lava flow or small cone is the volcanic landform; on a planetary scale, the Big Island of Hawaii, and the Ontong Java Plateau, one of the largest oceanic LIPs, could be considered as volcanic landforms (de Silva and Lindsay 2015). Large, long-lived volcanic clusters and fields are collections of spatially, temporally and magmatically connected volcanic landforms, which represent the interplay between eruptive and erosional histories. Clusters, fields and complexes that form LIPs represent the most impressive expression of volcanism and the most obvious volcanic landform at the planetary scale. These often include the entire range of individual landforms.

The variable scale of the same geological object suggests some primary controls on the characteristics of volcanic landforms. Magmatism on the Earth is fundamentally basaltic, but the variety of volcanic landforms reflects a wide range of compositions (from mafic to felsic) and resulting eruptive styles (effusive, extrusive and many explosive styles). On Earth, differentiation of the basaltic magma from the mantle while ascending or being stored in the lithosphere is responsible for the diversity of compositions, probably unique on the rocky planets in our solar system. The volcanic landforms are the expression of the magma composition and, ultimately, are the results of its modifications in the uppermost lithosphere, the plumbing system, and finally the central conduit (de Silva and Lindsay 2015).

2.4 The 'Volcano-Magma System'

Volcanoes have roots that reach deep inside Earth. In geology, a volcano is not only the geomorphic expression at the Earth's surface, but the plumbing system (magma generation, storage and transport). A volcanic construct, being just the surface part of a larger system developed at depth, can be considered as part of a *volcano-magma system* (Schmincke 2004). A volcano can be thought of as the surface expression of a deeper and complex magma reservoir and intrusions, i.e. the plumbing system. The system includes the shallow and deeper parts of the volcanic substructure, i.e. the subsurface feeding system linking the magma chamber(s) to the vent through the central conduit (Fig. 1). When Walker (2000) described the basaltic volcanoes, he provided a systemic definition of the term volcano: "When describing volcanoes, it is of great merit to recognize the existence of a larger entity, namely, the volcanic or magmatic system. A system may embrace the melting anomaly at the magma source, the conduits through which magma rises toward the surface, the magma chambers, intrusions, and geothermal zones, and the volcano itself".

Geomorphology studies emphasize the terrestrial or subaqueous landforms at the Earth's surface, but several criteria for understanding and ranking volcanic landforms rely on the plumbing system and on eruption styles. Magma ascent and eruptive processes play a role on the texture of volcanic rocks, therefore on many landforms made up of lavas or pyroclastic or both, and on landforms that involve more volatiles than solid material (e.g. maars). Furthermore, erosion can reveal some deep structures of volcanoes, such as their intrusions and roots, or even, following long-term erosion, plutons, exposed in large, deeply eroded calderas or in a cauldron.

The volcano-magma system is divided into four zones for simplicity (Schmincke 2004): (1) In the root zones, magma is generated by partial melting of pre-existing rocks. (2) Detachment and rise of magma. (3) Stagnation and differentiation. (4) At the Earth's surface, magma erupts either effusively as degassed lava flows, or extrusively, as viscous lava domes, or explosively, as tephra and volatiles. (5) Construction of an edifice or excavation of a depression, and in the case of high-rising eruptive columns, gas input into the atmosphere and troposphere. Whether a magma erupts effusively, extrusively, or explosively, depends on parameters such as composition, viscosity and rise speed, interaction with external water, and in particular, on the formation, expansion and bursting of bubbles formed when a magma becomes saturated with volatile compounds.

Following Walker's and Schmincke's schemes, Szakács (2010) advocated for a systemic definition of a volcano as a collection of 'subsystems', and offered two possible definitions of the term volcano: (1) an igneous system that reaches Earth's surface', and (2) the uppermost part of the volcanic system, that includes the shallow magma reservoir that feeds eruptive activity, all magma plumbing pathways and conduits tapping that reservoir, and the volcanic edifice at the surface, also including intrusions within such an edifice'. This second definition, which is similar to what Schmincke (2004) called a volcano-magma system, is represented by Fig. 1, a scheme of a volcano, consisting of the volcanic edifice, the central conduit, and the shallow plumbing system made up of different types of intrusion (Acocella 2021).

We note that the modern view on volcanoes and magma systems includes the systemic approach, represented by models of transcrustal plumbing systems with a non-permanent hydraulic connection, e.g. Cashman and Sparks (2013); Cashman et al. (2017), Sparks and Cashman (2017), or, following an earlier approach by Wilson et al. (1992), Castruccio et al. (2017). In geology and/or geophysics, a volcano is only the upper part of a system, which can be interpreted as both the subaerial and the shallow and/or deep subsytems, which are connected during eruptions. Modern quantitative research uses models in which the key plumbing system properties can be inferred from the topography of the volcanic constructs. In other words, dimensions such as height and basal radius and topographic profiles of edifices help invert the depth and size of the eruption-feeding magma chamber, as the structure of the magmatic system can determine the morphology of the volcanic edifice (for an application to arc volcanoes see Castruccio et al. 2017).

3 Towards a Geomorphological Definition

3.1 Construction and Destruction

In geomorphology, a volcano is the expression of an eruption on the Earth's surface, either a constructional landform (volcanic construct such as a cone) or a destructional landform (such as a caldera) due either to tectonic/magmatic processes or erosion. The geomorphological definition of a volcano that is adopted in this book, in line with the above arguments, can also include the shallow plumbing system as well as the erosional volcanic landforms: fast erosional processes such as flank failures or long-term decay are intrinsically associated with volcanoes on Earth, and take place during quiescence intervals of the eruptive activity and as soon as it ceases.

Volcanism directly creates and degrades landforms, and indirectly provides an age for both the landsurface over which the erupted material lies and the succession with which it is intercalated. In contrast to erosional landforms in basement rocks that can survive for long geologic periods on Earth, volcanoes usually have a relatively short-term existence, commonly less than one million years for individual constructs and a few tens of million years for volcanic fields (Thouret 1999). Volcanic landforms, in contrast to other types of landforms, result from near-simultaneous constructive and destructive processes. Finally, volcanic landforms are the subaerial expression of growth (constructional landforms) and excavation (explosively-driven or destructive landforms), but also erosional landforms (involving weathering, mass wasting and river expenditure) that may lead to inverted relief.

3.2 A Volcanic Construct Involves Several Components

A geomorphological definition of a volcano can be based on several criteria such as geometry, shape, size, tectonic setting, magma composition, and eruption style (Figs. 2 and 3). A holistic definition is required that can apply from the scale of a simple monogenetic cone to the most complex constructs of long-lived volcanic clusters and volcanic fields. As many as seven components or factors can be taken into account to create a holistic definition:

1. The tectonic setting, which governs how the magma composition is evolving. The tectonic and structural context (extensional, compressional, strike slip, etc.) also controls the magma path towards the surface.
2. The plumbing system plays a fundamental role, including the location and depth of reservoir, as the upper conduit to the vent feeds the volcano system. Volcanoes reflect the compositional system that feed them, exhibiting widely varying magma supply rates and resulting eruption styles.
3. Perhaps the most influential factor in shaping volcanic landforms is the manner in which gas escapes the magma (Oppenheimer 2005; Siebert et al. 2015), which controls whether the resultant volcanic construct will consist of lava or pyroclastic rocks. Both gas content and viscosity vary widely in magmas, and the more gas-rich, viscous, and lower-temperature compositions common at continental margins tend to explode violently, fragmenting the melt into tephra. This process can form small or large cones of ash and scoria, as well as widely dispersed ash layers from

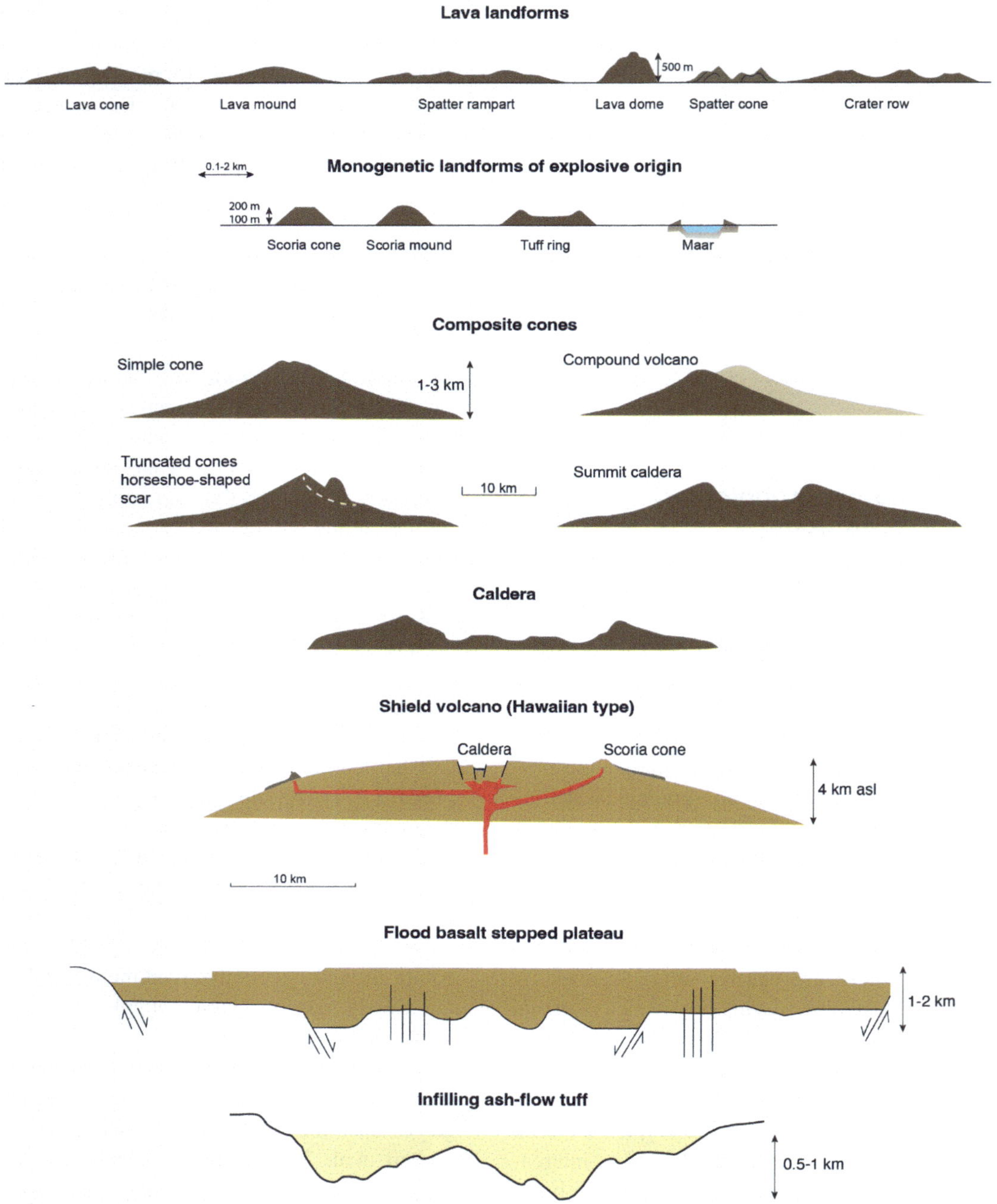

Fig. 2 Sketch of the principal types of volcanoes and volcanic landforms, including the size parameter. Modified after Ollier (1988), Siebert and Simkin (2002)

composite cones or calderas. In contrast, gas separates more easily from the less viscous, hotter, and less gas-rich magmas of both spreading centres and oceanic islands, where lava flows passively from vents to form low slope-angle oceanic shields.

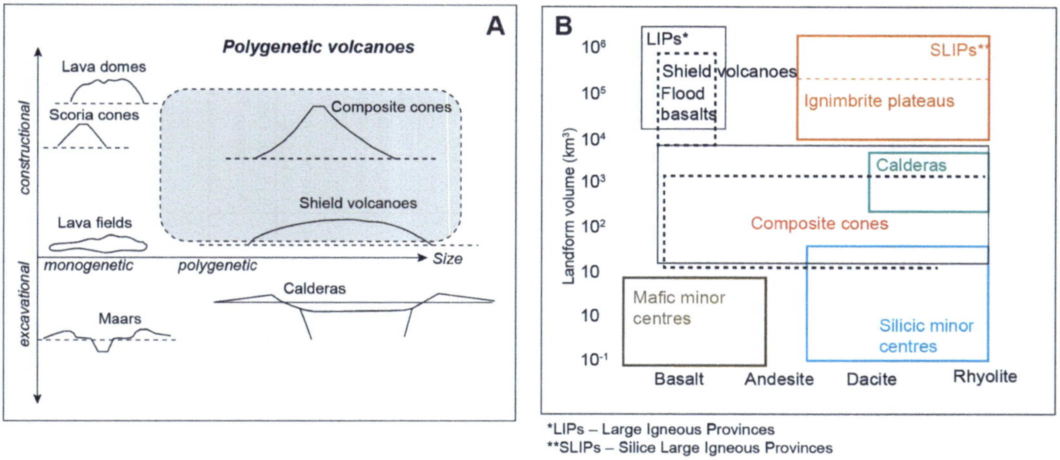

Fig. 3 **A** Diagram showing how volcanoes can be distributed according to two criteria: (1) geomorphic processes of construction and excavation, and (2) complexity of eruptive history, from monogenetic to polygenetic edifices (adapted from Davidson and de Silva 2000). **B** Diagram showing a similar distribution of volcanoes based on (1) volume and (2) magma composition, here the approximate silica content (adapted from de Silva and Lindsay 2015). Solid lines represent the characteristic range of composition and volume. Dashed lines indicate subordinate ranges

4. Such difference in gas emission and volatile content is controlled by magma composition (mafic, intermediate, or felsic) and physical properties (viscosity, temperature, shear strength, bubble content). Magma composition and properties in turn govern the eruptive style, from effusive (Hawaiian) and extrusive (Pelean) to explosive activity (related to the volatile content and eruption rate), e.g. from Strombolian to Vulcanian and Plinian. Eruption style is also determined by the presence of internal or external water, which contributes to either hydroclastic or phreatomagmatic eruptions.
5. Volcano geometry: shape and size (Fig. 2) are determined and/or controlled by several factors, namely: (1) the volume of volcanic products, (2) the interval length between eruptions, (3) the composition of volcanic products (4) the type of volcanic eruption style, (5) the geometry of the vent, and (6) the medium (subaerial, underwater or subglacial) on or into which the volcanic products are erupted.
6. Eruptive processes of volcanic landform development, which results in either constructional or excavational landforms (Fig. 3, modified after Davidson and de Silva (2000). Volcanoes reflect the main processes that shape them. Thus, the degree of complexity of their eruptive history enables us to distinguish, for example, monogenetic from polygenetic landforms.
7. Exogenous agents can play a role in the variety of volcanic landforms: weather (rainfall, temperature) and climate shifts over long periods, including glaciations. This drives erosion (weathering, removal and transport, downcutting) which alters the morphology of the edifices.
8. In certain circumstances human activities can alter the morphology of volcanic constructs, either by agriculture, infrastructure, or tourism.

To summarise, our geomorphological definition of volcanoes is as follows: a volcano is a geomorphic construct on the Earth's surface associated with volcanic eruption(s), resulting from the action of a vent, including a subsurface and deep plumbing system, and the surface edifice. This geomorphic construct results from the eight components or factors above: tectonic setting,

Table 1 Classification of elementary landforms

Landform Process Eruptive activity	Volcano and Landform type	Sub-type, Individual landforms	Geodynamic or tectonic setting	Magma composition, Deposit texture	Area, km^2 Volume, m^3 Aspect ratio Ar (av. average*)	Shape. Size* H, L, D, W (km) Ratios* H/W$_{co}$, H$_{co}$/W$_{co}$, D$_{cr}$/W$_{co}$. Slope angle α (°)	Growth, duration and prevailing factors	Examples and locations
ELEMENTARY CONSTRUCTIONAL LADNFORMS MADE UP OF LAVA AND CREATED BY EFFUSIVE ERUPTIONS	Lava flows Terrestrial, constructional landforms	Pāhoehoe flows, S and P-type; Sheet, shelly, rubbly, entrail, ropy	Mid-Ocean Ridges (MOR) Shield volcanoes	Mafic (basalt) lava, coherent	A 1–5 km^2 V 0.01–0.05 km^3 Ar ≤ 1/100	Flattish, smooth, undulating. L 1–10 km H < 0.01 W < 1 km. α < 5°	Hours to days Plug, high effusion rate	MOR (MAR, EPR). Hawai'i, Big Island
		'A'ā lava flows autobreccia	All, in particular volcanic arcs, shield volcanoes	Intermediate (andesite) Clastic lava Clinker, breccia	A 0.05–2.5 km^2 V 0.1–0.07 km^3 Ar 1/4–1/20	Relatively thick (tens of m), brecciated. L 1–5 km W < 0.5 km; α 5–20°	Days to months. Shear stress, Yield strength. Low effusion rate	Cumbre Vieja, La Palma, Canary Islands, 2021
		Transitional lava flows	In particular on shield volcanoes	Basalt	A 0.05–2 km^2 V 0.1–0.06 km^3 Low Ar	Between pāhoehoe and 'a'ā α 5–15°	Days to months Break in slope	Hawai'i, Big Island
		Block-lava flows	All, in particular volcanic arcs	Intermediate to silicic lava	A 0.12 – 0.25 km^2 V 0.006–0.03 km^3 Ar 1/4–1/20	Relatively thick, blocky. L 0.5–1 km W 0.25 km H 0.05–0.1 km	Weeks to years Yield strength Low effusion rate	Llaima volcano, Chile
		Silicic lava flows Coulees	Arc volcanoes, domes	Rhyolite, obsidian	A 0.2–0.125 km^2 V 0.1–0.25 km^3 Ar 1/2–1/20	Thick, stubby L < 1 km W < 0.25 H 0.1–0.5 km Max 14 km^3 Chao dome	Months to years High viscosity	Chao dome complex, Chile
	Subaqueous	Mid-Ocean Ridge with pillow lavas & volcanoes; lobate and sheet lavas & volcanoes	Submarine MOR	Basalt; associated with hyaloclastites	A thousands km^2 V hundreds km^3 Elongated	Length 70,000 km Lobe, pillow L 1 km, W 0.2–0.5 km; thick. Sheet L 2–3 km, W < 2 km; thin	Hours to days High effusion rate, spreading rate	All MORs
	Lava fields	Single lava flows	All	Mafic to felsic	A ≤ 1 km^2 V ≤ 1 km^3 Low Ar	Flow, levee, channel, tube, tumulus. L 1–5 km W 0.05–0.5 km H 0.01–0.05 km	Days Variable effusion rate	Etna, 1998. Chaos Crags south, Lassen, California
		Compound lava flows	All, in particular hot spots, LIPs, CFBs	Mafic to intermediate	A > 100 km^2 V > 2 km^3 Low Ar	Multiple, overlapping and intertwined. L 10–100 km W 2–10 km H ≥ 0.1 km	Months to years Long, sustained effusion rate	Kilauea, Hawai'i, NE rift zone 1983-2018
	Lava lakes	Summit lakes, Flank lakes. Active; inactive	Mid-ocean ridges and rifts	Basalt	A 0.01–0.04 km^2 V < 0.02 km^3	W ~120 m Av.D floor 60m W ~225 m Av.D 500 m	Persistent	Erta Ale, Ethiopia Niyragongo, RDC

Table 1 (continued)

LANDFORMS		Small lava edifices Constructional	Lava cones Lava mounds Lava discs	Hot spots, Intraplate	Basalt, andesite, trachyte	A 20 km^2 V 15 km^3. A <5 km^2 V 1 km^3. A 1.3 km^2 V 0.04 km^3	H 0.75 km, D 5 km H 0.2 km, D 1–5 km H 0.02–0.04 km, D 1.3 km	Low effusion rate	Beerenberg, Jan Mayen; Mt. Cotterill, SE Aust Vulcanello, Italy
		Fissure vents	Eruptive fissures Spatter ramparts	Hot spots, rift zone	Basalt; Lava, spatter, sometimes welded	A 0.05–0.015 km^2 V lava flow several km^3	Linear and narrow. L 0.5–3 km. H 0.01–0.05 km. Width $\leq$0.1 km.	Days to weeks Magma supply rate	Lakagigar, Fagradalsfjall, Iceland Tarawera, NZ
			Spatter cones, usually breached, feed lava flows and/or fields	In particular oceanic shields	Agglutinated and welded spatter (fallout)	A 0.01–0.04 km^2 V 0.001–0.01 km^3 Ar 5	Small mound H 0.05–0.2 km D 0.08–0.2 km; W_{cr}/W_{co} 0.36 Hco/Wco 0.22. α 30–35°	Hours to weeks	Pu'u0'o' East Rift zone, Hawaï'i 1983-2018
	MONOGENETIC LANDFORMS CREATED BY EXTRUSIVE ACTIVITY	Lava domes Exogenous and constructional landforms	Upheaved plugs or spines	Volcanic arcs Continental rifts	Dacite, rhyolite, trachyte	A < 0.1 km^2 V~ 0.03 km^3 Ar 1.8	Steep piton (spine) H 0.18 km D 0.5 km	Weeks. Yield strength, viscosity	Chopine, Chaîne des Puys, France
			Pelean or composite domes	Volcanic arcs Continental rifts	Dacite, rhyolite	A ~1.8 km^2 V~ 0.5 km^3 Ar ~0.3	Broad and high with talus H 0.5 km D 1–2 km	Months to years	Soufrière, FWI Santiaguito, Guat
			Low lava domes	Volcanic arcs Continental rifts	Dacite	A ~0.2 km^2 V < 0.01 km^3 Ar 0.25	Low and craggy H 0.05 km D 0.5 km	Weeks to months	Novarupta, Alaska
			Dome coulees	Intraplate Continental rifts	Andesite, dacite rhyolite, obsidian	A ~5 km^2 V < 1 km^3 Ar 0.06	Broad, feeding stubby flow H 0.3 km D 2.5 km	Months	Glass Mountain, California
		Endogenous domes	Cumulo-domes Dome complexes	Volcanic arcs Continental rifts	Andesite to dacite trachyte	A ~2 km^2 V 0.5-1 km^3 Ar 0.25	Overlapping dome H 0.5 km D 1.5-2 km	Months to years	Puy de Dôme, Chaîne des Puys
			Crypto-domes (forming bulges)	Volcanic arcs Continental rifts	Dacite	Destructed	Detected by bulge on volcano flank	Days to months, but transient	Mt. St. Helens March-18 May 1980, USA
	MONOGENETIC VOLCANOES CREATED BY EXPLOSIVE ERUPTIONS	Pyroclastic cones Constructional landforms	Scoria or cinder cones Scoria mounds: without crater	All, in particular continental rifts and shield volcanoes	Mafic to intermediate. Scoria lapilli, blocks and bombs (fallout)	A 0.6–4.9 km^2 V 0.1–0.3 km^3 0.4–3 km^3 with flows Ar 0.4–2	Conical hill, twin, nested. H= 0.05–0.4 km D= 0.25–2.5 km (av. 0.9) H_{co}/W_{co} 0.18. D_{cr}/W_{co} 0.4 H/W_{co} 0.26. α 25–35°	A few weeks to < 12 months (but 9 years at Paricutín)	Pariou and Côme, Chaîne des Puys, France; Cones on Mauna Kea, Hawai'i
			Pumice cones	All, in particular volcanic arcs	Intermediate to slicic (fallout)	A 0.03–0.8 km^2 V 0.03–0.3 km^3 Ar 0.4–3	Single hill. H 0.1–0.3 km D 0.2–1 km. H_{co}/W_{co} 0.3–0.5 D_{cr}/W_{co} 0.05–0.25 H/W_{co} 0.1–0.5 α 25–30°	A few weeks	Newberry, Oregon. Monte Pilato, Lipari, Italy
		Hydroclastic volcanoes Constructional	Tuff cones	Volcanic arcs, Intraplate	Ash and lapilli often cemented Fallout and surge	A 0.2–1.7 km^2 V 0.01–0.25 km^3 Ar 0.18–0.27	Small hill. H 0.05–0.3 km D 0.1–1.5 km W_{cr} 0.1–1.5 km. H_{co}/D_{cr} 0.2–0.5 km α 20–30°	Weeks to a few months	Udo, south Korea Tagus and Beagle, Isabela Islands, Galápagos

Table 1 (continued)

		landforms	Tuff rings with low rim	All, in particular intraplate	Ash and lapilli Fallout and surge	A 0.8–3.2 km^2 V 0.04–0.2 km^3 Ar 0.02–0.07	Shallow, crater and low rim. H_{rim} 0.05 km. D 1–2 km. W_{cr} 0.2–3 km, av. 0.75 H/D_{cr} 0.05–0.13 km. α 2–20°	Days to a few weeks	Diamond Head, Oahu, Hawai'i Volcán Xico, Mexico
		(Maars : destructive and construc-tional)	Maars, with diatreme, high rim, and crater often with lake	All, in particular intraplate	≥ 90 % non-juvenile ash, lapilli and blocks. Fallout and surge	A 0.13–0.8 km^2 Rim ejecta 0.003–0.045 km^3 (Av. 0.02 km^3). Ar 0.08–0.2	Subcircular crater and rim H_{rim} 0.01–0.15 km. D_{rim} 0.4–1.0 km (av.0.6). H_{co}/D_{cr} 0.2–0.5 km. D_{rim}/Depth 0.10. α 10–30°	A few days	Ukinrek, 1977, Alaska Lake Pavin, Chaîne des Puys, France
			Littoral cones	All Islands. Created by lava flow entering sea	Ash, lapilli, a few bombs Fallout	A 0.04–0.20 km^2 V 0.001–0.02 km^3 Ar 2.5–5	Small hill. H 0.05–0.2 km D 0.2–0.5 km H_{co}/W_{co} 0.33. α 20–30°	Days to weeks	Pu'u Hou, Mauna Loa coast, Hawai'i; Canary Islands
	Subaqueous construct	Submarine effusive erup-tions (constru-ctional)	Seamounts (guyots when eroded and flat topped)	Mid-ocean ridges and near slopes	Pillow lava Sheet lava	A 0.8–3.2 km^2 V 0.4–4.8 km^3 Ar 0.12–0.95	Single mountain. 100,000 > 1 km high; 25 million > 0.1 km high. H 0.1–3 km. D 1–2 km. α 10–30°	Weeks to months	East Pacific Rise
	Explosive activity and collapse	Craters (Destructive landforms)	Explosion craters	On cones and composite volc-anoes	All compositions	A 0.1–0.2 km^2	Symmetrical, assymetrical Elongated: cleft. Average W ≤ 1 km, D ≤ 0.5 km	Hours to days Recurrent	e.g. Vesuvius, Italy
			Crater rows	Along an eruptive fissure	Mostly intraplate and hot spot	A 2.5–12.5 km^2	En echelon, Length 5–25 km. W 0.1–0.5 km	Days to months	Laki, Skaftár fires 1783, Iceland
			Pit craters	Local collapse	No deposit. Hole in lava flows	A 0.002 km^2	50-100 m deep and narrow av. D 20–25 m	Instantaneous to hours	Halemaumau floor, Kilauea

Table 2 Classification of hybrid, polygenetic and other landforms

Landform Process Eruptive activity	Volcano and Landform type	Sub-type, Individual landform	Geodynamic or tectonic setting	Magma composition Deposit texture Process	Area, km² Volume, km³ Aspect ratio (all: average)	Shape. Size* H, D (km) Ratios* H/W_{co}, H_{co}/W_{co}, D_{cr}/W_{co}, Slope angle α (°)	Growth duration Prevailing factor	Examples Locations
HYBRID LANDFORMS constructional and erosional, created by interactions	Cones created by lava-water inter actions	Rootless cones on lava flows entering a lake or shallow sea	All, in particular hot spots and intraplate	Mafic composition. Lava ejecta, breccia	A < 0.001 km^2 V < 0.0001 km^3 Low Ar	Tiny cones. H 0.01–0.02 km. D 0.01–0.05 km	Instantaneous Quenching	Lake Myvatn, Iceland
	Intra- or sub-glacial edifices from ice ad magma interactions	Tuyas or Table mountains. Tindars (ridges)	All, in particular arcs and subaerial spreading ridges	Hydroclastic deposits, breccia, hyaloclastites, lava cap	A 0.8–1 km^2 V 0.6–1 km^3 Ar 0.7–0.8 A 5 km^2 V 2.5km^3	Table mountain: H 0.7 km D or W 1 km Ridge: H 0.5 km L 1–5 km	Months to year. Intraglacial explosive activity Inverted relief	Herðubreið, Hlöðufell; Klukkutindar, Iceland
		Lava deltas in sea (constructional)	Arc volcanoes, rifts	Breccia, lava, pillow lava; diamicts	A 0.15–0.45 km^2 V 0.04–0.23 km^3	H 0.3–0.5 km L 1.5 km W 0.1–0.3 km	Months to years Shallow water	James Ross Island Antarctica
POLYGENETIC SHIELD AND	Large and medium-sized oceanic shield volcanoes	Hawaïan type, Earth's largest construct. Summit caldera, rift zones, horseshoe scar	Oceanic hot spots	Basalt, picrite, hawaiite lava < 1% pyroclastic deposits	A 2,000–10,000 km^2 V 6,000 max 95,000 km^3 Ar 1–4/1000	Huge shield. H 9 km, D 25 –100 km. H/W ≤ 0.04 Shallow caldera D 3–4 km. W_{base}/W_{cald} > 10. α ≤ 7°	10^5-10^6 years High magma output. Caldera magma drainage.	Mauna Loa, Kilauea, Hawai'i. Piton de la Fournaise
		Galápagos type, medium shield with steep summit and deep caldera, horseshoe scar	Oceanic hot spots Active western group, older eastern group	Basalt, picrite lava < 5% pyroclastic deposits	A 300–1,300 km^2 V 10– 250 km^3 (subaerial) Ar 1–4/1000	Bowl shape, steep summit H 1.3 km, D 20–40 km. D_{cald} 3.5–9 km. W_{base}/W_{cald} 2.3–4. α 3-9 (max 25°)	10^5-10^6 years Repeated caldera collapse, ring dykes & fissures	Fernandina, Galápagos; Gran Canaria
	Small oceanic and continental shields	Icelandic type, low, small lava shield, with crater	Oceanic hot spots and thin continental crust	Basalt, picrite lava	A 80–500 km^2 V 1–20 km^3 Ar 1– 6/1000	Low and broad. H 0.1–0.6 km, D 10–20 km. Vent D_{cr} 0.1–2 km. α 1-6°	≥10^5 years Lava effusion	Skjalbreidur, Iceland; Erta Ale, Ethiopia
	Medium-sized continental shields	Turkana subtype Newberry subtype ring fissures, summit caldera, horseshoe scar	Intraplate, continental rifts Back-arc, subduction zones	Alkali basalt, basanite; trachyte, phonolite Calc-alkaline magma, andesite	A 20–300 km^2 V 20–100 km^3 (Austr 3,000 km^3) A 300–700 km^2 V 200–1,000 km^3	H < 1 km D 7–20 km. α 5–9°. Summit caldera 3-8 km H 1, D 20–30 km. α 4–6° $Area_{shield}/A_{lava\ complex}$ 36–62	≥10^5 years Caldera, ring fissures	Silali, Suswa, EARS (Kenya). Newberry, Medicine Lake
	Other shield-like volcanoes	Ignimbrite shield-like edifices with summit caldera and lava dome(s)	Intraplate Older hot spots	Bimodal (basalt-rhyolite), undersaturated magma Voluminous ignimbrite	A 1,500–5,000 km^2 V 150–300 km^3 Ar 1/1000	Large edifices H 1–1.5 km D 70–100 km. D_{cr} 7–15 km. α 2–5° flanks, 10° upper complex	≥10^6 years Uplifted, domed basement	Voon, Emi Koussi, Tibesti

Table 2 (continued)

LANDFORMS	COMPOSITE EDIFICES	Composite cones, and constructs from simple to compound, and clusters	Simple cones or stratocones: regular, irregular	Continental arcs, Island arcs, Continental rifts	Basaltic andesite to dacite Lava, pyroclastics	A 40–300 km^2 V 50–200 km^3 Larger in rifts Ar 0.01–0.02	High conical edifice. H 2 km, D 10–20 km. Dcr 0.3–1 km. Caldera 1–6 km. H/W 0.15–0.33. α 5–30°	10^4-10^5 years Simple cones	Licancabur, Lascar, Chile Mt. Adams, Cascades
			Truncated cones with summit caldera or horsehoe scar	Arc volcanoes Subduction zones	Andesite, dacite Trachy-andesite, basalt	A 80–200 km^2 V 5–150 km^3	High, broad cone with collapse scar. H 1.5 km D 10–20 km. α 5–30°	3-5x10^4 years Collapse scar	Mt. St. Helens Tambora
			Somma cones	Arc volcanoes	Phonolite, basalt, andesite. Lava, pyroclastics	A 80–200 km^2 V 10–400 km^3	Broad cone with summit caldera. H 1 km D 10–15 km. α 5–33°	0.35-23x10^5 years Explosive calderas	Vesuvius Hakone
			Compound cones	Continental arcs, and hybrid settings	Basaltic andesite, andesite, dacite, trachyte, phonolite Lava, pyroclastics	A 300–500 km^2 V 20–1,000 km^3	Two or more edifices. H 2 km, D 20–25 km. α 5–33°	10^5 – 10^6 years Massifs. Cones, lava fields, domes	Etna, Sicily; Ampato-Sabancaya, Peru; Ruapehu, NZ
			Large, long-lived volcanic clusters or volcano complexes	Continental margins with thick crust	Intermediate and silicic Domes, cones, lava fields, pyroclastics	A 100–600 km^2 V 200–2,000 km^3	Vast clusters or massifs. H 2– 3 km D 20–50 km. α 5–33°	>10^6 (max. 10^7) years. Collection of edifices	Chachani, Peru Aucanquilcha, Chile
		Calderas Excavational landforms truncating summit edifices	Mafic, collapse calderas on shields	Oceanic and terrestrial shields	Basalt; Lava, rare pyroclastics (Hydrothermal caldera formation, La Réunion)	A 13–50 km^2 V Collapse, lava, ejecta 10–40 km^3	Ellipitic shape. D_{cal} 4–8 km (max. 11). Depth 0.3–0.8 km. α 1–10°	Rapid, repeated collapse (magma drainage) and lava infilling	Halemaumau, Kilauaea Masaya, Nicaragua
			Explosive calderas (somma) on composite cones	Volcanic arcs Island arcs, subduction zones	Intermediate to silicic Pyroclastic deposits	A 1–50 km^2 V_{cal} <15 km^3 V PDCs and fallout 5–50 km^3	Small, subcircular. D_{cal} < 2 km, max 8 km, Depth ≥ 1 km. α 20–35°	Hours to days Catastrophic collapses, large ignimbrites	Vesuvius, Italy Ubinas, Peru Crater Lake, Oregon
	CALDERAS	Calderas on shield-like landforms	Explosive calderas on shield-like complex cones	Rifting Continental rifts	Peralkaline composition	A 80–120 km^2 Vcal 10–20 km^3 Vejecta 5–10 km^3	Relatively large, shallow. D 12 km H 0.3 km	Repeated collapses and lava infilling	Pantelleria, Italy Kone, Ethiopia
			Explosive calderas on low 'ignimbrite-shield-like' edifices	Volcanic arcs	Silicic composition Dacite to rhyolite	A 1500–2000 km^2 V 200–1,000 km^3 V Ignimbrites 100–600 km^3	Extensive, low. H 0.3–1.3 km D 45–50 km. Area $_{landform}$/$Area_{collapse}$ 3–4. α 2 –10°	Ignimbrite, day(s) but recurrent (flare ups) over several Myr	Panizos, Altiplano-Puna Volcanic Complex
		Caldera volcanoes	Resurgent ash-flow calderas, with central and peripheral domes.	Intraplate Uplifted or domed basement	Dacite to rhyolite Trachyte, trachyandesite Inflow and outflow ignimbrites.	A 200–2,500 km^2 V 100–1,000 km^3	Vast, nested depressions with central dome(s). D_{cal} 15–80 km. H_{rim} 0.5–2 km.	Catastrophic, incremental 10^4 - 10^6 years. Resurgence modes and	Valles, USA; Cerro Galán, Argentina;

Table 2 (continued)

	Excavational and constructional landforms	Nested caldera complexes and rhyolitic centres	Hot spot and extensional settings	Resurgent lava dome, block or graben	V Ignimbrites 100–1,000 km^3 (max 6,000 km^3)	D_{basal} 100–300 km. α 1–15°	duration (15-80 ka after collapse)	Toba, Sumatra; Yellowstone; Campi Flegrei, San Juan, Col.
		Volcano-tectonic depressions, 'inverse volcanoes'	Arc volcanoes Subduction zones Extensional	Dacite to rhyolite	A 2,000–35,000 km^2 V 100–1,000 km^3	Vast collection of calderas and edifices H 1–2 km D 50–200 km D_{cal} 50–80 km	Catastrophic and incremental	La Primavera, Mexico. Taupo Volcanic Zone, NZ
LARGE-SCALE PLATEAUS Earth's largest landforms Constructional and erosional	Large igneous provinces LIPs (Constructional)	Oceanic and continental flood basalt plateaus (CFBs)	Intraplate, plume heads, hot spots. Back-arc, plate margins	Basalt lava flows fed by long fissures and dykes. La Roza Member flow 40,000 km^2 V 2000 km^3	A 160,000 2-3 (max 7) x 10^6 km^2 V 1–44 x 10^6 km^3	Extensive, thick basaltic plateaus. H 1–3.5 km.W, L 350–750 km. Very high effusion rate 0.1-0.8km^3/yr	1 max 50 x 10^6 years. Pulses 1-5 Myr. Very high effusion rate	Columbia River Flood Basalt province; Otong Java
		Silicic Large Igneous Provinces (SLIP)	Continental rifts, compressional-extensional	Silicic to felsic. Bimodal: basalt and rhyolite	A 350,000 km^2 V 200,000 km^3	Extensive and thick (1–3 km) plateau of ignimbrite sheets	1 - 2 x 10^6 years High output 2000 -8000 km^3 DRE	Sierra Madre Occidental; Paraná-Etendeka
	Traps with structural scarps	Eroded continental flood basalt plateaus	LIPs	Basalt. Lava flows and interbedded breccia, paleosols and sediments	A 500,000 km^2 V 1,000,000 km^3	Succession of thick lava flows H 2–3.5 W 100–300 km	Four periods 66.3 -65.5 Ma. Short pulse 0.25-1 Myr	Deccan, India; CRFB USA; NE Australia
	Ignimbrite plateau with scarps	Ignimbrite or ash-flow tuff sheets	Continental arcs, collision to extensional settings	High-K dacite-rhyolite. Interbedded tephra fall deposits	A 1,000–10,600 km^2 V25–180 km^3 each ignimbrite	Succession of thick (10–110 m) welded and unwelded ignimbrites	At least 10 sheets 9-2.5 Ma. Short explosive pulses.	Capadoccia, Turkey
LAND-FORMS from HYDROT-HERMAL ACTIVITY	Geyser Hot spring	Hot springs with intermittent water eruption and steam	Caldera floor Hot spots	No rock. Sinter deposit (dissolved silica). Silica terraces. CO_2 travertine	A ≤ 0.001 km^2 V ≤ 0.002 km^3	Periodic jets 1.5–5 min., 14,000 to 32,000 l boiling water to height of 30–50 m	Neutral pH, chloride Deep seated	Old Faithful, USA; Lake Rotomahana, NZ
	Hot pools, bubbling mud pools	Fumaroles, hot turbid pools, explosion craters	Caldera floor Hot spots	Solfataras, sulphur mounds. Pits, small phreatic events	A ≤ 0.01 km^2 Minor V	Geothermal heated groundwater (30–100°)	Acid sulphate waters	Mammoth Springs, WY. Waiotapu, NZ
EXPOSED ROOTS, SHALLOW STRUCT-URES, AND INVERTED RELIEF	Cauldron: eroded caldera	Caldera roots and substructure; ring structures	Continental arcs Intraplate Pre-Tertiary rocks	Rhyolite, ignimbrite. Sinking of roof rocks into a large underlying magma reservoir	A 500–1,200 km^2 V 100-1250 km^3	Oval-shaped ring dykes D 20–50 (max 100) km H 1–2 km	Subsidence. Ring dyke, cone sheet. Carboniferous Devonian	Ardnamurchan, Mull, Scotland; Sabaloka, Soudan
	Eroded plugs: Neck; Culot	Eroded chimneys Eroded lava	All settings, in particular intraplate	Breccia (basalt) Lava, phonolite	A 0.001 km^2 V 0.008 km^3 A 0.045 km^2 V 0.012 km^3	H 82 m D 35 m H 264 m D 240 m	Removal of the central conduit	Aiguilhe, Le Puy, Auvergne; Devil's Tower, Wyoming
LAND-FORMS	Mesa Planeze	Eroded lava cap Thick, triangular lava flow remnants	In particular intra-plate; monogenetic fields	Lava, basalt, trachy-basalt, trachyandesite	A 0.5–1 km^2 V 0.05–0.25 km^3 A 0.5–10 km^2 V 0.1–5 km^3	L < 1 km H 0.1–0.5 km W ≤ 0.5 km. α < 2° L 1–5 km W 0.5–2 km H 0.2–0.5 km. α 2–5°	Removal of sides of the lava cap above substrate.	Cantal volcano, Auvergne, France Hopi buttes, AR

plumbing system, gas release, volatile content, which depends on magma composition, structural conditions, eruptive processes acting on landform development, exogenous agents, and human activities in certain circumstances.

It is not surprising that a wide range of volcanic features emerge from such a definition, based on characteristics listed in Tables 1 and 2. Furthermore, all these components can change in space and time. Eruptive processes include many that destroy and modify earlier constructs, adding to the variety of forms that fall under the name 'volcano'.

4 Principles of Existing Volcano Classifications

As stated by Siebert and Simkin (2002), various 'volcano shapes reflect the wide range of variations in lava chemistry, vent orientation, gas content and eruptive style'. As a result, the traditional classifications are linked to effusive, extrusive and explosive volcanic processes that build up volcanoes with various dimensions: 'volcanoes come in a variety of shapes and sizes' (Siebert et al. 2010). The British Geological Survey (https://www.bgs.ac.uk/discovering-geology/earth-hazards/volcanoes/how-volcanoes-form/) lists four types of volcanoes: stratovolcano (composite volcano), shield volcano, lava dome and caldera. Alternatively, the website of the University of Oregon (https://volcano.oregonstate.edu/types-volcanoes) asserts that six types can account for as much as 90% of all volcanoes: shield volcanoes, composite volcanoes, silicic caldera complexes, monogenetic fields, flood basalts, and mid-ocean ridges—this being a practical classification covering all settings on Earth.

First, let us consider some classical ways to categorise volcanoes:

1. Volcanoes can be either monogenetic or polygenetic, depending on their structure as well as their eruptive behaviour. The former are generally small, simple volcanic structures (up to a few hundred metres high), built up over a short period of time (one eruptive episode, usually a few days to a few years): for example, scoria cones and tuff rings. The short lifetime may be due to a decrease in magma supply (from relatively small magma chambers), plugging of the vent, or changes in tectonic setting. Polygenetic volcanoes are the opposite: they are typically complex, larger landforms of lava and pyroclastic material, formed over long periods of time, sometimes involving several cycles (in rare cases up to millions of years); they include most types of shield volcanoes and the majority of composite volcanoes. These are commonly fed by a larger magma chamber that remains for a long period of time, or forms over a short geological time before large eruptions. The actual volcanic eruption rate is determined by processes within the magmatic body (magma mixing/mingling and recharge, gas release, etc.) and related to the regional and local stress field (e.g. Walker 2000; Acocella 2021).
2. Based on the predominant eruption products that build up the given landform, there are (1) effusive volcanoes (producing mostly lava), including small landforms: lava domes, lava cones, lava mounds, and large landforms: shield volcanoes and simple composite volcanoes made up mostly of lava, and (2) pyroclastic volcanoes (typically small pyroclastic cones, but also ignimbrite shield-like volcanoes).
3. According to the size and external steepness of the volcano (Fig. 2), a graphic classification of the well-known types of lava dome, scoria cone, composite volcano, shield volcano, lava field, maar and caldera, was proposed and correlated with the magma composition from mafic to silicic end members (Davidson and de Silva 2000). However, this qualitative classification does not allow more precise distinctions to be made.

   Volcanoes can comprise large (area range of 1–1000 km^2) landforms and small (areal range of 0.01–1 km^2) landforms. The former include (1) volcanic edifices, volcanic mountains as constructs composed of one or more volcanic centres, (2) LIPs: large igneous provinces, flood basalt plateaus or basalt traps resulting from effusive activity, and (3) ignimbrite sheets forming plateaus that result from large explosive caldera-forming eruptions. These large forms can also encompass a variety of small landforms, e.g. those related to lava flows. However, the division into small and large landforms is not clear-cut. Small landforms can also be relatively large, e.g. the largest spatter cones (0.01 km^3) can reach the size of smaller scoria cones (Dóniz-Paez et al. 2012). A composite volcano can be a relatively modest landform, similar to a large scoria cone, or transitional between polygenetic landforms such as composite volcanoes and long-lived volcanic clusters. Size should also be considered in terms of overlapping volcanic fields such as caldera complexes and silicic multi-vent complexes. Volcanic mountains are not created in one go as large landforms, but instead grow gradually. As a result, these categories are only helpful for descriptive purposes.

4. One important geomorphic feature is highlighted by the grouping based on the general slope profile of the volcanic types (Fig. 2; San Diego State University, Department of Geological Sciences). According to this, scoria cones, which are the most common pyroclastic volcano type, are characterised by a straight slope, shield volcanoes by a convex slope, and composite volcanoes by a concave slope. In fact, models have worked with slope and profile of volcanoes, for example taking the remarkably symmetric geometrical form of stratovolcanoes (e.g. Fuji; Lacey et al. 1981) as a model of uniform shape for all volcanoes.
5. Simple and complex volcanoes can be distinguished based on the complexity of their shape. Simkin and Siebert (1994) apply these categories mainly to composite volcanoes and but also partly to pyroclastic cones (Fig. 3): simple and complex (e.g. twin) volcanoes, and simple scoria cones and complex landforms located along a fissure. Here, simple volcanoes form over a relatively short period of time, typically as a result of the activity of one vent. However, the classification from this point of view obviously overlaps with the classification according to internal structure (point 1) and size (point 3). It is not easy to tell which volcanoes are simple or complex. For example, effusive volcanoes may be simple, modest shield-like volcanic cones (e.g. Rangitoto, New Zealand); complex shield volcanoes of mature stage, with varied activity (e.g. Mauna Kea, Hawaii); or extensive lava fields with scoria cones (elongated: e.g. Erta Ale Range, Ethiopia; or lava fields extending in all directions: e.g. Columbia River basalt plateau, USA). Similarly, the term complexity is far from being defined.
6. Depending on the surface location and setting or medium of the eruption, volcanoes can be described as subaerial, subaquaeous (submarine and lacustrine) or subglacial.
7. Several authors have attempted to apply statistical methods to classify volcanoes. Here we present two of them.

Bishop (2009) presented a generic three-tiered classification using a dataset of the Newer Volcanic Province in southeast Australia, as a basic description of the complexity of the landform regardless of scale, tectonic setting, climatic regime, state of degradation, or local definition. The description of volcanic morphology using a generic threefold taxonomy consisting of simple, compound, and complex, is a benchmark from which attributes, including locality-specific traits, can be more easily incorporated, interpreted and understood. The nominal classification of landforms using this taxonomy is particularly useful within a GIS investigation as it offers a basic description of the complexity of landforms and the analysis of point pattern fields using spatial statistical techniques. Through the inclusion of attribute data, the taxonomic scheme improves the description of events and processes. Importantly, when coupled with a large spatial coverage, uniform scale and appreciable resolution of modern satellite images within a GIS environment, comparative and systematic spatial studies of global volcanic systems can readily be employed. Although mainly used to characterise monogenetic volcanoes, the scheme can also be used for other volcano systems. Furthermore, the classification provides insight into landscapes for which field data are missing and where access is difficult impractical, including extra-terrestrial surfaces.

Hone et al. (2007) explored the application of cladistics as a way to classify volcanoes. Cladistics, also known as parsimony analysis, is a technique widely applied in biology to classify organisms and identify evolutionary trends, reducing bias and subjectivity, and taking into account all the information provided in descriptive databases. Here, a dataset of 59 Quaternary volcanoes and 129 volcanic edifices from the Tohoku region of northeast Japan was obtained for volcano edifice characteristics such as volcano size, chemical composition, dominant eruptive products, volcano morphology, dominant landforms, volcano age, and eruptive history. The utility of cladistics lies in identifying groupings of volcanoes with similar characteristics: analysis of the study volcanoes both with and without chronology yielded three groups based on composition, eruptive products, and geomorphic characters. Volcano clusters had already been identified in northeast Japan (Tamura et al. 2002), and the cladistic groupings were unevenly distributed between the clusters. This result indicates that each cluster has a distinctive but coherent style of volcanism and, for groupings with time characteristics, distinctive evolution as magmatic systems. Cladistic analysis could be a useful tool for elucidating dynamic igneous processes that could be applied to other regions and globally.

Nevertheless, none of these conceptually attractive classification schemes, which have been adapted to real spatio-temporal relationships, can be taken as a basis for the objectives of our book. There are three reasons for this: (1) are not simple or staightforward, (2) most are based on descriptive features rather than quantitative characteristics of volcanic structures/constructs, and (3) they often result in overlapping groups or sets of volcanic landforms.

To consider the features of individual volcanoes of Earth to be classified, the reader is referred to the Global Volcanism Program (https://volcano.si.edu/), run by the Smithsonian Institution, and to the Volcano Hazards Program managed by the U.S. Geological Survey. This complete database was launched in 2013 as a catalogue of Holocene and Pleistocene volcanoes and eruptions, and replaced the three previous editions of Volcanoes of the World.

5 Principles of Classification of Volcanic Landforms in Our Book

Classifications of volcanic landforms need to be improved to incorporate the complexity in volcanic landform generation controlled by magma composition, eruption style, erupted material, and the medium in which they are created (terrestrial, subaqueous and subglacial), all of which influence the morphology of volcanic constructs.

We have compiled a table of classifications using the hierarchy based on the key criteria referred to in the definition of a volcano (sections 1–3). Firstly, we divide elementary, monogenetic landforms (Table 1) and non-elementary or polygenetic landforms (Table 2). Secondly, we differentiate the elementary landforms based on the type of eruptive activity: effusive, extrusive and explosive. The second subheading displays the volcano and landform type, and the subtype or individual landform. The following subheadings are criteria such as the geodynamic/tectonic setting, the magma composition and deposit texture, then the size parameters (surface area, volume and aspect ratio, shape and size ratios). This is completed by the duration and prevailing factor of growth. We then go on to provide examples.

Table 2 is based on a similar organization to that using the previous criteria. We start with hybrid landforms created by interactions between lava and water, and ice and magma.

This is followed by the polygenetic landforms, which are generally larger and more complex: (1) shield volcanoes with three types, mostly oceanic (Hawaiian, Galápagos), a small continental (Icelandic), and two medium-sized continental (Turkana and Newberry subtypes) ones, followed by other shield-like edifices; (2) the composite constructs with four types (simple, truncated, somma, and compound cones), followed by the large, long-lived volcanic clusters or volcano complexes, and (3) a wide range of calderas, i.e. excavational landforms, contrasting with the previous constructional edifices, based on magma composition and collapse mechanism as well as on size: (i) mafic collapse caldera, (ii) explosive caldera on a composite cone; and (iii) on a shield like-complex cone; (iv) explosive caldera on a low 'ignimbrite shield-like' edifice; (v) large resurgent ash-flow caldera, and (vi) volcano-tectonic depression or inverse volcano.

Table 2 also includes Earth's largest volcanic landforms: large-scale plateaus that are both constructional and erosional: these are Large Igneous Provinces (LIPs), first the continental flood basalt plateaus and second the silicic Large Igneous Provinces (SLIPs), followed by traps and ignimbrite plateaus.

Finally, Table 2 displays a range of exposed landforms that result from deep, long-term erosion, such as cauldrons, eroded plugs, and 'inverted relief' landforms such as volcanic mesas and planèzes. Eventually, we added a few manifestations of hydrothermal activity in volcanic areas such as geysers, etc.

6 Objectives and Rationale of Volcano Geomorphology

Publishing a comprehensive book on volcano geomorphology can be justified for at least two reasons. First, several books have been published on geomorphology (e.g. Summerfield 1991; Anderson and Anderson 2010; Goudie and Viles 2010; Ritter et al. 2011; Huggett and Shuttleworth 2022), but books specifically on volcano geomorphology are scarce or outdated and out of print (Cotton 1944, 1952; Ollier 1969, 1988). Second, many books of volcanology have been published over the past decade, but they do not describe landforms and geomorphic processes in sufficient detail. To our knowledge, only six modern books on volcanology include even one or a few chapter(s) on volcanoes as landforms and/or landscapes (Francis 1993; Chester 1993; Scarth 2000; Francis and Oppenheimer 2004; Lockwood and Hazlett 2010, 2022; Cas et al. 2024).

In summary, our book on 'Volcano geomorphology: landforms, processes and hazards' aims to:

(1) review the most recent outcomes of new research topics in geomorphology and physical volcanology, e.g. an improved classification and understanding of volcanic landforms, both monogenetic and polygenetic, with respect to structural settings, magmatism, and climate-driven erosion;
(2) improve our knowledge of processes and rates of growth, decay and erosion of volcanic landforms by integrating the results of the expanding subject of volcanology both in the field and in the laboratory, and;
(3) consider how volcanic landforms, landscapes, and processes can be studied by reviewing a wide array of classical and modern methods. The book includes a specific chapter on field, laboratory, and numerical methods applied to volcanic geomorphology.

This book is designed for a broad academic audience, such as university lecturers and researchers, undergraduate and graduate students, as well as those involved in nature conservation and projects related to the land.

7 Content of This Book

The book is laid out in eleven chapters, including contributions and case studies.

We first set the stage by describing the distribution of volcanoes within five principal tectonic settings (Chap. 1).

Chapters 2 and 3 portray elementary volcanic landforms, based on two principal textural components, i.e. lava and pyroclasts (tephra).

Chapters 4 and 5 are devoted to polygenetic volcanic landforms comprising shield volcanoes and strato- or composite volcanoes.

Chapter 6 showcases another kind of polygenetic, complex landforms: caldera and associated landforms, a negative (excavational) type of volcano.

Chapters 7 and 8 describe the degradational processes and erosion taking place on volcanoes over the short and long term.

Chapter 9 focuses on the role of geomorphology in assessing volcanic hazards.

Chapter 10 describes five recent case studies related to volcanic hazards.

Finally, Chap. 11 presents several classical and modern methods and techniques used to study active volcanoes and processes.

References

Acocella V (2021) Volcano-Tectonic processes. Advances in volcanology, IAVCEI. Springer, Switzerland, 552 pp. ISBN 978-3-030-65967-7

Anderson RS, Anderson SP (2010) Geomorphology: the mechanics and chemistry of landscapes. Cambridge Univ Press. Xvi + 637 p. https://doi.org/10.1017/S0016756810000932

Bishop MA (2009) A generic classification for the morphological and spatial complexity of volcanic (and other) landforms. Geomorphology 111(1–2):104–109. https://doi.org/10.1016/j.geomorph.2008.10.020

Borgia A, Aubert M, Merle O, van Wyk de Vries B (2010) What is a volcano? pp 1–10. In: Cañon-Tapia E, Szakacs A (eds) (2010) What is a volcano? Geol Soc Amer Spec Paper 470, Boulder, 140 pp. ISBN 978-0-8137-2470-6

Cas R, Giordano G, Wright JV (2024) Volcanology: Processes, Deposits, Geology and Resources. Springer International Publishing, 2 vol., 1833 pp. ISBN: 978-3-319-66612-9

Cashman KV, Sparks RSJ (2013) How volcanoes work: a 25 year perspective. Geol Soc Amer Bull 125 (6–6):664–690. https://doi.org/10.1130/B30720.1

Cashman KV, Sparks RSJ, Blundy JD (2017) Vertically extensive and unstable magmatic systems: a unified view of igneous processes. Science 355(6331):eaag3055. https://doi.org/10.1126/science.aag3055

Castruccio A, Diez M, Gho R (2017) The influence of plumbing sytem structure on volcano dimensions and topography. J Geophys Res Solid Earth 122:8839–8859. https://doi.org/10.1002/2017JB014855

Chester DK (1993) Volcanoes and society. Edward Arnold (Seve noaks), London, 356 p. ISBN 0-340-51761-1

Cotton CA (1944) Volcanoes as landscape forms. Whitcombe and Tombs, London, p 416

Cotton CA (1952) Volcanoes as landscape forms. Wiley, New York, p 416

Daly RA (1911) The nature of volcanic action. In: Proceedings of the American Academy of Arts and Science, vol 47, no 3, pp 47–122. American Academy of Arts and Sciences. https://doi.org/10.2307/20022712

Davidson J, de Silva S (2000) Composite volcanoes. In: Sigurdsson H, Houghton B, McNutt SR, Rymer H, Stix J (eds) Encyclopedia of volcanoes. Academic Press, San Diego, pp 663–68. ISBN 978-0-12-385938-9

Davies AG (2007) Volcanism on Io, A comparison witrh Earth, vol 7. Cambridge University Press, 355 p. ISBN 978-0-521-85003-2

de Silva S, Lindsay JM (2015) Primary volcanic landforms. In: Sigurdsson H, Houghton B, McNutt SR, Rymer H, Stix J (eds) The encyclopedia of volcanoes, 2nd ed. Academic Press Elsevier, pp 273–297. ISBN 978-0-12-385938-9

Dóniz-Paez J, Romero-Ruiz C, Sánchez N (2012) Quantitative size classification of scoria cones: the case of Tenerife (Canary Islands, Spain). Phys Geogr 33(6):514–535. https://doi.org/10.2747/0272-3646.33.6.514

Escher BG (1933) On the formation of Calderas. Leidsche Geologische Mededeelingen, III, pp 183–219, 1929

Fisher RV, Heiken G, Hulen JB (1977) Volcanoes. Princeton University Press, Princeton, NJ, Crucibles of change. https://doi.org/10.1515/9780691238210

Fornaciai A, Favalli M, Karátson D, Tarquini S, Boschi E (2012) Morphometry of scoria cones, and their relation to geodynamic setting: a DEM-based analysis. J Volcanol Geoth Res 217–218:56–72. https://doi.org/10.1016/j.jvolgeores.2011.12.012

Francis PW (1993) Volcanoes. A Planetary perspective. Clarendon Press, Oxford University Press, 443 pp. ISBN 0-19-854452-9

Francis PW, Oppenheimer C (2004) Volcanoes. Oxford University Press, 536 p. ISBN 978-0-19-925469-9

Geikie A (1903) Textbook of Geology. London, Macmillan, 4th ed., 343 p.

Jackson JA, Bates RL (ed) (1997) Glossary of Geology. American Geological Institute, Alexandria, Virginia, 769 p. ISBN 0-922152-34-9

Goudie A, Viles H (2010) Landscapes and geomorphology: a very short introduction. Oxford University Press, 160 p. ISBN 978-0-19-956557-3

Graettinger AH (2018) Trends in maar crater size and shape using the global Maar Volcano Location and Shape (MaarVLS) database. J Volcanol Geoth Res 357:1–13. https://doi.org/10.1016/j.jvolgeores.2018.04.002

Gregg TK, Lopes RM, Fagents SA (eds) (2021) Planetary volcanism across the solar system. In: Series comparative planetology, vol 1. Elsevier, 354 p. ISBN 978-0-12-813988-2

Green J, Short NM (1971) Volcanic landforms and surface features. A photographic Atlas and Glossary. Springer, New York, 519 p. ISBN 978-3-642-65152-6

Grosse P, Euillades P, Euillades L, Wyk V, de Vries B (2014) A global database of composite volcano morphometry. Bull Volcanol 76(1):784. https://doi.org/10.1007/s00445-013-0784-4

Gudmundsson A (2020) Volcanotectonics: understanding the structure, deformation, and dynamics of volcanoes. Cambridge University Press, Cambridge. ISBN 978-11-0702495-3

Gudmundsson A, Drymoni K, Browning J, Acocella V, Amelung F, Bonali FL, Elshaafi A, Galindo I, Geyer A, Heap MJ, Karaoglu O, Kusumoto S, Marti J, Pinel V, Tibaldi A, Thordarson T, Walter TR (2022) Volcanotectonics: the tectonics and physics of volcanoes and their eruption mechanics. Bull Volcanol 84(8):72. https://doi.org/10.1007/s00445-022-01582-4

Head JW, Wilson L (2022) Planetary volcanology: progress, problems, and opportunities. Bull Volcanol 84(3):23. https://doi.org/10.1007/s00445-022-01527-x

Hone DWE, Mahony SH, Sparks RSJ, Martin KT (2007) Cladistic analysis applied to the classification of volcanoes. Bull Volcanol 70:203–220. https://doi.org/10.1007/s00445-007-0132-7

Huff WS, Owen LA, Soldati A (2022) Volcanic landforms. In: Treatise on geomorphology, 2d edn, Chap 2.13. Elsevier, pp 340–375. https://doi.org/10.1016/B978-0-12-818234-5.00098-5

Huggett R, Shuttleworth E (2022) Fundamentals of geomorphology, Routledge, 5th edn, London, 682 p. ISBN 9781032169637

Karátson D, Favalli M, Tarquini S, Fornaciai A, Wörner G (2010) The regular shape of stratovolcanoes: a DEM-based morphometrical approach. J Volcanol Geotherm Res 193(3–4):171–181. https://doi.org/10.1016/j.jvolgeores.2010.03.01

Lacey A, Ockendon J, Turcotte D (1981) On the geometrical form of volcanoes. Earth Planet Sci Lett 54(1):139–143. https://doi.org/10.1016/0012-821X(81)90074-1

Lockwood JP, Hazlett RW (2010) Volcanoes, global perspectives. Wiley-Blackwell, UK, 539 pp. ISBN 978-1-4051-6249-4

Lockwood JP, Hazlett RW, de la Cruz-Reyna S (2022). Volcanoes, global perspectives, 2nd edn, Wiley-Blackwell, Oxford, 480 pp. ISBN 978-1-1194-7885-0

Lopes RMC, Mitchell KL, Williams D, Mitri G (2010) Beyond Earth: how extra-terrestrial volcanism has changed our definition of a volcano, pp 11–30. In: Cañon-Tapia E, Szakacs A (eds) (2010) What is a volcano? Geol Soc Amer Spec Paper 470, Boulder, 140 pp. ISBN 978-0-8137-2470-6

MacDonald GA (1972) Volcanoes. Prentice-Hall, New Jersey, 509 pp. ISBN 0-13-942219-6

Mazzini A, Etiope G (2017) Mud volcanism: an updated review. Earth-Sci Rev 168:81–112. https://doi.org/10.1016/j.earscirev.2017.03.001

NASA Science—Space Place. https://spaceplace.nasa.gov/volcanoes2/en/

Ollier C (1969) Volcanoes. Australian National University Press. ISBN 978-0-70-810532-0

Ollier C (1988) Volcanoes. B Blackwell, 228 p. ISBN 978-0-631-15664-X

Ollivier M, Roques F, Casoli F, Encrenaz T, Selsis F (2009) Planetary systems. In: Planetary systems. Astronomy and astrophysics library, Springer, Berlin, pp 1–20. https://doi.org/10.1007/978-3-540-75748-1_1

Oppenheimer C (2005) Volcanic degassing. In: Rudnick RL (ed) The crust, treatrise on geochemistry, vol 3. Elsevier, Amsterdam, pp 123–166 pp 123–166. ISBN 0-08-044847-X

Pike RJ, Chow GD (1981) Revised classification of terrestrial volcanoes and catalog of topographic dimensions, with new results on edifice volume. US Geol Survey Open-File Report 81-1038, 40 p

Ritter DF, Craig Kochel R, Miller JR (2011) Process geomorphology. Waveland Press Inc., 652 p. ISBN 13: 978-1-57766-669-1

Rittmann A (1936) Vulkane und Ihre Tätigkeit. Ferdinand Enke Verlag, Stuttgart, 188 p. Translated into English (by Vincent EA) in 1962: volcanoes and their activity. Interscience Publishers, 305 p. ISBN-13: 978-0471724506

Scarth A (2000) Volcanoes. UCL Press, London. ISBN 1-85728-223-X

Schmincke H-U (2004) Volcanism. Springer, Berlin, 324 pp. ISBN 3-540-43650-2

Siebert L, Simkin T (2002) Volcanoes of the world. An illustrated catalogue of Holocene volcanoes and their eruptions. Smithsonian Institution, Global Volcanism Prog. Digital Inf Serv GVP-3 http://www.volcano.si.edu.world/

Siebert L, Simkin T, Kimberley P (2010) Volcanoes of the world, 3rd edn. Smithsonian Institution, University of California Press. ISBN 978-0-520-26877-7

Siebert L, Cottrell E, Venzke E, Andrews B (2015) Earth's volcanoes and their eruptions: an overview. In: Sigurdsson H, Houghton B, McNutt SR, Rymer H, Stix J (eds) The Encyclopedia of Volcanoes, Chap 12, 2nd edn, Academic Press Elsevier, pp 239–255. ISBN 978-0-12-385938-9

Sigurdsson H, Houghton BF, McNutt SR, Rymer H, Stix J (eds) (2000) The encyclopedia of volcanoes, 1st edn, Academic Press, San Diego, 1417 p. ISBN 978-0-08-054798-5

Sigurdsson H, Houghton BF, McNutt SR, Rymer H, Stix J (eds) (2015) The encyclopedia of volcanoes, 2d edn, Academic Press Elsevier, Amsterdam, 1421 p. ISBN 978-0-12-385938-9

Simkin T, Siebert L (1994) Volcanoes of the World, 349 pp. Geoscience Press, Tucson, Smithsonian Institution, 349 p

Sparks RSJ, Cashman KV (2017) Dynamic magma systems: Implications for forecasting volcanic activity. Elements 13(1):35–40. https://doi.org/10.2113/gselements.13.1.35

Summerfield MA (1991) Geomorphology and global tectonics. Wiley, 367 p. ISBN 0-471-97193-6

Szakács A (2010) From a definition of volcano to conceptual volcanology, pp. 67–76, In: Cañon-Tapia E, Szakacs A (eds) (2010) What is a volcano? Geol Soc Amer Spec Paper 470, Boulder, 140 pp. ISBN 97-80-8137-2470-6

Tamura Y, Tatsumi Y, Zhao D, Kido Y, Shukuno H (2002) Hot fingers in the mantle wedge: new insights into magma genesis in subduction zones. Earth Planet Sci Lett 197: 105–106. https://doi.org/10.1016/S0012-821X(02)00465-X

Thouret J-C (1999) Volcanic Geomorphology. Earth-Sci Rev 47: 95–132. S0012-8252_99.00014-8

Thouret J-C (2004) Geomorphic processes and hazards on volcanic mountains, Chapter 11, pp 242–273, In: Owens PN, Slaymaker O (eds) Mountain Geomorphology, 313 p. Arnold, London. ISBN 0-340-76417-1

Tyrrel GW (1931) Volcanoes. Thornton Butterworth Ltd, 252 p.

Van Loon AJ (2010) Sedimentary volcanoes: Overview and implications for the definition of a volcano on Earth, pp 31–42, In: Cañon-Tapia E, Szakacs A (eds) (2010) What is a volcano? Geol Soc Amer Spec Paper 470, Boulder, 140 pp. ISBN 978-0-8137-2470-6

Walker GPL (2000) Basaltic volcanoes and volcanic systems, pp 283–289. In: Sigurdsson H, Houghton BF, McNutt SR, Rymer H, Stix J (eds) The encyclopedia of volcanoes, 1st edn, Academic Press, San Diego, 1417 p. ISBN 978-0-08-054798-5

Williams H, McBirney AR (1979) Volcanology. Freeman, Cooper and Co., San Francisco, 397 pp. ISBN 0-87735-321-2

Wilson L (2009) Volcanism in the solar system. Nature Geosci 2(6):389–397. https://doi.org/10.1038/ngeo529

Wilson L, Head JW, Parfitt EA (1992) The relationship between the height of a volcano and the depth to its magma source zone: a critical reexamination. Geophys Res Lett 19(13):1395–1398. https://doi.org/10.1029/92GL01073

Chapter 1 Spatial Distribution and Tectonic Setting of Volcanoes

Volcanism does not occur randomly on the Earth's surface. About 95% of the active volcanoes are located at or near plate boundaries (Figs. A and B in Appendix). Most active volcanoes develop in fold belts that coincide with earthquake zones. A large number of recent and extinct volcanoes form alignments or are located in narrow bands, along tectonically active zones of the Earth, such as mid-ocean ridges, continental margins (e.g. the western coast of the Americas), and island arcs (e.g. the Aleutians, Lesser Antilles; Fig. A in Appendix). Other aligned edifices are closely associated with fault systems along rift valleys (e.g. East African Rift, West Antarctic Rift).

There are exceptions to this general picture. Several volcanoes are located within plates, either in the ocean away from mid-ocean ridges, or deep within continents (Fig. B in Appendix). Most of these are 'hot-spot' volcanoes that develop where plumes rise up from the asthenosphere. This leads to magma generation, and hot magma bursts through the lithosphere to form large edifices both in oceans, such as the Hawaiian Islands, the Galápagos Islands, and La Réunion Island, and on continents, such as the Hoggar and Tibesti Mountains of the Sahara. Here we will consider active volcanoes (those which have erupted at least once over the past 11,000 years), recent edifices (with an eruption in the past 1 million years) and touch on volcanic fields or provinces that were emplaced millions of years ago.

1 Spatial Distribution of Volcanoes at the Earth's Surface

The spatial distribution of volcanoes is linked to different plate tectonic settings. The coincidence between the distribution of most volcanoes and the plate boundaries suggests that the processes governing plate motion also control the generation of the magma feeding the volcanoes. Magmatism and volcanism are processes resulting from heat generation in the core and mantle, and plate tectonics. The outer portion of our planet consists of seven major plates and several minor plates, which are in slow constant motion with regard to each other (Fig. 1A). A cross section of the 'layered Earth' together with a global digital elevation map (Fig. 1B) illustrates the tectonic and volcanic activity of the Earth over the last one million years, highlighting the types of plate boundaries and the distribution of volcanic fields. The rigid lithospheric plates of variable thickness (50 to $\leq$200 km) and width (tens to hundreds of kilometres) involve not only the Earth's crust but also the underlying upper mantle. The asthenosphere is a plastic, partially molten layer beneath the lithosphere in which mantle convection currents induce movements and interactions of the plates. The relative motion between neighbouring plates defines three types of boundaries (Acocella 2021): (1)

D. Karátson and J.-C. Thouret, *Volcano Geomorphology*, Advances in Volcanology,
https://doi.org/10.1007/978-3-031-77693-9_2

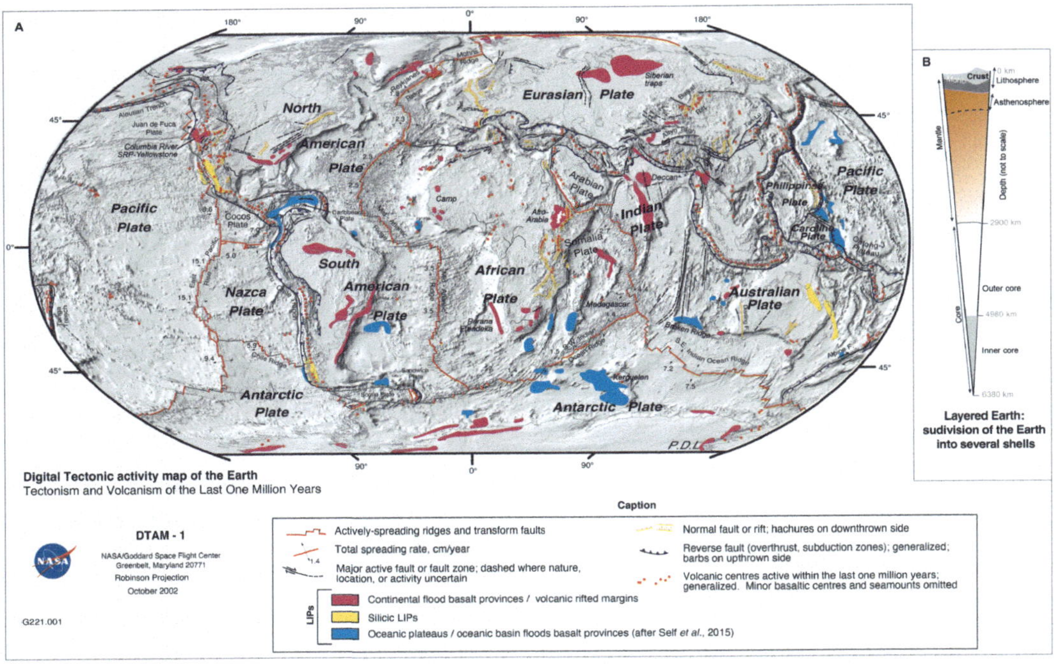

Fig. 1 A. Tectonic and volcanic activity of the Earth over the last one million years on a SRTM-Digital Elevation Model, highlighting the plate boundaries and the distribution of volcanoes (modified after Lowman et al. 1999; NASA 1999). Three different types of volcano and volcanic fields include: composite volcanoes (red triangles), monogenetic fields (yellow areas), and oceanic and continental shield volcanoes above hot spots (open circles). Large Igneous Provinces (LIPs, modified after Self et al. 2015) of different ages are shown, including continental flood basalt provinces, oceanic plateaus, and silicic LIPs. **B**. Cross-section on the right-hand side shows the layers within the Earth; note the position of the lithosphere and asthenosphere (after Acocella 2021, Fig. 1.2, p. 5)

divergent or constructive boundaries, where oceanic crust is formed (red lines, Fig. 2); (2) convergent or destructive margins, where crust is consumed and most active subaerial volcanoes form above the subduction zones (blue lines, Fig. 1); and (3) transform or strike-slip boundaries, where margins slide laterally past one another (yellow lines, Fig. 1). In addition, a number of volcanoes and extensive volcanic fields are located away from plate boundaries, in intraplate settings.

Table 1 provides an overview of the distribution of volcanoes and volcanic fields based on a number of criteria: tectonic setting and regime, abundance (volume and area), geometry and diversity of landforms, including the main magma composition, and finally the links with magmatic systems beneath the volcanic constructs (after Cas and Wright 1988; Thouret 1999, 2004).

2 Volcanoes and Volcanic Fields in Five Tectonic Settings

Figure 2 distinguishes four tectonic settings, ranked according to the annual magma production rate (Crisp 1984; Wohletz and Heiken 1992; Schmincke 2004; White et al. 2006; Papale et al. 2022; Galetto et al. 2023): (1) divergent plate boundaries including mid-ocean ridges; (2) continental rifts; (3) convergent plate boundaries or subduction zones near continents and island arcs, and; (4) intraplate setting, including hot spots and large igneous provinces. Following Schmincke (2004), we added a fifth, 'hybrid' setting for volcanoes in triple junction areas. Diagram Ba below the cross section of Fig. 2 shows that volcanoes at mid-ocean ridges (together with submarine volcanoes) and

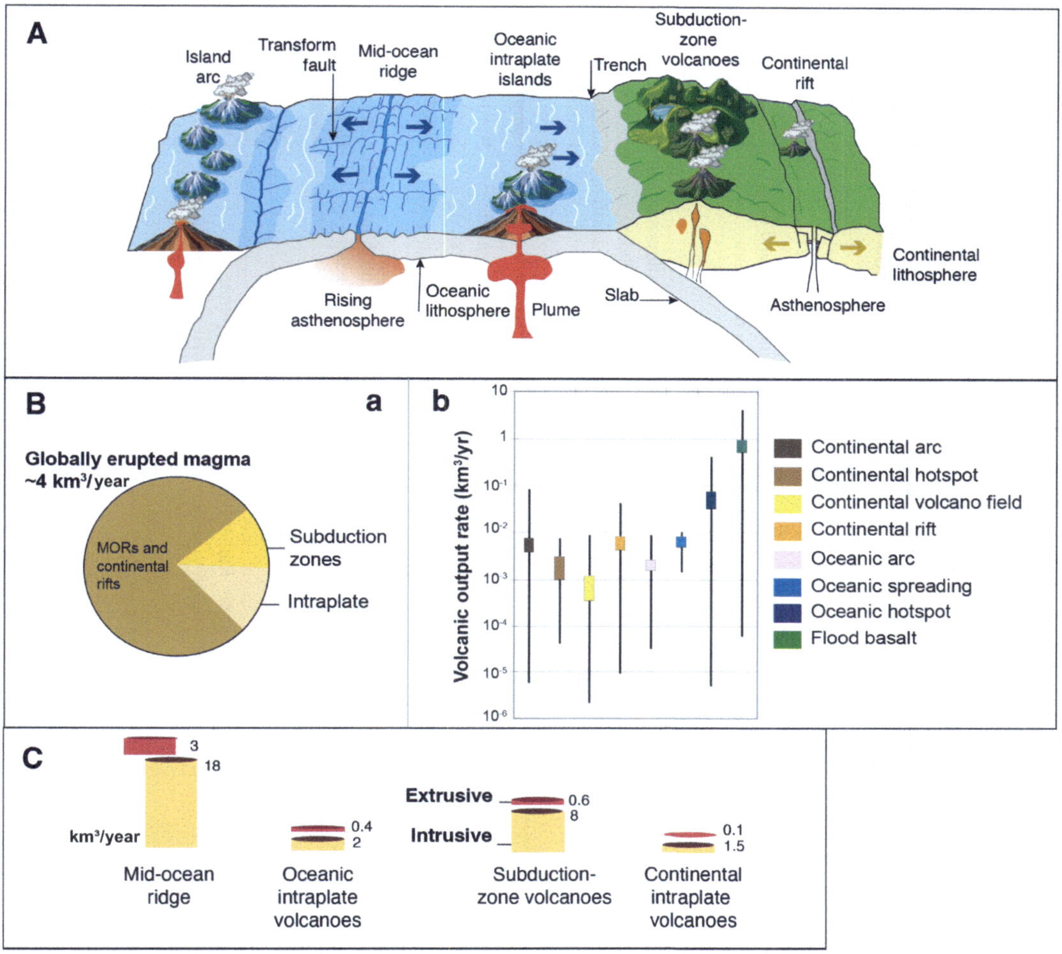

Fig. 2 A. Principal tectonic settings where volcanism and magmatism take place: divergent (mid-ocean ridges), convergent (subduction zones with continental and island arcs) plate margins. In addition, oceanic islands and continental rift volcanoes are shown in an intraplate setting (After Schmincke 2004, Fig. 2.2, p. 14). **Ba**. On the left-hand side, proportion of estimated mass of annual volcanic production in three tectonic settings (After Schmincke 2004, Fig. 2.5, p. 17. See also Siebert et al. 2010). **Bb**. Volcanic output rate (km^3/year) for eight categories of petro-tectonic settings (after White et al. 2006, Fig. 2, p. 7). Coloured boxes represent the range of one standard deviation from the mean rate. The black bars show the minimum and maximum rates for each setting. **C.** Annual magma production rate (km^3/yr) in different settings, distinguishing extrusive and intrusive emplacements (Wohletz and Heiken 1992; after Schmincke 2004, Fig. 2.6, p. 17)

those near subduction zones produce one order of magnitude more magma than their intraplate counterparts. The volumes of magma erupted at mid-ocean ridges represent 62% of the global Earth annual magma output, almost twice as much as that erupted at volcanic arcs (26%) and hot spots (12%) combined (Schmincke, 2004). Magma output rates (km^3/year) have been broken down into eight tectonic and petrological settings by White et al. (2006) (Diagram Bb, Fig. 3), revealing again that flood basalts (LIPs) produce the most magma, followed by oceanic hot spots, and then ocean-mid ridges, continental rifts and continental arcs at two orders of magnitude less magma. Regarding subaerial volcanism, Papale et al. (2022) constrained volume data using various records, observations and statistical models, and estimated the annual magma

Table 1 Distribution of volcanoes and volcanic fields related to tectonic settings, ranked in terms of abundance (volume and area) (after Cas and Wright 1988; Thouret 1999, 2004)

Tectonic setting and regime	Abundance (volume and area) on Earth	Volcanoes and volcanic fields	Geometry, diversity of landforms, magma type	Links with magmatic systems	Examples of volcanic field or province
1. Mid-ocean ridge and associated submarine volcanism. Extensional regime	Largest magmatic system on Earth. Largest number of seamounts	Central graben, rift zone, pillow and sheet flow volcanoes, seamounts	Linear and narrow. Rather uniform Tholeiitic basalts (Mid-ocean ridge basalts MORB)	Direct link with shallow magma chambers Dyke swarms	Mid-Atlantic Ridge, East Pacific Rise, Marianas Trough
2. Continental rift systems and passive margins Extensional regime	Abundant in volume	Rift volcanoes, shield volcanoes, composite cones, volcanic clusters	Narrow and linear or broad and long. Relative diversity. Bimodal compositions	Medium-depth and deep magma reservoirs	Eastern branch, East Africa Rift; Basin and Range, USA
2a. Intraplate setting within or near graben or rift Extensional regime	Less abundant in volume and area. Rarer in continental margins	Monogenic fields: scoria cones, domes, maars, tuff rings, lava flow fields	Elongate. Limited diversity of single volcanoes. Alkali basalts and other alkali rocks	Relatively shallow magma reservoirs and dyke swarms. Role of aquifer	Chichinautzín (Mexico), Chaîne des Puys (France), Eifel (Germany)
3. Subduction zone; Continental margin arc associated with trench Compressive regime	Abundant in area. Wide and voluminous	Composite cones, large volcanic clusters, calderas, complexes, multi-vent centres	Linear and wide, regular spacing. Diversity of landforms. Alkali series, Intermediate-silicic	Medium-depth and deep magma reservoirs	North, Central, South Andean Volcanic Zone, Cascades
3a. Islands arcs, Micro-continental arc; Marginal sea-back arc and basin behind island arc. Compressive regime	Abundant in volume but narrow areas	Composite cones, dome complexes, calderas	Linear and narrow. Diversity of landforms Ocean island basalts (OIBs) and high-silica magmas	Shallow and medium depth magma reservoirs	Lesser Antilles, Hellenic arc, Aleutians; Japan, New Zealand
4. Large igneous provinces (LIPs), Continental flood basalts (CFBs) Extensional regime	Very extensive and voluminous Submarine> >continental	Flood basalt plateaus and plains, traps, volcanic provinces	Large areas. Rather uniform Mostly basaltic, some silicic LIPs	Mantle plume Oceanic and continental hot spots	Deccan traps Columbia River Flood Basalts

(continued)

Table 1 (continued)

Tectonic setting and regime	Abundance (volume and area) on Earth	Volcanoes and volcanic fields	Geometry, diversity of land-forms, magma type	Links with magmatic systems	Examples of volcanic field or province
4a. Hot spot, Intraplate oceanic volcanism Extensional regime	Largest volume of single edifice type	Basaltic oceanic shield volcanoes, chain and sea-mounts	Large, but relatively uniform: shields. OIBs	Oceanic hot spots	Hawai'i, Mascarene Islands, Canary Islands
5. Hybrid settings: triple junction, slab edges Mixed regimes or stress fields	Relatively large, single edifices, but isolated examples	Large, compound volcanoes including adventitious scoria cones	Broad but isolated. Relative diversity Alkali basalts and series	Medium-depth and deep magma reservoirs	Etna (Sicily, Italy); Fuji (Japan)

output rate to be 4 km^3 over the last one million years, while Galetto et al. (2023) calculated the mass of magma erupted subaerially to range between 1.1 and 4.9×10^{13} kg over three decades (1980–2019). However, shorter time intervals in the order of 1000–10,000 years display significant fluctuations in output rate. If we consider active subaerial volcanoes, about 80% occur within convergent and 15% in divergent settings, while about 5% occur at intraplate margins (Fisher and Schmincke 1984), while Galetto et al. (2023) show that arcs erupted ~65% of the total mass over the three analysed decades. Figure 3C also shows that magmas intruded into the crust exceed those extruded in each of the tectonic settings by five to fifteen times. We describe the distribution of volcanoes and the structure of volcanic fields in five principal tectonic settings (Table 1).

2.1 Divergent Plate Boundaries: Geomorphology and Volcanism at Spreading Mid-Ocean Ridges

The mid-ocean ridges (MORs: Mid-Atlantic Ridge, East Pacific Rise, Juan de Fuca Ridge in the Pacific Ocean, and Carlsberg in the Indian Ocean) that extend over a combined length of 56,000 km in the Atlantic Ocean, Eastern Pacific and Indian Oceans are the volcanoes that produce the most magma, with two thirds of the Earth's annual volcanic output. MORs are by far the most frequent type of divergent plate boundary. They develop along the contact-line between plates that are moving away from each other. This motion creates lithospheric thinning at the contact-line, which in turn promotes decompression and magma generation. Variable rates of divergence (a few mm per year to ~15 cm per year; Püthe and Gerya 2014; Vithana et al. 2019) are proportional to the flux of magma, determining the structure, magmatic and volcanic features of the MOR (Acocella 2021). The Mid-Atlantic Ridge has the slowest spreading rate (2–4 cm/year), the Indian Ocean Ridge an intermediate rate (7 cm/year),

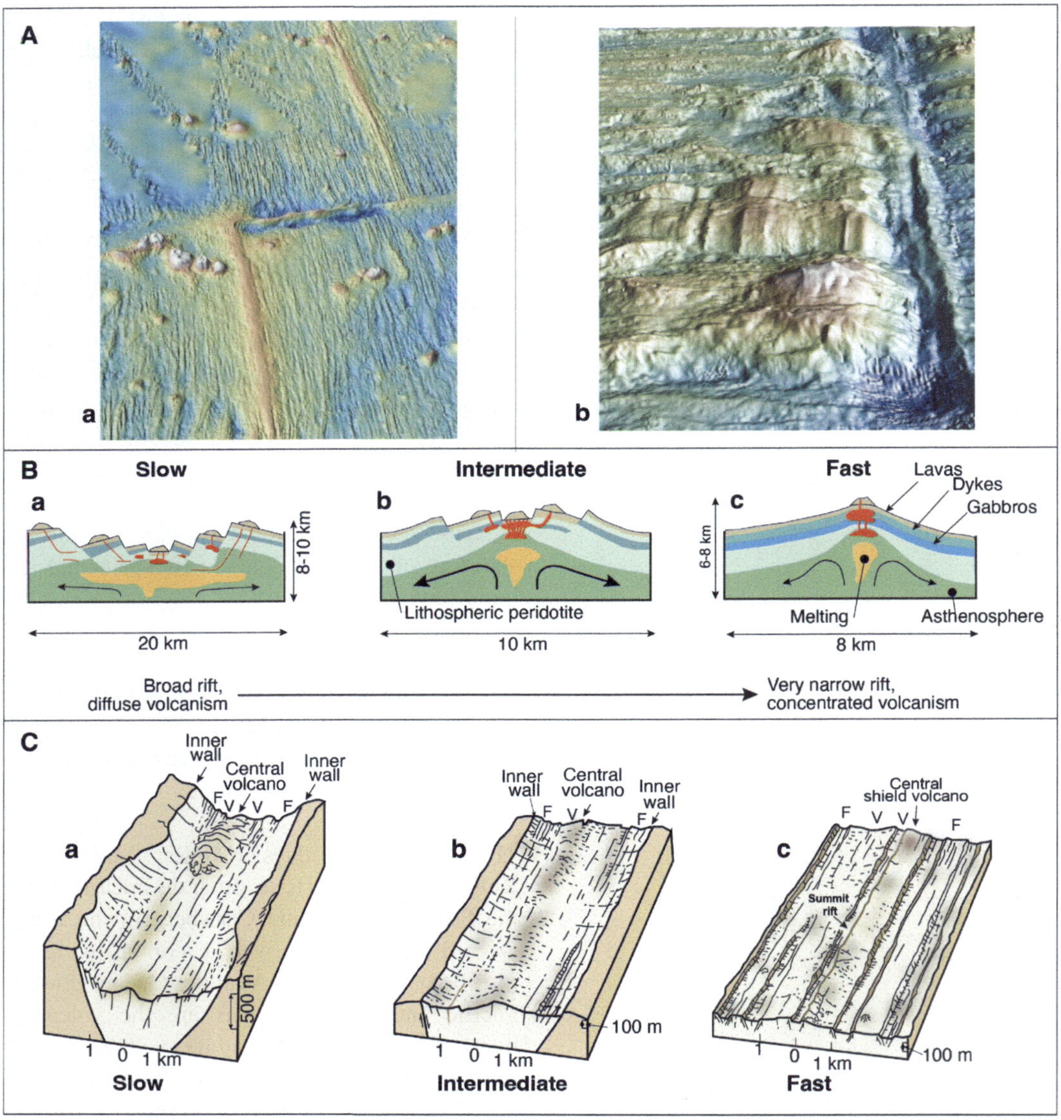

Fig. 3 Mid-Ocean Ridges. **A**. Images of Mid-Ocean Ridges (free for use from GeoMapApp). 1. Fast-spreading East Pacific Rise (https://www.geomapapp.org/gallery/index.html). Location: East Pacific Rise, Pacific Ocean. Latitude: 9° to 11° N, Longitude: 104° W. Image width: 250 km. Shipboard multibeam bathymetry data. This is a perspective view looking north along the East Pacific Rise spreading centre toward a prominent (dark blue = deeper) transform fault running eastward. The North American plate is to the right (east) of the rise, and the Pacific plate to the left (west). (The seafloor appears to be out of focus in areas where there is no high-resolution bathymetry data available). 2. Slow-spreading Mid-Atlantic Ridge (https://www.geomapapp.org/gallery/index.html). Location: Kane fracture zone, North Atlantic Ocean. Latitude: 23° N, Longitude: 45° W. Image width: 33 km. Shipboard multibeam bathymetry data. The image is looking in a west-north-westerly direction along the transform fault in the Kane fracture zone, where the African plate (upper right) and the North Atlantic plate (left) are being formed. This fracture zone is located in the central Atlantic Ocean, on the slow-spreading Mid-Atlantic ridge between Africa and the Caribbean Sea. (The checkerboard effect seen in part of the image is an artifact of the pattern made by the ships' tracks.) **B**. Cross sections of slow-, intermediate- and fast-spreading mid-ocean ridges based on spreading rate (modified after van Wyk de Vries 2018). Upper images illustrate the fast (11 cm/year) East Pacific Rise and the slow (3.5 cm/year) Mid-Atlantic ridge. **C**. Along-axis morphology of the central volcanoes rift showing their characteristics (modified after Macdonald 1982; Cas and Wright 1988). (a) Discontinuous volcano construction resulting from a succession of pillow lavas extruded at slow rates, (b) moderately continuous with en echelon offsets at intermediate eruption rates, and (c) almost continuous eruption creating a shield-like volcano with a summit rift at fast rates

and the East Pacific Rise the fastest (8–15 cm/year). In the Pacific, the highest rate (~15 cm/year) occurs along the length of the Pacific-Nazca plate boundary, from the Easter to the Juan Fernandez microplates. The largest number of active submarine volcanoes also occur along these oceanic axial zones. Magma generated along mid-ocean ridges and marginal basins is predominantly basaltic, including tholeiites, alkali basalts and basanites. Mid-Oceanic Ridge Basalts (MORBs) are typically tholeiitic basalts in the form of pillow lavas.

Mid-ocean spreading ridges, the primary sites of oceanic crust construction, form linear structures offset by transform faults (Fig. 3A; Soule 2015). A transform fault is a fault along a plate boundary where the motion is predominantly horizontal (Wilson 1965). Such strike-slip faults also form part of the plate boundary, along which lateral offset between segments of spreading ridges is accommodated. Axial volcanic ridges can be up to several hundreds of metres high, 1.5 km wide, and several to tens of kilometres long. Mafic magma extrusion generally occurs in a narrow 1–2-km-wide zone and ranges from linear fissures, through elongate shields, to discontinuous chains of central volcanoes constructed of pillow lavas and associated sheet or lobate lavas. Central volcanoes, or seamounts, also form outside the rift itself, particularly near large offsets in the ridge system and near topographic highs of the faster spreading ridges. Small magma pockets with slow eruption rates produce seamounts; small magma bodies with higher eruption rates produce hummocky fissure-fed flows.

The morphology of MORs, notably the spatial density and height of volcanic constructs, including axial topographic highs, abyssal hills, and seamount populations, is largely determined by the spreading rate (Fig. 3A; Soule 2015). The crest of the slow-spreading part of the Mid-Atlantic Ridge (MAR, Smith et al. 1997) is marked by an axial rift valley, a graben bounded by extensional faults (Fig. 3B). The rift valley floor is distinguished by networks of subparallel fractures from which volatiles, hot mineralized water and lava escape (Lockwood and Hazlett 2010). The basaltic lavas spread out slowly but continuously, so that few distinct volcanoes can form, and instead elongate broad structures less than a few hundred metres high are built up and deformed. Two types of submarine volcanoes have been found within the central rift zone of the slow-spreading MAR (Fisher and Schmincke 1984; Schmincke 2004): (1) steep-sided pillow volcanoes 50–200 m high and several tens to hundreds of metres wide, and (2) sheet flows representing a significant contribution, with at least 20–30% of all submarine lava extrusions, forming 3–8 m-thick sheets up to several km-long (see Figs. 5.10 and 5.11. *in*: Schmincke 2004). Submarine sheet lavas are thought to result from high eruption rates at the beginning of eruptive episodes, whereas pillow lavas form at lower rates at the end of these episodes. Where spreading is much faster (e.g. the East Pacific Rise), axial valley development is reduced or non-existent, so no substantial volcanic edifices have a chance to grow (Fig. 3A–C).

MORs are principally out of sight, at between one to five thousand metres below sea level. The only exceptions are islands where MORs emerge above sea level, as in the large island of Iceland and in the Afar region of Ethiopia, coinciding with areas of high magma production and varied volcanoes (Fig. A in Appendix). Ridges coincident with hot spots (e.g. Iceland in the North Atlantic, Tristan da Cunha in the South Atlantic), characterized by high magma output rates, are also emergent. Broadly elongate shield volcanoes cut by fissure systems parallel to the plate boundary are active, exemplified by Krafla and Askja in Iceland, and Erta Ale in the Afar. Iceland is one of the prime volcanic fields on Earth with twenty-two active volcanoes, showing edifices both along and off the mid-oceanic ridge. Iceland is renowned for having produced about 25% of the global volume of erupted lava flows since 1500 AD/CE (MacDonald 1972). Recent eruptions in 2014 (Holuhraun near Bárðarbunga beneath the Vatnajökull ice cap) and 2021–2022 (Fagradalsfjall, Reykjanes peninsula) exemplify fissure vents where prolonged effusive activity emplaced extensive lava-flow fields. To the

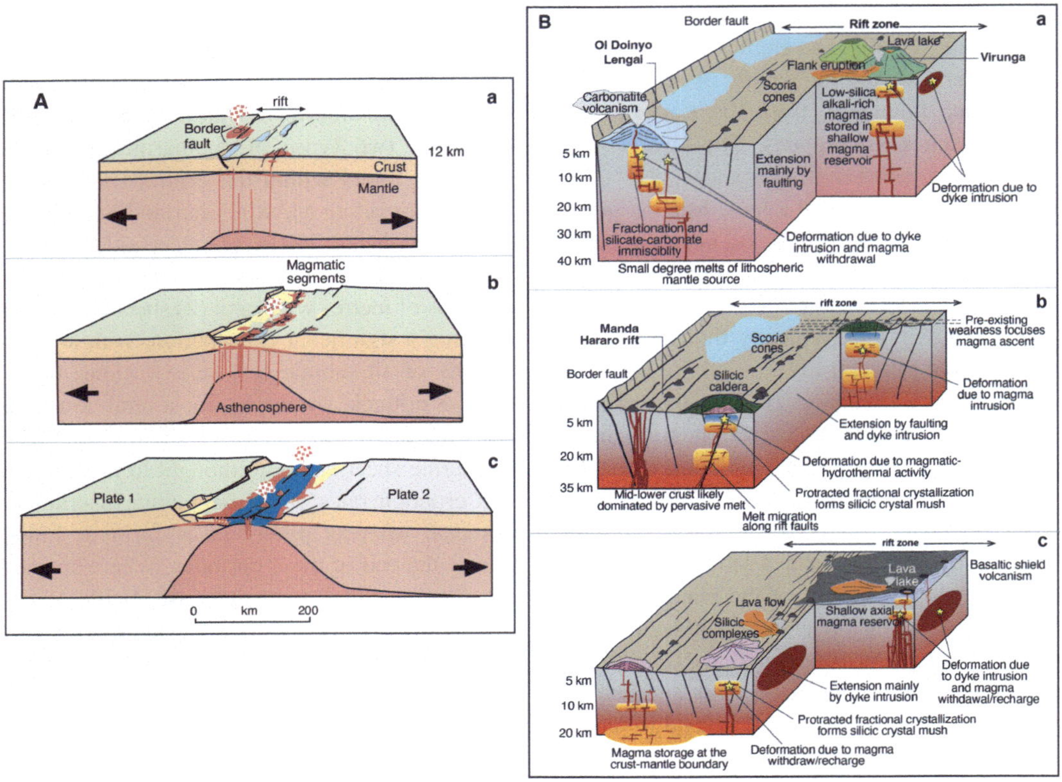

Fig. 4 Continental rift and magmatic systems linked to volcanoes. **A**. Three-stage model based on the Main Ethiopian Rift (MER) example for continental break-up leading to the formation of a volcanic passive margin (Ebinger 2005, Fig. 2.4, p. 2.19). (a): 0–5 Myr after onset of rifting at the surface; (b): 10–15 Myr after; (c): Seafloor spreading. **B**. Types of volcano constructs and relationships with magmatic systems at three different stages of rift evolution (Biggs et al. 2021, Nature Com., Fig. 3, p. 4). **a**. Early-stage rifting, based on Ol Doinyo Lengai and the Natron Rift in Tanzania (left), and the Virunga volcanoes on the western branch of the EARS (right). **b**. Mature continental rifting, based on the MER, with the bimodal association of silicic calderas above mid-lower-crust reservoirs and scoria cones located on fault systems. **c**. Nascent sea-floor spreading, based on Afar, Ethiopia, with the Manda Hararo Rift and silicic complexes on the left-hand side, and the Erta Ale shield and lava lake on the right-hand side.

south of Iceland along the Mid-Atlantic Ridge, a cluster of active volcanoes which form the Azores islands, and further south still are less studied volcanic islands near the MAR: Cape Verde, Ascension, St Helena, Tristan da Cunha, etc. (Fig. A in Appendix).

2.2 Volcanoes Along Continental Rift Valleys

Divergent plate boundaries are not restricted to mid-ocean ridges, as active volcanism, which is similar to that at MORs, is found within continents, related to extension of continental plates by mantle upwelling. These volcanoes have different and mixed magma compositions, and the continental rifts are characterized by high heat flow. Volcanoes develop far away from the plate boundaries within continents, but a closer inspection reveals that these edifices are often associated with tectonic features. For example, large, composite volcanoes occur in two tectonic regions of Africa: the East African Rift System and isolated massifs within and to the south of the Sahara (Fig. A in Appendix).

2.2.1 Geomorphology, Structure, and Magmatism of Continental Rifts

Continental rifts are prominent topographic features on the Earth, whose relief can exceed 2000 m. Fully developed rifts are 40–60 km wide and have subsided 2–3 km into uplifted plateaus. Volcanic edifices characterize both the rifts and their shoulders, the uplifted blocks extending for hundreds of kilometres on either side of many rifts. Rifts are narrow graben structures that range from a few hundred kilometres in length, such as the Rhine Rift (Germany), to broad rifted areas characterized by many horst and graben structures, such as the Basin and Range Province in the western USA. Other narrow grabens include the complex East African Rift System (EARS), the Baikal Rift (Russia), and the Rio Grande Rift (USA), which is part of the Basin and Range Province.

Continental rifts represent divergent plate boundaries or thinning of plate interiors in continental lithosphere (LaFemina 2015, his Fig. 3.11). Extensional stresses within the lithosphere are caused either by relative plate motions, or uplifting and doming of the lithosphere induced by a mantle plume. 'Passive' and 'active' rifts are traditionally distinguished on the basis of topography and genesis (Condie 1982; see Merle 2011 for a discussion on this terminology). Many of these grabens start as 'passive' rifts, with no initial asthenospheric convection; they produce no magma until stretching thins the lithosphere sufficiently for the asthenosphere to undergo decompression melting. 'Active' rifts, which are associated with asthenospheric convection, begin with large-scale volcanism and uplift followed by rifting. Merle (2011) simplified the early classification of rifts by placing them in a workable genetic tectonic scheme, in which he distinguished four types, termed subduction-, plume-, mountain-, and transform-related rifts. The rifts described here come under the plume- and transform-related types, rather than the subduction case, which is limited to early rifting processes.

The archetypal continental rift is the East African Rift System (EARS; inset map, Fig. 4A). The magmatic EARS runs along a divergent plate boundary between the Nubian and Somalian plates, extending nearly 5000 km inland, from Mozambique to the triple junction of Afar situated on a hot spot, where the EARS meets the Gulf of Aden and Red Sea oceanic rifts (inset map, Fig. 4A). The Main Ethiopian Rift (MER, Fig. 4A) and the adjacent Afar illustrate over a 1000 km-stretch how incipient continental rifting (southern MER) evolves through mature continental rifting (northern MER) and continental break-up (central Afar) into proto-oceanic rifting (northern Afar) (Medynski et al. 2016). Nowadays, the only example of a continental lithosphere that has completely rifted apart along faults is the Red Sea Rift that opened about 20 Ma ago and developed into a divergent oceanic plate boundary producing oceanic tholeiitic basalts. The degree of E–W extension increases northward from 0.2 to 0.7 cm/year. This corresponds to a northward decrease in crustal thickness, an increase in magmatism and volcanic activity, and shallowing earthquakes. In contrast, the ages of the erupted products indicate a southward migration of volcanic activity, from Ethiopia to Tanzania, suggesting that the most extended northern rift is also the oldest (Koptev et al. 2016; Acocella 2021).

EARS has two branches including a deep valley, several large lakes, and broad stratovolcanoes generally located on their shoulders (Fig. 4B). The western branch hosts Lakes Malawi, Tanganyika, and Albert, and Nyamuragira and Nyiragongo volcanoes (3470 m, Democratic Republic of the Congo) which are among the most active in the world. The permanent lava lake of Nyiragongo is infamous for lava outflows, which wreaked havoc in the city of Goma in 1977 and 2003. The eastern branch, also marked by smaller lakes (Natron, Turkana), has famous volcanoes on both shoulders, such as Kilimanjaro (5892 m), Kenya, Elgon and Ol Doinyo Lengai, with its rare carbonatite lavas (the single occurrence in historical times). The eastern branch opens up into the Danakil depression, known for its active Erta Ale lava lake, and terminates in the Red Sea. Rising only 600 m, with a diameter of 50 km at the base, Erta Ale is one of the most active basaltic volcanoes on Earth, with continuous eruptions and a lava lake since at least 1967.

Based on the Main Ethiopian Rift (MER) example, Ebinger (2005) provided a three-stage model for continental break-up leading to the formation of a passive volcanic margin (Fig. 4A). We combine Ebinger's model with the three-stage model of magmatic systems feeding volcanoes by Biggs et al. (2021). Volcanic constructs and magmatic feeding systems can be nicely linked to each of the three evolutionary stages of EARS. Biggs et al. (2021) further illustrates relationships between volcanic constructs, magmatic systems, storage and transport of magma, and processes feeding volcanoes (Fig. 4B).

1. During the first stage, 0–5 Myr after the onset of rifting at the surface, the layered lithosphere begins to thin through brittle and ductile deformation. The rise of asthenosphere to replace the thinning lithosphere transfers heat, and some decompression melting may occur. Volcanic constructs such as Ol Doinyo Lengai and the Natron Rift in Tanzania (left, Fig. 4Ba), and the Virunga volcanoes on the western branch of the EARS (right, Fig. 4Ba) both represent early-stage rifting at depth (40 km). Faulting exerts most of the extension, while deformation is due to dyke intrusion.
2. During the second stage, 10–15 Myr after the onset of rifting, the lithosphere continues to thin by faulting and ductile deformation. The asthenosphere beneath rises to shallower levels, leading to more melting. This melt rises through the heated and weakened mantle lithosphere along fissures parallel to the faults at the surface. Strain localizes to a narrow zone marked by magmatic intrusions into the crust, creating a magmatic segment. Mature continental rifting is characterized by a mid-lower crust of intermediate depth (35 km) dominated by pervasive melt, above which form silicic calderas, with monogenetic volcanoes located above pre-existing crustal weakness, such as at the Main Ethiopian Rift (Biggs et al. 2021) (Fig. 4Bb). Extension is due to faulting and dyke intrusion, while deformation is due to magma intrusion.
3. The third stage corresponds to seafloor spreading. The tectonically and magmatically thinned lithosphere finally ruptures in the heavily intruded zones, and new oceanic lithosphere is created. The thick piles of lavas in the magmatic segments load the weak plate, flexing it towards the new ocean basin to form seaward-dipping lavas. Afar, Ethiopia, with the Manda Hararo Rift (left), and the Erta Ale segment (right, Fig. 4Bc), represent basaltic shield volcanism of the nascent seafloor spreading, with lava lakes occurring alongside silicic complexes and lava flows. Rift extension is mainly due to dyke intrusion filling shallow axial magma reservoirs. Magma is stored at the crust-mantle boundary at a depth of about 20 km. Deformation is due to magma withdrawal and recharge.

Fig. 5 Magmatic systems feeding volcanoes, dykes and faults. **A**. Map of the Main Ethiopian Rift (MER) between the Nubian and Somalian plates, highlighting the rift boundary faults and magmatic systems along oblique fault systems, the Quaternary volcanic fields, central volcanoes, and calderas elongated W-E or WNW-ESE (Corti 2009, Fig. 9, p. 12). **Inset a**) the location of MER with respect to the entire EARS and south of the Afar triple junction, with extension rates of the three major plates. MER: Main Ethiopian Rift, C for Central, S for Southern, and N for Northern. **Inset b**) the most recent time interval (3-0 Ma) of the EARS after Bonini et al. 2005, *in*: Corti 2009, Fig. 42, p. 41). Black arrow indicates the direction of propagation of the extensional deformation. Red numbers indicate the approximate timing of onset of extensional activation, or reactivation, of structures. SA: Southern Afar; YTVL: Yerer Tullu–Wellel volcano-tectonic Lineament. **B**. Interpretative three-dimensional block diagram of the Central MER evolution (from Corti et al. 2003 *in:* Bonini et al. 2005, Fig. 47, p. 226). Quaternary magmatic systems along the sinistral Wonji fault system feeding volcanoes that are oblique to the rift axis. UC, upper crust; LC, lower crust; UM, upper mantle; m, magma; rf, rift fill; WFB, Wonji Fault Belt; 1st, first orthogonal rifting phase; 2nd, second oblique rifting phase. **C**. Magmatic system feeding monogenetic and polygenetic volcanoes (Ebinger and Casey 2001, Fig. 3, p. 529). Dyke intrusion is driven by magma pressure and tectonic extensional stress. Dykes intrude perpendicular to axis of extensional stress, in this case oblique to rift bounding faults. Lower crust deforms in ductile manner.

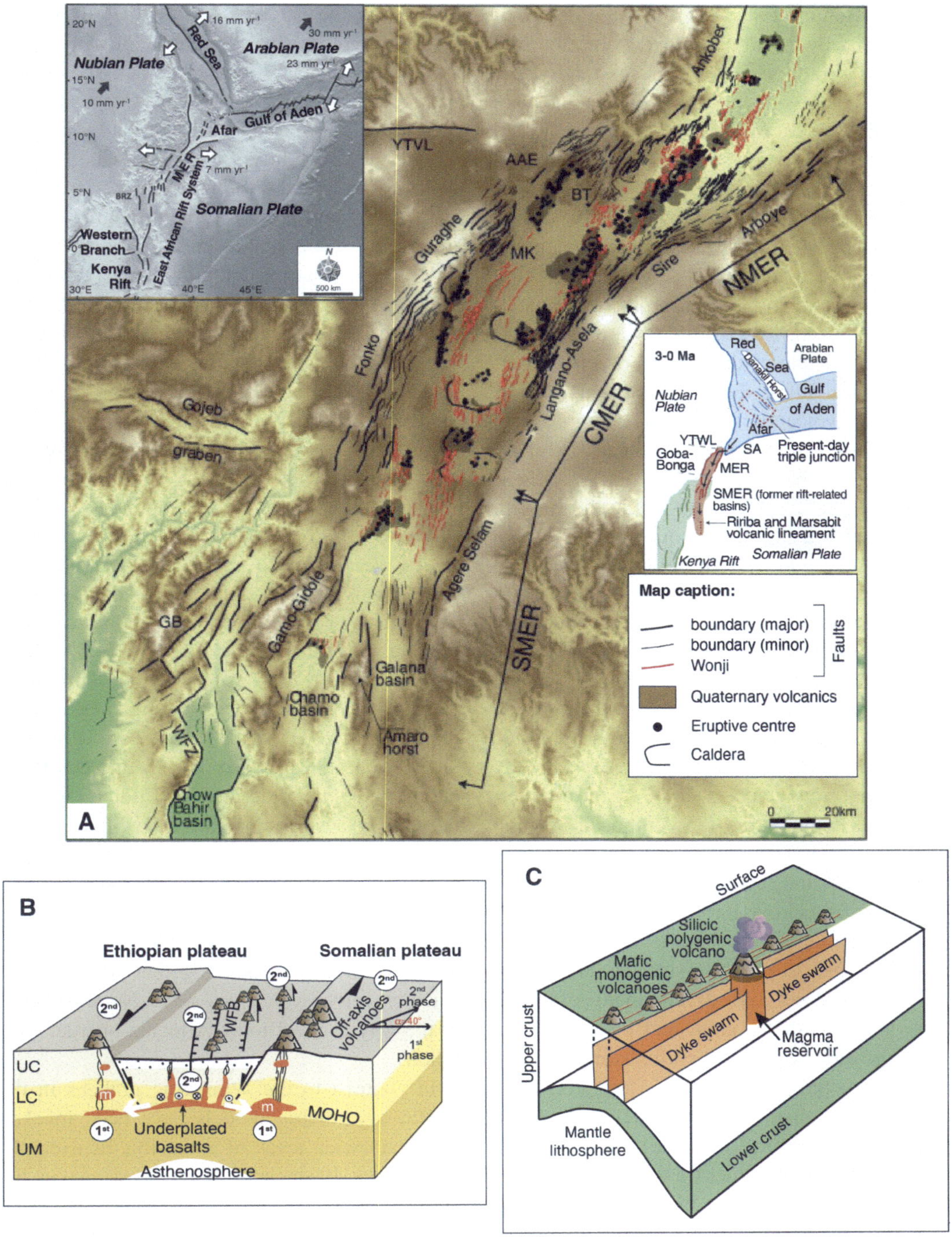

A map of the MER rift (Fig. 5A), featuring Quaternary-recent volcanic and tectonic activity, is characterized by NNE-SSW-striking, right stepping en-echelon zones about 20 km wide and 60 km long (Bonini et al. 2005; Corti 2009; Wadge et al. 2016). Figure 5B shows an interpretative three-dimensional block diagram of the evolution of the Central MER (Corti et al.

2003, *in:* Bonini et al. 2005), and the deformation induced by a two-phase extension: the first phase was orthogonal, while the second phase is oblique and sinistral. Oblique faults such as the NNE-SSW Wonji fault system have fed magmatic systems and concentrated deformation since the Quaternary-Pleistocene, whereas the NE-SW border faults along the rifts are of Mio-Pliocene age. Figure 5C shows a magmatic system in which magma is fed from a central magma chamber by dykes propagating laterally in the strong upper crust towards a central composite volcano of dacitic and rhyolitic composition. Secondary dykes propagated towards monogenic volcanoes of mafic composition, hence creating a bimodal volcanism (Ebinger and Casey 2001).

The magmatic systems lie within the rift, but are oriented at a slightly oblique angle to its NE axis. Volcanic edifices within a magmatic system generally comprise a predominantly composite volcano of felsic composition, often with a summit caldera, together with several mafic monogenic volcanoes, broadly aligned parallel to the system (Fig. 5C). The magma within the reservoir of the composite volcano may intrude laterally along the rift through dykes, eventually feeding the monogenic fissures, or intrude vertically and erupt within the composite volcano (Ebinger 2005). The magmatic systems, focusing magmatic and tectonic activity, are responsible for rift-opening through dykes mainly at depth and through normal faults and extensional fractures principally at the surface. The calderas accompanying the development of felsic reservoirs are usually elongated parallel to the extension direction or to pre-existing, oblique regional structures (Fig. 5A, B).

MER volcanism is bimodal, with mafic (basaltic) and felsic (rhyolitic) compositions, whereas intermediate rocks are scarce. The primary magmas are mafic: alkaline and, rarely, tholeiitic basalts. Phonolites and trachytes are also frequent, developed from the primary mafic melts. The widespread off-rift volcanism, building elongated volcanic edifices up to 1.5 km high, is structurally controlled by the unloading of the rift depression, directing the magma away from the rift axis toward the rift flanks. Trachytic volcanism is focused off-rift, likely related to a deeper source. The activity of normal fault scarps within the rift, responsible for local unloading on their hanging wall, may also explain the clustering of the monogenic vents at their footwall, a few hundred metres from the scarp (Acocella 2021). When magmas accumulate in crustal magma reservoirs, they may undergo assimilation of continental crust and magma differentiation. This results in the formation of rhyolitic magmas. These two different ways of producing magmas explain why intermediate, andesitic magmas are very scarce at continental rifts.

Volcanism associated with rift valleys or grabens is not restricted to East Africa. Volcanism is also found in rifts in all continents such as Lake Baikal in Central Asia, and in grabens across Western and Central Europe. Volcanoes have been active since the Oligocene to Pleistocene in rift valleys or grabens in France (Chaîne des Puys; e.g. André et al. 2013; Boivin et al. 2017), Germany (Eifel; Schmincke 2004) and the Czech Republic (Bohemia, Ulrych et al. 2011), with volcanoes of less than 11,000 years old found in the Chaîne des Puys, and just slightly older in the Eifel (Germany) and Garrotxa volcanic fields, NE Spain (Bolós et al. 2015). Volcanoes along strike-slip faults (e.g. the Anatolian fault) in an extensional setting produced widespread ignimbrite sheets in Capadoccia (Central Turkey; Le Pennec et al. 2005) in Miocene times, together with pre-historically active volcanoes (e.g. Nemrut Dagi) all the way to the Turkey-Armenia boarder (Mt. Ararat).

In contrast to the symmetric grabens in the EARS, asymmetric models have been invoked to explain wide areas of extension, as exemplified by the Basin and Range Province in the western USA. Asymmetric grabens used to be interpreted as having been formed by a gently dipping detachment fault that cuts at a low angle through the crust from one side of the graben down to the bottom of the lithosphere. However, Christie-Blick et al. (2007) cast doubt on the interpretation of specific regional detachment faults in the Basin and Range Province and the large extensional strains with which such faults are commonly associated. Volcanism associated with wide areas of extension are now interpreted as a passive response to tectonic processes, with no implication of mantle plume processes (Putirka and Platt 2012). These authors suggest that the Basin and Range volcanism was initiated by the transition from a subduction to a transform boundary (now manifest as the San Andreas fault), which caused a slab window to open as the subducting Farallon plate fell away.

2.3 Volcanism at Subduction Zones Along Convergent Plates

2.3.1 Continental Arcs and Islands Arcs: Location, Structure and Magmatism

The total global length of volcanic arcs is approximately 27,000 km, but they vary in length individually from tens to thousands of km. The most notorious volcanic arcs and island arcs form the so-called 'Ring of Fire' around the Pacific Ocean (Fig. B in Appendix).

The eastern edge of the Pacific Ocean coincides with the volcanic arcs down the length of the Americas. The arcs begin with the Scotia and South Sandwich Islands, and extend 7000 km to the north through the four volcanic zones termed Austral, Southern, Central, and Northern Andes on the continental margin of South America. They contain at least one hundred active volcanoes and up to one thousand are potentially active in the Cordillera of the Andes (Francis 1993). Among them, a few volcanoes are the highest and/or most iconic stratocones of the world such as Ojos del Salado (6879 m asl) in Chile and Nevado Coropuna (6377 m) in Peru. The historical (since 1600 AD/CE), second deadliest eruption worldwide occurred at the snow-clad Nevado del Ruiz in Colombia, which produced lahars killing 23,000 people on 13 November 1985. Active volcano arcs resume in Central America to become most extensive in Mexico (Popocatepetl, El Chichón). Active volcanoes are distributed throughout the western states of the USA and British Colombia (Canada), all the way through Alaska and the Aleutians, up to the Kurile Islands. Among them, Mt. St. Helens and Novarutpa—Katmai famously erupted explosively in 1980 and 1912, respectively.

Island arcs host most of the active volcanoes in the western edge of the Pacific Ocean (Fig. B in Appendix). South of the Kamchatka peninsula in easternmost Russia, Japan hosts many iconic volcanoes, the most renowned being Fuji, Asama, Unzen, and Aso, notorious for their catastrophic eruptions. Island arc volcanoes continue through the Philippines (Pinatubo being the site in 1991 of the third largest eruption of the twentieth century), Sulawesi, New Guinea (Lamington), the New Hebrides (Ambrym and Aoba), and Melanesian islands and the Tonga-Kermadec Archipelago. The latter is the site of an eruption which took place on 15 January 2022 at the shoaling island of Hunga Tonga-Hunga Ha'Apai (Allenvi 2022), the most powerful explosion since the eruptions of Krakatau in 1883 and Tambora in 1815. The Indonesian arcs, including Sulawesi, host the largest number (~125, or 14%) of active arc volcanoes. Krakatau was the site of the deadliest eruption in the modern world with a far-reaching tsunami that caused 36,000 deaths in 1883. Overall, three quarters of the Earth's active volcanoes lie along the Pacific "Ring of Fire". Farther southeast, the North Island of New Zealand is home to famous active volcanoes (e.g. Mt. Ruapehu-Tongariro-Ngauruhoe and Mt. Egmont-Taranaki) and was the site of the largest historical eruption in the caldera of Taupo ~232 AD/CE. Other destructive eruptions of volcanoes have occurred on island arcs such as the Lesser Antilles (Mount Pelée, whose eruption triggered a massive disaster in 1902 with 29,000 fatalities in the city of Saint-Pierre, Martinique). It is worth mentioning the volcanoes in the subduction zones between the convergent African and Eurasian plates, such as the Aeolian archipelago in Italy, and the Cyclades Islands in Greece, site of the Minoan (Late Bronze Age) eruption at Santorini ~1600 years BC/BCE.

Active arc volcanoes are typically concentrated in narrow bands 20–60 km wide located just over 100 km above the surface of the descending slab; however, both arc length and width can vary substantially due to the subducting plate geometry (from flat to steep), rate of descent, and other factors. The location of the volcanic arc results from the location of the thermal anomaly within the wedge above the slab, which in turn is constrained by the depth of release of volatiles from the slab, itself a function of the slab dip. Therefore, steeper slabs develop volcanic arcs nearer to the trench and shallower slabs develop volcanic arcs farther from the trench. Flat slabs may not reach the temperature and pressure conditions to release volatiles, thus inhibiting magmatism (e.g. northern Peru; e.g. Gutscher et al. 2000).

Volcanic arcs typically host a range of composite volcanoes and calderas (Chaps. 5 and 6), while some of them contain monogenetic fields above a thin overriding plate and/or along tensional fractures oblique or perpendicular to the arc. Arc volcanoes in general do not show frequent unrest, although the eruptive activity can be highly variable from rare persistently active volcanoes (e.g. Semeru, Sangay) to edifices commonly erupting once in several centuries (e.g. Pinatubo, Mt. Baker). The volume of magma erupted by arc volcanoes comprises less than 10% of the global production (Fig. 2B), but more than 80% of the roughly 5350 eruptions recorded in historical times (Schmincke 2004). Most of the volcanic rocks generated in arcs are collectively termed calc-alkaline due to the presence of calcium and alkaline elements, which come from the sediments on top of the subducting slab. Volcanism at island arcs formed on oceanic plates tends to produce mostly mafic and intermediate (i.e. andesitic) rocks, whereas volcanism at continental arcs may also produce felsic (i.e. dacitic and rhyolitic) rocks, in addition to intrusive rocks such as granite batholiths. The main factor is that sediments transported to depths contribute significantly to dewatering in the mantle wedge. Water released in this wedge causes significant changes in the overlying mantle, reducing the solidus temperature of the rocks of the asthenosphere from around 1400 to 1000 °C. As a result, arc magmas, in particular islands arc ones, typically contain a high proportion of water (as much as 6%). In the subduction zone, magma generation occurs at depths of up to ~200–250 km, creating volcanism at the surface at distances of around 100 km from the deep-sea trench.

2.3.2 Categories of Volcanic Arcs

Volcanic arcs consist of continental or island arcs, sometimes associated with back-arc volcanism. (1) Continental arcs grow where an oceanic plate is subducting beneath continental lithosphere, as in the Andes or the Cascades arc. In the case of the ongoing subduction zone west of the Central Andes, the light continental lithosphere (South America) is pushing upward and westward above the eastward subducting Nazca oceanic plate. (2a). Island arcs form where a denser oceanic plate is subducted beneath a lighter oceanic one (e.g. the Marianas). The subducted plate can roll back and cause migration of the surface volcanism. (2b) Island arcs also form where the volcanic arc grows on a sliver of continental crust that has separated from a continent (e.g. Japan, the Philippines), also known as microcontinental arcs. (3) In some regions, volcanic activity additionally occurs behind the arc on sea floor of coastal margins (e.g. Japan Sea) and within continental interiors (e.g. to the east of the Central Andes in Peru). This back-arc volcanism in an extensional setting tends to cover a broader surface area relative to the narrowly confined volcanism within arcs. However, back-arc volcanism is less vigorous as well as less dense than arc volcanism.

In addition to the type of convergent plates, arc volcanoes can be grouped on the basis of the nature of the convergence between plate boundaries. The convergence characteristics are angle, rate and the type of tectonic regime: extensional, strike-slip, compressional or contractional, with transtensional tectonics in hybrid cases. Acocella (2021) distinguished three categories of volcanic arcs:

(1) extensional arcs are characterized by predominantly arc-perpendicular extension, such as the Cascades arc where there is a low extension rate, the Taupo volcanic zone of New Zealand where there is a high extension rate, and the Tyrrhenian margin of central Italy, which developed on the margin of an extended back-arc zone;
(2) strike-slip arcs are characterized by predominantly horizontal motion expressed through a principal displacement zone: the South Andean Volcanic Zone, which is experiencing minor obliquity, and Sumatra (Indonesia), which is subject to major obliquity. The obliquity of the convergence between two plates is expressed by the angle between the convergence vector and the trench-perpendicular component; the higher the angle, the higher the obliquity.
(3) contractional arcs are characterized by dominant shortening, resulting from overall arc-orthogonal compression with a frequent but minor strike-slip component: Northeast Honshu, Japan, and the complex example of the Central Volcanic Zone (CVZ) of the Andes in South Peru and Northern Chile.

The CVZ illustrates the evolution of a contractional arc over extended geological periods, the most recent of which, from the Late Tertiary to the Quaternary, experienced voluminous volcanism in the thickest (60–70 km) portion of the orogen. The CVZ is characterized by a switch from steady-state volcanism creating large, andesite-dacite composite volcanoes to 'flare-up' volcanism (episodic higher flux magmatism), producing a province of caldera complexes and voluminous ignimbrites of dacitic to rhyolitic compositions (Chap. 6). The switch is related to the southward migration and subduction of the oceanic Juan Fernandez Ridge, which lies on the Nazca plate, over the past 25 Ma. Soon after the passage of the ridge the slab increased its dip, switching volcanism eastward. The steepening slab started to roll back, promoting decompression melting in the overlying, enlarged mantle wedge. This condition created an elevated thermal input, encouraging crustal melting and generating a significant amount of magma. As a result, volcanism migrated to the west from ~24 Ma, coinciding with an ignimbrite flare-up in the CVZ. A distinct flare-up within the Neogene Central Andean ignimbrite province produced the most voluminous ignimbrite plateau in the southern half of the Altiplano-Puna Volcanic Complex (APVC, 22° to 24° S), characterized by a magma production rate of nearly 20 km^3/km/Ma. Approximately 15,000 km^3 of magma have been erupted in the last ~ 10 Ma, concentrated in pulses occurring every ~2 Myr, with eruption rates one order of magnitude higher than the average rate for the flare-up of continental silicic provinces (de Silva et al. 2006; Salisbury et al. 2011). Magma emission in the APVC has been focused on ~20 large complex calderas, along five NW–SE trending magmatic belts aligned on NW–SE trending fault zones in the Altiplano (Chap. 6).

2.4 Hybrid Settings of Volcanoes Near Subducting Plates

Other, rarer volcanoes are located outside these subducting plates. Schmincke (2004) called them 'slab-edge volcanoes', such as Mt. Etna, Sicily (the largest active volcano in Europe) which lies at the junction of four plates or microplates (Fig. 6), Mt. Fuji located at a triple junction at the southern end of the subducting Pacific plate, and the Klyuchevskoy−Tolbachik complex, the highest and most active volcanoes in Kamchatka, Russia. Such highly productive volcanoes occur in a tensional environment between subducting plates or near the edge of a subducting slab. Etna is a complex volcano at the meeting point of three microplates (Adriatic, Ionian, and Tyrrhenian) with the African plate (Fig. 6), whose unusual form is the result of the superposition of various edifices, calderas, summit vents and rift zones. The chemical signatures of the mafic magmas of some of the slab edge volcanoes are most similar to intraplate mafic magmas, but they also share some characteristics with subduction zone-related, more

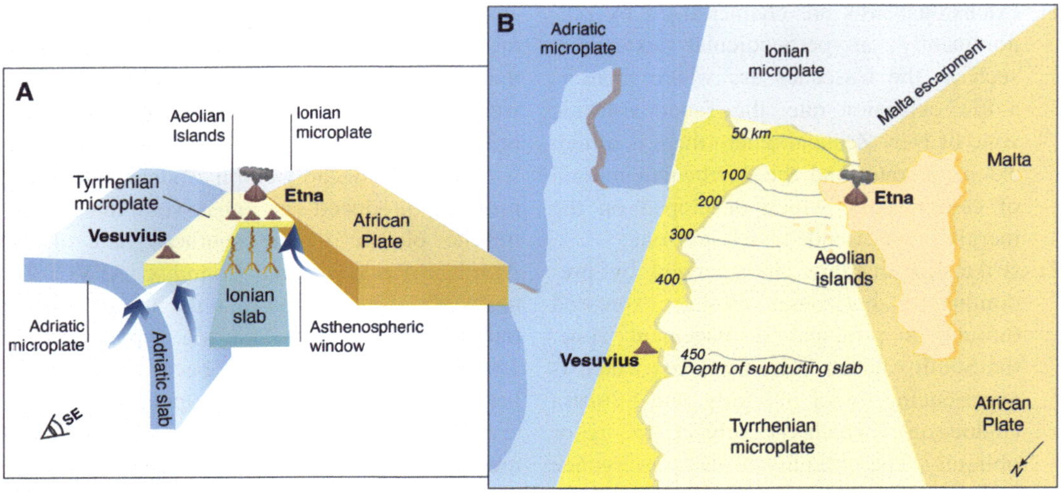

Fig. 6 Etna composite volcano as an example of a hybrid tectonic setting (after Schmincke 2004, Fig. 2.8 and 2.9, p. 19). Etna owes its structure, eruptive styles, and magma compositions to its unusual tectonic setting at a triple junction and as a slab-edge edifice. Left-hand side: 3-D model of the south Tyrrhenian subduction zone looking SE from a point above the Tyrrhenian sea. Right-hand side: Sketch map looking at southern Italy and Sicily from the northwest. Arrows represent local patterns of asthenospheric mantle flow driven by slab motion. Etna volcano is located outside the Aeolian arc and above asthenospheric mantle rising sideways from beneath the African plate

evolved magmas with high volatile content. A major reason for foci of high mafic magma production rates in slab-edge settings may be the fact that fertile asthenosphere is forced to rise sideways of the edges of down-going slabs (Fig. 6; Schmincke 2004), partially melting as it rises and decompresses. Thus, volcanoes located in hybrid settings can also produce hybrid magmas.

Other volcanoes, rarer than triple junction ones, form in hybrid tectonic settings along strike-slip faults and near transform plate boundaries. The horizontal motion along transform boundaries does not usually involve any major magma generation. Nevertheless, volcanism may occur in regions of transtension. These are regions where relative plate motions are oblique to the transform faults allowing extension of the crust, e.g. in the Southern Andes (Lara et al. 2006; Tibaldi et al. 2017), in the Basin and Range province (western USA), and along strike-slip fault systems in Sumatra, Anatolia, etc. Transpression and transtension are strike-slip deformations that deviate from simple shear because of a component of shortening or extension, respectively, orthogonal to the deformation zone (Dewey et al. 1998). Volcanism in transtensional regions can vary from monogenetic basaltic volcanic fields to the formation of composite cones (see Sect. 3, Chaps. 3 and 5).

2.5 Intraplate Hot-Spot Volcanoes and Large Igneous Volcanic Fields

Intraplate oceanic islands and volcanic fields of different sizes are situated away from the plate boundaries, although a few of them lie near mid-ocean ridges. The largest volcanic fields form plateaus of extraordinary size (thousands of km^2) on both continents and submarine floors. Intraplate volcanism, which includes hot-spot volcanoes and large igneous provinces termed LIPs, is generated by mantle upwelling, and is responsible for more than 10% of the global magma output (Fig. 2).

2.5.1 Hot spot volcanoes

Most authors estimate that the total number of hot spots ranges between 45 and 70 (e.g.

Ballmer et al. 2015). Hot spots have irregular but non-random distribution over the Earth's surface. They primarily lie in oceans, often near oceanic ridges, but are rare near subduction zones. The best-defined hot spots appear relatively stationary over time, as they move slowly with respect to the overriding plate. Typically, the direction of plate motion corresponds to the direction of migration of the aligned hot spot volcanoes. The migration results from age-progressive chains of dead and dormant volcanoes emanating from the loci of active volcanism at the present plume location, whether it be on land or at sea. The volume of plume magmatism is highly variable, ranging from mid-sized erupted volumes at oceanic islands like La Réunion and the associated Mascarenes plateau, and the Azores oceanic plateaus, to the massive thermal anomalies and volcanism of the African Super Plume under the EARS (O'Connor et al. 2019; Acocella 2021).

The Pacific Ocean contains several volcanic islands and archipelagoes associated with hot spots (Fig. B in Appendix). The highly studied Big Island, Hawai'i is formed of five shield volcanoes (Kīlauea, Mauna Loa, Mauna Kea, Kohala, and Hualalai: Chap. 4), attached to the still submerged seamount of Loihi southeast of Kīlauea. Mauna Loa is the largest volcano on Earth (up to 95,000 km^3), dwarfing all composite volcanoes, while Kīlauea is the most active volcano worldwide with an average magma extrusion rate of 0.10–0.12 km^3/ka (Clague and Sherrod 2014). Big Island lies at the southeastern end of a 3500 km-long chain of submerged seamounts, atolls and basaltic volcanic islands termed the Hawai'i-Emperor chain, which consists of more than 100 individual volcanoes. Other chains of hot spot oceanic islands exist in the South Pacific Ocean, such as the Tahiti-Tuamotu-Society, Marshall-Samoan and Louisville. All of them show abrupt bends halfway between the NNW-trending segment and SSE-tending trend (Fig. B in Appendix), indicating a dramatic change in motion of the Western Pacific lithosphere beneath them ~47 Myr ago (Torsvik et al. 2017). In the Indian Ocean, La Réunion Island, east of Madagascar, is a broad high island that lies above a hot spot, together with Mascarenes to the north. Two basaltic volcanoes (Chap. 4), the frequently active Piton de la Fournaise volcano and the deeply eroded Piton des Neiges which contains conspicuous cirques, form a basaltic oceanic shield on La Réunion Island (Defos du Rau 1959; Oehler et al. 2008; Salvany et al. 2012; Villeneuve et al. 2013).

Fracture zones offset the oceanic Canary Islands to the east near the African coast. These broad shield volcanoes illustrate the waning of eruptive activity from west to east, with Mt. Teide on the central island of Tenerife mentioned by Columbus in 1492 and the most recent (2021) scoria cone-building and lava-flow eruption at Tajogaite-Cumbre Vieja on the westernmost Island of La Palma (Carracedo et al. 2022; Chaps. 3 and 10). Isolated volcanic massifs are located in central and north Africa, which may coincide with intracontinental hot spots, which have created active rifts. Three such volcanic complexes in the Sahara are Jebel Marra in Sudan, Tibesti in Chad (Chap. 6) and Hoggar in Algeria. All contain tens of young volcanoes of different sizes and shapes, up to the size of the caldera of Jebel Marra. Another active volcano complex is the huge, elongated Mt. Cameroon along the NNE-trending Cameroon volcanic line, hosting the infamous Lake Nyos, a maar that had a lethal limnic eruption in 1986 (Chap. 3).

The Oceanic Island Basalts (OIBs) produced by the hot spots are geochemically distinct from MORBs. This suggests that the lower mantle, the inferred source of plumes, is compositionally different from the upper mantle, the source of MORBs. Hot spots may be the surface expression of mantle plumes, as detected by slow seismic velocity anomalies under major hot spot regions in Hawai'i, Iceland, the South Pacific, and Africa. Hot spot volcanoes are believed to be formed by long-lived plumes of hot upwelling mantle that originate from relatively stationary anomalies at thermal boundary layers such as the mantle transition zone at 400 to 600 km in depth or the core-mantle boundary at a depth of 2900 km. When the mantle plume reaches the lithosphere, its uppermost part

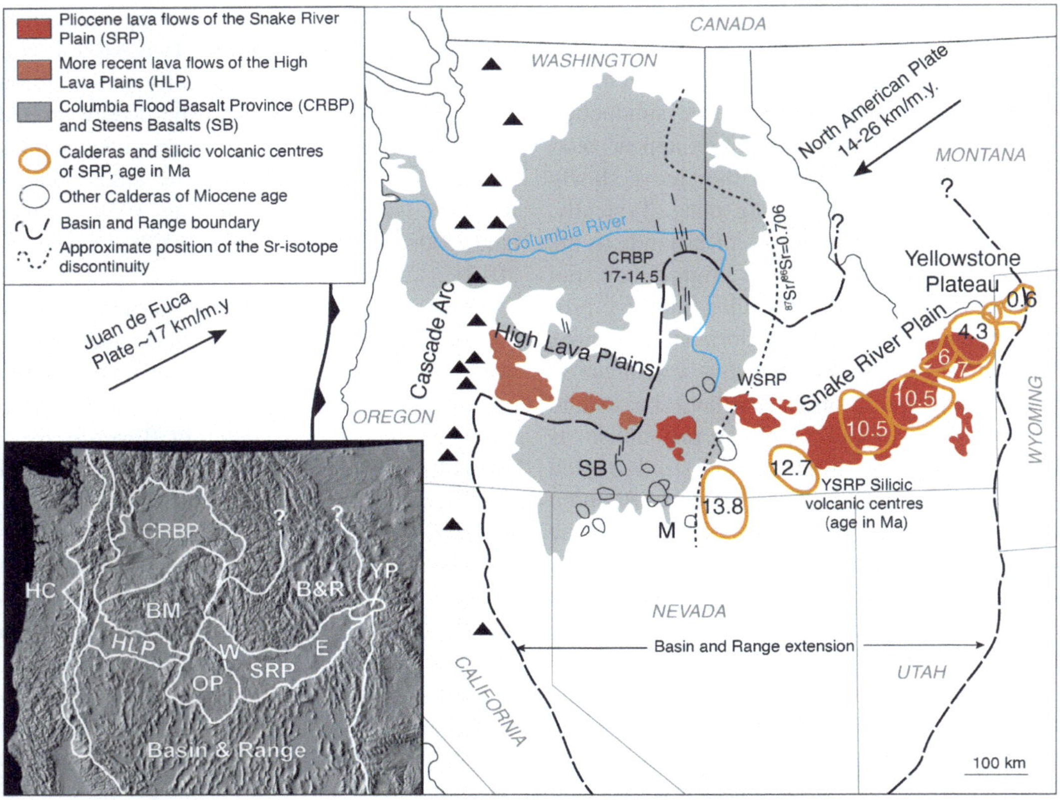

Fig. 7 Linear trends of age-progressive volcanism, plate motion over a hot spot or a mantle plume, as illustrated by the Yellowstone–Snake River Plain (YRSP) in the western USA (modified after Jordan 2005; Smith et al. 2009). This map shows the tectonic setting of the YSRP and the Oregon High Lava Plains (HLP) volcanic systems with respect to the Columbia River Basalt province. Pliocene and younger basalts of the YSRP and HLP are dark and light red, respectively. Circles are volcanic centres of the YSRP and Owyhee region; ages are given for some calderas to indicate age progression. The bold dashed line shows the limit of Basin and Range extension. The light dashed line indicates the approximate position of the Sr-isotope discontinuity: $^{87}Sr/^{86}Sr$ value of 0.706 is thought to broadly delineate the craton margin to the east and the 'oceanic' crust to the west. Also shown are dyke swarms which fed middle Miocene flood basalts; CRB, Columbia River basalt dykes; SB, Steens Basalt dykes. The inset map shows the extent of the geological and physiographic provinces on a digital topographic base: HLP = High Lava Plains, HC = High Cascades, CRBP = Columbia River Basalt Plateau, OP = Owyhee Plateau, SRP = Snake River Plain (E = Eastern, W = Western), B&R= Basin and Range, and YP = Yellowstone Plateau

widens out to form a mushroom-like head with a width of up to 1000 km and causes bulging of the overlying lithosphere. Partial melting of the mantle peridotite occurs when the plume reaches a depth of 100–150 km beneath the Earth's surface (White and Mackenzie 1995). As tectonic plates move over these stationary upwellings, magma production fuels active volcanism at the surface, which typically forms chains of shield volcanoes. Therefore, long-lived hot spot volcanism may last for a hundred million years, e.g. the ~100 Ma-old Hawaiian-Emperor chain in the Pacific Ocean.

In the same way, the role of a hot spot and plume head can be seen as a linear WSW-ENE trend of lava plateaus and calderas, from the Columbia River Basalt province of early-middle Miocene age to the Snake River Plain lava fields and Yellowstone volcanic centre and calderas of Quaternary age to Present (Fig. 7; Jordan

2005; Smith et al. 2009a, b). The Yellowstone hot spot resulted from a mantle plume interacting with the overriding North America plate, a process that has highly modified the continental lithosphere through magmatic and tectonic processes and produced the 16–17 Ma, 700-km-long YSRP silicic volcanic system (Hooper et al. 2007; Smith et al. 2009a, b). These authors consider the contemporary southwest extension of 2–3 mm/year, measured by GPS across the Yellowstone Plateau, which is a quarter of the overall Basin and Range opening rate, to be part of Basin-Range intraplate extension. Smith et al. (2009a, b) extrapolate the plume source southwest to its original location at 17 Ma and 600 km southwest and 200 km north of the YSRP, respectively, based on the geometry and its assumed depth of 660 km. Importantly, this location is beneath the southern part of the Columbia River Plateau of the same age and implies that the Yellowstone mantle plume may be the common source for both of these large volcanic fields (as shown in Fig. 7).

Hot-spot volcanism is commonly thought to derive from decompression melting of hot low-density mantle material ascending in mantle plumes. The persistent hot-spot magmatism is considered to result from mantle plume tails creating anomalous heat transfer to the lithosphere, whereas the large-volume magmatism during LIP formation is often attributed to mantle plume heads reaching the crust.

2.5.2 Intraplate Large Igneous Provinces (LIPs)

The Earth's largest provinces and the highest order grouping of volcanoes are known as LIPs, dominated by basaltic volcanic features (Saunder 2005; de Silva and Lindsay 2015). Bryan and Ernst (2008), followed by Ernst (2014, 2021), have revised the definition of LIPs: "Large Igneous Provinces are mainly mafic, magmatic provinces with areal extents >0.1 million km^2, igneous volumes >0.1 million km^3 and maximum lifespans of ~ 50 Myr that have intraplate tectonic settings or geochemical affinities. Despite their gigantic size, LIPs have been emplaced by igneous pulses of short duration (~ 1–5 Myr), at a high rate (0.1–1 km^3/year) that occurred on average every 20–30 million years". Some 35 LIPs have been emplaced in the past 250 million years (Bryan and Ernst 2008).

LIPs represent significant volcanic outputs in modern Earth Science for at least two reasons:

(1) LIPs play an essential role in the thermal evolution of Earth, as they potentially record the transport of material from the lower mantle or core-mantle boundary to the Earth's surface and are a fundamental component in convection models involving the whole mantle (Burke and Torsvik 2004; Bryan and Ferrari 2013). The mafic rocks, mainly flood basalts, of LIPs are associated with the occurrence of a mantle plume at the base of the lithosphere.
(2) Biospheric, hydrospheric, and atmospheric impacts of LIPs are now widely regarded as key trigger mechanisms for mass extinctions, although the exact mechanisms that bring about the deaths are still under discussion (Kelley 2007; Keller and Kerr 2014). Many of these LIPs seem to coincide with mass extinction events (e.g. the Deccan LIP, India, at the end of Cretaceous) and continental break-up. The Paraná-Etendeka flood basalts, for example, crop out in SE America and SW Africa and were coeval with the opening of the south Atlantic.

Some authors classify LIPs by the composition of their dominant products: bimodal (basalt and rhyolite), andesite and silicic provinces, or by their tectonic environments: continental interior, back arc, rift or plate margin, and oceanic settings (Bryan et al. 2002; Sheth 2007). Magmatism of the majority of LIPs is mafic, but some are dominated by silicic magmatism (Fig. 1). The silicic large igneous provinces (SLIPs), silicic-rock equivalents of LIPs (Bryan et al. 2002), like the Sierra Madre Occidental in Mexico, are smaller (~350,000 km^2), but massive (200,000 km^3) at the scale of volcanic landforms (de Silva and Lindsay 2015). Some of the LIPs show bimodality, producing flood basalts and silicic (effusive and explosive) eruptions, such as the Paraná-Etendeka SLIP (Deligne and Sigurdsson 2015).

LIPs occur both on continents and in oceans: they include continental flood basalts in the former, and volcanic passive margins, oceanic plateaus, and ocean-basin flood basalts in the latter. By far the largest LIPs occur on the sea floor, because oceanic lithosphere is thinner than that beneath continents and thus rifts more easily, permitting the eruption of vast amounts of mantle-derived partial melts. Thus, one can distinguish: (1) broad, flat-topped oceanic volcanic plateaus, such as the submarine Ontong Java (Lower Cretaceous) in the South Pacific, Kerguelen (Lower Cretaceous, 2.3 million km^2) in the Indian Ocean, and Caribbean-Colombian (Upper Cretaceous, 1.1 million km^2) in the Caribbean Sea. Ontong Java, with an extent of 3 million km^2 and a volume of 44 million km^3, is one of the largest of any type of landform on the Earth, and exhibits greater relief (rising 2 km above the surrounding seafloor) than most terrestrial counterparts. (2) The on-land plateaus of continental flood basalts often show km-thick traps (from the Swedish 'trappa' term meaning stairs), such as the Siberian traps (7 million km^2) of Late Permian age, the Deccan traps of Late Cretaceous age (500,000 km^2), and the smaller (164,000 km^2) Columbia River Flood Basalts (CRFB) emplaced 16.5–14.5 million years ago in Oregon, USA.

Based on composition, volcanic features, mode of emplacement, and source, we divide LIPs into two types: Flood basalt provinces and Ignimbrite plateaus.

Continental Flood Basalt provinces, plateaus and traps

Plateaus and traps of flood basalts are the expressions of large igneous provinces. Continental Flood Basalts (CFBs) consist of extensive (tens of thousands of km^2) sheets of lava flows, representing hundreds of thousands of km^3 of magma, which erupted from scattered fissure vents or dyke swarms (Rampino and Stothers 1988; de Silva and Lindsay 2015). The term 'plateau basalt' was introduced by Sir Archibald Geikie after his visit to the Columbia River plateau (Williams and McBirney 1979). Tyrrell (1937) coined the term 'flood basalts' to encompass all widespread basalts discharged from swarms of fissures and shield volcanoes, and the term multiple-vent basalts to denote extensive lava fields formed by the confluence of flows discharged from scattered vents active at about the same time. The word 'flood' appears in 'flood basalt provinces' because of the way rapid and voluminous effusions of lava cover pre-existing topography, creating a new landscape of subdued relief or tableland (Lockwood and Hazlett 2010). For example, the Columbia River Basalt (CRB) lavas buried mountainous landscapes with a relief of over one kilometre and spread out across broadly subsiding basins: the CRB contains approximately 300 individual lava flows increasing in cumulative thickness to as much as 3500 m toward the centre of the plateau. The underlying continental crust may have subsided as flows accumulated, or perhaps the crust was rifting open as eruptions took place. While the base of the volcanic pile sank, the top of the flows formed a low-level plain or basin floor that was frequently resurfaced by freshly erupted lavas (e.g. the High Lava Plain of Oregon).

Volumes of erupted lavas and eruption rates

The large volume of magmas erupted in short geologic periods (1–5 Myr) and the subsequent high eruption rates are the hallmarks of continental flood basalt provinces and distinguish them, by an order of magnitude, from the products of other types of volcanic activity (Thordarson and Self 1998). The eruptive rate (0.1–0.8 km^3/year on average) corresponds to that of the entire present mid-ocean ridge system and surpasses that of volcanic hot spots by a factor of 100–1000 (Schmincke 2004; Bryan and Ferrari 2013; Ernst 2014). Age dating has shown that these huge magma volumes formed within 1–5 Myr. The output of lava spikes dramatically during the period of volcanism, usually at the beginning. In the CRB, eruptions took place over an interval of 11–12 million years, but about 87% of the total volume was erupted in only about one and a half million years (Fig. 8). Six major groups of flows were emplaced, each representing numerous individual eruptions (Swanson et al. 1975, 1989; see Table in Fig. 8). The volume of magma erupted during each of the individual eruptions is also exceptional: La Roza Member flow of the CRB plateau covers an area of about 40,000 km^3 and represents a volume of about 2000 km^3 (Swanson et al. 1975, 1989; Thordarson and Self 1998). This flow was likely emplaced in less than 100,000 years during prolonged eruptions, each lasting years to decades, with the largest flow fields fed by tens of km-long fissures.

Such large erupted volumes and relatively frequent eruptions led to the geologically rapid construction of extensive lava plateaus, up to several km thick. For example, the size of the CRB in Oregon and Washington is 164,000 km^2 and its volume 174,300 km^3 (Fig. 8; Photo 1) (Tolan et al. 1989; Reidel et al. 2013). Together with the High Lava Plains of Oregon (~20,000 km^2) and the Snake River Plain of Idaho (~40,400 km^2), these volcanic fields are all thought to be related to the interaction of the Yellowstone plume with the North American lithosphere above it and Basin and Range extension in Miocene times (Fig. 7; Smith et al. 2009a, b). The more massive Deccan Flood Basalt Province, with a current areal extent of 500,000 km^2 and an average thickness of 1 km, is the product of intense magmatism, with 1.5 million km^3 over a period of almost one million years between ~66.3 and 65.6 Ma (Duncan and Pyle 1988; Sheth 2007; Schoene et al. 2019; Mittal et al. 2021a, b). The Deccan results from the Indian lithosphere interacting with the Réunion hot spot, now under La Réunion Island. Oceanic plateaus have even larger volumes, about 10–60 million km^3, while their eruptive rate is higher: 2–20 km^3/year, perhaps because they erupted through a thinner oceanic lithosphere.

LIPs have not occurred uniformly throughout the Earth's history. Statistical studies by Self et al. (2015) suggest periodicities in LIP production of approximately 30–35 and 60 million years, although their size and recurrence may have decreased since the Archean (Davies 2001; Herzberg et al. 2010). In basalt-dominated provinces, large magnitude basaltic eruptions have shorter recurrence intervals of 10^3–10^4 years than similar magnitude silicic eruptions, of up to 10^5 years. Such periodicity probably reflects variations in rates of mantle circulation, and there is a clear clustering of events at times of supercontinent breakup; LIP events are thus becoming an increasingly important tool in reconnecting dispersed continental fragments (Bryan and Ferrari 2013). One example is the South Atlantic, where the small remnants of flood basalts of the Etendeka plateau in Namibia are the same age as the Paraná flood basalts in South America. The split occurred between 132 and 125 Ma, following the rapid extrusion of flood basalts over a period of 1.6–3 Myr (Gomes and Vasconcelos 2021).

Morphology, architecture and physical emplacement of the continental flood basalts

Geomorphic features of continental flood basalt plateaus (CFBPs) rely on the architecture of the volcanic fields and the types of lava flows. Erosional dissection of the basaltic plateaus commonly results in the formation of a step-like topography, termed traps, on the sides of the valleys, such as the Deccan in central NW India (Photo 2). Such traps are caused by the variable resistance to erosion of different parts of the lava flows due to backweathering of the less stable scoria tops of flows or interbedded paleosols or sediments (Photo 3, e.g. the Photographic Atlas of Flood basalts by Sheth 2018a, b). Other typical trap landscapes have been described in basaltic successions in Yemen (Chiesa et al. 1989), Ethiopia (Hofman et al. 1997) and remains of the Central Atlantic Magmatic Province, e.g. in Morocco (Hachimi et al. 2011).

White et al. (2009) draws our attention to the physical emplacement of LIPs, based on descriptions of similar volcanic architecture from each flood basalt province. Flood basalt interpretation has been based upon the Columbia River Basalt province, the smallest, youngest,

A

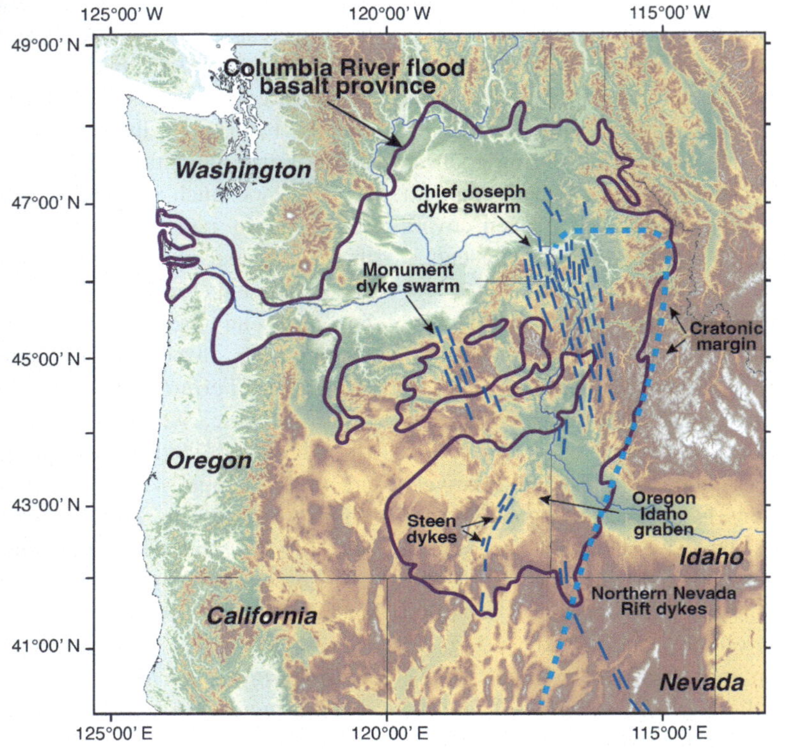

B

Stratigraphic subdivision of Columbia River Basalt Group (CRBG)

Series		Group	SubGroup	Formation (Age, Volume, % of CRBG)	Member	Mag*
Miocene	Upper	Columbia River Basalt Group	Yakima Basalt SubGroup	**Saddle Montain Basalt** (14-6 Ma, 2,400 km³ volume, 1.5% of CRBG)	Lower Monumental Member	N
					Ice Harbor Member	N,R
					Buford Member	R
					Elephant Mountain Member	R,T
	Middle				Pomona Member	R
					Esquatzel Member	N
					Weissenfels Ridge member	N
					Asotin Member	N
					Wilbur Creek Member	N
					Umatilla Member	N
				Wanapum Basalt (15.5-14.5 Ma, 10,800 km³ volume, 6.0% of CRBG)	Priest Rapids Member	R3
					Roza Member	T,R
					Frenchman Springs Member	N2
					Eckler Mountain Member	N2
	Middle / Lower			**Grande Ronde Basalt** (17-15.5 Ma, 151,700 km³ volume, 87% of CRBG)		N2
						R2
	Lower			**Picture Gorge Basalt**		N1
						R1
				Imnaha Basalt (17.5-17 Ma, 9,500 km³ volume, 5.5% of CRBG)		R1
						T
						N0
						R0

*Mag: Magnetic Polarity
N: Normal; R: Reversed; T: Transitional; subscripts denote magnetostratigraphic units
USGS, Swanson *et al.* (1989)

Fig. 8 Columbia River Basalt Group and Province: SRTM DEM-based map and Table for stratigraphic subdivisions, after Swanson et al. (1989, U.S. Geol. Survey, AGU Field Trip Guidebook T106. Science 238, pp 1105–1) and map modified from Reidel et al. (2013, Fig. 1 p. 2, GSA Spec Paper 497). Note dyke swarms in blue; age, volume and proportion of CRBG formations in Table Column 4. Contour of CRBG adapted from Public Domain map, US Geological Survey, modified and arguably the most intact CFBP (e.g. Reidel et al. 2013). Flood basaltic lava flows are emplaced as pāhoehoe-type lava-flow fields that advance and thicken by inflation during which time they reach great extents over periods of years or decades during sustained eruptions (see definitions in Chap. 2) (Swanson et al. 1975; Self et al. 1996, 2015). Thus, flood basalt eruptions are thought to have durations of years to decades, even perhaps centuries (Self et al. 2015). The flows are mainly pāhoehoe, though minor amounts of 'a'ā may also be present (see Chap. 2). Each flow field consists of several major lava flows, which in turn consist of multiple flow lobes. These extensive lobes, with aspect ratios (length/thickness) ranging from ~50 to ~500, give the provinces their 'layer-cake' or, when eroded, step-like, appearance. The major sheet lobes, containing the majority of the lava volume, are commonly 20–30 m thick, several kilometres wide, and show features consistent with in situ flow thickening by inflation (Self et al. 1996; see Fig. 9, Chap. 2).

These flow fields fed by very long (≥ 100 km) fissures and eruptive styles of flood basalt were Hawaiian-like in nature at the vent, featuring lava ponds (White et al. 2009). The longest and largest eruptive units on Earth are confined to CFBPs; some huge-volume flows are known to have advanced more than 750 km from their source. Individual flow units are thickest on shallow slopes, where they commonly exceed tens of metres in thickness and are spectacularly columnar jointed (Photo 1). The Snake River Plain flood basalt province in Southern Idaho is so young (a few thousand years) that its original topography is nearly intact. The surface was actually formed by many flows from numerous, widely separated vents. Around each vent, the outpouring of highly fluid lava built broad (several km at the base), nearly flat lava shields or mounds (30–60 m high) with slopes of less than 1°. Numerous inter-flow horizons, called boles, are most commonly red or brown. These have been variably interpreted as altered glassy bases of basalt flows, paleosols developed on exposed flow tops, altered volcanic ash beds, or local inter-flow sediments, to which Duraiswami et al. (2020) added the weathering of flow-top and flow-bottom breccias. Flood lava flows generally postdate and overlie mafic volcaniclastic deposits, and most of them are commonly underlain by beds of shattered, glassy volcanic debris called hyaloclastites. These glassy debris represent the filling of pre-existing river canyons with lava, termed intracanyon flows, which commonly show well developed entablature jointing (Lockwood and Hazlett 2010).

The length of flow fields was supply-limited rather than landscape- or cooling-limited (see Chap. 2). Lava flows longer than tens of kilometres are favoured by a pāhoehoe emplacement style (thermal insulation), sustained eruptions of years to tens of years, favourable slopes, unhindered flow conditions (e.g. dry river beds), and an insulated conduit system (lava tubes). Lava flow size is ultimately limited by supply. Thus, the key to these floods of lava is the immense volume of magma released during one eruption, rather than the lava viscosity, eruption rate or environmental conditions. Postulated eruption rates of ~4000 m^3/s are similar to the maximum sustained eruptions rates of the largest historic flood lava eruptions: the Laki event with a volume of 14 km^3 of basaltic magma which erupted over eight months in 1783–1784 AD/CE (Thordarson and Self, 1993), and Eldgjá 939–940 AD/CE in Iceland (Oppenheimer et al. 2018). It should be noted that ten years of effusion at the maximum sustained Laki eruption rate, estimated at ~4000 m^3/s by Thordarson and Self (1993), would yield a flood basalt flow field of ~1250 km^3.

Long fissure-type systems, near-vent associations, and plumbing system

Vents are poorly known, but some are long fissure-type systems. The total fissure system for the Roza flow is ~180 km in length at the effusion rate of 1000 m^3/s (Self et al. 2015). The fissures appear to cluster in time and space and are often erupted from a group of sub-parallel linear-vent systems, represented by dyke groupings or swarms (Fig. 8). The occurrence of dyke systems that can be traced laterally for tens to hundreds of kilometres implies that many flood basalt eruptions are fed by linear vent systems. In the CRB, associations of fountain-fed flows, spatter and lapilli scoria units, and pāhoehoe define complex linear vent systems that bear a resemblance to the cone complexes formed by modern-day fissure eruptions, e.g. the Tolbachik fissure eruptions, 1975–1976 and 2012–2013, in Kamchatka (Fedotov and Markhinin 1983; Gordeev et al. 2013).

Some authors (White et al. 2009; Self et al. 2015) have pointed out similarities between flood basalts and the largest historical effusive eruption in SE Iceland in terms of vent morphology with rows of craters along swarms of dykes (Fig. 9). The near-vent succession of fountain-fed lavas, scoria beds and lavas of the preserved 14.98 Ma-old Roza Member is identical to that found in the near vent successions of the 1783–1784 AD/CE Lakagigar and 939–940 AD/CE Eldgjá fissure-fed flood lava eruption (Self et al. 1996; Thordarson and Self 1993, 1998; White et al., 2009). The pyroclastic cones along the crater row are characterized by low to moderate slope angles (<20°) and minimum heights of up to ~160 m (Photo 4). They are composed of agglutinated and welded spatter and scoria that extend up to 750 m away from the vent. Thick, well-sorted fall deposits composed of scoria lapilli extend away from some vents and exhibit some characteristics comparable to the proximal deposits of violent Strombolian or subplinian eruptions (Brown et al. 2014).

Continental flood basalt provinces consist of intrusive systems in the crust, extensive flood lavas, and mafic volcaniclastic deposits in varying proportions (White et al. 2009). Intrusive rocks are inferred to represent the solidified remnants of a plumbing system that fed eruptions at the surface, as well as representing substantial accumulations of magma in the subsurface (Mittal et al. 2021; Mittal and Richards 2021). The vast majority of intrusive rocks within the upper crust is in widespread sills as well as spawning dyke assemblages of complex geometry. The CFBP magmatic system consists of some combination of interconnected sills or magma reservoirs throughout the crust (Ernst and Youbi 2017). According to Mittal et al. (2021) and Mittal and Richards (2021), the most plausible magmatic structure able to explain all observations made in the Deccan is a CFBP magmatic system that likely consists of a number of medium-sized magma bodies (< a few thousand km^3) instead of a couple of large

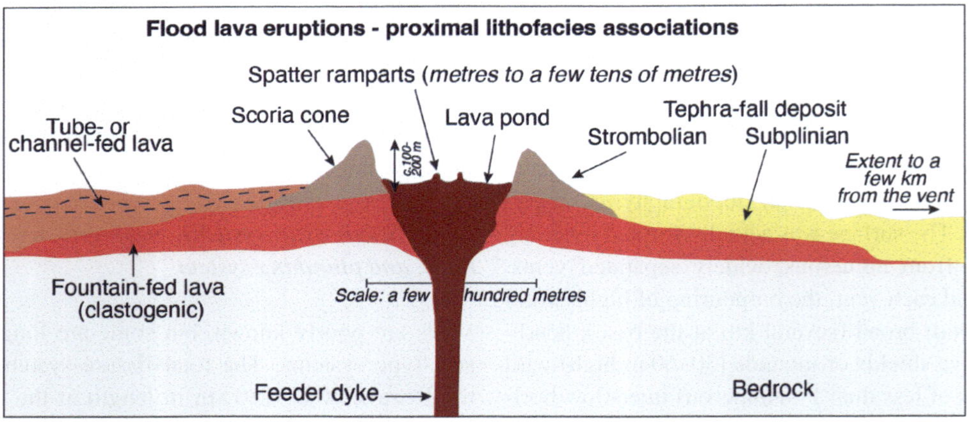

Fig. 9 Flood lava eruptions: proximal lithofacies associations (modified after White et al. 2009). Schematic illustration showing the stratigraphy of the near-vent successions produced by a single eruption episode during the Laki and Eldgjá eruptions, Iceland, as analogues to the Columbia River Flood Basalt, in particular near-vent Roza member eruption products. Approximate sizes have been added to the unscaled, original diagram of White et al. (2009, Fig. 8, p. 300)

magma chambers. These magma bodies have to be spatially separated by at least a few kms if not more, in order not to become thermally merged over hundreds of kyr (Mittal et al. 2021; Mittal and Richards 2021).

In the case of bimodal volcanism, a large-scale explosive style prevailed, such as Paraná-Etendeka, in which nine silicic eruptions ranging from VEI 8.7–9.3 produced as much as 2000–8000 km^3 DRE around 132 Ma (Deligne and Sigurdsson 2015). Silicic pyroclastic eruptions have emplaced deposits that are as voluminous, if not more so, than individual flood-basalt units, and with much greater dispersal. In addition, significant volumes of mafic volcaniclastic deposits in many CFBPs have revealed evidence for extensive phreatomagmatic eruptions that preceded flood basalt emplacement. Interaction of dykes and shoaling sills with near-surface aquifers favours the production of mafic volcaniclastic deposits, which comprise large vent complexes and substantial pyroclastic deposits (White et al. 2009). Thus, the dominance of explosive volcanism in LIPs is a result of either a silicic magma composition and/or environmental conditions at eruptive sites conducive to large-scale phreatomagmatism (see Chap. 3 for a definition).

Ignimbrite plateaus from SLIPs

Caldera-studded ignimbrite plateaus represent large volcanic fields such as the well-exposed Sierra Madre Occidental in Mexico (Ferrari et al. 2002), the Great Basin in the western United States, and continental hot spot tracks such as the Yellowstone-Snake River Plain that trends across southern Idaho (Fig. 7; Jordan 2005). The distinctive features are stacks of ignimbrite sheets defining a plateau several tens to hundreds of thousands of square kilometres in extent in southern Peru and northern Chile (Photo 5). Ignimbrite plateaus are the broad, gently sloping, thick landforms forming aprons surrounding large caldera complexes or clusters (de Silva and Lindsay 2015). These ignimbrites can be superimposed and can overlap each other, while individual silicic eruptive units can have thicknesses, areal extents, and volumes that are comparable to, or exceed, interbedded flood basalt lavas (for a definition of ignimbrites, see Giordano and Cas 2021). At the core of the plateau are the calderas and other source vents for the ignimbrites such as broad shield-like ignimbrites (de Silva and Lindsay 2015; see Chap. 6). Caldera complexes represent the main eruptive sources for the large volumes of silicic magma, and range from regional downsag structures to complex volcanic-tectonic collapse structures controlled by tectonic stresses and preexisting crustal architecture (Bryan et al. 2002).

Ignimbrite plateaus are found in a variety of tectonic environments where continental crust is present, either at subduction zones or at intraplate hot spots resulting in continental rifting. The largest volume silicic igneous provinces occur along accreted continental margins. The requirement of continental crust points to crustal melting and assimilation by mantle-derived basalt being involved in the production of the large volumes of silicic magma. The coherence of volcanic style, magmatic composition, and temporal transience of these episodes have been suggested to be the result of a 'flare-up' in magma production caused by an anomalous thermal input from the mantle (de Silva 1989). Detailed studies of continental flood basalt provinces are revealing complex spatial–temporal relationships between silicic and mafic volcanism. The simultaneous existence of mafic and silicic magmas in some eruptive centres (e.g. Messum Igneous Complex in Namibia), suggests that large calderas could be the site of both basaltic lava and rhyolitic ignimbrite-forming eruptions (Bryan et al. 2002).

Origin and Source of LIPs

Relative to MORBs, flood basalt provinces include tholeiitic basalts with higher MgO and more diverse major-element compositions. Contamination by either continental crust or subcontinental lithospheric mantle has been proposed to be the source of the distinct isotopic signatures in the basalts. However, the origin and source of flood basalt magma is highly debated (Coffin and Eldholm 1994; Schmincke 2004). Most authors think that flood basalts are generated when a large plume head impinges on the base of the lithosphere (Ernst and Buchan 2001; Campbell 2005). These mantle plumes are probably hotter than the mantle that rises beneath mid-ocean ridges and thus generate greater amounts of partial melting. For CFBPs that were erupted over a short geologic timescale (1–3 Myr, e.g. Deccan traps), flood basalts represent melts from a rapidly rising plume. Other authors (White and McKenzie 1995) think that lithosphere rifts above abnormally hot asthenosphere, which occurs when the head of a plume intersects a slowly drifting plate. In this model, the hydrated sub-continental lithosphere is thought to become partially melted above a hot rising plume, the heat being supplied by the plume, while the magma has its source in the much older lithosphere. This may be the case where the lithosphere is thicker than 100 km and flood basalts were erupted over a long period (10 Myr), as for the Paraná basalts in South America.

Associated silicic volcanic rocks in many provinces demonstrate the importance of crustal melting associated with the generation and emplacement of CFBPs (Self et al. 2015). Large-volume silicic igneous provinces ultimately reflect large-scale crustal melting processes in response to lithospheric extension and high thermal input from underlying hot mantle (Bryan et al. 2002). When plume heads impinge on continental crust, LIPs can be more silicic and produce the most violent forms of volcanism on the planet, with magnitudes exceeding VEI 9. Such eruptions can produce huge volumes of tephra that mantle thousands of km^2, such as the Cenozoic volcanoes of the San Juan Mountains in Colorado, USA or the Snake River Plain and Yellowstone volcanic systems (Fig. 7; see Chap. 6) (Christiansen 2001; Jordan 2005; Smith et al. 2009a, b; Foulger et al. 2015).

Correlation of continental flood basalt provinces with mass extinction events

Flood basalt sequences have recently received much attention because some of them are associated in time with major mass-extinctions (Fig. 10A; Rampino and Stothers, 1988; Courtillot and Renne 2003; Saunders 2005; Kelley 2007; Bryan et al. 2010; Rampino et al. 2019). The four largest mass extinctions in the last 260 Ma coincide with four traps (Fig. 10A), making a causal connection between the two, with some form of catastrophic climatic perturbation being the most likely hypothesis. The hottest debate centres around the eruption of the Deccan flood basalts (Fig. 10B), whose age (66 Ma) exactly coincides with that of the Cretaceous/Tertiary (K/T) boundary. The Chicxulub asteroid impact (Mexico) and the eruption of the massive Deccan flood basalt province are two proposed causes of the end-Cretaceous mass extinction, which included the demise of non-avian dinosaurs.

Searching for sound statistical links, Rampino et al. (2019) found that the ages of 10 of the 11 well-documented extinction episodes from the last 260 Myr show correlations, with a very high

Fig. 10 LIPs and mass extinctions. The Deccan traps case study. **A**. Diagram showing potential correlations between LIPs, namely continental flood basalts (CFBs) and oceanic plateaus, and mass extinctions and/or oceanic anoxia events (adapted from, and modified after Rampino et al. *in:* Koeberl C, Bice DM, eds, 2019). Strongly correlated events with age-scale boundaries of geologic periods are highlighted in red. **B.** Stratigraphy, sampling transects, and U/Pb age model for the Deccan traps (after Schoene et al. 2019, Fig. 1, p. 863). **a**. Elevation map of study location in the Western Ghats, India. A black segmented line denotes cross section X–X′ as shown in B. Sampling transects are located by colored dots. **b**. Geologic cross section through the field area, with sample locations indicated. Different basalt formations in the Deccan Traps are colour-coded to the stratigraphic column in C. Cross section is based on previous work, modified according to our geochronology. **c**. Volumetric stratigraphic column and magnetic chrons of the major formations of the Deccan Traps. The plotted sample heights are based on the composite stratigraphic section. The age model for the Deccan Traps, based on U-Pb geochronology, is shown with 95% credible intervals. Horizontal gray bars indicate eruption ages derived from populations of zircon dates from each horizon; black horizontal bars show dates refined from the stratigraphic Bayesian model. The vertical gray-shaded bar shows an age for the Chicxulub impact (all references *in*: Schoene et al.'s 2019)

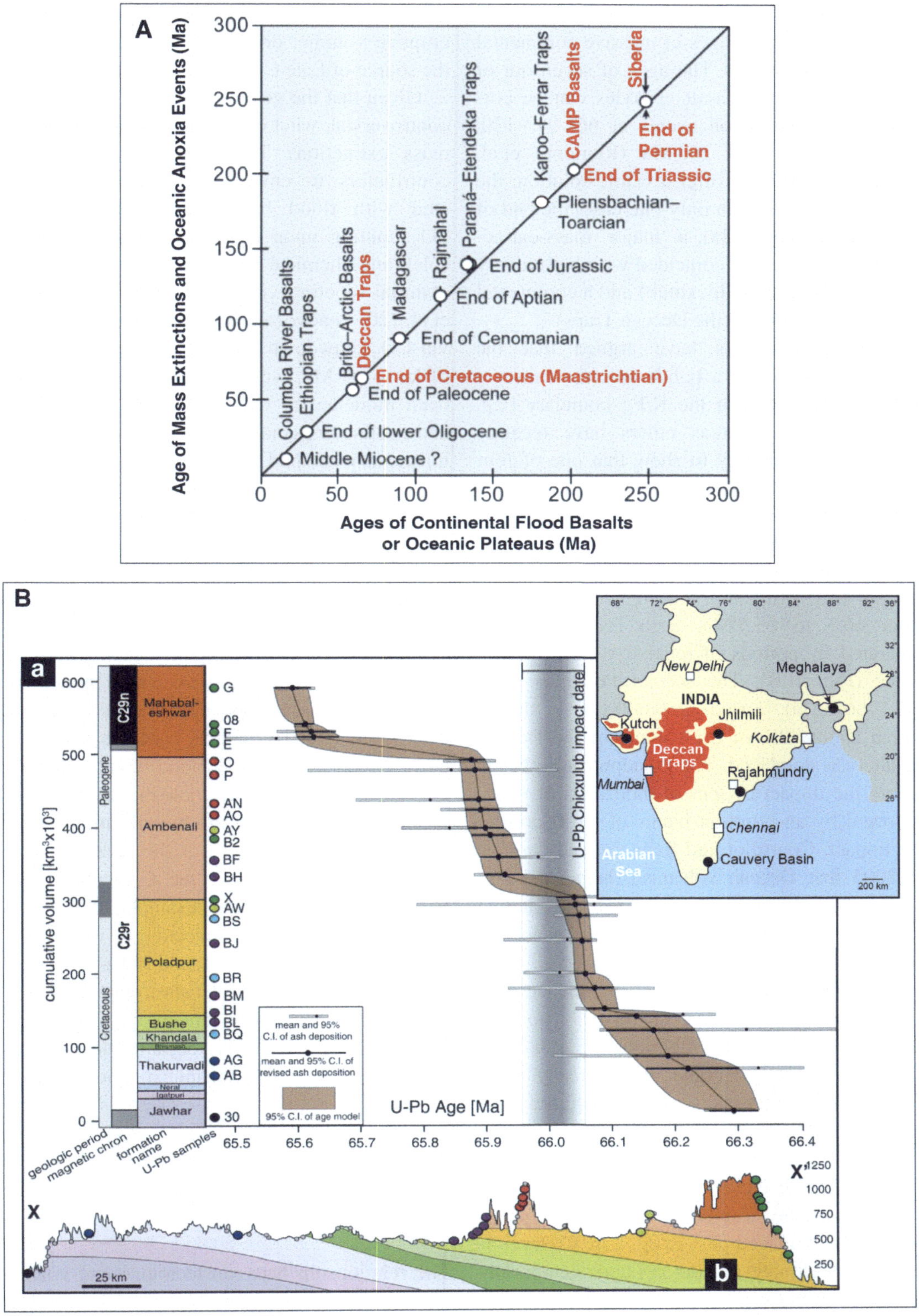
A
Age of Mass Extinctions and Oceanic Anoxia Events (Ma)
Ages of Continental Flood Basalts
or Oceanic Plateaus (Ma)
0
50
100
150
200
250
300
Siberia
End of
Permian
CAMP Basalts
End of Triassic
Karoo–Ferrar Traps
Pliensbachian–
Toarcian
Paraná–Etendeka Traps
End of Jurassic
Rajmahal
End of Aptian
Madagascar
End of Cenomanian
Deccan Traps
End of Cretaceous (Maastrichtian)
Brito-Arctic Basalts
End of Paleocene
Ethiopian Traps
End of lower Oligocene
Columbia River Basalts
Middle Miocene ?
B
a
cumulative volume [km³x10³]
geologic period
magnetic chron
formation name
U-Pb samples
Paleogene
Cretaceous
C29n
C29r
Mahabaleshwar
Ambenali
Poladpur
Bushe
Khandala
Thakurvadi
Neral
Igatpuri
Jawhar
G
08
F
E
O
P
AN
AO
AY
B2
BF
BH
X
AW
BS
BJ
BR
BM
BI
BL
BQ
AG
AB
30
mean and 95% C.I. of ash deposition
mean and 95% C.I. of revised ash deposition
95% C.I. of age model
U-Pb Age [Ma]
U-Pb Chicxulub impact date
65.5
65.6
65.7
65.8
65.9
66.0
66.1
66.2
66.3
66.4
New Delhi
Meghalaya
INDIA
Jhilmili
Kutch
Kolkata
Deccan
Traps
Mumbai
Rajahmundry
Chennai
Arabian
Sea
Cauvery Basin
200 km
X
X'
25 km
b
1250
1000
750
500
250

level of confidence, with the ages of the largest impact craters or the ages of massive continental flood-basalt eruptions. The ages of seven out of eleven major flood-basalt episodes can be correlated with extinction events at 66, 94, ~120, 183, 201, 252, and 260 Ma (Rampino et al. 2019) (Fig. 10A). To offer a viable solution, the authors stated that: "in only one case, the end of the Cretaceous (66 Ma), a 'major' mass-extinction event apparently coincided with both a very large impact crater (Chicxulub) and a continental flood-basalt eruption (the Deccan Traps)".

Earlier researchers have argued that the Chicxulub impact of a bolide was the cause of the mass extinction at the K/Pg boundary (e.g. Alvarez 2003), whereas others have recently used U/Pb chronology to show that one of four volcanic pulses in Deccan, with a high time-scale resolution of 40,000 years, preceded the Chicxulub impact (e.g. Schoene et al. 2015, 2019). These authors' results showed that the Deccan Traps erupted in four high-volume events from 66.3 to 65.5 Ma, each lasting ≤ 100 ka, separated by periods of relative volcanic quiescence (Fig. 10B). The second Poladpur pulse (66.1–66.0 Ma) lasted 250,000 years, during which about 80% of the volume of the Deccan basalts was emplaced. The Poladpur pulse began before the impact by tens of thousands of years, followed by an eruption hiatus of ≤ 100 ka after the impact. Courtillot and Renne already argued in 2003 that Deccan volcanism began prior to the impact and straddled it in time, excluding the invoked possibility of a causal connection between the bolide impact and volcanism.

By contrast, the same issue of Science contains a paper by Sprain et al. (2019) who used argon-argon dating of the volcanic ash from the Deccan traps to argue that a steady eruption of the flood basalts mostly occurred after the Chicxulub impact. Sprain et al. (2019) found that the Deccan traps did not erupt in three large, discrete pulses and that >90% of their volume erupted in < 1 million years, with ~75% emplaced post-K/Pg. Late Cretaceous records of climate change coincide temporally with the eruption of the smallest Deccan trap phases, suggesting that either the release of climate-modifying gases is not directly related to eruptive volume, or Deccan volcanism was not the source of Late Cretaceous climate change.

Given that the geochronological data remains controversial, what could be the mechanisms for mass extinction? The most commonly cited contributors to environmental change associated with flood basalts are CO_2 (warming), SO_2 (cooling upon conversion to sulfate aerosols), and chemical weathering of fresh basaltic material (cooling via CO_2 drawdown). Schoene et al. (2019) advocate for a cooling contribution via CO_2 drawdown: at time scales of hundreds of ka to >1 Ma, weathering of fresh basalt has been modelled to result in net CO_2 drawdown and cooling, especially if the basalt is at low latitudes, as were the Deccan traps.

Much remains to be discovered as to how flood basalt magmatism contributes to mass extinctions. An update has been provided by Bond et al. (2014) in the textbook 'Volcanism, Impacts, and Mass Extinctions: Causes and Effects' (Keller and Kerr 2014). The authors claim that there is no direct correlation between total volume of lava and extinction magnitude because there is always sufficient recovery time between individual eruptions to negate any cumulative effect of successive flood basalt eruptions. Bond et al. (2014) further argue that the best-constrained examples of death-by-volcanism record the main extinction pulse at the onset of often explosive volcanism (e.g. the Capitanian, end-Permian, and end-Triassic examples, Fig. 10A), suggesting that the rapid injection of vast quantities of volcanic gases (CO_2 and SO_2) is the trigger for a truly major biotic catastrophe. However, many voluminous large igneous province eruptions (e.g. the Cretaceous oceanic plateaus) are not associated with significant extinction loss.

3 Tectonic Interactions with Magmatism at Volcanoes

The relationship between tectonics and volcanism is traditionally thought to be the result of interaction between the basement structures and the regional stress field in which the volcanic edifice is located. Recent research has suggested a new perspective, as the magmatic activity and volcano load can in turn influence the tectonic environment. Interactions are thus twofold: (1) tectonics play a role in the location, alignment and spacing of volcanoes, the magma's path to the surface, and ultimately the growth of the volcanic edifices; (2) at rift zones and grabens, dyke and sill intrusions are largely independent of the regional tectonic context. Furthermore, magma intrusions can also deform the volcanic edifice, while volcano growth can modify local faulting.

3.1 Broad Scale Interactions

Considering interactions at the broadest scale possible, Cas and Wright (1988) classified areas of modern volcanic activity into one of the following three tectonic regimes:

(1) Divergent regime, characterized by: (a) mid-ocean ridge volcanism; (b) marginal seas or back-arc basins of spreading origin, volcanism behind oceanic island arcs (e.g. Lau-Havre Trough, Marianas Trough); (c) Continental rift volcanism: narrow, linear zones (e.g. East Africa rift zone), including aulacogens (rift zones that are no longer active, or failed arms of a triple junction; Burke 1977), and wide zones (e.g. Basin and Range Province, western USA);
(2) Convergent regime, found in: (a) continental margin arc volcanism associated with oceanic trench subduction zones (e.g. Andes, Cascades volcanic belt, western USA); (b) micro-continental arc volcanism associated with oceanic trench subduction zones (e.g. Japan, New Zealand, Indonesia), and; (c) young island arc volcanism associated with oceanic trench subduction zones (e.g. Tonga-Kermadec, Marianas, Aleutians, Hellenic, Scotia arcs);
(3) Passive regime, characterized by: (a) intraplate continental (flood) volcanism (e.g. eastern Australia Cenozoic volcanism); (b) intraplate oceanic islands volcanism (e.g. Hawaiian chain) and other oceanic volcanic seamounts.

3.1.1 Faulting, Stress Field and Magma Paths

The passage of magma through the crust and lithosphere is controlled by the crustal lithospheric stress-field and shallower stress-field configurations, and resulting fractures, i.e. normal faults, thrust faults, and strike–slip faults (Fig. 11, after Cas and Wright 1988). The rise of magma to the surface is dependent on the existence of vertical or subvertical fractures, or pathways in the crust. Such fractures are most likely to form when σ_1, the maximum principal stress, is orientated vertically or horizontally, and σ_3, the minimum principal stress, is orientated horizontally. If σ_1 is high enough to exceed the tensile strength of the rock then fractures will propagate parallel to this stress axis. Magma can move along these newly-created fractures only if the magma pressure is greater than σ_3. In a situation where σ_1 is vertical and σ_3 is horizontal, equivalent to normal faulting, the magma will move through subvertical dyke-like fractures toward the surface (Fig. 11-1). Where σ_1 is horizontal, the resultant fractures will be sub-horizontal if σ_3 is vertical, and could lead to low-angle thrusting (Fig. 11-2). If σ_1 is horizontal and σ_3 is also horizontal, vertical fractures will be produced and strike-slip movement will occur (Fig. 11-3). It is suggested that voluminous volcanism will generally only occur in extensional, strike-slip or passive (with associated pre-existing vertical crustal fractures) tectonic regimes, but not in highly compressional tectonic regimes, except in very rare cases, e.g. Ambrym, Vanuatu (Shreve et al. 2019). This exception, a counterintuitive caldera location in compressional regime, explains why active volcanic arcs may be created far away from the trench, at a distance of about 200–250 km.

3.1.2 Strike-Slip or Transform Faults and Oblique Convergence

Strike-slip or transform faults do not correspond to favourable locations for magma generation

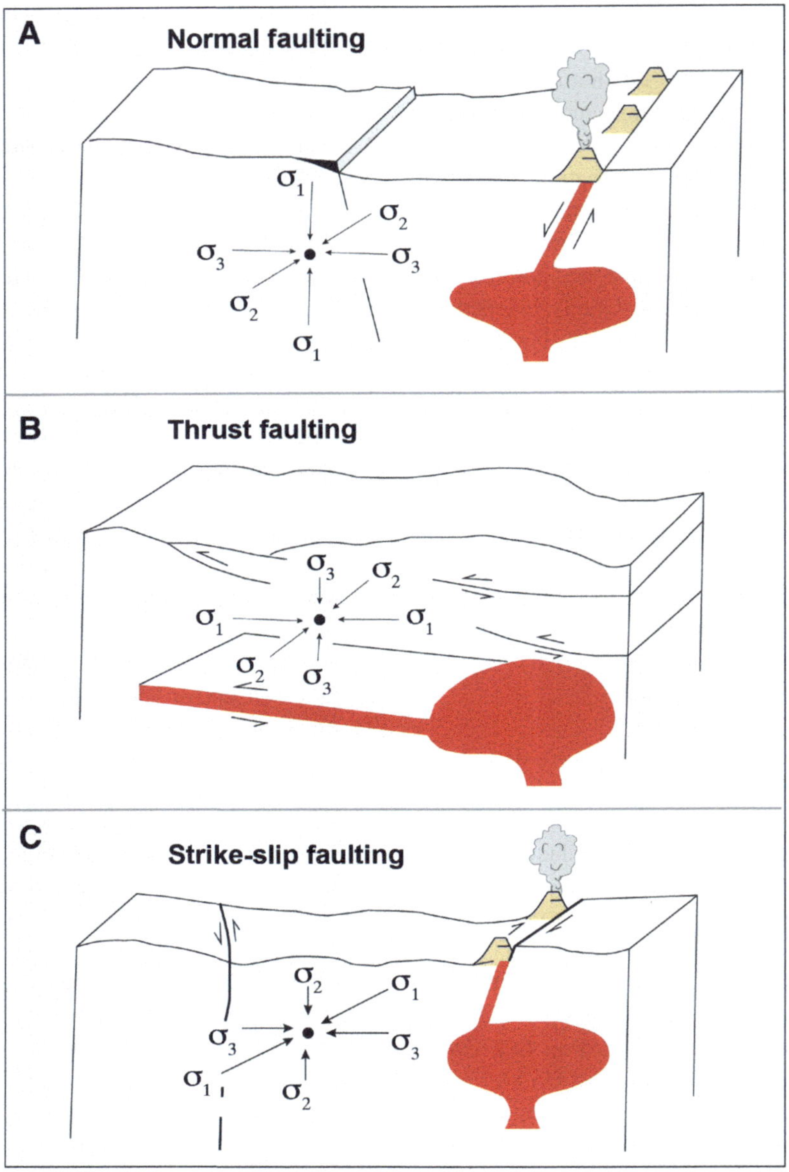

Fig. 11 Sketch showing how fault or fracture orientations through the crust and lithosphere control the path of magma feeding volcanoes and may result from three stress-field configurations: (a) normal faulting, (b) thrust faulting, and (c) strike-slip faulting (Courtesy of Cas and Wright 1987, Fig. 15.11, p. 463, see also Fig. 17.63 *in*: Cas et al. 2024). Voluminous volcanism generally only occurs in extensional, strike-slip or passive (atectonic) regimes, but not in highly compressional tectonic regimes

and edifice growth. However, composite volcanoes can grow along strike-slip faults in highly oblique convergent arcs such as in Sumatra, and along transform faults that offset ocean ridges such as the Mid-Atlantic Ridge (MAR) at the Azores hot spot.

The structure of the Azores archipelago, located on a hot spot next to the MAR (Fig. B in Appendix), is mainly controlled by the transtensional motion along the transform plate boundary between the Eurasian and African plates. The Azores archipelago straddles the MAR at a triple junction that includes an oblique and recently formed ultraslow rift, a zone of diffuse seafloor deformation, a major fracture zone and a hot spot (O'Neil and Sigloch, *in*: Kueppers and Beier 2018). The archipelago is mainly located to the east of the MAR, with seven islands to the east and only two islands to the west of it. As a result of the activity of the transtensional plate boundary, several of the Azores volcanoes show distinctive morphological, structural and volcanological features (Acocella 2021). At least three islands appear to be strongly elongated (Sao Miguel, Sao Jorge and Pico) parallel to the plate boundary. The elongation is highlighted by transtensional fault systems, fissure eruptions forming focused rift zones and aligned composite and monogenetic volcanoes. Several volcanoes, such as Sao Miguel, Terceira, Faial and Corvo, currently host calderas which are usually associated with the eruption of more evolved magmas or the emptying of large magma reservoirs.

Arc volcanoes often show evidence of regional convergence conditions; many edifices are elongated parallel to the direction of the maximum principal stress σ_1 and several arc volcanoes show a local alignment parallel to the σ_1 direction (Acocella 2021). These features, for example along the Andean and Japanese arcs, result from the preferred development of feeder dykes parallel to the regional σ_1 direction. Convergence angle, oblique or orthogonal, also has a role on the location of volcanoes on the overriding plate. When it is oblique, the most common case, volcanoes grow between en echelon strike-slip faults and within grabens in the strike-slip domain inland, away from the contractional domain. When the convergence is orthogonal, volcanoes grow between thrust faults produced by contraction and along secondary faults that accommodate the thrust faults (Acocella 2021, his Fig. 10.12).

3.2 Structural Setting and Stress Field Control on Volcano Distribution

The interactions between tectonics and volcanic structures can be considered at different scales from that of the volcano edifice, either monogenetic or polygenetic, to that of the entire volcanic field. As first-order interactions, the structural setting and stress field exert some control on the distribution, spacing and alignment of both composite and monogenetic volcanoes (Tibaldi et al. 2017).

3.2.1 Alignment and Spacing of Active Polygenetic and Monogenetic Volcanoes

The alignment of edifices along tectonic features has long been recognized at different regional scales, such as the twofold alignment of volcanoes in the Galápagos archipelago or the Boset-Bericha-Gudda volcanic centres along the NE-SW Wonji fault system (Ollier 1972; Corti 2009; Fig. 12A, B), and the Cameroon Volcanic Line to the north of the Gulf of Guinea (Lockwood and Hazlett 2010). Aligned composite volcanoes in divergent plate boundaries form as a consequence of localized decompression along the axis of rift zones, whereas those in convergent plate boundaries result from the availability of magma above the zone of partial melting of the slab, with the location of the volcanic arc being ultimately controlled by the slab dip. At a local scale, the alignment of monogenetic volcanoes along a direction determined by eruptive fissures commonly results from the shallow emplacement of a dyke, whose direction is largely controlled by the stress field, but partly by pre-existing structures in bedrock.

Most monogenetic fields display a statistically clustered vent distribution, with the volcanic alignment depending on the tectonic environment as well as other factors, e.g. topography, pre-existing structures, and local stress changes due to older intrusions (Le Corvec et al. 2013). However, some monogenetic fields like the Auckland Volcanic Field do not show any obvious alignment. The overall distribution of vent clusters within volcanic fields suggests that magma propagation is the result of interplay between deep magma source(s) and the shallow level stress field or pre-existing crustal structures (Acocella 2021).

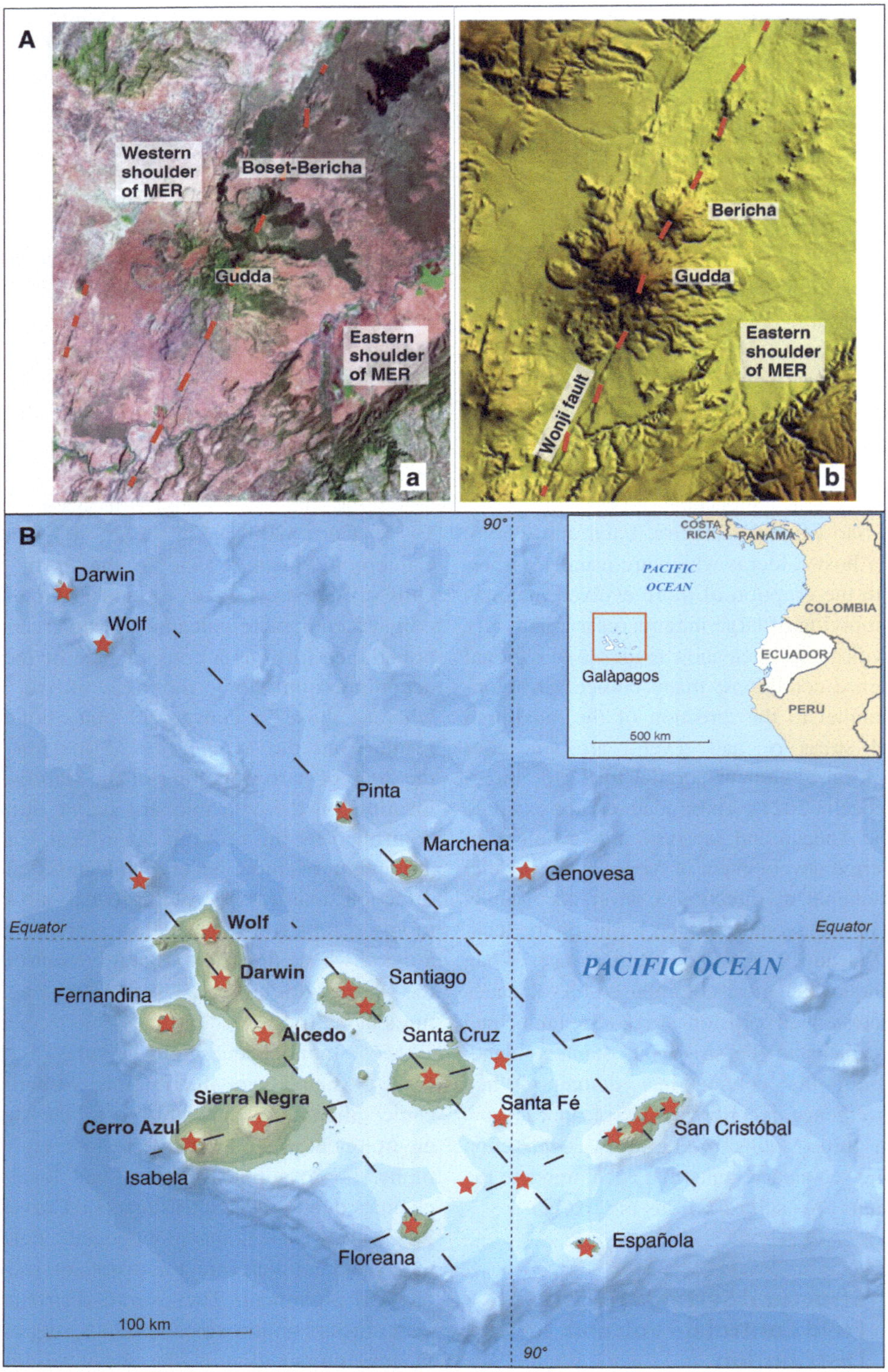

Fig. 12 Alignment and spacing of volcanoes. **A.** Alignment of the Boset-Bericha-Gudda volcanic centres along the NE-SW trending Wonji fault system in the Northern Main Ethiopian Rift (1: satellite image and 2: DEM, after Corti 2009, Fig. 9, p. 12). **B.** Alignment and spacing of volcanoes and islands in the Galapagos archipelago (modified after Ollier 1972). Dashed lines are apparent lineaments

The distance or spacing between polygenetic volcanoes within an associated group may be related to several first-order factors. In general, the thinner the crust or the lithosphere, the more closely spaced the volcanoes (e.g. northern branch of the EARS, Fig. A in Appendix). On similar grounds, the spacing of monogenetic vents has been related to the thickness of the crust beneath various volcanic fields in different extensional tectonic settings. This highlights a relationship between the distribution of monogenetic vents and the depth of origin of the magma, apart from at convergent plate boundaries, where the spacing of the composite arc volcanoes does not follow a pattern and is more difficult to interpret. Other studies (Muller et al. 2001; Pansino and Taisne 2019) relate complex volcano spacing to the load of a volcanic edifice: larger volcanoes impose a higher load on the crust, thus inducing a stronger stress field locally that is capable of attracting magma below it in the form of dykes directed toward the volcano from progressively wider regions. This ability to attract magma over wider areas may explain why larger volcanoes are also more widely spaced (e.g. the northern Cascades arc).

The flexural response of the lithosphere due to the load of a volcanic edifice may also control the distance between adjacent polygenetic volcanoes that develop successively. In particular, compressive stresses caused by crustal flexure due to volcanic loading prevent new upwellings nearby, forcing a new volcano to develop at a minimum distance equal to that at which the radial stresses switch from compressional to tensile. This change occurs at the inflection point of a flexed lithosphere, which is proportional to the thickness of the plate (Acocella 2021).

3.2.2 Far-Field Plate Motion and Local Stress Field Interactions in Continental Rifts

These interactions can be observed in continental rifts such as EARS. A volcano in the EARS setting should produce dykes and fissures parallel to the rift, as the direction of minimum horizontal stress is perpendicular to the rift. However, Wadge et al. (2016) found that only seven of the 21 volcanoes that had historical eruptions actually matched this geometrical ideal. The orientation of the eruptive fissures and dykes and/or the axis of the host rift segments at the other 14 volcanoes is oblique to the theoretical values. To explain the oblique orientation of these volcanoes to the rift axis, the authors invoke local (non-plate tectonic) variations in the stress field caused by a combination of the following: crustal heterogeneities and anisotropies (from structures in the Proterozoic basement), transfer zone tectonics at the tips of offset rift segments, gravitational loading by the volcanic edifices with 1–2 km relief, and magmatic pressure in central reservoirs. Wadge et al. (2016) found that the more oblique volcanoes tend to have large edifices, large eruptive volumes, and evolved and mixed magmas that might trigger explosive behaviour. Nine of the volcanoes have calderas of varying ellipticity, six of which are large, reservoir-collapse types mainly elongated across-rift (e.g. Kone) and three are smaller, elongated parallel to the rift and contain active lava lakes (e.g. Erta Ale, see Figs. 3B and 4A), suggesting different mechanisms of formation and stress fields.

These interactions between the stress field and pre-existing (basement) structures can be more complex at the regional scale in volcanic arcs. The complex setting of the Central Andes Volcanic Zone (CAVZ) provides striking examples of mutual feedback between tectonism and magmatism in a contractional volcanic arc (Fig. 13). This feedback stems from a statistical analysis of the distribution of 1518 monogenetic and polygenetic volcanoes of Miocene to Quaternary and Holocene age, and from the analysis of the morphostructural characteristics of 315 craters and cones to reconstruct the magma paths (Tibaldi et al. 2017). The tectonic setting involves normal faults, thrust and strike-slip faults, as well as folds in a complex contractional arc with tens of km-long strike-slip faults. The regional Holocene stress state involves an E–W horizontal principal stress (Fig. 13). During Pliocene and Pleistocene times, the main direction of shortening was also probably E–W, consistent with the development of N–S and NNE–SSW reverse faults and folds,

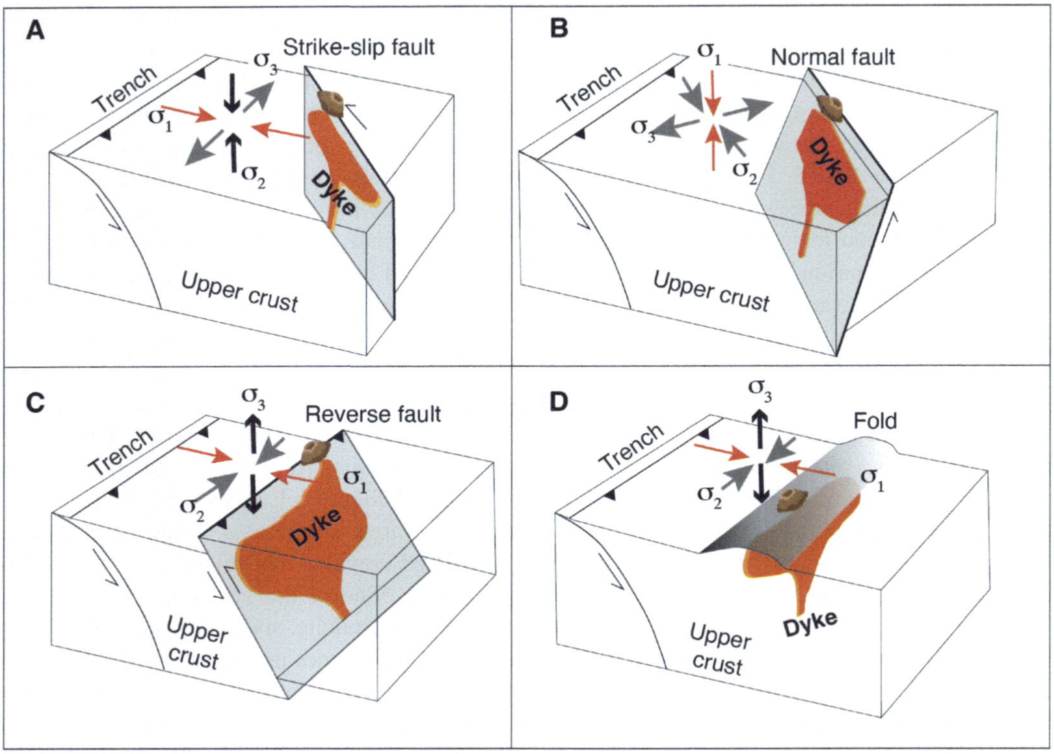

Fig. 13 Sketch (not to scale) of magma path orientation in the Central Andes (North Chile–Bolivia–Argentina) based on outcropping dykes and volcano morphometric parameters (after Tibaldi et al. 2017, Fig. 21, p. 37). In A and B magma paths are oblique to the trench strike and to the plate convergence direction. In A, magma path is also oblique to the principal stress axes. In C and D, magma path strike is perpendicular to the greatest principal stress σ_1 and to convergence. In C, magma paths sometimes use the reverse fault planes, while in D they use the hinge zone of folds

and NW–SE left-lateral strike-slip faults. Most magma paths trend N–S, NW–SE, and NE–SW, in decreasing order of frequency.

North–south normal and reverse faults and N–S folds guide volcano emplacement and magma paths (Fig. 13; Tibaldi et al. 2017). Fold hinges were particularly influential in controlling magma paths and location of volcanoes east of the Salar de Atacama. NW–SE faults also guided magma paths in the northern part of the CVZ. Zones of vertical strata, with bedding striking NE–SW, also acted as preferential magma paths in the north-eastern part of the region. These data suggest that at convergence zones, shallow magma paths can be more sensitive to the presence and geometry of upper crustal weaknesses than to the regional stress regime.

3.2.3 Tectonic Effects on Caldera Location, Morphology, and Formation

To show the relationships between volcanic complex or caldera location, morphology, and tectonics, three approaches have been undertaken based on structural and remote sensing data (e.g. shape of large calderas), and on laboratory experiments and modelling (Thouret 1999 and references therein; see Chap. 6). Classical morphotectonic studies include the statistical analysis of the geometry of drainage patterns and stream directions controlled by tectonics and the identification of morphotectonic features. For example, the main morphological characteristics of the volcanic complexes of Latium in Italy (Buonasorte et al. 1991; Trigila 1995) are strongly controlled by four prevailing tectonic directions for Vulsini, and by three for the Sabatini and Colli Albani areas. These studies allow us to infer where and how the structural setting of the sedimentary units and the recent tectonic activity of the area 'control' the location and shape of the calderas.

The influence of the structural framework on the shape and the formation of large caldera complexes has been inferred from remote sensing and ground-based structural analysis. Based on 2D-images produced by draping SPOT satellite images over a DEM, the relationships between the geodynamic setting, the regional faults, and the calderas of Toba and Tondano, Indonesia, have been described in terms of their evolution within a pull-apart basin (Lecuyer et al. 1997). The links between the main morphologic and structural features on the submerged portions of volcanic edifices such as Panarea and Stromboli in the Aeolian Islands (Gabbianelli et al. 1993), have been carried out through electroacoustic and high-resolution seismic profiles in the SE Tyrrhenian Sea. Both complexes show preferential development along NE– SW lineaments, which coincides with the regional structural trend of this sector of the Aeolian structure. Faulting, caldera collapse, and tectonic tilting were interrelated and were all fundamentally influenced by activation of the NE–SW fractures.

3.3 Volcanic Edifices Can Affect Their Tectonic/Magmatic System and Subsequent Growth

The tectonic effect of volcanic constructs on faulting has been tested through field observations, laboratory experiments and modelling. The effect of volcanic constructs on rift fault patterns is exemplified by three case studies (van Wyk de Vries and Merle 1996, and van Wyk de Vries 2018): (1) Fieale volcano (Asal Rift, Djibouti), showing inward-curving faults; (2) Axial and Brown Bear seamounts, showing Juan de Fuca Ridge curving into the volcanoes; and (3) Maderas volcano (Nicaragua), which is spreading and creating rifts predominantly normal to regional extension. Analogue models show increased fault throw at the volcano, caused by interaction of the regional stress field with that created up by the volcano mass itself. In order for faults to be reoriented there must be a ductile layer below the volcano, such as hot crust at a mid-ocean ridge, weak sedimentary strata, etc. Increased volcano mass and size and lower brittle-ductile ratios lead to increased fault curvature, potentially resulting in volcanoes on one side of a rift capturing the fault and forming a new rift axis. The fact of concentrating extension means that magma is more easily erupted. The resulting increased extension and magma eruption rate leads to rift thinning, which can favour the formation of oceanic crust (van Wyk de Vries and Merle 1996). In addition to volcanic constructs and gradual volcano spreading influencing faulting and rifting (Merle and Borgia 1996; van Wyk de Vries and Merle 1996), they also affect the slope instability of the edifice (van Wyk de Vries and Francis 1997). A volcano of sufficient size induces stresses that may deform its substratum. Huge oceanic shields on thin oceanic lithosphere cause the lithosphere to sag downward into the asthenosphere. As much as half or two-thirds of the growth of the shield volcanoes of Hawai'i may be countered by lithospheric subsidence (Peterson and Moore 1987; Lipman 1995; see also Chap. 4). In turn, this subsidence creates a feed-back of stresses which deform the edifice. Continued sinking well below sea level will lead to guyots, atolls, development of fringing coral reefs and/or to volcanic islands with shoaling phases (see Chap. 3).

3.4 Magmatism Can Modify the Morphology, Structure and Development of Volcanoes

Magma plays an important role in the development of plate boundaries. Older research emphasized the control of regional tectonics on rift zones and convergent plate boundaries, but recent work and modern case studies on seismo-tectonic and volcanic events in Afar and Iceland suggest that magmatism itself can modify the morphology, structure and development of volcanic edifices (Gudmundsson 2020; Acocella 2021).

3.4.1 Magmatism Shapes the Structure of Rifts at Divergent Plate Boundaries

Recent rifting episodes have been linked to the emplacement of tens of kilometre-long dykes along a single rift portion in continental, transitional and oceanic rifts (for example Krafla 1975–1984 and Bárðarbunga 2014 in Iceland; Afar 1978–79 and 2004, and East Pacific Rise 2005–2009). These rifting events led to a rapid and significant opening of a portion of a magmatic system along the plate boundary, but were characterized by a lack of faulting due to regional tectonics. The observed generation of faults and graben-like structures parallel to the rift axis resulted from the emplacement of dykes and was thus magma-induced (Acocella 2021). The overall graben-like and depressed morphology of the plate boundary at the surface with symmetric, normal faults defining the edges of the axial rift zone, results from repeated dyke emplacement at depth. It appears that while regional tectonic processes control the far-field plate motion and divergence, magmatic processes control the near-field evolution of mature plate boundaries and shape their structure and morphology (Gudmundsson 2020).

3.4.2 Magmatism Controls Extensional Arcs

Magma also plays an important role along convergent plate boundaries characterized by extension, such as the Taupo Volcanic Zone of New Zealand (Fig. 14, after Holden et al., 2015). These have tectono-magmatic features similar to those of continental rifts, suggesting that magmatic activity plays an important role in controlling the evolution of the extensional arc and its structure. Two features highlight the non-negligible role of magmatism in strike-slip and contracting arcs (Fig. 15) (Acocella 2021): (1) In strike-slip arcs with an oblique convergence, the magma intruded into the crust can thermally weaken it, focusing along strike-slip deformation induced by strain partitioning along the volcanic arc. The magma may control the structural architecture of the overriding plate and the location of volcanoes. (2) In compressional arcs with a perpendicular convergence, magma has been found to intrude from sills into stacked tabular intrusions, and even to rise to the surface, by locally increasing the principal vertical stress, which may switch from the minimum principal stress σ_3 to the maximum principal stress σ_1 (Fig. 15).

3.4.3 Shallow Magma Intrusions Can Deform Volcanoes

Where magma is close to the Earth's surface and uplift creates some space, shallow plumbing systems are strongly controlled by the way they uplift the surface. This was described long ago by early geologists, for example von Buch's theory of 'craters of elevation' (e.g. van Wyk de Vries et al. 2014). The full range of structures related to intrusive uplift includes extensional fracturing and faulting above dykes and sills, tilting of strata, marginal thrusting and strike-slip faulting. Shallow magma tectonics is illustrated by the conceptual Fig. 16, showing typical features of folding associated with magma intrusion within, below, and in the vicinity of a volcanic edifice. These intrusion-related folds lead to uplift and bulging of the surface or volcano flanks (van Wyk de Vries et al. 2014). Uplift also permits lateral movement, and large sections of crust can slide away and be transported via this process. Examples of this can be found on the Tertiary volcano on Mull (Scotland) and the south-west side of Etna (Italy). Such deformation can happen at all scales, and is common in monogenetic volcanoes. This is the case of the uplift, bulges and grabens of the Puy de Dôme-Petit Puy de Dome-Grosmaneaux, in the Chaîne des Puys, France (Fig. 17).

All these features indicate that the plumbing system of volcanoes and magmatic activity, mostly in the form of dyke propagation and sill emplacement, are largely, but not entirely, independent of the regional tectonic context.

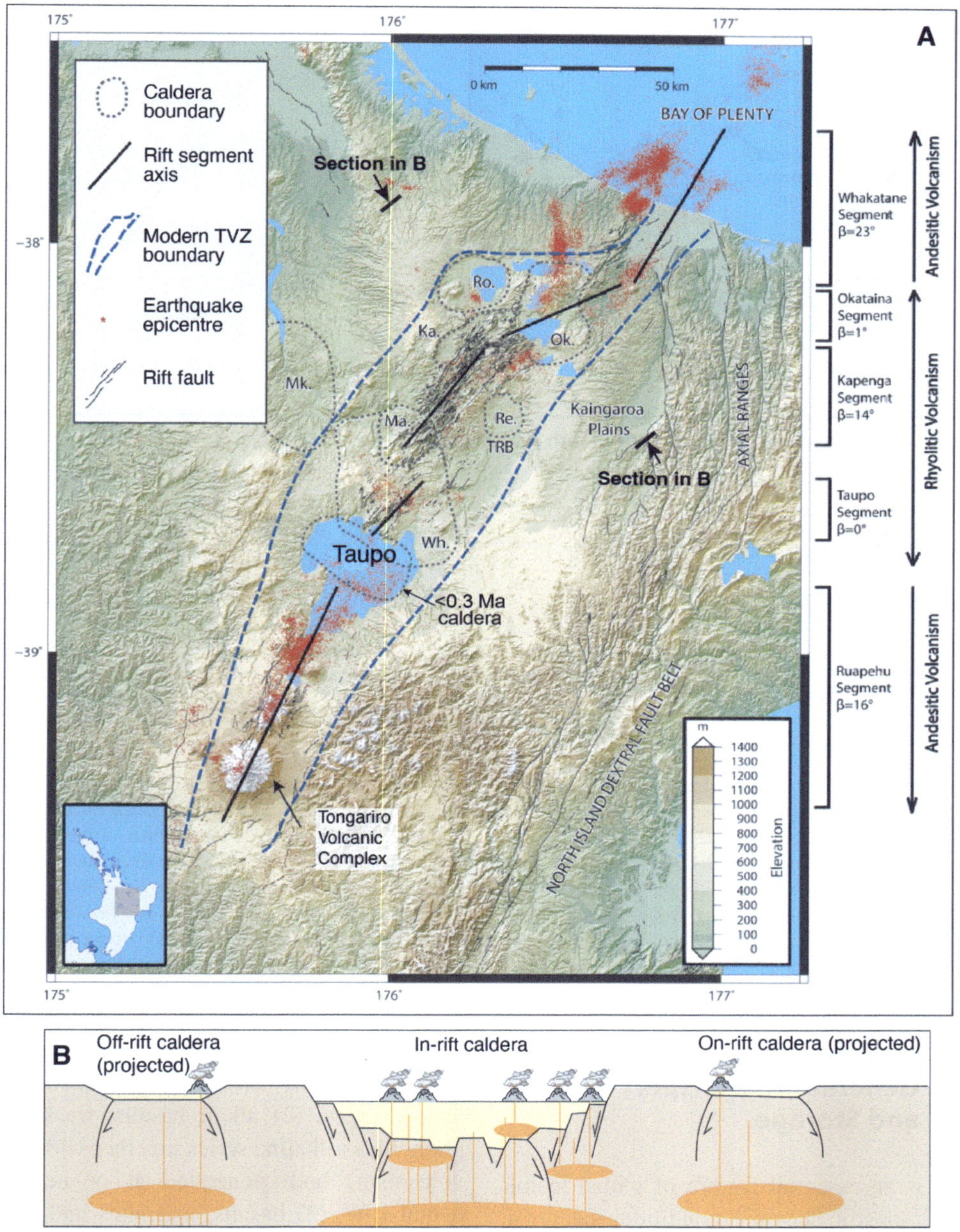

Fig. 14 Alignment and structural setting of in-rift and off-rift calderas in the Taupo Volcanic Zone (after Holden et al. 2015, Fig. 1, p. 2084). **A** Rift segment axis and dextral shear diagrams: light black lines are active rift faults from the GNS Science Active Faults Database. *B* (°) is the stretching factor with positive values signifying extension. Caldera boundary locations are approximate and shown as dashed grey lines. Ro. Rotorua Caldera, Ok. Okataina Caldera, Mk. Mangakino Caldera, Ma. Maroa Caldera, Re. Reporoa Caldera, Ta. Taupo Caldera, To. Tongariro Volcanic Complex, Wh. Whakamaru Caldera. Dashed blue lines represent modern rift boundaries. Bottom left inset shows location of study area. Arrow to Section in B in bold font indicates the axis of the cross section shown below the map. **B** Cross section depicts across-rift differences in caldera structure: the more productive in-axis 'polygenic' calderas are deeper and shaped by regional tectonics, whereas off-axis 'monogenic' calderas are less developed and subcircular (after Spinks et al. 2005, Fig. 3, p. 11)

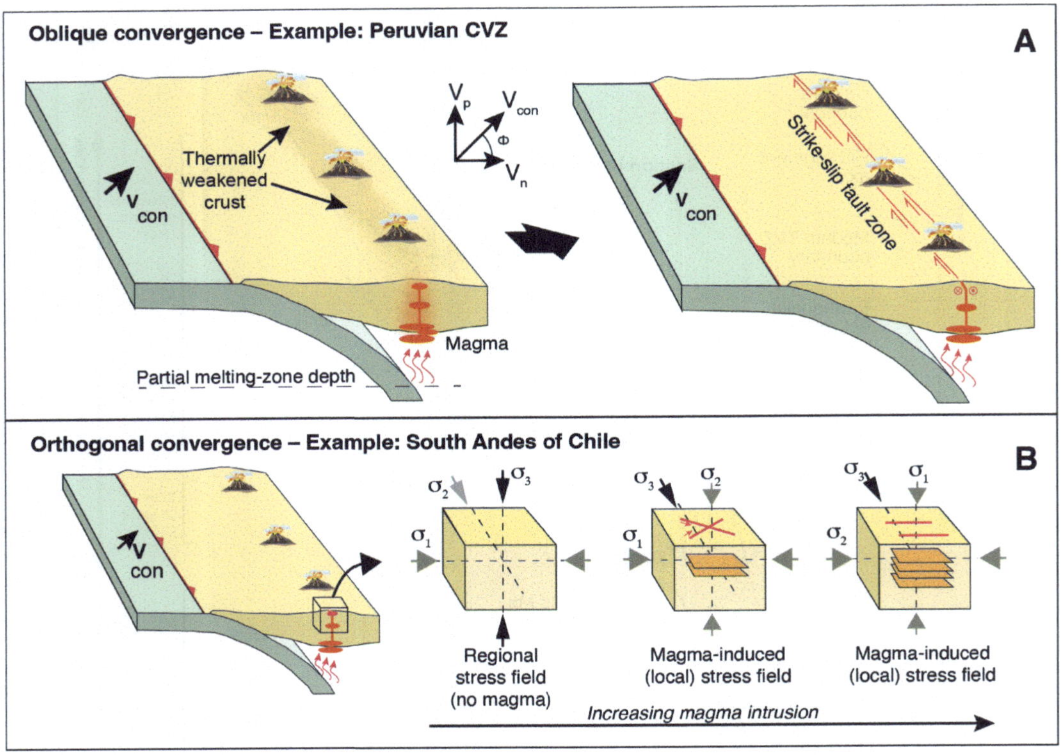

Fig. 15 Oblique and orthogonal convergence along portions of convergent plate boundaries experiencing subduction (map view after Acocella 2021, Fig. 10.24, p. 400). Convergence angles result in distinct structural locations of volcanoes on the overriding plate. **A**. In oblique convergence, the motion resulting from the two plates V_{con} makes an angle ϕ (angle of obliquity) with respect to the direction orthogonal to the trench, defining a trench-parallel V_p and trench-orthogonal V_n component of convergence. Volcanoes are located in the strike-slip domain away from the contractional domain in oblique convergence. **B**. In orthogonal convergence, the motion resulting from both plates defines the convergence V_{con}, which is orthogonal to the contact line between the plates, or the trench. Volcanoes are located between small normal faults and along strike-slip faults in orthogonal orientation

4 Volcanoes and Magma Generation, Transport and Storage

Magma is an essential feature of global tectonics. The specific tectonic conditions occurring in the different settings described above are responsible for different magma compositions and properties, which ultimately influences the type of volcanic edifice and eruptive activity. The position and behaviour (eruptive style) of volcanoes in each environment are linked to the nature of the rising magma.

Figure 18A, B shows a nomenclature of the principal magma types related to four tectonic settings, as summarised above in section 1: (1) tholeiitic basalts (MORBs) produced at mid-ocean ridges; (2) alkali basalts, trachytes and phonolites (alkaline series together with bimodal volcanism) both generated at or near continental rifts; (3) basalts, andesites and rhyolites (andesitic and calc-alkaline series) produced at volcanic arcs in subduction zones; and (4) alkali basalts corresponding to oceanic island hot spots. Decompression melting is the driving mechanism for the divergent boundary-related magmas, while partial melting is the driving mechanism for subduction-related and hot-spot magmas. Volcanoes in hybrid settings produce hybrid magmas similar to those of both

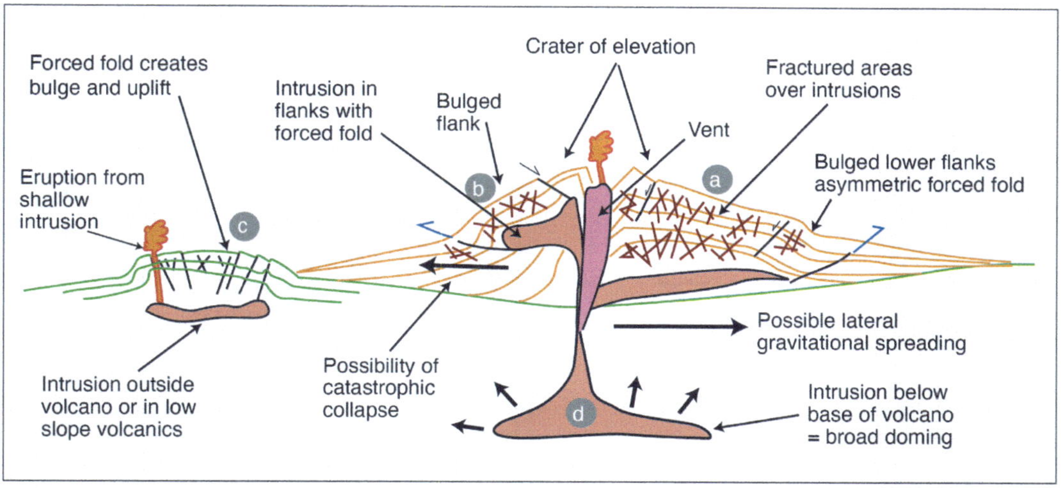

Fig. 16 Summary diagram showing the typical features of forced folding related to intrusions within, below, and outside of a volcanic edifice (modified after van Wyk de Vries et al. 2014). The intrusion-related forced folds lead to uplift and bulging of the surface or volcano flanks. Four cases are shown: (a) fractured areas on flank due to laccolith-like intrusion below the edifice (e.g. Momotombo); (b) when the flank becomes bulged (e.g. Mount St. Helens, USA or Petit Puy de Dôme, Chaîne des Puys) and may trigger the collapse of the unstable flank; (c) forced fold created by uplift of a previously flat-lying area (e.g. Grosmanaux, Chaîne des Puys) before any eruption of material; and (d) deep intrusion below base of volcano, inducing a broad doming of the above edifice, but where no clear surface faulting is observable. In all cases uplift, fracture and tilted strata produce an upper depression, the main criteria of 'craters of elevation'. Compare this conceptual model with the following geological cross-section (Fig. 17A)

continental rifts and subduction zones. A simple nomenclature of volcanic rocks is displayed in a geochemical Total Alkali Silica diagram, showing five lava series based on the international IUGS classification (Le Maitre 2002), as developed for example by Briot (1999).

Processes of magma production are combined in the conceptual Fig. 18C (after van Wyk de Vries 2018), together with the ways in which magma can be stored and transported to the surface at the different tectonic settings.

1. Mafic magma (MORB) at mid-ocean ridges is produced by mantle plume rise and decompression, and is influenced by melting rate and mantle composition. MORB can be identified by their chemical composition as they originate from the shallow asthenosphere, which is depleted in incompatible elements. Basaltic magmas in hot-spot shield volcanoes and continental LIPs are generated by decompression melting in mantle plume rising under oceanic or continental lithosphere. As the mantle within the plume is hotter than the surrounding mantle, it can start melting at a greater depth than for MORB beneath ocean ridges. In the case of hotspots, the top of the melting zone is determined by the thicker overlying lithosphere, and these alkali basalts are richer in alkalis and iron and poorer in silica than MORB.
2. Alkali basalts may also form along continental divergent plate boundaries, such as the East African Rift System. The lithosphere experiences a lower amount of extension with respect to oceanic ridges, and mantle melting occurs at depths similar to those beneath oceanic islands, dominated by alkaline compositions and their evolved derivatives, such as trachytes and phonolites (Acocella 2021). Where mantle plumes melt thicker, silicic continental crust, they produce highly silicic (>70% S_iO_2) rhyolites (e.g. Yellowstone USA).

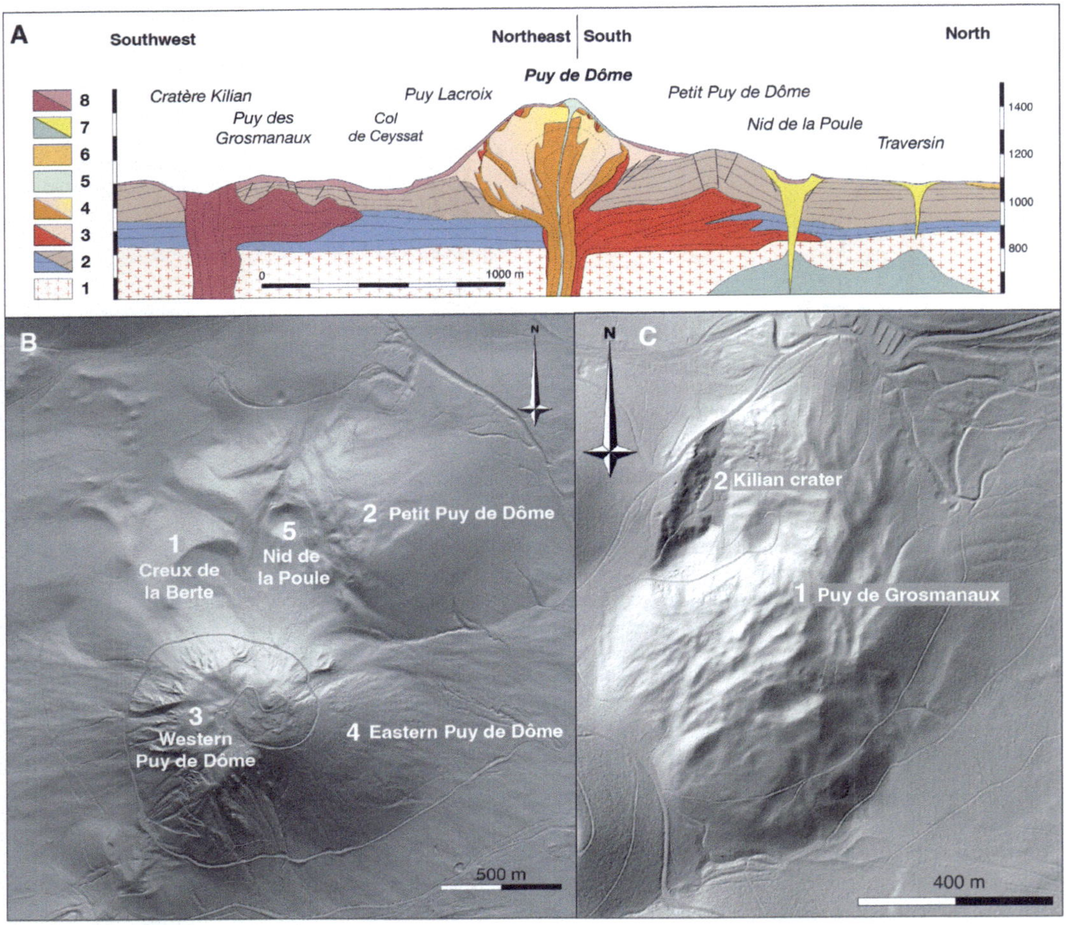

Fig. 17 Shallow magma intrusion creating uplift and deformation, as exemplified by the Puy de Dôme area, Chaîne des Puys (French Massif central), after Boivin et al. (2017). Extracted from the guide "Volcanologie de la Chaîne des puys", 6th edition 2017. **A**. Cross section of the Puy de Dôme and neighbouring edifices showing an *earlier* interpretation of the Puy de Dome structure based on field observations and geology (Boivin et al. 2009). 1. Basement formed by the Plateau des Dômes. 2. 'Old' strombolian cones and lava flows. 3. Trachytic cryptodome below the Petit Puy de Dôme and the first dome of the Puy de Dôme. 4. Second dome and cogenetic breccia with fumarolic and silicified material above the conduit. 5. Most recent pyroclastic density current deposit. 6. Strombolian tephra-fall deposit from the Puy Pariou. 7. Dyke of trachybasaltic composition and associated phreatomagmatic craters of Nid de Poule and phreatic vent of Traversin. 8. Trachytic cryptodome below the Puy des Grosmanaux, Kilian crater and associated tephra-fall and pyroclastic density current deposits. **B**. LIDAR-based DEM (from CRAIG 2017) shows the Petit Puy de Dôme area to the north, from the oldest to the most recent: 1 Strombolian cone of Creux de la Berte, 2 Fractured surface of Petit Puy de Dôme, 3 Western dome of Puy de Dôme, 4 Eastern flank and dome of Puy de Dôme, 5 Maar of Nid de la Poule (interpreted as a "crater of elevation": see Fig. 16). **C**. LIDAR-based DEM of the Kilian-Grosmanaux area to the south: 1 fractured surface of Puy de Grosmanaux, 2 Kilian crater. Copyright from Parc Natural Regional des Volcans d'Auvergne

3. Volcanoes that develop in arcs, either on continental margins or on islands, are fed by magmas derived from slab dehydration, volatile release and melting of slab and mantle. Back-arc extension magmas are produced by mantle wedge circulation caused by subduction and roll back of the slab. Most of the volcanic rocks generated in continental arcs are collectively termed calc-alkaline due to the presence of calcium and alkaline elements,

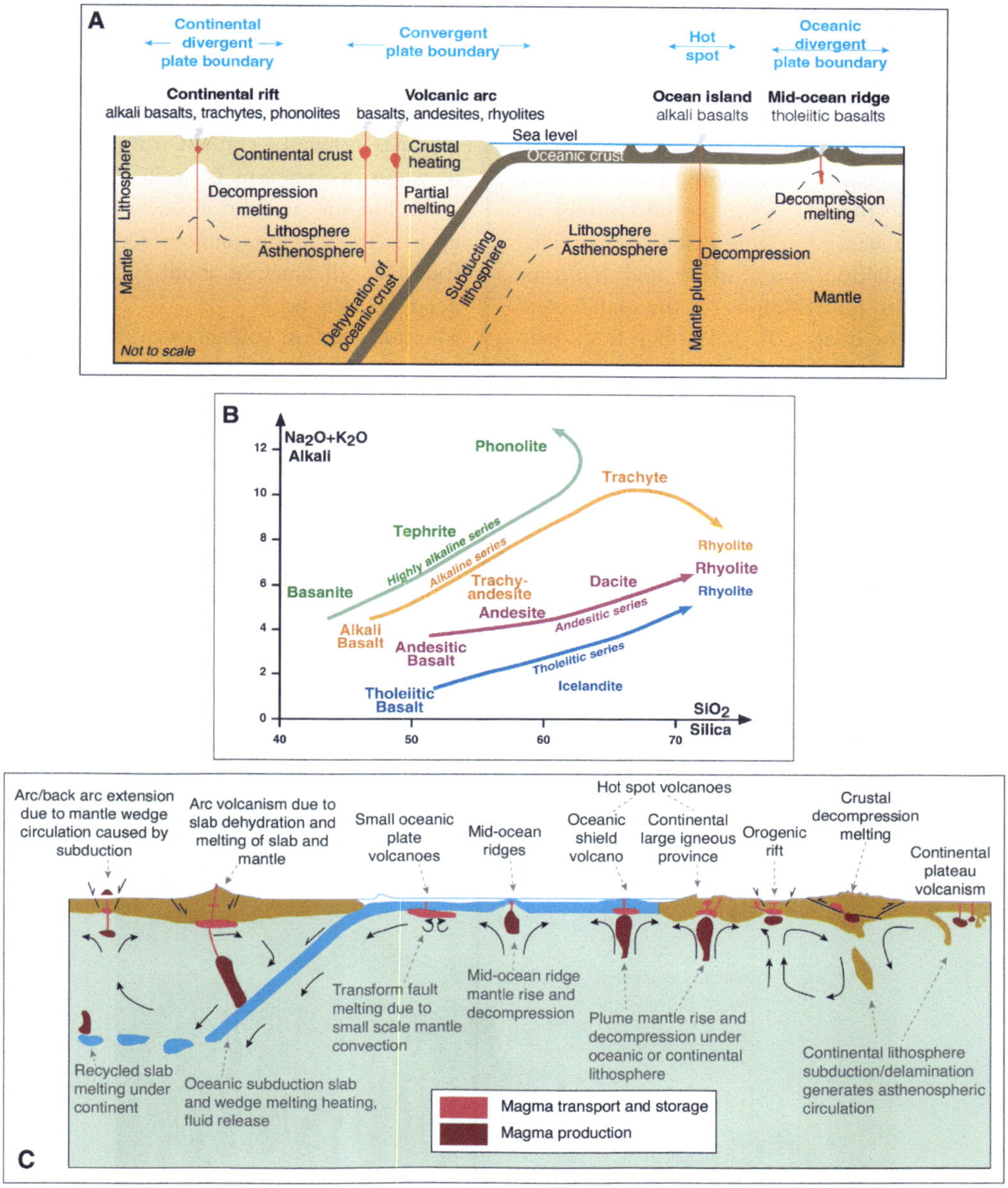

Fig. 18 Magma and lava compositions; magma generation, storage and transport. **A**. Schematic section through continental and oceanic lithosphere showing the major processes responsible for magma generation and the related main eruption compositions as a function of the tectonic plate setting (blue) (after Schmincke 2004; Rogers 2015, Fig. 4.18, p. 111 *in*: Acocella, 2021, Fig. 1.5, p. 8). Note that melting mechanisms give rise to either Mid-Ocean Ridge Basalts (MORBs) at spreading ridges, Ocean Island Basalts (OIBs) at oceanic shield volcanoes, Arc Island Basalts (AIBs) at island arc volcanoes (mostly explosive), or continental flood basalts (CFBs, mafic LIPs) creating plateaus and traps. **B**. Total Alkali-Silica diagram (after Le Maitre 2002) showing the four principal lava series (highly alkaline, alkaline, andesitic and tholeiitic) as found in different tectonic settings such as continental rifts, volcanic arcs, and mid-ocean ridges or oceanic hot spot volcanoes (modified after Briot 1999). **C**. Conceptual figure showing where magmas are generated, transported and stored, within the framework of the plate tectonic settings. Magma can be produced and volcanoes can develop in as many as eight areas: (1) arc-back arc island volcanoes, (2) arc volcanoes on continental margins, (3) small oceanic plate volcanoes, (4) mid-ocean ridges, (5) hot spot volcanoes, either oceanic shield volcanoes or LIPs, (6) continental orogenic rifts, (7) collision in orogenic belts, and (8) continental plateaus (modified after van Wyk de Vries 2018)

which come from the sediments riding on the subducting slab. Mixing of basaltic and rhyolitic magmas is the main process responsible for the formation of andesitic magma, with andesite being the most common rock type along volcanic arcs. Another process involved in andesitic magma is the partial melting of the mantle wedge, contaminated by a felsic melt of adakitic composition and hydrous fluids from dehydration of the subducted slab (Rivera et al. 2017). Further fractional crystallization of andesitic magma, possibly combined with the assimilation of crustal material, leads to even more silica-rich, dacitic and rhyolitic compositions.

4. Magmatism at island arcs is related to the processes taking place in the asthenosphere above the subducting slab. The main factor is that sediment transported to depths contributes significantly to dewatering in the region called the mantle wedge. Water released in this wedge causes marked changes in the overlying mantle, reducing the solidus temperature of the rocks of the asthenosphere from around 1400 to 1000 °C. As a result, arc magmas typically contain a high proportion of water (as much as 6%). In the subduction zone, magma generation occurs at depths of up to ~200–250 km, and surface volcanism is found at a distance of about 100 km from the deep-sea trench.
5. Magmas feeding volcanoes in rifts and continental plateaus are stored beneath the continental lithosphere and produced by repeated asthenospheric upwelling. In fold belts, melting of the crust and mantle lithosphere occurs by heating (from basaltic magma underplating), decompression (by removal of load through erosion or rifting/crustal stretching) or from the addition of volatiles that lower the melting point.

Photo 1 Columbia River Basalt Group and Plateau from Three Devil's Grade, Washington, USA. The Moses Coulee canyon has cut down into the succession of thick colonnades of the Frenchmen Springs Member basalt (15.5 Ma), which is capped by an upper basalt of the Roza Member (14.9 Ma). For scale, note the cattle feed lot, bottom left (*photo* Williamborg, 2006; public domain)

Photo 2 Deccan traps, India. Panorama of the Deccan flood basalt sequence in the Western Ghats escarpment, from the Arthur's Seat viewpoint (1347 a.s.l.) near the hill resort of Mahabaleshwar (1436 m). The section of sequence seen in the photo is ~1000 m thick, and comprises the Poladpur Formation, the Ambenali Formation, and the Mahabaleshwar Formation. The basalts are overlain by a regional ferricrete cap, tens of metres thick and densely forested (*photo* courtesy of H. Sheth)

Photo 3 Deccan traps, India. Detail of the Deccan flood basalt sequence exposed in the Western Ghats escarpment, same site as photo 2. The section of sequence shown here is ~400 m thick, and comprises the Mahabaleshwar Formation. The individual lava flows are extensive, 30–70 m-thick sheet lobes with columnar jointing, and are often separated by "red bole" beds, some of which are palaeosols, while others are altered glassy chilled margins of the flows, volcanic ash, interflow sediments, or flow-top and flow-bottom breccias (*photo* courtesy of H. Sheth)

Photo 4 Fissure, crater row, eruptive fissures and pyroclastic cones along the Laki (Lakagigar or Skaftar Fires) 1783–1784 AD/CE fissure eruption in southeastern Iceland (*photo* courtesy of Sz. Kósik)

Photo 5 Succession of thick ignimbrite sheets forming a high plateau, cut into by a canyon, Río Cotahuasi region, southern Peru. The cliff shows a ~ 1000 m-thick succession of welded and unwelded ignimbrites termed Alpabamba (Upper Miocene) overlain by unwelded Sencca ignimbrites (mid Pliocene to early Quaternary) that form high plateaus at 4300–4500 m a.s.l. (*photo* J.-C. Thouret)

References

Acocella V (2021) Volcano-Tectonic processes. Advances in Volcanology, IAVCEI, Springer, Switzerland, 552 pp. ISBN 978-3-030-65967-7

Acocella V, Rivalta E (2019) Calderas: structure, unrest, magma transfer and eruptions. Reference Module, Earth Syst. Environ. Sci. Elsevier, 15 p. https://doi.org/10.1016/B978-0-12-409548-9.10904-2

Allenvi (Alliance nationale de Recherche pour l'Environnement) (2022) L'éruption du volcan Hunga Tonga-Hunga Ha'apai le 15 janvier 2022: un ébranlement du système Terre à l'échelle planétaire. Note Inter-organismes, Groupe Thématique Atmosphère, Coordination: Lac C (CNRM), Le Pichon A (CEA) and Listowski C (CEA), 47 pp

Alvarez W (2003) Comparing the evidence relevant to impact and flood basalt at times of major mass extinctions. Astrobiol 3(1):153–161. https://doi.org/10.1089/153110703321632480

André M-F, Boivin P, Peiry, J-L, Thouret J-C (2013) Volcanoes, Basins and Cultural Heritage of the Auvergne Region. International Conference on Geomorphology, Paris Field-trip P5 and International Symposium in Auvergne, 1–5 September 2013, 128 pp., PRES Clermont Université Blaise Pascal, GEOLAB CNRS and Laboratoire Magmas et Volcans CNRS, OPGC et IRD, Clermont-Ferrand

Ballmer MD, van Keken PE, Ito G (2015) Hotspots, Large Igneous Provinces, and Melting Anomalies. In: Schubert G (ed) Treatise on geophysics, 2nd edn, Elsevier, pp 393–459. ISBN 9780444538024. https://doi.org/10.1016/B978-0-444-53802-4.00133-0

Biggs J, Ayele A, Fischer TP, Fontijn K, Hutchison W, Kazimoto E, Whaler K, Wright TJ (2021) Volcanic activity and hazard in the East African Rift Zone. Nature Comm 12:6881. https://doi.org/10.1038/s41467-021-27166-y

Boivin P, Besson J-C, Briot D, Camus G, Goër de Herve A, Gourgaud A, Labazuy P, Langlois E, de Larouzière F-D, Livet M, Mergoil J, Miaillier D, Morel J-M, Vernet G, Vincent P-M (2009) Volcanologie de la Chaîne des Puys, 5è ed, Carte 1/25000, notice 196 pp. Parc naturel régional des Volcans d'Auvergne

Boivin P, Besson J-C, Briot D, Camus G, Deniel C, Gourgaud A, Labazuy P, Langlois E, de Larouzière F-D, Livet M, Médard E, Merceccia C, Mergoil J, Miaillier D, Morel J-M, Thouret J-C, Vernet G (2017) Volcanologie de la Chaîne des Puys, 6è ed., Parc naturel régional des Volcans d'Auvergne, Carte 1/25000 couleur, notice bilingue, 199 pp

Bolós X, Marti J, Becerril L, Planagumá L, Grosse P, Barde-Cabusson S (2015) Volcano-structural analysis of La Garrotxa Volcanic Field (NE Iberia): Implications for the plumbing system. Tectonophysics 642:58–70. https://doi.org/10.1016/j.tecgto.2014.12.013

Bond DP, Wignall PB, Keller G, Kerr AC (2014) Large igneous provinces and mass extinctions: an update. In: Keller G, Kerr AC (eds) Volcanism, impacts, and mass extinctions: causes and effects. Geol Soc Amer Spec Paper 505, Chapter 2, pp 29–55. https://doi.org/10.1130/SPE505

Bonini M, Corti G, Innocenti F, Manetti P, Mazzarini F, Abebe T, Pecskay Z (2005) Evolution of the Main Ethiopian Rift in the frame of Afar and Kenya rifts propagation. Tectonics 24, TC1007. https://doi.org/10.1029/2004TC001680

Briot D (1999) From one lava to another. In: de Goër de Herve A (ed) Volcanism and volcanoes of the Auvergne. La Dépêche Scientifique du Parc, pp 13–14, Parc des Volcans d'Auvergne, Montlosier

Brown RJ, Blake S, Thordarson T, Self S (2014) Pyroclastic edifices record vigorous lava fountains during the emplacement of a flood basalt flow field, Roza Member, Columbia River Basalt Province, USA. Geol Soc Amer Bull 126:875–891. https://doi.org/10.1130/B30857.1

Bryan SE, Ernst R (2008) Revised definition of Large Igneous Provinces (LIPs). Earth-Sci Rev 86(1–4):175–202. https://doi.org/10.1016/j.earscirev.2007.08.008

Bryan SE, Ferrari L (2013) Large igneous provinces and silicic large igneous provinces: progress in our understanding over the last 25 years. Geol Soc Amer Bull 125(7–8):1053–1078. https://doi.org/10.1130/B30820.1

Bryan SE, Riley TR, Jerram DA, Stephens CJ, Leat PT (2002) Silicic volcanism: an undervalued component of large igneous provinces and volcanic rifted margins. In: Menzies MA, Klemperer SL, Ebinger CJ, Baker J (eds) Volcanic rifted margins. Geol Soc Amer Spec Paper 362, pp 99–120. https://doi.org/10.1130/SPE362

Bryan SE, Peate IU, Peate D, Self S, Jerram DA, Mawby MR, Marsh JS, Miller JA (2010) The largest volcanic eruptions on Earth. Earth-Sci Rev 102(3–4): 207–229. https://doi.org/10.1016/j.earscirev.2010.07.001

Buonasorte G, Ciccacci S, De Rita D, Fredi P, Lupia Palmieri E (1991) Some relations between morphological characteristics and geologic structure in the Vulsini Volcanic Complex, Northern Latium, Italy. Z Geomorph NF Suppl-Bd 82:59–71. 0044-2798/91/0082-0059

Burke K (1977) Aulacogens and continental breakup. Ann Rev Earth Planet Sci 5: 371–396. https://doi.org/10.1146/annurev.ea.05.050177.002103

Burke K, Torsvik TH (2004) Derivation of large igneous provinces of the past 200 million years from long-term heterogeneities in the deep mantle. Earth Planet Sci Lett 227(3–4):531–538. https://doi.org/10.1016/j.epsl.2004.09.015

Campbell IH (2005) Large igneous provinces and the mantle plume hypothesis. Elements 1:265–268. https://doi.org/10.2113/gselements.1.5.265

Carracedo JC, Troll VR, Day JMD, Geiger H, Aulinas M, Soler V, Deegan FM, Perez-Torrado FJ, Gisbert G, Gazel E, Rodriguez-Gonzalez A, Albert H (2022) The 2021 eruption of the Cumbre Vieja volcanic ridge on La Palma, Canary Islands. Geology Today 38(3):94–107. https://doi.org/10.1111/gto.12388

Cas RAF, Wright JV (1988) Volcanic successions, modern and ancient. Unwin Hyman, London, 528 pp. ISBN 0-04-552021-6

Cas RAF, Giordano G, Wright JV (2024) Volcanology—processes, deposits, geology and resources, 2 volumes. Springer Cham, 1833 p. ISBN 978-3-319-66612-9

Chiesa S, Civetta L, De Fino M, La Volpe L, Orsi G (1989) The Yemen trap series: genesis and evolution of a continental flood basalt province. J Volcanol Geoth Res 36(4):337–350. https://doi.org/10.1016/0377-0273(89)90078-4

Christiansen RL (2001) The Quaternary and Pliocene Yellowstone Plateau volcanic field of Wyoming, Idaho, and Montana. US Geol Surv Prof Paper 729-G, 120 pp. US Geol Surv. ISBN 0-607-95346-2

Christie-Blick N, Anders MH, Wills S, Walker CD, Renik B (2007) Observations from the Basin and Range Province (western United States) pertinent to the interpretation of regional detachment faults. Geol Soc London, Spec Publ 282, 1: 421–441. https://doi.org/10.1144/SP282.17

Clague DA, Sherrod DR (2014) Growth and degradation of Hawaiian volcanoes. In: Poland MP, Takahashi TJ, Landowski CM (eds) Characteristics of Hawaiian volcanoes, US Geol Surv. Prof. Paper 1801, pp 97–146. https://doi.org/10.3133/pp18013

Coffin MF, Eldholm O (1994) Large igneous provinces: crustal structure, dimensions, and external consequences. Rev Geophys 32:1–36. https://doi.org/10.1029/93RG02508

Condie KC (1982) Plate tectonics and continental drifts. Pergamon Press, Oxford

Corti G (2009) Continental rift evolution: From rift initiation to incipient break-up in the Main Ethiopian Rift, East Africa. Earth-Sci Rev 96:1–53. https://doi.org/10.1016/j.earscirev.2009.06.005

Courtillot VE, Renne PR (2003) On the ages of flood basalt events. CR Geosci 335:113–140. https://doi.org/10.1016/S1631-0713(03)00006-3

Crisp JA (1984) Rates of magma emplacement and volcanic output. J Volcanol Geoth Res 20(3–4):177–211. https://doi.org/10.1016/0377-0273(84)90039-8

Davies GF (2001) Dynamic Earth: plates, plumes and mantle convection. Cambridge University Press, Cambridge. ISBN 10:0521599334

Defos du Rau J (1959) Le relief de l'île de la Réunion, étude de morphologie volcanique. Institut de Géographie, Université de Bordeaux, Thèse, 320 pp

Deligne NI, Sigurdsson H (2015) Global rates of volcanism and volcanic episodes. In: Sigurdsson H, Houghton B, McNutt SR, Rymer H, Stix J (eds) The encyclopedia of volcanoes, 2nd edn, Chap 14. Academic Press Elsevier, pp 261–272. ISBN 978-0-12-385938-9

de Silva SL (1989) Altiplano-Puna volcanic complex of the central Andes. Geology 17(12):1102–1106. https://doi.org/10.1130/0091-7613(1989)017

de Silva SL, Lindsay J (2015) Primary Volcanic Landforms. In: Sigurdsson H, Houghton B, McNutt SR, Rymer H, Stix J (eds) The encyclopedia of volcanoes, 2nd edn, Chap 13, pp 273–297, Academic Press Elsevier. ISBN 978-0-12-385938-9

de Silva S, Zandt G, Trumbull R, Viramonte JG, Salas G, Jimenez N (2006) Large ignimbrite eruptions and volcano-tectonic depressions in the Central Andes: a thermo-mechanical perspective. Geol Soc London, Spec Publ 269:47–63. https://doi.org/10.1144/GSL.SP.2006.269.01.04

Dewey JF, Holdsworth RE, Strachan RA (1998) Transpression and transtension zones. Geol Soc London, Spec Publ 135(1):1–14. https://doi.org/10.1144/GSL.SP.1998.135.01.01

Duncan R, Pyle D (1988) Rapid eruption of the Deccan flood basalts at the Cretaceous/Tertiary boundary. Nature 333:841–843. https://doi.org/10.1038/333841a0

Duraiswami RA, Sheth HC, Gadpallu P, Youbi N, Chellai EH (2020) A simple recipe for red bole formation in continental flood basalt provinces: weathering of flow-top and flow-bottom breccias. Arab J Geosci 13:953. https://doi.org/10.1007/s12517-020-05973-9

Ebinger C (2005) Continental break up: the East African perspective. Astron Geophys 46(2):2.16–2.21. https://doi.org/10.1111/j.1468-4004.2005.46216.x

Ebinger CJ, Casey M (2001) Continental breakup in magmatic provinces: an Ethiopian example. Geology 29(6):527–530. https://doi.org/10.1130/0091-7613(2001)029<0527:CBIMPA>2.0.CO;2

Ernst RE (2014) Large Igneous provinces. Cambridge University Press, Cambridge, 641 pp. ISBN 978-0-521-87177-8

Ernst RE (2021) Large igneous provinces. In: Alderton D, Elias SA (eds) Encyclopedia of geology, vol 2, Academic Press, Elsevier, pp 60–68. ISBN 978-1-611974-02-7

Ernst RE, Buchan KL (2001) Mantle plumes: their identification through time. Geol Soc Amer Spec Paper pp 143, 145, 146, 147, 148, and 259. ISBN 978-0-8137-2352-5

Ernst RE, Youbi N (2017) How Large Igneous Provinces affect global climate, sometimes cause mass extinctions, and represent natural markers in the geological record. Palaeogeogr Palaeoclim Palaeoecol 478:30–52. https://doi.org/10.1016/j.palaeo.2017.03.0140031-0182

Fedotov SA, Markhinin Ye K (1983) The great Tolbachik fissure eruption: geological and geophysical data 1975–1976. Cambridge University Press, New York, 341 pp (translated from Russian)

Ferrari L, López-Martinez M, Rosas-Elguera J (2002) Ignimbrite flare-up and deformation in the southern Sierra Madre Occidental, western Mexico: implications for the late subduction history of the Farallón plate. Tectonics 21(4):17–24. https://doi.org/10.1029/2001TC001302

Fisher RV, Schmincke H-U (1984) Pyroclastic rocks. Springer, Berlin, 472 pp. ISBN 3-540-12756-9

Foulger GR, Christiansen RL, Anderson DL (2015) The Yellowstone "hot-spot" track results from migrating basin-range extension. In: Foulger GR, Lustrino M, King SD (eds) The interdisciplinary Earth: in Honor of Don L Anderson. Geol Soc Amer Spec Paper 514 AGU 71, pp 215–238. ISBN 978-0-8137-2514-7

Francis PW (1993) Volcanoes. A Planetary perspective. Clarendon Press, Oxford University Press, 443 pp. ISBN 0-19-854452-9

Gabbianelli G, Romagnoli C, Rossi PL, Calanchi N (1993) Marine geology of the Panarea-Stromboli area Aeolian Archipelago, Southeastern Tyrrhenian sea. Acta Vulcanol 3:11–20

Galetto F, Pritchard ME, Hornby AJ, Gazel E, Mahowald NM (2023) Spatial and temporal quantification of subaerial volcanism from 1980 to 2019: Solid products, masses, and average eruptive rates. Rev Geophys 61(3):e2022RG000783. https://doi.org/10.1029/2022RG000783

Giordano G, Cas R (2021) Classification of ignimbrites and their eruptions. Earth-Sci Rev 220:103697. https://doi.org/10.1016/j.earscirev.2021.103697

Gomes AS, Vasconcelos PM (2021) Geochronology of the Paraná-Etendeka large igneous province. Earth-Sci Rev 220:103716. https://doi.org/10.1016/j.earscirev.2021.103716

Gordeev EI, Murav'ev YD, Samoilenko SB, Volynets AO, Mel'Nikov DV, Dvigalo VN (2013) The Tolbachik fissure eruption of 2012–2013: preliminary results (translated from Russian). Doklady Earth Sci 452(2):1046–1050. Springer Nature https://doi:10.1134/S1028334X13100103

Gudmundsson A (2020) Volcanotectonics: Understanding the Structure, Deformation and Dynamics of Volcanoes. Cambridge University Press, 575 pp. ISBN 978-1-107-02495-3

Gutscher M-A, Spakman W, Bijwaard H, Engdahl ER (2000) Geodynamics of flat subduction: Seismicity and tomographic constraints from the Andean margin. Tectonics 19(5):814–833. https://doi.org/10.1029/1999TC001152

Hachimi HE, Youbi N, Madeira J, Bensalah MK, Martins L, Mata J, Medina F, Bertrand H, Marzoli A, Munhá J, Bellieni G (2011) Morphology, internal architecture and emplacement mechanisms of lava flows from the Central Atlantic Magmatic Province (CAMP) of Argana Basin (Morocco). Geol Soc London, Spec Publ 357(1):167–193. https://doi.org/10.1144/SP357

Herzberg C, Condie K, Korenaga J (2010) Thermal history of the Earth and its petrological expression. Earth Planet Sci Lett 292(1–2):79–88. https://doi.org/10.1016/j.epsl.2010.01.022

Hofmann C, Courtillot V, Féraud G, Rochette P, Yirgu G, Ketefo E, Pik R (1997) Timing of the Ethiopian flood basalt event and implications for plume birth and global change. Nature 389:838–841. https://doi.org/10.1038/39853

Holden H, Wallace L, Beavan J, Fournier N, Cas R, Ailleres L, Silcock D (2015) Contemporary ground deformation in the Taupo Rift and Okataina Volcanic Centre from 1998 to 2011, measured using GPS. Geophys J Intern 202, 3:2082–2105. https://doi.org/10.1093/gji/ggv243

Hooper PR, Camp VE, Reidel SP, Ross ME (2007) The Columbia River Basalts and their relationship to the Yellowstone hotspot and Basin and Range extension. In: Foulger GR, Jurdy JM (eds) Plumes, plates, and planetary processes. Geol Soc Amer Spec Paper 430, pp 635–668. https://doi:10.10.1130(2007).2430(30)

Jordan BT (2005) Age-progressive volcanism of the Oregon high lava plains: overview and evaluation of tectonic models. In: Foulger GR, Natland JH, Presnal DC, Anderson DL (eds) Plates, plumes and paradigms, 863 pp. Geol Soc Amer Spec Paper 388. ISBN 0-8137-2388-4

Karátson D, Veres D, Gertisser R, Magyari EK, Jánosi C, Hambach H (eds) (2022) Ciomadul (Csomád), The Youngest Volcano in the Carpathians. Volcanism, Palaeo-environment, Human Impact. Springer, 282 pp. ISBN 978-3-030-89139-8

Keller G, Kerr AC (eds) (2014) Volcanism, impacts, and mass extinctions: causes and effects. Geol Soc Amer Spec Paper 505, 455 pp. ISBN 978-0-8137-2505-5

Kelley SP (2007) The geochronology of large igneous provinces, terrestrial impact craters, and their relationships to mass extinctions on Earth. J Geol Soc 164:923–936. https://doi.org/10.1144/0016-76492007-026

Kim S, Wessel P (2011) New global seamount census from altimetry-derived gravity data. Geophys J Intern 186(2):615–631. https://doi.org/10.1111/j.1365-246x.2011.05076.x

Koptev A, Burov E, Calais E, Leroy S, Gerya T, Guillou-Frottier L, Cloething S (2016) Contrasted continental rifting via plume-craton interaction: applications to Central East African Rift. Geosc Front 7(2):221–236. https://doi.org/10.1016/j.gsf.2015.11.002

Kueppers U, Beier C (eds) (2018) Volcanoes of the Azores. Revealing the geological secrets of the northern central Atlantic Islands. Active Volcanoes of the World. Springer, Berlin, 355 pp. ISBN 978-3-662-58546-7

LaFemina PC (2015) Plate tectonics and volcanism. In: Sigurdsson H, Houghton B, McNutt R, Rymer H, Stix J (eds) The encyclopedia of volcanoes, 2nd edn, Chap. 3, pp 65–92, Academic Press Elsevier. ISBN 978-0-12-385938-9

Lara LE, Lavenu A, Cembrano J, Rodriguez C (2006) Structural controls of volcanism in transversal chains: resheared faults and neotectonics in the Cordón Caulle–Puyehue area (40.5° S), Southern Andes. J Volcanol Geoth Res 158:70–86. https://doi.org/10.1016/j.jvolgeores.2006.04.017

Le Corvec N, Menand T, Lindsay J (2013) Interaction of ascending magma with pre-existing crustal fractures in monogenetic basaltic volcanism: an experimental approach. J Geophys Res, Solid Earth 118(3):968–984. https://doi.org/10.1002/jgrb.50142

Lécuyer F, Bellier O, Gourgaud A, Vincent PM (1997) Tectonic control of the Tondano caldera North Celebes, Indonesia. CR Acad Sci Paris, Serie IIA 8(325):607–613

Le Maitre RW (ed) (2002) Igneous rocks—A classification and glossary of terms. In: Recommendations of the IUGS subcommission on the systematics of igneous rocks, 2nd edn, Cambridge University Press, 236 pp. ISBN-13 978-0-521-662215-4

Le Pennec J-L, Temel A, Froger J-L, Sen S, Gourgaud A, Bourdier J-L (2005) Stratigraphy and age of the Cappadocia ignimbrites, Turkey: reconciling field constraints with paleontologic, radiochronologic, geochemical and paleomagnetic data. J Volcanol Geoth Res 141(1–2):45–64. https://doi.org/10.1016/j.jvolgeores.2004.09.004

Lipman PW (1995) Declining growth of Mauna Loa during the last 100,000 years: rates of lava accumulation versus gravitational subsidence. In: Rhodes JM, Lockwood JP (eds) Mauna Loa revealed: Structure, Composition, History, and Hazards. Geophys Monogr 92, AGU, pp 45–78. ISBN 0-87590-049-6

Lockwood JP, Hazlett RW (2010) Volcanoes, Global perspectives. Wiley-Blackwell, UK, 539 pp. ISBN 978-1-4051-6249-4

Lowman PD, Yates J, Masuoka P, Montgomery BC, Salisbury DO, Yates J (1999) A digital tectonic activity map of the Earth. J Geosci Edu 47:428–437. https://doi.org/10.5408/1089-9995-47.5.428

McDonald GA (1972) Volcanoes. Prentice-Hall, New Jersey, 509 pp. ISBN 0-13-942219-6

MacDonald KC (1982) Mid-oceanic ridges: fine-scale tectonic, volcanic and hydrothermal processes within the plate boundary zone. Ann Rev Earth Planet Sci 10:155–190. https://doi.org/10.1146/annurev.ea.10.050182.001103

Medynski S, Pik R, Burnard P, Dumont S, Grandin R et al (2016) Magmatic cycles pace tectonic and morphological expression of rifting (Afar depression, Ethiopia). Earth Planet Sci Lett 446:77–88. https://doi.org/10.1016/j.epsl.2016.04.014

Merle O (2011) A simple continental rift classification. Tectonophys 513(1–4):88–95. https://doi.org/10.1016/j.tecto.2011.10.004

Merle O, Borgia A (1996) Scaled experiments of volcanic spreading. J Geophys Res 101(B6):13805–13817. https://doi.org/10.1029/95JB03736

Mittal T, Richards MA, Fendley IM (2021a) The magmatic architecture of continental flood basalts: I. Observations from the Deccan Traps. J Geophys Res, Solid Earth 126:e2021JB 021808. https://doi.org/10.1029/2021JB021808

Mittal T, Richards MA (2021b) The magmatic architecture of continental flood basalts: II. A new conceptual model. J Geophys Res Solid Earth 126:e2021JB021807. https://doi.org/10.1029/2021JB021807

Muller JR, Ito G, Martel SJ (2001) Effects of volcano loading on dyke propagation in an elastic half-space. J Geophys Res 106:11101–11113. https://doi.org/10.1029/2000JB900461

NASA (Goddard Space Flight Center) (1999) Global tectonic activity map of the Earth. Tectonism and volcanism of the last one million years. Greenbelt, Maryland. https://doi.org/10.5408/1089-9995-47.5.428

O'Connor JM, Jokat W, Regelous M, Kuiper KF, Miggins DP, Koppers AAP (2019) Superplume mantle tracked isotopically the length of Africa from the Indian Ocean to the Red Sea. Nature Com 10:5493. https://doi.org/10.1038/s41467-019-13181-7

Oehler J-F, Lénat J-F, Labazuy P (2008) Growth and collapse of the Reunion Island volcanoes. Bull Volcanol 70:717–742. https://doi.org/10.1007/s00445-007-0163-0

Ollier C (1972, 1988) Volcanoes. Basil Blackwell, Oxford, 288 pp. ISBN 0-631-15664-X

Oppenheimer C, Orchard A, Stoffel M, Newfield TP, Guillet S, Corona C, Sigl M, Di Cosmo N, Büntgen U (2018) The Eldgjá eruption: timing, long-range impacts and influence on the Christianisation of Iceland. Clim Change 147:369–381. https://doi.org/10.1007/s10584-018-2171-9

Pansino S, Taisne B (2019) How magmatic storage regions attract and repel propagating dykes. J Geophys Res Solid Earth 124(1):274–290. https://doi.org/10.1029/2018JB016311

Papale P, Garg D, Marzocchi W (2022) Global Rates of subaerial volcanism on Earth. Front Earth Sci 10:922160. https://doi.org/10.3389/feart.2022.922160

Peterson DW, Moore RB (1987) Geologic history and evolution of geologic concepts, Island of Hawai'i. In: Decker RW, Wright TL, Stauffer PH (eds) Volcanism in Hawai'i. US Geol Surv Prof Paper 1350, vol 1, pp 149–189

Poland MP, Takahashi TJ, Landowski C (eds) (2015) Characteristics of Hawaïan volcanoes. US Geol Surv Prof Paper 1801, 427 pp. ISSN 1044-9612

Püthe C, Gerya T (2014) Dependence of mid-ocean ridge morphology on spreading rate in numerical 3-D models. Gondwana Res 25(1):270–283. https://doi.org/10.1016/j.gr.2013.04.005

Putirka K, Platt B (2012) Basin and Range volcanism as a passive response to extensional tectonics. Geosphere 8(6):1274–1285. https://doi.org/10.1130/GES00803.1

Rampino MR, Stothers RB (1988) Flood basalt volcanism during the past 250 Million years. Science 241(4866):663–668. https://doi.org/10.1126/science.241.4866.663

Rampino MR, Caldeira K, Prokoph A (2019) What causes mass extinctions? Large asteroid/comet impacts, flood-basalt volcanism, and ocean anoxia—Correlations and cycles. In: Koeberl C, Bice DM (eds) 250 Million Years of Earth History in Central Italy: celebrating 25 Years of the Geological Observatory of Coldigioco, Chap 14, vol 542. Geol Soc Amer. https://doi.org/10.1130/2019.2542(14)

Reidel SP, Camp VE, Ross ME, Wolff JA, Martin BS, Tolan TL, Wells RE (eds) (2013) The Columbia River flood basalt province: stratigraphy, areal extent, volume, and physical volcanology, The Columbia River flood basalt province, vol 497. Geol Soc Amer Spec Paper pp 1–43. https://doi.org/10.1130/2013.2497(01)

Rivera M, Martin H, Le Pennec J-L, Thouret J-C, Gourgaud A, Gerbe M-C (2017) Petro-geochemical constraints on the source and evolution of magmas at El Misti volcano (Peru). Lithos 268–271:240–259. https://doi.org/10.1016/j.lithos.2016.11.009

Salisbury MJ, Jicha BR, de Silva SL, Singer BS, Jimenez NC, Ort MH (2011) $^{40}Ar/^{39}Ar$ chronostratigraphy of Altiplano-Puna volcanic complex ignimbrites reveals the development of a major magmatic province. Geol Soc Amer Bull 123(5–6):821–840. https://doi.org/10.1130/B30280.1

Salvany T, Lahitte P, Nativel P, Gillot P-Y (2012) Geomorphic evolution of the Piton des Neiges volcano (Réunion Island, Indian Ocean): competition between volcanic construction and erosion since 1.4 Ma. Geomorph 136:132–147. https://doi.org/10.1016/j.geomorph.2011.06.009

Saunders AD (2005) Large igneous provinces: origin and environmental consequences. Elements 1(5):259–263. https://doi.org/10.2113/gselements.1.5.259

Schoene B, Samperton KM, Eddy MP, Keller G, Adatte T, Bowring SA, Khadri SF, Gertsch B (2015) U-Pb geochronology of the Deccan Traps and relation to the end-Cretaceous mass extinction. Science 347(6218):182–184. https://doi.org/10.1126/science.aaa0118

Schoene B, Eddy MP, Samperton KM, Keller B, Keller G, Adatte T, Khadri SFR (2019) U-Pb constraints on pulsed eruption of the Deccan Traps across the end-Cretaceous mass extinction. Science 363(6429):862–866. https://doi.org/10.1126/science.aau2422

Schmidt R, Schmincke H-U (2002) From seamount to oceanic island, Porto Santo, central east Atlantic. Int J Earth Sci 91:594–614. https://doi.org/10.1007/s00531-001-0243-x

Schmincke H-U (2004) Volcanism. Springer, Berlin, 324 pp. ISBN 3-540-43650-2

Self S, Thordarson Th, Keszthelyi L, Walker GPL, Hon K, Murphy MT, Long P, Finnemore S (1996) A new model for the emplacement of Columbia River basalts as large, inflated pahoehoe lava flow fields. Geophys Res Lett 23(19):2689–2692. https://doi.org/10.1029/96GL02450

Self S, Coffin MF, Rampino MR, Wolff JA (2015) Large igneous provinces and flood basalt volcanism. In: Sigurdsson H, Houghton B, McNutt SR, Rymer H, Stix J (eds) The encyclopedia of volcanoes, 2nd edn. Elsevier Academic Press, Chap. 24, pp 441–458. ISBN 978-0-12-385938-9

Sheth HC (2007) Large Igneous Provinces (LIPs): definition, recommended terminology, and a hierarchical classification. Earth-Sci Rev 85(3–4):117–124. https://doi.org/10.1016/j.earscirev.2007.07.005

Sheth H (2018a) Flood basalt landscapes. In: Sheth H (ed) A photographic atlas of flood basalt volcanism. Springer, Cham, pp 7–31. https://doi.org/10.1007/978-3-319-67705-7_3

Sheth H (2018b) Morphology and architecture of Flood Basalt Lava Flows and sequences. In: Sheth H (ed) A photographic atlas of flood basalt volcanism. Springer, Cham, pp 33–79. https://doi.org/10.1007/978-3-319-67705-7_3

Shreve T, Grandin R, Boichu M, Garaebiti E, Moussallam Y, Ballu V, Delgado F, Leclerc F, Vallée M, Henriot N, Cevuard S (2019) From prodigious volcanic degassing to caldera subsidence and quiescence at Ambrym (Vanuatu): the infuence of regional tectonics. Nat Sci Rep 9:18868. https://doi.org/10.1038/s41598-019-55141-7

Siebert L, Simkin T, Kimberly P (2010) Volcanoes of the world, 3rd edn. Smithsonian Institution, University of California Press, Berkeley, 551 pp. ISBN 978-0-520-26877-7

Simkin TL, Tilling RI, Vogt PR, Kirby SH, Kimberly P, Stewart DB (2006) This dynamic planet: world map of volcanoes, earthquakes, impact craters and plate tectonics. US Geol Surv IMAP 2800. https://doi.org/10.3133/i2800

Smith DK, Humphris SE, Tivey MA, Cann JR (1997) Viewing the morphology of the Mid-Atlantic Ridge from a new perspective. EOS Trans AGU, 78(26):265 and 269. https://doi.org/10.1029/97EO00173

Smith RB, Jordan M, Steinberger B, Puskas CM, Farrell J, Waite GP, Husen S, Chang W-L, O'Connell R (2009) Geodynamics of the Yellowstone hotspot and mantle plume: Seismic and GPS imaging, kinematics, and mantle flow. J Volcanol Geoth Res 188:26–56. https://doi.org/10.1016/j.jvolgeores.2009.08.020

Smith RB, Jordan M, Steinberger B, Puskas CM, Farrell J, Waite GP, Husen S, Chang W-L, O'Connell R (2009) Geodynamics of the Yellowstone hotspot and mantle plume: seismic and GPS imaging, kinematics, and mantle flow. J Volcanol Geoth Res 188(1–3): 26–56. https://doi.org/10.1016/j.jvolgeores.2009.08.020

Soule A (2015) Mid-Ocean ridge volcanism. In: Sigurdsson H, Houghton B, McNutt SR, Rymer H, Stix J (eds) The encyclopedia of volcanoes, 2nd edn, Chap. 21. Academic Press Elsevier, pp 395–421. ISBN 978-0-12-385938-9

Spinks K, Acocella V, Cole J, Bassett K (2005) Structural control of volcanism and caldera development in the transtensinal Taupo Volcanic Zone, New Zealand. J Volcanol Geoth Res 144:7–22. https://doi.org/10.1016/j.jvolgeores.2004.11.014

Sprain CJ, Renne PR, Vanderkluysen L, Pande K, Self S, Mittal T (2019) The eruptive tempo of Deccan volcanism in relation to the Cretaceous-Paleogene boundary. Science 363(6429):866–870. https://doi.org/10.1126/science.aav1446

Swanson DA, Wright TL, Helz RT (1975) Linear vent systems and estimated rates of magma production and eruption for the Yakima Basalt on the Columbia Plateau. Amer J Sci 275:877–905. https://doi.org/10.2475/ajs.275.8.877

Swanson DA, Cameron KA, Everts RC, Pringle PT, Vance JA (1989) Cenozoic volcanism in the Cascade Range and Columbia plateau, Southern Washington, and northernmost Oregon. AGU field trip guidebook T106. Science 238:1105–1111

Thordarson T, Self S (1993) The Laki (Skaftar Fires) and Grimsvötn eruptions in 1783–1785. Bull Volcanol 55:233–263. https://doi.org/10.1007/BF00624353

Thordarson T, Self S (1998) The Roza Member, Columbia River Basalt Group; a gigantic pahoehoe lava flow field formed by endogenous processes? J

Geophys Res 103:27411–27445. https://doi.org/10.1029/98JB01355

Thouret J-C (1999) Volcanic geomorphology. Earth-Sci Rev 47:95–132. S0012-8252_99.00014-8

Thouret J-C (2004) Geomorphic processes and hazards on volcanic mountains. In: Owens P, Slaymaker O (eds) Mountain geomorphology, Chap. 11, pp 242–273, Routledge, London. ISBN 0-340-76417-I

Tibaldi A, Bonali FL, Corazzato C (2017) Structural control on volcanoes and magma paths from local- to orogen-scale: the central Andes case. Tectonophys 699:16–41. https://doi.org/10.1016/j.tecto.2017.01.005

Tolan TL, Reidel SP, Beeson MH, Anderson JL, Fecht KR, Swanson DA (1989) Revisions to the estimates of the areal extent and volume of the Columbia River Basalt Group. In: Reidel SP, Hooper PR (eds) Volcanism and tectonism in the Columbia River Flood-Basalt Province. Geol Soc Amer Spec Paper, p 239. https://doi.org/10.1130/SPE239-p1

Torsvik T, Doubrovine PV, Steinberger B, Gaina C, Spalman W, Domeier M (2017) Pacific plate motion change caused the Hawaiian-Emperor bend. Nature Com 8:15660. https://doi.org/10.1038/ncomms

Trigila R (ed) (1995) The Volcano of the Alban Hills. A contribution to: geomorphology, stratigraphy and volcano-tectonics, mineralogy, petrology, geochronology, crustal seismology, gravimetry, hydrogeology, fluid geochemistry and thermalism, geochemical monitoring, and volcanic hazard. Tipografia Sgs, Roma, p 283

Tyrrell GW (1937) Flood basalts and fissure eruptions. Bull Volcanol 1:89–111. https://doi.org/10.1007/BF03028044

Ulrych J, Dostal J, Adamovič J, Jelínek E, Špaček P, Hegner E, Balogh K (2011) Recurrent Cenozoic volcanic activity in the Bohemian Massif (Czech Republic). Lithos 123(1–4):133–144. https://doi.org/10.1016/j.lithos.2010.12.008

van Wyk de Vries B, Merle O (1996) The effect of volcanic constructs on rift fault patterns. Geology 24(7): 643–646.https://doi.org/10.1130/0091-7613(1996)024<0643:TEOVCO>2.3.CO;2

van Wyk de Vries B, Francis PW (1997) Catastrophic collapse at stratovolcanoes induced by gradual volcano spreading. Nature 387:387–390. https://doi.org/10.1038/387387a0

van Wyk de Vries B., van Wyk de Vries M (2018) Tectonics and volcanic and igneous plumbing systems. In: Burchardt S (ed) Volcanic and igneous plumbing systems: understanding magma transport, storage and evolution in the Earth's Crust, Chap. 7. Elsevier, pp 167–189. https://doi.org/10.1016/C2015-0-06837-X, ISBN 978-0-12-809749-6

van Wyk de Vries B, Marques A, Herrera R, Granjas JL, Llanes P, Delcamp A (2014) Craters of elevation revisited: forced folds, bulges and uplift of volcanoes. Bull Volcanol 76:875. https://doi.org/10.1007/s00445-014-0875-x.

Wadge G, Biggs J, Lloyd R, Kendall J-M (2016) Historical volcanism and the state of stress in the East African Rift System. Front Earth Sci 4:86. https://doi.org/10.3389/feart.2016.00086

Villeneuve N, Bachèlery P, Kemp J (2013) La Réunion Island: a typical example of a basaltic shield volcano with rapid evolution. In: Fort M, André M-F (eds) Landscapes and landforms of France, Chap 25. World Geomorphological Landscapes, Springer, pp 261–270. https://doi.org/10.1007/978-94-007-7022-5-25

Vithana MVP, Xu M, Zhao X, Zhang M, Luo Y (2019) Geological and geophysical signatures of the East Pacific Rise 8–10° N. Solid Earth Sci 4(2):66–83. https://doi.org/10.1016/j.sesci.2019.04.001

White RS, McKenzie D (1995) Mantle plumes and flood basalts. J Geophys Res Solid Earth 100(B9):17543–17585. https://doi.org/10.1029/95JB01585

White JDL, Bryan SE, Ross PS, Self S, Thordardson (2009) Physical volcanology of continental large igneous provinces: update and review. In: Thorardson T, Self S, Larsen G, Rowland SK, Hoskulddsson A (eds) Studies in volcanology, the legacy of GPL walker, Spec Publ IAVCEI no 2, pp 291–321. ISBN 978-1-86239-280-9

White SM, Crisp JA, Spera FJ (2006) Long-term volumetric eruption rates and magma budgets. Geochem Geophys Geosyst 7(3):Q03010. https://doi.org/10.1029/2005GC001002

Williams H, McBirney AR (1979) Volcanology. Freeman, Cooper and Co., San Francisco, 397 pp. ISBN 0-87735-321-2

Wilson JT (1965) A new class of faults and their bearing on continental drift. Nature 207(4995):343–347. https://doi.org/10.1038/207343a0

Wohletz K, Heiken G (1992) Volcanology and geothermal energy. University of California Press, Berkeley, 432 pp. ISBN 0-05-201914-0

Chapter 2 Elementary Landforms and Processes I: Lava Flows, Lava Lakes, and Lava Domes

Lava flows, lava fields and lava lakes in craters are elementary landforms produced by effusive eruptions, in addition to lava domes created by extrusive activity. We look first in Sect. 1 at lava flows, the most widespread volcanic landforms on Earth as well as on planetary landscapes of Earth's moon, Mars, Venus, Mercury and Io. Section 2 deals with lava lakes and their dynamics, and 3 covers lava domes and their emplacement.

1 Lava Flows

Lava flows, produced by effusive eruptions, are more volumetrically important than volcaniclastic formations in the Earth's volcanic areas. Lava flows of basalt composition (MORBs) have formed the present-day oceanic crust since the Jurassic (~160 million years, and back to Archean times for the old oceanic crust), and they are still ongoing at Mid-Ocean Ridges and active seamounts. They dominate the surface of many basaltic centres such as oceanic shield volcanoes above hot spots (Hawai'i, Galápagos, La Réunion Island), and in monogenetic fields (e.g. Newer Volcanic province in Australia, Chaîne des Puys in France, and Chichinautzín in Mexico). Lava flows with a wide range of compositions have by and large built up the composite cones and domes in volcanic arcs (Cascades Range, Andean volcanic zones, etc.). Much more extensive lava flows (10^5–10^6 km^2) of basaltic composition are found in continental flood basalt plateaus such as the Deccan traps (India) and Columbia River Basalt Group (Oregon, USA).

1.1 Definition of a Lava Flow

Active lava is a mixture of molten rock (liquid), crystals (solids), and vesicles (voids due to gas bubbles) (Harris and Rowland 2015). Compositions of lava range from ultramafic (e.g. komatiite, albeit absent on Earth since the Mesozoic) and mafic (basalt) to felsic (e.g. rhyolite). Different lava flows can be simply divided into low viscosity (low silica and high crystal content) and high viscosity (high silica and low crystal content) types. Other factors (flow geometry), surface appearance and field forms, or physical (temperature), rheological properties (yield strength) and cooling rate provide criteria for distinguishing lava flow types.

A *flow unit* is a separate cooling unit according to Walker (1970): "a lava flow top which cooled significantly and solidified before another flow unit was superimposed on it". A *simple lava flow* corresponds to relatively short eruptions with sustained effusion rates, so that its length (L) greatly exceeds its width (w). For example, Harris (2015a) described a single lava flow at Etna (October 1998), with a L/w ratio of

D. Karátson and J.-C. Thouret, *Volcano Geomorphology*, Advances in Volcanology,
https://doi.org/10.1007/978-3-031-77693-9_3

22, which was produced over 2–6 h with a mean output rate (VDRE/d) of 2–7 m^3/s.

Lava flow fields or *compound lava flows* are composed of more than one flow unit. Walker (1970) suggested that effusion rate was a criterion for distinguishing simple from compound lava flows. Basaltic lavas extruded at high effusion rates (typically tens to a few thousands of m^3/s) form far-reaching flows composed of simple lava flows, whereas those extruded at low effusion rates (one to a few tens of m^3/s) produce small flow units, which pile up close to the vent and form compound lava-flow fields. Compound lava flows have comparable length and width (ratio ~2), and effusion rates of tens of m^3/s over a period of weeks. The compound lava-flow field around the 2021 Tajogaite – Cumbre Vieja cone (La Palma, Canary Islands) contains at least nine lava flow units or lobes (Carracedo et al. 2022; see Chaps. 4 and 10). The lava-flow field mantled a surface of 12.35 km^2 and emplaced a volume of 194 million m^3 between 19 September and 13 December 2021. This yields a time-average discharge rate of 27 m^3/s, although within this there were notable variations in effusion rate.

1.2 Four Types of Lava Flow

Effusive eruptions on dry land produce three traditionally recognized types of lava flow: pāhoehoe, 'a'ā and block lava. Here we adopt the spelling of Hawaiian lava flows proposed by several authors, we thus refer to the nomenclature, origin, and spreading of Hawaiian terms according to Harris et al. (2017). We also describe a fourth type: pillow lavas, associated lavas and hyaloclastite deposits emplaced in subaqueous environments, either underwater or in marshy areas. Harris et al. (2017) wrote: "The most commonly used terms to describe lava surfaces today are not descriptive but, instead, are merely words, specifically the Hawaiian words 'a'ā (rough, brecciated basalt lava) and pāhoehoe (smooth, glassy basalt lava), plus block lava, i.e. thick brecciated lavas that are typically more silicic than basalt. 'A'ā and pāhoehoe were introduced into the Western geological vocabulary by American geologists working in Hawai'i during the 1800s. In 1933, Ruy Finch added the third lava type, block lava, to the classification scheme, while G.A. MacDonald formalized the tripartite system in 1953. A number of subtypes and transitional forms of all three major lava types have been defined, particularly since the 1970s and based on in-situ measurements of lava flow interiors" (e.g. Swanson 1973, 1979; Kilburn and Guest 1993; Cioni and Funedda 2005).

Lockwood and Hazlett (2010) summarized the main surface types of lava flows with respect to their composition and source vent. The common features of lava flow surfaces have been described in more detail by Kilburn (1981, 2015) and listed in his Table 55.3, for 'a'ā, block lava, pāhoehoe, and toothpaste lavas. Table 1 provides a summary of descriptions made by these authors.

Over the past 25 years, several authors have provided insight into lava flow morphology, structure, and mechanisms of formation (Calvari and Pinkerton 1999; Walker 2009; Chevrel et al. 2013; Gregg 2017; Harris et al. 2022) as well as into specific lava flow-related hazards (Kilburn 2015; Calvari 2019). There is a transitional spectrum from pāhoehoe to 'a'ā lavas, as well as between block lavas of silicic composition and dome coulees, with many shared characteristics within each pair (Cas and Wright 1988). We will describe both cases in Sects. 1.2.3 and 1.3.

Table 2 (modified after Harris et al. 2017) provides a classification of, and descriptive terminology for, five flow types (i.e. pillow lava, pāhoehoe, 'a'ā, block lava and coulee) and their sub-types. The classification is based on a number of descriptors: lava type and composition, surface state, vesicle state, breccia type (clasts for 'a'ā), and most importantly for the purposes of this book: surface morphology and flow-field form. In simple terms, the final lava flow morphology is formed by the interplay of those parameters that enhance flow advance and those that hinder it (Gregg 2017), which are related to factors that affect the rate of growth of the lava flow's surface crust. The behaviour

Table 1 Simple description of four main surface types of lava flows related to their composition and source vent (Lockwood and Hazlett 2010)

Surface and roughness	Common composition and SiO_2 content	Source vent structures	Typical length (L) Mean thickness (T) Mean discharge rate (MDR)
Pāhoehoe, smooth, hummocky and continuous surface, often with a millimetre-scale texture of interwoven lava threads	Basalt 45–55% SiO_2	Shield volcanoes, cinder cones, composite cones	L 10–50 km T 3–20 m MDR 10–1000 m^3/s
'A'ā, covered with loose, spiny rubble, and a jumble of irregular fragments	Basalt, basaltic andesite ≤55% SiO_2	Shield volcanoes, cinder cones, composite cones	L < 10 km T 5–25 m MDR 10–100 m^3/s
Block-lava flow, blocky, lithic. Covered by broken lava, fragments up to meters across with smooth, planar, and angular surfaces	Basaltic andesite, dacite 55–63% SiO_2. Alkaline rocks: trachyte 58–69% SiO_2 and phonolite 51–61% SiO_2	Composite cones, domes	L < 5 km; some < 15 km T 20–100 m MDR 1–10 m^3/s
Short and stubby coulee, blocky, glassy. Thick and steep extrusions	Dacite 63–71% SiO_2, rhyolite > 71% SiO_2. Alkaline rocks: trachyte 58–69% SiO_2 and phonolite 51–61% SiO_2	Dome coulees and obsidian flows	L 0.5–3 km T 50–300 m MDR 1–10 m^3/s

of the solidifying crust plays an essential role in lava flow morphology, as it thickens with time and deforms under stress (Gregg 2017; see Sect. 1.4).

1.2.1 Basaltic Lava Flows: Pāhoehoe, 'a'ā and Subaqueous Lava

Pāhoehoe: Definition, Types and Sub-types

Pāhoehoe is a Hawaiian term for basaltic lava having a smooth surface. It is the least viscous of all common lavas and most abundant on shield volcanoes and at subaerial mid-ocean ridges. Such lava flows are smooth, undulating, sometimes ropy, with sheets and oval lobes, and spherical vesicles. In contrast to 'a'ā, pāhoehoe lacks both surface and basal breccias (Harris and Rowland 2015). Pāhoehoe lavas can be observed in intraplate settings other than hot spots, such as Italy, the French Massif central, Morocco, Saudi Arabia, the East African rift, Réunion Island and the neighbouring Comoro Islands, Queensland, Australia, Japan, South Korea, and northern China.

Relatively flat pāhoehoe surfaces exhibit a number of typical topographic features a few metres wide and long (Fig. 1): small mounds called tumuli, uplifted blocks called lava rise terraces, broad depressions, and steep-walled holes, termed inflation pits (Lockwood and Hazlett 2010). Lava channels are present in the upstream reaches of a flow. After the initial emplacement, lava flows inflate (Hon et al. 1994; Walker 2009). Gradual inflation can thicken a flow by as much as several metres. In addition, the build-up of pressure from fresh lava injected beneath the crust can cause it to split open, releasing a lobe of molten lava, which quickly develops its own crust and repaves the flow surface to form a greater thickness. Multiple breakouts and lobe extrusions from a flow front can continue to advance the flow for days after the vent upslope has ceased activity. The smaller lobes, called pāhoehoe toes, may measure from a few centimetres to tens of metres in width. While some surfaces of stationary pāhoehoe flows inflate, other portions will deflate. Most flows show evidence of simultaneous flow deflation near their vents

Table 2 Classification of, and description terminology for, four lava flow types and their sub-types based on a series of descriptors (modified and adapted after Harris et al. 2017)

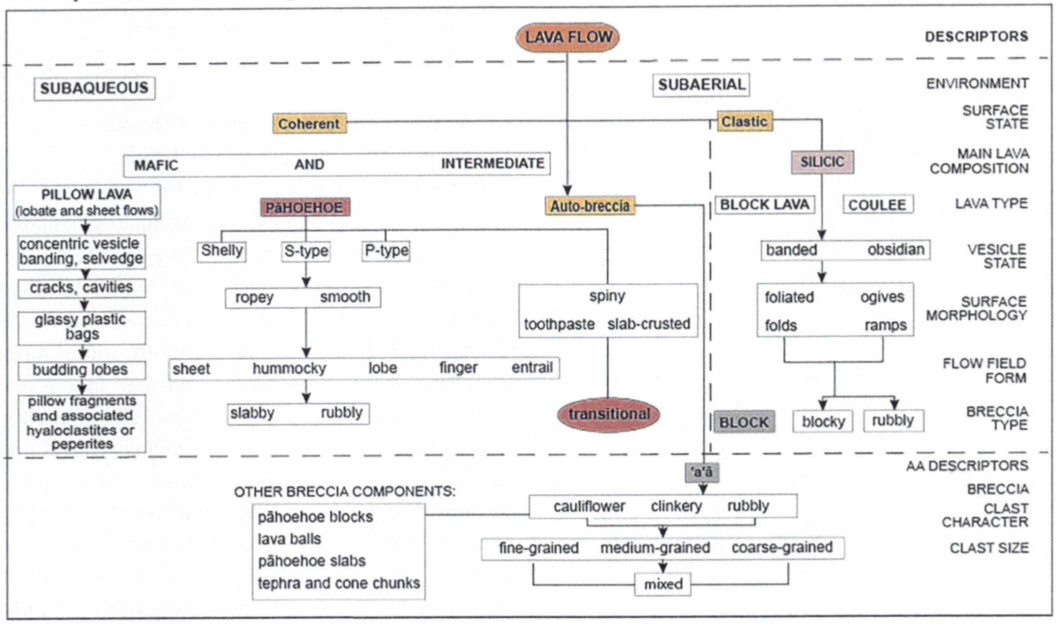

with corresponding flow inflation at their fronts (Lockwood and Hazlett 2010).

Basalt lava can be produced from vents as: (1) coherent flows from small *boccas* (openings) or from the overspill or breaching of a lava lake ponded in a crater; or (2) fire fountains of lava that rise up around the vent and then flow away (Cas and Wright 1988). Many eruptions of basalt lava flows begin with a phase of fire-fountaining of gas-rich magma, followed by the extrusion of coherent flows of relatively gas-poor magma. Flows formed from agglutinated lava spatter are associated with two vent structures: *spatter cones* occur as discrete mounds surrounding a lava fountain, whereas *spatter ramparts* are elongate structures along the trace of an eruptive fissure (Fig. 2; Photo 1). Highly gas-charged magmas giving rise to intense lava fountaining are likely to generate high-viscosity basaltic flows with high yield strengths. Magmas with low initial gas and crystal contents form fluid lavas from less vigorous fire fountains or lava lakes. These changes in gas and crystal content can also occur during eruptions such as the lavas erupted at Askja, Iceland in 1961 (Cas and Wright 1988) and at Piton de la Fournaise (Réunion Island) in 1985 (Boivin and Bachèlery 2009).

Harris and Rowland (2015) distinguish two common types of pāhoehoe based on surface features (Photo 2):

(1) *S-type* is a spongy, gas-rich 'fresh' lava, with a sponge-like appearance due to a high number of bubbles. S-type contains very abundant (up to 40% of the rock), small (mm-scale) spherical vesicles. The full bubble population is retained, as S-type pāhoehoe tends to be erupted without storage and degassing.
(2) *P-type*, less common than S-type, is a pipe vesicle-bearing or a blue glassy, late-stage, degassed lava. It is less vesicular but with larger individual vesicles than S-type lava, and is emplaced when underlying surface slopes are $<4°$. P-type pāhoehoe results from the eruption of lava that has undergone prolonged degassing, but minimal cooling, because it resided longer in the system.

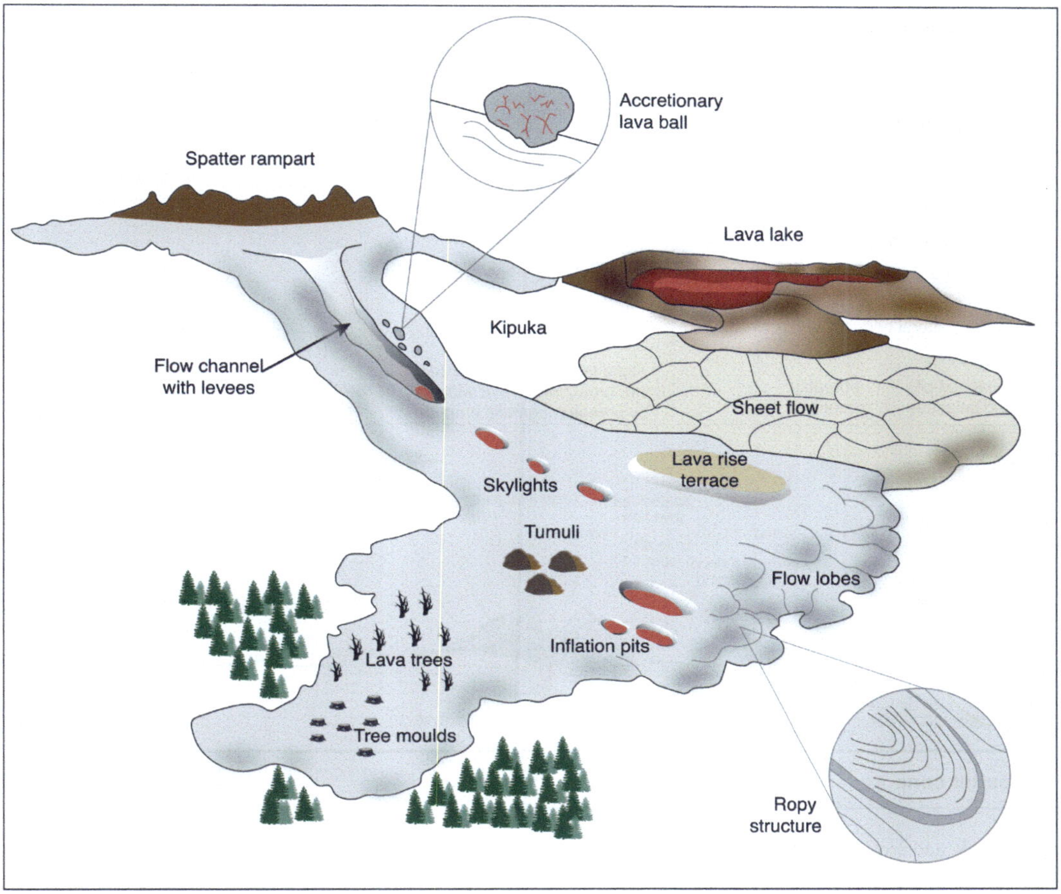

Fig. 1 Distinctive surface features of pāhoehoe flows (Lockwood and Hazlett 2010, Fig. 6.22, p. 147). See text for the explanation of minor landforms

Harris and Rowland (2015) also distinguish a number of pāhoehoe sub-types that range from gas-rich pāhoehoe to lavas increasingly degassed or emplaced at lower effusion rates:

(1) Pāhoehoe *sheet flows* are extensive, relatively flat and correspond to a very high effusion rate (hundreds of m^3/s), while their flow fronts can advance at several kilometres per hour. Sheets can be hundreds of metres across and very thin. Most sheet flows are highly vesicular S-type pāhoehoe.

(2) *Shelly* pāhoehoe are gas-rich, near vent lavas with a frothy surface and large voids inside. The metre-high voids may be due to segregation of gas and lava within a still fluid pāhoehoe flow, from which the gas is unable to escape because of the overlying quenched cm-thick crust.

(3) *Slabby, rough or rubbly* pāhoehoe can form if a flow stalls for a period of time. When the crust fails, it breaks into 'a jumble of jagged slabs' several cm-thick and up to one metre in length. Rough or rubbly pāhoehoe is composed of broken, angular to rounded fragments, similar to some 'a'ā lava flow fields.

(4) *Finger or entrail* pāhoehoe is formed by an individual lobe extruded at an extremely low effusion rate. Locally, multiple short but narrow pāhoehoe lobes, termed entrail lava, can form intricate drapery on cliffs and fault scarps.

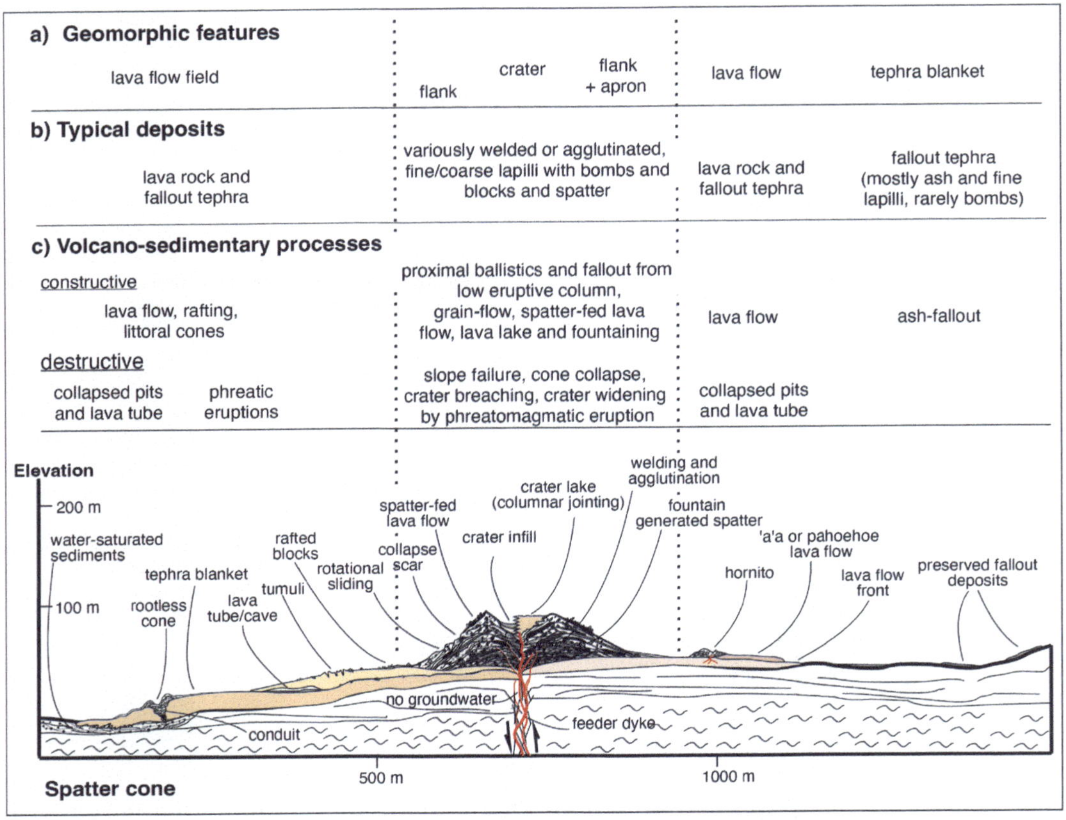

Fig. 2 Table and cross section illustrating spatter cones and other small landforms associated with basaltic lava flows (modified after Kereszturi and Németh 2013)

(5) Common *ropy pāhoehoe* characterize areas of the crust that are wrinkled and twisted into forms resembling lengths of coiled rope (Fink and Fletcher 1978; Lockwood and Hazlett 2010). The ropes are bent into convex curves in the direction of the flow movement (Fig. 1). Ropy forms result from the dragging and twisting of the thick, hot, plastic flow crust by movement of the liquid lava underneath. Friction results in the edges of narrow lava streams moving less rapidly than the middle.

'A'ā lava flow

'A'ā is a Hawaiian term for lava with a broken, brecciated surface. It is more common on composite volcanoes, but can also be found on shield volcanoes adjacent to pahoehoe (Photo 3). 'A'ā consists mainly of clinkers (Dutton, 1893) or auto-breccia, and exhibits sheared vesicles. Individual clinkers range in size from cm to m across, and together they form a rough, spiny and jagged surface on which walking and sampling are challenging (Fig. 3A; Photo 4). The surface of an 'a'ā flow is often a mix of clinker, accretionary lava balls, slabs of pāhoehoe from farther upflow, scoria fallen from fire fountaining at the source vent and, occasionally, large chunks of cone material collapsed onto the flow. The torn surfaces are the *spinose* protrusions, characteristic of 'a'ā clinker (Harris and Rowland, 2015). The principal model for clinker formation invokes a velocity profile, which increases from zero at the flow edge to highest at the flow centre, creating rotation within the molten flow. This rotation tears open the lava surface, continually exposing hotter core lava and causing individual clinker blocks to rotate relative to each other. At the flow front, there is a similar, but vertical velocity profile, causing an over-steepening, forward toppling and rolling, 'caterpillar tread' motion, as well as near continuous flow front collapse.

The structure of a typical 'a'ā flow in section, usually a few metres thick, consists of two distinct zones (Harris and Rowland 2015): an upper, rubbly part, and a lower, massive core of solid lava, which cooled slowly (Fig. 3A; Photo 5). Beneath the core, a reddened, baked layer of pre-existing soil is often present, separated from the massive bed by a thin layer of rubbly lava, which was surmounted by the advancing flow. The massive lava contains abundant, mostly spherical vesicles, sometimes deforming into pipe vesicles in the flow direction due to a high degree of shearing. There are occasional gas blisters which form large cavities tens of cm across. Fractures and flow-parallel shear structures are often apparent in the interior (Fig. 3A). Ramping can occur, causing blocks of solid core lava, with striated faces, to move upwards and possibly be extruded from the flow surface or front.

Lipman and Banks (1987) distinguished four 'a'ā subtypes among Mauna Loa's 1984 flows: (1) Slabby 'a'ā, which has a crust comprised of broken pāhoehoe slabs; (2) Scoriaceous or spiny 'a'ā, which comprises clasts of frothy, spiny, highly vesicular lava; (3) Clinkery 'a'ā, which consists of denser clots of cooled lava that have ground against each other when riding atop the flow; and (4) Blocky 'a'ā, which has a coarse grain size and subangular forms mixed with clinkery 'a'ā to form a heterogeneous rubble.

Transition from pāhoehoe to 'a'ā

Pāhoehoe and 'a'ā lavas often erupt from the same vent. Both are chemically identical, with their differences lying in the physical structure of the silicate melt and other factors. While changes between the two types are subtle, transition can only occur from pāhoehoe to 'a'ā and never the inverse (Francis 1993). Pāhoehoe surfaces develop only in low-viscosity basaltic lavas, whereas 'a'ā surfaces are found in lavas of a broader compositional range. Higher effusion rates, greater than 5–10 m^3/s, facilitate 'a'ā formation, while slower rates favour pahoehoe formation (Rowland and Walker 1990). Changes in slope are another factor: when pāhoehoe encounters steep slopes and flows faster downhill, it can divide into streams of 'a'ā lava in response to abrupt changes in shearing stress.

Several authors have defined the transition from pāhoehoe to 'a'ā lava flows, in particular on Mauna Loa and Kilauea volcanoes, Hawai'i (e.g. Cashman and Mangan 2014). Coupling of field and laboratory measurements with the results of analogue experiments shows that this transition can be calibrated as a threshold zone between shear-strain rate and apparent viscosity (Fig. 4A). This occurs at a threshold value of apparent viscosity apart from at very low strain rates, where pāhoehoe morphologies can be maintained to higher viscosities (Cashman and Mangan 2014). The rheological basis of the transition from pāhoehoe to 'a'ā flows can also be illustrated by three paths according to temperature and viscosity versus shear rate or flow speed (Fig. 4B): For path A–B, pāhoehoe flows come to rest while in a fluid state and continue to cool until solidification. For path A–C, pāhoehoe flows continue to move as their temperature drops below a critical point (determined largely by composition); at point X, they can no longer flow as a fluid and cooler parts of the flow convert to the 'a'ā form through rupturing. For path A-D, the pāhoehoe maintains a constant temperature, but converts to 'a'ā at point Y where the flow's velocity suddenly increases, as shown by flow over a cliff (Lockwood and Hazlett 2010).

Two pāhoehoe types form the transition to 'a'ā in Hawai'i (Harris and Rowland 2015):

(1) *Toothpaste* lava has the rheology of 'a'ā, but it flows very slowly on subhorizontal slopes, and thus does not break into clinker (Rowland and Walker 1987). Flow units are thicker (60–150 cm) than primary pāhoehoe. Degassed tongues or lobes of toothpaste lava, containing a few

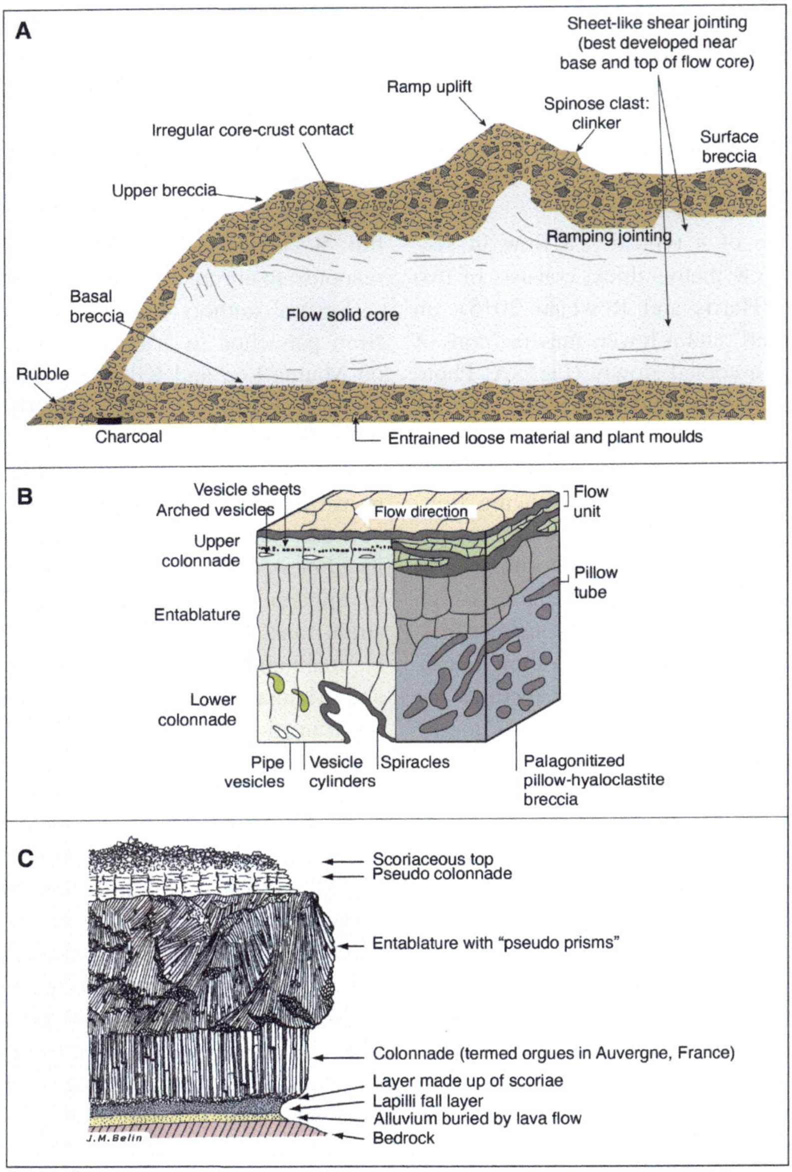

Fig. 3 Cross section of lava flows. **A** Cross section of 'a'ā flow showing the three layers (upper breccia, massive core, basal breccia), steep front as well as ramps and shear jointing (modified after Cas and Wright 1988; Lockwood and Hazlett 2010). **B** Cross section through a basaltic lava flow (after Schmincke 2004, Fig. 9.9, p. 131): the thick flood basalt lava flow that entered the Miocene Columbia River canyon. Pillow lava in the lower part of the flow dipping down towards the former transport direction to the left consists of thin lava tubes, and some of the quickly cooled lava became brecciated. The water was so shallow that the lava flow produced its own delta with a later subaerial flow unit travelling over dry ground. **C** Basaltic lava flow layers, after de Goër, 1985 *in:* Gourgaud et al. 2009: lower colonnade, entablature, and upper (pseudo) colonnade. The sharp boundary between the larger basal and the smaller upper column marks the meeting point of the solidification fronts during cooling. The front rises slowly upward from below and relatively rapidly downward into the interior of the lava flow

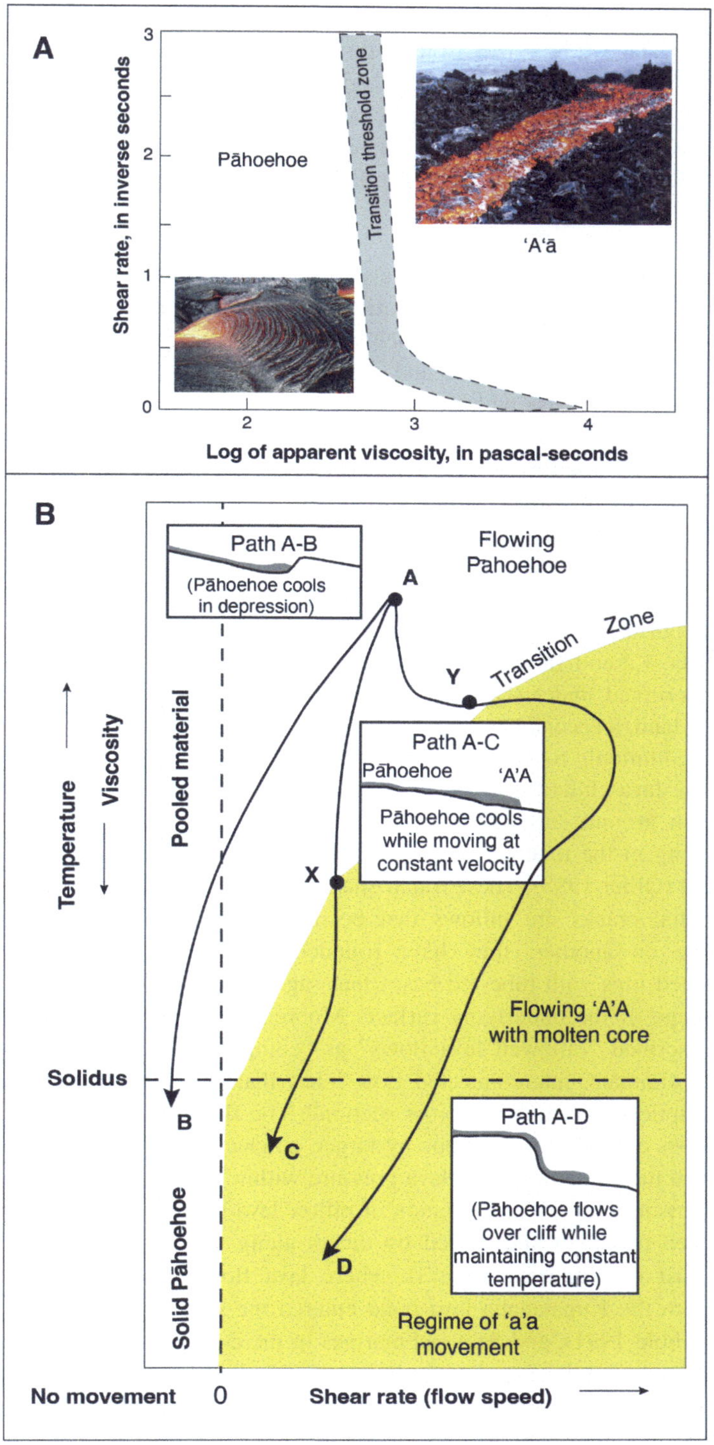

Fig. 4 Pāhoehoe-to-'a'ā transition. **A** Plot of shear rate versus apparent viscosity fields for two flow surface morphologies (pāhoehoe and 'a'a), drawn from a combination of field observations and laboratory experiments that represent the best approximation, to date, of the pāhoehoe-to-'a'ā transition in Hawaiian lava flows (Cashman and Mangan, 2014, A Century of Studying effusive eruption in Hawai'i, Fig. 5, p. 362, Chap. 9, Prof. Paper 1801, U.S. Geol. Survey). **B** The rheological basis for the pāhoehoe–'a'ā transition (Lockwood and Hazlett, 2010, Fig. 6.13, p. 138). Theoretical aspects of the pāhoehoe–'a'ā transition are discussed by Hulme (1974) and field aspects by Peterson and Tilling (1980). See text for further explanation

large vesicles, can be formed when late-stage flow oozes from within the primary pāhoehoe field where it had been stored. The surface may be characterized by small, transverse undulations, and grooves and ridges orientated parallel to the flow direction.

(2) *Transitional pāhoehoe* results from emplacement of relatively degassed and crystallized lava at relatively high local effusion rates and velocities. The surface is rough and dull, and the interior relatively dense with large, sheared vesicles.

1.2.2 Pillow Lavas, Associated Lava Types and Deposits

Basalt lavas created at MORs are much more extensive on the ocean floor than on land. Hence, the classification of lava flows includes a fourth, underwater type, mentioned above in Sect. 1.2 and Table 2. If lava flows into water or is erupted under seawater or into water bodies on land, a special structure known as *pillow lava* is commonly formed (Fig. 3B; Photo 6A and B). The lava chills rapidly to form a glassy, plastic skin around still low viscosity lava, and rolls along in the manner of plastic bags full of liquid (Ollier 1988). These round-shaped bags with radial cracks are pillows that become heaped one on another; they have rounded or flattened tops with tube-fed bases that sag into the shape of the underlying surface. Moore (1975) described “pillowed lava flows” as “composed of elongate, interconnected flow lobes that are elliptical or circular in cross section”. The flow lobes are fed from upslope by larger, connected lava tubes that maintain lava pressure within the growing lobes. The formation of pillow lava has been observed and filmed by divers along the coast of Kilauea (Hawai’i), where lava flows from the Kupaianaha lava pond entered the sea (Tibble 1991), and from submarines in the deep sea (e.g. on the East Pacific Rise in 2005–2006, Soule 2015). Channels were about a metre wide, and the channelized flows moved at a speed of 1–3 m/s. Volumetric flow rates of about 0.7 m^3/s were estimated. Pāhoehoe lava flowing into the sea was accompanied by continuous, minor explosions and implosions. Much of the hot lava was abruptly chilled, but extensive fragmentation took place, producing fine-grained, angular debris. Pillows formed and the lava advanced by the development of re-entrant selvedges (quenched glassy margins) around budding subaqueous lava tubes, a process analogous to the formation of toes of advancing, subaerial pahoehoe (Cas and Wright 1988, their Fig. 4.14), reproduced in laboratory experiments (Gregg and Fink 1995). Pillow lavas are well known for their thick selvedges, their radial and concentric cooling cracks and their concentric vesicular banding (Staudigel and Koppers 2015; Photo 6A and B). Cross sections of pillows show both intra-pillow and inter-pillow cavities. The latter can be filled by quenched palagonite glass.

Along mid-ocean ridges, pillow lava flows occur alongside *lobate flows* towards the tips of the eruptive fissure, emplaced as ridges comprised of mounds or hummocks, whereas the central fissure is occupied by *sheet flows* (Soule 2015). Pillow lavas predominate on seamounts, where they are tubular in shape, often with circular or oval cross sections (see Chap. 3, Sect. 3.5). Pillows decrease in size from the base of a pillow sequence to the top, and with increasing distance from the vent. Pillow lavas have very gently sloping flow surfaces when they are extruded at relatively high effusion rates, but they may also form pillow-walls along-strike of fissures with a relatively small supply of magma. Pillow lavas do not form massive lava flows at seamounts, whereas they form tabular lava bodies several tens of metres thick at mid-ocean ridges with very high lava eruption rates.

Submarine lavas also include lobate lavas and sheet lavas together with *hyaloclastites,* as glass quenches on contact with water (Table 2; White et al. 2015). When 'a'ā flows enter the sea, the flow may be fragmented into debris consisting of small glassy fragments known as hyaloclastite breccia. Huge amounts of hyaloclastites may form, over which lava can advance and construct a lava delta (Fig. 3B; Mattox and Mangan 1997). Such hyaloclastite deposits form part of a pillow lava sequence, along with lava debris and quenched lava fragments, collectively belonging to the typical succession of deposits at shoaling volcanoes, as described in Chap. 3, Sect. 7.

1.2.3 Block-Lava Flows

Block-lava flows are mainly silicic lava flows (dacite and rhyolite), with some andesitic flows, and are commonly associated with domes and composite cones. Andesite and dacite, being more viscous and exhibiting higher yield strength than basalt, flow less far and build up thicker flows. Block-lava flows can be observed in several tectonic settings characterized by intermediate and silicic magma compositions, such as island arcs and intraplate volcanism with the exception of hot spots (e.g. Etna in Sicily, Kameni Islands in the Greek Cyclades, the East African Rift, Saudia Arabia, northern China, etc.).

Block-lava flows of intermediate to silicic composition

Block-lava flows exhibit a blocky surface with large, angular, smooth-sided blocks up to several metres in size. Clasts forming the breccia are "relatively smooth polyhedral blocks bounded by dihedral angles, lacking the exceedingly rough and spinose character of typical 'a'ā" (MacDonald 1953). Block lavas generally contain a larger amount of autobreccia relative to the solid core component than spiny 'a'ā flows do, and the blocks tend to be notably less vesicular. Autobrecciated textures are common in andesitic lavas, with angular blocks banded by regular chisel marks left by the splitting of lava during cleft formation. Near the vent, movement of the centre of the flow often continues after consolidation of the flow margins, forming levees as the level of the flow subsides upslope while continuing to spread downslope. The vent may be marked by a towering lava dome. Pāhoehoe surfaces are never found on blocky andesite and dacite flows, while lava tubes are also less common in andesites than in basalts (Lockwood and Hazlett 2010). Kilburn (2015) noted that another transition existed between lava flows, from 'a'ā to block-lava. The 'a'ā-blocky transition occurs when the lava interior becomes too crystalline to move forward by flow alone, at which stage its upper levels, including the surface, also fracture as they deform. However, a blocky surface never evolves into 'a'ā.

Having a higher viscosity than their basalt counterparts (in the range of 10^5–10^{11} Pa s; Harris and Rowland 2015), silicic lava flows erupted from many continental composite cones are much thicker and shorter than 'a'ā flows, and move much more slowly. Flows typically reach only a few km long and are tens of m thick, even up to a few hundred metres in thickness (e.g. the Chao dacite flow in Northern Chile: de Silva et al. 1994). These highly viscous flows exhibit steep fronts, commonly an order of magnitude higher than typical basaltic 'a'ā flows, and consist of piles of huge angular blocks (Lockwood and Hazlett 2010. Photo 7A and B). The internal structure of a block-lava flow, similar to an 'a'ā flow, comprises a basal and surface breccia with a coherent core (Harris and Rowland 2015). The core is dense, vesicle-free to vesicle-poor and often crystal-rich. Block lavas may move as a true viscous flow close to their source vents; MacDonald (1953) wrote that "examination of old flows suggests that block-lava flows more commonly advance by sliding or gliding over the surface of the underlying material in much the same manner as some glaciers". Flow front advances are extremely slow over long-time intervals, e.g. 2–13 m per day at Santiaguito, Guatemala (Harris and Rowland 2015).

1.2.4 Coulees or High-Silica Lava Flows

Coulees are thick extrusions emplaced on slopes steep enough for shear stresses to exceed the yield strength, thus permitting the flow to ooze down slope. Dacitic and rhyolitic lava flows are extremely sluggish, and form thick, steep extrusions: the thickness ranges from 50 to 500 m, with an average of about 100 m (Cas and Wright 1988). The Chao dacite coulee in northern Chile, for example, has a volume of about 24 km^3 and was erupted in three distinct flow units. The flow front of the largest unit is over 300 m high. Such coulees form a transition to dome coulee (see Sect. 3.3). In rhyolite flows a variety of lithologies and textural features can be found: obsidian, layers containing spherulites (radiating aggregates of alkali feldspar, which are found in the glassy carapace), pumiceous layers, horizons of stony rhyolite, and perlite (hydrated obsidian). The principal lithological component of most rhyolitic lavas, especially domes, is foliated stony rhyolite. High-silica lava flows tend to be glassy, and many rhyolite flows, also termed obsidian *coulees*, consist almost entirely of glass (Photo 8A and B). This is because the "high viscosity of siliceous melts suppresses the diffusion of ions and growth of microlites as degassing occurs prior to eruption" (Lockwood and Hazlett 2010). The most common type of high-silica volcanic glass is obsidian, a black, bubble-free glass that readily fractures to develop conchoidally-ribbed surfaces. The compositions of obsidian range from high-silica dacite to rhyolite.

Fink (1983) first described the stratigraphy of an obsidian flow: overlying an ash-fall layer (onset of the eruption phase), the succession consists of a basal lava breccia, coarsely vesicular pumice, an obsidian core, finely vesicular pumice, and a surface breccia. Each of the layers has different rheological properties: the most remarkable being the large difference in viscosity and density between the surface carapace compared to the main mass of obsidian (Francis 1993). The obsidian flow is subject to compression or extension as it moves over topographical highs and lows. Because the carapace is much more rigid than the obsidian core, the deformation causes the carapace to fold and fracture, producing a complex surface topography, while the core responds plastically. The wavelength of these folds may be tens of metres. Spectacular flow folds are often developed in the low-viscosity obsidian layer itself, expressed in contorted layers of pumiceous material.

Rhyolite coulees often contain well-developed flow foliations (Cas and Wright 1988; Francis 1993). These consist of interbanded and foliated layers of varying crystallinity (obsidian and stony rhyolite), spherulite content, and vesicularity. The prominent foliations are generated during stretching, shearing and attenuation during flow. The foliations often develop ramp structures, steep at the top of the flow, curving down to be asymptotic against the base in the upflow direction. The flow is usually compressive as lava builds up behind the brecciated flow-front, which retards the advancing flow. If the lava becomes extensional (following a sudden increase in slope) shear planes dip in the opposite direction, as seen in Fig. 5A. Ramping and fold structures are common in dacitic lavas containing obsidian and pumice in the Aeolian Islands and Sardinia, where pervasive folding, foliation, thrusts and fractures affect the whole flow thickness (Fig. 5B).

The rough and blocky surface of a rhyolitic lava flow commonly wrinkles into cross-flow pressure ridges termed ogives (Fig. 5A; Photo 8A and B), many metres high and spaced 10–20 m apart, that are convex down-flow with limbs running close and parallel to the flow margins. The ogives result from the freezing of the flow front while melt continues to drain from the vent. Compression forces the surface of the flow near its terminus to buckle and warp as the additional flow mass accumulates behind the stagnated, dam-like flow terminus. Other surface features of rhyolite lavas include pumice diapirs. These rose from the basal pumiceous layer of the flow, due to gravity instability caused by the inherent density inversion in the flow (Francis 1993).

1.3 Characteristic Landforms of Lava Flows

An active lava flow system can be split into two principal components: a feeder system, which is either a channel or a tube, and a zone of dispersed lava supplied by the feeder system (Lipman and Banks 1987). The most common distribution system is made up of lava channels that develop because the lava margins cool faster and experience more drag than the centre (Harris and Rowland 2015). Hulme (1974) demonstrated early on that a Bingham fluid placed on a slope would naturally form a channel because the material at the flow margins is

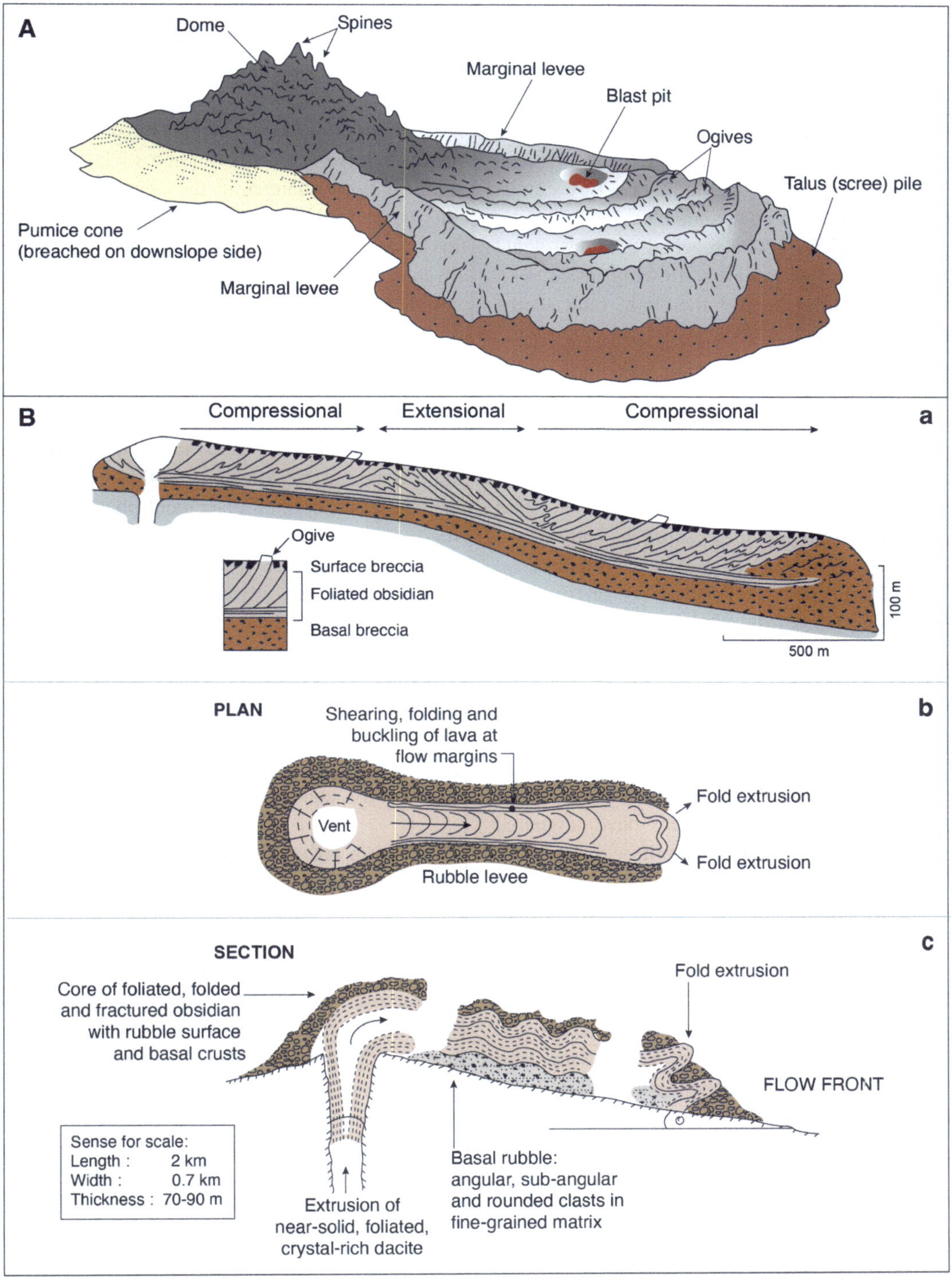

Fig. 5 Felsic lava flow, coulee: **A** Typical features and surface landforms of a high-silica (rhyolite or dacite) lava flow (Lockwood and Hazlett 2010, Fig. 6.13, p. 138). **B a** Cross section along the length of the Rocche Rosse coulee, Lipari Island, illustrating the general foliation patterns of a rhyolite obsidian coulee (modified after Cas and Wright 1988, in turn modified from Hall 1978). **b** Schematic plan and **c** cross-section and details of the key structures in the San Pietro Island silicic flow units, Sardinia (Harris and Rowland 2015, Fig. 17.7, p. 330)

too thin to provide enough force to cause flow. Alternatively, channels may form as a result of preexisting topography within which the lava is confined. Lava tubes are important volcanological phenomena because they extend the distance which a flow can reach, as the lava remains insulated. For example, lava flows emitted at Pu'u' Ō'ō and Kupaianaha vents (Kilauea) between 1983 and 2003 reached the coast some 12 km down valley after flowing as a system of lava tunnels over the 'Pali' fault scarps (Kauahikaua et al. 2003).

1.3.1 Lava Flow Channels and Levees

Channels are molten lava streams whose surface is exposed to the sky. The stream is contained between two banks of stationary lava, termed levees. Observations on the Mauna Loa 1984 effusive eruption helped Lipman and Banks (1987) to define four channel zones in pahoehoe flows (Fig. 6A and B): (1) *flow toe* or *front*; (2) *zone of dispersed flow* where some parts of the lava start to slow relative to other parts and shear zones develop between parts of the lava moving at different velocities; (3) *transitional channel zone*, marked by a distinct faster-flowing central stream, bounded by lava that is beginning to stagnate, and early-stage levees; and, (4) *stable channel zone*, where levees have stagnated and all motion is confined to the central channel. Four types of channel levees were described by Sparks et al. (1976): (1) *Initial* levees, which are the first zone of marginal stationary lava to form, and will eventually determine the width of the channel; (2) *Rubble* levees, a form of initial levee, comprised of flow front breccia; (3) *Overflow* levees, formed by overflowing lava that builds on top of existing levees, which tend to form overhanging levees. (4) *Accretionary* levees where lava adheres onto the channel from its outer margin.

Dietterich and Cashman (2014) investigated the origin of lava channel networks with modern tools such as 2D network analysis developed for river systems and neural networks, 3D LiDAR geomorphological analysis, as well as historical records of flow emplacement at Mauna Loa, Hawai'i. They found that the channel network complexity is a function of underlying slope and that the degree of flow branching, network connectivity, and longevity of individual channels all influence the final flow morphology: flow and channel width and levee height. Because channel networks govern the distribution of lava supply within a flow, changes in the channel topology can dramatically alter the effective volumetric flux in any branch, affecting both flow length and advance rate. Specifically, branching will slow and shorten flows, while merging can accelerate and lengthen them. The result also has implications for forecasting in lava flow-prone areas.

Harris et al. (2022) described the anatomy of a channel-fed 'a'ā lava flow system at Kilauea, Hawai'i. Mapping, analysis and sampling showed that a 5.5 km-long, solidified channel resulted from three emplacement phases: initial, free-flowing, and late-stage ponded. These three emplacement phases were coupled with variations in underlying substrate, slope, and volume flux. These temporal and spatial variables can be combined to define four channel types and geometries: tubed, stable, ponded, and braided; five lava facies: smooth pāhoehoe, rough/spiney, slabby, transitional, and 'a'ā; and four levee types: initial-rubble, surge-fed overflow, pond-fed overflow, and accretionary.

1.3.2 Lava Tubes or Pyroducts

Lava tubes were defined by Greeley (1987) as "a conduit beneath the surface of solidified lava through which molten lava flows". Lava tubes were termed *pyroducts* by Greeley (1987), although Lockwood and Hazlett (2010) noted that pyroduct was a term first coined by Reverend Titus Coan, when he witnessed active subterranean 'rivers of fire' during the 1843 eruption of Mauna Loa. Coan specifically mentioned their role in insulating lava from heat loss. Lockwood and Hazlett (2010) prefer the term pyroduct to "lava tubes, tunnels or pipes, as these are rarely tubular in geometry". They defined pyroduct as "any internal lava conduit in a flow, irrespective of shape and size, regardless of whether it contains molten lava during eruptive activity or is preserved as an elongate cave after eruptive activity ends and molten

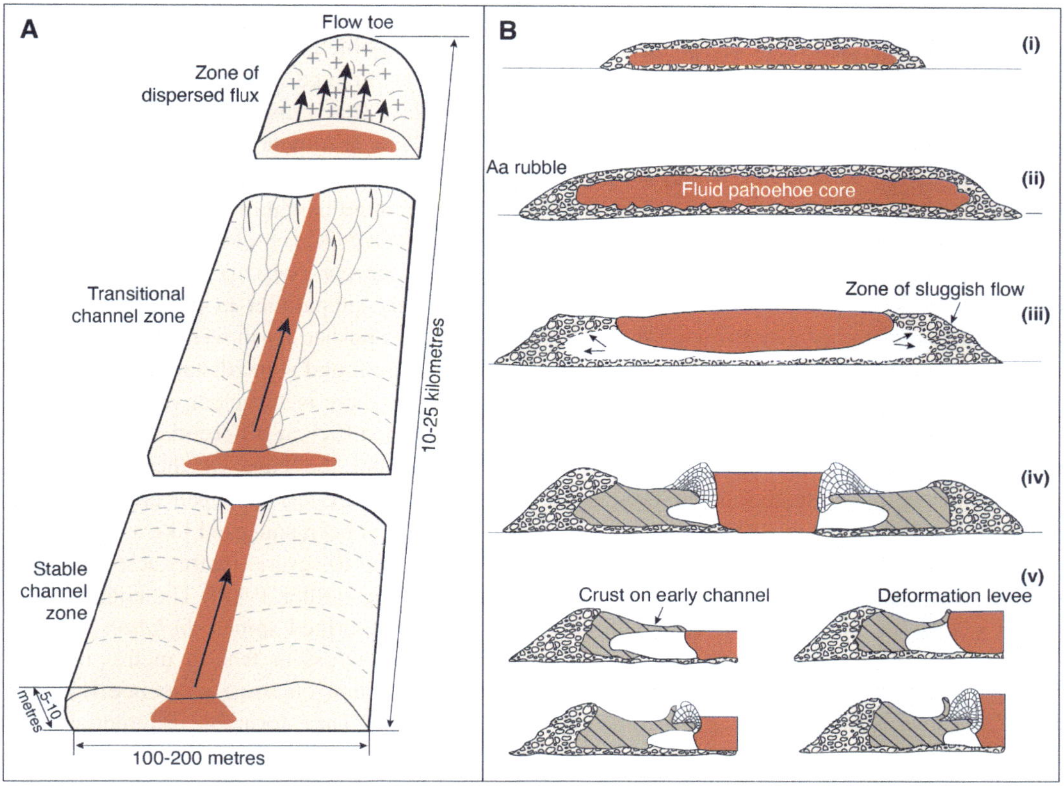

Fig. 6 Lava channel, example from Mauna Loa, 1984. **A** 3-D view showing the structures along the channel zone (modified after Lipman and Banks 1987). **B** Development of 'a'ā channel and levees from emplacement of (i) an initial flow, through (ii–iii) formation of transitional (rubble) levees, finishing with formation of (iv) overflow levees, and finally (v) deformation levees. Each zone is thus characterized by an increasingly complex levee morphology (modified after Lipman and Banks 1987, Fig. 57.9, and Harris and Rowland 2015, Fig. 17.12, p. 336)

rock drains away". Lava tubes or pyroducts can branch and braid, undergo dramatic changes in height and width over very short distances, and are subject to collapse both during and after eruptions. A collapsed roof opens up skylights, or holes in the tube roof (Cotton 1944). Tube coalescence occurs when lava breaks through the roof or wall of an older lava tube, which can result in the unexpected appearance of vents several kilometres downstream (e.g. Pu'u'O'o-Kupaianaha at Kilauea: Kauahikaua et al. 2003).

The formation of pyroduct roofs involves two processes: (i) narrowing of the channel rims by freezing of lava levees along channel walls, and (ii) the accretion of plates of crust that are skimmed off channel surfaces where flow obstructions are encountered. Peterson et al. (1994) listed five mechanisms for creating a tube: (1) Crustal growth by a stationary, hanging levee extending entirely across the channel, and forming a roof; (2) Repeated channel overflow, with increasingly overhanging levees at the sides. When the rims join, an arched roof is created and the light-bulb or triangular tube is formed; (3) Jamming of solidified crust plates at channel constrictions. Lava that oozes up between crust plates welds them into a solid roof; (4) The crust of a stalled flow behind a breakout forms a bulbous tube, with each bulge being a zone of stalling and inflation, and; (5) When lava enters the ocean, waves can abruptly chill the crust across the entire surface of a molten stream, thus forming a tube across the surf zone.

Lava tubes are much more common in pāhoehoe than 'a'ā, and their length varies from a few metres to 100 km or even more, as on the Undara lava plateaus of North Queensland, Australia (Ollier and Brown 1965; Atkinson et al. 1975; Photo 9A–C). The Undara volcano erupted 0.19 Ma ago and formed lava fields covering 1550 km^2 with a volume of approximately 23 km^3. One of the flows extended 160 km on a gradient that averaged only 0.3°, and contains long wide caves (Ollier and Brown 1965). The exceptional length results from a combination of very high effusion rates, high-temperature hawaiites, a favourable topography, and lava tube efficiency (Atkinson et al. 1975). A prominent geomorphological feature of the flow is the 40 km-long, 5–20 m-high, 200 m-wide "Wall" (a pressure ridge) that is thought to lie above one of the lava tubes which occupied a former stream course at this location. Although less numerous and often shorter, lava tubes also exist in 'a'ā lava flows. Examination of several compound 'a'ā lava flow fields on Etna confirmed that a complex network of tubes forms at successively higher levels within the flow field, and that tubes generally advance by processes that include flow inflation and tube coalescence (Calvari and Pinkerton 1999).

1.3.3 Minor Surface Landforms Associated with Lava Flows

A number of minor landforms can be found on the lava flow surface, in particular on pāhoehoe but also on 'a'ā flows (e.g. on Icelandic shield volcanoes; Rossi 1997; Rossi and Gudmundsson 1996). Most features form in response to pressure build-up within a tube system and to inflation processes (Fig. 1).

R.A. Daly (1911 *in:* Lockwood and Hazlett 2010) introduced the term *tumulus* (plural *tumuli*) to refer to the steep-sided mounds observed on surface of many pāhoehoe flows. Most are elliptical in ground plan and range in size up to 15 m in diameter and are 1.5–3 m in height. The tumulus crust commonly splits apart at the top during uplift of the original surface due to lava overpressure, and molten lava drains away laterally, leaving grottos or holes (e.g. Rossi and Gudmundsson 1996). *Lava blisters* on pāhohoe, which resemble tumuli but are less common, are created by the surficial swelling of a plastic lava flow crust in response to the puffing up of gas or vapour from beneath the flow. They are hollow structures 1–150 m in diameter, with a maximum height of 30 m.

Pressure ridges are elongate (up to 100 m long) and narrow (<10 m) uplifts of the lava crust, mostly subparallel to the flow direction. These ridges pushed by movement of the underlying lava are found some distance above bottlenecks where there is sufficient width for a ridge to form, and sufficient differential flow between the centre and the edge of the flow to create drag. *Squeeze-ups* are open cracks through which lava has been squeezed like a dyke. Some of them are up to several metres across and one kilometre-long (Ollier 1988). The equivalent on 'a'ā flows are termed spines. *Inflation pits,* less than a metre across or tens of metres in diameter, are common on some pāhoehoe flow surfaces (Fig. 1). They form by differential uplift of the surrounding crust rather than collapse processes.

Spatter cones, shatter rings and *hornitos* are often built up over adventitious vents where active lava breaks out at the surface after a long subterranean passage. Orr (2011) describes *shatter rings* as "circular to elliptical volcanic features, typically tens of meters in diameter, which form over active lava tubes. They are characterized by an upraised rim of blocky rubble and a central depression". *Hornitos* are small, steep, rootless spatter cones up to several metres high, which develop when lava is forced up through a skylight and then accumulates around the opening, sometimes accompanied by explosions induced by trapped groundwater (Fig. 2). Examples of well-preserved hornitos and other minor landforms of an exceptional size can be found in the Wudalianchi Geopark in Northern China (Xiao and Wang, 2009).

Accretionary lava balls are typically egg-shaped, smooth-sided boulders 1–3 m in diameter that consist of low-density internal cores enclosed by glassy rims. Most originate within spatter cones formed by lava fountains during the early phase of long-lived eruptions. Because the bulk density of these porous, lava-coated spatter masses is much less than the surrounding melt, they immediately rise back up to the surface, where they are rafted or tumbled downstream in flow channels (Lockwood and Hazlett 2010). Fine examples can be seen in the Vesuvius (Italy) lavas erupted in 1944, in which spectacular lava balls occur in phenocryst-rich (30–50%) tephrite and phonotephrite flows with otherwise typical blocky surface morphologies (Vilardo and Ventura 2008; Bisson et al. 2021).

More extensive or peculiar landforms are created when flowing lava enters the sea, a lake or marsh. *Lava deltas* build up on the shores of volcanic islands, such as on the coast of La Palma, fed by the 2021 Tajogaite-Cumbre Vieja effusive eruption. When 'a'ā lavas enter the sea, they do so explosively, fragmenting the lava and forming piles of debris known as littoral cones. Lava deltas can mantle pillow lavas and hyaloclastites (Fig. 3B) and in turn can be covered by late flows advancing from the land, extending the coastline. When lava encroaches into lakes or marshes, ground-water is trapped and if sufficiently heated it can flash into steam, which bursts through the lava to form *pseudocraters* or rootless vents (e.g. Lake Myvatn, Iceland: Photo 10).

Fluid basalt lavas commonly bury tilled land and trees in tropical lowlands and islands like Hawai'i. Where relatively low viscosity lava ($<10^3$ Pa s, Chevrel et al. 2019) sweeps through forests, it can solidify around living, cold, damp trees. Fast-flowing lava, on the other hand, sets them on fire, incinerating the wood, while also solidifying around the outside of the former tree trunks, forming casts or *tree moulds* (Finch 1931; Lockwood and Hazlett 2010). Tree moulds are more common around the Hawaiian shield volcanoes, but they can also occur at composite cones (e.g. Fuji: Ogawa 1980). Tree moulds remain standing as the surrounding, more liquid lava drains away, forming hollow pillars of basalt, or standing *lava trees* (Chevrel et al. 2019; Biren et al. 2020).

1.4 Physical and Rheological Parameters Driving Lava Flows

Lava flows owe their dynamics to the effusion rate, the topography on which they expand, and the cooling history and rheological parameters. To these primary factors controlling how lava flows develop, Kilburn (2015) added the local environment, namely slope angle and whether lava extrusion occurs on land, underwater or beneath ice.

1.4.1 Aspect Ratio and Rheological Parameters

Lava flows have different morphologies based on variations in viscosity and temperature. A simple diagram (Fig. 7A) expresses such variations in morphology based on the aspect ratio (AR: thickness/area) of principal lava types according to their composition and viscosity (Walker, 1973). Basalt lavas are thin and extensive (AR > 1/100), whereas rhyolitic lavas are thick and compact (AR = 1/4 to 1/10). The thickness and width of a flow are related to effusion rate (F or Q), the slope (α) and three fluid properties: Bingham viscosity (η), yield strength (σ_y) and specific weight ($g\rho$). The aspect ratio of a lava flow can be predicted from the equation:

$$\sigma_0/F\eta)^{0.25}(g\rho)^{0.75} \quad (1)$$

Aspect ratio therefore depends mainly on yield strength. Basalt lavas with low yield strength give rise to low AR flows, whereas more silicic lavas with higher yield strengths will be emplaced as higher AR flows. AR may be insensitive to effusion rate, but a higher effusion rate would increase the temperature, and therefore reduce the yield strength, lowering the aspect ratio (Cas and Wright 1988).

A few rheological parameters help understand lava flow dynamics: shear stress and yield strength related to flow thickness, viscosity related to temperature and shear rate (Fig. 4A), and finally cooling rate. Pioneer studies include those by Jaggar (1917), Nichols

(1939), Wentworth et al. (1945), Hulme (1974), Chester et al. (1985) and Moore (1987), but laboratory experiments combined with field measurements have provided more insight into the role of rheological parameters and understanding lava flow dynamics, e.g. Chevrel et al. (2013), Cashman and Mangan (2014), Harris and Rowland (2015), Stasiuk and Jaupart (1997) and Gregg (2017). The main parameters include shear stress τ that needs to be reached before flow will occur, which is determined by the yield strength σ_y (or τ_0) of the fluid: flow only takes place if the downslope force is larger than the yield strength. Yield strength in turn is related to a critical flow thickness, as lavas tend to be Bingham fluids (Hulme 1974). Basaltic lava has a yield strength of ~10^3 Pa, whereas that of silicic lava is around 10^5 Pa. Viscosity, yield strength and crystallinity control the lava flow dynamics. The dynamic viscosity η of lava will increase as temperature and density decrease. Viscosities measured on lava flows at eruption temperature are in the range of 10^2–10^3 Pa s for basaltic flows (density ρ 2600–2800 kg/m^3 at T = 1050–1200 °C), increasing to 10^4–10^7 for andesite (ρ 2450 kg/m^3 at T = 950–1170 °C) and up to 10^9–10^{13} Pa s for rhyolite (ρ 2200 kg/m^3 at T = 700–900 °C) (Kilburn 2015). The crystal content or crystallinity of the lava will also increase the viscosity.

In modern volcanology, the complex rheological behaviour of lavas is measured and modelled as a function of the chemical composition of the liquid phase, the fraction of crystals and bubbles, and the temperature and strain rate (Chevrel et al. 2013; Greggs 2017). Using field measurements on a lava flow in Central Iceland, rheological models, and thermodynamic calculations, Chevrel et al. (2013) showed that the flow viscosity was transient and quickly became non-Newtonian as the groundmass crystallised. They showed that rheological properties, as inferred from the morphology of the flow, must be distinguished from the lava behaviour at the eruption onset and during most of its flow: viscosity increased by more than six orders of magnitude during emplacement from the vent to the flow front. This is due to the fact that viscosity is highly dependent on the crystallisation sequence and crystal shapes, meaning that the bulk chemical composition of the erupted material cannot be used alone to determine viscosity. Thus, while the bulk chemistry can be used to characterize lavas of basaltic composition, it may not apply to silicic lavas in which vitrification resulting from degassing and cooling ultimately causes lava flows to halt.

1.4.2 Factors Controlling Flow Length and Flow-Front Advance Rate

In general, the length of short duration flows is limited by magma supply, whereas the length of long-lived flows is limited by cooling (Cashman and Mangan 2014). The length of a simple channelized lava flow will increase with lava effusion rate if the maximum flow length is controlled by cooling, if cooling rates are constant, and if effusion rate controls flow velocity. However, there

Fig. 7 Aspect ratio of lava flows. Controls on lava flow length and flow-front advance. **A** Aspect ratio and dimensions of lava flows of different compositions (modified after Walker 1973; Cas and Wright 1988). Y-axis: thickness (T, in m), X-axis: covered area (A, in km^2) or diameter (D, in km). Dotted line: volume (in km^3). Dashed line: T/A ratio. **B** Plot of the length of channelized Hawaiian lava flows as a function of average volumetric flow (effusion) rate. Lines are original trends identified by Walker (1973). 'Ocean-limited' flows reached the ocean, and therefore the flow distances are minimum. Unfilled red diamonds represent short-duration (<12 h) lava flows erupted from Kilauea in July (short flow) and December 1983 (long flow). Blue squares represent flows from 3-week long Mauna Loa eruptions in 1984 (short flow) and 1859 (long flow) (Cashman and Mangan 2014, A Century of Studying effusive eruption in Hawai'i, Fig.12, p. 371, Chap. 9, Prof. Paper 1801, U.S. Geol Survey, pp. 357–394). **C.1** Volumetric flow rate versus flow-front velocity estimated for historical flows (diamonds and squares: Rowland and Walker 1990) and constraints from well-observed recent flows (colored boxes: data from Kauahikaua et al. 2003). P = pāhoehoe, A = 'a'ā. **C.2** Plot showing distance versus time for recent flows in Hawai'i, according to different volumetric fluxes and front velocities (Kauahikaua et al. 2003). Individual fields are labelled for volumetric flow rates; dashed lines are flow-front velocities. Colours in 1 and 2 refer to the same data (both plots in: Cashman and Mangan 2014, A Century of Studying effusive eruption in Hawai'i, Fig. 13, p. 372, Chap. 9, Prof. Paper 1801, U.S. Geol Survey, pp. 357–394)

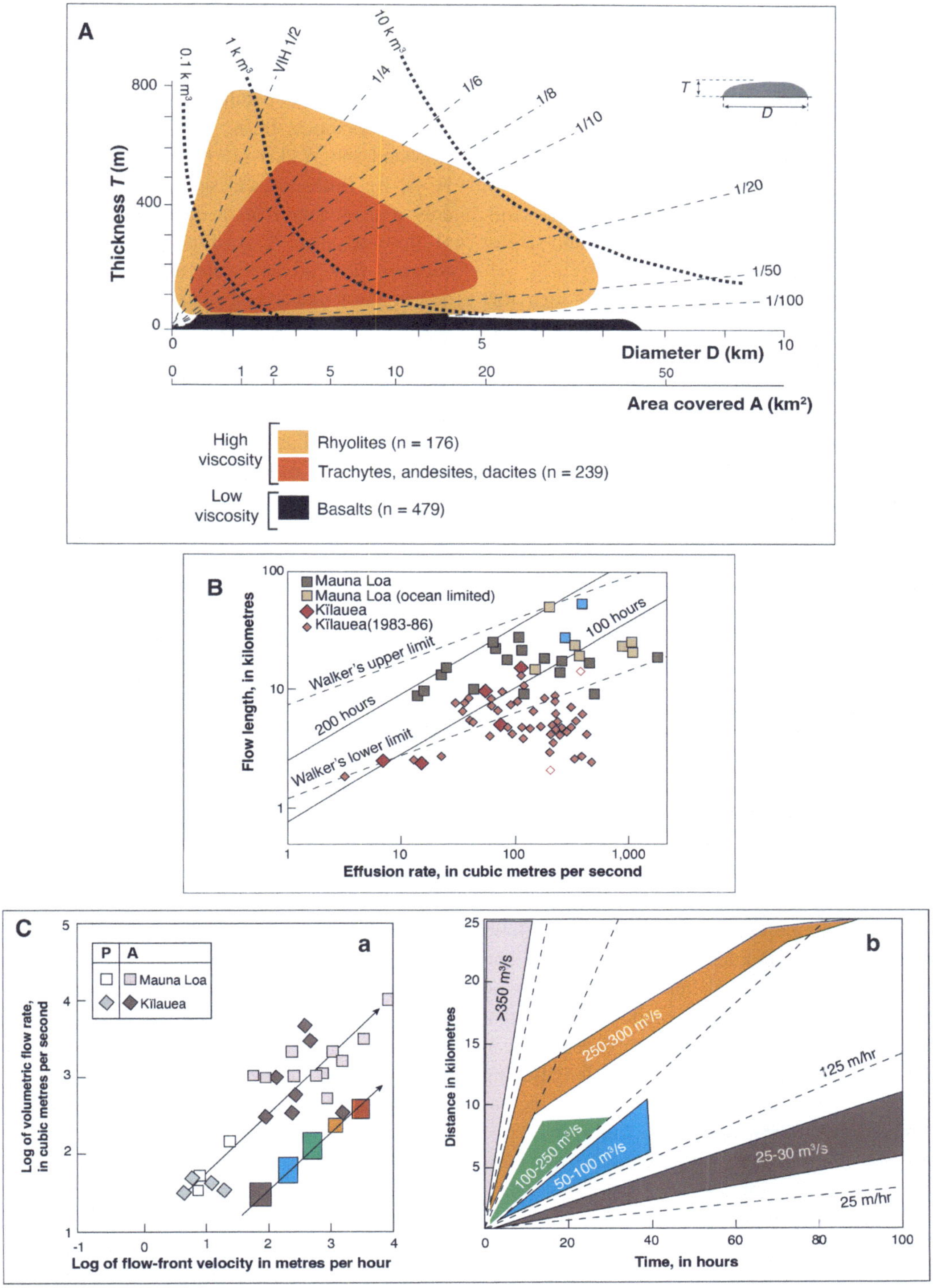
A
0.1 k m³
1 k m³
VIH 1/2
10 k m³
1/4
1/6
1/8
1/10
1/20
1/50
1/100
T
D
800
400
0
Thickness T (m)
0
5
10
Diameter D (km)
0
1
2
5
10
20
50
Area covered A (km²)
High viscosity
Rhyolites (n = 176)
Trachytes, andesites, dacites (n = 239)
Low viscosity
Basalts (n = 479)
B
Mauna Loa
Mauna Loa (ocean limited)
Kīlauea
Kīlauea(1983-86)
Walker's upper limit
200 hours
100 hours
Walker's lower limit
100
10
1
Flow length, in kilometres
1
10
100
1,000
Effusion rate, in cubic metres per second
C
a
P
A
Mauna Loa
Kīlauea
5
4
3
2
1
Log of volumetric flow rate, in cubic metres per second
-1
0
1
2
3
4
Log of flow-front velocity in metres per hour
b
>350 m³/s
250-300 m³/s
100-250 m³/s
50-100 m³/s
125 m/hr
25-30 m³/s
25 m/hr
25
20
15
10
5
Distance in kilometres
0
20
40
60
80
100
Time, in hours

is no clear correlation between channelized-flow lengths and effusion rates (Fig. 7B). Effusion rate is measured and defined in three different modes (Harris et al. 2007): instantaneous effusion rate IER; time-averaged discharge rate (weekly, daily) TADR; and mean output rate MOR, i.e. volume extruded from the start to the end of the eruption. TADR and MOR are metrics commonly used to quantify the timescale and size of lava flow eruptions (e.g. Calvari 2019).

Modern studies have changed the idea that long lava flows formed when low-viscosity lava is erupted at high-effusion rates (Cashman et al. 1998). In contrast with simple channelized flows proportional to the volumetric flux, long-lived flows are more complex because lava channels evolve in space and time as levees form, flow fronts freeze, flows split around topographic obstacles and lava spillovers create new channel branches (Dietterich and Cashman 2014). These modifications to channel networks affect both the flow advance rate and the ultimate length achieved by individual flow lobes (Cashman and Mangan 2014). In addition to changes in effusion rate occurring at the vent, the high fluidity of pāhoehoe lavas makes them susceptible to topographic confinement that promotes flow lengthening, whereas channel bifurcations caused by topographic obstacles create multiple parallel channels that can limit individual flow-lobe lengths. Finally, flow fields generally widen rather than lengthen (ratio $w/l > 2$) when the magma supply is sporadic and flows are emplaced as discrete events (Cashman and Mangan 2014). Detailed field measurements on active lava flows on Kilauea have shown that inflation of lava flows after emplacement can result in reactivation and continued lengthening of these flows. Inflated sheets can thus act as insulated lava tubes that permit the transport of lava over great distances at near isothermal conditions (e.g. Self et al. 1996).

Nevertheless, effusion rate exerts a fundamental control on the initial advance rates of channelized lava flows (Fig. 7C-a). This relation is illustrated using data from well observed eruptions, where initial effusion rates for channelized flows ranged from 25 to 1000 m^3/s and corresponding initial rates of flow advance varied from <0.02 to 3–4 m/s (Fig. 7C-b; Kauahikaua et al. 2003). These initial flow advance velocities form a trend that parallels, but is offset toward higher velocities from, the data compiled by Rowland and Walker (1990) on older flows (Fig. 7C–a, b). Both datasets suggest that slope plays only a secondary role in controlling initial rates of flow advance. The observed correspondence between flow advance rate and volumetric flow rate suggests that advance rates inferred from flow features (e.g. runup height on tree moulds, superelevation on channel bends, and lava crystallinity) can be used to estimate effusion rates from older eruptions. These data also highlight the importance for hazard assessment of obtaining accurate estimates of effusion rates during early stages of eruptive activity. As lava flows typically advance more slowly when farther from their source vents, eruptions from low-elevation flank vents pose much greater hazards than those from vents high on the rift zones of shield volcanoes (Cashman and Mangan 2014). This was illustrated by the December 2022 Mauna Loa lava flows which extruded from the Mokuaweoweo summit caldera (above 4100 m.a.s.l.) far from any settlement, as opposed to the 2018 Kilauea lava flows at low elevation that erupted along the lower East rift zone containing residential areas (below 1000 m.a.s.l.).

Cooling also operates as lava moves away from the vent, because of heat loss due to radiation, conduction through the ground and levees, and convection into the atmosphere above the flow surface. Another cooling factor is the incorporation of levee material and erosion of the bedrock by the hot moving lava. The temperature of the lava decreases with distance from the vent, due to radiative heat loss per metre advanced, and heat loss due to convection. Harris and Rowland (2015) provided an example of an initial surface temperature of 1000 °C cooling by 4.5 °C per kilometre, decreasing to 0.13 °C per km once the surface temperature decreases to 250 °C. However, the temperature decline is not linear downflow, as it increases rapidly near the flow front. At the same time, the crystal content increases with distance from the vent, due to cooling-related crystallization.

In channels, cooling per unit distance appears to increase gradually along most of the flow with a rapid increase near the flow toe. Crystallinity rates were measured at 1.2% per degree cooling, and were higher at Etna than at Kilauea (Harris and Rowland 2015). The presence of a tube roof greatly reduces heat loss from flowing lava. As a result, tube-fed lavas can extend great distances even at quite low volume fluxes because the cooling rates are so low (0.7 °C/km, and down to 0.35 °C/km distally). Finally, the rheology changes down flow as temperature decreases and crystallinity increases with distance. Viscosity and yield strength increase with distance, which means that the flow velocity declines with distance. These parameters related to distance help define the cooling-limited flows.

1.5 Thermal Structure and Inflation Mechanism of Lava Flows

The thermal structure of active pāhoehoe flows helps understand how active flows will inflate. Inflation is a process by which a fixed and stationary cooling surface, which has developed sufficient strength to prevent the escape of fluid lava, becomes lifted upwards due to the influx of new lava (Walker 1991; Hon et al. 1994). Inflation is a ubiquitous process in an active pāhoehoe flow field and causes single lobes, multiple coalesced lobes, or entire flow fields to uplift (Hon et al. 1994). If broad areas of pāhoehoe are uplifted or inflated, lava rises (Walker 1991), also termed inflated pāhoehoe sheet lobes (Thordarson and Self 1998), will result. These are broad areas, hundreds of metres across with roughly horizontal surfaces, that have been inflated by a few metres to tens of metres and which are bounded by a steep escarpment.

1.5.1 Thermal Structure of a Lava Flow

Hon et al. (1994) carried out measurements on active pāhoehoe flows at Kilauea. A lava flow in section shows five thermally and rheologically distinct layers, from base to top: (1) A basal crust characterized by a low to moderate temperature (600–900 °C) and a high viscosity. The crust cools on contact with bedrock surface, while its thickness increases with time and cooling rate. (2) A lower visco-elastic, partially molten layer. (3) A molten core with the highest temperature: 1150 °C (Hawai'i), 1080 °C (Etna), and lowest viscosity: 10^2 to 10^3 Pa s in Hawai'i to 10^3 to 10^4 Pa s at Etna. The core thickness decreases with time and cooling rate. (4) An upper visco-elastic, partially molten layer is characterized by relatively lower temperatures, with its top defined by an isotherm at 800 °C and the base defined by an isotherm at 1070 °C. (5) Finally, the surface crust (with a lower temperature of around 850 °C), which is solid but brittle, where rapid heat loss results in a reduction in surface temperature of 50–300 °C almost immediately upon exposure (quenching effect).

If the lava supply remains constant, both the basal and uppermost crustal layers increase in thickness (Hs) over time (t) at a rate controlled by conductive cooling, that is, $H_s = C\sqrt{t}$, C being a constant (Hon et al. 1994; Cashman and Mangan 2014). The basal crustal layer is assumed to grow at 70% of the rate of the uppermost crustal layer because of contact with the underlying rock (Hon et al. 1994). The thickness of the molten core appears to stabilize after about 100 h, but it may take years to solidify a thick lava flow, as suggested by modelling carried out on the La Roza member of the Columbia River Basalt Group (CRBG, Thordarson and Self 1998, in: Harris et al. 2017) (Fig. 8A).

1.5.2 Inflation Mechanism

For a flow to inflate internally, the outer crust must either deform or break. Successive fractures at either the sheet flow margins or central axial crack thus record the process of lava flow inflation (Cashman and Mangan 2014). Mapping of structures within inflating pāhoehoe flows has also proven important for understanding active lava transport systems (Hon et al. 1994). For example, elongate tumuli form over primary lava feeder networks, and hummocky tumuli characterize areas where tubes have slowed or stalled. These inflation features generate a distinctive surface morphology that has made it possible to

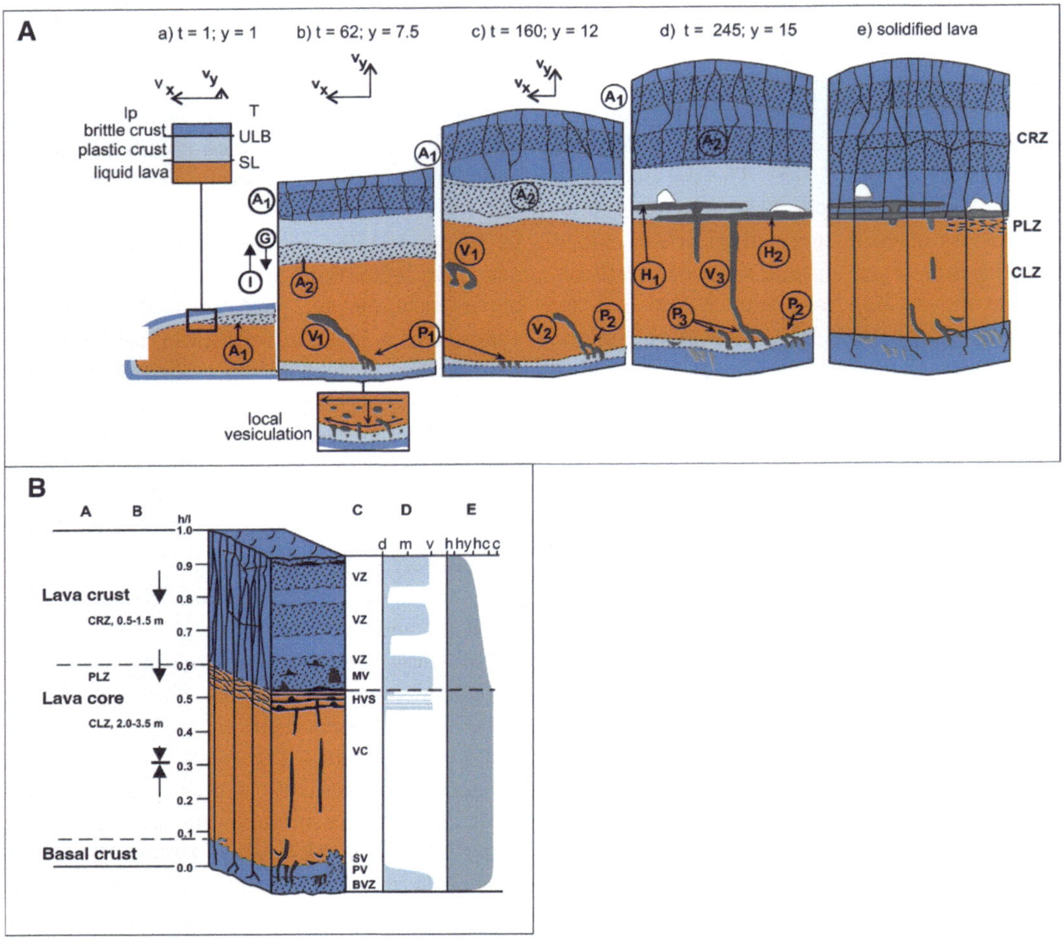

Fig. 8 Inflation mechanisms of a pāhoehoe flow. **A** Idealized cross sections of Thordarson and Self (1998, Fig. 14, p. 27431) showing the growth stages of inflated sheet lobe pāhoehoe, based on the La Roza member, CFRB (t = time in days, y = thickness in metres, and vx and vy give the lateral and vertical expansion vectors, respectively; the length of line is proportional to the amount of expansion). Direction of inflation (I) and growth direction of the upper crust (G) are indicated by arrows (left side of section b). The liquid core is orange, the brittle crust is dark blue, and the plastic crust, which is above the temperature of brittle behaviour (ULB) and below the solidus temperature (SL), is light blue. Joints in the crust zone (CRZ), columnar zone (CLZ), and platy zone (PLZ) are indicated with black lines and dashed horizontal lines, respectively. Other features are A_1, A_2 (dotted) = vesicular zones, P_1, P_2, P_3 = pipe vesicles, V_1, V_2, V_3 = vesicle cylinders, H_1, H_2 = continuous horizontal vesicle sheets or dome-shaped features; megavesicles in white; the subscripted numbers of P, V, H provide the sequence of formation. **B** Composite log of Thordarson and Self (1998, Fig. 21, p. 27442) illustrating the characteristic structures of inflated sheet lobe pāhoehoe. Left side breaks out the characteristic three-part division of sheet lobes (A) and jointing styles (B), and the right side shows the characteristic distribution of vesicle structures (C), vesiculation (D), and crystallinity (E). In D, d = dense (0–5 vol% vesicles), m = moderately vesicular (10–20 vol%), and v = vesicular (30–40 vol%). In E, h = hyaline, hy = hypohyaline, hc = hypocrystalline, and c = holocrystalline lava textures

identify flow inflation in many other terrestrial and submarine basaltic lava-flow fields as well as in some LIPs (e.g. Roza member in CRBG, Thordarson and Self 1998).

Inflation also produces characteristic structures and textural sequences (vesiculation and crystallinity) in the flow interior (Fig. 8A). Thick, inflated units will remain molten for weeks, even months to years, so that vesicles can organize themselves into distinctive patterns and sequences. Vesicles tend to increase in size, but decrease in number, away from the top and basal surfaces, with the flow centre being dense and massive (Thordarson and Self 1998). A composite log by these authors illustrates the characteristic structures of inflated sheet lobe pāhoehoe (Fig. 8B). Inflation and forward flow of pāhoehoe, together with injection of new material into the flow interior, result in a three-part division of the unit with a dense lava core bounded by vesicular upper and lower crust containing cyclic vesicle horizons, hanging contacts, and segregation structures (Thordarson and Self 1998).

1.6 Columns and Columnar Jointing in Lava Flows

Columnar jointing is more common in mafic lava flows than in silicic ones, but they occur in a wide range of lava flows (see below). However, basalt lava flows a few to several tens of metres thick develop spectacular stacks of tiers (sets or rows) of cooling columns (Fig. 9A; Photo 11).

1.6.1 Columns, Colonnade and Entablature

Columnar-jointed lavas usually exhibit elongated, polygonal columns (several decimetres wide, 1–10 m high), with a height-to-width ratio in general between 10 and 100 (Hetényi and Milazzo 2014) (Fig. 9A and B). Above a thin basal zone lies the *lower colonnade* (Tomkeieff 1940), a collection of well-developed, adjacent, vertical columns from less than a metre to as much as 20 m in height, which occupies up to 30% of the flow thickness. In the next zone up, the columns are much narrower and show a variety of orientations, with closely spaced cooling cracks. Termed *entablature* by Tomkeieff (1940), this irregular pattern comprises about 60% of the flow mass. At the top of the flow is a thinner zone, the *upper colonnade*, in which the columns are again wide but much less well developed than below (Lockwood and Hazlett 2010). Entablature is lacking in some flows and in others there may be multiple sets (tiers) of entablature and colonnade jointing (Fig. 9).

As lava flows cool, they shrink, setting up significant internal stresses, and form columnar jointing. The most prominent joints form at approximately right angles to cooling surfaces, which are usually the top and bottom of the flow, but may be at the sides as well, for example where a flow cools against the wall of a former gorge or a crater. The joints tend to develop in three directions at roughly 60° to each other and form multi-sided columns (Lockwood and Hazlett 2010). The orderly, prismatic columns produced by jointing commonly possess five or six faces, each face being a separate joint, and are therefore polygonal in cross section: on average, half of the columns are hexagonal and one third are pentagonal, with the remainder being polygonal (O'Reilly 1879; Budkevitsch and Robin 1994). For example, hexagons outnumber pentagons at the The Giant's Causeway (Northern Ireland), as well as Devil's Postpile (California, USA), but pentagons predominate over hexagons in the Craters of the Moon basalt (Idaho, USA) (Budkevitsch and Robin 1994). Cross sections of the columns are usually equant polygons, hence column diameter is proportional to joint spacing and column face width (Fig. 9B). The diameter of the columns usually ranges from ~0.2 to > 2m, while the height can range into tens of metres, as at lava columns.

1.6.2 Columnar Jointing: A Complex Set of Thermodynamic Processes

Columnar joints can be defined as "a planar, extensional rupture surface of a larger network of polygonal fractures that delimits a set of columns, the entire system resulting from contraction of the material" (Hetényi and Milazzo 2014). Columnar joints form when thermal stresses due to cooling cannot be relaxed by the viscous behaviour of the material, and their difference overcomes the tensile stress of the material, resulting in rupture. This set of processes is strongly dependent on temperature and cooling rate. The development of columnar joints was

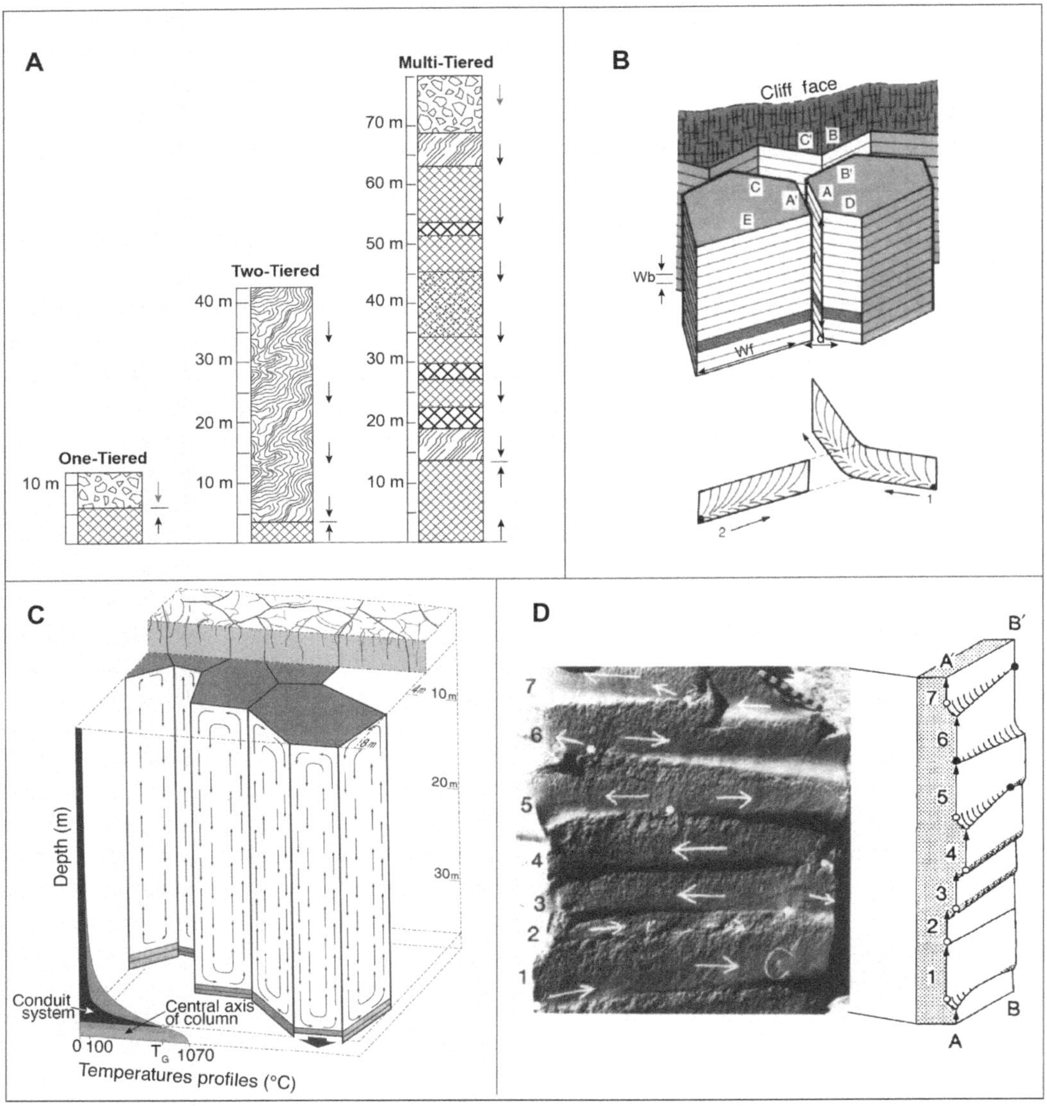
A
Multi-Tiered
Two-Tiered
One-Tiered
70 m
60 m
50 m
40 m
30 m
20 m
10 m
B
Cliff face
Wb
Wf
C
Depth (m)
10m
20m
30m
Conduit system
Central axis of column
0 100
T_G 1070
Temperatures profiles (°C)
D
A
B
A′
B′

highly studied through modelling and experiments over the years 1980–2000 (e.g. Ryan and Sammis 1981; Degraff and Aydin 1987; Budkevitsch and Robin 1994; Allain and Limat 1995; Grossenbacher and McDuffie 1995; Manga and Ventura 2005).

Almost all cooling lava bodies (including lava flows, lava lakes and lava infills in maars or tuff rings) are cut by cracks known as *joints*, which originate almost entirely during lava flow emplacement. The growth of individual cracks proceeds perpendicular to the cooling surface and the initial, often irregular network of joints becomes polygonal below this surface (Fig. 9C). Its characteristic spacing depends on the cooling rate, which is determined by two factors: chemical composition of the magma and geological setting of the emplacement. (1) There is a continuous transition in column width from large in highly alkaline lavas (phonolite), through intermediate in intermediate compositions (andesite), to narrow columns in mafic lavas (basalt). (2) The geological emplacement of the igneous body is highly important in influencing column width, as it determines the boundary conditions for cooling and the thickness of the material to cool. The cooling rate at the surface can be accelerated by direct contact with water or ice and hence yield narrower columns than might otherwise be expected.

A key issue in columnar joint formation is the control on column diameter. Column diameter, proportional to fracture width, depends on the rate of cooling, slower cooling resulting in wider columns and faster cooling in narrower columns (Degraff and Aydin 1987). In two-tier and multi-tier lava flows, downward-growing columnar joints usually meet upward-growing joints well below the middle of the flows, which indicates relatively more rapid cooling of upper portions compared to lower portions. This supports the idea that convection of water in columnar joints connected to the surface may be an important mechanism for cooling the upper portion of these flows, whereas conduction is probably the dominant cooling mechanism at the base (Degraff and Aydin 1987).

Cooling surfaces commonly generate their own sets of columnar joints, which work their way stepwise into slowly solidifying flow interiors of all flows (Fig. 9A and C). Over a period of days, weeks or months, joint sets propagating

◀ **Fig. 9** Development of polygonal columnar jointing in basaltic lava flow. **A** Configuration and overall growth directions of columnar joints in three structural types of lava flow. Relative spacing and regularity of lines schematically show changes of column diameter and regularity of column faces from tier to tier. Solid arrows show growth directions of columnar joints. Open arrows show probable growth direction of irregular subvertical joints (not shown) in the heterogeneous flow-top regions (pebbly pattern). One-tiered flow has two sets of vertical joints, but only one set is properly columnar. Two-tiered flow was examined at a location where the flow-top (not shown) is mostly missing. Multi-tiered flow with an internal vesicular zone (stippled) (after DeGraff and Aydin 1987, Fig. 14, p. 614). **B** Idealized vertical columns with horizontal bands on joint surfaces. Symbols represent column length (l), column diameter (d), column-face width (Wf), band width (Wb), and aspect ratio (l/d) (after Aydin and DeGraff 1988, Fig. 3, p. 472). **C** The downward propagation of columnar joints in a cooling basalt flow, modified after Ryan and Sammis (1981) *in:* Budkevitsch and Robin (1994, Fig. 3, p. 223). Cellular convection along the joint planes is simplified with small arrows. In the actual network of cracks, fluid convection is not confined to within a single joint face and the flow pattern is likely to be more complex. At the leading edge of the crack tip, the growth of the columns by incremental crack advance is illustrated by the newly formed crack segments (shaded). Crack propagation (large filled arrow) is punctuated and stops where the temperature is at the upper limit of the glass transition, T_G (~750 °C). Two depth-temperature profiles are shown: one for the crack-conduit system (light grey) and one through the core of a column (black). For some distance behind the crack front (about twice the column diameter), the interior of a column is warmer than, and loses heat by conduction to, the cooler joint planes. **D** Interpretation of joint-surface morphology of a column face near a flow base (Scale in inches). **a** Horizontal bands of the column face. Each band has a single plumose structure and represents an individual crack that formed during a discrete cracking event. **b** Sketch of the column face showing the origins of the cracks (dots), hackles of the plumose structures (short thin lines), and crack terminations or outlines (continuous thick lines). Arrows are on horizontal plume axes and show local propagation directions. Numbers give the sequence of crack formation. **c** Oblique view of cracks on cutout section A-A'-B'-B. Overhangs (with stippling) and ledges of column face are shown by parallel curved lines. Profile of cracks along the cut A-A' shows plume axes (open circles) and vertical components of crack propagation (arrows) (after DeGraff and Aydin 1987) from opposite cooling surfaces may meet to form single, through-going columns (Budkevitsch and Robin 1994). Extension begins as stress builds up at the tip end of a partly developed joint, followed by sudden fracturing of the rock a few centimetres farther inside. The rock is momentarily relaxed after each spasm of propagation, but tension steadily accumulates with ongoing cooling until the breaking point is once again reached. The cyclical nature of this process results in panelled joint surfaces (Fig. 9D). Where exposed by later erosion, the face of each panel typically features a feathery structure, termed a *plumose* pattern, the fine lines of which spread apart in the direction the panel opened over a period of microseconds and terminate in sharp, wedge-shaped edges termed *hackles*, the inception line for fresh cracking (Fig. 9D).

Joints in the colonnade may exhibit surface markings known as banded striae or '*chisel marks*' oriented normal to column axes and plumose structures (DeGraff and Aydin 1987). Each stria consists of a smooth band and a rough band, each containing a single plumose structure, while both bands correspond to a joint growth increment (Fig. 9D). The striae form by fracturing, which occurs when thermal contraction-induced stresses exceed the yield strength of the rock. The fracture propagates inward (to hotter regions) until it reaches the glass transition temperature. This temperature interval is thought to be a constant, with a measured value of 53 °C for a Hawaiian basalt (Ryan and Sammis 1981). Analysis of plumose structures and intersections of cracks led to a kinematic model of columnar jointing, which provided the first direct proof that columnar joints grow incrementally from external to internal regions of solidifying magma bodies (DeGraff and Aydin 1987). Columnar joints form by nucleation and growth of new cracks on the edges of older cracks. Each new crack begins at a point and propagates mostly normal to column axes and along the leading edge of a developing column face, where thermal stress is concentrated (Grossenbacher and McDuffie 1995).

Lava flows provide field and laboratory tools to study the compositional evolution of magma derived from active volcanoes. In a complementary way, albeit less accessible and rarer, lava lakes in craters offer a window into the upper part of active magmatic systems.

2 Volcanic Lava Lakes

Lava lakes are large volumes of exposed molten lava, usually basaltic, contained in a vent, crater, or other broad depression termed a lava pond (Davies 2014). Most of them are located within summit calderas on active shield volcanoes (e.g. Halemaumau crater at Kilauea), but lava lakes can occur at composite volcano summits (e.g. Nyiragongo, DRC) and more rarely in maars (Hopi Buttes, Arizona, USA). Lava lakes are confined by pit walls or levees produced by repeated overflow episodes, often with a mobile surface crust under which magma circulates. Active lava lakes represent the exposed, uppermost part of convecting magma systems and provide windows into the dynamics of magma convection, degassing and cooling (Spampinato et al. 2008). As such, they form the end-member of magma degassing mechanisms. Degassing also occurs persistently at open vents of large cones (e.g. Villarrica, Chile; Etna, Sicily), and temporarily as fumaroles on crater flanks (e.g. Vulcano, Italy), in crater water lakes (e.g. Poas, Costa Rica; Kelut, Indonesia; Ruapehu, New Zealand) as well as through soils (e.g. Mammoth Mountains, California; Vulcano).

2.1 Active Lava Lakes

Active lava lakes represent a variety of open-vent volcanism in which a sizable body of lava accumulates at the top of the magma column, constrained by the crater geometry (Lev et al. 2019). The convection currents stir molten, gas-rich magma within the uppermost level of a conduit (Tilling 1987; Tazieff 1994). Conduit convection, driven most likely by density contrasts resulting from degassing, is a conceptual model to explain the longevity of lava lakes. Heat loss from the lake surface is balanced by the enthalpy difference between ascending and sinking magma in the conduit (Francis et al. 1993; Oppenheimer and Francis 1998; Harris et al. 1999, 2005). The longevity of lava lakes has been explained by a two-way exchange between the lava lake and the underlying reservoir via a conduit carrying hot, gas-rich magma upwards to the lake, and returning cooled, degassed magma downwards. Witham and Llewellin (2006) proposed a mechanism emphasizing the role of the gas flux in driving flow, and argued that the stability of lava lakes is strictly dependent on lake-conduit geometry, solubility and gas expansion, and magma volatile content. Several authors have examined lava lake dynamics, focusing on gas and thermal budgets, seismicity, description of surface features, and eruptive style. Others have modelled the surficial features of lava lakes, in particular patterns of the crust movements on lake surfaces (Spampinato et al. 2008).

2.2 Typical Features

Lava lakes are rare, yet they are found at volcanoes across the tectonic environment spectrum (subduction zones, rifts, and hotspots) (Lev et al. 2019). Lava lakes are often circular or subcircular in appearance, and typically tens to hundreds of metres across. Active lava lakes, which typically persist for decades, are different to lava ponds (or passive lava lakes) or lava pools (or inactive lava lakes), where lava has flowed away from an active vent and collected in a topographic depression, but has no local source of heat, so it eventually solidifies. Features common to all lava lakes, regardless of composition (e.g. tholeiitic basalt at Erta' Ale, phonolite at Erebus, and nephelinite at Nyiragongo), include variable lake surface level, variations in the surface velocity over a time scale of minutes to days, superficial currents, localized explosive activity, a solidified crust covering a portion of the lake surface, persistent yet variable gas emission rates, and convection (Tazieff 1994).

Craters hosting lava lakes often contain platforms or terraces due to variations in surface level for the most active lakes, while overflow from lava lakes or flank eruptions leaves a series of residual thin lava slabs on the flattish calderas (e.g. Erta' Ale). Unmanned aerial systems for image acquisition have revolutionized the study and monitoring of volcanic activity and active craters where field access is unsafe or impossible (Fig. 10A–C). For example, the morphology of Nyiragongo's crater has been surveyed by an array of techniques including monitoring and topographic measurements (Burgi et al. 2014), unmanned aerial systems (Smets et al. 2019), and satellite imagery and SAR interferometry (Barrière et al. 2022).

A map of the Nyiragongo's 1.2 km-wide crater (Paris and Grünewald 2015) shows three platforms related to variations in lake level since 1948 and three fissure eruptions in 1977, 2002 and 2021: P1 at 3270 m.a.s.l (prior to 1977), P2 at 3185 m (1995–2002), and P3 at 3100 m (2003 to present) (Fig. 10A and B; Photo 12A). The sequential increase in lava lake volume and rapid fissure eruptions probably gave rise to Nyiragongo's peculiar crater morphology (Tazieff 1984; Durieux 2002). The 1977 eruption of Nyiragongo produced very fluid pāhoehoe that moved at several tens of km/h down steep slopes and killed and trapped elephants, leaving behind well preserved moulds, while the 2002 flank eruption wreaked havoc in the city of Goma. Barriere et al. (2022) and Burgi et al. (2014), among others, used historical field records from 1948 and recent expeditions to reconstruct the evolution of the crater and the morphology of the pit containing the lava lake over a 20-year period (2002–2021). Figure 10A and C shows the striking variations in platform collapse, with a height difference of 650 m between the lowest and highest platform (P1) in 1977, and of 300 m in 2002. This reconstruction and further estimates of the variations between 2008 and the 2021 flank eruption, based on satellite radar images and photogrammetry, have helped Smets et al. (2019) and Barrière et al. (2022) appraise the evolution of the lava lake conduit and the volcano plumbing system. These authors estimated the lava flow accumulation based on high-spatial resolution DEMs: at least

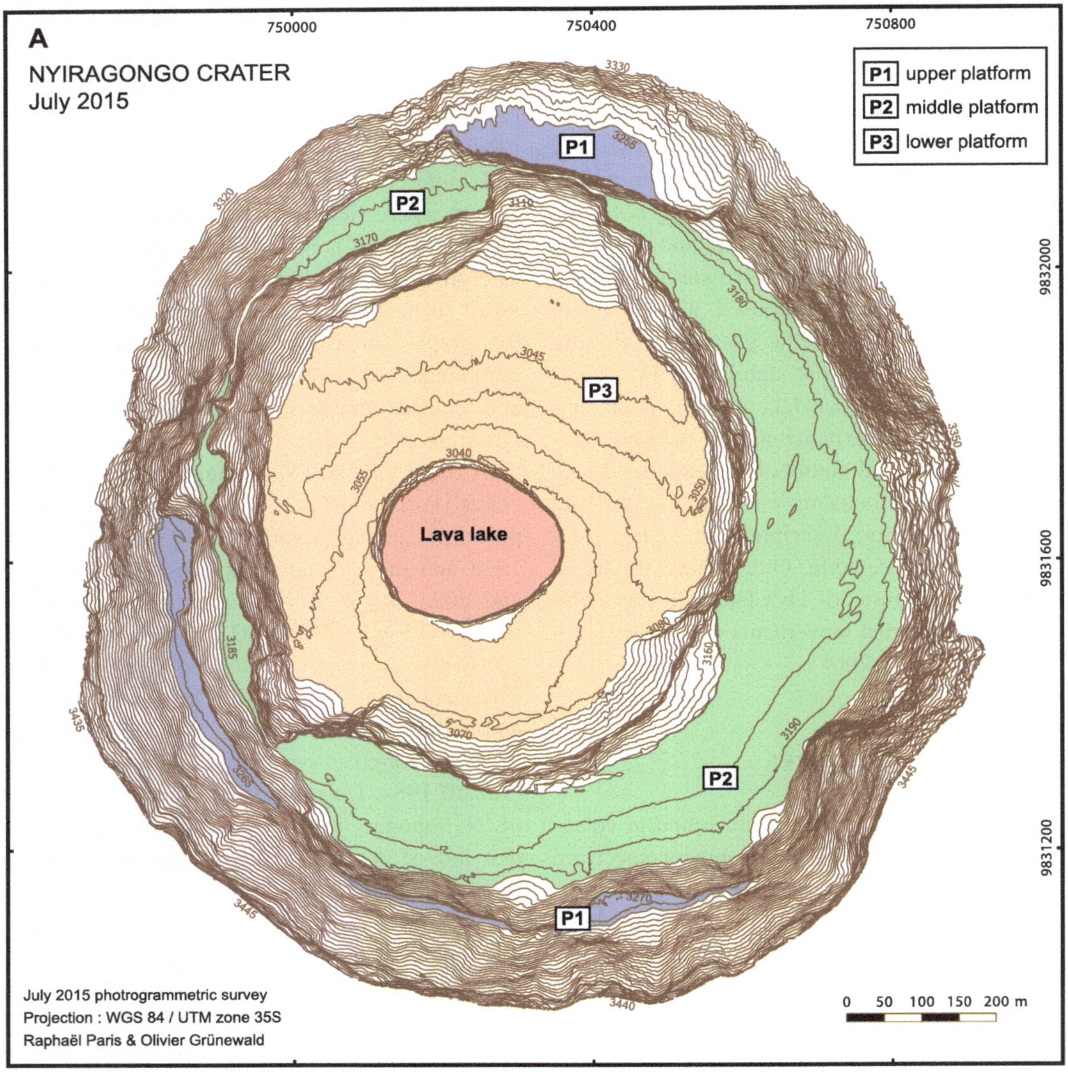

Fig. 10 Lava lakes: crater, platform, pit and conduit at Nyiragongo, DRC. **A** Map showing the Nyiragongo volcano crater (3470 m.a.s.l.), with its lava lake enclosed in a c. 300 m-deep pit crater surrounded by cliffs and three platforms or terraces left by lava lake rise and overflow (Virunga Volcanic Province, DRC; courtesy of Paris and Grünewald 2015). P1 (upper platform): lake level from 1971 to 1977, followed by a collapse in the range of 650 m during the 1977 fissure eruption. P2 (middle platform): lake level from 1996 to 2002, and renewed collapse of around 300 m during the 2002 fissure eruption. P3 (upper platform): lake level from 2008 to 2015. See also Smets et al. (2019). **B**. Topography of the volcanic crater of Nyiragongo, and conduit model. Schematic representation of the lava lake with the eruption vents of January 2002 marked. The height difference from the SE rim (altitude 3425 m.a.s.l.) to the crater floor (P3) is ~400 m, with P2 and P1 corresponding respectively to the 2002 and 1977 lava lake levels. The shallow reservoir (hc in the range of 1000–4000 m) is connected to the lava lake through a vertical conduit (of diameter dc with mean height roughness εc), open to the atmosphere with PA = ~0.071 MPa, with Pl indicating pressure at the bottom of the lava lake. Conduit and reservoir pressures are indicated by Pc and Pr, respectively (Burgi et al. 2014, Fig. 2, p. 4108). Photograph 1: Lava lake in pit crater. Photograph 2: Pit crater and three platforms (both pictures are courtesy of R. Paris). **C** Changes in lava lake elevation (in m.a.s.l.) from historical field records and recent expeditions (1948–2020) and from the continuous estimates obtained in the present study (red line). Details on the pre-2016 references can be found in Smets et al. (2019). The new estimate of the crater floor after the 2002 eruption (~ 2880 m.a.s.l.) is taken into account. Remnants of former crater floor left after the 1977 eruption are marked as P1 platform (~ 3270 m.a.s.l), while levels reached before the 2002 eruption are marked as P2 platform (~ 3185 m.a.s.l). P3 platform is the crater floor that hosts the lava lake of 2022 (Barrière et al. 2022, Fig. 1b, p. 2)

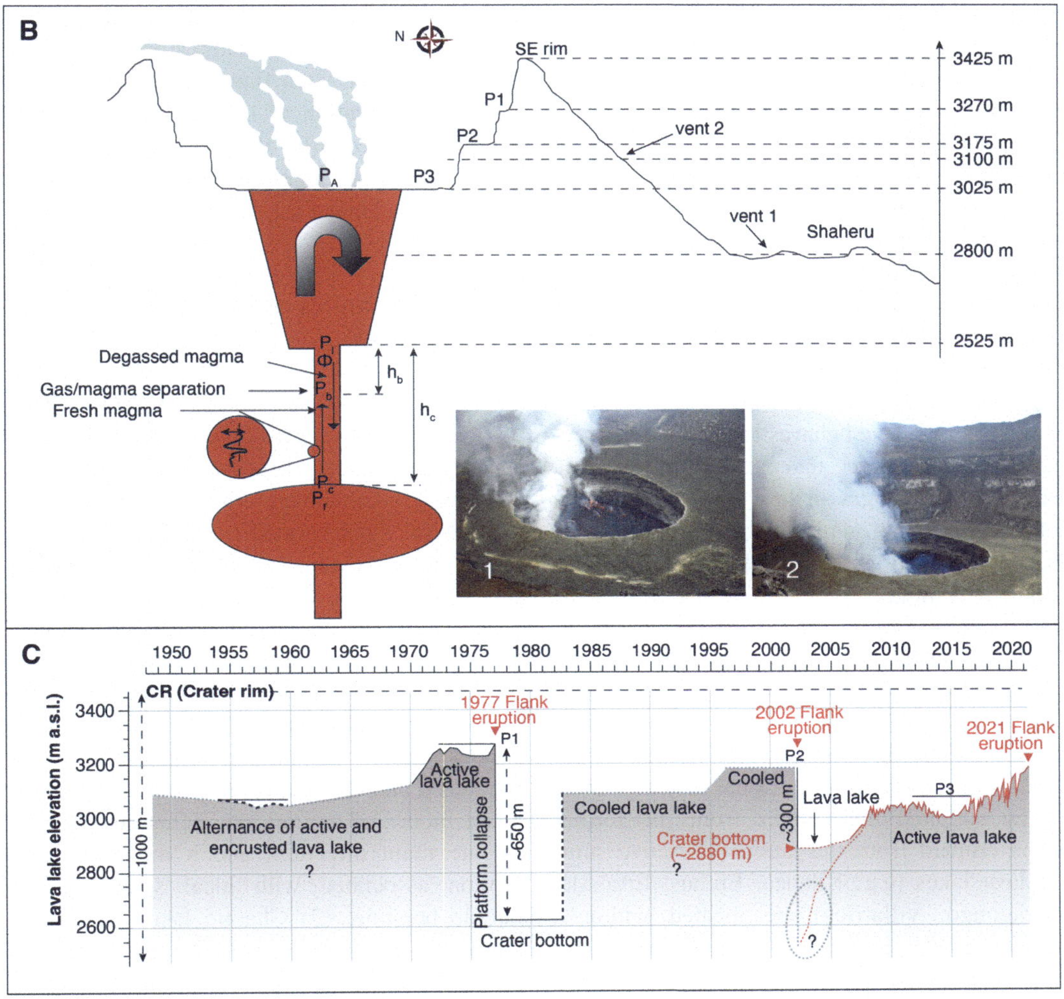

Fig. 10 (continued)

10.5×10^6 m^3 of lava accumulated in Nyiragongo crater between March 2016 and March 2018. This volume corresponds to half of the lava flow volume emitted during the disastrous January 2002 flank eruption of the volcano.

2.3 Types of Lava Lakes

Persistent lava lakes, which provide vivid and splendid images to volcanologists and tourists alike (e.g. Erta 'Ale, Photo 12B) are rare on Earth, whereas inactive, solidified lava lakes are more abundant. Subtypes have been distinguished on the basis of magma supply mechanism. Swanson et al. (1979) identified two types of lava lakes: *active*, if the lake is directly connected to a magma conduit, or *inactive*, if it results from confinement of a flow in depressions (e.g. Kilauea Iki, Alae, and Makaopuhi craters on Kilauea) or behind natural levees. Such lakes are stagnant, cooling and solidifying slowly in place over a period of years (Wilson and Parfitt 1993).

In an active lava lake, circulation within the lake, conduit, and reservoir enables the lake to remain molten and allows thermal, gas, and mass fluxes to be maintained over tens of years (e.g. Le Guern et al. 1979; Le Guern 1987). Lava lakes at open-vent basaltic volcanoes provide a window into the uppermost magma system to gain more direct insight into changes in reservoir pressure (Tilling 1987). At Kilauea Volcano (Hawai'i) the surface height of the lava lake in Halemaumau Crater fluctuates with surface deformation over short (hours to days) and long (weeks to months) time scales. This correlation implies that the lake behaves as a simple piezometer of the subsurface magma reservoir (Patrick et al. 2015). Changes in lava level and summit deformation scale with, and shortly precede, changes in eruption rate from Kilauea's East Rift Zone, indicating that summit lava level can be used for short-term forecasting of rift zone activity and associated lava flow hazards at Kīlauea.

Harris et al. (1999) further distinguished two subtypes of active lava lakes: summit lava lakes and flank lava lakes, with the key difference being that during normal activity most of the supplied mass sinks from the former but is erupted from the latter. (1) Active summit lava lakes (e.g. Kilauea, Erebus, Erta'Ale, Nyiragongo, Villarrica) are located within a summit crater above the magma reservoir, and remain active for tens of years. They are characterized by maintaining a stable volume and having minor overflows. Mass balance is maintained by circulation, where cooled and/or degassed dense magma sinks down into the lake to be replaced by rising hot and/or volatile saturated, buoyant magma. (2) Flank lava lakes are associated with effusive eruptions and form at eruptive vents, e.g. on Kilauea (at Mauna Iki during the 1901–1902 eruption, Mauna Ulu and Alae during the 1969–1974 eruptions and at Kupaianaha and Pu'u 'O'o between 1986 and 1997), as well as at some Icelandic shield volcanoes. The input of magma is balanced by extrusion of lava onto the flanks of the volcano or from fissure eruptions, rather than by circulation, leading to mass loss by lake overflows or through vents that feed surface flows or tube systems. They thus lead to the building of lava shields and lava-flow fields around the lake. The lifetimes of such flank lava lakes are limited to the duration of the eruption, becoming established soon after the beginning and draining at the end of an eruption (e.g. lava pond at Kupaianaha, Hawai'i).

2.4 Dynamics of Persistent Lava Lakes

In their review of lava lakes, Lev et al. (2019) analysed seven active lakes: Erebus (Ross Island, Antarctica), Erta'Ale (Afar, Ethiopia), Kilauea (Halemaumau crater, disappeared by May 2018, but reformed in 2022), Marum crater, Ambrym (Vanuatu), Masaya (Nicaragua), Nyiragongo (Democratic Republic of Congo), and Villarrica (Chile). These lakes span all the tectonic environments and a range of magma compositions. Lev et al. (2019) suggest that these active lava lakes be divided into two main classes based on their circulation patterns: organized and chaotic. (1) Organized lakes have a few modes of motion, one being slow motion in one main direction, and a mode of faster motion associated with local spattering and bubble bursting. (2) Chaotic lakes have many styles of motion and appearance, where the loci of upwelling, bursting, and crustal coverage change both spatially and temporally over short timescales. The surface of chaotic lakes has little crustal coverage, while the surface of organized lakes is mostly crusted over. The authors categorize Kilauea, Nyiragongo, and Erta'Ale as 'organized', and the lakes of Masaya, Marum (and likely Villarrica), as 'chaotic'. Erebus is an exception and falls between these two types.

Is there a common mechanism that links the behaviour of the different lakes and the two classes of lake? Lev et al. (2019) placed the seven lakes in a continuous spectrum based on a linear relationship between lake area, outgassing flux, and average surface speed. This global relationship suggests that all lakes are controlled by common processes. The controlling physical mechanism is actually a combination of lava lake size and gas flux, exercising a balance between opposing forces: (i) a resisting force imposed by the lake's crust and the viscosity of the lava, both acting to slow circulation and gas emission, and (ii) a driving force governed by rising gases pushing through the lake and up to the surface to be released into the atmosphere and the sinking of cooled, degassed magma. This scenario predicts a global correlation between all three observations—lake size, circulation speed, and gas emission—with an influence of rheology (including the combined effect of crystal content, melt viscosity, and vesicularity), and the thermal flux from the lake.

Overall circulation is driven by the upwelling of fresh, gas rich, highly vesicular magma, which is extremely buoyant compared with degassed, cooler magma sinking from the surface. The gas then accumulates beneath the crust, and is released when the skin is broken—due to rock falls or when the pressure of accumulated gas overcomes the tensile strength of the crust. This modulation leads to a two-regime pattern of flow. In general, large lakes (e.g. Nyiragongo, Kilauea) are more likely to develop a skin compared to smaller ones, but it is the relationship between gas flux and lake area that controls the behaviour, as shown by Erta'Ale, a small lake that behaves in a placid and organized manner, stemming from its low gas emission.

As mentioned earlier, lava lakes can be found in all tectonic settings, but they may demonstrate distinctive behaviour in convergent margins and in plate interiors or divergent margins. The arc volcano lakes (Villarrica, Marum and Masaya) have a high volatile flux, small surface area, and therefore high gas/area ratios, which create a highly vigourous circulation (average velocities >1 m/s). Lake size plays a role in controlling circulation: faster velocities are observed at small lakes, while slower velocities are observed at both small and large lakes. The geometry of the conduit-lake system affects flow. For small lakes, it is reasonable to assume that the widening from the conduit to the lake is gradual. In contrast, in the large lakes, a marked break in diameter between the conduit and the lake bounds has been documented or inferred (e.g. Halemaumau at Kilauea). The eight-fold increase in surface area of the lake at Erta'Ale over a modest 20 m rise in lake level is the result of a similar step-wise increase in diameter between the conduit and the lake. Therefore, the three lakes that experience organized circulation are those where there is a sharp increase in diameter from the conduit to the lake. This observation suggests that this sudden increase in width helps mask the deep, upwelling circulation noted at the smaller lakes, and separates a shallow, more organized, circulation from the deep, more chaotic, circulation pattern.

Lev et al. (2019) used image velocimetry to analyse lake surface motion and identify the spatial distribution of surface velocity fields. They were able to quantify the type of flow at each lake as either 'radial' (Nyiragongo), 'side-to side' (Kilauea, Erta Ale, Erebus) or 'multi-radial' (Marum, Masaya, and Villarrica, which has an intermittent lava lake that resumed activity in 2016). Considering the geometry of conduits and lava lakes, radial and multi-radial motion can point to a core-annular flow within the conduit, while side-to-side flow can be the surface expression of a side-to-side conduit flow regime. Organized lakes seem to experience both core-annular and side-to-side conduit flow regimes, while chaotic lakes all show core-annular flow.

To sum up, Lev et al. (2019) found that lava lakes follow a linear relationship between average surface speed and the total gas flux/lake surface area ratio. This relationship points to the combined importance of both flux and lake size in addition to the total volume of gas emitted, and suggests that a shared deep mechanism controls the supply of heat and gas to all lakes. On the other hand, the differences between chaotic and organized lakes highlight the important role of the geometry of the conduit-lake transition, which superimposes a shallow signal onto that of the deep circulation. The spatial patterns of surface motion suggest that the release of gas bubbles at chaotic lakes is more efficient (i.e. bubbles are less likely to be retained and recycled) compared to that at organized lakes. In addition, the data presented in Lev et al.'s review (2019) indicate that the solidified crust of organized lakes plays a role in regulating convection and outgassing in lava lakes.

2.4.1 Lava Lake Convective Regime: Case Study of Erta'Ale

Erta'Ale volcano located within the Danakil Depression in Ethiopia hosts one of the few permanent convecting lava lakes, which has probably been active for more than a century. The volcano has a 0.7 × 1.6 km elliptical summit crater hosting a steep-sided pit crater about 150 × 160 m containing a lava lake of about ~ 6200 m^2. Harris et al. (2005) surveyed the Erta 'Ale lava lake in February 2002 and modelled its dynamics, emphasizing convection which allows the lava lake to remain active. Buoyant magma ascending into the lake replaces cooled magma dropping back down into the lake, allowing the lake to remain fluid while maintaining a small and stable (non-overflowing) volume. This implies circulation between the lake and a deeper system to which it is connected via an open conduit, as well as within the lake itself. Such motions involve the smooth and steady movement of crust away from upwelling zones towards sinks where the crust is consumed. Rates of motion and vigour of convection can, however, vary over a scale of minutes to hours. Thermal data collected at Erta 'Ale acquired in 2002 (Harris et al. 2005; Harris 2008) revealed phases of both vigorous and sluggish convection characterized by high (0.1–0.4 m/s) and low (<0.08 m/s) surface velocity, respectively. The surface temperature varied from 280 to 650 °C, and the heat flux was 5–25 MW. The lava volume flux was estimated to be 0.1–0.8 m^3/s. Lake activity was first reported in 1906, giving a calculated volume of 1.28 km^3 over the period (90 years × 0.45 m^3/s).

Figure 11 illustrates a model to explain variable lava lake convective regimes (Harris 2008). Several processes were involved at Erta 'Ale: (1) low density (buoyant), gas-rich, hot magma ascending into the lake; (2) surface degassing and heat loss causing magma to decrease in bubble content, cool and crystallize; (3) magma becomes denser and more viscous and sinks, through density-driven convection; (4) highly viscous and dense magma eventually sinks from the lake to be replaced by an equal flux of fresh magma. Harris et al. (2005) found that, while convection requires only very small density differences, variations in the rate of convection, as expressed by surface motion, can easily be driven by cooling induced by lake surface heat loss. Transitions in the vigour of lava lake convection may thus be explained in terms of surface cooling and crystallization which increase the density and viscosity of convecting layers.

3 Lava Domes Created by Extrusive Activity

In contrast to lava lakes, a dome will form at the vent when crystal-rich, degassed lava cools as soon as it reaches the Earth's surface. There has been a move over the past forty years or so away from descriptive studies of volcanic domes to quantitative analyses, provided by monitoring of active domes using an array of new techniques, field measurements following dome eruptions, and laboratory experiments. This combination of methods has resulted from detailed studies of dome eruptions such as Merapi, Indonesia (1994, 2006, 2010, 2018 to present), Unzen, Japan (1991–1994), extrusions inside the crater of Mt. St-Helens (2004–2008), Chaitén, Chile (2008–2009), and the spectacular episodes of growth and destruction of Soufriere Hills (Montserrat Island) between 1995 and 2015. Domes are no longer just considered as landforms; their growth is increasingly seen as one specific eruption type.

3.1 Definition

MacDonald (1972) defined domes as 'a steep-sided heap of viscous lava formed over vents'. A circular, steep-sided mound is formed at the Earth's surface because this type of lava is so viscous that it cannot flow away readily and accumulates around the vent. The resultant dome may stand alone, form in the crater of a composite cone, or more rarely in a scoria

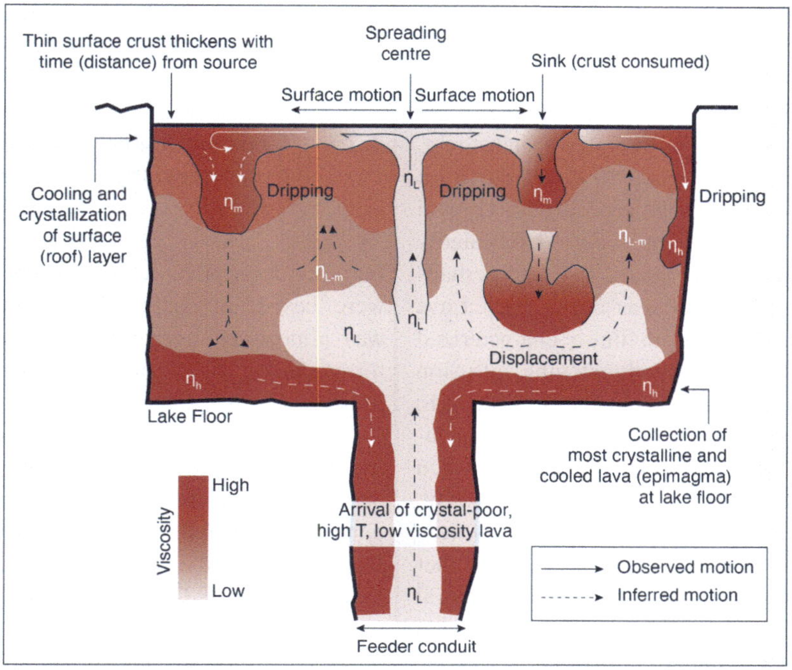

Fig. 11 A cross-section model to explain variable lava lake convective regimes in active lava lakes based on Erta' Ale (modified from Harris 2008). Low viscosity (high temperature, crystal-poor) lava at *hL* enters the lake. Material at *hL* ascends to the surface where it experiences fast motion. Radiative heat loss drives cooling and crystallization, causing the viscosity and density of the surface layer to increase, and the velocity at which the layer circulates to decrease. Eventually negative buoyancy causes dripping/sinking of the highest viscosity material at *hh* which descends to the lake floor. Displacement of moderate viscosity material by the descending, crystal-rich plume causes ascent of moderate (*hm*) and low-to-moderate (*hL-m*) viscosity lava (modified from Harris 2008, Fig. 2)

cone, or form part of a cluster or dome complex (Fink 1990). Dome-building eruptions are common: around 6% of eruptions on Earth are lava dome-forming, particularly in convergent plate boundary settings, such as island arcs (Calder et al. 2015). The majority of domes are of intermediate (andesite and dacite, e.g. Santiaguito, Guatemala), or alkaline composition (trachyte, e.g. in the Chaîne des Puys, France; phonolite, e.g. in Velay, French Massif central), while rhyolitic domes form specific landforms such as 'mesas' (table) or 'tortas' (cake), with asymmetric coulees (e.g. Purupurini dome cluster, southern Peru, Photo 13). This characteristic torta shape is attributed to high viscosity that prevents the lava from flowing very far. Such high viscosity can be obtained in two ways: by high silica and crystal content in the magma, or by degassing of low-viscosity magma (Cas and Wright 1988; Francis 1993).

A simple distinction between two end-member domes is based on the emplacement of *exogenous* domes or extrusions onto the Earth surface, or at shallow levels beneath the Earth's surface, i.e. *endogenous,* such as cryptodomes, when intruded magma never reaches the surface (Francis et al. 1993; Lockwood and Hazlett 2010). One such cryptodome was emplaced inside the Mt. St. Helens composite cone before the 18 May 1980 eruption. One rare class of volcanic domes results from the upheaval of the plug filling the upper part of a pipe-like volcanic conduit, the material being pushed up like a cork. These are called *plug domes* (e.g. the 1912 Novarupta plug, Katmai, Alaska) or spines, such as the one erected before the 8 May 1902 Mt. Pelée eruption.

Lava domes can form relatively rapidly, over days or weeks (e.g. Chaitén, Chile, 2008), but protracted growth and repeated, partial destruction during explosive phases or collapses may alternate over a period of years to decades (e.g. Soufrière Hills, Montserrat, 1995–2015, Druitt and Kokelaar 2002; Wadge et al. 2014). Growth and collapse of domes has occurred repeatedly at Merapi, Java since at least 1984, and perhaps as far back as 1931 (Gertisser et al. 2023). Lava dome eruptions vary from those characterized by relatively passive, unthreatening effusion (e.g. flat-topped *tortas* in the Central Andes), to others with varied, unpredictable, and threatening explosive activity, such as Unzen between 1991 and 1995 (Nakada et al. 1999; Calder et al. 2015). Effusion rates vary widely from as little as 0.01 m^3/s to over 100 m^3/s, with the higher rates having major implications on the structural stability of the dome (Voight 2000). Sudden collapse of lava domes may generate *pyroclastic density currents* (PDCs), which can result in significant numbers of fatalities, such as the worst volcanic disaster of the twentieth century at Mount Pelée, Martinique, which caused about 29,000 fatalities.

Dome morphology can vary from steep-sided to tabular in cross-section and circular or elliptical to irregular in plan view. The morphology of lava domes also depends on the underlying topography: domes which are restricted to crater areas have a nearly circular shape, while those erupted on the flank of volcanic edifices tend to be elongated downslope. Lava domes may also develop as complexes along arcuate chains up to 20 km long (e.g. Nevado Chachani, Peru; Aguilar et al. 2022). Dome lavas cover a wide compositional range from basaltic at mid-ocean ridges through to rhyolitic (Chaitén, Chile), although the majority are of intermediate composition (Calder et al. 2015), or include alkaline trachytes (e.g. Puy de Dôme and Sarcoui, Chaîne des Puys, France, see Chap. 10) and phonolites (Mt. Gerbier de Jonc, Velay, France) in continental intraplate settings. However, the morphology and eruptive behaviour of domes owe more to the high viscosity of the magma, its volatile content, and limited ability to flow rather than to their overall composition (Fink and Anderson 2017).

3.2 Geometric and Rheological Characteristics of Lava Dome Types

Earlier, descriptive classifications of domes were based on shape, hence dome typology was almost as diverse as the range of landform shapes. For example, Scarth (1994) or Moriya (1978) distinguished 'simple' domes (e.g. Sarcoui, Auvergne), 'pitons' or spines (Puy Chopine, Auvergne), 'bristling' domes (Soufrière de Guadeloupe, French West Indies), 'nested' domes (Montaña Rajada, Tenerife, Canary Islands), 'cracked and displaced' domes (Kameni Vouno, Santorini, Cyclades Islands), and 'asymmetrical' domes (Paulina Peak, Newberry volcano, USA). In fact, dome morphology is controlled by a combination of magma rheology and ascent dynamics, substrate topography and vent geometry, and endogenous/exogenous growth. An additional distinction stems from petrology, as crystal-poor, obsidian domes of rhyolitic or rhyodacitic composition, can be separated from crystal-rich domes of basaltic andesite to dacite compositions (Calder et al. 2015).

The average volume of lava domes is about 10^7 m^3, globally (Calder et al. 2015). The morphology of domes can be described by statistical geometrical characteristics such as the height/diameter ratio (Fig. 12A and B). Domes range in diameter from a few tens of metres to a few kilometres, and their height can reach 0.5 to 1 km. Lava domes have a H (Height)/ D (Diameter) ratio greater than 0.1 (Moryia 1978; Stasiuk and Jaupart 1997): for example, H/D 0.22 for the Puy de Dôme in the Chaîne des Puys, France (Boivin et al. 2017). Aguirre-Díaz et al. (2006) calculated H/D ratios of 21 edifices from 0.18 for dome-coulees, 0.17 for Pelean (composite) domes, down to 0.09 for low domes in the Valle del Bravo volcanic field, Mexico. Slope angle is also characteristic: based on the DEM analysis of 29 lava domes worldwide, Karátson et al. (2013) found a correlation between the mean

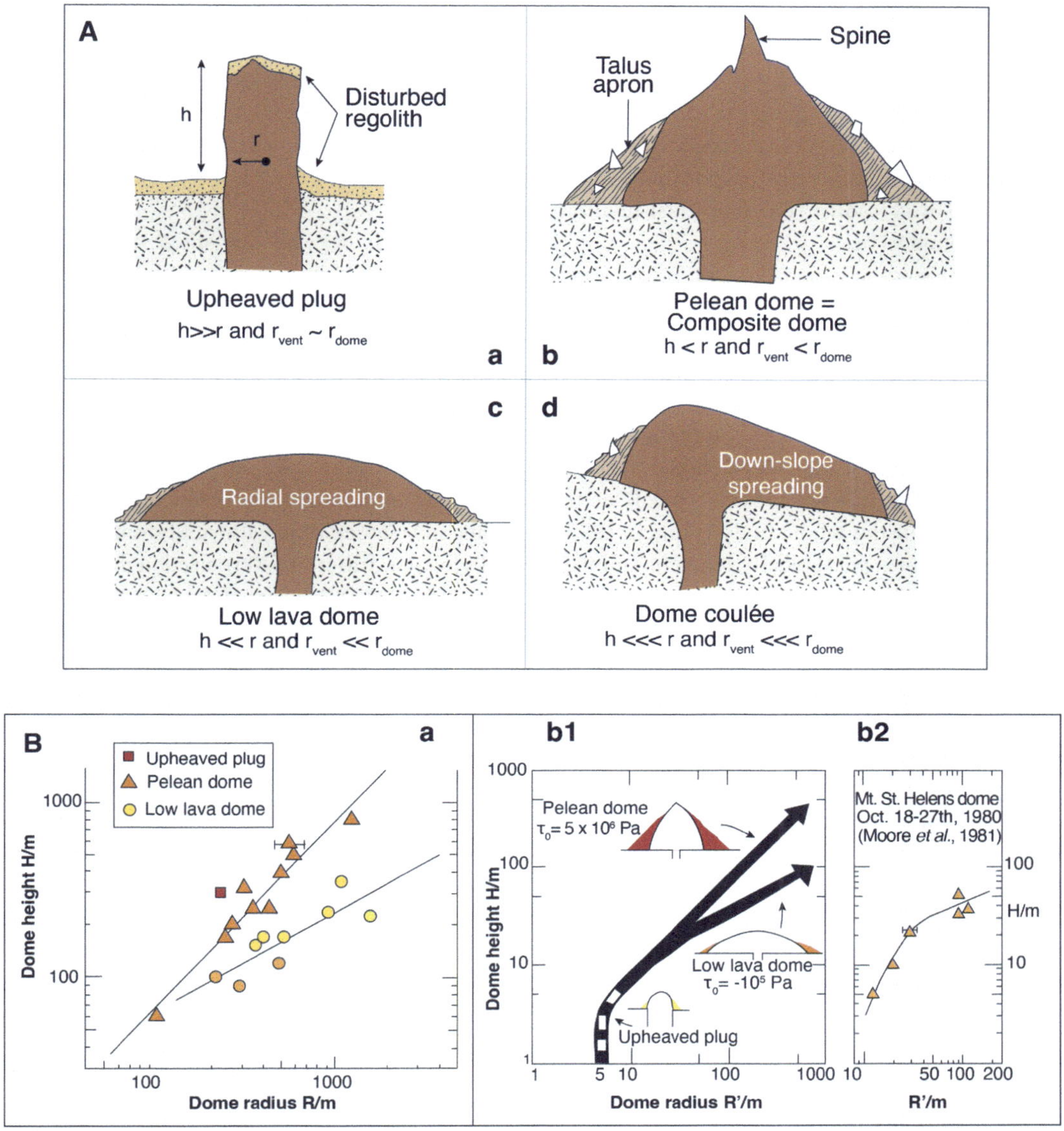

Fig. 12 Lava dome types, geometry and rheology. **A** Four types of lava domes (modified after Blake 1989) with corresponding sketch and ratios (height/radius and radius vent/radius dome): Upheaved plug, Pelean or composite dome, Low lava dome, and Dome coulee. **B** Two plots showing dome types based on: **a** Geometry, i.e. the height/radius ratio from measurements and statistics (modified after Blake 1989; Fink 1990). **b** Lava rheology (yield strength, shear stress and viscosity) based on experiments (b1) and measurements (b2), conducted by the U.S. Geological Survey, Mt. St. Helens Observatory (modified after Blake 1989; Fink 1990)

slope value of the upper dome flanks and the age of the domes.

Based on laboratory experiments, Blake (1989) found that the height and radius of a dome were related by a simple equation:

$$H = 1.76(\tau_0 R/\rho g)^{1/2} \qquad (2)$$

where H = height, R = radius, τ_0 = yield strength, ρ = density, and g = acceleration due to gravity.

It follows that if H and R can be measured, then the yield strength can be determined. Measurements made during the growth of a basaltic andesite lava dome at Soufrière St. Vincent in 1979 showed that the height of the lava dome was consistently proportional to the square root of its radius. The empirical relationship was found to be:

$$H = 5.75R^{1/2} \tag{3}$$

This relationship suggests that the yield strength of the Soufriere dome was about 2.6×10^5 Pa, about an order of magnitude greater than would be expected if the dome were composed of plastic lava. In fact, the carapace of chilled lava surrounding the hot interior provides much of the yield strength of the whole body (Francis 1993). Active lava domes are typically made of an inner, somewhat ductile lava core and a cooler, brittle outer rock carapace (Fink 1990). At steep dome margins, or at the fronts of active lobes, loose or unsupported portions of the carapace commonly collapse to form extensive talus or aprons of pyroclastic material around the composite domes (e.g. Soufrière Hills and Mt. Pelée).

Lava domes can be best separated into two main classes on a rheological basis, primarily based on yield strength τ_0. The yield strength is the critical threshold stress that must be attained before deformation, i.e. lava movement will occur: (1) viscous lava of silicic composition piles up in domes and cannot spread if rheological conditions are $h<h_0$ so $\tau<\tau_0$; (2) Lava of intermediate composition spreads from dome coulees if rheological conditions are $h>h_0$ so $\tau>\tau_0$.

If the stress applied to the lava:

$$\tau = h\rho g \sin(\theta), \tag{4}$$

(where h is dome height or lava depth, ρ is density, g is gravity and sin(θ) is the slope of the underlying surface), exceeds the threshold stress that must be attained before deformation can occur:

$$\tau_0 = h_0\rho g \sin(\theta), \tag{5}$$

where h_0 is the critical height. Deformation will occur, so flow will occur if:

$$\tau > t_0 = h\rho g \sin(\theta) > h_0\rho g \sin(\theta), \tag{6}$$

meaning that lava thickness h exceeds the critical thickness h_0.

A transition from a dome pile to a 'dome coulee', driven by a gravitational force or shear stress, occurs in the down slope direction when:

$$\tau = h\rho g \sin(\theta) > \tau_0 \tag{7}$$

Yield strength depends on silica content (high silica induces higher τ_0), temperature (higher temperature induces lower τ_0), and crystallinity (higher crystal content induces higher τ_0). Yield strength thus varies from 1 to 10^3 Pa for basalts to ~10^5 Pa for dacites, and $\geq 10^5$ to ~10^7 Pa for rhyolites. A study of the dacite lava flows and domes of the intra-caldera Kameni Islands, Santorini, in Greece emphasized the relationship of dome height and elapsed time with yield strength.

Pyle and Elliott (2006) used a high-spatial resolution LiDAR-based DEM, aerial photos and archived reports, to map dacite 'a'ā lava flows and domes in great detail. By combining maps, aerial photos and archived reports, the authors were able to reinterpret the dome-forming eruptions of the Kameni Islands in 1866–1870 and 1939–1940, showing similar patterns of behaviour, with progressive increases in dome height with time. The dependence of dome height on elapsed time for the Kameni and other domes is close to $t^{1/4}$, which is the expected value derived from analysis of domes with a crustal yield strength (Fink and Griffiths 1998; Griffiths 2000): the best fit curve through the Kameni data with a $t^{1/4}$ dependence is height (H) ~ 1.2 $t^{1/4}$. Pyle and Elliott (2006) estimate that such behaviour is consistent with a model of slow inflation of a dome with a strong crust. Lava domes on the Kameni Islands have a crustal yield strength (4×10^7 Pa) that is lower by one order of magnitude than that of the domes at Mt. St. Helens (1981–1986: 1.3×10^8 Pa) and St. Vincent (1979: 1.5×10^8 Pa).

The models in Blake (1989), Fink and Griffiths (1998) and Griffiths (2000) showed that even simple domes grow in rather complex ways. Material moves radially outwards, stretching the carapace, such that a single point approaches the margin of the dome linearly, but never quite reaches it—the margin of the dome undergoes simultaneous radial compression and circumferential extension. Pelean domes present slightly different problems (Francis 1993). They are constructed of such viscous magmas that their yield strengths can reach 10^5 to 10^8 Pa, so they are barely able to deform under their own weight. They typically have steep, pyramidal profiles, whose shape is influenced by the heaps of talus accumulating around the flanks of the growing dome. The high viscosity of the associated magma means that Pelean domes are also more likely to be violently disrupted by explosive activity than low, flat lava domes. This was the case with Mt. Lamington (Papua New Guinea, 1951, Taylor 1958) and the Soufrière Hills (Montserrat) domes between 1995 and 2010 (Wadge et al. 2014).

3.3 Principal Classes of Domes

Blake (1989) and Fink (1990) used a classification that includes four types of dome based on their morphology and rheological properties (Fig. 12A): upheaved plugs (spines), Pelean domes (now termed *composite* domes), low lava domes (also called pancakes), and 'dome-coulees' related to silicic flows (i.e. on the spectrum between a dome and a lava flow). This variety of domes suggests a decreasing ratio of height/diameter and a decreasing yield strength from spines to 'dome coulees' (Fig. 12B). A plot showing yield strength versus shear stress shows a distinction in the behaviour between low domes and upheaved plugs: this distinction has been supported by measurements conducted on Mt. St. Helens extrusions following the 1980 eruption (Fig. 12b2). We add a fifth class here, of cumulo-domes.

(1) *Upheaved plug (or piton)*

Plugs composed of dacite/rhyolite lava with very high crystal content and very high yield strength have a high aspect ratio. If the magma is so viscous that it cannot deform on emerging from conduit, it will form an upheaved plug. The plug radius equals the vent radius (e.g. Puy Chopine, Chaîne des Puys, Photo 14). Height (h) exceeds radius (r), hence $h \gg r$ and $r_{vent} \sim r_{dome}$. However, true upheaved plugs are rare because they are prone to collapse, such as the 230 m-high spine that grew at Mt. Pelée summit for about nine months before the 8 May 1902. It was a vertically grooved rock pinnacle, with a shark-fin form and both vertical and curved faces, which underwent constant crumbling. The spine grew at a velocity of about 10 m per day, but it collapsed into a big pile of rubble only forty days after its formation. This was also the case with lava spine intrusions in the Mt. St. Helens crater between 2004 and 2008 (Sherrod et al. 2008). Lava spines extruded through and deformed glaciers in the constrained crater space. Some lava plugs may carry a layer of country rock sediments or granite on top of the piston. This happened twice at Usu volcano, Hokkaido in Japan, once in 1910 and again in 1943 (Francis 1993). The second event, called "Showa Sin-Zan" (or New Roof Mountain), is well known thanks to the diagram by Masao Mimatsu, the village postmaster, who recorded its growth by drawing a series of profiles on the paper covering his window from May 12 to November 1944. This was the first detailed record of the growth of a lava dome, which reached 900 m along its longest axis and 300 m in height. In November 1944, a plug of lava eventually punched its way through the elevated dome, which was still carrying a cap of baked clay, rising until it was about 100 m above the top of the first dome, and nearly 300 m above ground level.

(2) *Pelean or composite domes*

Pelean dome morphology is dominated by a pyramidal or conical shape, topped by crags and spines protruding from a steep collar of debris and a talus apron due to collapse, with $h < r$ and $r_{vent} < r_{dome}$. Pelean domes are characterized by craggy topography and lava spines which form jagged battlements along the crest of the dome (e.g. Lassen Peak, USA). Steep collars of debris formed by collapse of unstable pinnacles often encircle the dome. These collapses usually take place as ordinary 'cold' avalanches, but when collapses involve large volumes of the dome, incandescent rocks within the core may be exposed, and collapse-driven pyroclastic density currents can be released. This was the case at Lamington (Papua New Guinea) in 1951 (Francis 1993), Santiaguito (Guatemala) in 1972 and 2002 (Rose et al. 1976; Harris et al. 2003), Merapi (Indonesia) in 1984, 1992 and 2006 (Gertisser et al. 2023), and Chaitén (Chile) in 2008–2009 (Pallister et al. 2013).

Repeated cycles of growth and destruction over many years lead to a jumbled edifice a few hundred metres high. Spines are extruded (exogenous growth) and mass is also injected (endogenous growth). The eponymous name comes from Mount Pelée, whose dome-building activity began around May 1902. A period of spine growth accompanied the dome-building, spanning the appearance of the first spine at this time to the total disintegration of the main spine by September 1903. This prolonged spine extrusion and crumbling led to a block-covered surface. By September 1903, after 17 months, it reached a height of 305 m and a width of 915 m. Regular dome collapses from Pelean domes feed extensive fans of block-and ash-flow deposits. The first eruption of Soufrière Hills volcano produced over 1 km^3 of magma, but only between 100 and 200×10^6 m^3 of that volume currently sits in the dome, which is surrounded by low-angle pyroclastic fans (Photo 15).

(3) *Low lava domes (or pancakes)*

These domes have a smooth, gently rounded profile, and flat top, contrasting with steep and conical composite domes. Two examples are given by the trachytic Cliersou and Aumône domes of early Holocene age in the Chaîne des Puys, France ($H/\sqrt{A} = 0.15$; Boivin et al. 2017). They lack brecciated slopes and jagged summits. A lower yield strength and viscosity allows this type of dome to spread/flow under its own weight, with:

$$h \ll r \quad \text{and} \quad r_{vent} \ll r_{dome} \tag{8}$$

A low lava dome formed at the bottom of the Popocatepetl crater, Mexico, in December 1997, which was elongate and relatively thin, with a pancake shape, 350 m long × 150 m wide and 10–15 m high.

(4) *Dome-coulees*

This form of dome consists of short (<2 km), thick (hundreds of metres) lava flows with a blocky surface. It exemplifies the transition between a dome and a high-silica 'coulee' (see Sect. 1.2.5). Rheological properties and/or slope are such that they allow flow in a preferential direction. Dome coulees have two associated landforms: the dome is roughly circular in plan view ($L \sim W$), whereas the coulee is elongate in a preferential direction ($L > W$). If a block-lava flow is produced, then it is elongate ($L \gg W$), e.g. the Little Glass Mountain coulee, Medicine Lake (California, USA) with H = 30–40 m has a ratio L (1550 m)/W (860 m) = 1.7.

Rhyolite domes are distinctive due to their shape, lithology, surface features and internal structure (Cas and Wright 1988). (1) Tholoids, which are circular in plan with a small surface area and grow within craters, may belong to the class of low domes. "Mesa" or "torta" domes are rhyolite or dacite lavas with an approximately circular plan forming asymmetric, biscuit-shaped bodies with a flat summit delimited by cliffs (Purupurini, Southern Peru; Photo 13). The Inyo domes in California, and Big Obsidian Flow in Newberry Caldera, Oregon (see Chap. 5), as well as many of the large-volume dacite domes of the Central Andes in South Peru and North Chile, can be classified as dome coulees since the flow is elongate in one direction on the underlying gentle slope. Purupurini domes are crystal-rich dacitic dome coulees of middle Holocene age, with margins 200–400 m high and a volume of 2–5 km^3, while the enormous Chao dacite coulee in Chile is around 15 km^3. They do not produce the pyroclastic fans associated with Pelean domes, with their volume being restricted to the dome and the flow, surrounded by screes. Coulees with high levees and ogives display the morphological features typical of silicic lava flows.

(5) *Cumulo-domes*

Cumulo-domes may form a fifth class of domes. Large (several km) and high (0.5–1 km), complex domes result from several intrusions and may hide previous structures. The term cumulo-dome was coined by Fouqué (1879) and expounded on by Jaggar (1920) to describe "hills of accumulation", squeezed up as "cylindric plugs in craters and irregularly dome-like elevations over them" (Cotton 1944). Ollier (1988) termed the rhyolitic domes at Tarawera, New Zealand, which were formed during the 1886 eruption, as cumulo-domes. Cumulo-domes show a broad compositional range of lavas from andesite to rhyolite, also including trachyte and phonolite.

The complex edifice of the Puy de Dôme in the centre of the Chaîne des Puys, France, may exemplify a cumulo-dome (Photo 16), characterized by multiple intrusions and a low explosivity. Its volume of ~0.37 km^3 (Deniel et al. 2020), computed on the basis of a LiDAR DEM and field observations, conceals four distinct volcanic constructs (Fig. 13A and B). Its inner structure has been unravelled by means of an array of geophysical techniques, including electrical resistivity tomography (ERT), high-resolution gravity and magnetic survey, and muon tomography (Portal et al. 2013, 2019; Barnoud et al. 2021). The new interpretation of the inner structure reveals strong differences to the previous interpretation based on field observations and laboratory measurements (Fig. 13B). Early authors (*in*: Boivin et al. 2009) described the Puy de Dôme edifice as the result of two buttressed structures crowned by a flat summit area (Fig. 13Ba; see also Fig. 17A, Chap. 1). The smooth structure to the east was built inside the scar of the collapse structure of the more craggy, western edifice (Boivin et al. 2017; Deniel et al. 2020). The volcanic edifice is not composed of one dome alone, but has grown inside, and has covered a cluster of large, pre-existing scoria cones. Barnoud et al. (2021) inverted gravimetry and muography data to image the 3D-density structure of the Puy de Dôme (Fig. 13Bc). The gravimetry data shows that the low-density, porous trachytic rocks of the Puy de Dôme are dry and permeable, while the muography data is used to delineate the dense trachytic core surrounded by a less dense talus. In addition, the Puy de Dôme cumulo-dome belongs to a complex of edifices in the central part of the monogenetic range whose morphology is due to shallow magmatic intrusions that have deformed and uplifted the neighbouring volcanoes of the Vasset-Grosmanaux and the Petit Puy de Dôme-Nid de la Poule area (Fig. 13C; see Fig. 17BC, Chap. 1).

3.4 Lava Dome Structures and Textures

A lava dome is characterized by a wide range of structures and textures (Anderson et al. 1995; Calder et al. 2015). These control the dome's permeability and thus its ability to lose gas—the buoyant driving force—thus determining whether it erupts passively or explodes catastrophically (Sparks 1997; Druitt and Kokelaar 2002). A lava dome consists of an inner lava core and an outer rock carapace, further flanked by a brecciated talus, all having different rheological properties. The core is characterized by ductile deformation, whereas the carapace is subject to brittle fracture, and the talus is composed of blocky fragments resulting from brittle fracture. Eruptive stress conditions, degassing, and thermal contraction enhance the brittle response of the lava (Sparks 1997), inducing brecciation and occasionally polygonal jointing (e.g. prismatically jointed blocks).

Externally, lava domes are associated with a field of non-sintered to poorly sintered blocks ranging in size from centimetres to tens of metres. Away from the vent area, the carapace sometimes exhibits compressional ridges, generated by folding of the dome surface due to internal shear traction from a spreading lava core. Lava dome surfaces and surroundings are also characterized by small pit craters created by gas and ash explosions (Calder et al. 2015). Internally, a lava dome may occasionally develop an 'onion-skin' foliation caused by pulses of intruded magma, which are successively pushed to the outermost dome margin to form concentric layers in a repeated process (Sherrod et al. 2008).

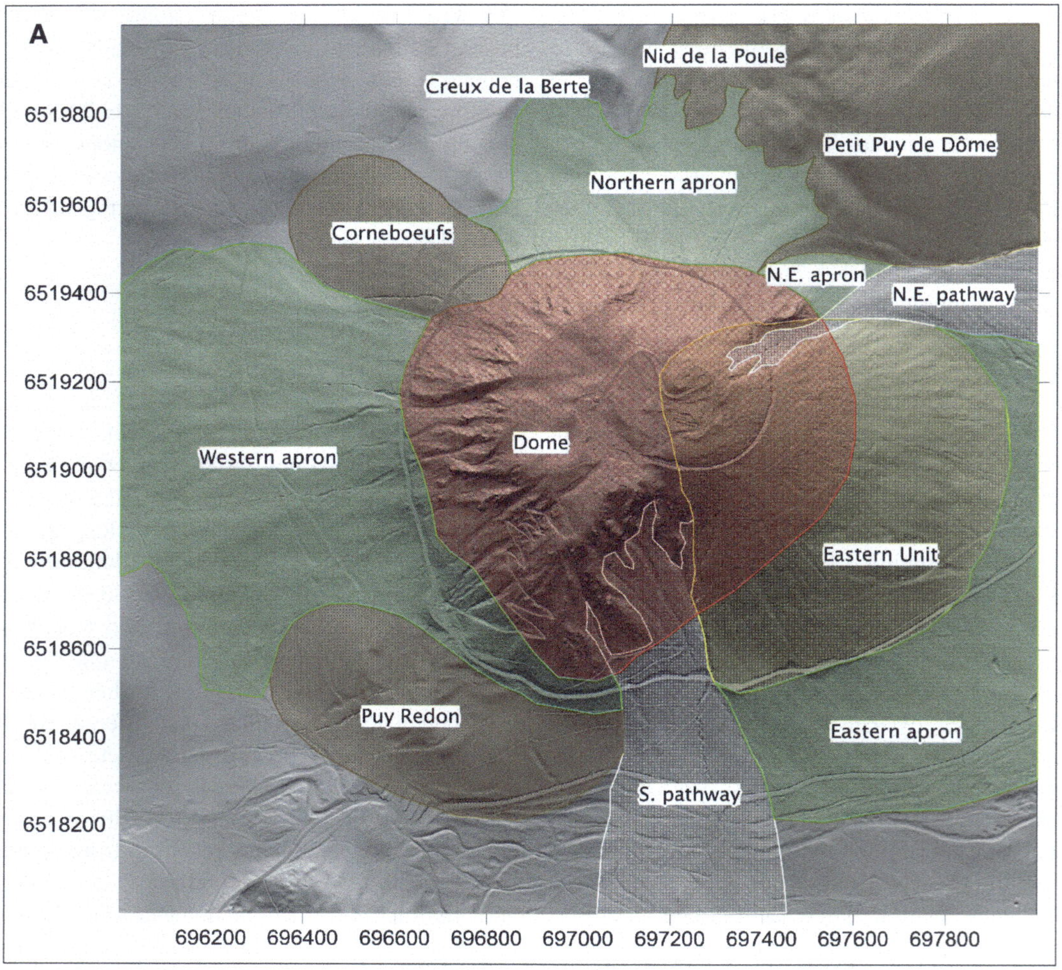

Fig. 13 Inner structure and growth of the Puy de Dôme cumulo-dome in the Chaîne des Puys (Auvergne, France): **A** Geomorphological units forming the Puy de Dôme cumulo-dome marked on a high-spatial resolution (0.5 m) Lidar-based DEM (LidArverne/CRAIG-GeoPhenix, 2011) (after Boivin *in:* Deniel et al. 2020, Suppl. Inf., Fig. 3, p. 8). **B** Cross sections of the Puy de Dôme trachytic cumulo-dome based on geophysical surveys including **a** Electrical Resistivity Tomography, **b** gravimetry and **c** muon tomography (courtesy of A. Portal et al. 2013; see also Barnoud et al. 2021). The muographic data acquired at one location bring a slight improvement to the model compared to the gravimetric data alone. The trachytic core of the dome and its surrounding less dense talus are better delineated. Results show that the Puy de Dôme is dry and permeable, with low density porous trachytic rocks. **C.** NNE-SSW sections illustrating the four phases of the Puy de Dôme eruption (11.0 ± 0.2 ka), after Deniel et al. (2020, Fig. 23, p. 47). 1: Granito-gneissic basement, 2: older volcanic formations (lava flows and cinder cones), 3: cpx-trachyte and associated block-and-ash flow (BAF) deposits (light colour), 4: bi-trachyte and associated BAF (light colour), 5: terminal pyroclastite, 6: fumarolised and silicified zones, 7: fumaroles. *Phase 1*: intrusion of cpx-trachyte encounters a mechanical discontinuity, forming a cryptodome below the Petit Puy de Dôme. *Phase 2*: following changes in the local stress field, cpx-trachyte follows a new ascent path and reaches the surface to build the dome. *Phase 3*: the bi-trachytic magma that constitutes the main volume of the magma chamber reaches the surface through the same conduit, penetrating the first dome as successive intrusions and breaking it up, embedding the resulting fragments. Fumarolised zones develop along fractures. *Phase 4*: toward the end of the bi-trachyte eruption, an explosive event occurs east of the summit producing small-volume deposits. Fluids still released from the dome feed fumaroles around the summit

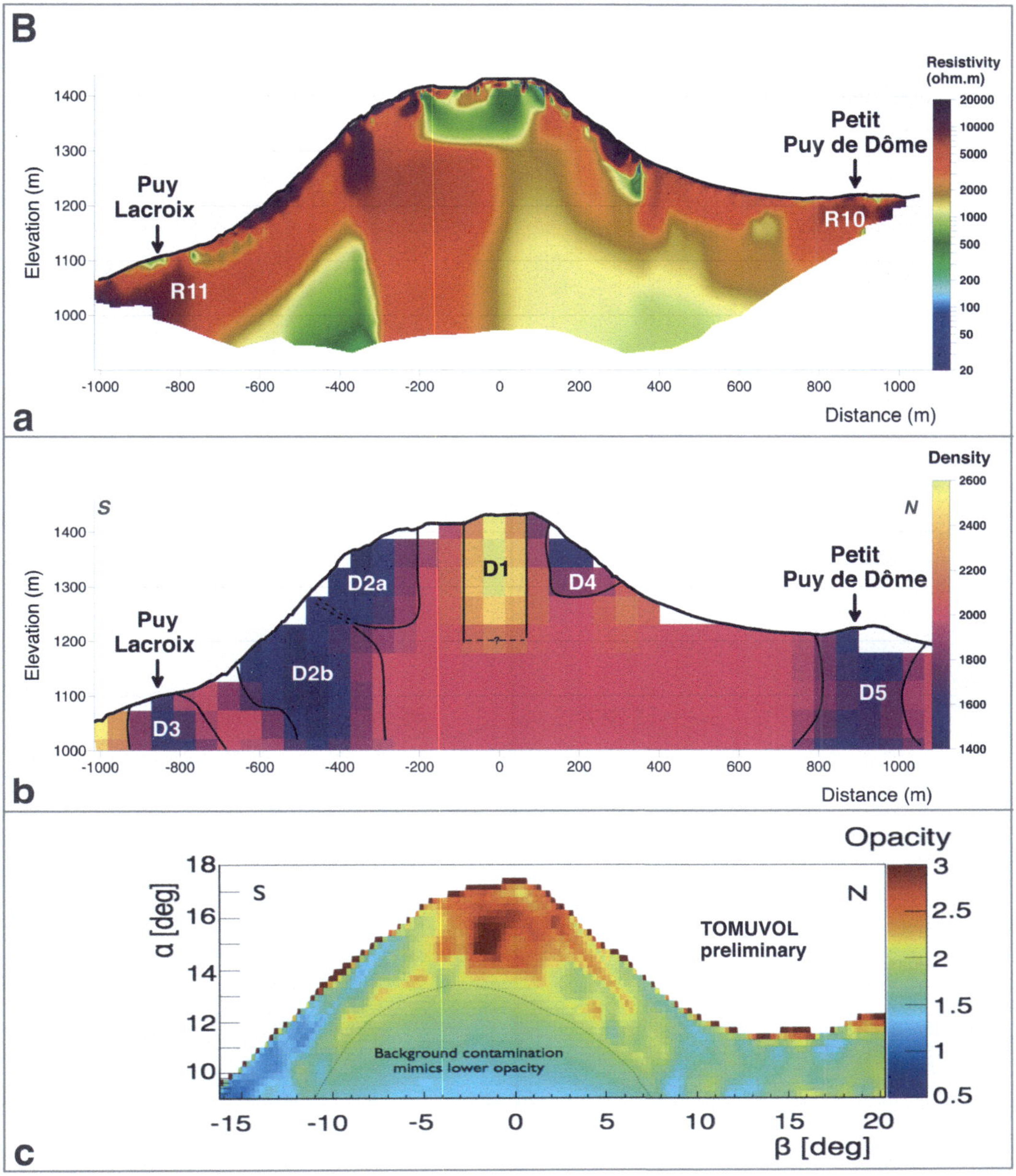

Fig. 13 (continued)

Lava domes exhibit diverse textures, but two main types can be distinguished (Calder et al. 2015): (1) Crystal-rich domes, which tend to be uniformly crystalline, are composed of dense, micro-vesicular lavas, and their porosity is generally lower than 35 vol%, with the carapace generally being more porous than the core. Despite their low porosity, dome rocks are also characterized by high permeability due to pore connectivity in the form of microfractures. Thus, these lava domes tend to outgas via localized permeable structures.

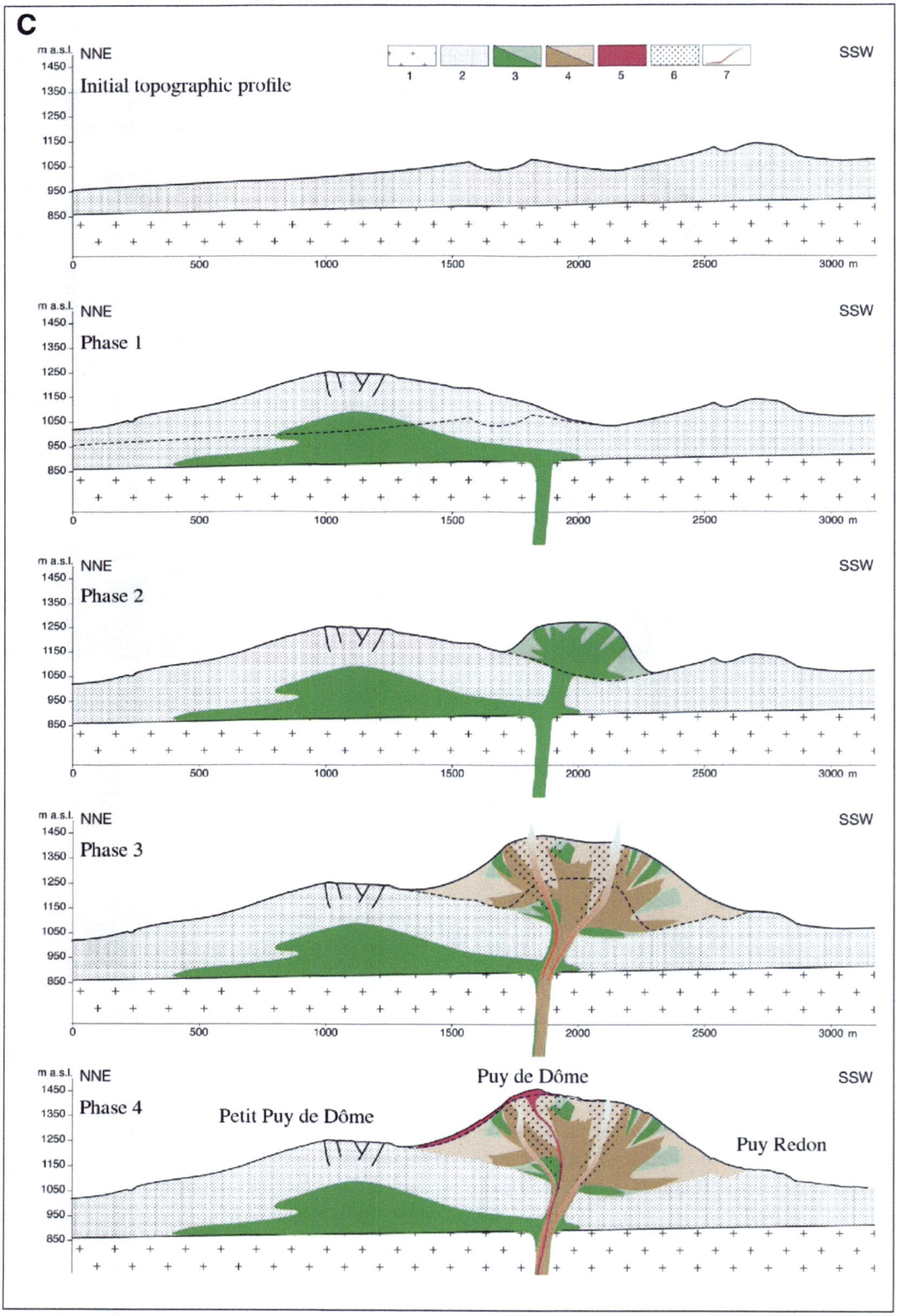

Fig. 13 (continued)

(2) Obsidian domes, which display stark textural contrasts with respect to crystal-rich domes (Calder et al. 2015). The carapace generally comprises highly porous or vesicular bands around a dense core. Obsidian contains mostly rounded pores ranging from nanometres to centimetres in diameter. Upon eruption, outgassing occurs and the pores subsequently collapse. Upon cooling, the obsidian glass is subjected to intense alteration, which takes the form of spherulites formed by devitrification and the crystallisation of anhydrous minerals.

3.5 Lava Dome Eruptions: Dynamics, Morphology, and Timescale

Over the past three decades, research on dome-building eruptions has provided rich contributions to our understanding of the magma transport, eruptive styles, and morphology of lava domes. Several authors (e.g. Sparks 1997; Giachetti et al. 2010; Rusiecka and Martel 2022; Utami et al. 2022) have carried out modern textural and geochemical studies of clasts produced by key dome eruptions. For example, Boudon et al. (2015) suggested that superficial explosions of a growing lava dome may be promoted through porosity reduction caused by both vesicle flattening due to gas escape and syn-eruptive cristobalite precipitation. Both processes generate an impermeable and rigid carapace allowing overpressurisation of the inner parts of the lava dome by the rapid input of vesiculated magma batches. The relative thickness of the cristobalite-rich carapace has an inverse relationship to the external lava dome surface area. Explosive activity is thus more likely to occur at the onset of lava dome extrusion, in agreement with observations, as the likelihood of superficial lava dome explosions depends inversely on lava dome volume. Such studies of dome-forming eruptions include the Mt. Pelée 1902–1905 and 1929–1932 catastrophic episodes (Boudon et al. 2015), extrusions at Mount St. Helens (1980–1986 and 2004–2008; e.g. Sherrod et al. 2008), Mt. Redoubt, Alaska (1989–1990; Miller 1994), Unzen Volcano, Japan (1991–1995; Nakada et al. 1995, 1999), and Chaitén, Chile (2008–2009; Pallister et al. 2013).

Chaitén and Mt. St. Helens can be used to illustrate short dome eruptive episodes (Calder et al. 2015), either associated with explosive activity, or restricted to endogenous and exogenous growth, respectively.

(1) Chaitén volcano in the Southern Volcanic Zone of the Andes was the shortest eruption (~18 months) and one of only three rhyolitic lava eruptions observed in historic time (the other two are Katmai-Novarupta in 1912, and Puyehue-Cordón Caulle, Chile in 2011). On 1 May 2008, Chaitén erupted suddenly and violently. The very rapid growth of a large (500×10^6 m^3) compound lava dome (Photo 17) was associated with synchronous explosive activity: the initial Plinian events were followed by a transitional explosive-effusive phase in May 2008 (Calder et al. 2015). The exogenous lava dome phase produced several large flow lobes that nearly covered the preexisting lava dome from June through to September 2008. In October 2008, a spine extrusion phase began, accompanied by continuing endogenous growth. This phase ended on 19 February 2009 when a large dome collapse removed about 10% of the dome volume and generated block-and-ash flows, which entered the Chaitén River valley. Dome growth soon resumed within the collapse scar forming a large endogenous lobe on the west side of the lava dome complex, which continued to grow until late 2009/early 2010. The volume of lava added to the lava dome complex was about 800×10^6 m^3, while that emitted during the explosive phase was about 300×10^6 m^3, giving a total output of about 1.1 km^3 (Pallister et al. 2013). The low viscosity magma explains the very high extrusion rate that peaked at 66 m^3/s during the first two weeks of the lava extrusion and averaged 45 m^3/s over the first four months.

(2) Mount St. Helens illustrates repeated dome extrusions, with almost no explosive eruptions, which took place in 1980–1986 and 2004–2008 within the horseshoe-shaped depression left by the 18 May 1980 eruption (Calder et al. 2015). Firstly, several episodes of endogenous and exogenous lava dome growth eventually built up a 1 km-diameter, 350 m-high Pelean lava dome with a volume of 74×10^6 m^3 from October 1980 through to the end of the eruption in 1986. Secondly, the volcano reawakened on 23 September 2004 after 18 years of near-quiescence. Magma rose above the crater floor in a series of seven spines or shear lobes up to 460 m above the crater floor over a four-year period with almost no explosive activity. The early spines extended southwards forming a whaleback morphology; as a consequence, the glaciers present in the crater area became deformed and shrank (Walder 2007). The last spine emerged more slowly, forming a dome-shaped rubbly mass. High initial extrusion rates of 4.9 m^3/s declined to 0.7 m^3/s. The eruption ended in 2008, having built a 92×10^6 m^3 lava dome (Sherrod et al. 2008; Gabrielli et al. 2020, their Fig. 3).

Protracted dome-building eruptions interrupted by periods of quiescence characterize the recurrent growth, collapse and block-lava flow cycles at the Soufrière Hills Volcano, Montserrat (1995–2010; Wadge et al. 2014), the PDCs and rockslide avalanches at Merapi from at least 1984 (Gertisser et al. 2023), and the Santiaguito dome complex (Guatemala) from 1922 to present (Harris et al. 2003; Massaro et al. 2022).

(1) The long-lived eruption at the Soufrière Hills (Photo 15) consisted of the growth and collapse of a sequence of andesite lava domes reaching volumes of 200–300×10^6 m^3, at an average extrusion rate of 3–5 m^3/s, and peak extrusion rates >15 m^3/s, punctuated by five distinct pauses in extrusion (Wadge et al. 2014). The pauses in extrusion, which lasted between 1.5 and 4.5 years, fall within the duration ranges of pauses recorded at other andesitic to dacitic lava dome eruptions (Calder et al. 2015). Accompanying lava dome growth throughout the eruption, rockfalls and PDCs were generated by dome and shear lobe collapse. Three periods of Vulcanian explosions generated fountain-collapse PDCs, frequent and vigorous ash-venting episodes, a debris avalanche (26 December 1997) and many mudflows (Druitt and Kokelaar 2002; Giachetti et al. 2010). Frequent collapses of the Soufrière Hills lava domes involved a total of 210×10^6 m^3 of material. PDCs formed coastal deltas and submarine fans extending tens of kilometres offshore.

(2) The dome complex of Santiaguito (Fig. 14A) illustrates a sequence of dome growth and collapses or explosions with distinct morphologies, eruption styles and silica content over the time-period 1922–2021 (Harris et al. 2003; Wallace et al. 2020; Massaro et al. 2022). Santiaguito domes first grew on the flank of Santa Maria volcano in 1922, and formed an alignment of four edifices in which Harris et al. (2003) recognized three episodes 10 to 30 years apart (Fig. 14A): (1) spines from 1922 to 1929, (2) Pelean or composite domes between 1939 and 1972, (3) low lava domes with thick and stubby block-lava flows (4.4 km) from 1972 to 2002, the growth of which is still ongoing at the time of writing in 2024. The sequence of events implies that τ_0 and h_0 are decreasing, implying that SiO_2 content is decreasing with time, as has the extrusion rate (Fig. 14A and B). Massaro et al. (2022) pointed out three orders of periodicity over the 1922–2021 period: (i) a nearly constant periodicity of ~10 years over the whole period; (ii) an intermediate periodicity of ~3.5 years from 2007 to 2017, and (iii) a short-term periodicity detected in 2001–2003, 2005–2007, 2011–2015 (~1 year), in 2011–2014 (~6 months) and in 2001–2003 (3–4 months).

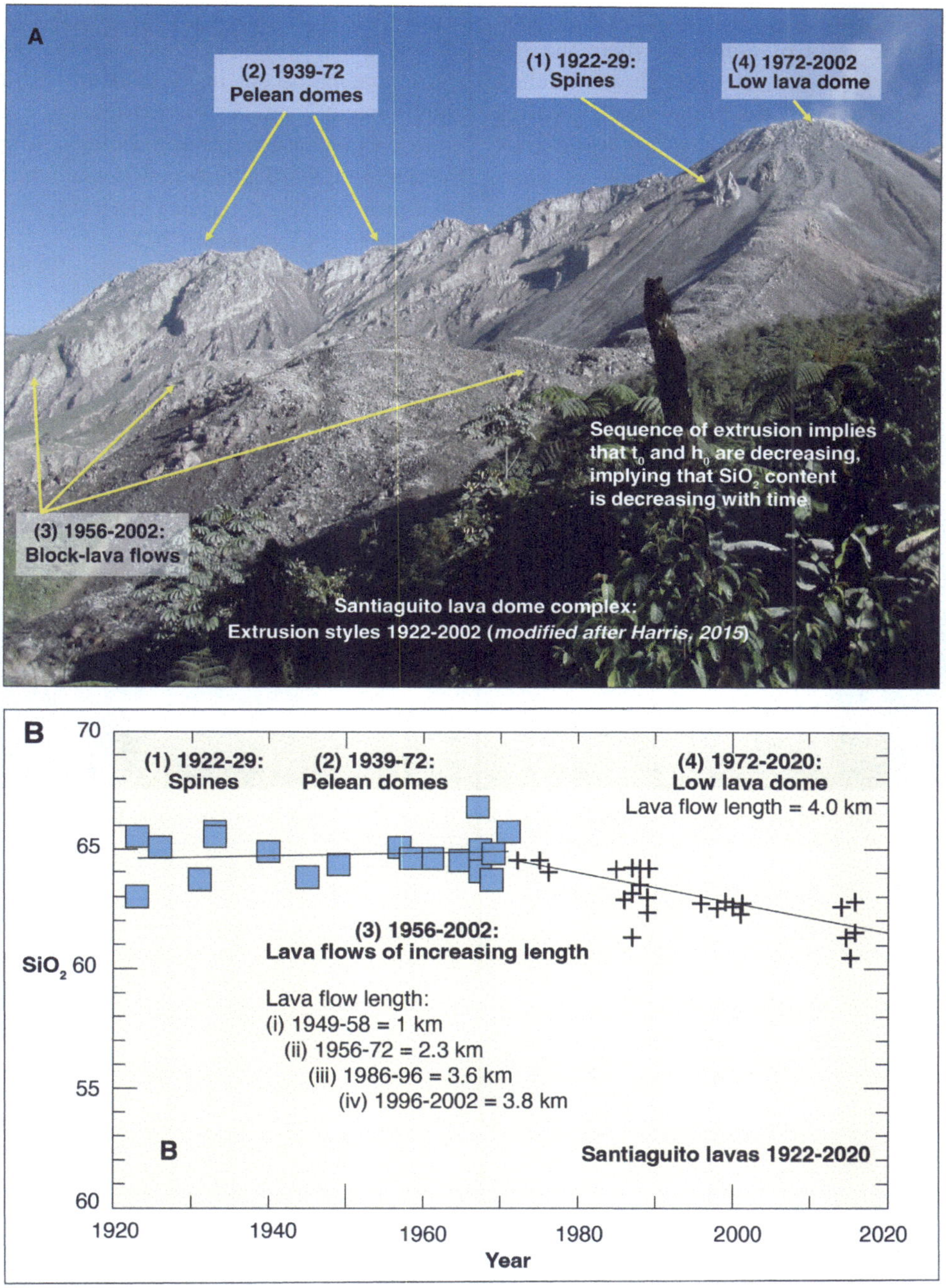

Fig. 14 Santiaguito dome complex, Guatemala: **A** Picture showing the Santiaguito dome complex and the time evolution of different dome types over the period 1922–2022 (modified after Harris 2015b). The dome range includes, from left to right: El Brujo, La Mitad, El Monje and Caliente domes. The dome complex volume is between 1.1 and 1.3 km^3. **B** The compositional diagram shows a decrease in SiO_2 through time until 2002–2021, which corresponds to the most recent Caliente dome-coulee and its block-lava flows (modified after Harris 2015a, b; Wallace et al. 2020; Massaro et al. 2022). Decrease in SiO_2 corresponds to lower viscosity and subsequent increase in block-lava flow length over time. Extrusion rate declined over the same period to an average of 0.44–0.53 ± 0.1 m^3/s (modified after Harris et al. 2003)

Overall, lava domes are increasingly recognized as a specific class of eruption, where the gas phase competes with a relatively slow ascending magma whose rheology progressively evolves from a viscous liquid to that of brittle rock (Calder et al. 2015). Dome eruptions are characterized by pulsatory magma ascent at multiple timescales (e.g. the Santiaguito dome complex, Fig. 14A and B). The cyclicity and complexity of lava dome eruptions, involving the dynamics of magma ascent in the conduit, extrusion, emplacement, and destruction-related hazards, represents a challenge in volcanology, as shown by Merapi in 2010 and its new dome, which has been growing and collapsing from August 2018 up to the time of writing in 2024.

Photo 6 Spatter rampart and active spatter cones feeding a lava flow, Piton de la Fournaise, La Réunion Island, December 2020 (*photo* courtesy of M.O. Chevrel)

Photo 7 Fresh P-type (blue glassy) and S-type (ropey) pāhoehoe, Pu'u 'O'o lava flow field, Kilauea, Hawai'i (*photo* courtesy of A. Harris)

Photo 8 Loose, autobrecciated 'a'ā at the contact with smooth, undulating sheet pāhoehoe lava flow, Kilauea, Hawai'i (*photo* courtesy of A. Harris)

Photo 9 Unweathered, young 'a'ā lava flow, about 5 m in thickness, Llaima volcano, Chile. Clinker autobreccia forming a rough, spiny surface with steep edges (*photo* J.-C. Thouret)

Photo 10 Structure and subunits of 'a'ā lava flow, Llaima volcano, Chile, about 4 m in thickness. From base to top: oxydised tephra substrate (below the hammer), basal breccia, massive core, and autobrecciated lava (*photo* J.-C. Thouret)

Photo 11 **A** Pillow lava forming underwater on the San Juan de Fuca Ridge, Eastern Pacific Ocean (*photo* M.R. Perfit and the Monterey Bay Aquarium Institute). **B** Pillow lava surrounded by selvedges of basaltic glass (hyaloclastites). Rupe di Aci Castello near Catania, Sicily, Italy (*photo credit* Mindat ID 52417, https://www.mindat.org/min-52417.html)

A

B

Photo 12 **A** Stubby, andesitic block-lava flows of Holocene age, Nevado Sabancaya, southern Peru (*photo credit* Google Earth image). Maximum length is 5 km from the vent. Note the steep front >50 m in height, minor surface landforms, levees and flow subunits (bottom right): Ashfalls from current eruptions obscure minor landforms on top and at the edges of the proximal lava flows. **B** Block-lava flow, Colima volcano, Mexico, showing an autobreccia with large, angular, smooth-sided blocks, decimetres to a few metres in size (*photo* J.-C. Thouret)

Photo 13 **A** Glass Mountain obsidian flow (coulee) at Medicine Lake Volcano, California, USA. Spectacular, steep-sided rhyolite and dacite obsidian flows with ogives, cumulatively about 6 km long, 3 km wide and 200–300 m thick, which erupted outside of the eastern caldera rim of Medicine Lake (*photo* J. Donnelly-Nolan; Public domain). **B** Coulee of Little Glass Mountain, Medicine Lake, showing ogives This coulee is 1550 m long, 860 m wide, and 30–40 m thick (*photo* courtesy of A. Harris)

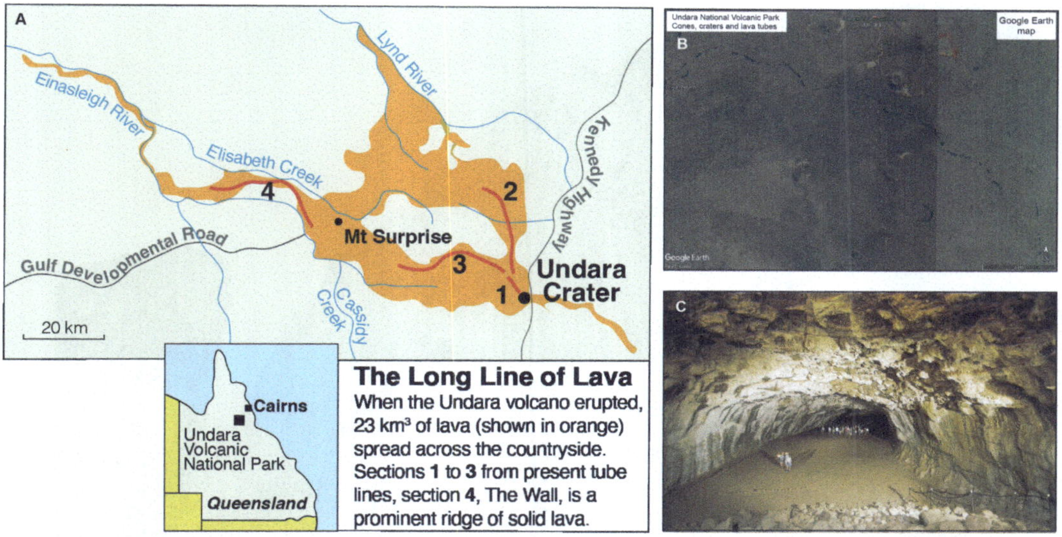

Photo 14 Undara plateau in NE Queensland, Australia. **A** Orange colour on map represents the 5500 km^2 extent of young (0.19 Ma) basalt lava flows, with a maximum length of 160 km and a volume of 23 km^3. **B** Lava flow on a very gentle slope can be tracked over tens of km through the line of collapse features and skylights above lava tubes. **C** Example of a lava tube (pyroduct) about 30 m high and 40 m wide (*photo credit* Undara Volcanic National Park, Queensland)

Photo 15 One of the pseudocraters or rootless cones (about 200 m wide and 30 m high) on a lava flow which flowed into the Lake Myvatn ~2300 years ago, northern Iceland (*photo* D. Karátson)

Photo 16 Colonnade and columnar jointing, basaltic lava flow 1.8 ± 0.2 Ma, Chilhac, Velay, Auvergne, France showing a 7 m-high colonnade overlain by a 15-m high entablature. The average diameter of columns is 30–40 cm (*photo* courtesy of P. Boivin)

A

B

Photo 17 **A** Active lava lake with pit crater in Nyiragongo volcano, Democratic Republic of the Congo. Note the overbank pahoehoe lava flow surrounding the 220 m-wide pit crater (*photo* courtesy of R. Paris). **B** Lava lake in the pit crater of Erta Ale shield volcano, Ethiopia. The lava lake is about 120 m in diameter. The high caldera rim (light-coloured cliff) is in the background (*photo* courtesy of A. Gourgaud)

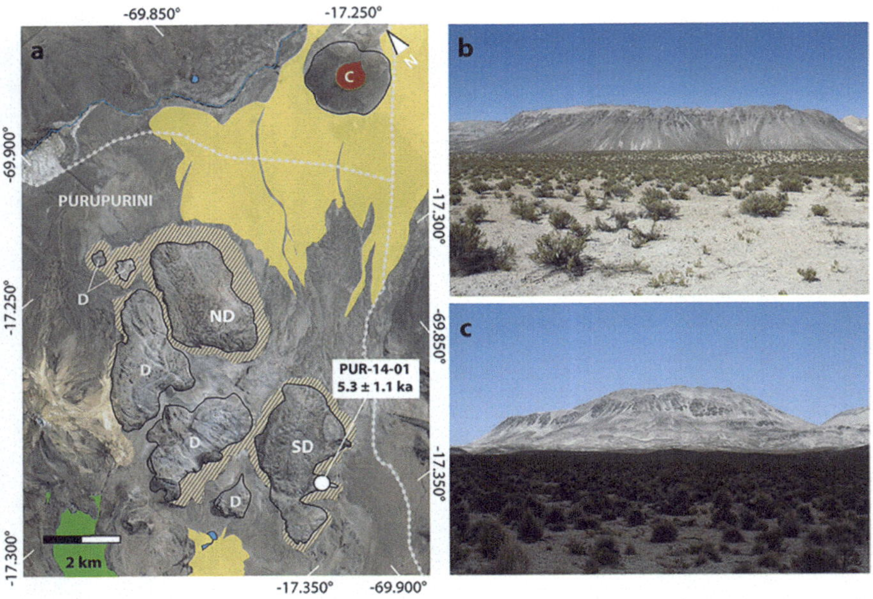

Photo 18 **A** Map and Google Earth image of the cluster of Purupurini dome-coulees of Holocene age, southernmost Peru. **B** 'Torta' type, dacitic dome coulee about 4.4 km long, 2.4 km wide and 400 m thick (ND in map). C. Dome coulee (SD in map) with one of located cosmogenic dated samples. Reference of the map and photos in: Bromley GRM, Thouret JC, Schimmelpfennig I, Mariño J, Valdivia D, Rademaker K, Vivanco Lopez SdP, ASTER Team (Aumaître G, Bourlès D, Keddadouche K) (2019): In situ cosmogenic 3He and 36Cl and radiocarbon dating of volcanic deposits refine the Pleistocene and Holocene eruption chronology of SW Peru. Bull Volcanol 81:64. https://doi.org/10.1007/s00445-019-1325-6

Photo 19 Puy Chopine, upheaved plug-type of dome, located in the Strombolian cone of the Puy des Gouttes, Chaîne des Puys, France. The trachytic protrusion (160 m high and 500 m in diameter at the base) was formed by a highly explosive eruption about 10,000 years ago, while the Puy des Gouttes is about 30,000 years old (*photo* courtesy of J. Damase/CD63).

Photo 20 Soufrière Hills Volcano, Montserrat Island. Composite or Pelean dome (1995–1999) surrounded by thick aprons of volcaniclastic debris and pyroclastic density current deposits channelled down radial valleys (*photo* courtesy of T. Druitt)

Photo 21 Puy de Dôme, Chaîne des Puys, France. Cumulo-dome looking northwest, showing the steep, craggy western flank, the eastern apron of volcaniclastic debris, and the flattish summit (see text for the internal structure, Chap. 2, Sect. 3.4) (*photo* courtesy of P. Boivin)

Photo 22 Oblique aerial photograph of the compound Chaitén lava (rhyolite) dome, southern Chile. Various lava lobes and pyroclastic density current deposits are marked. Caldera is about 3 km in diameter (*photo* J.S. Pallister, US Geological Survey, 24 January 2010; see (Pallister et al. 2013) for details)

References

Aguilar CR, Thouret J-C, Samaniego P, Wörner G, Jicha B, Paquette J-L, Suana E, Finizola A (2022) Growth and evolution of long-lived, large volcanic clusters in the Central Andes: the Chachani volcanic cluster, southern Peru. J Volcanol Geoth Res 426:107539. https://doi.org/10.1016/j.jvolgeores2022.107539

Aguirre-Díaz GJ, Jaimes-Viera MC, Nieto-Obregón J (2006) The Valle del Bravo volcanic field: geology and geomorphometric parameters of a Quaternary monogenetic field at the front of the Mexican Belt. In: Siebe C, Macias JL, Aguirre-Diaz GJ (eds) Neogene-Quaternary Continental Margin Volcanism: a perspective from Mexico. Geol Soc Amer, Spec Paper 402. Penrose Conference Series, pp 139–154. ISBN 10 0-8137-2402-3

Allain C, Limat L (1995) Regular patterns of cracks formed by directional drying of a colloidal suspension. Phys Rev Lett 74(15):2981–2984. https://doi.org/10.1103/PhysRevLett.74.2981

Aydin A, DeGraff JM (1988) Evolution of polygonal fracture patterns in lava flows. Science 239:471–476. https://doi.org/10.1126/science.239.4839.471

Anderson SW, Fink JH, Rose WI (1995) Mount St. Helens and Santiaguito lava domes: the effect of short-term eruption rate on surface texture and degassing processes. J Volcanol Geoth Res 69:105–116. https://doi.org/10.1016/0377-0273(95)00022-4

Atkinson A, Griffin TJ, Stephenson PJ (1975) A major lava tube system from Undara Volcano, North Queensland. Bull Volcanol 39:266–293. https://doi.org/10.1007/BF02597832

Barnoud A, Cayol V, Lelièvre PG, Portal A, Labazuy P, Boivin P, Gailler L (2021) Robust Bayesian joint inversion of gravimetric and muographic data for the density imaging of the Puy de Dôme Volcano (France). Front Earth Sci 8:575842. https://doi.org/10.3389/feart.2020.575842

Barrière J, d'Oreye N, Smets B, Oth A, Delhaye L, Subira J et al (2022) Intra-crater eruption dynamics at Nyiragongo (DR Congo), 2002–2021. J Geophys Res, Solid Earth 127. https://doi.org/10.1029/2021JB02385

Biren J, Harris A, Tuffen H, Chevrel MO, Gurioli L, Vlastélic I, Schiavi F, Benbakkar M, Fonquernie C, Calabro L (2020) Chemical, textural and thermal analyses of local interactions between lava flow and a tree—case study from Pahoa, Hawai'i. Front Earth Sci 8:233. https://doi.org/10.3389/feart.2020.00233

Bisson M, Tadini A, Gianardi R, Angioletti A (2021) The use of historical cartography and ALS technology to map the geomorphological changes of volcanic areas: a case study from Gran Cono of Somma-Vesuvius volcano. Geomorphology 380:107624. https://doi.org/10.1016/j.geomorph.2021.107624

Blake S (1989) Viscoplastic models of lava domes. In: Fink H (ed) Lava flows and domes. Emplacement mechanisms and hazard implications. IAVCEI Proceedings in Volcanology, vol 2. Springer, pp 88–126. ISBN: 3-540-50684-5

Boivin P, Bachèlery P (2009) Petrology of 1977 to 1998 eruptions of Piton de la Fournaise, La Réunion Island. J Volcanol Geoth Res 184(1–2):109–125. https://doi.org/10.1016/j.jvolgeores.2009.01.012

Boivin P, Besson J-C, Briot D, Camus G, Goër de Herve A, Gourgaud A, Labazuy P, Langlois E, de Larouzière F-D, Livet M, Mergoil J, Miaillier D, Morel J-M, Vernet G, Vincent P-M (2009) Volcanologie de la Chaîne des Puys, 5è ed, Carte 1/25000, notice 196 pp.

Boivin P, Besson J-C, Briot D, Camus G, Deniel C, Gourgaud A, Labazuy P, Langlois E, de Larouzière F-D, Livet M, Médard E, Merceccia C, Mergoil J, Miaillier D, Morel J-M, Thouret J-C, Vernet G (2017) Volcanologie de la Chaîne des Puys, 6è ed, Parc naturel régional des Volcans d'Auvergne, Carte 1/25000 couleur, notice bilingue 199 p. ISBN: 978-2-910357-09-2

Boudon G, Balcone-Boissard H, Villemant B, Morgan DJ (2015) What factors control superficial lava dome explosivity? Nat Sci Rep 5:14551. https://doi.org/10.1038/srep14551

Budkevitsch P, Robin P-Y (1994) Modelling the evolution of columnar joints. J Volcanol Geoth Res 59:219–239. https://doi.org/10.1016/0377-0273(94)90092-2

Burgi P-Y, Darrah TH, Tedesco D, Eymold WK (2014) Dynamics of the Mount Nyiragongo lava lake. J Geophys Res Solid Earth 119:4106–4122. https://doi.org/10.1002/2013JB010895

Calder ES, Lavallée Y, Kendrick JE, Bernstein M (2015) Lava dome eruptions. In: Sigurdsson H, Houghton B, McNutt AR, Rymer H, Stix J (eds) The encyclopedia of volcanoes, chap 18, 2nd edn. Academic, pp 343–362. ISBN: 978-0-12-385938-9

Calvari S (2019) Understanding basaltic lava flow morphologies and structures for hazard assessment. Ann Geophys 62(2):1–17. https://doi.org/10.4401/ag-8048

Calvari S, Pinkerton H (1999) Lava tube morphology on Etna and evidence for lava flow emplacement mechanisms. J Volcanol Geoth Res 90:263–280. https://doi.org/10.1016/S0377-0273(99)00024-4

Carracedo JC, Troll VR, Day JMD, Geiger H, Aulinas M, Soler V, Deegan FM, Perez-Torrado FJ, Gisbert G, Gazel E, Rodriguez-Gonzalez A, Albert H (2022) The 2021 eruption of the Cumbre Vieja volcanic ridge on La Palma, Canary Islands. Geol Today 38(3):94–107. https://doi.org/10.1111/gto.12388

Cas RAF, Wright JV (1988) Lava flows, chap 4. Volcanic successions. Chapman & Hall, London, pp 59–92. ISBN: 0-04-552021-6

Cashman KV, Mangan MT (2014) A century of studying effusive eruptions in Hawai'i. In: Poland MP, Takahashi TJ, Landowski CM (eds) Characteristics of Hawaiian volcanoes. US Geol Survey Prof Paper 1801, chap 9, pp 357–394. ISSN: 1044-9612

Cashman C, Pinkerton H, Stephenson J (1998) Introduction to special section: long lava flows. J Geophys Res Solid Earth 103(B11):27281–27289. https://doi.org/10.1029/98JB01820

Chester DK, Duncan AM, Guest JE, Kilburn CRJ (1985) Mount Etna: the anatomy of a volcano. Chapman and Hall, London. ISBN: 978-94-010-8309-6

Chevrel MO, Platz T, Hauber E, Baratoux D, Lavallée Y, Dingwell DB (2013) Lava flow rheology: a comparison of morphological and petrological methods. Earth Planet Sci Lett 384:109–120. https://doi.org/10.1016/j.epsl.2013.09.022

Chevrel MO, Harris AJL, Ajas A, Biren J, Gurioli L, Calabro L (2019) Investigating physical and thermal interactions between lava and trees: the case of Kīlauea's July 1974 flow. Bull Volcanol 81:6. https://doi.org/10.1007/s00445-018-1263-8

Cioni R, Funedda A (2005) Structural geology of crystal-rich, silicic flows: a case study from San Pietro Island (Sardinia, Italy). Geol Soc Amer Spec Paper 396:1–14. https://doi.org/10.1130/2005.2396(01)

Cotton CA (1944) Volcanoes as landscape forms, Part II. Whitcombe and Tombs Ltd., New Zealand, pp 70–399

Davies AG (2014) Lava lakes. In: Encyclopedia of planetary landforms. Springer, New York. https://doi.org/10.1007/978-1-4614-9213-9_57-1

Degraff JM, Aydin A (1987) Surface morphology of columnar joints and its significance to mechanics and direction of joint growth. Geol Soc Amer Bull 99(5):605–617. https://doi.org/10.1130/0016-7606(1987)

Deniel C, Boivin P, Miallier D, Gerbe MC (2020) Multi-stage growth of the trachytic lava dome of the Puy de Dome (Chaîne des Puys, France). Field, geomorphological and petro-geochemical evidence. J Volcanol Geoth Res 396:106749. https://doi.org/10.1016/j.jvolgeores.2019.106749

de Silva SL, Self S, Francis PW, Drake RE, Ramirez CR (1994) Effusive silicic volcanism in the Central Andes: the Chao dacite and other younger lavas of the Altiplano-Puna Volcanic Complex. J Geophys Res 99(B9):17805–17825. https://doi.org/10.1029/94JB00652

Dietterich HR, Cashman KV (2014) Channel networks within lava flows: Formation, evolution, and implications for flow behavior. J Geophys Res Earth Surf 119:1704–1724. https://doi.org/10.1002/2014JF003103

Druitt TH, Kokelaar BP (eds) (2002) The eruption of Soufriere Hills Volcano, Montserrat, from 1995 to 1999, Geol Soc London Memoir, London, vol 21, 645 pp. ISBN: 1-86239-098-3

Durieux J (2002/2003) Volcano Nyiragongo (DR Congo): Evolution of the crater lava lakes from the discovery to the present. Acta Vulcanol 14(15):137–144. https://doi.org/10.1400/19087. Casalini id: 2231613

Dutton CE (1893) Hawaiian volcanoes (2005 edition). University of Hawai'i Press, Honolulu, p 235

Finch RH (1931) Lava tree casts and tree molds. The Volcano Letter (Hawaiian Volcano Observatory), vol 316, pp 1–3

Finch RH (1933) Block lava. J Geol 41(7):769–770

Fink JH (1983) Structure and emplacement of a rhyolitic obsidian flow: Little Glass Mountain, Medicine Lake Highland, northern California. Geol Soc Amer Bull 94:362–380. https://doi.org/10.1130/0016-7606(1983)94<362:SAEOAR>2.0.CO;2

Fink JH (1990) Lava flows and domes. Emplacement mechanisms and hazard implications. IAVCEI Proceedings in Volcanology, vol 2, Springer, 249 pp. ISBN: 3-540-50684-5

Fink JH, Fletcher RC (1978) Ropy pahoehoe: surface folding of a viscous fluid. J Volcanol Geoth Res 4(1–2):151–170. https://doi.org/10.1016/0377-0273(78)90034-3

Fink JH, Griffiths RW (1998) Morphology, eruption rates, and rheology of lava domes: insights from laboratory models. J Geophys Res 103(B1):527–545. https://doi.org/10.1029/97JB02838

Fink JH, Anderson SW (2017) Emplacement of Holocene Silicic Lava Flows and Domes at Newberry, South Sister, and Medicine Lake Volcanoes, California and Oregon. US Geol Survey Scient Investig Rep 2017-5022-I, 41 pp. https://doi.org/10.3133/sir20175022I

Fouqué F (1879) Santorin et ses éruptions. G Masson Editeur, Paris, p 370

Francis PW (1993) Volcanoes. A planetary perspective. Clarendon Press, Oxford University Press, 443 pp. ISBN: 0-19-854452-9

Francis PW, Oppenheimer C, Stevenson D (1993) Endogenous growth of persistently active volcanoes. Nature 366:554–557. https://doi.org/10.1038/366554a0

Gabrielli S, Spagnolo M, De Siena L (2020) Geomorphology and surface geology of Mount St. Helens Volcano. J Maps 16(2):585–594. https://doi.org/10.1080/17445647.2020.1790048

Giachetti T, Druitt RH, Burgisser A, Arbaret L, Galvet C (2010) Bubble nucleation, growth and coalescence during the 1997 Vulcanian explosions of Soufrière Hills Volcano, Montserrat. J Volcanol Geoth Res 193(3–4):215–231. https://doi.org/10.1016/j.jvolgeores.2010.04.001

Gertisser R, Troll VR, Walter TR, Nandaka IGMA, Ratdomopurbo A (eds) (2023) Merapi volcano: geology, eruptive activity, and monitoring of a high-risk volcano. Springer, Active Volcanoes of the World. ISBN: 978-3-031-15039-5

Gourgaud A, Boivin P, Briot D, Livet M, Mergoil J, Miallier D, Vernet G, Vincent PM (2009) Planète volcan, l'Auvergne pour comprendre le volcanisme. Ed Focus, CRDP d'Auvergne, 175 p avec CD-Rom. https://www.cndp.fr/crdp-clermont/crdp/LireArticle-760-librairie-crdp-htm

Greeley R (1987) The role of lava tubes in Hawaiian volcanoes. In: Decker RW, Wright TL, Stauffer PH (eds) Volcanism in Hawai'i, vol 2. US Geol Survey Prof Paper 1350, chap 59, pp 1589–1601

Gregg TKP (2017) Patterns and processes: subaerial lava flow morphologies: a review. J Volcanol Geoth Res 342:3–12. https://doi.org/10.1016/j.jvolgeores.2017.04.022

Gregg TKP, Fink JH (1995) Quantification of submarine lava-flow morphology through analog experiments. Geology 23(1):73–76. https://doi.org/10.1130/0091-7613(1995)023<0073:QOSLFM>2.3.CO;2

Griffiths RW (2000) The dynamics of lava flows. Annu Rev Fluid Mech 32:477–518. https://doi.org/10.1146/annurev.fluid.32.1.477

Grossenbacher KA, McDuffie SM (1995) Conductive cooling of lava: columnar joint diameter and stria width as functions of cooling rate and thermal gradient. J Volcanol Geoth Res 69:95–103. https://doi.org/10.1016/0377-0273(95)00032-1

Harris AJL (2008) Modeling lava lake heat loss, rheology, and convection. Geophys Res Lett 35:L07303. https://doi.org/10.1029/2008GL033190

Harris AJL (2015a) Lava flows: structures and morphologies. Master Course presentation, LMV, Université Clermont-Auvergne (Unpublished)

Harris AJL (2015b) Persistent degassing, lava lakes and lava domes. Master Course presentation, LMV, Université Clermont-Auvergne (unpublished)

Harris AJL, Rowland SK (2015) Lava flows and rheology. In: Sigurdsson H, Houghton B, McNutt SR, Rymer H, Stix J (eds) The encyclopedia of volcanoes, chap 17, 2nd edn. Academic, pp 321–342. ISBN: 978-0-12-385938-9

Harris AJ, Rose WI, Flynn LP (2003) Temporal trends in lava dome extrusion at Santiaguito 1922–2000. Bull Volcanol 65:77–89. https://doi.org/10.1007/s00445-002-0243-0

Harris AJL, Carniel R, Jones J (2005) Identification of variable convective regimes at Erta'Ale lava lake. J Volcanol Geoth Res 142(3):207–223. https://doi.org/10.1016/j.jvolgeores.2004.11.011

Harris AJL, Dehn J, Calvari S (2007) Lava effusion rate definition and measurement: a review. Bull Volcanol 70(1):1–22. https://doi.org/10.1007/s00445-007-0120-y

Harris AJL, Rowland SK, Chevrel MO (2022) The anatomy of a channel-fed ʻaʻā lava flow system. Bull Volcanol 84:70. https://doi.org/10.1007/s00445-022-01578-0

Harris AJL, Flynn L, Rothery DA, Oppenheimer C, Sherman SB (1999) Mass flux measurements at active lava lakes: implications for magma recycling. J Geophys Res 104:7117–7136. https://doi.org/10.1029/98JB02731

Harris AJL, Rowland SK, Villeneuve N et al (2017) Pāhoehoe, ʻaʻā, and block lava: an illustrated history of the nomenclature. Bull Volcanol 79:7. https://doi.org/10.1007/s00445-016-1075-7

Hetényi G, Milazzo MP (2014) Columnar joints. In: Encyclopedia of planetary landforms. Springer, New York. https://doi.org/10.1007/978-1-4614-9213-9_57-1

Hon K, Kauahikaua J, Denlinger R, McKay K (1994) Emplacement and inflation of pähoehoe sheet flows: observations and measurements of active lava flows on Kilauea volcano, Hawai'i. Geol Soc Amer Bull 106:351–370. https://doi.org/10.1130/0016-7606(1994)106<0351:EAIOPS>2.3.CO;2

Hulme G (1974) The interpretation of lava flow morphology. Geophys J Roy Astron Soc 39(2):361–383. https://doi.org/10.1111/j.1365-246X.1974.tb05460.x

Jaggar TA (1917) Thermal gradient at Kilauea lava lake. J Washington Acad Sci 7:397–405

Jaggar TA (1920) Seismometric investigation of the Hawaiian Lava column. Bull Seism Soc Am 10(4):155–275. https://doi.org/10.1785/BSSA0100040155

Karátson D, Telbisz T, Harangi SZ, Magyari E, Dunkl I, Kiss B, Jánosi CS, Veres D, Braun M, Fodor E, Biró T, Kózik SZ, von Eynatten H, Lin D (2013) Morphometrical and geochronological constraints on the youngest eruptive activity in East-Central Europe at the Ciomadul (Csomád) lava dome complex, East Carpathians. J Volcanol Geoth Res 157–158:56–72. https://doi.org/10.1016/j.jvolgeores.2013.01.013

Kauahikaua JP, Sherrod DR, Cashman KV, Heliker C, Hon K, Mattox TN, Johnson JA (2003) Hawaiian lava-flow dynamics during the Pu'u 'O'o-Kupaianaha eruption: a tale of two decades. In: Heliker C, Swanson DA, Takahashi TJ (eds) The Pu'u O'o-Kupaianaha eruption of Kilauea volcano, Hawai'i: the first 20 years. US Geol Survey Prof Paper 1676, pp 63–87. QE523.KSP89.2003

Kereszturi G, Németh K (2013) Monogenetic basaltic volcanoes: genetic classification, growth, geomorphology and degradation. In: Németh K (ed) Updates in volcanology—new advances in understanding volcanic systems. Intech, pp 3–88. ISBN: 978-953-51-0915-0

Kilburn CR (1981) Pahoehoe and aa lavas: a discussion and continuation of the model of Peterson and Tilling. J Volcanol Geoth Res 11(2–4):373–382. https://doi.org/10.1016/0377-0273(81)90033-0

Kilburn CRJ (2015) Lava flow hazard and modeling. In: Sigurdsson H, Houghton B, McNutt SR, Rymer H, Stix J (eds) The encyclopedia of volcanoes, chap 55, 2nd edn. Academic, Elsevier, pp 957–969. ISBN: 978-0-12-385938-9

Kilburn CRJ, Guest JE (1993) 'A'a lavas of Mount Etna, Sicily. In: Kilburn CRJ, Luongo G (eds) Active lavas. Monitoring and modelling, Part 1, chap 3. UCL Press, Routledge, London, pp 73–106, 388 pp. ISBN: 978-1-032-35070-7

Le Guern F (1987) Mechanism of energy transfer in the lava lake of Nyiragongo (Zaire), 1959–1977. J Volcanol Geoth Res 31(1–2):17–31. https://doi.org/10.1016/0377-0273(87)90003-5

Le Guern F, Carbonnelle J, Tazieff H (1979) Erta'Ale lava lake: heat and gas transfer to the atmosphere. J Volcanol Geoth Res 6(1–2):27–48. https://doi.org/10.1016/0377-0273(79)90045-3

Lev E, Ruprecht P, Oppenheimer C, Peters N, Patrick M, Hernández PA, Spampinato L, Marlow J (2019) A global synthesis of lava lake dynamics. J Volcanol Geoth Res 381:16–31. https://doi.org/10.1016/j.jvolgeores.2019.04.010

Lipman PW, Banks NG (1987) 'A'ā flow dynamics, Mauna Loa. In: Decker RW, Wright TL, Stauffer PH (eds) Volcanism in Hawai'i, US Geol Survey Prof. Paper 1350, chap 57, pp 1527–1567

Lockwood JP, Hazlett RW (2010) Volcanoes. Global perspectives. Wiley-Blackwell, UK, 539 pp. ISBN: 978-1-4051-6249-4

MacDonald GA (1953) Pähoehoe, 'a'ä and block lava. Amer J Sci. 251(3):169–191

MacDonald GA (1972) Volcanoes. Prentice-Hall, New Jersey, 509 pp. ISBN: 0-13-942219-6

Manga M, Ventura G (2005) Kinematics and dynamics of lava flows. Geol Soc Amer Spec Paper 396:219 pp. ISBN: 0-8137-2396-5

Massaro S, Costa A, Sulpizio R, Coppola D, Soloviev A (2022) Detecting multiscale periodicity from the secular effusive activity at Santiaguito lava dome complex (Guatemala). Earth Planets Space 74:107. https://doi.org/10.1186/s40623-022-01658-7

Mattox TN, Mangan MT (1997) Littoral hydrovolcanic explosions: a case study of lava-seawater interaction at Kilauea Volcano. J Volcanol Geoth Res 75:1–17. https://doi.org/10.1016/S0377-0273(96)00048-0

Miller TP (1994) Dome growth and destruction during the 1989–1990 eruption of Redoubt volcano. J Volcanol Geoth Res 62:197–212. https://doi.org/10.1016/0377-0273(94)90034-5

Moore JG (1975) Mechanism of formation of pillow lava: Pillow lava, produced as fluid lava cools underwater, is the most abundant volcanic rock on Earth, but only recently divers observed it forming. Amer Sci 63:269–277. https://www.jstor.org/stable/27845462

Moore JG (1987) Preliminary estimates of the rheological properties of 1984 Mauna Loa lava. In: Decker RW, Wright TL, Stauffer PH (eds) Volcanism in Hawai'i, US Geol Survey Prof Paper 1350, pp 1569–1588

Moriya I (1978) Morphology of lava domes. Bull Dept Geography, Komazawa University of Japan 14:55–69 (in Japanese)

Nakada S, Miyake Y, Sato H, Oshima O, Fujinawa A (1995) Endogenous growth of dacite dome at Unzen volcano (Japan), 1993–1994. Geology 23(2):157–160. https://doi.org/10.1130/0091-7613(1995)023<0157:EGODDA>2.3.CO;2

Nakada S, Shimizu H, Ohta K (1999) Overview of the 1990–1995 eruption at Unzen Volcano. J Volc Geoth Res 89(1–4):1–22. https://doi.org/10.1016/S0377-0273(98)00118-8

Nichols RL (1939) Viscosity of lava. J Geol 47(3):290–302

Ogawa T (1980) The lava caves and lava tree moulds of Mt. Fuji. J Assoc Jpn Cavers 2(3):3–50

Ollier (1988) Volcanoes. Basil Blackwell, Oxford, 228 pp. ISBN: 0-631-15664-X

Ollier CD, Brown MC (1965) Lava caves of Victoria. Bull Volcanol 28:215–229
Oppenheimer C, Francis P (1998) Implications of longeval lava lakes for geomorphological and plutonic processes at Erta 'Ale volcano, Afar. J Volcanol Geoth Res 80:101–111. https://doi.org/10.1016/S0377-0273(97)00041-3
O'Reilly JP (1879) Explanatory notes and discussion on the nature of the prismatic forms of a group of columnar basalts, Giant's Causeway. Trans Royal Irish Acad 26:641–734 with plates XV–XVIII. https://www.jstor.org/stable/30079106
Orr TR (2011) Lava tube shatter rings and their correlation with lava flux increases at Kīlauea Volcano, Hawai'i. Bull Volcanol 73:335–346. https://doi.org/10.1007/s00445-010-0414-3
Pallister JS, Diefenbach AK, Burton WC, Muñoz J, Griswold JP, Lara LE, Lowenstein JB, Valenzuela CE (2013) The Chaíten rhyolite lava dome: eruption sequence, lava dome volumes, rapid effusion rates and source of the rhyolite magma. Andean Geol 40:277–294
Paris R, Grünewald O (2015) Map of Nyiragongo crater (based on stereo-photogrammetry) with the lava lake and three platforms. In: Smets et al (2019) EGU Geophys Res Abstracts, Vienna, vol 21, p 1-1, 1 pp.
Patrick MR, Anderson KR, Poland MP, Orr TR, Swanson DA (2015) Lava lake level as a gauge of magma reservoir pressure and eruptive hazard. Geology 43(9):831–834. https://doi.org/10.1130/G36896.1
Peterson DW, Tilling RI (1980) Transition of basaltic lava from pāhoehoe to 'a'ā, Kilauea Volcano, Hawai'i: field observations and key factors. J Volcanol Geoth Res 7(3–4):271–293. https://doi.org/10.1016/0377-0273(80)90033-5
Peterson DW, Holcomb RT, Tilling RI, Christiansen RL (1994) Development of lava tubes in the light of observations at Mauna Ulu, Kilauea volcano, Hawai'i. Bull Volcanol 56:343–360. https://doi.org/10.1007/BF00326461
Portal A, Labazuy P, Lénat J-F, Béné S, Boivin P, Busato E, Cârloganu C, Combaret C, Dupieux P, Fehr F, Gay P, Laktineh I, Miallier D, Mirabito L, Niess V, Vulpescu B (2013) Inner structure of the Puy de Dôme volcano: cross-comparison of geophysical models (ERT, gravimetry, muon imaging). Geosci Instrum Method Data Syst 2:47–54. https://doi.org/10.5194/gi-2-47-2013
Portal A, Fargier Y, Labazuy P, Lénat J-F, Boivin P, Miallier D (2019) 3D electrical imaging of the inner structure of a complex lava dome, Puy de Dôme volcano (French Massif Central, France). J Volcanol Geoth Res 373:97–107. https://doi.org/10.1016/j.jvolgeores.2019.01.019
Pyle D, Elliott JR (2006) Quantitative morphology, recent evolution, and future activity of the Kameni Islands volcano, Santorini, Greece. Geosphere 2(5):253–268. https://doi.org/10.1130/GES00028.1
Rose WI, Pearson T, Bonis S (1976) Nuee ardente eruption from the foot of a dacite lava flow, Santiaguito Volcano, Guatemala. Bull Volcanol 40:23–38. https://doi.org/10.1007/BF02599827
Rossi MJ (1997) Morphology of the 1984 open-channel lava flow at Krafla volcano, northern Iceland. Geomorphology 20(1–2):95–112. https://doi.org/10.1016/S0169-555X00007-X
Rossi M-J, Gudmundsson A (1996) The morphology and formation of flow-lobe tumuli on Icelandic shield volcanoes. J Volcanol Geoth Res 72:291–308. https://doi.org/10.1016/0377-0273(96)00014-5
Rowland SK, Walker GPL (1987) Toothpaste lava: characteristics and origin of a lava structural type transition between pāhoehoe and 'a'ā. Bull Volcanol 49:631–641. https://doi.org/10.1007/BF01079968
Rowland SK, Walker GPL (1990) Pāhoehoe and 'a'ā in Hawai'i: volumetric flow rate controls the lava structure. Bull Volcanol 52:615–628. https://doi.org/10.1007/BF00301212
Rusiecka MK, Martel C (2022) Nucleation delay in water-saturated rhyolite during decompression in shallow volcanic systems and its implications for ascent dynamics. Bull Volcanol 84(61):56–76. https://doi.org/10.1007/s00445-022-01569-1
Ryan MP, Sammis CG (1981) The glass transition in basalt. J Geophys Res 86(part B):9519–9535. https://doi.org/10.1029/JB086iB10p09519
Scarth A (1994) Volcanoes. UCL Press, London, 273 pp. ISBN: 1-85728-223-X
Schmincke H-U (2004) Volcanism. Springer, 324 pp. ISBN: 3-540-43650-2
Self S, Thordarson Th, Keszthelyi L, Walker GPL, Hon K, Murphy MT, Long P, Finnemore S (1996) A new model for the emplacement of Columbia River basalts as large, inflated Pahoehoe lava flow fields. Geophys Res Lett 23(19):2689–2692. https://doi.org/10.1029/96GL02450
Sherrod DR, Scott WE, Stauffer PH (eds) (2008) A volcano rekindled: the renewed eruption of Mont St. Helens, 2004–2006. US Geol Surv Prof. Paper 1750, 865 pp. https://doi.org/10.3133/pp1750
Smets B, Wauthier C, Dille A, Paris R, Samyn D, d'Oreye N, Kervyn F (2019) Unmanned aerial system observations of Nyiragongo crater and lava lake. EGU Geophys Res Abstracts, Vienna, vol 21, p 1-1, 1 p.
Soule A (2015) Mid-ocean ridge volcanism. In: Sigurdsson H, Houghton B, McNutt SR, Rymer H, Stix J (eds) The encyclopedia of volcanoes, chap 21, 2nd edn. Academic, pp 395–421. ISBN: 978-0-12-385938-9
Spampinato L, Oppenheimer C, Calvari S, Cannata A, Montalto P (2008) Lava lake surface characterization by thermal imaging: Erta 'Ale volcano (Ethipia). Geoch Geophys Geosyst 9(12):Q12008. https://doi.org/10.1029/2008GC002164
Sparks RSJ (1997) Causes and consequences of pressurization in lava dome eruption. Earth Planet Sci Lett

150(3–4):177–189. https://doi.org/10.1016/S0012-821X(97)00109-X

Sparks RSJ, Pinkerton H, Hulme G (1976) Classification and formation of lava levees on Mt Etna, Sicily. Geology 4:269–271. https://doi.org/10.1130/0091-7613(1976)4<269:CAFOLL>2.0.CO;2

Stasiuk MV, Jaupart C (1997) Lava flow shapes and dimensions as reflections of magma system conditions. J Volcanol Geoth Res 78(1–2):31–50. https://doi.org/10.1016/S0377-0273(97)00002-4

Staudigel H, Koppers AAP (2015) Seamounts and island building. In: Sigurdsson H, Houghton B, McNutt SR, Rymer H, Stix J (eds) The encyclopedia of volcanoes, chap 22, 2nd edn. Academic, pp 405–421. ISBN: 978-0-12-385938-9

Swanson DA (1973) Pāhoehoe flows from the 1969–1971 Mauna Ulu eruption, Kilauea volcano, Hawai'i. Geol Soc Amer Bull 84:615–626. https://doi.org/10.1130/0016-7606(1973)84<615:PFFTMU>2.0.CO;2

Swanson DA, Duffield WA, Jackson DB, Peterson DW (1979) Chronological narrative of the 1969-71 Mauna Ulu eruption of Kilauea Volcano, Hawai'i. US Geol Surv Prof Paper 1056, 55 pp.

Taylor GA (1958) The 1951 eruption of Mount Lamington, Papua. Austr Bur Miner Resour Geol Geophys Bull 38:1–117

Tazieff H (1984) Mt. Nyiragongo: renewed activity of the lava lake. J Volcanol Geoth Res 20:267–280. https://doi.org/10.1016/0377-0273(84)90043-X

Tazieff H (1994) Permanent lava lakes: observed facts and induced mechanisms. J Volcanol Geoth Res 63: 3–11. https://doi.org/10.1016/0377-0273(94)90015-9

Thordarson T, Self S (1998) The Roza member, Columbia River Basalt Group: a gigantic pāhoehoe lava flow field formed by endogenous processes? J Geophys Res 103(Bll):27411–27445. https://doi.org/10.1029/98JB01355

Tibble GW (1991) Underwater observations of active lava flows from Kilauea volcano, Hawai'i. Geology 19:633–666. https://doi.org/10.1130/0091-7613(1991)019<0633:UOOALF>2.3.CO;2

Tilling RI (1987) Fluctuations in surface height of active lava lakes during 1972–1974 Mauna Ulu eruption, Kīlauea volcano, Hawai'i. J Geophys Res 92:13721–13730. https://doi.org/10.1029/JB092iB13p13721

Tomkeieff SI (1940) The basalt lavas of the Giant's Causeway district of Northern Ireland. Bull Volcanol 6(1):89–143. https://doi.org/10.1007/BF02994875

Utami SB, Andújar J, Costa F, Scaillet B, Humaida H, Carn S (2022) Pre-eruptive excess volatiles and their relationship to effusive and explosive eruption styles in semi-plugged volcanoes. Front Earth Sci 10:882097. https://doi.org/10.3389/feart.2022.882097

Vilardo G, Ventura G (2008) Geomorphological map of the 1944 Vesuvius lava flow (Italy). J Maps 4(1):225–234. https://doi.org/10.4113/jom.2008.1003

Voight B (2000) Structural stability of andesite volcanoes and lava domes. Philos Trans R Soc London 358:1663–1703. https://doi.org/10.1098/rsta.2000.0609

Wadge G, Robertson REA, Voight B (eds) (2014) The eruption of Soufriere Hills volcano, Monserrat from 2000 to 2010. Geol Soc London Memoir 39, 501 pp. https://doi.org/10.1144/M39

Walder JS (2007) Emplacement of a silicic lava dome through a crater glacier: Mount St. Helens, 2004–2006. Ann Glaciol 45:14–20. https://doi.org/10.3189/172756407782282426

Walker GPL (1970) Compound and simple lava flows and flood basalts. Bull Volcanol 35:579–590. https://doi.org/10.1007/BF02596829

Walker GPL (1973) Mount Etna and the 1971 eruption-lengths of lava flows. Philos Trans Royal Soc London Ser A Math Phys Sci 274(1238):107–118. https://doi.org/10.1098/rsta.1973.0030

Walker GPL (1991) Structure and origin by injection of lava under surface crust, of tumuli, "lava rises", "lava rise pits", and "lava-inflation clefts" in Hawai'i. Bull Volcanol 53(7):546–558. https://doi.org/10.1007/BF00298155

Walker GPL (2009) The endogenous growth of pāhoehoe lava lobes and morphology of lava-rise edges. In: Thordarson T, Self S, Larsen G, Rowland SK, Höskuldsson (eds) Studies in volcanology: the legacy of Georges Walker. Spec Publ IAVCEI, no 2. Geol Soc London, pp 17–32, 413 pp. ISBN: 978-1-86239-280-9

Wallace PA, Lamb OD, De Angelis S, Kendrick JE, Hornby AJ, Díaz-Moreno A, González PJ, Felix W, von Aulock FW, Lamur A, Utley JEP, Rietbrock A, Chigna G, Lavallée Y (2020) Integrated constraints on explosive eruption intensification at Santiaguito dome complex, Guatemala. Earth Planet Sci Lett 536(116139):16 pp. https://doi.org/10.1016/j.epsl.2020.116139

Wentworth CK, Carson MH, Finch (1945) Discussion on the viscosity of lava. J Geol 53(2):94–104. https://doi.org/10.1086/625252

White JDL, McPhie J, Soule SA (2015) Submarine lavas and hyaloclastite. In: Sigurdsson H, Houghton B, McNutt SR, Rymer H, Stix J (eds) The encyclopedia of volcanoes, chap 19, 2nd edn. Academic, pp 363–373. ISBN: 978-0-12-385938-9

Wilson L, Parfitt EA (1993) The formation of perched lava ponds on basaltic volcanoes: the influence of flow geometry on cooling-limited lava flow lengths. J Volcanol Geoth Res 56(1–2):113–123. https://doi.org/10.1016/0377-0273(93)90053-T

Witham F, Llewellin EW (2006) Stability of lava lakes. J Volcanol Geoth Res 158:321–332. https://doi.org/10.1016/j.jvolgeores.2006.07.004

Xiao L, Wang C (2009) Geologic features of Wudalianchi volcanic field, northeastern China: implications for Martian volcanology. Planet Space Sci 57:685–698. https://doi.org/10.1016/j.pss.2008.08.005

Chapter 3 Elementary Landforms and Processes II: From Simple, Terrestrial Landforms to Hydroclastic, Glaciovolcanic and Underwater Landforms

Pyroclastic and hydroclastic cones, and glaciovolcanic landforms are relatively small, so-called monogenetic, simple volcanic edifices that result from short-term, commonly low-magnitude, explosive eruptions. They comprise the most common elementary volcano types on Earth that occur in all tectonic settings. We have grouped them under a common umbrella, as these landforms and deposits are created by short-lived eruption(s) of magma, usually from single vents, and a full range of interactions with external water. Other subaqueous (underwater) landforms are also presented here. Hence, in this chapter we consider dry (with no external addition of water), hydroclastic, glaciovolcanic and subaqueous edifices and deposits formed by negligible to substantial, even excessive amounts of water (or ice), as depicted in Fig. 1.

1 General Features of Simple Volcanoes in Terrestrial Environments

As pyroclastic and hydroclastic cones are much smaller than shield volcanoes (Chap. 4) and composite cones (Chap. 5), some authors refer to them as minor landforms (e.g. Lockwood

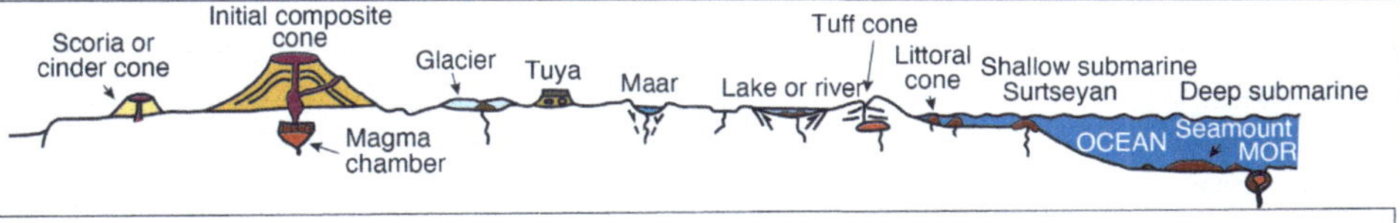

Role of external water	Little to none		Substantial	Major	Subordinate	Optimum	Excessive
Eruptive phenomena	Hawaiian, Strombolian	Effusive; Vulcanian-Plinian; dome growth	Subaqueous englacial, lacustrine	Phreatomagmatism Molten fuel-coolant interactions	Phreatic steam	Surtseyan Phreatoplinian	Effusive Extrusive, dome
Dominant products	Scoria, ballistics, tephra fall	Tephra fall, PDCs	Pillow lava hyaloclastites, breccia	Ballistics, bombs tephra fall (PDCs) accretionary lapilli, vesicular tuff	Lithic ballistics	Hyaloclastites Tephra fall, ballistics	Lava flows; pillow, lobate & sheet flows
Fragmentation	cm-sized	cm to metre-sized	mm- to cm sized to metre-sized	cm to metre-sized	dm to m-sized	μm- to mm sized superheated steam	mm- to m-sized, condensing steam, liquid water

Fig. 1 Illustration of volcanic environments indicative of magma and (external) water interactions for monogenetic terrestrial and subaqueous volcanoes: from on-land, dry explosive cones and other edifices (on the left) to hydroclastic cones and glaciovolcanic landforms, all the way to subaqueous and submarine landforms (on the right). Table below shows role of external water, eruptive phenomena, dominant products, and fragmentation for each category of volcanoes. Adapted and modified from Wohletz (1993) in: Wohletz et al. (2013)

D. Karátson and J.-C. Thouret, *Volcano Geomorphology*, Advances in Volcanology,
https://doi.org/10.1007/978-3-031-77693-9_4

and Hazlett 2010). However, this is not entirely appropriate as large scoria cones can reach heights of 400–500 m, exceeding the size of small composite volcanoes, and a maar or tuff ring crater can be up to 1.5 km wide. The volume of erupted magma controls the size of any edifice. Thus, the term minor should be reserved for describing landforms that are up to a few tens of metres across, such as the products related to effusive activity (minor landforms associated with lava flows, see Chap. 2). Lava domes, also discussed in Chap. 2, vary in height from tens to hundreds of metres, showing that size cannot be the only criterion for falling into the major or minor landform categories. Nevertheless, explosive cones are relatively small volcanoes—this term is used, for example, by White and Ross (2011)—and their volume ranges between 0.1 and 5 km^3, while composite volcanoes usually reach tens of km^3 (Chap. 5).

Cones of explosive origin presented in this chapter are considered in the literature to be *monogenetic* in origin. This implies one eruption or one eruption phase (Walker 2000; Valentine and Connor 2015), corresponding to a short lifetime (<1 year, rarely a few years and at most up to 100 years), commonly composed of <1 km^3 of mafic magma (Kereszturi and Németh 2013). Less frequently, explosive cones may have a longer lifetime, with a more complex evolutionary history, and may evolve into a small composite volcano (e.g. Cerro Negro, Nicaragua; Puy Pariou in the Chaîne des Puys, France; Photo 23). Therefore, there is a wide spectrum of monogenetic landforms from the simplest scoria cones to more complex edifices that lie halfway between monogenetic and polygenetic volcanism (Németh and Kereszturi 2015; Becerra-Ramírez et al. 2022). Truly monogenetic explosive cones are predominantly built up of pyroclastic/phreatomagmatic fall and minor surge/flow deposits, but if the cone evolves, the role of lava flows may increase significantly over a period of a few years (e.g. Paricutín, Mexico: Fig. 2 showing selected scoria cone images). While pyroclastic and hydroclastic products are mostly cone-building, i.e. deposited from a low (<1 km) eruption column over a short distance, the associated lava flows can reach longer distances, up to several kilometres, filling palaeo-valleys and developing a more complex landform. Thus since explosive monogenetic volcanic landforms can be associated with effusive landforms, it is possible to integrate them into a single system (e.g. Murcia and Németh 2020).

The genetic categories of explosive cones, their subdivision, the evolution and complexity of the individual forms remains a highly debated issue in volcanology. It has been a major topic since the 1990s, fostered by the analysis and interpretation of high-resolution DEMs as well as geophysical surveys (see Chap. 11). Thanks to comparative analysis and physical–mathematical modelling, the study of explosive cones has experienced a renaissance, with dozens of publications in international journals on the subject each year. (For one of the comprehensive works on the topic, see Kereszturi and Németh 2013.)

Categorisation of explosive cones can be based on their shape, composition and the primary processes that built them up. In general, they are created by low-magnitude explosive eruptions (mainly Strombolian, and its submarine counterpart, Surtseyan; or other relatively low-intensity phreatomagmatic activity), producing mixed, i.e. both juvenile and non-juvenile, fragments, whose size ranges from fine ash to bombs. In this book, explosive cones are divided into pyroclastic cones, including *scoria cones* and *pumice cones,* i.e. cones resulting from magmatic explosions; and hydroclastic cones, which encompass *maars, tuff rings* and *tuff cones,* i.e. cones resulting from phreatomagmatic eruptions. Some of these cones may be genetically related to each other, with one form evolving into another. In the following sections of this chapter we firstly describe the different

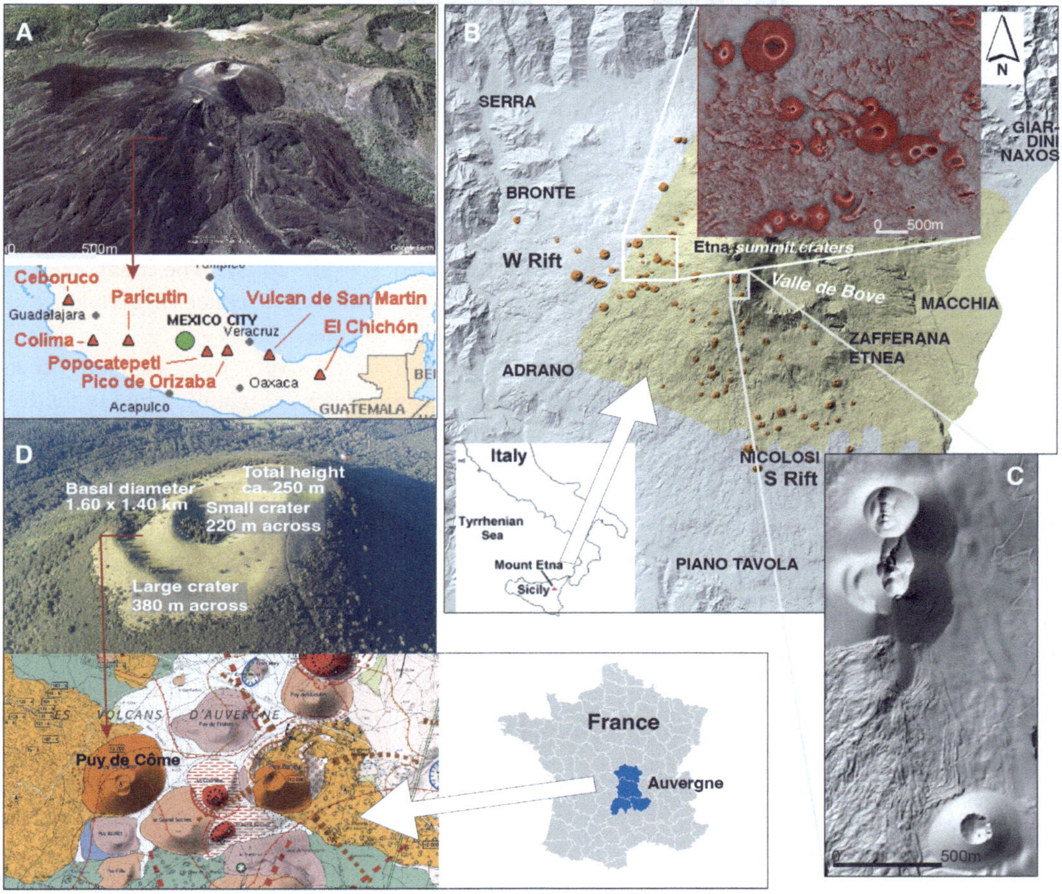

Fig. 2 Scoria cones represent the most common volcano type on Earth. **A** Oblique satellite view of Paricutín scoria cone and lava flow (source: Google Earth). **B** and **C** Scoria cones on west and south flank of Mt. Etna, respectively (Favalli et al., unpublished material). **D** Aerial photo of the nested craters of Puy de Come scoria cone, with size parameters, and location on the volcanological map of the Chaîne des Puys (courtesy P. Boivin)

elementary types—their occurrence, morphology, material, and formation—followed by the factors and processes of formation, development and transformation. In the last section we consider other monogenetic or small compound volcanoes that result from interaction between magma and ice: *glaciovolcanic* landforms.

Taking into account the degree of consolidation of the material, some authors classify *ash cones* separately from *tuff cones* (e.g. Lockwood and Hazlett 2010): the former are cones of looser material, the latter of material that has hardened into tuff. However, this distinction is not necessarily justified, as the resulting shape does not change simply because of the consolidation of the material. Another cone type, sometimes referred to in the literature as a pyroclastic cone, but also discussed in the context of Hawaiian eruptions and small-volume lava flows, is the *spatter cone* (see

Chap. 2). Although this landform is generally an order of magnitude smaller than monogenetic cones (<200 m in diameter and 10–20 m in height: Vespermann and Schmincke 2000), it can be rightly included among them (e.g. Wood 1979; Kereszturi and Németh 2013), because its changing explosive activity produces mixed Hawaiian and Strombolian eruption phases, i.e. spattering and scoria cone-building activity sequentially construct a cone shape. However, we cannot define a spatter cone as a volcano based on the definition given in the Introduction of this book. Although a spatter cone may be formed from a Hawaiian lava fountain during the activity of a true vent, it is often formed from a rootless or pseudo-vent by lava spattering. Thus, the small size, along with the genesis, are both criteria for identifying spatter cones as lava-flow-associated minor landforms, hence their inclusion in Chap. 2 (i.e. elementary landforms made up of lava).

2 Pyroclastic Cones

2.1 Scoria Cones

The most common and regular type of volcanic landform on Earth is the scoria (or in US English, cinder) cone (Fig. 3, Photo 24). It is formed from low-viscosity magma (typically basalt and basaltic andesite) that generally erupts with low explosive activity, either Hawaiian or Strombolian, or, more rarely, with high explosive activity, either Vulcanian to violent Strombolian, or even Subplinian or phreatomagmatic (Valentine et al. 2005; Pioli et al. 2008; Jordan et al. 2016). However, the style can be variable even within a single eruption. The predominant constituent of scoria cones, as its name implies, is scoria, but they also contain fine or coarse ash, lapilli, and/or a mixture of all these, often with bombs, which may be due to a fluctuation in explosive energy (McGetchin et al. 1974; Valentine et al. 2005; Mannen and Ito 2007). For example, unusually violent Strombolian eruptions (Pioli et al. 2008), caused by external water, increase the ash content in the cone layers. In addition, these eruptions can be related either to a single vent creating a simple cone landform, or to a few focal points along a fissure resulting in an elongated or twin-cratered cone (Riedel et al. 2003). The final primary shape may also lack a crater in certain cases (e.g. collapsed cones).

Apart from relatively rare, isolated occurrences, scoria cones are mainly found in the following volcanic settings: (1) widely distributed in low relief extensional intraplate or subduction-related volcanic fields (e.g. Springerville, Arizona, and Cima volcanic field, California, both in USA; Michoacán-Guanajuato, Mexico; Southeast Australia; Chaîne des Puys, France; Middle Atlas, Morocco; Cappadocia, Turkey; Bakony-Balaton Uplands, Hungary; Wuhlandada field, NE China); (2) dotting the slopes of large shield volcanoes (e.g. Mauna Kea, Hawaii, USA; Mt. Cameroon, Cameroon); (3) on mainly mafic composite or other polygenetic volcanoes (e.g. Etna, Italy; Fuji, Japan), and (4) clustered in caldera volcanoes (e.g. Aso, Japan: Photo 25). Scoria cones may also occur in shallow seas in rare cases, produced by hydrovolcanic eruptions. They are also frequent landforms of the rocky planets of Mars and Venus, and given that good resolution DEMs are available for Mars, their morphometry there has been extensively studied (Brož et al. 2015).

Scoria cones on shield or other large polygenetic volcanoes are usually associated with rift zones (e.g. Mauna Kea, Etna; see Chap. 4). In addition to rifts, it is also common that scoria cone distribution is tectonically influenced, i.e. located along fissures (e.g. twin cones, longitudinally elongated cones: see Fig. 2). Topography exerts the greatest control on cone distribution for volcanoes with slopes of greater than 10° (Tibaldi 1995; Corazzato and Tibaldi 2006), while the scoria cone location on a more gentle slope as well as the direction of crater breaching (by lava flow or subsequent erosion) are determined by faults or fractures.

2.1.1 Structure, Formation and Growth

The internal structure and the regular shape of scoria cones are the result of predominantly Strombolian-type pyroclasts, in particular scoriae. The scoria fragments and bombs are ejected on a ballistic trajectory of up to a few hundred metres, before piling up on the ground. Once the pile reaches a certain height (Fig. 4), the scoriae begin to move down slope, building up a cone around the vent, which is emplaced on steep slopes at the angle of friction for scoriae (of about 30–33°) (Vespermann and Schmincke 2000). This results in smooth, uninterrupted slopes for the cone, whose summit is truncated by a crater. In some cases the material composing scoria cones may contain a higher percentage of ash fraction than previously thought, in the event of unusually violent Strombolian eruptions. Therefore, Riedel et al. (2003) also incorporated the laterally propagating, swirling ash fallout from such eruptions in their model of scoria cones.

In addition to the eruptive processes, the shape of the cone is affected by associated syn-eruptive and secondary processes, such as sintering/welding of material near the vent, long-term mass movement of loose scoriae down the slope, sliding of scoriae, and the occurrence of occasional lava flows. As a result, the structure of the scoria cone, as viewed in cross-section, can be divided into several characteristic facies (Fig. 3).

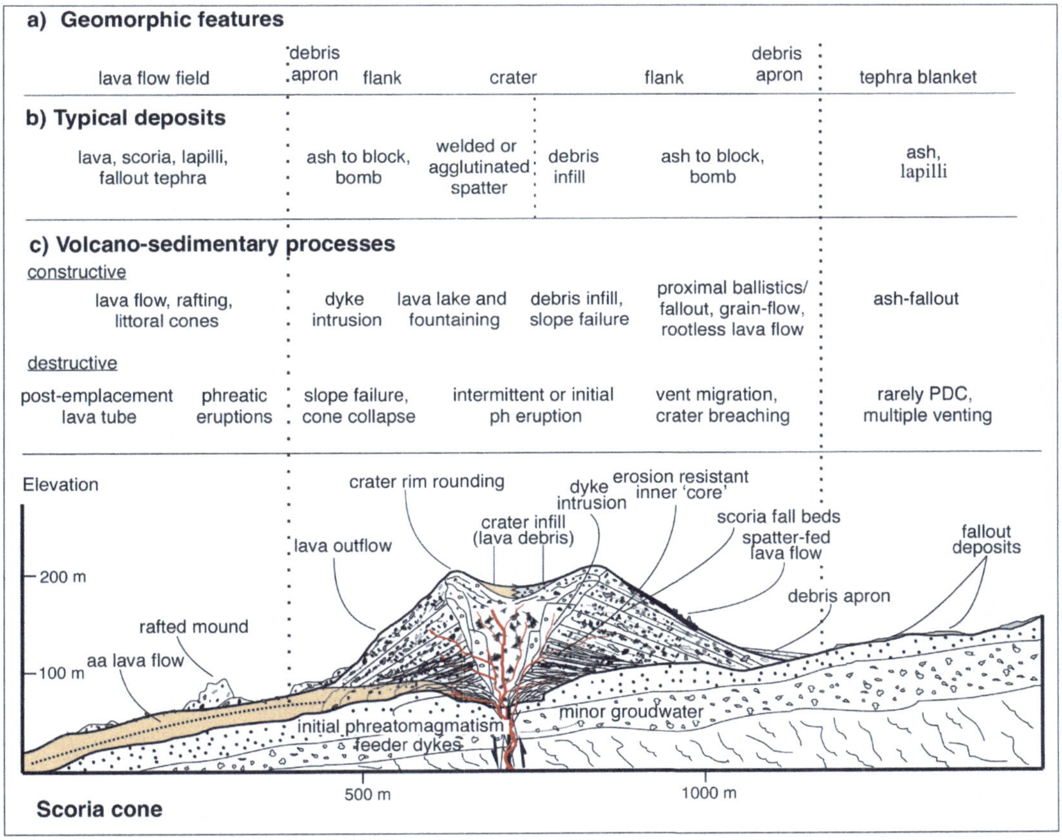

Fig. 3 Anatomy of a scoria cone: structure, geomorphological features, deposits, and volcano-sedimentary processes (modified after Kereszturi and Németh 2013)

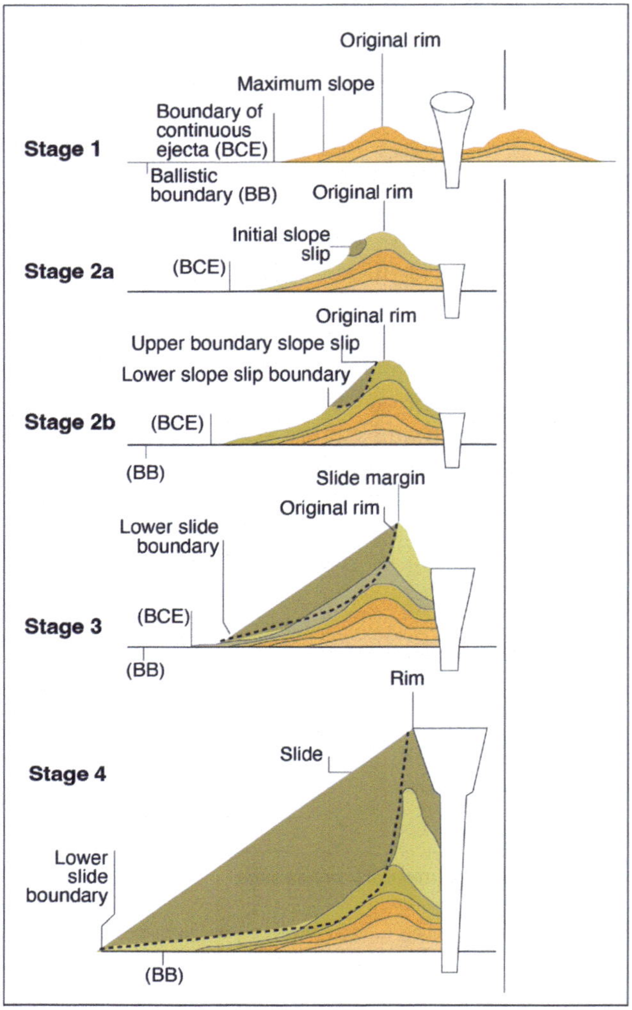

Fig. 4 Four growth stages of a scoria cone (Vespermann and Schmincke 2000; Schmincke 2004, Fig. 9.28, p. 141)

The *vent* or *crater facies* is the inner part of the cone that is layered during the initial activity and is formed by lava-like aggregated scoria (agglutinate) from Hawaiian lava flows or Hawaiian and Strombolian eruptions. This facies is the most resistant to erosion. Upwards or outwards within the cone, aggregation is less effective, and the clasts (flattened bombs several tens of centimetres or often several metres in diameter, and lapilli) become more visible. The *crater wall facies* is made up of bombs (often streamlined spindle bombs and breadcrust bombs), lapilli, and ash. The core facies at the interior of the vent and crater wall remains black or reddish black owing to quenching, while the slope facies is usually red due to slow and limited contact with air leading to oxidation of the iron content. The slopes are formed by a repetition of scoria layers. Due to the variable intensity of these eruptions and notably the pulsating nature of the predominant Strombolian eruptions, these facies are well stratified. The layers often exhibit grading, partly due to the variability of the eruption (e.g. increasing energy–reverse grading) and partly due to the grain flow on a steep slope. The apron of the cone is represented by the *talus* (debris-slope) *facies*, which consists of layers of scoriae that have been intercalated by slope processes.

Scoria cone development is frequently accompanied by lava flows originating from the crater that cut across its rim, or originating from a deeper part of the vent at or above the contact of the cone with the bedrock. The volume of the lava material often exceeds that of the scoria cone. The lava outflow often tears off and transports a section of the scoria cone (Fig. 3; e.g. La Vache and Lassolas cones, and Puy Pariou, Chaîne des Puys, France, Photos 26 and 27).

One of the best examples of scoria cone growth and associated lava flows is provided by the eruptive activity of Paricutín in Mexico (Fig. 2). It formed along a fissure in a farmer's cornfield in 1943 and grew to a 336 m-high cone in one year by violent explosions that produced scoriae and ash fall (Segerstrom 1960, 1966; Luhr et al. 1993; Becerril et al. 2021 and references therein). It then continued to erupt, at a lower intensity, until 1952, reaching a height of 460 m after 9 years. Such a lengthy lifetime is exceptional: 50% of the scoria cones studied by Wood (1980) were formed within 30 days and 95% within 1 year. Another example of recent scoria cone growth is the compound Cerro Negro scoria cone in Nicaragua (McKnight and Williams 1997). Since its formation in 1850, it has undergone more than 20 Strombolian to subplinian eruptions that formed a 250 m-high cone and associated lava field (for more details, see next section and Fig. 7).

The growth rate of Paricutín's scoria cone was very fast at the beginning: after one day it was a 35 m-high mound, and after one week it reached a height of 167 m due to the production of several million m^3 of scoria per day (Larrea et al. 2017). After a few weeks the activity gradually slowed down, showing that cone growth rate is not uniform. Many other scoria cones globally are known to have built up quickly (e.g. most recently in 2021, Tajogaite at Cumbre Vieja, La Palma, Canary Islands, see Sect. 3.2.2 and Chap. 10), reaching a height of 100–200 m in days. An increase in crater diameter, in particular in the case of rapidly-formed cones, may be due not simply to its widening by the eruption(s), but also to associated small or large scoria slides triggered by associated tremors.

Craters and/or vents on a given cone can vary not only in size but also in number, e.g. at Lemptégy volcano, Chaîne des Puys (France), two monogenetic scoria cones of different compositions were formed by eruptions from a pair of closely spaced vents that were active at different times (de Goër de Hervé 1999; Delcamp et al. 2014). At Paricutín two initial vents merged into one, and this eruption almost moved beyond the scoria cone stage towards the end of its growth, not only because of its size but also because of frequent lava flows and evolving magma composition (Fig. 2a).

2.1.2 Morphometry

Average scoria cone height is around 100–150 m, but they may reach 300 or exceptionally 500 m. Their steep outer slopes are around 30°. They have a relatively large crater, usually with a diameter equal to or greater than their height, some tens of metres deep, although craterless scoria mounds do exist (e.g. Victoria, Australia: Ollier 1988). Scoria cones have a cone diameter of 0.5–1 km, but some are as wide as 2 km. Their volume is usually less than 0.1 km^3, but can reach up to 0.2–0.3 km^3 (see morphometric data for Paricutín and Cerro Negro in Fig. 7). Fornaciai et al. (2012) investigated the morphometry of more than 500 predominantly North and Central American scoria cones, mostly based on DEMs with a 10 m spatial resolution. The cones had an average height of 106 m (with a maximum of 300 m), a diameter of 813 m (with a maximum of 1995 m), and a volume of 0.03 km^3 (with a maximum of 0.28 km^3).

The increase in size of scoria cones is regular, following a morphometric pathway, during which it preserves its shape. After initial qualitative approaches (e.g. Colton 1937), the quantitative morphometric studies of scoria cones—particularly the pioneer works of S.C. Porter, M. Settle and C.A. Wood in the 1970–1980s—are considered as representing major milestones in volcano geomorphology. The quantitative approach of these authors characterising the evolution of scoria cones is still widely used (for a historical account, see Vörös et al. 2022). Porter (1972), Settle (1979) and Wood

(1980) determined the typical dimensions of fresh, regular-shaped scoria cones worldwide on the basis of topographic maps to derive well-defined ratios: cone height/cone diameter (H/W_{co}) and crater width/cone diameter (W_{cr}/W_{co}) (Fig. 5). Examining these morphometric ratios, Wood (1980), using data from 910 scoria cones on Earth, obtained a height versus cone diameter (H/W_{co}) ratio of about 1/5 (0.18) and a crater diameter versus cone diameter ratio of 2/5 (0.40) for active cones. These basic morphometric parameters increase proportionally, thus for a given scoria cone growth, the three main parameters generally increase continuously according to the given ratios.

Wood (1980) also found the volume to be an important parameter: his studies showed that cones with larger volumes tend to be more intact and more regular than their smaller counterparts. It was suggested that a smaller volume (= less magma erupted) equates to less pre-eruption vesiculation, which means that less energy is required to erupt. This also yields a larger grain size and may result in more frequent clastogenetic lava flows from the coalescence of scoriaceous and lava clasts, and volcanic debris falling to the surface, which may make the slope more resistant to erosion. The output of lava flows on one flank may result in irregular cone shapes. In contrast, small-volume cones are formed in a shorter period of time, which means that the resulting irregularities and deformations to the shape (e.g. due to swift changes in wind direction) are less likely to be compensated for, unlike those caused by multiple or prolonged eruptions.

Following other topographic map- or stereogrammetry-based morphometric studies (e.g. Hasenaka and Carmichael 1985; Moriya 1986; Tibaldi 1995), a new research phase began in the 2000s, based on high-spatial resolution digital elevation model (DEM) data, which has refined or modified the earlier findings. Favalli et al. (2009) showed that a different way of calculating the height value is required for scoria cones rising on the gentle slopes of flanks of composite and shield volcanoes (Fig. 5). In fact, the calculation used by Settle, Porter, Wood and others for a scoria cone located on a horizontal substrate (e.g. flat volcanic field) yields a higher value than reality, as compared to a scoria cone located on a slope (typically, on the flanks of a large volcano). In addition, in the latter case, there is a natural tendency for the lava to flow around the cone, which partially buries the cone and thus reduces its height. As a result of these considerations, Favalli et al. (2009) obtained an H/W ratio of less than 0.18 for young scoria cones on Mt. Etna ($H/W = 0.137$) (Fig. 6a), leading to the suggestion that the real morphometric ratios for terrestrial scoria cones are somewhat lower than the previously accepted values.

In terms of morphometric evolution, Paricutín showed a generally steady increase in size for the first five years of activity (1943–1947) in terms of cone diameter and height, but this regular growth was modified towards the end of activity by two factors. Firstly, the volcano sank under its own weight and its foot became buried by lava flows, similar to many Etnean cones, and thus its relative height and diameter were reduced (Fig. 6b). Secondly, the crater diameter also decreased towards the end, caused by the emergence of a parasitic cone within the crater. Lava flows were present from the beginning of eruptive activity at Paricutin, and they remained intermittent throughout (the lava flow that breached the crater rim caused a sudden increase in its diameter). Eruption of more voluminous lava flows resulted from a change in magma composition from basaltic to basaltic andesite (Erlund et al. 2010). The lava flows after the first year were much more voluminous than the pyroclastic-fall material (see Fig. 2), flooding a total area of about 10 km^2 and destroying two villages.

Some scoria cones can evolve even further and become relatively large ($\geq$ 500 m high). However, evolution is not only about size: a scoria cone may evolve into a composite cone based on eruptive characteristics. Lines of evidence for this scenario are provided by Cerro Negro (Nicaragua; Fig. 7). It is a relatively small (240 m-high) but highly productive scoria cone, one of four, belonging to the El Hoyo (or Las Pilas)

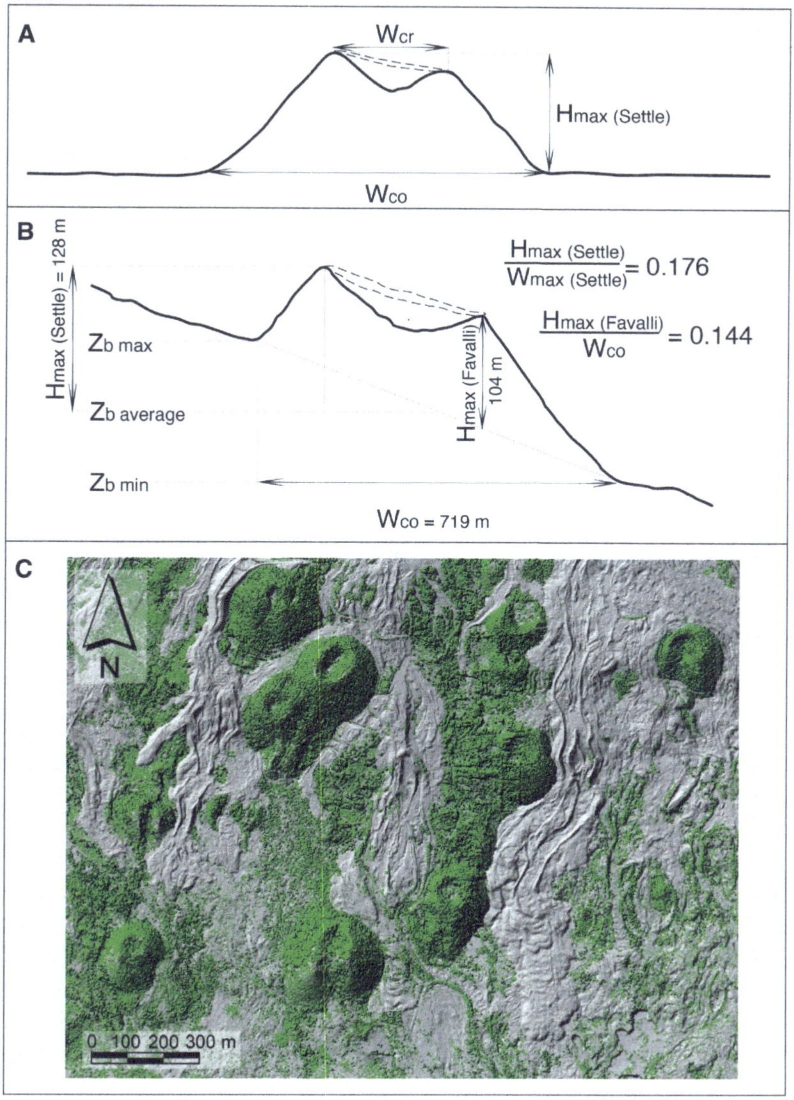

Fig. 5 Morphometric parameters of scoria cones. **A** Calculation of the height of scoria cones on a horizontal plane (Settle 1979) or a dipping plane (e.g. on a large composite volcano; Favalli et al. 2009). On a large volcano flank, the lava flows surround cones, thus typically reducing their original size. Examples from the southern slope of Etna (Favalli et al. 2009). The classical method of Settle defines Hmax as the maximum elevation above the average basal elevation of the cone. In the improved method (Favalli et al. 2009), Hmax is the maximum elevation above the fitted basal plane. The two methods give significant differences in the case of a dipping plane. **B** On a dipping plane, e.g. that of a large volcano, Hmax may be further reduced by lava overflows: examples from the southern flanks of Mount Etna

volcanic complex or massif (see Chap. 6), and is fed by its most recent conduit (Venugopal et al. 2016). Since 1850, when eruptions started, it has been characterized by a 1.5 km^3/kyr magma production rate on average. Its morphometric evolution (Fig. 7a, b; e.g. Thouret 1999) shows similarities and differences with isolated composite volcanoes in Central America (see Fig. 7d). Morphometrically, regular-shaped composite cones can be distinguished from

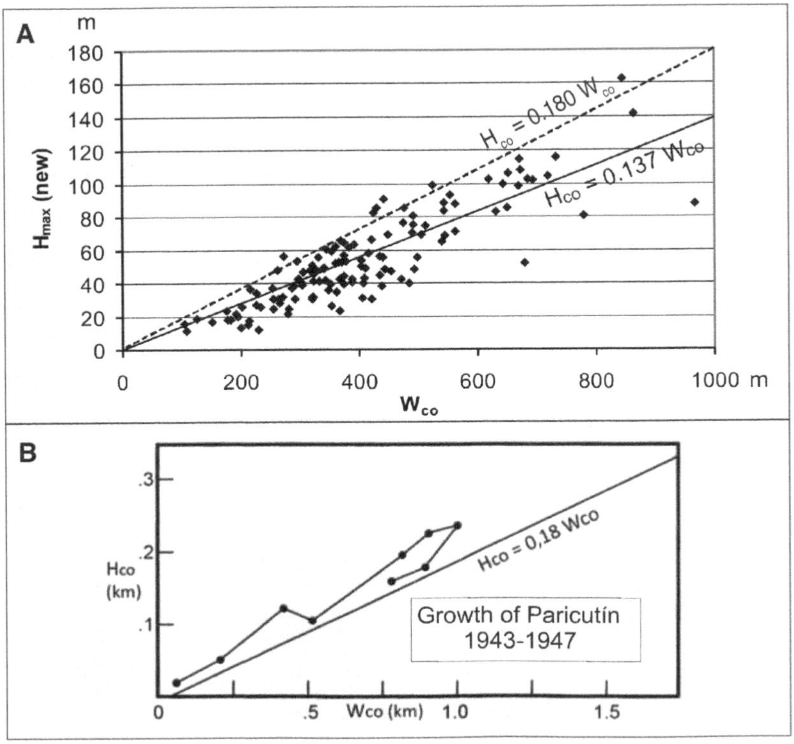

Fig. 6 Scoria cone morphometry. **A** H/Wco (cone height vs. cone width) plot for Etna's scoria cones using LIDAR DEM data (Favalli et al. 2009). Hmax is defined as the improved definition in Fig. 5. **B** Evolution of Paricutín scoria cone in the light of H/Wco

scoria cones by their smaller W_{cr}/W_{co} ratio: 0.029 (for details, see Chap. 5) compared to the 0.40 value of the scoria cones, respectively, based on a large number of examples (Wood 1980; McKnight and Williams 1997). Cerro Negro has a transitional shape between a scoria and composite cone, mainly because its crater sometimes widens due to very violent eruptions. Overall, during its short life time, Cerro Negro seems to have followed the morphometric trend of Arenal composite volcano (Costa Rica), which reflects an even higher variability in eruption style. While Paricutín also had energetic eruptions in comparison to a typical scoria cone, those of Cerro Negro have been even more energetic, fed by more evolved, water-rich magmas (Portnyagin et al. 2012) that produced greater column heights than those of Parícutin.

Such anomalous scoria cones might also be exemplified by the aforementioned Lemptégy scoria cone, Chaîne des Puys, France, not only in terms of eruptive style, but also tectonic evolution (Petronis et al. 2013). Sunset Crater, San Francisco Volcanic Field, is also a well-known example of variable eruptive styles, including highly energetic subplinian eruptions during its short-lived activity in ~1085 AD/CE (Alfano et al. 2019: Photo 28). Another example is Tavurvur scoria cone, New Britain, which forms a twin volcano with its slightly more violent counterpart, Vulcan tuff cone (see Sect. 3.2.2) within Rabaul caldera (Bernard and Bouvet de Maisonneuve 2020).

In conclusion, variability of the morphometric ratios and the less regular shape of scoria cones may be due more to the complex nature of their evolution than previously thought (Valentine et al. 2005; Kereszturi et al. 2012). A great diversity of eruptive processes characterised the complex Tajogaite scoria cone on the

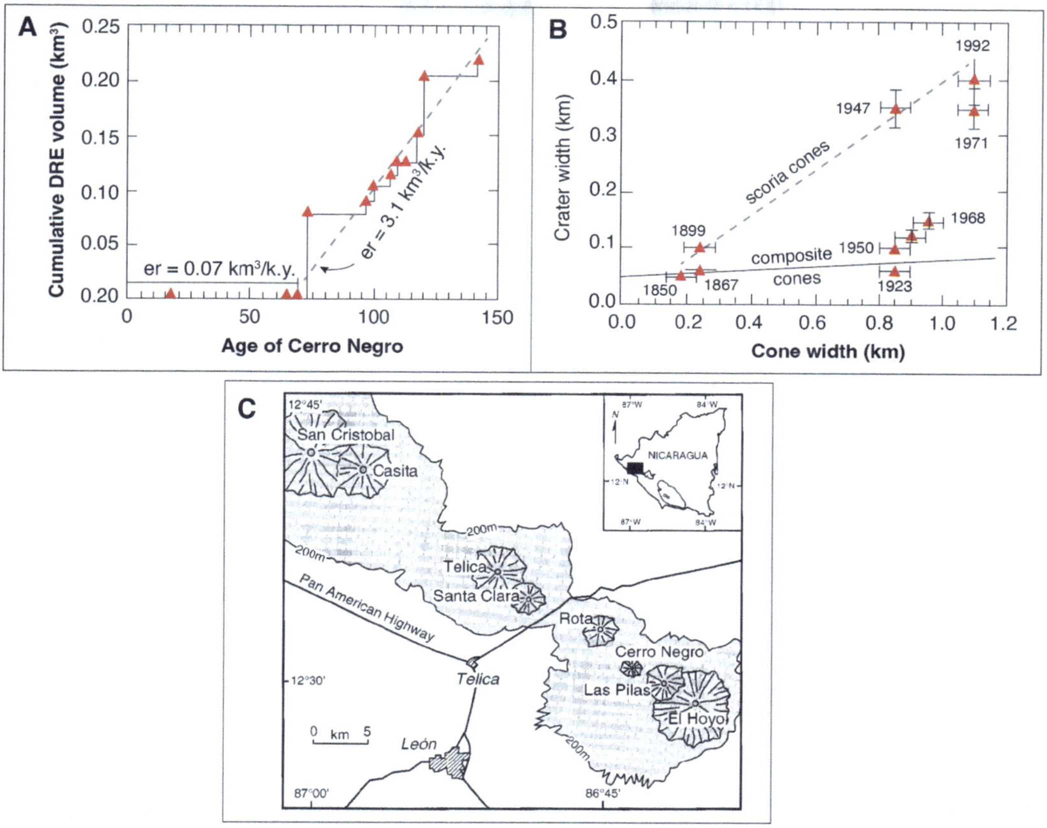

D

Volcano	Eruptivity (km³/ka)	Crater width/ cone width ratio	Eruption style	Total (vol. km³) DRE	Pyrocl. vol./ total vol. (%)	Cone height (km)	Eruption rate (kg/s)	Column height (km)
Paricutin	0.12	0.31	St	2.1	67	0.42	10^1 - 10^4	0.1 - 0.6
Jorullo	0.12	0.25	St, V	~2.0	~60	0.35	n.d.	n.d.
Cerro Negro	1.6	0.08 - 0.36	St, V, Sp	0.22	77	0.25	10^3 - 10^5	0.1 - 8.0
Izalco	8.8	0.07	St, V	2.1	~18	0.65	10^{3} [a]	≤ Plinian
Arenal	1.3	0.06	St, V, Sp, P	5.0	~20	1.10	10^4 - 10^5 [b]	≤ Plinian
Strombolian	n.d.	n.d.	St	–	1-20	–	10^1 - 10^6	0.1 - 5

Fig. 7 Characteristics of the Cerro Negro (Nicaragua) scoria cone, including volume, compared with that of composite cones (after McKnight and Williams 1997, *in:* Thouret 1999 with permission Earth-Sci. Rev., Fig. 2, p. 98, and Table 1, p. 99, with permission from Geological Society of America). **A** Diagram showing the cumulative DRE volume age Cerro Negro. **B** Diagram of crater width versus cone width for cinder cones and composite cones. **C** Map showing location of Cerro Negro in western Nicaragua. **D** Table displaying several geometric and eruptive characteristics for distinguishing scoria (cinder) cones from composite cones. Data from several authors, all referenced in McKnight and Williams 1997)

Cumbre Vieja ridge (La Palma, Canary Islands), formed during the activity between September December 2021 (Carracedo et al. 2021; Romero et al. 2022; Bonadonna et al. 2022; Chap. 10). Its final structure, reaching 187 m above the previous ground level, was associated with six additional vents, a number of hornitos, and a 12 km^2 lava-flow field (Civico et al. 2022).

2.2 Pumice Cones

Pumice cones are a rare, albeit interesting, type of pyroclastic cone. Pumice is typically deposited during high-energy explosive eruptions of dacitic-rhyolitic magma over long distances, and is therefore not a cone-forming product. However, rare cases to the contrary result in the formation of small cones. Although early works attempted to distinguish them (e.g. Cotton 1944, 1952), they are commonly presented in conjunction with tuff cones, which demonstrate the influence of external water (see Sect. 3.3). Examples of pumice cones in a strict sense include the rhyolitic cone within Newberry caldera volcano (Fig. 8: Oregon, USA), the Nabeyama tuff cone (Sakurajima volcano, Japan: Photo 29), the pumice hills of Monte Pilato (Island of Lipari, Italy), and the Puu Wa'awa'a trachytic pumice cone of the Hualalai shield volcano (Hawaii, USA).

The origin of pumice cones is puzzling. They may be formed by a special interaction of gas- and SiO_2-rich magma and water either during subaqueous eruptions or when lava enters a water body. In this case, even if the contact between magma and water leads to fragmentation, it does not necessarily cause phreatomagmatism and the formation of tuff cones and tuff rings (see Sects. 3.2 and 3.3). Instead, the presence of water prevents major dispersion of pyroclasts and eventually a pumice cone forms (Moriya 1994). In another scenario, the arrival of a new rhyolitic magma batch into an existing one may result in contrasting fragmentation producing a wide range of pyroclastic textures, and magmatic explosions pile up pumice and obsidian fragments to form a pumice cone (Davi et al. 2011).

Magma-water interaction may account for pumice cone formation at Newberry caldera (MacLeod et al. 1995; Jensen and Donnelly Nolan 2017). There, six major eruptions from the early Holocene to about 1300 years ago have included basaltic lava flows, rhyolitic explosive events, and the extrusion of obsidian flows within the caldera. Around 7300 years ago, eruptions from a central fissure created a ridge in the middle of caldera, and divided it into two parts: present-day Paulina Lake and East Lake (Fig. 8). Between these lakes, Interlake Flow emerged along the ridge, and Central Pumice Cone was formed by small-scale explosions when hot viscous rhyolitic lava erupted up through lake water and shattered. The pumice cone rises ~230 m above East Lake.

Nabeyama tuff cone, with a wide crater, is associated with Sakurajima volcano on Kyushu Island (Japan). It was formed during the Tenpyo-Hoji phreatomagmatic eruption in AD/CE 764 (Kobayashi et al. 2013; Nishihara et al. 2022). Ejecta of the cone, which emerged on the beach, are composed of pumice fragments and ash, and are also rich in obsidian flakes as a result of rapid cooling, and the eruption mechanism is thought to be similar to that of Newberry cone (Moriya 1994). Half of the Nabeyama cone has already been removed by wave action, and its foot was partly buried by subsequent lava flows (Kobayashi et al. 2013).

Unusual rhyolitic magmatic explosions have been identified at Monte Pilato in the case of its AD/CE 776 eruption (Davi et al. 2011). Juvenile fragments show a repetition of strong textural heterogeneities from highly vesicular pumice to dense black obsidian. Moreover, agglomerated and welded pumice is thought to represent a portion of magma heated by a batch of hotter melt that entered the magma chamber and triggered the eruption. Spatter accretion may have been due to the rhyolitic pumice fragments emplaced from pyroclastic fountaining. This kind of explosive magmatic eruption was followed ~400 years later by the phreatomagmatic then effusive eruption of the Roche Rosse obsidian lava flow from the same rhyolite (Dellino and La Volpe 1995).

The morphometry of pumice cones is similar to that of scoria cones, except that their crater is generally larger: 620 m for the Central Pumice Cone at Newberry, ~600 m for the breached half-crater at Nabeyama, and 1000 m for the breached half-crater at Monte Pilato.

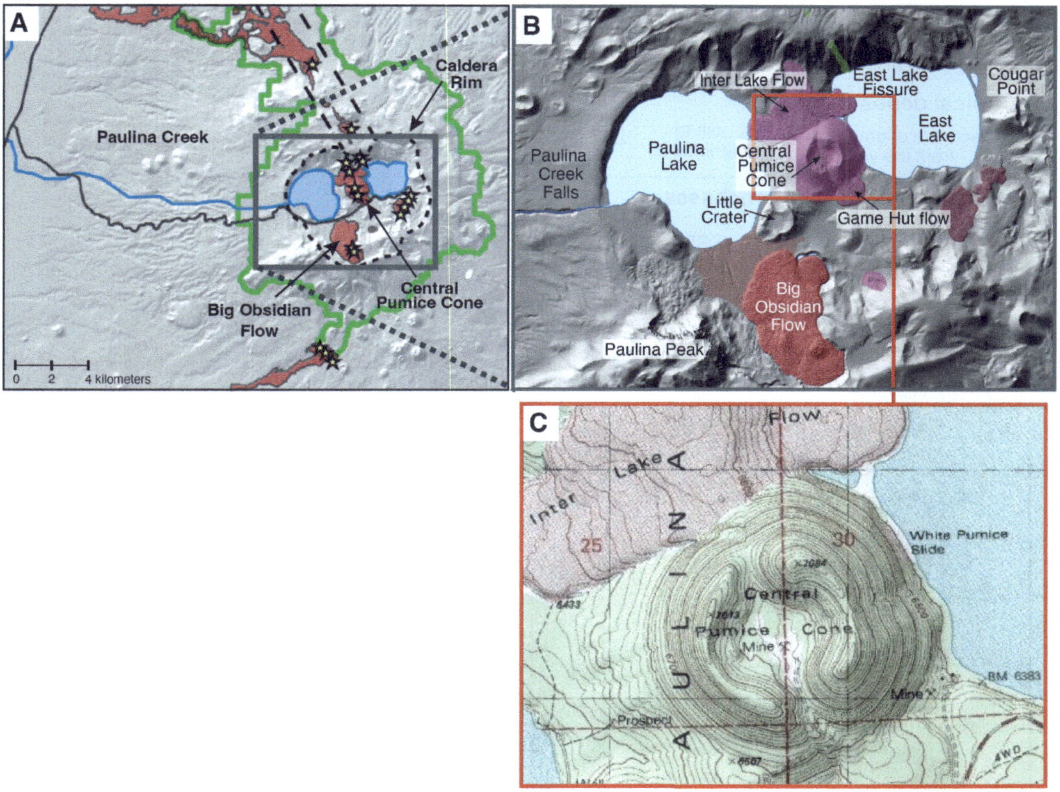

Fig. 8 Pumice cone of Newberry volcano. **A** Overview: shaded relief image based on 10 m US DEM. **B** Six major Holocene eruptions within the Newberry caldera produced basaltic lava flows, rhyolitic explosions, and the extrusion of obsidian flows. Around 7300 years ago, fissure eruptions divided the caldera into Paulina Lake and East Lake. Interlake Flow was then emplaced between the lakes, and Central Pumice Cone (230 m tall) was formed by small explosions when viscous rhyolitic lava erupted up through lake water and shattered. (Map source: Jensen and Donnelly Nolan 2017). **C** USGS topographic map of Central Pumice Cone (scale 1:25,000)

3 Hydroclastic Cones

3.1 Maars

Maars (Fig. 9) represent the second most common terrestrial volcanic landform after scoria cones (Sheridan and Wohletz 1983). Together with tuff rings and tuff cones, they are formed by phreatomagmatic eruptions (Ollier 1967; Lorenz 1973, 2003; White and Ross 2011; Németh et al. 2011; de Silva and Lindsay 2015; Valentine et al. 2017). A maar is a broad, shallow, subcircular, bowl-shaped volcanic crater whose sides cut into the country rocks. Its name originates from the German term *maar* (meaning pond or tarn), because the closed crater is usually filled by a lake. The crater floor is typically located below the surrounding surface. This feature, among others, distinguishes maars from the similar *tuff rings* (see next section).

Maar craters are commonly surrounded by a low rim of tephra-ring deposits, composed of a mixture of ejecta from subsurface explosions resulting from the interaction of magma, gas and groundwater (Zimanowski et al. 1997; Lorenz 2003; see below). Maars usually have simple eruptive histories (Palladino et al. 2015), typically comprising a series of short-lived (hours to

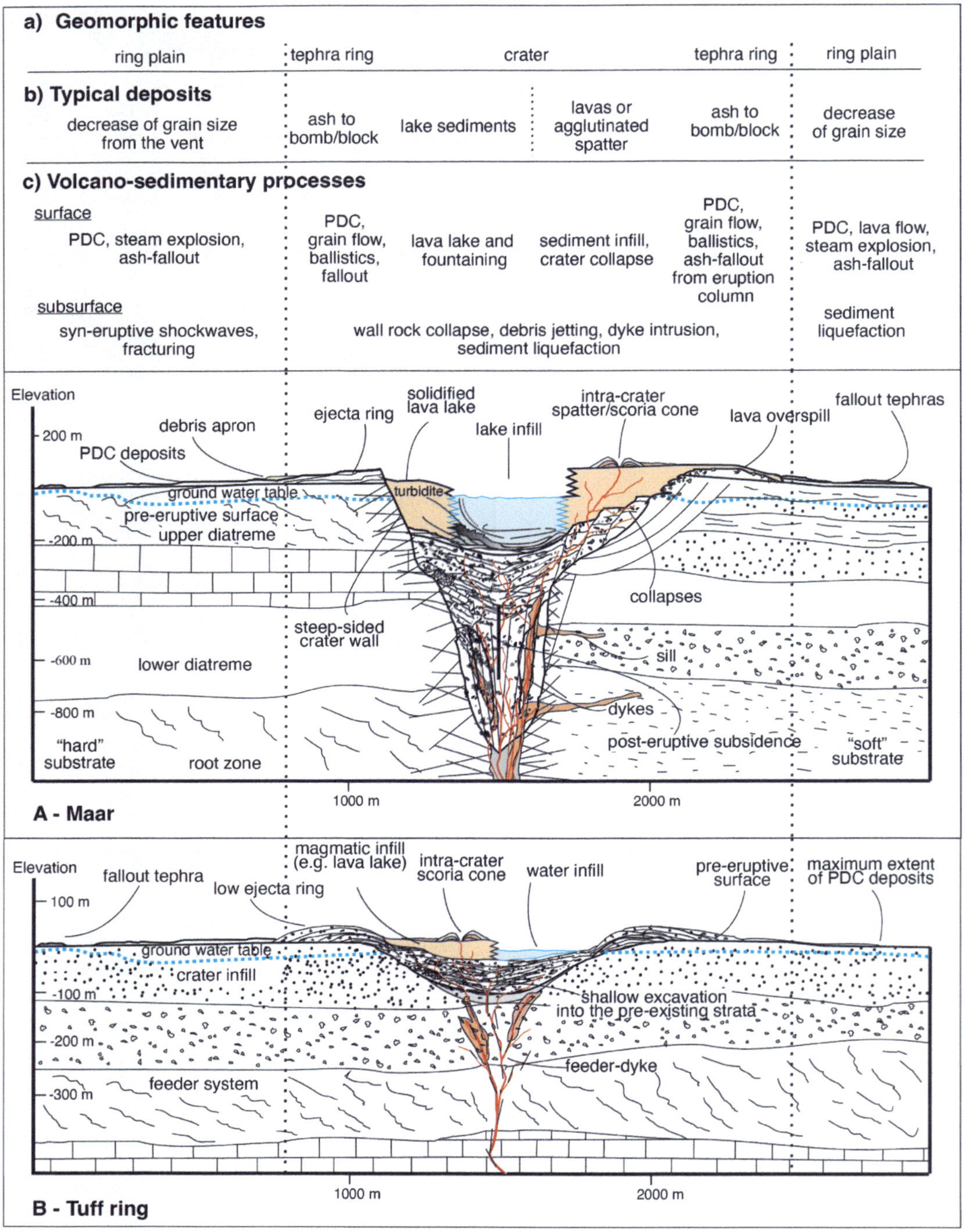

Fig. 9 Principal types of hydroclastic volcanoes. **A** Maar: geomorphological features, deposits and volcano-sedimentary processes (modified after Kereszturi and Németh 2013). **B** Tuff ring with characteristics shown in Table A (modified after Kereszturi and Németh 2013). **C** Tuff cone: geomorphological features, deposits and volcano-sedimentary processes (modified after Kereszturi and Németh 2013)

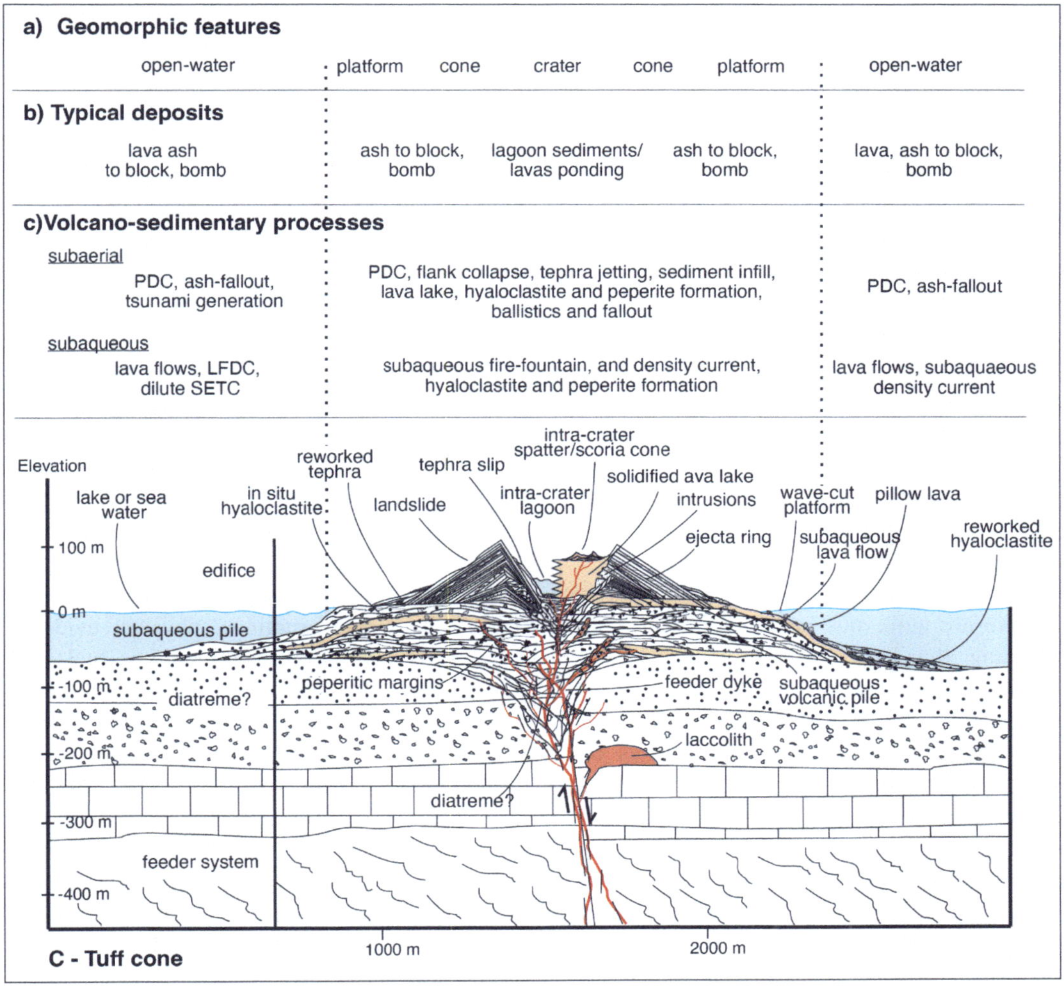

Fig. 9 (continued)

a few days) phreatomagmatic explosive events accompanied by annular collapses. Studies of maar deposits and laboratory experiments have sought to understand how powerful magma-water interactions create wide and relatively deep craters (Wohletz 1986; Németh et al. 2008; White and Ross 2011; Graettinger and Valentine 2017). These interactions however are temporally limited the hot magma quickly boils off available water, shutting down the reaction. Therefore, the host maar is often succeeded by scoria cones, for example monogenetic volcanic fields associated with scoria cones, such as Chaîne des Puys, France, e.g. Lake Pavin maar, Photo 30; Michoacán-Guanajuto, Mexico; Springerville, Arizona and Hopi Buttes, New Mexico, both in USA; Bakony–Balaton Uplands, Hungary (Aranda Gómez et al. 1992; Mergoil and Boivin 1993; Martin and Németh 2007; Boivin et al. 2017; Thouret et al. 2021). Lava flows from the central vent can also sometimes be trapped in the maar (e.g. Hopi Buttes or Bakony-Balaton Uplands). The relationship between pyroclastic and hydroclastic cone types is discussed later.

In addition to occurring in monogenetic fields, maars can form anywhere where there is sufficient near-surface water. For a comprehensive database of the 240 Quaternary maars in a wide range of geodynamic settings, see the Maar Volcano Location and Shape (MaarVLS) database (Graettinger 2018). To mention a few examples of landform complexity, maars can be found on the shield volcanoes of the Canary Islands (e.g. Lanzarote, Gran Canaria, Spain), and at the base of the Halla shield volcano on Jeju Island (South Korea), where maars, tuff rings and tuff cones are found in the coastal zone beneath the higher-lying scoria cones. Intraplate basaltic volcanic areas with well-known maars include the Bakony-Balaton Uplands Volcanic Field (Hungary, 8–2 Ma: Wijbrans et al. 2007) and the Newer Volcanic Province (Southeastern Australia, 8–1 Ma: Heath et al. 2020).

One of the most diverse volcanic settings is connected to the Cameroon tectonic line. In addition to the large volcanoes associated with it, there are numerous maars of different sizes, including the largest—the Debundscha maar (Cameroon), located on the coastal area at the western base of Mt. Cameroon. Sudden gas or limnic eruptions from maar crater lakes filled with CO_2 gas are a major hazard in this area: an overflow from the Monoun crater in 1984 and one from the Nyos crater in 1986 caused major catastrophes and casualties (35 and 1700 victims, respectively) (Baxter and Kapila 1989; Kling et al. 2005; Aka and Yokoyama 2013; Kusakabe 2017).

Other recent maar-forming eruptions that have been observed and scientifically described include the Rotohana eruption associated with the 1886 eruption of Tarawera volcano in New Zealand, the Niahue eruption in Chile (1956), and the Ukinrek maar eruption in Alaska (1977).

An unusual environment characterised Alaska's Espenberg volcanic field maars (Devil Mountain Lakes) in the Seward Peninsula (Begét et al. 1996). They were formed during the last Ice Age, when a thick layer of permafrost provided a large reserve of water to fuel the explosion a little at a time. A series of large to very large maars were created between ~100 and 20 ka, the largest of them being 5 km wide. And as in many other settings, volcanism did not end with the maar explosions, but produced lava flows; the youngest lava, the Lost Jim Lava Flow, is only 3000 years old. Ukinrek maar, which is located in a similar setting, was formed as a double maar in the course of a 10-day eruption in 1977 resulting from rising and ponded magma interacting with small amounts of shallow, initially frozen groundwater within glacial till and underlying silicic pyroclastic deposits (Kienle et al. 1980; Ort et al. 2018).

3.1.1 Structure, Formation and Growth

Although some of the early literature invoked the involvement of magmatic gases in maar-forming eruptions (see discussion in Cotton 1952 and later in White and Ross 2011), since the beginning of the twentieth century they have been attributed predominantly to the interaction of external (subsurface) water and erupting magma. This finding has been supported by observations of hydrovolcanic eruptions, physical evidence, as well as laboratory experiments on fuel–coolant interactions, although alternating magmatic, volatile-driven explosions can also occur (e.g. Jaggar 1949; Thorarinsson 1966; Boivin et al. 1982; Sheridan and Wohletz 1983; Wohletz 1986; Zimanowski et al. 1997; Morrissey et al. 2000; Stix and Marten de Moor 2018). Because maars typically form during violent phreatomagmatic eruptions, the most striking features of the sequence composing the crater rim are the proportion of quenched, magmatic ejecta (lapilli, bombs) and the amount of bedrock ejecta, often non-volcanic rocks. This is a consequence of sudden deep-seated, high-intensity eruptions, creating ripped ballistic blocks or bombs of more than 3 m across. The frequent occurrence of accretionary lapilli and vesicular tuff/lapilli tuff in the stratigraphic sequence is evidence of the presence of subsurface water. Even very small amounts of water are sufficient to cause explosions, as proven by the abundance of maars in arid to semi-arid climates. The subordinate igneous component of the maar is mostly basalt, but there are also basaltic andesites, andesites (e.g. Hokkaido, Japan), rhyolites (San Francisco Volcanic Area, Arizona), dacites (Papua New Guinea) and even carbonatites (Tanzania) (Austin-Erickson et al. 2008; Ross et al. 2017).

Phreatomagmatic processes that generate maars and tuff rings typically comprise base surges and fallout. Tuff and lapilli tuff of pyroclastic-fall origin may be intercalated with the surge deposits, and explosion breccias are also common in and around the vent. Characteristic depositional facies may distinguish magmatic from phreatomagmatic eruptions (see the structure of tuff rings, Fig. 9B). Maar eruptions are rarely accompanied by lava flows (e.g. Ubehebe craters, California, USA; Mt. Porndon, Victoria, Newer Volcanic Province, Australia). There are examples of concomittant neighbouring maar and scoria cone activity: e.g. Narse d'Espinasse and Puy de l'Enfer in the Chaîne des Puys, and La Sauvetat, Devès massif, France (Boivin and Gourgaud 1978; Boivin et al. 2017).

The violence of phreatomagmatic explosions and the velocity of the rising magma do not allow the xenoliths that are ripped up to sink back down, so they are carried away by the gases released from the magma when it erupts. A good example of this type of bedrock sampling is the Ukinrek maar in Alaska, which is composed of 75% non-juvenile components, with silicic pyroclastics building up the wider area (Pirrung et al. 2007; Ort et al. 2018). Sudan's Malha maar is even richer in xenoliths, with a thin basalt ring and material mainly composed of granite, gneiss and sandstone debris from the surrounding area.

The explosion can extract country rock from varying depths from a few tens of metres to as much as 500 m below the surface. The well-known Pliocene maar and tuff ring successions of Tihany or Szigliget (Bakony–Balaton Uplands, Hungary) contain xenoliths of Permian red sandstone, Triassic limestone, clay shales, etc., excavated from a depth of several hundred metres (e.g. Martin and Németh 2004). Peridotite and lower-crust granulite inclusions from the mantle, which occur in many maar deposits worldwide, yield very important information on the crustal region of the Earth. The substrate controls the focus of the explosions. In Hungary, there is a classification into soft and hard substrate (i.e. loose sediments vs. carbonate rocks), with the latter causing disruption to deep-seated karstic groundwater (Auer et al. 2007). A mixed hard-soft setting was found for two large coalescent maars at the Pali Aike volcanic field near the Argentina-Chile border (Ross et al. 2011).

The deepest, subvolcanic part of the maar is called a *diatreme* (Lorenz 1970, 1973, 1986). Explosions propagating downward fragment and erupt wall rock at deeper and deeper horizons, enlarging the conduit (i.e. creating a diatreme) to the depth of the aquifer. Based on experimental data, maar-diatreme eruptions are likely sourced by explosions in the uppermost ~200 m, and even shallower ones (<100 m) are likely to dominate deposition onto tephra rings (Valentine et al. 2014). Most explosions below ~200 m will not affect the surface but contribute to the formation of, and to the vertical mixing of materials within, a diatreme. Diatremes are chimney-like vents, widening at the top, and filled by vent breccias, including xenoliths and slip blocks, down to a depth of 2 km. There are well-known diatremes in the Swabian Alb (Germany), but the most famous are the Kimberley (South Africa) ones, which are filled with kimberlite, a diamond-rich, peridotite-bearing rock, and have been excavated to depths of several hundred metres (Lorenz 1975; Kurszlaukis et al. 2009). Crater- and vent-filling deposits showing contrasting degradation, and affected by subsequent tectonic movements, are common in the Bakony–Balaton Uplands, Hungary (Németh et al. 2003).

3.1.2 Morphometry

Maar craters (Table 1) are considered as being subcircular landforms, but the recent global survey of their morphometry, Maar VLS (Graettinger 2018) indicates that they frequently have compound shapes with overlapping circles resulting from a few to tens of discrete explosions. Moreover, some maars display an elongated shape: in extreme cases, such as the Nejapa and Ticomo maars in Nicaragua, they have elongation values of <0.45.

Given the low rim, the height of maar craters relative to the surroundings is commonly <30 m (an exception is the Joya Honda maar, ~100 m high: Vespermann and Schmincke 2000), whereas the crater diameter shows great variability from several tens of m to >1 km, the mean value being 600–800 m (Graettinger 2018). The H/W_{co} ratio varies between 0.01 and 0.013 (Graettinger 2018). Lorenz (1973) considers that the width of a maar is equal to half of the depth of its diatreme. Rarely, maars can be 2.5–3 km or more across; in addition to the exceptional Espenberg volcanic field maars, examples include the maars of Kamchatka (Russia), and Kilbourne Hole, New Mexico (USA). Kilbourne Hole, a 50 m-high maar, has a crater 2500 m wide, and hence a H/W_{cr} ratio of only 0.02 (Wohletz and Sheridan 1983). Given the commonly large crater relative to the basal cone diameter, Wood (1979) determined a W_{cr} / W_{co} ratio (including tuff rings) of around 0.6 as opposed to the smaller (0.4) ratio of scoria cones.

The inner crater wall is always very steep, and may continue to deform through slides and possibly small collapses during and immediately after the explosion, filling the interior with debris or post-eruptive sediment. Thus, crater depth of maars (commonly <100 m, less commonly up to 300 m: White and Ross 2011) is difficult to constrain due to maar lakes and/or sedimentary infill; crater depth/crater diameter ratios that have been obtained are about 0.1 (White and Ross 2011; Graettinger 2018).

Whereas the inner rim of maars is sharp (with dips up to 30°), outer tephra deposits extend away from the rim at low slope angles (5–10°). Slopes angles depend on the volume ejected, the pre-eruptive topography, and the moisture content of the tephra (Lorenz 2003).

3.2 Tuff Rings

Eruptions with a higher contribution from a magmatic explosion create more substantial landforms, tuff rings and tuff cones, which are higher and have steeper flank slopes.

From a geomorphological point of view, tuff rings are not very different to maars. Tuff rings are low ring-shaped forms with a large crater, similar in occurrence to subaerial maars. However, the crater floor is at the same level as, or higher than, its surroundings (e.g. Xico volcano near Mexico City, Photo 31; Hverfjall tuff ring near Myvatn Lake, Iceland, Photo 32). Another essential geomorphic difference is the elevated crater wall due to the increased tuff and lapilli tuff production that is deposited from phreatomagmatic eruptions. These minor differences control the morphometry of tuff rings, which in turn can overlap with that of maars. The height difference with the surrounding ground can be up to 50–100 m (for example the 250 m tuff ring sequence at Capelinhos, see below).

The accumulation of loose pyroclastics often occurs asymmetrically. The final emplacement of the loose tuff material makes the inner crater wall more gently sloping than the inner rim of maars, while the outer crater wall is steeper (often >10°, but typically less than 25°: Vespermann and Schmincke 2000; White and Ross 2011). The rim of the Diamond Head tuff ring near Honolulu (Oahu, Hawaii) is about 100 m high, but rises to 232 m in the direction of the prevailing wind due to the consistent trade wind.

3.2.1 Formation and Structure

The magmatic component predominates in a tuff ring sequence, comprising as much as 90%, which contrasts with the non-juvenile componentry of maar deposits. It is mostly basalt or basanite, but can be more differentiated or silica-rich, with tephrite, phonolite, andesite, and rarely dacite.

Tuff rings are typically formed by Surtseyan eruptions in lake or shallow marine environments, or phreatomagmatic eruptions on land. The violent explosions produce base surges and ejection of so-called cocktail jets. Vesicular clasts and accretionary lapilli are common in a tuff ring sequence due to the presence of the wet medium. The vesicles may be formed both by water vapour absorption or by secondary mudflows, which are preserved as small cavities once the pyroclastic deposit dries out. They may subsequently be filled by secondary minerals, helping to cement and harden the deposit. It should be noted that the presence of water does not necessarily mean the absence of dry surges, as temperatures up to 200–300 °C are not uncommon.

Table 1 Geometric and morphometric characteristics of maars – and other hydroclastic cones – and comparison with statistical parameters from the maar VLS worldwide dataset (completed and modified after Graettinger 2018; Thouret et al. 2021)

Maar	Rim diameter (km) Catchment area (km^2)	Mean and maximum rim height (m)	Rim height/ diameter ratio	Excavation depth (m): mean, maximum	Depth diameter ratio; depth ratio[b] $(V/S)/D$	Aspect ratio AR $D_{minor}D_{major}$	Elongation *EL* $A/\pi(D_{maj}/2)^2$	Isoperimetric circularity *IC* $4\pi A/P^2$	Surface *S* (km^2) Volume *V* ($10^6\ m^3$)	Mean/ max slope bedding dip (°)
Maar VLS[a]	1.122 ± 0.833	10–150	0.009–0.13	5–400	Average 0.10	0.81 ± 0.13	0.80 ± 0.12	0.90 ± 0.08	*S* 0.17—2.75 *V* 17.30–447.5	Outer slope 10–25° Inner slope 30–45°
Mode	0.603	n.a.	n.a.	n.a.	n.a.	0.84	0.62	0.93	S ~ 1.17 V ~ 86.8	Outer 12° Inner 35°
50th percentile	0.905	n.a.	n.a.	n.a.	n.a.	0.83	0.82	0.92	S ~ 7 V ~ 251	n.a.

[a] Maar Volcano Location and Shape (Maar VLS) after Graettinger (2018) and case studies *in:* Thouret et al. (2021)
[b] DR Depth Ratio as proposed by Rouwet et al. (2019)

The structure of the tuff rings is divided into different facies depending on the thickness, grain size, and lithoclast content of the pyroclastic-surge and associated, interbedded pyroclastic-fall deposits. Pyroclastic fall may be related to phases when magmatic activity (i.e. violent Strombolian) prevails over phreatomagmatic activity. Notably, the order of deposit facies, i.e. antidune, massive and planar-bedded, as previously observed by Wohletz and Sheridan (1979) for a number of examples in North America and northern Mexico, does not correspond to the order described by Sohn and Chough (1989) for Korean examples, namely the Suwolbong and Songaksan tuff ring. In the latter case, massive surge deposits preceded the antidune facies. This change can be explained by differences in initial conditions (e.g. particle concentration, turbulence rate) and the complexity of the respective facies. In both facies models, the layers become thinner as they move away from the vent, and the wavelength decreases in the dune facies. For comparison of complex facies relationships observed at maars and tuff rings, see Valentine and Connor (2015).

We point out that the deeper structure of tuff rings, unlike maars, does not always have a deep diatreme. This is due to the fact that the external source of phreatomagmatic explosions of tuff rings is supplied by surface water, such as large lakes (Belousov and Belousova 2000). Where a diatreme does co-exist with a tuff ring, it usually extends only to shallow depths. In the case of interaction of deeper groundwater (aquifers, hydrothermal system) with upwelling magma, diatreme facies similar to maars can result.

3.2.2 Spectacular Examples of Tuff Rings

Spectacular and well-documented examples of water-magma interaction processes and the emergence of the tuff rings from seawater include Capelinhos near Fayal in the Azores (1957–1958), Surtsey Island near Iceland (1963–1967), Vulcan Island in the Taal caldera (Philippines, 1865, 1937, 1994), and the eruption of Miyakeyima volcano in a shallow sea environment (Izu Island, Japan, 1983). Specifically, a new magma batch injected into an existing hydrothermal system led to the eruption of Víti crater in 1724 (Montanaro et al. 2021), introducing the famous Myvatn Fires eruptions within Krafla caldera (Iceland), and creating a tuff ring with a 300 m-wide crater filled by a lake (Photo 33).

The Surtsey eruptions were scientifically documented at first hand (Thorarinsson 1967), occurring on a volcanic island that grew up out of the sea. This island, together with the larger Vestmannaeyjar off the southern coast of Iceland and the surrounding islands and seamounts, is linked to a system of fissures that cross the country and continue into the ocean. Surtsey was formed on a 130 m-deep undersea shelf area over a period of four years from November 1963, by recurrent eruptions of alkaline basaltic magma. As magma reached sea level, the characteristic phreatomagmatic activity, later termed 'Surtseyan-type' (Kokelaar 1983), created the southeastern tuff ring by steam eruptions, ash clouds and several hundred metre-high cocktail jet clouds. Six months after the first eruptions, in early 1964, activity shifted to the northwest, and a second tuff ring was built up by lava fountaining and lava flows. After a year of dormancy, the older vent reactivated between 1966 and 1967, mainly producing lava flows. The total mass of material exposed was about 1 km^3, only 9% of which is above sea level (Moor 1985). The irregular-shaped tuff-ring complex, which rises to 110 m and is largely filled with lava, has subsided by more than 1 m in the last 50 years, mainly due to compaction of the strata. Its coastal zone has been heavily eroded by wave abrasion. During the eruption, two other small islands formed from separate vents (Syrtlingur and Jolnir), but these were subsequently washed away by wave action, while the most distant active vent, Surtla, did not rise above sea level.

Another, much older but very well studied tuff ring is the crater of Laacher See in the Eifel Mountains (Germany: Photo 34). It was formed 12,900 years ago, bringing more than 6 km^3 of phonolitic magma to the surface over a period of a few days. In the resulting succession there is thick tuff from pyroclastic-fall that mantled the topography, valley-filling ignimbrites, and a variety of phreatomagmatic base surge deposits (van den Bogaard and Schmincke 1984). Following extremely violent Plinian eruptions, the initial pressure supplied by magmatic gases was replaced by groundwater incorporated from the surrounding valley, reinforcing the phreatomagmatic character of the eruption. Repeated explosions widened, eroded and deepened the crater vent. The high-sulphur ash released into the atmosphere during the eruption could have triggered regional or global cooling in the Younger Dryas (Baldini et al. 2018).

3.3 Tuff Cones

A tuff cone is a landform reflecting a further evolution of a tuff ring, when a massive volcanic construct, which may become as large as, or larger than, a scoria cone, is created. Tuff cones record weaker—i.e. less energetic (see Fig. 9)—hydrovolcanic explosive activity than tuff rings, where erupted material is not ejected very far, but instead builds a cone. For example, the Koko crater near the Diamond Head tuff ring (Hawaii) is 368 m high (Photo 35). This is considered very large, as most of the tuff cones are <300 m in height (Head et al. 1981; White and Ross 2011; Brand and Brož 2015). In the case of Surtseyan eruptions (Capelhinos, The Azores; Surtsey, Iceland; and Velay, France), hyaloclastite tuffs, instead of magmatic scoriae, constitute the tuff cone.

Tuff cone craters are often smaller in diameter than tuff rings because of the lower intensity explosions, but they are usually larger than those of scoria cones. The Udo tuff cone in Korea, for example, has a crater diameter of 1.5 km. As a result, the H/W_{cr} ratio varies between 0.2 and 0.5 and H/W_{co} between 0.09 and 0.11 (Brand and Brož 2015). The accumulated tuff and lapilli tuff sequence is steeper than that of tuff ring layers: the slope of the outer flank is usually 20–30°, and reaches as much as 45° near the vent. The steepness of these slopes triggers regular slides both on the outer slope as well as into the crater interior (e.g. Lanzarote, Canary Islands).

One of the best-explored tuff cones (Sohn and Chough 1992; Sohn et al. 2012; Brenna et al. 2012) is Ilchulbong (Sunrise Peak), a fortress-like construct on the east coast of the South Korean

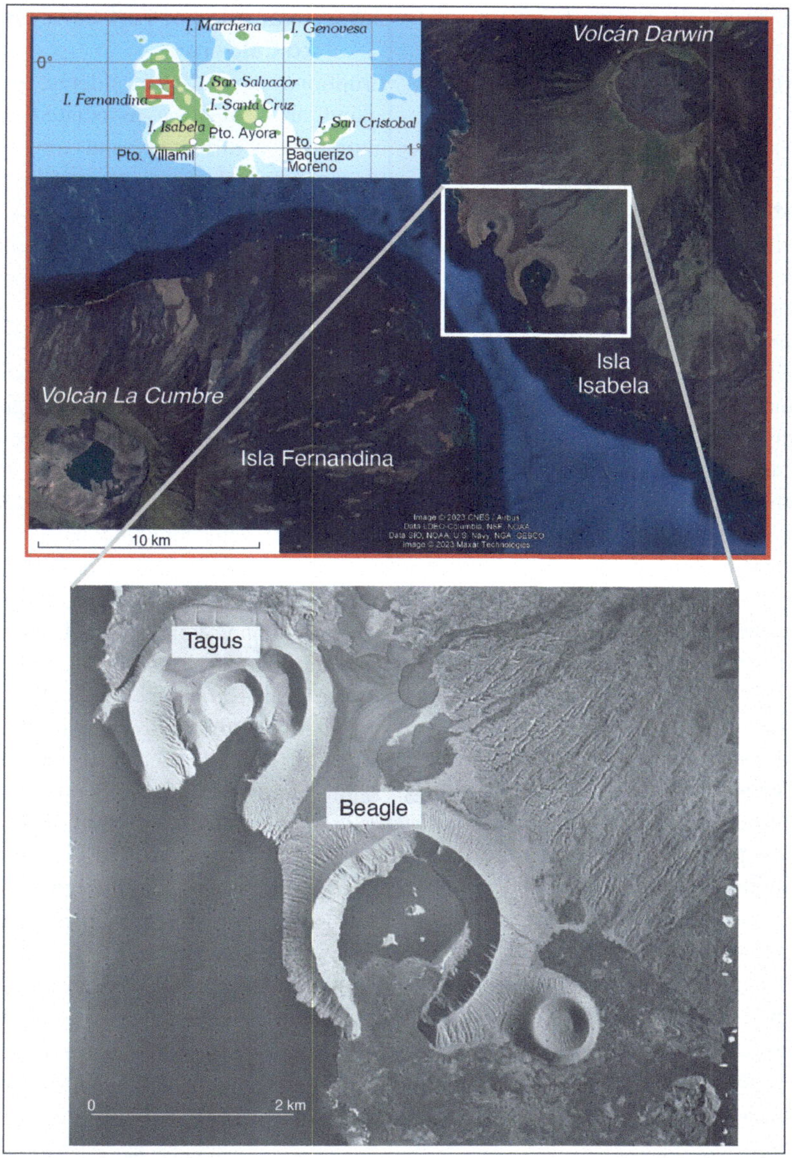

Fig. 10 Tuff cones on Isabela, largest of the Galápagos Islands: Tagus and Beagle are two legendary tuff cones located in the foreground of the Darwin shield volcano (see text). (*source* Smithsonian Institution US Air Force photograph, GVP 04731, public domain)

shield-volcano island of Jeju. Its crater has a diameter of 600 m, a height of 179 m and its floor is 90 m above sea level. The cone was created by Surtseyan eruptions ~5000 years ago. Since then its outer slopes have been partly eroded by sea wave abrasion, providing spectacular exposures that reveal the internal structure of the cone, with its steeply outward-dipping phreatomagmatic tuff layers at an angle of 45°. The island, recognised and protected as a geopark, is visited by millions of tourists every year.

A well-documented pair of tuff cones is located in the Galápagos Islands, near the Darwin shield volcano (Fig. 10). Charles Darwin's ship, the Beagle, anchored in the sheltered cove of a complex twin crater, comprising

Tagus, with a crater diameter of 1.5 km, and another, wider tuff cone, with a crater diameter of about 1.8 km and a height of 220 m, named after the ship. Between them are fresh basalt lavas from the shield volcano.

3.3.1 Formation and Structure

Tuff cones build up more slowly than maars and tuff rings, and can reach a large size. The short-lasting phreatomagmatic explosions in the sea or a lake result in a series of wet ash clouds with pyroclastic fallout and surges spreading over a short distance. Compared with the formation of tuff rings, the less violent eruptions are dominated by tephra fall, so that the tuff cones are mainly composed of fallout and a small fraction of surge deposits. Vesiculated clasts and accretionary lapilli are common. The increase in the proportion of fallout layers is traditionally thought to be due to an increase in the water/magma ratio (see Cas and Wright 1988; Wohletz and Heiken 1992; Fig. 11), although studies on Jeju (e.g. Sohn and Chough 1992; Sohn et al. 2012) showed that the dominance of fallout material is independent of this ratio. At the same time, an important difference compared with tuff rings is that primary pyroclastics are intercalated with debris-flow, mudflow and slipped layers during synvolcanic redeposition, both in the steep crater walls and on the outer slopes. Sohn et al. (2012) clarified that the Ilchulbong cone was formed in three phases separated by significant pauses, during which compaction of tuff and significant erosion occurred along with seismicity-induced slumping and slipping of deposits. Despite these processes, tuff cones can retain their steep slopes and height because the loose sediment, if unaffected by sliding, dries out quickly due to its high temperature, and does not move downslope.

Tuff cones generally do not have a deep diatreme structure, with a few exceptions: e.g. Sæfell tuff cone, Iceland (Mattsson et al. 2005); Costa Giardini diatreme, Iblei Mountains, Sicily (Calvari and Tanner 2011).

3.3.2 Examples of Tuff Cones Related to Calderas

Eruptions in shallow calderas, particularly in island arcs, may be associated with the formation of a tuff cone, and some extraordinarily large ones have been cited in the literature. However, the geological evidence for these giant tuff cones is ambiguous, and requires further investigation.

During the 3600 BP Late Bronze Age (Minoan) eruption of Santorini (Chap. 6), there appears to have been a tuff construct or large tuff cone in the caldera with a height of >500 m, based on the current elevated position of the Minoan tuffs on the caldera cliffs, along with the down-faulted masses in the caldera (Johnston et al. 2014). This cone may have been built up to the point of a complete caldera infill, eventually blocking access to sea water. No traces of this supposed cone remains, but if this was the case, then the topography would have become higher and conical (Karátson et al. 2020).

At Ambrym shield volcano, New Hebrides, Vanuatu (Fig. 12; see Chap. 4), which has one of the largest basaltic calderas on Earth (12 km in diameter), caldera formation was previously thought to have been related to the collapse of an unusual, very large tuff cone during highly explosive phreatomagmatism (Robin et al. 1993). In contrast, more recent studies (e.g. Cronin and Németh 2005) have argued that there was no large caldera-forming eruption or large tuff cone, and that the observable volcaniclastic sequences can be accounted for by smaller-scale events emplacing phreatomagmatic fall and surge sequences as well as lahar and fluvial deposits. In the caldera, two relatively large tuff cones (300–400 m in height), Marum and Benbow—the latter filled by a lava lake until 2018—have been shaped by phreatomagmatic eruption-induced pit crater subsidence (see multiple craters of Marum, Fig. 12).

The island of Aoba (Ambae), Vanuatu, is also a basaltic shield volcano, with a NE-SW-trending rift zone dotted with scoria cones. The

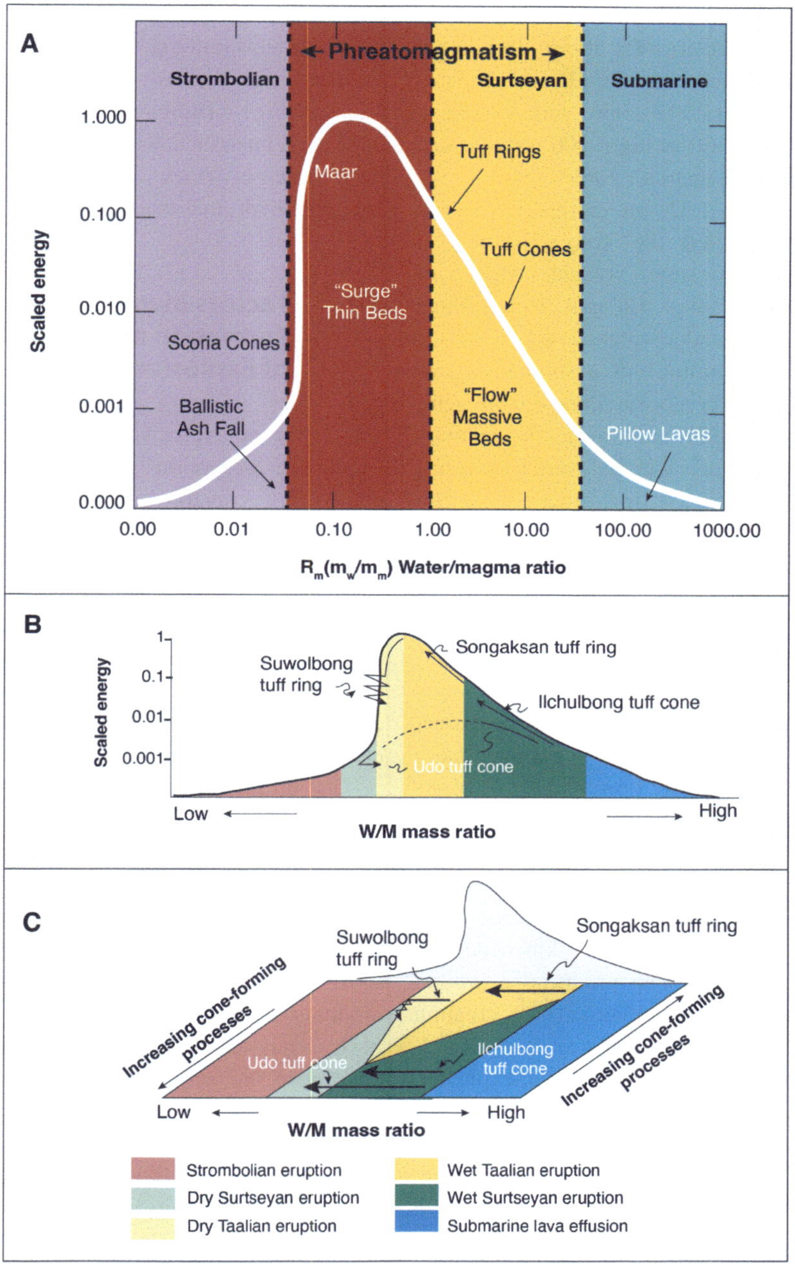

Fig. 11 The type of pyroclastic and hydroclastic cone formed is largely determined by the water/magma ratio (modified after Wohletz and Sheridan 1983). A dry magmatic explosive eruption will produce a pyroclastic cone (e.g. scoria cone), whereas too much water will prevent any explosive cone from forming. In between, phreatomagmatic activity commonly produces hydroclastic cones: the most violent explosions (about 1:3 water/magma ratio) result in maars and tuff rings, while an increasing ratio results in a tuff cone. **A** Scaled kinetic energy relative to the initial melt thermal energy is shown as a function of water/magma mass ratio. Ranges are indicated for Strombolian, phreatomagmatic (maar) or shallow hydrovolcanic (Surtseyan), and submarine activity as well as for corresponding landforms and deposit bedforms (modified after Wohletz and Heiken 1992, in turn adapted from Wohletz and McQueen 1984). **B** and **C** Hydroclastic landforms and magma-water interactions in South Korea. **B** Interactions depicted as scaled energy versus Water (W)/Magma (M) mass ratio. **C** Ranges of hydroclastic eruptive styles and resulting landforms based on water/magma ratio (Sohn 1996 in: Thouret 1999, Fig. 1, p. 97 with permission from Geol Soc Amer)

summit region is occupied by three crater lakes within a nested caldera. Of them, the large, active Voui lake underwent phreatic and phreatomagmatic activity in 2005–2006, with Surtseyan eruptions creating a 50 m-high, 500 m-wide tuff cone in the crater (Bani et al. 2009).

On 15 January 2022, an extremely powerful explosion affected the Tonga Islands. It occurred at the two small volcanic islands of Hunga Tonga and Hunga Ha'apai in the Tonga Islands, Tonga-Kermadec volcanic arc. The two islands—emerged summits of a submarine caldera of an andesitic composite cone—were actually joined by a largely underwater island with a volume of ~1.8 km^2, which formed as a tuff cone over the course of hydromagmatic eruptions in 2014/2015 (Garvin et al. 2018; Allenvi 2022: see photo 41). The composite volcano has been active for more than 110 years, and an even older pyroclastic succession related to caldera formation is also exposed (Cronin et al. 2017; Brenna et al. 2022). After seven years of inactivity, the latest eruptions started in December 2021. Following intermittent activity, on January 13–14, 2022, an unusually powerful set of blasts sent ash surging into the stratosphere. Then on 15 January, another explosion launched material as high as 50 km, triggering destructive tsunami waves (Yuen et al. 2022). Within moments an eruption cloud more than 200 km-wide was formed. The explosion also produced a ground surge that moved parallel to the surface, creating a shock wave, and a prodigious amount of volcanic lightning. The extremely large explosion was captured by satellite images. A sonic explosion also occurred, possibly when the volcano collapsed underwater and seawater rushed in, causing a huge displacement of seawater (Nam et al. 2023). The sound effect was recorded as far as 9600 km away in Alaska, and the atmospheric shockwave circumnavigated the Earth. The central landmass—the new tuff cone, and most of the two previous islands—was destroyed by the giant (phreatoplinian) eruption, with only the highest ground remaining above water.

The answer to why this eruption was so energetic lies in the explosive interaction of magma and water. This interaction is not a linear relationship, i.e. there is a critical range in the magma/water ratio at which maximum energy releases (e.g. Wohletz et al. 2013). During the Hunga Tonga—Hunga Ha'apai eruption, magma may have interacted with water close to the surface, which provides ideal conditions for highly energetic hydromagmatic processes to take place.

3.4 Factors Involved in the Formation and Evolution of Explosive Cones

The types of explosive cones described above (with the exception of the unusual pumice cones) are constructed as a function of the water/magma ratio (Fig. 11a, b). Namely, as early experimental works (e.g. Sheridan and Wohletz 1981; Wohletz 1983) showed, the explosive efficiency—measured as the ratio of eruptive kinetic energy to magmatic thermal energy—is related to the mass ratio of water and magma and the confining pressure (Wohletz et al. 2013).

Dry eruptions occur with low water/magma ratios. Then, depending on the amount of volatiles dissolved in the magma, low-intensity, explosive (Strombolian) eruptions take place, during which a *scoria cone* is formed. The most efficient water/magma ratio required for the most violent explosions is 1:3. More significant amounts of water are associated with Taalian phreatomagmatic or hydrovolcanic eruptions, creating a *tuff ring* while a high-energy explosion at subsurface or shallow depth leads to *maar* formation with associated disruption of the bedrock. Note that the scaled energy is roughly the same for tuff rings and maars.

The question of the efficiency of the water/magma ratio may be complicated by the physical (e.g. temperature) and chemical (e.g. salinity) conditions of the available water and by a number of other factors that determine the mechanism of phreatomagmatic eruptions, which makes the 1:3 ratio far from being a general one. More available water, i.e. a high water/magma ratio that characterises shallow marine or lake environments, is more likely to result in

Fig. 12 Tuff cones at Ambrym Island, New Hebrides, Vanuatu. Ambrym, a basaltic shield volcano, hosts two large tuff cones, Marum (left) and Benbow (right), formed by small-scale phreatomagmatic eruptions and pit crater subsidence (*source* Google maps and Google Earth)

Surtseyan eruptions, that is, pyroclastic surges and ash fallout building up a *tuff cone*. In the case of submarine conditions under considerable hydrostatic pressure (depths of>2–3 km), fragmentation may be limited and submarine lava flows appear with accompanying pillow lava formation: see Sect. 3.6. For a detailed physical assessment of this question, in particular the critical pressure, see Wohletz et al. (2013) and Cas and Giordano (2014).

One given cone type can evolve into another if the water/magma ratio changes over a short period of time. A common sequence is the formation of a maar or tuff ring, first by phreatomagmatic eruption, with high-energy explosions, within which, or from which, a scoria cone is later formed by gentler magmatic explosive eruptions as water is depleted. This was the case at Surtsey in 1963, and Capelinhos, Faial in 1957–1958, for example. This characteristic

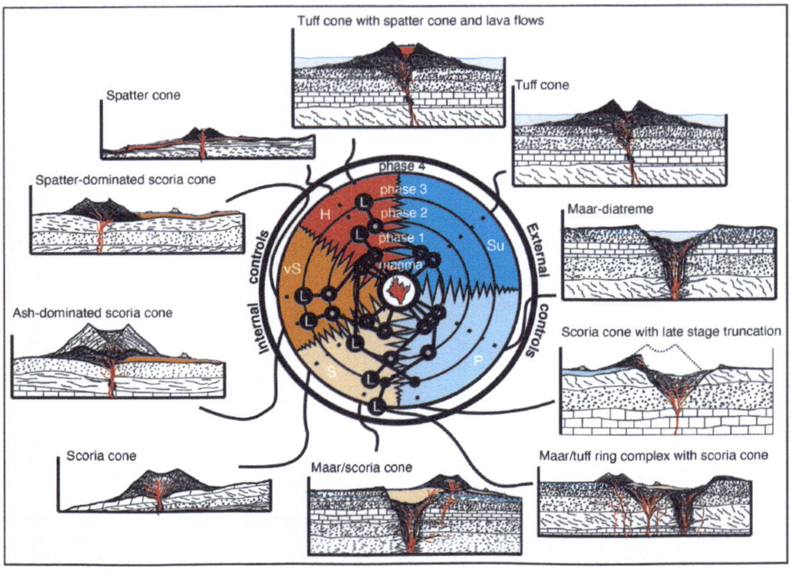

Fig. 13 Types and evolutionary paths of pyro- and hydroclastic cones (courtesy of Kereszturi and Németh 2013). Eruption history (E) is defined by a spectrum of constant or changing eruptive processes during activity. The initial magma (in the centre of the circle in red) is fragmented by means of internal and external processes, which determine the magma fragmentation mechanism and eruption style (phase 1). If any change (e.g. shift in vent position, exhaustion of ground-water, etc., see main text) occurs, it will trigger a new phase (phase 2, 3, 4, n); successive rings moving away from the centre until the eruption stops. Colour code for the eruptive style: Hawaiian (red, H), Strombolian (light orange, S), violent Strombolian (dark orange, vS), Surtseyan (dark blue, Su), Phreatomagmatic (light blue, P). 'L' stands for lava effusion

may also occur following the arrival of a new, dry magma batch. The resulting landform is called a nested tuff ring or a maar-scoria cone complex.

However, the eruptive processes can be more complicated or may be inverted. For example, frequent changes in the water/magma ratio can cause magmatic and phreatomagmatic explosive events and deposits to alternate or become mixed within a given eruptive succession (e.g. Houghton and Schmincke 1989). The water/magma ratio may increase in the process (e.g. by tapping an aquifer), often resulting in a tuff cone built on a tuff ring. Complex landforms can also be created by vent migration, as seen in the evolution of Surtsey Island, during which lava flows were also produced. The role of different factors, such as the primary magmatic explosions and the presence and amount of external water, in the formation of more complex landforms is illustrated in a circular diagram (Fig. 13, Kereszturi and Németh 2013), in which subsequent eruptive styles responsible for creating a given landform or a combination of landforms are also indicated.

4 Basaltic Volcanic Fields

As indicated by the name, these fields are compositionally basaltic in a broad sense, and consist of a collection of small volume, short-lived, monogenetic volcanoes of explosive origin, as described above, together with a variety of landforms, in particular lava flow fields (Chap. 2). They may also include lava domes of more evolved composition (Valentine and Connor 2015). Basaltic volcanic fields occur in subduction, rift, and intraplate tectonic settings. They can also form on the flanks of major edifices (composite cones, shield volcanoes and calderas) or in the areas surrounding these long-lived systems. Although individual edifices are short-lived, the basaltic fields can be active for several million years.

The number of volcanoes (scoria cones and their attendant lava flow fields, maars, tuff rings, and domes) can range from tens to more than one thousand edifices in basaltic volcanic lava fields (Valentine and Connor 2015). The list given by these authors describes 17 extensive (>1000 km^2) volcanic fields, including nine basaltic volcanic fields comprising more than 100 edifices, ranked here on the basis of abundance, from most to least (as many as 1040 edifices down to 100): Michoacán-Guanajuato, Mexico; San Francisco, USA; Springerville, USA; Newer Volcanic Province, Australia; Pinacate and Camargo, Mexico; East and West Eiffel, Germany; Pali-Aike, Chile-Argentina; and Lunar Crater, USA. Medium-sized basaltic volcanic fields (a few hundred km^2) still contain dozens of volcanoes; for example the Auckland Volcanic Field, a cluster of ~55 small basaltic volcanoes spread over 350 km^2, and the Chaîne des Puys (Auvergne, France), granted UNESCO World Heritage status in 2018, includes about 80 scoria cones, maars, tuff rings, and domes in a linear zone 30 km north south by 4 km west to east above the Limagne fault. Near the centre of the Trans-Mexican Volcanic Belt, directly south of Mexico City, no less than 146 Quaternary and Holocene scoria cones were counted in an area of ~1000 km^2, yielding a vent density of 0.15 per km^2. The Sierra Chichinautzín, also located in the Trans-Mexican Volcanic Belt, about 20 km west of Popocateptl and 15 km south of Mexico City, is even larger (5945 km^2), with at least 227 volcanic constructs, including youthful scoria cones close to the city of Mexico (Siebe and Macías 2004). A recent study shows that the Sierra Chichinautzin, previously thought of as a single volcanic field, can be separated into four groups on the basis of radiometric dating, between 1.3 Ma and 7 ka old, and other data (Jaimes-Viera et al. 2018).

Regional tectonic settings play a critical role in the distribution of basaltic volcanic fields, either forming vent clusters or vent alignments. Clusters of 10–100 individual vents, with a total collective diameter of ~50 km, appear in convergent margins (e.g. Chukogu, southwest Honshu, Japan). The location of clusters along the NE–SW trending volcanic front has apparently moved from Pliocene to Quaternary times closer to the subduction trench, with the changing geometry of the back-arc tectonic setting (Kamata 1998 in: Valentine and Connor 2015). More common are numerous vents aligned in long (75–150 km) and relatively narrow (<50 km) zones along major faults or parallel to subduction zones. For instance, the volcanic field of Baja California, Mexico, displays several groups, totalling as many as 900 scoria cones and related volcanoes. They are distributed along the principal NW–SE direction of the peninsula and slightly oblique NNW–SSE striking faults, reflecting changes in the paleo-subduction zone in the region (Germa et al. 2013 in: Valentine and Connor 2015).

5 Glaciovolcanic Landforms and Magma-Ice Interactions

Glaciovolcanic landforms are mostly monogenetic edifices resulting from interactions between water or ice and magma in a cold environment. Termed *intraglacial, subglacial* and *glaciovolcanic,* these landforms and interaction processes have been recognized for the past 80 years in Iceland and British Columbia. Iceland, a country with the greatest concentration of glaciovolcanic edifices and the highest frequency of subglacial eruptions, has long been the hub of glaciovolcanology (e.g. Gudmundsson 2005). Interest in glaciovolcanism has increased over the past 25 years, with a growing body of publications about Antarctica and Mars, whose deposits can be used as proxies for determining the distribution of ice sheets in space and time, thus providing clues to palaeoclimate reconstructions (e.g. Skilling 2002; Smellie et al. 2008). This is possible since glaciovolcanic deposits are far more long-lived than the ice that forms them.

The terms subglacial and intraglacial are often used to classify the volcanic eruptions and the resulting landforms. Strictly speaking, the term *subglacial* only applies to processes occurring under ice cover without direct contact with the atmosphere. The term *intraglacial* is more general since it also applies to eruptions and volcanoes that have broken through glaciers (Jones 1969; Jakobsson and Gudmundsson 2008). Subglacial and intraglacial volcanic rocks are especially prominent in the Móberg (Palagonite) Formation of Iceland, which is a chronostratigraphical unit (Kjartansson 1960 *in:* Jakobsson and Gudmundsson 2008), comprising all strata formed during the Brunhes geomagnetic epoch towards the end of the Pleistocene (0.78–0.01 Ma).

Magma-ice interactions can occur in a number of ways and produce a variety of landforms and products (Lescinsky and Fink 2000; Larsen 2003; Mee et al. 2006; Komatsu et al. 2007; Jakobsson and Gudmundsson 2008; Smellie 2006, 2008; Edwards et al. 2009, 2010, 2015; Wilson and Head 2002, 2007; Wilson et al. 2013). Ice, in the form of glaciers, snow-fall or frozen lakes, commonly occurs as a sheet emplaced onto a silicate rock surface, and magma rising through the shallow crust always travels in dykes through brittle fractures. Thus, possible initial geometries of the interaction between rising magma along dykes (or sills) and ice are: (a) injection of a magmatic dyke into a layer of ice; (b) formation of a sill-like magma intrusion beneath ice at the ice-rock interface, which may ultimately develop into a subglacial lava flow; and (c) emplacement of a lava flow or a pyroclastic deposit on top of glacier ice from a vent outside the ice layer (Wilson and Head 2002, 2007; Wilson et al. 2013).

5.1 Main Glaciovolcanic Landforms

The three most common glaciovolcanoes (in Iceland: Fig. 14) are flat-topped tuyas, elongate tindars, and conical tuyas (Russell et al. 2014; Pedersen et al. 2020). We note again that móberg, which is sometimes used to describe hyaloclastite ridges of consolidated, mafic to intermediate hyaloclastite, is a chronostratigraphical unit, in which pillow lavas, hyaloclastites and cap lavas are included.

Glaciovolcanic landforms show three geomorphic types (Edwards et al. 2015):

(1) Flat-topped *tuyas* have distinctive morphologies in plan view and profile and generally have a length-to-width ratio of 2:1 or less. The flat top generally comprises subhorizontal lava flows that were erupted above the level of the surrounding englacial lake (above the passage zone, Fig. 15A; Photo 36). The tuya morphology can form edifices produced during one single eruption or volcanoes built over longer (thousands to tens of thousands of years) timescales (Fig. 15B).
(2) *Tindars* are elongate, steep-sided ridges that generally have a length-to-width ratio of >2:1, with the length ranging from 2 km to more than 20 km. These ridges are the most common landforms in Iceland, where their north–south elongation broadly records the stress field of the extensional setting (Pedersen et al. 2020) (Fig. 14; Photo 37).
(3) *Conical landforms* are built by eruptions similar to tuya ones, but fail to produce lava caps. However, other types of deposits have been documented, including extensive sheets, often composed of pillow lavas, but sometimes comprising an edifice with columnar jointed lava at the base overlain by pillow lavas and hyaloclastites. Individual units can be 200–300 m thick and have been interpreted as being subglacially erupted lava flows.

Pedersen et al. (2020) distinguished three morphometric groups from a database of 155 glaciovolcanic edifices within the Icelandic neovolcanic zones, formed during the last 0.78 Ma (Fig. 14). Three planimetric variables (basal length, average basal width and average summit plateau width) and their ratios define three groups: (1) flat-topped edifices (subdivided into equidimensional and elongated), (2) linear ridges, and (3) conical edifices with no or small summit plateaus. These morphometric groups correspond to the common terminology for tuya, tindar and cone/mound, respectively.

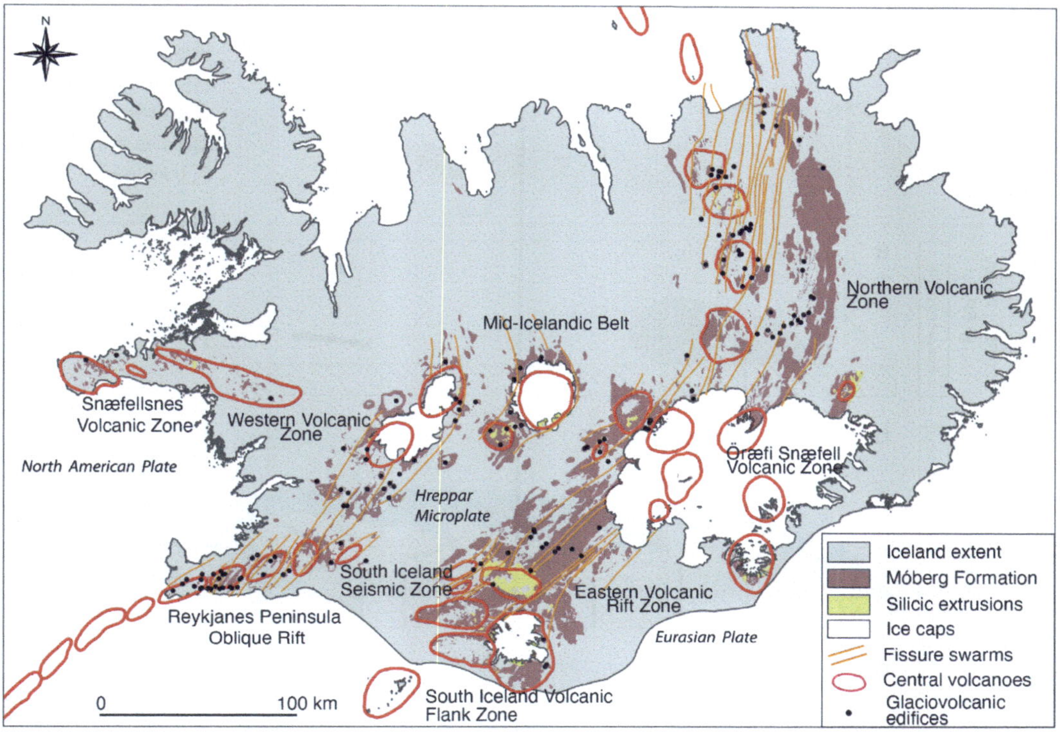

Fig. 14 Map of Icelandic glaciovolcanic edifices (after Pedersen et al. 2020, Fig. 1, p. 3). Location of Icelandic glaciovolcanic edifices (black dots) analyzed geomorphometrically. Volcanic systems (fissure swarms in orange, central volcanoes in red) are outlined. The background shows the outline of Iceland in grey and the outline of the largest glaciers in white (after the National Land Survey of Iceland). The Móberg Formation is outlined in brown and silicic extrusives are in yellow. The neovolcanic zones are: the Reykjanes Peninsula Oblique Rift (RPOR), the Western Volcanic Zone (WVZ), the Mid-Icelandic Belt (MIB), the South Iceland Seismic Zone (SISZ), the South Iceland Volcanic Flank Zone (SIFZ), the Eastern Volcanic Rift Zone (EVRZ), the Northern Volcanic Zone (NVZ), the Öræfi Snæfell Volcanic Zone (ÖSVZ), and the Snæfellsnes Volcanic Zone (SVZ). All references to the background volcanism in Iceland are included in Pedersen et al. (2020)

The glaciovolcanic edifice volumes range from 0.15 to 32 km^3; conical edifices are the smallest (<0.1 km^3) while the flat-topped edifices are the largest (>1 km^3).

Tuyas, called table mountains by Icelanders, are characterized by pillow lavas at their bases and hyaloclastites at their summit reflecting a change in depth of ice-water cover during edifice growth (Fig. 15A and B, Photo 36). The steep flanks of most table mountains, buttressed by breccias, are due to the original confinement by ice, while the flat tops represent eruption into the floors of overlying meltwater lakes. The summit is often capped by a subaerially erupted lava flow, which forms a resistant top to the table mountain, and may be crowned by basaltic scoriae. However, due to potential confusion with the more common table mountains that are formed by the erosion of flat-lying resistant sedimentary beds or lava flows in arid terrains around the world, the term tuya is now preferred to describe these distinctive features after Tuya Mountain in British Columbia (Mathews 1947).

Werner and Schmincke's (1999) detailed study of lithofacies (mostly hyaloclastites, tuffs and breccia forming the edifice, and scoria and lava topping the sequence: Fig. 15B) contrasts with a rapid, continuous, entirely subglacial, monogenetic evolution. Many tuyas and similar volcanoes were constructed over several

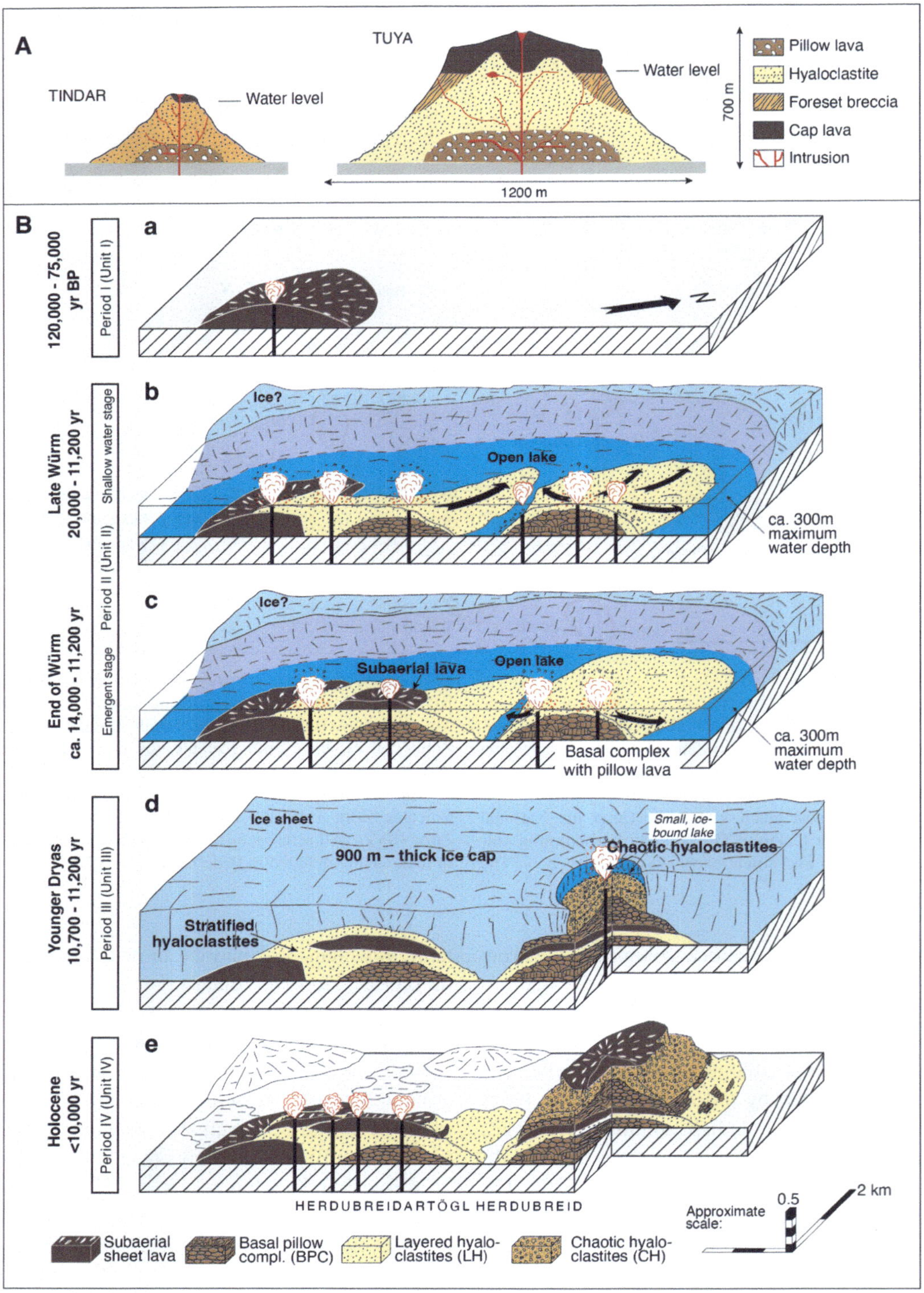
A
TINDAR
Water level
TUYA
Water level
700 m
1200 m
Pillow lava
Hyaloclastite
Foreset breccia
Cap lava
Intrusion
B
a
120,000 - 75,000 yr BP
Period I (Unit I)
N
b
Late Würm 20,000 - 11,200 yr
Shallow water stage
Ice?
Open lake
ca. 300m maximum water depth
c
End of Würm ca. 14,000 - 11,200 yr
Period II (Unit II)
Emergent stage
Ice?
Subaerial lava
Open lake
Basal complex with pillow lava
ca. 300m maximum water depth
d
Younger Dryas 10,700 - 11,200 yr
Period III (Unit III)
Ice sheet
900 m – thick ice cap
Small, ice-bound lake
Chaotic hyaloclastites
Stratified hyaloclastites
e
Holocene <10,000 yr
Period IV (Unit IV)
HERDUBREIDARTÖGL HERDUBREID
Approximate scale:
0.5
2 km
Subaerial sheet lava
Basal pillow compl. (BPC)
Layered hyaloclastites (LH)
Chaotic hyaloclastites (CH)

eruptive periods in changing environments characterized by contrasting volcanic and sedimentary processes. Werner and Schmincke (1999) described five stages of tuya growth, illustrated in Fig. 15B, since the Last Interglacial period (between ~130,000 and 115,000 years BP). The majority of the exposed hyaloclastite deposits formed in large, stable lakes, as indicated by widespread, up to 300-m-thick, continuous sections of deep water, shallow water and emergent facies. In contrast, chaotic assemblages of coarser-grained, more poorly sorted and more strongly palagonitized hyaloclastite tuffs and breccias, as well as scoria and lava, are interpreted to have formed under sub- or englacial conditions in small, chimney-like ice cavities or ice-bound lakes (Fig. 15B, diagram d).

Ridges are the most morphometrically distinct glaciovolcanic edifice due to their extreme elongation (several kilometres), which can form topographic divides a few tens of km long (Lockwood and Hazlett 2010). These are long, steep-sided ridges of hyaloclastite, commonly altered to brownish palagonite, such as Helgafell, southwest Iceland (Schopka et al. 2006; Photo 37). The dataset of Pedersen et al. (2020) reveals a morphological evolution from an eruptive fissure to a ridge or tindar: (a) an initial eruptive fissure concentrates into one vent, generating an equidimensional conical edifice, which progressively grows a larger plateau as the eruption continues; (b) the edifice maintains its elongation with growth, suggesting that a fissure was the dominant vent structure, while also developing a significant flat top during protracted eruption; and (c) the fissure lengthens during the eruption and the edifice does not reach a plateau-building stage.

A large proportion of glaciovolcanic edifices have been built by mafic magma, in particular of basaltic composition; however, several case studies have also been published on intermediate to silicic glaciovolcanism in Iceland, the western United States, the Andes, and Antarctica. Intermediate composition lavas appear to produce much less fragmental material than basaltic lava (e.g. McGarvie 2009). In Iceland, effusive deposits are predominant over explosive deposits in intermediate and silicic lava fields, and deposit morphology and orientations of cooling joints are key to recognising glaciovolcanism. The edifices and deposits produced by glaciovolcanic eruptions of more evolved magma compositions can be more difficult to identify. For example, the relatively low viscosity of basaltic magmas means that, in subaerial environments, mafic lava flows have low aspect ratios, whereas confinement by ice generates anomalously high aspect ratios. Sideromelane is not a common constituent of subaerial basaltic lavas, thus glass-rich basaltic deposits can be a clear indicator of 'water-enhanced cooling'. Edwards et al. (2015 and references therein) present characteristic properties of non-basaltic glaciovolcanism, such as confined lava flows, radial jointing, phreatomagmatic tephra and peperite in rhyolitic edifices in Iceland; confined dacitic lava domes at Garibaldi volcano

◀ **Fig. 15** Tuyas and tindars. **A** Tindar (linear ridge) versus Tuya (table mountain) with indication of former water level (modified after Jakobsson and Gudmundsson 2008; Courtesy of Gudmundsson). 'Passage zone' is a transition zone between a subaerial lava and its subaqueous flow-foot breccia (Jones 1969 *in:* Jakobsson and Gudmundsson 2008). **B** Formation of a tuya or table mountain in englacial and lacustrine environments, and evolution of explosive activity due to interaction with ice cap and open lake from the Interglacial, based on the Herdubreid area in central Iceland (modified after Werner and Schmincke 1999). **a** Interglacial stage, with small lava flow shield. **b** Last glaciation: explosive activity in deep water, forming pillow lavas, associated hyaloclastites, and volcaniclastic sedimentation in a deep, open lake surrounded by ice. **c** End of last glaciation: subaerial lava caps shield volcanoes and hyaloclastites above the level of ice meltwater and lake. **d** Cold spell 'Younger Dryas' in early Holocene helped cover small volcanoes with a thick ice cap. Sustained, englacial explosive activity formed chaotic hyaloclastites (and lava deltas) in ice-bound water bodies. **e** Post-glacial (Holocene) volcanic activity in ice-free landscape. On the left-hand side, several small volcanoes can coalesce to form a tindar (linear, constructional ridge made up of pillow lavas and hyaloclastites). On the right-hand side, subaerial lava flow forms a resistant cap over the underlying fragmental rocks so that a flat-topped mountain table is left when the ice cap disappears (mimicking an 'inverted relief' landform)

(British Columbia); confined dactic lava flows with radial jointing and hydrothermally altered tephra deposits, and pillow lava deltas at Hoodo (British Columbia) and Tahake (Antarctica). Ice-clad volcanoes can also comprise stacked andesitic lava flows, e.g. ice-bound flows with pseudo-pillow fractures at Garibaldi, and in the Cascades.

5.2 Volcano-Ice Interactions and Processes

Evidence of confinement and the presence of water are key to identifying glaciovolcanic lithofacies. The link between confinement and presence of water is ice. Coherent volcanic lithofacies that are important for recognition of glaciovolcanism include pillow lavas, massive lava, and intrusions. Subglacial effusive eruptions occurring beneath thick ice fields produce large volumes of meltwater that create a watery interface around the erupting lava. Lava morphologies can also sometimes suggest an ice-dominated eruption environment. Thickened individual lava flows or stacks of lava flows featuring vertical cooling surfaces provide evidence for ephemeral confinement of lava or water, or both (Lescinsky and Fink 2000).

Volcano-ice interactions are illustrated in Fig. 16A, depicting the three main classes of historical glaciovolcanic eruptions and their products (Edwards et al. 2015):

The first class (1A and 1B) of *subglacial* eruptions occurs beneath ice, either thick or thin. Two end-member styles exist for this first class: (A) Subglacial eruptions beneath a >500 m-thick ice-sheet show simultaneous melting deformation of ice and growth of volcanic edifice. (B) Eruption beneath thin ice <200 m: melting of ice cover produces a cauldron with steep walls and little ice flow. The first class of subglacial eruptions is common in Iceland, including the Gjálp eruption in 1996, several Grísmvötn eruptions, and the first part of the Eyjafjallajökull eruption in 2010 (see below). The majority of Pleistocene tuyas found globally are of this type.

The second and third classes of *supraglacial* eruptions occur beside and on ice or snow cover. The second class includes supraglacial eruptions where the ice surrounding the volcano exerts considerable influence on the properties and distribution of the eruption products (Edwards et al. 2015). These eruptions are typical of a large number of snow-capped and ice-clad volcanoes, including many in the Andes, Alaska, and Kamchatka, during which *pyroclastic density current* (PDC)-forming eruptions produce meltwater and floods, and cause regional, sometimes protracted hazards. This class includes the eruptions of Nevado del Ruiz in 1985, Mt. Redoubt in Alaska (1989–1990 and 2009), as well as the effusive portion of the 2010 Eyjafjallajökull eruption. The third class induces meltwater between lava flow and snow and ice cover. Small-volume lavas originate above the ice surface but flow over and/ or beneath the snow/ice surface. This class involves minor volcano-ice interactions that produce identifiable glaciovolcanic deposits, and is the most widepread type of glaciovolcanism worldwide. The deposits generated can record the presence of snow/ice across the full range of latitudes, occurring above all composite cones

Fig. 16 Glacio-volcanic eruptions, resulting landforms and processes. **A** Schematic diagrams illustrating the three main types of historical glacio-volcanic eruptions (Edwards et al. 2015, Fig. 20.10, p. 389): Class 1A. Subglacial eruption beneath a >500 m-thick ice sheet showing simultaneous melting (eventually leading to jökuhlhaups), and deformation of ice and growth of volcanic edifice. Class 1B. Eruption from beneath thin ice, where melting of ice cover produces a cauldron with steep walls and little ice flow. Class 2. Supraglacial eruption common to ice-clad composite cones where the resulting pyroclastic density currents can induce rapid melting and produce lahars. Class 3. Small-volume lavas originate from above the ice surface but flow over and/ or beneath the snow/ice surface. **B** Diagram representing heat transfer efficiency (f%) plotted as a function of degree of fragmentation (%) for two types of eruptions: (1) effusive glacio-volcanic eruptions where quench fragmentation dominates (f <30), and (2) explosive glacio-volcanic eruptions (f >60) driven by magmatic fragmentation (modified from Gudmundsson 2003, in: Edwards et al. 2015, Fig. 20.8, p. 388)

▶

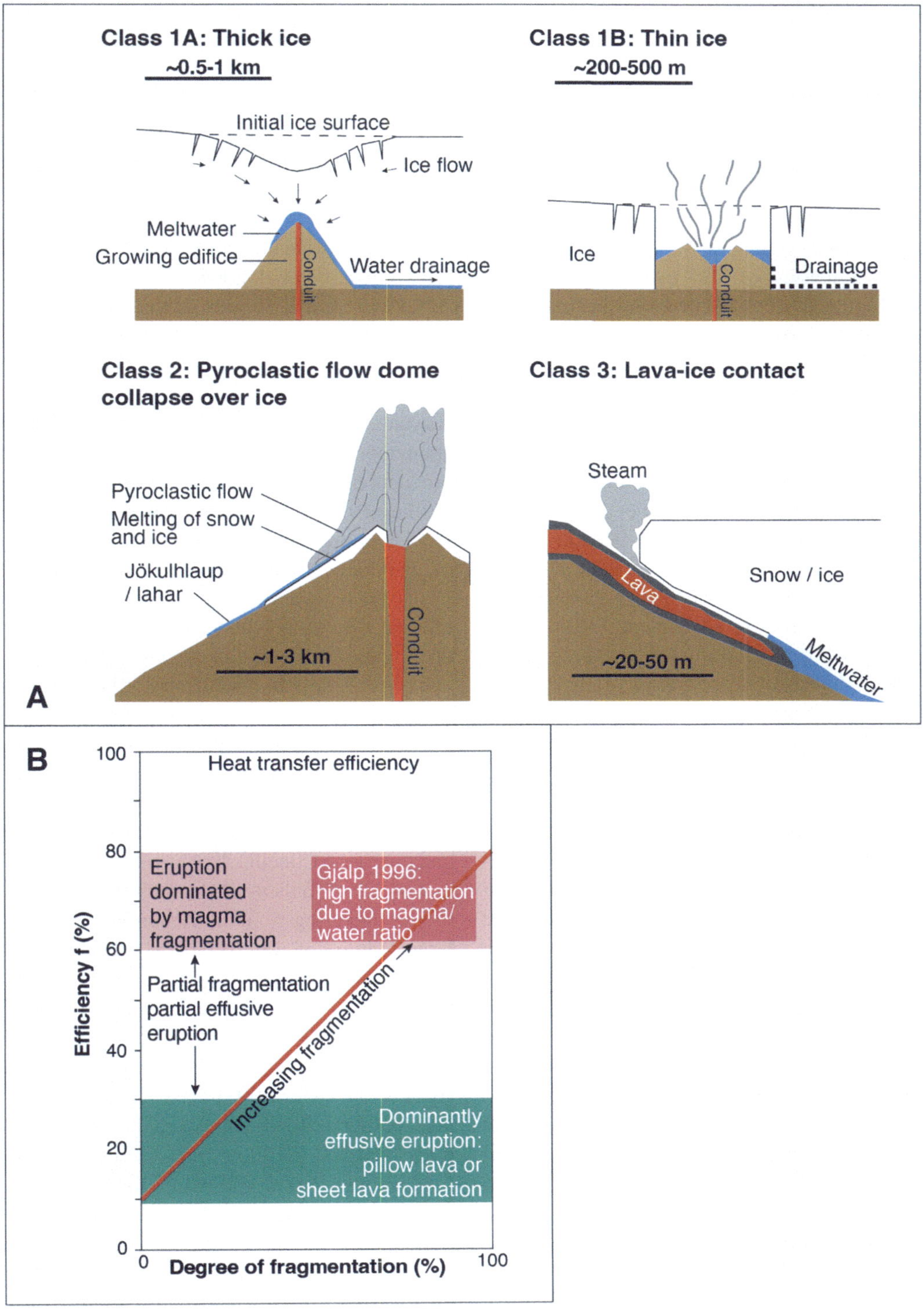
Class 1A: Thick ice
~0.5-1 km
Initial ice surface
Ice flow
Meltwater
Growing edifice
Conduit
Water drainage
Class 1B: Thin ice
~200-500 m
Ice
Conduit
Drainage
Class 2: Pyroclastic flow dome collapse over ice
Pyroclastic flow
Melting of snow and ice
Jökulhlaup / lahar
~1-3 km
Conduit
Class 3: Lava-ice contact
Steam
Lava
Snow / ice
~20-50 m
Meltwater
A
B
Heat transfer efficiency
Eruption dominated by magma fragmentation
Gjálp 1996: high fragmentation due to magma/ water ratio
Partial fragmentation partial effusive eruption
Increasing fragmentation
Dominantly effusive eruption: pillow lava or sheet lava formation
Efficiency f (%)
0
20
40
60
80
100
Degree of fragmentation (%)
0
100

that have had a significant snow/ice cover during past glaciations.

5.2.1 Three Generic Scenarios for Volcano-Ice Interactions

Wilson et al. (2013) described three generic scenarios for volcano-ice interactions, two being subglacial and the third being supraglacial in terms of environment and processes. Possible initial geometries of the interaction between ice and magma are: (1) injection of a magmatic dyke into a layer of ice; (2) formation of what is initially a sill-like magma intrusion beneath ice at the ice-rock interface, which may ultimately develop into a subglacial lava flow, and; (3) emplacement on top of ice of a lava flow or pyroclastic material from a source elsewhere (Wilson and Head 2002; Wilson et al. 2013).

Case 1 of Wilson et al.: injection of magma into an ice layer as a dyke provides an inherent instability in the resulting geometry (see Fig. 16A), resulting in a fragmental deposit surrounding the vent or fissure at the base of the ice sheet. Subsequent evolution of the subglacial volcanic mass may involve construction of a pillow mound, then an explosively generated subaqueous tuff cone and ultimately a subaerial edifice with or without lava-fed deltas, as found in Antarctica (Smellie 2000; Smellie and Skilling 1994; Skilling 2002). The rifting episode associated with the Bárðarbunga-Holuhraun eruption in 2014–2015 provided the first observations of major dyke propagation under ice (Reynolds et al. 2017). Three shallow ice cauldrons with volumes ranging from 1 to 18 million m^3 formed in Dyngjujökull glacier above the 48-km-long lateral path of the magma. A subglacial eruption is the only plausible mechanism to explain the observed rapid heat transfer at a location far from known geothermal areas (Reynolds et al. 2017).

As shown by Fig. 16B, efficiency of heat transfer is controlled by effusion rate and hydrostatic pressure, which help distinguish two cases of fragmentation produced by either effusive or explosive eruptions (Gudmundsson et al. 2004). If lava supply rates are very high, and hydrostatic pressures relatively low, extensive fragmentation of quenched lava will occur and extensive subglacial hyaloclastite deposits can be formed. This eruption is dominated by magma fragmentation, as occurred at Gjálp beneath the Vatnajökull ice cap in 1996. If effusion rates are slower and the hydrostatic pressure is higher, pillow lavas similar to those in submarine environments, where effusive eruptions dominate, will be formed instead (Gudmundsson et al. 2004).

Glaciovolcanic sequences thus differ depending on their location under thick or thin <200 m ice (Fig. 16A: class 1a and b). Distinctive glaciovolcanic sequences, known as sheet-like sequences of Mount Pinafore type (after the eponymous Antarctic edifice) appear to be the products of eruptions under thin ice (Smellie and Skilling 1994). They are outflow deposits and the sequences are typically composed, from base up, of diamictite (tillite), tephra-sourced sandstones, hyaloclastite, water-cooled lava, and finally subaerial lava. Eruptions under much thicker ice produce quite different lithofacies. The Icelandic outcrops of sheet-like sequences of Dalsheidi type are the products of voluminous subglacial fissure eruptions individually up to >30 km^3 in volume. The Dalsheidi sequence of four major lithofacies, totalling 300 m in thickness, shows, from base up, diamictite, lava (sill), pillow lavas, hyaloclastite, and mudstone (Smellie 2008).

Case 2 of Wilson et al.: melting of the base of ice overlying a sill intrusion can produce a large volume of water (Wilson and Head 2002, 2007; Tuffen 2007). The water may accumulate relatively slowly under a glacier, but if it reaches a large body of water, it may eventually escape very rapidly in the form a *jökulhlaup* (Björnsson 1992, 2003). The latter is a glacial lake outburst flood with exceptional discharge rates (10,000–50,000 m^3/s), as estimated on 5 November 1996 following the Gjálp eruption north of Grímsvötn beneath the 8100 km^2 Vatnajökull ice cap. The 1996 Gjálp eruption lasted for 13 days and started as a 4 km-long fissure eruption under 600 m-thick ice in the northwest part of the Vatnajökull ice cap (Edwards et al. 2015). The Gjálp eruption eventually created a 6 km-long, 0.5 to 2 km-wide, and 450 m-high basaltic andesite hyaloclastite ridge (tindar) of 0.8 km^3 in volume. The eruption melted more than 2.5 km^3 of ice at the base of the glacier, at an initial rate of about 500,000 m^3/day according to the estimated volume of a large down sag and cauldron collapse in the overlying glacial surface (Gudmundsson et al. 2004; Jarosch et al. 2008). The meltwater drained away continuously from the eruption site, 15 km to the south into the Grímsvötn caldera, where it accumulated in a subglacial lake. Although the Gjálp eruption stopped on 13 October, the meltwater did not reach the edge of Vatnajökull, 50 km away, until 5 November, when it burst forth, creating a jökulhlaup that destroyed bridges, roads and powerlines while also transporting house-size blocks of ice several kilometres from the glacier's ruptured margin. The torrential floods and lahars added several km^3 of new land to the coast of Iceland (Lockwood and Hazlett 2010).

Case 3 of Wilson et al.: in *supraglacial* eruptions, the presence of the ice plays no part in the magmatic eruption that emplaces a lava flow or pyroclasts. The consequences depend on the rate of thickening of the tephra-fall deposit and its final total thickness. Immediate heating of underlying ice is minimal except for very close to the vent. If the volcanic material is hot on emplacement (lava flow or PDC), the effects on the underlying ice will depend on both the rate of advance and the thickness of the flowing material, and the rate at which water can migrate beneath the advancing deposit to escape at its edges.

The March–May 2010 Eyjafjallajökull eruption can be used to illustrate Case 3. The explosive eruption is best known for having disrupted air traffic in Europe and across the Atlantic Ocean, but it also displayed a large variety of interactions between magma, ice and snow (Magnússon et al. 2012; Edwards et al. 2015; Oddsson et al. 2016). During the first effusive phase, basaltic lava flowed on top of the snow. However, the rate of melting was too slow to remove the snow ahead of the advancing lava, leading to gradual melting of snow over a period of weeks to months. The main explosive eruption started in the shallow ice-filled summit caldera on 14 April and melted the 200 m-thick ice in about 4 h through vertical-walled ice cauldrons. Ice melting was rapid in the first 24 h (500–1000 m^3/s), slowing a little on the second day (200 m^3/s). Meltwater drained northward along the steep outlet glacier Gígjökull, subglacially for the first 1–1.5 km, then emerging in several jökulhlaups with discharges of the order of 1,000 to 10,000 m^3/s. On the eighth day of the eruption, lava started to flow subglacially toward the north, melting through the glacier (Photo 38). The lava advanced about 2 km in a few days, forming a 20 m-thick lava flow by melting the ice above the flow front. Magma-ice interaction and melting became minimal after the lava stopped flowing in early May (Edwards et al. 2015).

The Eyafjallajökull eruption is a rare example of subglacial eruptions, another being the Gjálp 1996 event which remained subglacial throughout. Instead, most historically eruptions within glaciers have broken through the ice cover with an emergent, subaerial, usually Surtseyan phase, producing airborne phreatomagmatic tephra (Gudmundsson 2005). In some cases the subaerial explosive eruption is a major part of the overall activity with a sizeable part of the volcanic material being transported out of the vent area through an eruption plume and dispersed as tephra over surrounding areas and beyond (e.g. to Europe from Eyajfjalljökull in 2010).

5.2.2 Lava Flows and Pyroclastic Material from Supraglacial Eruptions

Supraglacial eruptions may involve either a lava flow advancing onto ice from a vent located off the glacier, or the deposition of pyroclasts onto ice from an explosive eruption. Wilson et al. (2013) distinguishes two cases:

(1) Lava flowing onto glacial ice is rare, compared with descriptions of secondary generation of lahar and meltwater flood events (e.g. Major and Newhall 1989). Mee et al. (2006) published a rare *in situ* case of an andesite lava from a winter eruption in

Chile that flowed over snow. The evidence comprised a flow front 5 m in width and 5 m in height composed of blocky glassy breccia that formed as a talus apron during post-emplacement gravitational instability, and a 20 m-wide zone behind the flow-front forming a basal layer several metres thick that consisted of glassy andesite lava showing distinctive, cross-cutting, curviplanar pseudo-pillow fractures and abundant perpendicular secondary fractures (Wilson et al. 2013).

(2) Pyroclasts deposited on glaciers, whether derived from fallout or PDCs, are reworked and redeposited by either aeolian activity or lahars generated by PDC-generated melting (Waitt et al. 1983; Thouret 1990; Pierson et al. 1990). They will also be advected *en masse* to marginal locations by ice flow, where they will be deposited in moraines and/or extensively reworked as further mass flows (Manville et al. 2000). Off-ice deposits are mixtures of pyroclasts, ice, snow, and water. The snow-ice components will melt out, leaving behind a lag deposit comprising porous ashy material that is likely to be washed away by rain or winnowed by wind (Manville et al. 2000).

This second case is exemplified by the 13 November 1985 eruption of Nevado del Ruiz in Colombia. Despite the moderate size of the eruption (VEI 2; 10–20 million m^3 of tephra: Pierson et al. 1990), interactions of hot pyroclasts with snow and ice released 30–50 million m^3 of meltwater in 30–90 min, feeding lahars with a volume of 10^8 m^3 that travelled at least 100 km (see Chap. 7). The eruption caused a 16% loss in area and 9% loss in volume of snow, firn and ice (Photo 39A–C). Turbulent PDCs mixed with snow mechanically to produce meltwater at a rate of 0.5–1.6 mm/s (Thouret et al. 2007). Such a melting rate is compatible with the removal of 2.3–8.2 m in thickness on average of snow and firn from the most affected 18 km^2 of the ice cap, based on isopach maps (Thouret 1990; Pierson and Janda 1994; Pierson et al. 1990; Thouret et al. 1995, 2007). The Nevado del Ruiz case study and other eruptions on snow- and ice-capped composite cones (Mt. Redoubt and Mt. Spurr in Alaska in 1990 and 1993, respectively: Waitt 1995; Ruapehu in New Zealand in 1995: Cronin et al. 1996; Tokachi-Dake in Japan in 1926: Katsui 1986; and Mt. St. Helens in Washington in 1980: Waitt et al. 1983) show that eruption-induced multi-phase gravity flows result from hot pyroclastics interacting with snow and ice, as summarized in a ternary diagram. Primary lahars and floods can result from eight interaction processes on ice-clad composite volcanoes (Thouret et al. 2007, and references therein): (i) snow avalanches, snowmelt and slushflows triggered by dry high-energy pyroclastic surges, (ii) wet-surge transformation to lahars; (iii) melting and incorporation of metre-thick snow-slab avalanches (Photo 39A); (iv) scouring and gullying effect of pyroclastic flows on steep glaciers (Photo 39B), transforming to ice-rich mixed diamictites; (v) turbulent pyroclastic flows and dilute pyroclastic surges that trigger dynamic mixing and fluid drag within a thick cover of loose snow and low-density firn (Photo 39C); (vi) mass failure of glacier ice caused by eruptive activity and seismic shaking; (vii) breakouts of glacier-dammed lakes by volcanic debris avalanches, and (viii) tephra-fall deposits interacting with snow, ice and liquid water.

The complex glacier-volcano interactions encompass mechanical and thermal effects (Fig. 17B), accounting for efficient and rapid heat transfer from PDCs to snow, firn and ice (Walder 2000a, b). All processes that generate meltwater are thermally driven. Snow becomes partially melted and vaporized, and the vapour flux fluidises the pyroclastic flow and causes it to convect through vapour-induced bubbling (Walder 2000a). The vapour and ingested air fluidize the flow, which can engulf all the meltwater produced. Vigorous bubbling and convection at the interface between PDCs and snow is favoured by high temperatures, small grain size and a relatively thin PDC layer. The unstable vapour-driven fluidization and convective motion scours and mixes snow into the PDC, resulting in a slurry (Walder 2000b). These PDCs include hot (>400°C) thin (<2 m) flows,

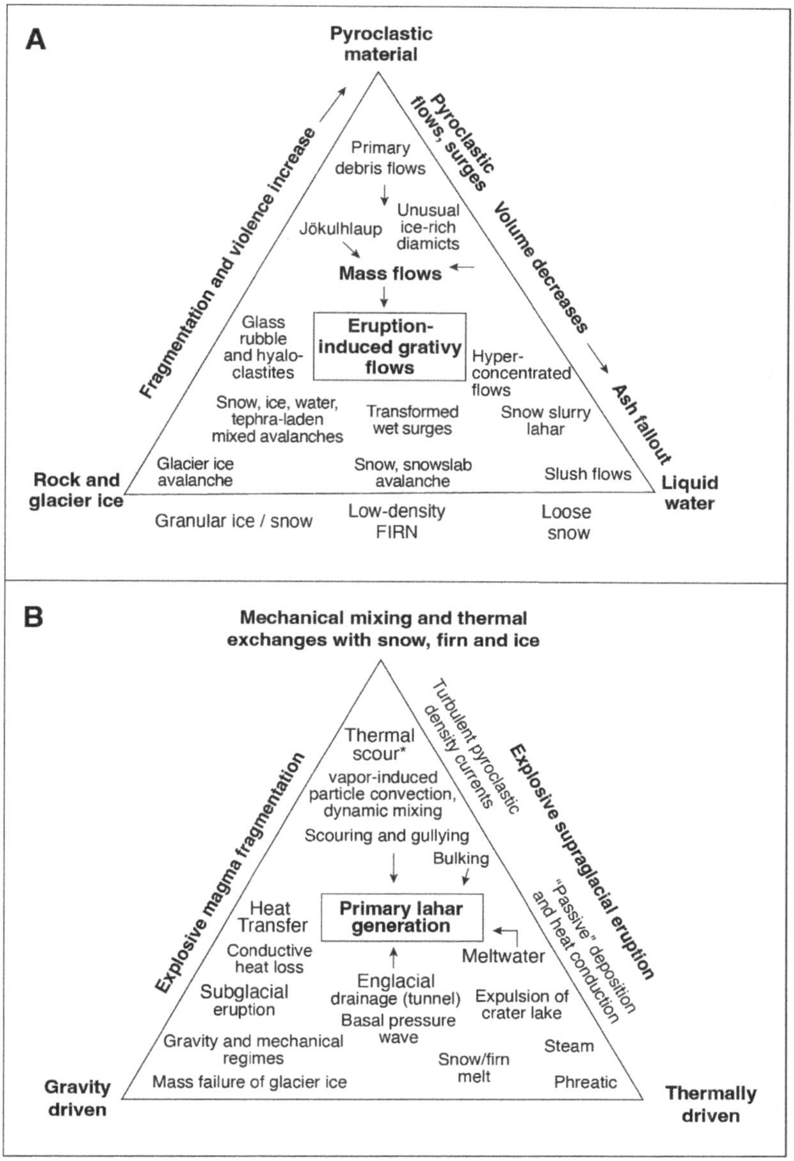

Fig. 17 Ternary diagrams showing deposits and processes linked to volcano-glacier interactions. **A** Multiphase gravity flows and deposits linked to volcano-glacier interactions on composite cones. This ternary diagram modified from Manville et al. (2000) *in:* Thouret et al. 2007, Fig. 1, p. 116 and Fig. 10, p. 124 is based on the Mt. St. Helens (1980), Nevado del Ruiz (1985), Mt. Redoubt (1990), Mt. Spurr (1992) and Ruapehu (1995) case studies. **B** Dynamic transformations between multiple gravity flows and ruption-triggered lahars on snow-capped composite cones ('thermal scour', 'vapour-induced particle convection', after Walder 2000b *in:* Thouret et al. (2007)

high-energy dry surges, as well as wet surges that transform into lahars owing to the segregation of their dense, wet part (Thouret et al. 2007 and references therein).

In summary, the complex interactions of hot tephra with snow and ice on composite cones include: the thermal erosion of snow, firn and ice; the mechanical abrasion and incorporation of snow and ice particles into PDCs; the turbulent regime of the PDC; and the englacial drainage producing tunnels or increasing their dimensions and scouring the ice margins at their outlets. All these processes contribute to the production of meltwater that eventually feeds large-volume lahars (up to 10^8 m^3) that can travel as far as 100 km from the ice cap (Thouret et al. 2007).

6 Elementary Volcanic Landforms in Underwater Environments

Subaqueous volcanism represents about 85% of the estimated annual, global volcanism (White et al. 2003, 2015a, b). Submarine volcanism is more widespread and voluminous than its subaerial counterpart, for the simple reason that sea floor occupies 70% of the Earth, and mid-ocean ridges (together with flood basalts at LIPs) produce the highest magma output of all tectonic settings (see Chap. 1). Table 2 after White et al. (2015b) shows the dominance of subaqueous volcanism over subaerial volcanism, yet it largely takes place unseen in deep sea environments.

The subaqueous environments are dominated by the lava-forming eruptions at MORs, and to a lesser extent by explosive submarine ones, either deep or shallow, typically in other tectonic settings such as island arcs, although explosive volcanism can also occur at or near MORs. In their review of the constraints, processes and products of submarine volcanism, Cas and Giordano (2014) state that the interaction between magma and ambient water can produce a variety of interaction styles, processes and deposits. The interaction may be *passive*, producing coherent lava because a vapour film at the interface insulates the magma from more dynamic interaction. The interaction may be *active*, when collapse of such an insulating vapour film leads to direct contact between liquid water and magma. Direct contact causes either instantaneous cooling of the magma to glass and a significant contraction in volume, leading to *in situ* quench

Table 2 Abundance and significance of explosive submarine eruptions (by White et al. 2015b)

Volcanism and composition	Subaerial	Subaqueous
Estimated % of global volcanism	15% (0.6 km^3/year)	85% (3.5 km^3/year)
• Basaltic volcanism	0.19 km^3/year (30%)	2.8 km^3/year (80%)
• Non-basaltic volcanism	0.43 km^3/year (70%)	0.6–0.7 km^3/year (20%)
Sulphur as non-water volatiles	~2.4 Mt/year	~14.6 Mt/year
Basalt ~ 5×10^6 tons/km^3	~0.93 Mt/year	~13.9 Mt/year
Other ~ 10^6 tons/km^3	1.4 Mt/year	~0.68 Mt/year
Volcanic heat released as percentage of global total	14% (30% at 1200 °C + 70% at 800 °C)	86% (80% at 1200 °C + 20% at 800 °C)
Estimated proportion of explosive eruptions	80% (maximal estimate) of 15% 12% of all eruptions are explosive subaerial ones	5% (minimal estimate) of 85% 4% of all eruptions are explosive subaqueous ones

Subaqueous eruptions, quantitatively dominated by spreading ridge events, are more commonly basaltic than subaerial ones, and make up a much larger share of total global volcanism. Some subaqueous eruptions are explosive. A small proportion of these explosive eruptions is enough to contribute significantly to the global transfer of heat and volatiles (sulphur here is representative of non-water volatiles) to the oceans. Volumes are derived from Crisp (1984), with a sixth of subduction volcanism considered subaqueous; basalt versus non-basalt proportions are estimates (White et al. 2015b) fragmentation, or thermal detonation of the liquid water, which is instantaneously superheated and expands explosively.

6.1 Constraints and Processes

All authors highlight how complex submarine volcanic processes are compared with subaerial processes, largely due to the very different properties of the ambient water mass compared with the atmosphere, and the way that erupting magma can interact with the water mass. Both water and magma have particular properties that can interact in subaqueous environments.

At least five properties of water affect volcanic eruption processes (Cas and Giordano 2014):

(1) The confining, hydrostatic pressure from an overlying water column is calculated at ~1 MPa per 100 m of water depth. The water depth corresponding to the critical pressure of pure water is about 2200 m, whereas in seawater it is about 3000 m. Explosions cannot occur at depths near to or greater than the critical point depth equivalent.
(2) Water has a high thermal conductivity and heat capacity. Heat transfer to water causes submarine melts to quench quickly, which helps preserve the syn-fragmentation microtextures in submarine pyroclastics (e.g. hyaloclastites), while glasses provide information about the depth of fragmentation.
(3) For water to behave explosively, it must transform from a supercritical state to a subcritical gas, which can expand explosively. This occurs when the ambient pressure is relatively low.
(4) Specific volume of water under pressure at magmatic temperatures. Since explosions represent instantaneous volume expansions, they can only occur under relatively low pressures.
(5) The bulk modulus, i.e. deformability/ compressibility of liquid water is 2340 MPa for seawater, and it increases with increasing pressure, which in subaqueous environments is equivalent to increasing water depth.

At least four properties of magma affect its eruption behaviour in submarine environments (Cas and Giordano 2014):

(1) Magma composition is mostly basaltic in mid-ocean ridge (MORB), ocean-island (OIB), and other intraplate settings. However, more evolved, rhyolitic compositions are also involved in island-arc magmas, which are more viscous and contain more volatiles.
(2) Volatile contents and their solubility in silicate magmas; oceanic basaltic magmas have low volatile contents compared with more evolved magmas, with carbon dioxide and water contents being significant in most cases.
(3) Thermal energy of magma. Basaltic phreatic/phreatomagmatic explosive eruptions are generally more energetic than rhyolitic ones. Conversely, when magma is supercooled by direct contact with liquid water, it is instantaneously chilled through the glass transition temperature to form glass.
(4) Viscosity of magmas erupted in submarine environments. The viscosity of magma erupted in deep water, where hydrostatic pressure is high, will be significantly lower than for a compositionally identical one erupted in shallow water, or one erupted in a subaerial setting from which volatiles can readily exsolve. This will affect the rheology of erupting magmas and their ability to vesiculate (Cas and Giordano 2014).

6.2 Abundant Submarine Landforms: Abyssal Hills, Seamounts, and Guyots

Three distinct submarine landforms on the Earth's seafloor are recognised:

(1) Topographic features known as *abyssal hills* (10–20 km long, 2–5 km wide, 50–300 m high) are commonly found near

spreading ridges. They are oriented perpendicular to the spreading direction, and they characterize ≥30% of the ocean floor (Macdonald et al. 1996).

(2) Larger than abyssal hills, underwater mountains called *seamounts* rise above the seafloor at a depth of 1000–5000 m bsl. They are the most abundant type of volcano, and yet the least explored physiographic feature on Earth (Photo 40A). Seamounts occur as isolated elevations on the seafloor with a circular or elliptical plan view, comparatively steep slopes, and a relatively small summit area (Menard 1964). Staudigel et al. (2010) defined a seamount as any geographically isolated topographic feature on the seafloor of more than 100 m in height including ones whose summit regions may temporarily emerge above sea level. This 100 m height threshold is relevant for seamounts built on a smooth abyssal terrain, while a height of 1000 m is more appropriate for seamounts in rough terrain (Staudigel and Koppers 2015). Seamount shape is conical or elongated with a pointy summit. They can be found in nearly all seafloor tectonic settings, in particular in intraplate settings. Seamounts outnumber subduction zone and hot spot volcanoes by several orders of magnitude (LaFemina 2015). Wessel (2001) and Wessel et al. (2010) estimated that there are at least 100,000 seamounts >1 km high and probably 25 million seamounts >100 m high, scattered over the ocean floor as well as along mid-oceanic ridges. A modern study of global distribution and morphology of seamounts 0.4–2.5 km-high identified 19,325 new seamounts (Gevorgian et al. 2023), expanding a previously published catalogue reporting 24,643 seamounts (Kim and Wessel 2011). They are generally extinct volcanoes composed of piles of lava that sometimes break the ocean surface. They resemble moderate-sized, subaerial shield volcanoes in shape, slope and height (McDonald 1972; Mendel and Sauter 1997; Staudigel and Koppers 2015). Seamounts include pillows as the most common lava-flow type. They also include volcaniclastic rocks, the most common type being fine-grained hyaloclastites, autochthonous pillow breccias, and allochthonous pillow-fragment breccias. The most widespread and voluminous material is the volcaniclastics deposited in the aprons of seamounts (Fig. 19; Schmincke 2004). The upper sections of seamounts show feeder dykes and horizontal sills as most important intrusive rock types, while plutonic rocks play a larger role in their deep structure (Fig. 19).

One of the best studied seamounts is Loihi, lying about 54 km south of Kilauea caldera (Big Island, Hawai'i: see Chap. 4), and 975 m below sea level. Its summit is about 2.5 km above the surrounding sea floor and its volume is currently about 1.7 km^3 (Clague and Sherrod 2014). Another example is the recently discovered active seamount termed 'Fani Maoré' to the east of Mayotte, one of the four volcanoes forming the Comoros Archipelago (North Mozambique channel, Fig. B in Appendix). Eruptive activity began here on 10 May 2018, and by May 2019 the volcano had grown to 820 m above the seafloor and ~5 km^3 in volume (Feuillet et al. 2021; Berthod et al 2022): this is the largest active submarine eruption ever documented. The new volcanic edifice lies at the tip of a 50-km-long ridge composed of many other recent edifices and lava flows. This volcanic ridge is an extensional feature inside a wide transtensional boundary that transfers strain between the East African and Madagascar rifts.

(3) A variation of the seamount is the *guyot,* which has a nearly flat top (Smith and Cann 1992; Smoot 1995). Named after the Swiss-American geographer A.H. Guyot, this term was given by the father of the ocean floor drift theory, H. Hess, who explained the occurrence of these flat-bottomed mountains at a distance from their associated reefs by their break-up and subsequent drift away from their place of origin.

Guyots look like seamounts, but their slopes are steeper, as they are composed of lava mixed with tephra, while the flattish form of their summit is due to wave erosion. They may have undergone processes such as crater and caldera formation or sector collapse. Their height from the seafloor can be up to 5500 m, with a basal diameter of 10–20 km. The guyots of the Pacific Ocean have been best studied. The Geisha seamounts in the Northwest Pacific (Photo 40B) are a good example of the type of submarine topography typical of this landform. Due to mass movements, the upper part of the guyots is steeper (20–25°), while the lower part develops a wide, flat debris apron. When the summit level first reaches sea level, the combined effects of wave abrasion, continued downslope mass movements, and the thermal subsidence of the oceanic lithosphere, result in a broad, flat-topped shape. Shoaling atolls have been created by fringing reefs above the subsiding guyot top. Although the corals that build up on the sinking volcano keep it elevated above sea level for some time, it eventually sinks. Darwin's point is the threshold at which the coral reef can no longer hold the atoll above the surface.

6.3 Types of Submarine Eruptions and Products

As with subaerial eruptions, submarine eruptions range from effusive and lava-forming to explosive ones, but effusive eruptions producing lava flows are much more common, in particular at and near MORs as well as on seamounts. In this section we consider processes that create underwater deposits and landforms.

6.3.1 Submarine Effusive Eruptions and Lava Flows

Effusive basaltic volcanism is the hallmark of subaqueous volcanism, while explosive eruptions are limited due to higher confining pressure underwater (see Sect. 6.1). The ambient pressure at the depth of MORs about 2500 m is 250 MPa. Most MOR basalts do not contain sufficient volatiles to erupt explosively: typical vesicle content for MOR basalts is <2 vol% and commonly less than 0.5 wt% (Gregg 2013), therefore MOR eruptions are effusive. In Chap. 1, we learned that submarine eruptions at MORs produce pillow lavas, lobate and sheet lavas of mafic composition. Submarine lava flow morphologies are similar to, but distinct from, subaerial basaltic lava morphologies because of the different ambient conditions (Gregg 2013). A number of terms describe pillowed, lobate, lineated, ropy (folded), or jumbled (hackly) lava flows. The corresponding landforms at and near MORs are: (1) *pillow lava cones* with lateral pillow ridges; (2) *pillow volcanoes* similar to small (hectometre-sized) shields; and (3) broad (kilometre-sized), tapered *sheet volcanoes* (Schmincke 2004).

However, an abundance of water does not produce lava pillows alone. *Lobate lava*, very common on the Pacific Ocean floor, is the dominant morphology on many fast and intermediate spreading-rate ridges, and reflects higher eruption rates than those at which pillow lavas are produced. Single lobes of lobate lavas are much wider than they are thick. Lobe surfaces are smooth, lacking the external features of pillow lavas, but many have ropy folds on the surface of the largest lobes (White et al. 2015a). Sheet lavas are produced at the highest eruption/flow rates. The surfaces may be smooth, ropey, lineated in the direction of flow, or jumbled. Sheet lavas are typically restricted to near-vent regions and areas where flow is constrained by topographic barriers or flow channelization (White et al. 2015a). Almost all submarine lava flows are inflated to some extent, which can result in submarine lava pillars. These are hollow cylinders of basalt commonly <20 m high and <2 m in diameter, morphologically similar to subaerial lava trees or tree moulds (see Chap. 2). They form during the emplacement, inflation, and drain-back of lobate sheet flows. Pillar height reflects the maximum thickness of the lava flow (Chadwick 2008 in: Gregg 2013).

Head and Wilson (2003) concurred with most earlier workers that for magma compositions typical of spreading centres and their vicinities, the most likely circumstance is the quiet effusion of magma with minor gas exsolution, and the production of somewhat vesicular pillow lavas or sheet flows, depending on effusion rate. Coherent lava flows such as basaltic pillow lavas and even coherent silicic lava flows and domes can be erupted on the seafloor without explosive interaction because of the efficient, slow cooling and insulation effects provided by 'film boiling' (Head and Wilson 2003). However, the vapour film, being a dynamic fluid, may experience natural instabilities such as Taylor and Kelvin–Helmholtz ones (e.g. Wohletz 1986), leading to collapse of the film and allowing direct contact between liquid water and magma. This would increase the cooling rate of the magma at the interface. High cooling rates cause rapid volume contraction, and very high contractional strain rates in the magma/glass. As a result, brittle contractional fracturing occurs and a network of contractional cooling cracks propagate inwards from the surface of the magma into the body of the magma as the magma-glass phase transition front propagates inward. The magma thus fragments in a process called *quench fragmentation*, or cooling contraction granulation/fragmentation, due to the thermal shock that the magma is subjected to. Magma may also vesiculate to a high degree in submarine environments without explosive fragmentation if the vesicle growth rate is slow, caused by a low volatile overpressure. Highly vesicular pillow lava can form at depths of 4 km, and pumiceous silicic lavas (with a pumice carapace) can occur.

Submarine lavas range from low aspect ratio basaltic pillow and massive sheet lavas to high aspect ratio rhyolitic lava flows and domes. As the SiO_2 content and viscosity increase, the tendency for submarine lavas to experience autobrecciation and quench fragmentation increases. Basaltic lavas vary from completely coherent to variably quench-fragmented pillows to massive sheet flows. Almost all andesitic to silicic lavas have marginal hyaloclastites and autobreccias, noted by transitions from the coherent core, to an *in situ* jigsaw-fit breccia and to variably clast-rotated breccia textures. Lava-forming eruptions can produce contemporaneous or post-eruptive aprons or deltas of debris. *Peperite*, which is a specific type of hyaloclastite (see Chap. 2), can form in any setting where magma intrudes into, or lava is flowing over, water-saturated, unconsolidated sediments (McPhie et al. 1993; Skilling 2002). Peperite results from the quench fragmentation of the magma by the pore water in the sediments, and the simultaneous heating of the pore water, which begins to convect and then dynamically mixes the juvenile clasts and the host sediment.

6.3.2 Explosive Submarine Environments and Products

Submarine explosive activity is much more common than previously assumed, in particular in island arc environments. Water depth for explosive eruptions is usually less than 1 km, although this remains contentious (White et al. 2015b), as indicated by explosive deposits found at depths of up to 4 km (Newland et al. 2022). Prior to the 1990s, the high ambient pressures exerted at most MORs were thought to preclude explosive eruptions. However, researchers have found bubble glass shards known as *limu o Pele* or *Pele's seaweed* along the Pacific MOR and at Loihi seamount (Chadwick et al. 2008). They formed from the violent expansion of magmatic gases and correspond to a Strombolian eruption style (Kokelaar 1986). Direct observations, using submersibles or remotely operated vehicles, of submarine explosive eruptions have revealed cyclic activity of mafic magma with pyroclastic material at depths <600 m at NW Rota-1 volcano, Marianas Arc (from 2004 through 2010), from two intermittently active vents about 1200 m bsl. at West Mata Volcano, Lau Basin in 2008, and probably between 0.7 and 1.5 km for explosive silicic eruptions in 2012 at Havre volcano, Kermadec Arc, New Zealand (Chadwick et al. 2008; Gregg 2013). The weakly explosive submarine eruption of NW Rota V volcano in the Marianas arc in 2006 at water depths of 550 m was similar in style to Strombolian explosions from subaerial volcanoes (Clague et al. 2009). Fragmentation probably resulted from both explosive fragmentation of very large gas bubbles, and by cooling-quenching granulation.

Explosive submarine eruptions involve two separate processes that contribute to the fragmentation of lava and to the formation of deep-sea pyroclastic rocks (Lockwood and Hazlett 2010). Simple quenching of hot lava results in differential stresses and fracturing that produce hyaloclastites (see Chap. 2). Recent studies of submarine pyroclastic deposits also indicate that actual lava fountains have opened on the deep seafloor, producing primary vitric ejecta that erupts as globules of molten rock, as fragments of glass bubbles and as Pele's tears. Beneath about 3000 m, water at high temperature becomes super-critical and two fluid phases can coexist at high temperature (Wohletz 2003), although neither of these fluids have the potential for the massive volume expansion that drives shallow-water and subaerial explosive volcanism. Explosive eruptions at these depths must be driven by exsolved magmatic gases contained within the erupting lava. The concentration of these gases (mainly CO_2) is not normally sufficient to drive explosive activity, but pockets of such gases may be formed by magmatic processes before they are erupted (Head and Wilson 2003).

Figure 18A (after Koyaguchi and Woods 1996 in: White et al. 2003) illustrates the effects of variations in water mass fraction and mass flux of basaltic magma on the dynamics of subaqueous eruption plumes. The diagram also shows two resulting types of eruption-fed deposits from: (1) tall, buoyant plumes with increasing mass fraction of water, and (2) low dense fountain-like plumes with a small mass fraction of water (Fig. 18A). These explosive eruptions involving mafic magma do not result in specific landforms apart from near-vent pyroclastic aprons, but instead create two types of pyroclastic deposit: (1) massive PDC beds, inversely-graded fallout deposits from high, buoyant plumes, and (2) either cold, wet PDC beds, or hot, dry PDC agglomerates, sometimes welded, which originate from low, dense plumes.

6.3.3 Two Environments, Deep and Shallow Water, for Explosive Eruptions

Explosive submarine volcanism can be divided into two environments in terms of depth below sea level (above and below the 300 m bsl threshold), i.e. deep and shallow water volcanism (McPhie and Allen 2008) (Fig. 18B).

The first type of deep subaqueous volcanism is driven by magmatic volatiles and can be broken down into three styles: (a) *Neptunian* (a term introduced by McPhie and Allen 2008), which includes a collapsing eruption column producing pumice-rich density currents and water-settled fall deposits, usually silicic in composition; (b) *Fountains*, usually basaltic in composition, producing fluidal lapilli and bombs, dynamic stressing, steam explosions and quench fragmentation, with near-vent fallout and a minor amount of resedimented facies, and; (c) *Strombolian-like*, usually basaltic in composition, producing *limu*, glassy vesicular ash and scoria lapilli.

The second deep subaqueous type is steam-driven. Steam explosions occur where water and wet sediment are trapped by low-viscosity lava, producing *limu*, with small volume and very limited dispersal.

Volcanism in shallow water occurs as phreatomagmatic explosive eruptions, with two eruptive styles: either Surtseyan, usually basaltic in composition, or phreatoplinian, usually silicic in composition (Fig. 18B).

Allen and McPhie (2009) described the Neptunian pyroclastic deposits in subaqueous volcanic successions as comprising massive to graded, very thick (several to tens of metres), laterally extensive (several kilometres) beds of nonwelded pumice lapilli with volumes up to tens of km^3. This facies may be overlain by laminated ash or bimodal ash and giant (>1 m) pumice clasts, and underlain by coarse lithic breccia. The association is inferred to be typical of sustained magmatic volatile-driven explosive eruptions from vents at water depths of ~1300–200 m. Eruption magnitude, fragmentation mechanisms, and juvenile pyroclast characteristics, especially vesicularity, are very similar in Neptunian and Plinian eruptions, but column behaviour differs primarily because of the contrasting physical properties of the ambient medium (water versus air).

Allen and McPhie (2009) considered the differences between Neptunian eruptions in subaqueous

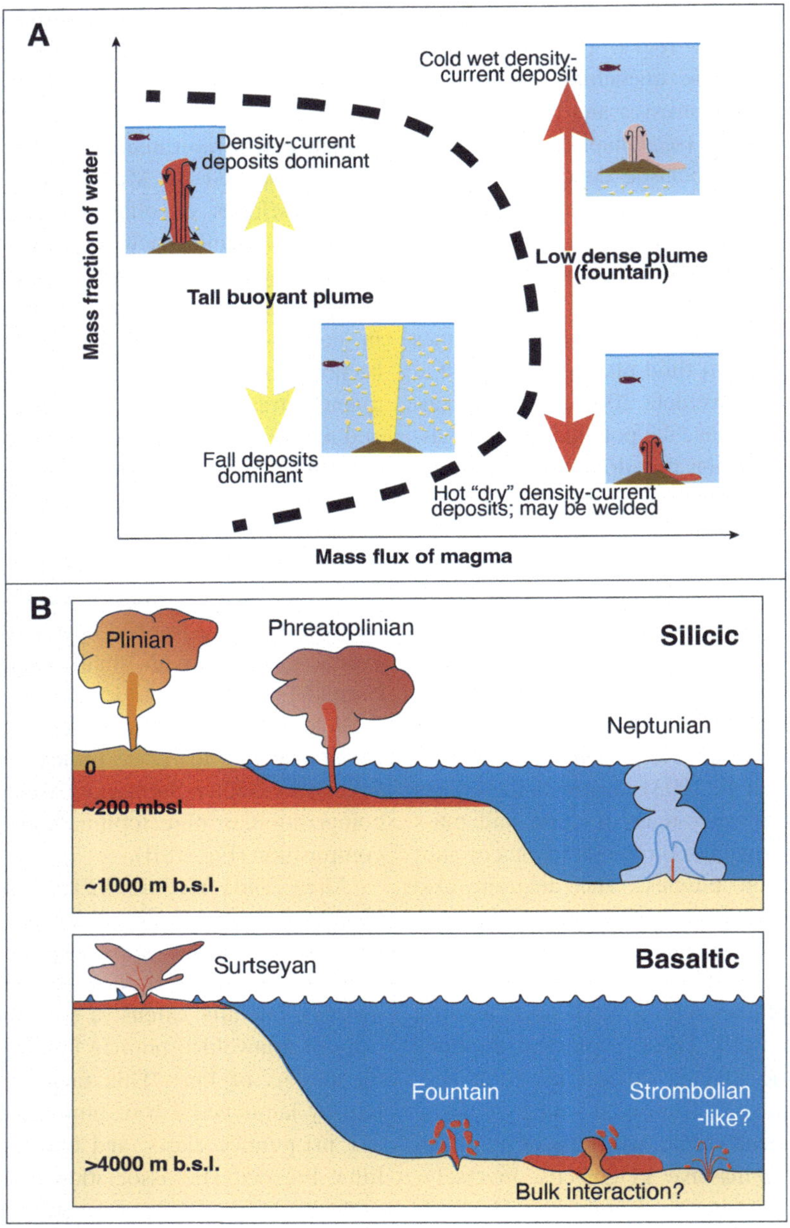

Fig. 18 Explosive submarine eruptions. **A** Effects of varying water mass fraction and mass flux of basaltic magma on the dynamics of subaqueous eruption plumes, and resulting eruption-fed deposits (modified after White et al. 2003, Fig. 4, p. 11). **B** Sketch depicting the principal modes of explosive submarine eruptions: shallow and deep submarine environments, with silicic and basaltic magma compositions (modified after Allen and McPhie 2009)

environments versus Plinian eruptions, their subaerial counterpart, as being constrained by hydrostatic pressure and the cooling effects of the ambient medium. They proposed that pyroclast dispersal in deep water involves a combination of flotational buoyant rise of some pyroclasts, especially large pumiceous blocks with hot, gas-rich interiors, fallout and collapse of columns due to the cooling effects of water on pyroclasts and condensation of their internal gases, making them dense and prone to collapse. Deposits include large rafted and settled pumiceous blocks, submarine fallout deposits, and hot to cold syn-eruptive pumiceous mass-flow deposits.

Head and Wilson (2003) showed that the most likely mechanism of production of pyroclastic deposits in deep submarine environments is Strombolian activity (Fig. 18A). This eruptive style is due to the localized concentration of volatiles in magma that have a low rise rate; magmatic gas collects by bubble coalescence, and ascends in large isolated bubbles which disrupt the magma surface in the vent, producing localized blocks, bombs, and pyroclastic deposits. Head and Wilson (2003) also stressed that a special case of the Hawaiian eruption style is likely to occur in submarine environments if magmatic gas buildup occurs in a magma reservoir by the upward drift of gas bubbles. In this case, a layer of foam will build up at the top of the reservoir in a sufficient concentration to exceed the volatile content necessary for disruption and Hawaiian-style activity.

The deposits and landforms are predicted to be somewhat different from those of a typical primary magmatic volatile-induced Hawaiian eruption. Typical pyroclast sizes might be smaller; and fountain heights may exceed those expected for the purely magmatic Hawaiian case. Cooling of descending pyroclasts would be more efficient, leading to different types of proximal deposits; and runout distances for density flows would be greater, potentially leading to submarine pyroclastic deposits surrounding vents up to distances of tens of metres to a kilometre.

Many examples of submarine fire fountain spatter deposits are recorded in the literature from both the modern seafloor and ancient submarine rock successions, with known occurrences at depths down to at least 2000 m. With high to moderate-height fountains, the larger surface area of the fountain, and the higher degree of expansion out of the vent results in greater interaction with the water mass leading to higher degrees of quench fragmentation of the magma and ejected spatter clasts. With lower fountains there is a high proportion of large spatter clasts to matrix because there is less interaction with the water mass. Submarine fire fountains are likely to behave in a similar manner to subaerial ones because both involve a dense jet of magma liquid injecting a less dense ambient fluid. Head and Wilson (2003) have, however, shown that deep-water fire fountains are likely to be lower than their subaerial counterparts because of the properties of the water mass. Deposition from subaqueous eruption columns will involve diminished ballistic dispersal, variations in settlement of fallout, and mass flow collapse of the eruption column once water has been ingested, generating hot to cold mass flows of pyroclastic debris (Fig. 18B).

If hot vesicular pumice is released into water, cooling and contraction of volatiles in the vesicles due to the cooling effects of the water mass will cause water to be drawn into the inter-connected vesicles, leading to the pumice becoming water-logged and less buoyant. Allen and McPhie (2009) suggested that this would make subaqueous eruption columns more prone to gravitational collapse than their subaerial counterparts. On the other hand, pyroclasts are more buoyant in water than in air because of the lower density contrast between solids and water than between solids and air. Water-settled fallout deposits should be better hydraulically sorted than subaerial counterparts.

6.3.4 Explosive Submarine Eruptions Involving Silicic Magma

Studies of submarine eruptions involving silicic magma compositions showed that they can also produce domes and cryptodomes (White et al. 2015b), for example at Milos Island in Greece (Stewart and McPhie 2006). Felsic magmas are subordinate to mafic magmas in ocean basins and much less frequently detected by ocean floor surveys. Silicic lavas and domes are common in modern submarine arc- and back-arc settings, while they form an integral component of numerous ancient submarine volcanic successions (White et al. 2015b). High-viscosity silicic magmas produce thicker and less extensive lavas than basaltic ones. They generate tabular lavas, equant domes and dome complexes, and lobe-hyaloclastite lavas, with coherent and autoclastic facies. Silicic tabular lavas and domes erupted under water are similar in size and shape to their subaerial counterparts. They are tens to hundreds of metres thick but shorter; most tabular lavas reach less than 10 km, and domes cover tens to a couple of hundred metres (White et al. 2015b). Some silicic submarine domes are entirely glassy, and some have outer zones of pumiceous hyaloclastite and coherent pumice facies. Quenched margins, cooling joints, and traces of hydrothermal alteration on the surface of lava domes can be characteristic, as in the case of the 7.3 ka Kikai-Akahoya submarine post-caldera rhyolite lava dome (Hamada et al. 2023).

'Giant' pumice clasts up to 10 m may spall from the upper margins of active submarine silicic lavas and domes (Fig. 18). Sometimes the spalled clasts rise through the water column and reach the surface. Once waterlogged, the clasts settle out of suspension to the seafloor. However, Cas and Giordano (2014) questioned this interpretation in their review. Based on a study by Rotella et al. (2013) around Macauley volcano in the Kermadec Arc, non-explosive rafted pumice fallout deposits found up 25 km around this volcano originated from the non-explosive detachment or spalling of highly vesiculated rafts of pumice from vesiculating magma emerging from a submarine vent. This hot pumice raft deposit highlights the importance of caution in the interpretation of pumice deposits found on the seafloor—they are not always in situ pyroclastic fallout deposits resulting from a significant explosive submarine eruption.

Large-scale underwater felsic explosive eruptions have formed calderas. Many ancient submarine explosive calderas and their deposits are well documented in the literature (e.g. Shinjima Island in Kagoshima Bay, Japan), with large-volume caldera-fill ignimbrites, some highly welded, making up the majority of the deposits. The mechanism for fragmentation and eruption of caldera-forming ignimbrites appears to be similar to that of subaerial calderas (see Chaps. 5 and 6). To ensure heat retention and welding in the erupting pyroclastic deposits, they must be shielded from direct interaction with the ambient water mass. This could be achieved by high mass and volumetric eruption rates, and the development of a gaseous and or steam cupola above the vent and the discharging pyroclastic mass filling the caldera. Submarine calderas are of similar size to subaerial calderas and shaped by similarly large and explosive eruption processes (White et al. 2015b). Modern submarine calderas are circular to elliptical and 2–10 km across, with flat or slightly concave floors 1–7 km wide underlain by pyroclastic deposits. Most explosive calderas appear to occur in depths of up to 0.1 km, although some deeper ones have been interpreted. The southern Kermadec Arc between the Australian and Pacific plates has 33 submarine volcanoes, with ~800 km^3 of silicic and rhyolitic centres, which includes calderas. One such caldera is Havre volcano whose rim lies between 600 and 1000 mbsl, and which erupted explosively in 2012 forming an eruption plume that was captured by satellite, and a pumice raft spreading from an initial 400 km^2 to cover ~20,000 km^2 within days (Carey et al. 2014 in: White et al. 2015b).

6.3.5 Shallow-Water Surtseyan Explosive Volcanism

This style of eruption has been recognised since 1962–1964 when the monogenetic tephra-cone volcano offshore Iceland was created (see Sect. 3.3.2). It is recognised as a typical phreatomagmatic explosive eruption style in relatively shallow water, such as marine shelves and in lakes. The deposits that make up Surtseyan tuff cones (see Sect. 3.3) are bedded, include subaerial and subaqueous fallout deposits, base surge deposits, abundant ballistic deposits, lateral tephra jet lobate deposits, and sometimes small lava flows, all reflecting the cone-building eruption stage. Phreatic and phreatomagmatic explosions are highly unlikely at depths of more than a few hundred metres because of the effects of increasing hydrostatic pressure in eliminating the stable film-boiling regime that is a prerequisite to thermo-hydraulic detonation of explosions.

7 Shoaling and Emergent Volcanic Islands

A broad description of monogenetic and compound submarine edifices would be incomplete without mentioning the *emergent volcanic islands*, which help in the interpretation of features and deposits of the *shoaling* phase during the growth of now emerged islands, such as the Madeira archipelago, Mid-Atlantic (Fig. A in Appendix) (Schmidt and Schmincke 2002).

Shoaling islands due to shallow submarine eruptions are the most known in the Pacific. Perhaps the best studied of them is Lateiki (previously known as Metis Shoal), a highly active volcano in the Tofua Arc with a number of temporary island-building eruptions and one submarine eruption in the last 55 years (Yeo et al. 2022). Several of its eruptions reportedly built ephemeral shoaling islands in the eighteenth and nineteenth centuries. The most recent eruption in 2019–2020 resulted in the construction of a short-lived island that was quickly eroded by wave action. A few more examples documented at volcanic islands that were shoaling volcanoes at some stage include Ba volcano, Viti Levu, Fiji (McPhie 1995), Tok Island, South Korea (Sohn 1995), and the evolution from a seamount to an emergent island at Porto Santo, Madera archipelago (Schmidt and Schmincke 2002). The evolution of the shoaling island of Hunga Tonga-Hunga Ha 'apai and its powerful phreatoplinian explosion in January 2022 provides another case study (see Sect. 3.3.3; Photo 41).

Six stages, interpreted from a succession of lithofacies, help reconstruct the growth of island volcanoes from subaqueous to shoaling and eventually subaerial environments (Sohn 1995; McPhie 1995; Schmidt and Schmincke 2002): (1) effusion of pillow lava in deep water, with autoclastic breccia and quenched fragments, and hyaloclastite formation below the pressure compensation level (PCL); (2) explosive, hydroclastic eruptions above the PCL but below sea level, producing clastic aprons with resedimented breccias and fragmented pillow lavas, (3) intercalation of bio-sedimentary lenses in relatively deep sea during periods of eruptive quiescence; (4) tephra from pyroclastic eruptions, redeposited when the shoaling volcano becomes eroded by sea waves; (5) phreatomagmatic deposits from eruptive phases building a tuff ring and tuff cone during an explosive transition from subaqueous to emergent environments, suggesting changes in the mixing ratio of water to magma; (6) after the volcano emerges above sea level, either subaerial lava flows cap the island, protecting the edifice against wave erosion, or pyroclastic or phreatomagmatic eruptions can persist if external water enters the volcano conduit by seepage through a permeable substrate. In addition, syn-volcanic intrusions from shallow magma batches may cut the entire succession of deposits.

Figure 20 shows a schematic cross section of a seamount that became an emergent ocean island near Gran Canaria in the Canary Islands (Schmincke 2004). The cross section shows an intrusive core made of plutonic rocks crosscut by dykes and sills, buttressed by lateral extrusive cores made up of pillow lavas, and clastic fans in deep water surrounding the seamount, the summit of which became covered by subaerial volcanics in turn when it emerged. The lower flanks and ring plain are mantled by aprons of distal, fine-grained deposits once the island shoals and emerges. Note in Fig. 20 that the entire island has subsided some time ago because subsidence rate has outcompeted growth due to the down-bent plate under the load of the Canary Islands oceanic shields. The oceanic to intermediate plate has been thinned by thermal melting from the Canary hot spot. Above the submarine stage indicated at the clastic fan, the emergent stage starts when volcaniclastic aprons are built on the flanks of the shoaling island. Widespread sediments may extend for more than 100 km away from the steep-sided volcanic edifice. The shoaling volcano grows and emerges above the sea surface when the eruptive rate is significantly higher than the erosional rate. Most clastic material produced is reworked and redeposited in adjacent sedimentary basins, forming volcanic aprons of fragmented pillow lava breccia, tephra layers, distal turbidites intercalated with products of landslides or debris avalanches (Fig. 20).

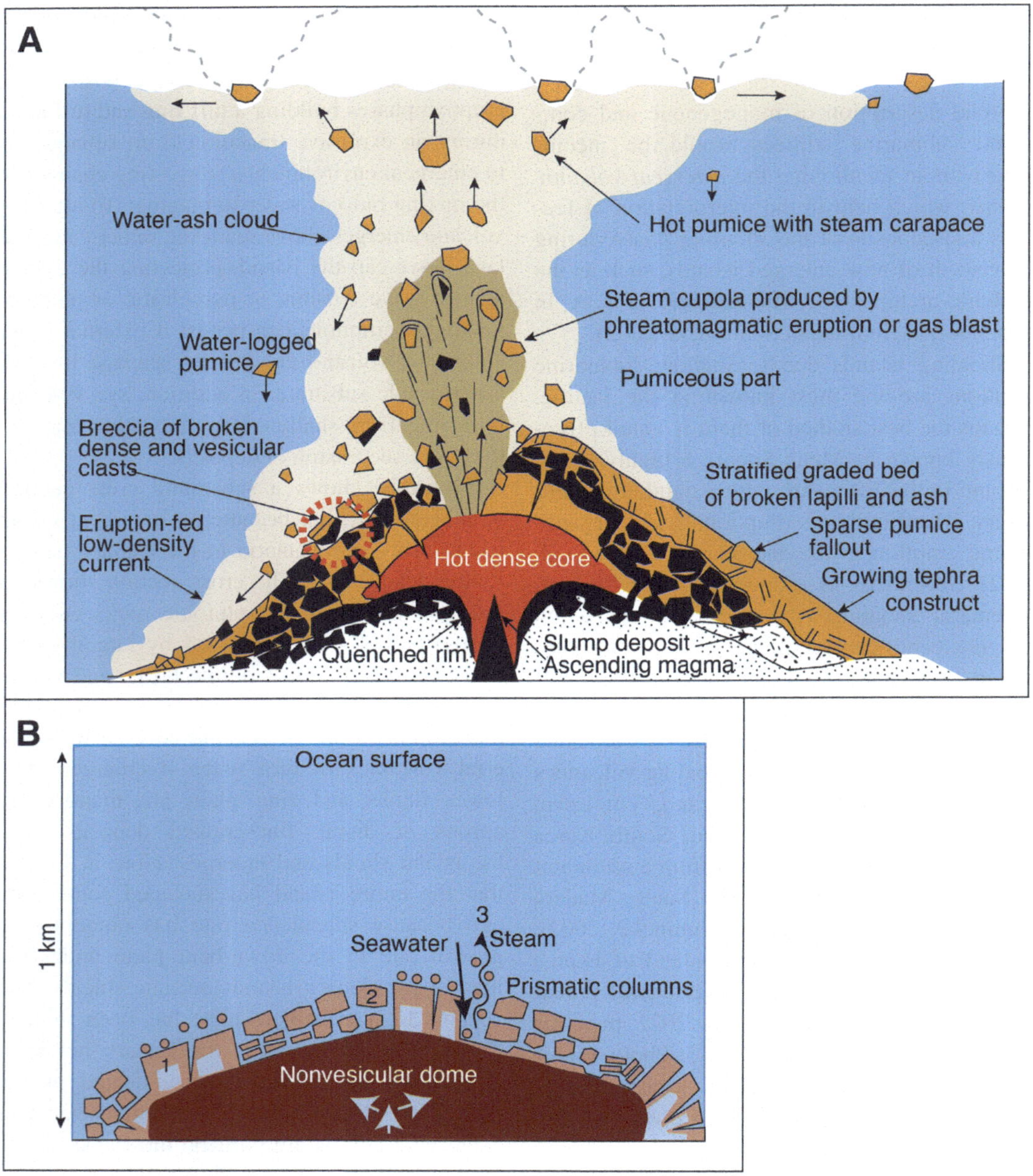

Fig. 19 Explosive submarine eruption and emplacement of dome (White et al. 2015b, Fig. 22.8, p. 419). **A** Eruption from a submarine pumiceous dome with phreatomagmatic or gas explosion and generation of a density current (Kano 2003). Fragments from quench-mechanical fragmentation and/or explosion accumulate as breccias around the dome. Hot pumice blocks may ascend to the water surface, drift, then settle separately when saturated with water. **B** Emplacement of deep-water silicic domes involves frothing of magma at the seafloor, chilling-in of vesicles in an outer carapace, and degassing of the lava core though interconnected vesicles that subsequently collapse. Numbers 1–3: 1. In situ pumice carapace, 2. Giant pumice slab, 3. Possible magma water explosions (Allen et al. 2010; White et al. 2015b)

Seamount growth and lifetime can be divided into six stages (Fig. 21) (Staudigel and Koppers 2015): (1) small seamounts (100–1000 m in height) are by far the most abundant type of seamount, made up of 80% pillow lavas; (2) mid-size seamounts (>1000 m in size) are high enough to begin establishing an internal magma plumbing system inside the volcano itself and above the oceanic crust; (3) the transition to shallow seamounts is mostly characterized by an increase in the production of submarine volcaniclastic rocks, while the transition to hydromagmatic activity becomes dominant at a depth of about 700 m; (4) islands and reefs are both formed when the summits of very large seamounts emerge above the sea level and there is a shift to subaerial volcanic activity. The islands produce substantial amounts of volcaniclastics (see Fig. 19). Coral reefs grow to compensate for the subsidence of the volcanic tropical islands caused by cooling of the lithosphere on which they were formed. Islands are first surrounded by a fringing reef, which develops into a barrier reef and finally into an atoll. Coral reefs ultimately drown, as their growth does not keep pace with the subsidence rate of the island; (5) volcanically extinct seamounts do not grow any further. Once completely drowned, seamounts do not erode any more, and as a result some of the older seamounts are among the oldest landforms on Earth (e.g. the 145 My-old Look Seamount in the Western Pacific), and: (6) subducting seamounts. The transport of seamounts towards subduction zones typically ends with their descent into the trench.

Submarine volcanic processes may be much more complex compared to subaerial processes, largely due to the way in which erupting magma can interact with the water mass. There is still ample scope for investigation of submarine landforms and processes, as a large number of underwater volcanic landforms remain hidden or unaccesible, and understanding of submarine eruptions lags behind that of their subaerial counterparts.

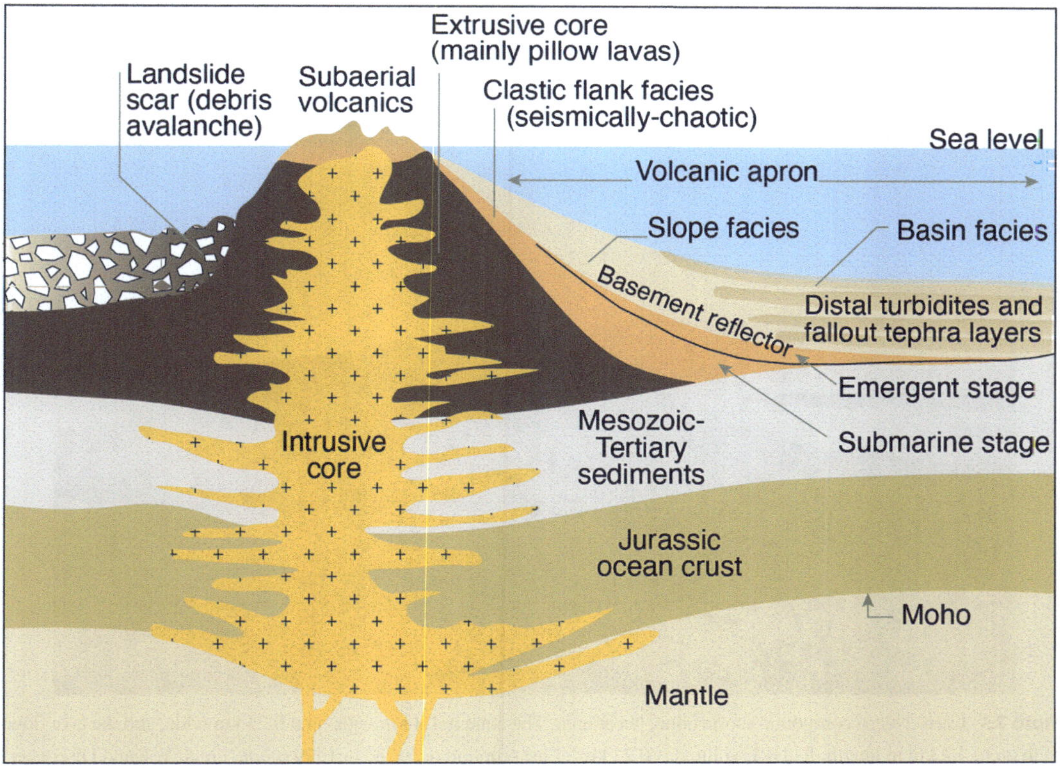

Fig. 20 Schematic cross section of a seamount/volcanic island and the clastic fan surrounding an island, based on drilling into the submarine flanks of Gran Canaria (after Schmincke 2004, Fig. 6.8, p. 76)

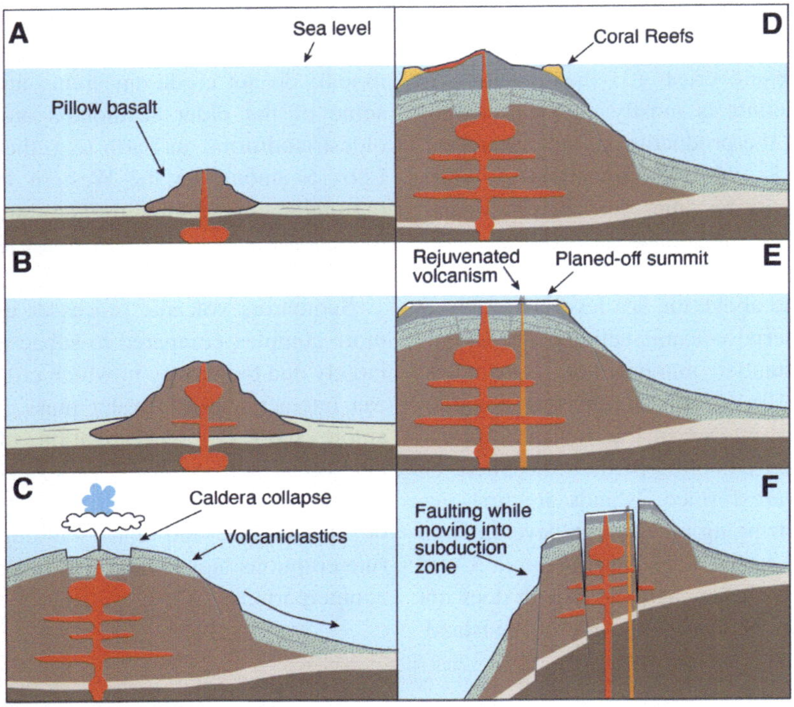

Fig. 21 Seamount evolution (after Staudigel and Clague 2010, *in:* Staudigel and Koppers 2015, Fig. 22.8, p. 419). Cycle of the birth, rise, caldera formation, emergence, subsidence, and the collapse of a seamount, ending with fragmentation of the seamounts in subduction zones

Photo 23 Cerro Negro compound scoria cone, Nicaragua. The cone is 1.4 km wide, and 0.25 km wide, and the lava flows reach up to 4.2 km in length. Its DRE volume is 0.22 km^3. Last eruption occurred in 1999 (*photo credit* Volcano Discovery)

Photo 24 Stútur scoria cone, Norðurnámshraun lava field (AD/CE 1477), Landmannalaugar, Iceland (*photo* D. Karátson). Cone width is 150 m, cone height is c. 60 m, crater width is 50 m. Note the greenish-yellowish moss cover on outer flanks

Photo 25 Kometsuka scoria cone, located on the northwest side of the Nakadake (to the right), a postcaldera cone complex in Aso caldera (Kyushu, Japan, *photo* D. Karátson). The cone formed c. 1800 years ago and its formation was accompanied by lava flows. The Aso caldera walls are visible in the background

Photo 26 Compound, monogenetic edifice of Puy Pariou, Chaîne des Puys, France 1 Initial tuff ring; 2. Ring segment pushed by lava flow from former lava lake; 3. Lava flow; 4. Strombolian cone. The Puy Pariou cone formed about 10,000 years ago (*photo* courtesy of P. Boivin)

Photo 27 Twin cones of Lassolas and La Vache, the youngest (~8600 years-old) cones of the Chaîne des Puys, France. The two horseshoe-shaped craters are largely breached and open to the south. Their highly explosive activity produced scoria, ash and pumice-fall deposits of trachybasaltic composition that covered an area in excess of 100 km^2 (*photo* courtesy of P. Boivin)

Photo 28 Sunset Crater, formed ~ AD/CE 1085 during the latest eruption of the San Francisco Volcanic Field, Arizona (USA; *photo* D. Karátson). The cone is located at about 2500 m a.s.l., with a height of about 300 m, a width of 1000–1200 m, has an uneven, slightly oval crater with a diameter of 400–600 m and a depth of about 120 m

Photo 29 Nabeyama pumice cone (Japan) as viewed from the north (*photo* D. Karátson). It was created during the Tenpyo/Hoji phreatomagmatic eruption in 764 AD/CE. Irregularly shaped on its western side (not visible), the cone has a basal diameter of c. 700 m, a crater diameter of almost 600 m (breached to the east) and a crater depth, which is equal to the height of the cone, of 200 m. Facing us, its northern side is strongly eroded by parasol ribbing

Photo 30 Lake Pavin maar looking south, located at the southern end of the Chaîne des Puys, France. It formed by a violent phreatomagmatic eruption ~6730 years ago at the foot of the Strombolian Montchal cone (in the background). The maar is about 905 m in diameter with a 58–110 m-high rim, while the lake is 770 m in diameter and 92 m deep (*photo* courtesy of M. Sagot)

Photo 31 Xico Volcán tuff ring, Valle de Chalco, 28 km southeast of Mexico City, looking north (Mexico). The tuff ring is ~1000 m in diameter, while its rim is 150 m in diameter and 25 to 50 m high. Note another slightly older, subdued tuff ring to the northeast of Xico (*photo* courtesy of C. Siebe)

Photo 32 The Hverfjall tuff ring (Iceland), part of the Krafla volcanic system, was created c. 2500 years ago by phreatomagmatic explosions at the southern end of a more than 20-km-long fissure (*photo* courtesy of A. Karátson)

Photo 33 The Víti Crater (Krafla caldera, Iceland), a tuff ring formed by explosive eruption in 1724 as part of the five-year-long period of eruptions known as the Myvatn Fires). Its diameter is 350 m, the lake is 150 m across (*photo* D. Karátson)

Photo 34 Laacher See in East Eifel Volcanic Field, Eifel Mountains (Germany; *photo* D. Karátson). Erupted 12,900 years ago, the low, elongated tuff ring, up to 140 m high and 1.2×1.6 km across (to the left), was formed by a range of preatomagmatic processes including pyroclastic fall, flow and base surge (to the right; geologist gives scale)

Photo 35 Koko tuff cone, eastern coast of Oahu (Hawai'i, USA), located on the eastern flank of the Koolau shield volcano. Known as Koko Crater, it is part of the Honolulu Series, and erupted during the early Holocene (*credit*: Google Earth). The elongated cone base is 2×1 km, while its crater is 600 m across, with the highest point at 368 m a.s.l

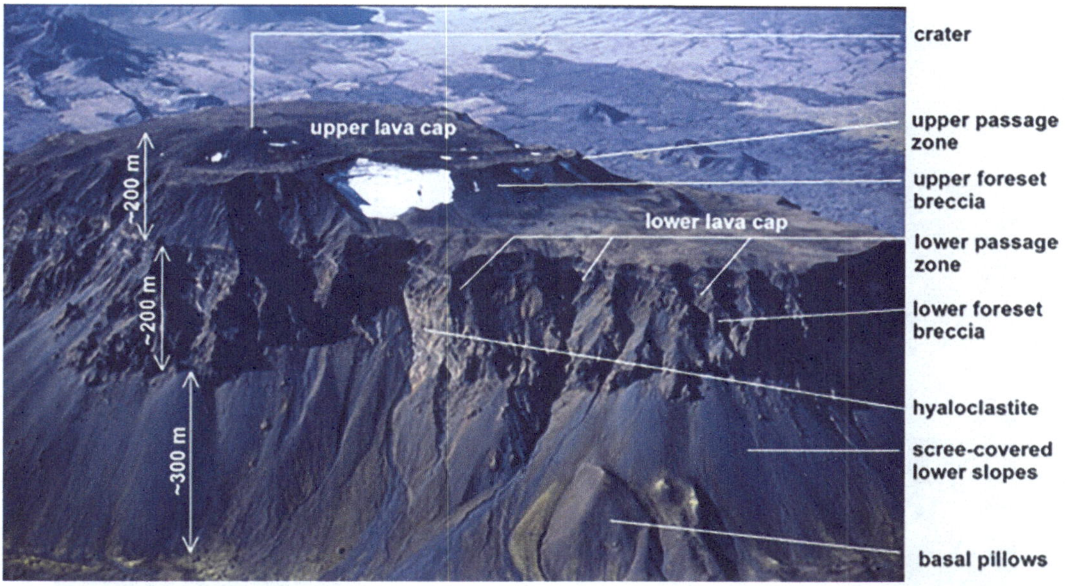

Photo 36 The tuya Hlöðufell in the Western Volcanic Zone, Iceland; aerial view from the northeast. The summit rises 700 m above the surrounding Holocene lava fields. The two lava caps indicate a rise in the level of the intraglacial lake during the formation of the table mountain (*photo* courtesy of M.T. Gudmundsson; see Jakobsson and Gudmunsson, 2008)

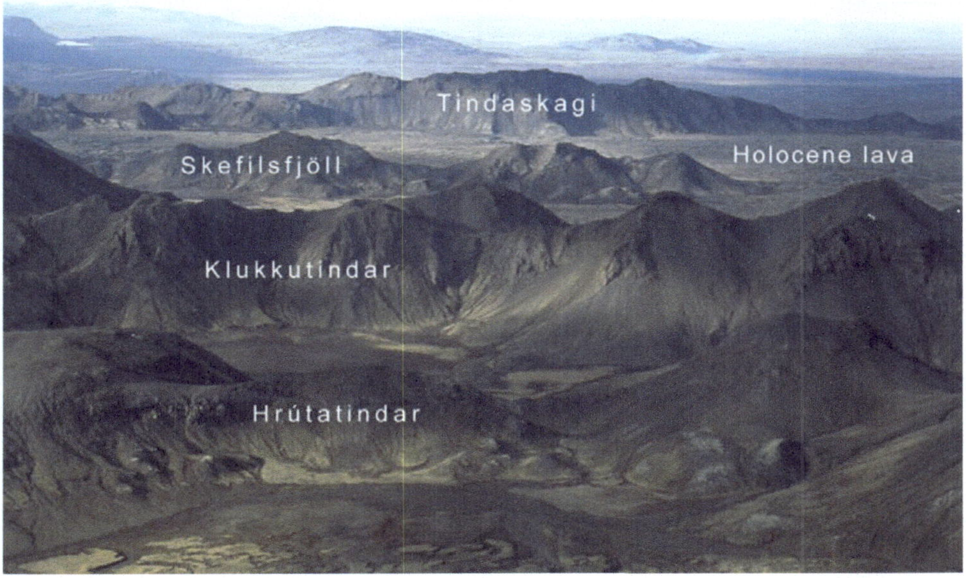

Photo 37 Tindar morphology (linear ridges of hyaloclastites and breccia), southwest of Hlöðufell tuya, SW Iceland. Aerial view from the southeast (*photo* courtesy of M.T. Gudmundsson; see Jakobsson and Gudmundsson, 2008)

Photo 38 Eyjafjallajökull, Iceland: 30–50 m-deep canyon carved in the Gígjökull glacier, fractured and covered by thick tephra during the 2010 eruption (*photo* courtesy of M.T. Gudmundsson, April 2010)

Photo 39 Transient landforms resulting from the effects of tephra and PDC interactions with the ice cap of Nevado del Ruiz Volcano, Colombia, 13 November 1985. **A** Septum of Arenas crater, Nevado del Ruiz: unstable, fractured ice cap and scar of ice avalanche slab (*photo* J.-C. Thouret). **B** Deep, radial gullies on tephra and snow at the ice cap margin (*photo* J.-C. Thouret). **C**. Mixed (ice, tephra and snow) avalanche deposit below the Nereidas glacier (*photo* J.-C. Thouret)

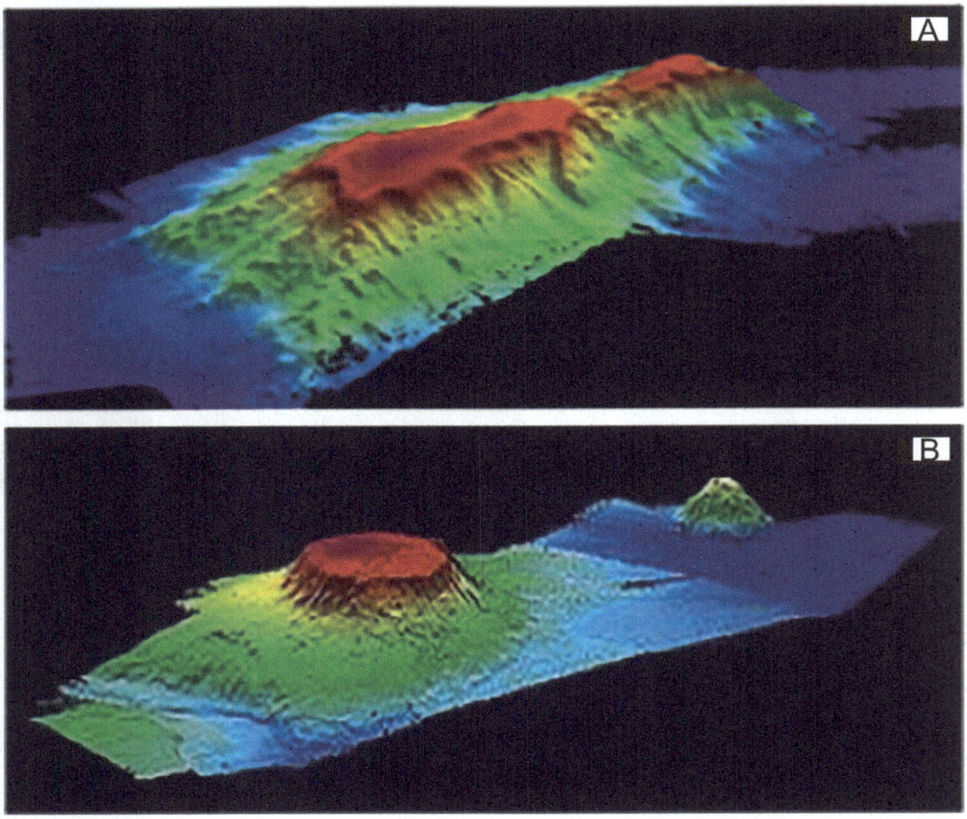

Photo 40 Seamount (**A**) and guyot (**B**), Geisha chain (94–102 Ma), northwest Pacific Ocean (Smoot, 1995). Seamount and guyot volumes range between 700 and 1500 km^3 each. Guyots rise 4.5–4.9 km above the marine basin floor; plateau areas, at about 1400 to 1500 m below seal level, range from 14 to 146 km^2 (*photo credit* Woods Hole Oceanographic Institution)

Photo 41 Honga Tonga-Hunga Ha'apai shoaling volcano, Tonga Island (above a submarine caldera) before and after the large 15 January 2022 eruption. **A** 7 January 2022; **B** 15 January 2022. *photo credit* Allenvi (2022)

References

Aka FT, Yokoyama T (2013) Current status of the debate about the age of Lake Nyos dam (Cameroon) and its bearing on potential flood hazards. Nat Hazards 65:875–885. https://doi.org/10.1007/s11069-012-0401-4

Alfano F, Ort MH, Pioli L, Self S, Hanson SL, Roggensack K, Allison CM, Amos R, Clatke AB (2019) Subplinian monogenetic basaltic eruption of Sunset Crater, Arizona, USA. Geol Soc Am Bull 131(3–4):661–674. https://doi.org/10.1130/B31905.1

Allen SR, McPhie J (2009) Products of neptunian eruptions. Geology 37(7):639–642. https://doi.org/10.1130/G30007A.1

Allen SR, Fiske RS, Tamura Y (2010) Effects of water depth on pumice formation in submarine domes at Sumisu, Izu-Bonin arc, western Pacific. Geology 38(5):391–394. https://doi.org/10.1130/G30500.1

Allenvi, Alliance nationale de Recherche pour l'Environnement (2022) L'éruption du volcan Hunga Tonga-Hunga Ha'apai le 15 janvier 2022 : un ébranlement du système Terre à l'échelle planétaire. Note Inter-organismes, Groupe Thématique Atmosphère, Coordination : Lac C. (CNRM), Le Pichon A (CEA) and Listowski C (CEA), 47 p.

Aranda-Gómez JJ, Luhr JF, Pier G (1992) The La Breña – El Jagüey Maar Complex, Durango, México: I. Geological evolution. Bull Volcanol 54:393–404. https://doi.org/10.1007/BF00312321

Auer A, Martin U, Németh K (2007) The Fekete-hegy (Balaton Highland Hungary) "soft-substrate" and "hard-substrate" maar volcanoes in an aligned volcanic complex—implications for vent geometry, subsurface stratigraphy and the palaeoenvironmental setting. J Volcanol Geoth Res 159(1–3):225–245. https://doi.org/10.1016/j.jvolgeores.2006.06.008

Austin-Erickson A, Büttner R, Dellino P, Ort MH, Zimanowski B (2008) Phreatomagmatic explosions of rhyolitic magma: experimental and field evidence. J Geophys Res Solid Earth 113(B11):B11201. https://doi.org/10.1029/2008JB005731

Baldini JUL, Brown RJ, Mawdsley N (2018) Evaluating the link between the sulfur-rich Laacher See volcanic eruption and the Younger Dryas climate anomaly. Clim past 14(7):969–990. https://doi.org/10.5194/cp-14-969-2018

Bani P, Oppenheimer C, Varekamp JC, Quinou T, Lardy M, Carn S (2009) Remarkable geochemical changes and degassing at Voui crater lake, Ambae volcano, Vanuatu. J Volcanol Geoth Res 188(4):347–357. https://doi.org/10.1016/j.jvolgeores.2009.09.018

Baxter PJ, Kapila M (1989) Acute health impact of the gas release at Lake Nyos, Cameroon, 1986. J Volcanol Geoth Res 39(2–3):265–275. https://doi.org/10.1016/0377-0273(89)90064-4

Becerra-Ramírez R, Dóniz-Páez J, González E (2022) Morphometric Analysis of Scoria Cones to Define the 'Volcano-Type' of the Campo de Calatrava Volcanic Region (Central Spain). Land 11(6):917. https://doi.org/10.3390/land11060917

Becerril L, Larrea P, Salinas S, Mossoux S, Ferrés D, Widom E, Siebe C, Martí J (2021) The historical case of Parícutin volcano (Michoacán, México): challenges of simulating lava flows on a gentle slope during a long-lasting eruption. Nat Hazards 107:809–829. https://doi.org/10.1007/s11069-021-04607-x

Begét JE, Hopkins DM, Charron SD (1996) The largest knownmaars on earth, Seward Peninsula, Northwest Alaska. Arctic, 49, 1, Calgary, AB, Canada, Arctic Institute of North America, 62–69.

Belousov A, Belousova M (2000) Formation of partially submerged tuff ring during the 1996 sublacustrine surtseyan eruption in Karymskoye lake, Kamchatka, Russia. In: Terra Nostra 2000/6: international Maar conference, pp 41–52

Bernard O, Bouvet de Maisonneuve C (2020) Controls on eruption style at Rabaul, Papua New Guinea—insights from microlites, porosity and permeability measurements. J Volcanol Geoth Res 406:107068. https://doi.org/10.1016/j.jvolgeores.2020.107068

Berthod C, Komorowski JC, Gurioli L, Médard E, Bachèlery P, Besson P, Verdurme P, Chevrel O, Di Muro A, Peltier A, Devidal JC, Nowak S, Thinon I, Burckel P, Hidalgo S, Deplus C, Loubrieu B, Pierre D, Bernell S, Pitel-Roudaut M, Réaud Y, Fouchard S, Bickert M, Le Friant A, Paquet F, Feuillet N, Jorry SL, Fouquet Y, Rinnert E, Cathalot C, Lebas E (2022) Temporal magmatic evolution of the Fani Maoré submarine eruption 50 km east of Mayotte revealed by in situ sampling and petrological monitoring. CR Géoscience 354(S2):195–223. https://doi.org/10.5802/crgeos.155

Boivin P, Gourgaud A (1978) Synchronisme des dynamismes de type "maar" et "strombolien" près de la Sauvetat (Sud-Devès, Massif-Central français). CR Som Soc Géol France 1:24–27

Boivin P, Bourdier J-L, Camus G, Goër de Herve A de., Gourgaud A, Kieffer G, Mergoil J, Vincent PM, Auby R (1982) Influence des magmas sur l'activité phréatomagmatique: approche volcanologique et thermodynamique. Bull Volcanol 45, 1: 25–39. https://doi.org/10.1007/BF02600387

Boivin P, Besson J-C, Briot D, Camus G, Deniel C, Gourgaud A, Labazuy P, Langlois E, de Larouzière F-D, Livet M, Médard E, Merceccia C, Mergoil J, Miaillier D, Morel J-M, Thouret J-C, Vernet G (2017) Volcanologie de la Chaîne des Puys, 6è ed, Parc naturel régional des Volcans d'Auvergne, Carte 1/25000 couleur, bilingual notice, 199 pp

Björnsson H (1992) Jökulhlaups in Iceland: prediction, characteristics and simulation. Ann Glaciol 16:95–106. https://doi.org/10.3189/1992AoG16-1-95-106

Björnsson H (2003) Understanding jökulhlaups: from tales to theory. J Glaciol 56(200):1002–1010. https://doi.org/10.3189/002214311796406086

Bonadonna C, Pistolesi M, Biass S, Voloschina M, Romero J, Coppola D, Folch A, D'Auria L, Martín-Lorenzo A, Dominguez L, Pastore C, Hardy M-PR,

Rodríguez F (2022) Physical characterization of long-lasting hybrid eruptions: the 2021 Tajogaite Eruption of Cumbre Vieja (La Palma, Canary Islands). J Geophys Res 127(11):e2022JB025302. https://doi.org/10.1029/2022JB025302
Brand BD, Brož P (2015) Tuff cone. In: Hargitai H, Kereszturi Á (eds) Encyclopedia of planetary landforms, vol 3. Springer, New York, pp 2197–2204. ISBN: 978-1-4614-3133-6
Brenna M, Cronin SJ, Smith IEM, Pontesilli A, Tost M, Barker S, Tonga'onevai S, Kula T, Vaiomounga R (2022) Post-caldera volcanism reveals shallow priming of an intra-ocean arc andesitic caldera: Hunga volcano, Tonga, SW Pacific. Lithos 412–413:106614. https://doi.org/10.1016/j.lithos.2022.106614
Brenna M, Cronin SJ, Smith IEM, Sohn YK, Maas R (2012) Spatio-temporal evolution of a dispersed magmatic system and its implications for volcano growth, Jeju Island Volcanic Field, Korea. Lithos 148:337–352. https://doi.org/10.1016/j.lithos.2012.06.021
Brož P, Čadek O, Hauber E, Rossi AP (2015) Scoria cones on Mars: Detailed investigation of morphometry based on high-resolution digital elevation models. J Geophys Res Planets 120(9):1512–1527. https://doi.org/10.1002/2015JE004873
Calvari S, Tanner LH (2011) The Miocene Costa Giardini diatreme, Iblean Mountains, southern Italy: model for maar-diatreme formation on a submerged carbonate platform. Bull Volcanol 73:557–576. https://doi.org/10.1007/s00445-010-0436-x
Carracedo JC, Troll VR, Day JMD, Geiger H, Junca MA, Soler V, Deegan F, Pérez Torrado FJ, Pinto GG, Gazel E, González AR, Minguez HA (2021) The 2021 eruption of the Cumbre Vieja Volcanic Ridge on La Palma, Canary Islands. Geol Today 38(3):94–107. https://doi.org/10.31223/X5D06D
Cas RAF, Wright JV (1988) Modern volcanoes and volcanic centres. In: Cas RAF, Wright JV (eds) Volcanic successions: modern and ancient. Springer, Dordrecht, Netherland, pp 362–410. ISBN-13: 978-0-412-44640-5
Cas RAF, Giordano G (2014) Submarine volcanism: a review of the constraints, processes and products, and relevance to the Cabo de Gata volcanic succession. It J Geosci 133(3):362–377. https://doi.org/10.3301/IJG.2014.46
Chadwick Jr WW, Cashman KV, Embley RW, Matsumoto H, Dziak RP, de Ronde SEJ, Lau TK, Deardorff ND, Merle SG (2008) Direct video and hydrophone observations of submarine explosive eruptions at NW Rota-1 volcano, Mariana arc. J Geophys Res 113:B08S10. https://doi.org/10.1029/2007jb005215
Civico R, Ricci T, Scarlato P, Taddeucci J, Andronico D, Del Bello E, D'Auria HPA, Pérez NM (2022) High-resolution Digital Surface Model of the 2021 eruption deposit of Cumbre Vieja volcano, La Palma, Spain. Sci Data 9:435. https://doi.org/10.1038/s41597-022-01551-8
Clague DA, Sherrod DR (2014) Growth and degradation of Hawaiian volcanoes. In: Poland MP, Takahashi TJ, Landowski CM (eds) Characteristics of Hawaiian Volcanoes, U.S. Geol Survey Prof Paper 1801, pp 97–146. http://dx.doi.org/10.3133/pp1801
Clague DA, Paduan JB, Davis AS (2009) Widespread Strombolian eruptions of mid-ocean ridge basalt. J Volcanol Geoth Res 180(2–4):171–188. https://doi.org/10.1016/j.jvolgeores.2008.08.007
Colton SH (1937) The basaltic cinder cones and lava flows of the San Francisco Mountain volcanic field. Mus N Ariz Bull 10:1–49
Corazzato C, Tibaldi A (2006) Fracture control on type, morphology and distribution of parasitic volcanic cones: an example from Mt. Etna, Italy. J Volcanol Geotherm Res 158(1–2):177–194. https://doi.org/10.1016/j.jvolgeores.2006.04.018
Cotton CA (1944) Volcanoes as landscape forms. Whitcombe and Tombs Ltd, Christchurch, New Zealand, 416 pp. ISSN: 0028-0836
Cotton CA (1952) Volcanoes as landscape forms. Wiley, NY, 416 pp. ISSN: 0028-0836
Crisp JA (1984) Rates of magma emplacement and volcanic output. J Volcanol Geoth Res 20(3–4):177–211. https://doi.org/10.1016/0377-0273(84)90039-8
Cronin SJ, Neall VE, Lecointre J, Palmer AS (1996) Unusual 'snow slurry' lahars from Ruapehu volcano, New Zealand, September 1995. Geology 24(12):1107–1110. https://doi.org/10.1130/0091-7613(1996)024
Cronin SJ, Brenna M, Smith IEM, Barker S, Tost M, Ford M, Tonga'onevai S, Kula T, Vaiomounga R (2017) New volcanic island unveils explosive past. EOS 98(1)
Cronin SJ, Németh K (2005) Where are the giant tuff cones and ignimbrites of Ambrym? A more conventional story of mafic volcanism at Ambrym Volcano, Vanuatu. In: Pettinga J, Wandres AM (eds) Abstract volume of the 50th annual conference, Kaikoura, New Zealand. Geol Soc New Zealand Misc Publ 119A, pp 21–22
Davì M, de Rosa R, Donato P, Sulpizio R (2011) The Lami pyroclastic succession (Lipari, Aeolian Islands): a clue for unravelling the eruptive dynamics of the Monte Pilato rhyolitic pumice cone. J Volcanol Geoth Res 201(1–4):285–300. https://doi.org/10.1016/j.jvolgeores.2010.09.010
De Goër de Hervé A, Camus G, Lavina P, Michelin Y, Montel J-L, Tort M (1999) Lemptégy volcan à ciel ouvert pour comprendre la Chaîne des Puys (The Lemptéguy: an open volcano to understand the Chaîne des Puys). Volcan de Lemptéguy sl, France, 56 pp. ISBN: 2-9514344-0-5
De Silva S, Lindsay JM (2015) Primary volcanic landforms. In: Sigurdsson H, Houghton B, McNutt SR, Rymer H, Stix J (eds) The The Encyclopedia of Volcanoes, 2nd edn. Academic, Elsevier, pp 273–297. ISBN: 978-0-12-385938-9
Delcamp A, van Wyk de Vries B, Stéphane P, Kervyn M (2014) Endogenous and exogenous growth of the monogenetic Lemptégy volcano, Chaîne des Puys, France. Geosphere 10(5):998–1019. https://doi.org/10.1130/GES01007.1

Dellino P, la Volpe L (1995) Fragmentation versus transportation mechanisms in the pyroclastic sequence of Monte Pilato-Rocche Rosse (Lipari, Italy). J Volcanol Geotherm Res 64(3–4):211–231. https://doi.org/10.1016/0377-0273(94)00084-T

Edwards BR, Tuffen H, Skilling IP, Wilson L (2009) Introduction to special issue on volcano-ice interactions on Earth and Mars: the state of the science. J Volcanol Geoth Res 185:247–250. https://doi.org/10.1016/j.jvolgeores.2009.06.003

Edwards BR, Magnusson E, Thordarson T, Gudmundsson MT, Höskuldsson A, Oddsson B, Haklar J (2012) Interactions between snow/firn/ice and lava/tephra during the 2010 Fimmvörduháls eruption, south-central Iceland. J Geophys Res 117:B04302. https://doi.org/10.1029/2011JB008985

Edwards BR, Gudmundsson MT, Russel JK (2015) Glaciovolcanism. In: Sigurdsson H, Houghton B, McNutt SR, Rymer H, Stix J (eds) The Encyclopedia of Volcanoes, chap 20, 2nd edn. Academic, Elsevier, pp 377–394. ISBN: 978-0-12-385938-9

Erlund EJ, Cashman KV, Wallance PJ, Pioli L, Rosi M, Johnson E, Granados HD (2010) Compositional evolution of magma from Parícutin Volcano, Mexico: the tephra record. J Volcanol Geotherm Res 197(1–4):167–187. https://doi.org/10.1016/j.jvolgeores.2009.09.015

Favalli M, Karátson D, Mazzarini F, Pareschi MT, Boschi E (2009) Morphometry of scoria cones located on a volcano flank: a case study from Mt. Etna (Italy), based on high-resolution LiDAR data. J Volcanol Geoth Res 186(3–4):320–330. https://doi.org/10.1016/j.jvolgeores.2009.07.011

Feuillet N, Jorry S, Crawford WC, Deplus C, Thinon I, Jacques E, Saurel JM, Lemoine A, Paquet F, Satriano C, Aiken C, Foix O, Philippe K, Laurent A, Rinnert E, Cathalot C, Donval J-P, Guyader V, Gaillot A, Scalabrin C, Moreira M, Peltier A, Beauducel F, Grandin R, Ballu V, Romuald D, Pelleau P, Gomez J, Besançon GL, Bernard P, Bachélery P, Fouquet Y, Bertil D, Lemarchand A, van Der Woerd J (2021) Birth of a large volcanic edifice offshore Mayotte via lithosphere-scale dyke intrusion. Nat Geosci 14(10):787–795. https://doi.org/10.1038/s41561-021-00809-x

Fornaciai A, Favalli M, Karátson D, Tarquini S, Boschi E (2012) Morphometry of scoria cones, and their relation to geodynamic setting: a DEM-based analysis. J Volcanol Geoth Res 217–218:56–72. https://doi.org/10.1016/j.jvolgeores.2011.12.012

Graettinger AH (2018) Trends in maar crater size and shape using the global Maar Volcano Location and Shape (MaarVLS) database. J Volcanol Geoth Res 357:1–13. https://doi.org/10.1016/j.jvolgeores.2018.04.002

Graettinger AH, Valentine GA (2017) Evidence for the relative depths and energies of phreatomagmatic explosions recorded in tephra rings. Bull Volcanol 79:88. https://doi.org/10.1007/s00445-017-1177-x

Garvin JB, Slayback DA, Ferrini V et al (2018) Monitoring and modeling the rapid evolution of earth's newest volcanic island: Hunga Tonga Hunga Ha'apai (Tonga) using high spatial resolution satellite observations. Geophys Res Lett 45(8):3445–3452. https://doi.org/10.1002/2017GL076621

Gevorgian J, Sandwell DT, Yu Y, Kim S-S, Wessel P (2023) Global distribution and morphology of small seamounts. Earth Space Sci 10:e2022EA002331. https://doi.org/10.1029/2022EA002331

Gregg TKP (2013) Deep-sea eruptions. In: Fagents SA, Gregg TKP, Lopes RMC (eds) Modeling volcanic processes. The physics and mathematics of volcanism, chap 12. Cambridge University Press, pp 258–274. ISBN: 978-0-521-89543-9

Gudmundsson MT (2003) Melting of ice by magma-ice-water interactions during subglacial eruptions as an indicator of heat transfer in subaqueous eruptions. In: White JDL, Smellie JL, Clague D (eds) Explosive subaqueous volcanism. Am Geophys Union Geophys Monograph 140, pp 61–72

Gudmundsson MT (2005) Subglacial volcanism in Iceland. In: Caseldine C, Russell A, Hardardóttir J, Knudsen O (eds) Iceland: modern processes, past environments, chap 6. Developments in quaternary science, vol 5. Elsevier, pp 127–151. ISBN: 0-444-50652-7

Gudmundsson MT, Sigmundsson F, Björnsson H, Högnadóttir T (2004) The 1996 eruption at Gjálp, Vatnajökull ice cap, Iceland: efficiency of heat transfer, ice deformation and subglacial water pressure. Bull Volcanol 66:46–65. https://doi.org/10.1007/s00445-003-0295-9

Hamada M, Hanyu T, McIntosh IM, Tejada MLG, Chang Q, Kaneko K et al (2023) Evolution of magma supply system beneath a submarine lava dome after the 7.3-ka caldera-forming Kikai-Akahoya eruption. J Volcanol Geoth Res 434:107738. https://doi.org/10.1016/j.jvolgeores.2022.107738

Hasenaka T, Carmichael ISE (1985) The cinder cones of Michoacán-Guanajuato, central Mexico: their age, volume and distribution, and magma discharge rate. J Volcanol Geoth Res 25(1–2):105–124. https://doi.org/10.1016/0377-0273(85)90007-1

Head JW, Wilson L (2003) Deep submarine pyroclastic eruptions: theory and predicted landforms and deposits. J Volcanol Geoth Res 151:155–193. https://doi.org/10.1016/S0377-0273(02)00425-0

Head JW, Sparks RS, Bryan WB, Walker GPL, Greely R, Whitford-Stark JL, Guest JE, Wood CA, Shultz PH, Carr MH (1981) Distribution and morphology of basalt deposits on planets. Basaltic volcanism on the terrestrial planets. Pergamon Press Inc., New York, pp 701–800

Heath M, Phillips D, Matchan EL (2020) Basalt lava flows of the intraplate Newer Volcanic Province in south-east Australia (Melbourne region): $^{40}Ar/^{39}Ar$ geochronology reveals ~8 Ma of episodic activity. J Volcanol Geoth Res 389:106730. https://doi.org/10.1016/j.jvolgeores.2019.106730

Houghton BF, Schmincke H-U (1989) Rothenberg scoria cone, East Eifel: a complex Strombolian and phreatomagmatic volcano. Bull Volcanol 52:28–48. https://doi.org/10.1007/BF00641385

Jaggar TA (1949) Steam blast volcanic Eruptions. A study of Mount Pelée in Martinique as type volcano. The Hawaiian Volcano Research Association, Honolulu, Hawaii, 137 pp

Jaimes-Viera MC, Martin Del Pozzo AL, Layer PW, Benowitz JA, Nieto-Torres A (2018) Timing the evolution of a monogenetic volcanic field: Sierra Chichinautzin, Central Mexico. J Volcanol Geoth Res 356:225–242. https://doi.org/10.1016/j.jvolgeores.2018.03.013

Jakobsson SP, Gudmundsson MT (2008) Subglacial and intraglacial volcanic formations in Iceland. Jökull Spec Issue: the Dynamic Geology of Iceland 58:179–196

Jarosch A, Gudmundsson MT, Högnadóttir T, Axelsson G (2008) Progressive cooling of the hyaloclastite ridge at Gjálp, Iceland, 1996–2005. J Volcanol Geoth Res 170:218–229. https://doi.org/10.1016/j.jvolgeores.2007.10.012

Jensen RA, Donelly Nolan JM (2017) Field-trip guide to the geologic highlights of Newberry Volcano, Oregon. Scient Investig Rep 2017-5022-J2, 30 pp. https://doi.org/10.3133/sir20175022J2

Johnston EN, Sparks RSJ, Phillips JC, Carey S (2014) Revised estimates for the volume of the Late Bronze Age Minoan eruption, Santorini, Greece. J Geol Soc 171:583–590. https://doi.org/10.1144/jgs2013-113

Jones JG (1969) Intraglacial volcanoes of the Laugarvatn region, south-west Iceland-1. J Geol Soc London 124:197–211. https://doi.org/10.1144/gsjgs.124.1.0197

Jordan SC, Le Pennec JL, Gurioli L et al. (2016) Highly explosive eruption of the monogenetic 8.6 ka BP La Vache et Lassolas scoria cone complex (Chaîne des Puys, France). J Volcanol Geoth Res 313:15–28. https://doi.org/10.1016/j.jvolgeores.2015.12.006

Kano K (2003) Subaqueous pumice eruptions and their products: a review. In: White JDL, Smellie JL, Clague DA (eds) Explosive subaqueous volcanism. Geophys Monogr Ser 140:203–229. https://doi.org/10.1029/140GM14

Karátson D, Telbisz T, Gertisser R, Strasser T, Nomikou P, Druitt T, Vereb V, Quidelleur X, Kósik S (2020) Constraining the landscape of Late Bronze Age Santorini prior to the Minoan eruption: Insight from volcanological, geomorphological and archaeological findings. J Volcanol Geotherm Res 401:106911. https://doi.org/10.1016/j.jvolgeores.2020.106911

Katsui Y et al (1986) The 1985 eruption of Nevado del Ruiz volcano, Colombia, and associated mudflow disaster (in Japanese). Rep Natural Disaster Scientific Res, Japan, No B-60-7, 102 pp

Kereszturi G, Jordan G, Németh K, Dóniz-Páez JF (2012) Syn-eruptive morphometric variability of monogenetic scoria cones. Bull Volcanol 74:2171–2185. https://doi.org/10.1007/s00445-012-0658-1

Kereszturi G, Németh K (2013) Monogeentic basaltic volcanoes: genetic classification, growth, geomorphology and degradation. In: Németh K (ed) Updates in volcanology—new advances in understanding volcanic systems. Intech, pp 3–88. ISBN: 978-953-51-0915-0

Kienle J, Kyle PR, Self S, Motyka RJ, Lorenz V (1980) Ukinrek Maars, Alaska, I. April 1977 eruption sequence, petrology and tectonic setting. J Volcanol Geoth Res 7(1–2):11–37. https://doi.org/10.1016/0377-0273(80)90018-9

Kim S, Wessel P (2011) New global seamount census from altimetry-derived gravity data. Geophys J Int 186(2):615–631. https://doi.org/10.1111/j.1365-246x.2011.05076.x

Kling GW, Evans WC, Tanyileke G, Kusakabe M, Ohba T, Yoshida Y, Hell JV (2005) Degassing Lakes Nyos and Monoun: defusing certain disaster. Proc Natl Acad Sci 102(40):14185–14190. https://doi.org/10.1073/pnas.0502274102

Kobayashi T, Miki D, Sasaki H, Iguchi M, Yamamoto T, Uto K (2013) Geological map of Sakurajima volcano. 2nd edn. Geological Survey of Japan

Kokelaar BP (1983) The mechanism of Surtseyan volcanism. J Geol Soc 140:939–944. https://doi.org/10.1144/gsjgs.140.6.0939

Kokelaar BP (1986) Magma-water interactions in subaqueous and emergent basaltic volcanism. Bull Volcanol 48:275–289. https://doi.org/10.1007/BF01081756

Komatsu G, Arzhannikov SG, Arzhannikova AV, Ershov K (2007) Geomorphology of subglacial volcanoes in the Azas plateau, the Tuva Republic, Russia. Geomorphology 88:312–328. https://doi.org/10.1016/j.geomorph.2006.12.002

Kurszlaukis S, Mahotkin I, Rotman AY, Kolesnikov GV, Makovchuk IV (2009) Syn- and post-eruptive volcanic processes in the Yubileinaya kimberlite pipe, Yakutia, Russia, and implications for the emplacement of South African-style kimberlite pipes. Lithos 112(1):579–591. https://doi.org/10.1016/j.lithos.2009.05.016

Kusakabe M (2017) Lakes Nyos and Monoun gas disasters (Cameroon)—limnic eruptions caused by excessive accumulation of magmatic CO_2 in crater lakes. Geochem Monogr Ser 1(1):1–50. https://doi.org/10.5047/gems.2017.00101.0001

LaFemina PC (2015) Plate tectonics and volcanism. In: Sigurdsson H, Houghton B, McNutt SR, Rymer H, Stix J (eds) The Encyclopedia of Volcanoes, chap 3, 2nd edn. Academic, Elsevier, pp 65–92. ISBN: 978-0-12-385938-9

Larrea P, Salinas S, Widom E, Siebe C, Abbitt RJF (2017) Compositional and volumetric development of a monogenetic lava flow field: the historical case of Parícutin (Michoacán, Mexico). J Volcanol Geotherm Res 348:36–48. https://doi.org/10.1016/j.jvolgeores.2017.10.016

Larsen G (2003) A brief overview of eruptions from ice-covered and ice-capped volcanic systems in Iceland during the past 11 centuries: frequency, periodicity and implications. In: Smellie JL, Chapman MG (eds) Volcano-ice interaction on Earth and Mars, vol 202. Geol Soc London Spec Pub, pp 81–90, 431 pp. ISBN: 1-86239-121-1

Lescinsky DT, Fink JH (2000) Lava and ice interaction at stratovolcanoes: use of characteristic features to determine past glacial extents and future volcanic hazards. J Geophys Res 105:23711–23726. https://doi.org/10.1029/2000JB900214

Lockwood JP, Hazlett RW (2010) Volcanoes, global perspectives. Wiley-Blackwell, UK, 539 pp. ISBN: 978-1-4051-6249-4

Lorenz V (1970) Some aspects of the eruption mechanism of the Big Hole Maar, Central Oregon. Geol Soc Amer Bull 81(6):1823–1830. https://doi.org/10.1130/0016-7606(1970)81[1823:SAOTEM]2.0.CO;2

Lorenz V (1973) On the formation of maars. Bull Volcanol 37(2):183–204. https://doi.org/10.1007/BF02597130

Lorenz V (1975) Formation of phreatomagmatic maar-diatreme volcanoes and its relevance to kimberlite diatremes. In: Physics and chemistry of the Earth. First international conference on Kimberlites, Cape Town, 24–28 September 1973, pp 17–27. https://doi.org/10.1016/B978-0-08-018017-5.50006-7

Lorenz V (1986) On the growth of maars and diatremes and its relevance to the formation of tuff rings. Bull Volcanol 48:265–274. https://doi.org/10.1007/BF01081755

Lorenz V (2003) Maar-diatreme volcanoes, their formation, and their setting in hard-rock or soft-rock environments. Geolines 15:72–83

Luhr JF, Simkin T, Cuasay M (1993) Parícutin: the volcano born in a Mexican cornfield. US Geoscience Press, Arizona, 427 pp

Macdonald KC, Fox PJ, Alexander RT, Pockalny R, Gente P (1996) Volcanic growth faults and the origin of Pacific abyssal hills. Nature 380:125–129. https://doi.org/10.1038/380125a0

MacLeod NS, Sherrod DR, Chitwood LA, Jensen RA (1995) Geologic Map of Newberry volcano, Deschutes, Klamath, and Lake Counties, Oregon: US Geol Survey Miscell Geol Investig Map I-2455, scales 1:62,500 and 1:24,000

Magnússon E, Gudmundsson MT, Sigurdsson G, Roberts M, Höskuldsson F, Oddson B (2012) Ice-volcano interactions during the 2010 Eyjafjallajökull eruption, as revealed by airborne radar. J Geophys Res 117:B07405. https://doi.org/10.1029/2012JB009250

Major JJ, Newhall CG (1989) Snow and ice perturbation during historical volcanic eruptions and the formation of lahars and floods. A global review. Bull Volcanol 52:1–27. https://doi.org/10.1007/BF00641384

Manville V, Hodgson KA, Houghton BF, Keys JR, White JDL (2000) Tephra, snow and water: complex sedimentary responses at an active snow-capped stratovolcano, Ruapehu, New Zealand. Bull Volcanol 62:278–293. https://doi.org/10.1007/s004450000096

Mannen K, Ito T (2007) Formation of scoria cone during explosive eruption at Izu-Oshima volcano, Japan. Geophys Res Lett 34(18):L18302. https://doi.org/10.1029/2007GL030874

Martin U, Németh K (2004) Mio/Pliocene phreatomagmatic volcanism in the western Pannonian Basin. Geological Institute of Hungary, Budapest, 193 pp. ISBN: 963 671 238 7

Martin U, Németh K (2007) Blocky versus fluidal peperite textures developed in volcanic conduits, vents and crater lakes of phreatomagmatic volcanoes in Mio/Pliocene volcanic fields of Western Hungary. J Volcanol Geotherm Res 159(1–3):164–178. https://doi.org/10.1016/j.jvolgeores.2006.06.010

McDonald GA (1972) Volcanoes. Prentice-Hall, New Jersey, 509 pp. ISBN: 0-13-942219-6

McGarvie DW (2009) Rhyolitic volcano-ice interactions in Iceland. J Volcanol Geoth Res 185:367–389. https://doi.org/10.1016/j.jvolgeores.2009.06.003

McGetchin TR, Settle M, Chouet BA (1974) Cinder cone growth modeled after Northeast Crater, Mount Etna, Sicily. J Geophys Res 79(23):3257–3272. https://doi.org/10.1029/JB079i023p03257

McKnight SB, Williams SN (1997) Old cinder cone or young composite volcano? The nature of Cerro Negro, Nicaragua. Geology 25(4):339–342. https://doi.org/10.1130/0091-7613(1997)025<0339:OCCOYC>2.3.CO;2

McPhie J (1995) A Pliocene shoaling basaltic seamount: Rakiraki, Fiji. J Volcanol Geoth Res 64:193–210. https://doi.org/10.1016/0377-0273(94)00050-Q

McPhie J, Allen SR (2008) Explosive submarine volcanism (Unpubl). Course of the CEV Commission of Explosive Volcanism, IAVCEI General Assembly, Reykjavik

McPhie J, Doyle M, Allen R (1993) Volcanic Textures. A guide to the interpretation of textures in volcanic rocks. CODES Key Centre, University of Tasmania, 196 p. ISBN 0-85901-522-X

Mathews WH (1947) 'Tuyas', flat-topped volcanoes in northern BC. Am J Sci 245(9):560–570

Mattsson HB, Höskuldsson Á, Hand S (2005) Crustal xenoliths in the 6220 BP Saefell tuff-cone, south Iceland: evidence for a deep, diatreme-forming, Surtseyan Eruption. J Volcanol Geotherm Res 145(3–4): 234–248. https://doi.org/10.1016/j.jvolgeores.2005.01.009

Mee K, Tuffen H, Gilbert J (2006) Snow-contact volcanic facies and their use in determining past eruptive environments at Nevados de Chillán volcano, Chile. Bull Volcanol 68:363–376. https://doi.org/10.1007/s00445-005-0017-6

Menard HW (1964) Marine geology of the Pacific. McGraw Hill (No. QE39. M46 1964)

Mendel V, Sauter D (1997) Seamount volcanism at the super slow-spreading Southwest Indian Ridge between 57°E and 70°E. Geology 25(2):99–102. https://doi.org/10.1130/0091-7613(1997)025<0099:SVATSS>2.3.CO;2

Mergoil J, Boivin P (1993) Le Velay, son volcanisme et les formations associées. Géologie de la France 3: 3-96, 1 carte couleurs

Montanaro C, Mortensen AK. Weisenberger TB, Dingwell DB, Scheu B (2021) Stratigraphic reconstruction of the Víti breccia at Krafla volcano (Iceland): insights into pre-eruptive conditions priming explosive eruptions in geothermal areas. Bull Volcanol 83(11):81. https://doi.org/10.1007/s00445-021-01502-y

Moore JG (1985) Structure and eruptive mechanisms at Surtsey Volcano, Iceland. Geol Mag 122(6):649–661. https://doi.org/10.1017/S0016756800032052

Morgan JK, Moore GF, Clague DA (2003) Slope failure and volcanic spreading along the submarine south flank of Kilauea volcano, Hawaii. J Geophys Res Solid Earth 108(B9):2415. https://doi.org/10.1029/2003JB002411

Moriya I (1994) Types of Quaternary Volcanoes along the Circum-Pacific Mobile Belt. J Geogr 103(7):730–748. https://doi.org/10.5026/jgeography.103.7_730

Moriya I (1986) Morphometry of pyroclastic cones in Japan. Bull Soc Vol Jpn 3:58–76

Morrissey M, Zimanowski B, Wohletz K, Büttner R (2000) Phreatomagmatic fragmentation. In: Sigurdsson H, Houghton BF, McNutt SR, Rymer H, Stix J (eds) The Encyclopedia of Volcanoes. Academic, San Diego, pp 431–445

Murcia H, Németh K (2020) Effusive monogenetic volcanism. In: Németh K (eds) Updates in volcanology—transdisciplinary nature of volcano science. IntechOpen, London, pp 1–15. https://doi.org/10.5772/intechopen.94387

Nam K, Wang F, Yan K, Zhu G (2023) Characteristics and time-series monitoring by GOES-17 of volcano flume on 15 January 2022 from Tonga submarine volcanic eruption. Geoenviron Disasters 10:2. https://doi.org/10.1186/s40677-023-00232-x

Németh K, Haller MJ, Siebe C (eds) (2011) Maars and scoria cones: the enigma of monogenetic volcanic fields. J Volcanol Geoth Res 201(1–4):V–VIII. https://doi.org/10.1016/j.jvolgeores.2011.03.001

Németh K, Goth K, Martin U, Csillag G, Suhr P (2008) Reconstructing paleoenvironment, eruption mechanism and paleomorphology of the Pliocene Pula maar, (Hungary). J Volcanol Geotherm Res 177(2):441–456. https://doi.org/10.1016/j.jvolgeores.2008.06.010

Németh K, Kereszturi G (2015) Monogenetic volcanism: personal views and discussion. Int J Earth Sci 104: 2131–2146. https://doi.org/10.1007/s00531-015-1243-6

Németh K, Martin U, Csillag G (2003) Erosion rate calculation based on eroded monogenetic alkaline basaltic volcanoes of the Mio/Pliocene Bakony-Balaton Highland Volcanic Field, Hungary. Geolines 15:93–97

Newland EL, Mingotti N, Woods AW (2022) Dynamics of deep-submarine volcanic eruptions. Sci Rep 12, 3276. https://doi.org/10.1038/s41598-022-07351-9

Nishihara A, Geshi N, Naruo H (2022) Long-term change of the eruption activities of Sakurajima volcano, Japan, inferred from the fallout tephra deposits. Front Earth Sci 10:988373. https://doi.org/10.3389/feart.2022.988373

Oddsson B, Gudmundsson MT, Edwards BR, Thordarson T, Eyjólfur Magnússon E, Sigurðsson G (2016) Subglacial lava propagation, ice melting and heat transfer during emplacement of an intermediate lava flow in the 2010 Eyjafjallajökull eruption. Bull Volcanol 78:48. https://doi.org/10.1007/s00445-016-1041-4

Ollier CD (1967) Maars. Their characteristics, varieties and definition. Bull Volcanol 31:45–73

Ollier CD (1988) Volcanoes. Basil Blackwell, Oxford, p 228

Ort MH, Lefebvre NS, Neal CA, McConnell VS, Wohletz KH (2018) Linking the Ukinrek 1977 maar-eruption observations to the tephra deposits: new insights into maar depositional processes. J Volcanol Geoth Res 360:36–60. https://doi.org/10.1016/j.jvolgeores.2018.07.005

Palladino DM, Valentine GA, Sottili G, Taddeucci J (2015) Maars to calderas: end-members on a spectrum of explosive volcanic depressions. Front Earth Sci 3:36. https://doi.org/10.3389/feart.2015.00036

Pedersen GBM, Grosse P, Gudmundsson MT (2020) Morphometry of glaciovolcanic edifices from Iceland: types and evolution. Geomorphology 370:107334. https://doi.org/10.1016/j.geomorph.2020.10733 40169-555X

Petronis MS, Delcamp A, van Wyk de Vries B (2013) Magma emplacement into the Lemptégy scoria cone (Chaîne des Puys, France) explored with structural, anisotropy of magnetic susceptibility, and paleomagnetic data. Bull Volcanol 75:753. https://doi.org/10.1007/s00445-013-0753-y

Pioli L, Erlund E, Johnson E, Cashman K, Wallance P, Rosi M, Granados HD (2008) Explosive dynamics of violent Strombolian eruptions: the eruption of Parícutin Volcano 1943–1952 (Mexico). Earth Planet Sci Lett 271(1–4):359–368. https://doi.org/10.1016/j.epsl.2008.04.026

Pirrung M, Büchel G, Lorenz V, Treutler H-C (2007) Post-eruptive development of the Ukinrek East Maar since its eruption in 1977 A.D. in the periglacial area of south-west Alaska. Sedimentology 55(2):305–334. https://doi.org/10.1111/j.1365-3091.2007.00900.x

Porter SC (1972) Distribution, morphology, and size frequency of cinder cones on Mauna Kea Volcano, Hawaii. Geol Soc Am Bull 83(12):3607–3612. https://doi.org/10.1130/0016-7606(1972)83[3607:DMASFO]2.0.CO;2

Portnyagin MV, Hoernle K, Mironov NL (2012) Contrasting compositional trends of rocks and olivine-hosted melt inclusions from Cerro Negro volcano (Central America): implication for decompression-driven fractionation of hydrous magmas. Int J Earth Sci 103:1963–1982. https://doi.org/10.1007/s00531-012-0810-3

Pierson TC, Janda RJ (1994) Volcanic mixed avalanches: a distinct eruption-triggered mass-flow process at snow-clad volcanoes. Geol Soc Amer Bull 106(10):1352–1358. https://doi.org/10.1130/0016-7606(1994)106

Pierson T, Janda R, Thouret J-C, Borrero C (1990) Perturbation and melting of snow and ice by the 13 November 1985 eruption of Nevado del Ruiz, Colombia, and consequent mobilization, flow, and deposition of lahars. J Volcanol Geoth Res 41(1–4):17–66. https://doi.org/10.1016/0377-0273(90)90082-Q

Reynolds HI, Gudmundsson MT, Högnadottir T, Magnússon E, Pálsson F (2017) Subglacial volcanic activity above a lateral dyke path during the 2014-2015 Bárðarbunga-Holuhraun rifting episode, Iceland. Bull Volcanol 79:28. https://doi.org/10.1007/s00445-017-1122-x

Riedel C, Ernst GGJ, Riley M (2003) Controls on the growth and geometry of pyroclastic constructs. J Volcanol Geotherm Res 127:121–152. https//doi.org/10.1016/S0377-0273(03)00196-3

Robin C, Eissen J-P, Monzier M (1993) Giant tuff cone and 12-km-wide associated caldera at Ambrym volcano (Vanuatu, New Hebrides Arc). J Volcanol Geoth Res 55:225–238. https://doi.org/10.1016/0377-0273(93)90039-T

Romero JE, Burton M, Cáceres F, Taddeucci J, Civico R, Ricci T, Pankhurst MJ, Hernández PA, Bonadonna C, Llewellin EW, Pistolesi M, Polacci M, Solana C, D'Auria L, Arzilli F, Andronico D, Rodríguez F, Asensio-Ramos M, Martín-Lorenzo A, Hayer C, Scarlato P, Perez NM (2022) The initial phase of the 2021 Cumbre Vieja ridge eruption (Canary Islands): products and dynamics controlling edifice growth and collapse. J Volcanol Geoth Res 431:107642. https://doi.org/10.1016/j.jvolgeores.2022.107642

Ross P-S, Delpit S, Haller MJ, Németh K, Corbella H (2011) Influence of the substrate on maar-diatreme volcanoes—an example of a mixed setting from the Pali Aike volcanic field, Argentina. J Volcanol Geoth Res 201(1–4):253–271. https://doi.org/10.1016/j.jvolgeores.2010.07.018

Ross P-S, Núñez GC, Hayman P (2017) Felsic maar-diatreme volcanoes: a review. Bull Volcanol 79:20. https://doi.org/10.1007/s00445-016-1097-1

Rotella MD, Wilson CJN, Barker SJ, Wright IC (2013) Highly vesicular pumice generated by buoyant detachment in subaqueous volcanism. Nat Geosci 6:129–132. https://doi.org/10.1038/NGEO1709

Rouwet D, Chiodini G, Ciuccarelli C, Comastri A, Costa A (2019) Lago Albano, the "anti-Nyos-type" lake: the past as a key for the future. J Afr Earth Sci 150:425–440. https://doi.org/10.1016/j.jafrearsci.2018.09.019

Russell JK, Edwards BR, Porrit LA, Ryane C (2014) Tuyas: a descriptive genetic classification. Quaternary Sci Rev 87:70–81. https://doi.org/10.1016/j.quascirev.2014.01.001

Schmidt R, Schmincke H-U (2002) From seamount to oceanic island, Porto Santo, central east Atlantic. Int J Earth Sci 91:594–614. https://doi.org/10.1007/s00531-001-0243-x

Schmincke H-U (2004) Volcanism. Springer, Berlin, 324 pp. ISBN: 3-540-43650-2

Schopka HH, Gudmundsson MT, Tuffen H (2006) The formation of Helgafell, southwest Iceland, a monogenetic subglacial hyaloclastite ridge: sedimentology, hydrology and volcano–ice interaction. J Volcanol Geoth Res 152:359–377. https://doi.org/10.1016/j.jvolgeores.2005.11.010

Segerstrom K (1960) Erosion and related phenomena at Parícutin in 1957. US Geol Surv Bull 1104-A, pp 1–18

Segerstrom K (1966) Parícutin, 1965—aftermath of eruption. US Geol Surv Prof Paper 550-C, US Government Printing Office, Washington, pp C93–C101

Settle M (1979) The structure and emplacement of cinder cone fields. Am J Sci 279:1089–1107

Sheridan MF, Wohletz KH (1981) Hydrovolcanic explosions: the systematics of water-pyroclast equilibration. Science 212(4501):1387–1389. https://doi.org/10.1126/science.212.4501.1387

Sheridan MF, Wohletz KH (1983) Hydrovolcanism: basic considerations and review. J Volcanol Geoth Res 17(1–4):1–29. https://doi.org/10.1016/0377-0273(83)90060-4

Siebe C, Macías JL (2004) Volcanic hazards in the Mexico City metropolitan area from eruptions at Popocatépetl, Nevado de Toluca, and Jocotitlán stratovolcanoes and monogenetic scoria cones in the Sierra Chichinautzin Volcanic Field. In: Siebe C, Macías JL, Aguirre-Díaz GJ (eds) Neogene-Quaternary continental margin volcanism: a perspective from Mexico. Geol Soc Amer Spec Paper 402, pp 253–329. https://doi.org/10.1130/2004.VHITMC.PFG-p1

Skilling IP (2002) Basaltic pahoehoe lava-fed deltas: large-scale characteristics, clast generation, emplacement processes and environmental discrimination. In: Smellie JL, Chapman MG (eds) Volcano-ice interaction on Earth and Mars. Geol Soc London Spec Publ. 202, pp 91–113, 431 pp. ISBN: 1-86239-121-1

Smellie JL (2000) Subglacial eruptions. In: Sigurdsson H, Houghton B, McNutt SR, Rymer H, Stix J (eds) The Encylopedia of Volcanoes. Academic, Elsevier, pp 403–418. ISBN: 978-0-12-385938-9

Smellie JL (2006) The relative importance of supraglacial versus subglacial meltwater escape in basaltic subglacial tuya eruptions: an important unresolved conundrum. Earth-Sci Rev 74:241–268. https://doi.org/10.1016/j.earscirev.2005.09.004

Smellie JL (2008) Basaltic subglacial sheet-like sequences: evidence for two types with different implications for the inferred thickness of associated ice. Earth-Sci Rev 88:60–88. https://doi.org/10.1016/j.earscirev.2008.01.004

Smellie JL, Skilling IP (1994) Products of subglacial volcanic eruptions under different ice thickness: two examples from Antarctica. Sedim Geol 91:115–129. https://doi.org/10.1016/0037-0738(94)90125-2

Smellie JL, Johnson JS, McIntosh WC, Esser R, Gudmundsson MT, Hambrey M, Wyk de Vries B (2008) Six million years of glacial history recorded in volcanic lithofacies of the James Ross Island volcanic group, Antarctica peninsula. Paleogeogr Paleoclim Paleoeco 260:122–148. https://doi.org/10.1016/j.palaeo.2007.08.011

Smith DK, Cann JR (1992) The role of seamount volcanism in crustal construction at the Mid-Atlantic Ridge (24–30°N). J Geophys Res 97(B2):1645–1658. https://doi.org/10.1029/91JB02507

Smoot NC (1995) Mass wasting and subaerial weathering in guyot formation: the Hawaïan and Canary Ridges as examples. Geomorphology 14:29–41. https://doi.org/10.1016/0169-555X(95)00035-4

Sohn YK (1995) Geology of Tok Island, Korea: eruptive and depositional processes from a shoaling to emergent island volcano. Bull Volcanol 56:660–674. https://doi.org/10.1007/BF00301469

Sohn YK, Chough SK (1989) Depositional processes of the Suwolbong tuff ring, Cheju Island (Korea). Sedimentology 36(5):837–855. https://doi.org/10.1111/j.1365-3091.1989.tb01749.x

Sohn YK, Chough SK (1992) The Ilchulbong tuff cone, Cheju Island, South Korea. Sedimentology 39(4): 523–544. https://doi.org/10.1111/j.1365-3091.1992.tb02135.x

Sohn YK, Cronin SJ, Brenna M, Smith IEM, Németh K, White JDL, Murtagh RM, Jeon YM, Kwon CW (2012) Ilchulbong tuff cone, Jeju Island, Korea, revisited: a compound monogenetic volcano involving multiple magma pulses, shifting vents, and discrete eruptive phases. Geol Soc Amer Bull 124(3–4):259–274. https://doi.org/10.1130/B30447.1

Staudigel H, Clague DA (2010) The geological history of deep-sea volcanoes. Biosphere, hydrosphere, and lithosphere interactions. Oceanography 23:58–71. https://www.jstor.org/stable/24861062

Staudigel H, Koppers AAP (2015) Seamounts and Island building. In: Sigurdsson H, Houghton B, MacNutt SR, Rymer H, Stix J (eds) The Encyclopedia of Volcanoes, chap 22, 2nd edn. Academic, Elsevier, pp 405–421. ISBN: 978-0-12-385938-9

Staudigel H, Koppers AAP, Lavelle JW, Pitcher TJ, Shank TM (2010) Defining the word seamount. Oceanography 23(1, Spec Issue on Mountains in the Sea, Box 1):20–21

Stewart AL, McPhie J (2006) Facies architecture and Late Pliocene-Pleistocene evolution of a felsic volcanic island, Milos, Greece. Bull Volcanol 68:703–726. https://doi.org/10.1007/s00445-005-0045-2

Stix J, Maarten de Moor J (2018) Understanding and forecasting phreatic eruptions driven by magmatic degassing. Earth Planets Space 70:83. https://doi.org/10.1186/s40623-018-0855-z

Thorarinsson S (1966) The Surtsey eruption. Course of events during the year, pp 84–90

Thorarinsson S (1967) The Surtsey eruption and related scientific work. Polar Rec 13(86):571–578. https://doi.org/10.1017/S0032247400058113

Thouret J-C (1990) Effects of the 13 November eruption on the snow pack and ice cap of Nevado del Ruiz (Colombia). J Volcanol Geoth Res 41(1–4):177–201. https://doi.org/10.1016/0377-0273(90)90088-W

Thouret J-C, Vandemeulebrouck J, Komorowski J-C, Valla F (1995) Volcano-glacier interactions: field survey, remote sensing, and modelling—case study: Nevado del Ruiz, Colombia. In: Slaymaker O (ed) Steepland geomorphology. Wiley, pp 63–88. ISBN: 978-0-340-76417-6

Thouret J-C (1999) Volcanic geomorphology—an overview. Earth Sci Rev 47(1–2):95–131. https://doi.org/10.1016/S0012-8252(99)00014-8

Thouret J-C, Ramirez J, Naranjo JL, Gibert-Malengreau B, Vandemeulebrouck J, Funk M, Valla F (2007) The Nevado del Ruiz ice cap, Colombia, 21 years after: volcano-glacier interactions, meltwater generation, and lahar hazards. Ann Glaciol 45:115–127. https://doi.org/10.3189/172756407782282589

Thouret J-C, Boivin P, Miallier D, Donnadieu F, Dumoulin J-P, Labazuy P (2021) Post-eruption evolution of maar lakes and potential instability: the Lake Pavin case study, French Massif Central. Geomorphology 382:107663. https://doi.org/10.1016/j.geomorph.2021.107663

Tibaldi A (1995) Morphology of pyroclastic cones and tectonics. J Geophys Res 100(B12):24521–24535. https://doi.org/10.1029/95JB02250

Tuffen H (2007) Models of ice melting and edifice growth at the onset of subglacial basaltic eruptions. J Geophys Res 112:B033203. https://doi.org/10.1029/2006JB004523

Valentine GA, Graettinger AH, Sonder I (2014) Explosion depths for phreatomagmatic eruptions. Geophys Res Lett 41(9):3045–3051. https://doi.org/10.1002/2014GL060096

Valentine GA, Connor CB (2015) Basaltic volcanic fields, chap 23. In: Sigurdsson H, Houghton B, McNutt SR, Rymer H, Stix J (eds) The Encyclopedia of Volcanoes, 2nd edn. Academic, Elsevier, 423–439 pp. ISBN: 978-0-12-385938-9

Valentine GA, Krier D, Perry FV, Heiken G (2005) Scoria cone construction mechanisms, Lathrop Wells volcano, southern Nevada, USA. Geology 33(8):629–632. https://doi.org/10.1130/G21459AR.1

Valentine GA, White JDL, Ross P-S, Graettinger AH, Sonder I (2017) Updates to concepts on phreatomagmatic maar-diatremes and their pyroclastic deposits. Front Earth Sci 5:68. https://doi.org/10.3389/feart.2017.00068

Van den Bogaard P, Schmincke H-U (1984) The Eruptive center of the Late Quaternary Laacher See Tephra. Geol Rundsch 73:933–980. https://doi.org/10.1007/BF01820883

Venugopal S, Moune S, Williams-Jones G (2016) Investigating the subsurface connection beneath Cerro Negro volcano and the El Hoyo Complex, Nicaragua. J Volcanol Geoth Res 325:211–224. https://doi.org/10.1016/j.jvolgeores.2016.06.001

Vespermann D, Schmincke HU (2000) Scoria cones and tuff rings. In: Sigurdson H, Houghton BF, McNutt SR, Rymer H, Stix J (eds) The Encyclopedia of Volcanoes. Academic, New York, pp 683–694

Vörös F, van Wyk de Vries B, Guilbaud M-N, Görüm T, Karátson D, Székely B (2022) DTM-based comparative geomorphometric analysis of four scoria cone areas—suggestions for additional approaches. Remote Sens 14(23):6152. https://doi.org/10.3390/rs14236152

Valentine GA, Connor CB (2015) Basaltic volcanic fields. In: Sigurdsson H, Houghton B, McNutt SR, Rymer H, Stix J (eds) The Encyclopedia of Volcanoes, chap 23, 2nd edn. Academic, Elsevier, pp 423–439. ISBN: 978-0-12-385938-9

Waitt RB (1995) Hybrid wet flows formed by hot pyroclasts interacting with snow, during the 1992 eruption at Crater Peak, Mount Spurr volvano, Alaska. In: Keith TEC (ed) The 1992 Eruptions of Crater Peak Vent, Mount Spurr Volcano, Alaska. US Geol Surv Bull 2139, pp 107–118. QE75.B9 no. 2139

Waitt RB, Pierson TC, MacLeod NS, Janda RJ, Voight B, Holcomb RT (1983) Eruption-triggered avalanche, flood and lahar at Mount St.Helens—effects of winter snowpack. Science 221(4618):1394–1397. https://doi.org/10.1126/science.221.4618.1394

Walder JS (2000a) Pyroclast/snow interactions and thermally driven slurry formation. Part 1: Theory for monodisperse grain beds. Bull Volcanol 62(2):105–118

Walder JS (2000b) Pyroclasts/snow interactions and thermally driven slurry formation. Part 2: Experiments and theoretical extension to polydisperse tephra. Bull Volcanol 62(2):119–129

Walker GPL (2000) Basaltic volcanoes and volcanic systems. In: Sigurdson H, Houghton BF, McNutt SR, Rymer H, Stix J (eds) The Encyclopedia of Volcanoes. Academic, New York, pp 283–289

Werner R, Schmincke HU (1999) Englacial versus lacustrine origin of volcanic table mountains: evidence from Iceland. Bull Volcanol 60:335–354. https://doi.org/10.1007/s004450050237

Wessel P (2001) Global distribution of seamounts inferred from gridded Geosat/ERS-1 altimetry. J Geophys Res 106(B9):19431–19441. https://doi.org/10.1029/2000JB000083

Wessel P, Sandwell DT, Kim SS (2010) The global seamount census. Oceanography 23(1, Special Issue on Mountains in the Sea):24–33. https://www.jstor.org/stable/24861056

White JDL, Smellie JL, Clague DA (2003) Introduction: a deductive outline and topical overview of subaqueous explosive volcanism. In: White JDL, Smellie JL, Clague DA (eds) Explosive subaqueous volcanism. AGU Monograph 140, pp 1–20, 379 pp. ISBN: 0-87590-999-X

White JDL, McPhie J, Soule SA (2015a) Submarine lavas and hyaloclastite. In: Sigurdsson H, Houghton B, McNutt SR, Rymer H, Stix J (eds) The Encyclopedia of Volcanoes, chap 19, 2nd edn. Academic, Elsevier, pp 363–375. ISBN: 978-0-12-385938-9

White JDL, Schipper CI, Kano K (2015b) Submarine explosive eruptions. In: Sigurdsson H, Houghton B, McNutt SR, Rymer H, Stix J (eds) The Encyclopedia of Volcanoes, chap 31, 2nd edn. Academic, Elsevier, pp 553–569. ISBN: 978-0-12-385938-9

White JDL, Ross P-S (2011) Maar-diatreme volcanoes: a review. J Volcanol Geoth Res 201(1–4):1–29. https://doi.org/10.1016/j.jvolgeores.2011.01.010

Wijbrans J, Németh K, Martin U, Balogh K (2007) $^{40}Ar/^{39}Ar$ geochronology of Neogene phreatomagmatic volcanism in the western Panonnian Basin, Hungary. J Volcanol Geoth Res 164(4):193–204. https://doi.org/10.1016/j.jvolgeores.2007.05.009

Wilson L, Head JW (2002) Heat transfer and melting in subglacial basaltic volcanic eruptions: implications for volcanic deposit morphology and meltwater volumes. In: Smellie JL, Chapman MG (eds) Volcano-ice interaction on Earth and Mars, vol 202. Geol Soc London Spec Pub, pp 5–26, 431 pp. ISBN: 1-86239-121-1

Wilson L, Head JW (2007) Heat transfer in volcano-ice interactions: synthesis and applications to processes and landforms on Earth. Ann Glaciol 45:83–86. https://doi.org/10.3189/172756407782282507

Wilson L, Smellie JL, Head JW (2013) Volcano-ice interactions. In: Fagents SA, Gregg TKP, Lopes RMC (eds) Modeling volcanic processes. The physics and mathematics of volcanism, chap 13. Cambridge University Press, pp 275–299. ISBN: 978-0-521-89543-9

Wohletz KH (1983) Mechanisms of hydrovolcanic pyroclast formation: grain-size, scanning electron microscopy, and experimental studies. J Volcanol Geoth Res 17(1–4):31–63. https://doi.org/10.1016/0377-0273(83)90061-6

Wohletz K (1986) Explosive magma-water interactions: thermodynamics, explosion mechanisms, and field studies. Bull Volcanol 48:245–264. https://doi.org/10.1007/BF01081754

Wohletz KH (1993) Atmospheric dust dispersal analysed by granulometry of the Misers Gold Event. J Geophys Res 98(B1):557–566. https://doi.org/10.1029/92JB01888

Wohletz K (2003) Water/magma interaction: physical considerations for the deep submarine environment. In: White JDL, Smellie JL, Clague DA (eds) Explosive subaqueous volcanism. Amer Geophys Union Monograph 140, pp 25–49, 379 pp. ISBN: 0-87590-999-X

Wohletz KH, Heiken G (1992) Volcanology and geothermal energy. University of California Press, Berkeley, 432 pp

Wohletz KH, McQueen (1984) Volcanic and stratospheric dustlike particles produced by experimental water-melt interactions. Geology 12(10):591–594. https://doi.org/10.1130/0091-7613(1984)12<591:VASDPP>2.0.CO;2

Wohletz KH, Sheridan MF (1979) A model of pyroclastic surge. Geol Soc Amer Spec Paper 180:177–193

Wohletz KH, Sheridan MF (1983) Hydrovolcanic explosions II. Evolution of basaltic tuff rings and tuff cones. Am J Sci 283:385–413

Wohletz K, Zimanowski B, Büttner R (2013) Magma-water interactions. In: Fagents SA, Gregg TKP, Lopes RMC (eds) Modeling volcanic processes. The physics and mathematics of volcanism, chap 11. Cambridge University Press, pp 230–254. ISBN: 978-0-521-89543-9

Wood CA (1979) Monogenetic volcanoes of the terrestrial planets. Proc Lunar Planet Sci Conf 10th, Houston, TX, March 19–23, 1979, vol 3. Pergamon Press Inc, New York, pp 2815–2840

Wood CA (1980) Morphometric evolution of cinder cones. J Volcanol Geoth Res 7(3–4):387–413. https://doi.org/10.1016/0377-0273(80)90040-2

Yeo IA, McIntosh IM, Bryan SE et al (2022) The 2019–2020 volcanic eruption of Late'iki (Metis Shoal), Tonga. Sci Rep 12:7468. https://doi.org/10.1038/s41598-022-11133-8

Yuen DA, Scruggs MA, Spera FJ, Zheng Y, Hu H, McNutt SR, Thompson G, Mandli K, Keller BR, Wei SS, Peng Z, Zhou Z, Mulargia F, Tanioka Y (2022) Under the surface: pressure-induced planetary-scale waves, volcanic lightning, and gaseous clouds caused by the submarine eruption of Hunga Tonga-Hunga Ha'apai volcano. Earthq Res Adv 2(3):100134. https://doi.org/10.1016/j.eqrea.2022.100134

Zimanowski B, Büttner R, Lorenz V (1997) Premixing of magma and water in MFCI experiments. Bull Volcanol 58:491–495. https://doi.org/10.1007/s004450050157

Chapter 4 Polygenetic Landforms I: Shield Volcanoes

1 Introduction, Definition and General Features

Following the small terrestrial and submarine volcanoes discussed in the previous chapter, we now turn to *polygenetic* volcanoes. These have in common a complex plumbing system with a number of conduits, and produce a range of eruptive products, often with various lithologies (Cas and Wright 1988), and build up over tens to hundreds of kyr up to a few Myr. The two main groups of polygenetic volcanoes are shield volcanoes and strato- or composite volcanoes.

Shield volcanoes result from edifice-building effusive eruptions, when an outpouring of basaltic lava creates a positive central landform. The term 'shield' volcano comes from the German *Schildvulkan*, introduced by German scientists at the beginning of the twentieth century, and became used in English *(shield volcano)* in the 1950s (e.g. Stearns and Macdonald 1946; Cotton 1952). The name derives from most shield volcanoes' similarity in shape to a warrior's shield lying face down, i.e. a low, convex topographic profile with a flat top.

Whereas only about one seventh of the subaerial volcanoes on Earth are shield volcanoes (Suzuki 1977), this volcano type is considered the most common in the Solar System, found on all terrestrial planets and some moons. This is due to the common association of shields with hot spots (Chap. 2). On Mars, some shields, such as Olympus Mons (which at 22 km in height is the highest mountain in the Solar System), grew very large over millions or tens of millions of years due to the thick, load-bearing crust and the non-moving plates. For comparison, the highest mountain on our planet, Mauna Loa (Photo 42), also a shield volcano, rises more than 9 km from the sea floor, which way surpasses the elevation of Mount Everest.

On Earth, the largest shields are not only high but are also the most voluminous volcanoes, reaching a total (submarine + subaerial) volume of up to tens of thousands of km^3. However, the size range of shield volcanoes varies widely. The smallest lava-built volcanoes considered in some literature as shields are < 1 km^3 in size (e.g. Reykjanes Peninsula, Iceland, or Michoacán-Guanajuato Volcanic Field, Mexico). Wood (1979) termed the smallest basalt shields 'lava cones', while the term '(low) lava shield' was given by other authors (Pike 1978; Greeley 1982; Andrew and Gudmundsson 2007).

Fundamental research on shield volcanoes include works by Cotton (1944, 1952), Whitford-Stark (1975), Williams and McBirney (1979) and Ollier (1988) on typology, and Wood (1979), Pike and Clow (1981), Mark and Moore (1987), Hasenaka (1994) and Rowland and Garbeil (2000) on topography-based morphometry. In addition, several comprehensive books on shield volcanoes have been published, in particular on the Hawaiian Islands, such as those by

D. Karátson and J.-C. Thouret, *Volcano Geomorphology*, Advances in Volcanology,
https://doi.org/10.1007/978-3-031-77693-9_5

Poland et al. (2014) and Carey et al. (2015), or on La Réunion and the Comore Islands, such as Michon and Staudacher (2009) and Bachèlery et al. (2016). From research nowadays, when DEMs have become available world-wide, and with constantly expanding methodology, we highlight the work of Grosse and Kervyn (2018) who compiled a comprehensive morphometric classification of 158 shield volcanoes on Earth using SRTM DEM data.

Shield volcanoes have several features, but they share at least two common characteristics:

(1) They all consist predominantly of monotonous lava flows erupted from central vents or flank fissures (e.g. rift zones: Walker 1999). In addition to the lava flows, shields often display low-explosivity lava fountaining, similar to that of scoria cones and spatter landforms, which remain minor in terms of volume. In Hawaii, for example, 90–95% of a shield volcano mass is built up of lava. Eruptions at shield volcanoes are less likely to be explosive, but this can occur when differentiated magmas erupt, as in the case of some oceanic but mainly continental shield volcanoes. Occasionally, if water gets into the vent—e.g. when groundwater is perched in summit or rift dike swarms—, phreatomagmatism can also occur (see later; e.g. McPhie et al. 1990; Clague and Dixon 2000).

(2) They have gentle slopes whose average value ranges between 5° and 15° (e.g. Macdonald 1971; Hasenaka 1994; Walker 2000; Francis and Oppenheimer 2003). Benefitting from precise, comparative DEM analyses, the mean threshold slope of the whole edifice which distinguishes shields from composite volcanoes has been determined at 12° (Grosse and Kervyn 2018). However, this is only an average value, and the slopes may be much steeper on some parts of the flanks. Thus, mean slope is just one of many other important parameters, implying that a number of characteristics (not just morphometric) are required for an accurate classification.

2 Occurrence, Constraints on Morphology, and Main Types

Shield volcanoes can develop where prolonged, voluminous and fluid pāhoehoe and 'a'ā lava flows build up a central cone. Therefore, most of them occur at hot spots where an extended time span as well as an abundant supply of basaltic melt is ensured. The largest shield volcanoes are found on the Hawaiian Islands with subaerial volumes up to 6000 km^3. Another well-known hot spot occurrence is the Galápagos Islands, which have smaller shield volcanoes, in the order of a few 100 km^3 (subaerial volume). Even smaller lava shields of Holocene age can be found in the west and north rift zones of Iceland where hot spot activity and on-land spreading are combined with subglacial conditions to form various volcanic landforms. Shield volcanoes also occur in continental rift zones (e.g. East African Rift System) and arc or back-arc settings (e.g. Cascade Range, Central America, New Hebrides). We will see how a given geodynamic context can modify the magma composition and eruptive style, resulting in shape variations.

In general, the broad shape and low profile of shield volcanoes is attributed to the thickening of basalt lava flows with distance, which in turn is related to the rheology of readily-flowing lava (Francis and Oppenheimer 2003). Moreover, the elongated shape and the broad, flat summit of large shields seems to be constrained by five factors, which are not necessarily related to each other, nor are they all always present: (1) development of rift zones (a single, extended rift, e.g. Oahu, La Palma; two opposite trending rifts, e.g. Kīlauea, Mauna Loa; or 'stellate-type' rift zones, e.g. Mauna Kea, Tenerife, see Sect. 3.4); (2) the far-reaching movement of lava along rifts in lava tubes/lava tunnels; (3) occasional volcano spreading (promoted by preferential intrusive activity along rift zones); (4) gravitational deformation above a weak, hydrothermally altered volcano core that might also contribute to flattening of the summit region (Merle et al. 2010); (5) the appearance of collapse caldera(s) and/or summit pit craters, which make the uppermost part flattish and excavated. Note that not all shields exhibit calderas, e.g. Eastern Galápagos Islands, or Mount Cameroon; and rift zones do not form on smaller-sized shield volcanoes (e.g. simple lava cones) on Iceland, which in itself puts them in a subtype apart. Moreover, these features can also be absent on larger shields. Marsabit (Kenya)—a huge, continental off-rift shield volcano, not very high (1707 m a.s.l.) but very broad (6100 km^2 in area)—lacks a central caldera, but displays scattered vents (almost 200 scoria cones and maars) that follow a local NE-SW structural trend (van Hazinga 2023).

The formation of steeper morphology on the upper flanks (of both larger and smaller shields, i.e. the Western Galápagos, or Michoacán–Guanajuato Volcanic Field, see Sects. 4.2 and 5.2) can be caused by at least four factors that constrain slope gradient involved: (1) more differentiated (e.g. trachytic) magma compositions (Hasenaka 1994; Wolfe et al. 1997); (2) low effusion rates producing relatively short and thick ('a'ā) flows (Chadwick and Howard 1991; Hasenaka 1994); (3) accumulation of pyroclastic material, along rift zones and fissures, leading to oversteepening (Hasenaka 1994; Rowland and Garbeil 2000), and (4) dyke intrusions, ring dykes along rift zones or ring (circumferential or arcuate) fissures (Chadwick and Dieterich 1995; Annen et al. 2001; Geist et al. 2014) that raise the substructure. Dyke intrusion patterns are controlled by the local stress field, which in turn controls the spatial distribution of faults/fissures or adventive vents (e.g. Rowland and Garbeil 2000; Bagnardi et al. 2013; Tibaldi et al. 2014). The above listed factors are interrelated, and may change over the long lifetime (in the order of 600–800 kyr) of a shield volcano, which can give rise to the complexity of landforms (e.g. Lipman 1995, 2000; Clague and Sherrod 2014).

Benefitting from the morphometric data obtained by Grosse and Kervyn (2018), we show a schematic grouping to demonstrate the wide variety of shields (Fig. 1). It is based on edifice volume versus elongation, which are correlated with magma supply rate and vent distribution, respectively, while also taking into account the complexity of the plumbing system and volcanic construct. Applying such criteria can indicate a number of shield groups, but these are either overlapping or transitional, and hence cannot define separate types. In turn, Fig. 2 presents the (subaerial) height versus basal width of shields. Although these two parameters simplify the geomorphic complexity of shields, such a plot allows a classification which is well-known in the literature, and which we follow in this chapter (Table 1). We distinguish four shield volcano types based on the principle geomorphic features and geodynamic setting. Selected examples categorised into one of the four types, showing characteristic slope patterns, are shown in Fig. 3.

Traditionally, three important types of shields, made up predominantly of basalt, have been described based mainly on size and slope features. Two of them occur above oceanic hot spots, and are distinguished by the different eruptive activity (principally the magma supply) of the underlying mantle plume. Accordingly, *large oceanic hot spot shields* and *medium-sized oceanic hot spot shields* are classified, termed *Hawaiian* and *Galápagos* types in the literature, which have gentler and steeper uppermost slopes, respectively. Due to the different slope pattern, the Galápagos type (which is exemplified by the western islands) resembles an 'overturned soup bowl' (Williams and McBirney 1979) rather than a shield, even if the geomorphic difference applies only to the subaerial part. Both the Hawaiian and Galápagos types show a central caldera; however, differences in structure, rate of plate motion, and position relative to the hot spot may cause variations in shape within both groups.

In addition to these larger structures, small shields are also well-known in various geodynamic settings. On the basis of the height/diameter ratio of shields, the *Icelandic* type, along with the even smaller '*scutulum type*', both composed of basalt, have long been distinguished (see pioneering morphometry in Whitford-Stark 1975). Highlighted occurrences include the shield volcanoes of Iceland, the Faroe Islands, and those in

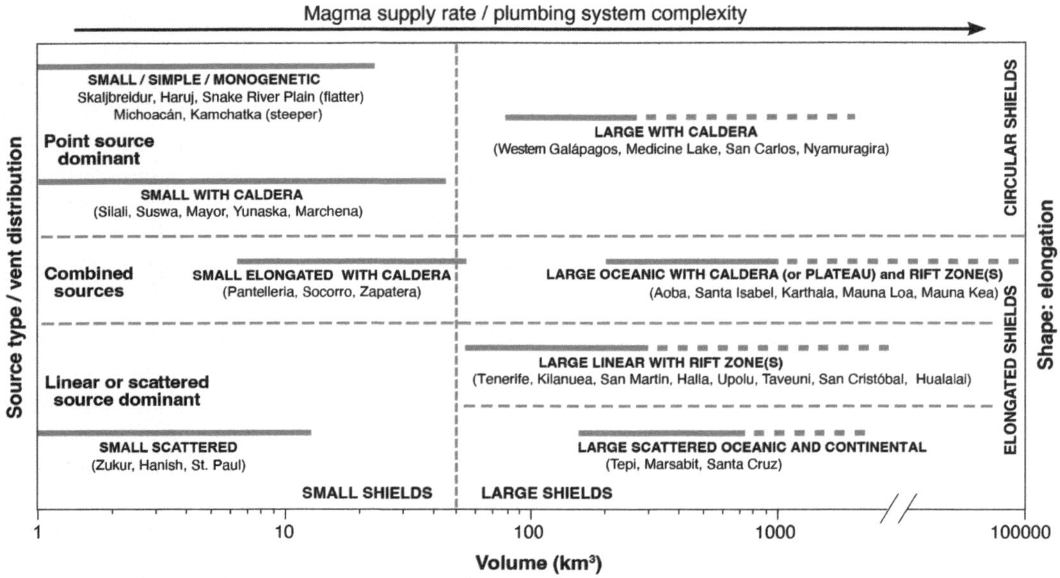

Fig. 1 Schematic grouping of shield volcanoes with examples based on a combination of criteria: (1) volume in km^3 (lower axis); (2) size, magma supply rate, complexity increasing towards the right side (upper axis); (3) source type and vent distribution (left axis); (4) elongation, either circular or elongated (right axis). Volume of 50 km^3 is an approximate threshold separating small (left-hand side) versus large (right-hand side) shield types. Horizontal bars indicate the subaerial volume range of each shield type, while dashed bars predict maximum values taking into account the underwater edifice volume (modified and adapted, in cooperation with P. Grosse, from Grosse and Kervyn 2018)

the Michoacán-Guanajuato Volcanic Field (Mexico), and the Snake River Plain (Idaho, USA). However, in this book, we do not separate the scutulum but group it together with the Icelandic type as *small oceanic and continental shields*, because there are no real geomorphic differences between them, and their sizes overlap.

At the same time, in continental intraplate (in particular, rift) settings, the medium-sized '*Turkana type*' was proposed by Webb and Weaver (1975) for some large volcanoes of the East African Rift System (EARS). These, in addition to basalt, are composed of more evolved (e.g. trachytic but also rhyolitic) magma. In other continental areas, hot spot-related, ancient, eroded shield volcanoes in Eastern Australia (Oligocene) or Canada (Late Miocene to Pliocene) in some cases with larger dimensions than those in the EARS have also been considered as belonging to the Turkana type (e.g. Lockwood and Hazlett 2010). These shields, in addition to lava flows, display explosive eruptions, late-stage lava domes, and unusual caldera formation. However, the Turkana type remains poorly studied in particular due to the significant degradation of the Australian and Canadian examples, which prevents a clear inference of their primary shape being made. Therefore, their classification is ambiguous, and they share morphometric similarities with the Galápagos type (except for the upper steep part which is absent). In addition, there are arc-related shields with similar dimensions to the Turkana (and partly Galápagos) type either with more mafic or more silicic compositions (e.g. Lipman 2000; de Silva and Lindsay 2015). Predominantly mafic examples are Okmok (Alaska, USA), Tsirk (Kamchatka), Ambrym (New Hebrides, Vanuatu); while silicic examples commonly have a bimodal composition and include Newberry and Medicine Lake, Oregon (USA). Both of these subtypes can produce explosive eruptions, and are typically associated

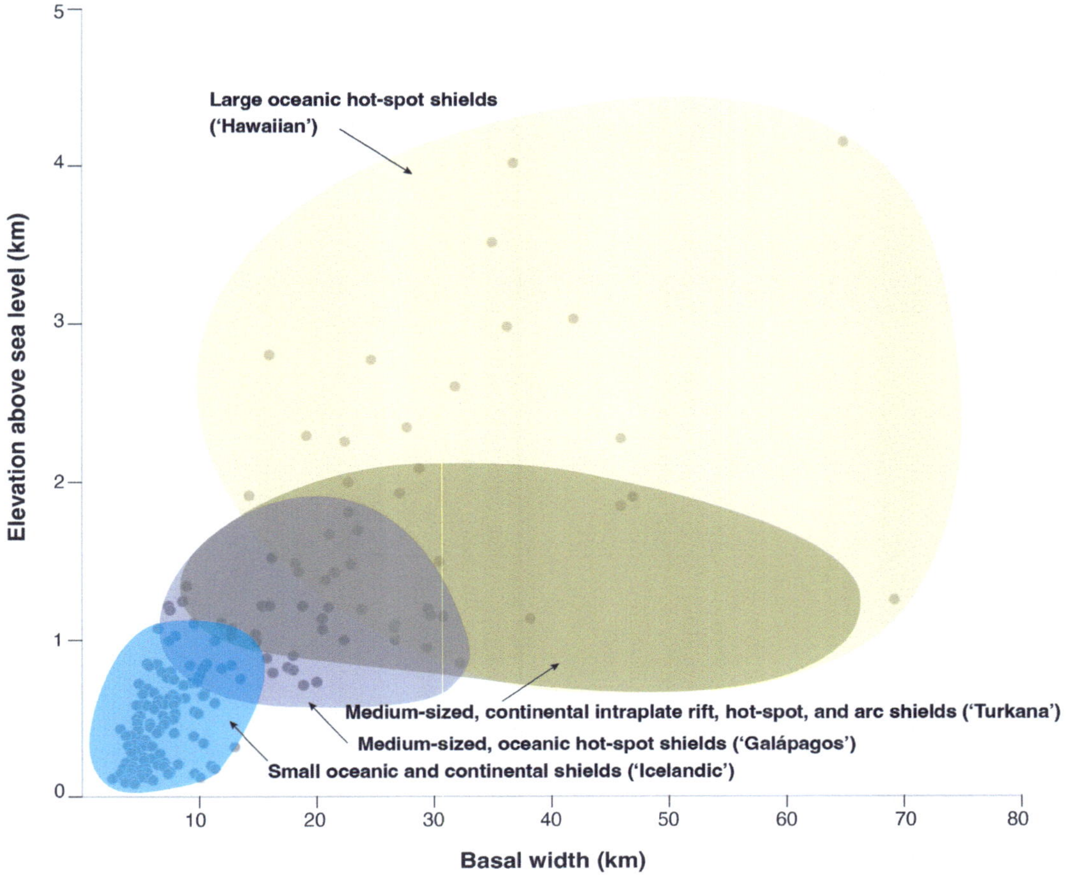

Fig. 2 Height (relative elevation)/basal width of 158 shield volcanoes with coloured fields referring to the four main type of shields presented in this book. Individual data points are from Grosse and Kervyn (2018). For the Hawaiian shields, only the subaerial part is considered for morphometry

with calderas, as at the Turkana shields. Because of the similar shapes and the wide spectrum in lithology, as well as their frequent explosivity, we do not distinguish the Turkana and Newberry examples as separate types: they are considered under a common umbrella termed *medium-sized, continental intraplate and arc shields*. However, the subtypes can differ in terms of effusive versus explosive eruptions and respective products, and will be described separately.

Finally, we note that for arc-related shields that are composed mainly of ignimbrites, the term *'Masaya type'* was proposed by Williams and McBirney (1979), named later *ignimbrite shield* by Baker (1981). Although arc shields can also be explosive (see Sect. 6.2), ignimbrite shields in our book are considered a separate class, and are discussed together with calderas in Chap. 6.

In the following sections, specific features of the four shield volcano types defined above are described, ending with constraints on possible evolutionary pathways.

3 Hawaiian Shields

3.1 General Geomorphic Features

The basic shape of this shield volcano type is mainly shown by the visible, subaerial shape of the large Hawaiian volcanoes, which represents only a relatively small portion of the edifices in

Table 1 Types of shield volcanoes with general characteristics

Shield type and subtype		Volume, km^3	Diameter (subaerial), km	Height (subaerial)/width	Mean slope	Structural features	Growth (long-term effusion) rates	Magma composition
Large oceanic hot-spot shields (Hawaiian)		1000–6000[a]	15–100	0.04–0.1	7-9°	Rift zones; submarine rifts; radial vents along eruptive fissures; summit caldera	2–8 m^3/s (extrapolated Myr average) 0.1 km^3/yr (Kīlauea, 200 yr average)	Picrite, basalt (tholeiitic and alkali), (shield); hawaiite to trachyte (post-shield)
Medium-sized oceanic hot-spot shields (Galápagos)		150–250 (Western islands) 10–150 (Eastern islands)	10–30	0.04–0.07 (Western islands) 0.03–0.06 (Eastern islands)	5–9° (Western islands) 3–6° (Eastern islands)	Submarine rifts; ring and radial fissures on upper flanks; summit caldera (Western islands); single rift (Eastern islands)	0.01 km^3/yr at Sierra Negra, 2005– (but in the long-term, only 1% of Hawaiian shield-building)	Picrite, basalt (both tholeiitic and alkali); rarely, trachyte, dacite, rhyolite
Small oceanic and continental shields (Icelandic)		<1–20[b]	3–15	0.02–0.05	2–6°	Monotonous lava flow from central crater; lava tubes; tumuli and lava rises	n.d. (eruptive rates vary up to > 50 m^3/s)	Picrite (small shields), tholeiitic basalt (large shields)
Medium-sized continental intraplate and arc shields	Turkana subtype	20–100 Up to 3000[c]	10–20 Up to 100[c]	0.03–0.07	5–9°	Ring fissures; summit caldera	n.d.	Alkali basalt, basanite, trachyte, phonolite, rhyolite
	Newberry subtype	200–1000	20–30	0.03–0.04	5–9°	Ring fissures; summit caldera	n.d.	Calc-alkaline basalt, basaltic andesite, minor andesite, dacite, rhyolite

[a] 10,000–95,000 including the submarine portion
[b] largest shields in Michoacán-Guanajuato volcanic field can reach 50 km^3
[c] In Australia, poorly constrained

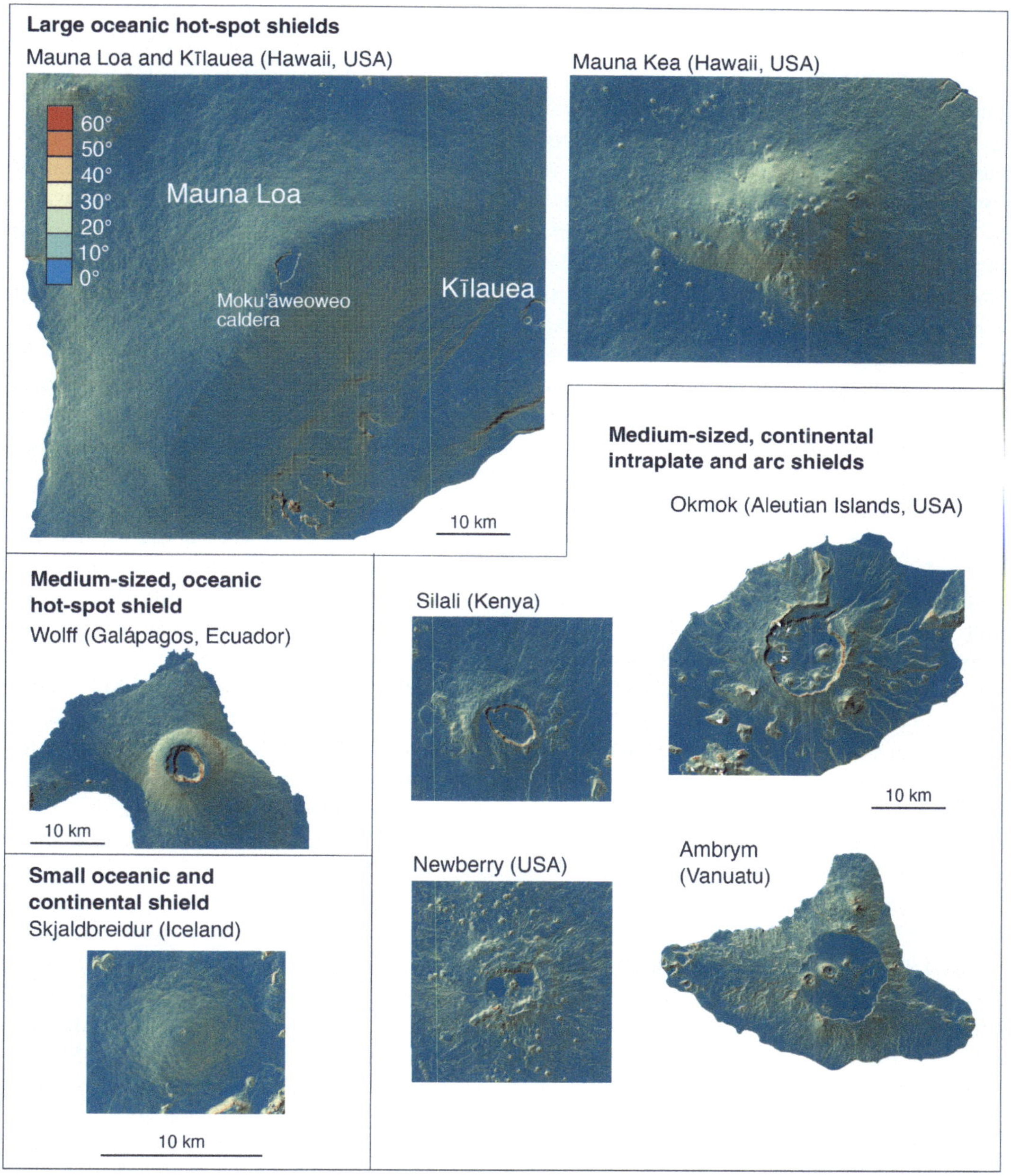

Fig. 3 Slope pattern of selected shield volcanoes representing the four to five main types in Fig. 2 based on SRTM DEM data. Note the different dimensions of shield types, and even the different scale for Icelandic shields

terms of area, volume, and slope characteristics. The 'shield' shape, with an elevation of up to 3000–4000 m a.s.l., refers to a profile characterized by 6–8° slopes on the lower flank, slightly steeper 8–10° on the main and upper flanks, resulting in ~ 7–9° mean slopes; by contrast the top of the shield is characterized by a flattish summit plateau (for morphometric parameters, see Grosse and Kervyn 2018). The uppermost flank slopes can be steeper at postshield volcanoes, e.g. Mauna Kea. The area of fault scarps, like at the Hilina Fault system of Kīlauea, can also be characterised by steep slopes (> 10°: Mark and Moore 1987).

Although the characteristic subaerial shape of these volcanoes with gentle lower flanks gave the name of 'shield' to these structures, bathymetric studies at Hawaii have revealed that the submarine flanks are always steeper (up to 15–18°). This is a consequence of a number of interrelated factors: subaqueous chilling of lava flows by seawater; fragmentation of lava at the shoreline; a greater angle of repose (of moving lava flows) in water; and the decrease in lava density due to the buoyant effect of water, all result in the lava not being able to flow under water as easily, causing it to thicken and spread laterally (Mark and Moore 1987; Fialko and Rubin 1999; Clague and Sherrod 2014). Therefore, even though precise morphometry can be determined for the subaerial part of the volcanoes, the obtained values do not constrain the whole morphology of the edifice. Importantly, on old (> 0.8 Ma) Hawaiian volcanoes this slope change (subaerial lower flanks versus submarine ones) has been used to identify submerged shorelines (Clague and Moore 1991; Lipman 1995; Faichney et al. 2010).

In the Hawaiian Islands, shield volcanoes can be found on eight islands. The island of La Réunion, in particular the active Piton de la Fournaise, also belongs to the Hawaiian type, although it is significantly smaller (~ 500 km^3 subaerial volume) and shows some transitional morphometric features to the Galápagos shields. Some of the Canary Islands (e.g. La Palma, El Hierro, the basal part of Gran Canaria and Tenerife) as well as other hot spot volcanoes like Mauritius, Mount Karthala (Comore Islands), Mount Cameroon (West Africa), Cape Verde, Samoan Islands etc., most of which are smaller in size, are also examples of active to dormant Hawaiian shields. Karthala and Cameroon belong to the steeper shields (up to 12–15° in some areas of the upper flanks) due to local tectonics, lower flank spreading, and/or sliding (Kervyn et al. 2014; Bachèlery et al. 2016). Most of the Canary Islands shield volcanoes, more modest in size, are also steeper; the reason for this is the typically lower eruption rates and the more evolved and thus lower temperature magma, which result in more viscous 'a'ā lava flows that seldom create far-reaching lava tunnels (Carracedo and Day 2002).

Taking a closer look at the Hawaiian Islands (Fig. 4), they are final products of the continuous, long-lasting magma supply of the Hawaiian hot spot over the past 80 Ma, as discussed in Chap. 2. Even though the magma output is very high and large shield volcanoes are built up, (i) island subsidence due to the load on the crust, (ii) flank instability resulting in sliding, and (iii) high-rate mass wasting and erosion in the prevailing tropical climate, degrade them in a couple of million years. In fact, the Hawaiian Ridge and Emperor Seamounts towards the NW comprise submerged, relict landforms stretching almost to Japan. The youngest volcano above the hot spot, still under water, is Lo'ihi or Kama'ehuakanaloa (35 km offshore of the Big Island); the youngest emergent one is Kīlauea (which started 0.3 Ma); whereas the oldest is Ni'ihau (6 Ma), but with rejuvenated activity at around 0.35 Ma (Clague and Sherrod 2014). Such an impressive set of progressively younger volcanoes but with different evolutionary stages (see Sect. 3.6) provides insight into their long-term formation.

The fifteen emerged Hawaiian shield volcanoes plus three submerged ones at the westernmost end form a SE-NW trend of progressively degraded shields to the NW, ideal for study. Notably, the maximum size a well-developed shield volcano can reach during its active period is not the same: in the normal life time, typically less than 1 Myr (Lipman 1995), only eight volcanoes reached an elevation >1500 m a.s.l. (Clague 1996). The largest shield volcano, Mauna Loa, has an area of 5100 km^2 and volume of 95,000 km^3 including its submarine portion. This extremely large value is due to the fact that a significant mass, assigned to Kīlauea in previous calculations, has now been added to Mauna Loa (Clague and Sherrod 2014); also note that the subaerial portion is only 3000 km^3 (Grosse and Kervyn 2018). Such a cumulative volume is two orders of magnitude larger than that of one of the largest, iconic composite volcanoes, Mount Fuji, 400–500 km^3 in

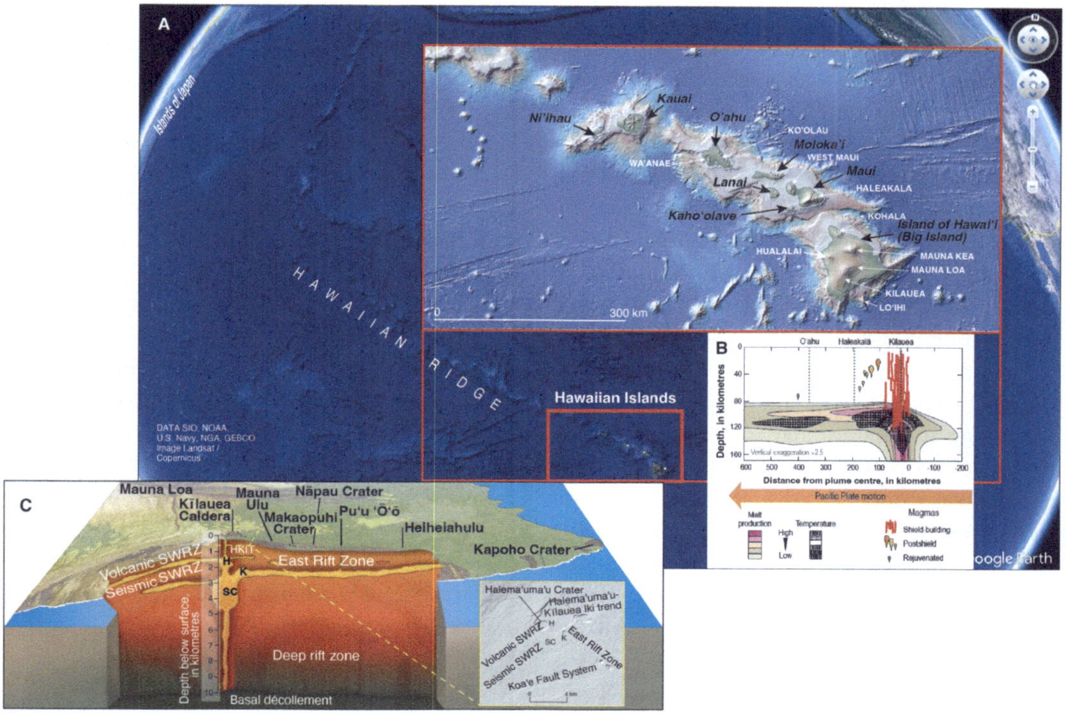

Fig. 4 Setting of the Hawaiian–Emperor volcanic chain in the East Pacific Ocean and the Hawaiian mantle plume. **A** The Hawaiian Islands with volcano names. **B** Cross-section diagram showing the evolving position of the mantle plume (plate motion is from right to left) and its inferred structure below the Hawaiian hot spot. Note isotherms and melt-production isopleths. (From Ribe and Christensen 1999 *in:* Clague and Sherrod 2014). **C** Updated interpretation showing a summit magma system consisting of two long-term reservoirs (a seismic Southwest Rift and an East Rift Zone with a molten core, both at a depth of 3 km and connected to the south caldera magma reservoir), a volcanic Southwest Rift zone within 1 km of the surface that is connected to the shallower magma reservoir; and a Halema'uma'u-Kilauea Iki trend within 1 km of the surface, also connected to the shallower summit magma reservoir (Poland et al. 2014). The model proposes that the East and Southwest Rift Zones host molten cores at 3–5 km depth

total volume (see Chap. 5). Haleakalā (island of Maui), Kauai, and Wai'anae (island of Oahu), in decreasing order, are 70,000–50,000 km^3 in volume, whereas Mauna Kea is slightly over 40,000 km^3; the young Kīlauea has a volume of ~ 10,000 km^3 (Robinson and Eakins 2006; Clague and Sherrod 2014).

3.2 Constraints on Subsidence and Uplift

Size of the largest shield volcanoes can also be assessed in the light of island uplift versus subsidence. Significant mantle upwelling that would cause topographic uplift in the vicinity of the current Hawaiian hot spot are suppressed by volcanic loading (Zhong and Watts 2002). Combined with the fact the Hawaiian Islands are located on a relatively thin (30–60 km) and rapidly moving (up to 10 cm/year) part of the Pacific plate (Acosta et al. 2003), the enormous weight of the large shields (Mauna Kea and Mauna Loa) of the Big Island causes them to sink, depressing the seafloor by a further 4–5 km (Moore 1987). The Big Island is currently subsiding at a rate of about 0.35 mm/year (Zhong and Watts 2002), which is twice the rate of height addition in historical eruptive activity. Thus, the total thickness of erupted basalt succession of these largest shields is 13–14 km. By contrast, older, inactive Hawaiian islands (e.g. Lanai, Molokai) tend to rise due to mantle upwelling and plume-plate interactions (Zhong and Watts 2002).

Due to the subsidence history and the fast drift over the hot spot, rocks of up to only 6 million years old can be studied on land in the Hawaiian Islands. This contrasts with the 23 million year-old exposed history (the oldest rocks are in Fuerteventura) in the slow-moving Canary Islands (van den Bogaard 2013). The shield volcanoes of the Canary Islands are not sinking in general, and some of the islands have even experienced a rise in basement as a result of subsequent intrusions. The reason for this is their smaller size and therefore smaller mass, and also the stable oceanic plate margin connected to the older (Jurassic) and thicker (60–100 km) slow-moving (2 cm/year) African lithospheric plate.

3.3 Productivity

Mauna Kea and Loa, as well as Kīlauea volcano, are also among the most productive volcanoes on Earth, with a long-term average growth rate of 8 km^3 per hundred years (although the main shield-building phase at Mauna Kea is now over). Their surfaces are renewed very rapidly, with the most active today, Kīlauea, having a surface area of 30 × 60 km, 95% of which has been formed within the last 1500 years. It erupts almost continuously, with the most recent effusive periods being 2013–2018, 2020–2021, 2021–ongoing (Global Volcanism Program: https://volcano.si.edu/volcano.cfm?vn=332010). Its long-term eruptions mostly from the East Rift Zone between 1983 and 2018 have produced lava flows covering more than 100 km^2, which destroyed hundreds of houses, and created a new coastline.

Similarly, 90% of the surface of the even more extensive (3300 km^2) Mauna Loa is less than 4000 yrs old (Lockwood and Lipman 1987; Lipman 1995). The time interval required for creating such a large shield volcano was about 800 kyr (Lipman 1995); most of the huge amount of lava that flowed out during this time is below sea level, constituting the majority of the volcanic edifice. A smaller fraction of Mauna Loa is terrestrial pāhoehoe and 'a'ā lavas; the main island has grown by an average of 0.02 km^2 per year over the last 600 kyr as a result of on-land lava flows (Lipman 1995). Mauna Kea was last active about 4500 yr ago, and Mauna Loa in 2022, after 38 years of dormancy. The site of this last eruption was the Moku'āweoweo summit caldera (Fig. 3), which is filled by lava flows from historical times; subsequently, the eruption was fed by fissures in the Northeast Rift Zone. The third largest shield volcano on the main Hawaiian island, Hualālai, like Mauna Kea, is now extinct and covered with scoria cones, albeit smaller ones (2500 m a.s.l.); its activity declined in historical times (last eruption occurred in 1800–1801).

3.4 Structure-Controlled Landforms: Rift Zones, Radial Vents, and Summit Calderas

Shield volcanoes on Hawaii tend to display well-developed rift zones. Most of the largest volcanoes (e.g. Mauna Loa) are characterised by long-term activity along two or three rifts and/or radial vents on the lower flanks. In the rift zones, the effusive activity produces essentially lava flows, and during the postshield stage, spatter cones, spatter ramparts and scoria cones, which may be more scattered. The deeper-seated part of rift zones hosts feeder dyke intrusions. These can number up to several thousand on a given shield volcano, and the material comprising the dykes and their deep roots solidifies as huge, intrusive masses of gabbro as they cool.

Fissure eruptions may take place at the beginning, where an isolated shield volcano first grows on the seafloor, but the feeder dykes develop in all directions without preferred, well-defined rift zones. This situation is exemplified by West Maui's shield (e.g. Lockwood and Hazlett 2010). However, rifts can develop even in the pre-shield stage, as is the case with Loi'hi

(Malahoff 1987; Clague et al. 2019). In a mature stage, e.g. considering the shape of Mauna Loa, the shield volcano becomes slightly elongated along two developed and one less developed rift zone, each more than 50 km in length. Orientation of the rifts is controlled by gravitational stress; preferably, in the case of three rift arms, they diverge at 120° angle from each other at the summit; however, Kīlauea and Mauna Loa display two rifts running at an angle of at least 150° to each other (Fig. 6). If a new volcano is born next to an existing one, its buttressing effect will create rifting parallel to the older volcano (see the case of Kīlauea abutting Mauna Loa). Triple rift zones are found on East Maui and Lanai, while less distinct and double or single rift zones occur on other, older Hawaiian islands such as Molokai, Oahu and Kaua'i. Kaua'i does not have rifts on land, but below sea level some ridges appear to be rift zones (Eakins et al. 2003).

The islands of Tenerife and El Hierro, in the Canary Islands, display three rift zones (Fig. 5). These stellate-type rifts do not show any substantial differences from those in Hawaii, except they are shorter (up to a few tens of kilometres). Occasionally, only a single rift is active due to the smaller magma supply and structural setting, exemplified by La Palma (Acosta et al. 2003). Its historical and recent rift zone eruptions (see Chaps. 2 and 3) have occurred in the southern part, along the linear ridge of a N-S oriented rift called Cumbre Vieja, which is 1800–1900 m high, (Carracedo et al. 2022). Further examples of single-rift shields include Mount Cameroon (Cameroon, West Africa) and Karthala (Comore Islands), shields of Samoa (Savaii, Upolu and Manu'a Islands), or Taveuni (Fiji). Note that these shields are not exclusively of hot-spot origin; for instance the younger (< 5 Ma) Samoan volcanism shows an additional subduction-related mantle source (Strak and Schellart 2018).

It is not fully understood whether rifting occurs actively or passively. When the volcano grows high, its flanks become unbuttressed and unstable, and may extend by gravitational spreading. This favours the 'passive' development of rift zones, i.e. by extensional fractures and injection of dykes. On the other hand, compressive uplift detected at Kīlauea may imply forceful, 'active' intrusion during rift-zone growth (for a discussion, see Poland et al. 2014).

Volcano spreading was also proposed, for example, at Mount Cameroon. This volcano is four times the height of Kīlauea above sea level, but only a tenth of its total volume (~10,000 km^3). The elongated, axisymmetrical shield volcano shape (Favalli et al. 2014) extends in a single NE-SW-trending rift zone, aligned along the Cameroon Tectonic Line (Mathieu et al. 2011; see Chap. 2). Emplacement of vertical dykes following the rift zone may contribute to the steep flanks of the volcano (Annen et al. 2001). The elongated shape might also be accounted for by enhanced lateral spreading of the lower flanks relative to the summit (Kervyn et al. 2014).

In addition to rift zones, the presence of radial vents (linear flank fissures) can also characterize the Hawaiian shields. This is most obvious on Mauna Loa, where eruptive fissures occur mostly on the southwestern, western and northern flanks, and are oriented radially to the summit caldera. The western flank also displays submarine occurrences. These vents are likely due to spreading in the arcuate Northeast and Southwest Rift Zones, which triggers extension on the west and north sides of the volcano. Scattered across the flanks of the volcano, these vents pose a more serious potential lava-flow hazard than the summit or rift eruptions, located farther away from settlements. On the older Hawaiian volcanoes, e.g. Wai'anae and Kaua'i, erosion has eroded surface vents, but radial dykes have been exposed (Macdonald et al. 1960). Radial vents on other shield volcanoes worldwide are rare, but La Réunion's Piton de la Fournaise is an example of frequent eruptions of this kind, even in a submarine setting (e.g. Dumont et al. 2022).

The summit part of Hawaiian shield volcanoes (e.g. at the intersection of the rift zones) typically forms a plateau, excavated in the centre by a caldera (Fig. 6). Outpouring of large-volume lava flows, as well as magma movement and withdrawal from shallow depths, trigger caldera formation by gradual subsidence. Such

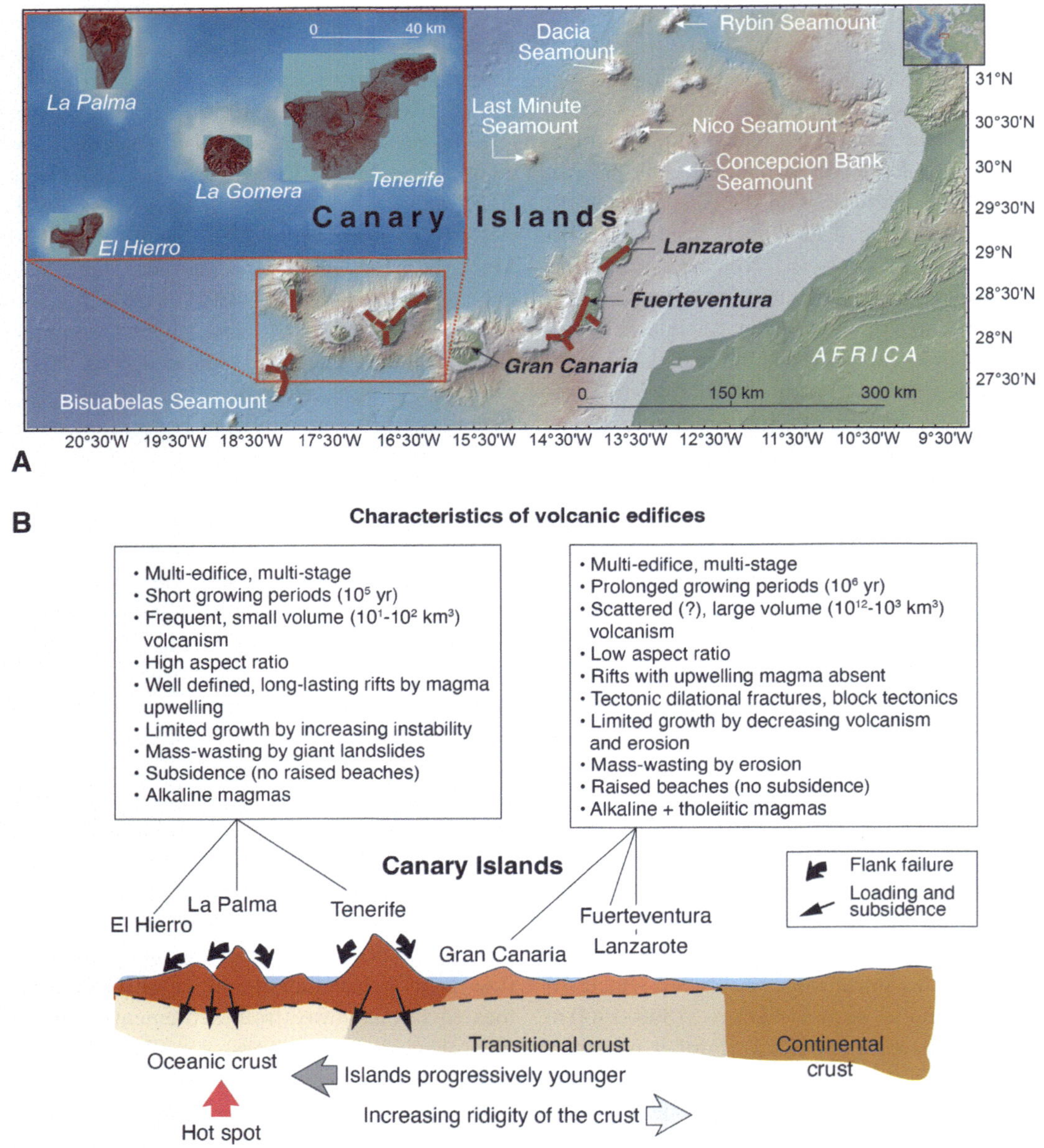

Fig. 5 Bathymetry and topography of the Canary Islands, displaying the rift zones (after Carracedo 1999). **A** Submarine and subaerial topography (GeoMapApp; topography of the western islands is enhanced by red relief images (RRI) based on SRTM DEM); seamount names (after Rivera et al. 2016) are also shown. **B** Characteristics of the western and eastern islands (after Carracedo 1994)

caldera formation has been termed 'magmatic calderas' (e.g. Merle et al. 2010) or 'drainage calderas' (Lockwood and Hazlett 2010; Chap. 7). 'Drainage' refers to the drainage of magma from a reservoir, which may explain why the caldera volume can be much larger than the volume of erupted products (Merle et al. 2010). Pit craters and repeated infill by lava flows are typical associated with such calderas. The calderas are 5–6 km in diameter on Big Island,

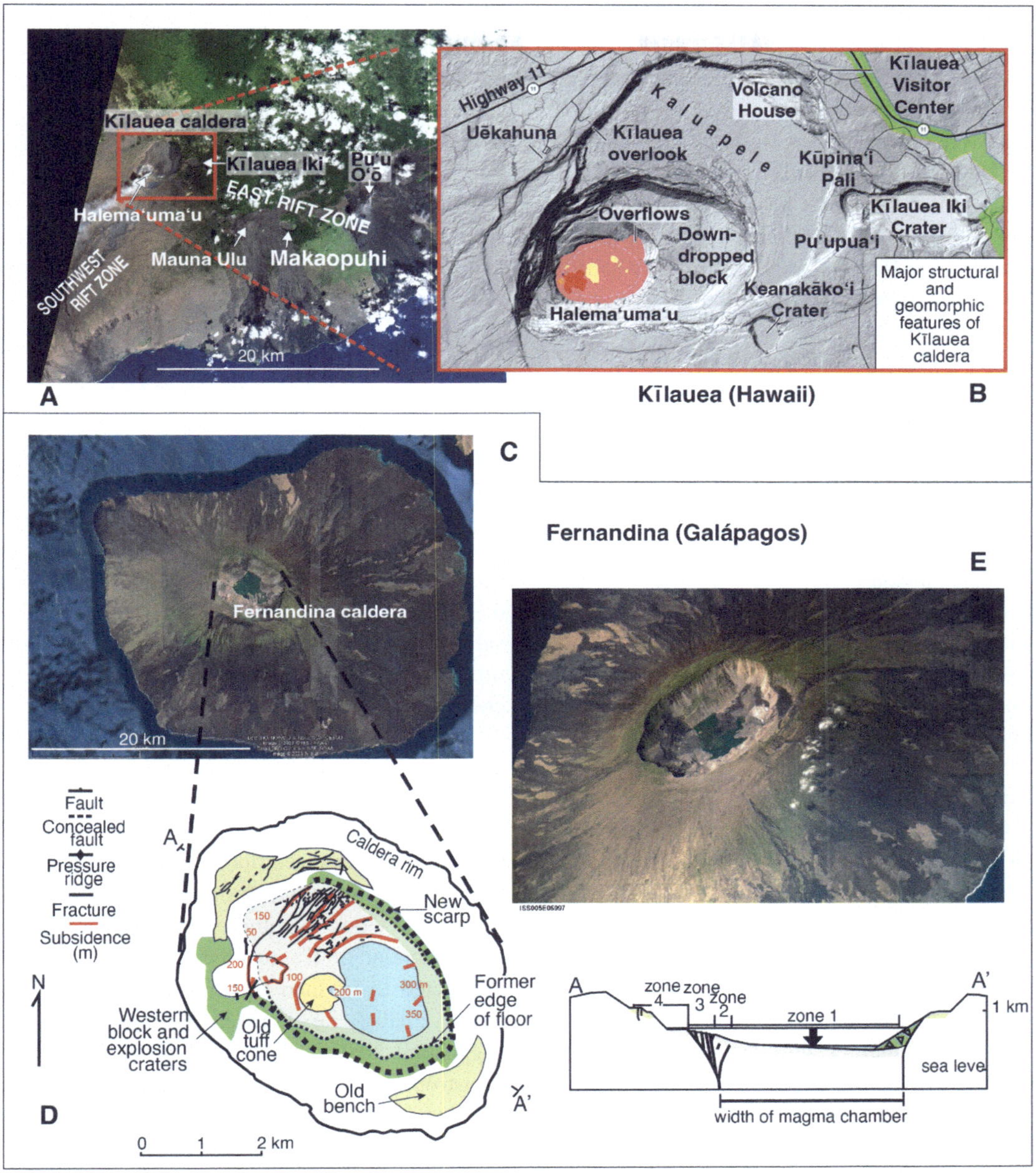

Fig. 6 Shield volcano caldera of Kīlauea, Hawaii Islands, and Fernandina, Galápagos Islands. Kīlauea. **A** satellite image, 2012 (credit: *NASA Earth Observatory,* ISS005-E-6997*)*; **B** DEM showing major structural and geomorphic features (credit: Global Volcanism Program/USGS, https://volcano.si.edu/volcano.cfm?vn=332010). Fernandina; **C** Google Earth image; **D** Topographic and structural features (Howard 2010); **E** Oblique satellite image, 2002 (credit: NASA Earth Observatory, *Image ISS005-E-6997*)

relatively little affected by erosion (see e.g. Moku'āweoweo caldera at Mauna Loa' summit in Fig. 3). Their depth, varying with time due to subsequent collapses and lava infill, is a couple of hundred metres, and they are the sites of generally modest activity compared to rift activity.

The most active part of Hawaii's main island today is the 3×5 km Kīlauea caldera and its East Rift Zone (Figs. 4 and 6; Photo 43). Kīlauea's Late Pleistocene shield volcano sits on the southeast slope of Mauna Loa above the Hawaiian hot spot. It is fed by a complex magma reservoir containing both deep and shallow magma storage zones (Flinders et al. 2020). Its summit caldera first formed 1500 yrs ago and was rejuvenated in the eighteenth century, up to a depth of 600 m in historical times, which has now been significantly infilled by lavas, e.g. from Halema'uma'u pit crater within the caldera (Neal and Lockwood 2003). The 2018 rift zone eruption drained a significant amount of magma from the volcano's summit reservoir, leading to subsidence, and withdrawal of the active Halema'uma'u lava lake. Eventually, this process caused the rock above the reservoir to fail, triggering the onset of episodic caldera collapse (Anderson et al. 2018). It should be noted that not all huge lava flow eruptions have immediately resulted in caldera subsidence, e.g. the 'Ailā'au 'au eruptions of Kīlauea in the fifteenth century (Clague et al. 1999). However, it is thought that such eruptions could initiate deformations eventually leading to caldera formation, which may take several years.

Caldera formation has already occurred on the youngest feature, Lo'ihi, indicating that calderas can be created at almost any time during the history of the shield volcano. Without a clear topographic rim, caldera structures are also visible in Kaho'olawe, West Maui, East Moloka'i, Ko'olau, Wai'anae, as the location of calderas commonly coincides with positive gravity anomalies centered on rift zone intersections, probably above the dense magma column that extends downward through the crust and into the mantle (Fig. 4B; Kauahikaua et al. 2000).

In contrast to Hawaii, a number of Hawaiian shield volcanoes worldwide, e.g. in the Canary Islands and La Réunion, display arcuate, open depressions. These are basically horseshoe-shaped landforms of flank failure origin due to increased flank steepness, partly transformed by subsequent erosion (for the Canarian shields: Carracedo 1996, Carracedo et al. 2007a, b). However, the formation of Las Cañadas caldera on Tenerife, for example, is more complex as it may also be due to caldera collapse (Martí 2019). At Piton de la Fournaise, La Réunion, the large size of its breached summit depression, the Enclos Fouqué 'caldera' (which is a horseshoe-scar open eastward towards the sea: Photo 44), has been attributed to the underlying, well-developed hydrothermal system leading to slow flank subsidence (formation of a so-called *hydrothermal caldera:* Merle et al. 2010, see Chap. 6).

3.5 Landslides and Fault Scarps

Huge landslides whose deposits have mainly been detected in submarine settings (see Fig. 4) are a typical process that affect the Hawaiian shield volcanoes in Hawaii, La Réunion, and the Canary Islands. Here, we refer to the voluminous deposits of past giant rockslides that removed a significant part of the shield volcano in a short time, a process first recognized by Moore (1964), and discussed in detail in Chap. 8.

Submarine bathymetric surveys carried out offshore of the Hawaiian Islands by the U.S. Geological Survey from the mid-1980s have revealed extensive debris fields and rotational slumps in the deep sea floor hundreds of kilometres from the islands (Moore et al. 1989), proving the previous hypotheses on landslides. The large amount of fragmented material could be connected to on-land morphostructures, in particular, 'palis', i.e. long cliffs along the coast, suggesting that the debris was emplaced by catastrophic rock debris avalanches (see Chap. 8). One of the most obvious structures, likely a precursor of such a process, is Kīlauea's Hilina pali and Hilina slump in the southern margin of the Big Island, the latter showing recent seaward sliding of 10 cm/yr (Morgan et al. 2003). The Nuuanu slide, related to the flank failure of Koolau volcano in Oahu 2 Ma, spread 160 km across the sea floor, most of which was uphill. The hummocky debris field emplaced from the slide is more than 35 km in width, with megablocks of the shield volcano tens of kilometres long and up to 1.5 km in thickness. Such catastrophic events must have created gigantic tsunamis in the Pacific Ocean. On the island of Lāna'i and Moloka'i, chaotic coral and lava breccia, inferred to be of tsunami origin, were recovered as high as 155 m above sea level (Moore et al. 1994), although island uplift may also explain (or contribute to) the high altitude of these deposits.

In total, seventeen slides have been identified around the Hawaiian Islands, and dozens more along the Hawaiian Ridge, all of which occurred in the past 30 Myr (Moore et al. 1989; Holcomb and Robinson 2004; Denlinger and Morgan 2014). One-third of the landslides might have had the capacity to generate megatsunamis (Clague and Sherrod 2014). Also, almost all of the Canary Islands have experienced giant landslides during their evolution (Masson et al. 2006; Hunt and Jarvis 2017). Likewise, submarine fans around La Réunion have also been interpreted as partly related to large landslides (Oehler et al. 2008; Le Friant et al. 2011), although small-scale, long-term mass-wasting leading to removal of island material has also played a significant role (see Chap. 8; Salvany et al. 2012; Gayer et al. 2021).

In Hawaii, submarine landslides seem to take place during the most significant volcano growth phase, during the transition from the pre-shield to post-shield stage. This means that tectonics of the active volcanoes triggers the landslides. However, landslides revealed along the entire chain of islands imply that the process can take place much later, when the volcano has already become extinct (Clague and Sherrod 2014). The exact reasons for landslides are not fully understood—for example, questions arise as to why large landslides can occur even when subaerial slopes are relatively gentle. Moreover, the seafloor in many places slopes inward (towards the island) as the growing island causes the oceanic plate to be depressed (Denlinger and Morgan 2014). At the same time, recent studies have clarified the role of magma pathways in flank instability—e.g. the stresses are generated in the south flank of Kīlauea by magma intrusion along the East Rift Zone. Interestingly, as already mentioned, Mauna Loa and Kīlauea have cyclic variations of rapid flank movements. Decades-long growth and slow slumping of the summit and the rift zone is punctuated by brief periods of rapid slope failure during which the seaward flank of the growing rift zone slides down in seconds to minutes (Denlinger and Morgan 2014). Stability of the volcanic edifice can also be weakened by hydrothermal weathering which is facilitated by repetitive intrusions. This can lead to ductile deformation of basal layers below the volcano, favouring flank spreading and collapse (e.g. for Hawaii: Morgan et al. 2007; for La Réunion: Michon and Saint-Ange 2008).

3.6 Evolutionary Scheme

Theoretically, a Hawaiian shield volcano will evolve through four eruptive stages: the preshield, shield, postshield, and rejuvenated stage (Fig. 7; Lipman 1995; Lipman et al 2000; Clague and Sherrod 2014). These stages are roughly the consequence of a plate drifting over a hot spot. The evolutionary process is basically the same for the Canary Islands, although postshield stage has not been distinguished (Carracedo et al. 1998, 2007a, b). A similar evolution with two stages was also evoked for La Réunion (Upton and Wadsworth 1972; Oehler et al. 2008; Peltier et al. 2012), while three stages have been proposed for the Galápagos shields, see Sect. 4.4 (Harpp and Geist 2018).

Stages observed in the Hawaiian Islands are gradational, and can be further subdivided. Between the postshield and rejuvenated stage, an erosional period of variable length may occur (Sect. 3.6.4), and of course there is a (long) period of erosion after the eruptive stages (as long as the volcano is above sea level). The four eruptive stages, in addition to the sometimes deviating growth history, reflect compositional variations in the magma, which are constrained principally by the amount and rate of heat supplied to the lithosphere as it moves over the hot spot (Haase et al. 2019). In the preshield stage, a small amount of partial melting occurs in a relatively deep magma reservoir (40–90 km: Grove and Till 2015) under high pressure, producing alkali basalts (Clague and Dixon 2000).

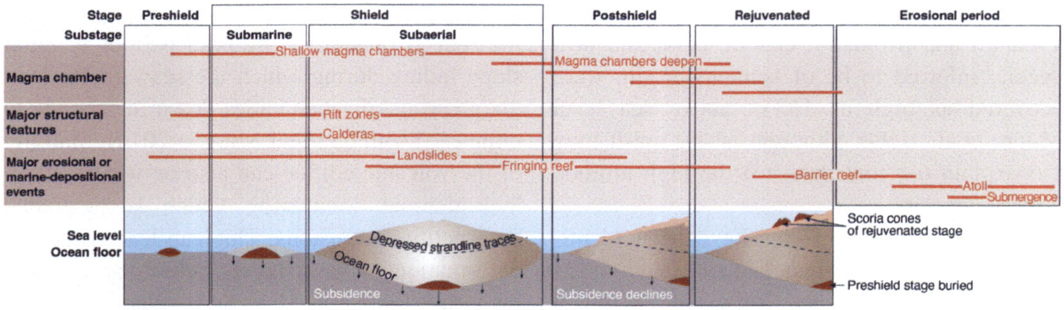

Fig. 7 Evolution of Hawaiian shield volcanoes showing five growth stages: preshield, shield, postshield, rejuvenated, followed by an erosional period, with major structural features and events represented temporally by red horizontal bars (Clague and Sherrod 2014)

The main stage is the shield stage, exemplified by Mauna Loa. It occurs when the shield volcano is directly above the hot spot, and intense magma supply results in the formation of a shallower (only a few-km-deep), high-volume magma chamber, producing large quantities of tholeiitic basalt by intense partial melting of the crust. Next, in the postshield stage, a decreased magma flux is caused by a lack of significant magma storage system; sporadically supplied alkaline magmas are generated directly from the upper mantle (Clague 1987; Sinton et al. 2017) and commonly erupt at parasitic or adventive vents, e.g. scoria cones. Mauna Kea and Hualālai in the Big Island of Hawaii are examples of this stage. With regard to evolution of the main shields, of further interest are the so-called Kea (compositionally depleted) and Loa (relatively enriched) geochemical trends that run geographically subparallel along the Hawaiian Islands and include a number of adjacent shields (Williamson et al. 2019). The two trends may reflect two plume heads: the Kea trend representing the original Hawaiian plume tail, while the more recent Loa trend—that has shown increased eruption rates in the past 5 Ma—represents an emerging plume head southeast of the Hawaiian tail (Zhang et al. 2023).

3.6.1 Preshield Stage

A Hawaiian volcano starts its life as a seamount in the deep sea when it passes over the hot spot. Alkali basalts erupt mainly as pillow and sheet lava flows. The preshield erupts tholeiitic and transitional lavas early on, showing a compositional trend that was first observed at Lo'ihi Seamount (Clague and Sherrod 2014). The preshield stage has also been identified on land at Kīlauea, Kohala, and perhaps Hualālai volcano.

The slopes of Lo'ihi are north–south striking and relatively steep, 10–15°. It has rift zones and flank failures, and its submarine summit is crowned already by a complex caldera structure and pit craters (Clague et al. 2019). Although Lo'ihi is growing rapidly and 3000–3500 m high, it is still almost 1000 m below the sea level. Average lava accumulation (vertical growth) rate for the preshield stage was estimated at about 5 m/kyr (Guillou et al. 1997). On this basis, Lo'ihi is predicted to eventually emerge from the sea within 50,000–200,000 years. Submarine preshield stage seamounts at the Society and Marquesas hot spots resemble those observed at Hawaii (Haase et al. 2019). In the Canary Islands, long-term erosion has exposed seamount rocks in Fuerteventura and La Palma that may correspond to the preshield stage (Carracedo et al. 1998).

At the end of the preshield stage, the volcano enters a zone of shallow water, resulting in phreatomagmatism. However, explosive eruptions can occur even before the water surface is reached. At Lo'ihi, lava fountains and even Strombolian activity were observed which took place due to the high magmatic volatile content of the alkalic magmas (Clague et al. 2003; Schipper et al. 2010).

3.6.2 Shield Stage

The volumetrically largest part of a shield volcano is created in the shield stage, over about 300 kyr (Lipman 1995). In fact, this is the period when the shield shape is constructed, associated with the appearance of rift zones, occasionally with radial vents, and a flat top commonly with a summit caldera.

Directly above the hot spot, increased magma supply results in the production of large amounts of mostly tholeiitic lava flows. On the main island of Hawaii, Mauna Loa is in the mature stage, whereas Kīlauea is the early stage of shield-building. Mauna Loa, despite its 2022 activity, shows evidence of decline. It has an age of 700 ka and rose above sea level ~ 300 ka (Lipman 1995; Jicha et al. 2012), although, notably, it still produced submarine eruptions from radial vents after emergence (Wanless et al. 2006). Both Mauna Loa and Kīlauea also display minor alkaline lava flows around their periphery (Coombs et al. 2006; Wanless et al. 2006).

Two relatively young Canary islands currently also represent the shield stage: El Hierro (1.1 Ma) and La Palma (1.7 Ma). These have shown continuous activity during their lifetime, with early-stage basalts as well as evolved alkaline rocks.

The island of La Réunion has two shield volcanoes: the older Piton des Neiges (1.4 Ma to 30 ka), and the younger, active Piton de la Fournaise (560 ka to present), which displays ongoing volcanism, with at least 150 eruptions since the seventeenth century. In its late-stage evolution, Piton des Neiges erupted differentiated alkaline products (trachytes, trachyandesites) probably from an edifice that had evolved into a composite volcano (Salvany et al. 2012), so must have surpassed the shield stage. Piton de la Fournaise shows many similarities to Kīlauea, including the shallow magma plumbing system, occurrence of flank instability (sector collapses, block displacements) as well as structural and compositional evolution, which may define it as a Hawaiian-type shield-stage volcano (Michon et al. 2009; Peltier et al. 2021). However, there are also differences: edifice size, life span (Piton de la Fournaise has been active for twice as long), and different flank instability mechanisms (Oehler et al. 2008; Lénat et al. 2012; Peltier et al. 2021). These are attributed to the higher magma supply and faster overriding plate at Hawaii compared to La Réunion. We have already referred to the large caldera size relative to the erupted products at Piton de la Fournaise, possibly related to subsidence due to hydrothermal alteration.

As noted above, the bulk of a shield volcano, 80–95% according to Clague and Sherrod (2014), is created during the shield stage. Although only a smaller fraction of the total volume is emplaced subaerially, on-land exposures provide a perfect opportunity to see how shield volcanoes build up. At the shields on the Big Island of Hawaii, the individual size of the lava flows is generally not significant, but their frequency and distribution are widespread. They comprise a monotonous sequence of pāhoehoe lavas, fed above all by the far-reaching rift zones. A very important and characteristic feature is the previously-mentioned fluidity of basalt that often moves along lava tunnels, which maintain the high temperature and thus allow the lavas to flow further. The fluidity of the lavas and their movement via tunnels explains the relatively gentle slopes and convex cross-section of the shield volcano profile. An interesting difference between the shields of the Big Island, reflecting their evolutionary stage (Bleacher and Greeley 2008), is that only 30% of the surface of Mauna Loa was formed by a network of lava moving in tunnels; 60–70% of the surface of Kīlauea has been created via tunnel-fed lava flows within the past 1500 years; and no lava tunnels are seen on the surface of Mauna Kea, which is now in the postshield stage.

The fast growth during the shield stage is represented by the high accumulation rates of superimposed lava flows, which in the long term can be measured as the rate between vertical increment (m)/duration (kyr) determined from a precisely dated stratigraphic column. Values obtained can reach up to 10–15 m/kyr (Guillou et al. 1997; Sharp and Renne 2005).

The most comprehensive data are available for Mauna Kea, using dated cores from the 2.7-km-deep Hawaii Deep Scientific Drilling project. Accumulation rates on Mauna Loa's submarine Southwest Rift Zone were pointed out to have been extremely high (25 m/kyr) in the most intense growing period of the shield stage at around 470 ka; later on, the rate decreased significantly to ~ 1 m/kyr for the next 200 ka (Jicha et al. 2012).

During the subaerial shield stage, the coalescence of numerous shelves aggraded by far-reaching lava flows forms a coastal plain. In this period, the frequent eruptions ensure that the coastal plain is able to keep pace with the subsidence of the volcano, and indeed Kīlauea and Mauna Loa, which have the youngest flows, are characterized by well-developed coastal terraces (Lipman 1995).

Phreatomagmatic eruptions may also occur during the shield stage, chiefly due to the replenishment of hydrothermal fluids by rainfall. In Hawaii, when the top of the magma chamber rises above the orographic rain belt, the likelihood of these eruptions decreases. Mauna Loa has already grown to this stage, whereas Kīlauea has shown phreatomagmatic activity in recent times (e.g. the 1790 AD/CE Keanakakoï Ash Member eruption: McPhie et al. 1990). Phreatomagmatic activity can also be related to volcanic aquifers, exemplified by Piton de La Fournaise (Join et al. 2005; Michon et al. 2013).

When the significant magma supply ceases, typically due to migration of the plate farther away from the hot spot, the shield growth stage terminates (in Hawaii, 600 kyr to 1 Myr on average). The most active phase of shield building in the Canary archipelago is currently taking place on La Palma and El Hierro islands (Carracedo et al. 1998).

3.6.3 Postshield Stage

In the postshield stage of the Hawaiian shield volcanoes, tholeiitic lavas give way to more alkaline lavas. The compositional shift is transitional. Many Hawaiian volcanoes (Wai'anae on O'ahu, East Moloka'i, West Maui, Haleakalā, Kohala, Mauna Kea, and Hualālai) have undergone the postshield stage. However, there are two—Ko'olau and Lāna'i—where the postshield stage is missing, and volcanism has not been renewed since 1.3 Ma on the latter. The postshield stage follows the shield stage immediately, and may last for hundreds of kyr; for example, on Mauna Kea, its duration was 260 kyr (Wolfe and Morris 1996), while on Haleakalā, it lasted an exceptional 1 Myr (Sherrod et al. 2003).

During the postshield stage, volumetrically minor eruptions occur that do not significantly change the shield shape except for the summit region. The caldera may be filled with scoria cones and lava flows originating from small-volume eruptions, which can raise the summit region and at the same time cover the outer slopes possibly in spatial relationship to the rift zone system (Photo 45). In fact, distribution of postshield scoria and spatter cones follow pre-existing rift zones on West Moloka'i, Haleakalā, and Hualālai (Photo 46). By contrast, the postshield centres are more dispersed on Kohala, East Moloka'i, and Wai'anae. The summit and the upper slopes of Manua Kea became covered by hundreds of scattered scoria cones from 250 ka. In particular, their appearance in the summit region explains why Mauna Kea is relatively steep—a morphometric characteristic which was further enhanced by glacial erosion: the late Pleistocene glaciation fed small glaciers, which steepened the slopes below the summit (Porter 1972). In the Canary Islands, only La Gomera is in the postshield stage, namely, in a 'gap' between the shield and rejuvenated stage (Carracedo et al. 1998).

3.6.4 Rejuvenated Stage

After the postshield activity, a shorter or longer erosional period (as much as ~2 Myr in Ni'ihau) followed in most of the Hawaiian Islands, in particular the older ones, manifesting itself by sliding processes, coastal abrasion (wave action), and mostly fluvial erosion. However, out of the fifteen emergent volcanoes, five major Hawaiian Islands: Ni'ihau, Kaua'i, Ko'olau (O'ahu), East Moloka'i, and West Maui have experienced rejuvenation. At

Kaua'i, no erosional period occurred between the postshield and rejuvenation stage, while east of Moloka'i, the islands are either still in the erosion period or in the postshield stage, up to Mauna Loa.

Activity on the rejuvenated-stage shield volcanoes recurred repeatedly for short intervals, but as this stage is considered very modest in terms of volume, it has not changed the general trend of long-term, effective erosion that has significantly reduced the height and extent of most islands. The older volcanoes, Ni'ihau and Kaua'i, show the largest surface area coverage of rejuvenated volcanism (~35%: Sherrod et al. 2007), whereas it is minor on the younger shields. Chemically, the erupted volcanics are very similar to those of the postshield stage (i.e. alkaline).

The origin and cause of this late-stage magmatism in the Hawaiian Islands is not yet well understood. The source of rejuvenated magmatism may be the upper part of the plume that undergoes a low degree of partial melting (Garcia et al. 2010). The oceanic plate is already moving away from the hot spot, and perhaps the cooling, shrinking crust provides the means for the magma to rise up, in particular when landslides induce fracturing favourable to small-scale volcanic eruptions.

In the Canary Islands, the eastern islands of Fuerteventura, Lanzarote, Gran Canaria and Tenerife also show a rejuvenated stage (Fig. 5). However, in terms of magma composition and volume of eruptive products, these islands show no real contrast between the shield/postshield and the rejuvenated stage. More importantly, the rejuvenated volcanism can be volumetrically significant. Magma compositions are much more diverse at all stages (Carracedo et al. 1998), although the Holocene eruptions are typically basaltic in composition.

One of the best examples of rejuvenated-stage volcanism can be seen on Gran Canaria, because this island has reached the end of its active life time. Its eruptive activity started 14.6 Ma and consisted mainly of a complex shield-building and a rejuvenated stage, separated by a 4 Myr-long erosional gap (van den Bogaard and Schmincke 1998; Schneider et al. 2004). Poorly constrained, buried Late Miocene volcanic edifice(s) of the shield-stage were followed by rejuvenated activity—in particular, the emplacement of a huge Pliocene volcano called Roque Nublo made up of products with an evolved composition. Due to an equilibrium between long-term volcanism and erosion, all stages on the island are adequately represented in the present-day topography; the higher levels are intensely eroded but often form remnants or outliers (Carracedo and Day 2002), highly useful for geomorphological reconstruction (Karátson et al. 2016). Hence, although much of the Roque Nublo edifice has been removed by erosion, based on the spatial relationships of its preserved lava and breccia successions it has been possible to reconstruct the former composite volcano (e.g. Pérez-Torrado et al. 1995). Moreover, using a morphometric approach to its paleo-surfaces, the original areal extent, volume (up to 1000 km^3), and elevation (>3000 m) of the original cone could be determined (Karátson et al. 2016). Figures imply that Roque Nublo showed dimensions comparable to the current Teide composite cone on Tenerife (3718 m).

This latter island roughly displays the same evolutionary progression as Gran Canaria (Ancochea et al. 1999; Carracedo et al. 2007a, b; Carracedo and Troll 2013). After a very long, 5.5 Myr erosional stage, rejuvenated activity on Tenerife at around 3.5 Ma produced the 10 × 17 km Las Cañadas caldera and the long eruptive history ended with the formation of the nested Late Pleistocene composite volcanoes of Pico de Teide (200 ka) and the smaller Pico Viejo (30 ka; Carracedo et al. 2007a, b). To conclude, Gran Canaria's and Tenerife's rejuvenated to recent evolution is significant compared to their counterparts in the Hawaiian islands, and they have undergone less erosion.

Other shield volcanic islands worldwide similar to Gran Canaria or Tenerife also show significant rejuvenation. At Fogo, one of the Cape Verde Islands, the rejuvenated stage produced the Fogo composite volcano (with ongoing eruptions) hosted in a 9 km-wide half-caldera of an early-stage basaltic shield volcano (for more details, see Chap. 5).

In Hawaii, coral colonies and reefs start to grow along the coast during the course of long-term erosion (e.g. Lāna'i, Kaua'i and Ni'ihau). Eventually an atoll forms, surrounded by a coral ring (Fig. 7). Atolls are typical of the Hawaiian Islands west of Ni'ihau, in particular west of a rocky reef called Gardner Pinnacle, but are not present in the Canary Islands, where colder waters only allow deeper coral reefs to form.

The submerged islands can survive for long periods (several Myr). The actual coastline is controlled by the long-term erosion, reef growth, sedimentation (e.g. carbonate deposition), and also sea-level change. Interestingly, while coral reefs protect the coast, living organisms established on coastlines may act as erosive agents (Naylor et al. 2013).

4 Galápagos Shields

4.1 General Features of a Complex Tectonic Setting

The evolution of the Galápagos shield volcanoes, in the East Pacific Ocean, ~1000 km off the west coast of Ecuador, is related to their specific tectonic setting. They are connected to a hot spot (the Galápagos Hot Spot, Fig. 8; also see Chap. 2) in the same way as the Hawaiian volcanoes. However, their formation has been controlled by complex spatial relationships surrounding the Galápagos Spreading Centre (a fault zone separating the Cocos and Nazca plates), and also by the fact that the edifices grew on a transitional type of crust with greater thickness (Mittelstaedt et al. 2014). The Galápagos Islands have been emplaced on the eastward-drifting Nazca Plate over the past ~ 3 Ma (Argus et al. 2011; Harpp and Geist 2018), with Fernandina, the westernmost island, currently lying above the hot spot (Villagomez et al. 2014).

The coalesced base of the Galápagos Islands forms a large, submarine plateau (the Galápagos Platform), which represents the westernmost extension of the Carnegie Ridge stretching eastward from the Galápagos Islands. This plateau may have been created by thick sequences of episodic, large-volume lava flows, and displays flat submarine terraces up to 40 km in length and 14 km in width, and individual terrace scarps up to 300 m in height (Geist et al. 2008). In addition to the emergent island shields on this plateau, a few more submarine volcanoes are also found in the surrounding deeper seafloor, connected to the Cocos Ridge. Possibly, some submerged part of the Carnegie Ridge was also composed of islands like the Galápagos several millions of years ago (Ali and Aitchison 2014).

Both the large plates and the microplates have moved in different directions and at different rates, resulting in a complex, changing regional stress pattern (Klein et al. 2005; Schouten et al. 2008; Schwartz et al. 2022). The combined tectonic setting (i.e. hot spot and mid-ocean ridge) along with the thickened crust are the main reason why the composition of the erupted magma covers both tholeiitic and alkali-rich compositions (White and Hofmann 1978; Vicenzi et al. 1990; Geist et al. 2005), the latter being more common than in Hawaii. Moreover, unlike in Hawaii, some shield volcanoes have also erupted more evolved products (dacite, rhyolite), e.g. Alcedo volcano on the western Isabela Island, or the Santiago, Pinzón, and Rábida shields in the eastern islands (Geist et al. 1994; Harpp and Geist 2018).

Particularly, in the past 1 Myr, intense volcanic activity occurred at the western Galápagos shields, manifested by calderas, along rift zones, and scattered around the shield volcano flanks, producing several times greater magma output than that of the eastern islands (Cullen et al. 1987; Glass et al. 2007). Thus, the Galápagos Islands are characterized by historical to current activity. More than 65 confirmed eruptions have taken place since 1800, most of which were in the westernmost islands (i.e. closer to the hot spot). Wolf, Cerro Azul and Fernandina are the most active. Wolf last erupted in 2015 and 2022, Cerro Azul in 1959, 1979, 1998, 2008, while Fernandina has eruptions every few years (most recently in 2020), and it had one of the largest (VEI 4) detected eruptions in the Galápagos Islands in 1968, associated with caldera formation (see Sect. 4.2.3).

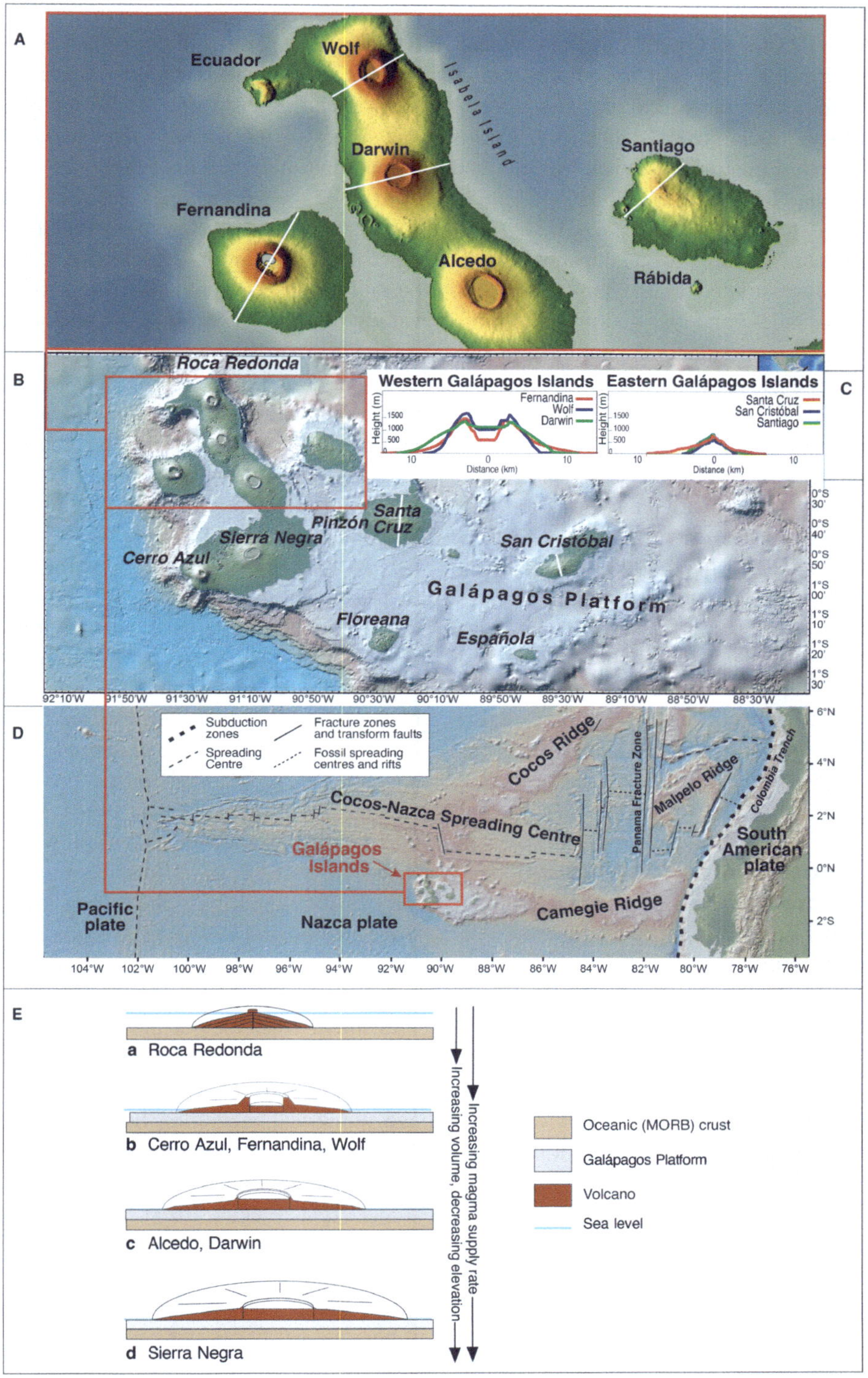

Fig. 8 Features of the Galápagos Islands. **A** Topography; **B** Large-scale topography and bathymetry; **C** Representative topographic profiles of the Western and Eastern Galápagos Islands (for the position of profiles, see white lines in A and B); **D** Geodynamic features shown on submarine topography; **E** Ideal cross-sections of selected shields in relation to increasing magma supply (after Naumann and Geist 2000). All maps and the profiles are based on SRTM DEM data; A was compiled using Open Street Map, B and D with GeoMapApp (www.geomapapp.org)

The Galápagos shield type has an overturned bowl-shape (Photo 47) with circular symmetry, steep upper slopes, and a well-developed, wide, central caldera with ring fissures. In addition to the Galápagos Islands, San Carlos (Bioko Island, Equatorial Guinea, Africa) and Karkar (Papua New Guinea) also have a similar morphology.

At the same time, the 'Galápagos' shield shape is only restricted to the larger, western, younger volcanoes, and the geomorphic variations in the eastern islands are not easy to link to tectonic and magmatic processes (Harpp and Geist 2018). The characteristics of the western shields and the differences to the eastern ones have been long recognized (e.g. McBirney and Williams 1969), and various hypotheses have been put forward to explain them (e.g. Nordlie 1973; Munro and Rowland 1996; Naumann and Geist 2000; Rowland and Garbeil 2000; Geist et al. 2014). Most recently, Harpp and Geist (2018) synthetized the morphometric and geochemical differences and proposed a general model, based on the changing tectonic regime over a Myr-long time scale, which is presented below.

4.2 Geomorphic and Structural Characteristics of the Western and Eastern Galápagos Shields

4.2.1 Morphometry

The Galápagos shield volcanoes are clearly different in size from the Hawaiian type, being two magnitudes smaller in volume, with a basal (subaerial) diameter of < 30 km, and a height up to 1700 ma.s.l. (set on a ca 4000 m-deep seafloor). This smaller scale is due to the lower magma supply rate.

The volcanoes of the western and eastern Galápagos can be divided into two groups based on many aspects. The younger (< 0.5 Ma to present) western Galápagos shield volcanoes are much larger than the eastern ones (150–250 versus 10–150 km^3, these figures referring to the subaerial constructs). Not only the mean slopes (5–9°) but also the upper slopes (9–11°) of their flanks are steeper than those of the eastern counterparts (for detailed morphometric data, see Grosse and Kervyn 2018). Within this, summit slopes of the western shields, e.g. Wolf or Fernandina, may reach 15–18° and even rarely 20° (while the eastern ones are < 10–15°), which contrasts with the Hawaiian shields. On the other hand, the lower (subaerial) slopes are gentle on both the western and eastern shields, with values similar to Hawaii (3–7°, and even flatter on the eastern shields). Lastly, there are extensive submarine slopes (i.e. on the Galápagos Platform), again with somewhat higher gradients (4–9°: Nordlie 1973), similar to Hawaii.

When considering slope features, a further important factor in the shape of the flatter eastern Galápagos shield volcanoes is that they are located further from the edge of the Galápagos Platform, and therefore they may have formed a more extensive apron in the surrounding shallow sea. In contrast, the steeper western volcanoes are at the edge of the platform, where abrasion by the sea and mass movements are more significant, enhancing the slope gradient (as suggested e.g. for Volcán Ecuador during prolonged erosional gaps: Rowland et al. 1994). Offshore of Fernandina, over-steepened, shallow submarine flanks consist of lava flows that transgressed the coastline, and have since been truncated by wave erosion and slumping (Geist et al. 2006).

4.2.2 Structural Characteristics

The shield volcanoes of the Western Galápagos Islands have ring or circumferential eruptive fissures around their summit caldera. Analogue and numerical modelling have shown that calderas induce a stress field that favours the formation of ring dykes around the edge of the caldera (Gaete et al. 2019). The ring fissures of the summit regions shift to radial fissures down the flanks (Chadwick and Howard 1991; Xu et al. 2023), and, further away, continue into submarine rift zones, which are most developed on Fernandina, Cerro Azul, Ecuador, Wolf, and Darwin volcanoes. On the basis of recent eruptions (1995, 2005, 2009) at Fernandina, the circumferential and radial eruptions were all fed by subhorizontal sill intrusions initiated from the shallow (1 km-deep) magma reservoir (Bagnardi et al. 2013). Ring fissures on the calderas and radial fissures on the lower flanks are attributed to significant volcano load in the central part, which takes precedence over the tectonic control.

Whereas the lower flanks of the western Galápagos shields are composed mostly of extensive lava flows from radial vents, the morphology of the summit region, in addition to small-volume lava flows, is controlled by a greater proportion of pyroclastic material including spatter ramparts and scoria cones. These products are derived from the ring fissures and create steeper summits (Rowland 1996; Naumann and Geist 2000). This material control on the overturned bowl-shape provides an alternative to the older model of intrusive doming of the central steep part (McBirney and Williams 1969), although recent studies have confirmed the role of intrusions and caldera uplift (e.g. Bagnardi et al. 2013).

Ring-like faults and related fissures can be observed not only in the Galápagos, but also, for example, on the shield volcano of Deception Island (Antarctica) with a caldera harbour. Other shield types can also exhibit ring faults and ring fissures, like Silali and Suswa in the EARS, or the shoaling shield volcano of Niuafo'ou atoll (Tonga Islands). These belong to the Turkana and Newberry subtype shields, respectively, within the medium-sized continental intraplate and arc shields (see Sect. 6).

4.2.3 Calderas

The western shield volcanoes are topped by large calderas (4–8 km in diameter), whereas none of the eastern ones have a caldera, apart from Pinzón, with a small summit caldera 1.5 km-wide, and perhaps Rábida, whose arcuate faults could be related to an infilled caldera (Harpp and Geist 2018). Presence or lack of calderas is supported by recent gravity measurements (e.g. Vigouroux et al. 2008). Within the large western calderas, Sierra Negra, Alcedo, and Darwin have shallower calderas (250–400 m: Photo 48), whereas Wolff, Cerro Azul, and Fernandina have deeper ones (500–900 m) (Munro and Rowland 1996). The deeper calderas also have a greater volume (10–12 km^3), except Cerro Azul which, despite its great depth (400 m), has the smallest diameter (3.8 km), and thus a caldera volume of only 3 km^3 (Grosse and Kervyn 2018). Caldera formation was well studied during Fernandina's activity in 1968, which started with VEI 4 hydromagmatic eruptions. Its existing caldera (Fig. 6) experienced 350 m of coherent but asymmetric (trapdoor-like) subsidence (Chap. 6), adding to its 700 m depth prior to 1968. Old benches at each end of the elliptical caldera provide evidence of earlier partial basalt filling, testifying that Fernandina has experienced repeated collapse and lava filling episodes, similar to many other western Galápagos calderas (Howard 2010).

The presence of calderas in the Western Galápagos is direct evidence of shallow magma chambers (Geist et al. 2014), similar to, or even shallower than, those in the mature shield-building stage at Hawaii. By contrast, shields of the Eastern Galápagos Islands have no calderas and tend to be elongated along a single rift zone (Grosse and Kervyn 2018, see also Fig. 12 in Chap. 2). Those with the most elongate shape display structural alignments of eruptive cones, pit craters, fissures and faults, whereas there is no evidence of the ring or radial fissure structures found on the western shields (Chadwick and Howard 1991). Exemplified by Santa Cruz, the elongated shields experience faulting, flank spreading, and intrusions parallel to the rift zone, accompanied by small-scale volcanic activity (Schwartz et al. 2022).

4.3 Explaining the Contrasting Features of the Galápagos Shields

Early hypotheses (e.g. Nordlie 1973) stated that the older, eastern Galápagos shields were smaller (and lacked have calderas) because of long-term erosion, and hence represented a late-stage development. However, the observed structural and morphometric differences are more likely due to the different tectonic situation and magma supply over time (Geist et al. 2014; Harpp and Geist 2018). This is true even though the easternmost islands (e.g. San Cristóbal) are more significantly degraded.

Most of the Western Galápagos shield volcanoes, which are younger (or even active), erupt evolved basalts owing to a high magma supply due to them being located directly over the Galápagos mantle plume (Harpp and Geist 2018). The magma flux from the Galápagos plume seems to have increased over the past 1 Myr, despite the thicker oceanic crust beneath the western islands. This is explained by the negative correlation of the magmatic flux of the Galápagos volcanoes with distance from the Galápagos Spreading Centre, and the western islands currently fed by the hot spot are now furthest from the mid-ocean ridge (Harpp and Geist 2018).

By contrast, the much smaller, older (1–3 Ma) Eastern Galápagos volcanoes were different from the beginning. They erupted relatively primitive magmas probably from isolated, ephemeral magmatic plumbing systems that were supplied by smaller magmatic fluxes over their lifetimes. Thus, the eastern volcanoes do not represent an evolved (i.e. long-term erosional) stage of the young western volcanoes but, instead, were formed by a different, less intense constructional mechanism controlled by the proximity to the Galápagos Spreading Centre and a lower degree of plume activity (Harpp and Geist 2018).

4.4 Evolution of the Galapagos Shields

Taking into account the different evolution of the western and eastern Galápagos shield volcanoes outlined above, a fourfold scheme can be put forward (based on Nordlie 1973; Geist et al. 2014; Harpp and Geist 2018):

(1) *construction of the submarine base* and coalescence of thick lava flows over the hot spot and a relatively thin crust for the Eastern Galápagos volcanoes;

(2a) *subaerial stage* at Eastern Galápagos: close to the mid-ocean ridge between 1 and 3 Ma, these volcanoes were characterized by a small, deep magma reservoir and relatively low magma supply showing limited chemical variability, and affected by strong tectonic control typically resulting in one rift zone;

(2b) *subaerial stage* at Western Galápagos: here, the present terrestrial shield volcanoes are created further away from the mid-ocean ridge, emplaced on a thickened crust by increased magma supply, with relatively homogenous composition; a thick intrusive (sill) system has developed which generates nested ring fissures, central caldera formation, and an inverted 'soup-bowl' shape (e.g. Fernandina, Wolff, Darwin);

(3) *'dying' subaerial stage* at Western Galápagos: transported away from the plume, the magma reservoir cools and fractional crystallization produces more evolved magma compositions (dacites, trachytes, rhyolites: e.g. Alcedo, Santiago, Pinzón);

(4) *erosional stage* (observed only at some of the Eastern Galápagos Islands): results from significant erosion of the edifice, characterized by reduced slope angles in the central part; occasionally, late-stage adventive vents develop (e.g. San Cristóbal).

5 Icelandic Shields

5.1 General Features

This shield type comprises relatively small (Table 1), flattish, concentric edifices consisting predominantly of lava flow successions. Whitford-Stark (1975) and Pike and Clow (1981)—on the basis of dozens of examples worldwide—characterized the Icelandic shield as a symmetric cone or mound with or without a crater (Photo 49). The Icelandic shields are typically monogenetic: they build up by subsequent effusions of lava over a short period of time (a few to several tens of years: Cas and Wright

1988; Andrew and Gudmundsson 2007). They have near-circular plan shapes, and the low central cone is commonly surrounded by an apron (Jakobsson et al. 1978). However, there are significant differences in size and occasionally in shape (e.g. slope pattern), as presented below.

The Icelandic shields occur mostly in the West and North Volcanic Zone of Iceland, although well-known examples are also found in the Snake River Plain (USA); the Michoacán–Guanajuato Volcanic Field (Mexico); the Haruj (Libya) and Ithnayn (Saudi Arabia) volcanic fields; the Newer Volcanic Field (Australia); scattered occurrences in the volcanic areas of Kamchatka; and the Faroe Islands (deeply eroded). At the same time, small shields can occur anywhere in the world either associated with monogenetic volcanic fields or alone, fed by successive, short-lived effusive eruptions from fissures or central vents.

The birth of shield volcanoes in Iceland is thought to be related to effusive fissure eruptions. Subsequently, the eruption becomes concentrated at a reduced number of vents, producing a few small, overlapping shields. Then, the eruption gradually becomes more focused, and one vent generates the main shield which buries the previous smaller shields (Gudmundsson 2000). This sequence of events may last for decades or centuries (Thordarson and Höskuldsson 2008). A kind of shield volcano formation, reshaping a previous tuya, may have started during the recent effusive activity of Fagradalsfjall volcano in the Reykjanes Peninsula (Halldórsson et al. 2022: Photo 50).

The lithology of Icelandic shield volcanoes is overwhelmingly basaltic, although basaltic andesites and andesites also occur outside Iceland (e.g. Michoacán-Guanajuato). In Iceland, basalts can be grouped into two categories. Most of the shields are composed of olivine tholeiite basalt, but some are composed of more primitive picrite basalt. As the latter lava is denser, its effusion results in very small, but slightly steeper shields, many of which may have been buried by subsequent olivine tholeiite shields (Jakobsson et al. 1978). These are called 'compound shield volcanoes' by Andrew and Gudmundsson (2007).

5.2 Morphometric Features as Constrained by Shield Formation

The Icelandic shield is a small-sized edifice bounded by gentle slopes. The usually straight slopes can be slightly steeper near the vent (Andrew and Gudmundsson 2008; Grosse and Kervyn 2018). In the eponymous Iceland there are about twenty shield volcanoes. They have contrasting sizes, several are only 0–1 km^3 in volume, with the majority between 1 and 10 km^3, and a few reach 20 km^3 with up to 15 km as the largest diameter. Skjaldbreiður, the iconic exemplary type of Icelandic shield, is located in the central part of the West Zone (Photo 49). The mean slope angle of its flanks is 5°, but slopes can reach 8°, especially near the central crater which is 300×400 m in width and 60 m in depth (Andrew and Gudmundsson 2008). Skjaldbreiður is considered to be the largest Holocene shield in Iceland, with an estimated volume of 12 km^3 (Grosse and Kervyn 2018), a value that might be only half of the full value if the buried part is taken into consideration (Andrew and Gudmundsson 2007).

In Iceland, there are somewhat flatter and steeper shield volcanoes ranging between two end-member values (< 2° to 6–9°: Thordarson and Höskuldsson 2008). The low profile of the former is exemplified by Þeistareykjarbunga, the steeper, more conical by Skjaldbreiður or Trölladyngja. The flat shields can be formed by low-rate, steady effusion of lava that may last for decades to centuries in the case of the largest edifices. Steeper, conical shields are built by prolonged, relatively high-discharge (>1000 m^3/s) and fountain-fed surface lava flows (e.g. from sustained lava lakes) which create a symmetric lava cone from shelly pāhoehoe lava. Subsequently, when the discharge is lower, lava is fed through lava tubes from the central vent to the propagating flow fronts, creating an extended lava apron. In the latter, slabby pāhoehoe and 'a'ā lava are most common. Among the most prominent features of a typical lava shield are tumuli and, toward the margin of the apron, lava rises (Andrew and Gudmundsson 2007).

The formation of relatively large-volume shield volcanoes in Iceland was constrained by the Pleistocene glaciation and then the disappearance of the ice cap (Andrew and Gudmundsson 2007). During the most extensive and thickest ice cover, the glacial load suppressed magma accumulation and associated volcanism. At the end of the Pleistocene glaciation, melting of a significant amount of ice resulted in decreased lithostatic pressure, which in turn resulted in increased size of deep-seated magma reservoirs. This process provided excess magma, and, as a consequence, resulted in the formation of lava shields of relatively large size. Indeed, most of the Icelandic shields are post-glacial, i.e. of Early Holocene age (10,000–5000 yr). Later on, the complete disappearance of glacial load reduced the amount of eruptible magma in later Holocene times; eventually the divergent plate movements of Eurasia and Africa relieved the ice-induced stresses and lava-shield activity died out (Andrew and Gudmundsson 2007). On the other hand, the neoglaciation (started about 5000 years ago) resulted in the formation of present-day glacier caps, covering a number of shields in Iceland. The second largest ice cap, Langjökull in the Western Volcanic Zone that developed at that time has associated shield volcanoes covered by glaciers (e.g., Þórisjökull/Thórisjökull, Geitlandsjökull: Photo 51).

Of other Icelandic type shields elsewhere, we highlight the Snake River basaltic area, Idaho (USA), where long-term volcano-tectonic evolution (Yellowstone hot spot and Basin and Range crustal extension: Fig. 6 in Chap. 2) was associated with multivent volcanism (Hughes et al. 1999). Low, monogenetic Quaternary shield volcanoes were formed by outpourings of a number of individual tube-fed pāhoehoe lavas. Single and coalesced shield volcanoes, e.g. Wapi Lava Field or Hell's Half Acre, have small volumes (in the order of 1 km^3) and very flat slopes (< 1°).

In the Michoacán–Guanajuato monogenetic field, with over 1000 vents of different sizes, there are hundreds of shields similar in size to those in Iceland. However, there is a subtype which is definitely steeper. In the light of detailed DEM studies (Grosse and Kervyn 2018), this specific group displays a mean slope of 10–12°, increasing to 15–20° towards the summit. Such a deviation can be explained by the basaltic andesitic or andesitic magma composition related here to the subduction-related setting: firstly, the erupted lavas are more viscous, have a higher yield strength, and therefore have a greater thickness, shorter length, and overall solidify at a steeper angle; second, these shields do not have a crater, but instead have a summit lava dome or scoria cone, which again makes the uppermost flanks steeper. The slightly different lithology and the different summit morphology are also reflected by a certain amount of pyroclastic material—i.e. the appearance of explosive eruptions, which play a minor but not negligible role in the geomorphic characteristics.

This steeper subtype of the Icelandic shield may also include the conical Rangitoto Island volcano (off Auckland, North Island, New Zealand: Lowe et al. 2016). Being part of the Auckland Volcanic Field, Rangitoto is a 4 km-wide, 240 m-high island with a complex, circular, flattish lava construct formed only 650–550 years ago on lavas that predated the shield by 6 kyr. Shortly after, ~ 500 yrs ago, large scoria cones were emplaced in the central part producing tephra deposits. In contrast with the young, short shield-building effusive stage, older lavas as well as tephras indicate renewed activity lasting for thousands of years.

We can also find relatively steep shields in the western part of the continental USA, perhaps best represented by the subduction-related Belknap volcano (Oregon). Belknap (2095 m a.s.l.) and Little Belknap (1922 m a.s.l.) are two summits of a 10–12 km-wide, 'double' shield volcano with 500–600 m relative height and 5–6° flank slopes. Their huge basalt and basaltic andesite lava flows, which erupted between ~ 3000 and 1500 years ago, were crowned by a basaltic scoria cone (Belknap) and a smaller effusive vent (Little Belknap). Other examples of such small-volume shields include Table and Badger Mountains near Lassen volcano, Goosenest close to Medicine Lake, and Whaleback near Mount Shasta (Lockwood and Hazlett 2010), all constituted of basalt and basaltic andesite.

6 Medium-Sized Continental Intraplate and Arc Shields

As presented above, the medium-sized (a few 100, and rarely up to a few 1000 km^3 volume) shield volcanoes of Earth, which commonly have explosive activity in addition to the monotonous, effusive shield-building, show a mixed geomorphic picture. Many of them are similar in size to the Galápagos type, without the steeper upper flanks, while they often have a central caldera that is larger than those in the Galápagos Islands due to the explosive character of the volcano. Importantly, explosive activity in their wider environment (in almost all terrestrial occurrences) always produces other volcano types as well (composite volcanoes, or calderas of larger size) in addition to strictly shield volcanoes. Moreover, in the case of these mixed fields, it can be difficult to distinguish the shield volcanoes from other types, all the more so because the composite volcanoes themselves can be of various types or transitional (see Chap. 5). Here, we distinguish continental intraplate rift and hot-spot shields, termed the Turkana subtype; and arc shields, either on continents, microcontinents, or island arcs, which are not specifically recognised in the literature, but for which we use the term Newberry subtype.

6.1 Intraplate Rift and Hot-Spot Shields (Turkana Subtype)

This subtype, based on occurrences in the Kenya Rift in the EARS, was originally proposed by Webb and Weaver (1975). It was not well established in the literature as a distinct type, until it was re-introduced by Lockwood and Hazlett (2010), who added other examples.

The Turkana shields share some basic features with the Western Galápagos shields, such as concentric ring fissures in their summit caldera, but they have a different lithology and related eruptive characteristics, due mostly to having developed on continental crust rather than oceanic crust. Most importantly, they are composed of evolved alkali eruptives (trachyte, phonolite) in addition to basalt, which are responsible for occasional explosive eruptions and caldera formation. Evolved magma can also result in the emplacement of lava domes. On Silali volcano, dome activity was responsible for the so-called summit trachytes on the western flank (Mutonga 2012). Post-caldera rhyolite lava domes and flows are also frequent on Ethiopian shield volcanoes (Fontijn et al. 2018).

The eponymous EARS edifices comprise, among others, Emuruangogolak, Silali, Kiplombe, Paka, Longonot, and Suswa volcanoes (Fig. 9), which are shields with a broad diameter (30–40 km) and medium- to large-sized central calderas, rising south of Lake Turkana and sitting in the rift valley. For example, the Holocene Silali volcano (1528 m a.s.l.), which has an elevation of 700–800 m relative to its surroundings and is dotted on its flanks by a few monogenetic cones and pit craters, has a flat summit level with a 300-deep caldera 5×8 km in diameter. Basalts erupted from flank fissures are common, whereas trachytic lavas and pyroclastics occur both on the flanks, and around and inside the caldera. The Late Pleistocene Emuruangogolak (1285 m a.s.l.), north of Silali, has a smaller caldera (5×3.5 km) filled and dotted with post-caldera lava flows and cones, with well-preserved flank cones and maars erupted along rift-parallel faults cutting the volcano.

These shields have average slope values (5–9°) similar to those of the Western Galápagos, but lack the overturned bowl profile. This implies that the slope pattern is different when looked at in detail: based on SRTM DEM-based morphometry (Grosse and Kervyn 2018), the lower flanks have a gentle slope (< 6° on average), the main flank is steeper (~8°), and the upper slopes are even steeper (11°), but well below the Galápagos upper slope values (i.e. 15–18°). The similar areal extent but the somewhat lower summit slope values relative to the Western Galápagos shields explains the relatively smaller volume of the Turkana shields in Kenya.

There are other, larger edifices worldwide that may also belong to the Turkana subtype (Lockwood and Hazlett 2010), such as the massive Oligocene shields in eastern Australia. These

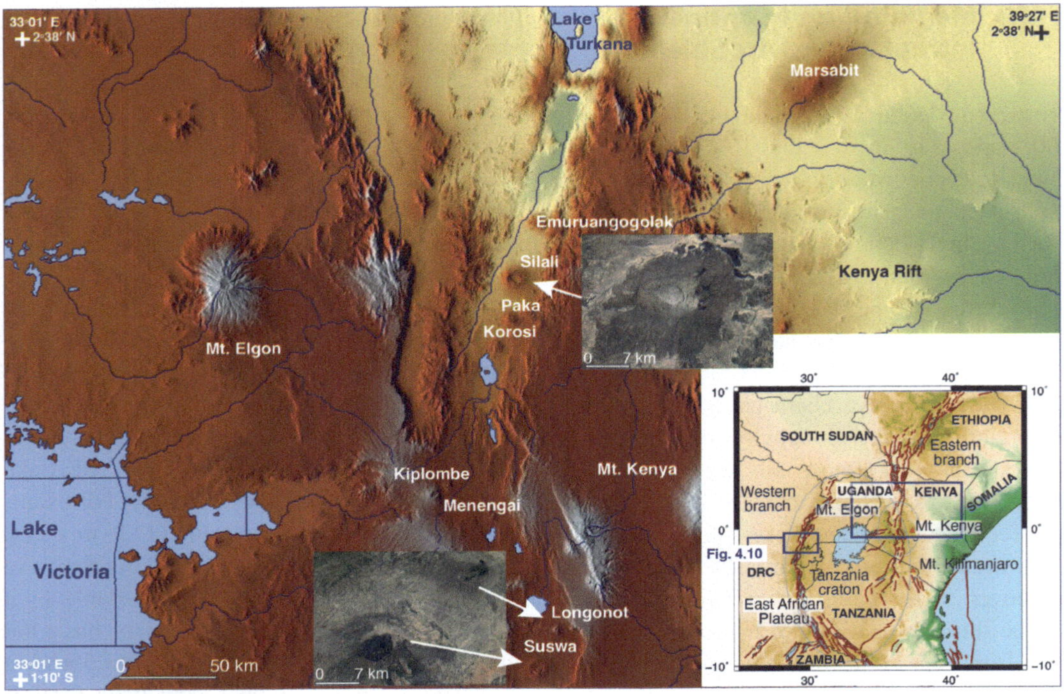

Fig. 9 Shaded and coloured relief image of the region south of Lake Turkana (the Kenya Rift) in the eastern branch of the East African Rift System (EARS), showing examples of Turkana shields. Map is based on SRTM DEM data and compiled using Open Street Map. Inserts displaying Silali, Longonot and Suswa are taken from Google Earth. Map is based on SRTM DEM data and compiled using Open Street Map

volcanoes, similar to the EARS ones, erupted more evolved magmas in addition to basalt (i.e. trachyte, trachyandesite, rhyolite), and the lavas were accompanied by a significant proportion of pyroclastics. Unfortunately, there is no detailed morphometric study on them, making it difficult to infer the original geomorphic characteristics. Located along the eastern margin of the continent, examples from 20–100 km in diameter include Hillsborough, Springsure, Buckland and Nebo volcano in Queensland, and Tweed, Nandewar, Ebor, Canobalas and Warrumbungles volcano in New South Wales (Johnson et al. 1989; Tapu et al. 2023).

However, these Eastern Australian shields have a different origin to the EARS ones as they were formed by hot-spot activity. When the northward-drifting Australian plate ploughed into the submarine, hotspot-related Ontong-Java Plateau, ongoing plume activity of the East Australian (or Crosgrove) hot spot fed a track of well-developed shield volcanoes from ~ 35 Ma to 6 Ma (Knesel et al. 2008; Davies et al. 2015). This chain of volcanoes forms the longest continental hot-spot track on Earth, 2000 km in length. The changing magma flux resulted in variable magma types, eruptive style, and volcano size (Tapu et al. 2023), which is similar to the differences between Hawaiian shields. Namely, during the period 35–27 Ma, in the area of present-day Queensland, intense magma supply created large shields (> 800 km^3) mostly with a bimodal composition and lifetimes of up to 7 million years. The largest edifice of this stage is Tweed volcano, with an assumed original volume of ~ 4000 km^3 (see Chap. 8). Subsequently, magma output reduced, and thus the southern, younger shields in New South Wales were smaller (< 300 km^3) and shorter-lived (< 1.5 Myr; Tapu et al. 2023).

Continental hot-spot shields somewhat resembling their Australian counterparts can also be found in the Anahim Range, Canada, where the hypothesized Anahim hot spot may have created a 300 km-long track of volcanoes formed on the westward-drifting North America plate. The volcanic chain consists of a number of shields, domes, and other volcano types, collectively called the Anahim Volcanic Belt (Souther 1986). They show an age progression from 14 Ma to historical times. Three large shields include, from west to east, Rainbow Range (8.7–7.2 Ma), Ilgachuz Range (6.1–4.0 Ma), and Itczha Range (3.5–1.1 Ma) (Bevier 1981; Souther 1994). These heavily eroded, dissected shields are 20–30 km in diameter, with well-preserved lower slopes at 5–8°, and consist of trachyte, rhyolite, and subordinately basalt lavas and pyroclastics.

Some eroded volcanoes of the Upper Rhine Graben (Germany) within the European Cenozoic Rift System, which are 10–20 Ma in age, also represent medium-sized rift- (and, perhaps, also hot spot)-related shields (Dèzes et al. 2004; Wilson and Downes 2006; Rotstein and Schaming 2011). The largest edifice is Vogelsberg, a deeply eroded, symmetric shield volcano with a diameter of 50 km and a possible total original volume of 600 km^3, which erupted a wide spectrum of tholeiitic and alkali basalts and basanites to trachyte, phonolite, and rhyolite (Bogaard and Wörner 2003).

As referred to above, caldera formation at the Turkana shields can be associated not only with large-volume lava outpourings but also with ignimbrite eruptions. The Silali caldera, 6×4 km in diameter (Smith et al. 1995) is thought to have been formed by gradual subsidence, following the growth of an early shield volcano (400–200 ka). Caldera-forming mechanisms included downsagging, propagation of a circumferential fissure zone, and collapse of the summit area associated with explosive activity; both basalt and trachyte eruptions continued until Early Holocene times. The nested caldera of the trachytic-phonolitic Suswa volcano, one of the largest shields (2356 m a.s.l.: Fig. 9), may have been created in the past 250 ka (Skilling 1993; Guth and Wood 2014). Growth of an early shield volcano ended with an ignimbrite eruption and lava flows associated with incremental formation of a large (8×12 km) outer caldera. This was followed by construction of a post-caldera cone on the southern caldera rim. Then, a new caldera was formed north of this cone, truncating its crater by an unusual intracaldera graben. In some other cases, explosivity has even more severely truncated the original shield volcano, best exemplified by the Late Pleistocene–Early Holocene trachytic Menengai volcano (see Fig. 9). Its early-stage broad shield, constructed by monotonous lava flows, was truncated at 35.6 ka by the caldera-forming ignimbrite eruption of the Menengai Tuff, covering an area > 115,000 km^2 (Blegen et al. 2019). The caldera, 12×18 km across, has subsequently been the site of at least one more ignimbrite eruption and prolonged effusive activity, probably until historical times; today, it is covered by a number of post-caldera lava flows and scoria cones (Macdonald et al. 1994; Riedl et al. 2020).

Emplacement of the Turkana shields and formation of their calderas seem to be strongly controlled by the prevailing tectonic regime (Wadge et al. 2016). The Menengai caldera might have been formed by subsidence, but post-caldera tectonic activity also occurred, including faulting and resurgent doming, which provide insights into the interaction of rift-scale extension and local upwarping due to magmatic processes (Riedl et al. 2020). Bosworth et al. (2003) pointed out that the long axes of elongated Turkana calderas are parallel to the upper crustal minimum horizontal stress direction. During the Late Pleistocene, formation of these calderas was accompanied by a 45° rotation of the minimum stress direction, and hence their long axes (from E–W to NW–SE), and at the same time an increase in the magnitude of the maximum horizontal stress.

6.2 Arc Shields (Newberry Subtype)

Shield volcanoes related to arc volcanism – either on continental margins or island arcs – are considered another subtype. These are also made up of differentiated, but in this case, typically silicic and less alkaline magma. In this geodynamic setting, i.e. subduction zone, the most common volcano types are composite volcanoes (Chap. 5) and calderas (Chap. 6). However, almost all subduction zone segments worldwide (e.g. the Aleutian Islands, the Cascade Range, the Kamchatka-Kurile arc, the New Hebrides) include shield volcanoes of basalt, basaltic andesite to andesite composition as well. At these volcanoes, in addition to effusive activity making up an (early-stage) shield shape, explosive eruptions (mostly in a mature stage) can also occur, similar to the Turkana-type shields. Calderas are also common, frequently associated with the highly explosive activity. We note that in the case of a broad, low shield with a caldera, the classification of the original landform is often controversial (see e.g. the Global Volcanism Program), partly because the lower flank of a truncated shield is not very different from a truncated composite volcano, so if the upper cone is missing, it is difficult to deduce its shape without detailed studies.

The best known continental-margin examples are Newberry and Medicine Lake (Photo 52). They are located in an extensional back-arc setting east of the High Cascade volcanic arc. Both display a low, broad shield morphology complicated by evolved lavas, with a central caldera at the top (e.g. Donnelly-Nolan et al. 2016). They seem to have been fed by multiple, small magma reservoirs during their extended life-time, and eruptive products show a bimodal composition similar to many of their neighbouring calc-alkaline High Cascade composite volcanoes (Chap. 5). Basalt or basaltic andesite lava flows of Newberry and Medicine Lake cover an area of 3000 and 2000 km^2, respectively, whereas the shields themselves are ~ 30 km in diameter, have a volume of 400–600 km^3, and display gentle slopes similar to the Turkana type at rather constant values (4–6°: Grosse and Kervyn 2018). However, the uppermost slopes appear to be a bit steeper owing to stubby lava flows and domes, terminating in a flattish, uneven top level (due to the presence of the summit caldera).

In more detail, Newberry (Jensen and Donnelly-Nolan 2017) has a basalt/basaltic andesite shield volcano base that formed 400 ka (see Fig. 8, Chap. 3). The outer flanks are dotted by some 400 scoria cones, many of them only a few thousand years old. Its central caldera hosts Paulina and East Lake, separated by Central Pumice Cone and Interlake Rhyolite Flow. The 1300 year-old Big Obsidian Lava flow, in the southern part of the caldera, is the most recent material (see Chap. 3; Photos 13 and 52). The northern and southern flanks of the volcano are covered by basalt/basaltic andesite lava flows and associated vents. Approximately 200 km southward, the even larger shield of Medicine Lake had a similar evolution, and was active from 500 ka to historical times, the last eruption producing Glass Mountain obsidian flow only 900 years ago.

As for caldera formation, the present-day 8 × 6 km-wide Newberry caldera is thought to have been created at 75 ka when about 25 km^3 rhyolitic tephra erupted (Jensen and Donnelly-Nolan 2017). Moreover, there were at least two more previous caldera-forming explosive eruptions, which followed the growth of the early shield construction. At Medicine Lake volcano, its 12 × 7 km-wide central caldera seems to have formed gradually as a result of subsidence in response to repeated effusive basaltic eruptions. Thus, unlike at Newberry, a large explosive eruption responsible for caldera formation cannot be identified (although the eruption of a dacite tuff ~80 ka may have initiated caldera formation: Donnelly-Nolan 2010). Subsidence and uplift of the caldera may have been associated with flank eruptions as well (Donnelly-Nolan 2010).

A more complex example from the Central Andes is the trachytic Payún Matrú volcano (Argentina). It has an impressive morphology: a broad, circular shield volcano base (with 8.5° mean flank slope) is truncated by a regular-shaped elliptical caldera, 9 × 7 km in diameter, which was created by an ignimbrite eruption between 168 and 82 ka (Germa et al. 2010). Post-caldera activity up to 7 ka included lava flows and explosive eruptions of trachyandesitic to rhyolitic composition. The original edifice has been reconstructed as a broad, flattened shield 2300 m in height, 20 km in width, with a volume of ~ 250 km^3 (Germa et al. 2010). However, more detailed studies are required to decipher the structural and geomorphic characteristics of such explosive shields to produce a well-founded typology.

A good example of explosive shield volcanoes in the realm of island arcs is Okmok, USA, which has an impressive, partly glacier-covered 10 km-wide central caldera (Fig. 3, Photo 71). The volcano is located in the Aleutian arc which is constructed on continental crust to the east and oceanic crust to the west, and Okmok is found just west of the ocean-continent transition on Umnak Island (Finney et al. 2008). Predominantly mafic magma (of basaltic to andesitic composition) is erupted along the Aleutian Islands, typically forming composite volcanoes and lava domes (see Chap. 5). However, shield volcanoes like Okmok also occur. Like Newberry, its early edifice was a basaltic shield volcano, but possibly more symmetric, which was subsequently truncated by two highly explosive caldera-forming eruptions at 12,000 and 2050 years ago (the latter affected people in Europe by causing severe cooling in the northern hemisphere: McConnell et al. 2020). As a result, a nested caldera 12 × 10 and 10 × 10 km in diameter, respectively, was formed (Byers 1959; Larsen et al. 2007). During these explosive eruptions, a small amount of dacitic/rhyolitic magma was ejected, followed by a much larger amount of andesite and basaltic andesite (Larsen et al. 2007). The preserved lower flanks of the original edifice have a mean slope < 10° (Grosse and Kervyn 2018). Such a value is somewhat steeper than those of Newberry and Medicine Lake, and implies that the shape of upper, truncated edifices cannot be clearly inferred.

The island volcanoes of Ambrym in the New Hebrides, Vanuatu (Fig. 3) alongside its northern neighbour, Ambae (or Aoba), show some structural and geomorphic deviations. They have already been described under the tuff cones section (Chap. 3). They are part of the New Hebrides Island arc that formed due to subduction of the Australian-Indian Plate beneath the Pacific Plate from approximately 10 Ma (Haase et al. 2020). Lithology of the Ambrym and Ambae shield volcanoes is dominated by calc-alkaline basalt (i.e. more evolved rocks are subordinate or missing), while the structural control is enhanced relative to their continental counterparts. Ambrym has a double rift (oriented N 105° S) ending in an inactive northern tip, and Ambae has a single rift. In the latter, the ENE-WSW-trending rift results in anomalously steep (10–15°) flanks (Grosse and Kervyn 2018) perpendicular to the rift zone in the central part of the edifice, and formation of a summit plateau that hosts a nested caldera (the larger of which is 4.5 km across). By contrast, Ambrym has a broad, low-profile shield encompassed by flank slopes similar to those of Okmok (around 7°: (Grosse and Kervyn 2018). We have already evoked its debated caldera formation in Chap. 3; gradual caldera subsidence has recently been linked to rifting episodes, dyke propagation, and ring fault activation (Shreve et al. 2021), i.e. unrelated to highly explosive eruptions. At the less known Ambae shield, although explosive eruptions were frequent during its history, subsidence of its caldera might have also been gradual, possibly due to massive outpourings of basalt lava (Warden 1970). However, a recent alternative model suggests that its caldera formation has also been affected by summit loading due to the hydrothermally altered rocks in the volcano core (Merle et al. 2010).

Arc shields also occur in Kamchatka and the Kuril Islands, formed in relation to the subduction of the Pacific plate, and in the north, by the young Komandorsky plate, under the Okhotsk plate (Portnyagin et al. 2005). Along the volcanic belt, the dominantly calc-alkaline and also alkaline rocks constitute two arcs: Central Kamchatka (Sredinny Range) and Southern Kamchatka–Kuril Islands (eastern volcanic front). During the Holocene, active volcanism has created mostly composite volcanoes and calderas (Chap. 5), but there are a few exceptions that can be classified as shield volcanoes (Ponomareva et al. 2007). Examples include Snegovoy and Maly Palyalpan (without a caldera) in Kamchatka, and Tsirk (with an eroded 6 km-wide caldera, probably of Lower Pleistocene age: Erlich 1986) on Iturup (Etoforu) in the Kuril Islands. Furthermore, there are other large calderas in Kamchatka that are referred to as shield volcanoes in the literature: e.g. Tao-Rusyr (Erlich 1986) and Gorely (Ponomareva et al. 2007). However, they are more complex in shape, have significant silicic magmatism and post-caldera activity—Gorely, for instance, hosts five overlapping cones—, and are therefore more likely to be classified as Somma-type composite volcanoes (see Chap. 5).

Finally, one more island arc volcano belonging to this subtype should be mentioned: Mayor Island (New Zealand), which is a peculiar volcano built up of pantellerite (high-alkaline or peralkaline rhyolite; Houghton et al. 1992). Its submarine diameter is 15 km in diameter, and the island which rises 300 m above sea level is only 4 km in width. The rhyolite lava construct makes the subaerial part relatively steep-sloped (13°: Grosse and Kervyn 2018). The island has a small central caldera (2.5 km in width), filled by late-stage post-caldera lava flows and domes. The caldera seems to have been formed gradually; apart from likely subsidence episodes, the last caldera-forming event 6300 yr ago was a Plinian eruption that produced tephra on mainland North Island (Houghton et al. 1992).

7 The Wide Diversity of Shield Volcanoes—Concluding Remarks

7.1 Shape Variations of Shield Types and Subtypes

In this chapter, we have documented the general shield types, which show a rich variety of variation in shape to create ‘subtypes’. As was shown, the wide spectrum of shields and shield-like edifices has transitional shapes often with overlapping eruptive and morphometric characteristics. However, even though some subtypes might be difficult to place in one of the four classical types, many aspects of the general evolutionary pattern are well-constrained.

Shields are born as small lava constructs from point-like or fissure vents. Creation of small-sized shields or cones seems independent of geodynamic setting. Lava constructs can evolve into small lava-dominated edifices, which are called by various terms in the literature. The smallest shield volcano is often referred to as scutulum, but it can quickly evolve into the Icelandic type. Small and flat Icelandic shields of basaltic composition can increase in size if there is an increased or long-lasting magma supply, but the edifice still remains below about 10 km^3 in size. Steeper landforms can be built up where, for example in a subduction setting (e.g. Mexico), they are fed by basaltic andesitic magma, accompanied by some explosive activity. Overall, the above scenarios apply to cases of modest or moderate magma supply.

On the contrary, significant hot-spot activity leads to the construction of sizeable shields, up to tens of thousands of km^3. In the case of emplacement on an oceanic plate, the resultant landform will be the larger Hawaiian shield, or the medium-sized Galápagos shield (found in the western islands of this archipelago). The type that is created depends on plume intensity (magma output rate), overriding plate speed, and tectonic control. Both types are dominated by basaltic lavas, although while the Hawaiian Islands are almost exclusively basaltic, the composition of shields in the Galápagos Islands, as well as in the Canaries, may be more evolved. In both types, symmetric or elongated shields can form, depending on the prevailing or changing tectonic control. Symmetric versus elongated shapes may even appear side by side: examples include Mauna Loa versus Mauna Kea in the Hawaiian Islands; Santa Cruz versus San Cristóbal in the Eastern Galápagos Islands; and La Gomera versus La Palma (southern rift zone) and Gran Canaria versus Tenerife in the Canary Islands.

Medium- to large-sized continental shields perhaps show an even richer variety of landforms. Continental intraplate shields can be hot spot- and rift-related (collectively called the

Turkana subtype), and are fed by magmas with a wide range of geochemical compositions, typically of alkaline affinity. The same applies to the subduction-related continental margin- and island arc-shield occurrences (termed the Newberry subtype): the eruptive products are mainly characterized by calc-alkaline magma, often with quite a diverse lithology. Rifts can also form at island arcs. On all these shield volcanoes, in addition to the volumetrically dominant, predominantly mafic lava flows, explosive activity can also play a significant role, which can lead to the formation of a central caldera several km large. However, in these cases, it must be recognised that the shape is transitional, and further studies are needed to gain a better understanding of their structural and geomorphic features.

It is also poorly known how precisely any of the geodynamic settings contribute to the specific kind of shield. To take an example, Turkana-type alkali shields may be associated not only with rifts but also with continental hot spots (Australia, Canada), where, in terms of size variability, they are similar to the Hawaiian shields, whereas their explosive character overlaps with the arc-related Newberry subtype. Another issue is the appearance of rift zones: they may evolve in almost all settings (less frequently in continental plates), but the development of a triple, double or single rift, often next to each other, is influenced by the local stress field.

7.2 Coexistence of Shield- and Composite Volcanoes in Similar Settings

Finally, creating a bridge to Chap. 5, which deals with strato- or composite volcanoes, we mention the possible relationship of explosive shields (fed by evolved magma) to other, more composite volcano types. There are several examples of shield volcanoes from both rift (e.g. EARS) and subduction settings (e.g. New Hebrides) which have edifices classified as calderas or stratovolcanoes by the Global Volcanism Program in their vicinity. In particular, the EARS has the largest range of magma types on Earth (Rooney 2020), which implies a multitude of magmatic processes (Biggs et al. 2021). Here, factors relating to the complex geodynamic setting (Fig. 4B in Chap. 1)—i.e. a combination of rifting and plume or 'super-plume' (Hansen et al. 2012), the variable thickness of the crust, and the evolution stage of the rift including changes in local stress field (for details, see Chap. 1)—are all responsible for creating different magmatic sources and magma compositions, which in turn results in diverse volcanic eruptions and landforms (Fontijn et al. 2018; Hunt et al. 2019; Biggs et al. 2021).

We illustrate the coexistence of shield- and composite volcanoes by taking two edifices from the Virunga Volcanic Province, located along the western branch of the EARS. In particular, there are two famous active volcanoes located next to each other, Nyamuragira (Nyamulagira) and Nyiragongo, which are classified as shield- and composite volcano, respectively (Fig. 10). Their setting provides an excellent example of why distinctive geomorphic evolutions can occur in the same setting and from an apparently similar magma plumbing system. In addition to Nyiragongo's current history and lava lake, presented in Chap. 2, we present here a brief overview comparing the evolution of the two volcanoes.

Nyamuragira (3058 m) is a highly productive, effusive, basanitic-tephritic shield volcano, which is difficult to put into one of the four main shield types. Its summit is truncated by a small, circular caldera, 1.6×-2.3 km across, formed in historical times by multiple collapses. It has steeper mean slopes (8–10°) than the Turkana subtype, possibly indicating a geomorphic transition toward a composite volcano. This is shown by occasional Strombolian explosions in addition to common Hawaiian lava fountaining, as well as fissure effusions (Smets et al. 2015). About 13 km to the south, Nyiragongo (3470 m) is a steep-sloped (30°) compound composite volcano (Demant et al. 1994) with a 1.2 km-wide, terraced summit crater. The name Nyiragongo refers to the

FIG 4.10 - 05/02/24

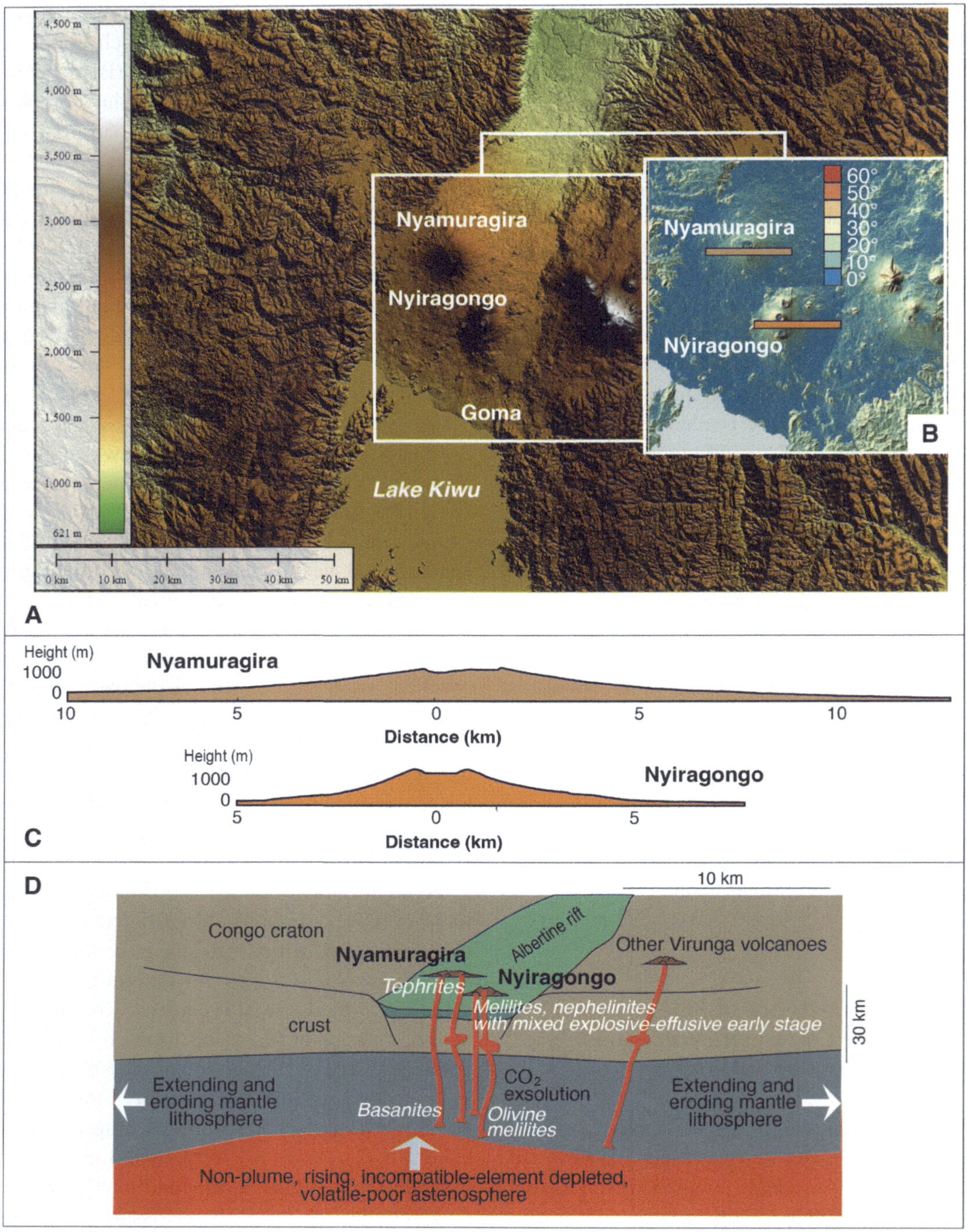

Fig. 10 Features of Nyamuragira (3058 m) shield- and Nyiragongo (3470 m) composite volcano north of Lake Kiwu in the western branch of the East African Rift System (EARS). **A** Coloured SRTM DEM image with inset map (**B**) showing slope patterns; **C** Topographic cross-section of Nyamuragira and Nyiragongo based on SRTM DEM data (position of cross sections is marked with orange bar in B); **D** proposed magma generation model after Minissale et al. (2022)

central, highest cone, which lies between (and partially overlapping) two older, smaller cones (Fig. 10) that collectively make up a compound edifice. Compositionally, Nyiragongo erupts ultra-alkaline products belonging to the foidite field. During the construction of this composite volcano, which ended a few centuries ago with a crater collapse, the lava composition changed from melilitite to leucite, then to nephelinite (Pouclet and Bram 2021). In historical times, the latter has come to the surface in the form of the most fluid lava flows on Earth. However, the early cone-building stage, fed by carbonated melilitic magma, also produced lava fountain-related tephras (exposed in the stepped terraces within the crater) in addition to short summit lava flows, and such a mixed explosive and effusive activity may have created the steep upper flanks (Pouclet and Bram 2021).

To summarise, we observe that different/changing eruptive styles can result in different volcano types at a short distance from each other and possibly within a short time scale. The different upper cone morphology of Nyamuragira and Nyiragongo could be related to their slightly contrasting eruptive styles mainly during their early evolution, which in turn would have been controlled by slightly different magma compositions reflecting some structural and geochemical complexities of the lithosphere (see Fig. 10: Minissale et al. 2022).

Photo 42 Mauna Loa (4170 m), Hawai'i (USA), the largest shield volcano on Earth. It has an area of 5000 km^2 and volume of 95,000 km^3 including the much more substantial submarine part. View from Saddle Road that runs between Mauna Loa and Mauna Kea, showing the 1843 pahoehoe lava flows from Mauna Loa (*photo* courtesy of Sz. Kósik)

Photo 43 Kīlauea caldera, Hawai'i (USA), viewed from the north, showing limited lava lake activity in the *Halema'uma'u* pit crater (2015); crater diameter of 800 m (*photo* courtesy of S. Kósik)

Photo 44 Interior of Enclos Fouqué 'caldera', La Réunion Island (France), a horseshoe-scar within Piton de La Fournaise). Lava flows come from the central Dolomieu crater (where the photo was taken from), which formed in 2007 (*photo* courtesy of L. Gourbet)

Photo 45 Mauna Kea (4205 m), the highest volcano on Hawaii's Big Island (USA). Now in the postshield stage, it is raised by summit scoria cones. Mauna Loa is mainly hidden in clouds in the distance (*photo* courtesy of S. Kósik)

Photo 46 The central depression of Haleakalā shield volcano (Maui, Hawai'i Islands, USA), is called the Haleakalā Crater. However, this negative landform, created in the intersection of two huge fault-controlled valleys, was in fact formed by landslides and erosion. The depression has subequently been filled by lava flows and scoria cones (visible at the bottom) belonging to the Hāna volcanics, which represents the waning phase of postshield volcanism (*photo* courtesy of S. Kósik)

Photo 47 Darwin volcano (1330 m-high, Galápagos Islands, Ecuador), one of the most iconic Galápagos-type shields, is located in the northern, narrow part of Isabela Island. It last erupted in 1813, and has fresh, unvegetated lava flows on all flanks, which reach both the eastern and western coasts (*photo* H. Jiménez, Galápagos Eco-Lodge, http://blog.galapagosecolodge.net/2012/09/galapagos-island-geography-and-geology_1590.html)

Photo 48 The Darwin volcano caldera in the foreground, with Wolf volcano (1710 m) in background to the right (Galápagos Islands). Aerial view from the southeast). The bottom of the symmetrical Darwin caldera, 5-km wide and 200 m deep, is covered by young lava flows (*photo* P. Ramón, 2003, Global Volcanism Program, https://volcano.si.edu/gallery/ShowImage.cfm?photo=GVP-11650)

Photo 49 The Skjaldbreiður shield volcano (1066 m high and 6 × 8 km across in diameter), southwest Iceland. It is an archetypal Icelandic shield and one of the largest, and probably formed over a period of only decades at the beginning of the Holocene. Thingvallavatn Lake in the foreground (*photo* D. Karátson)

Photo 50 Fagradalsfjall (Iceland), a 40-km-long tuya, formed during the Last Glacial. Comprising eruptive fissures, cones and lava fields, it belongs to the Reykjanes Volcanic Zone. After a long, quiet period of 6000 years, Fagradalsfjall has begun reshaping into a shield during recent (2021- ongoing) eruptive activity. Photo shows a 2021 eruption of the spatter cone (*photo* courtesy of Gábor Nagy)

Photo 51 Examples of glacier-clad broad shield volcanoes belonging to the Langjökull ice cap (Iceland): Thorisjökull in the foreground, and Geitlandsjökull in the background (aerial *photo* D. Karátson)

Photo 52 The Newberry Caldera (Oregon, USA) is representative of arc shields. It has a broad, low morphology 3000 km^2 in area, and its central part is occupied by the Paulina Lake and East Lake, divided by the Newberry pumice cone. Big Obsidian Lava flow is visible within the caldera in the foreground (*photo* courtesy of S. Kósik)

References

Acosta J, Uchupi E, Smith D, Muñoz A, Herranz P, Palomo C, Llanes P, Ballesteros M, ZEE Working Group (2003) Comparison of volcanic rifts on La Palma and El Hierro, Canary Islands and the Island of Hawaii. In: Clift P, Acosta J (eds) Geophysics of the Canary Islands. Mar Geophys Res 24:59–90. https://doi.org/10.1007/s11001-004-1162-6

Ali JR, Aitchison JC (2014) Exploring the combined role of eustasy and oceanic island thermal subsidence in shaping biodiversity on the Galápagos. J Biogeogr 41(7):1227–1241. https://doi.org/10.1111/jbi.12313

Ancochea E, Huertas MJ, Cantagrel JM, Coello J, Fúster JM, Arnaud N, Ibarrola E (1999) Evolution of the Cañadas edifice and its implications for the origin of the Cañadas Caldera (Tenerife, Canary Islands). J Volcanol Geotherm Res 88(3):177–199. https://doi.org/10.1016/S0377-0273(98)00106-1

Anderson KR, Johanson IA, Patrick MR, Gu M, Segall P, Poland MP, Montgomery-Brown EK (2018) Magma reservoir failure and the onset of caldera collapse at Kīlauea Volcano in 2018. Science 366(6470). https://doi.org/10.1126/science.aaz1822

Andrew REB, Gudmundsson A (2007) Distribution, structure, and formation of Holocene lava shields in Iceland. J Volcanol Geotherm Res 168(1–4):137–154. https://doi.org/10.1016/j.jvolgeores.2007.08.011

Andrew REB, Gudmundsson A (2008) Volcanoes as elastic inclusions: their effect on the propagation of dykes, volcanic fissures, and volcanic zones in Iceland. J Volcanol Geotherm Res 177(4):1045–1054. https://doi.org/10.1016/j.jvolgeores.2008.07.025

Annen C, Lénat J-F, Provost A (2001) The long-term growth of volcanic edifices: numerical modelling of the role of dyke intrusion and lava-flow emplacement. J Volcanol Geotherm Res 105(4):263–289. https://doi.org/10.1016/S0377-0273(00)00257-2

Argus DF, Gordon RG, DeMets C (2011) Geologically current motion of 56 plates relative to the no-net-rotation reference frame. Geochem Geophys Geosyst 12(11). https://doi.org/10.1029/2011GC003751

Bachèlery P, Lénat J-F, Di Muro A, Michon L (2016) Active volcanoes of the southwest Indian Ocean: Piton de la Fournaise and Karthala. Active volcanoes of the world. Springer, 428 p. https://doi.org/10.1007/978-3-642-31395-0

Bagnardi M, Amelung F, Poland MP (2013) A new model for the growth of basaltic shields based on deformation of Fernandina volcano, Galápagos Islands. Earth Planet Sci Lett 377–378:358–366. https://doi.org/10.1016/j.epsl.2013.07.016

Baker MCW (1981) The nature and distribution of upper Cenozoic ignimbrite centres in the Central Andes. J Volcanol Geotherm Res 11:293–315. https://doi.org/10.1016/0377-0273(81)90028-7

Bevier ML (1981) The rainbow range, British Columbia: a miocene peralkaline shield volcano. J Volcanol Geotherm Res 11(2–4):225–251. https://doi.org/10.1016/0377-0273(81)90025-1

Biggs J, Ayele A, Fischer TP, Fontijn K, Hutchison W, Kazimoto E, Whaler K, Wright TJ (2021) Volcanic activity and hazard in the East African Rift Zone. Nat Commun 12:6881. https://doi.org/10.1038/s41467-021-27166-y

Bleacher JE, Greeley R (2008) Relating volcano morphometry to the developmental progression of Hawaiian shield volcanoes through slope and hypsometric analyses of SRTM data. J Geophys Res 113(B9). https://doi.org/10.1029/2006JB004661

Blegen N, Brown FH, Jicha BR, Binetti KM, Faith JT, Ferraro JV, Gathogo PN, Richardson JL, Tryon CA (2019) The Menengai Tuff: a 36 ka widespread tephra and its chronological relevance to Late Pleistocene human evolution in East Africa. Quaternary Sci Rev 152:152–168. https://doi.org/10.1016/j.quascirev.2016.09.020

Bogaard PJF, Wörner G (2003) Petrogenesis of Basanitic to Tholeiitic Volcanic Rocks from the Miocene Vogelsberg, Central Germany. J Petrol 44(3):569–602. https://doi.org/10.1093/petrology/44.3.569

Bosworth W, Burke K, Strecker M (2003) Effect of stress fields on magma chamber stability and the formation of collapse calderas. Tectonics 22(4). https://doi.org/10.1029/2002TC001369

Byers FM Jr (1959) Geology of Umnak and Bogoslof Islands Aleutian Islands Alaska. Investigations of Alaskan Volcanoes. US Geol Surv Bull 1028-L, US Government Printing Office, Washington, USA

Carey R, Cayol V, Poland M, Weis D (2015) Hawaiian volcanoes, from source to surface. Geophysical Monograph Series 208. AGU, Washington DC, USA, 582 p. https://doi.org/10.1002/9781118872079

Carracedo JC (1994) The Canary Islands: an example of structural control on the growth of large oceanic-island volcanoes. J Volcanol Geotherm Res 60(3–4):225–241. https://doi.org/10.1016/0377-0273(94)90053-1

Carracedo JC (1996) Morphological and structural evolution of the western Canary Islands: hotspot-induced three-armed rifts or regional tectonic trends? J Volcanol Geotherm Res 72(1–2):151–162. https://doi.org/10.1016/0377-0273(95)00080-1

Carracedo JC (1999) Growth, structure, instability and collapse of Canarian volcanoes and comparisons with Hawaiian volcanoes. J Volcanol Geotherm Res 94(1–4):1–19. https://doi.org/10.1016/S0377-0273(99)00095-5

Carracedo JC, Badiola ER, Guillou H, Paterne M, Scaillet S, Torrado FJP, Paris R, Fra-Paleo U, Hansen A (2007a) Eruptive and structural history of Teide Volcano and rift zones of Tenerife, Canary Islands. Bull Geol Soc Am 119(9–10):1027–1051. https://doi.org/10.1130/B26087.1

Carracedo JC, Day S (2002) Canary Islands: classic geology in Europe 4. Terra Publishing, Harpenden, UK, 294 p. ISBN: 1-903544-07-6

Carracedo JC, Day S, Guillou H, Badiola ER, Canas JA, Torrado FJP (1998) Hotspot volcanism close to a passive continental margin: the Canary Islands. Geol Mag 135(5):591–604. https://doi.org/10.1017/S0016756898001447

Carracedo JC, Rodriguez-Badiola E, Guillou H, Paterne M, Scaillet S, Perez-Torrado FJ, Paris R, Fra-Paleo U, Hansen A (2007b) Eruptive and structural history of Teide Volcano and rift zones of Tenerife, Canary Islands. Bull Geol Soc Am 119(9–10):1027–1051

Carracedo JC, Troll VR (2013) Teide volcano: geology and eruptions of a highly differentiated oceanic stratovolcano. Springer, Heidelberg, 279 p. ISBN: 978-3-642-25892-3

Carracedo JC, Troll VR, Day JMD, Geiger H, Aulinas M, Soler V, Deegan FM, Perez-Torrado FJ, Gisbert G, Gazel E, Rodriguez-Gonzalez A, Albert H (2022) The 2021 eruption of the Cumbre Vieja volcanic ridge on La Palma, Canary Islands. Geol Today 38(3):94–107. https://doi.org/10.1111/gto.12388

Cas RAF, Wright JV (1988) Modern volcanoes and volcanic centres. In: Cas RAF, Wright JV (eds) Volcanic successions: modern and ancient. Springer, Dordrecht, Netherlands, pp 362–410. ISBN-13: 978-0-412-44640-5

Chadwick WW Jr, Dieterich JH (1995) Mechanical modeling of circumferential and radial dyke intrusion on Galapagos volcanoes. J Volcanol Geotherm Res 66 (1–4):37–52. https://doi.org/10.1016/0377-0273(94)00060-T

Chadwick WW Jr, Howard KA (1991) The pattern of circumferential and radial eruptive fissures on the volcanoes of Fernandina and Isabela Islands, Galapagos. Bull Volcanol 53:259–275. https://doi.org/10.1007/BF00414523

Clague DA (1987) Hawaiian alkaline volcanism. Geol Soc Spec Publ 30(1):227–252. https://doi.org/10.1144/GSL.SP.1987.030.01.10

Clague DA (1996) The growth and subsidence of the Hawaiian-Emperor volcanic chain. In: Keast A, Miller SE (eds) The origin and evolution of Pacific island biotas, New Guinea to eastern Polynesia: patterns and processes. SPB Academic Publishing, Amsterdam, pp 35–50. ISBN: 90-5103-136-X

Clague DA, Batiza R, Head JWIII, Davis AS (2003) Pyroclastic and hydroclastic deposits on Loihi Seamount, Hawaii. Geophys Monogr Ser 140:73–95. https://doi.org/10.1029/140GM05

Clague DA, Dixon JE, (2000) Extrinsic controls on the evolution of Hawaiian ocean island volcanoes. Geochem Geophys Geosyst 1(4). https://doi.org/10.1029/1999GC000023

Clague DA, Hagstrum JT, Champion DE, Beeson MH (1999) Kīlauea summit overflows: their ages and distribution in the Puna District, Hawai'i. Bull Volcanol 61:363–381. https://doi.org/10.1007/s004450050279

Clague DA, Moore JG (1991) Geology and petrology of Mahukona Volcano. Hawaii. Bull Volcanol 53:159–172. https://doi.org/10.1007/BF00301227

Clague DA, Paduan JB, Caress DW, Moyer CL, Glazer BT, Yoerger DR (2019) Structure of Lo'ihi Seamount, Hawai'i and lava flow morphology from high-resolution mapping. Front Earth Sci 7. https://doi.org/10.3389/feart.2019.00058

Clague DA, Sherrod DR (2014) Growth and degradation of Hawaiian volcanoes. In: Poland MP, Takahashi TJ, Landowski CM (eds) Characteristics of Hawaiian volcanoes. US Geol. Surv. Prof. Paper 1801, Reston, VA, pp 97–146. https://doi.org/10.3133/pp18013

Coombs M, Eakins B, Cervelli P (2006) Growth and collapse of Hawaiian volcanoes. J Volcanol Geotherm Res 151(1–3). https://doi.org/10.1016/j.jvolgeores.2005.07.022

Cotton CA (1944) Volcanoes as landscape forms. Whitcombe and Tombs Ltd, Christchurch, New Zealand, 416 p. ISSN: 0028-0836

Cotton CA (1952) Volcanoes as landscape forms. Wiley, NY, 416 p. ISSN: 0028-0836

Cullen AB, McBirney AR, Rogers RD (1987) Structural controls on the morphology of Galapagos shields. J Volcanol Geothermal Res 34(1–2):143–151. https://doi.org/10.1016/0377-0273(87)90099-0

Davies DR, Rawlinson N, Iaffaldano G, Campbell IH (2015) Lithospheric controls on magma composition along Earth's longest continental hotspot track. Nature 525:511–514. https://doi.org/10.1038/nature14903

de Silva S, Lindsay JM (2015) Primary volcanic landforms. In: Sigurdsson H, Houghton B, McNutt SR, Rymer H, Stix J (eds) Encyclopedia of volcanoes. Academic Press, San Diego, CA, pp 273–297. ISBN: 978-0-12-385938-9

Demant A, Lestrade P, Lubala RT, Kampunzu AB, Durieux J (1994) Volcanological, and petrological evolution of Nyiragongo volcano, Virunga volcanic field, Zaire. Bull Volcanol 56:47–61

Denlinger RP, Morgan JK (2014) Instability of Hawaiian volcanoes. In: Poland MP, Takahashi TJ, Landowski CM (eds) Characteristics of Hawaiian volcanoes. US Geol Surv Prof Paper 1801, Reston, VA, pp 149–176. ISBN: 978-1-4113-3872-2

Dèzes P, Schmid SM, Ziegler PA (2004) Evolution of the European Cenozoic Rift System: interaction of the Alpine and Pyrenean orogens with their foreland lithosphere. Tectonophysics 389:1–33. https://doi.org/10.1016/j.tecto.2004.06.011

Donnelly-Nolan JM (2010) Geologic Map of Medicine Lake Volcano, Northern California. US Geol Surv Scientific Investigations Map 2927, 48 p

Donnelly-Nolan JM, Champion DE, Grove TL (2016) Late Holocene volcanism at Medicine Lake Volcano, northern California Cascades. US Geol Surv Prof Paper 1822, Reston, VA, 59 p. https://doi.org/10.3133/pp1822

Dumont Q, Cayol V, Froger JL, Peltier A (2022) 22 years of satellite imagery reveal a major destabilization structure at Piton de la Fournaise. Nat Commun 13:2649. https://doi.org/10.1038/s41467-022-30109-w

Eakins BW, Robinson JE, Kanamatsu T, Naka J, Smith JR, Takahashi E, Clague DA (2003) Hawaii's volcanoes revealed. US Geol Surv Geologic Investigation Series Map I-2809, Reston, VA. ISBN: 0-607-95494-9

Erlich E (1986) Geology of the calderas of Kamchatka and Kurile Islands with comparison to calderas of Japan and the Aleutians, Alaska. US Geol Surv Open-File Rep 86-291, Vancouver, WA, 300 p. https://doi.org/10.3133/ofr86291

Faichney IDE, Webster JM, Clague DA, Paduan JB, Fullagar PD (2010) Unraveling the tilting history of the submerged reefs surrounding Oahu and the Maui-Nui Complex, Hawaii. Geochem Geophys Geosyst 11(7). https://doi.org/10.1029/2010GC003044

Favalli M, Karátson D, Yepes J, Nannipieri L (2014) Surface fitting in geomorphology—examples for regular-shaped volcanic landforms. Geomorphology 221: 139–149. https://doi.org/10.1016/j.geomorph.2014.06.009

Fialko YA, Rubin A (1999) What controls the along strike slopes of volcanic rift zones? J Geophys Res 104(B9):20007–20020. https://doi.org/10.1029/1999JB900143

Finney B, Turner S, Hawkesworth C, Larsen J, Nye C, George R, Bindeman I, Eichelberger J (2008) Magmatic differentiation at an island-arc Caldera: Okmok Volcano, Aleutian Islands, Alaska. J Petrol 49(5):857–884. https://doi.org/10.1093/petrology/egn008

Flinders AF, Caudron C, Johanson IA, Taira T, Shiro B, Haney M (2020) Seismic velocity variations associated with the 2018 lower East Rift Zone eruption of Kīlauea, Hawai'i. Bull Volcanol 82:47. https://doi.org/10.1007/s00445-020-01380-w

Fontijn K, McNamara K, Tadesse AZ, Pyle DM, Dessalegn F, Hutchison W, Mather TA, Yirgu G (2018) Contrasting styles of post-caldera volcanism along the Main Ethiopian Rift: implications for contemporary volcanic hazards. J Volcanol Geotherm Res 356:90–113. https://doi.org/10.1016/j.jvolgeores.2018.02.001

Francis P, Oppenheimer C (2003) Applications of satellite remote sensing techniques to volcanology. In: Mather MP (ed) TERRA-1: understanding the terrestrial environment. The role of Earth observations from space. Taylor & Francis Ltd, London, UK, pp 37–52. ISBN: 0-7484-0044-3

Gaete A, Kavanagh JL, Rivalta E, Hazim SH, Walter TR, Dennis DJC (2019) The impact of unloading stresses on post-caldera magma intrusions. Earth Planet Sci Lett 508:109–121. https://doi.org/10.1016/j.epsl.2018.12.016

Garcia MO, Swinnard L, Weis D, Greene AR, Tagami T, Sano H, Gandy CE (2010) Petrology, Geochemistry and geochronology of Kaua'i Lavas over 4.5 Myr: implications for the origin of rejuvenated volcanism and the evolution of the Hawaiian Plume. J Petrol 51(7):1507–1540. https://doi.org/10.1093/petrology/egq027

Gayer E, Michon L, Villeneuve N (2021) Volcanic island multi-stage construction inferred from a simple geometrical approach: example of Réunion Island. Geomorphology 392. https://doi.org/10.1016/j.geomorph.2021.107900

Geist DJ, Bergantz G, Chadwick WWJr (2014) Galápagos Magma Chambers. In: Harpp KS, Mittelstaedt E, d'Ozouville N, Graham DW (eds) The Galápagos: a natural laboratory for the Earth sciences. Geophysical Monograph 204. Wiley, Hoboken, New Jersey, p 69. ISBN: 978-1-118-85241-5. https://doi.org/10.1002/9781118852538

Geist DJ, Diefenbach BA, Fornari DJ, Kurz MD, Harpp K, Blusztajn J (2008) Construction of the Galápagos platform by large submarine volcanic terraces. Geochem Geophys Geosyst 9(3). https://doi.org/10.1029/2007GC001795

Geist DJ, Fornari DJ, Kurz MD, Harpp KS, Soule SA, Perfit MR, Koleszar AM (2006) Submarine Fernandina: magmatism at the leading edge of the Galápagos hot spot. Geochem Geophys Geosyst 7:Q12007. https://doi.org/10.1029/2006GC001290

Geist DJ, Howard KA, Jellinek AM, Rayder S (1994) The volcanic history of Volcán Alcedo, Galápagos Archipelago: a case study of rhyolitic oceanic volcanism. Bull Volcanol 56:243–260. https://doi.org/10.1007/BF00302078

Geist DJ, Naumann TR, Standish JJ, Kurz MD, Harpp KS, White WM, Fornari DJ (2005) Wolf Volcano, Galápagos Archipelago: melting and magmatic evolution at the margins of mantle plume. J Petrol 46(11):2197–2224. https://doi.org/10.1093/petrology/egi052

Germa A, Quidelleur X, Gillot PY, Tchilinguirian P (2010) Volcanic evolution of the back-arc Pleistocene Payun Matru volcanic field (Argentina). J South Amer Earth Sci 29:717–730. https://doi.org/10.1016/j.jsames.2010.01.002

Glass JB, Fornari DJ, Hall HF, Cougan AA, Berkenbosch HA, Holmes ML, White SM, de la Torre G (2007) Submarine volcanic morphology of the western Galápagos based on EM300 bathymetry and MR1 side-scan sonar. Geochem Geophys Geosyst 8(3): https://doi.org/10.1029/2006GC001464

Greeley R (1982) The Snake River Plain, Idaho: representative of a new category of volcanism. J Geophys Res 87(B4):2705–2712. https://doi.org/10.1029/JB087iB04p02705

Grosse P, Kervyn M (2018) Morphometry of terrestrial shield volcanoes. Geomorphology 304:1–14. https://doi.org/10.1016/j.geomorph.2017.12.017

Grove TL, Till CB (2015) Melting the Earth's upper mantle. In: Sigurdsson H, Houghton B, McNutt SR, Rymer H, Stix J (eds) Encyclopedia of volcanoes. Academic Press, San Diego, CA, pp 35–47. ISBN: 978-0-12-385938-9

Gudmundsson A (2000) Dynamics of volcanic system in Iceland: example of tectonism and volcanism at Juxtaposed Hot Spot and Mid-Ocean Ridge Systems. Annu Rev Earth Planet Sci 28:107–140. https://doi.org/10.1146/annurev.earth.28.1.107

Guillou H, Garcia MO, Turpin L (1997) Unspiked K-Ar dating of young volcanic rocks from Loihi and Pitcairn hot spot seamounts. J Volcanol Geotherm Res 78:239–249. https://doi.org/10.1016/S0377-0273(97)00012-7

Guth A, Wood J (2014) Geology of the Suswa Region, Kenya. Geologic map of the Southern Kenya Rift. Digital Map and Chart Series DMCH016. US Geol Surv, Boulder, Colorado, USA, Sheet 1. https://doi.org/10.1130/2014.DMCH016.S1

Haase KM, Beier C, Kemner F (2019) A comparison of the magmatic evolution of Pacific intraplate volcanoes: constraints on melting in mantle plumes. Front Earth Sci 6. https://doi.org/10.3389/feart.2018.00242

Haase KM, Gress MU, Lima SM, Regelous M, Beier C, Romer RL, Bellon H (2020). Evolution of magmatism in the new Hebrides island arc and in initial back-arc rifting, SW Pacific. Geochem Geophys Geosyst 21:e2020GC008946. https://doi.org/10.1029/2020GC008946

Halldórsson SA, Marshall EW, Caracciolo A, Matthews S, Bali E, Rasmussen MB, Ranta E et al (2022) Rapid shifting of a deep magmatic source at Fagradalsfjall volcano, Iceland. Nature 609:529–534. https://doi.org/10.1038/s41586-022-04981-x

Hansen SE, Nyblade AA, Benoit MH (2012) Mantle structure beneath Africa and Arabia from adaptively parameterized P-wave tomography: Implications for the origin of Cenozoic Afro-Arabian tectonism. Earth Planet Sci Lett 319–320:23–34. https://doi.org/10.1016/j.epsl.2011.12.023

Harpp KS, Geist DJ (2018) The evolution of Galápagos volcanoes: an alternative perspective. Front Earth Sci 6. https://doi.org/10.3389/feart.2018.00050

Hasenaka T (1994) Size, distribution, and magma output rate for shield volcanoes of the Michoacán-Guanajuato volcanic field, Central Mexico. J Volcanol Geotherm Res 63(1–2):13–31. https://doi.org/10.1016/0377-0273(94)90016-7

Holcomb RT, Robinson JE (2004) Maps of Hawaiian Islands Exclusive Economic Zone interpreted from GLORIA sidescan-sonar imagery. US Geol Surv Scientific Investigations, Map 2824, Reston, VA. ISBN: 0-607-98618-2

Houghton BF, Weaver SD, Wilson CJN, Lanphere MA (1992) Evolution of the Quaternary peralkaline volcano: Mayor Island, New Zealand. J Volcanol Geotherm Res 51(3):217–236. https://doi.org/10.1016/0377-0273(92)90124-V

Howard KA (2010) Caldera collapse: perspectives from comparing Galápagos volcanoes, nuclear-test sinks, sandbox models, and volcanoes on Mars. GSA Today 20:4–10. https://doi.org/10.1130/GSATG82A.1

Hughes SS, Smith RP, Hackett WR, Anderson SR (1999) Mafic Volcanism and Environmental Geology of the Eastern Snake River Plain, Idaho. In: Hughes SS, Thackray GD (eds) Guidebook to the geology of eastern Idaho: Idaho Museum of Natural History, pp 143–168

Hunt JA, Pyle DM, Mather TA (2019) The geomorphology, structure, and lava flow dynamics of peralkaline rift volcanoes from high-resolution digital elevation models. Geochem Geophys Geosyst 20(3):1508–1538. https://doi.org/10.1029/2018GC008085

Hunt JE, Jarvis I (2017) Prodigious submarine landslides during the inception and early growth of volcanic islands. Nat Commun 8(2061). https://doi.org/10.1038/s41467-017-02100-3

Jakobsson SP, Jónsson J, Shido F (1978) Petrology of the Western Reykjanes Peninsula, Iceland. J Petrol 19(4):669–705. https://doi.org/10.1093/petrology/19.4.669

Jensen RA, Donnelly-Nolan JM (2017) Field-trip guide to the geologic highlights of Newberry Volcano, Oregon. US Geol Surv Scientific Investigations Report 2017-5022-J2, Reston, Virginia, 30 p. https://doi.org/10.3133/sir20175022J2

Jicha BR, Rhodes JM, Singer BS, Garcia MO (2012) $^{40}Ar/^{39}Ar$ geochronology of submarine Mauna Loa volcano, Hawaii. J Geophys Res 117.https://doi.org/10.1029/2012JB009373

Johnson RW, Knutson J, Taylor SR (1989) Intraplate volcanism: in eastern Australia and New Zealand. Cambridge University Press, Sydney, 408 p. ISBN: 0-521-38083-9

Join JL, Folio JL, Robineau B (2005) Aquifers and groundwater within active shield volcanoes. Evolution of conceptual models in the Piton de la Fournaise volcano. J Volcanol Geotherm Res 147:187–201. https://doi.org/10.1016/j.jvolgeores.2005.03.013

Karátson D, Yepes J, Favalli M, Rodríguez-Peces MJ, Fornaciai A (2016) Reconstructing eroded paleovolcanoes on Gran Canaria, Canary Islands, using advanced geomorphometry. Geomorphology 253:123–134. https://doi.org/10.1016/j.geomorph.2015.10.004

Kauahikaua J, Hildenbrand T, Webring M (2000) Deep magmatic structures of Hawaiian volcanoes, imaged by three-dimensional gravity models. Geology 28(10): 883–886. https://doi.org/10.1130/0091-7613(2000)28<883:DMSOHV>2.0.CO;2

Kervyn M, van Wyk de Vries B, Walter TR, Njome MS, Suh CE, Ernst GGJ (2014) Directional flank spreading at Mount Cameroon volcano: evidence from analogue modeling. J Geophys Res 119(10):7542–7563. https://doi.org/10.1002/2014JB011330

Klein EM, Smith DK, Williams CM, Schouten H (2005) Counter-rotating microplates at the Galapagos triple junction. Nature 433:855–858. https://doi.org/10.1038/nature03262

Knesel KM, Cohen BE, Vasconcelos PM, Thiede DS (2008) Rapid change in drift of the Australian plate records collision with Ontong Java plateau. Nature 454:754–757. https://doi.org/10.1038/nature07138

Larsen JF, Neal C, Schaefer J, Beget J, Nye C (2007) Late Pleistocene and Holocene caldera-forming eruptions of Okmok caldera, Aleutian Islands, Alaska. In: Eichelberger J, Gordeev E, Izbekov P, Kasahara M, Lees J (eds) Volcanism and subduction:

the Kamchatka Region, Amer Geophys Union, Geophys Monogr 172, pp 343–364. https://doi.org/10.1029/172GM24

le Friant A, Lebas E, Clément V, Boudon G, Deplus C, de Voogd B, Bachèlery P (2011) A new model for the evolution of La Réunion volcanic complex from complete marine geophysical surveys. Geophys Res Lett 38:L09312. https://doi.org/10.1029/2011GL047489

Lénat J-F, Bachèlery P, Merle O (2012) Anatomy of Piton de la Fournaise volcano (La Réunion, Indian Ocean). Bull Volcanol 74:1945–1961. https://doi.org/10.1007/s00445-012-0640-y

Lipman PW (1995) Declining growth of Mauna Loa during the last 100,000 years: rates of lava accumulation vs. gravitational subsidence. In: Rhodes JM, Lockwood JP (eds) Mauna Loa Revealed: structure, composition, history, and hazards. AGU Geophys Monogr Ser 92, Washington, DC, pp 45–80. https://doi.org/10.1029/GM092p0045

Lipman PW (2000) Calderas. In: Sigurdsson H, Houghton B, McNutt SR, Rymer H, Stix J (eds) Encyclopedia of volcanoes. Academic Press, San Diego, CA, pp 643-662. ISBN: 978-0-12-385938-9

Lipman PW, Sisson TW, Ui T, Naka J (2000) In search of ancestral Kīlauea volcano. Geology 28(12):1079–1082. https://doi.org/10.1130/0091-7613(2000)28<1079:ISOAKV>2.0.CO;2

Lockwood JP, Hazlett RW (2010) Effusive volcanic eruptions and their products. In: Lockwood JP, Hazlett RW, de la Cruz-Reyna S (eds) Volcanoes—global perspectives. Wiley, Chichester, UK, pp 127–172. LCCN 2021031545

Lockwood JP, Lipman PW (1987) Holocene eruptive history of Mauna Loa volcano. In: Decker RW, Wright TL, Stauffer PH (eds) Volcanism in Hawaii. US Geol Surv Prof Paper 1350, US Government Printing Office, Washington, pp 509–535. https://doi.org/10.1017/S0016756800022421

Lowe DJ, Shane PA, de Lange PJ, Clarkson BD (2016) Guidebook for Rangitoto Island AQUA Field Trip, Auckland, 2016. School of Science, University of Waikato, in association with the Australasian Quaternary Association (AQUA) Biennial Conference, School of Science, University of Waikato, 35 p. ABN 78 458 664 047

Macdonald GA (1971) Geologic map of the Kipuka Quadrangle, Hawaii. US Geol Surv, Washington, 1:24000, Map GQ-957. https://doi.org/10.3133/gq957

Macdonald GA, Davis DA, Cox DC (1960) Geology and ground-water resources of the Island of Kauai, Hawaii. Hawaii Division of Hydrography 13, Honolulu, Hawaii, 212 p

Macdonald R, Navarro JM, Upton BGJ, Davies GR (1994) Strong compositional zonation in peralkaline magma: Menengai, Kenya Rift Valley. J Volcanol Geotherm Res 60:301–325. https://doi.org/10.1016/0377-0273(94)90057-4

Malahoff A (1987) Geology of the summit of Loihi submarine volcano. In: Decker RW, Wright TL, Stauffer PH (eds) Volcanism in Hawaii. US Geol Surv Prof Paper 1350, US Government Printing Office, Washington, pp 133–145. https://doi.org/10.1017/S0016756800022421

Mark RK, Moore JG (1987) Slopes of the Hawaiian ridge. In: Decker RW, Wright TL, Stauffer PH (eds) Volcanism in Hawaii. US Geol Surv Prof Paper 1350, US Government Printing Office, Washington, pp 101–107. https://doi.org/10.1017/S0016756800022421

Martí J (2019) Las Cañadas caldera, Tenerife, Canary Islands: a review, or the end of a long volcanological controversy. Earth Sci Rev 196. https://doi.org/10.1016/j.earscirev.2019.102889

Masson DG, Harbitz CB, Wynn RB, Pedersen G, Løvholt F (2006) Submarine landslides: processes, triggers and hazard prediction. Philos Trans R Soc A 364:2009–2039. https://doi.org/10.1098/rsta.2006.1810

Mathieu L, Kervyn M, Ernst GGJ (2011) Field evidence for flank instability, basal spreading and volcano-tectonic interactions at Mt Cameroon, West Africa. Bull Volcanol 73:851–867. https://doi.org/10.1007/s00445-011-0458-z

McBirney AR, Williams H (1969) Geology and petrology of the Galápagos Islands. Mem Geol Soc Am 118:197 p. ISBN: 9780813711188

McConnell JR, Sigl M, Plunkett G, Burke A, Kim WM, Raible CC, Wilson AI, Manning JG, Ludlow F, Chellman NJ, Innes HM, Yang Z, Larsen JF, Schaefer JR, Kipfstuhl S, Mojtabavi S, Wilhelms F, Opel T, Meyer H, Steffensen JP (2020) Extreme climate after massive eruption of Alaska's Okmok volcano in 43 BCE and effects on the late roman republic and ptolemaic kingdom. Proc Nat Acad Sci (PNAS) 117(27):15443–15449. https://doi.org/10.1073/pnas.2002722117

McPhie J, Walker GPL, Christiansen RL (1990) Phreatomagmatic and phreatic fall and surge deposits from explosions at Kilauea volcano, Hawaii, 1790 AD: Keanakakoi Ash Member. Bull Volcanol 52:334–354. https://doi.org/10.1007/BF00302047

Merle O, Barde-Cabusson S, van Wyk de Vries B (2010) Hydrothermal calderas. Bull Volcanol 72:131–147. https://doi.org/10.1007/s00445-009-0314-6

Michon L, Di Muro A, Villeneuve N, Saint-Marc C, Fadda P, Manta F (2013) Explosive activity of the summit cone of Piton de la Fournaise volcano (La Réunion island): a historical and geological review. J Volcanol Geotherm Res 264:117–133. https://doi.org/10.1016/j.jvolgeores.2013.06.012

Michon L, Saint-Ange F (2008) Morphology of Piton de la Fournaise basaltic shield volcano (La Réunion Island): characterization and implication in the volcano evolution. J Geophys Res 113(B3). https://doi.org/10.1029/2005JB004118

Michon L, Staudacher T (2009) Recent advances on the geodynamics of Piton de la Fournaise volcano. J Volcanol Geotherm Res 184(1–2):1–224

Michon L, Villeneuve N, Catry T, Merle O (2009) How summit calderas collapse on basaltic volcanoes: new insights from the April 2007 caldera collapse of Piton de la Fournaise volcano. J Volcanol Geotherm Res 184(1–2):138–151. https://doi.org/10.1016/j.jvolgeores.2008.11.003

Minissale S, Casalini M, Cucciniello C, Balagizi C, Tedesco D, Boudoire G, Morra V, Melluso L (2022) The geochemistry of recent Nyamulagira and Nyiragongo potassic lavas, Virunga Volcanic Province, and implications on the enrichment processes in the mantle lithosphere of the Tanzania-Congo craton. Lithos 420–421. https://doi.org/10.1016/j.lithos.2022.106696

Mittelstaedt E, Soule AS, Harpp KS, Fornari D (2014): Variations in crustal thickness, plate rigidity, and volcanic processes throughout the Northern Galápagos Volcanic Province. In: Harpp KS, Mittelstaedt E, d'Ozouville N, Graham DW (eds) The Galápagos: a natural laboratory for the Earth sciences. Geophys Monogr 204. Wiley, Hoboken, New Jersey, pp 263–284. https://doi.org/10.1002/9781118852538

Moore JG (1964) Giant submarine landslides on the Hawaiian Ridge. US Geol Surv Prof Paper 501:95–98

Moore JG (1987) Subsidence of the Hawaiian ridge. In: Decker RW, Wright TL, Stauffer PH (eds) Volcanism in Hawaii. US Geol Surv Prof Paper 1350, US Government Printing Office, Washington, pp 85–100. https://doi.org/10.1017/S0016756800022421

Moore JG, Clague DA, Holcomb RT, Lipman PW, Normark WR, Torresan ME (1989) Prodigious submarine landslides on the Hawaiian ridge. J Geophys Res 94(B12):17465–17484. https://doi.org/10.1029/JB094iB12p17465

Moore JG, Normark WR, Holcomb RT (1994) Giant Hawaiian landslides. Annu Rev Earth Planet Sci 22:119–144. https://doi.org/10.1146/annurev.ea.22.050194.001003

Morgan JK, Clague DA, Borchers DC, Davis AS, Milliken KL (2007) Mauna Loa's submarine western flank: Landsliding, deep volcanic spreading, and hydrothermal alteration. Geochem Geophys Geosyst 8(5):Q05002. https://doi.org/10.1029/2006GC001420

Morgan JK, Moore GF, Clague DA (2003) Slope failure and volcanic spreading along the submarine south flank of Kīlauea volcano, Hawaii. J Geophys Res 108(B9). https://doi.org/10.1029/2003JB002411

Munro DC, Rowland SK (1996) Caldera morphology in the western Galápagos and implications for volcano eruptive behavior and mechanisms of caldera formation. J Volcanol Geotherm Res 72(1–2):85–100. https://doi.org/10.1016/0377-0273(95)00076-3

Mutonga M (2012) Geothermal geology of Silali volcano, Kenya. Geotherm Resour Council Trans 36: 943–950

Naumann T, Geist DJ (2000) Physical volcanology and structural development of Cerro Azul Volcano, Isabela Island, Galápagos: implications for the development of Galápagos-type shield volcanoes. Bull Volcanol 61:497–514. https://doi.org/10.1007/s004450050001

Naylor LA, Coombes MA, Viles HA (2013) Reconceptualising the role of organisms in the erosion of rock coasts: a new model. Geomorphology 157–158:17–30. https://doi.org/10.1016/j.geomorph.2011.07.015

Neal CA, Lockwood, JP (2003) Geologic map of the summit region of Kīlauea Volcano, Hawaii, US Geol Surv Geol Invest Map, I-2759

Nordlie BE (1973) Morphology and structure of the Western Galápagos volcanoes and a model for their origin. Geol Soc Amer Bull 84(9):2931–2956. https://doi.org/10.1130/0016-7606(1973)84<2931:MASOTW>2.0.CO;2

Oehler J-F, Lénat J-F, Labazuy P (2008) Growth and collapse of the Reunion Island volcanoes. Bull Volcanol 70:717–742. https://doi.org/10.1007/s00445-007-0163-0

Ollier CD (1988) Volcanoes. Basil Blackwell Publication, Oxford, 228 p

Peltier A, Ferrazzini V, di Muro A, Kowalski P, Villeneuve N, Richter N, Chevrel O, Froger JL, Hrysiewicz A, Gouhier M, Coppola D, Retailleau L, Beauducel F, Gurioli L, Boissier M, Brunet C, Catherine P, Fontaine F, Lauret F, Garavaglia L, Lebreton J, Canjamale K, Desfete N, Griot C, Harris A, Arellano S, Liuzzo M, Gurrieri S, Ramsey M (2021) Volcano Crisis Management at Piton de la Fournaise (La Réunion) during COVID-19 lockdown. Seism Res Lett 92(1):38–52. https://doi.org/10.1785/0220200212

Peltier A, Massin F, Bachèlery P, Finizola A (2012) Internal structure and building of basaltic shield volcanoes: the example of the Piton de La Fournaise terminal cone (La Réunion). Bull Volcanol 74(8):1881–1897. https://doi.org/10.1007/s00445-012-0636-7

Pérez-Torrado FJ, Carracedo JC, Mangas J (1995) Geochronology and stratigraphy of the Roque Nublo Cycle, Gran Canana, Canary Islands. J Geol Soc Lond 152:807–818. https://doi.org/10.1144/gsjgs.152.5.0807

Pike RJ (1978) Volcanoes on the inner planets: some preliminary comparisons of gross topography. In: Proc. lunar planet sci. conf. 9th, vol 3, pp 3239–3273

Pike RJ, Clow GD (1981) Martian volcanoes in a classification of central edifices. Abstracts of Papers presented to the Third International Colloquium on Mars, pp 199–201

Poland MP, Takahashi TJ, Landowski CM (2014) Characteristics of Hawaiian volcanoes. US Geological Survey Professional Paper 1801, US Government Printing Office, Washington, 429 p. https://doi.org/10.3133/pp1801

Ponomareva VV, Churikova T, Melekestsev IV, Braitseva OA, Pevzner M, Sulerzhitskii L (2007) Late Pleistocene-Holocene volcanism on the Kamchatka Peninsula, Northwest Pacific Region. In: Eichelberger J, Gordeev E, Izbekov P, Kasahara M, Lees J (eds) Volcanism and subduction: the Kamchatka Region, Amer Geophys Union, Geophys Monogr 172, pp 165–198. https://doi.org/10.1029/172GM15

Porter SC (1972) Distribution, morphology, and size frequency of cinder cones on Mauna Kea Volcano, Hawaii. Geol Soc Am Bull 83(12):3607–3612. https://doi.org/10.1130/0016-7606(1972)83[3607:DMASFO]2.0.CO;2

Portnyagin M, Hoernle K, Avdeiko G, Hauff F, Werner R, Bindeman I, Uspensky V, Garbe-Schönberg D (2005) Transition from arc to oceanic magmatism at the Kamchatka-Aleutian junction. Geology 33(1):25–28. https://doi.org/10.1130/G20853.1

Pouclet A, Bram K (2021) Nyiragongo and Nyamuragira: a review of volcanic activity in the Kivu rift, western branch of the East African Rift System. Bull Volcanol 83:10. https://doi.org/10.1007/s00445-021-01435-6

Ribe NM, Christensen UR (1999) The dynamical origin of Hawaiian volcanism. Earth Planet Sci Lett 171(4):517–531. https://doi.org/10.1016/S0012-821X(99)00179-X

Riedl S, Melnick D, Mibei GK, Njue L, Strecker MR (2020) Continental rifting at magmatic centres: structural implications from the Late Quaternary Menengai Caldera, central Kenya Rift. Geol Soc Lond 177:153–169. https://doi.org/10.6084/m9.figshare.c.4647818.v1

Rivera J, Canals M, Lastras G, Hermida N, Amblas D, Arrese B, Martín-Sosa P, Acosta J (2016) Morphometry of Concepcion bank: evidence of geological and biological processes on a large volcanic seamount of the Canary Islands Seamount Province. PLoS ONE 11(5):e0156337. https://doi.org/10.1371/journal.pone.0156337

Robinson JE, Eakins BW (2006) Calculated volumes of individual shield volcanoes at the young end of the Hawaiian Ridge. J Volcanol Geotherm Res 151(1–3):309–317. https://doi.org/10.1016/j.jvolgeores.2005.07.033

Rooney TO (2020) The Cenozoic magmatism of East Africa: Part II. Lithos 360–361. https://doi.org/10.1016/j.lithos.2019.105291

Rotstein Y, Schaming M (2011) The Upper Rhine Graben (URG) revisited: Miocene transtension and transpression account for the observed first-order structures. Tectonics 30(3):TC3007. https://doi.org/10.1029/2010TC002767

Rowland SK (1996) Slopes, lava flow volumes, and vent distributions on Volcán Fernandina, Galápagos Islands. J Geophys Res 101(B12):27657–27672. https://doi.org/10.1029/96JB02649

Rowland SK, Garbeil H (2000) Slopes of oceanic basalt volcanoes. Geophys Monogr Ser 116:223–247. https://doi.org/10.1029/GM116p0223

Rowland SK, Munro DC, Perez-Oviedo V (1994) Volcán Ecuador, Galapágos Islands: erosion as a possible mechanism for the generation of steep-sided basaltic volcanoes. Bull Volcanol 56:271–283. https://doi.org/10.1007/BF00302080

Salvany T, Lahitte P, Nativel P, Gillot P-Y (2012) Geomorphic evolution of the Piton des Neiges volcano (Réunion Island, Indian Ocean): competition between volcanic construction and erosion since 1.4Ma. Geomorphology 136(1):132–147. https://doi.org/10.1016/j.geomorph.2011.06.009

Schipper CI, White JDL, Houghton BF, Shimizu N, Stewart RB (2010) Explosive submarine eruptions driven by volatile-coupled degassing at Lo'ihi Seamount, Hawai'i. Earth Planet Sci Lett 295(3–4):497–510. https://doi.org/10.1016/j.epsl.2010.04.031

Schneider J-L, Torrado FJP, Torrente DG, Wassmer P, del Carmen Cabrera Santana M, Carracedo JC (2004) Sedimentary signatures of the entrance of coarse-grained volcanoclastic flows into the sea: the example of the breccia units of the Las Palmas Detritic Formation (Mio-Pliocene, Gran Canaria, Eastern Atlantic, Spain). J Volcanol Geotherm Res 138:295–323. https://doi.org/10.1016/j.jvolgeores.2004.07.007

Schouten H, Smith DK, Montési LGJ, Zhu W, Klein EM (2008) Cracking of lithosphere north of the Galapagos triple junction. Geology 36(5):339–342. https://doi.org/10.1130/G24431A.1

Schwartz DM, Harpp K, Kurz MD, Wilson E, van Kirk R (2022) Low-volume magmatism linked to flank deformation on Isla Santa Cruz, Galápagos Archipelago, using cosmogenic ^{3}He exposure and ^{40}Ar/^{39}Ar dating of fault scarps and lavas. Bull Volcanol 84. https://doi.org/10.1007/s00445-022-01575-3

Sharp WD, Renne PR (2005) The ^{40}Ar/^{39}Ar dating of core recovered by the Hawaii Scientific Drilling Project (phase 2), Hilo, Hawaii. Geochem Geophys Geosyst 6(4). https://doi.org/10.1029/2004GC000846

Sherrod DR, Nishimitsu Y, Tagami T (2003) New K-Ar ages and the geologic evidence against rejuvenated-stage volcanism at Haleakalā, East Maui, a post-shield-stage volcano of the Hawaiian island chain. Geol Soc Amer Bull 115(6):683–694. https://doi.org/10.1130/0016-7606(2003)115<0683:NKAATG>2.0.CO;2

Sherrod DR, Sinton JM, Watkins SE, Brunt KM (2007) Geologic Map of the State of Hawai'i. US Geol Surv Sci Investigations Map 3143, Reston, VA, 72 p. https://doi.org/10.3133/sim3143

Shreve T, Grandin R, Smittarello D, Cayol V, Pinel V, Boichu M, Morishita Y (2021) What triggers caldera ring-fault subsidence at Ambrym volcano? Insights from the 2015 dike intrusion and eruption. J Geophys Res Solid Earth 126:e2020JB020277. https://doi.org/10.1029/2020JB020277

Sinton JM, Eason DE, Duncan RA (2017) Volcanic evolution of Moloka'i, Hawai'i: implications for the shield to postshield transition in Hawaiian volcanoes. J Volcanol Geotherm Res 340:30–51. https://doi.org/10.1016/j.jvolgeores.2017.04.011

Skilling IP (1993) Incremental caldera collapse of Suswa volcano, Gregory Rift Valley, Kenya. J Geol Soc 150(5):885–896. https://doi.org/10.1144/gsjgs.150.5.0885

Smets B, Kervyn M, d'Oreye N, Kervyn F (2015) Spatio-temporal dynamics of eruptions in a youthful extensional setting: insights from Nyamulagira Volcano (D.R. Congo), in the western branch of the East African Rift. Earth Sci Rev 150:305–328. https://doi.org/10.1016/j.earscirev.2015.08.008

Smith M, Dunkley PN, Deino A, Williams LAJ, McCall GJH (1995) Geochronology, stratigraphy and structural evolution of Silali volcano, Gregory Rift, Kenya. J Geol Soc 152(2):297–310. https://doi.org/10.1144/gsjgs.152.2.0297

Souther JG (1986) The western Anahim Belt: root zone of a peralkaline magma system. Can J Earth Sci 23(6):895–908. https://doi.org/10.1139/e86-091

Souther JG (1994) The Ilgachuz Range and Adjacent Parts of the Interior Plateau, British Columbia. Geol Surv Canada. ISBN 0–660–15500–1.

Strak V, Schellart WP (2018) A subduction and mantle plume origin for Samoan volcanism. Sci Rep 8.https://doi.org/10.1038/s41598-018-28267-3

Stearns HT, Macdonald GA (1946) Geology and groundwater resources of the island of Hawaii. US Geol Surv, Honolulu, Hawaii, 363 p

Suzuki T (1977) Volcano types and their global population percentages. Volcanol Soc Jpn 22(1):27–40

Tapu AT, Ubide T, Vasconcelos PM (2023) Increasing complexity in magmatic architecture of volcanoes along a waning hotspot. Nat Geosci 16:371–379. https://doi.org/10.1038/s41561-023-01156-9

Thordarson T, Höskuldsson Á (2008) Postglacial volcanism in Iceland. Jökull 58:197–228

Tibaldi A, Bonali FL, Corazzato C (2014) The diverging volcanic rift system. Tectonophysics 611:94–113. https://doi.org/10.1016/j.tecto.2013.11.023

Upton BGJ, Wadsworth WJ (1972) Aspects of magmatic evolution on Réunion Island. Philos Trans R Soc Lond A 271:105–130. https://doi.org/10.1098/rsta.1972.0002

van den Bogaard P (2013) The origin of the Canary Island Seamount Province—new ages of old seamounts. Sci Rep 3(1). https://doi.org/10.1038/srep02107

van den Bogaard P, Schmincke H-U (1998) Chronostatigraphy of Gran Canaria. In: Weawer PPE, Schmincke H-U, Firth JV, Duffield W (eds) Proceeding of the ocean drilling program. Scientific Results 157, College Station, TX, pp 127–140

van Hazinga C (2023) Using remote sensing and GIS to identify magmatic strain accommodation: the case study of Mt Marsabit, Kenya. MSc Thesis, Salem University, Massachusetts, USA, 56 p

Vigouroux N, Williams-Jones G, Chadwick WW Jr, Geist DJ, Ruiz A, Johnson D (2008) 4D gravity changes associated with the 2005 eruption of Sierra Negra volcano, Galápagos. Geophysics 73(6):WA29–WA35. https://doi.org/10.1190/1.2987399

Villagomez DR, Toomey DR, Geist DJ, Hooft EEE, Solomon SC (2014) Mantle flow and multistage melting beneath the Galápagos hotspot revealed by seismic imaging. Nat Geosci 7:151–156. https://doi.org/10.1038/ngeo2062

Vicenzi EP, McBirney AR, White WM, Hamilton M (1990) The geology and geochemistry of Isla Marchena, Galapagos Archipelago: an ocean island adjacent to a mid-ocean ridge. J Volcanol Geotherm Res 40(4):291–315. https://doi.org/10.1016/0377-0273(90)90110-2

Wadge G, Biggs J, Lloyd R, Kendall JM (2016) Historical volcanism and the state of stress in the East African rift system. Front Earth Sci 30(4). https://doi.org/10.3389/feart.2016.00086

Walker GPL (1999) Volcanic rift zones and their intrusion swarms. J Volcanol Geotherm Res 94(1–4):21–34. https://doi.org/10.1016/S0377-0273(99)00096-7

Walker GPL (2000) Basaltic volcanoes and volcanic systems. In: Sigurdsson H, Houghton B, McNutt SR, Rymer H, Stix J (eds) Encyclopedia of volcanoes. Academic Press, San Diego, CA, pp 283–289. ISBN: 978-0-12-385938-9

Wanless VD, Garcia MO, Rhodes JM, Weis D, Norman MD (2006) Shield-stage alkalic volcanism on Mauna Loa Volcano, Hawaii. J Volcanol Geotherm Res 151(1–3):141–155. https://doi.org/10.1016/j.jvolgeores.2005.07.027

Warden AJ (1970) Evolution of Aoba caldera volcano, New Hebrides. Bull Volcanol 34:107–140. https://doi.org/10.1007/BF02597781

Webb PK, Weaver SD (1975) Trachyte shield volcanoes: a new volcanic form from South Turkana, Kenya. Bull Volcanol 39:294–312. https://doi.org/10.1007/BF02597833

White WM, Hofmann AW (1978) Geochemistry of the Galápagos Islands: implications for mantle dynamics and evolution. Year Book Carnegie Institution Washington 77:596–606

Whitford-Stark JL (1975) Shield volcanoes. In: Fielder G, Wilson L (eds) Volcanoes of the Earth, Moon, and Mars. St Martin's Press, NY, pp 66–74

Williams H, McBirney AR (1979) Volcanology. Freeman Cooper Co, San Francisco, CA, p 397

Williamson NMB, Weis D, Scoates JS, Pelletier H, Garcia MO (2019) Tracking the geochemical transition between the Kea-dominated Northwest Hawaiian Ridge and the bilateral Loa-Kea trends of the Hawaiian Islands. Geochem Geophys Geosyst 20(9):4354–4369. https://doi.org/10.1029/2019GC008451

Wilson M, Downes H (2006) Tertiary-Quaternary intraplate magmatism in Europe and its relationship to mantle dynamics. In: Gee DG Stephenson R (eds) European lithosphere dynamics. Geol Soc Lond Mem 32:147–166. https://doi.org/10.1144/GSL.MEM.2006.032.01.09

Wolfe EW, Morris J (1996) Geologic map of the Island of Hawaii. US Geol Surv Map I-2524-A. https://doi.org/10.3133/i2524A

Wolfe EW, Wise WS, Dalrymple GB (1997) The geology and petrology of Mauna Kea Volcano, Hawaii—a study of postshield volcanism. US Geol Surv Prof Paper 1557, US Government Printing Office, Washington, 129 p. https://doi.org/10.3133/pp1557

Wood DA (1979) Dynamic partial melting: its application to the petrogenesis of basalts erupted in Iceland, the Faeroe Islands, the Isle of Skye (Scotland) and the Troodos Massif (Cyprus). Geochim Cosmochim Acta 43(7):1031–1046. https://doi.org/10.1016/0016-7037(79)90092-9

Xu W, Xie L, Bürgmann R, Liu X, Wang J (2023) The 2022 eruption of Wolf volcano, Galápagos: the role of caldera ring-faults during magma transfer from InSAR deformation data. Geophys Res Lett 50:e2023GL103704. https://doi.org/10.1029/2023GL103704

Zhang L, Cao Z, Zartman RE, Li C, Sun S, Liu L, Sun W (2023) An emerging plume head interacting with the Hawaiian plume tail. Innovation 4(2):100404. https://doi.org/10.1016/j.xinn.2023.100404

Zhong S, Watts AB (2002) Constraints on the dynamics of mantle plumes from uplift of the Hawaiian Islands. Earth Planet Sci Lett 203:105–116. https://doi.org/10.1016/S0012-821X(03)00289-9

Chapter 5 Polygenetic Landforms II: Composite Cones: From Individual Constructs to Clusters

1 Terminology: Strato- or Composite?

Besides shield volcanoes, another kind of polygenetic volcano is the strato- or composite volcano. A *stratovolcano* (from the Latin *stratum*) is a traditional term for the most familiar 'textbook' volcano type to the laymen, with the concave cone being the most depicted form in history, and the one most frequently drawn by kids.

Many such cones are constituted of a succession of lava and pyroclastic rocks (tephra), which appear to be the stratified eruptive products of a conical geometry. However, in reality, the edifice structure is often far from a 'layer-cake', being complicated by adventive vents, multiple summits etc., so it tends to be 'composite'. This has led to the use of an alternative term in the literature for stratovolcanoes: that of *composite volcano,* dating back to H. Williams and A. McBirney's Volcanology (Williams and McBirney 1979), a term which has become popular in the past few decades. As these authors wrote, "it seems preferable (…) to speak of cones built partly of effusive and partly of pyroclastic layers as composite". In accordance, a modern definition of composite volcanoes is "relatively large, long-lived constructional volcanic edifice comprising lava and volcaniclastic products from one or more vents" (Davidson and de Silva 2000).

On the other hand, the term 'composite' can also typify some large shield volcanoes—either composed of stratified beds, often predominantly successive, low-angle lava flows, or of superimposed edifices—, as well as composite monogenetic cones. As for large volcanoes, Etna in Italy (Photo 53), or Kirishima in Japan are good examples of compound, amalgamated edifices composed of shield(s), stratocones, and a number of monogenetic centres. Thus both terms are relevant, as shown by the fact that 'stratovolcano' is still also used in high-profile scientific forums and publications, such as the volcano classification system of Global Volcanism Program (https://volcano.si.edu/), the LaMEVE (Large Magnitude Explosive Volcanic Eruptions, https://www2.bgs.ac.uk/vogripa/view/controller.cfc?method=lameve) database of the British Geological Survey (Crosweller et al. 2012), and the appendix titled Catalog of Earth's Documented Holocene Eruptions (Siebert et al. 2015) in the 2nd edition of the Encyclopedia of Volcanoes. However, here in this book, to avoid terminological confusion, we prefer the term composite volcano—or, more specifically, in order to highlight the simplest conical landform which is a distinctive element, *composite cone.* By opting for this, we follow Francis (1993), who divided volcanic edifices into monogenetic and polygenetic, and considered *simple cones, compound volcanoes* and *volcano complexes* as three successive groups of composite

D. Karátson and J.-C. Thouret, *Volcano Geomorphology*, Advances in Volcanology,
https://doi.org/10.1007/978-3-031-77693-9_6

polygenetic volcanoes. Several other compilations (e.g. Walker 2000; Lockwood and Hazlett 2010; de Silva and Lindsay 2015) also discuss composite volcanoes as a group of polygenetic edifices.

It is worth briefly referring to how and when mankinds' long-standing contact with 'stratovolcanoes' evolved into an understanding of them. Because of their beauty and majesty, explorers and people who visited volcanoes were immediately drawn to composite volcanoes. Many cultures regard them as iconic or sacred mountains. Composite volcanoes worldwide with both cultural and artistic significance include Popocatépetl, Mexico; Cotopaxi, Ecuador; Mayon, The Philippines; Ararat (Ağri Daği), Turkey; etc. (see also a compilation on Volcanoes in Art in the Encyclopedia of Volcanoes by Sigurdsson 2015, and his book Melting the Earth: the history of ideas about volcanoes, Sigurdsson 1999). In more modern times, a memorable event occurred in 1767, when the English diplomat William Hamilton sketched the changing shape of Vesuvius as it was erupting. Perhaps the most notable volcano painter, William Turner, captured the eruption of Mount Vesuvius in 1817 under this very title *(Vesuvius in eruption)*. Other world-famous artwork includes the series of woodcuts of Fuji by Japanese painters Hiroshige and Hokusai in the eighteenth and nineteenth centuries, consisting of dozens of images.

The first scientific studies of the concave, regular-shaped cones were published in the nineteenth century by Milne (1878) and Becker (1885). Mostly descriptive classifications followed from Sapper (1927), Cotton (1944, 1952), then MacDonald (1972). The first morphometric studies included those by Wood (1978) and Pike and Clow (1981), using topographic data. Pike and Clow (1981) were also the first to propose a volcano classification based on volcano size, the presence of caldera, and general lithology (see later). Modern reviews on composite volcanoes were published by Francis (1993), Thouret (1999), Davidson and de Silva (2000), Lockwood and Hazlett (2010), and de Silva and Lindsay (2015).

Composite volcanoes are also a legitimate source of concern worldwide, as—in addition to the common lava flows and small to moderate explosions—they are occasionally or periodically characterised by violent explosive eruptions (Volcanic Explosivity Index, VEI$\geq$4). Notably, because significant populations worldwide exploit the fertile soils associated with composite volcanoes, often in favourable climates (e.g. Central America, Lesser Antilles, Northern Andes, Japan, Java: see Chap. 8), the nearby villages or towns even nowadays are frequently affected by lethal eruptions, either directly (via pyroclastic density currents: PDCs: see Chap. 10) or by accompanying phenomena (e.g. lahars: see Chap. 7). While moderately powerful to violent explosive eruptions at composite volcanoes (including lava dome groups) are relatively common (selected examples of VEI 3 include: Tajogaite/Cumbre Vieja, 2021; Merapi, 2014; Montagne Pelée, 1929–32; Etna, 1669; VEI 4: Merapi, 2010; Eyafjallajökull, 2010; M. Pelée, 1902; Bandai, 1988; VEI 5: Mount St. Helens, 1980; Vesuvius, AD/CE 79), highly explosive (VEI>6) eruptions are less frequent—these are mainly associated with the formation of truncated cones, that is, caldera collapse (see Sect. 2, 5.2, and Chap. 6). It is important to know, however, that even if the more powerful eruptions are rare (see Chap. 9, Sect. 5 for a probabilistic assessment), many significant eruptions have occurred at composite volcanoes in historical times alone. Table 1 highlights some examples of VEI 7 and VEI 6 eruptions since the Middle Ages, along with basic volcano geomorphic features.

From a geomorphological point of view, this list alone illustrates the rich variety of composite landform types, telling us that both regular, symmetrical (e.g. Santa María) and complex (e.g. Krakatau) edifices can produce highly explosive eruptions. However, in general, regular-shaped cones that tend to erupt basaltic/andesitic magma are characterized by a less explosive eruptive behaviour (with no caldera formation). On the contrary, more evolved composite cones often, although not exclusively, erupt more differentiated magma, i.e. dacite, rhyolite, associated with small- to medium-sized (summit) caldera formation.

Table 1 Main features of VEI 7 and 6 historical eruptions of composite volcanoes

	Volcano (elevation m a.s.l.)	Volume (km^3)[a]	Geographic setting, country	Geodynamic setting	Year of eruption	Volcano geomorphic features/changes	Erupted volume (km^3 bulk)[b]
VEI 7	Changbaishan/Paektu-san (2744)	76	China/North Korea	Intraplate, back-arc	946 AD/CE	Central edifice truncated by a circular, 5×5 km summit caldera	100
	Rinjani (3726)	263	Lombok Island, Indonesia	Island arc subduction, front	1257	Large, central edifice truncated by an elliptical, 6×9 km summit caldera	91
	Tambora (2850)	640	Sumbawa Island, Indonesia	Island arc subduction, rear arc	1815	Large, central edifice truncated by a circular, 6×6 km summit caldera	110
VEI 6	Pago/Witori 724)	n. d.	New Britain, Papua New Guinea	Island arc subduction, front	1370 ± 160 BC/BCE	5.5×7.5 km multiple caldera hosting post-caldera cones	30
	Ceboruco (2280)	21	Mexico	Continental margin subduction	1005 ± 15 AD/CE	Central edifice truncated by a nested double caldera (4 and 1.5 km across, respectively)	11
	Huynaputina (4679)	n. d.	Central Andes, Peru	Continental margin subduction	1600	Irregular edifice with a half-open depression hosting overlapping tuff cones	26
	Krakatau (283)	? (partly sub-marine)	Sunda Strait, Indonesia	Island arc subduction, front	1883	Cones (islands), remnants of older caldera, disappeared/truncated in 1883; post-caldera cone	48
	Santa María (3745)	16	Guatemala	Continental margin subduction	1902	Regular cone with asymmetric flank crater hosting a new lava dome complex	20
	Pinatubo (1486)	123	Luzon Island, the Philippines	Island arc subduction, front	1991	Central edifice truncated by a circular, 2.5 km-wide summit caldera	13
	Hunga Tonga–Hunga Ha'apai (114)	? (mostly submarine)	Tonga Islands	Island arc subduction, front	2022	Emergent islands of a 6 km-wide submarine caldera, destroyed by the 2022 eruption	10[c]

Volcano volume estimates after Grosse et al. (2012), dates and eruptive volumes from Crosweller et al. (2012) and, for Hunga Tonga–Hunga Ha'apai (2022), from Poli and Shapiro (2022)

[a] Data from Grosse et al. (2012); n. d.: no data

[b] Based on the LaMEVE database (Crosweller et al. 2012)

[c] Based on Poli and Shapiro (2022)

2 Classification of Landforms from Simple Cones to Compound Edifices

In the past few decades, the 'boom' of DEM-based work, thanks to the progressively higher resolution and the increasing availability of source DEMs, has resulted in the linking of topography to classification, landform evolution, and even eruptive style (e.g. Grosse et al. 2009, 2012, 2014; Karátson et al. 2010a, 2012). Castruccio et al. (2017)—building on earlier work by Lacey et al. (1981)—applied physical modelling and volcano morphometry to infer the depth of the magma plumbing system, whereas Cosburn and Ray (2020) performed mathematical modelling looking at dyke distribution below the volcano. These works, in addition to many others not listed here, have highlighted the diversity of composite volcanoes. An evolved composite volcano can be complicated by summit or flank lava domes, cryptodomes, adventive vents, e.g. scoria cones (the latter typically located on the flanks), even another composite volcano (e.g. forming a twin or compound edifice), along with modifications to the central vent area (e.g. nested or coalesced craters (Fig. 1). Moreover, the upper cone can experience sector or flank failure or summit collapse, hence becoming truncated. Presenting this complexity in detail is in line with the growing knowledge of regional volcano geomorphology, with the new findings confirming the validity of 'composite' landforms.

During the growth of a simple cone—often over a lifetime of only < 10 kyr, although sometimes tens of 10 kyr, up to c. 500 kyr—, it can be subjected to a multitude of endogenic and exogenic changes. The most important of these include: changes to the magma plumbing system, changes to the magma composition, the role of aquifer(s) on eruptive style, vent position (e.g. vent migration), structural geology conditions, volcano spreading, and flank failure. Most changes are due to the eruptive and tectonic activity, i.e. related to endogenous forces, although some processes, e.g. flank failure, can also have exogenic factors (see Chap. 7); and erosion, commonly at a lower rate, can equally play a role in forming the edifice

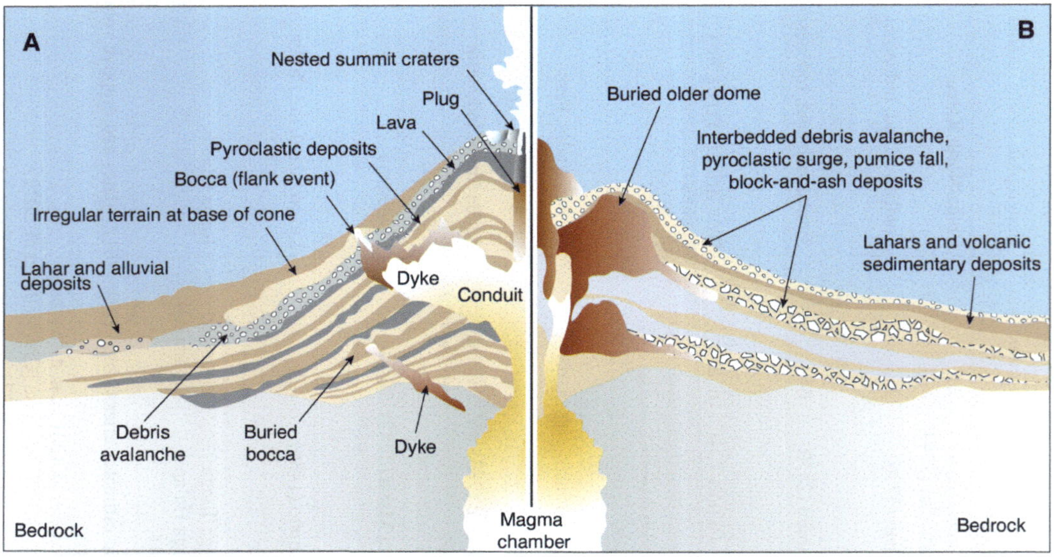

Fig. 1 Cross-section of an ideal composite cone (**A**) and a lava dome complex (**B**) illustrating the major features (after Lockwood and Hazlett 2010, modified). Composite volcanoes can show transitional shapes (even from their initial stage). The shape may become complicated, but not systematically. Compound cones/volcanoes are usually a combination of these two landforms (**A, B**), stacked side by side or on top of each other

shape during its active lifetime (see Chap. 8). As a consequence, the primary shape of a composite volcano may be modified, truncated or destroyed on various time scales, and the resulting geomorphic changes need to be included in any landform classification. It is important to emphasize the temporal aspect on geomorphic evolution since both endo- and exogenic factors can change over time. Moreover, there can be changes and breaks in eruption rate as well as long repose times without any eruption. Such changes over time will strongly affect the morphology at any given stage of evolution (see Sect. 6).

Here, four composite volcano types are distinguished based mainly on topographic and morphometric criteria (Fig. 2), following (1) Moriya's (1979) classification for Japanese volcanoes, (2) the terminology of Francis (1993), (3) modified and adjusted according to the recent DEM-based findings. These types are characterized by a rough transition from simple cones (regular or irregular) to a more complex shape in terms of increasing size (basal width, volume, but not necessarily elevation), summit size, complexity (e.g. increased ellipticity and number of peaks), and decreasing steepness (Grosse et al. 2012). Note that the main types are not fully constrained by these morphometric parameters since there is a geomorphic continuum (Pike 1978; Grosse et al. 2014). Moreover, as we will demonstrate, there is not necessarily any common edifice evolution from 'simple' to 'complex' landforms.

(A) *Simple cones.* Simple cones have two subtypes: *regular cone* and *irregular cone.* A regular cone is considered to be the elementary landform of composite cones which is usually the first to develop (Sect. 5.1), and displays commonly concave or linear slopes. Formation of an irregular cone can be caused by modification of a regular cone due to a number of factors, although it is not infrequent for an irregular cone to form at the outset. However, the general cone shape remains largely circular and relatively symmetrical.

(B) *Truncated cones.* These may evolve from either of the former, with the formation of (1) a smaller or larger (commonly 1–8 km wide) summit caldera typically due to large explosive eruption(s); or (2) an asymmetric, horseshoe-shaped collapse scar due to flank failure(s).

(C) *Somma volcanoes.* This landform is preceded by significant cone truncation due to the processes referred to above, after which subsequent activity may build up new eruptive centres (lava domes, intracaldera cones) inside the caldera. This results in a geomorphically more complex shape, which generally remains uniform, and often symmetric, in plan view. In some literature (e.g. Moriya 1979; Mannen et al. 2018), *caldera volcano* has been used for such an edifice, but in this book we follow the terminology of the Encyclopedia of Volcanoes (e.g. Davidson and de Silva 2000) in restricting the latter term to *calderas,* that is, 'negative' volcanoes (whose constructive, positive preceding landform, if any, has been destroyed: see Chap. 6). At Somma volcanoes, even though the caldera landform may dominate the present-day morphology, remnants of a previous, central, truncated composite volcano are still conspicuous (e.g. the ring-shaped Mount Somma enclosing the now higher Vesuvius cone: Photo 54).

(D) *Compound cones.* This is an amalgamation of edifices built up by the diverse or prolonged activity of multiple vents, shifting vents on composite cones. As a result, a compound volcano can include a simple regular cone, a twin volcano, or a massif (with juxtaposed or superimposed centres) etc., with or without a caldera. Overall, the edifice can become elongated, irregular and asymmetric, although it sometimes retains a common base.

We can speak of a 'single' volcano in so far as the shape is uniform in terms of plan view and (in general) in terms of the common magma plumbing system. This is a significant difference

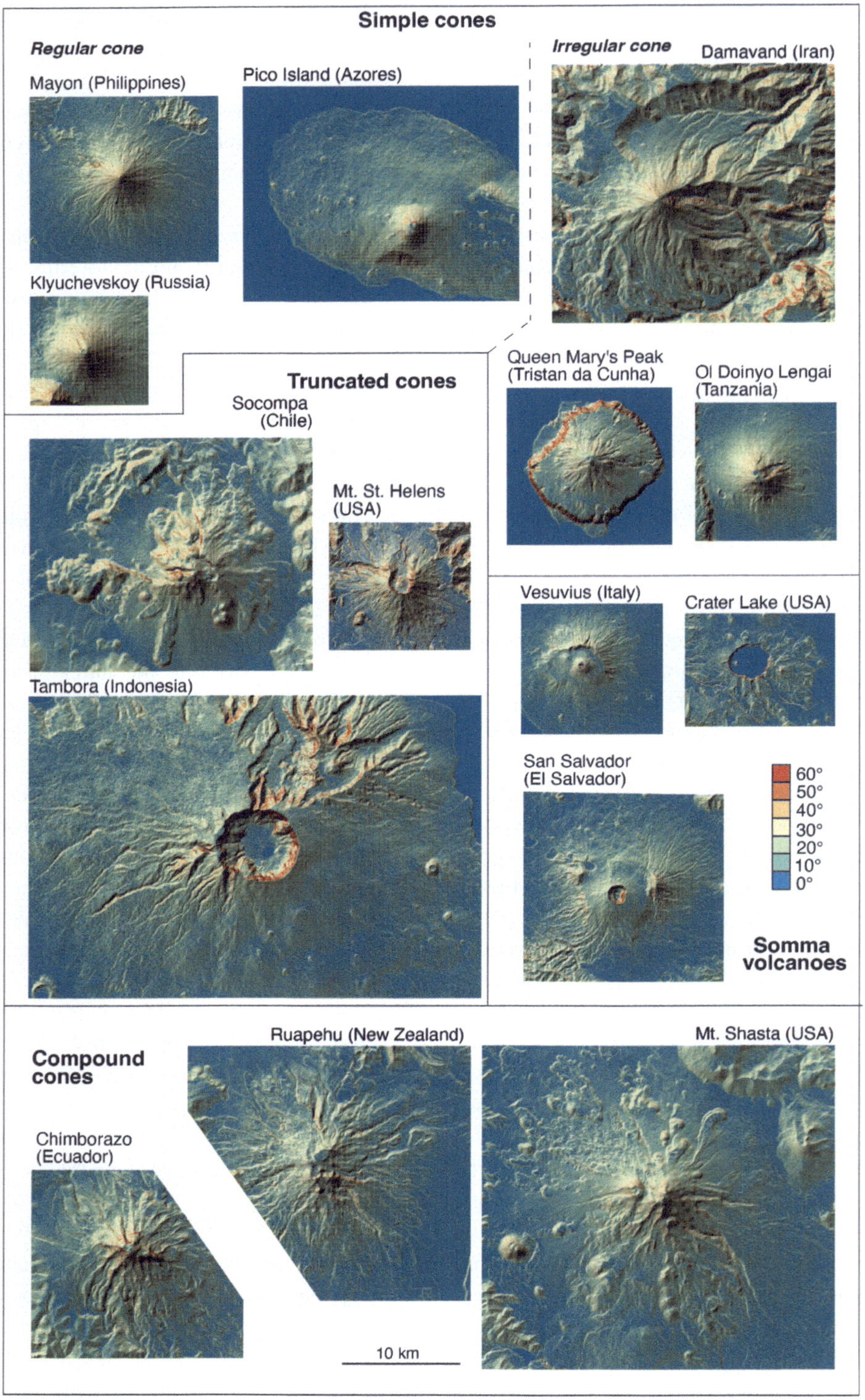

Fig. 2 Four main types of composite cones showing selected examples via slope maps (based on SRTM DEM data)

from *large volcanic clusters (LVC),* also termed volcano complexes, which comprise a number of composite volcanoes and other kinds of centres (see Sect. 7).

Table 2 gives the volume and lifetime of selected composite cones based on the above geomorphological classification. The lack of a general trend within these types is reflected, for example, in the lack of any relationship between size and age. Even long-lived volcanoes can remain as simple cones. An example of such is Fuji (Fig. 3), whose oldest structure is ~270 ka, on which composite cones of the older (100 ka), then present-day (10 ka) Younger Fuji developed (Photo 55). Despite the extended lifetime, after a smaller shift in magma composition (from dacite/andesite to basaltic andesite/basalt at ~160 ka) the volcanic system has remained as a regular cone (Aoki et al. 2019). Another intraplate example is Damávand, which has an even longer lifetime (<1.8 Ma), with the present edifice (the irregular cone of Young Damávand) dating back as far as ~550 ka (Davidson et al. 2004).

In the following, we address the questions of eruptive activity and lithofacies associations of composite cones (Sect. 3), and present their diversity by geodynamic setting (Sect. 4). Then, based on the categorisation, the four main types in terms

Table 2 Volume and lifetime estimates of some composite volcanoes divided into types used in this book (modified from Wood 1978 and Wadge 1982; most volumetry data updated from Grosse et al. 2012)

Type		Name	Volume (km^3)	Lifetime (ka)	Lithology
Simple cones	Regular cone	Santa María	16	<32[1]	Andesite, basaltic andesite
		Fuego	20	30[2]	Basaltic andesite, basalt
		Klyuchevskoy	50	7[3]	Basalt, basaltic andesite
		Fuji	400–500 (60)	270 (modern Fuji 10)[4]	Basalt, basaltic andesite
	Irregular cone	Damávand	400 (55)	< 1.8 Ma (Young Damávand 550)[5]	Trachyandesite
Truncated cones		St. Helens	16	28[6]	Dacite, andesite, basalt
		Tambora	640	43[7]	Trachyandesite, trachybasalt
Somma volcanoes		Somma-Vesuvius	16	35[8]	Phonolite, phono-tephrite, trachyte, latite
		Oshima- Ōshima	20	30–40[9]	Basalt, andesite
		Hakone	132	230[10]	Andesite, basalt, dacite/rhyolite
Compound cones		Sakurajima	12	<30[11]	Andesite, dacite
		Kirishima	104	22[12]	Andesite, basalt, dacite
		Asama	17	30[13]	Andesite, basaltic andesite, dacite
		Avachinsky	75	60[14]	Andesite, basaltic andesite, basalt
		Shiveluch	116	80[15]	Andesite, basaltic andesite, basalt
		Ruapehu	150	230[16]	Andesite, basaltic andesite, dacite

Ages from [1]Conway et al. (1994), [2]Vallance et al. (2001), [3]Ponomareva et al. (2007), [4]Aoki et al. (2019), [5]Davidson et al. (2004), [6]Clynne et al. (2008), [7]Gertisser et al. (2012), [8]Cioni et al. (2008), [9]Yamamoto (1984), [10]Mannen et al. (2018), [11]Nishihara et al. (2022), [12]Imura (1992), [13]Wadge (1982), [14]Krashennikov et al. (2020), [15]Pevzner et al. (2014), [16]Leonard et al. (2021); lithology from Global Volcanism Program

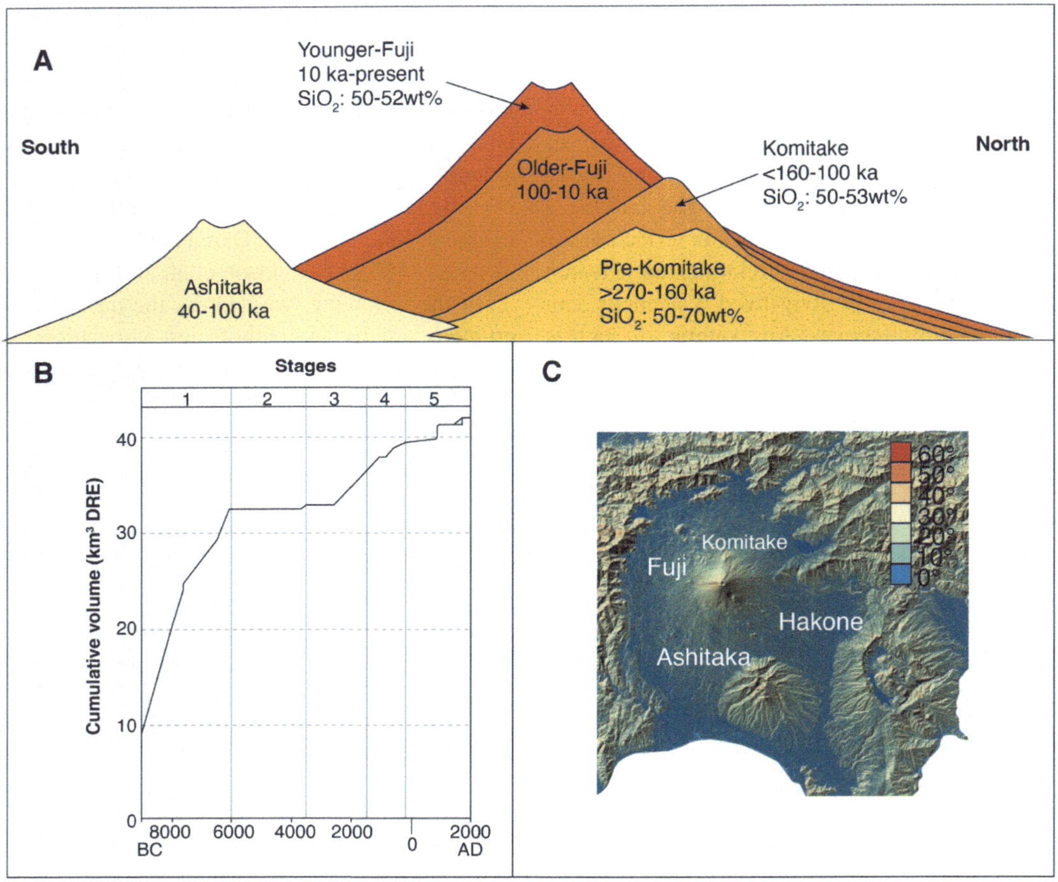

Fig. 3 Edifice evolution of Fuji composite cone. **A** Superimposed edifices, **B** Volumetric growth in the past ten thousand years (after Yoshimoto et al. 2010; Aoki et al. 2019), **C** Slope pattern map of the area around Fuji (based on SRTM DEM data). Observe the intact flanks of active Fuji, the dissected flanks of Late Pleistocene Ashitaka volcano (see Chap. 8), and those of Hakone caldera, the collapse of which truncated a previous composite volcano (see text)

of geomorphic features and morphometry are analysed, along with possible evolutionary pathways (Sect. 5). Spatio-temporal constraints are presented in Sect. 6, focussing on what is meant by 'relatively large' or 'long-lived' according to the Davidson and de Silva definition (2000), and finally, we present the details of LVCs (Sect. 7).

3 Eruptive Characteristics and Lithofacies Associations

Composite volcanoes are built up by material from small-scale explosive (e.g. Strombolian, less commonly Hawaiian) summit eruptions and shorter or longer lava flows, punctuated by products from less frequent moderate (e.g. Vulcanian) or highly explosive (Plinian) activity. Flank eruptions (e.g. from satellite vents, radial dykes) are also common. As a result, the proximal part of the edifice is composed of a succession of lava flows intercalated with pyroclastic deposits, e.g. tephra fallout and agglutinate. Hence, exposed inner crater or caldera rims can often be used for establishing an overall volcano stratigraphy (e.g. Nyiragongo: Demant et al. 1994; Tambora: Sigurdsson and Carey 1992).

3.1 The Role of Eruptive Style on Cone Shape

The existence, magnitude (VEI), and frequency of explosive eruptions during the lifetime of a composite volcano depends on a number of factors: (1) tectonic setting, which controls (2) the magmatic plumbing system (related to the crust type, structure, thickness) and (3) related magma generation (e.g. simple deep-seated, or shallow multi-layered magma reservoirs); (4) the physical and chemical properties of the magma (including the amount of volatiles); and (5) environmental factors such as available external water.

One of the most important factors constraining explosivity is the magma viscosity. In contrast to shield volcanoes, most composite volcanoes erupt higher-viscosity magmas (andesite, dacite) which flow more slowly and remain closer to the source (Giordano et al. 2008; Lev 2018). In addition, because of the commonly less viscous magma of shield volcanoes, any gases contained within the magma have an easier time escaping, while in the case of the higher-viscosity magmas producing composite volcanoes, gases are put under greater pressure and can lead to occasional more violent eruptions.

Explosive eruptions are common at composite volcanoes, but their role in building up the volcanic edifice is largely related to the VEI value. Low-VEI eruptions (e.g. Hawaiian, Strombolian, Vulcanian activity) produce lava fountaining, tephra fountaining, scoria fall, or small ash plumes, which are cone-building. On the contrary, for the rarer high-VEI explosions, a significant volume of explosive material (several km^3 or more) is deposited from PDCs, usually at a distance of < 10 km and exceptionally 15–20 km from vent. Hence, although they also play a role in cone-building along valleys in proximal to medial areas, they mostly contribute to the succession of the lower flanks and more distal areas. However, identification of the eruptive processes (within this, explosivity) does not always provide unambiguous constraints on the main volcano type. The most comprehensive available source of volcano classification, the Global Volcanism Program, assigns volcano type, taking into account the main features (including main geomorphic ones) of both the volcanic edifice and the eruptive products, but this still does not preclude uncertainties with regard to classification due to the existence of numerous transitional landforms.

Apart from the higher viscosity of magmas at composite cones, there is another major factor behind their steep, >20–25° (and commonly conical) shape which relates to eruption style rather than magma composition (Wood 1978). We described the case of Nyiragongo and Nyamulagira in Chap. 4, showing that cone-building eruptions at Nyiragongo (i.e. fountain-fed tephras in addition to short, stubby summit lava flows, prior to its present-day lava lake activity) may have been responsible for creating its steep upper slopes. In particular, the gas-rich magmas could have resulted in 1–2 km-high lava- and tephra fountains at Nyiragongo, and this might be the case at many other composite cones as well. Paroxysmal explosive events of lava and ash fountaining at Etna (e.g. from 2011 to 2015: Calvari et al. 2018) or Villarrica (in 2015: Romero et al. 2018) may have led to a similar mode of cone-building. At these volcanoes, basaltic lava flows predominate in the upper cone. Similarly, in the case of the historical eruptions of tephritic-phonotephritic magmas at Somma-Vesuvius, high, even subplinian lava fountaining may have occurred (e.g. in 1631: Mastrolorenzo et al. 1993; 1906: Cole and Scarpati 2010; see also the historical overview by Principe et al. 2004). Therefore, at least in these cases, the steep composite cone landform has nothing to do with the eruption of more viscous magma, i.e. andesite or dacite in volcanic arcs. In fact, a great number of regular-shaped composite cones over the world are of mafic composition (down the American Cordilleras: e.g. Herbert, Cleveland, Shishaldin, Pavlof, Pacaya, Santa María, Fuego, Agua, Villarrica, see Sect. 5), whose steep upper slopes may form by lava/tephra fountaining and short, stubby lava flows.

While the simple, conical landform is the starting point, increasing complexity of the edifice can be introduced during a volcano's evolution. However, although shape variations may follow an evolutionary trend, and geomorphic changes are common within the lifetime of many composite volcanoes, an initial (regular) volcano shape does not necessarily evolve over time. In other words, a general trend in shape evolution from simple to complex cannot be inferred: "there is no standard sequence, no unidirectional progression, and nothing predetermined in the evolution of stratovolcanoes" (Hildreth and Lanphere 1994). For instance, in addition to active composite volcanoes in the CVZ of the Central Andes, there are a large number of relatively simple cones and cone remnants (Karátson et al. 2012) dating back as far as 15 Ma—the oldest ones being deeply eroded but with no evidence of changes in shape (see also Chap. 8). Overall, the presence or absence of geomorphic evolution is ultimately related to whether the factors of eruptive activity listed above remain constant for long periods of time, or whether they change frequently, resulting in modifications to the volcanic landforms.

3.2 Lithofacies Associations

The successive phases of active and dormant periods during the eruptive activity of composite volcanoes create distinctive sedimentary environments or *facies*. Their primary characteristics are briefly summarised below, and their further evolution/transformation is discussed in Chaps. 7 and 8 on volcano degradation. The characteristic *lithofacies pattern* (Fig. 4) tells us that the primary or syn-eruptive products from effusive and explosive activity, and secondary or epiclastic products from continuous redeposition, do not occur randomly away from the vent, but in a defined order and character. The proportion of epiclastic sediments increases gradually, while the proportion of lavas decreases sharply in favour of volcaniclastics with distance from the vent (Davidson and de Silva 2000). In other words, a given facies is a function of distance from vent, and is controlled by the prevailing eruptive style and also the erosional processes.

1. *Main vent association* (including the core facies and proximal facies). This is the material derived from the eruptive centre, comprising one or several vents, of a composite volcano. In an active volcano, with a crater or summit caldera, the core facies consists of vent-filling breccia, with the proximal facies constituted of a succession of lava flows interlayered with fallout deposits, which may be supplemented by minor redeposited volcaniclastics. Lava dome(s) with massive lavas or talus slopes are also common within the crater/caldera interior or outer rims. In the case of extinct, degraded volcanoes, once the deeper levels have been exhumed they reveal facies composed of subvolcanic vent fill (e.g. shallow intrusive rocks, radially oriented dykes and sills, and roots of lava flows). It should be stressed that these rocks may, depending on hydrothermal alteration, become resistant or, on the contrary, loose, which determines their preservation or (erosional) removal, respectively. Hydrothermal ore bodies and veins may also be present, and coarse-grained fragmental material (e.g. megablocks from debris-avalanches, coarse-grained material from collapsed lava domes) can also be found in the facies association. This facies can be identified at eroded volcanoes by the rapid alternation of lavas and pyroclastics, the absence of epiclastic deposits, the variability in grain size, and the overall presence of coarse-grained clasts.

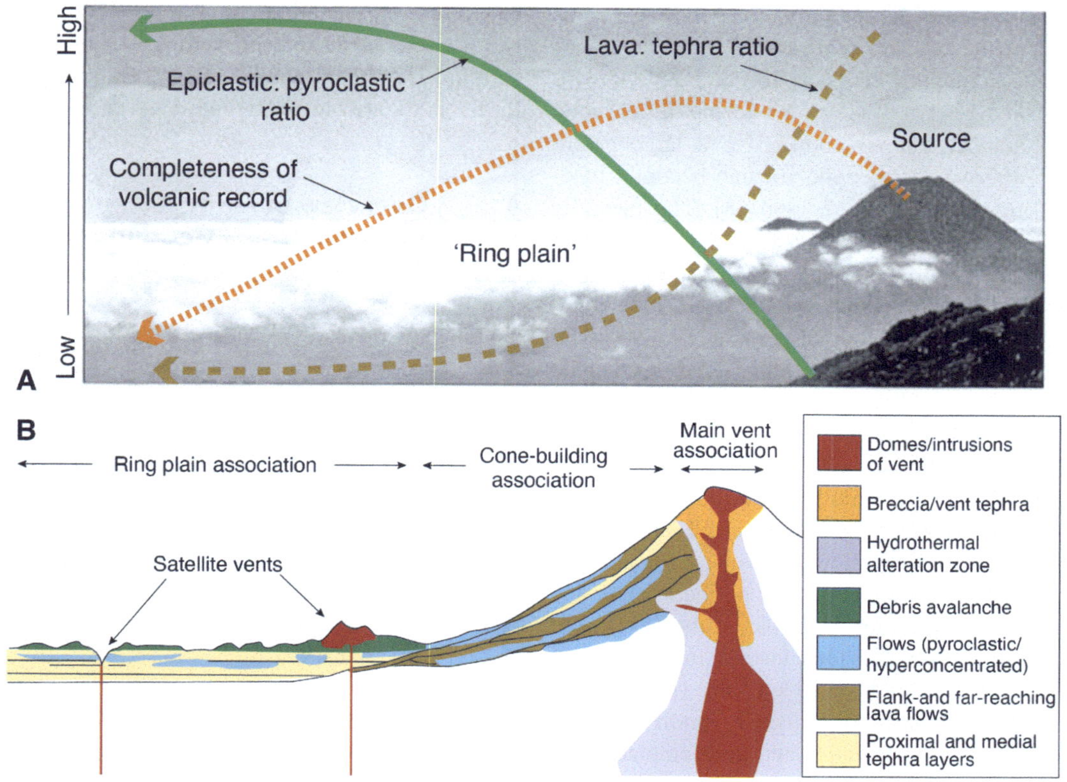

Fig. 4 The facies of composite volcanoes are determined by the primary, syn-eruptive products deposited during eruption, and the secondary volcaniclastics transported during and after eruption (modified after Davidson and de Silva 2000)

2. *Medial or cone-building facies.* This comprises the diverse material constituting the main flanks of the composite volcano: lava-flow tongues, pyroclastic-density current (PDC) deposits on the mid- and lower slopes, tephra-fall deposits, and debris-flow deposits on the medial valley banks and the more gentle lower flanks. In a study of 700 composite volcanoes, slope values of main flanks vary highly (14–32°: Grosse et al. 2012) depending on the lithology and the complexity of the volcano edifice. Downslope, in the presence of a well-developed drainage network, geomorphological inversion is common: fresh lava flows are emplaced in valleys, while the inter-valley ridges are composed of older material. In the case of a less entrenched drainage network, the lava flows and PDC deposits tend to cover the medial and lower flanks of the volcano by filling valleys and depressions. The flow direction and accumulation area of lava flows is often determined by the fact that the crater or a summit caldera is open in one direction, creating an outlet. By contrast, the emplacement of tephra-fall deposits is not primarily influenced by morphology but by the prevailing wind direction. This variability can lead to a complex 3D structure due to spatial and temporal changes. As a result, looking at the succession in cross-section represents a dynamic pattern, reflecting the differential emplacement of tephra-fall and pyroclastic-flow deposits and their redeposited counterparts (see Chaps. 7 and 8).
3. *Distal or ring-plain facies.* This is built up of the material of the outward-dipping,

low-slope debris apron surrounding a composite volcano (with lower flank slope values ranging between 10° and 22°: Grosse et al. 2012). It is formed during the volcanic activity but is modified to a large extent afterwards, once the volcano becomes inactive (see Chaps. 7 and 8). The facies is composed of a small portion of tephra-fall and PDC layers, and a majority of epiclastic material redeposited from the upper flanks. The latter also includes occasional but widespread volcanic debris-avalanche deposits (Chap. 7). The fallout layers comprise fine-grained material deposited from high eruption columns, while the PDC deposits are mainly pumiceous pyroclastic-flow deposits (ignimbrites) and far-reaching block-and-ash flow deposits. Although the stratigraphic record may be incomplete due to distal setting and erosion, the observable succession provides insight into the characteristic eruptive style of the volcano. The distal deposits can mantle, modify, or create new topography, and their emplacement and subsequent reworking have an impact on the drainage network (Major 2022; see Chaps. 8 and 9). The channel geometry of gullies and valleys controls the deposition processes, with the exception of large debris avalanches. Both PDCs and debris flows have high yield strength, so the resultant deposit will often be slightly convex in cross-section.

It can be very difficult, and sometimes impossible, to separate redeposited volcanogenic materials from primary pyroclastics in the distal apron area, since sooner or later almost all pyroclastic material will be partially or completely redeposited. In submarine, lagoonal, or lake environments, e.g. offshore island arcs, the distal facies may not always be present, and/or the typical facies pattern may be modified by underwater mass movements (see Chap. 7). A further modifying factor may be the overlapping of debris aprons of aligned composite volcanoes: an example in a modern setting is the triple edifice of Taranaki-Pouakai-Kaitake on North Island, New Zealand (Cronin et al. 2021); and another one in an ancient setting is Călimani-Gurghiu-Harghita chain, East Carpathians, Romania (Szakács and Seghedi 1995).

4 Overview of Main Occurrences by Geodynamic Setting

Composite volcanoes occur in all settings except mid-ocean ridges. Here, we present composite volcanoes with typical and atypical occurrences by geodynamic setting, also looking at their characteristic lithology; the focus is on well-known examples, but also on specific features.

4.1 Occurrences in Subduction and Collision Zones

These comprise the most important tectonic settings, hosting the greatest number of the Earth's composite volcanoes. The majority of active arc volcanoes (e.g. in the North, Central and South American Cordilleras, Kamchatka, Java, the islands of the Southwest Pacific: Gill 2016) can be found within 150–350 km of convergent plate margins, i.e. subduction zones with an oceanic trench (see Chap. 1). The remainder of the volcanoes are found at atypical subduction zones or in collision settings.

4.1.1 Active Subduction Zones

Spatial distribution and magma generation

Volcanoes usually form linear ranges up to a few hundreds of kilometers long at subduction zones. In terms of their distribution, volcanoes in arc segments are often evenly spaced, a few tens of km apart (see Chap. 1). In the Cascades, for example, the mean distance between volcanoes is ~45 km, but ranges from 35 to 170 km (Lockwood and Hazlett 2010). Volcanoes sometimes form overlapping edifices, as in the Guatemalan segment of the Central American Arc (Reynolds 1987; Carr et al. 2003). The changing distribution and the existence of volcano clusters may be accounted for by the variable structure and thickness of the lithosphere within the volcanic arc (Hildreth 2007). Volcano distance also depends on the plate convergence rate and melt production (e.g. Tatsumi et al. 2020), and along-arc locations of composite volcanoes may additionally be affected by temperature variations in the mantle wedge (Lee and Wada 2017).

As described in Chap. 1, composite volcanoes in subduction settings commonly erupt calc-alkaline magmas forming volcanic arcs. Higher-K (i.e. alkaline) and less Fe-enriched magmas tend to erupt in the generally older rear arc (Turner et al. 2003). However, calc-alkaline and alkaline affinities can often be found close together spatially, for example in the Sunda arc: calc-alkaline andesites occur at the volcanic front (e.g. Merapi, Semeru, Sumbing), contrasting with alkaline rocks in the rear arc (e.g. Tambora, Muriah, Watukubu; Gertisser et al. 2012). Likewise, in the ~1000 km-long, discontinuous Trans-Mexican Volcanic Belt, the typical calc-alkaline rocks are accompanied by intraplate alkaline and adakitic rocks (Ferrari et al. 2012).

Based on the overall lithology, Pike and Clow (1981) distinguished geomorphic classes of composite volcanoes by size and calc-alkaline versus alkaline composition. Of their seven classes, five were connected to island- or continental arcs. These authors reported that, on average, alkaline magmas create larger volcanoes than calc-alkaline ones. However, even a schematic distinction between calc-alkaline and alkaline volcanoes is not easy due to the frequent spatial-temporal coexistence of compositional diversity. Evolved composite volcanoes can erupt dacite and rhyolite in addition to andesite and basaltic andesite (and often with a bimodal composition). These are mostly found in continental arcs (e.g. the Cascades: Humphreys and Grunder 2022; the Northern or Central Andean Volcanic Zone: Stern 2004), while exclusively mafic composite volcanoes are typically found along microcontinental arcs (e.g. Kamchatka, Japan) and island arcs (e.g. Lesser Antilles, Kuril Islands). Note, however, that individual volcanoes might show compositional variations: in the Cascades, Mount St. Helens has been characterised by a complex lithology from andesitic and basaltic eruptives to dacites over the past ~40 ka (Smith and Leeman 1993). On the contrary, North Sister produced compositionally homogeneous low-K basaltic andesites over its <400 ka history (Schmidt and Grunder 2009).

In the following, we highlight selected occurrences of the Circum-Pacific Belt, focussing on the main composite volcano types and assemblages.

North America

The northernmost occurrence is the Aleutian Arc (Fig. 5; Miller et al. 1998; Schaefer and Nye 2008; Alaska Volcano Observatory website, http://www.avo.alaska.edu/volcanoes/). Most of the composite volcanoes are snow- or ice-clad regular-shaped cones or ones truncated by calderas (Miller and Smith 1987). Kanaga, Herbert, Carlise, Cleveland (Fig. 5A–C), Vsevidof, Shishaldin, and the twin peaks of Pavlof and Pavlof Sister represent the most regular, conical composite volcanoes. Vsevidof (whose most recent eruption was in 1878) and its eastern neighbour, the dormant Rechesnoi (Fig. 5D and E) are also interesting in terms of what a cone looks like when its morphology is constantly modified, and when it is subject to millennia of glacial erosion. Shield volcanoes with calderas (Okmok: Fig. 5E) or without (Westdahl) can also coexist with composite volcanoes exceptionally. Prominent calderas which truncate the original composite volcano(es) include circular landforms such as Yunaska or Aniakchak, or more irregular ones like Fisher or Katmai. In Alaska, Augustine and Redoubt are famous for their debris avalanches and the resulting horseshoe scars (Chap. 7).

The Cascade Arc includes three dozen major volcanic composite volcanoes (Photo 56). The highest and one of the largest composite edifices is Mount Rainier (Sisson and Vallance 2009), whose dormant young cone completely covered a sector collapse scar formed by the

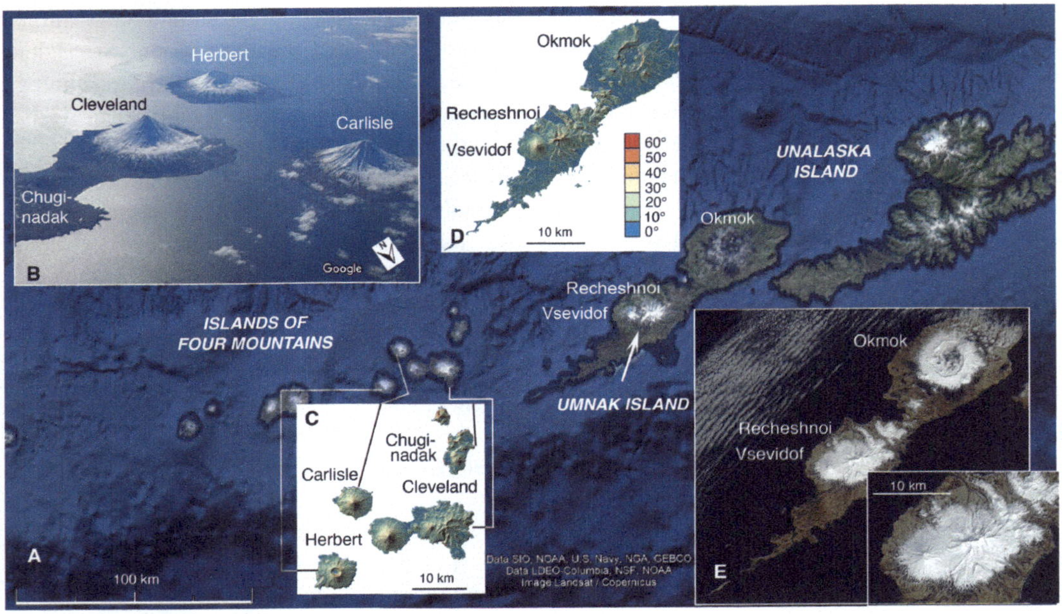

Fig. 5 Satellite imagery, topography, and slope pattern of the central Aleutian Islands. **A** Google Earth image, **B** Oblique satellite view of the Islands of Four Mountains showing regular composite cone morphologies (and a 2 km-wide summit caldera on Herbert volcano; credit: Kim/Google Earth, 2023); **C** and **D** slope pattern maps of the Islands of Four Mountains and Umnak Island, respectively (based on SRTM DEM data); **E** LANDSAT-8 satellite image of Umnak Island showing a regular cone (Vsevidof), a heavily glaciated dormant composite cone (Rechesnoi), and a shield volcano caldera (Okmok; credit: NASA Earth Observatory, Joshua Stevens, 2014)

large Osceola Mudflow (a debris avalanche) following a major explosive eruption ~5600 years ago. One of its famous neighbours, Mount St. Helens is well-known for its 1980 eruption and sector collapse (Chaps. 7 and 9; Lipman and Mullineaux 1981) with post-caldera lava dome activity until 2008. Mount Shasta, the second highest but the most voluminous of the Cascade volcanoes, is constituted of four main edifices emplaced over the past 600 ka. It also experienced one of the largest Quaternary subaerial debris avalanches (Crandell et al. 1984; Tortini et al. 2015).

Central America

Along the Trans-Mexican Volcanic Belt (e.g. Varley 2012), out of almost 40 large Holocene centres, a dozen are classified as composite volcanoes by the Global Volcanism Program. The current morphology of the highest volcanoes has been influenced by glaciation: small tropical glaciers are still present on some of the highest (>5000 m) volcanoes (e.g. *Citlaltépetl* / Pico de Orizaba: Soto 2023). The most known, Popocatépetl, is a regular-shaped composite edifice, which is the volcano's third rejuvenated cone, each time constructed in the past 23 kyr and following a significant flank collapse (Espinasa-Pereña and Martín-Del Pozzo 2006). The volcano also hosted a glacier on its north flank, which has now melted due to the coupled effect of climate change and eruptions from the crater (Delgado-Granados et al. 2007). Toluca, a broad composite volcano with summit lava domes is truncated by a 1.5 km-wide Late Pleistocene horseshoe scar (Norini et al. 2008). Interestingly, its growing lava domes showed specific, repeated interaction between the hydrothermal system and the volcano glaciers. During dome growth, a large amount of melting water produced thick fluvioglacial sequences, subsequently covered by block-and-ash flow deposits from dome collapses (Capra et al. 2015). Of the other volcanoes in Mexico, twin-peaked Colima consists of an extinct >1 Ma, cone and a currently active, regular-shaped composite cone: Photo 57) developed in a horseshoe scar, clearly visible in the current morphology (Roverato et al. 2011); El Chichón is a relatively small lava dome complex with a summit crater, and is famous for its VEI 5 1982 eruption which exploded the most recent dome, producing devastating PDCs (De la Cruz-Reyna and Martín-Del Pozzo 2009). Southward, the Central American Volcanic Arc—particularly in Guatemala, El Salvador, Nicaragua and Costa Rica—has an unparalelled wealth of composite volcanoes. At the time of writing (end of 2024), Santiaguito and Fuego have shown continuous eruptions (ongoing), while San Miguel and San Cristóbal were active until 2023/2024. In the past 30 Myr the volcanism has migrated from east to west (Saginor et al. 2013). Interestingly, in Guatemala, the older rear arc or 'volcanoes behind the front' (Cameron and Walker 2006) has shown a systematic temporal geomorphic evolution from large composite volcanoes to small monogenetic ones. This is explained by potentially short residence times of magmas in crustal reservoirs and small degrees of crustal assimilation over areas behind the front, which accompanied smaller scale, deeper magmatic systems feeding monogenetic volcanoes.

Figure 6 highlights two areas in Guatemala and El Salvador that illustrate the coexistence of volcanoes of different types and ages.

In Guatemala, the active dacitic Santiaguito lava dome complex (Harris et al. 2003: Chap. 2) started to grow at the base of the regular-shaped cone of Santa María (Table 1) in 1922. The nearby Lake Atitlán is of multiple interest: its caldera, formed in the Miocene, hosts, or is surrounded by, a series of impressive, steep, regular-shaped partly overlapping active to dormant composite cones: San Pedro, Atitlán, Tolimán, and further east, Acatenango, Fuego, and Agua, one of the most regular-shaped cones in the world. Due to the tropical climate, the cone surfaces quickly become vegetated, thus satellite images of the cone flanks give a good indication of the approximate date of the last eruption, i.e.

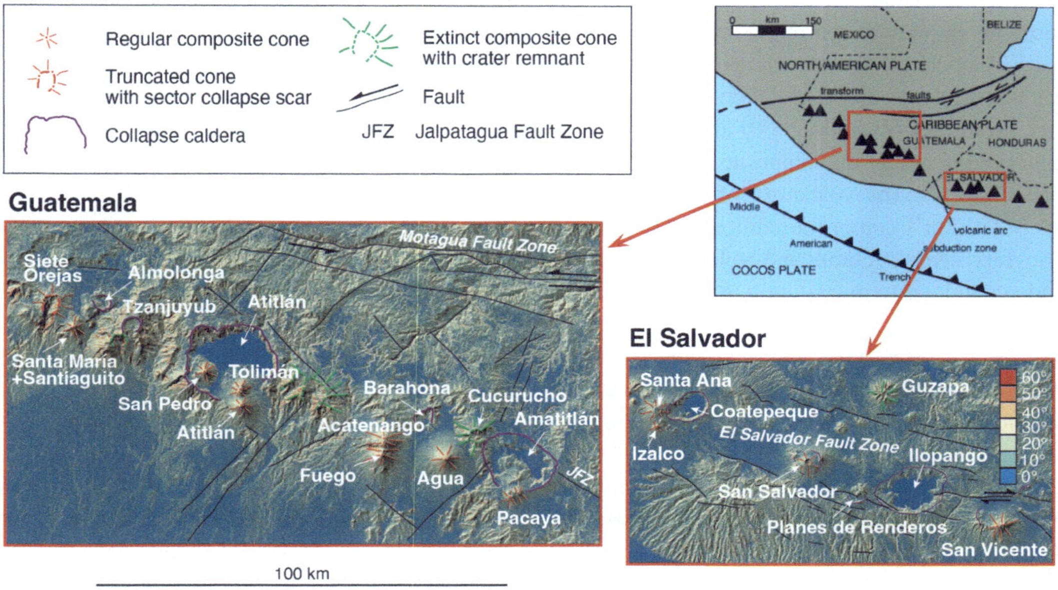

Fig. 6 An example of the high diversity of the coexistence of composite volcanoes and calderas in subduction settings: major volcano geomorphic features of two selected areas in Guatemala (left) and El Salvador (right). Intact active, and dormant or extinct, eroded volcanic landforms are both visible. The Motagua Fault belongs to a complex transform fault zone that separates the North American plate from the Caribbean Plate. Fault patterns after Bonis (1993), Canora et al. (2014), and Suñe-Puchol et al. (2019). Slope pattern maps based on SRTM DEM data

whether they are covered with brownish scoria or ash, or already green with vegetation. Pacaya sits in the 16 × 14 km-wide Amatitlán caldera, and is a complex of post-caldera composite volcanoes and domes, the most recent being the active Mackenney basaltic cone which has had almost 50 major eruptions since the Spanish conquest (Bardintzeff and Deniel 1992; Cisneros de León et al. 2023).

Most volcanoes in El Salvador are found in the Median Trough of the El Salvador Fault Zone. Close to the Guatemalan border, the elliptical, 7 × 9 km Coatepeque caldera was created by collapse of composite volcanoes east of Santa Ana volcano, partially truncating the flanks of the latter in the process. The caldera margins and interior are dotted by pyroclastic cones, lava domes and, on the southern flank of Santa Ana, the 650 m-high Izalco, a small perfectly-shaped, straight-sided youngest cone created in 1770 and which has been active since then (Carr and Pontier 1981; Lechner et al. 2013).

In Nicaragua, in addition to San Cristóbal and Momotombo, which are regular-shaped cones, Telica compound volcano consists of a number of juxtaposed cones and nested craters. In Costa Rica, we draw attention to two composite volcanoes. The perfectly symmetric, young Arenal cone has frequently produced VEI 4 eruptions in its history (Soto and Alvarado 2006) and represents an 'embryonic' stage in composite cone evolution (see Chap. 3). The volcanism of Irazú, the largest (~250 km^3) composite edifice in the country, also represents a compound volcano which has undergone long-term evolution (850 ka). Recently characterised by effusive, Strombolian and phreatomagmatic (e.g. Vulcanian) activity, historical eruptions of Irazú are controlled by ubiquitous magma mixing between two distinct magma batches (Alvarado et al. 2006).

The Lesser Antilles Arc is characterised by a low subduction rate that has resulted in low magma production rates and, as a consequence, only small islands. Indeed, the volcanoes on the islands are mostly lava domes or lava dome complexes below 1500 m a.s.l.—e.g. Montagne Pelée, Martinique, and Soufrière Hills, Montserrat (Chap. 2).

South America

South of Panama Bay, the Andean Volcanic Belt (Chap. 1) consists of four segments: the Northern (NVZ), Central (CVZ), Southern (SVZ), and Austral (AVZ) Volcanic Zones. Between them, inactive gaps are due to a shallowing in dip of the subducting plate which prevents sufficient magma generation. While composite volcanoes predominate, the largest eruptions (including many VEI 8 super-eruptions in the Earth's history) occurred in calderas and caldera systems (see Chap. 7).

In the NVZ, Nevado del Ruiz (Colombia), a lava dome complex which grew up in an older caldera, had frequent lahars due to melting of its ice cap during its history. In 1985 there was a VEI 3 eruption, the deadliest in South America with 23,000 victims drowned or washed away in the mud (Williams 1990; Chap. 3). Regular-shaped Sangay (5286 m) in Ecuador is one of the most active volcanoes on Earth; it also had huge, far-reaching Pleistocene debris avalanches (Valverde et al. 2021). Ecuador still has four glacier-capped volcanoes—Chimborazo, Cotopaxi, Cayambe and Antisana, all around 6000 m a.sl.—, but their ice covers are quickly disappearing. Of the two most famous, Chimborazo is a huge, complex, Plio/Pleistocene edifice, whose present cone was built after a 35 ka debris avalanche, and also erupted a few times in the Holocene (Barba et al. 2008). Cotopaxi, one of the most regular-shaped (Sect. 5.4) and at the same time most highly active composite cones in the world, shares a similar evolution to Chimborazo but is just 4000 yr old (Hall and Mothes 2008).

Southward, beyond the Peruvian flat-slab segment, is the CVZ. It hosts a magnificent chain of Pliocene–Quaternary volcanoes—the highest in the world—which has formed in the western Cordillera and on the western margin of the Altiplano–Puna Plateau (Kay and Coira 2009; Scott et al. 2018; Giambiagi et al. 2022), a semi-arid upland with an average altitude of c. 4000 m (i.e. the elevated base level of the volcanoes). Its complex evolution makes it an ideal subject for tectonic, volcanological and erosion studies. Along the CVZ, in addition to the commonly occurring andesites—which are obviously abundant in the eponymous Andes, building up generations of composite volcanoes—there are also large amounts of rhyolites and dacites. The latter are less associated with composite volcanoes than with calderas and complex caldera fields, which were created during voluminous ignimbrite eruptions (Chap. 6; for summary papers, see de Silva and Kay 2018; Wörner et al. 2019).

Two of the most interesting areas within the CVZ are shown in Fig. 7. In the north, Nevado Coropuna, Peru's highest and largest volcano, is an ice-covered compound edifice with an elongated east–west oriented morphology. It consists of half a dozen summit cones, so should be termed a large volcano cluster. Like Coropuna, its neighbour, Solimana (6093 m) has only fumarolic activity, but Solimana's morphology is completely different, truncated by a huge horseshoe scar. These two large volcanoes, as well as many others to the south, are situated between huge, deep canyons (e.g. Cotahuasí and Colca) that cut into an impressive succession of ignimbrites dating back 20–25 Ma (Thouret et al. 2007, 2016), interstratified with coeval, and generally overlain by young andesites, the latter derived from the Quaternary or currently active composite volcanoes. El Misti, a historically active volcano (most recent phreatic event in 1985) near the city of Arequipa, had a 'classic' composite volcano history: its young (>10 ka) regular-shaped cone in the caldera of an older volcano is the third to form here (Thouret et al. 2001). Further eastward is Ubinas, Peru's most active volcano (with explosive summit eruptions ongoing in 2024). Its complex cone is truncated by a small summit caldera 1.2–1.4 km across, which hosts a double tuff cone; and its structural evolution has been determined

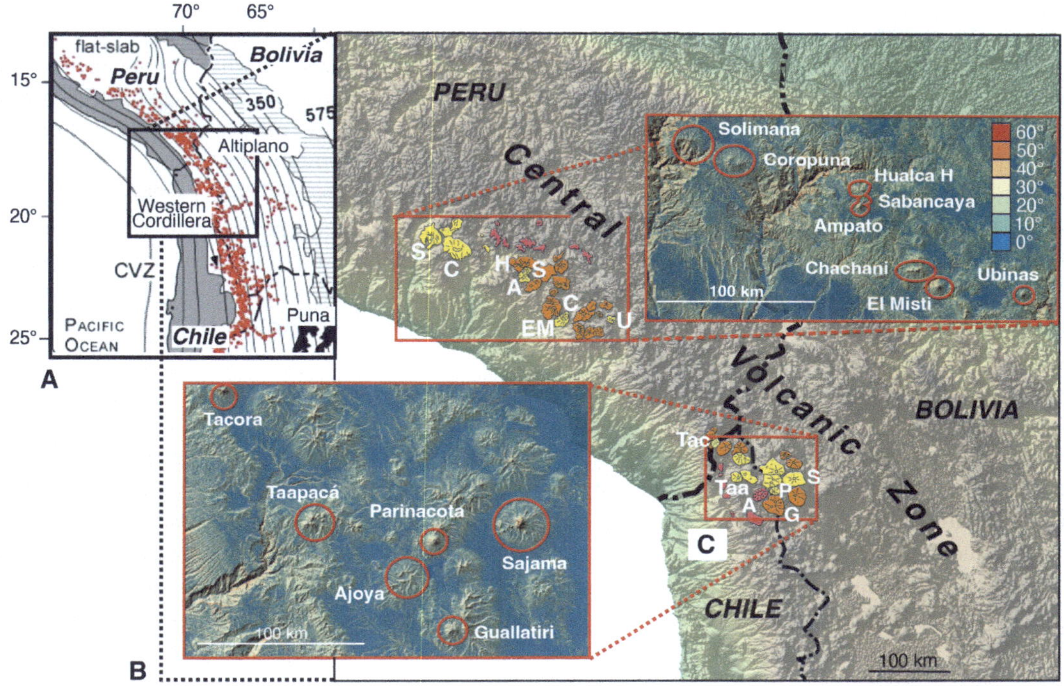

Fig. 7 Two areas of particular volcano geomorphological interest in the Central Volcanic Zone (CVZ) of the Central Andes. **A** Location of the border region of Peru-Bolivia-Chile (the Arica Bend), showing contours to the Wadati-Benioff zone with depths in km, and volcanic centres in red (Kay et al. 1999). **B, C** Major composite volcanoes of Miocene (purple) through Pliocene/Quaternary (orange) age to recent or active (yellow) in Peru (**B**) and Bolivia-Chile (**C**), respectively (modified after M. Mamani and G. Wörner, unpublished). Enlarged are slope pattern maps based on SRTM DEM data; the same volcanoes circled in red

both by the asymmetric, sloping basement of the volcano and the current inner hydrothermal system (Gonzales et al. 2014). Southward, the historically largest (VEI 6) eruption in the Andean Cordilleras occurred at Huaynaputina (Peru) in 1600 AD/CE (Table 1), associated with significant ash cover, and resulting in devastation of large areas by lahars (Thouret et al. 2002). The volcano is a large but geomorphologically inconspicuous edifice with lava dome clusters and craters in a horsehoe scar of debris avalanches.

The Chilean-Bolivian border region has a number of composite volcanoes that reach over 6000 m a.s.l.: Nevados de Payachata (Parinacota and Pomerape: Photo 58), Taapacá, Huallatiri, and other, older volcanic edifices, some of which deeply eroded, such as Ajoya with an erosional caldera (Fig. 7). Taapacá and Huallatiri are complex edifices consisting mostly of lava domes, while the others are composite cones. Taapacá, a large volcanic cluster, and neighbouring Parinacota, a composite volcano, are examples of contrasting eruptive behaviour, lifetime, and morphology. The former erupted homogenous dacites for ~1.5 Ma, while the latter erupted variable products with a wide range of composition from basalt to rhyolite in only 60 ka (Clavero et al. 2004a; Hora et al. 2007; Fig. 8). The different compositional range alongside the highly variable eruption rates for the two volcanoes implies distinct magmatic regimes in the subvolcanic plumbing system. Taapacá has experienced slow and steady production of hybrid dacites from 1.5 Ma to the present at an eruption rate of less than 0.02 km^3/kyr. In contrast, the magma at Parinacota ranged in composition from basalt to rhyolite over 150 kyr, produced at a rate of 0.2 km^3/kyr; following a large flank failure 9 ka a new cone has been constructed of homogenous andesite at very high eruption rates (10 km^3/kyr) (Blum-Oeste and Wörner 2016; Wörner et al. 2019, also see Sect. 6). East of these volcanoes is Sajama—again a younger cone built upon older domes—which is Bolivia's highest mountain. Its huge size is all the more remarkable because it probably became extinct towards the end of the Pleistocene and has since been carved and eroded by glaciers. Further southward lie the highly symmetric, regular cone of Licancábur; the overlapping twin volcano of San Pedro–San Pablo; the amalgamated edifice of Láscar, the most active volcano in the Chilean CVZ; and the huge, compound 'Nevados' Ojos del Salado (6893 m), which is the highest volcano in the world (Photo 59). The latter is not glaciated due to the extremely dry, cold-desert climate, and serves as an analogue for Martian landscapes (see the special issue of Astrobiology: Kereszturi 2020).

Beyond the Pampean flat slab segment in the SVZ lies the volcano cluster of San Pedro–Tatara (or –Pellado), consisting of two successive edifices built within a 0.5 Ma caldera 12×6 km in size (Singer et al. 1997). Further to the south, the volcanoes that rise around the glacial lakes of the Chilean Lake District are below 3000 m in altitude, but their upper cones are snow- or ice-capped for most of the year (Photo 60). They include the regular-shaped, active cone of Villarrica (whose lava lake is now continuously monitored, see Chap. 2). In addition, recent, highly explosive (VEI 4 or 5) eruptions of other centres to the south, such as the lava dome complex of Chaitén (2008–2012, see Chap. 2), or the composite cones of Puyehue–Cordón Caulle (2011–2012) and Calbuco (2015), have brought about increased volcano monitoring in Chile (Amigo 2021).

From Kamchatka to Japan

The microcontinental arc of Kamchatka extending down to the Kuril Islands was formed by the westward subduction of the Pacific Plate (Plechov and Kamenetsky 2023). Dominated by composite volcanoes (Ponomareva et al. 2007), its specific geological feature is the interplay between glaciations and volcanic activity during the Pleistocene, which resulted in voluminous caldera-forming eruptions, glacial advances, and rapid sea level changes, affecting the local and even the global climate (Bindeman et al. 2004). Klyuchevskoy, one of the most regular-shaped, highly productive basaltic cones in the

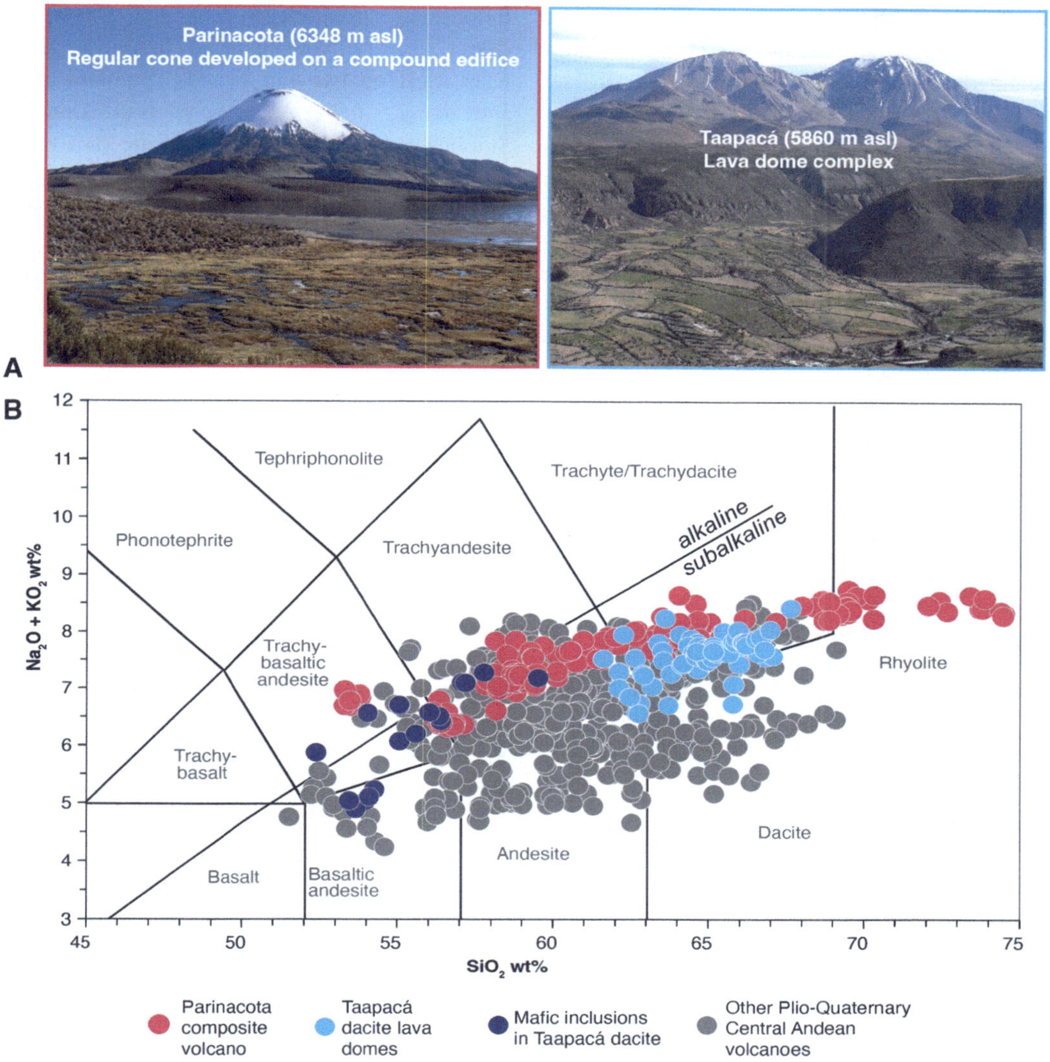

Fig. 8 Geochemical comparison of two adjacent volcanic edifices in the CVZ of the Central Andes: Parinacota and Taapacá, after Rout et al. (2021). CVZ reference data from Mamani et al. (2010). **A** Typical morphology of a regular cone versus a lava dome cluster (*photo*: D. Karátson). **B** TAS diagram of eruptive products of the two volcanoes. Mafic enclaves from Taapacá dacites are shown in dark blue, and are indicative of magma mixing. See text for explanation

world, has an edifice of 50 km^3 which was constructed in only 6 kyr (Table 2; see Sect. 6.2). Shiveluch, which is more than twice as big and ten times as old, has a double cone: the 65 ka Stary Shiveluch was truncated by a 9-km-wide late-Pleistocene flank failure, and the Molodoy Shiveluch lava dome complex was constructed within the half-open scar in the Holocene (see Chap. 7). The third highly active volcano, Bezymianny, has a modern edifice grown on an older lava dome complex, and produced a flank failure with a horseshoe scar accompanied by a lateral blast in 1956 during activity similar to the 1980 Mount St. Helens eruption (also see Chap. 7).

South of the Kuril Islands, the volcanism in Japan is due to two subduction zones: that of the Pacific plate in the north and the Philippine plate in the south. The country hosts more than one hundred active volcanoes (for geomorphological descriptions, see Moriya 1983; Koarai et al. 1996). On the island of Honshū, Fuji (or Fujisan), the iconic, highly symmetric, regular-shaped basaltic composite cone (Aoki et al. 2019) is superimposed onto an older edifice, the Late Quaternary Komitake, and itself consists of the Older Fuji and the recent Younger Fuji (Fig. 3; Photo 55), which has grown as fast as Klyuchevskoy in only 10 kyr (Table 2). Fuji is adjacent to the much older Ashitaka composite cone with an eroded crater remnant (see Chap. 8), and is 20 km from Hakone somma caldera which truncated a former composite volcano (see Sect. 5.3). Further north, still on Honshū island, is Bandai-san—a complex of overlapping composite cones—known for its 1888 flank collapse. Two more Japanese volcanoes should be mentioned since they have produced the two deadliest eruptions in recent decades. The irregular-shaped cone of Ontake (also on Honshū) has been continuously monitored following constant unrest since its 1991 eruption. Unfortunately the seismometers, GPS stations and tiltmeter were unable to predict the phreatic activity which led to an eruption on 27 September 2014 (Yamaoka et al. 2014). Ontake is a popular tourist attraction for hikers, and on that day dozens of people were climbing the volcano, 60 of whom were killed or went missing during the eruption. Unzen or Unzendake (on Kyushu)—a lava dome complex (also see Chap. 2) known for having caused a tsunami in 1792 due to gravitational dome collapse which killed almost 15,000 people—reawoke in the autumn of 1990, and displayed extensive lava dome activity up to 1996 (Nakada et al. 1999; Photo 61). In 1991, an unusually large PDC killed 43 people including three volcanologists (Katia and Maurice Krafft, and Harry Glicken).

Indonesia and the Philippines

In the equatorial region, there are two prominent, highly volcanically active areas: the Indonesian archipelago and the Philippines (see Chap. 1). Volcanism in Indonesia manifests itself on two big islands (Sumatra, Java) and many smaller ones (e.g. Bali, Lombok, Sumbawa, Flores), all of which are part of the Sunda Arc (Fig. 9). Magmatism is due to the northward subduction of the Indo-Australian Plate beneath the Eurasian Plate, creating the Java Trench, with some influence of extensional tectonics due to the rotation of Sumatra in the Sunda Strait (Schlüter et al. 2002). There are differences along the island arc: while subduction shears off large amounts of sediment in the Sumatra foreland (originating mainly from the Ganges and Brahmaputra rivers in India) and pushes it to form an outer, non-volcanic island chain, the sea off Java is much deeper and no outer island chain has formed. The active volcanic chain in Sumatra and Java extends some 30–50 km from the subduction front.

The majority of volcanoes in Sumatra are older edifices, although there are active and even erupting volcanoes as well (e.g. Sinabung, Kerinci). Krakatau, which is situated in the Sunda Strait between Sumatra and Java, is famous for its 1883 eruption, resulting in a somma caldera formation (see Sect. 5.3). Java, to the east, is one of the most volcanically diverse spots on Earth. Figure 9 shows some regular-shaped cones, horseshoe scars, and post-caldera edifices of different ages. Perhaps the best known from a volcanological point of view is the active Merapi volcano in central Java (Gertisser et al. 2023; Photo 62). It has had a complex geomorphic evolution: the oldest (<170 ka) volcanic rocks have been covered by Old Merapi (dated at 30 to 5 ka) which experienced one or more Mount St. Helens-type sector collapse (see Chap. 7). This was eventually overlain by the regular cone, a lava dome complex, of the youngest New Merapi, showing almost continuous lava dome activity with devastating block-and-ash flows and lahars over the last two thousand years (Newhall et al. 2000; Gertisser et al. 2012). Among the many other important volcanoes in the Indonesian archipelago, Tambora (Figs. 2 and 10) stands out, whose

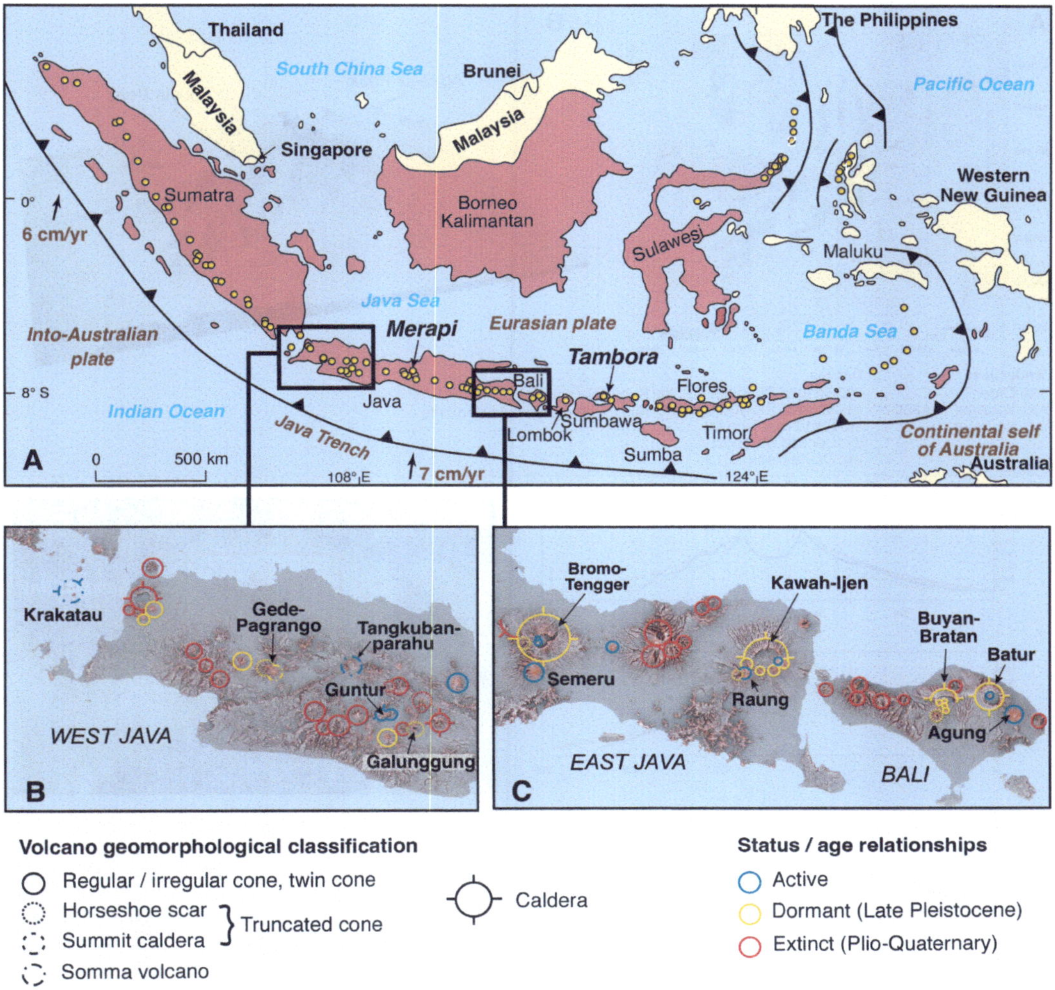

Fig. 9 Volcano geomorphic features of Java. **A** Geodynamic and geographic setting of Java within Indonesia (red colour). Yellow dots mark active or dormant volcanoes (after Gertisser and Keller 2003). **B, C** Enlarged areas of West Java and East Java and Bali, respectively, showing selected composite cone types as classified in this book. Status and age of volcanoes are after Global Volcanism Program. Background maps are the RRI (red relief image) of Java based on SRTM DEM (Favalli, M. and Karátson, D., unpublished)

circular summit caldera was formed in the 1815 VEI7 eruption.

In the Philippine archipelago, opposing subduction zones are responsible for the Philippine tectonics, resulting in two quasi-parallel deep-sea trenches. Large composite edifices, compound volcanoes and amalgamated clusters occur typically on thickened crust (Paguican et al. 2021; see Sect. 7). Of its active composite volcanoes, we highlight the iconic, highly symmetric andesite cone of Mayon (Sect. 5.1.1).

The Southwest Pacific

In the southwest Pacific, active subduction-related arc volcanism is represented by New Britain, the Solomon Islands, then the curve of the Vanuatu (New Hebrides) arc to Fiji, and finally the south-trending Tonga–Kermadec arc (Taylor et al. 2023). On some of the larger islands (e.g. New Britain: active to dormant; Fiji: extinct, heavily eroded), sizeable composite volcanoes occur (e.g. the Ulawun and Bamus

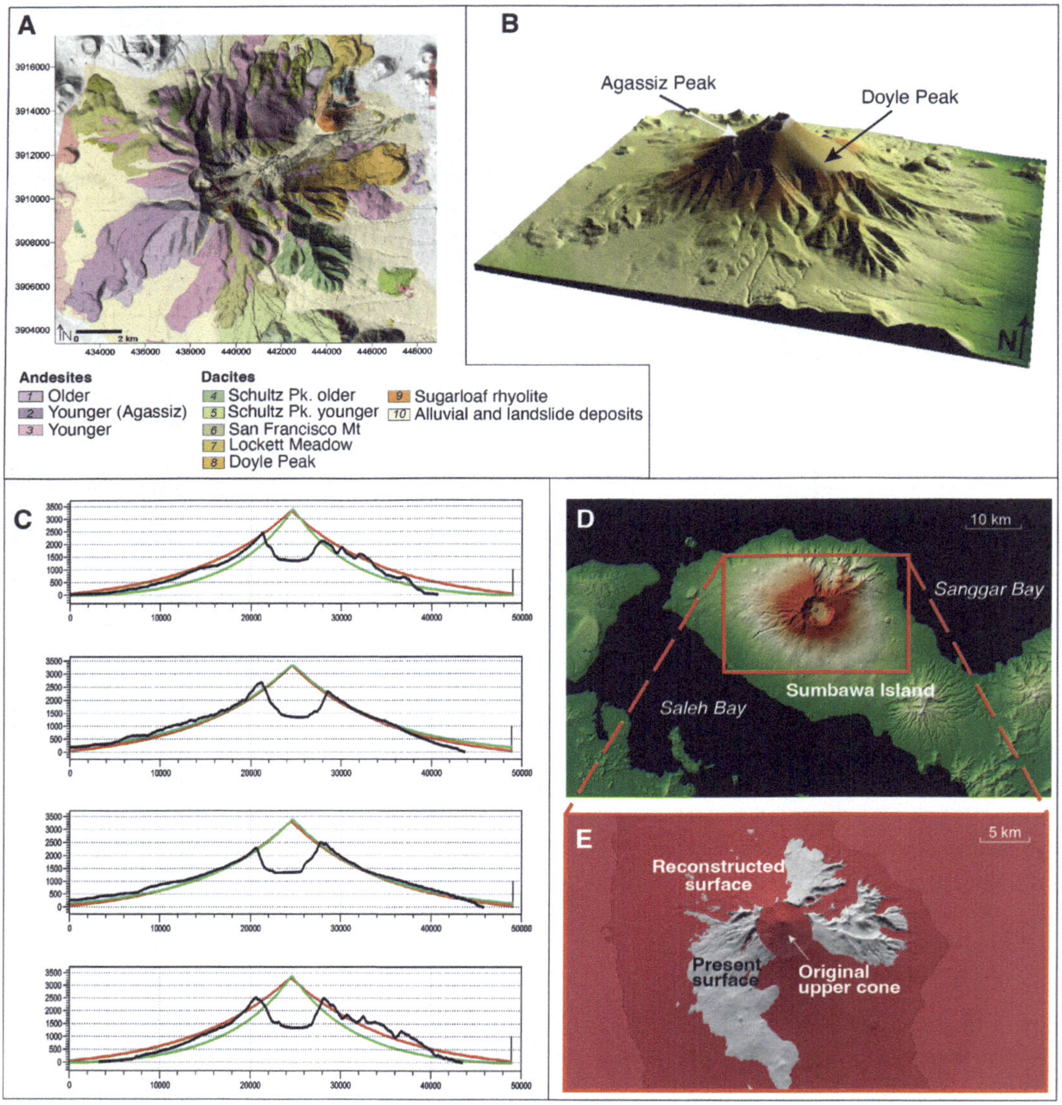

Fig. 10 Geomorphological reconstruction of truncated composite cones. **A, B** Geology and geomorphological reconstruction of San Francisco Peak (modified after Karátson et al. 2010b). **B, C** Reconstructed cone of Tambora in cross sections (**B** green line for elliptic base, red line for circular base), and 3D view (**C**) of best-fitting cone with an elliptic base (Favalli et al. 2016)

regular-shaped cones, see Fig. 14 in Chap. 7), but elsewhere volcano morphology is typically restricted to small- to medium-sized composite cones, lava domes, and monogenetic edifices. In the Fiji Islands, active volcanism is only found on the small neighbouring islands.

The Tonga–Kermadec subduction zone represents the fastest convergent plate boundary on Earth. In the absence of continental crust, the island arc has served as an excellent example for studying subduction magmatism (for a summary, see Smith and Price 2006). Volcanoes of the Tonga Islands are mostly emergent portions of submarine composite edifices, such as Hunga Tonga–Hunga Ha'apai presented in Chap. 3.

Finally, at the southern continuation of the Kermadec Trench, the superb volcanism of New Zealand is the result of subduction of a thick oceanic plate beneath the continental Indo-Australian Plate in the shallow Hikurangi Trough. Basically, the volcanism in the North Island is twofold. Typical subduction-related andesites from deeper-seated sources erupt from regular-shaped composite cones (Taranaki/formerly Egmont) or compound edifices (Tongariro, Photo 63; Ruapehu, Photo 64); whereas, closer to the subduction zone, shallow crustal magma reservoirs erupt silicic products in the form of caldera volcanism at the Taupō Zone (see Chap. 6).

4.1.2 Atypical Subduction Zones

Mixed or irregular subduction settings also contain active or dormant composite volcanoes. The Mediterranean subduction system (see Chap. 1) includes, among others, the Aegean archipelago, whose most prominent composite volcano, Stromboli, rises to almost 3000 m from the seafloor (Favalli et al. 2005; Zamboni et al. 2016). In the archipelago, atypical subduction of the Ionian Plate is characterised by both arc- and intraplate geochemical signatures, and different types of volcanoes (lava domes, calderas) coexist with composite cones. Some active volcanoes of the Appennine Peninsula, which has significant internal extension and high-K eruption products (Brancolini et al. 2019) also fit in here (e.g. Somma-Vesuvius, see Sects. 5.3 and 6). As mentioned in Chap. 1, Etna in Sicily has formed in an unusual, hybrid tectonic setting as a volcano complex (Photo 53). Although voluminous lava flows along extended rift zones mean that Etna resembles a 'shield-like' volcano, explosive eruptions (commonly Strombolian but sometimes even Plinian) have also occurred in its history, i.e. showing a transition toward a composite landform.

4.1.3 Collision Zones

Volcanism linked to collision- or extinct subduction-related settings is perhaps best known in southwest Asia, in the Alpine-Himalaya orogenic belt. Magmatism in Anatolia occurred due to the convergence and collision between the African, Arabian and Eurasian plates in Miocene times. Its famous composite volcanoes lie next to extensive ignimbrites (Ilbeyli et al. 2004). For instance, Ercyies Dağı and Hasan Dağı are large compound volcanoes with both calc-alkaline and alkaline compositions that have been active in the past <1 Ma (Şen et al. 2003). Eastward, the even larger, double-peaked calc-alkaline Ararat (Ağri Daği)—the highest volcano in Turkey, close to the Armenian border—still shows fumarolic activity; the higher Greater Ararat is a compound edifice, while the Little Ararat has a regular-shaped cone. Northeastward, the Causasus Mountains comprise at least 70 Quaternary volcanic centres, with two dozen well-preserved composite volcanoes, mostly effusive cones (Lebedev and Vashakidze 2014), of which the two-peaked, calc-alkaline dacitic Elbrus, the highest point in Europe (5642 m), stands out.

South of the Caspian Sea, Damávand, Iran (5670 m: Photo 65), is not only even higher, but is the most voluminous (400 km^3) volcanoes in the collision zone of Alborz Mountains (Fig. 2; Momeni and Madariaga 2022). The compound, trachyandesitic edifice is currently dormant: it has ongoing fumarole activity but there is no evidence of any eruption since the emplacement of ~7 ka lava flows. The present-day cone, the irregular-shaped Young Damávand, has been constructed in the past ~550 ka on the partially collapsed edifice of Old Damávand (1.2–0.8 Ma: Davidson et al. 2004).

Southward of the Alborz, the Zagros Mountains also represent collision belts, e.g. Saray, Sahand, and Sabalan, which are deeply eroded Miocene–Quaternary composite volcanoes a few hundred km south of the Causasus (Ghalamghash et al. 2019). The Makran continental arc, close to the Arabian sea, still displays active subduction but with no active volcanoes: e.g. the symmetric Bazman, and the twin-peaked, irregular-shaped, more eroded Taftan, which are composite volcanoes with Late Quaternary activity (Delavari et al. 2022).

The Inner Carpathians contain some still recognisable Mid-Miocene to Late Quaternary calc-alkaline andesitic-dacitic volcanoes (Szakács and Seghedi 1995; Karátson 1996; Lexa et al. 2010). This region represents typical collision volcanism where the original subduction pattern is no longer clearly defined, as the initial deep-sea trench has long disappeared. Its latest eruptive period created the prolonged volcanism from 850 to 28 ka of Ciomadul (Csomád) dacitic lava dome complex in Romania (Karátson et al. 2022: Photo 66).

Of other collision settings worldwide, Papua New Guinea, located in a complicated convergence zone involving the Australian and a number of micro-Pacific Plates, has volcanoes in a peculiar tectonic setting. In addition to subduction-related volcanoes (e.g. Lamington with its breached summit crater and post-1951 lava dome), which include nearby New Britain, the inner highlands of Papua New Guinea contain extinct (Late Pleistocene?) composite volcanoes which may have been emplaced in a microplate collision setting (Holm et al. 2016; Liu et al. 2023). Bosavi—a heavily dissected but relatively well preserved, truncated regular cone with a 35–40 km basal diameter and a 5 km-wide erosionally modified caldera (see Chap. 8)—is probably the youngest of more than a dozen slightly or deeply eroded composite volcanoes scattered across the Papuan Thrust and Fold Belt (Löffler 1977).

4.2 Occurrences in Intraplate and Rift Zone Settings

Another major realm where composite volcanoes occur is in the interior of lithospheric plates. Typically associated with grabens and deep faults, they can be found on almost all continents as well as in oceans, often in relation to hot spots or other mantle magma sources. As hot spots can be responsible for continental breakup, i.e. rifting (see Chap. 1), composite volcanoes are often found at continental rift zones (e.g. EARS) and mid-ocean ridges associated with hot spots (e.g. Iceland).

Similarly to subduction zones, the magma that makes up the composite volcanoes may be also differentiated in this continental setting, often being bimodal, and commonly alkali-rich. Differentiated magmas can produce both explosive eruptions as well as lava flows and domes. As we saw in the case of Nyiragongo (Chap. 4), the formation of a composite cone shape as opposed to a shield morphology depends on whether the conduit is capable of producing, in addition to lava flows, cone-building explosive products from a single (or focused) central vent. As with the composite volcanoes in subduction zones, intraplate and rift volcanoes also frequently undergo caldera formation, hence the original conical shape is often truncated.

4.2.1 Intraplate Settings Associated with Hot Spots and/or Mid-Ocean Ridges

In the Atlantic region, from north to south, Iceland, the Azores, the Canary and Cape Verde Islands, and Tristan da Cunha are worth noting. They represent the full range of hot-spot settings, some of which are associated with mid-ocean ridges (for details on geodynamics, see Chap. 1). Here, we look at the link between tectonic and magmatic activity to see how composite edifices, which may coexist with shield volcanoes, are produced.

Volcanism on Iceland is due to the Icelandic hot spot coinciding with the Mid-Atlantic Ridge. In comparison with Iceland, the Azores is a larger, more diffuse submarine plateau, with elevated WNW-ESE trending ridges that host nine, mostly Quaternary age islands (Kueppers and Beier 2018). These have been fed by the Azores hot spot, which is overall characterized by a smaller magma production rate than Iceland, and is less well defined. As a consequence, the Azores Islands comprise mostly typical composite volcanoes, whereas in Iceland, as we saw in Chap. 4, fissure eruptions and shield volcanoes of all sizes predominate.

However, composite volcanoes and calderas also exist on Iceland. Among them, we find steep (up to 30°) regular-shaped edifices: e.g. Öræfajökull, the highest volcano (2119 m:

Photo 67); or elongated ones, e.g. Hekla (see below). Apart from these, quite a number of other, large 'central volcanoes' (Thordarson and Larsen 2007) which have less steep slopes (4–8°) and are truncated by a broad summit with 7–15 km-wide calderas, represent a specific class of calderas (e.g. Grímsvötn, Torfajökull, Krafla: see Chap. 6). One of the most known composite volcanoes of the eastern rift zone of Iceland, renowned for its 2010 eruption, is Eyjafjallajökull. It has an elongated ice-covered edifice with a 2.5 km-wide summit caldera. Following historical eruptions of little significance, increased activity from lava fountaining to phreatomagmatic explosions between 20 March and 23 June 2010 induced melting of portions of the ice cap (Höskuldsson et al. 2010). The summit crater released fine-grained ash that drifted eastward, and remained suspended in the atmosphere for weeks. The ash passed directly over Europe, halting flights for nearly a week in a shutdown situation, resulting in significant economic loss. 40 km northward, Hekla erupts basaltic andesites, in contrast to the tholeiitic basalts that are common to the rift-zone volcanoes. However, it has also had numerous dacitic explosive eruptions during historical times that have deposited tephra throughout Iceland, creating valuable marker horizons. The two magma types behind these eruptions are unrelated: basaltic andesite evolved from basalt through crystal fractionation, and dacite formed by crustal melting, the two mixing together to produce the Hekla andesites (Gudmundsdóttir et al. 2011).

To the north of Iceland, the basaltic composite cone of Beerenberg on Jan Mayen Island (Norway), the northernmost volcano on Earth, is located in an unusual tectonic setting at the intersection of the Jan Mayen Fracture Zone (a transform fault) and the mid-ocean ridge. It has had historical flank eruptions, the last of which was in 1985 (Gjerløw et al. 2016).

The existence of composite volcanoes (with or without calderas) in the Iceland region, as well as in the Azores, is due to evolved, silicic magmatism. However, the magma generation mechanisms at the two places are somewhat different. In Iceland, which has a thickened crust, silicic magmas can erupt due to either assimilation of hydrothermally altered upper crust or magma differentiation in the lower crust (Schattel et al. 2014). In contrast, silicic magmatism in the Azores has been fed by a heterogeneous and less hot, hydrous mantle source, in other words, a 'wetspot' (Beier et al. 2015).

The volcanoes of the Azores are mostly of Quaternary age (2 Ma to present, except Santa Maria Island which is 6 Ma). They have erupted a wide range of magmas from basalt to trachyte. The diversity of the islands is marked by sixteen large composite or shield volcanoes (in addition to basaltic fissure vents), of which nine are considered to be active or dormant. Most volcanoes are relatively small and a number of them are truncated by calderas. Out of the nine islands (with one or more central volcano each), six have erupted evolved silicic, typically alkaline magmas—often explosive and caldera-forming—in addition to mafic (basaltic) ones that produced a predominance of lava flows. Corvo, one of the two islands located to the west of the Mid-Atlantic Ridge, is a small composite volcano with a basal (submarine) diameter of ~8 km, showing shield volcano evolution followed by explosive activity and collapse of a 2 km-wide caldera (Kueppers and Beier 2018; Bossis et al. 2023). Faial or Fayal is the westernmost of the six eastern islands: it has an andesitic/trachytic composite cone shape and is truncated by a central caldera of the same size as that of Corvo. Faial is famous for its submarine Capelinhos eruption (1957–58, Chap. 3) that created a new island which merged with the western peninsula of the island. To the east is the island of Pico which hosts the highest and the most regular-shaped composite cone of the Azores, a predominantly basaltic, symmetric edifice with a 600 m-wide summit crater (Fig. 2). Further eastward on São Miguel Island, Sete Cidades has a 5 km-wide caldera; although it is classified as a 'stratovolcano' in the Global Volcanism Program, it has never had a prominent composite edifice. Its first-stage, long-lived (200 kyr) shield volcano started to produce trachytic explosive eruptions, and eventually fresh basaltic injections that resulted in large caldera-forming eruption(s) (Beier et al. 2006).

In the Canary Islands, as presented in Chap. 4, the lack of subsidence and the prolonged volcanism (>10 Ma in the central and eastern islands) may be responsible for the additional generation of silicic magmatism (Carracedo and Troll 2016) which led to the creation of composite volcanoes during late-stage (rejuvenated) island evolution.

The Cape Verde Islands comprise ten islands ranging from Miocene to present in age, of which only Fogo (a 25 km-wide island) is active. The island is constituted of alkaline basalts and carbonatites of a basal, submarine to subaerial shield volcano of Pliocene–Quaternary age, overlain by a Late Quaternary shield (which collapsed to the east at ~70 ka), capped by the tephritic-basanitic to phonotephritic present-day edifice which has been active from 30 ka (Cornu et al. 2021). Current eruptions occur at this latter, 'rejuvenated' volcano of the same name: Fogo, which is a symmetric composite cone that has grown up within the 9-km-wide horseshoe scar (Cha Caldera) related to the above-mentioned flank failure. Interestingly, since the island is affected by north-easterly trade winds, most precipitation falls on the eastern flanks and thus those volcano slopes are steeper and more intensely gullied (Martínez-Moreno et al. 2018). Studies on the high volatile content of Fogo's magma led to the finding that magma storage below the volcano is ~20 to 30 km, and thus explosive eruptions are fed directly from the mantle. This way, mafic eruptions are sustained by the high volatile concentrations, and the deep CO_2 exsolution drives magma ascent (DeVitre et al. 2023). These eruptions are frequent but not very explosive (i.e. Hawaiian, Strombolian), so thousands of people live in the fertile caldera. However, the lava flows can lead to disruption of the land use; for example, in 1995, 5000 people were evacuated or left their homes, while the 2014–2015 eruption destroyed two villages.

In the South Atlantic Ocean, the circular, 12 km-wide island of Tristan da Cunha (British Overseas Territory), some 400 km to the east of the Mid-Atlantic Ridge, is the most recent volcano related to the Walvis Hot Spot (Geissler et al. 2020). This symmetrical island is a conical edifice, crowned by Queen Mary's peak (2060 m) with a 300 m-wide summit crater (Fig. 2), but rising to a total of ~5500 m from the seafloor. Classification of the volcano is ambiguous. The Global Volcanism Program includes it among the shield volcanoes; on the contrary, the British Geological Survey and the Oregon State University websites (as well as Cas and Wright 1988) classify Tristan as a strato- or composite volcano. Hicks et al. (2012) described the edifice as "a complex volcano in its postshield phase consisting of interlayered lava and pyroclastic flows". In fact, the conical upper part of the volcano has a cone shape, bounded by 25–30° slopes above ~800 m, in contrast to the lower flanks which flatten out to 8° (Dunkley 2002). This morphology would be in line with a composite volcano in the rejuvenation stage, as seen at Tenerife or Fogo. Tristan da Cunha's eruptions have lasted from >1 Ma to the present day, producing basanite, tephrite and trachyandesite. Most of the subaerial activity occurred in the past ~100 ka (Hicks et al. 2012), but there have only been two historical eruptions: around 1700 and a more recent, minor flank eruption in 1961–62.

As for the Pacific Ocean, apart from the subduction-related occurrences, volcanic intraplate islands do not have composite volcanoes due to the fast plate movement over hot spots. However, there are a few examples in the Indian Ocean. The active Heard Island (Mawson Peak: 2345 m), along with the McDonald Islands, lies on the southwestern part of the Kerguelen Plateau. Heard Island is the emergent portion of two volcanoes: the glacier-clad basaltic-trachytic Big Ben composite volcano that makes up most of the island, and the smaller Mount Dixon at the northwestern peninsula. Volcanism started within the past 0.5 Ma, and Big Ben's historically active Mawson Peak (which has had ongoing eruptions since 2010) rises up within a 5–6 km-wide horseshoe scar (Duncan et al. 2016).

There are a number of intraplate occurrences on the continents as well, including those two related to the European Cenozoic Rift System:

(1) The Massif Central (Auvergne, France) erupted alkaline rocks (from basanite and phonolite to basalt, trachyte, latite and rhyolite) from Miocene to Late Quaternary times. In addition to the Late Pleistocene and Holocene Chaîne des Puys (Chap. 1), it comprises two large composite volcanoes. Mont Cantal, a large, symmetric edifice is a volcanic cluster, with a basal diameter of 50×70 km and an area of 2500 km^2, and is considered as being the largest volcano ever in Europe, which went extinct in the Late Pliocene. Originally, it may have been twice as large as Etna with a volume of >400 km^3 and an altitude of 3500 m (Nehlig et al. 2003), although it has now been reduced by flank failures and glaciations. The Monts Dore consists of two adjacent, partly overlapping edifices, the older Guéry (3.2–1.5 Ma) and the younger (1.1–0.2 Ma) and higher Sancy volcano (Nomade et al. 2012).
(2) To the northeast, the Late Miocene Kaiserstuhl composite volcano is part of the diverse pattern of rift-related volcanoes in the Upper Rhine Graben (see Chap. 4).

There are both shield and composite volcanoes in the South China/North Korea border region, of which three have undergone historical eruptions. These and a number of other centres in Korea can be found in a back-arc region related to the subduction of the Pacific Plate (Fan et al. 2020). The most famous of them, Changbaisan (in Chinese; Paektusan in Korean), is a composite volcano containing a caldera lake named Lake Tianchi (Heaven Lake), whose present form was established during a VEI 7 explosive eruption in AD/CE 946 (Table 1; Pan et al. 2017; Yang et al. 2021).

Looking at the interior of other continents, we can highlight two occurrences of composite volcanoes in North America:

(1) In Canada, the extension- or rift-related Northern Cordillera Volcanic Province (Edwards and Russell 2000) comprises two compound edifices. Of them, the basaltic-trachytic-rhyolitic Edziza has a complex structure, truncated by a 2 km-wide summit caldera and complicated by lava domes, stretching over a shield volcano plateau (Souther 1992). Volcanism has lasted from the late Miocene to subglacial eruptions in the Holocene (with last eruption at ~AD/CE 950).
(2) Far to the south, associated with the Colorado Plateau in Utah, Colorado, Arizona, and New Mexico (USA)—an area that experienced extensive Neogene volcanism—, is a series of monogenetic volcanic fields including composite volcanoes and lava domes of various sizes (e.g. San Francisco, Springerville, Zuni, Mt. Taylor, Mt. Baldy volcanic fields). The andesitic-dacitic San Francisco Peaks (Arizona) erupted between 0.6 and 0.4 Ma (Karátson et al. 2010b), and is the largest of five composite volcanoes in the San Francisco Volcanic Field. Truncated to the east by its Inner Basin—a horseshoe-shaped depression originating from either flank failure or glaciation—, its original dimensions have been significantly reduced. It was originally a ~4400 m-high, two-peaked cone with a volume of about 80 km^3 (Fig. 10A and B), of which ~7.5 km^3 was removed by the flank collapse and/or subsequent glacial erosion, creating the present-day enlarged basin, and a further ~2 km^3 has been removed from the outer valleys by fluvial erosion (Karátson et al. 2010b). Mount Taylor is a Late Pliocene, even more deeply eroded, symmetric, lava-dominated composite volcano (Goff et al. 2021). It has a peculiar, elongated, erosionally enlarged central depression shaped by long-term fluvial erosion (see Chap. 8).

4.2.2 Continental Rift Setting

This setting also allows magma differentiation to occur, permitting the creation of composite volcanoes. As presented in Chaps. 1 and 4, in particular, the East African Rift System (EARS) is a magnificent model area for the process of continental rifting, diverse magma formation, and the emplacement of various volcano types.

In the following, peculiarities of two of the composite volcanoes are highlighted.

Ol Doinyo Lengai (Tanzania) is the only active volcano on Earth to emit carbonatites. Its conical edifice (Fig. 2) rises above Lake Natron, in the Gregory Rift Valley to the south. As the alkali-rich natrocarbonatite magma (450 °C) contains little silica, it has a relatively low viscosity, hence carbonatite lavas are among the fastest flowing. However, Lengai had a long history (dating back to 350 ka: Sherrod et al. 2013) with changing composition. A cone-building phonolitic stage occurred prior to 11 ka, terminated by a sector collapse, after which the younger cone has grown up, composed mainly of nephelinite. Hence, the superimposition of the two edifices makes the cone shape slightly irregular. Historical eruptions (recorded since 1880) consist of ~30 years-long cycles, characterized by carbonatite effusion punctuated by subplinian explosions. These latter produce magma of both nephelinitic/phonolitic and carbonatitic composition, and are due to the rise in H_2O in the melt over a 30-yr time scale, ultimately resulting in H_2O vapour saturation that triggers explosive eruption (Lundstrom et al. 2022). The most recent eruptive cycle which began in 1983 involved natrocarbonatite lava flows and culminated with explosive eruptions in 2007–2008; since then, lava effusion has been practically continuous. Lengai has two craters, of which the northern, active one has varied in form during historic times, either being surrounded by 200 m-deep crater walls or completely filled by lava.

Meru (also in Tanzania) has a conical shape, too, but it is breached by a horseshoe-shaped scar resulting from a flank failure that produced the Momella debris avalanche, spreading over an area of 1250 km^2 as far as Kilimanjaro (Delcamp et al. 2017; also see Chap. 7). The volcano has had at least three explosive eruptions up to VEI 5 over the past 40 ka (Kisaka et al. 2021), in addition to a more recent minor explosion in 1910.

Finally, rift-related volcanoes also occur in Antarctica. Off the coast of West Antarctica, Ross Island is composed of four composite volcanoes, all related to the West Antarctic Rift System (Chap. 1), which is comparable in size to the EARS. Of the Ross Island volcanoes, Erebus (3794 m) is the only active one, and is the southernmost erupting volcano on Earth. It had an early submarine stage at ~1.3 Ma, producing basal submarine basanite lavas, followed by a subaerial composite edifice of more differentiated phonolitic composition (Oppenheimer and Kyle 2008; Iverson et al. 2014). Due to the relatively higher viscosity of phonolite, the upper part is steeper. Erebus has had two caldera collapses in the past 100 ka, possibly associated with large explosive eruptions. The late-stage caldera is marked by a summit plateau at 3200 m, on which the active cone has grown. This cone has a double summit crater, the inner one containing a long-term phonolitic lava lake (Chap. 2), unusually rich in CO_2 which produces Strombolian and phreatomagmatic eruptions (Hill et al. 2022). Another volcano, Mount Melbourne, is situated about 300 km away. It is a slightly elongated but overall more regular-shaped composite cone about a thousand m lower in altitude than Erebus, which evolved from subglacial to subaerial activity <123 ka. Typified by mafic effusive activity, it also erupts trachyte and rhyolite during occasional Plinian eruptions (Giordano et al. 2012).

5 Characterization of Main Landform Types

Here, based on the lessons learned from various settings and composite volcano characteristics presented in the previous section, we describe the four main composite volcano types in detail (Table 3). We focus on their geomorphic evolution, and incorporate information from typical and atypical examples. We also present a comparative plot (Fig. 11) which displays volcano volume, mean slope of main flank, and roughly indicates the trend in shape complexity and geochemical affinity (calc-alkaline versus alkaline). Overall, alkaline volcanoes have a gentler slope due to the lower viscosity of alkaline/peralkaline magmas (see discussion in Cas and Wright 1988), but slope pattern is also controlled by

Table 3 Morphometric, geomorphic and structural features of the four main types of composite cones/volcanoes presented in this chapter

Composite cone type or subtype	Volume (km^3)	Diameter (subaerial), (km)	H/W	Mean (flank) slope (°)	Major features	Magma composition
Simple cones						
Regular cone *Examples*: <u>Small</u>: Irruputuncu, Momotombo <u>Large</u>: Kronotsky, Sangay	1–70	4–30	>0.15 (up to ~0.22)	>20 (up to ~30°)	Concave to linear slopes; central crater common; adventive flank vents; calc-alkaline rocks tend to produce steeper flanks than alkaline ones	Calc-alkaline basalt to andesite; minor dacite-rhyolite; less frequent alkaline rocks (e.g. trachyte, phonolite)
Irregular cone *Examples*: Lascar, Damávand	5–100	7–35	0.10–0.15	15–20°	Concave to irregular slopes; central crater(s) common; elongated/deformed slopes common; adventitive flank vents	Calc-alkaline basalt (rare) to andesite; additionally, dacite, rhyolite; less frequent alkaline rocks (e.g. trachyte, phonolite)
Truncated cones						
Examples: <u>Small</u>: Augustine, Zarechny <u>Large</u>: Shiveluch, Tambora	5–150	10–40	not differentiated		Concave to irregular slopes; truncated central crater or caldera; adventitive flank vents	Same as above
Somma volcanoes						
Small: Vesuvius, Cosigüina **Large**: Veniaminof	10–400	10–40	not differentiated		Concave to irregular slopes; central caldera; post-caldera vents (cones, lava domes); adventitive flank vents	Same as above
Compound cones						
Small: Las Pilas, Telica **Large**: Elbrus	20–1000	10–60	<0.10	<15°	Concave to irregular slopes; amalgamated cones common; central craters, caldera(s); summit plateau; post-caldera cones, lava domes; adventitive flank vents	Same as above

Morphometry values are taken (or averaged) from Grosse et al. (2009, 2012), lithology from Global Volcanism Program and a number of references specified in main text

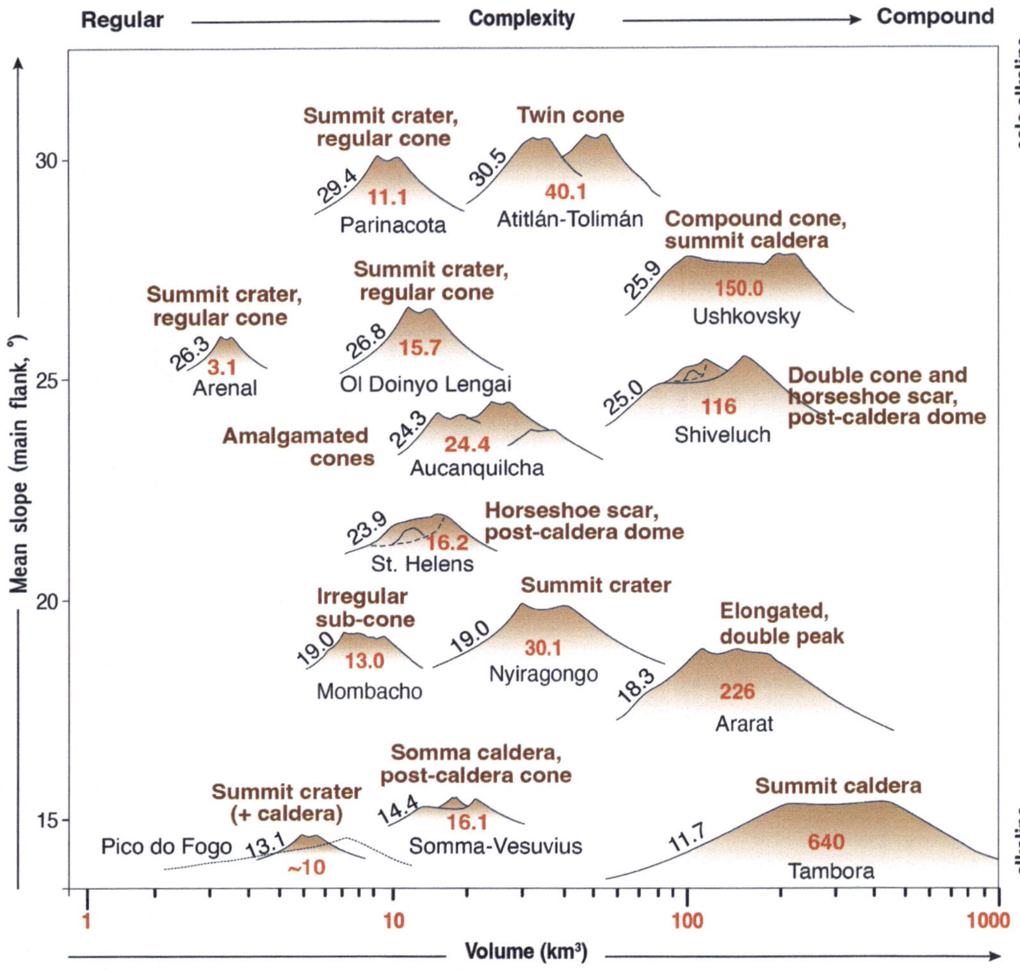

Fig. 11 Composite cone types with selected examples defined by edifice volume, mean slope of main flank, overall complexity of shape, and geochemical affinity (shown in general). Morphometric data after Grosse et al. (2014)

eruptive activity. For example, enhanced explosivity can result in steep upper flanks like those of the nephelinitic/carbonatitic Ol Doinyo Lengai or the ultra-alkaline Nyiragongo. The same-side axes of the plot are not necessarily related in other respects: for example, increasing volume does not always correlate with increasing complexity. A good example is Tambora (see Sect. 5.2).

5.1 Simple Cones

5.1.1 Regular Cone

Composite cones may initially evolve from unusual scoria cones, which start to effuse relatively large amounts of lava, and most characteristically produce violent explosions (Davidson and de Silva 2000). Good examples are Izalco (El Salvador) and the compound Cerro Negro

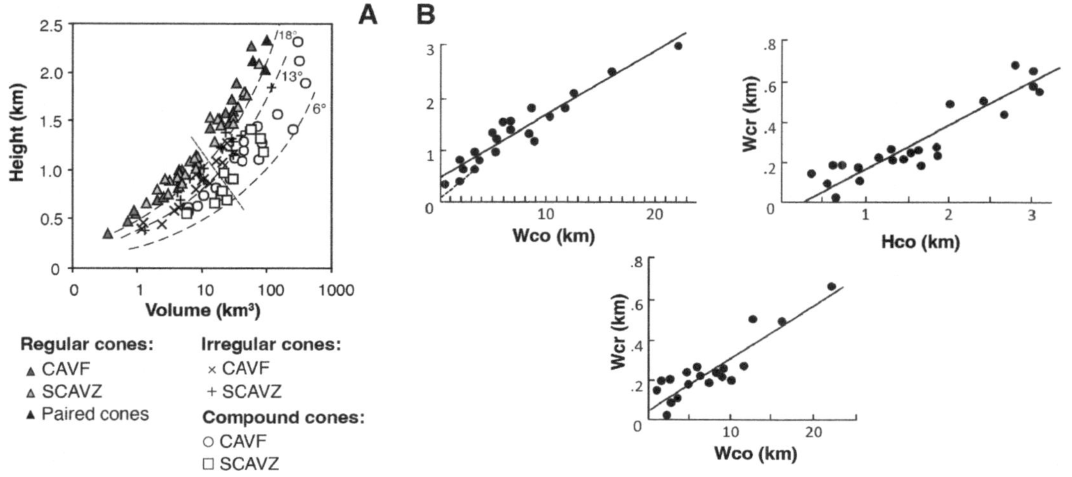

Fig. 12 Morphometry of composite cones. **A** Height (H_{co}) versus volume (V) of regular, irregular and compound cones from the Central American Volcanic Arc (CAVF) and the southern CVZ of the Andes (SCAVZ) (after Grosse et al. 2009). **B** H_{co}/W_{co}, W_{cr}/H_{co}, and W_{cr}/W_{co} ratios of selected Circum-Pacific composite cones (after Wood 1978)

(Nicaragua; see Chap. 3). Mount McLoughlin, a dissected cone in the Cascades, seems to encapsulate a scoria cone (Harris 2005). As discussed in Chap. 3, the key factor, apart from issuing frequent lava flows, is probably the change in explosive eruptive style toward Vulcanian or even subplinian due, for example, to an increased volatile content in the magma, as for Cerro Negro. In addition to the changing eruptive style, eruptive rates may also increase. At the same time, the maintenance of a conical shape, as demonstrated previously, is due to the small-scale but intense lava and tephra fountaining alternating with short cone-building lava flows. This early cone-building stage is in line with the fact that the overwhelming majority of conical composite volcanoes are typically made up of basalt, basaltic andesite, and andesite.

However, the size of cones formed this way is highly variable, ranging from the 'embryonic' ones to overgrown giants such as Fuji (see Fig. 3). Grosse et al. (2009) analyzed ~40 regular-shaped composite cones from the Central American Volcanic Arc and the southern CVZ (Andes), which have relative heights ranging between 350 and 2250 m and volumes of between <1 and 75 km^3 (Fig. 12A). The highest Andean cones, like Parinacota and Licancábur, with a relative height of 2–2.5 km above the Altiplano, are smaller than many others either along arcs (e.g. Klyuchevskoy) or within plates (e.g. Damávand), the latter approaching 3.5 km in height (Grosse et al. 2012). Recently, Castruccio et al. (2017) studied composite volcano morphometry with regard to the location and size of magma reservoir, and proposed that the reservoir geometry and physical conditions (size, depth, overpressure, and effusion rate) control the size of the volcano that builds up above it. Large volcanoes (>2000 m relative height, >10 km cone radius, i.e. corresponding to the largest composite cones) are usually basaltic systems with overpressured magma sources >15 km deep, while smaller volcanoes (<2000 m height, <10 km cone radius) are fed by more evolved but shallower (<10 km) magma reservoirs.

Basal diameter of regular composite cones ranges from a few to 20–30 km (Table 3; Wood 1978; Grosse et al. 2012). The main area of activity is the summit region, usually with one or more summit craters. Fissure vents and flank pyroclastic cones or lava domes are not uncommon. Crater diameter is generally less than 1 km, typically 300–700 m, but the crater size and morphology often changes during the active period, as new craters may open, merge and nest into each other.

The increase in size of simple composite cones, similar to that of scoria cones, can be demonstrated by morphometric ratios. Based on the basic shape parameters (cone height: H_{co}, cone width: W_{co}, crater width: W_{cr}) of 26 Circum-Pacific composite volcanoes of different sizes, Wood (1978) found linear relationships (Fig. 12B). The H_{co}/W_{co} ratio is about 1:8 (= 0.125), and the W_{cr}/W_{co} is only 1:35 (= 0.029). Grosse et al. (2009) obtained similar H_{co}/W_{co} values for regular-shaped cones (>0.15, Table 3). These constant ratios imply that once the volcano reaches a characteristic 'composite cone' shape, it increases proportionally. As shown in Table 3, fast-growing cones can reach a maximum size within 5–10 kyr.

Both above ratios are significantly lower than the corresponding ratios for scoria cones ($H_{co}/W_{co}=1{:}5$, $W_{cr}/W_{co}=2{:}5$). According to Wood (1978), the morphometric differences—due to increased complexity at the summit, e.g. appearance of twin craters, nested craters (see later)—become observable beyond a cone diameter of only about 2 km, i.e. well under the size range of the largest scoria cones. The differences are mostly due to increased magma productivity: at composite cones, all additional volume has to be distributed across the growing edifice (Davidson and de Silva 2000). Hence, for a given, relatively small addition to the height, there will be an accompanying increase in width. The W_{cr}/W_{co} ratio is even more different to that of scoria cones. This is because the crater diameter does not change for a long time (or at all), so for an enlarging, wide cone with the same crater diameter, the ratio will become progressively lower.

Likewise, the average slope of the growing composite cone gradually decreases with the flattening of the lower flanks and widening of the apron, from mean values of >30° for the smallest (2 km cone diameter) to 15° for the largest (cone diameter exceeding 20 km) for the 26 Circum-Pacific volcanoes studied by Wood (1978). Grosse et al. (2009) obtained similar values for regular cones in the range of 21–34°. However, the upper part of the cone always remains very steep: for the most regular-shaped 19 composite cones in the world, the average slope of the upper flank (about one third of the cone height) is 28°, and the highest slope values occasionally reach 40° (Karátson et al. 2010a).

Vertical growth of the cone generally slows over time. Davidson and de Silva (2000) attribute this decrease to two factors. First, the magma can only be driven up to the vent by the overpressure within the magma reservoir system, and the lithostatic pressure of the increasing weight of the volcano obviously acts against this pressure. The slower upward movement of magma may be reinforced by magma differentiation (leading to increased viscosity). On the other hand, because magma has to travel further to reach the top of the edifice, summit lava flows become rarer, and, in many cases, activity shifts to adventive (mostly flank) cones or fissures. We presented a number of composite cones in the Circum-Pacific belt where this is the case. A good example is the most recent eruption of Fuji (1707 AD/CE, Hōei eruption) which occurred not from the summit crater but from a newly formed, huge flank crater.

Another important aspect of composite cone growth is that while a scoria cone builds up a straight slope due to the friction repose angle of fallout scoria, the composite cone slope profile becomes concave. Such a slope, which may persist for a long period, is called an 'equilibrium slope' by Davidson and de Silva (2000), and represents a balance between construction (mass addition by eruptions) and degradation (mass redistribution through erosion). One reason for the concave profile is simply the greater travel distance of material deposited or re-deposited along the flank, as the redistributing eruptive and erosional processes that accumulate material downslope are more efficient when the cone is higher and has a wider base. Ultimately, the significant increase in height and the efficiency of mass movements downslope (as on other mountain slopes) result in a concave profile for most symmetrical composite cones as they develop.

Formation of the concave profile can also be studied in terms of morphometry. A long time ago, Milne (1878) concluded that the slope of regular cones is best described by a logarithmic function. The volume of such a cone can be integrated by its radius expressed as:

$$r = Be^{Mh}$$

where e is Euler's number, B and M are constant values, and h is the relative height of the volcano (also see Francis 1993). Concentrating on the most regular-shaped examples, Karátson et al. (2010a) used SRTM DEM data to show that the nineteen simple cones selected had a very high degree of circular symmetry (Fig. 13). If we consider the average slope of these extremely regular volcanoes, a relationship between the height and radius of the volcano for the lower half to two-thirds of all cones is similar to that found by Milne; the slope profile of the composite cone does indeed follow a logarithm in the wider, lowermost part.

However, looking at the upper part of the average slope profile of the most regular cones, two groups can be distinguished (Fig. 14A). One group of volcanoes (e.g. Mayon in the Philippines, Villarrica in Chile: Photo 68, Kronotsky in Kamchatka) has a concave slope for the upper third of the flank, but the slope profile is close to a parabolic curve, and has therefore been termed a 'P-type' volcano by Karátson et al. (2010a). The second group (e.g. Parinacota, Licancábur in the Central Andes) displays, on the contrary, a straight slope for the upper third of the cone flank up to the summit, i.e. a true cone ('conical' shape), termed a 'C-type' volcano. Notably, the magma composition of these regular cones is very similar (predominantly basalt, basaltic andesite, andesite). The slight but possibly systematic difference between the two groups may be that C-type composite volcanoes have a slightly more evolved magma composition with higher SiO_2 and/or higher water content (Fig. 14B), probably resulting in a higher frequency of small explosions (in addition to large ones). Accordingly, the upper third of the cone flank seems to made up of more frequent lava fountaining and/or scoria fall: this could be interpreted as a small, straight-sided scoria cone growing on top of the edifice. In contrast, for P-type volcanoes, the frequency of small explosions seems to be lower and the frequency of lava flows is higher. As a consequence, the upper cone is shaped more by lava flows, whose material is emplaced onto the downslope region of the cone, resulting in an overall concave ('equilibrium') slope.

Of course, there are transitions between these two geomorphic subtypes, both in terms of slope pattern and volcanic activity; and the eruptive characteristics can vary over the lifetime of a volcano: the morphometry observed at a given moment reflects the prevailing activity during the period that has just taken place. Although this model needs further refinement, it does point out that subtle differences in the eruptive activity of volcanoes can be captured successfully by morphometry.

5.1.2 Irregular Cone

Longer lifetimes (tens of 10 kyr) of composite cones often result in a more irregular shape, although a cone can be asymmetric from the very beginning. Modification to the shape might be induced by edifice growth on an uneven or tilted substrate, shifting and multiplication of summit vents, appearance of lava domes or cryptodomes, distortion of shape due to the changing stress field and so on (Davidson and de Silva 2000; Grosse et al. 2009; Aoki et al. 2019). The simplest geomorphic change is due to the appearance of minor adventive (or satellite) centres, e.g. scoria cones, typically scattered around the flanks of the composite volcano. These might erupt different material from the eruptive products of the main vent in terms of magma composition, possibly fed by distinct, sometimes independent magma batches. While satellite vents have little effect on the regular conical shape of the composite edifice (Fig. 4), lava domes and cryptodomes can make it more irregular. Another obvious way of deforming a regular cone is by tectonism, as exemplified by the Damávand volcano. The asymmetry of its younger cone can be explained by deformation due to pronounced, gravity-driven spreading toward the east, along with subsidence of the summit region, in the order of 5 mm/year (Shirzaei et al. 2011).

Obviously, the geomorphic complications result in morphometric changes as well. Grosse et al. (2009) found that approximately 20

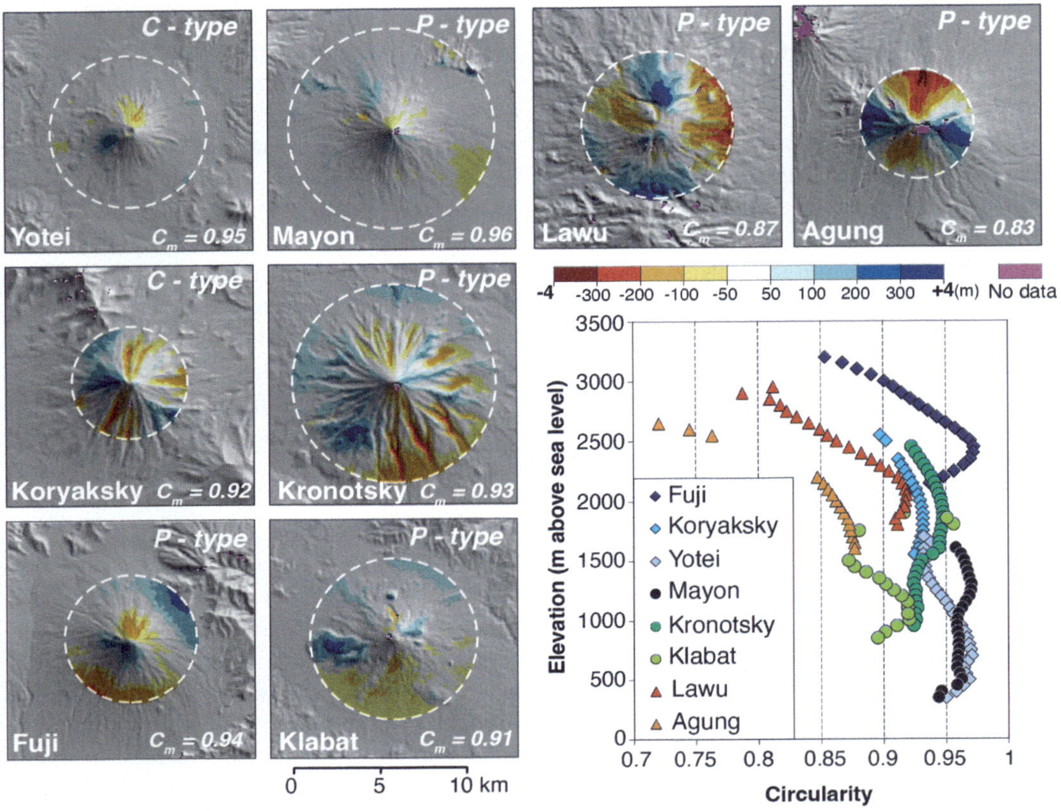

Fig. 13 Circular symmetry of selected composite cones. The most regular volcanoes (e.g. Mayon, Yotei, Kronotsky) have a circular symmetry with a value ≥ 0.9. Irregular, but still simple cones can show values of 0.7–0.8 (for an improved definition and assessment of circularity, see Karátson et al. 2010a)

studied irregular composite volcanoes along the Central American Volcanic Arc have somewhat lower H_{co}/W_{co} ratios (0.10–0.15) than regular cones (Table 1, Fig. 12A). Likewise, irregular cones have slightly lower mean slope values (15–20°) relative to those of regular cones (Grosse et al. 2009). Note that all these are average values, and the upper flanks can remain steep despite an often wider summit area and the geomorphic complications in shape.

5.2 Truncated Cones

During growth, composite volcanoes can be truncated by one of two mechanisms.

One occurs when the edifice becomes unstable due either to the arrival of a new magma, hydrothermal activity, the appearance of a (crypto)dome, or to structural/geomorphic changes like earthquakes, faulting or simply the oversteepening of the growing cone, to the point

Fig. 14 Minor but apparently systematic shape and geochemical differences proposed for regular composite cones (Karátson et al. 2010a). **A** While the average slope profile of the lower two-thirds of all highly symmetric cones follows a logarithmic curve, a parabolic curve delimits the upper flank of P-type volcanoes, and a straight slope characterizes the upper part of C-type volcanoes; examples shown for both groups. **B** SiO_2 vs molar MgO/(MgO + FeO) plot for lavas of nineteen regular composite cones. P-type volcanoes are characterized by lower SiO_2 and H_2O content, which could be the cause of a higher incidence of effusive events, whereas higher SiO_2 and/or higher H_2O content are observed at C-type volcanoes, which could result in a higher incidence of mildly explosive (e.g. Strombolian) eruptions ▶

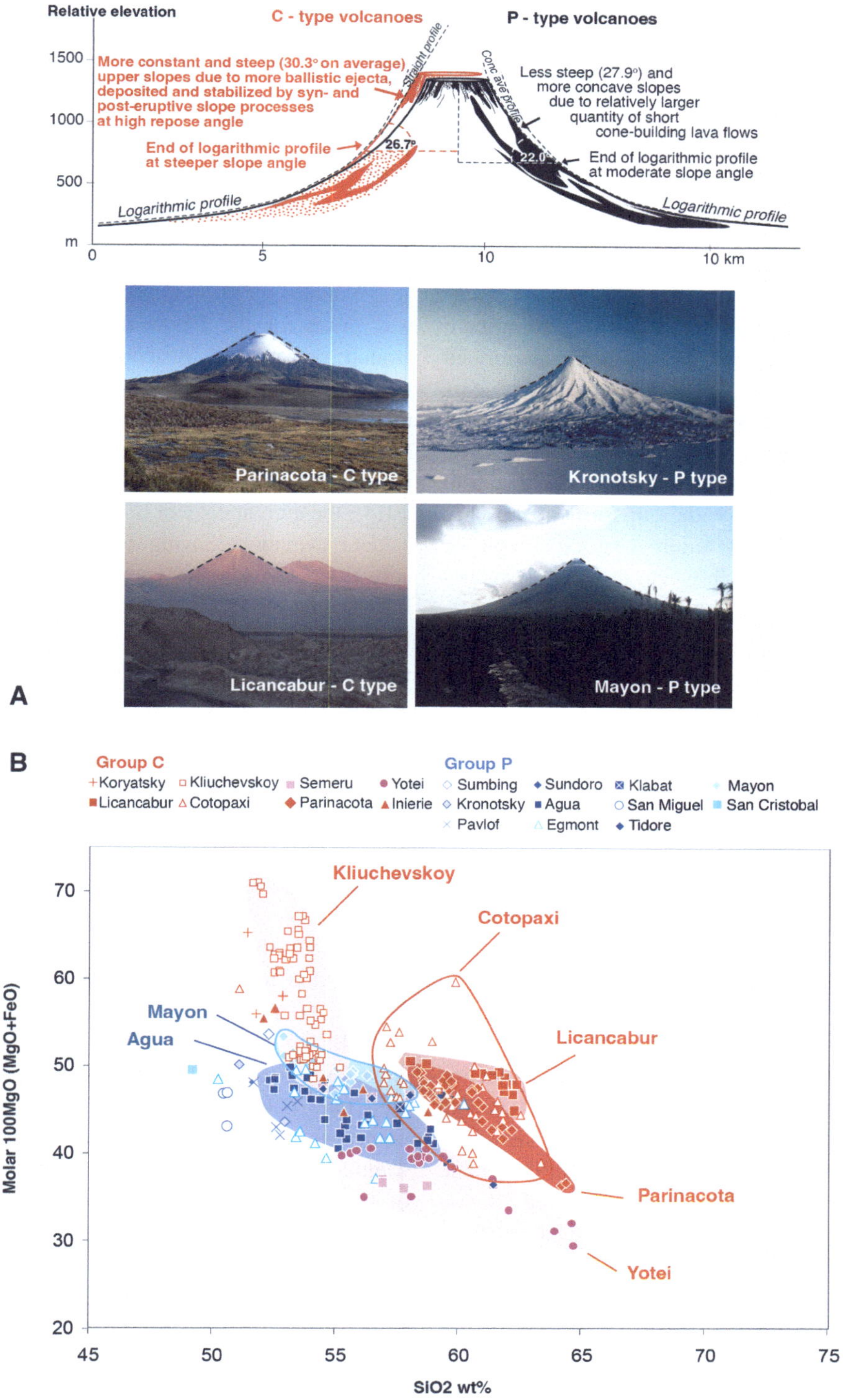
Relative elevation
C - type volcanoes
P - type volcanoes
1500
1000
500
m
More constant and steep (30.3° on average) upper slopes due to more ballistic ejecta, deposited and stabilized by syn- and post-eruptive slope processes at high repose angle
End of logarithmic profile at steeper slope angle
26.7°
Straight profile
Concave profile
Less steep (27.9°) and more concave slopes due to relatively larger quantity of short cone-building lava flows
22.0°
End of logarithmic profile at moderate slope angle
Logarithmic profile
Logarithmic profile
0
5
10
10 km
Parinacota - C type
Kronotsky - P type
Licancabur - C type
Mayon - P type
A
B
Group C
+Koryatsky
Kliuchevskoy
Semeru
Yotei
Licancabur
Cotopaxi
Parinacota
Inierie
Group P
Sumbing
Sundoro
Klabat
Mayon
Kronotsky
Agua
San Miguel
San Cristobal
Pavlof
Egmont
Tidore
Kliuchevskoy
Cotopaxi
Mayon
Agua
Licancabur
Parinacota
Yotei
Molar 100MgO (MgO+FeO)
70
60
50
40
30
20
45
50
55
60
65
70
75
SiO2 wt%

at which the upper flank of the edifice fails and partially collapses producing a volcanic debris avalanche (Chap. 7; for a recent summary, see Roverato et al. 2021). In this case, truncation is asymmetrical. The summit region can become significantly lower: at Mount St. Helens, for example, the 1980 flank failure reduced the volcano height by 400 m. The failure typically results in the formation of a horseshoe-shaped scar. This is an asymmetric depression commonly 1–6 km in diameter, with a significant opening or breach. Of the examples mentioned in Sect. 4, some that show collapse scars include: Redoubt, Augustine, St. Helens (USA); Toluca, Colima (Mexico); Socompa (Chile); Bandai-san (Honshū, Japan). Larger scar diameters up to 6–8 km wide can form in the case of complete summit failure (e.g. Socompa, or Reventador, Ecuador) when the two branches of the horseshoe can be set at an obtuse or perpendicular angle. Subsequently, renewed activity from the same vent often restores the conical shape by filling the scar (e.g. Socompa, Fuji, Taranaki, Bezymianny), and the new cone may even completely 'heal' the scar and cover up the depression (Rainier, Chimborazo, Parinacota).

(2) The other mechanism is summit caldera formation commonly during a subplinian or Plinian eruption, when the (partial) emptying of the magma chamber triggers the (symmetric) collapse of the upper cone. The resulting landform is a medium-sized (2–10 km-wide), generally circular caldera (e.g. Aniakchak, El Chichón, Tambora, etc., see also Fig. 2). Aniakchak, for instance, before its Holocene caldera-forming eruption(s), was formed of the so-called Ancestral Aniakchak edifice, of which only remnants of a glacially sculpted Pleistocene composite volcano have been preserved (Bacon et al. 2014). Tambora has one of the largest summit calderas, with its regular, circular depression 6 km across (Fig. 2). The caldera was formed at the end of the climactic phase of the 1815 eruption (Gertisser et al. 2012), truncating an earlier, slightly elongated composite volcano. Fitting an exponential profile to the volcano flanks with an elliptic edifice base allows reconstruction of the cone to be carried out, indicating a ~3500–3700 m elevation of the original cone (Fig. 10: Favalli et al. 2016).

5.3 Somma Volcanoes

Caldera collapse of a composite volcano can affect larger parts of, or the whole, edifice. In this case, a somma caldera (e.g. Cotton 1944; McBirney and Williams 1979) is created (Fig. 2). The name was given for the toponymous Somma caldera around present-day Vesuvius (Fig. 2), referring to the originally highest point, 'summa' (1281 m a.s.l.: Photo 54). Prior to the eruption ~18 ka, Somma-Vesuvius was preceded by a composite volcano originally 1900 m high (Favalli et al. 2014); the current caldera is result of a multicyclic caldera evolution, having been built up repeatedly in response to the emptying of large, shallow magma reservoirs during four prehistoric and historical Plinian eruptions, the latest being the AD/CE 79 Pompeii eruption (Cioni et al. 1999). Krakatau is another well-known example. In this case, outer islands (Panjang, Sertung) represent an older caldera, whereas the central Rakata is a remnant island—truncated by several hundred metres—left behind after the 1883 explosive eruption as a result of gradual caldera collapse (Madden-Nadeau et al. 2021). However, vertical elevation changes can be more significant. For instance, the regular, circular Crater Lake was formed due to the collapse of Mount Mazama (Photo 73), a composite cone originally ~3700 m high (Bacon and Lanphere 2006; also see Sect. 7 and Chap. 6). Given the current average altitude of the Crater Lake caldera rim (about 2200 m), this represents a cone lowering of 1500 m. Importantly, somma volcanoes can also form from a cluster of pre-existing lava domes. At Kaguyak caldera (Alaska, USA), the original edifice consisted of nine clusters of lava domes, five of which were truncated by a Holocene collapse (Fierstein and Hildreth 2008).

In the case of an irregular original cone shape, the caldera will not be as well-rounded. Irregular-shaped examples include Fisher and Katmai calderas (Alaska, USA). Katmai was formed by the VEI 6 Novarupta eruption (1912) from a complex volcano with four summits (Hildreth and Fierstein 2003). It is more difficult to calculate the original height for these irregularly shaped calderas. Hakone (Honshū, Japan), with a current diameter of 12×8 km and average altitude of ~1000 m, may have been formed by the collapse of an original composite volcano, roughly constrained to a height of 2700 m (Kuno et al. 1970). Further shape variations in calderas include double depressions: the regular cone of Ceboruco (Mexico), for instance, is truncated by two concentric summit calderas.

Following caldera formation at active volcanoes, significant geomorphic changes often occur due to post-caldera rejuvenation, such as the growth of new lava domes or small composite volcanoes. Post-caldera features can be relatively small like those in Crater Lake (e.g. Wizard Island, plus mainly underwater scoria cones and lava domes), but more significant landforms—conspicuous cones—are present in some other well-known occurrences such as Anak Krakatau within the 1883 Krakatau caldera (Indonesia), or Usu at the rim of Toya caldera and Miharayama within Izu-Oshima caldera (both in Japan).

5.4 Compound Cones

Some of the largest composite edifices on Earth are extremely complex in shape. One way of increasing complexity is if the cone is doubled. Twin systems, that is, two neighbouring composite volcanoes, include Gareloi, Pavlof and Pavlov Sister, Fuego and Acatenango, Colima, San Pedro–San Pablo, Parinacota–Pomerape (Nevados de Payachata), Tupungato-Tupungatito, Atitlán-Tolimán, and Ararat–Little Ararat (see Fig. 11). Twin volcanoes often reflect a vent shift such that the younger of the two is active, leaving its older twin extinct. Bezymianny and Klyuchevskoy in Kamchatka, however, are twins which are concurrently active, and form part of the larger Klyuchevskoy Volcanic Group (Ponomareva et al. 2007). Propagation of vents—more than just two—and formation of juxtaposed cones with nested craters is exemplified by Telica compound volcano, Nicaragua. As a result, compound cones are generally more voluminous than simple regular cones, which they may occur directly next to. For example, regular-shaped Cotopaxi (5897 m) is just 350 m lower than its irregular neighbour, the compound Chimborazo (6268 m), but it is much 'slimmer': its volume is 66 km^3, less than half that of Chimborazo (145 km^3; Grosse et al. 2012).

As volcanoes become even more compound, they are represented by amalgamated edifices termed multiple volcanoes (Francis 1993; Grosse et al. 2009) resulting from vent migration over short distances. Examples include Citlaltépetl, Irruputuncu, Lascar and Ojos del Salado in the CVZ (Andes); Ruapehu (New Zealand); Ushkovsky (Kamchatka, Russia); Ercyies Dağı and Hasan Dağı (Turkey). The highest volcano on Earth, Ojos del Salado, occupies an area of ~ 70 km^2 with dozens of cones, craters and lava domes, which can still, however, be considered as a single volcanic edifice, i.e. a uniform volcanic structure, probably with the same magma reservoir or at least shared magmatic evolution (Francis 1993). The compound edifice may have grown within an earlier caldera. Ruapehu, a long-lived (~230 ka) edifice, is made up of four subsequent formations, their distribution being constrained by intense glaciations, and its broad summit displays at least six Holocene centres (Leonard et al. 2021).

The shape of a compound edifice, which tends to be wider relative to the simple cone type, is also characterized by a different morphometry (see Fig. 12A). They have a reduced H_{co}/W_{co} ratio (<0.10) and are less steep (<20°) compared to the regular or irregular composite cones Grosse et al. 2009).

6 Evolution, Lifetime, and Growth Rate of Composite Cones

Two important questions can be asked about the time–space evolution of composite cones. The first concerns the volcano's lifetime, i.e.

how long it remains active; and the second the rhythm of activity, which can be continuous or discontinuous. Accordingly, we can establish time intervals (duration and recurrence) of eruptive periods that occur between repose intervals. Reconstruction of long-term activity and identification of eruptive phases of a volcano can be done by carrying out detailed stratigraphy coupled with geochronology (see Chap. 10) to help determine eruption rates (e.g. for Mount Adams: Hildreth and Lanphere 1994; for Damavánd: Davidson et al. 2004; for Crater Lake/Mount Mazama: Bacon and Lanphere 2006; for Parinacota: Hora et al. 2007).

6.1 Lifetime: Fast Growth of Composite Cones

The question of the lifespan of composite cones can be answered using the information from detailed case studies. Based on the examples referred to above, their lifetime is generally less than 1 Myr, and mostly less than 0.5 Myr (Hildreth and Lanphere 1994); longer lifetimes of some volcanoes results from subsequent activity of long-lived large volcano clusters (Sect. 7). However, as seen above (Table 3), the growth of a composite cone in many cases takes only a few kyr or a few tens of kyr (lifetime of Mount St. Helens: 28 ka; Young Fuji: 10 ka; Parinacota: 9 ka; Klyuchevskoy: 7 ka; Cotopaxi: 4 ka), and in the case of cone collapse, many cones (like Parinacota) 'recover' fairly fast in geological terms to re-build a cone of similar height.

On the other hand, many composite volcanoes have prolonged lifetimes with repeated cone-building phases. If we take the example of Mt. Adams (the eastern neighbour of Mt. St. Helens), its oldest rocks are 940 ka; subsequently, three active cone-building periods (or 'spurts') have punctuated the volcanic evolution, at 520–490, 460–425, and 40–10 ka (Fig. 15; Hildreth and Lanphere 1994). This implies that the volcanic cone was built and evolved over short intermittent stages during the last 0.5 Ma. In particular, in the cone-building stages, the eruption rate could reach 1.6–5 km^3 per kyr ('peak rate'), compared to slow evolution over several millennia when the volcano grew by only 0.05–1 km^3/ka ('background rate'). Another well-studied volcano, Parinacota, dates back to 167 ka for the earliest andesite lavas (Fig. 15; Clavero et al. 2004b; Hora et al. 2007). This was followed by a period of inactivity in the interval from 120 to 60 ka, before there was a build-up of rhyodacite domes and an early composite volcano (52–20 ka). At 9 ka a large part of this volcano was destroyed by a flank failure three times the size of that at Mount St. Helens, and present-day Parinacota was constructed on the remnants of the failed cone. The cone-building rate of the two superimposed volcanoes averages out at <1 km^3/kyr, but the young cone grew ten times faster (Hora et al. 2007); the two cones have a total volume of about 40 km^3 of which the young cone accounts for only a quarter (Karátson et al. 2012).

6.2 Variability in Eruptive Activity

Although only a few composite volcanoes have yet been subjected to detailed geochronological studies, obtained ages show a high variability. Klyuchevskoy has been growing over the past 7 kyr at a remarkably high eruption rate of 27–35 km^3/year (Crisp 1984; Fedotov et al. 2010), creating a corresponding, large edifice volume of 50 km^3. At present, it erupts on average 60 million metric tons of basalt a year, which is 25% of the total annual amount of magma produced by all seventy volcanoes in the Kuril-Kamchatka volcanic belt (Fedotov et al. 2010). Tongariro, on the contrary, is an example of a slow-productivity compound volcano. It has a volume of 90 km^3, and a lifetime of 350 ka, which indicates a long-term average of ~0.25 km^3/kyr, i.e. two orders of magnitude lower than that of Klyuchevskoy (Leonard et al. 2021). Long-term eruptive rates calculated for another compound volcano, San Pedro–Tatara (Chile), are quite similar at 0.2–0.3 km^3/kyr (Singer et al. 1997).

The findings imply that composite volcanoes can operate at quite different eruption rates: at

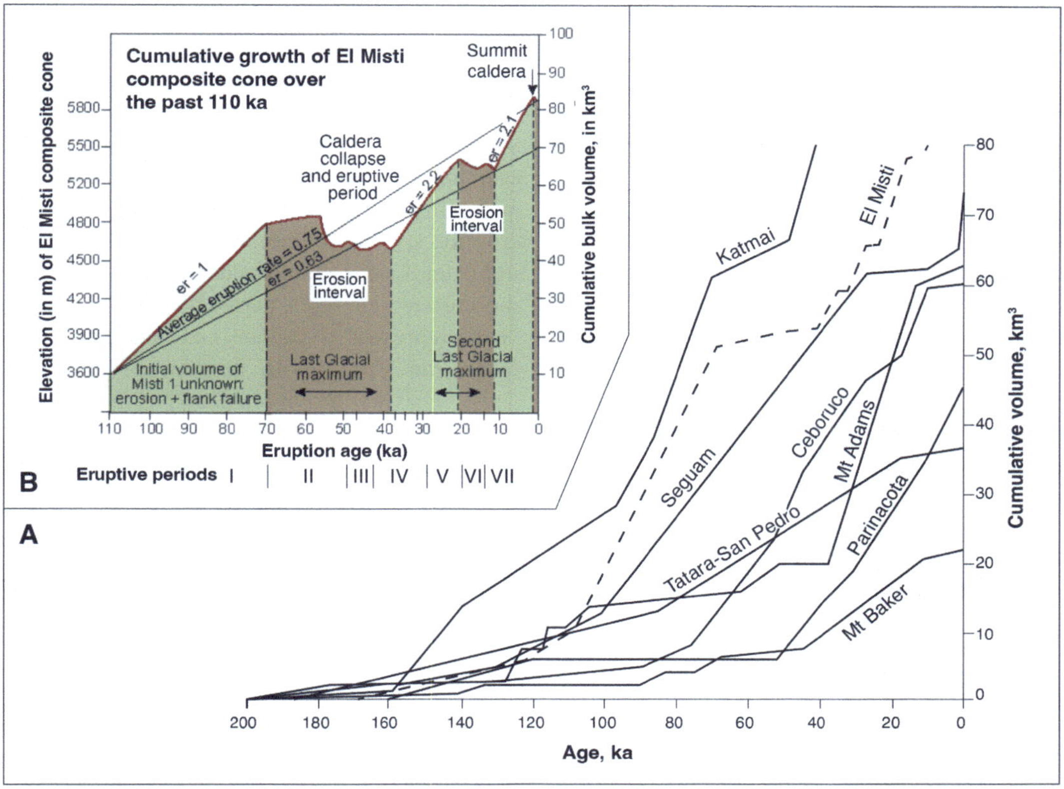

Fig. 15 Magma output at composite cones. **A** During the lifetime of a composite volcano, activity may fluctuate between two end-members: high cone-building 'peak' eruptive rates alternating with low to modest 'background' rates. Examples from Hora et al. (2007). **B** Cumulative growth of El Misti composite cone over the past 110 kyr, showing average eruptive rate, pauses or declining growth due to erosional and flank failure events (Thouret et al. 2001, Fig. 14, p. 1609)

one end those which are an order(s) of magnitude higher during active cone-building stages (spurts); and at the other those which have (often prolonged) periods of quiescence in between these growth periods. The cone-building spurts may occur cyclically but also irregularly during the lifetime of a volcano. As Hildreth and Lanphere (1994) emphasize, growth of an imposing cone may cover only 1–5% of the active lifetime of the composite volcano.

We can also learn a lot from the historical record of eruption frequencies of active volcanoes that may characterise their long-term behaviour:

1. *Persistent eruptive activity.* Sakurajima in Japan (persistently active since 1955), Semeru (Java, Indonesia), and Sangay (Ecuador) have been active on a daily basis for several decades or even centuries.
2. *Frequent eruptive activity.* More common is a periodicity of a few months to ten years (e.g. Fuego, Izalco, Arenal, Lascar, Pavlof, Vesuvius).
3. *Infrequent activity.* Volcanoes with an eruption frequency on the order of a hundred years include Pinatubo, Nevado del Ruiz, Osorno, Turrialba, Fuji, and St Helens.
4. *Rare activity with long-lasting quiescence intervals.* A thousand-year eruption frequency is common among the High Cascadian volcanoes like Mounts Adams, Baker, Hood, and Rainier, but the observational record may not be long enough to make inferences about the favoured modes

of eruptive behaviour (Hildreth and Lanphere 1994). Long-term variations and fluctuations in eruptive activity can be compared for Parinacota, Mount Adams, San Pedro–Tatara and other volcanoes (Hora et al. 2007) in Fig. 15.

The issue of frequency, and the variation in rate of eruptive activity are commonly described using terms such as *pulse, phase* (or simply *eruption), episode, stage,* and *period* (Simkin et al. 1981; De la Cruz-Reyna 1991). The latter three are usually used to describe increasingly longer time intervals, comprising one or more of the previous eruptive intervals.

6.3 Periods or 'Cycles' of Eruptive Activity

Eruptive periods and longer periods of repose in between may form repeatedly. These may happen regularly ('cycles') or irregularly. At Mount Adams, there have been irregular active periods lasting for 30–50 kyr (Hildreth and Lanphere 1994). Fuego in Guatemala has been active for about 100 kyr. During this time, the following four phases have been repeated regularly on the order of 100 kyr: (1) short, annual periods of volcanic activity with high-energy eruptions; (2) a period of debris flows, lahars and coarse-grained stream flows redepositing eruptive products; (3) a period of decreasing resedimentation in the drainage system; (4) a quiet period of the dormant volcano with fluvial erosion (Vessell and Davies 1981; Martin and Rose 1981). Since 1999, Fuego has been back to its eruptive cycle following two decades of quiescence, characterized by persistent Strombolian activity punctuated by occasional paroxysmal eruptions of greater magnitude (Naismith et al. 2019).

More or less regular eruptive periods have also been identified for Somma-Vesuvius. From ~35 to 18 ka, the Somma composite volcano grew up through modest, mostly effusive activity. Thereafter, a violent, explosive cycle began with Plinian eruptions triggering repeated caldera collapses. Up to the present day there has been an overall trend (Cioni et al. 2008): (1) the volume of erupted magma of the past Somma-Vesuvius eruptions has been decreasing, in contrast to an increase in their frequency from 3900 years BP; (2) apart from the recurrent Plinian eruptions, intermediate activity shows a large variability in terms of magnitude, intensity and eruption style, and a range of repose times varying from dozens to hundreds of years, including the current quiescent period (3) the past 2000 years represent a calmer cone-building cycle (i.e. Gran Cono) with smaller, but frequent, active phases. Strombolian and effusive eruptions occurred in the first-third, fifth-seventh, tenth-thirteenth centuries and between 1631 AD/CE and 1944.

7 Large, Long-Lived Volcanic Clusters or Volcano Complexes

Large and long-lived volcanic clusters (LVCs) are assemblages of composite cones and lava domes, sharing a common eruptive history and a similar tectonic domain with several closely-spaced conduits that are active more or less contemporaneously. Volcano clusters can also be termed *volcano complexes* (Francis 1993) to denote the prevailing landform characteristics. They differ from the composite cone types presented above in that they are not individual constructs that one can easily identify on a map: volcanic clusters represent a form of 'distributed volcanism', as opposed to individual, central volcanism (Francis 1993). An LVC has no clear outer limit and many constructs and their products are overlapping, while the sources (vents, fissures) are not easy to trace back.

Studies of individual composite cones across the Andes, the Cascades (USA) and the EARS over the past 30 years have provided a fundamental understanding of their eruptive behaviour, timescale, and magmatic evolution which can be used to explain the variability of erupted products. By contrast, there are few studies on the eruptive and compositional history and thermal evolution of long-lived volcanic clusters (>1 Ma), such as the Aucanquilcha cluster (Klemetti and Grunder 2007; Grunder et al. 2008), Taápaca in northern Chile (Clavero et al. 2004b), the Nevado Chachani and Nevado Coropuna clusters in southern Peru (Table 4; Aguilar et al. 2022). However, other volcano clusters have been thoroughly studied, particularly in the Cascades (USA), for example the volcanic fields of Mount Adams (Hildreth and Lanphere 1994), Mount Baker (Hildreth and Fierstein 2003), and the Mount Mazama and Crater Lake region (Bacon and Lanphere 2006).

7.1 LVC Definition and Differences from Other Composite Cones

Large volcanic clusters (LVCs) are the most voluminous volcanic landforms (Table 4) (Francis 1993; Grosse et al. 2009, 2012). An LVC is defined as an extensive (250–700 km^2) and voluminous (200–600 km^3) assemblage of spatially, temporally and genetically related, clustered volcanic edifices and eruptive centres. It comprises composite cones, compound cones with associated lava flows and pyroclastic deposits, overlapping lava dome complexes and their coulees, as well as fissure-fed lava-flow fields (Francis 1993). As such, LVCs are larger than typical composite cones in the Central Andes ($\leq$140 km^2, e.g. El Misti, Peru; Parinacota, Chile; Table 4) or even compound cones ($\leq$200 km^2, e.g. Ampato-Sabancaya, Peru; Taápaca, Chile; Table 4).

LVCs remain active over a long (>1 Ma) period, roughly twice as long as the growth period of typical individual composite or compound volcanoes in the Cascades, Southern Andes or Aleutian Islands (e.g. Hildreth and Lanphere 1994; Singer et al. 1997; Hildreth and Fierstein 2003; Coombs and Jicha 2021; Table 4). Exceptionally, the eruptive activity of an LVC may extend to as much as 11 Ma in the same area, as is the case with the Aucanquilcha cluster in Chile (Klemetti and Grunder 2007; Grunder et al. 2008; Table 4). Beyond differences in size and periods of activity, we argue that LVCs are specific in that they develop inside, or overlap, calderas surrounded by medium-sized aprons of ignimbrite sheets. Field and lithological relationships indicate that pulses of silicic ignimbrite-forming magmas are followed by the slow 'bleeding' of magmas with a variable range in composition from silicic to mafic, again followed by a regime of more uniform, hybrid intermediate andesite lava flows and domes. So, their volume, lifetime, and the overall evolution of the magmatic regimes, distinguish LVCs from other polygenetic volcanoes (Table 4; Fig. 16, after Aguilar et al. 2022).

7.2 Case Studies of LVCs

A striking example of the Pleistocene and Holocene evolution of a bimodal cluster is represented by the Boset-Bericha-Gudda volcanic complex (BBGVC, see Fig. 9A, Chap. 1) in the Northern Main Ethiopian Rift (MER, Siegburg et al. 2018). The BBVC has been divided into four main eruptive stages, comprising: (1) early rift floor emplacement, (2) formation of the Gudda Volcano (cones, domes and lava flows) during two main cycles, separated by caldera formation, (3) formation of the Bericha Volcano, with cones, domes, and lava flows, and (4) sporadic fissure eruptions. $^{40}Ar/^{39}Ar$ geochronology provides evidence for episodic activity at the BBVC from ~120 ka to the present-day (Siegburg et al. 2018). Low-volume mafic episodes are more frequent (~10 kyr cyclicity) than felsic episodes (~100 kyr cyclicity), but the latter are more voluminous. Felsic episodes have on average a higher eruption rate (2–5 per 1000 years) and productivity (0.563 km^3/kyr) at Gudda, compared with Bericha (1–2 per 1000 years, with 0.063 km^3/kyr). The voluminous lava flow eruptions (~65 km^3) from the BBGVC at ~120 ka and since ~16 ka are coincident with dated activity at nearby volcanic complexes in the MER. The young age of lavas and current fumarolic activity along the fault system suggest that the BBGVC is still potentially active. (Siegburg et al. 2018).

Table 4 Characteristics of large volcano clusters (LVC) of the Andean Central Volcanic Zone (CVZ), and comparison with some compound and individual composite cones elsewhere

Large volcanic clusters (LVC) and compound cones	Surface area (km^2)	Volume (km^3) ± 10%	No. of eruptive centres	Lifetime (Ma)	Eruptive pulses/stages (Ma)	DRE magma production (km^3)	Bulk magma output rate (km^3/kyr)	Compositional range	Caldera: associated or buried	References
Chachani LVC, CVZ, South Peru	580–600	289–346 ± 29/35	12	1.27	2: 'old' 1.1–0.6, 'young' 0.4–0.05	198–231 (80% lava flows)	Aver. 0.14–0.22 Peak 0.28–0.3 'Old' group: ~0.14 'Young' group: ~0.22	Mafic andesite to dacite	Probably buried	Aguilar et al. (2022)
Coropuna LVC, CVZ, South Peru	400–450	135–200	5	1.1	4: 'old' 1.1–0.4, 'young' 0.2–0.0014	130–180 (90% lava flows)	Aver. 0.2 Peak 0.3	Mafic andesite to dacite	Probably buried	Mariño et al. (2020)
Ampato (A) – Sabancaya (S) compound, CVZ, South Peru	c. 240–260	A 38–42 S 6–10 Total 44–52	3 1	A 0.4–0.45 S 0.005–0.01	5 Ampato 0.45–0.11 Sabancaya 0.01–0.005	40–50	A 0.10–0.12 S 0.6–1.7 (max) Aver. 0.08–0.09	Andesite and dacite	Probably buried	Samaniego et al. (2016), Thouret et al. (1994), Bromley et al. (2019)
Aucanquilcha compound cone, CVZ, North Chile	c. 80	37	1	1	4	70	Aver. 0.04–0.10 Peak 0.16	Dacite		Klemetti and Grunder (2007)
Aucanquilcha LVC, CVZ, N Chile	700	327 ± 20	19	11	3 (1.5–3 Myr)	300	Aver. 0.03 Peak 0.11–0.22	Mafic andesite to dacite		Grunder et al. (2008) , Walker et al. (2013)
Lastarria - Cordón del Azufre LVC, CVZ Chile -Argentina	LVC 500 Lastarria 90	45 Lastarria 200 cluster	7	0.6 – <0.3	4 Lastarria 3 Cordón del Azufre	40 180	0.1–0.15 Max 0.33	Andesite and dacite	Yes Current inflation	Naranjo (1992), Froger et al. (2007)
Nevados de Payachata twin, compound cones, CVZ, N Chile	190 50	Parinacota 46 Pomerape 40	2	0.163	4 Parinacota	40 35	Parinocota 0.25–0.31 (max 0.5–1.2)	Dacite and rhyolite		Wörner et al. (1988), Davidson et al. (1990), Hora et al. (2007)
Ollaguë compound volcano, CVZ, N Chile	480	80–90	>7	<1	4	70–80	Aver. 0.07–0.08	Dacite		Feeley and Davidson (1994)

(continued)

Table 4 (continued)

Large volcanic clusters (LVC) and compound cones	Surface area (km^2)	Volume (km^3) ± 10%	No. of eruptive centres	Lifetime (Ma)	Eruptive pulses/stages (Ma)	DRE magma production (km^3)	Bulk magma output rate (km^3/kyr)	Compositional range	Caldera: associated or buried	References
Llullaillaco compound volcano, CVZ, N Chile	150	50–60	4	1.5–0.048	2	40–50	Aver. 0.04–0.06	Dacite		Richards and Villeneuve (2001)
Taapacá lava dome cluster, CVZ, N Chile	180–250	35–73	4	1.5	4	30–65	0.05 domes; 0.07 comp. volc.	Dacite		Clavero et al. (2004b)
Lascar twin, compound cone CVZ, N Chile	55–185	50–60	3	<0.43	4	30–40	Aver. 0.02–0.03 Peak 0.14	Dacite		Gardeweg et al. (1998), Matthews et al. (1994)
For the purpose of comparison										
Mt. Mazama and volcanic field, Cascades, USA	500	58–112	Many	0.42	Many	100–120	0.42 Peak 0.8 Volc field 0.07	Mafic andesite to rhyodacite	Crater Lake caldera	Bacon and Lanphere (2006)
Mt. Adams and volcanic field, Cascades, USA	1250	230–400	Many	0.94	Many		Aver. 0.24–0.42	Basalt to dacite		Hildreth and Lanphere (1994)
El Misti individual composite cone	105–130	70–83	2	<0.83 max <0.112 min	4	55–65	Aver. 0.63 Peak 2.1	Andesite to rhyolite	Summit caldera	Thouret et al. (2001), Rivera et al. (2017)
Ubinas individual composite cone	65–90	55–60	1	0.376 to present	2	45–55	Aver. 0.17–0.22	Mafic andesite to rhyolite	Summit caldera	Thouret et al. (2005), Rivera et al. (2014)

All references appear in Aguilar et al. (2022)

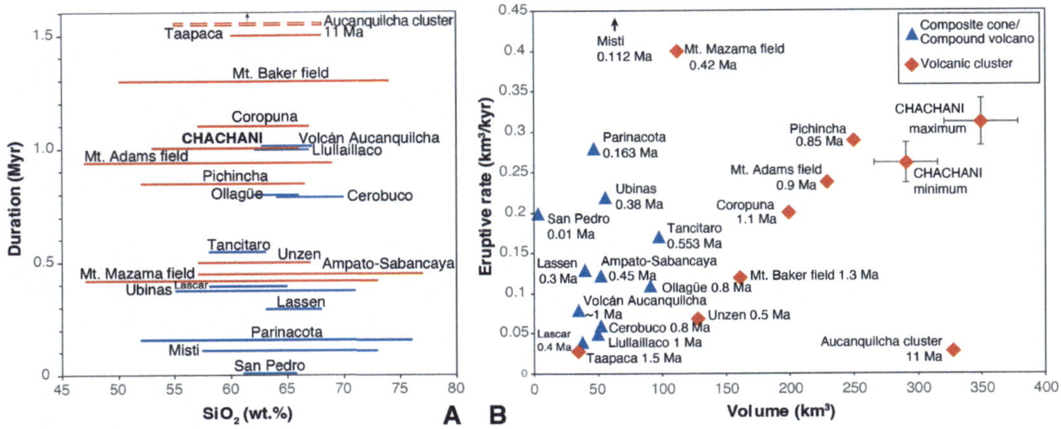

Fig. 16 **A** SiO_2 content versus duration of activity, and **B** volume versus magma eruptive rate plotted for the Chachani volcanic cluster and other clusters, compared to compound volcanoes and individual composite volcanoes in the CVZ, and other volcanic clusters worldwide. Individual composite cones and compound volcanoes (red triangles) of the CVZ include: San Pedro, Ollagüe, Lascar, Llullaillaco, El Misti, Ubinas, Tancitaro, Parinacota, Auncanquilcha, Uturuncu, and Ampato-Sabancaya. Large volcanic clusters, fields or complexes (blue triangles) encompass: Pichincha volcano complex, compound Unzen, compound Taápaca, and Coropuna cluster for the purposes of comparison, although the crust is thinner than in the CVZ: Mount Baker field, Mount Adams field, and Mount Mazama field. Average eruptive rate has been calculated using bulk volume of erupted material for each of the volcanic systems. Two black dots with associated error range indicate the minimum and maximum volume estimates for Chachani-LVC. All references in: Aguilar et al. (2022), Fig. 1, p. 4

Paguican et al. (2021) studied volcano 'massifs' in the Philippines which include small flat cones, small steep cones, and large composite cones. (Note however, that in the context of composite volcano classification in this book, we consider massifs as a class of compound cones.) These authors provided geometric characteristics of the massifs, distinct from other smaller edifices, such as volume in the range of 76–675 km^3 (average ~267 km^3), height ranging from 1012 to 2175 m (average ~1533 m), and a range of slopes between 11° and 22° (mean 15°). On average, the volume of massifs reaches ten times that of the large cones, and their total height is twice that of the cones. The slope angle parameter turns out to be indistinguishable for massifs and large cones. Paguican et al. (2021) further analysed morphometric parameters to define massifs among other categories of volcanoes, such as size and irregularity, and to a lesser extent the size of the summit region and edifice truncation and edifice elongation. Results of principal component and clustering analyses show that a small number of variables describing the edifice size provide the largest contribution to the analysis: the area, average width and maximum axis of the base and summit, and the height and volume of the edifice. In contrast, the profile shape of the volcano (height/basal width ratio, mean and standard deviation of slope of whole edifice) turns out to weakly contribute to a distinction between edifices. Overall, 'massifs' are large, irregular, and have more peaks than most of the smaller edifices. This suggests that, in addition to climate factors, volcano age correlated with edifice height must also be considered.

In the Philippines, thirteen of the 'massifs' are situated above the thickened crust of the Bicol–Negros–Panay, Central Mindanao, and Central Luzon regions; the distribution of the large cones and massifs suggests that the thickness of the crust may influence the evolution of arc volcanoes. Thicker crust than average makes magma more likely to stall during ascent, forming long-lived magma chambers that focus volcanic activity (Karlstrom et al. 2010).

One of the best studied examples is illustrated by the Quaternary eruptive history of the volcanic region of Crater Lake caldera and the former Mount Mazama edifice (Bacon and Lanphere 2006) which can together be considered as a volcano cluster. The large Mazama edifice (see Sect. 5.3) was constructed on a Pleistocene silicic lava field, amongst monogenetic and shield volcanoes ranging from basalt to andesite, similar in composition to parental magmas for Mazama. Long-lived volcanic activity focused on this large field over half a million years or more could potentially have led to a thermal input affecting the regional magma flux, the establishment of climactic magma chambers and powerful eruptions. Bacon and Lanphere (2006) quantified the Quaternary eruptive history of the large Mazama–Crater Lake field to define processes and conditions that led to voluminous explosive eruptions. The average eruption rate for all Mazama products was ~0.4 km^3/kyr, but major edifice construction episodes had rates of ~0.8 km^3/kyr. The long-term eruption rate for regional monogenetic and shield volcanoes was only ~0.07 km^3/kyr.

Photo 53 Mount Etna (3357 m, Sicily, Italy), the largest present-day compound polygenetic volcano in Europe. Eruptive activity in its summit region, comprising a number of craters such as Voragine and Bocca Nuova, has also created adventive scoria cones like those from 2001/2002 in the foreground (*photo* D. Karátson)

Photo 54 An iconic strato- or composite volcano in Italy: the Somma-Vesuvius edifice **A** *photo credit* Google Earth; **B** the volcano with Herculaneum in the foreground, *photo* D. Karátson). The truncated caldera of Mount Somma has been formed by repeated collapses, the latest one of which was in AD/CE 79, and hosts the youngest, post-caldera "Gran Cono", the present-day Vesuvius (1281 m)

Photo 55 Mount Fuji (3776 m), the sacred mountain of Japan. One of the most regular-shaped composite cones in the world, Young Fuji, the present-day edifice, has developed on an older one over the past 10,000 years (*photo* D. Karátson)

Photo 56 Three adjacent, irregular-shaped composite cones in the central Oregon Cascade Ranges: from left to right, South Sister (highest: 3157 m), Middle Sister, and North Sister. All of them ceased activity during the late Pleistocene, but adventive vents produced late-stage lava flows as recently as about 1600 years ago (*photo* D. Karátson)

Photo 57 West flank of historically active Volcán de Colima volcano (3839 m, Mexico), whose latest eruption was in 2019. Somewhat lower than its northern twin, the Late Pleistocene Nevado del Colima has been constructed within a 5 km-wide horseshoe-shaped scar. Note the block-and-ash flow fan on its mid-slope (*photo* courtesy of Sz. Kósik)

Photo 58 The twin composite volcanoes of Nevado de Payachata in Chile-Bolivia. **A** the highly symmetrical, Holocene, 6336 m-high Parinacota (to the right) and the Late Pleistocene, 6222 m-high Pomerape. Note the debris-avalanche deposits damming Laguna Cotacotani. **B** Plateau domes on the shore of Chungara Lake, on which the youngest Parinacota cone grew (*photos* D. Karátson)

Photo 59 Ojos del Salado, the world's highest (6893 m) compound volcano in Chile-Argentina, last erupted c. 1000 years ago. Its elongated summit complex comprises a number of craters, pyroclastic cones, and lava domes (**A**). The 3.9 Ma Laguna Verde ignimbrite is in the foreground on the right hand side. Ojos is one of half a dozen other cones towering well above 6000 m on the 5000 m-high Chilean altiplano (**B**) (*photo* courtesy of B. Nagy)

Photo 60 Volcanoes in the Chilean Lake District: Lonquimay (2832 m), a simple, symmetrical composite cone of dominantly Holocene age (centre left, highest) with an ice-filled summit crater; red-coloured Cona Navidad (Christmas Cone) in the foreground (crater diameter c. 200 m), formed in 1988; and Tolhuaca (or Tolguaca) composite cone (2739 m), last erupted c. 4000 years ago, in the distance to the right of Lonquimay (*photo* courtesy of B. Nagy)

Photo 61 After 200 years of dormancy, the activity of Unzen lava dome complex (Kyushu, Japan) in 1990–1996 significantly reshaped the morphology of the summit region. During dome growth, thousands of block-and-ash flows were generated; the summit region comprises a stack of new lobes that have raised the edifice by more than 100 m (*photo* D. Karátson, taken in spring 1997 when the summit lava spine, shown in the inset, still contained sulphurous fumaroles)

Photo 62 The lava dome complex of Merapi (Java, Indonesia): the light brownish-coloured, 'smoking' New Merapi (<5000 years) towers above Old Merapi (in the foreground, covered by forest) (*photo* D. Karátson)

Photo 63 Tongariro, the northern, lower compound composite edifice in the North Island of New Zealand, is composed of half a dozen overlapping volcanoes with Red Crater in the foreground, and the highest, regular-shaped Ngauruhoe cone (2291 m) in the background to the right (these two features formed in the last 3000 years) (*photo* courtesy of S. Kósik)

Photo 64 Ruapehu, the largest active volcano in New Zealand (highest point is Tahurangi, 2797 m), is a compound edifice south of Tongariro (**A**), elongated in a NNE-SSW direction and surrounded by an extensive ring plain. It has a deep, acidic crater lake, named *Te Wai ā-moe* (meaning Crater Lake in Māori) with an active hydrothermal system (**B**) (*photo* courtesy of Sz. Kósik)

Photo 65 Damávand, the highest, collision-related irregular composite cone (5670 m) in Iran, southwest Asia. The present cone represents Young Damávand (c. 1.5 Ma old), while the ridge with rockfall lobes to the right is remnant of an old cone. Photo taken from the north (*photo* courtesy of K. Ghahraman)

Photo 66 The Late Pleistocene Ciomadul (Csomád) lava dome complex, Romania. Activity of the adjacent domes covered almost 1 million years, with the last eruption less than 30,000 years ago. Behind the lava domes on the right hand side is Sf. Ana (St. Anne) crater lake (*photo* D. Karátson)

Photo 67 The broad central volcano of Öræfalljökull composite cone, Iceland, with a 4×5 km subglacial caldera clearly visible on the left hand side. The glacier-clad volcano belongs to the largest ice cap of Vatnajökull in Iceland (*photo* courtesy of G. Nagy)

Photo 68 Simple, regular composite cones around the world with parabolic (P-type) slope profiles: **A** Osorno, Chile (*photo* courtesy of B. Nagy), **B** Kaimon, Kyushu, Japan (*photo* D. Karátson), **C** Villarrica, Chile (*photo* D. Karátson), **D** Mayon, The Philippines (*photo* courtesy of N. Irapta)

References

Aguilar R, Thouret J-C, Samaniego P, Wörner G, Jicha B, Paquette J-L, Suaña E, Finizola A (2022) Growth and evolution of long-lived, large volcanic clusters in the Central Andes: the Chachani Volcano Cluster, southern Peru. J Volcanol Geotherm Res 426:107539. https://doi.org/10.1016/j.jvolgeores.2022.107539

Alvarado GE, Carr MJ, Turrin BD, Swisher CC III, Schmincke H-U, Hudnut KW (2006) Recent volcanic history of Irazú volcano, Costa Rica: alternation and mixing of two magma batches, and pervasive mixing. In: Rose WI, Bluth GJS, Carr MJ, Ewert JW, Patino LC, Vallance JW (eds) Volcanic hazards in central America. GSA Special Papers 412, pp 259–276. https://doi.org/10.1130/2006.2412(14)

Amigo A (2021) Volcano monitoring and hazard assessments in Chile. Volcanica 4(S1):1–20. https://doi.org/10.30909/vol.04.S1.0120

Aoki Y, Tsunematsu K, Yoshimoto M (2019) Recent progress of geophysical and geological studies of Mt. Fuji Volcano, Japan. Earth-Sci Rev 194:264–282. https://doi.org/10.1016/j.earscirev.2019.05.003

Bacon CR, Lanphere MA (2006) Eruptive history and geochronology of Mount Mazama and the Crater Lake region, Oregon. GSA Bull 118(11–12):1331–1359. https://doi.org/10.1130/B25906.1

Bacon CR, Neal CA, Miller TP, McGimsey RG, Nye CJ (2014) Postglacial eruptive history, geochemistry, and recent seismicity of Aniakchak volcano, Alaska Peninsula. US Geol Surv Prof Paper 1810, Reston, VA, USA, 74 p. ISSN: 1044-9612

Barba D, Robin C, Samaniego P, Eissen J-P (2008) Holocene recurrent explosive activity at Chimborazo volcano (Ecuador). J Volcanol Geotherm Res 176(1):27–35. https://doi.org/10.1016/j.jvolgeores.2008.05.004

Bardintzeff J-M, Deniel C (1992) Magmatic evolution of Pacaya and Cerro Chiquito volcanological complex, Guatemala. Bull Volcanol 54:267–283. https://doi.org/10.1007/BF00301482

Becker GF (1885) The geometrical form of volcanic cones and the elastic limit of lava. Am J Sci 30(178):283–293. https://doi.org/10.2475/ajs.s3-30.178.283

Beier C, Haase KM, Abouchami W (2015) Geochemical and geochronological constraints on the evolution of the Azores Plateau. In: Neal CR, Sager WW, Sano T, Erba E (eds) The origin, evolution, and environmental impact of oceanic large Igneous Provinces. GSA Special Papers 511, pp 27–55. https://doi.org/10.1130/2015.2511(02)

Beier C, Haase KM, Hansteen TH (2006) Magma evolution of the Sete Cidades Volcano, São Miguel, Azores. J Petrol 47(7):1375–1411. https://doi.org/10.1093/petrology/egl014

Bindeman IN, Ponomareva VV, Bailey JC, Valley JW (2004) Volcanic arc of Kamchatka: a province with high-$\delta^{18}O$ magma sources and large-scale $^{18}O/^{16}O$ depletion of the upper crust. Geochim Cosmochim Acta 68(4):841–865. https://doi.org/10.1016/j.gca.2003.07.009

Blum-Oeste M, Wörner G (2016) Central Andean magmatism can be constrained by three ubiquitous endmembers. Terra Nova 28(6):434–440. https://doi.org/10.1111/ter.12237

Bonis SB (1993) Mapa Geológico Guatemala, 1:250,000 ND 15-8-G, Instituto Geográfico Nacional, Guatemala

Bossis R, Regard V, Carretier S (2023) Initial shape reconstruction of a volcanic island as a tool for quantifying long-term coastal erosion: the case of Corvo Island (Azores). Earth Surf Dyn 11(3):529–545. https://doi.org/10.5194/esurf-11-529-2023

Brancolini G, Civile D, Donda F, Tosi L, Zecchin M, Volpi V, Rossi G, Sandron D, Ferrante GM, Forlin E (2019) New insights on the Adria plate geodynamics from the northern Adriatic perspective. Mar Pet Geol 109:687–697. https://doi.org/10.1016/j.marpetgeo.2019.06.049

Calvari S, Cannavó F, Bonaccorso A, Spampinato L, Pellegrino AG (2018) Paroxysmal explosions, lava fountains and ash plumes at Etna Volcano: eruptive processes and hazard implications. Front Earth Sci 6. https://doi.org/10.3389/feart.2018.00107

Cameron BI, Walker JA (2006) Diverse volcanism in southeastern Guatemala: the role of crustal contamination. In: Rose WI, Bluth GJS, Carr MJ, Ewert JW, Patino LC, Vallance JW (eds) Volcanic hazards in central America. GSA Special Papers 412, pp 121–139. https://doi.org/10.1130/2006.2412(07)

Canora C, Martínez-Díaz JJ, Villamor P, Staller A, Berryman KR, Álvatez G, José A, Capote R, Díaz M (2014) Structural evolution of the El Salvador Fault Zone: an evolving fault system within a volcanic arc. J Iber Geol 40(3):471–488. https://doi.org/10.5209/rev_JIGE.2014.v40.n3.43559

Capra L, Roverato M, Groppelli G, Caballero L, Sulpizio R, Norini G (2015) Glacier melting during lava dome growth at Nevado de Toluca volcano (Mexico): evidences of a major threat before main eruptive phases at ice-caped volcanoes. J Volcanol Geotherm Res 294:1–10. https://doi.org/10.1016/j.jvolgeores.2015.02.005

Carr MJ, Feigenson MD, Patino LC, Walker JA (2003) Volcanism and geochemistry in Central America: progress and problems. Geophysical Monograph-American Geophysical Union 138:153–174

Carr MJ, Pontier NK (1981) Evolution of a young parasitic cone towards a mature central vent; Izalco and Santa Ana volcanoes in El Salvador, Central America. J Volcanol Geotherm Res 11(2–4):277–292. https://doi.org/10.1016/0377-0273(81)90027-5

Carracedo JC, Troll VR (2016) The geology of the canary islands. Elsevier, Amsterdam, Netherlands, 621 p. ISBN: 978-0-12-809663-5

Cas RAF, Wright JV (1988) Modern volcanoes and volcanic centres. In: Cas RAF, Wright JV (eds) Volcanic Successions: Modern and Ancient. Springer, Dordrecht, Netherlands, pp 362–410. ISBN-13: 978-0-412-44640-5

Castruccio A, Diez M, Gho R (2017) The influence of plumbing system structure on volcano dimensions and topography. J Geophys Res 122(11):8839–8859. https://doi.org/10.1002/2017JB014855

Cioni R, Bertagnini A, Santacroce R, Andronico D (2008) Explosive activity and eruption scenarios at Somma-Vesuvius (Italy): towards a new classification scheme. J Volcanol Geotherm Res 178(3):331–346. https://doi.org/10.1016/j.jvolgeores.2008.04.024

Cioni R, Santacroce R, Sbrana A (1999) Pyroclastic deposits as a guide for reconstructing the multi-stage evolution of the Somma-Vesuvius Caldera. Bull Volcanol 61:207–222. https://doi.org/10.1007/s004450050272

Cisneros de León A, Danišík M, Schindlbeck-Belo JC, Kutterolf S, Schmitt AK, Freundt A, Kling J, Wang K-L, Lee H-Y (2023) Timing and recurrence intervals for voluminous silicic eruptions from Amatitlán caldera (Guatemala). Quaternary Sci Rev 301. https://doi.org/10.1016/j.quascirev.2022.107935

Clavero JE, Sparks RSJ, Polanco E, Pringle MS (2004a) Evolution of Parinacota volcano, central Andes, northern Chile. Revista Geologica de Chile 31(2):317–347. https://doi.org/10.4067/S0716-02082004000200009

Clavero JE, Sparks RSJ, Pringle MS, Polanco E, Gardeweg MC (2004b) Evolution and Volcanic Hazards of Taapaca Volcanic Complex, Central Andes of Northern Chile. J Geol Soc 161:603–618. https://doi.org/10.1144/0016-764902-065

Clynne MA, Calvert AT, Wolfe EW, Evarts, RC, Fleck RJ, Lanphere MA (2008) The Pleistocene eruptive history of Mount St. Helens, Washington, from 300, 000 to 12, 800 years before present. In: Sherrod DR, Scott WE, Stauffer PH (eds) A volcano rekindled: the renewed eruption of Mount St. Helens, 2004–2006. US Geol Surv Prof Pap 1750, Reston, VA, pp 593–627. https://doi.org/10.3133/pp1750

Cole PD, Scarpati C (2010) The 1944 eruption of Vesuvius, Italy: combining contemporary accounts and field studies for a new volcanological reconstruction. Geol Mag 147(3):391–415. https://doi.org/10.1017/S0016756809990495

Conway FM, Diehl JF, Rose WI, Matias O (1994) Age and Magnus flux of Santa Maria, Guatemala: correlation of paleomagnetic waveforms with the Mono lake excursion. J Geol 102(1):11–24. http://www.jstor.org/stable/30065708

Coombs ML, Jicha BR (2021) The eruptive history, magmatic evolution, and influence of glacial ice at long-lived Akutan volcano, eastern Aleutian Islands, Alaska, USA. GSA Bull 133(5–6):963–991. https://doi.org/10.1130/B35667.1

Cornu M-N, Paris R, Doucelance R, Bachélery P, Bosq C, Auclair D, Benbakkar M, Gannoun A-M, Guillou H (2021) Exploring the links between volcano flank collapse and the magmatic evolution of an ocean island volcano: Fogo, Cape Verde. Sci Rep 11. https://doi.org/10.1038/s41598-021-96897-1

Cosburn K, Ray M (2020) Analysing the topographic form of stratovolcanoes. J Volcanol Geotherm Res 407.https://doi.org/10.1016/j.jvolgeores.2020.107123

Cotton CA (1944) Volcanoes as landscape forms. Whitcombe and Tombs Ltd, Christchurch, New Zealand, 416 p. ISSN: 0028-0836

Cotton CA (1952) Volcanoes as landscape forms. Wiley, NY, 416 p. ISSN: 0028-0836

Crandell DR, Miller CD, Glicken HX, Christiansen RL, Newhall CG (1984) Catastrophic debris avalanche from ancestral Mount Shasta volcano. Calif Geol 12(3):143–146. https://doi.org/10.1130/0091-7613(1984)12<143:CDAFAM>2.0.CO;2

Crisp JA (1984) Rates of magma emplacement and volcanic output. J Volcanol Geotherm Res 20(3–4):177–211. https://doi.org/10.1016/0377-0273(84)90039-8

Cronin SJ, Zernack AV, Ukstins IA, Turner MB, Torres-Orozco R, Stewart RB, Smith IEM, Procter JN, Price R, Platz T, Petterson M, Neall VE, McDonald GS, Lerner GA, Damaschcke M, Bebbington MS (2021) The geological history and hazards of a long-lived stratovolcano, Mt. Taranaki, New Zealand. N Z J Geol Geophys 64(2–3):456–478. https://doi.org/10.1080/00288306.2021.1895231

Crosweller HS, Arora B, Brown SK, Cottrell E, Deligne NI, Guerrero NO, Hobbs L, Kiyosugi K, Loughlin SC, Lowndes J, Nayembil M, Siebert L, Sparks RSJ, Takarada S, Venzke E (2012) Global database on large magnitude explosive volcanic eruptions (LaMEVE). J Appl Volcanol 1:4. https://doi.org/10.1186/2191-5040-1-4

Davidson J, de Silva S (2000) Composite volcanoes. In: Sigurdsson H, Houghton B, McNutt SR, Rymer H, Stix J (eds) Encyclopedia of volcanoes. Academic Press, San Diego, CA, pp 663–681. ISBN 978-0-12-385938-9

Davidson J, Hassanzadeh J, Berzins R, Stockli DF, Bashukooh B, Turrin B, Pandamouz A (2004) The geology of Damavand volcano, Alborz Mountains, northern Iran. GSA Bull 116(1–2):16–29. https://doi.org/10.1130/B25344.1

De la Cruz-Reyna S (1991) Poisson-distributed patterns of explosive eruptive activity. Bull Volcanol 54:57–67. https://doi.org/10.1007/BF00278206

De la Cruz-Reyna S, Martín Del Pozzo AL (2009) The 1982 eruption of El Chichón volcano, Mexico: eye-witness of the disaster. Geofísica Internacional 48(1):21–31. ISSN: 2954-436X

de Silva S, Kay SM (2018) Turning up the Heat: High-Flux Magmatism in the Central Andes. Elements 14(4):245–250. https://doi.org/10.2138/gselements.14.4.245

de Silva S, Lindsay JM (2015) Primary volcanic landforms. In: Sigurdsson H, Houghton B, McNutt SR, Rymer H, Stix J (eds) Encyclopedia of volcanoes. Academic Press, San Diego, CA, pp 273–297. ISBN 978-0-12-385938-9

Delavari M, Mehrabi B, Zelenski M, Chaplygin I, Nekrylov N, Shakeri A, Taran Y (2022) The Bazman and Taftan volcanoes of southern Iran: implications for along-arc geochemical variation and magma storage conditions above the Makran low-angle subduction zone. J Asian Earth Sci 233.https://doi.org/10.1016/j.jseaes.2022.105259

Delcamp A, Kervyn M, Benbakkar M, Kwelwa S, Peter D (2017) Large volcanic landslide and debris avalanche deposit at Meru, Tanzania. Landslides 14(3):833–847. https://doi.org/10.1007/s10346-016-0757-8

Delgado Granados H, Miranda PJ, Huggel C, Ortega del Valle S, Ibargüengoitia MAA (2007) Chronicle of a death foretold: extinction of the small-size tropical glaciers of Popocatépetl volcano (Mexico). Global Planet Change 56(1–2):13–22. https://doi.org/10.1016/j.gloplacha.2006.07.010

Demant A, Lestrade P, Lubala RT, Kampunzu AB, Durieux J (1994) Volcanological and petrological evolution of Nyiragongo volcano, Virunga volcanic field, Zaire. Bull Volcanol 56:47–61. https://doi.org/10.1007/BF00279728

DeVitre CL, Gazel E, Ramalho RS, Venugopal S, Steele-MacInnis M, Hua J, Allison CM, Moore LR, Carracedo JC, Monteleone B (2023) Oceanic intraplate explosive eruptions fed directly from the mantle. Earth Atmos Planet Sci 120(33). https://doi.org/10.1073/pnas.2302093120

Duncan RA, Quilty PG, Barling J, Fox JM (2016) Geological development of Heard Island, Central Kerguelen Plateau. Aust J Earth Sci 63(1). https://doi.org/10.1080/08120099.2016.1139000

Dunkley PN (2002) Volcanic hazard assessment of Tristan da Cunha. British Geological Survey, Nottingham, UK, 57 p

Edwards BR, Russel JK (2000) Distribution, nature and origin of Neogene-Quaternary magmatism in the northern Cordilleran volcanic province, Canada. GSA Bull 112(8):1280–1295. https://doi.org/10.1130/0016-7606(2000)112<1280:DNAOON>2.0.CO;2

Espinasa-Pereña R, Martín-Del Pozzo AL (2006) Morphostratigraphic evolution of Popocatepetl volcano, Mexico. In: Siebe C, Macías JL, Aguirre-Díaz GJ (eds) Neogone-Quaternary continental margin volcanism: a perspective from Mexico. GSA Special Papers 402, pp 101–123. https://doi.org/10.1130/2006.2402(05)

Fan X, Chen Q-F, Legendre CP, Guo Z (2020) Intraplate volcanism and regional geodynamics in NE Asia revealed by anisotropic Rayleigh-wave tomography. Geophys Res Lett 47(1). https://doi.org/10.1029/2019GL085623

Favalli M, Karátson D, Gertisser R, Fornaciai A (2016) Volcanic geomorphology of Tambora (Sumbawa island, Indonesia) on the basis of SRTM DEM data. Geophys Res Abstracts, EGU 18, Paper: 8916

Favalli M, Karátson D, Yepes J, Nannipieri L (2014) Surface fitting in geomorphology—examples for regular-shaped volcanic landforms. Geomorphology 221:139–149. https://doi.org/10.1016/j.geomorph.2014.06.009

Favalli M, Pareschi MT, Neri A, Isola I (2005) Forecasting lava flow paths by a stochastic approach. Geophys Res Lett 32(3). https://doi.org/10.1029/2004GL021718

Fedotov SA, Zharinov NA, Gontovaya LI (2010) The magmatic system of the Klyuchevskaya group of volcanoes inferred from data on its eruptions, earthquakes, deformation, and deep structure. J Volcanol Seism 4:1–33. https://doi.org/10.1134/S074204631001001X

Ferrari L, Orozco-Esquivel T, Manea V, Manea M (2012) The dynamic history of the Trans-Mexican Volcanic Belt and the Mexico subduction zone. Tectonophysics 522–523:122–149. https://doi.org/10.1016/j.tecto.2011.09.018

Fierstein J, Hildreth W (2008) Kaguyak dome field and its Holocene caldera, Alaska Peninsula. J Volcanol Geotherm Res 177(2):340–366. https://doi.org/10.1016/j.jvolgeores.2008.05.016

Francis P (1993) Volcanoes. A planetary perspective. Clarendon Press, Oxford, UK, 452 p. ISBN: 0-19-854033-7

Geissler WH, Wintersteller P, Maia M, Strack A, Kammann J, Eagles G, Jegen M, Schoemer A, Jokat W (2020) Seafloor evidence for pre-shield volcanism above the Tristan da Cunha mantle plume. Nat Commun 11. https://doi.org/10.1038/s41467-020-18361-4

Gertisser R, Charbonnier SJ, Keller J, Quidelleur X (2012) The geological evolution of Merapi volcano, Central Java, Indonesia. Bull Volcanol 74:1213–1233. https://doi.org/10.1007/s00445-012-0591-3

Gertisser R, Keller J (2003) Trace element and Sr, Nd, Pb and O isotope variations in medium-K and high-K volcanic rocks from Merapi Volcano, Central Java, Indonesia: evidence for the involvement of subducted sediments in Sunda Arc Magma Genesis. J Petrol 44(3): 457–489. https://doi.org/10.1093/petrology/44.3.457

Gertisser R, Troll VR, Walter TR, Gusti I, Nandaka MA, Ratdomopurbo A (eds) (2023) Merapi Volcano—geology, eruptive activity, and monitoring of a high-risk volcano. Active Volcanoes of the World, Springer, Cham, 572 p. https://doi.org/10.1007/978-3-031-15040-1_1

Ghalamghash J, Schmitt AK, Chaharlang R (2019) Age and compositional evolution of Sahand volcano in the context of post-collisional magmatism in northwestern Iran: evidence for time-transgressive magmatism away from the collisional suture. Lithos 344–345:265–279.https://doi.org/10.1016/j.lithos.2019.06.031

Giambiagi L, Tassara A, Echaurren A, Julve J, Quiroga R, Barrionuevo M, Liu S, Echeverría I, Mardónez D, Suriano J, Mescua J, Lossada AC, Spagnotto S, Bertoa M, Lothari L (2022) Crustal anatomy and evolution of a subduction-related orogenic system: Insights from the Southern Central Andes (22–35°S). Earth-Sci Rev 232.https://doi.org/10.1016/j.earscirev.2022.104138

Gill J (2016) Island arc volcanism, volcanic arcs. In: Harff J, Meschede M, Petersen S, Thiede J (eds) Encyclopedia of marine geosciences. Encyclopedia of earth sciences series. Springer, Dordrecht, Netherlands, pp 379–383. https://doi.org/10.1007/978-94-007-6238-1_20

Giordano D, Russell JK, Dingwell DB (2008) Viscosity of magmatic liquids: a model. Earth Planet Sci Lett 271: 123–134. https://doi.org/10.1016/j.epsl.2008.03.038

Giordano G, Lucci F, Phillips D, Cozzupoli D, Runci V (2012) Stratigraphy, geochronology and evolution of the Mt. Melbourne volcanic field (North Victoria Land, Antarctica). Bull Volcanol 74:1985–2005. https://doi.org/10.1007/s00445-012-0643-8

Gjerløw E, Haflidason H, Pedersen RB (2016) Holocene explosive volcanism of the Jan Mayen (island) volcanic province, North-Atlantic. J Volcanol Geotherm Res 321:31–43. https://doi.org/10.1016/j.jvolgeores.2016.04.025

Goff F, McIntosh W, Peters L, Wolff JA, Kelley SA, Goff CJ, Osburn GR (2021) Volcanic evolution of Mount Taylor Stratovolcano, New Mexico: facts and misconceptions. In: Frey BA, Kelley SA, Zeigler KE, McLemore VT, Goff F, Ulmer-Scholle DS (eds) The geology of the Mount Taylor Area. New Mexico Geological Society Special Publication 14:17–28. ISBN: 978-1-58546-012-0

Gonzales K, Finizola A, Lénat J-F, Macedo O, Ramos D, Thouret J-C, Fournier N, Cruz V, Pistre K (2014) Asymmetrical structure, hydrothermal system and edifice stability: the case of Ubinas volcano, Peru, revealed by geophysical surveys. J Volcanol Geotherm Res 276:132–144. https://doi.org/10.1016/j.jvolgeores.2014.02.020

Grosse P, Euillades PA, Euillades LD, van Wyk de Vries B (2014) A global database of composite volcano morphometry. Bull Volcanol 76.https://doi.org/10.1007/s00445-013-0784-4

Grosse P, van Wyk de Vries B, Euillades PA, Kervyn M, Petrinovic IA (2012) Systematic morphometric characterization of volcanic edifices using digital elevation models. Geomorphology 136(1):114–131. https://doi.org/10.1016/j.geomorph.2011.06.001

Grosse P, van Wyk de Vries B, Petrinovic IA, Euillades PA, Alvarado GE (2009) Morphometry and evolution of arc volcanoes. Geology 37(7):651–654. https://doi.org/10.1130/G25734A.1

Grunder AL, Klemetti EW, Feeley TC, McKee CM (2008) Eleven million years of arc volcanism at the Aucanquilcha Volcanic Cluster, northern Chiean Andes: implications for the life span and emplacement of plutons. Royal Soc Edinb 97(4):415–436. https://doi.org/10.1017/S0263593300001541

Gudmundsdóttir ER, Larsen G, Eiríksson J (2011) Two new Icelandic tephra markers: the Hekla Ö tephra layer, 6060 cal. yr BP, and Hekla DH tephra layer, ~6650 cal. yr BP. Land-sea correlation of mid-Holocene tephra markers. Holocene 21(4):629–639. https://doi.org/10.1177/0959683610391313

Hall M, Mothes P (2008) The rhyolitic–andesitic eruptive history of Cotopaxi volcano, Ecuador. Bull Volcanol 70:675–702. https://doi.org/10.1007/s00445-007-0161-2

Harris AJ, Rose WI, Flynn LP (2003) Temporal trends in lava dome extrusion at Santiaguito 1922–2000. Bull Volcanol 65:77–89. https://doi.org/10.1007/s00445-002-0243-0

Harris SL (2005) Fire Mountains of the West: the Cascade and Mono Lake Volcanoes, 3rd edn. Mountain Press, Missoula, 454 p

Hicks A, Barclay J, Mark DF, Loughlin S (2012) Tristan da Cunha: constraining eruptive behavior using the $^{40}Ar/^{39}Ar$ dating technique. Geology 40(8):723–726. https://doi.org/10.1130/G33059.1

Hildreth W (2007) Quaternary magmatism in the cascades—geologic perspectives. US Geol Surv Prof Paper 1744, Reston, VA, USA, 125 p. ISBN: 1-4113-1945-1

Hildreth W, Fierstein J (2003) Geologic map of the Katmai Volcanic Cluster, Katmai National Park, Alaska. US Geol Surv Geologic Investigations Series I-2778, 43 p. https://doi.org/10.3133/i2778

Hildreth W, Lanphere MA (1994) Potassium-argon geochronology of a basalt-andesite-dacite arc system: the Mount Adams volcanic field, Cascade Range of southern Washington. GSA Bull 106(11):1413–1429. https://doi.org/10.1130/0016-7606(1994)106<1413:PAGOAB>2.3.CO;2

Hill GJ, Wannamaker PE, Maris V, Stodt JA, Kordy M, Unsworth MJ, Bedrosian PA, Wallin EL, Uhlmann DF, Ogawa Y, Kyle P (2022) Trans-crustal structural control of CO_2-rich extensional magmatic systems revealed at Mount Erebus Antarctica. Nat Commun 13.https://doi.org/10.1038/s41467-022-30627-7

Holm RJ, Rosenbaum G, Richards SW (2016) Post 8 Ma reconstruction of Papua New Guinea and Solomon Islands: microplate tectonics in a convergent plate boundary setting. Earth-Sci Rev 156:66–81. https://doi.org/10.1016/j.earscirev.2016.03.005

Hora JM, Singer BS, Wörner G (2007) Volcano evolution and eruptive flux on the thick crust of the Andean Central Volcanic Zone: $^{40}Ar/^{39}Ar$ constraints from Volcán Parinacota, Chile. GSA Bull 119(3–4):343–362. https://doi.org/10.1130/B25954.1

Höskuldsson Á, Larsen G, Gudmundsson MT, Oddsson B, Magnússon E, Sigmarsson O, Óskarsson N, Jónsdóttir I, Sigmundsson F, Einarsson P, Hreinsdóttir S, Pedersen R, Högnadóttir Þ, Thordarson T, Hayward C, Hartley M, Meara R, Arason Þ, Karlsdóttir S, Petersen GN (2010) The Eyjafjallajökull eruption April to May 2010: magma fragmentation, plume and tephra transport, and course of events. Abstract of presentation in Program of the Eyjafjallajökull and Aviation Conference (15–16 Sept 2010) and associated Eyjafjallajökull Eruption Workshop (17–19 Sept 2010)

Humphreys ED, Grunder AL (2022) Tectonic controls on the origin and segmentation of the Cascade Arc, USA. Bull Volcanol 84:102. https://doi.org/10.1007/s00445-022-01611-2

Ilbeyli N, Pearce JA, Thirlwall MF, Mitchell JG (2004) Petrogenesis of collision-related plutonics in Central Anatolia, Turkey. Lithos 72(3–4):163–182. https://doi.org/10.1016/j.lithos.2003.10.001

Imura R (1992) Eruptive history of the Kirishima volcano during the past 22000 years. Geographical reports of Tokyo Metropolitan University 27, pp 71–89. ISSN: 03868710

Iverson NA, Kyle PR, Dunbar NW, McIntosh WC, Pearce NJG (2014) Eruptive history and magmatic stability of Erebus volcano, Antarctica: insights from Englacial tephra. Geochem Geophys Geosyst 15(11):4180–4202. https://doi.org/10.1002/2014GC005435

Karátson D (1996) Rates and factors of stratovolcano degradation in a continental climate: a complex morphometric analysis for nineteen Neogene/Quaternary crater remnants in the Carpathians. J Volcanol Geotherm Res 73(1–2):65–78. https://doi.org/10.1016/0377-0273(96)00016-9

Karátson D, Favalli M, Tarquini S, Fornaciai A, Wörner G (2010a) The regular shape of stratovolcanoes: a DEM-based morphometrical approach. J Volcanol Geotherm Res 193(3–4):171–181. https://doi.org/10.1016/j.jvolgeores.2010.03.012

Karátson D, Telbisz T, Singer BS (2010b) Late-stage volcano geomorphic evolution of the Pleistocene San Francisco Mountain, Arizona (USA), based on high-resolution DEM analysis and ^{40}Ar/^{39}Ar chronology. Bull Volcanol 72:833–846. https://doi.org/10.1007/s00445-010-0365-8

Karátson D, Telbisz T, Wörner G (2012) Erosional rates and erosion patterns of Neogene to Quaternary stratovolcanoes in the Western Cordillera of the Central Andes: an SRTM DEM based analysis. Geomorphology 139–140:122–135. https://doi.org/10.1016/j.geomorph.2011.10.010

Karátson D, Veres D, Lahitte P, Telbisz T, Wulf S, Gertisser R, Dibacto S, Kiss B, Magyari EK, Novothny Á, Jánosi C (eds) (2022) Evolution of the Ciomadul Volcanic Field—lava domes and explosive eruptions. In: Karátson D, Veres D, Gertisser R, Magyari EK, Jánosi C, Hambach U (eds) Ciomadul (Csomád), the youngest volcano in the Carpathians. Volcanism, palaeoenvironment, human impact. Springer, Switzerland, pp 39–63. https://doi.org/10.1007/978-3-030-89140-4_3

Karlstrom L, Dufek J, Manga M (2010) Magma chamber stability in arc and continental crust. J Volcanol Geotherm Res 190(3–4):249–270. https://doi.org/10.1016/j.jvolgeores.2009.10.003

Kay SM, Coira BL (2009) Shallowing and steepening subduction zones, continental lithospheric loss, magmatism, and crustal flow under the central Andean Altiplano-Puna Plateau. In: Kay SM, Ramos VA, Dickinson WR (eds) Backbone of the Americas: shallow subduction, plateau uplift, and ridge and terrane collision. GSA Memoirs 204, pp 229–259. https://doi.org/10.1130/2009.1204(11)

Kay SM, Mpodozis C, Coira B (1999) Magmatism, tectonism, and mineral deposits of the Central Andes (22–33 S latitude). In: Skinner BJ (ed) Geology and ore deposits of the Central Andes. Special Publication. Society of Economic Geology 7, pp 27–59

Kereszturi Á (ed) (2020) Unique and potentially mars-relevant flow regime and water sources at a high Andes-Atacama site. Astrobiology. Spec Collect: Atacama 20(6):723–740.https://doi.org/10.1089/ast.2018.2013

Kisaka M, Fontijn K, Shemsanga C, Tomašek I, Gaduputi S, Debaille V, Delcamp A, Kervyn M (2021) The late Quaternary eruptive history of Meru volcano, northern Tanzania. J Volcanol Geotherm Res 417.https://doi.org/10.1016/j.jvolgeores.2021.107314

Klemetti EW, Grunder AL (2007) Volcanic evolution of Volcán Aucanquilcha: a long-lived dacite volcano in the Central Andes of northern Chile. Bull Volcanol 70:633–650. https://doi.org/10.1007/s00445-007-0158-x

Koarai M, Tsuzawa M, Kumaki Y, Sekiguchi T (1996) Geomorphological maps for Japanese volcanoes by Geographical Survey Institute, Japan. GeoJournal 38:329–337. https://doi.org/10.1007/BF00204725

Krashennikov SP, Bazanova LI, Ponomareva VV, Portnyagin MV (2020) Detailed tephrochronology and composition of major Holocene eruptions from Avachinsky, Kozelsky, and Koryaksky volcanoes in Kamchatka. J Volcanol Geother Res 408:107088. https://doi.org/10.1016/j.jvolgeores.2020.107088

Kueppers U, Beier C (2018) Volcanoes of the Azores. Revealing the geological secrets of the northern central Atlantic islands. Active volcanoes of the world series. Springer, Berlin, Germany, 355 p. ISBN: 978-3-642-32225-9

Kuno H, Oki Y, Ogino K, Hirota S (1970) Structure of the Hakone caldera as revealed by drilling. Bull Volcanol 34:713–725. https://doi.org/10.1007/BF02596700

Lacey A, Ockendon JR, Turcotte DL (1981) On the geometrical form of volcanoes. Earth Planet Sci Lett 54(1):139–143. https://doi.org/10.1016/0012-821X(81)90074-1

Lebedev VA, Vashakidze GT (2014) The catalogue of Quaternary volcanoes of the Greater Caucasus based on geochronological, volcanological and isotope-geochemical data. J Volcanol Seismolog 8:93–107. https://doi.org/10.1134/S0742046314020043

Lechner HN, DeMets C, Hernandez D, Rose WI (2013) A pilot GPS study of Santa Ana Volcano (Ilamatepec) and Coatepeque caldera, El Salvador. In: Rose WI, Palma JL, Granados HD, Varley N (eds) Understanding open-vent volcanism and related hazards. GSA Special Paper 498, Boulder, Colorado, USA, pp 57–75. ISBN: 978-0-8137-2498-0

Lee C, Wada I (2017) Clustering of arc volcanoes caused by temperature perturbations in the back-arc mantle. Nat Commun 8.https://doi.org/10.1038/ncomms15753

Leonard GS, Cole RP, Christenson BW, Conway CE, Cronin SJ, Gamble JA, Hurst T, Kennedy BM, Miller CA, Procter JN, Pure LR, Townsend DB, White JDL, Wilson CJN (2021) Ruapehu and Tongariro stratovolcanoes: a review of current understanding, New Zealand. J Geol Geophys 64(2–3):389–420. https://doi.org/10.1080/00288306.2021.1909080

Lev E (2018) Why Guatemala's volcano of fire is so much deadlier than Hawaii's Kilauea. Interview with Einat Lev, CBC News. Posted: 6 June 2018. https://science.fas.columbia.edu/2018/06/06/

Lexa J, Seghedi I, Németh K, Szakács A, Konečný V, Pécskay Z, Fülöp A, Kovacs M (2010) Neogene-Quaternary Volcanic forms in the Carpathian-Pannonian Region: a review. Open Geosci 2(3):207–270. https://doi.org/10.2478/v10085-010-0024-5

Lipman PW, Mullineaux (1981) The 1980 eruptions of Mount St. Helens, Washington. US Geol Surv Prof Paper 1250, US Government Printing Office, Washington DC, USA, 844 p

Liu Y, Li S, Jiang S, Liu J, Chen Y, Jiang Z, Wang L, Dai L, Zhu J, Suo Y, Santosh M, Wang P (2023) Origin of microplates under oblique subduction system in New Guinea: inferences from gravity and magnetic data. Gondwana Res 120:175–189. https://doi.org/10.1016/j.gr.2022.09.001

Lockwood JP, Hazlett RW (2010) Effusive volcanic eruptions and their products. In: Lockwood JP, Hazlett RW, de la Cruz-Reyna S (eds) Volcanoes—global perspectives. Wiley, Chichester, UK, pp 127–172. LCCN 2021031545

Löffler E (1977) Geomorphology of Papua New Guinea. Australian National University Press, Canberra, 195 p. ISBN: 0 7081 0410 X

Lundstrom CC, Hervig R, Fischer TP, Sivaguru M, Yin L, Zhou Z, Lin X, Grossi-Diniz R (2022) Insight into differentiation in alkalic systems: nephelinite-carbonate-water experiments aimed at Ol nyo Lengai carbonatite genesis. Front Earth Sci 10.https://doi.org/10.3389/feart.2022.970264

MacDonald GA (1972) Volcanoes. Pentice-Hall, Englewood Cliffs, NJ, USA, 510 p

Madden-Nadeau AL, Cassidy M, Pyle DM, Mather TA, Watt SFL, Engwell SL, Abdurrachman M, Nurshal MEM, Tappin DR, Ismail T (2021) The magmatic and eruptive evolution of the 1883 caldera-forming eruption of Krakatau: integrating field- to crystal-scale observations. J Volcanol Geotherm Res 411.https://doi.org/10.1016/j.jvolgeores.2021.107176

Major JJ (2022) Subaerial volcaniclastic deposits—influences of initiation mechanisms and transport behaviour on characteristics and distributions. Geol Soc Lond Spec Publ 520:29–100. https://doi.org/10.1144/SP520-2021-142

Mamani M, Wörner G, Sempere T (2010) Geochemical variations in igneous rocks of the Central Andean orocline (13°S to 18°S): tracing crustal thickening and magma generation through time and space. GSA Bull 122(1–2):162–182. https://doi.org/10.1130/B26538.1

Mannen K, Yukutake Y, Kikugawa G, Harada M, Itadera K, Takenaka J (2018) Chronology of the 2015 eruption of Hakone volcano, Japan: geological background, mechanism of volcanic unrest and disaster mitigation measures during the crisis. Earth Planets Space 70. https://doi.org/10.1186/s40623-018-0844-2

Martin DP, Rose WI Jr (1981) Behavioral patterns of Fuego volcano, Guatemala. J Volcanol Geotherm Res 10(1–3):67–81. https://doi.org/10.1016/0377-0273(81)90055-X

Martínez-Moreno FJ, Santos FAM, Madeira J, Pous J, Bernardo I, Soares A, Esteves M, Adão F, Ribeiro J, Mata J, da Silveira AB (2018) Investigating collapse structures in oceanic islands using magnetotelluric surveys: the case of Fogo Island in Cape Verde. J Volcanol Geotherm Res 357:152–162. https://doi.org/10.1016/j.jvolgeores.2018.04.028

Mastrolorenzo G, Munno R, Rolandi G (1993) Vesuvius 1906: a case study of a paroxysmal eruption and its relation to eruptive cycles. J Volcanol Geotherm Res 58(1–4):217–237. https://doi.org/10.1016/0377-0273(93)90110-D

McBirney AR, Williams H (1979) Geology and petrology of the Galápagos Islands. Geol Soc Amer Mem 118, 197 p. ISBN: 9780813711188

Miller TP, McGimsey RG, Richter DH, Riehle JR, Nye CJ, Yount ME, Dumoulin JA (1998) Catalogue of the historically active volcanoes of Alaska. US Geol Surv Open File Report 98-582, Fairbanks, AK, USA, 104 p

Miller TP, Smith RL (1987) Late Quaternary caldera-forming eruptions in the eastern Aleutian arc, Alaska. Geology 15(5):434–438. https://doi.org/10.1130/0091-7613(1987)15<434:LQCEIT>2.0.CO;2

Milne J (1878) On the form of volcanoes. Geol Mag 5(8):337–345. https://doi.org/10.1017/S0016756800150678

Momeni S, Madariaga R (2022) Long-term triggered seismicity on the Mosha fault by Damavand volcano, Iran: implications on the seismic hazard of Tehran metropolis. Front Earth Sci 10:945297. https://doi.org/10.3389/feart.2022.945297

Moriya I (1979) Geomorphological developments and classification of the Quaternary volcanoes in Japan. Geogr Rev Jpn 52:479–501. https://doi.org/10.4157/grj.52.479

Moriya I (1983) Geomorphology of Japanese volcanoes. Tokyo University Press, Tokyo, Japan, 135 p

Naismith AK, Watson IM, Escobar-Wolf R, Chigna G, Thomas H, Coppola D, Chun C (2019) Eruption frequency patterns through time for the current (1999–2018) activity cycle at Volcán de Fuego derived from remote sensing data: Evidence for an accelerating cycle of explosive paroxysms and potential implications of eruptive activity. J Volcanol Geotherm Res 371:206–219. https://doi.org/10.1016/j.jvolgeores.2019.01.001

Nakada S, Shimizu H, Otha K (1999) Overview of the 1990–1995 eruption at Unzen volcano. J Volcanol Geotherm Res 89(1–4):1–22. https://doi.org/10.1016/S0377-0273(98)00118-8

Nehlig P, Boivin P, de Goër A, Mergoil J, Prouteau G, Sustrac G, Thiéblemont D (2003) Les volcans du Massif Central. Revue Géologues. Numéro spécial. Massif Central, Orleans, France, 41 p

Newhall CG, Bronto S, Alloway B, Banks NG, Bahar I, del Marmol MA, Hadisantono RD, Holcomb RT, McGeehin J, Miksic JN, Rubin M, Sayudi SD, Sukhyar R, Andreastuti S, Tilling RI, Torley R, Trimble D, Wirakusumah AD (2000) 10,000 Years of explosive eruptions of Merapi Volcano, Central Java: archaeological and modern implications. J Volcanol Geotherm Res 100(1–4):9–50. https://doi.org/10.1016/S0377-0273(00)00132-3

Nishihara A, Geshi N, Naruo H (2022) Long-term change of the eruption activities of Sakurajima volcano, Japan, inferred from the fallout tephra deposits. Front Earth Sci 10:988373. https://doi.org/10.3389/feart.2022.988373

Nomade S, Scaillet S, Pastre J-F, Nehlig P (2012) Pyroclastic chronology of the Sancy stratovolcano (Mont-Dore, French Massif Central): new high-precision $^{40}Ar/^{39}Ar$ constraints. J Volcanol Geotherm Res 225–226:1–12. https://doi.org/10.1016/j.jvolgeores.2012.02.006

Norini G, Capra L, Groppelli G, Lagmay AMF (2008) Quaternary sector collapses of Nevado de Toluca volcano (Mexico) governed by regional tectonics and volcanic evolution. Geosphere 4(5):854–871. https://doi.org/10.1130/GES00165.1

Oppenheimer C, Kyle P (eds) (2008) Volcanology of Erebus volcano, Antarctica. Spec issue J Volcanol Geotherm Res 177:531–754. https://doi.org/10.1016/j.jvolgeores.2008.10.006

Paguican EM, Grosse P, Fabbro GN, Kervyn M (2021) Morphometric classification and spatial distribution of Philippine volcanoes. J Volcanol Geotherm Res 418:107251. https://doi.org/10.1016/j.jvolgeores.2021.107251

Pan B, de Silva SL, Xu J, Chen Z, Miggins DP, Wei H (2017) The VEI-7 Millennium eruption, Changbaishan-Tianchi volcano, China/DPRK: New field, petrological, and chemical constraints on stratigraphy, volcanology, and magma dynamics. J Volcanol Geotherm Res 343:45–59. https://doi.org/10.1016/j.jvolgeores.2017.05.029

Pike RJ (1978) Volcanoes on the inner planets: some preliminary comparisons of gross topography. In: 9th lunar and planetary science conference, pp 3239–3273

Pike RJ, Clow GD (1981) Martian volcanoes in a classification of central edifices. Abstracts of Papers presented to the Third International Colloquium on Mars, pp 199–201

Plechov PJ, Kamenetsky VS (2023) Magmatism in Kamchatka and the Kurile Islands. Petrology 31:261–382. https://doi.org/10.1134/S0869591123030086

Poli P, Shapiro NM (2022) Rapid characterization of large volcanic eruptions: Measuring the impulse of the Hunga Tonga Ha'apai explosion from teleseismic waves. Geophys Res Lett 49:e2022GL098123. https://doi.org/10.1029/2022GL098123

Ponomareva VV, Melekestsev I, Braitseva O, Churikova T, Pevzner M, Sulerzhitsky L (2007) Late Pleistocene-Holocene Volcanism on the Kamchatka Peninsula, Northwest Pacific Region. In: Eichelberger J, Gordeev E, Izbekov P, Kasahara M, Lees J (eds) Volcanism and subduction: the Kamchatka Region, pp 165–198. https://doi.org/10.1029/172GM15

Pevzner MM, Tolstykh ML, Babansky AD, Layer P, Volynets AO (2014) First data on the isotope age and composition of the parental melts of the initial stage of the Shiveluch volcanic massif. Volcanism and associated processes, Materials of the annual conference celebrating the Volcanologist day. Petropavlovsk-Kamchatsky, Institute of Volcanology and Seismology, pp. 104–107 (in Russian). http://www.kscnet.ru/ivs/publication/volc_day/2014/art16.pdf

Principe C, Tanguy J-C, Arrighi S, Paiotti A, Le Goff M, Zoppi U (2004) Chronology of Vesuvius' activity from A.D. 79 to 1631 based on archeomagnetism of lavas and historical sources. Bull Volcanol 66:703–724. https://doi.org/10.1007/s00445-004-0348-8

Reynolds JH (1987) Timing and sources of neogene and quaternary volcanism in South-Central Guatemala. J Volcanol Geotherm Res 33(1–3):9–22. https://doi.org/10.1016/0377-0273(87)90052-7

Romero JE, Vera F, Polacci M, Morgavi D, Arzilli F, Alam MA, Bustillos JE, Guevara A, Johnson JB, Palma JL, Burton M, Cuenca E, Keller W (2018) Tephra from the 3 March 2015 sustained column related to explosive lava fountain activity at Volcán Villarrica (Chile). Front Earth Sci 6:98. https://doi.org/10.3389/feart.2018.00098

Rout SS, Blum-Oeste M, Wörner G (2021) Long-term temperature cycling in a shallow magma reservoir: insights from sanidine megacrysts at Taapaca volcano, Central Andes. J Petrol 62(9):1–32. https://doi.org/10.1093/petrology/egab010

Roverato M, Capra L, Sulpizio R, Norini G (2011) Stratigraphic reconstruction of two debris avalanche deposits at Colima Volcano (Mexico): insights into pre-failure conditions and climate influence. J Volcanol Geotherm Res 207(1–2):33–46. https://doi.org/10.1016/j.jvolgeores.2011.07.003

Roverato M, di Traglia F, Procter J, Paguican E, Dufresne A (2021) Factors contributing to volcano lateral collapse. In: Roverato M, Dufresne A, Procter J (eds) Volcanic debris avalanches. From collapse to hazard. Springer, Cham, pp 91–119. https://doi.org/10.1007/978-3-030-57411-6_5

Saginor I, Gazel E, Condie C, Carr MJ (2013) Evolution of geochemical variations along the Central American volcanic front. Geochem Geophys Geosyst 14(10):4504–4522. https://doi.org/10.1002/ggge.20259

Sapper K (1927) Vulkankunde. J Engelhorns nachf, Stuttgart, Germany, 474 p, maps 4

Schaefer JR, Nye C (2008) The Alaska Volcano Observatory—20 years of volcano research, monitoring, and eruption response. Alaska Division of Geological and Geophysical Surveys, Alaska GeoSurvey News NL 2008-001, 11(1):1–9

Schattel N, Portnyagin M, Golowin R, Hoernle K, Bindeman I (2014) Contrasting conditions of rift and off-rift silicic magma origin on Iceland. Geophys Res Lett 41(16):5813–5820. https://doi.org/10.1002/2014GL060780

Schlüter HU, Gaedicke C, Roeser HA, Schreckenberger B, Meyer H, Reichert C, Djajadihardja Y, Prexl A (2002) Tectonic features of the southern Sumatra-western Java forearc of Indonesia. Tectonics 21(5): https://doi.org/10.1029/2001TC901048

Schmidt ME, Grunder AL (2009) The evolution of North Sister: a volcano shaped by extension and ice in the central Oregon Cascade Arc. GSA Bull 121(5–6):643–662. https://doi.org/10.1130/B26442.1

Scott EM, Allen MB, Macpherson CG, McCaffrey KJW, Davidson JP, Saville C, Ducea MN (2018) Andean surface uplift constrained by radiogenic isotopes of arc lavas. Nat Commun 9:969. https://doi.org/10.1038/s41467-018-03173-4

Şen E, Kürkcüoğlu B, Aydar E, Gourgaurd A, Vincent PM (2003) Volcanological evolution of Mount Erciyes stratovolcano and origin of the Valibaba Tepe ignimbrite (Central Anatolia, Turkey). J Volcanol Geotherm Res 125(3–4):225–246. https://doi.org/10.1016/S0377-0273(03)00110-0

Sherrod DR, Magigita MM, Kwelwa S (2013) Geologic Map of Oldonyo Lengai (Olnyo Lengai) Volcano and surroundings, Arusha Region, United Republic of Tanzania. US Geol Surv Open-File Report 1306, Reston, VA, USA, 65 p

Shirzaei M, Walter TR, Nankali HR, Holohan EP (2011) Gravity-driven deformation of Damavand volcano, Iran, detected through InSAR time series. Geology 39(3):251–254. https://doi.org/10.1130/G31779.1

Siebert L, Cottrell E, Venzke E, Andrews B (2015) Earth's volcanoes and their eruptions: an overview. In: Sigurdsson H, Houghton B, McNutt SR, Rymer H, Stix J (eds) Encyclopedia of volcanoes. Academic Press, San Diego, CA, pp 239–255. ISBN: 978-0-12-385938-9

Siegburg M, Gernon TM, Bull JM, Keir D, Barfod DN, Taylor RN, Abebe B, Ayele A (2018) Geological evolution of the Boset-Bericha Volcanic Complex, Main Ethiopian Rift: 40Ar/39Ar evidence for episodic Pleistocene to Holocene volcanism. J Volcanol Geotherm Res 351:115–133. https://doi.org/10.1016/j.jvolgeores.2017.12.014

Sigurdsson H (1999) Melting the Earth: the history if ideas on volcanic eruptions. Oxford University Press, Oxford, UK, 260 p. ISBN: 978-0-19-510665-7

Sigurdsson H (2015) Volcanoes in Art. In: Sigurdsson H, Houghton B, McNutt SR, Rymer H, Stix J (eds) Encyclopedia of volcanoes. Academic Press, San Diego, CA, pp 273–297. ISBN: 978-0-12-385938-9

Sigurdsson H, Carey S (1992) The eruption of Tambora in 1815: environmental effects and eruption dynamics. In: Sigurdsson H, Carey S (eds) The year without a summer. National Museum of Canada, Canada, Ottawa, pp 16–45

Simkin T, Siebert L, McClelland L, Bridge D, Newhall CG, Latter JH (1981) Volcanoes of the world: a regional directory, gazetteer, and chronology of volcanism during the last 10,000 years. Hutchingson Ross, Stroundsburg, PA, USA, 240 p

Singer BS, Thompson RA, Dungan MA, Feeley TC, Nelson ST, Pickens JC, Brown LL, Wulff AW, Davidson JP, Metzger J (1997) Volcanism and erosion during the past 930 k.y. at the Tatara-San Pedro complex, Chilean Andes. GSA Bull 109(2):127–142. https://doi.org/10.1130/0016-7606(1997)109<0127:VAEDTP>2.3.CO;2

Sisson TW, Vallance JW (2009) Frequent eruptions of Mount Ranier over the last ~2,600 years. Bull Volcanol 71:595–618. https://doi.org/10.1007/s00445-008-0245-7

Smith DR, Leeman WP (1993) The origin of Mount St. Helens Andesites. J Volcanol Geotherm Res 55(3–4):271–303. https://doi.org/10.1016/0377-0273(93)90042-P

Smith IEM, Price RC (2006) The Tonga-Kerdamec arc and Havre-Lau back-arc system: their role in the development of tectonic and magmatic models for the western Pacific. J Volcanol Geotherm Res 156(3–4):315–331. https://doi.org/10.1016/j.jvolgeores.2006.03.006

Soto GJ, Alvarado GE (2006) Eruptive history of Arenal volcano, Costa Rica, 7 ka to present. J Volcanol Geotherm Res 157(1–3):254–269. https://doi.org/10.1016/j.jvolgeores.2006.03.041

Soto V (2023) Bedrock outcropping in the accumulation zone of the largest glacier in Mexico (Glaciar Norte of Citlaltépetl), as evidence of a possible accelerated extinction. J Mt Sci 20(2):338–354. https://doi.org/10.1007/s11629-021-7216-2

Souther JG (1992) The late Cenozoic Mount Edziza volcanic complex, British Columbia. Geol Surv Canada Memoir 420, 320 p

Stern CR (2004) Active Andean volcanism: its geologic and tectonic setting. Revista Geológica de Chile 31(2):161–206.https://doi.org/10.4067/S0716-02082004000200001

Suñe-Puchol I, Aguirre-Díaz GJ, Dávila-Harris P, Miggins DP, Pedrazzi D, Costa A, Ortega-Obregón C, Lacan P, Hernández W, Gutiérrez E (2019) The Ilopango caldera complex, El Salvador: Origin and early ignimbrite-forming eruption of a graben/pull-apart caldera structure. J Volcanol Geotherm Res 371:1–19. https://doi.org/10.1016/j.jvolgeores.2018.12.004

Szakács A, Seghedi I (1995) The Călimani-Gurghiu-Harghita volcanic chain, East Carpathians, Romania: volcanological features. Acta Vulcanol 7(2):145–153

Tatsumi Y, Suenaga N, Yoshioka S, Kaneko K, Matsumoto T (2020) Contrasting volcano spacing along SW Japan arc caused by difference in age of subducting lithosphere. Nat Sci Rep 10.https://doi.org/10.1038/s41598-020-72173-6

Taylor B, Benyshek EK, Goodliffe AM (2023) Subduction of submarine arc volcanoes beneath the Solomon islands arc. Geoscience 13(8):236. https://doi.org/10.3390/geosciences13080236

Thordarson T, Larsen G (2007) Volcanism in Iceland in historical time: volcano types, eruption styles and eruptive history. J Geodyn 43(1):118–152. https://doi.org/10.1016/j.jog.2006.09.005

Thouret J-C (1999) Volcanic geomorphology—an overview. Earth Sci Rev 47(1–2):95–131. https://doi.org/10.1016/S0012-8252(99)00014-8

Thouret J-C, Finizola A, Fornari M, Legeley-Padovani A, Suni J, Frechen M (2001) Geology of El Misti volcano near the city of Arequipa, Peru. GSA Bull 113(12):1593–1610. https://doi.org/10.1130/0016-7606(2001)113<1593:GOEMVN>2.0.CO;2

Thouret J-C, Jicha BR, Paquette J-L, Cubukcu EH (2016) A 25 Myr chronostratigraphy of ignimbrites in South Peru. Implications for the volcanic history of the Central Andes. J Geol Soc 173(5):734–756. https://doi.org/10.1144/jgs2015-162

Thouret J-C, Juvigné E, Gourgaud A, Boivin P, Dávila J (2002) Reconstruction of the AD 1600 Huaynaputina eruption based on the correlation of geologic evidence with early Spanish chronicles. J Volcanol Geotherm Res 115(3–4):529–570. https://doi.org/10.1016/S0377-0273(01)00323-7

Thouret J-C, Wörner G, Gunnell Y, Singer B, Zhang X, Souriot T (2007) Geochronologic and stratigraphic constraints on canyon incision and Miocene uplift of the Central Andes in Peru. Earth Planet Sci Lett 263(3–4):151–166. https://doi.org/10.1016/j.epsl.2007.07.023

Tortini R, van Wyk de Vries B, Carn SA (2015) Seeing the faults from the hummocks: tectonic or landslide fault discrimination with LiDAR at Mt Shasta, California. Front Earth Sci 3:48. https://doi.org/10.3389/feart.2015.00048

Turner S, Foden J, George R, Evans P, Varne R, Elburg M, Jenner G (2003) Rates and processes of Potassic Magma Evolution beneath Sangeang Api Volcano, East Sunda Arc, Indonesia. J Petrol 44(3):491–515. https://doi.org/10.1093/petrology/44.3.491

Vallance JW, Schilling SP, Matías O, Rose WI, Howell MM (2001) Volcano hazards at Fuego and Acatenango, Guatemala. US Geological Survey Open-File Report 01-431, Vancouver, WA, 23 p

Valverde V, Mothes PA, Beate B, Bernard J (2021) Enormous and far-reaching debris avalanche deposits from Sangay volcano (Ecuador): multidisciplinary study and modeling the 30 ka sector collapse. J Volcanol Geotherm Res 411:107172. https://doi.org/10.1016/j.jvolgeores.2021.107172

Varley N (2012) Volcanoes of Mexico. In: Meyers RA (ed) Encyclopedia of sustainability science and technology. Springer, New York, NY, pp 11613–11633

Vessell RK, Davies DK (1981) Non-marine sedimentation in an active fore-arc basin. In: Ethridge FG, Flores RM (eds) Recent and ancient non-marine depositional environments: models for exploration. SEPM Spec. Publ 31, pp 31–45

Wadge G (1982) Steady state volcanism: evidence from eruption histories of polygenetic volcanoes. J Geophys Res 87(B5):4035–4049. https://doi.org/10.1029/JB087iB05p04035

Walker GPL (2000) Basaltic volcanoes and volcanic systems. In: Sigurdsson H, Houghton B, McNutt SR, Rymer H, Stix J (eds) Encyclopedia of volcanoes. Academic Press, San Diego, CA, pp 283–289. ISBN: 978-0-12-385938-9

Williams H, McBirney AR (1979) Volcanology. Freeman Cooper Co, San Francisco, CA, 397 p

Wood CA (1978) Morphometric evolution of composite volcanoes. Geophys Res Lett 5(6):437–439. https://doi.org/10.1029/GL005i006p00437

Williams SN (ed) (1990) Nevado del Ruiz Volcano, Colombia. J Volcanol Geotherm Res Spec Issue 1:41(1–4):1–379; 11:42(1):1–224

Wörner G, Mamani M, Blum-Oeste M (2019) Magmatism in the Central Andes. In: Wörner G, Schildgen T, Reich M (eds) Central Andes: mountains, magmas, and minerals. Elements, vol 14, pp 237–244, https://doi.org/10.2138/gselements.14.4.237

Yamamoto M (1984) Origin of calc-alkaline andesite from Oshima-Ōshima volcano, North Japan. J Fac Sci Hokkaido Univ IV 21(1):77–131

Yamaoka K, Geshi N, Hashimoto T, Ingebritsen SE, Oikawa T (2016) The phreatic eruption of Mt. Ontake volcano in 2014. Special issue. Earth Planets Space 68:175. https://doi.org/10.1186/s40623-016-0548-4

Yang Q, Jenkins SF, Lerner GA, Li W, Suzuki T, McLean D, Derkachev AN, Utkin IV, Wei H, Xu J, Pan B (2021) The millennium eruption of Changbaishan Tianchi Volcano is VEI 6, not 7. Bull Volcanol 83:74. https://doi.org/10.1007/s00445-021-01487-8

Yoshimoto M, Fujii T, Kaneko T, Yasuda A, Nakada S, Matsumoto A (2010) Evolution of Mount Fuji, Japan: inference from drilling into the subaerial oldest volcano, pre-Komitake. Island Arc 19(3):470–488. https://doi.org/10.1111/j.1440-1738.2010.00722.x

Zamboni D, Gazel E, Ryan JG, Cannatelli C, Lucchi F, Atlas ZD, Trela J, Mazza SE, de Vivo B (2016) Contrasting sediment melt and fluid signatures for magma components in the Aeolian Arc: implications for numerical modeling of subduction systems. Geochem Geophys Geosyst 17(6):2034–2053. https://doi.org/10.1002/2016GC006301

Chapter 6 Calderas and Caldera Complexes

1 Definition and Early Classification

Calderas are subcircular depressions in volcanic areas, with a diameter of between about two and several tens of kilometres. The term has a late Latin origin, meaning 'a pot for boiling' (Shorter Oxford Dictionary), and was first introduced into the geological vocabulary ('a deep cauldron-like cavity on the summit of a volcano') when Leopold von Buch published his 1815 memoir of his visit to the Canary Islands. Large calderas are impressive volcanic landforms due to their size: 10–30 km in diameter, 500–1000 m in depth, surrounded by extensive aprons of ignimbrite covering hundreds to thousands of km^2, as seen in satellite images (e.g. Cerro Galán in Argentina; Valles in New Mexico, USA; Photos 69 and 70). The largest known calderas associated with silicic pyroclastic eruptions that produce ash-flow tuffs or ignimbrites can reach 60 km in diameter, while the associated pyroclastic deposits have volumes exceeding 10^3 km^3.

A caldera is the wide topographic basin that results from different types of collapse in response to the removal of several cubic kilometres of magma, and usually has inward-facing caldera walls surrounded by an elevated topographic rim (e.g. Crater Lake, Oregon USA; Santorini, Greece; Toba, Sumatra, Indonesia; Okmok, Aleutians, USA, Photo 71). In addition to their potential volcanic hazards, calderas host many of the Earth's ore deposits and geothermal energy resources, as well as providing key insights into the generation and evolution of large-volume silicic magma bodies (Hildreth 1981; Lipman et al. 1984; Lipman 1992, 1997; Cashman and Giordano 2014; Giordano et al. 2024).

In his classic publication on 'Calderas and their origin', Williams (1941) summarised the definitions and types used during the first part of the twentieth century. William's broad classification (Table 1) was based on type locality and included many different types of volcanic events. Further insight into the relationships between caldera formation and ignimbrites was provided by R. L. Smith (1960, 1979) who drew attention to the ash-flow sheets associated with voluminous caldera-forming eruptions. Williams and McBirney (1979) redefined calderas as 'large collapse depressions, more or less circular or cirque-like in form, the diameter of which is many times greater than any included vent'.

Williams and McBirney based their sixfold classification on eponymous caldera volcanoes (Table 1). They added two shield calderas, from Hawai'i and the Galápagos, created by drainage of magma from beneath the summit region and/or eruption of lavas from circumferential fissures near the summit, respectively (see Chap. 4). Walker (1984) distinguished five types based on the distribution of vents within the caldera. In

D. Karátson and J.-C. Thouret, *Volcano Geomorphology*, Advances in Volcanology,
https://doi.org/10.1007/978-3-031-77693-9_7

Table 1 Early definition of caldera categories and mechanisms of collapse proposed by four authors or groups of authors (after Williams 1941, 1942; Williams and McBirney 1979; Lipman 1984, 1997, 2000a)

Key process Authors	Geometric, geomorphic and structural features; eponymous volcano Williams (1941, 1942)	Chain of processes of subsidence and collapse; eponymous volcano Williams and McBirney (1979)	Distribution of post-caldera vents (volcano examples) Walker (1984)	End-members of caldera geometry, with inferred shape and depth of pluton Lipman (1984, 1997, 2000a)
Types or end-members, and eponymous volcanoes or examples	Explosion and collapse Krakatau Kilauea Katmai	Krakatau: foundering of the top of large composite volcanoes following eruptions of siliceous pumice falls and flows <100 km^3. The land simply sinks without rupturing	Type C: a single vent occupies a central or near-central position (Vulcano, Vico, Italy)	Plate or piston. Floor crust remains intact as it drops along enclosing ring faults One only, large eruption
	Cryptovolcanic Glencoe Miscellaneous	Katmai: drainage of a central magma reservoir to feed new volcanoes or fissure eruptions beyond the base of the cone	Type L: vents distributed in straight line or linear zone (Hakone, Japan; Tengger, Indonesia)	Piecemeal. Floor broken into blocks, which subsides to different degrees. Several cycles of eruption and collapse resulting in nested or compound calderas
	Erosional caldera (see Chap. 8)	Valles: foundering along arcuate fractures independent of pre-existing volcanoes as a consequence of, and simultaneously with, the discharge of colossal volumes of siliceous pumice falls and flows >100 km^3	Type M: a single vent occurring at or near the caldera margins (no example offered)	Downsag. The land, overlying magma bodies nearly too small or deep to create caldera, simply sinks without rupturing. Wide and/or deep pluton
	Volcanic graben Summit Sector	Masaya: piecemeal cauldron subsidence of a broad shallow depression occupying most of the central portion of a low inconspicuous shield	Type R: vents along an arcuate line paralleling the caldera margin (Masaya, Okmok, La Primavera, etc.)	Trapdoor. Collapse along one side of a ring fracture system, with floors sloping at an angle toward the location of maximum subsidence Asymmetric pluton
	Volcanic vents or fissure throughs	Atitlán: cauldron subsidence unrelated to an earlier cone, but associated with eruptions from volcanoes near the rim or from nearby fissures	Type S: widely scattered vents within the caldera (Newberry, USA; Phlegrean Fields and Pantelleria, Italy)	Funnel, possibly the least common type. Small caldera a few km across whose enclosing faults converge at a depth of several km into narrow pipe-like conduits. Small and deep pluton
	Major volcano-tectonic depression	Volcano-tectonic depression	Inverse volcano; e.g La Primavera (Mexico)	Caldera complex

Calderas on shield volcanoes are not included here (see Chap. 4) the 1980s, processes were regarded as the most significant means of explaining caldera collapse, and Walker (1984) noted the importance of downsag in many calderas. Lipman (1997, 2000a) followed up on this and completed the caldera classification based on their form, size, geometry and relationships with plutons beneath poly-featured depressions. Cole et al. (2005) reviewed the terminology and history of development of ideas on caldera shape, formation, origin and mechanisms. Branney and Acocella (2015) further developed mechanisms of caldera collapse and incorporated features and deposits around and within the large ash-flow calderas.

Calderas have been distinguished according to size, shape, single versus multiple collapse, composition, and geodynamic setting. Calderas and caldera complexes can be found in almost every volcanic context. Nevertheless, large calderas are more common near subduction zones, in active continental margins and islands arcs, as well as in back-arc setting. A useful distinction can be made into three principal types: (1) calderas at basaltic shield volcanoes and shield-like edifices (see Chap. 4), (2) calderas at the summit of intermediate-composition composite cones in continental or island arcs (see Chap. 5), and (3) silicic 'ignimbrite' or 'ash-flow calderas', which form during large-volume explosive eruptions (Branney and Acocella 2015), hence termed 'explosive calderas' by Lockwood and Hazlett (2010). Here we look at the second type to explain collapse mechanisms, but focus mainly on the third one, which not only forms on the summit of individual volcanoes, but also develops independently of any preexisting edifice on uplifted basement and in large volcanic fields.

2 Terminology

Cole et al. (2005) distinguished the following geomorphological and volcanological terms: caldera, caldera complex, cauldron, ring structure and associated features (Fig. 1).

- A *caldera* is a volcanic structure, generally large, which is principally the result of collapse or subsidence into the top of a magma chamber during or immediately following eruptive activity. The complexity of calderas lies in the fact that original boundaries can be blurred by subsequent erosion, including volcanic features that may no longer have any original topographic expression; the long-lasting erosion can reveal deeper levels of collapse, such as low-lying areas within the high and rugged San Juan Mountains, Colorado, after nearly 30 Myr of erosion (Lipman 1984).
- A *crater* may resemble a caldera in form but is much smaller (usually less than 1 km across) and differs genetically in being a constructional form rather than a product of destruction. Subsidence features less than a kilometre across are usually called *collapse pits*, which lack ejecta rims and have very steep walls, with talus-filled floors. Pits and other collapse craters commonly occur on the caldera floor and rift zones of young, oceanic shield volcanoes, especially near the summit where magma withdrawal from flank eruptions at lower elevations can easily remove support from the thin overlying crust (Holohan et al. 2011).
- A *caldera complex* consists of spatially and structurally associated nested or overlapping calderas of different ages. The size and volume of some caldera complexes is astonishing: for example, La Garita complex of six calderas (35 × 75 km) in San Juan Mountains, Colorado (Fig. 2) has an overall size of about 2000 km^2 (Lipman 2000a, b; see Sect. 6.3.1) and may belong to the plate/piston group (see Table 1), where the roof block dropped over 2.5 km during the 27.6 Ma eruption of the 5000 km^3 Fish Canyon Tuff (Hildreth 1979; Wilson and Hildreth 1997).
- *Cauldron* is an eroded caldera in which most of the eruptive deposits accompanying collapse have been removed by erosion, and older volcanic or sedimentary units and plutonic rocks are now exposed below the caldera floor, often termed caldera root (Fig. 1B). There is no sharp demarcation between calderas and cauldrons, which

simply reflect different parts of a caldera structure; surface eruptions and foundering into a shallow magma chamber contribute to the development of both. Cauldrons tend to occur in pre-Tertiary terrains and in the basement, and the principal difference is that cauldrons are thought to result from a passive foundering of the roof of a body of magma. Cole et al. (2005) suggested that eroded caldera structures can be called cauldrons or ring structures, depending on how much erosion has occurred.

- *Ring structure* is a magma reservoir, exposed by deep erosion (generally >2 km), beneath an inferred caldera structure (Fig. 1B). In ring structures evidence of a caldera structure has largely been removed by erosion, exposing a magma chamber or chambers, which are the plutonic equivalent of the caldera volcanics.

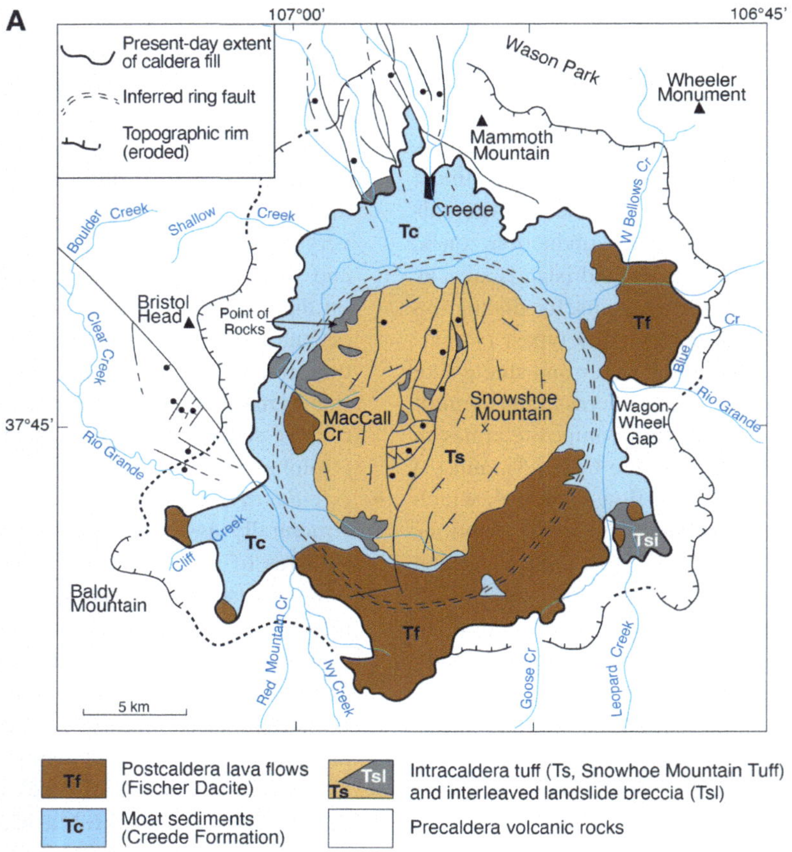

Fig. 1 Differences between (A) caldera, (Ba, b, c) cauldron and ring structure (after Cole et al. 2005, Fig. 1, p. 4) and related cross sections (from Williams and McBirney 1979, modified). The examples used are: **A** Creede caldera, Colorado, nested in La Garita caldera complex (Lipman 2000a, b; Bethke and Hay 2000, see location in Fig. 2). Geological map shows approximate location of eroded topographic caldera rim, present-day extent of moat-fill deposits, inferred ring fault, and late normal faults formed during resurgent doming and mineralization (after Lipman 2000a, Fig. 7, p. 659). Note the intra-caldera tuff and interleaved landslide breccia and post-caldera lava flows. **Ba** Ossipee cauldron, Caroll County, New Hampshire (Kingsley 1931) along with cross section (modified from Williams and McBirney 1979). **Bb** Kudaru ring structure, Nigeria (Bain 1934, all references in: Cole et al. 2005, Fig. 1, p. 4). **Bc** Cross section of the Sabaloka igneous complex, Sudan, about 700 million years old, showing ring dykes and plateau ignimbrites filling cauldron subsidence, and subsided roof of magma chamber (Almond 1977 *in:* Francis 1993, Fig. 14.3, p. 293)

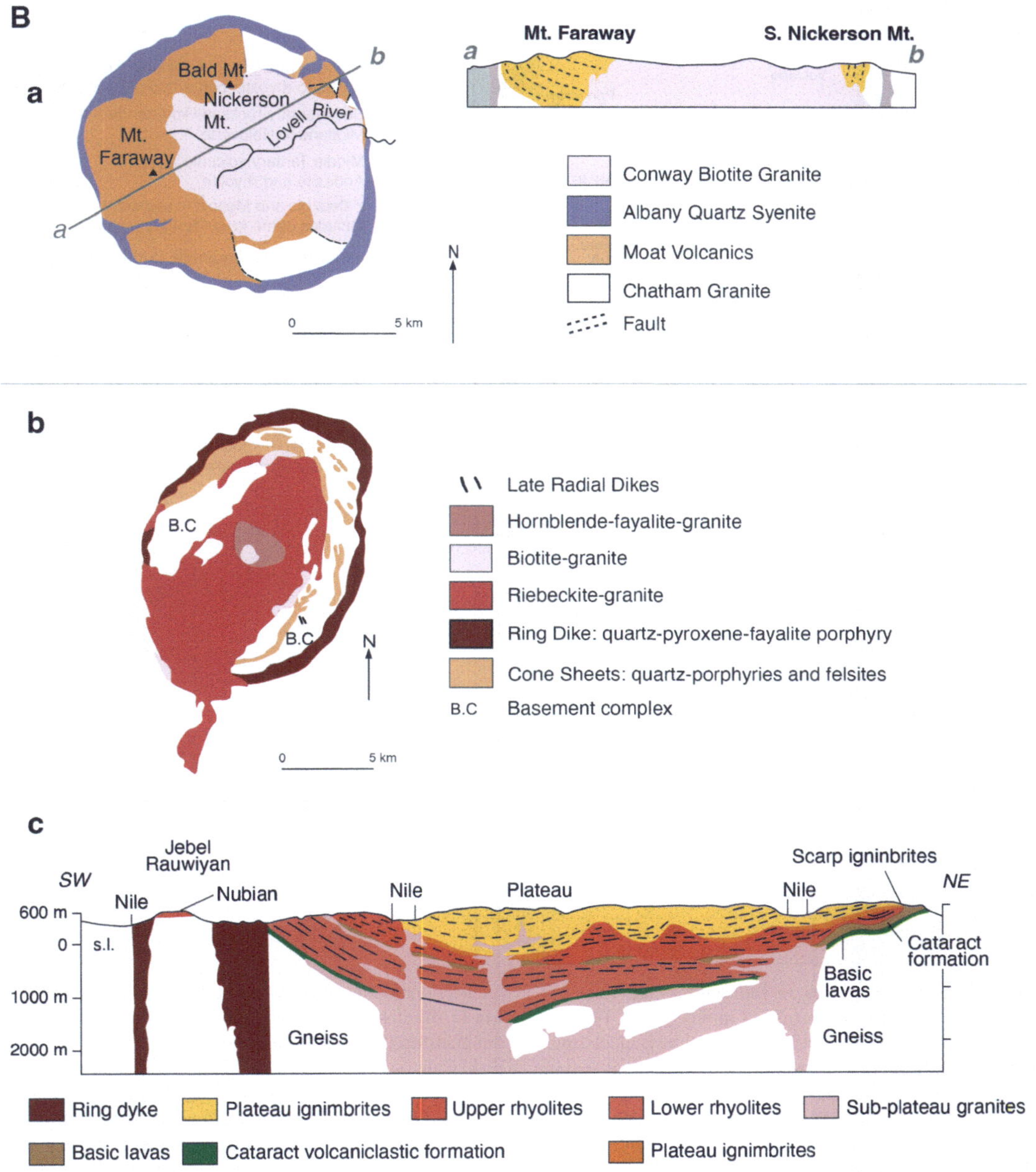

Fig. 1 (continued)

- The term *volcano-tectonic depression* was coined by Williams (1941) and later called 'inverse volcano' by Walker et al. (1981). Cas and Wright (1988) defined it as 'an extensive basin (tens to hundreds of km long), often bounded by tectonic faults, containing calderas and ignimbrites or a caldera complex nested in a large volcanic field that usually includes cones, domes, and other types of edifices'. Case studies of La Primavera volcanic field (Mexico) and the Taupo Volcanic Zone (New Zealand) are described in Sect. 11.

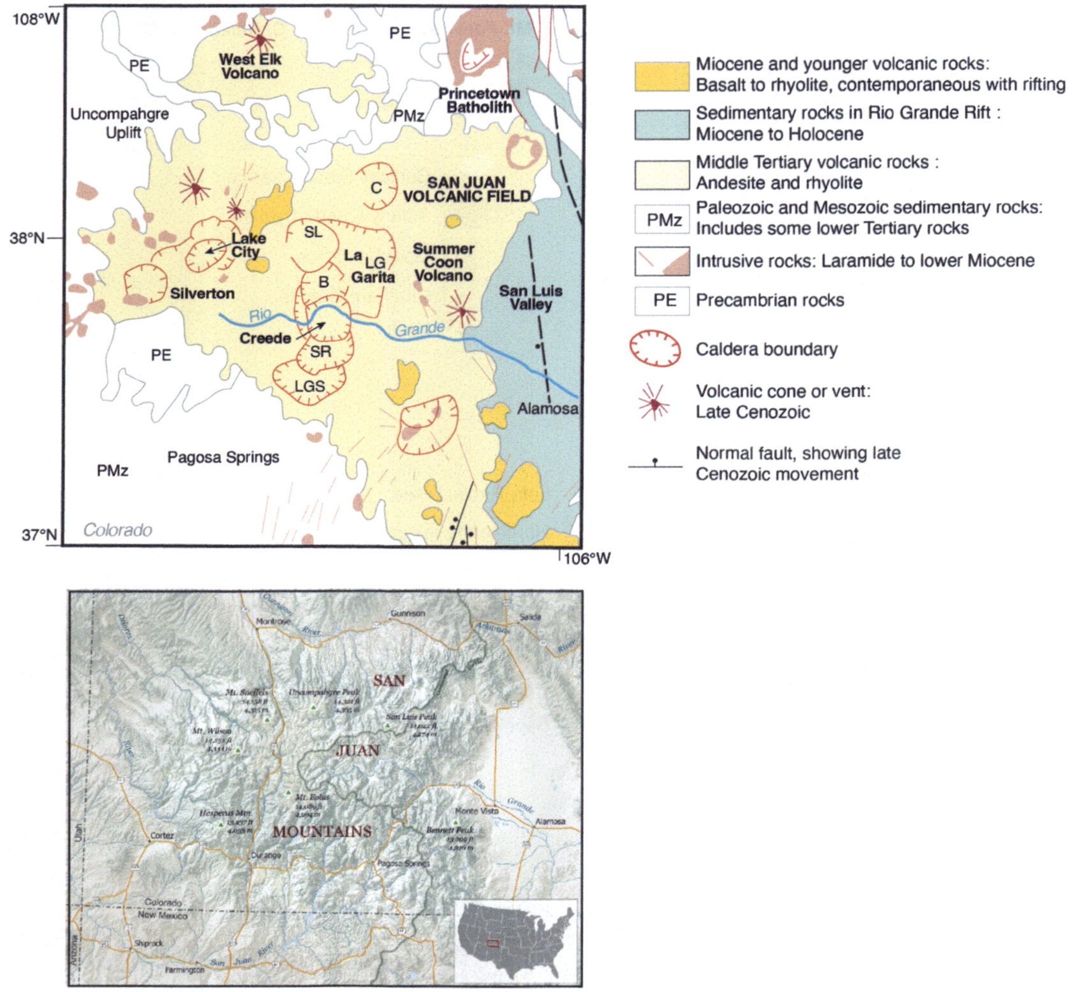

Fig. 2 Caldera complexes in the central San Juan volcanic field, Colorado, USA (after Lipman 2000b, Fig. 1, p. 11; Lipman 2006, Fig. 1, U.S. Geol. Survey, Geol Invest Series I-2799). **A** As many as ten exposed calderas varying in size from 10 to 75 km in maximum diameter, the largest calderas (La Garita, Platoro) being associated with the most voluminous eruptions. Eruption of at least 8800 km^3 of dacitic-rhyolitic magma as nine major ash-flow sheets (individually 150–5000 km^3) was accompanied by recurrent caldera subsidence between 28.3 and about 26.5 Ma in the central San Juan Mountains. After collapse of the giant La Garita caldera during eruption of the 27.6 Ma Fish Canyon Tuff, seven additional explosive eruptions and calderas (San Luis complex of 3 calderas, Bachelor, Creede, South River, and La Garita South) formed inside the La Garita depression within about 1 Myr. **B** Inset map of the San Juan Mountains in Colorado, USA based on SRTM DEM (Public Domain)

3 Geomorphic and Structural Boundaries of a Caldera

The shape and size of calderas is variable. Some calderas are near circular and have a simple history of collapse (e.g. Reporoa, New Zealand; Nairn et al. 1994; Beresford and Cole 2000; Creede, Colorado, USA; Bethke and Hay 2000), while others are more rectilinear (e.g. Taupo, New Zealand; Cole et al. 1998; Silent Canyon, Nevada, USA; Ferguson et al. 1994). Size can also vary from small pit-craters (e.g. in Masaya

Caldera, Nicaragua) to very large nested collapse structures up to 20–30 km in diameter (e.g. La Garita, Colorado; Lipman 2000b; Aso caldera, Japan), while many form part of caldera complexes. The largest Cenozoic structure is the 74 ka Toba caldera in Sumatra, which has an asymmetrical, elongated elliptical shape 40×105 km along the NW Great Sumatra strike-slip fault (Chesner et al. 1991; Chesner 2012; Bellier and Sébrier 1994; Photo 72). The largest calderas occur within continental crust, and many of these are composite, multicyclic, and provide evidence of sequential subsidence into geometrically complex magma chambers. However, calderas associated with explosive activity worldwide have a fairly similar range of dimensions despite their location in different tectonic settings (Lipman 2000a).

Figure 3 depicts the topographic and structural features of a large caldera, such as the rim, the inner topographic walls, the caldera floor, the subcircular ring fault, the collapse collar with annular arcuate fractures, as well as intracaldera ignimbrite and landslide breccia (Lipman 1984, 1997, 2000b; Cole et al. 2005; Branney and Acocella 2015; Acocella 2021). The most distinctive feature of a caldera consists of one or more ring faults, usually arcuate in map view, along which the roof of the chamber collapses (Acocella 2021); the ring fault represents the *structural boundary* of the caldera (Fig. 3). The ring fault may reach the surface if there is significant subsidence. Conversely, a broad flexure forms above the ring fault if the subsidence is limited, giving way to a downsag caldera with no topographic scarp. However, severe erosion during and after caldera collapse usually prevents the existence of an observable pristine ring fault, leaving an erosional scarp. This is the *topographic boundary* of the caldera, formed of a contour joining the highest points that surround the depression, and reaching up to 1 km in height above the caldera floor (Figs. 2 and 3). The topographic boundary is normally wider than the structural rim because landsliding modifies the caldera rims soon after or contemporaneously with caldera collapse by mass wasting, erosion and scalloping. Post-collapse erosion (sliding) of the inner wall first enlarges the area of the topographic caldera basin, but continued erosional dissection down to the structural boundary of caldera may reduce its size in some cases (Lipman 1984, 2000a; Nairn et al. 1994). The wide depression corresponding

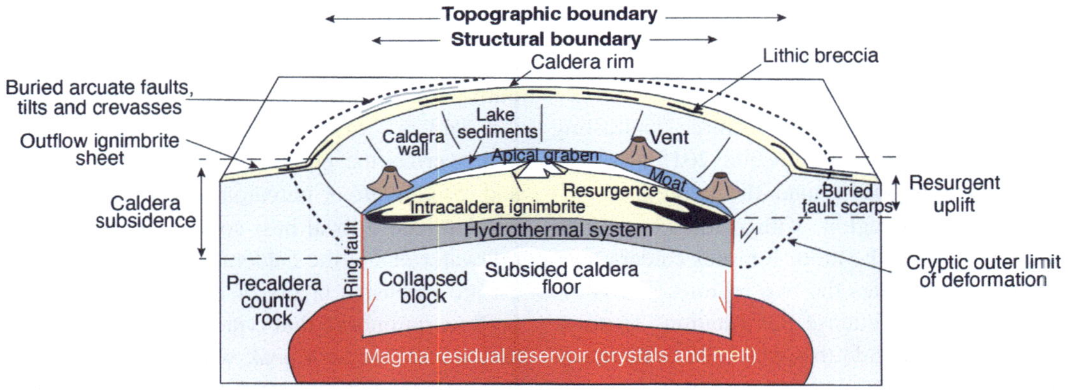

Fig. 3 Main topographic, geomorphic features and structural elements of a resurgent caldera (not to scale) (adapted and modified from Cole et al. 2005; Branney and Acocella 2015; Acocella 2021). Note the caldera rim, a partly buried caldera wall with scalloped scarps, a subsided caldera floor overlain by a fill of thick ignimbrite, landslide breccias, and caldera-lake sediments. The cryptic structural boundary lies outside the caldera rim. A thin outflow ignimbrite sheet contains proximal lithic breccias, and may partly conceal a conical-shaped zone of peripheral extension and slight inward tilting. The ring facture zone is largely denuded and buried. A central resurgent dome is flanked by post-collapse eruptive vents

to a caldera is not always a distinctive geomorphic feature because the rim may have been highly eroded (Fig. 1A) and/or the depression filled by post-caldera volcanism or the floor uplifted by resurgence. This occurs with old calderas >1 Ma and/or if incremental collapses form nested or overlapping calderas (e.g. La Garita, San Juan Mountains, Colorado; Fig. 2).

The resurgent Creede caldera, the youngest, nested depression in the La Garita caldera, Colorado, retains a variety of features (recognizable resurgent dome, surrounded by moat sediments) between the topographic and geomorphic depression and the caldera structure <20 km in diameter (Fig. 1A, after Lipman 1984, 2000b; Bethke and Hay 2000). Creede caldera is now best defined by the unconformity between the caldera infilling and wall rocks. This boundary is located between the inferred initial topographic rim and the concealed structural boundary that is approximately defined by the arcuate alignment of resurgent structures and postcaldera vents. The structurally-bounded subsided block is only about one-third of the area of the original topographic caldera, demonstrating the extent to which structural caldera size can be affected by erosional processes.

4 Geomorphic and Geological Features Within, Around, and Beneath Calderas

A caldera can be divided into three contrasting regions (Branney and Acocella 2015): (1) the extracaldera region around the depression, (2) the intracaldera region, with caldera vents and conduits, and (3) the more complex caldera margin, which separates the two regions, and is the site of the most intense deformation. To these three regions, we add that which lies underneath the depression.

4.1 Around the Caldera

During an explosive caldera-forming eruption, the region surrounding the caldera is covered by an extensive tephra blanket that may extend several tens of km from the caldera. This outflow blanket is no more than 100 m thick, and commonly thickens into basins, forming pyroclastic aprons or plateaus with a specific pattern of radial drainage network of ravines (e.g. around Valles, New Mexico, Photo 70 or around Aso caldera, Photo 73). Ignimbrite aprons exhibit unique surface textures and minor landforms (de Silva and Bailey 2017). These include pervasive joints and fractures that lead to yardangs in a dry climate (alternating ridges and furrows grooved and fluted by wind abrasion) as well as prominent mounds and fissure networks at deposit surface. Such features related to cooling and degassing processes are favoured by the involvement of external water buried by hot pyroclastic flows.

In vertical section, this pyroclastic sheet exhibits a distinctive caldera-forming eruption sequence, comprising up to five units that overlay the pre-caldera rocks (Branney and Acocella 2015): (1) phreatic ash layers from precursory explosions, (2) a Plinian pumice-fall layer which records the start of the caldera-forming explosive eruption, (3) an ignimbrite outflow sheet consisting of three subunits: (a) a lower part that records the pyroclastic fountaining phase generating radial and waxing PDC flows, (b) extensive heterolithic breccias emplaced by PDCs during the climactic, caldera-forming phase of the eruption, and (c) the ignimbrite outflow sheet that records waning PDCs from the caldera), (4) a coignimbrite ashfall layer that overlies the pumice-rich top of ignimbrite, and (5) phreatic or phreatomagmatic ashfall layers that record local post-collapse explosivity as rainwater enters the caldera. Of particular interest are extensive layers of lithic breccias in the outflow ignimbrite that typically increase in size and abundance to a peak with unit 3b (Fig. 3); these record the climactic phase of the caldera-forming eruption, a time when abundant blocks were entrained into the PDCs owing to conduit flaring, erosion, faulting, and rock avalanching, as the caldera subsided. However, this caldera-forming eruption sequence may not be present in all calderas.

4.2 Within the Caldera

A thick intracaldera-forming eruption sequence occurs within the caldera. The intracaldera region comprises the foundered caldera floor strata buried by a significant fill of massive ignimbrite, meso- and megabreccias, and overlying caldera-lake sediments. Intracaldera ignimbrites are more intensely welded, hydrothermally altered and contain larger blocks than the equivalent outflow ignimbrite (Branney and Acocella 2015; Giordano et al. 2024). Lenses of mesobreccia are abundant in most caldera fills and record frequent rock-fall avalanches from steepening caldera wall scarps. Megablocks of country rocks, hundred of metres to a couple of kilometres in size, reside within many calderas.

Volcaniclastic sediments, mass wasted or washed in from the caldera walls, commonly overlie the caldera fill ignimbrite. Some calderas (Campi Flegrei, Italy; Santorini, Greece) are partly or completely flooded with seawater, others have caldera lakes (Toba, Sumatra; Crater Lake, USA; Pinatubo, Philippines; Apoyo, Nicaragua; Ksudach, Kamchatka, Russia). These lakes, which often occur in the early post-caldera phase, gradually deepen to hundreds of metres and accumulate sedimentary successions tens to hundreds of metres thick, in which the reworked, erupted pyroclastic material is predominant. Resurgent doming of the central caldera fill may create an island with a surrounding lacustrine moat (Fig. 1A), such as at Creede (San Juan Mountains, Colorado) and Valles (New Mexico). Postcaldera volcanoes may grow and emerge from the caldera lake to form islands (Wizard Island, Crater Lake; Nea and Palea Kameni, Santorini: Photos 74 and 75A, B).

The caldera floor commonly shows deformation, intrusions and hydrothermal activity (Branney and Acocella 2021). The lithology and fracturation of the preexisting roof and the morphology and depth of the magma reservoir, which lies below the caldera floor, both control how much the caldera floor breaks up as it subsides. A hydrothermal system commonly lies between the caldera floor and the magma chamber at depths of between about 1 and 4 km. The floor of the caldera and, to a lesser extent, the fill are typically cut by dykes, sills and intrusive domes, particularly in calderas that have a protracted eruption history (e.g. Valles in New Mexico, Cerro Galán in Argentina, Yellowstone in Wyoming, USA). The visible depth of a caldera is usually less than the actual amount of structural subsidence, in some cases by an order of magnitude, owing to pyroclastic infilling synchronous with bedrock collapse (Fig. 3).

Caldera-forming eruptions may start at a point source, either at a central location or within the marginal fracture system of an incipient caldera. Stress patterns change as the caldera subsides and as successive parts of the magma reservoir are depressurized, and new fractures propagate and dilate, causing vents to shift and multiply (e.g. at Santorini, Greece and Long Valley, Nevada; Holohan et al. 2008). Compositional changes in the pyroclastic ring dyke match the compositional zoning of the caldera fill and reveal that an arcuate fissure conduit is unzipped progressively during the caldera subsidence to form a complete ring.

4.3 Caldera Margins

The caldera margin is the most complex and dynamic part of the landform. It is the site of intense deformation, intrusion, and topographic instability, susceptible to landslides and erosion, and it also coincides with the zone of eruption vents, proximal deposits, and intense hydrothermal alteration. Collapse often leads to deformation outside the topographic rim, particularly where downsag has occurred, but a collapse collar usually delimits the deformation zone, encompassing all areas that have been visibly affected by formation of the caldera (Fig. 3). The early-formed prominent caldera scarps are unstable and retreat outward by as much as several kilometres as a result of erosion, avalanches, and landslides that may leave embayment-shaped scarps as at Creede and La Garita (San Juan Mountains, Colorado, Figs. 1A and 2, respectively) and at Aso (Japan, Photo 76A and B). The distance of retreat of the wall varies with wall height, its erodibility, and the time elapsed since the subsidence.

The outer boundary, or limit of deformation during subsidence, typically projects upward and outward from the outer margin of the magma reservoir to the land surface at an angle of ~35°. Arcuate extensional faults and crevasses occur outside of the topographic caldera rim, typically accompanied by slight inward downsag of flanking strata (Fig. 3). The structure of a caldera margin is influenced by the thickness and rigidity of the subsiding block and the depth of subsidence relative to the caldera diameter. Small amounts of subsidence may involve significant inward tilt or peripheral downsag, which may extend several kilometres within the caldera. Steeper tilts within the caldera are associated with the dilation of spectacular arcuate crevasses around the caldera, some 100-m wide and 300-m deep, filled with ignimbrite and breccia. Dips of strata around some cauldrons may approach vertical, such as at the Devonian-aged Ossippee Mountains of Devonian age (New Hampshire, Fig. 1B).

Calderas tend to collapse along a linked system of interconnected arcuate faults. With increasing subsidence, the deformation may increasingly focus on a single fault. It is rare to see the actual fault system, which is generally buried or eroded, except at Miyakejima in Japan, where an outward-dipping fault formed during subsidence in 2000 (Fig. 4; Photo 77), and later movement was accompanied by, and then followed by, inward-dipping peripheral, concentric faulting on the outer wall (Geshi et al. 2002). This is consistent with analogue models of caldera collapse in which steep outward-dipping arcuate faults propagate upward near the caldera margins in response to the sudden stress reorientation caused by the decreased support from the magma reservoir (Lipman 2000a; Roche et al. 2000). With continued subsidence these connect to form an annular ring-fault system, leaving a crustal wedge in the hanging wall unsupported, so it collapses along a new set of concentric, arcuate normal faults outside of the initial outward-dipping faults (Fig. 5A; Lipman 2000a). The depth of the subsidence, for a given caldera diameter, is a factor in shaping the structural margin. Where one side of a caldera subsides trapdoor-wise more than the other, that margin may develop a pair of concentric arcuate faults, whereas the opposite, shallower side may be dominated by tilting in the case of downsag.

4.4 Beneath the Caldera: Subcaldera Intrusions, and Cauldron Roots

The structure of ignimbrite calderas can be revealed, especially if they are deeply eroded like in the San Juan Mountains of Colorado, allowing scientists to observe the roots of calderas and plutonic rocks. This is not the case for calderas on truncated composite cones, which become eroded away after a few Myr of existence. Bachmann et al. (2002) noted that the voluminous, dominantly dacitic ignimbrite sheets that occur in the San Juan Mountains act as 'a link between the plutonic and volcanic realms'. However, the ignimbrite sheets from volcanic magma (rhyolite, dacite) are not analogous to the residual magma of granitic or granodioritic plutons in terms of volume and crystal content.

4.4.1 Ring Dykes and Cone Sheets

Ring complexes are exposed by deep erosion in and around calderas, revealing ring dykes and cone sheets that describe arcs or circular outcrop patterns many km in diameter. The two intrusive structures typically coexist in a single complex, oriented around a common centre, and both appear to be related to caldera development, the ring dykes principally with magma drainage and collapse, and the cone sheets with build up of intrusive pressure. They can be divided into two groups (Lockwood and Hazlett 2010):

1. *Ring dykes* extend vertically in rocks or may flare outward with depth, away form the centres of the ring that they form at the surface. They can develop in response to the withdrawal of magma from underlying chambers. The Loch Ba ring dyke of Devonian age on the Isle of Mull, Scotland, the 'most perfect

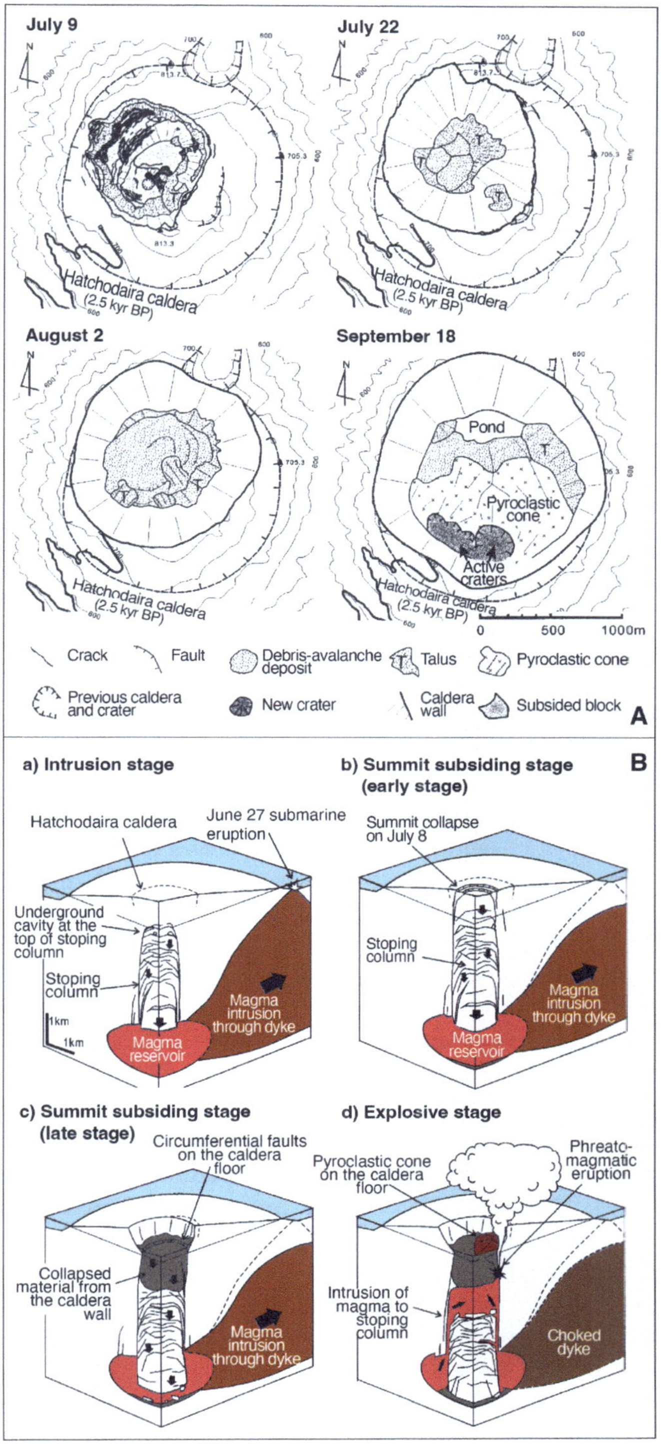

Fig. 4 Miyakejima caldera-forming eruption in 2000 (Geshi et al. 2002). **A** Structural evolution of the collapsed caldera on 9 July, 22 July, 2 August, and 18 September 2000, based on aerial photos by Asia Air Survey Co. Ltd. and Japan Air Self Defense Force. Talus has a slope of about 30–45° and consists of collapsed materials from the caldera wall. Subsided blocks are only visible on 9 July (after Geshi et al. 2002, Fig. 7, p. 62). **B** Schematic illustration of the development of the collapsed caldera (after Geshi et al., 2002, Fig. 11, p. 66). **a** Intrusion stage before the surface collapse. Evacuation of magma from the reservoir caused stoping of the roof rock of the reservoir. Underground stoping formed a cavity at the top of the stoping column. **b** The early stage of the summit subsidence. Surface subsidence occurred on 8 July when the cavity reached a critical depth of less than 1 km. The roof rocks of the cavity could no longer carry their own weight and collapsed into the cavity. Release of gas and steam filled the cavity and caused a phreatic eruption. **c** The late stage of the summit subsidence. Continuous evacuation of magma from the reservoir caused the subsidence of the reservoir roof. The top of the stoping column was filled with the collapsed materials from the outward migrating caldera wall. **d** Explosive stage. Invasion of magma to the stoping column caused phreatomagmatic eruption after the westward intrusion of magma ceased

known example' according to Bailey et al. (1924), forms a nearly continuous oval 5.8 by 8.5 km in diameter with a maximum width of about 300 m. Stratigraphic correlation indicates that the bedrock inside the ring complex subsided by about 150 m, suggesting that the ring dykes intruded along active faults concurrently with downfaulting. Ring dykes can form wide, complex networks, ranging in width from a few hundred m to over 1 km. The first-studied ring complex, Glencoe in Scotland, has basaltic, andesitic, and rhyolitic rocks and interbedded sediments within an oval area 8 km across and is enclosed by a rhyolitic ring intrusion (Clough et al. 1909).

2. *Cone sheet* dykes taper downward toward a common point, and develop above the roofs of shallow magma chambers. Several hundred cone sheets lie in Devonian ring complexes on the Isle of Mull and Ardnamurchan Peninsula in SW Scotland (Anderson 1936). A cone sheet tends to consist of only one or a few dykes around the ring, with individual dykes being fine-grained, basaltic and thin, seldom exceeding 5 m in width. The outermost sheets dip inward at angles of 35–45°, but those closer to the centre of the complexes dip more steeply, up to 75°. Projections of the dips of the cone sheets converge toward a focus at a depth of about 5 km beneath the middle of each complex.

4.4.2 Cauldron Subsidence

Many deeply eroded intrusive ring complexes, like those of Scotland, Ireland, New England, Norway, and Nigeria, represent the roots of cauldrons or calderas (Williams and McBirney 1979). The term cauldron subsidence has been used to denote the sinking of roof-rocks into a large underlying body of magma (Francis 1993). The Sabaloka cauldron has been cut through by the Nile in northern Sudan, exposing preserved ring dykes (Almond 1977; Fig. 1B). An oval-shaped ring dyke of porphyritic microgranite measuring 25 × 15 km encloses a vertical sequence of up to 2000 m of inward-dipping volcanic rocks, mostly rhyolitic ignimbrites, which foundered intermittently into a plutonic complex of granite and syenite. The ignimbrites bear a close chemical and petrographic resemblance to the rocks of the ring dyke from which they were probably extruded. In addition, several feeder-dykes that cut the interior of the subsided central block can be traced upward into rhyolitic ignimbrites and downward into fine-grained granite (Williams and McBirney 1979; Francis 1993).

Cauldron subsidence can be compared to a piston subsiding within its cylinder: subsurface ring dykes are the geological expression of the cylinder, while the subsided block and thick accumulations of welded ignimbrite form the piston. Cauldron subsidence and subsurface movement of magma are probably the more significant causes of collapse than eruptions. There are examples in which the collapse volume was much greater than that of the extruded magma (e.g. Santorini, Campi Flegrei) and others in which the caldera floors dropped with little if any eruption beyond the caldera boundary (e.g. mafic and peralkaline calderas, see below). Eruptions that accompany collapse may account for part of the volume of withdrawn magma, but they are not necessarily the cause of collapse; in many instances it is the collapse that triggers the eruption rather than vice versa (Francis 1993).

In some cases, a caldera forms when magma has been siphoned away from a magma reservoir to an eruptive site that may be several km away. A well-known example is the collapse of Katmai volcano in 1912, where the eruptive site was Novarupta, at a distance of 10 km (Hildreth and Fierstein 2000). Another case is a caldera located at the headwall of a former flank collapse amphitheatre, such as the Cañadas caldera in Tenerife (Marti 2019). Alternatively, calderas may also result from the lateral intrusion of magma through a dyke or sill, eventually feeding a minor to moderate distal effusive eruption. Thus, it should be remembered that calderas may be associated with effusive or explosive eruptions of any size (Druitt and Sparks 1984; Branney and Acocella 2015). There are, however, calderas that form in the absence of any eruption (Hawaiian shields) and others that result from the coalescence of craters formed by a series of small ($\leq 1\ km^3$) eruptions.

5 Relationships with Tectonics and Regional Structures: Elliptical and Graben Calderas

Regional faults commonly affect the structure of a caldera. Their shape can be polygonal (e.g. Ischia, Rotorua, Cole et al. 2005), elongate, as in MER, Ethiopia (see Fig. 5A; Chap. 1; Corti 2009; Wadge et al. 2016), a straight-sided graben (Las Cañadas, Tenerife, Marti 2019), or variable according to regional tectonics, such as in the Taupo volcanic zone, New Zealand (Wilson et al. 1984, 2009; Holden et al. 2015; see Fig. 15, Chap. 1) and along the rift in Iceland (Krafla, Askja: Acocella 2007, and Bárðarbunga: Sigmundsson 2019). Regional tectonics may affect some caldera features, like the shape in plan view, generating an elliptical collapse with the major axis parallel to the maximum horizontal principal stress σ_m. Commonly elliptical calderas reflect the shape of anisotropic, elongated or even circular chambers at depth, due to a regional tectonic contribution (Acocella 2021).

However, an elliptical caldera, elongated parallel to the maximum direction of extension, forms from an initially subcircular (in plan view) reservoir. Its expected maximum eccentricity: $E = Lmin/Lmax$ is $\sim$ 0.8, where *Lmin* is the length of the minor axis of the ellipse and *Lmax* the length of the major axis. The ellipticity of a caldera with a subcircular reservoir can be thus ascribed to regional tectonics (Acocella 2007). This explains why calderas are elongated parallel to the least compressive stress σ_3: e.g. Askja, Krafla across the slow MOR in Iceland, Suswa, Silali and Paka in EARS (Wadge et al. 2016), and Okataina and Taupo in the Taupo Volcanic Zone in New Zealand (Fig. 14, Chap. 1). Conversely, the control of pre-existing structures can cause calderas to become elongated along orientations that differ from the regional σ_3, related to elliptical shallow reservoirs and parallel to the direction of dykes (e.g. Grimsvötn, Iceland; Erta Ale, Afar; Pantelleria, Italy). In his analysis of the frequency of elliptic calderas as a function of the angle α (between the major axis of the caldera and the regional structural trend), Acocella (2007) considered 57 elliptic calderas worldwide with an eccentricity of $E<0.9$. Most (49%) calderas are (sub)perpendicular to the regional rift structures controlling volcanism; 42% are (sub) parallel to the rift structures; and only 5% of them are oblique to the rift structures, but always parallel to significant pre-existing structures. Thus, the structure and development of a caldera is almost independent of pre-existing regional structures or stresses, as also shown by experiments (Holohan et al. 2005).

Significant magma chamber withdrawal may reactivate regional normal faults, if present, forming graben calderas (Acocella 2021). *Graben calderas* are elongated rift segments, bounded by normal faults, which collapse during large eruptions under regional extension and whose width approximates that of the erupting reservoir below. Examples are the WNW-ESE striking grabens of Faial Island with the Caldeira do Cabeço Grande, and Pico Island with Caldeira Grande, both in the Azores archipelago. Graben calderas, inferred on continental (southern Basin and Range, western USA), transitional (central Afar) and oceanic rifts (East Pacific Rise), thus represent an end-member of strongly elongated calderas. Finally, regional tectonic stress may affect the anticipated and delayed formation of a caldera, for example regional extension decreases the stability of the roof rock overlying a magma chamber, promoting early onset of caldera collapse, whereas compression on relatively short timescales increase roof rock stability, delaying collapse.

6 Caldera Development and Evolution

An early sketch by Williams (1941), adapted by Bacon (1983) and shown in Cole et al. (2005) and Lockwood and Hazlett (2010) depicts caldera development, which involves the explosive eruption at the summit of a composite cone above a large magma reservoir, the climactic eruption with ignimbrites from ring-fault vents, and the concomitant step-by-step collapse of the edifice and foundering of the roof of the magma reservoir (Lipman 1984, 1997, 2000a, b; Wohletz and Heiken 1992). A typical cycle of caldera subsidence and ash-flow eruptions is shown in Fig. 5 based on natural examples of recent and Tertiary calderas which were formed without a pre-existing edifice (Lipman 1984, 2000b): (1) pre-collapse volcanism with initial tumescence, due to magma accumulation, (2) caldera subsidence concomitant with major eruption(s), (3) ignimbrite and sedimentary infill

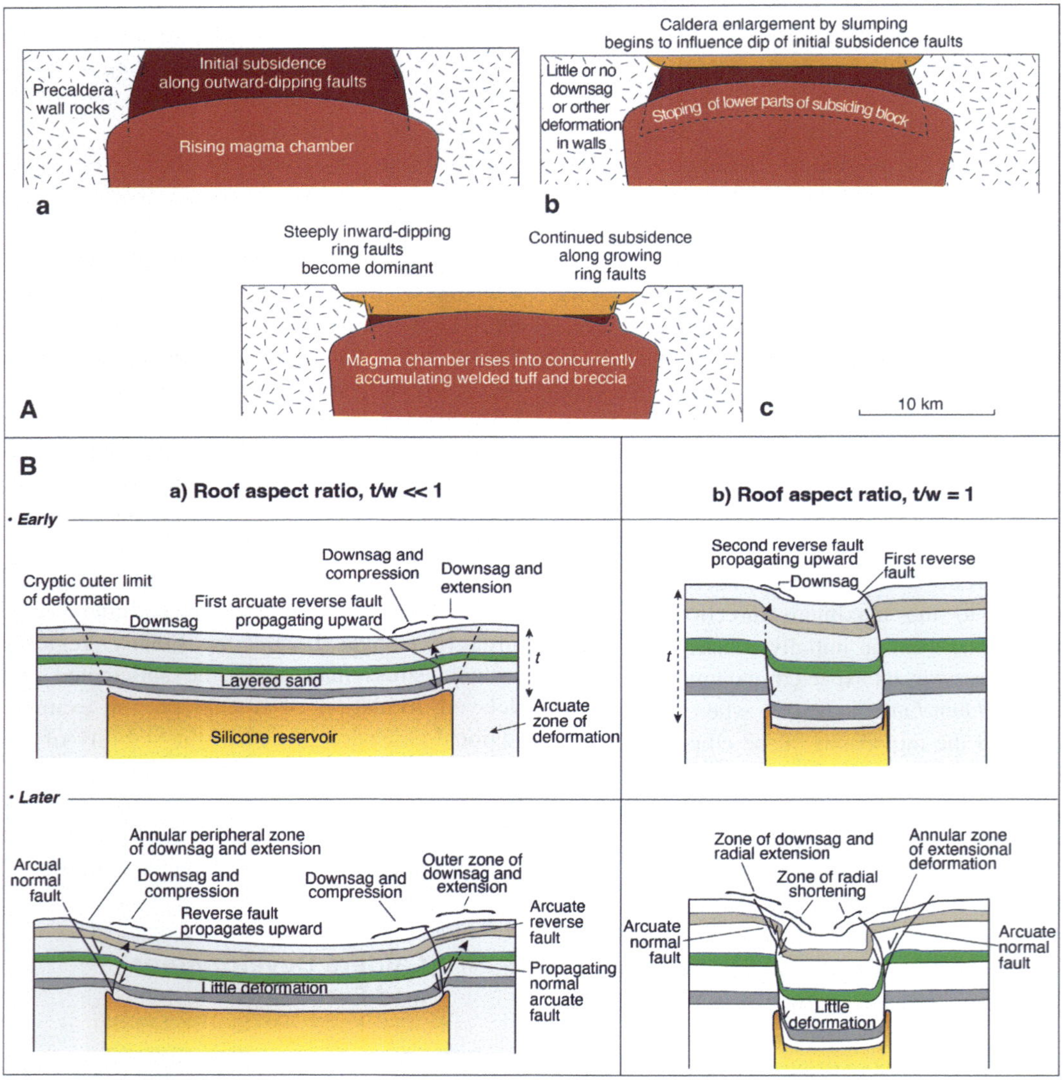

Fig. 5 **A** Possible evolution of ring-fault geometry, in which initial outward-dipping ring faults are replaced by inward dips, as progressive subsidence destabilizes the caldera walls, and continued upward rise of magma reduces the thickness of the subsiding plate or piston (after Lipman 1997, 2000a, Fig. 4, p. 650). Note how experimental analogue models (B) have been designed based on Lipman's theory. **B** Progressive structural development of calderas as inferred from two- and three-dimensional analogue models (Roche et al. 2000; Branney and Acocella 2015, Fig. 16.7, p. 308). In both (**a**) wide and (**b**) narrow calderas, initial downsag is followed by propagation of vertical to outward-dipping reverse faults that gradually link around the caldera. Slightly later, normal arcuate faults and tilting form beyond the reverse faults, creating an annular zone of intense deformation, which in real calderas involves multiple arcuate normal faults, crevasse dilation, brecciation, rapid denudation, and burial by caldera fill

on the caldera floor and cooling of the magma chamber, and (4) rejuvenation due to input of new magma, promoting post-caldera magmatism and resurgence, followed by hydrothermal activity and mineralisation.

6.1 Pre-caldera Volcanism

Prior to collapse, several small to medium-sized composite cones of intermediate composition may develop above smaller, near-surface magma reservoirs. These eventually coalesce to form a large magma body that feeds ignimbrite eruptions or heats the shallow crust sufficiently to allow the buoyant rise of a large pluton. Broad uplift related to the emplacement of plutons may lead to the development of arcuate ring and radial fractures.

The slow emplacement of large magma bodies causes ring fractures to form. Prior to many ash-flow eruptions, a compositionally zoned magma chamber develops by differentiation of the magma, magma mixing and exchange of material with wall rocks. Several calderas have distinct eruptive cycles with compositional zonation developing in the magma reservoirs during non-eruptive intervals. Eruptive periods are sometimes separated by periods of quiescence, lasting from 10,000 to several 100,000 years. At Mt. Mazama, a large composite cone at the site of the current Crater Lake caldera (Sect. 7.2), the magma reservoir reached a preclimatic zoned stage at least 27 ka, prior to the formation of the caldera 6845 ± 50 BP (Bacon and Lanphere 2006).

Evidence of a similar pre-caldera stage of development can also be found at many other caldera-related volcanic fields, as well as at collapsed composite cones like Mt. Mazama. The pre-caldera Glass Mountain obsidian domes along the eastern side of California's Long Valley caldera, source of the Bishop Tuff, are an especially visible example. Fisher Caldera in the Alaskan-Aleutian volcanic arc represents an example of the recent development of a large magma body (Lockwood and Hazlett 2010). Measuring 12 by 18 km and lying near the tip of the Alsaka peninsula, Fisher is the largest of the dozen Holocene calderas within the arc. Eight small composite cones developed between 40,000 and 9000 years ago in a ring defining the outline of the future caldera (Stelling et al. 2005). These signified incipient fracturing in the roof of a young, shallow magma body. The pre-caldera phase of volcanism at Fisher Caldera ended abruptly 9100 years ago. Collapse ensued concurrently with a violent Plinian eruption, and some 70 km^3 of material spewed out as pumice-rich ash flows that traveled northwest to the Bering Sea, more than double the volume of the 1883 Krakatau eruption (Stelling et al. 2005).

6.2 Syn-caldera Stage

Central area of previously formed, clustered volcanoes caves into the collapsed area. Following the eruption of many ash flows and syngenetic caldera collapses, the basin fills with ignimbrite, covering the floor of the caldera. Section 4.2 above describes intra-caldera ignimbrite being an order of magnitude thicker than the cogenetic ignimbrite outflow sheet. Initial collapse along bounding faults is followed by slumping of oversteepened caldera walls and accumulation of voluminous collapse breccias interleaved with ignimbrite in the caldera fill. The main magma body underlies the entire caldera and is compositionally zoned, becoming more mafic downward.

6.3 Post-caldera Resurgence, Renewed Eruptions and Hydrothermal Activity

During later stages of caldera collapse, a central area may be uplifted by emplacement of new magma (termed resurgence) and fragmented into smaller domains. Resurgence is asymmetrical, with greater uplift in the area of greatest prior collapse. Some resurgent uplift is accommodated by movement along ring faults in the opposite sense to that which occurred during caldera subsidence. Welded tuffs and other caldera deposits may be uplifted as a structural dome over the magma body. Extensional graben faults form over the crest of the dome. The original caldera floor is often almost obliterated by the rise of the magma chamber to near the level of the pre-caldera land surface. The caldera moat is partly filled by lava domes and volcaniclastic sediments, terminating a caldera cycle. Hydrothermal activity and mineralization are prevalent late in the cycle and develop within the caldera deposits and floor rocks, as well as along extensional faults that cross the resurgent dome (Lipman 1984; Wohletz and Heiken 1992; Cole et al. 2005; Lockwood and Hazlett 2010).

6.3.1 Renewed Eruptions

The development of a caldera by no means ends the life of the source magma chamber (Lockwood and Hazlett 2010). Influxes of new melt from below usually continue, re-pressurising the reservoir and forcing new eruptions during the post-caldera stage of development. After a period of tens to several hundreds of thousands of years other Plinian eruptions associated with renewed chamber collapse may occur from different magmatic sources in the same area, creating new calderas that overlap or lie within the older one. Nested and overlapping, compound calderas show that two or three caldera-forming events may originate from an individual magma reservoir during their lifetime, as shown by the well documented San Juan caldera cluster of Colorado (Lipman 2006). The La Garita caldera depicted in Fig. 2A formed during a single eruption pulse about 28 Ma, accompanied by multiple subsidence segments. Eruption of the resulting Fish Canyon Tuff, comprising at least 5000 km^3 of ash-flow tuffs and tephras, is the most voluminous pyroclastic eruption documented on Earth, even without the additional 2500 km^3 of downwind ash fallout. Over the next 1.1 Myr, seven more calderas formed within the giant basin of La Garita (Fig. 2). Much younger than the San Juan volcanics, the Yellowstone Plateau of Montana-Wyoming encloses a compound caldera some 60 km in diameter that formed over the course of three Plinian eruptions 1.8, 1.2 and 0.6 Ma (Christiansen 2001; Fig. 6A). The magma chamber is much larger than the depression and the modelled roots of this giant magma system can be followed to depths of more than 40 km (Fig. 6A and B; Lowenstern 2020; Lowenstern et al. 2006; Farrell et al. 2014).

6.3.2 Post-caldera Hydrothermal Activity and Mineralisation

Calderas host a hydrothermal system in which fluids from shallow or deep aquifers are heated, pressurized, and contaminated by magmatic volatiles (Branney and Acocella 2015). Spectacular surface manifestations include hot springs, geysers, fumaroles, and mud pools, for example at Campi Flegrei, Yellowstone, and Rotorua (New Zealand). The accumulation of magma beneath Yellowstone is the heat engine that drives the activity of the largest active geyser field on Earth, providing aesthetic and touristic highlights in Western USA, similar to those in New Zealand and Iceland.

Calderas make excellent ‘chimneys’ for the circulation of mineralizing fluids following their formation, while many of the world’s most important metallic ore deposits are associated with caldera-fill rocks (Lipman 1992). Convecting hydrothermal fluids may cause chemical and mineralogical changes to rocks locally or pervasively in the caldera floor and fill, and particularly in marginal zones. Ore mineralization may derive directly from the magma, magmatic fluid, or hot groundwater, and mostly occurs along permeable fracture networks. Mineralization may begin soon after caldera collapse, or millions of years later.

Resurgent activity results mostly from the replenishment of a gas-depleted magma reservoir with fresh batches of magma after caldera formation, although an additional factor is the increased water pressures in maturing, confined aquifers. This is of significance for the controversy concerning active resurgent calderas such as Campi Flegrei, whose historical episodes of unrest can be interpreted either in terms of magmatic or hydrothermal activity.

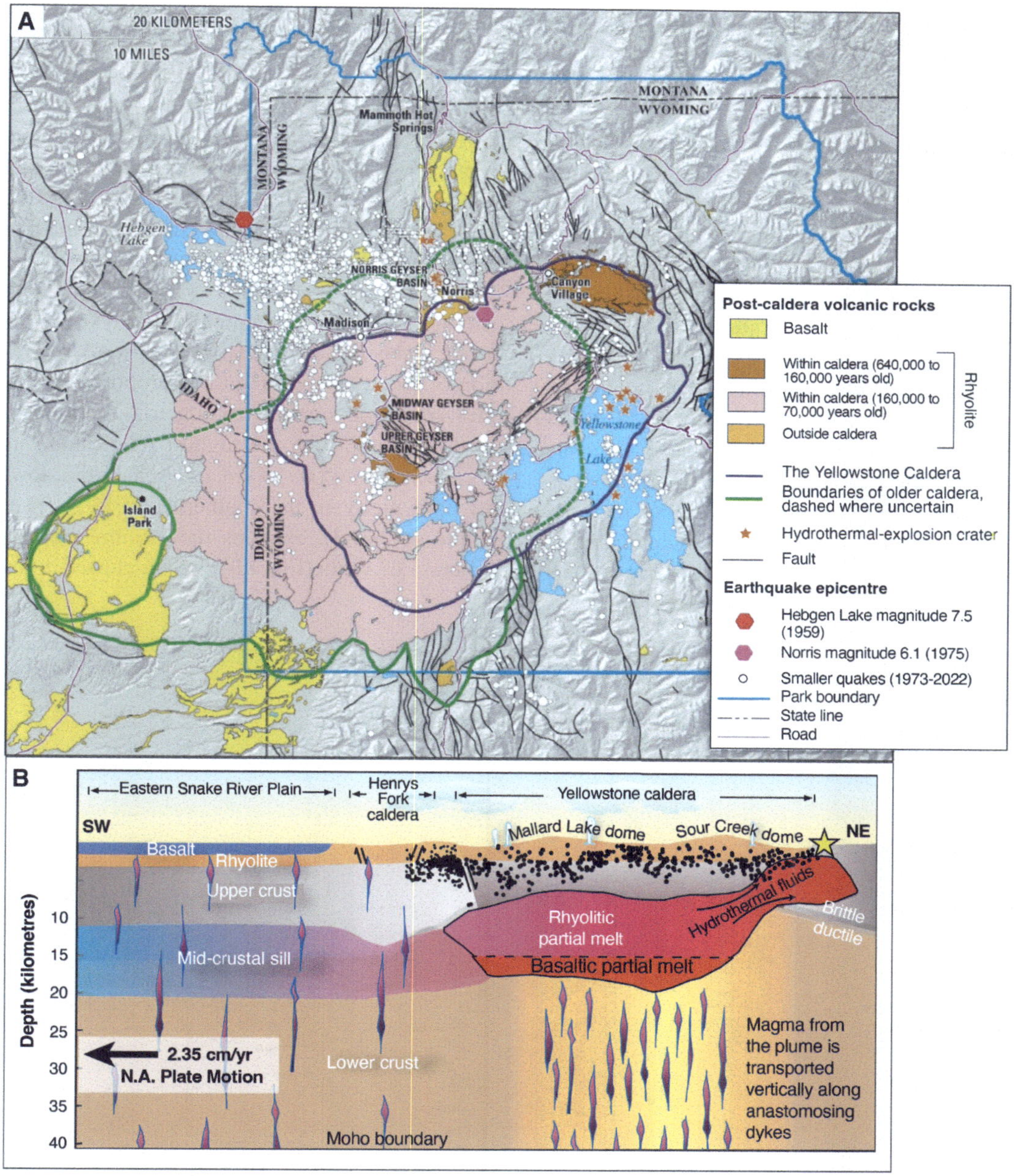

Fig. 6 The compound Yellowstone caldera. **A.** Simplified geological map of the compound Yellowtone caldera in the Yellowstone National Park (scale 1:63,360, Lowenstern 2020, USGS Volcano Hazards Program, Public domain) showing the boundaries of the older calderas and post-caldera volcanic rocks. Earthquake epicenters are also indicated. Map features are from Christiansen (2001, Fig. 7, page G18) and Lowenstern et al. (2006). **B** Plume-plate interaction model for the Yellowstone hot spot and evolution of the Yellowstone-Snake River Plain volcano system. The model shows how basaltic melt from the plume rises buoyantly through the upper mantle and lower crust, partially melting the silicic continental crust and creating a partially melted magma reservoir beneath Yellowstone and well beyond the caldera. This energetic magmatic system drives Yellowstone's hydrothermal system, extraordinarily high heat flow, and seismicity. As the lithosphere passes over the plume the plate cools and thermally contracts above the crystallizing fractionated crustal magma reservoir forming the characteristic bimodal rhyolite-basalt volcanic system of the Yellowstone-Snake River Plain (after Fig. 4, p. 3072 in: Farrell et al. 2014, Gephys. Res. Lett. 41, 9: 3068-3073; https://doi.org/10.1002/2014GL059588; Free Access)

6.3.3 Migration of Eruptive Activity and Unrest at Active Calderas: Campi Flegrei

Campi Flegrei (Barberi et al. 1991) is a trachytic caldera complex ~12 km across hosting two nested calderas that formed during eruptions ~39 ka (Campanian Ignimbrite, CI; Rosi et al. 1983; Scarpati et al. 2020) and ~15 ka (Neapolitan Yellow Tuff, NYT; Orsi et al. 1992; Scarpati et al. 1993; Deino et al. 2004) (Fig. 7). Continued collapse over the past ~4 ka means that Campi Flegrei must be regarded as an active caldera. The subsidence associated with both depressions is of the order of ~2 km. However, marine deposits and post-collapse volcanism have mostly filled the caldera depression.

Post-caldera volcanism has led to more than 70 monogenic vents located predominantly in the northeast sector of the caldera (Acocella 2021). Eruptive activity migrated progressively inward over Epochs 1 (15–9.5 ka), 2 (8.6–8.2 ka), and 3 (4.8–3.7 ka) filling most of the caldera (di Vito et al. 1999, 2016). After the last caldera collapse at ~15 ka, the eruptive vents progressively migrated from the caldera rim inwards (Rivalta et al. 2019) (Fig. 8A). This is due to a decrease in the unloading stresses within the caldera, owing to its progressive infilling over the last 15 ka. As unloading stresses became less dominant, the dyke trajectories steepened and led to the inward migration of the vents over the three eruptive epochs. Moreover, the overall asymmetric topography of the caldera, with the highest relief along its northern rim (200–300 m.a.s.l.) and the lowest bathymetry in the southern part (50–100 m below sea level), explains the asymmetric distribution of vents (Fig. 8B), which cluster in the

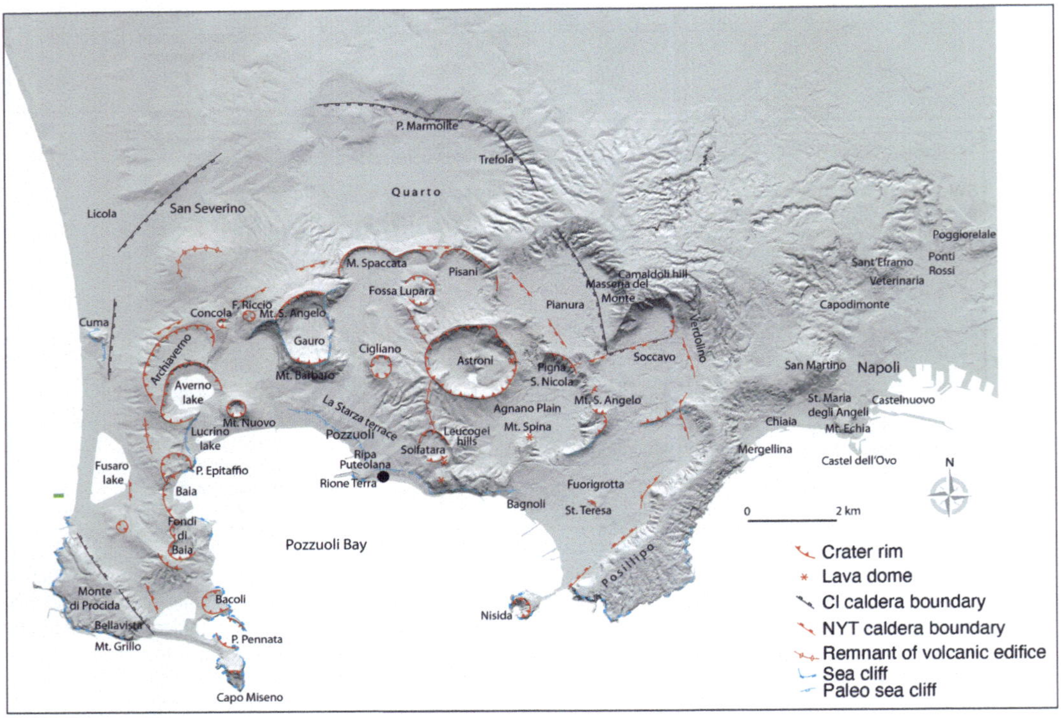

Fig. 7 DEM-based map of the Campi Flegrei nested caldera and the Campanian area (Courtesy of Costa et al. 2022). The main features are the two collapse structures that formed during the ~39 ka Campanian (CI) eruption (outer scarps) and the ~15 ka Neapolitan Yellow Tuff (NYT) eruption (inner scarps). Vent areas for some of the eruptions in the last 15 ka are marked, which include the Astroni, Nisida and Gauro cones, Lake Averno, and Monte Nuovo

Fig. 8 **A** Distribution of the post-caldera vents (coloured circles) at Campi Flegrei nested caldera as a function of three epochs during the last 15 kyr eruptive history. **a** Shaded relief map of the caldera with location of eruptive vents. Volcanism migrates inward with time, clustering in the northeast part of the caldera. **b** Boxplot showing the radial distribution of eruptive vents. The boxes have horizontal lines at the first quartile q1, median, and third quartile q3. The whiskers correspond to 99.3% data coverage. The horizontal size of the boxes represents the time interval of the epoch. **c** Volume of eruptions and cumulative (grey line) erupted volume (After Rivalta et al. 2019: courtesy of E. Rivalta et al. 2019, Fig. 2, p. 5, Science Advance, 5, 7: 1–11). **B** The post-caldera eruptive pattern at Campi Flegrei results from the progressive thickening of the marine and volcanic infill within the caldera (yellow), which gradually shifts the location of the eruptive vents (triangles) inward. The asymmetric distribution of the vents is explained by the highest relief to the northeast, which develops stronger gravitational loads and thus attracts magma (V. Acocella and E. Rivalta, Fig. 9A, p. 10, Reference module in Earth Syst. Envir. Sci., Elsevier). **C** Geophysical model of the shallow and deep caldera structure (Courtesy of M. Calo and A. Tramelli, 2018, Scient. Rep., Nature Portfolio), synthetizing the information retrieved from the P wave seismic velocities (Vp), Vp/Vs ratio, Quality factor of the P and S waves (Qp and Qs), and Scattering (12 Hz). The seismic layer (marked by low Vp/Vs and high Qp and Qs) separates the shallow part of the caldera from the deeper one. The deep body (marked by high Vp and Vp/Vs and low Qp and Qs) rises from depth and stops at about 2.3 km where it flattens out beneath the seismic layer. The most permeable part of the seismic layer allows the migration of magmatic fluids into the shallow part of the caldera where it mixes with the geothermal reservoir beneath the Pozzuoli region (marked by high Vp/Vs and low Qp and Qs and low scattering) by means of shallow active faults driving them toward the surface. The Pozzuoli reservoir is generally bounded by gas-saturated country rock (marked by low Vp/Vs, Qp and Qs, and high scattering) where very shallow water saturated reservoirs may also exist beneath the main craters of the caldera

currently onshore northeast part. Immediately before the last eruption, the intruded magma migrated laterally from an oblate chamber below the caldera centre to beneath Monte Nuovo and then propagated upward to feed a ring dyke. The vent for the most recent Monte Nuovo eruption in 1538 lies just outside the central part of the caldera but still within its rim (Rivalta et al. 2019; Acocella 2021).

Campi Flegrei caldera is a densely inhabited area and has suffered several periods of unrest over the last few centuries (Costa et al. 2022). The unrest process is highly debated because of conflicting interpretations about the magmatic versus hydrothermal mechanism driving the current caldera deformation. Using the microseismicity recorded during the 1984 unrest, seismic tomography models allowed Calò and Tramelli (2018) to image the shallow substructure (5 km in depth) of the Campi Flegrei caldera. Results show (Fig. 8C): (1) a plumbing system with a diameter of 1 km located at a depth of between 2.3 km and 4 km, (2) a 0.5 km-thick caprock located at 2 km depth, interpreted as the main structure regulating the fluid interchange between deep and shallow sectors of the caldera, (3) the shape and volume of a shallow reservoir beneath the city of Pozzuoli, which played a key role during the 1982–1984 unrest, and (4) several small reservoirs beneath the main craters of the caldera.

7 A Simple Descriptive Distinction of Calderas Based on Magma Composition

Caldera features and their variations can be distinguished based on composition for descriptive purposes, although caldera types and collapses are not directly linked to eruptive products (Cole et al. 2005). The first, simple distinction is between mafic calderas and felsic ones on the basis of the mechanisms by which the magma chamber can be emptied (Sigmundsson 2019; Acocella 2021).

7.1 Mafic Calderas on Shield Volcanoes

Shield calderas are large, shallow depressions at the summit (see Chap. 4). The roof of the basaltic oceanic shield volcanoes typically collapses owing to the withdrawal of the magma column beneath the summit and its drainage towards

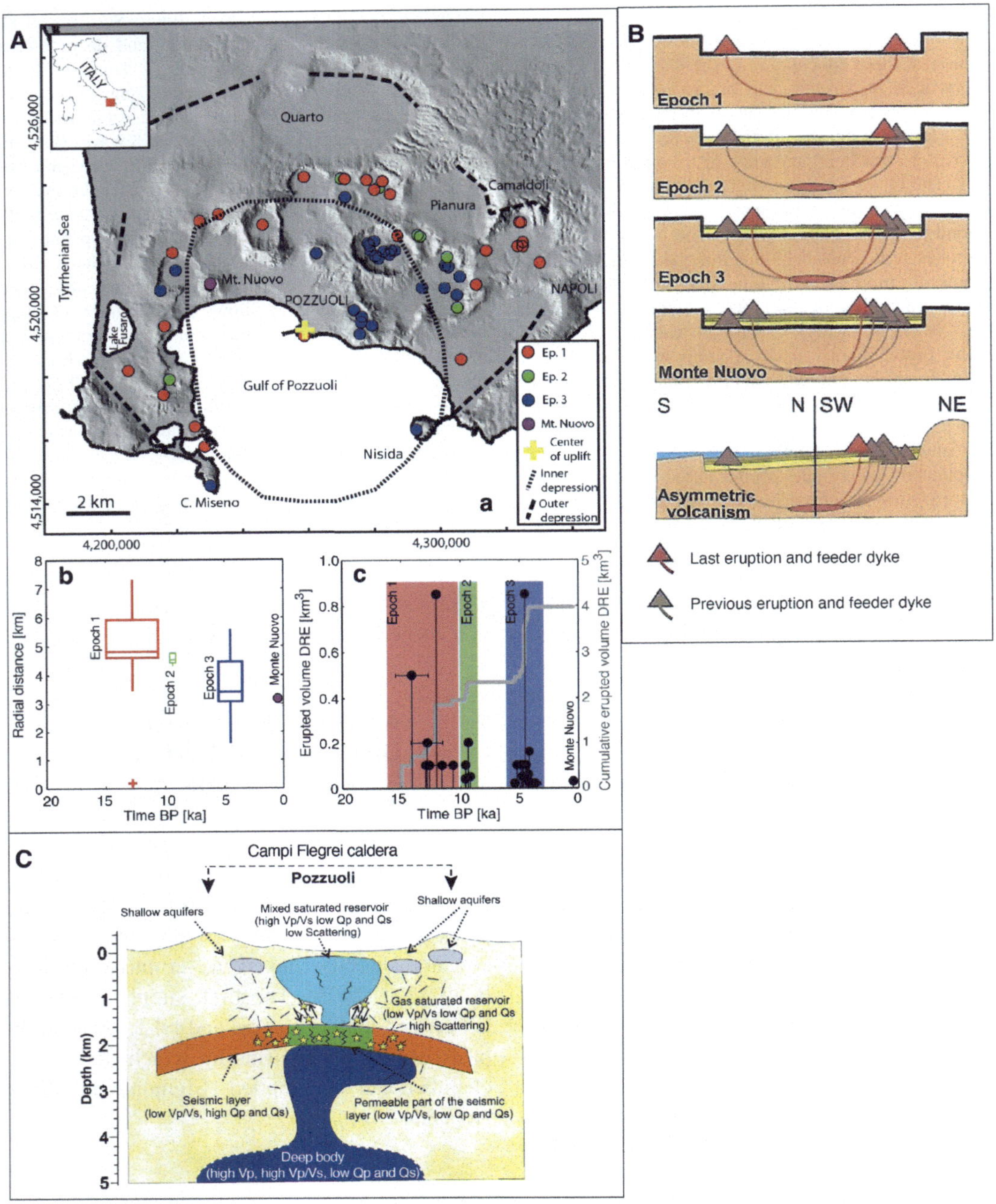

the rift zones (Fig. 9A). Lockwood and Hazlett (2010) call them 'drainage calderas' caused by the lateral drainage of magma from underlying magma chambers typically underneath oceanic shields. Three mechanisms can lead to summit caldera collapse at basaltic shields: (1) the intruding magma and sills can expand the shield thereby causing extension and collapse at the surface, (2) extension due to magma intrusion may lead to an unstable situation which, along with magma withdrawal, induces caldera formation, and (3) lateral propagation of magma along a distant rift zone may induce change in stress and slip of ring faults, hence a far-away

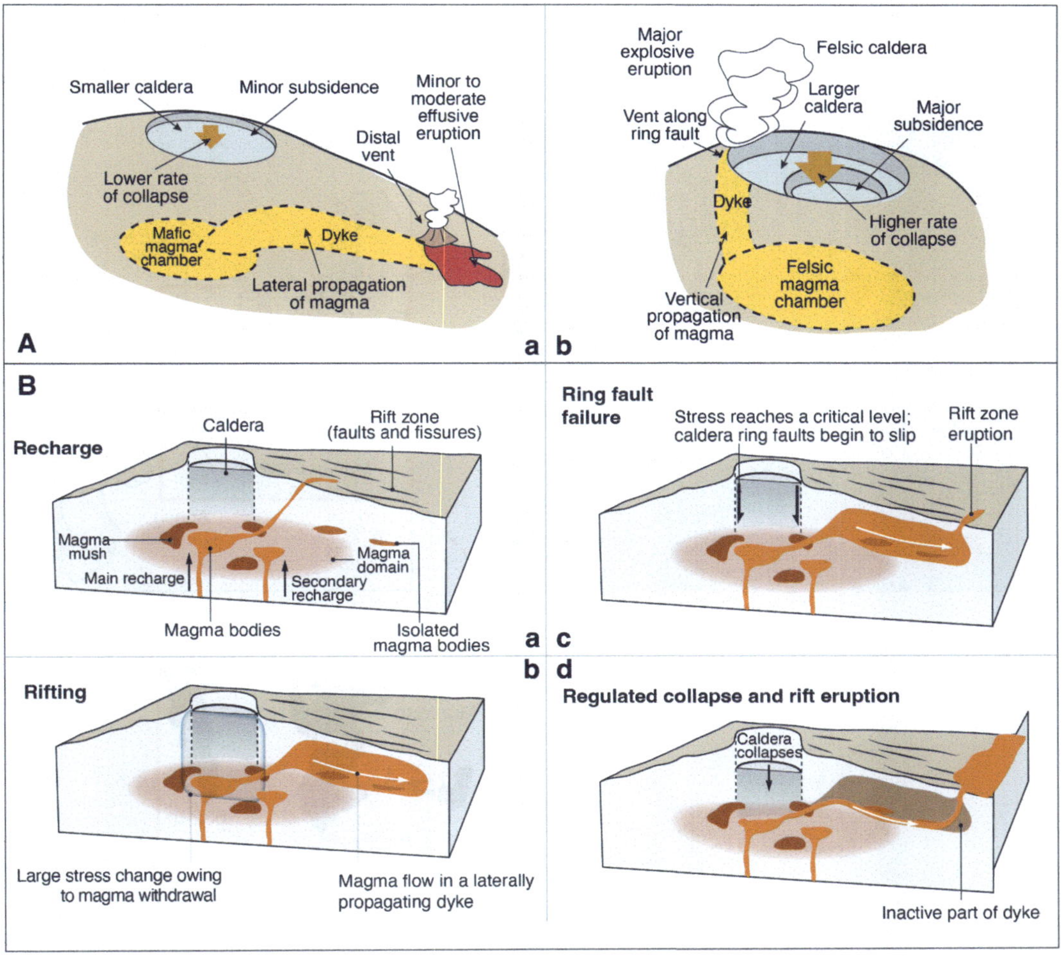

Fig. 9 **A** Recurrent features commonly associated with mafic (left) calderas versus felsic (right) calderas (not to scale; Acocella, 2021, Fig. 5.3, p. 166). **B** Generalized cross-section along a volcano rift illustrating how four steps of caldera formation can be coupled with lateral transfer of magma into rifts and distant rift zone eruptions (Sigmundsson 2019, Fig. 1, p. 1201): (**a**) Recharge into the magma domain (liquid magma bodies and crystal mush) occurs before an eruption. (**b**) Rifting is followed by (**c**) ring-fault failure and finally (**d**) caldera collapse at the centre of the volcano, regulated by a major eruption in the rift. Note that mafic calderas can also be formed by voluminous explosive eruptions involving mafic ignimbrites and Plinian pumice-fall, e.g. Masaya, Nicaragua (Sect. 7.1.1)

effusive eruption can lead to caldera collapse (Sigmundsson 2019) (Fig. 9B).

7.1.1 Explosive Calderas and Hydrothermal Calderas on Mafic Shields

Explosive caldera-forming eruptions also occur on mafic shields and may have been underestimated. Masaya caldera (11 × 6 km and 0.3 km in depth), Nicaragua, nested at the summit of a broad, low basaltic shield of Las Sierras (Fig. 10A), is an unusual example of collapse due to large explosive eruptions in a fashion typical of silicic calderas, thus contrasting with mechanisms invoked for Hawaiian and Galápagos shield calderas (see Chap. 4). The main collapse at Masaya was concurrent with explosive eruptions that produced basaltic

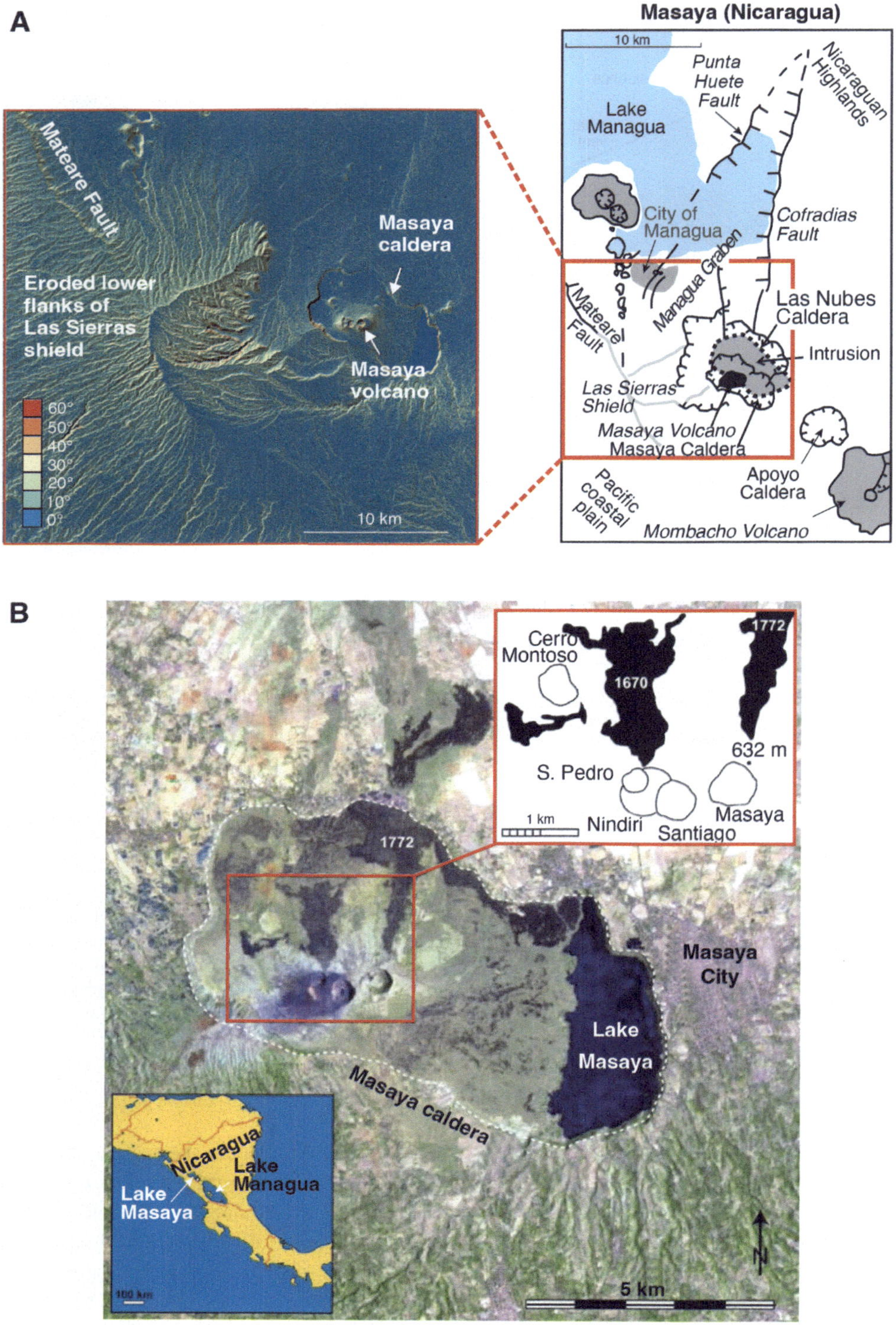

Fig. 10 Masaya caldera, Nicaragua, illustrating the development of a caldera on a mafic shield (Las Sierras) following explosive eruptions. The Managua Graben and neighbouring volcanoes are shown. **A** Structural map (modified after Williams and Stoiber 1983; Rymer et al. 1998) and DEM. **B** ETM+ Landsat satellite image of Masaya caldera. Outline of pit crater rims and historic lava flows is inset. Flow field marked with? is of unknown date, but has a similar (young, unvegetated) surface to those of 1670 and 1772. (After Harris 2009, Fig. 1, p. 542)

ignimbrites and widespread basaltic Plinian fall deposits (Williams and Stoiber 1983; Pérez et al. 2020). The latter authors report a volume of 14 km^3 DRE for the 9 ka San Antonio Tephra, the 2.1 ka Masaya Triple Layer and the 1.9 ka Masaya Tuff/Ticuantepe Lapilli. Masaya caldera contains a number of central cones cut by four pit craters that partially intersect: Masaya, Nindiri, Santiago and San Pedro (Fig. 10B; Harris 2009). Historical and ongoing activity has emplaced young lava flows and a lava lake in the active Santiago crater. Explosive eruptions occur after periods when the magma is unable to degas and become supersaturated with volatiles (Walker et al. 1993). This leads to explosive eruptions and caldera collapse. Based on gravity data and observed pit craters, the collapse is interpreted to have occurred on outward-dipping faults (Rymer et al. 1998).

Another collapse mechanism model has been invoked for large calderas on basaltic volcanoes such as Piton de la Fournaise, La Réunion, Island, Nuku-Hiva in French Polynesia, and Aoba in Vanuatu. As an alternative to the standard 'Hawaiian' model, a different model of hydrothermal caldera formation was applied to the Enclos caldera of Piton de la Fournaise (Merle et al. 2010) (see Chap. 4). Geophysical investigations reveal that some calderas on basaltic volcanoes host hydrothermal fluid circulation controlled by structural discontinuities. Analogue experiments show that the hydrothermally altered core of an edifice can slowly flow, thus causing the vertical collapse of the resistant roof and forming large calderas (Barde-Cabusson and Merle 2007). In the situation of caldera formation by ductile deformation of the hydrothermal system, experiments show that the flat floor forms early by summit flattening before the formation of the ring fault along which the collapse occurs. The process of this flat floor caldera-like formation is similar to that of a classical caldera, given that the rheology of the whole system is very similar, composed of a viscous reservoir that is able to flow, overlain by a sequence of brittle volcanic rocks.

7.1.2 Explosive Calderas on Shield-Like Complex Cones of Peralkaline Composition

Volcanic edifices of peralkaline composition, which belong to shield volcanoes, are also called shield-like complex cones (Houghton et al. 1992). They go through similar stages of caldera collapse and filling to explosive basaltic calderas. Their calderas are explosive in origin and they may include a post-caldera phase of resurgence. Peralkaline calderas typically occur in zones of rifting, such as Pantelleria, associated with a submerged continental rift (Mahood and Hildreth 1983, 1986; Orsi et al. 1991), and calderas of the K'one (Gariboldi) volcanic complex, Ethiopia, associated with the Main Ethiopian Rift (Acocella, et al. 2002). Circular peralkaline calderas show moderate sizes: most are ~12 km in diameter with collapses of ~300 m, in response to small volume pyroclastic eruptions. At Pantelleria, the caldera rim height is ~90 m, and the estimated volume of pyroclastic material is 7 km^3 (Green Tuff; Wolff and Wright 1981). The Gran Canaria caldera (see Chap. 4) is an exception, with a rim wall up to 1 km high, and erupted pyroclastic material in the order of 100 km^3 (Schmincke 1967), but this probably represents numerous collapse events, as have occurred at nearby Las Cañadas (Marti and Gudmundsson 2000; Marti 2019). At some peralkaline calderas, post-caldera lava extrusion is associated with magma intrusion into the root zone of the caldera that leads to uplift or resurgence of the central portion of the floor (e.g. Suswa, Johnson 1969; Pantelleria, Mahood and Hildreth 1983; Orsi et al. 1991).

7.1.3 Calderas at the Summit of 'Shield-Like' Ignimbrite Constructs

Distinct volcanic constructs, intermediate between ash-flow calderas and shield-like complex edifices on uplifted basement, may not fit in any of the afore-mentioned categories, as they are made up of ignimbrites forming aprons surrounding large calderas. Such shield-like edifices have been discussed in intraplate settings (e.g. Tibesti in Chad, Jebel Marra in Sudan) and volcanic arcs (e.g. Altiplano-Puna Volcanic Complex). They have been called 'ignimbrite shields' (de Silva and Lindsay 2015), but this locution may be confusing, as shield volcanoes (Chap. 4) are made up overwhelmingly of lava flows, with only a minority of pyroclastic products.

Ignimbrite shield-like volcanoes: case studies

Large ignimbritic constructs of peralkaline rhyolites abound in the Neogene Volcanic Field of Tibesti (VFT, Chad), 30,000 km^2 in size, which is probably of hot-spot origin (Gourgaud and Vincent 2004). Among the five recognized eruptive phases that built up these constructs, Deniel et al (2015) document a third phase that formed three large 'ignimbritic shield-like' volcanoes 70–100 km in diameter, with low angle slopes (2–3°). They do not coincide with any individual pre-existing volcano, but are associated with significant updoming of the basement, uplifted aprons of ignimbrites and/or unusual clusters of domes. The third phase ended with the collapse of three large calderas: Tarso Voon (about 7–5 Ma), Emi Koussi (2.4–1.33 Ma), and Yirrigué (0.43 Ma) in the Tarso Toussidé volcanic complex (Photo 78). The Voon ignimbrite covers an area of 2200–2600 km^2 with an estimated volume of about 130 km^3. The younger Yirrigué ignimbrite covers ~3200 km^2 with a volume of about 150 km^3.

The Voon, Emi Koussi and Yirrigué ash-flow calderas are smaller (~11–15 km in diameter) than silicic ash-flow calderas, and result from a single collapse, while post-caldera activity occurred without floor resurgence. Their well-defined rims and planar floors suggest plate (piston)-type caldera formation (Table 1; Elshaafi and Gudmundsson 2021). They also share geomorphic features with the Galápagos shield calderas, in particular with steepened slopes (≥ 10°) at their summits (Permenter and Oppenheimer 2007). The older ignimbritic sheets intercalated with the plateau basalts, usually with no exposed vents, are probably related to this type of caldera setting on horsts with significant basement uplift, which favour the accumulation of large volumes of magma in upper crustal magma chambers. Such accumulation of laccolithic bodies may have resulted in regional updoming prior to any eruption.

Emi Koussi (Photo 79), 60–70 km in diameter and the highest summit (3415 m a.s.l.) in the Sahara desert, is the largest volcano of Tibesti, albeit with less than 1000 m relative height (Deniel et al. 2015). The complex summit caldera of the volcano (13.5 × 11 km) results from the emission of three successive Quaternary (2.4–1.32 Ma) pyroclastic sequences, consisting of silicic products followed by small volume basaltic lavas (Gourgaud and Vincent 2004). With an area of ~2780 km^2 and an estimated volume of ~150 km^3, the Emi Koussi ignimbrite is the largest of the three above-mentioned volcanoes. It is classified as an 'ignimbritic volcano', although it differs from the others in several significant features. (1) Its slopes are less than 10° but steeper than the 2–3° slopes typical of ignimbritic sheets that form aprons around calderas. (2) The different units of its highly welded ignimbrites resemble lava flows, except in their upper and lower parts, and especially at their contact with the basement. Similar lava-like features are also observed in ignimbrites from other continental LIPs, such as Paraná-Etendeka and Karoo. (3) The chemical composition of the erupted products is also distinct, with predominantly undersaturated magmas (Deniel et al. 2015; Elshaafi and Gudmundsson 2021).

The existence of such unusual 'shield-like ignimbrite volcanoes' may be due to structural factors: (1) Doming of basement can induce shield-like landforms with relatively steep slopes, although these are not controlled by repose-angle slopes like other pyroclastic, monogenetic cones; (2) Physical characteristics, welding, and chemical composition of ignimbrites can make them analogous to gently sloping lava flows. Other volcanic constructs that do not fit easily into any of the so-called 'ignimbrite volcano' categories are also present in Tibesti: elevated rhyolite domes (Abéki, Photo 78) and ignimbrite feeder dykes (Kilehegé) such as those found at cauldrons (Sect. 4.4). Elevated domes are characterized by higher uplift, the preponderance of cone-sheet swarms and no obvious relation with any former basaltic shield. Tardso Abéki is an asymmetric dome 40 km in width, with no caldera but instead a central block forming a relief of 1200 m over an area of 17 × 11 km (Vincent 1970). This is also the case of the Shiribetsu volcanic complex (Hokkaido, Japan), which has an ignimbrite plateau and subsequent central lava domes, but does not display caldera collapse (Goto et al. 2020).

Do so-called 'ignimbrite shields' exist?

Lipman (2000a) aptly discussed the term 'ignimbrite shields'. Spectacular silicic centres of Neogene age in the Altiplano-Puna of the central Andes have been described as ignimbrite shields (e.g. Cerro Panizos, Laguna Colorada, both in Bolivia, and Purico in Chile) (de Silva and Lindsay 2015). In contrast to wide, deep calderas, these ignimbrite sheets form a thick, extensive apron dipping gently (8–10°) outward from a central lava dome complex, but they display a downsag feature at their summits with no significant collapse (Salisbury et al. 2011). Such constructs have been proposed (de Silva and Lindsay 2015) as an individual class of explosive centres or a transitional type of landform akin to that of ignimbrite sheets (see Silicic large igneous provinces, Chap. 1). However, the lack of ignimbrite shields in North America and elsewhere suggests possible interpretative problems with the Andean 'ignimbrite shields' resulting from their lack of pervasive dissection, inaccessibility and somewhat limited field studies. At least some 'ignimbrite shields' seem alternatively interpretable as conventional, structurally bounded, nonresurgent calderas, surrounded by their ignimbrite aprons, which were then filled and/or overflowed by intracaldera lavas and pyroclastics. Analogous large, non-resurgent calderas that were once completely buried by later flows and other volcanic deposits have been variably exposed by erosion in many volcanic areas (San Juan volcanic field in Colorado, Lipman and McIntosh 2008; Cappadocia field in Turkey, Froger et al. 1998; Le Pennec et al. 2005, Cotahuasi-Ocoña-Marán area in southern Peru, Thouret et al. 2016). The pre-erosion morphology of these deeply eroded and filled San Juan Mountains calderas would have looked strikingly like the so-called ignimbrite shields of the Andes.

7.2 Felsic Calderas

Calderas on composite cones (e.g. Somma on Vesuvius created by the 79 AD/CE Plinian eruption; Miyakejima, Japan formed in 2000, see Sect. 8.1) are due to a combination of fast subsidence and explosive activity, which results in the foundering of the pre-existing volcano owing to a climactic eruption of a voluminous magma reservoir. The roof of the dacite-rhyolitic or trachytic calderas founders owing to the emptying of the magma chamber due to intense explosive activity and high magma production. This also occurs on basement tumescence or doming independent of any single large volcano. Lockwood and Hazlett (2010) collectively call them 'explosive calderas' associated with violent pyroclastic eruptions, and caused by the subsidence of magma chamber roofs. Many authors call the large felsic calderas 'ash-flow' calderas or 'ignimbrite calderas' that formed in continental or island-arc areas independent of any single large volcano (Smith 1960; Smith and Bailey 1968; Sheridan 1979; Self et al. 1986; Wohletz and Heiken 1992; Schmincke 2004). Thus we choose to differentiate the two types of explosive calderas, firstly those at the summit of composite cones, secondly those on uplifted basement.

8 Two Categories of Explosive Calderas and Processes

Explosive calderas can be broken down into two main categories: (1) relatively small- or medium-sized depressions at the summit of composite cones or on truncated cones, and (2) vast, deep depressions on domed basement independent of any individual pre-existing volcano. Further distinctions will be discussed based on the composition of the caldera-forming eruption.

8.1 Small and Medium-Sized Calderas on Composite Cones

In the first case, relatively small calderas (<10 km) are formed by explosive silicic eruptions, typically sub-Plinian or Plinan. A caldera-forming eruption took place at Miyakejima volcano (Japan) in 2000, enabling scientists to observe the formation of a relatively small caldera (1.6 km across) on an active basaltic andesite composite cone (Schmincke 2004; Nakada and Druitt 2005). The Miyakejima edifice (814 m.a.s.l. + 1000 m.b.s.l) forms an island in the Izu-Bonin volcanic chain about 200 km south of Tokyo (Fig. 11). Observations and subsequent interpretations of this eruption were made possible through an array of techniques involving geophysical surveys, earthquake swarms, and satellite imagery (Geshi et al. 2002). From 26 June to 8 July 2000, magma migrated northwestward below the sea floor, starting from a reservoir below the summit. A very large volume of magma (1–2 km^3) had apparently escaped laterally from the main reservoir beneath the centre of the island. Collapse of the summit of the volcano began on 8 July and lasted until the middle of August. Smaller eruptions lasted until late September. The largest eruption occurred on 18 August after subsidence had stopped, producing 8×10^9 kg of tephra. The total volume of tephra erupted from the volcano in 2000 amounts to 1.1×10^7 m^3 DRE. The growth rate of the caldera was 1.4×10^7 m^3/d and the final volume of the caldera was about 6×10^8 m^3 (Geshi et al. 2002). Because the volume of erupted material was exceedingly small, a little more than 1% of the void created by caldera collapse, the eruption could not have triggered caldera collapse.

The collapse process itself appears to have happened basically in two parts (Geshi et al. 2002; Schmincke 2004) (Fig. 4B): (1) a central stoping rock column 600–700 m in diameter is thought to have subsided piston-like, bounded by outward-dipping faults; (2) at the same time as the central piston subsidence there was significant lateral enlargement of the surface basin by caving-in of the vertical walls along listric faults. The final caldera had a diameter of 1.6 km in September, the floor being 450 m deeper than the former land surface. Subsidence of the central rock column was calculated as ~40–53 m/d assuming 40 days of subsidence from July to August. The total subsidence then amounted to a record 1.6–2.1 km to account for the volume of the caldera depression (Schmincke 2004). The origin of the void that caused the summit collapse is easily accounted for by lateral drainage of magma at depth, as indicated by seismic evidence. The volume of fresh magma at depth was still large and convecting to account for the very high SiO_2 degassing that continued for years. This prevented the return of some 5000 people who were evacuated from the island until 2005.

Many composite cones never experience collapse calderas like Miyakejima, because the magma chambers underlying these edifices are generally too deep and small to collapse. The roof cannot founder if the reservoir is deep, hence the thickness of crust forming the roof must be 10 km maximum. A critical ratio relates the width of a magma chamber to the thickness of its roof (Lockwood and Hazlett 2010). If the ratio is <1, the roof will be strong enough to support long-lasting composite cones. It the ratio is >1, meaning that the magma chamber has a very thin roof, then a major volcanic edifice may not develop at all.

8.1.1 Andesitic-Dacitic Calderas

Calderas associated with andesitic–dacitic volcanism, typical of composite cones, usually involve the destruction of these constructs. The collapse of Mt. Mazama to form Crater Lake, USA (Bacon 1983; Bacon and Lanphere 2006), Krakatau, Indonesia (Self and Rampino 1981; Simkin and Fiske 1983), and Santorini, Greece

(Druitt et al. 1999) are the best studied examples of this type of caldera. These eruptions all resulted in the collapse of a composite cone of basaltic andesite/andesite composition to form a caldera (Self and Rampino 1981; Bacon 1983; Wheller and Varne 1986; Druitt et al. 1999). An important fact is that the material erupted in the climactic phases was either totally dacitic, or was from a zoned magma chamber in which there was a significant amount of dacitic material. This suggests that the magma resided in the magma chamber for long enough to evolve and become more silicic than the magma usually erupted. The climactic eruptions are also usually of much larger volume than usual. This may mean that a stage was reached that was favourable for magma accumulation.

A well-studied example is the collapse of Mt. Mazama, Oregon and formation of Crater Lake (Williams 1942; Bacon 1983; Bacon and Lanphere 2006). Mt Mazama is the name given to a group of shield and composite cones of intermediate composition that began to form a high massif in the central Cascades range about half a million years ago (see Chap. 5). The climactic caldera-forming eruption began 6845 ± 50 BP (Bacon and Lanphere 2006). About 50 km^3 of rhyodacitic magma erupted as tephra and ash flows from ring faults, as Mazama stratocone (40–52 km^3) sank in a piston-like manner into the underlying, partly emptied chamber, forming the present Crater Lake caldera (Photo 74). While the ash flows continued to propagate down preexisting canyons, extending to distances as great as 70 km, the composition of the magma changed from rhyodacite to andesite and gabbroic crystal mush, illustrated by the colour of the pumice changing from pale to dark grey, mirroring a zoned magma chamber (Photo 80). The late-stage eruption of low-silica magma, producing small post-collapse constructs of mafic composition in continental calderas that formed during powerful silicic eruptions, is a common occurrence and further evidence of the role of mafic melts in initiating these eruptions (Lockwood and Hazlett 2010).

Santorini: multi-phase caldera collapse, cyclic eruptive behaviour, and timescales

The multi-phase caldera of Santorini, in the Greek Cyclades Islands, is one of the best studied calderas worldwide (Druitt and Francaviglia 1992; Druitt et al. 1999; Druitt and Vougioukalakis 2019). Santorini has drawn the attention of scientists since the classic late nineteenth century study by Ferdinand Fouqué (Fouqué 1879), who was one of the first to show that calderas form largely by vertical collapse, not explosive evacuation. Santorini is located on a 25 km-thick continental crust in the Christiana–Santorini-Kolumbo volcanic field, within a NE-SW transtensional rift 100 × 45 km in size that is subject to stretching (Nomikou et al. 2019), situated obliquely to the Hellenic subduction zone in the South Aegean Sea (Fig. 11A). This rift connects the extinct Christiana centre to the SW, Santorini caldera with its intracaldera Kameni Volcano, Kolumbo Volcano, and 24 other submarine cones of the Kolumbo chain to the NE (Druitt and Vougiuokalakis 2019; Preine et al. 2022). Figure 11A shows an interpretation of the NE-SW bathymetric section through Christiana Volcano, Christiana basin, Santorini caldera, Anhydros basin and Kolumbo Volcano (Preine et al. 2022). Their study highlights the fundamental tectonic control on magma emplacement and shows that the Christiana-Kolumbo volcanic field, having produced more than 100 explosive eruptions in the last 400,000 years, evolved from a volcanic field with local centres that matured only recently to form the vast Santorini edifice.

Figure 12B shows the topography, morphology and setting of the caldera, including the main Thera Island, Therasia Island, and the modern Nea and Palea Kameni Islands (Karátson et al. 2018; Photo 75A and B). Two outer islands, Thera and Therasia predate the Late Bronze Age (LBA, also called the "Minoan", dated at 1630 BC/BCE) eruption, while two inner ones post-date the LBA eruption: Nea and Palea Kameni, formed between 197 BC/BCE and 1950 AD/CE (Druitt et al.

2019). Like many other explosive volcanoes, Santorini has a repetitive three-fold cyclic behaviour (Druitt et al. 2019): (1) Plinian eruption with or without caldera collapse; (2) inter-Plinian edifice construction with moderate explosive activity; (3) repose, leading up to the next Plinian eruption, with a return period of 10–30 kyr. From about 350 ka onwards, volcanism became highly explosive, with at least eleven Plinian eruptions between then and the LBA eruption (Druitt et al. 1999). Albeit poorly constrained at sea, the volume of each of these eruptions probably ranges from a few to a few tens of km^3 of intermediate to silicic magma.

Santorini is a multi-phase caldera where more than twelve Plinian eruptions have occurred over the last 350 ka, at least four of which have caused caldera collapses, namely the 172 ka Lower Pumice 2, the 80 ka Upper Scoria 1, the 22 ka Cape Riva, and the LBA eruption at ~3600 yr BP (Fig. 11C) (Druitt et al. 1999, 2019). The Plinian eruptions are fed from magma reservoirs at 4–8 km depth that developed over several centuries prior to eruption by the arrival of high-flux

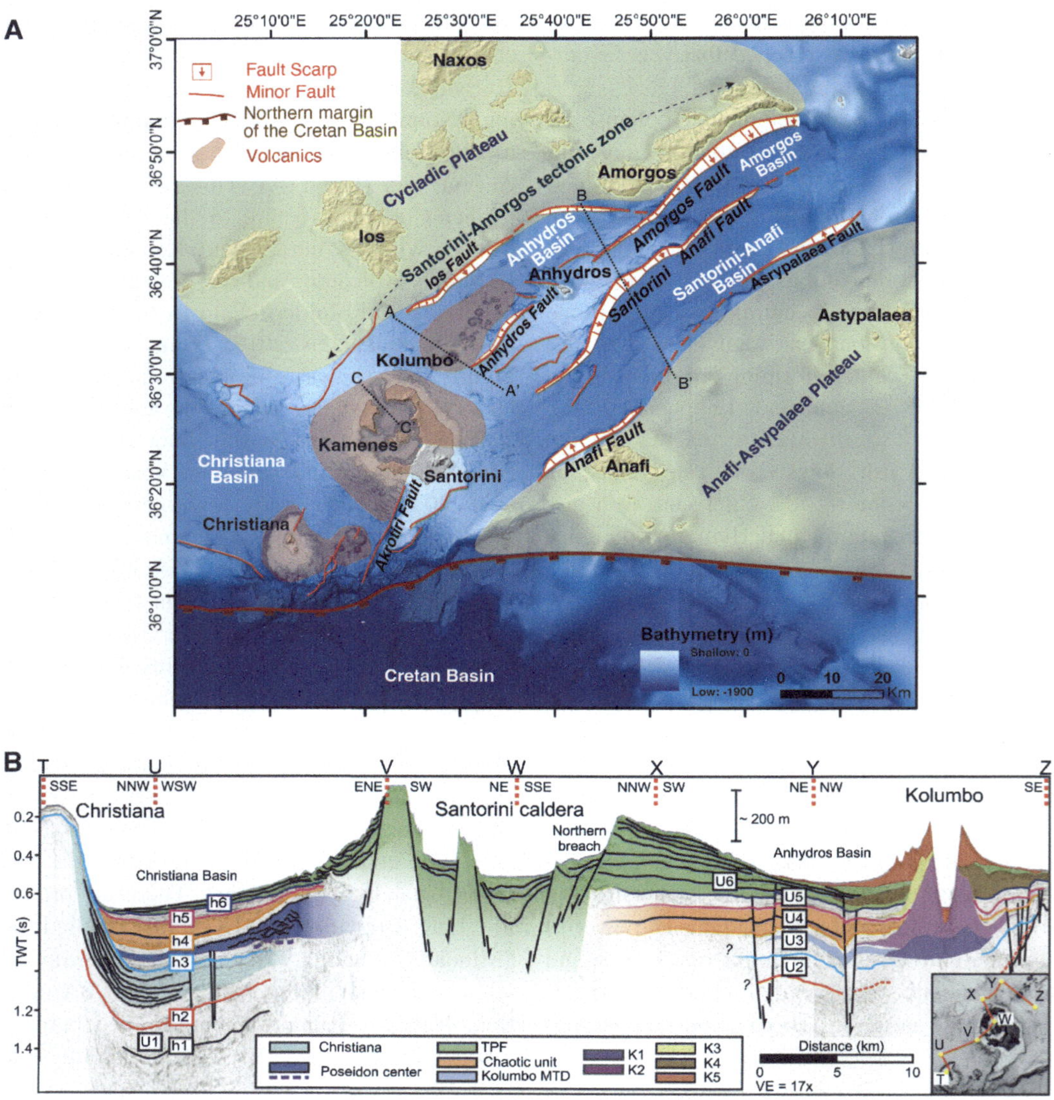

◀ **Fig. 11** Santorini multiphase caldera: tectonic setting, stratigraphy, and DEM-based geomorphic reconstruction. **A** Major tectonic features of the Christiana–Santorini–Kolumbo (CSK) volcanic field and the Santorini–Amorgos Tectonic Zone. Note bathymetry key at lower right (Nomikou et al. 2019; Fig. 1, p. 172). **B** Interpretation of the seismic profile connecting volcanic centres of the Christiana-Santorini-Kolumbo volcanic field (Preine et al., 2022, Fig. 2, p. 98). TWT—two-way travel time; h1–h6—marker horizons; U1–U6—volcano-sedimentary units; TPF—Thera Pyroclastic Formation; MTD—mass transport deposit; K1–K5—Kolumbo units; VE—vertical exaggeration. **C** Simplified geological map of Santorini, with location map inset. **D** Simplified stratigraphic column, showing the products of Plinian eruptions in blue (mainly magmas of intermediate compositions) and red (mainly silicic magmas), and the products of inter-Plinian periods in brown, five of which are named. The two earliest products of the volcanic field are shown in green. Asterisk labels are as follows: * = historically dated; ** = dated by radiometric methods, principally K-Ar and $^{40}Ar/^{39}Ar$; *** = dated by correlation with deep-sea ash layers. Abbreviations: BCE = before Common Era; CE = Common Era. (Druitt et al., 2019; Fig. 1, p. 178). **C** Reconstruction of the Late Bronze age caldera (Karátson et al. 2018, Fig. 4, p. 6). Digital elevation model (DEM) reconstruction of Santorini comparing (**a**) the present-day topography and (**b**) the proposed topography prior to the Minoan eruption. The latter shows the reduced size of a Pre-Kameni island, a smaller and shallower caldera harbour restricted to the north, a possibly smaller caldera outlet (dotted blue line) in the north, and a continuous southern caldera rim connecting Thera and Therasia through Aspronisi. **E** Multiphase caldera collapse at Santorini, geomorphic reconstruction (Druitt and Francaviglia 1992; Karstens et al. 2023, Fig. 4, p. 7). Interpreted topographic evolution of the Santorini caldera during the Minoan eruption. **a** Pre-Minoan topography with northern shallow marine caldera basin ('shallow lagoon') and pre-Minoan cliffs (dotted red lines). **b** Tuff cone formed during the third phase of the Minoan eruption. **c** Post-Minoan caldera based on mapping of the base of Unit 3 with caldera cliffs ('post-Minoan cliffs') which collapsed after the caldera collapse (marked in blue). **d** Present-day topography

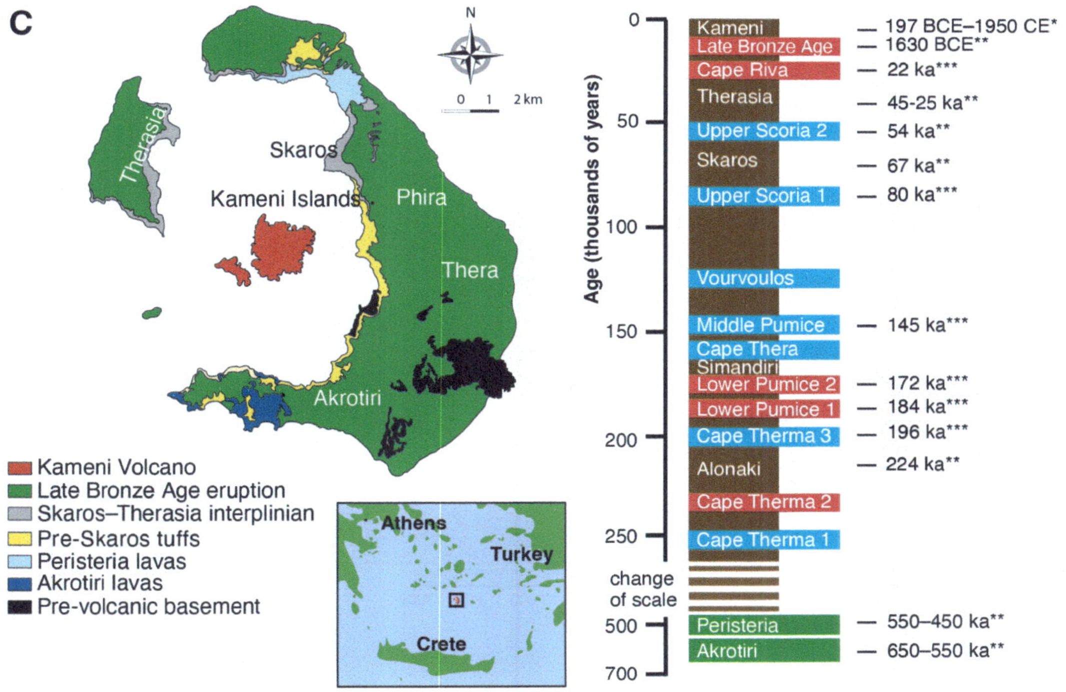

Fig. 11 (continued)

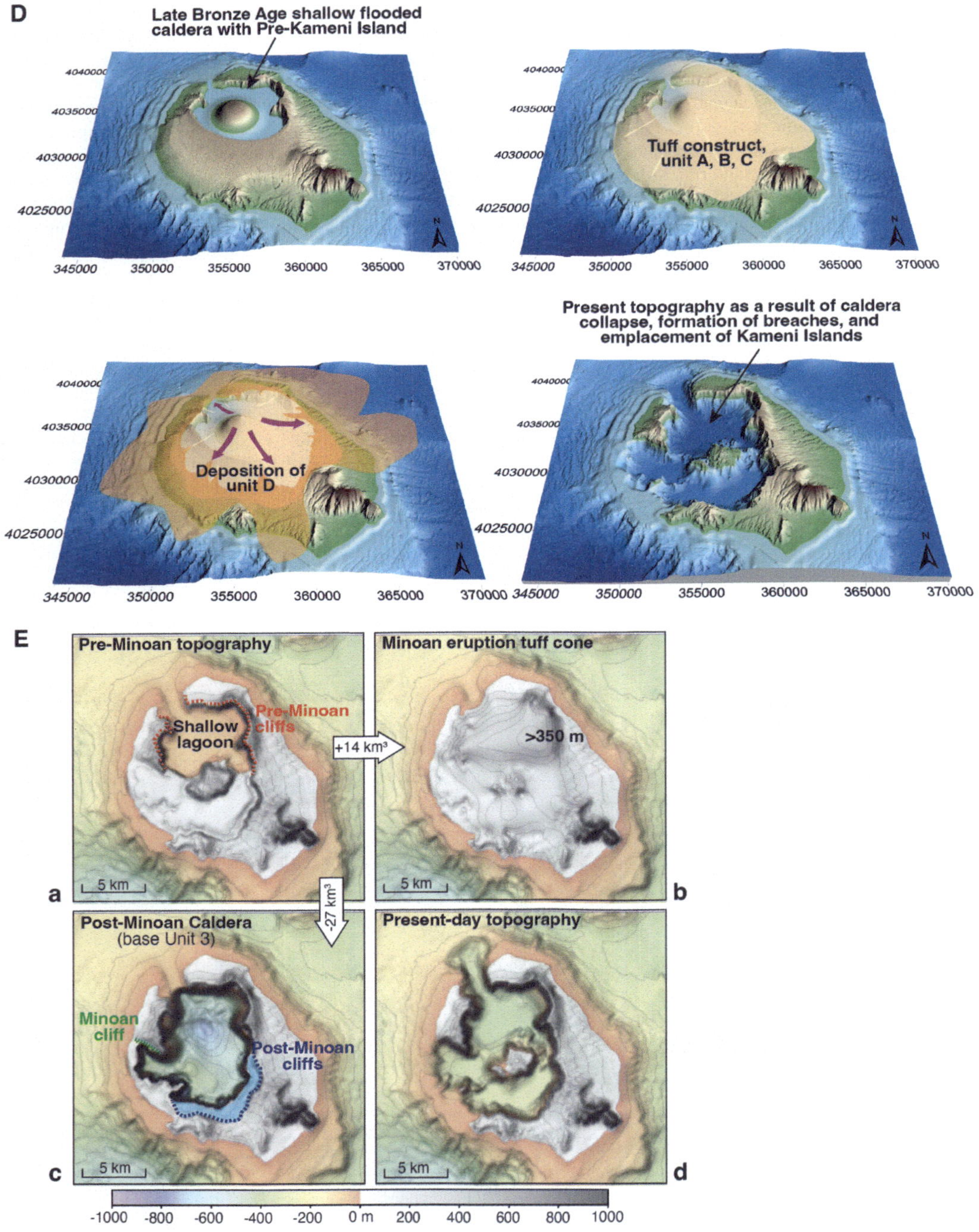

Fig. 11 (continued)

magma pulses from deeper in the sub-caldera reservoir. The LBA eruption, one of the largest on Santorini, may have discharged as much as 86 km^3 (~34.5 km^3 DRE) of rhyodacitic magma, forming the present-day 300–400 m-deep caldera (Johnston et al. 2014; Druitt et al. 2019; Karstens et al. 2023). Steep unconformities present in the cliff successions of Santorini have been interpreted as ancient caldera walls, with at least three events of caldera collapse (Lower Pumice 2, Upper Scoria 1, and Cape Riva) prior to the last one in the Late Bronze Age (Druitt and Francaviglia 1992).

The wealth of information acquired on this well-studied caldera has allowed reconstructions to be made of the topography prior to the LBA (Minoan) eruption. For example, Karátson et al. (2018) have used a photo-statistical method, complemented by granulometry and high-precision K–Ar dating to elaborate on the size and age of the intracaldera pre-Kameni Island and reconstruct the topography of the LBA caldera. Pre-Kameni and other parts of Santorini were destroyed during the 3.6 ka Minoan eruption, and their fragments were incorporated as lithic clasts in the Minoan pyroclastic deposits. Applying the Cassignol-Gillot K–Ar dating technique to the black glassy andesite lithics, Karátson et al. (2020) propose that the island started to grow at ~20.2 ka soon after the Cape Riva eruption. Figure 11C shows the proposed physiography of Santorini prior to the Minoan eruption using a combined DEM with 10 m resolution offshore and 15 m resolution onshore. The reconstructed DEM provides a 3D view of Minoan Santorini: a closed caldera rim in the southwest connecting Thera and Therasia, an unconstrained strait in the north, a smaller caldera bay, and a 1.5–3 km^3 Pre-Kameni island (with low dome shape) in the centre of the flooded bay.

To establish a benchmark for large volcanic eruptions, Karstens et al. (2023) have re-calculated the volume of the iconic Minoan eruption, by integrating seismic reflection and P-wave tomography datasets with computed tomography-derived sedimentological analyses. Their results reveal a total DRE eruption volume of 34.5 ± 6.8 km^3, which encompasses 21.4 ± 3.6 km^3 tephra fall deposits, 6.9 ± 2 km^3 ignimbrites, and 6.1 ± 1.2 km^3 intra-caldera deposits. Based on integration of geophysical and sedimentological data, the authors re-interpreted the DEM-based topographic evolution of the Santorini caldera during the Minoan eruption in four phases (Fig. 11D) starting with (a) the pre-Minoan topography, (b) a Minoan eruption tuff cone, (c) the post-Minoan caldera with post-Minoan cliffs that collapsed after the LBA caldera collapse (marked in blue), and ending with (d) the present-day topography.

New investigations on Santorini and similar calderas suggest two major models

(1) Large eruptions at Santorini are preceded by events on two very different timescales (Druitt et al. 2019): (i) prolonged generation of silicic magma deep in the sub-caldera pluton over many thousands of years, followed by (ii) rapid transfer of that magma to the top of the sub-caldera pluton to form a transient magma chamber, from which eruption takes place a few centuries to a few decades afterwards.
(2) Magma chambers are probably transient features on a geological timescale and they are assembled by high-flux pulses of magma from deeper in the sub-caldera pluton on geologically shorter timescales than previously thought. Progress in recent petrological and geochemical investigations on the kinetic growth of minerals contained in ignimbrites produced by large caldera-forming eruptions all point to a short period of time between the peak growth of large magma reservoirs and the onset of eruption (Druitt et al. 2019; Flaherty et al. 2018). Based on phenocryst diffusion chronometry data from the analysed pumice of the LBA deposit (Flaherty et al. 2018), phenocryst growth records one or more episodes of magma transfer from a source region of gabbroic to dioritic bulk composition in the 8–12 km depth range to a pre-eruptive magma chamber at 4–6 km in depth, within a few centuries prior to the eruption. High-flux transfer events may provide a way of reconciling a continuous supply of basaltic magma and heat to the base of the crust with the intermittency of large silicic eruptions at Santorini.

There is also evidence that voluminous eruptions at large calderas in general tap into multi-crustal intrusion sills and laccoliths (Cashman and Giordano 2014) that remain otherwise disconnected between eruptions. Magmas are now widely believed to be generated in bodies of crystal mush, from which they are extracted and accumulated into melt-dominated magma chambers prior to caldera forming-eruptions. These authors present key observations: (1) many large eruptions tap multiple melt sources, (2) large melt bodies are probably transient features, and (3) eruptions of crystal-rich magma are probably driven by roof collapse and fragmented by sudden decompression. Models invoked to explain such observations suggest that in many systems that erupt initially crystal-poor silicic magma, particularly those in extensional settings, melt may be stored within multiple isolated and/or partially connected lenses. This type of complex storage geometry could provide a mechanism for the rapid assembly of large melt bodies, as well as a local source of triggering recharge magma.

8.2 Large, Ash-Flow or Ignimbrite Calderas with Rhyolitic Composition

Ash-flow calderas are very large (10–30 km, maximum 60 km in diameter), and their form and eruptive products are generated during a few major events (Walker 1984; Lipman 1997). Such vast calderas typically form independently of any single large volcano. Many, if not all, appear to form in association with volcanic complexes and fields (see stage 1 in Fig. 12A and B, after Wohletz and Heiken 1992; Lipman 2000a). Some large calderas are actually the roof regions of larger granitic to granodioritic plutons that are exposed where the upper few kilometres have been stripped away by erosion. The area of subsidence corresponds roughly to the maximum diameter to which the underlying magma reservoir developed as it evolved, though it may be somewhat smaller.

Fig. 12 Resurgent caldera: **A** Generalized evolution of resurgent ignimbrite calderas based largely on studies of Cenozoic calderas in the western USA (after Lipman 1984, 1997, 2000a, Fig. 2, p. 648; Wohletz and Heiken 1992, building on a model by Smith and Bailey, 1968). **a** Precollapse volcanism. Clustered composite cones of intermediate-composition develop over isolated, small high-level plutons that eventually coalesce to form a large silicic magma body that feeds ignimbrite eruptions or heats the shallow crust sufficiently to allow the buoyant rise of a large pluton. The site of the subsequent caldera collapse is indicated by dotted lines; thick arrows indicate upward movement of magma. **b** Caldera structure geometry just after ash-flow eruptions and concurrent caldera collapse. Central area of older, clustered volcanoes caves into collapsed area. Intra-caldera tuff ponds during subsidence and is an order of magnitude thicker than the cogenetic outflow ignimbrite sheet. Initial collapse along bounding faults is followed by slumping of oversteepened caldera walls and accumulation of voluminous collapse breccias that interfinger with ignimbrite in the caldera fill. Caldera floor subsides asymmetrically and is tilted to the left in the diagram. Dashed lines approximate the compositional zones within the pluton. **c** Resurgence and post-caldera deposition. Resurgence is asymmetrical, with greater uplift in the area of greatest prior collapse. Extensional graben faults form over the crest of the dome. Some resurgent uplift is accommodated by movement along rift faults in the opposite direction to that during caldera subsidence. Welded tuffs and other caldera deposits are uplifted as a structural dome over the magma body. Magma body has risen into volcanic pile and intruded cogenetic intra-caldera ignimbrite. Original caldera floor has been almost entirely obliterated by the rise of the magma chamber to close to the level of the pre-caldera land surface. Caldera moat is partly filled by lava domes and volcaniclastic sediments. Hydrothermal activity and mineralization are prevalent late in cycle and develop within the caldera deposits and floor rocks, as well as along extensional faults that cross the resurgent dome (adapted from Lipman 1984; Wohletz and Heiken 1992). **B** Possible main evolutionary stages of a large felsic volcano with caldera: **a** initial tumescence due to magma accumulation; **b** major eruption(s) and caldera formation; **c** evanescence and cooling of the shallow chamber; **d** rejuvenation due to the input of new magma, promoting resurgence (after Smith and Bailey, 1968). **C** Main resurgence modes observed in experiments and nature mainly as a function of the Aspect ratio A_r (where Ar = thickness z/width L) of the crust overlying the magma chamber (Acocella et al. 2001, Fig. 4, p. 145 *in:* Acocella 2021, Fig. 5.12, p. 178). **a** High A_r develops a resurgent block with consistent layers within the caldera and peripheral volcanic activity (Ischia and Pantelleria type); dashed purple lines illustrate local configuration of the σ_1 trajectories controlling the development of the reverse ring fault. **b** Low A_r develops a resurgent dome with domed layers, summit depression and volcanic activity at the periphery and within the dome (Valles and Long Valley types) ►

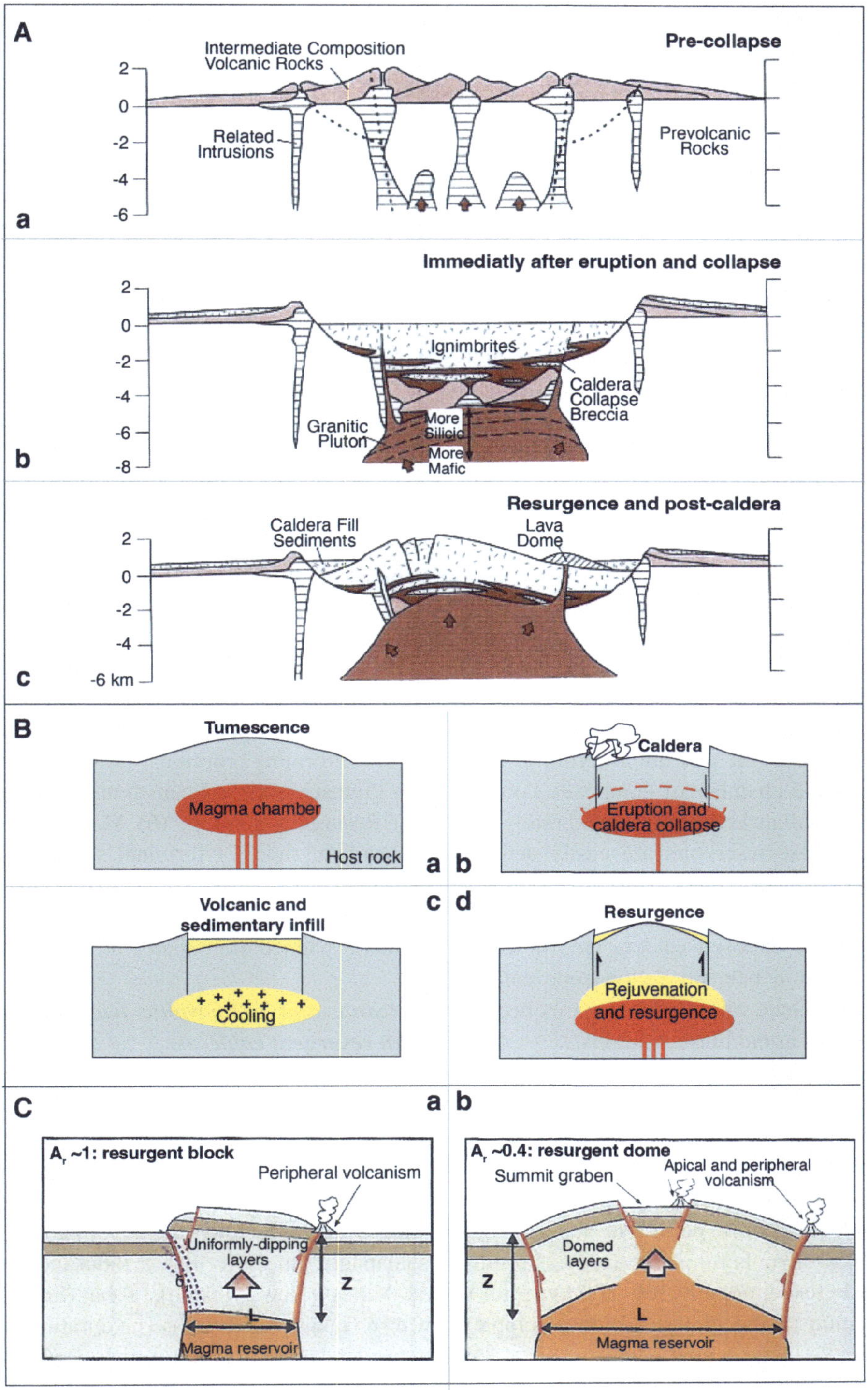
A
Pre-collapse
Intermediate Composition
Volcanic Rocks
Related
Intrusions
Prevolcanic
Rocks
a
Immediatly after eruption and collapse
Ignimbrites
Caldera
Collapse
Breccia
Granitic
Pluton
More
Silicic
More
Mafic
b
Resurgence and post-caldera
Caldera Fill
Sediments
Lava
Dome
c
-6 km
B
Tumescence
Magma chamber
Host rock
Caldera
Eruption and
caldera collapse
a b
c d
Volcanic and
sedimentary infill
Cooling
Resurgence
Rejuvenation
and resurgence
C
a b
A, ~1: resurgent block
Peripheral volcanism
Uniformly-dipping
layers
Z
L
Magma reservoir
A, ~0.4: resurgent dome
Summit graben
Apical and peripheral
volcanism
Domed
layers
Z
L
Magma reservoir

Ash-flow calderas have their own characteristics that distinguish them from calderas of composite cones (Vincent 1994; Bourdier 2021).

(1) Their formation is due to the high volume of Plinian eruptions, being typically greater by a few hundreds to thousands of km^3 than the Plinian eruptions (of a few km^3 to a few tens of km^3) on composite cones.
(2) They are at least as big on average as summit calderas of composite cones (8 km at Crater Lake), as the diameter of a caldera is roughly correlated to the volume of emitted magma (Cas and Wright 1988). The diameter usually exceeds 10 km (22 km at Valles) and subsidence of the caldera floor is generally >1 km. The actual depth of the caldera floor, buried or resurgent, is often deeper than the depth of summit calderas.
(3) Structurally, they are associated with the activity of intracrustal magma reservoirs, the volume and composition of which are distinct from those of composite cones (Cashman and Giordano 2014). The composition is dacitic and often rhyolitic, while the magma chamber of composite cones is of intermediate composition, and rarely rhyolitic. These reservoirs are easily detected by geophysical methods and are of the size of batholiths, sometimes wider than the size of the exposed caldera (see modelled cross section beneath Yellowstone caldera, Fig. 6B). Once solidified, reservoirs become granitic or granodioritic plutons.
(4) These huge collapse depressions are associated with the largest volume pyroclastic deposits recorded. The structures are usually, but not always, in areas that have experienced extensive volcanic activity over a considerable time period in the order of 350–450 kyr before the caldera-forming episode (exceptionally 300–800 kyr prior to formation of the Platoro Caldera complex, Colorado; Dungan et al. 1989).

8.2.1 Resurgent Calderas

Large felsic calderas often show features associated with resurgence (although other calderas can also show this phenomenon). A resurgent caldera forms a broad topographic depression with a central elevated massif. The post-caldera stage of volcanism is commonly characterized by uplift of caldera floors. This uplift, termed *caldera resurgence*, is typically accompanied by small eruptions that form minor cones and lava flows, as illustrated by Okmok (Larsen 2013) and Aso (Hasenaka et al. 2022) calderas (Photos 71 and 76). The term *resurgent caldera or cauldron*, in which the cauldron block, following subsidence, is uplifted, usually in the form of a structural dome, was coined by RL Smith and RS Bailey (US Geological Survey) in 1962, while they were mapping the Valles-Toledo caldera complex in the Jemez Mountains of New Mexico. Using the Valles caldera as a model, but with additional information from other cauldrons (Fig. 1), Smith and Bailey (1968) developed a model of resurgent calderas involving seven stages of volcanic, structural, sedimentary, and plutonic events, which led to Lipman's model (Fig. 12A and B): (1) Regional tumescence and generation of ring fractures; (2) Caldera-forming eruptions; (3) Caldera collapse; (4) Preresurgence volcanism and sedimentation; (5) Resurgent doming; (6) Major ring-fracture volcanism, and; (7) Terminal solfatara and hot-spring activity. These stages define the cycle of resurgent cauldrons, which in the Valles-Toledo caldera spanned more than 1 million years.

Volcanic and geomorphic features associated with resurgent calderas

The uplifted area is usually circular to elliptical, from a few km (Pantelleria, Italy) to a few tens of km (La Pacana, Chile), proportional to the size of the caldera. The amount of uplift ranges from a few tens of metres (Alcedo, Galápagos) to more than a thousand metres, as at Valles (New Mexico), Toba (Sumatra) and Cerro Galán (Argentina). The duration and uplift rate of resurgence are not well constrained, but from data acquired in Campi Flegrei, Toba, Siwi (Vanuatu) and Iwo-Jima (Japan), the duration of resurgence is very variable, lasting from 1 to 79 ka, and different uplift rates range from

Table 2 Main features of active resurgent calderas (Acocella 2021)

Active, resurgent caldera	AC (km^2)	AR (km^2)	Uplift (m)	Vr (km^3)	Vv (km^3)	Type	T (ka)	Tr (ka)	Composition	Supply	Regional extension (mm/year)
Ischia	46	~20	~1000	9±2	<1	B	22±5	<33	T	Yes	<2
Pantelleria	42	7	~350	1.5±0.3	0.5±0.2	B	27	<18	T, R	Yes	<2
Campi Flegrei	~200	~30	~250	3±0.5	4±0.5	D, G	<4	<15	T	Yes	<2
Long Valley	450	~80	>400	~15	<100	D, G	100	80	R, Da	Yes	0.6
Valles	283	~65	≥1000	32±5	<5	D, G	<40	27±27	R, RD	Yes	<4
Yellowstone SCD	2500	214	?	>27	0	D, G	<84	<123±9	R	Yes	<4.3
Yellowstone MLD	2500	164	?	20±7	<600	D, G	~440	?	R	Yes	<4.3
Toba	2350	950	1100	320±30	20±5	B	<40	<70	R, Da	Yes	0
Siwi	36	18	>250	4.5±1	0.5±0.3	D, G	<18	2	Ba, TA	Yes	0
Iwo-Jima	63	17	>120	>1.5	0	D	>2	<0.8	TA	inferred	<2

AC area of caldera, AR area of resurgence, Vr volume of resurgence, Vv DRE volume of syn-resurgence volcanism, Type B block, D dome, G graben, T time between caldera collapse and resurgence, Tr duration of resurgence. Composition: T trachyte, R rhyolite, Da dacite, RD rhyodacite, Ba basalt, TA trachyandesite, Supply evidence of magma supply, Reg. ext. rates of regional extension. Yellowstone SC Sour Creek dome, Yellowstone ML Mallard Lake dome. Values in parentheses are partly constrained and may not be representative

centimetres to tens of centimetres per year. Most resurgent uplift occurs episodically, within a few thousand years, interrupted by pauses or even subsidence. Table 2 displays the main features of active resurgent calderas (Acocella 2021).

Large, resurgent rhyolitic calderas include: Long Valley, California, USA (Bailey et al. 1976) which erupted the Bishop Tuff (Wilson and Hildreth 1997); Cerro Galán, Argentina (Francis et al. 1983; Francis and de Silva 1989) which erupted the Cerro Galán ignimbrite (Sparks et al. 1985; Folkes et al. 2011), and Valles-Toledo (New Mexico, USA), where resurgence was first fully discussed. Valles is one of the best studied of this type of caldera, located on the edge of the Rio Grande Rift, an area of lithospheric thinning and extension (Fig. 13; Smith et al. 1961; Smith and Bailey 1968; Heiken et al. 1986, 1990). The eruption and deposition of two major rhyolitic ignimbrites accompanied the collapse. The first (lower member of the Bandelier Tuff; Heiken et al. 1986, 1990; Self et al. 1986) occurred at 1.47 Ma, from ring fractures around the Valles caldera, and the second (upper member of the Bandelier Tuff) at 1.12 Ma, during which most collapse occurred. The ignimbrite is now exposed on all sides of the caldera with thicknesses ranging from 15 to 270 m, and occurs within the caldera as 400–1100 m of densely welded ignimbrite. Geophysical evidence suggests the collapse was asymmetric (Heiken et al. 1986), producing two calderas (Toledo and Valles; Fig. 13A) with trapdoor geometry. Resurgence of more than 1 km occurred during the next 100 kyr (Self et al. 1986), accompanied by radial fracturing, to form Redondo Dome (Fig. 13B). At the same time a caldera lake formed depositing sediments in the moat surrounding the resurgent dome. Between 1.12 and 0.13 Ma rhyolite domes and lava flows, and some airfall deposits and small ignimbrites, were erupted around the ring fracture system (Fig. 13A). Like at Long Valley caldera, the eruptions took place in clockwise succession around the area of resurgent uplift. Scattered basaltic eruptions also occurred around the fringe of resurgent silicic calderas, revealing the involvement of mafic magma in the reservoir system.

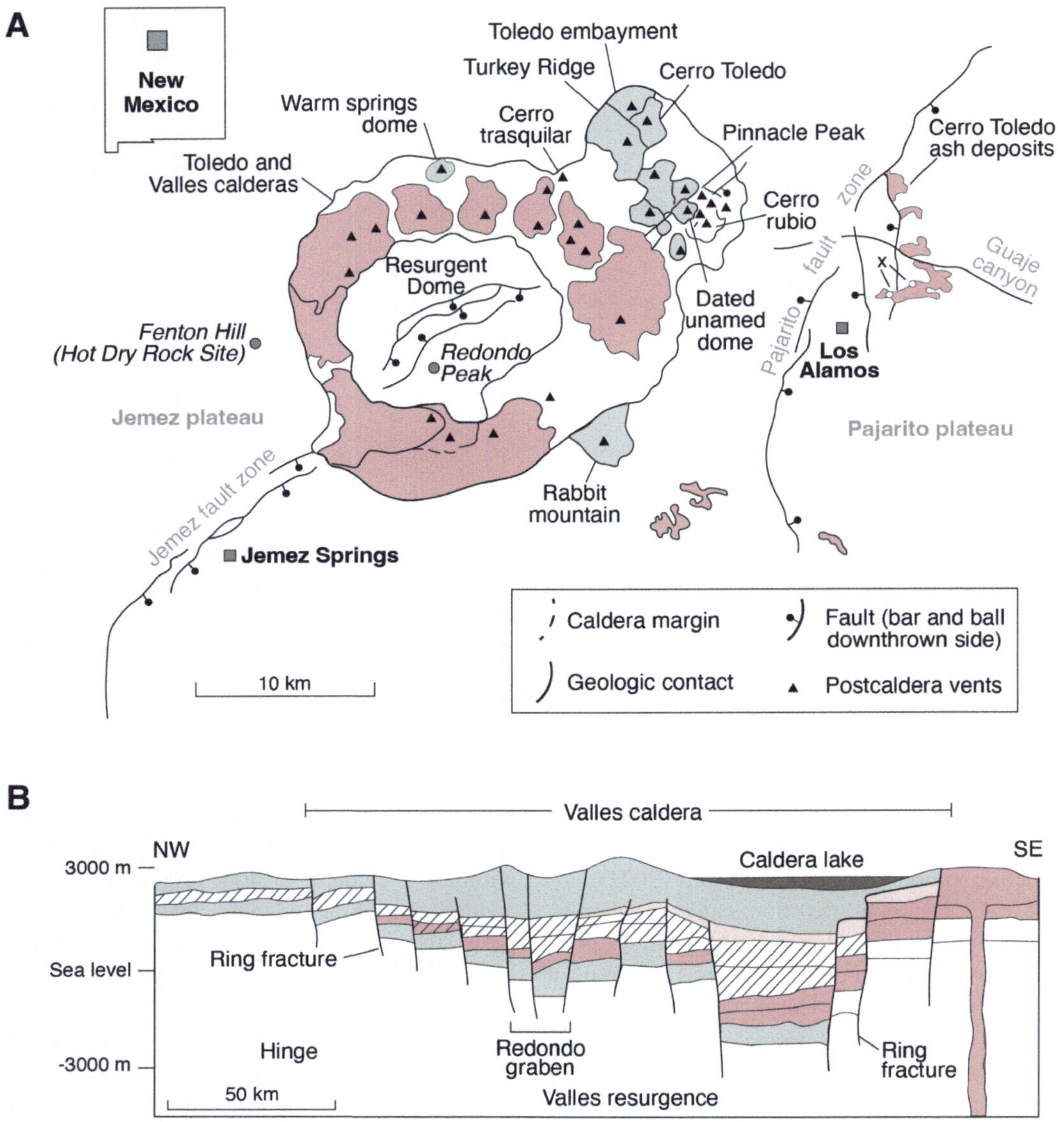

Fig. 13 Valles and Toledo resurgent caldera complex in New Mexico (Wohletz and Heiken 1992, Figs. 12, p. 157 and 15, p. 160. See also Cole et al. 2005). **A** Sketch map of the Valles/Toledo caldera complex, showing post-caldera rhyolite domes. The grey colour represents Toledo-age domes (1–1.4 Ma); the pink shading indicates Valles domes that range in age from less than 1 Ma to ~0.1 Ma. Triangles mark the vents of intracaldera volcanoes (Wohletz and Heiken 1992). **B** Schematic cross section of the Valles/Toledo calderas. The caldera complex, which appears to have a trapdoor shape, has a shallow dip on the northwest side where it overlies Paleozoic sedimentary rocks and Precambrian metamorphic and igneous basement. Most collapse has occurred along the eastern caldera margin where the caldera overlies the Rio Grande rift and a thick sequence of poorly consolidated rift sedimentary rocks (Wohletz and Heiken 1992)

During resurgence, eruptions can be highly effusive, often involving lavas of similar composition as the magma that erupted during caldera formation, as for example at Yellowstone caldera (Lockwood and Hazlett 2010). Yellowstone is the youngest of the large calderas in the USA (of which Valles and Long Valley are the largest), is located above a mantle hot-spot trace and last erupted about 640 ka (see Fig. 6, Chap. 1; Fig. 6A). A model elaborated by Farrell et al. (2014) depicts how the crustal magma reservoir 90 km long and 5–17 km deep extends well beyond the boundary of the nested compound calderas (Fig. 6B), indicating that the hot spot is now moving to the NE. The magma from the plume, transported vertically in dyke swarms drives Yellowstone's hydrothermal system, extraordinary high heat flow, and seismicity. Yellowstone is far from extinct, and the intense hydrothermal activity at the NE boundary of the caldera system attracts almost 4 million of tourists per year to the area. The model also shows the bimodal rhyolite-basalt melt zoned body that characterizes the Yellowstone and Snake River Plain system.

The resurgence phenomenon was first recognized by R. W. van Bemmelen in his work on the vast Toba caldera in northern Sumatra (van Bemmelen 1949). It is the youngest resurgent caldera in the world (Photo 72), formed only 74 ka by a massive eruption (Youngest Toba Tuff, YTT) which covered a surface of about 30,000 km^2, and has an estimated volume of about 2800 km^3 including co-ignimbrite ash (Chesner 2012; Storey et al. 2012). About two kilometres of subsidence took place initially at Toba to form a deep lake almost 100 km long and 30 km across (Fig. 14) (Francis 1993). The caldera covers about 2270 km^2, is elongated parallel to the major regional Sumatra Fault, and is not a simple structure, having undergone several episodes of subsidence. Chesner et al. (1991) and Chesner (2012) showed that four separate eruptions over the last 1.2 million years were involved in the formation of the compound Toba caldera (Haranggaol Dacite Tuff 1.2 Ma, Oldest Toba Tuff 0.840 Ma, Middle Toba Tuff 0.501 Ma, and Youngest Toba Tuff 0.074 Ma), with repose periods of 340,000–400,000 years between eruptions. The compound Toba caldera encompasses all previous calderas (MTT, OTT, and HDT) and most of the pre-existing volcano in the north. Following the 74 ka YTT eruption, the island of Samosir composed of rhyolitic lava domes covering 640 km^2, formed a tilted block (60 × 20 km) gently dipping towards the west. It has since been uplifted by at least 1100 m to its present position (Chesner 2012). Uplift also formed the Uluan peninsula, a symmetrical counter-part to Samosir on the SE side of the caldera (Fig. 14). As there is a Plinian fall deposit missing from the bottom of the YTT sections, it seems likely that the YTT ignimbrites were emplaced by fountain collapse at the onset of the eruption rather than being generated by collapsing Plinian eruption columns.

9 Resurgence Modes

Scientists have used a combination of analogue experiments, field observations of natural examples and theoretical calculations to propose resurgence modes.

9.1 Resurgence Mechanism

From analogue experiments and natural caldera cases, Acocella et al. (2001) postulated two modes of resurgence, namely resurgent blocks and resurgent domes, which are distinguished by the aspect ratio of magma reservoir width to crustal roof thickness (Fig. 12C). Note that each large ignimbrite eruption (shown in yellow) was followed by the eruption of basaltic and andesitic-trachyte material (shown in pink and grey, respectively): Aspect ratio, Ar, of the crust overlying the magma chamber, where $Ar = z$ thickness/L width). The ratio relating the width of a magma chamber to the thickness of its roof has been analyzed experimentally and theoretically in the laboratory (Roche et al. 2000; Acocella 2007, 2021).

1. At high Ar of around 1, where the depth of the magma chamber is similar to the width of the uplifted area, a resurgent block develops with coherent layers within the block, and peripheral volcanic activity. This occurred at the shallow, largely submerged calderas of Ischia and Pantelleria. Inward-dipping and steepening reverse faults define the base of the main block uplift, while normal faults fringe the summit rise. Figure 12C illustrates

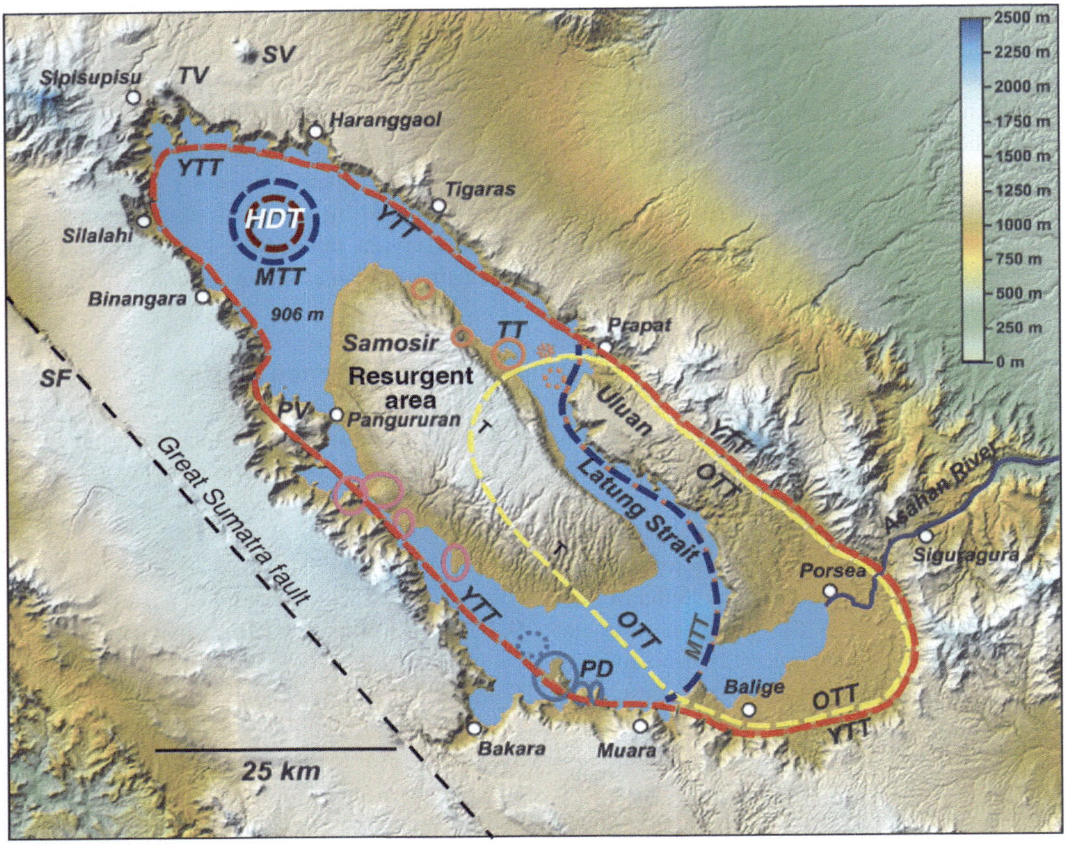

Fig. 14 Location map of the compound Toba Caldera plotted onto a DEM-based image (Chesner 2012, Fig. 2, p. 7). From lake bottom to caldera rim, total relief ranges from about 900 to 1700 m. Lake Toba covers about two thirds of the caldera floor and is up to 500 m deep. Caldera locations indicated by dashed lines: HDT Haranggaol Dacite Tuff in red brown, OTT Oldest Toba Tuff in yellow, MTT Middle Toba Tuff in dark blue, and YTT Youngest Toba Tuff in red. The Samosir Island resurgent block reaches 1630 m.a.s.l. Its upper surface dips gently (5–8°) to the west, while its eastern edge consists of a series of parallel fault scarps with considerable displacement. Samosir lava domes are outlined by orange circles and Pardepur lava domes are in light blue. Across the Latung Strait, lies the Uluan peninsula which appears to be a semi-symmetrical counterpart to Samosir Island with its steep western scarp and 10–15° eastward dips. Dotted orange and light blue circles are sub-lacustrine lava domes, identified by bathymetry, which belong to the post-resurgence Samosir and Pardepur lava dome groups respectively. Orange asterisk is an area of possible sub-lacustrine hydrothermal activity associated with another Samosir lava dome. Pink ovals are areas of recent uplift that may be underlain by hypabyssal intrusions. Abbreviations are: PV Pusikbukit volcano, PD Pardepur lava domes, TT TukeTuk lava domes, TV Tandukbenua volcano, SV Singgalang volcano. The Great Sumatra strike-slip Fault is shown. Towns are shown by white circles

the local configuration of the σ_1 trajectories controlling the development of the reverse ring fault. Melt may work its way up along the outer fault system, feeding small volcanic eruptions within the caldera moat.

2. At low *Ar* of~0.4, where the depth of the magma chamber is less than the width of the uplifted area and the chamber width is more than double the thickness of its roof crust, a resurgent dome with dome layers, and an apical depression can develop, associated with volcanic activity at the periphery and within the dome. This occurred at the Jemez, Valles and Long Valley calderas. In Fig. 12C, diagram b shows reverse faults on the dome, but normal faults bound a summit graben, while volcanic activity occurs at apical (graben fault) and peripheral (reverse dome ring fault)

locations (Acocella et al. 2001). The area of subsidence can be a simple circular depression, but if regional crustal extension prevails, this may modify the stress imposed by upwelling magma to create a graben whose orientation lies perpendicular to the direction of regional extension. Post-caldera volcanism not only occurs within the moat, but also in the graben or simple centre of the uplift as well.

In fact, experimental work by Walter and Troll (2001) suggests that a complete gradation of caldera structures exists between these two styles of resurgence. In the case of Ischia and Pantelleria, the rising melt was able to displace the block of crust upward without causing significant internal deformation. In the case of Jemez and Long Valley, the thin roof block fractured internally during uplift, permitting the ascent of magma through the centre of the caldera floor to the surface.

9.2 Resurgence Models

Resurgence is a notable feature of many large felsic calderas (Table 2). Early interpretations attributed resurgence to a build up in pressure in the magma chamber, or from shallow intrusions. Four mechanisms related to the dynamics of a magma chamber and triggering resurgence are now recognized: regional detumescence, magmatic rebound and pressurization (Marsh 1984; Kennedy et al. 2012), in addition to the viscosity contrast and vesiculation between the residual magma in the reservoir and that responsible for resurgence (Branney and Acocella 2015).

Regional detumescence implies that, upon caldera formation, an originally inflated crust relaxes, pushing downward on the magma chamber and squeezing magma upward against, around and through the fallen block, triggering resurgence. This is poorly documented in nature.

Magmatic rebound and/or decompression following the formation of a caldera is related to the sluggish response of the viscous magma to the sudden drawdown caused by the eruptions during caldera formation and/or to the vesiculation of resurgent magma due to decompression. However, the onset of resurgence often ranges from a few tens of thousands of years (Valles, Ischia, Pantelleria) to several hundred thousand years (Yellowstone), suggesting the lack of a direct relationship between the collapse and resurgence.

Magma pressurization includes an increase in the pressure within a magma chamber resulting from magmatic convection, vesiculation or the influx of new magma at the base of the caldera block, or as shallow laccolith-like intrusions. These have been observed in eroded, Miocene calderas (Lake City, Colorado: Kennedy et al. 2012; Kumano, Japan: Miura 1999). This process is feasible, but it still does not explain the crucial, anomalous accumulation of magma required to trigger dramatic uplift instead of feeding eruptions.

Viscosity contrast and vesiculation. There is an imbalance between the higher volume of the resurgent block or dome with regard to the volume of magma erupted during resurgence, suggesting some process that hinders the extrusion of the pressurized magma (Acocella 2021). This process may be related to the relatively low viscosity contrast (less than nine orders of magnitude) between the magma responsible for resurgence and the residual magma below the caldera. Such a viscosity contrast provides a rheological barrier, hindering the upward propagation of the new injected magma through dykes, which instead accumulates in sill- or laccolith-like intrusions, stagnates and promotes resurgence.

Three cases can be distinguished (Galetto et al. 2017): (1) A thin layer of viscous residual magma, contrasting with the low-viscosity new magma, promotes dyke propagation and eruption, but without resurgence. (2) A thick layer of medium viscosity residual magma, having a weak contrast with the new magma, hinders dyke propagation, promoting stagnation and resurgence, while peripheral eruptions may occur. Thicker and medium viscosity residual magma provides a less permeable barrier, leading to resurgence without eruptions in the area, as observed at Ischia, Yellowstone or Iwo-Jima. 3). A similar thick layer of medium viscosity propagates at shallower levels: dykes may come to a stop within the altered intracaldera tuff, developing one or more sills which then constitute a rheological barrier for successive dykes and stagnation in laccoliths, thus promoting resurgence (Galetto et al. 2017). These conditions provide a more permeable barrier, leading to resurgence accompanied by dyking and thus eruptions within the resurgent area, as observed at Campi Flegrei (see Fig. 8B).

10 Subsidence Mechanisms Inducing Distinct Caldera Geometries

The key element for caldera formation is the removal of magmatic support from the crust overlying the chamber, but caldera collapse mechanisms and geometries vary widely, and no two calderas form in exactly the same way (Lockwood and Hazlett 2010). The process of caldera collapse, final caldera morphology and type of associated eruption depend on many factors (Cole et al. 2005), such as the amount and rate of subsidence, the critical depth, size and shape of the magma reservoir, the roof thickness ratio and its strength, and the magma composition. The regional stress field and pre-existing tectonic structures also play a role. To gain access to this information, a combination of field data, experimental and numerical modelling is required.

10.1 Interpretations of Mechanisms of Caldera Collapse

As shown in Table 1, several authors have distinguished calderas on the basis of their morphology, visible structure, distribution of vents, and magma composition. However these characteristics do not provide information about the shape and behaviour of the magma reservoir and the physical conditions of the host rocks that lead to the formation of the ring faults, which ultimately determine the occurrence of caldera collapse.

By comparing analogue models with natural cases, Acocella (2007) used the s/d ratio (subsidence s versus diameter d of a caldera) to infer a progressive structural evolution of caldera collapse with increasing depth of subsidence (Fig. 15A). The evolution shows four stages, as follows: limited subsidence is mostly accommodated by downsag in stage 1, characterized by s/d <0.025. During stage 2, a steep outward-dipping ring fault propagates and subsidence progresses within the range 0.025 < s/d <0.055. With further subsidence, in stage 3, downsag develops towards the outer edge within 0.055 < s/d <0.07. With more subsidence in stage 4, with s/d >0.07, the outer zone develops arcuate normal faults and crevasses (Fig. 15A; Acocella 2007). Thus, the s/d ratio enables geologists to recognize diagnostic surface features to categorize a caldera within a precise structural and evolutionary context along a continuum mainly controlled by progressive subsidence.

Furthermore, this s/d ratio helps us understand that the depth of a magma chamber plays a major role, as illustrated by a number of calderas for which s has been correlated with diameter d of the magma chamber (Fig. 15B; Acocella 2021). This correlation is linked to the erupted volume fraction of caldera collapse (six fractions ranging from 60 to 100%) and the aspect ratio of the caldera $Ar=z/L$ of the roof (Fig. 16A; Geyer et al. 2006). It follows that a larger volume of magma must be extruded at greater depth to form a caldera, meaning that deeper and narrower magma chambers must evacuate a larger percentage of magma to reach a caldera structure similar to that of shallower and wider chambers. For very high aspect ratio, a caldera may not even form at the surface. Another factor related to the aspect ratio, Ar, regards the effective connectivity between the magma chamber and the surface during a caldera-forming eruption. Lower values of Ar (Ar <1.6) form coherent collapses, ensuring connection from the chamber to the surface with newly formed faults feeding

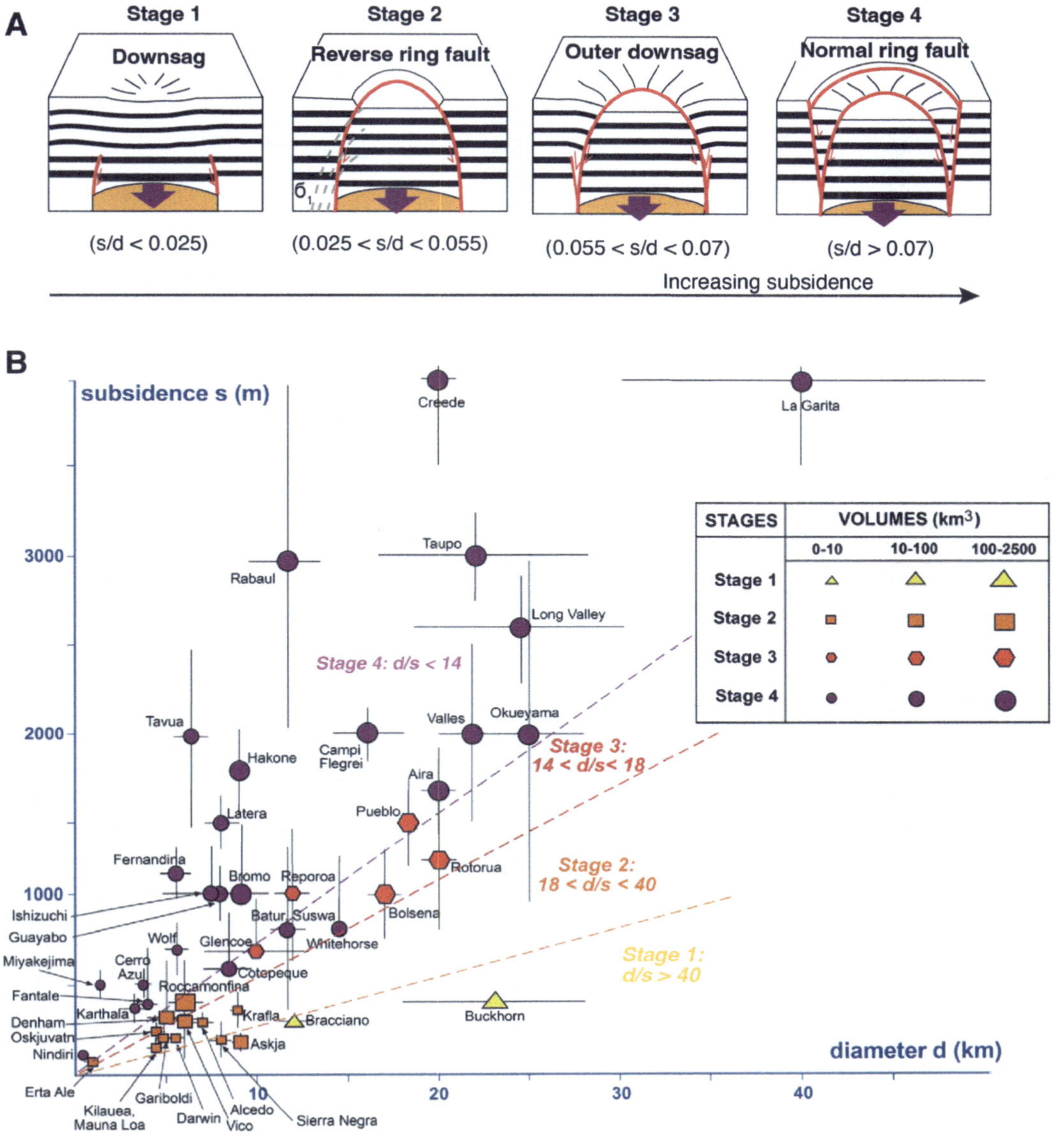

Fig. 15 Evolution of caldera geometries based on subsidence rate and diameter of calderas. **A** Evolution of natural caldera geometries, summarized into four stages from a combination of experimental data and actual examples (after Acocella 2007, Fig. 21, p. 144). To the right, caldera end-members (Lipman 1997; Cole et al. 2005) and related conditions. Four evolutionary stages of caldera collapse, based on the *s/d* ratio (amount of subsidence *s*/caldera diameter *d*), obtained from analogue models and measurements in nature. Stage 1: downsag; Stage 2: reverse ring fault (dashed grey lines illustrate local configuration of the σ_1 trajectories controlling the development of the reverse fault); Stage 3: peripheral downsag; Stage 4: peripheral normal ring fault. **B** Variation of subsidence (s) with the diameter (d) of 46 natural calderas. The calderas are classified according to the four experimental stages of Fig. 15A. Dashed lines define specific d/s ratios for each caldera stage. The size of the symbols is proportional to the DRE volume of erupted magma during the formation of the caldera (inset) (after Acocella 2007, Fig. 27, p. 150)

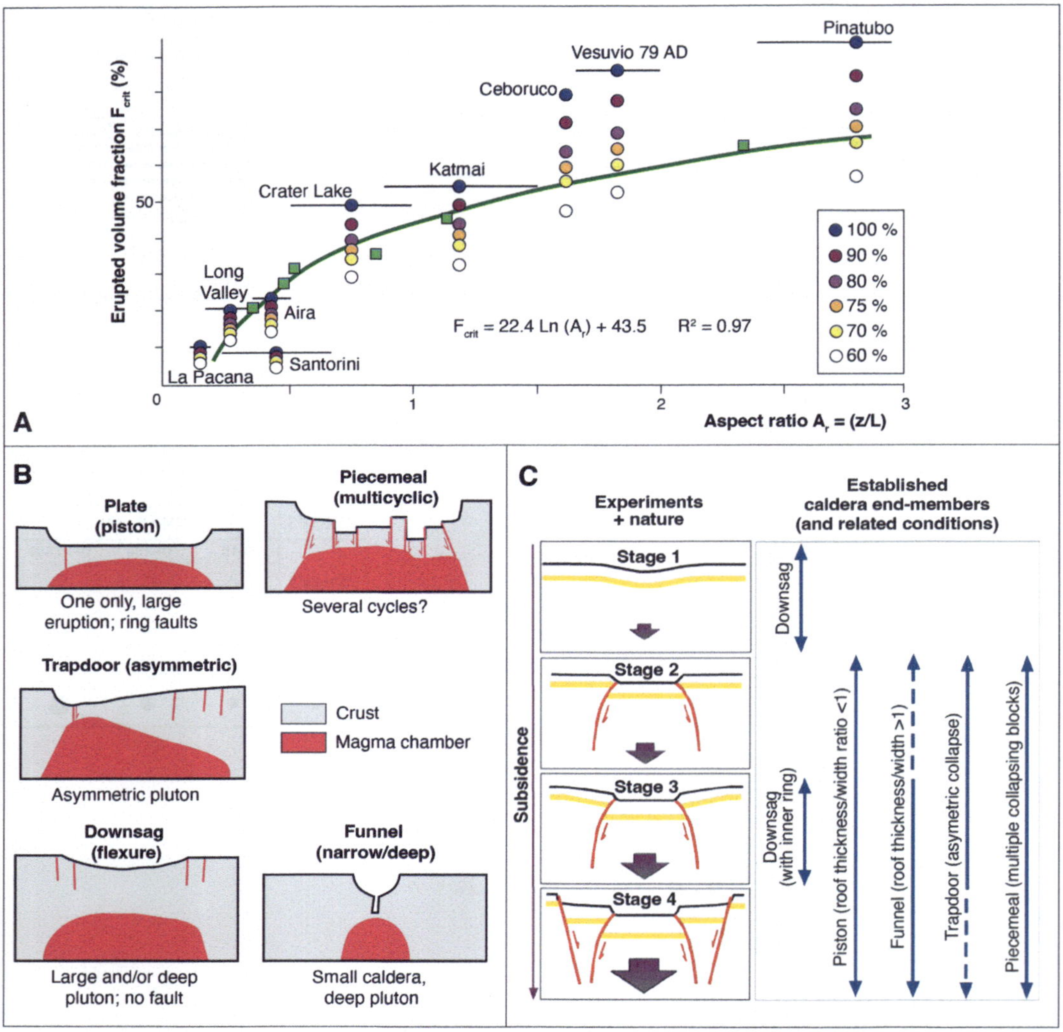

Fig. 16 Comparison of analogue models with natural calderas. **A** Erupted volume fraction (F_{crit}%) at the caldera onset as a function of the aspect ratio of the roof of the magma chamber *Ar* (where *Ar* = thickness *z*/width *L* of the roof). The erupted volume fraction is the portion of magma that needs to be erupted to trigger caldera collapse. Values of F_{crit} for natural examples are calculated considering different percentages (60–100%) of erupted magma; green squares indicate experimental values of F_{crit} at the collapse onset and green line indicates the related log-fit (after Geyer et al. 2006, Fig. 3, p. 380 *in:* Acocella 2021, Fig. 5.17, p. 186). The result implies that volume fractions required to trigger caldera collapse are lower for chambers with low aspect ratios (shallow and wide) than for chambers with high aspect ratios (deep and small). **B** Classical interpretation of five end-members of caldera collapse with geometrical relationships with plutons (Lipman 1997, 2000a). See text and Table 1 for explanation. **C** Reconciling analogue models with natural examples based on field studies (Acocella 2007, Fig. 31, p. 154). Analogue models stem from four stages of increasing subsidence while natural examples (end-members) stem from Lipman's types (2000a, b)

vents capable of sustaining the eruption. Higher values of *Ar* (*Ar* >1.6) form non-coherent collapses, which may hinder magma rise and thus the continuation of the eruption.

Also influencing caldera structure is the ratio between the thickness of the floor *t* (roof of magma reservoir) and the extraction width *w*: this is the roof aspect ratio *t*/*w* (Fig. 5B;

Roche et al. 2000; Branney and Acocella 2015). Mining subsidence and models show that the pattern of surface deformation varies: in subcritical extraction, t/w >0.7, a single point of compression develops in the centre of the subsided area, surrounded by a zone of extension, whereas in supercritical extraction, t/w <0.7, a platelike, non-deformed central zone forms, bounded by a narrow zone of compression that passes outward into a marginal zone of extension as seen as calderas (Fig. 5B).

Using the model of progressive structural evolution with increasing depth of subsidence (*s/d* ratio), the interpretation of caldera collapse can be reconciled with the popular, classical approach of Lipman's (1997), following Walker (1984), who defined five end-member structural types along with contrasting caldera collapse geometries (plate or piston, piecemeal, trapdoor, dowsag and funnel calderas: Fig. 16B) based on field data from active and eroded extinct calderas. Acocella (2007) compares the evolution of natural caldera geometries with Lipman's end-members by defining the constraints on the *s/d* ratio (Fig. 16C). The classical geometric classification carries some limitations concerning the structure and evolution of calderas because the end-members do not match some field observations: several of the proposed end-member structures may coexist in the same caldera as in models of rare examples where the caldera floor and structural margins are sufficiently exhumed and well exposed. For example, Scafell caldera in England possesses all the characteristics of trapdoor, piecemeal, or downsag types (Kokelaar and Branney 1999). In addition, the geometric classification cannot solve the imbalance between the volume of the large magma reservoir and that of the resurgent dome/block (Acocella 2021).

In fact, Lipman (2000a) was well aware of these limitations, noting that natural examples show a structural gradation between each end-member, and that some calderas of the San Juan Mountains involve not just a nested ring-fault plate structure but also significant downsag and piecemeal subsidence. He stated that 'Many well-studied calderas involve subsidence processes intermediate between idealized end-members; Many calderas have such transitional attributes that subclassification can be inherently ambiguous and subjective'. Lipman further added: 'Interpreting caldera structures in terms of a continuum of subsidence styles, rather than end-members types, can clarify relations between eruptive styles and structural processes' in correlation with the size of the eruption and the location and geometry of the cogenetic magma chamber.

Physical subsidence models have yet to assess the possible influence of subvolcanic crystal mushes with multiple melt lenses (Cashman and Giordano 2014). The depth of subsidence, for example, may be limited by the thickness and crystal content of a mush-filled sill-like reservoir, and the subsidence may migrate laterally during an eruption when partly interconnected melt lenses are erupted in rapid succession. Change in composition over the caldera-forming eruption sequence (from andesitic to ryhodacitic, reflecting the evacuation of different parts of a complex reservoir), subsidence rate, and the existence of preexisting faults or weak, hydrothermal zones also may influence the pattern of subsidence deformation.

10.2 Another Mechanism of Caldera Collapse

Marti et al.'s (2009) classification scheme is based on the conditions that lead to caldera formation and not on the morphology of the resulting depression. The authors have combined field data with analogue and numerical modelling to propose a genetic classification of calderas based on the stress conditions that permit the formation of ring faults and the pressure evolution in the magma reservoir during a caldera-forming eruption. A fundamental geophysical distinction of calderas relies on the pressure variation inside the magma chamber with regard to the eruption (Gottsmann and Marti 2008; Marti et al. 2009); thus two end-members, namely overpressure and underpressure calderas, develop from different initial conditions and

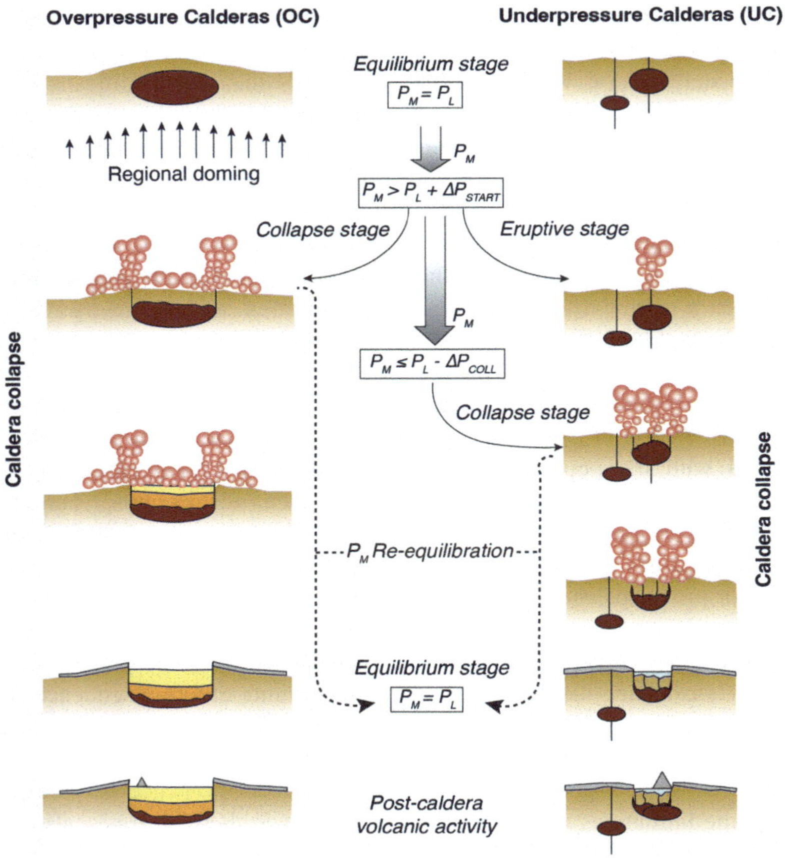

Fig. 17 Sketch of the evolution of 'overpressure' (OC) and 'underpressure' (UC) calderas. ΔP_{COLL} underpressure required to induce caldera collapse; ΔP_{START} overpressure required to trigger an eruption; P_L lithostatic pressure; P_M magmatic pressure. Note regional doming (not necessarily in the presence of an individual pre-existing volcano) and the sheer size of the magma reservoir at the onset of the process for OCs, whereas UCs develop on relatively small edifices, usually composite cones. Also note collapse stage starts early at OCs, but follows eruptive stage at UCs (Marti et al. 2009, Fig. 4, p. 260)

generate different sequences of caldera-forming deposits, as shown in Fig. 17.

1. *Underpressure calderas* (UC) result from ring-fault subsidence after significant decompression of the magma chamber following a pre-caldera eruptive episode. The initiation of collapse is a consequence of the eruption, as it is thought to occur in natural examples according to the classical model of caldera formation. UCs are the most common type of collapse caldera and usually occur associated with long-lived volcanic systems preferentially at composite cones (Marti et al. 2009). Magma withdrawal during the caldera-forming eruption in UC leads first to a decrease in overpressure in the magma chamber, often as a result of several eruption phases, so that the conditions within the chamber range from overpressure to underpressure when subsidence starts. This is reflected in the sequence of caldera-forming deposits that records Plinian eruptions that pre-date the syn-collapse deposits. These calderas may occur in any type of crust and tectonic setting, but they are more common in extensional settings. The deposits are generally felsic calk-alkaline and alkaline, but

can also be mafic calk-alkaline and alkaline in composition. Possible examples of UC are Crater Lake (Bacon 1983; Bacon and Lanphere 2006), Katmai (Hildreth 1991; Hildreth and Fierstein 2000; Hildreth et al 2003) and Santorini (Druitt et al. 1999; Druitt and Vougioukalakis 2019).

2. In *overpressure calderas* (OC), the stress conditions developing the caldera ring faults are achieved prior to initiation of the eruption when an overpressurized sill-like magma chamber is loaded by magmatic, regional doming or undergoes regional extension or large-scale magmatic underplating (Gudmundsson 1998, 2007). Here the initiation of collapse is coeval with the eruption onset and the resulting eruptions are often very large (Fig. 17). This simultaneous initiation has been noted at recent calderas in Iceland, for example at Bárðarbunga in 2014–2015 (Sigmundsson 2019; see also Pedersen and Gudmundsson, Chap. 10). Most of the caldera-forming sequence in OC corresponds to syn-collapse deposits, as there is no record of substantial eruptive phases prior to the formation of ring faults. The caldera-forming eruption (CFE) starts when the ring faults interesect the magma chamber, thus favouring a sudden decompression of magma and its fragmentation inside the chamber. This CFE tends to generate quite large plate/piston or trap-door collapse structures (diameter >25 km) related to significant volumes of erupted magma (>100 km^3). The associated magmas are primarily calk-alkaline, mainly rhyolitic or dacitic, and the calderas are associated with extensional and shear-extensional structures in continental subduction zones and continental rifting. Possible examples of OC are La Pacana (Lindsay et al 2001), Cerro Galán (Francis et al. 1978, 1983; Folkes et al. 2011), Aguas Calientes (Petrinovic 1999) and El Abrigo (Pittari et al. 2008).

A major difference between these two genetic types of calderas is that in OC the onset of subsidence occurs during the first episode of the caldera-forming eruption, while in UC subsidence is immediately preceded by a significant eruptive epidode not associated with any down-faulting of caldera blocks. Consequently, almost all the magma erupted in OC eruptions accompanies caldera subsidence, while in the case of UC a variable proportion of magma may be erupted without causing caldera collapse. In both cases, the eruption begins due to an excess pressure inside the magma chamber. Recently, using one deep (> 5 km) magma system at Aira caldera and one shallow (3 km) magma system at Kikai caldera, both in southwestern Japan, Geshi et al (2023) showed that caldera collapse was formed by rapid decompression of a magma chamber. These authors posit that the underpressure required for a magma chamber to collapse is proportional to the square of the depth to the magma chamber for calderas of the same horizontal size: large underpressure (61 MPa) at Aira and small underpressure (18 MPa) at Kikai before the onset of caldera collapse. Thus, distinct magma chamber underpressure thresholds correlate caldera collapse with magma chamber dimensions and the magnitude of the underpressure in the magma chamber may control the pattern of collapse and the eruption sequence of ignimbrite.

11 Caldera Complexes and Volcano-tectonic Depressions

Most multiple calderas or caldera complexes, also called compound or polygenetic calderas, occur within broadly extensional areas (Cole et al 2005). The Taupo Volcanic Zone, New Zealand (~1.6 Ma), for example, includes eight calderas (Wilson et al. 1995) (Fig. 14, see Chap. 1), of which the Taupo (Cole et al. 1998) and Okataina (Nairn 1989) calderas are the youngest. The largest silicic calderas often occur in clusters in continental arcs. In the Central San Juan Mountains, Steven and Lipman (1976), Lipman (2000a), and Bachmann et al. (2002) mapped at least 18 separate calderas nested within one another between 22 and 30 Ma, with multiple collapses associated with the eruption of nine major ignimbrites, which cover about 25,000 km^2 in total (Fig. 2). Long-term erosion has exposed some of the sub-volcanic batholiths. In southern Peru, Neogene and Quaternary ignimbrites cover an estimated 25,000 km^2 (Mamani et al. 2010; Thouret et al. 2007, 2016). In Bolivia, in the western Cordillera, there is an even more impressive complex termed the Altiplano-Puna volcanic complex (APVC), which is presented next together with Cerro Galán.

11.1 Caldera Complexes in the Central Andes

Widespread ignimbrites were recognized by Francis and de Silva (1989) in the APVC where northernmost Chile, Argentina and Bolivia meet in the Puna region. Mapping these ignimbrites has proven to be easier than identifying their source calderas, as erosion and post-caldera volcanism, or other pyroclastic deposits from other sources, have hidden the initial landforms (see Sect. 7.1.3). Several eruptive pulses have emplaced ignimbrite sheets over the past 30 Myr that cover at least 50,000 km^2 in the APVC. The most recent pulses took place around 9–10 Ma, 3–5 Ma and 1–2 Ma, coinciding with the most recent uplift of the Andean Cordillera. Intense magmatism and concomitant voluminous explosive eruptions have been called 'ignimbrite flare ups' by de Silva (1989) and de Silva et al. (2006).

For example, one extensive Pliocene ignimbrite (Atana) has been traced back to La Pacana caldera, a huge but structurally ill-defined landform 65 × 35 km in size, located in northernmost Chile near Argentina and Bolivia (Francis 1993). La Pacana is a good illustration of two aspects of large caldera complexes: the mismatch between the topographic and structural margins of a caldera, and the structure of a resurgent centre. The resurgent block (Cerros de la Pacana, 4905 m a.s.l.) is long and narrow (50 × 12 km), and exhibits antiform, welded ignimbrites and clearly defined NW–SE striking summit graben faults.

The APVC has been re-assessed with new geochronological techniques combined with field research. Salisbury et al. (2011) precisely dated and estimated the volumes of ignimbrites erupted from the Pastos Grandes and the Guacha large caldera complexes located in the Lipez area of Bolivia (Fig. 18A). The new age determinations demonstrate a distinct onset of APVC ignimbrite volcanism with modest output rates, an episodic middle phase with the highest eruption rates, followed by a decline in volcanic output (Fig. 18B). The cyclic nature of individual caldera complexes and the spatiotemporal pattern of the volcanic field as a whole are consistent with both incremental construction of plutons as well as a composite Cordilleran batholith (Salisbury et al. 2011).

An outstanding example of a caldera complex is Cerro Galán, located on the high-altitude Puna of northwest Argentina (Cas and Cashman 2011; Folkes et al. 2011). Despite its 35 km-diameter elliptical structure, Cerro Galán was

Fig. 18 Ignimbrites from the Altiplano-Puna Volcanic Complex and inferred location of caldera complexes (modified after Salisbury et al. 2011). **A** Location map showing sample localities and estimated original ignimbrite distributions (thick coloured outlines) of newly described ignimbrite units from the Bolivian Altiplano-Puna volcanic complex. Extents of previously described Vilama and Panizos ignimbrites are also shown (references in Salisbury et al. 2011). Outcrops from the SERGEOMIN geologic maps are colour-coded to demonstrate correlations to related ignimbrite units. Distribution of Tara and Guacha ignimbrites in Chile (references in Salibsury et al. 2011). $^{40}Ar/^{39}Ar$ sample locations from their study are shown as circles; stars are published isotopic age data used to aid in correlating ignimbrites in their study. Map numbers correspond to $^{40}Ar/^{39}Ar$ and K-Ar dates listed in Tables 1 and 2 of Salisbury et al. 2011). Dashed lines show approximate or inferred locations of caldera collapse scarps. Locations of characteristic remanent magnetism (ChRM) sample sites are shown as squares. Base map is a digital elevation model (DEM) compiled using 90 m Shuttle Radar Topography Mission (SRTM) data. **B** Cumulative volume versus time for ignimbrite eruptions in the Altiplano-Puna volcanic complex (APVC; thick line). Shown for comparison are the Southern Rocky Mountain (SRM), Southwest Nevada (SWN), Yellowstone, and Taupo volcanic fields (references in Salisbury et al., 2011, Fig. 10, p. 838). Altiplano-Puna volcanic complex ignimbrite volcanism has a distinct start at ~11 Ma, followed by waxing volcanism punctuated by three discrete pulses, and apparent waning activity during the last 3 Myr ▶

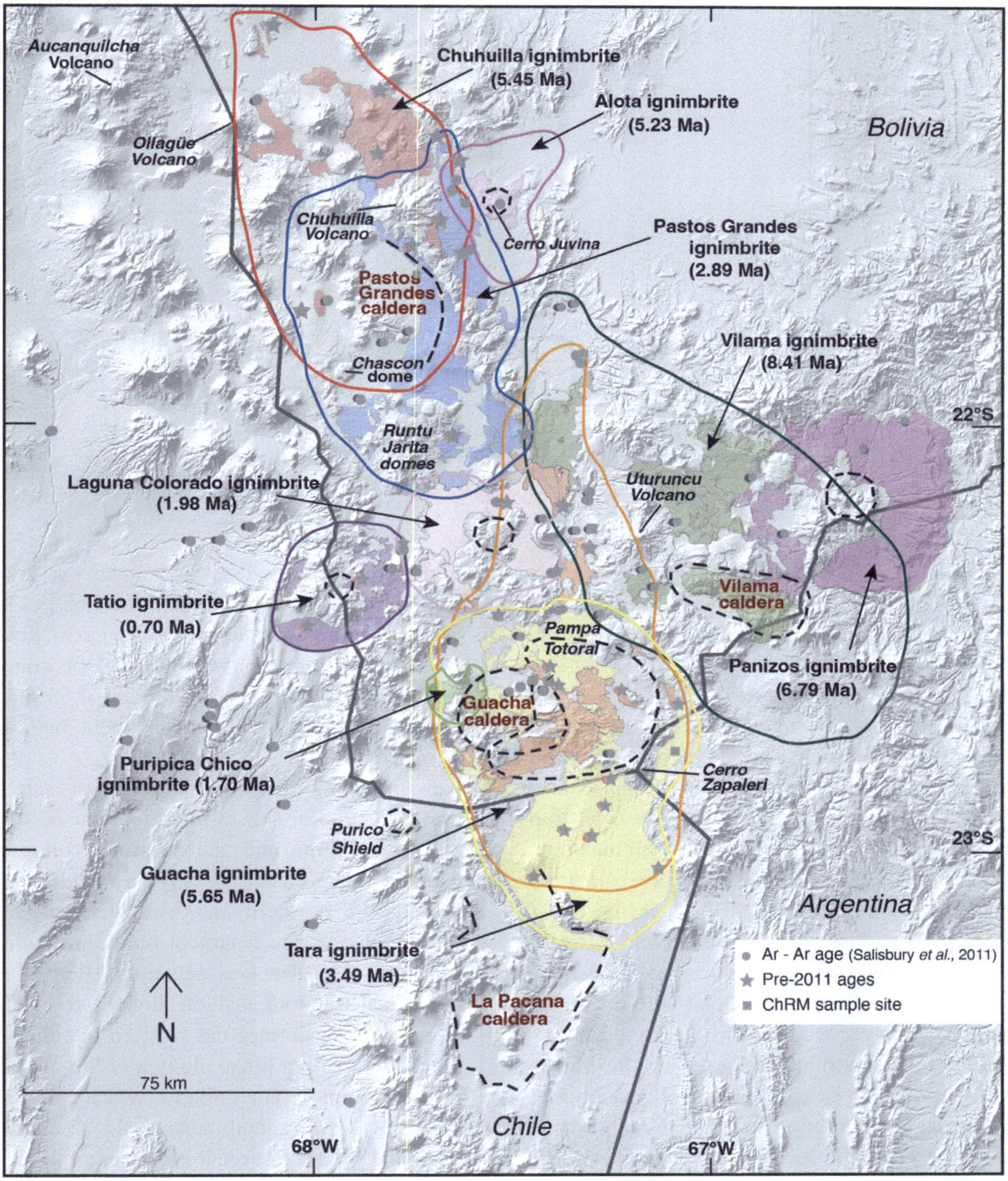
Aucanquilcha Volcano
Ollagüe Volcano
Chuhuilla ignimbrite (5.45 Ma)
Alota ignimbrite (5.23 Ma)
Bolivia
Chuhuilla Volcano
Cerro Juvina
Pastos Grandes ignimbrite (2.89 Ma)
Pastos Grandes caldera
Chascon dome
Vilama ignimbrite (8.41 Ma)
22°S
Runtu Jarita domes
Laguna Colorado ignimbrite (1.98 Ma)
Uturuncu Volcano
Tatio ignimbrite (0.70 Ma)
Vilama caldera
Pampa Totoral
Panizos ignimbrite (6.79 Ma)
Guacha caldera
Puripica Chico ignimbrite (1.70 Ma)
Cerro Zapaleri
Purico Shield
23°S
Guacha ignimbrite (5.65 Ma)
Argentina
Tara ignimbrite (3.49 Ma)
Ar - Ar age (Salisbury et al., 2011)
Pre-2011 ages
ChRM sample site
La Pacana caldera
N
75 km
Chile
68°W
67°W

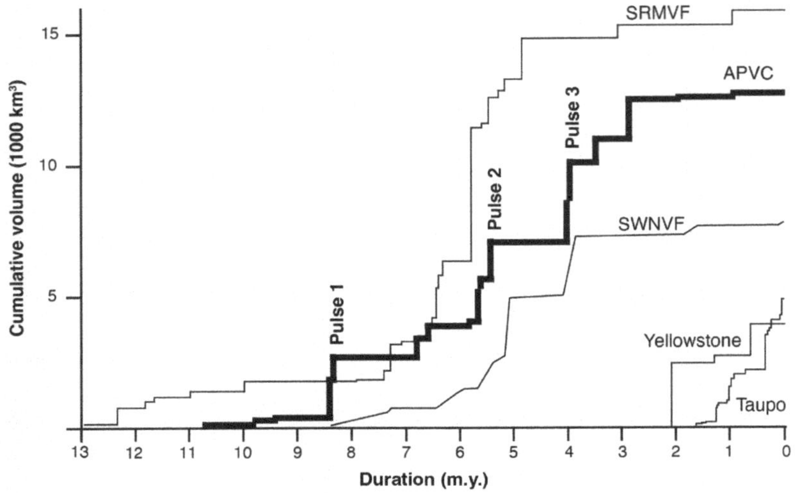

Fig. 18 (continued)

unidentified until the advent of Landsat satellite images in 1977 (Friedman and Heiken 1977 in: Francis 1993) and independently by Francis and Baker (1978) and Francis et al. (1978). The Cerro Galán system coincides with a piecemeal graben bounded by N-S trending basement faults of about 35 km in length. The central part of the graben-like form contains the Cerro Galán caldera sensu stricto (Photo 69), a roughly isosceles trapezoid shape, elongated east–west, with maximum dimensions of 27×16 km (Sparks et al. 1985; Francis 1993; Folkes et al. 2011, their Fig. 11).

The Cerro Galán caldera complex was reinvestigated more recently (Cas and Cashman 2011). Precise $^{40}Ar/^{39}Ar$ geochronology coupled with paleomagnetic data and field volcanology resulted in important revisions to the stratigraphy of ignimbrites, caldera collapse mechanism and the evolution of the caldera complex (Fig. 19) The main results can be summarised as follows (Folkes et al. 2011):

1. The Cerro Galán system has erupted nine major ignimbrites between ~6 Ma and ~2 Ma. The most recent Cerro Galán Ignimbrite (CGI) is ~2.08 Ma. This ignimbrite is found in a circumference of 40 km in all directions from the structural margins, with a maximum distal extent of ~80 km. The preserved areal extent of the CGI is ~2,400 km^2 and the volume is 630 km^3 DRE, approximately two thirds of the commonly quoted value (Francis et al. 1983; Sparks et al. 1985), of which about half is intracaldera fill.
2. New mapping and interpretation of satellite images allows a re-interpretation of the caldera margins and the form of collapse (Fig. 19). The caldera is strongly fault-controlled: the northern and southern margins on the edge of the resurgent block are fault-bounded. Collapse was accommodated in a piecemeal trapdoor fashion along a regional N-S fault that defines the western margin of the caldera with a hinge along a N-S regional fault to the east. The collapse was locally as at least 1.4 km in depth beneath the area now occupied by the resurgent block, equal to the thickness of Cerro Galán Ignimbrite within the resurgent centre.
3. Magma volumes increased with subsequent eruptions in the Galán volcanic system, perhaps reflecting the thermomechanical evolution of the upper crust in response to the older Toconquis Group ignimbrite magmas 'priming' the crust for accumulation of larger

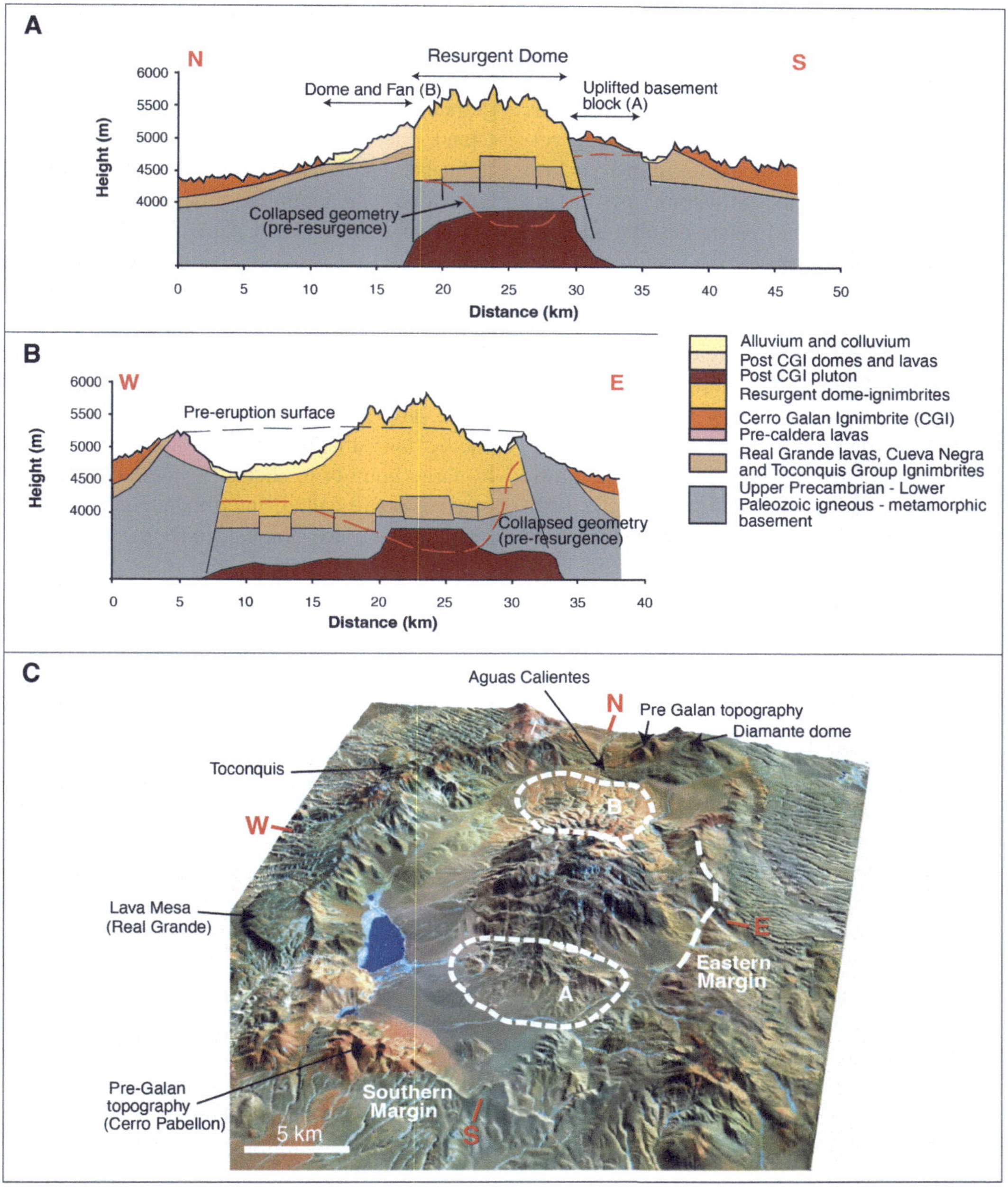

Fig. 19 Cerro Galán volcanic system (after Folkes et al. 2011, Fig. 9, p. 1445). Geological cross sections **a** N-S and **b** E-W across the Cerro Galán depression, with cross sections indicated in c. Vertical exaggeration is ×5. Profiles were generated from a 30 m-DEM image using ASTER data. Interpreted floor geometry shown as piecemeal blocks. Post-collapse resurgent geometry of blocks is compared with immediate post-collapse geometry shown in red. Approximate locations of major caldera bounding faults are shown (Folkes et al. 2011). **c** Aerial view produced by draping ASTER data over the ASTER-30 m DEM showing the location of main features at Cerro Galán. Location of the cross sections is shown by the compass directions in red (E-W and N-S). A is the southern basement-dominated portion of the resurgent dome, and B is the post resurgent dome and volcaniclastic apron complex to the north (Folkes et al. 2011)

batholith-sized chambers for the most recent ignimbrite-producing magmas.
4. The Cerro Galán system experienced significant volcanic activity in the time periods between the eruptions of the nine large-volume ignimbrites. Evidence for numerous lava flows, silicic dome extrusions, pyroclastic flows and surges, and young basaltic-andesite scoria cones suggests that magmatic and volcanic activity has been continuous since ~6 Ma to about 2 Ma. The large-volume ignimbrites represent increased thermal and magmatic fluxes to the upper crust in the period before the eruption of these crystal-rich silicic deposits.
5. After resurgence of Cerro Galán, dacitic lavas oozed quietly to the surface about 2.1 Ma from a vent on the northernmost part of the caldera ring fracture, piling up in thick coulées. There are no historical records of unrest at Cerro Galán, but hot springs are still present at several locations.

11.2 Volcano-tectonic Depressions

Very large calderas or caldera complexes (Toba, Yellowstone, Cerro Galán) should not be included in volcano-tectonic depressions, which are even bigger features than calderas and result from broader crustal stresses than those imposed by any single magma reservoir. Volcano-tectonic depressions are very large grabens or composite grabens in continental areas stretching along the axes of volcanic arcs (Lockwood and Hazlett 2010). Regional stresses associated with plate convergence and heating from deep-seated magma generation and emplacement lead to the development of these major landforms (between one and several thousand km^2). They enclose multiple volcanic edifices and centres, and allow for the build-up of thick sequences of volcanic and sedimentary material on their floors. Individual volcano-tectonic depressions range up to 800 km in length and may be as much as 150 km in width. Boundary faults generally involve both extensional and strike-slip movements. Large water bodies, such as lakes Managua and Nicaragua in Central America and Kagoshima Bay in Japan, result from flooding of depression floors. A large volcano-tectonic graben beneath Kagoshima Bay on Kyushu Island is at least 120 km long × 20–25 km wide. It hosts several large volcanoes (Kaimondake, Kirishima) and a number of large silicic calderas such as Ata to the south, Aira (20 × 20 km) in the middle, and the smaller Hakuto and Kobayashi calderas to the north, which are surrounded by extensive andesitic to dacitic and rhyolitic ignimbritee sheets.

Other volcanic collections of negative landforms are multivent complexes of andesitic-dacitic and alkaline compositions, surrounded by large ignimbrite sheets, which lack a central cone: they are called ‘intermediate-silicic multivent centres’, also known under the heading of volcano-tectonic depressions (Cas and Wright 1988). Most of the depressions are located in subduction zone (Mexico, The Philippines, New Zealand) and island arc (Lesser West Indies) settings, although they can also occur in continental intraplate settings (East African Rift System). Andesitic examples include the Laguna de Bay and Taal volcano-tectonic depressions with an area of between 300 and 1500 km^2 in the Philippines. Laguna de Bay has the appearance of a caldera with dimensions of 25 × 12 km, nested in a large collapse structure 35 × 45 km in diameter, of uncertain origin, which is one of the largest described. Major Pleistocene eruptions of alkali-rich andesitic ignimbrites from Laguna de Bay and Taal have joined together several Tertiary island volcanoes to form the present landmass of southwestern Luzon.

Rhyolitic multivent centres belong to the largest volcanic landforms on the Earth’s surface. Typically, they lack a topographically impressive cone, instead they form large, broad volcano-tectonic depressions which Walker et al. (1981) aptly termed *inverse volcanoes*. They are multivent complexes, and the low-profile volcanic centres are a collection of low rhyolitic hills, which may be scattered over a few hundred square kilometres, and situated within a shallow depression (e.g. La Primavera volcanic field, Mexico; Fig. 20A and B; Walker et al. 1981; Sourisseau et al. 2020). These rhyolite hills are composed of rhyolitic domes, coulees and pumice cones. The hills rise from gently sloping ignimbrite sheets, often containing more than one ignimbrite, together with interbedded pumice-fall and epiclastic deposits. The volcanic field comprises a flattish plain with hills in a shallow volcano-tectonic depression, and a broad shield-like landform (Fig. 20A; Sourisseau et al. 2020). The diameter of these shallow volcano-tectonic depressions varies from about 50–200 km, while the width of the volcanic multi-vent centre is 20–60 km. Gentle slopes vary from $<1°$ to $5°$ with a maximum height of only a few hundred metres. Total volume of these multi-vent centres of intermediate to silicic composition varies in the order of 100 km^3 to over 1000 km^3.

Numerous lava domes have also developed within the depression or along the circular ring fractures, covering the caldera rim. Individual rhyolite lavas do not travel far from the vent, but in time they may coalesce on some volcanoes to produce a nearly complete ring of rhyolite lava (Fig. 20A). La Primavera volcano illustrates somewhat complicated features. Some of the lavas are located around an almost complete ring, while some lavas occur on a transverse line crossing it. Most of these domes are monogenetic, with a vent-to-lava ratio of 1:1. However, there are three larger, polygenetic structures, which can be called independent volcanoes in their own right (Fig. 20B). These are younger, petrologically different, and may lie on an outer ring structure. La Primavera lacks a well-defined structural expression of a caldera because subsequent uplift of the central part of the volcano has masked original topographic relationships. A significant circular area of lake sediments 10 km in diameter and ~100 m thick is associated with rhyolite lavas erupted on a ring fracture (Cas and Wright 1988). The sedimentary succession consists of thinly bedded, white ashy sediments, ash turbidites, some beds of diatomite and an unusual bed of giant pumice. Finally, the volcano-tectonic depressions also contain active hot springs and sinter deposits.

11.2.1 Felsic Caldera Complexes: Earth's Most Frequently Active and Productive Systems

The Taupo Volcanic Zone shows a pronounced segmentation into northeastern and southwestern andesite-dominated extremities with composite cones and no calderas, and a central, 125 km-long rhyolite-dominated segment (Wilson 1993; Wilson et al. 1995; Davy and Caldwell 1998). Eight rhyolitic caldera centres have so far been identified in the central segment, of which two (Mangakino and Kapenga) are compound (Fig. 9B, Chap. 1). These centres account for 34 inferred caldera-forming ignimbrite eruptions, in the ~1.6-Myr lifetime of the central TVZ. The modern central Taupo Volcanic Zone is the most frequently active and productive silicic volcanic system on Earth, erupting rhyolite at ~0.28 m^3/s: Wilson et al. (1995) suggested this rate has been constant for at least the past 0.34 Ma. Rhyolite ($\geq$ 15,000 km^3 bulk volume in total) is the dominant magma type erupted in the TVZ, mostly as caldera-forming ignimbrite eruptions, as well as rhyolite lava domes and flows. Andesite lava flows are an order of magnitude less abundant, and basalt lava flows and dacite domes are minor in volume (<100 km^3 each).

The volumetric output of this type of rhyolitic centre has been estimated for several multivent centres. For La Primavera volcano, which has been active over the past 100,000 years, the average output is 0.06 km^3 DRE per 100 years (Walker et al. 1981). This output rate is at least an order of magnitude less than the average output of about 0.4 km^3 DRE per 100 years for both Taupo and Okataina, which are the most active rhyolitic centres in New Zealand and probably the most productive rhyolitic volcanoes known on Earth. Average central TVZ eruption rates from large caldera-forming events alone are 3.8 km^3/kyr, similar to those estimated at intracontinental silicic volcanic provinces: 3.0 km^3/kyr for Yellowstone (2 Ma to present) and 5 km^3/kyr for the San Juan Mountain volcanic field (29.4–26.9 Ma) (Wilson et al. 1995, 2009; Cole et al. 2005, 2014).

A striking difference between caldera complexes and composite cones, and their types of

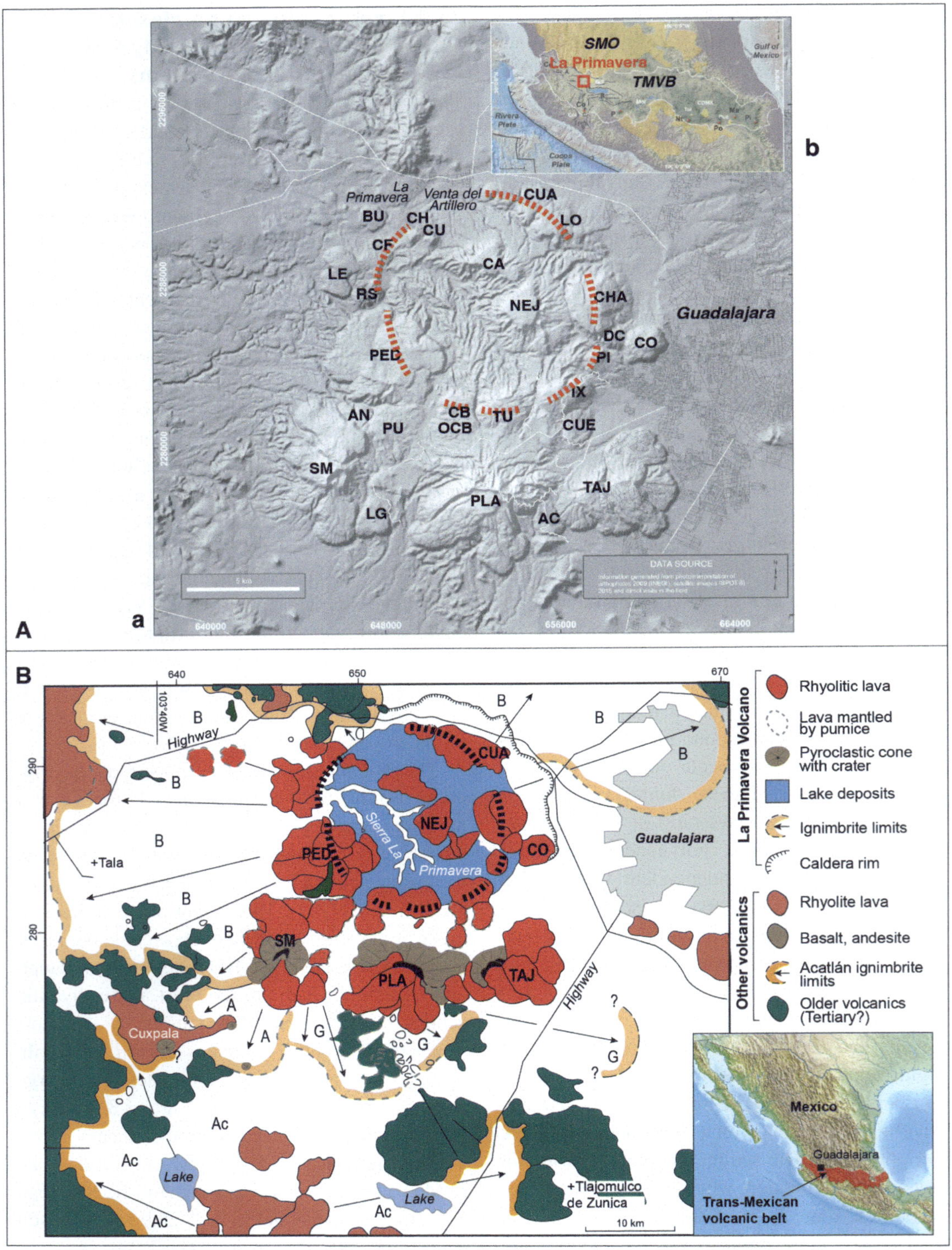
A
a
b
SMO
La Primavera
TMVB
Rivera Plate
Cocos Plate
Gulf of Mexico
La Primavera
Venta del Artillero
BU
CH
CU
CUA
LO
CF
CA
LE
RS
NEJ
CHA
Guadalajara
DC
CO
PED
PI
IX
CB
TU
AN
PU
OCB
CUE
SM
TAJ
PLA
LG
AC
DATA SOURCE
5 km
2296000
2288000
2280000
640000
648000
656000
664000
B
640
650
670
103°40W
290
280
Highway
+Tala
Sierra La Primavera
NEJ
PED
CUA
CO
SM
PLA
TAJ
Guadalajara
Cuxpala
Lake
Ac
A
G
+Tlajomulco de Zunica
10 km
La Primavera Volcano
Rhyolitic lava
Lava mantled by pumice
Pyroclastic cone with crater
Lake deposits
Ignimbrite limits
Caldera rim
Other volcanics
Rhyolite lava
Basalt, andesite
Acatlán ignimbrite limits
Older volcanics (Tertiary?)
Mexico
Guadalajara
Trans-Mexican volcanic belt

Fig. 20 Caldera complex and volcano-tectonic depression. **A** La Primavera caldera complex (Sourisseau et al. 2020, Fig. 1, p. 2). **a** Shaded relief model of La Primavera with the locations of lavas, domes, and volcanic centres. The continuous and dotted red lines indicate the traces of the visible and inferred caldera ring-fault, respectively. Grey lines represent the streets of main cities and towns, and white lines the main roads in the area. Abbreviations are: Lava: CF = Cañon de las Flores and LG = Llano Grande. Domes: AC = Arroyo Colorado, AN = Animas, BU = Burro, CA = Cerro Alto, CHA = Chapulin, CH = Chato, CO = Colli, CU = Cuate, CUA = Cuartilla, CUE= Cuesta, CB = Culebreado, DC = Dos Coyotes, IX = Ixtahuatonte, LE = León, LO = Lobera, OCB = Old Culebreado, PED = Pedernal, PI = Pilas, PV = Piñar de la Venta, PU = Puerta, RS = Rio Salado and TU = Tule. Volcanic centre: NEJ = Nejahuete, PLA = Planillas, SM = San Miguel and TAJ = Tajo. **b** Insert map that shows the present tectonic plate configuration and the extent of the Sierra Madre Occidental (SMO) and the Trans-Mexican Volcanic Belt (TMVB). La Primavera caldera (red square) is located on the western part of the TMVB (WGS 84 ellipsoid is used in the UTM 13 coordinate system at 1:100,000 scale). A: Tepic-Zacoalco Graben; B: Chapala Graben; C: Colima Graben. Volcanic centres are: Ce = Ceboruco; Co = Colima; P = Parícutín; Nt = Nevado de Toluca; Po = Popocatépetl; Ma = Malinche; and Pi = Pico de Orizaba. Cities: Gdl = Guadalajara; Col = Colima; Mor = Morelia; Tol = Toluca; CDMX = Ciudad de Mexico; Pue = Puebla. **B** La Primavera volcano-tectonic depression, containing the caldera complex shown in A (modified after Walker et al. 1981). Geology of La Primavera volcano and adjacent areas. Black dashed line around the blue area corresponds approximately to the inferred caldera rim as shown in A. The arrows represent the generalized flow directions of the ignimbrites. Rhyolite hills previously marked with numbers on the map by GPL Walker are: NEJ, CUA, PED, SM, PLA, TAJ and CO; these domes are mentioned in map A

activity, is the repose periods between eruptions (Davidson and de Silva 2000). In La Primavera, an eruption has occurred on average about every 2000 years for the past 100,000 years. In Taupo (Photo 81), the average repose period for the past 50,000 years is ~3000 years, but has varied from a few hundred years to 16,000 years, and in Okataina for the same period it is ~2500 years (Cas and Wright 1988; Wilson et al. 2009). As for larger rhyolitic fields, the example of the San Juan caldera complexes and associated ignimbrites shows that the eruption of at least 8800 km^3 of dacitic-rhyolitic magma as nine major ash-flow sheets (individually 150–5000 km^3) was accompanied by recurrent subsidence of at least seven large calderas over a time interval of 2.5 Myr (Steven and Lipman 1976; Lipman 2000b, 2006). The life expectancy of individual rhyolitic systems can therefore be well over 1 Myr; e.g. the Yellowstone compound caldera is at least 2 Myr old (Christiansen 2001).

12 Challenging Classification of Calderas and Caldera Complexes

The different collapse caldera types that have been proposed from field studies (Table 1) do not necessarily represent different genetic conditions (Cole et al. 2005; Branney and Acocella 2015). They may only represent variations in host rock mechanics during caldera collapse, slightly different magma chamber geometries, different regional structural controls or different magma rheologies. Therefore, any attempt toward classification can be challenging, and any simple structural classification raises questions for at least two reasons: (1) the internal structure of most calderas is poorly constrained and challenging to resolve even by drilling and geophysical studies; (2) Downsag, trapdoor asymmetry, caldera-floor faults, and ring-fault systems may be considered as common structural attributes of calderas that develop in nature as well as in analogue models (Branney and Acocella 2015).

A variety of parameters help define caldera types, but those linked to caldera morphology should be distinguished from the mechanisms of collapse or subsidence (Cole et al. 2005; Gottsmann and Marti 2008; Geyer and Marti 2008; Geshi et al. 2012).

(1) The surface morphology of a caldera, even prior to any resurgence, rarely provides sufficient evidence to place the caldera within a structural classification. First, the number of collapse events, rarely single, often repeated, and sometimes multiple, leads to

nested, polycyclic calderas and overlapping, compound calderas Second, the final caldera morphology and type of associated eruption depend on several factors (Cole et al. 2005): the distribution of vents within the caldera, caldera-floor coherency, the ring fault geometry, syn- and post-collapse deposits, the presence or absence of resurgence, and the dominant composition of the erupting magma.

(2) Mechanisms of caldera collapse or cauldron subsidence are complex as they involve many parameters: the amount and rate of subsidence, the ratio between subsidence and critical depth and shape of the magma reservoir, the roof thickness ratio, excess pressure at the onset of the eruption and rapid decompression of the magma chamber that helps distinguish overpressure from underpressure calderas (Marti et al. 2009).

Resurgence is the long-term surface expression of the repeated shallow emplacement of new magma, unable to reach the surface and erupt because of the relatively low-viscosity contrast with the residual magma. In the short term, caldera unrest is often characterized by deformation, seismicity and degassing, which usually results from the shallow emplacement of magma (Newhall and Dzurisin 1988). Comparison between short-term unrest and longer-term resurgence suggests a connection, with resurgence probably resulting from the cumulative net uplift produced by multiple unrest episodes. Therefore, uplift during unrest at a resurgent caldera may be the short-term expression of resurgence. The accumulated magma during unrest may subsequently be transferred, potentially leading to an eruption. Processes of resurgence may generate pathways for the emplacement of radial dykes and cone sheets, which may help forecast the location of future vents.

13 Concluding Remarks

1. The key element for caldera formation is the removal of magmatic support from the crust overlying the chamber, but the process of caldera collapse and the final caldera morphology depend on many factors. The modern approach to investigating caldera formation combines field observations and geophysical surveys acquired during active caldera unrest with structural criteria and analogue experiments.
2. A first-order description based on erupted magma composition (Cole et al. 2005) is useful but there is no direct link between compositional range and collapse mechanism. Calderas may be associated with effusive or explosive eruptions of any size. However, large explosive andesitic–dacitic and rhyolitic eruptions generally lead to the emplacement of high volume ignimbrites and creation of large, deep collapse depressions.
3. Even though several differences exist in the collapse of mafic and felsic calderas, the mechanism, structure and evolutionary stages of collapse may be the same and depend upon the amount of subsidence with regard to the caldera diameter. Thus, the *s/d* ratio allows four stages of collapse to be defined, with increasing subsidence, also linked to the aspect ratios of the roof and caldera (Acocella 2007, 2021).
4. The shape and growth of the underlying magma chamber plays a crucial role in the collapse process. Caldera collapse may be initiated if the chamber is shallow enough and has reached a critical width, and the size of the chamber plays an important role in the collapse geometry. However, end-members determined on the basis of collapse geometry and the depth and size of the magma chamber

(Lipman 2000a, b) must be considered as transitional and can be viewed in the light of progressive increase of subsidence (Acocella 2007).

5. Ring faults involved in the caldera collapse process may be primary and initiated during caldera collapse, or reactivated older tectonic or volcanic structures. The primary faults may dip steeply inward- or outward-dipping, or be vertical, and dip may also vary with depth. Depth to the magma chamber will affect the way collapse occurs.
6. Tumescence causes doming of the overlying strata. Resurgence is due to several factors, but the viscosity contrast (linked to vesiculation) between the magma responsible for resurgence and the residual magma below the caldera may play the most significant role.
7. The final morphology of a caldera will vary according to the way the caldera floor breaks up, how collapse proceeds during the eruption—in particular the symmetry and coherency of the caldera floor, and the presence of post-caldera deformation.

Several important questions remain (Branney and Acocella 2015), such as: Are most calderas cyclic, or do some simply undergo repeated caldera-forming eruptions, following recharge events? Does the caldera structure affect the frequency and style of eruptions? How common is pre-tumescence, and what controls whether a caldera undergoes resurgence? How do the magma reservoirs assemble, evolve, and erupt?

Photo 69 Cerro Galán caldera, northwest Argentina (27 × 16 km lying within a graben 35 km in length), with resurgent Cerro Galán dome, and aprons of thick ignimbrites reaching up to 100 km (*photo credit* Landsat 8 satellite image, Wikipedia)

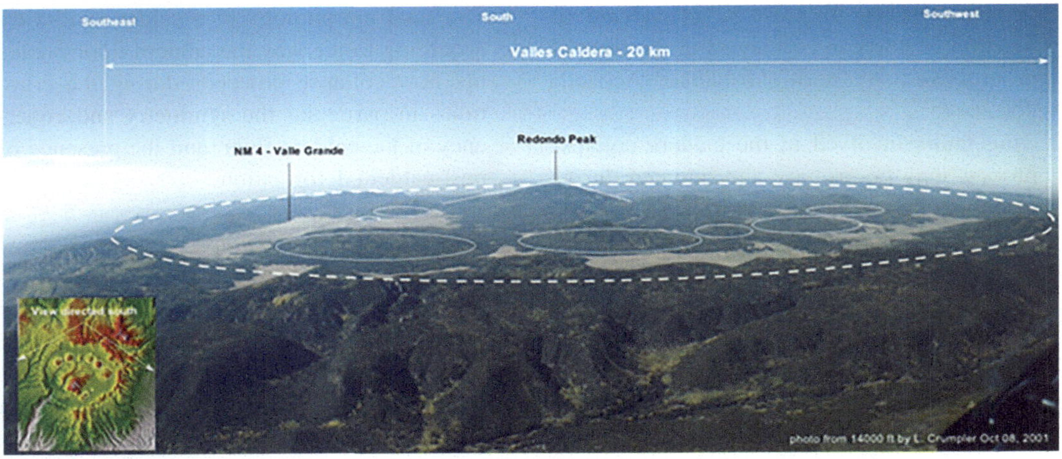

Photo 70 Valles–Toledo Caldera (type of ash-flow caldera, ~ 20 km in diameter), New Mexico, USA (Photo looking south, *credit* L. Crumpler; New Mexico Museum of Natural History and Science)

Photo 71 Okmok Caldera, 10 km in diameter, Aleutian Island arc, Alaska, USA, looking north. The caldera formed in the Holocene, and is dotted by several subsequent cones, with one of the largest eruptions occurring in 43 BC/BCE. Drainage of the caldera lake caused one of the largest floods in Alaska, breaching the Crater Creek to the northeast (*photo credit* Wikipedia)

Photo 72 Toba caldera, Sumatra, Indonesia (100 km long and 30 km wide). Note the resurgent Samosir Island in the middle and the extensive ignimbrite aprons surrounding the large caldera (*photo credit* Vulkane.net, 'Toba: Supervulkan-eruption auf Sumatra')

Photo 73 Example of a strongly welded ignimbrite, Aso caldera, Japan. The 5 to 6 m-thick cliff shows columnar jointing, while the foreground shows cut away hexagonal prisms about 0.5–1 m across; hammer in foreground for scale (*photo* D. Karátson)

Photo 74 Crater Lake caldera (Oregon, USA), containing post-caldera Wizard Island, a basaltic cone (*photo* courtesy of A. Harris)

Photo 75 **A** Aerial view of the Santorini caldera, north is to the right Cyclade Islands, Greece with the resurgent Nea and Palaea Kameni Islands (*photo credit* https://flic.kr/s/aHsiDnvBkC). **B** View of the eastern part of the caldera, looking north, showing the pyroclastic deposits of the Cape Therma eruption overlain by the Lower Tuff and Upper Tuff, the latter being the white Minoan deposits on top (Late Bronze Age: ~3600 BP), forming prominent cliffs. The eastern tip of the more recent Kameni Islands is visible on the left hand side (*photo* D. Karátson)

Photo 76 **A** Aso caldera, Kyushu, Japan. Panoramic view of Aso caldera, central Kyushu, from the northwestern caldera wall towards the resurgent Mount Aso in the background (*photo* K. Mannen *in:* Miura et al. (2009) Geol Soc Japan 115, 12. https://doi.org/10.5575/geosoc.115.12.XIX_XX). **B** Nakadake post-caldera cone in Aso caldera (*photo credit* IStock, gnoparus; public domain)

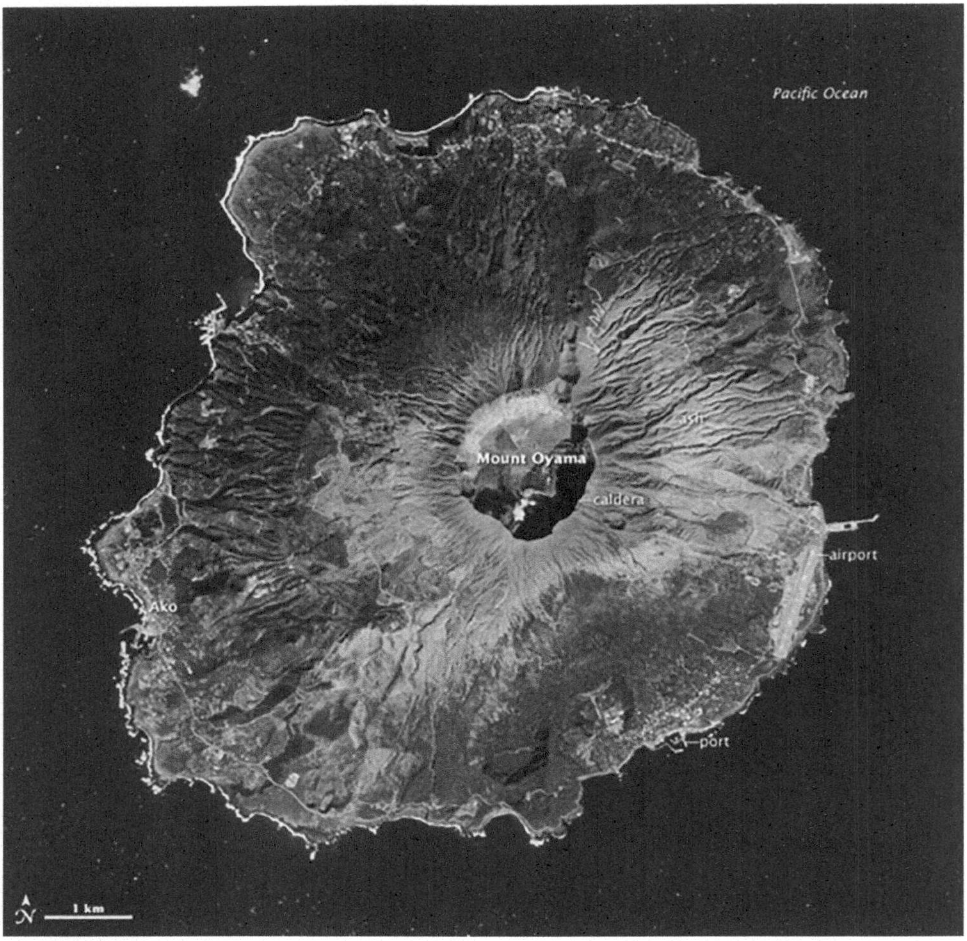

Photo 77 Miyakejima Island (Izu archipelago, Japan) and summit caldera formed by the eruption in 2000 (*photo credit* NASA Earth Observatory image, Landsat 8 OLI data from the U.S. Geological Survey; image taken on 11 February 2015)

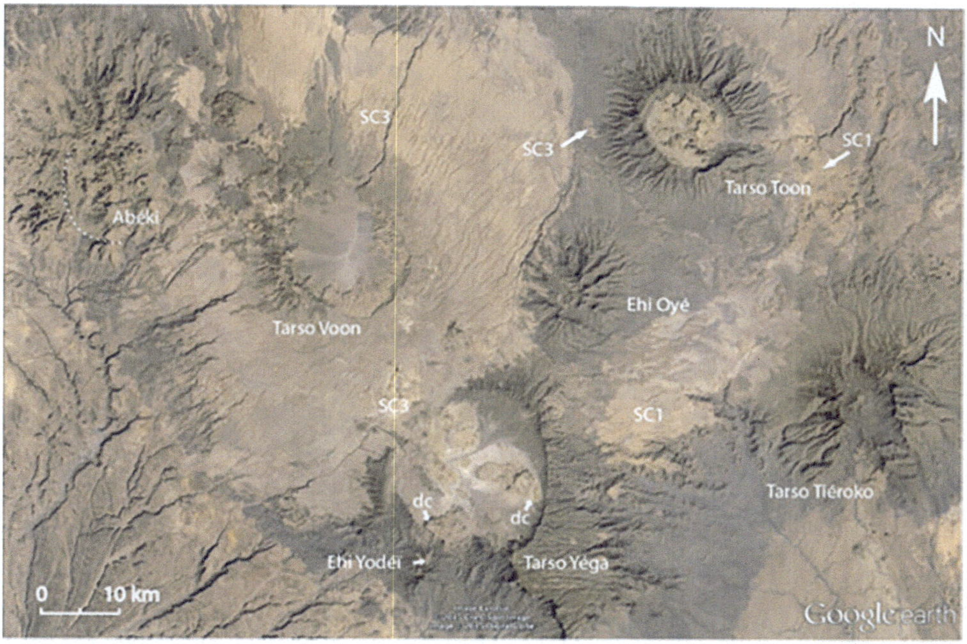

Photo 78 Shield-like ignimbrite volcanoes Tibesti, Chad. Calderas are surrounded by ignimbrites: Tarso Voon, Tarso Toon, Tarso Yega and Tarso Tiéroko, as well as elevated rhyolite domes, e.g. Abéki (*photo credtit* of C. Deniel, 2015)

Photo 79 Emi Koussi shield volcano, largest (60–70 km at the base) and highest (3415 m.a.s.l.) in the Tibesti Volcanic Field, Chad. Three calderas are contained within the 12×15 km summit depression (*photo credtit* ID ISS026-E-17074, Nasa Earth Observatory, ISS expedition 26 Crew; image taken 11 January 2011)

Photo 80 Two ignimbrite deposits totalling a thickness of ~ 80–90 m, Crater Lake, Oregon, USA. A clear change in composition occured during the 6845 ± 50 BP eruption, with a thick rhyodacitic, yellowish lower part overlain by an andesitic, grey upper part, reflecting a zoned magma chamber. The Pinnacles exposure of the unwelded ignimbrite displays spectacular erosional features (fairy chimneys, here called pinnacles) along the steep-sided canyon of Wheeler Creek (*photo* courtesy of Sz. Kósik)

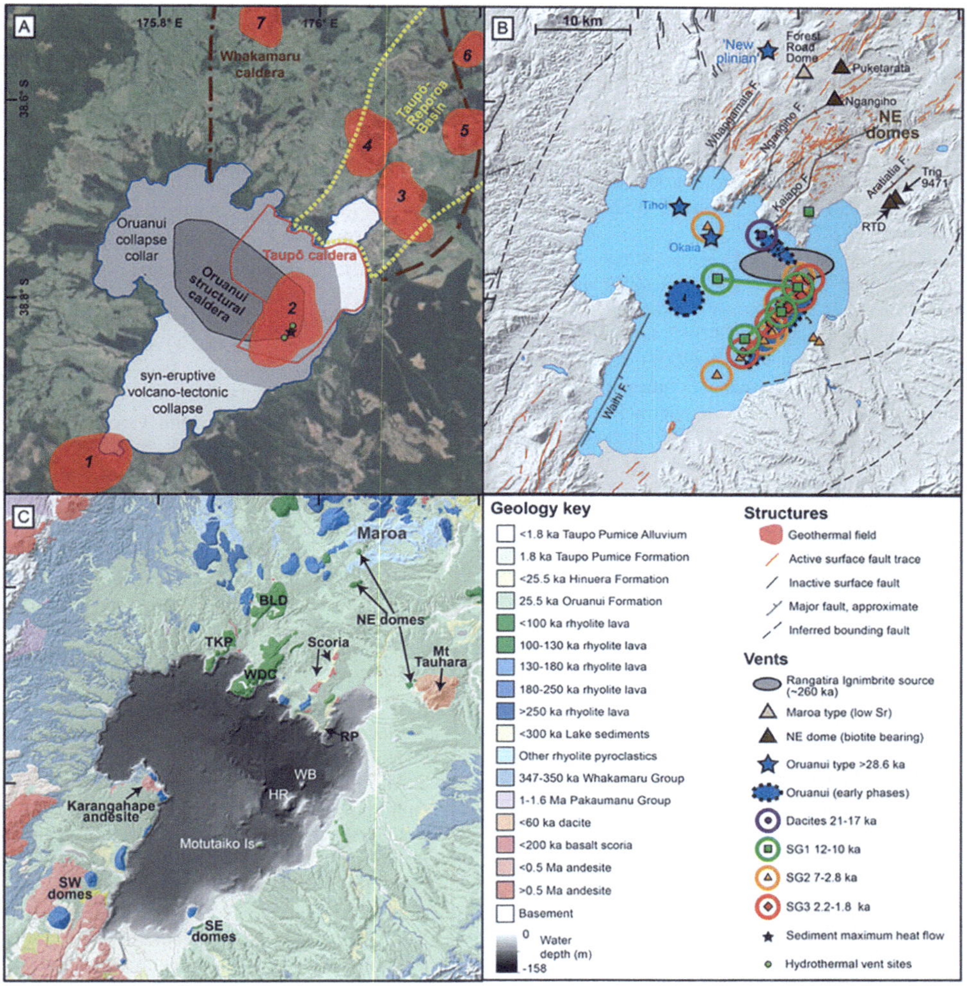

Photo 81 Taupō Volcano, North Island, New Zealand. Maps from the Taupō area, Figure 2 and DEM taken from Barker et al. (2020) highlighting key geological features, geothermal areas, and calderas. At least five caldera collapses (Rangatira, Maroa, two Oruanui, Hatepe) occurred between 260 ka and 1.8 ka (Wilson, 2001, Wilson et al., 2006; Lowe et al. 2021). Barker SJ, Wilson CJN, Illsley-Kemp F, Leonard GS, Mestel ERH, Mauriohooho K, Bruce LA, Charlier BLA (2020): Taupō: an overview of New Zealand's youngest supervolcano. New Zealand J Geol Geophys 27 p. https://doi.org/10.1080/00288306.2020.1792515

References

Acocella V (2007) Understanding caldera structure and development: an overview of analogue models compared to natural calderas. Earth-Sci Rev 85(3–4):125–160. https://doi.org/10.1016/j.earscirev.2007.08.004

Acocella V (2021) Volcano-tectonic processes. Advances in volcanology. IAVCEI, Springer, 552 pp. ISBN: 978-3-030-65967-7

Acocella V, Rivalta E (2019) Calderas; structure, unrest, magma transport and eruptions. Reference module in Earth Systems and Environmental Sciences, Elsvier, 15 pp. https://doi.org/10.1016/B978-0-12-409548-9.10904-2

Acocella V, Cifelli F, Funiciello R (2001) The control of overburden thickness on resurgent domes: insights from analogue models. J Volcanol Geotherm Res 111:137–153. https://doi.org/10.1016/S0377-0273(01)00224-4

Acocella V, Korme T, Salvini F, Funiciello R (2002) Elliptic calderas in the Ethiopian Rift: the control of pre-existing structures. J Volcanol Geotherm Res 119(1–4):189–203. https://doi.org/10.1016/S0377-0273(02)00342-6

Almond DC (1977) The Sabaloka igneous complex, Sudan. Philos Trans Roy Soc London A287:595–633. https://doi.org/10.1098/rsta.1977.0160

Anderson EM (1936) Ring dykes and cauldron subsidence: the dynamics of formation of cone sheets. Proc Royal Soc Edinb 56:128–157

Bachmann O, Dungan M, Lipman P (2002) The Fish Canyon magma body, Colorado: rejuvenation and eruption of an upper crustal near-solidus batholithic magma chamber upon voluminous mafic underplating. J Petrol 43(8):1469–1503. https://doi.org/10.1093/petrology/43.8.1469

Bacon CR (1983) Eruptive history of Mount Mazama and Crater Lake caldera, Cascade Range, USA. J Volcanol Geotherm Res 18(1):57–115. https://doi.org/10.1016/0377-0273(83)90004-5

Bacon CR, Lanphere MA (2006) Eruptive history and geochronology of Mount Mazama and the Crater Lake region, Oregon. Geol Soc Amer Bull 118(11–12):1331–1359. https://doi.org/10.1130/B25906.1

Bailey RA, Dalrymple GB, Lanphere MA (1976) Volcanism, structure and geochemistry of Long Valley Caldera, Mono County, California. J Geophys Res 81:725–744. https://doi.org/10.1029/JB081i005p00725

Bailey EB, Clough CT, Wright WB et al (1924) Tertiary and pre-Tertiary geology of Mull, Loch Aline and Oban: a description of parts of sheets 43, 44, 51, and 52 of the geological map. Memoirs Geol Survey of Great Britain, vol 43. HM Stationery Office, London, 445 pp

Bain ADN (1934) The younger intrusive rocks of the Kudaru Hills, Nigeria. Quaternary J Geol Soc London 90:201–223. https://doi.org/10.1144/GSL.JGS.1934.090.01-04.0

Barberi F, Cassano E, La Torre P, Sbrana A (1991) Structural evolution of Campi Flegrei caldera in light of volcanological and geophysical data. J Volcanol Geotherm Res 48:33–49. https://doi.org/10.1016/0377-0273(91)90031-T

Barde-Cabusson S, Merle O (2007) From steep-slope volcano to flat caldera floor. Geophys Res Lett 34(10):1–5. https://doi.org/10.1029/2007GL029784

Bellier O, Sébrier M (1994) Relationship between tectonism and volcanism along the Great Sumatran Fault Zone deduced by spot image analyses. Tectonophysics 233(3–4):215–231. https://doi.org/10.1016/0040-1951(94)90242-9

Beresford SW, Cole JW (2000) Kaingaroa Ignimbrite, Taupo Volcanic Zone, New Zealand: evidence for asymmetric caldera subsidence of the Reporoa Caldera. NZ J Geol Geophys 43:471–481. https://doi.org/10.1080/00288306.2000.9514903

Bethke PM, Hay RL (2000) Overview: ancient Lake Creede. Geol Soc Am Spec Pap 346:1–8. https://doi.org/10.1130/0-8137-2346-9.1

Bourdier J-L (2021) Géologie du volcanisme. Sciences sup, Dunod, 80 pp. ISBN: 978-2-10-081800-6

Branney M, Acocella V (2015) Calderas. In: Sigurdsson H, Houghton B, McNutt SR, Rymer H, Stix J (eds) The encyclopedia of volcanoes, Chap 16, 2nd edn. Academic Press, Elsevier, pp 299–315. ISBN: 978-0-12-385938-9

Calò M, Tramelli A (2018) Anatomy of the Campi Flegrei caldera using enhanced seismic tomography models. Sci Rep 8(1):16254. https://doi.org/10.1038/s41598-018-34456-x

Cas RAF, Wright JV (1988) Volcanic successions, modern and ancient. Unwin Hyman, London, 528 pp. ISBN: 0-04-552021-6

Cas RAF, Cashman K, eds (2011) The Cerro Galán Ignimbrite and Caldera: characteristics and origins of a very large volume ignimbrite and its magma system. Bull Volcanol 73(10):186 pp. https://doi.org/10.1007/s00445-011-0564-y

Cashman KV, Giordano G (2014) Calderas and magma reservoirs. J Volcanol Geotherm Res 288:28–45. https://doi.org/10.1016/j.jvolgeores.2014.09.007

Cas R, Giordano G, Wright JV (2024) Volcanology: processes, deposits, geology and resources. Earth sciences, geography and environment. Springer, Cham, 1833 p. eBook ISBN: 978-3-319-66613-6

Chesner CA, Rose WI, Deino R, Drake R, Westgate JA (1991) Eruptive history of Earth's largest Quaternary caldera (Toba, Indonesia). Geology 19:200–203. https://doi.org/10.1130/0091-7613(1991)019<0200:EHOESL>2.3.CO;2

Chesner CA (2012) The Toba Caldera complex. Quaternary Int 258:5–18. https://doi.org/10.1016/j.quaint.2011.09.025

Christiansen RL (2001) The Quaternary and Pliocene Yellowstone Plateau volcanic field of Wyoming, Idaho, and Montana. Geology of Yellowstone Park. US Geol Survey Prof Paper 729-G. ISBN: 0-607-95346-2

Clough CT, Maufe HB, Bauiley EB (1909) The cauldron subsidence of Glen Coe, and the associated igneous phenomena. Quaternary J Geol Soc London 65:611–678. https://doi.org/10.1144/GSL.JGS.1909.065.01-04.35

Cole JW, Milner DM, Spinks KD (2005) Calderas and caldera structures: a review. Earth-Sci Rev 69(1–2):1–26. https://doi.org/10.1016/j.earscirev.2004.06.004

Cole JW, Brown SJA, Burt RM, Beresford SW, Wilson CJN (1998) Lithic types in ignimbrites as a guide to the evolution of a caldera complex, Taupo volcanic centre, New Zealand. J Volcanol Geotherm Res 80:217–237. https://doi.org/10.1016/S0377-0273(97)00045-0

Cole JW, Deering CD, Burt RM, Sewell S, Shane PAR, Matthews NE (2014) Okataina Volcanic Centre, Taupo Volcanic Zone, New Zealand: a review of volcanism and synchronous pluton development in an active, dominantly silicic caldera system. Earth-Sci Rev 128:1–17. https://doi.org/10.1016/j.earscirev.2013.10.008

Corti G (2009) Continental rift evolution: from rift initiation to incipient break-up in the Main Ethiopian Rift, East Africa. Earth-Sci Rev 96:1–53. https://doi.org/10.1016/j.earscirev.2009.06.005

Costa A, Di Vito MA, Ricciardi GP, Smith VC, Talamo P (2022) The long and intertwined record of humans and the Campi Flegrei volcano (Italy). Bull Volcanol 84(5):1–27. https://doi.org/10.1007/s00445-021-01503-x

Davidson J, de Silva S.L (2000) Composite volcanoes. In: Sigurdsson H, Houghton B, McNutt SR, Rymer H, Stix J (eds) The encyclopedia of volcanoes, 1st edn. Academic Press, pp 663–681. ISBN: 978-0-12-385938-9

Deino AL, Orsi G, de Vita S, Piochi M (2004) The age of the Neapolitan Yellow Tuff caldera-forming eruption (Campi Flegrei caldera–Italy) assessed by ^{40}Ar/^{39}Ar dating method. J Volcanol Geotherm Res 133(1–4):157–170. https://doi.org/10.1016/S0377-0273(03)00396-2

Deniel C, Vincent PM, Beauvilain A, Gourgaud A (2015) The Cenozoic volcanic province of Tibesti (Sahara of Chad): major units, chronology, and structural features. Bull Volcanol 77:74. https://doi.org/10.1007/s00445-015-0955-6

de Silva SL (1989) Altiplano-Puna volcanic complex of the central Andes. Geology 17(12):1102–1106. https://doi.org/10.1130/0091-7613(1989)017<1102:APVCOT>2.3.CO;2

de Silva SL, Francis PW (1989) Correlation of large ignimbrites: two case studies from the Central Andes of north Chile. J Volcanol Geotherm Res 37:133–149. https://doi.org/10.1016/0377-0273(89)90066-8

de Silva SL, Lindsay J (2015) Primary volcanic landforms. In: Sigurdsson H, Houghton B, McNutt SR, Rymer H, Stix J (eds) The Encyclopedia of volcanoes, 2nd edn. Academic Press, Elsevier, Chap 15, pp 273–297. https://doi.org/10.1016/B978-0-12-385938-9.00015-8

de Silva SL, Bailey JE (2017) Some unique patterns on ignimbrites on Earth: a 'bird's eye' view as a guide for planetary mappers. J Volcanol Geotherm Res 342:47–60. https://doi.org/10.1016/j.jvolgeores.2017.06.009

de Silva SL, Zandt G, Trumbull R, Viramonte JG, Salas G, Jiménez N (2006) Large ignimbrite eruptions and volcano-tectonic depressions in the Central Andes: a thermomechanical perspective. In: Troise E, De Natale G, Kilburn CRJ (eds) Mechanisms of activity and unrest at large calderas. Geol Soc London, Spec Pub 269, pp 47–63. https://doi.org/10.1144/GSL.SP.2006.269.01.04

Di Vito MA, Isaia R, Orsi G, de Vita S, D'Antoniuo M, Pappalardo L et al (1999) Volcanism and deformation since 12,000 years at the Campi Flegrei caldera (Italy). J Volcanol Geotherm Res 91:221–246. https://doi.org/10.1016/S0377-0273(99)00037-2

Di Vito MA, Acocella V, Aiello G, Barra D, Battaglia M, Caransdente A et al (2016) Magma transfer at Campi Flegrei (Italy) before the last 1538 AD eruption. Sci Rep 6:32245. https://doi.org/10.1038/srep32245

Druitt TH, Sparks RSJ (1984) On the formation of calderas during ignimbrite eruptions. Nature 370:679–681. https://doi.org/10.1038/310679a0

Druitt TH, Francaviglia V (1992) Caldera formation on Santorini and the physiography of the islands in the late Bronze Age. Bull Volcanol 54:484–493. https://doi.org/10.1007/BF00301394

Druitt TH, Vougioukalakis G (Guest eds) (2019) South Aegean volcanic arc. Elements 15(3):153–196. ISSN: 1811-5209

Druitt TH, Pyle DM, Mather TA (2019) Santorini volcano and its plumbing system. Elements 15(3):177–184. https://doi.org/10.2138/gselements.15.3.177

Druitt TH, Edwards L, Mellors RM, Pyle DM, Sparks RSJ, Lanphere M, Davies M, Barriero B (1999) Santorini volcano. Geol Soc Memoir 19, 165 pp. ISBN: 1-86239-048-7

Dungan MA, Lipman PW, Colucci M, Ferguson K, Balsey S (1989) Southeastern (Platoro) Caldera Complex. In: Chapin CE, Zidek J (eds) Field excursion to volcanic terranes in the Western United States, vol 1. Southern Rocky Mountains Region. Memoir New Mexico Bur Mines Miner Resour, vol 46, pp 305–329. ISBN: 1-883905-15-X

Elshaafi A, Gudmundsson A (2021) Central volcanoes and caldera collapses in the late Miocene-Late Pleistocene Tibesti Volcanic Province, northwest Chad. J Geodyn 145:101846. https://doi.org/10.1016/j.jog.2021.101846

Farrell J, Smith RB, Husen S, Diehl T (2014) Tomography from 26 years of seismicity revealing that the spatial extent of the Yellowstone crustal magma reservoir extends well beyond the Yellowstone caldera. Geophys Res Lett 41(9):3068–3073. https://doi.org/10.1002/2014GL059588

Ferguson JF, Cogbill AH, Warren RG (1994) A geophysical–geological transect of the Silent Canyon complex, Pahute Mesa, Nevada. J Geophys Res 99B:4323–4339. https://doi.org/10.1029/93JB02447

Flaherty T, Druitt TH, Tuffen H, Higgins MD, Costa F, Cadoux A (2018) Multiple timescale constraints for high-flux magma chamber assembly prior to the Late

Bronze Age eruption of Santorini (Greece). Contrib Mineral Petrol 173:75. https://doi.org/10.1007/s00410-018-1490-1

Folkes CB, Wright HM, Cas RAF, de Silva SL, Lesti C, Viramonte JG (2011) A re-appraisal of the stratigraphy and volcanology of the Cerro Galán volcanic system, NW Argentina. Bull Volcanol 73:1427–1454. https://doi.org/10.1007/s00445-011-0459-y

Fouqué F (1879) Santorin et ses éruptions. Masson, Paris. Santorini and Its Eruptions (English translation by AR McBirney, 1999). John Hopkins University Press, Baltimore, 495 pp

Francis PW (1993) Volcanoes. A planetary perspective. Clarendon Press, Oxford University Press, 443 pp. ISBN: 0-19-854452-9

Francis PW, Baker MCW (1978) Sources of two large ignimbrites in the Central Andes: some LANDSAT evidence. J Volcanol Geotherm Res 4(1–2):81–87. https://doi.org/10.1016/0377-0273(78)90029-X

Francis PW, de Silva SL (1989) Application of the Landsat Thematic Mapper to the identification of potentially active volcanoes in the Central Andes. Rem Sens Environ 28:245–255. https://doi.org/10.1016/0034-4257(89)90117-X

Francis PW, Hammill M, Kretzschmar G, Thorpe RS (1978) The Cerro Galán Caldera, north-west Argentina and its tectonic setting. Nature 274:749–751. https://doi.org/10.1038/274749a0

Francis PW, O'Callaghan L, Kretzschmar GA, Thorpe RS, Sparks RSJ, Page RN (1983) The Cerro Galán ignimbrite. Nature 301:51–53. https://doi.org/10.1038/301051a0

Froger JL, Lénat JF, Chorowicz J, Le Pennec JL, Bourdier JL, Köse O, Gourgaud A (1998) Hidden calderas evidenced by multisource geophysical data; example of Cappadocian Calderas, Central Anatolia. J Volcanol Geotherm Res 85(1–4):99–128. https://doi.org/10.1016/S0377-0273(98)00052-3

Galetto F, Acocella V, Caricchi L (2017) Caldera resurgence driven by magma viscosity contrasts. Nat Commun 8:1750. https://doi.org/10.1038/s1467-017-01632-y

Geshi N, Acocella V, Ruch J (2012) From structure- to erosion-controlled subsiding calderas: evidence thresholds and mechanics. Bull Volcanol 74:1553–1567. https://doi.org/10.1007/s00445-012-0617-x

Geshi N, Shimano T, Chiba T, Nakada S (2002) Caldera collapse during the 2000 eruption of Miyakejima Volcano, Japan. Bull Volcanol 64:55–68. https://doi.org/10.1007/s00445-001-0184-z

Geshi N, Miyagi I, Saito G, Conway CE (2023) Caldera collapse thresholds correlate with magma chamber dimensions. Sci Rep 13(1):7463. https://doi.org/10.1038/s41598-023-34411-5

Geyer A, Marti J (2008) The new worldwide collapse caldera database (CCDB): a tool for studying and understanding caldera processes. J Volcanol Geotherm Res 175(3):334–354. https://doi.org/10.1016/j.jvolgeores.2008.03.017

Geyer A, Folch, Marti J (2006) Relationship between caldera collapse and magma chamber withdrawal: an experimental approach. J Volcanol Geotherm Res 157:375–386. https://doi.org/10.1016/j.jvolgeores.2006.05.001

Giordano G, Cas R, Wright JV (2024) Subaerial pyroclastic density currents (pyroclastic flows and surges). In: Cas R, Giordano, Wright JV (eds) Volcanology: processes, deposits, geology and resources. Textbooks in earth sciences, geography and environment. Springer, Cham. https://doi.org/10.1007/978-3-319-66613-6_10

Goto Y, Miyoshi M, Danhara T, Tomiya A (2020) Evolution of the Quaternary silicic volcanic complex of Shiribetsu, Hokkaido, Japan: an example of ignimbrite shield volcanoes in an island arc setting. Int J Earth Sci 109(8):2619–2642. https://doi.org/10.1007/s00531-020-01906-9

Gottsmann J, Marti J (eds) (2008) Caldera volcanism: analysis, modeling and response. Developments in volcanology, vol 10. Elsevier, 492 pp. ISBN: 978-0-444-53165-0

Gourgaud A, Vincent PM (2004) Petrology of two continental alkaline intraplate series at Emi Koussi volcano, Tibesti, Chad. J Volcanol Geotherm Res 129(4):261–290. https://doi.org/10.1016/S0377-0273(03)00277-4

Gudmundsson A (1998) Formation and development of normal fault calderas and the initiation of large explosive eruptions. Bull Volcanol 60:160–170. https://doi.org/10.1007/s004450050224

Gudmundsson A (2007) Conceptual and numerical models of ring-fault formation. J Volcanol Geotherm Res 164:142–160. https://doi.org/10.1016/j.jvolgeores.2007.04.018

Harris AJL (2009) The pit-craters and pit-crater-filling lavas of Masaya volcano. Bull Volcanol 71:541–558. https://doi.org/10.1007/s00445-008-0241-y

Hasenaka T, Kawaguchi M, Torii M, Yasuda A, Hokanishi N (2022) Volcanic events and magma interactions before, during and after the caldera-forming eruption of Aso, Kyushu island, Japan. In: Ohta T, Ito T, Osada M (eds) Rock mechanics and engineering geology in volcanic fields, 1st edn. CRC Press, pp 3–9. https://doi.org/10.1201/9781003293590

Heiken G, Goff F, Stix J, Tamanyu S, Shafiqullah M, Garcia S (1986) Intracaldera volcanic activity, Toledo Caldera and embayment, Jemez Mountains, New Mexico. J Geophys Res 91B:1799–1815. https://doi.org/10.1029/JB091iB02p01799

Heiken G, Goff F, Gardner JN, Baldridge WS (1990) The Valles/Toledo Caldera Complex, Jemez Volcanic Field, New Mexico. Annu Rev Earth Planet Sci 18:27–53. https://doi.org/10.1146/annurev.ea.18.050190.000331

Hildreth W (1979) The Bishop Tuff: evidence for the origin of compositional zonation in silicic magma chambers. In: Chapin CE, Elston WE (eds) Ash-flow tuffs. Geol Soc Amer, Spec Paper 180, pp 43–75. ISBN: 0-8137-2180-6

Hildreth W (1981) Gradients in silicic magma chambers: implications for lithospheric magmatism. J Geophys Res Solid Earth 86(B11):10153–10192. https://doi.org/10.1029/JB086iB11p10153

Hildreth W (1991) The timing of caldera collapse at Mount Katmai in response to magma withdrawal toward Novarupta. Geophys Res Lett 18:1541–1544. https://doi.org/10.1029/91GL01083

Hildreth W, Fierstein J (2000) Katmai volcanic cluster and the great eruption of 1912. Geol Soc Amer Bull 112:1594–1620. https://doi.org/10.1130/0016-7606(2000)112<1594:KVCATG>2.0.CO;2

Hildreth W, Lanphere MA, Fierstein J (2003) Geochronology and eruptive history of the Katmai volcanic cluster, Alsaka Peninsula. Earth Planet Sci Lett 214:93–114. https://doi.org/10.1016/S0012-821X(03)00321-2

Holden H, Wallace L, Beavan J, Fournier N, Cas R, Ailleres L, Silcock D (2015) Contemporary ground deformation in the Taupo Rift and Okataina Volcanic Centre from 1998 to 2011, measured using GPS. Geophys J Int 202(3):2082–2105. https://doi.org/10.1093/gji/ggv243

Holohan EP, Troll VR, Walter TR, Munn S, McDonnell S, Shipton ZK (2005) Elliptical calderas in active tectonic settings: an experimental approach. J Volcanol Geotherm Res 144:119–135. https://doi.org/10.1016/j.jvolgeores.2004.11.020

Holohan EP, Troll VR, van Wyk de Vries B, Walsh JJ, Walter TR (2008) Unzipping Long Valley: an explanation for vent migration patterns during an elliptical ring fracture eruption. Geology 36:323–326. https://doi.org/10.1130/G24329A.1

Holohan EP, Schopfer MPJ, Walsh JJ (2011) Mechanical and geometric controls of the structural evolution of pit crater and caldera subsidence. J Geophys Res Solid Earth 116:B07202. https://doi.org/10.1029/2010/JB008032

Houghton BF, Weaver SD, Wilson CJN, Lanphere MA (1992) Evolution of a Quaternary peralkaline volcano: Mayor Island, New Zealand. J Volcanol Geotherm Res 51:217–236. https://doi.org/10.1016/0377-0273(92)90124-V

Johnson RW (1969) Volcanic geology of Mt. Suswa, Kenya. Philos Trans R Soc Lond Ser a: Math Phys Sci 265:383–412

Johnston EN, Sparks RSJ, Phillips JC, Carey S (2014) Revised estimates for the volume of the Late Bronze Age Minoan eruption, Santorini, Greece. J Geol Soc 171:583–590. https://doi.org/10.1144/jgs2013-113

Karatson D, Thouret J-C, Moriya I, Lomoschitz A (1999) Erosion calderas: origins, processes, and climatic/structural control – an overview. Bull Volcanol 61(3):174–193. https://doi.org/10.1007/s004450050270

Karátson D, Gertisser R, Telbisz T, Vereb V, Quidelleur X, Druitt TH, Nomikou P, Kosik S (2018) Towards reconstruction of the lost Late Bronze age intra-caldera island in Santorini, Greece. Sci Rep 8:7026. https://doi.org/10.1038/s41598-018-25301-2

Karátson D, Telbisz T, Gertisser R, Strasser T, Nomikou P, Druitt T, Vereb V, Quidelleur X, Kósik S (2020) Constraining the landscape of Late Bronze Age Santorini prior to the Minoan eruption: Insights from volcanological, geomorphological and archaeological findings. J Volcanol Geotherm Res 401:106911. https://doi.org/10.1016/j.jvolgeores.2020.106911

Karstens J, Preine J, Crutchley GJ, Kutterolf S, van der Bilt WGM, Hooft EE, Druitt TH, Schmid F, Cederstrøm JM, Hübscher C, Nomikou P, Carey C, Kühn M, Elger J, Berndt C (2023) Revised Minoan eruption volume as benchmark for large volcanic eruptions. Nature Commun 14:2097. https://doi.org/10.1038/s41467-023-38176-3

Kennedy B, Wilcox J, Stix J (2012) Caldera resurgence during magma replenishment and rejuvenation at Valles and Lake City calderas. Bull Volcanol 74:1833–1847. https://doi.org/10.1007/s00445-012-0641-x

Kingsley L (1931) Cauldron subsidence in the Ossipee Mountains. Amer J Sci 5(128):139–168. https://doi.org/10.2475/ajs.s5-22.128.139

Kokelaar BP, Branney MJ (1999) Inside silicic calderas (Snowdon, Scafell, and Glencoe, UK): interaction of caldera development, tectonism and hydrovolcanism. CEV Field Workshop, Guidebook, 150 pp

Larsen JF (2013) Unraveling the diversity in arc volcanic eruption styles: examples from the Aleutian volcanic arc, Alaska. J Volcanol Geotherm Res 327:643–668. https://doi.org/10.1016/j.jvolgeores.2016.09.008

Le Pennec JL, Temel A, Froger JL, Sen S, Gourgaud A, Bourdier JL (2005) Stratigraphy and age of the Cappadocia ignimbrites, Turkey: reconciling field constraints with paleontologic, radiochronologic, geochemical and paleomagnetic data. J Volcanol Geotherm Res 141(1–2):45–64. https://doi.org/10.1016/j.jvolgeores.2004.09.004

Lindsay JM, de Silva S, Trumbull R, Emmermann R, Wemmer K (2001) La Pacana caldera, N Chile: a re-evaluation of the stratigraphy and volcanology of one of the world's largest resurgent caldera. J Volcanol Geotherm Res 106:145–173. https://doi.org/10.1016/S0377-0273(00)00270-5

Lipman PW (1984) The roots of ash flow calderas in western North America: windows into the tops of granitic batholiths. J Geophys Res 89B:8801–8841. https://doi.org/10.1029/JB089iB10p08801

Lipman PW (1992) Ash-flow calderas as structural controls of ore deposits—recent work and future problems. US Geol Surv Bull 2012:L1–L12

Lipman PW (1997) Subsidence of ash-flow calderas; relation to caldera size and magma-chamber geometry. Bull Volcanol 59:198–218. https://doi.org/10.1007/s004450050186

Lipman PW (2000a) Calderas. In: Sigurdsson H, Houghton B, McNutt SR, Rymer H, Stix J (eds) Encyclopedia of volcanoes, Part V, 1st edn. Academic Press, San Francisco, pp 643–652. ISBN: 978-0-12-385938-9

Lipman PW (2000b) The central San Juan caldera cluster: regional volcanic framework. Geol Soc Am Spec Pap 346:9–69. https://doi.org/10.1130/0-8137-2346-9.9

Lipman PW (2006) Geologic map of the central San Juan caldera cluster, Southwestern Colorado, including the Database for the Geologic map of the Central San Juan Caldera Cluster, Southwestern Colorado by JE Robinson, DR Dutton, DW Ramsey, PW Lipman and TJ Felger. Geol Invest Series I-2799, US Geol Survey Prof Paper

Lipman PW, McIntosh WC (2008) Eruptive and noneruptive calderas, northeastern San Juan Mountains, Colorado: where did the ignimbrites come from? Geol Soc Amer Bull 120(7–8):771–795. https://doi.org/10.1130/B26330.1

Lipman PW, Self S, Heiken G (1984) Introduction to special issue on calderas. J Geophys Res 89(B10):8219–8221. https://doi.org/10.1029/JB089iB10p08219

Lockwood JP, Hazlett RW (2010) Volcanoes. Global perspectives. Wiley, UK, 539 pp. ISBN: 978-1-4051-6249-4

Lowenstern JB (2020) Simplified geological map of the Yellowstone caldera. US Geol Survey, Volcano Hazards Program

Lowenstern JB, Smith RB, Hill DP (2006) Monitoring super-volcanoes: geophysical and geochemical signals at Yellowstone and other large caldera systems. Philos Trans Royal Soc a: Math Phys Eng Sci 364(1845):2055–2072. https://doi.org/10.1098/rsta.2006.1813

Mahood G, Hildreth W (1983) Nested calderas and trapdoor uplift at Pantelleria, Strait of Sicily. Geology 11:722–726. https://doi.org/10.1130/0091-7613(1983)11<722:NCATUA>2.0.CO;2

Mahood GA, Hildreth W (1986) Geology of the peralkaline volcano at Pantelleria, Strait of Sicily. Bull Volcanol 48:143–172. https://doi.org/10.1007/BF01046548

Mamani M, Wörner G, Sempere T (2010) Geochemical variations in igneous rocks of the Central Andean orocline (13°S to 18°S): tracing crustal thickening and magma generation through time and space. Geol Soc Am Bull 122:162–182. https://doi.org/10.1130/B26538.1

Marsh BD (1984) On the mechanics of caldera resurgence. J Geophys Res 89:8245–8251. https://doi.org/10.1029/JB089iB10p08245

Marti J (2019) Las Cañadas caldera, Tenerife, Canary Islands: a review, or the end of a long volcanological controversy. Earth-Sci Rev 102889. https://doi.org/10.1016/j.earscirev.2019.102889

Marti J, Gudmundsson A (2000) The Las Cañadas caldera (Tenerife, Canary Islands): an overlapping collapse caldera generated by magma-chamber migration. J Volcanol Geotherm Res 103:161–173. https://doi.org/10.1016/S0377-0273(00)00221-3

Marti J, Geyer A, Folch A (2009) A genetic classification of collapse calderas based on field studies, and analogue and theoretical modelling. In: Thordarson T, Self S, Larsen G, Rowland SK, Hoskuldsson A (eds) Studies in volcanology: the legacy of GPL Walker. Spec Publ IAVCEI 2, Geol Soc London, pp 249–266. ISBN: 978-1-86239-280-9

Merle O, Barde-Cabusson S, van Wyk de Vries B (2010) Hydrothermal calderas. Bull Volcanol 72:131–147. https://doi.org/10.1007/s00445-009-0314-6

Miura D (1999) Arcuate pyroclastic conduits, ring faults, and coherent floor at Kumano caldera, southwest Honshu, Japan. J Volcanol Geotherm Res 92(3–4):271–294. https://doi.org/10.1016/S0377-0273(99)00089-X

Nairn IA (1989) Sheet V16 AC, Mount Tarawera. Geological Map of New Zealand 1:50,000. Dep Sci Ind Res, Wellington

Nairn IA, Wood CP, Bailey RA (1994) The Reporoa caldera, Taupo volcanic zone: source of the Kaingaroa Ignimbrites. Bull Volcanol 56:529–537. https://doi.org/10.1007/BF00302833

Nakada, Druitt TH (2005) The 2000 eruption of Miyakejima volcano, Japan: introduction to special issue. Bull Volcanol 67:203–204. https://doi.org/10.1007/s00445-004-0403-5

Newhall CG, Dzurisin D (1988) Historical unrest at large calderas of the world. US Geol Survey Prof Paper 1855. Library of Congress Catalog no. 88600541, 1108 pp

Nomikou P, Hübscher C, Carey S (2019) The Christiana-Santorini-Kolumbo volcanic field. Elements 15:171–176. https://doi.org/10.2138/gselements.15.3.171

Orsi G, Gallo G, Zanchi A (1991) Simple-shearing block resurgence in caldera depressions. A model from Pantelleria and Ischia. J Volcanol Geotherm Res 47:1–11. https://doi.org/10.1016/0377-0273(91)90097-J

Orsi G, Antonio M, De Vita S, Gallo G (1992) The Neapolitan Yellow Tuff, a large-magnitude trachytic phreatoplinian eruption; eruptive dynamics, magma withdrawal and caldera collapse. J Volcanol Geotherm Res 53:275–287. https://doi.org/10.1016/0377-0273(92)90086-S

Pérez W, Freundt A, Kutterolf S (2020) The basaltic plinian eruption of the ~6 ka San Antonio Tephra and formation of the Masaya caldera, Nicaragua. J Volcanol Geotherm Res 401:106975. https://doi.org/10.1016/j.jvolgeores.2020.106975

Permenter JL, Oppenheimer C (2007) Volcanoes of the Tibesti massif (Chad, northern Africa). Bull Volcanol 69:609–626. https://doi.org/10.1007/s00445-006-0098-x

Petrinovic IA (1999) La caldera del colapso de Cerro Aguas Calientes, Salta, Argentina: evolucion y esquema estructural. Acta Geol Hispanica 34(2–3):243–253

Pittari A, Cas RA, Wolff JA, Nichols HJ, Marti J (2008) The use of lithic clast distributions in pyroclastic deposits to understand pre- and syn-caldera collapse processes; a case study of the Abrigo ignimbrite, Tenerife, Canary Islands. In: Gottsmann J, Marti J (eds) Caldera volcanism: analysis, modelling and response, Chap 3. Developments in Volcanology, Elsevier, Amsterdam, pp 97–142. https://doi.org/10.1016/S1871-644X(07)00003-4

Preine J, Karstens J, Hübscher C, Nomikou P, Schmid F, Crutchley GJ, Druitt TH, Papanikolaou D (2022)

Spatio-temporal evolution of the Christiana-Santorini-Kolumbo volcanic field, Aegean Sea. Geology 50(1):96–100. https://doi.org/10.1130/G49167.1

Rivalta E, Corbi F, Pasasarelli L, Acocella V, Davis T, Di Vito MA et al (2019) Stress inversions to forecast magma pathways and eruptive vent location. Sci Adv 5(7):1–11. https://doi.org/10.1126/sciadv.aau9784

Roche O, Druitt TH, Merle O (2000) Experimental study of caldera formation. J Geophys Res 105:395–416. https://doi.org/10.1029/1999JB900298

Rosi M, Sbrana A, Principe C (1983) The Phlegrean Fields: structural evolution, volcanic history and eruptive mechanisms. J Volcanol Geotherm Res 17:273–288. https://doi.org/10.1016/0377-0273(83)90072-0

Rymer H, van Wyk de Vries B, Stix J, Williams-Jones G (1998) Pit crater structure and processes governing persistent activity at Masaya Volcano, Nicaragua. Bull Volcanol 59:345–355. https://doi.org/10.1007/s004450050196

Salisbury MJ, Jicha BR, de Silva SL, Singer BS, Jiménez NC, Ort MH (2011) $^{40}Ar/^{39}Ar$ chronostratigraphy of Altiplano-Puna volcanic complex ignimbrites reveals the development of a major magmatic province. Geol Amer Soc Bull 123(5–6):821–840. https://doi.org/10.1130/B30280.1

Scarpati C, Cole P, Perrotta A (1993) The Neapolitan Yellow Tuff—a large volume mutiphase eruption from Campi Flegrei, southern Italy. Bull Volcanol 55:343–356. https://doi.org/10.1007/BF00301145

Scarpati C, Sparice D, Perrotta A (2020) Dynamics of large pyroclastic currents inferred by the internal architecture of the Campanian Ignimbrite. Sci Rep 10(1):22230. https://doi.org/10.1038/s41598-020-79164-7

Schmincke H-U (1967) Cone sheet warm, resurgence of Tejeda caldera, and early geologic history of Gran Canaria. Bull Volcanol 31:153–162. https://doi.org/10.1007/BF02597011

Schmincke H-U (2004) Volcanism. Springer, Berlin, 324 pp. ISBN: 3-540-43650-2

Self S, Rampino MR (1981) The 1883 eruption of Krakatau. Nature 294:699–704. https://doi.org/10.1038/294699a0

Self S, Goff F, Gardner JN, Wright JV, Kite WM (1986) Explosive rhyolitic volcanism in the Jemez Moutains: vent locations, caldera development and relation to regional structure. J Geophys Res 91:1779–1798. https://doi.org/10.1029/JB091iB02p01779

Sheridan MF (1979) Emplacement of pyroclastic flows: a review. In: Chapin CE, Elston WE (eds) Geol Soc Amer, Spec Paper 180, pp 125–136. ISBN: 0-8137-2180-6

Sigmundsson F (2019) Calderas collapse as magma flows into rifts. Recent caldera collapses show the importance of distant volcanic rift zones. Science 366(6470):1200–1201. https://doi.org/10.1126/science.aaz7126

Simkin T, Fiske RS (1983) Krakatau 1883: the volcanic eruption and its effects. Smithsonian Inst Press, Washington, DC, 464 pp. ISBN: 874748410

Smith RL (1960) Ash flows. Geol Soc Amer Bull 71(6):795–841. https://doi.org/10.1130/0016-7606(1960)71[795:AF]2.0.CO;2

Smith RL (1979) Ash flow magmatism. In: Chapin CE, Elston WE (eds) Ash flow tuffs. Geol Soc Amer Bull, Spec Paper 180, pp 5–27. ISBN: 0-8137-2180-6

Smith RL, Bailey RA (1968) Resurgent cauldrons. In: Coats RR, Hay RL, Anderson CA (eds) Studies in volcanology, Geol Soc Amer Memoirs 116, pp 613–662. https://doi.org/10.1130/MEM116-p613

Smith R L, Bailey RA, Ross CS (1961) Structural evolution of the Valles Caldera, New Mexico, and its bearing on the emplacement of ring dykes. US Geol Surv Prof Paper 424-D:145–149

Sourisseau D, Macías JL, Tenorio FG, Avellán DR, Girón RS, Bernal JP, Saldaña JLA, Murillo ZT (2020) New insights into the stratigraphy and 230Th/U geochronology of the post-caldera explosive volcanism of La Primavera caldera, Mexico. J South Amer Earth Sci 103:102747. https://doi.org/10.1016/j.jsames.2020.102747

Sparks RSJ, Francis PW, Hamer RD, Pankhurst RJ, O'Callaghan LO, Thorpe RS, Page R (1985) Ignimbrite of the Cerro Galán Caldera, NW Argentina. J Volcanol Geotherm Res 24:205–248. https://doi.org/10.1016/0377-0273(85)90071-X

Stelling P, Gardner JE, Begét J (2005) Eruptive history of Fisher caldera, Alaska, USA. J Volcanol Geotherm Res 139(3–4):163–183. https://doi.org/10.1016/j.jvolgeores.2004.08.006

Steven TA, Lipman PW (1976) Calderas of the San Juan volcanic field, south-western Colorado. US Geol Survey Prof Paper 958, 35 pp

Storey M, Roberts RG, Saidin M (2012) Astronomically calibrated $^{40}Ar/^{39}Ar$ age for the Toba supereruption and global synchronization of late Quaternary records. Proc Nat Acad Sci 109(46):18684–18688. https://doi.org/10.1073/pnas.1208178109

Thouret J-C, Wörner G, Gunnell Y, Singer B, Zhang X, Souriot T (2007) Geochronologic and stratigraphic constraints on canyon incision and Miocene uplift of the Central Andes in Peru. Earth Planet Sci Lett 263:151–166. https://doi.org/10.1016/j.epsl.2007.07.023

Thouret JC, Jicha BR, Paquette JL, Cubukcu EH (2016) A 25 myr chrono-stratigraphy of ignimbrites in south Peru: implications for the volcanic history of the Central Andes. J Geol Soc London 173(5):734–756. https://doi.org/10.1016/S0377-0273(01)00323-7

Van Bemmelen RW (1949) The geology of Indonesia. General Government Printing Office, The Hague, Netherlands 1A, 732 pp

Vincent PM (1970) The evolution of the Tibesti volcanic province, eastern Sahara. In: Clifford TN, Gass IG (eds) African magmatism and tectonics. Oliver and Boyd, Edinburgh, pp 301–319, 461 pp. https://doi.org/10.1017/S001675680005679X

Vincent PM (1994) Les calderas. In: Bourdier J-L (ed) Le volcanisme, Chap 6. Manuel et Méthodes 25, BRGM, pp 181–195, 420 pp

Wadge G, Biggs J, Lloyd R, Kendall J-M (2016) Historical volcanism and the state of stress in the East African Rift System. Front Earth Sci 4:86. https://doi.org/10.3389/feart.2016.00086

Walker GPL (1984) Downsag calderas, ring faults, caldera sizes, and incremental caldera growth. J Geophys Res 89B:8407–8416. https://doi.org/10.1029/JB089iB10p08407

Walker GPL, Wright JV, Clough BJ, Booth B (1981) Pyroclastic geology of the rhyolitic volcano of La Primavera, Mexico. Geol Rundsch 70:1100–1118. https://doi.org/10.1007/BF01820184

Walker JA, Williams SN, Kalamarides RI, Feigenson MD (1993) Shallow open-system evolution of basaltic magma beneath a subduction zone volcano: the Masaya Caldera Complex, Nicaragua. J Volcanol Geotherm Res 56:379–400. https://doi.org/10.1016/0377-0273(93)90004-B

Walter TR, Troll V (2001) Formation of caldera periphery faults: an experimental study. Bull Volcanol 63(2–3):191–203. https://doi.org/10.1007/s004450100135

Wheller GE, Varne R (1986) Genesis of dacitic magmatism at Batur Volcano, Bali Indonesia: implications for the origins of stratovolcano calderas. J Volcanol Geotherm Res 28:363–378. https://doi.org/10.1016/0377-0273(86)90031-4

Williams H (1941) Calderas and their origin. University of California, Bull Geol Sci 25:239–346

Williams H (1942) The geology of Crater Lake National Park, Oregon, with a reconnaissance of the Cascade range south to Mount Shasta. Carnegie Institute of Washington Publication 540, 162 pp

Williams H, McBirney AR (1979) Volcanology. Freeman, Cooper & Co., San Francisco, 398 pp. ISBN: 0-87735-321-2

Williams SN, Stoiber RE (1983) Masaya-type caldera—redefined as the mafic analogue of Krakatau-type caldera. EOS Trans Am Geophys Union 64(45):877

Wilson CJN (1993) Stratigraphy, chronology, styles and dynamics of late Quaternary eruptions from Taupo volcano, New Zealand. Philos Trans Royal Soc London A343:295–306. https://doi.org/10.1098/rsta.1993.0050

Wilson CJN, Hildreth W (1997) The Bishop Tuff: new insights from eruptive stratigraphy. J Geol 105:407–439. https://doi.org/10.1086/515937

Wilson CJN, Gravley DM, Leonard GS, Rowland JV (2009) Volcanism in the central Taupo Volcanic Zone, New Zealand: tempo, styles and controls. In: Thorardson T, Self S, Larsen G, Rowland SK, Hoskulddsson A (eds) Studies in volcanology, the legacy of GPL Walker, Spec Publ IAVCEI no 2, pp 225–247. ISBN: 978-1-86239-280-9

Wilson CJN, Rogan AM, Smith IEM, Northey DJ, Nairn IA, Houghton BF (1984) Caldera volcanoes of the Taupo volcanic zone, New Zealand. J Geophys Res Solid Earth 89(B10):8463–8484. https://doi.org/10.1029/JB089iB10p08463

Wilson CJN, Houghton BF, McWilliams MO, Lanphere MA, Weaver SD, Briggs RM (1995) Volcanic and structural evolution of Taupo Volcanic Zone, New Zealand: a review. J Volcanol Geotherm Res 68:1–28. https://doi.org/10.1016/0377-0273(95)00006-G

Wohletz K, Heiken G (1992) Volcanology and geothermal energy. University of California Press, 432 pp. ISBN: 0-05-201914-0

Wolff JA, Wright JV (1981) Formation of the Green Tuff, Pantelleria. Bull Volcanol 44:681–690. https://doi.org/10.1007/BF02597091

Chapter 7 Short-Term Degradation of Active Volcanoes

1 Overview of Degradation Processes

The different types of volcanoes on Earth have been presented in the previous chapters. Volcanic edifices are all created by the accumulation, or 'aggradation' of eruptive products (cf. Cas and Wright 1988), although the duration of the eruption and the time between periods of activity can vary widely. During inactive periods, volcanoes tend to undergo 'degradation', but this may also occur even during active periods. Newton's aphorism about gravity, "what goes up, must come down", can also be applied to volcanoes (e.g. Fink 1992) in the sense that what is built up by eruptions is subsequently removed by degradation, involving various erosional processes.

1.1 Definition, Scope, and Facies Relationships

Degradation (a general term referring to the overall lowering of the ground surface by erosion), associated with the transport of removed material, leads to the progressive destruction of a volcanic edifice. While this can also be brought about in a broader sense by eruptive processes (e.g., breaching of a crater by a lava flow, or summit caldera collapse of a composite volcano during an explosive eruption), this chapter and the next deal only with external, i.e. erosional processes. *Erosion* is a widely used specific term for the processes governed by *external* or *exogenic* forces (or agents) that mechanically wear away the land surface.

Erosional processes in volcanic terrains are often referred to as *epiclastic processes* (Cas and Wright 1988), which lead to the emplacement of *volcanogenic sedimentary deposits* (McPhie et al. 1993). Such deposits are considered as end-products, and can be formed even after a long time and multiple processes, as many phases of remobilisation and resedimentation may occur. In volcanic terrains, the most characteristic epiclastic processes are those which affect fresh, bare, non-vegetated surfaces and transport/remobilise volcaniclastic debris. These processes are essentially not different from those that transport debris in any other terrain; but because the transported clasts are typically more variable in density (from massive lava clasts to weathered rocks to scoria or pumice), volcaniclastic debris tends to be more poorly sorted and may differ texturally and structurally to that in sedimentary successions (Cas and Wright 1988; McPhie et al. 1993).

The characteristics of the degradation processes on any volcano or volcanic terrain depend on the following factors:

(1) the intensity of, and the time interval between, the eruptions if the volcano is still active;

D. Karátson and J.-C. Thouret, *Volcano Geomorphology*, Advances in Volcanology,
https://doi.org/10.1007/978-3-031-77693-9_8

(2) the lithological, sedimentological and physical properties of the erupted products (thickness, grain size, density, porosity, etc.);
(3) the type of sedimentary environment: terrestrial (including fluvial and lacustrine) or submarine (deep open-sea or shallow submarine);
(4) the size and topography (e.g. slope features) of the affected area;
(5) the prevailing climate (arid, semiarid, humid, etc.).

Given the variety of these factors, the time scale and efficacy of erosion processes can vary greatly (Fig. 1). On active volcanoes, due to the often unstable, unconsolidated materials constituting unvegetated surfaces, they tend to be rapid. The rate of erosion may subsequently slow down, but this is highly dependent on the climate. Climate conditions will determine the rate of rock weathering, soil formation and vegetation growth (Chap. 8), which constrain the general rate of erosion, even in the short term.

Eventually, all primary volcanic material is removed, transported away and redeposited by erosion. In a sedimentological context, this (often multiple) process is termed *resedimentation.* This broad term comprises two different temporal groups:

(a) *syn-eruptive* and *inter-eruptive processes,* which occur during the active or dormant period of a volcano;
(b) *post-eruptive processes,* which occur after the volcano has become extinct.

The syn- and inter-eruptive processes differ significantly from long-term post-eruptive erosion, particularly due to specific, high-energy mass movements that occasionally transport a large volume of material (Fisher and Smith 1991). Although some high-energy processes (e.g., debris flows, rockfalls) may also occur long after the volcanism has ceased, they are most common during the active lifetime of a volcano. The high flux of transported material is often associated with eruptive activity. Therefore, we present them separately: Chap. 7 is devoted to high-energy, short-term erosion processes, while Chap. 8 deals with common erosional processes, in particular fluvial erosion and slope evolution over a longer time scale.

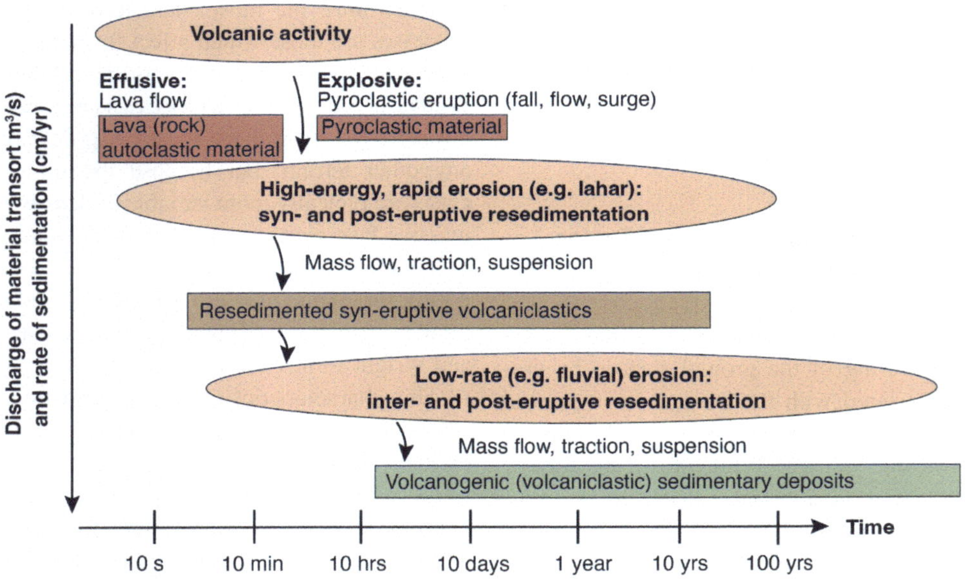

Fig. 1 Generation, sedimentation and resedimentation of eruptive products on volcanoes in terms of time (duration of process), and rate of material transport (modified after McPhie et al. 1993; Cas and Wright 1988; Manville et al. 2009)

Note, however, that fluvial erosion processes can act in the short term on active volcanoes as well. During volcanically active periods, especially in fresh, modified landscapes after eruptions, fluvial erosion, which is generally thought to be 'slow', can be significantly enhanced (Kataoka et al. 2009; Major 2022). Fluvial channels might become unstable, causing channel widening and channel migration (Photo 82). Dramatically increased fluvial erosion has been documented after the eruptions of Mount St. Helens in 1980, El Chichón in 1982, Pinatubo in 1991, and Unzen in 1990–95 (Inbar et al. 2001; Gran et al. 2011; Pierson and Major 2014; Gomez et al. 2021). Remobilisation of fresh volcanic deposits results in a high sediment yield, and occurs in association with rill development (Manville et al. 2009; see Chap. 8). Persistent channel instabilities (e.g., formation of braided rivers) and geomorphic adjustments can prolong significant channel erosion for years or decades (Major 2022).

Sediment yield (usually expressed as Mg/km^2/year or m^3/km^2/year) can also be viewed as a long-term, basin-wide, average erosion rate (Major et al. 2000; Pierson and Major 2014). Volcanic catchments with syn-eruption annual sediment yields of 10^4 to 10^7 t/km^2 are among the highest sediment producers on Earth (Gran and Montgomery 2005; Pierson and Major 2014) (Fig. 2). However, periods of activity triggering intense erosion alternate with intervals of repose (Manville et al. 2009), switching between short eruption phases, in the order of hours, days or rarely years, and long-term erosion over a kyr- or Myr time scale. Ultimately, the stratigraphic record of a volcano or volcanic field comprises both primary eruptive products and the syn/inter- and post-eruptive volcaniclastic deposits resulting from short- and long-term erosional processes.

Dynamic changes in degradation processes contribute significantly to the volcanic lithofacies (Smith 1991; Orton 1995). We have already shown the facies assemblages of composite cones in Chap. 5. Although there are fundamental differences on land and under water, for all volcanic environments a basic distinction can be made between *proximal* (near-vent), *medial* (typically, volcano flank) and *distal* (e.g., ring plain and beyond) facies in terms of prevailing processes. Given that eruptive processes affect mainly the proximal, and to some extent the medial facies, the proportion of pyroclastic material decreases away from the volcanic

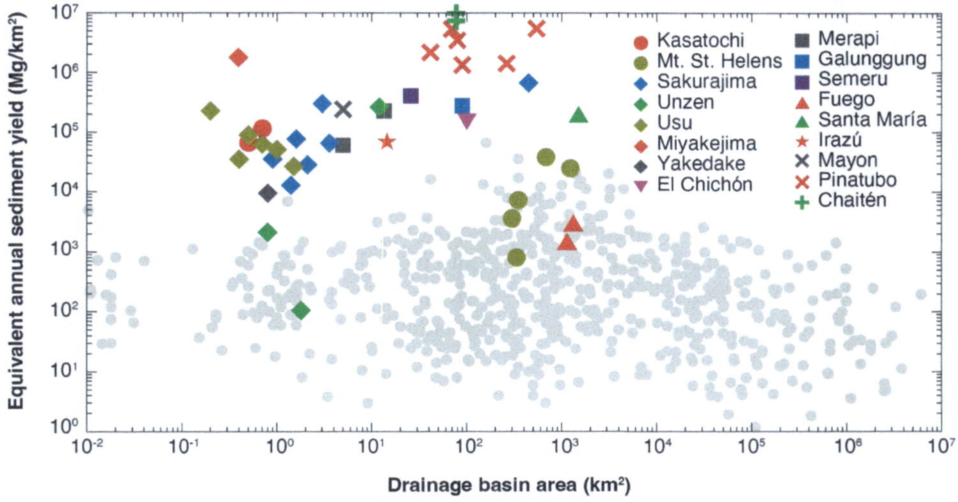

Fig. 2 Average annual (mostly suspended) sediment yield as a function of drainage basin area for yields from volcanically disturbed basins and nonvolcanic river basins worldwide (Pierson and Major 2014). Coloured data points represent drainage basins affected by recent volcanic eruptions. Grey data points are for basins in volcanic terrain where no recent eruptions have occurred and in nonvolcanic terrain

centre, accompanied by an increase in epiclastic material (Smith 1987, 1991; McPhie et al. 1993). On the other hand, proximal, medial and distal are relative concepts, whose deposits show a general decrease in thickness and grain size with distance from vent. They vary in size and distance from the centre, both in relative and absolute terms; facies dimensions depend on the size of the volcano, the amount and intensity of the supply of volcanic material, and the mode and possibility of material transport over a wide range of distances (a few km to hundreds of km).

We do not address the increasingly diverse sedimentological literature available on facies relationships, but recommend Cas and Wright (1988) and McPhie et al. (1993) for greater detail, and the review papers by Manville et al. (2009) and Martí et al. (2018) for further reading. In Volcanic Successions by Cas and Wright (1988), a number of facies models are described: terrestrial and submarine volcanoes; basaltic fissure volcanoes; basaltic oceanic seamounts; submarine silicic volcanoes; basaltic and silicic subglacial volcanoes; and Precambrian volcanic successions.

1.2 Differences Between Terrestrial and Underwater Environments

In terrestrial volcanic environments, the surface of an erupting or recently dormant volcano is usually composed of fresh, uncompacted lava or pyroclastic material. Under most climates, this surface is subject to syn- and inter-eruptive erosion processes. Deep, temporary gullies, in particular on volcano flanks, often form as a result of periodic fluvial erosion or slope mass movements (gully development will be addressed in Chap. 8). In Fig. 3, examples from Honshū, Japan, shows differences between erosional processes and sedimentation on an active and a dormant/extinct volcano, respectively, in a terrestrial environment involving a lake.

For subaqueous (commonly submarine) environments, there is a major difference between sedimentation in deep water and in shallow water offshore of volcanic islands or island arcs. In deep submarine settings, the material deposited from non-volcanic mass movements is occasionally supplemented by a small amount of pyroclastic fallout (Carey and Schneider 2011), whereas more significant volcaniclastic sedimentation occurs near the islands themselves (Fisher and Schmincke 1984). These differences are illustrated for subduction-related island arcs and oceanic intraplate volcanism in Fig. 4 (Carey 2000), and the specific related features are addressed in Sect. 4.

In the vicinity of island-arc volcanoes, in addition to the core area of an on-land eruption centre, we can distinguish four areas: (a) a coastal zone fringing the island, (b) shallow marine slopes, (c) canyons, (d) submarine alluvial fans (Kuenzi et al. 1979; Vessell and Davies 1981; Allen et al. 2007). Obviously, the intensity of volcanic activity is a key factor affecting sedimentation around an active arc, but compared to terrestrial settings, biogenic sedimentation generally increases at the expense of volcaniclastic sedimentation away from the active centre. Closer to the volcano, pyroclastic and hydroclastic debris are transported into the surrounding forearc- and back-arc basins during eruptions, while epiclastic sedimentation dominates during the lulls in volcanic activity. The amount of intercalated fallout tephra depends on the intensity of the eruption (the height and duration of the eruption column) and the prevailing wind direction.

Apart from studies on some emergent areas of the Canary Islands (e.g., Staudigel and Schmincke 1984), knowledge about the sedimentation at oceanic hotspot islands comes from deep-sea research (drilling, bathymetry) around the Hawaiian and Canary Islands. This has helped to clarify the volcanic evolution of these islands, as described in Chap. 4. However, compared to terrestrial settings, there is less information about either the debris transport processes (affecting depths to as much as

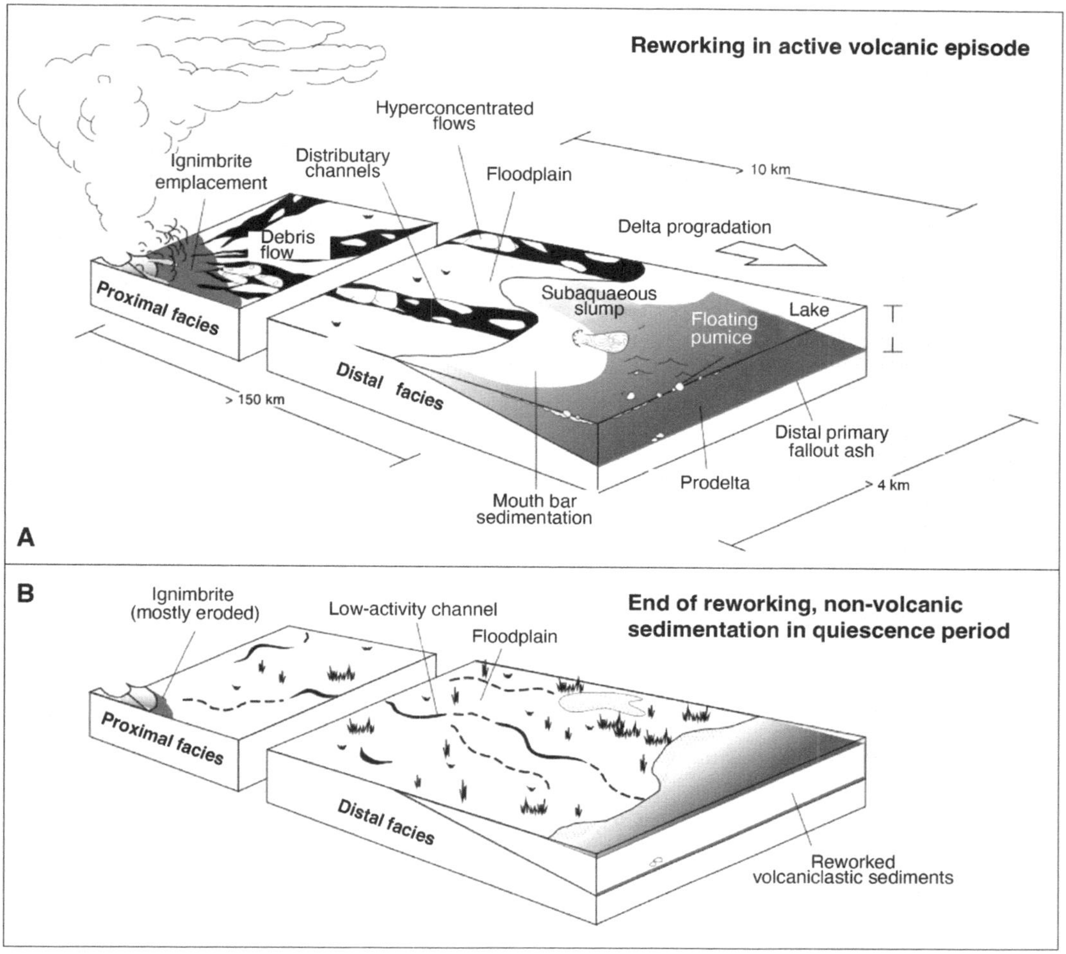

Fig. 3 Syn- and post-volcanic sediment transport and sediment deposition in terrestrial and lake environments (example from 2.4 Ma Mushono Tephra, Japan: Kataoka 2005)

thousands of metres) or the facies relationships. What can be said is that (1) rapid sedimentation in this setting is governed by large-scale, massive debris slumping rather than explosive activity, the effect of which is subordinate, and (2) in this case, too, as we move away from the islands, there is common increase in non-volcanic, including biogenic/bioclastic, sedimentation (Schmincke and Sumita 1998; Oehler et al. 2008; Trofimovs et al. 2013).

1.3 Short-Term Degradation Processes Specific to Volcanoes

Active volcanoes, especially the larger ones, are unstable edifices due to internal and external factors of both volcanic and non-volcanic origin. These factors may make the edifice unstable, and instability will commonly lead to the failure or lateral collapse of a bigger or smaller portion

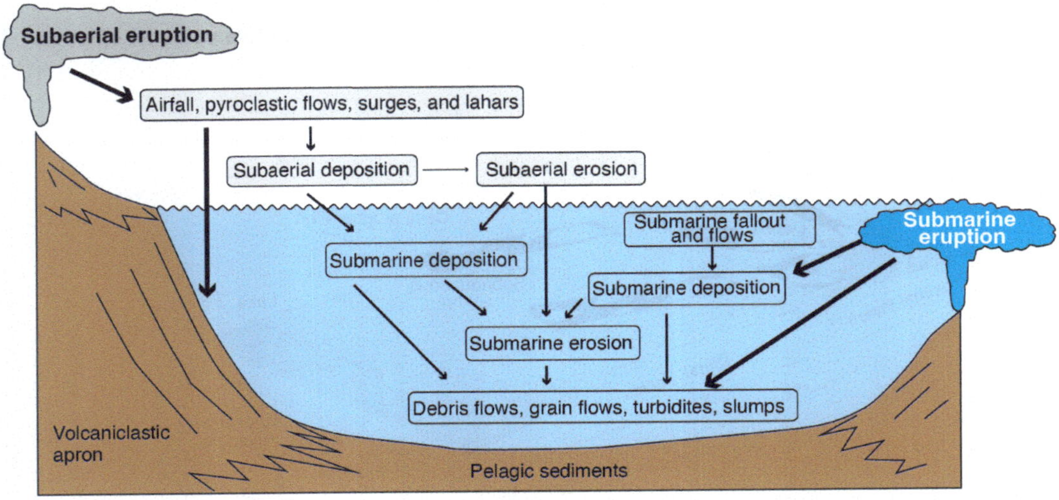

Fig. 4 Sources and transport processes of volcaniclastic sediments in convergent-margin island arcs (left) and oceanic intraplate volcanism (right). Thick arrows indicate direct transformations, thin arrows indicate reworking. (After Carey 2000; Manville et al. 2009)

of the edifice, and the formation of *volcanic debris avalanches* (Sect. 2; e.g., Voight et al. 1981; Siebert 1984). The resulting landform characteristics on both shield and composite volcanoes have already been presented in Chaps. 4 and 5, respectively: here, we focus on the characteristics of the processes and deposits. Debris avalanches typically characterize active and dormant periods, although they may also occur when the volcano becomes extinct, albeit less commonly. Thanks to extensive research over the past few decades, more than one thousand debris avalanches have now been identified from nearly six hundred volcanoes around the world (Siebert and Roverato 2021).

Alternatively, a bare volcanic surface covered with loose, volcaniclastic or lava debris often undergoes small-scale slope mass movements (Cas and Wright 1988). In dry climates (less than 200–250 mm rainfall per year), these mass movements can take place as virtually water-free bodies, but if debris flows incorporate more than 5% water, specific mass movements called *lahars* can occur (Sect. 3; Smith and Lowe 1991; Vallance 2000). This Javanese term has become widely used in volcanology for all flow phenomena from volcanoes that carry more debris or mud than a conventional river or stream. Given appropriate conditions (e.g., abundant water, remobilisable, ductile bedrock), debris avalanches can also be transformed into lahars. It should be stressed that debris avalanches, and more particularly lahars, can also occur in the absence of an eruption and at extinct volcanoes. In fact, lahars can be triggered up to a long time after the eruption, and hence pose a long-lasting threat on dormant volcanoes (Thouret et al. 2020).

Both debris avalanches and lahars can occur in subaqueous environments as well (either flowing into water, or triggered subaqueously), eventually transforming into turbidites. The subaqueous processes and deposits are described at the end of this chapter in Sect. 4.

The particular importance of debris avalanches and lahars and their impact on a volcanic edifice can be summarised as follows:

(1) they fundamentally determine the shape and geomorphic evolution of the volcano, because they constrain the growth of the edifice and greatly contribute to the formation of the volcaniclastic apron surrounding the volcano;

(2) they may radically affect and alter the drainage system, as they may supply a high volume of sediment load over a long period of time compared with the quiescence periods;
(3) together with the often increased fluvial erosion acting in the inter- and post-eruptive period(s) (see Chap. 8), they form an essential part of volcanic cycles, that is, the alternation of eruptive periods, rapid (accelerated) erosion processes including the associated, short-term mass flows, and then quiescent periods with subdued, 'average' fluvial erosion.
(4) As a consequence, debris avalanches and lahars pose an important hazard to the surrounding human settlements and wildlife, which is presented in detail in Chap. 9.

2 Volcanic Debris Avalanches

On 18 May 1980 the northern flank of Mount St. Helens volcano (Washington, USA) failed and collapsed to the north (Photo 83), as the prelude to a foreseen, well-monitored but still unexpected eruption in terms of the initial process. The giant landslide that occurred right after an earthquake on the morning of 18 May (see Sect. 2.4.1) was a real shock to volcanologists, as such events were largely unknown until then (except for Bandai-san, Honshū, Japan 1888: Sekiya and Kikuchi 1889; and Bezymianny, Kamchatka, Russia 1956: Gorshkov 1959). As a result of this lack of knowledge, the scenario, size, and frequency of volcanic debris avalanches from composite cones had been underestimated prior to the eruption of Mount St. Helens. However, in the years that followed, not only were the 1980 eruption products from Mount St. Helens analysed in extraordinary detail (first published immediately afterwards in a US Geological Survey Professional Paper: Lipman and Mullineaux 1981), but scientists also discovered previous debris-avalanche deposits from Mount St. Helens, as well as at hundreds of composite and shield volcanoes worldwide (Table 1 presents a selection of the largest deposits), and detailed the characteristics of the initiation and transport mechanisms in terrestrial and submarine settings (e.g., Crandell et al. 1984; Siebert 1984; Francis and Wells 1988; Moore et al. 1994; McGuire et al. 1996; Keating and McGuire 2000). In recent years, the literature on debris avalanches has grown exponentially, also resulting in fundamental summary studies and books (Ui et al. 2000; Inokuchi 2006; Ponomareva et al. 2006; van Wyk de Vries and Davies 2015; Yoshida 2016) as well as a volume on the topic in the Advances in Volcanology (Roverato et al. 2021).

2.1 General Features

The size and height of any volcano is constrained by the supply of volcanic material, the balance between the pressure in the magma reservoir and that of the crust, the physical properties of the rocks, and the load-bearing capacity of the underlying substrate. As a growing volcano is commonly unstable and contains structural weaknesses, it is prone to either slow sliding or, beyond a certain threshold, sudden flank failure (van Wyk de Vries and Davies 2015). In the latter case, a volcanic debris avalanche is generated. These avalanches are slope mass movements which usually remove a much greater rock mass than rockfalls, debris flows, and non-volcanic debris avalanches (e.g., Hsü 1975; Voight 1978): the total volume of material removed from large composite cones can reach tens of km^3 and hundreds of km^3 from shield volcanoes in submarine setting (Table 1).

Flank failure plays a crucial role in determining the construction threshold of an edifice, thereby hindering volcano growth. A considerable number of well-studied volcanoes have collapsed repeatedly over their lifetime (see Sect. 2.3). For instance, Taranaki (former Mount Egmont) volcano, New Zealand, experienced fourteen failures over its lifetime of c. 130 ka; Shiveluch, Kamchatka (Russia), thirteen failures; and Colima (Mexico) eight failures, respectively, over the past c. 30 ka. Whereas volcano construction occurs on a millennial to tens of millennia scale, the flank failure is a very

Table 1 The largest known debris avalanches (in chronologic order, going back in time)

Volcano	Location	Type	Age	Volume (km^3)	Runout (km)	Trigger (type)	References	Comment
Kamen	Kam-chatka, Russia	Composite	1.2 ka	4–6	30	Earth-quake(s)		
Fuji	Japan	Composite	2.9 ka	1.76	24	–	Miyaji et al. (2004)	
Chokai	Japan	Composite	2.6 ka	2.85	26	M	Yoshida (2016)	
Galunggung	Indonesia	Composite	4.2 ka	23	20	M	Bronto (1989)	
Rainier	USA	Composite	5.6 ka	3.8	120	M	Vallance and Scott (1997) and John et al. (2008)	Largest of 10 debris avalanches
Socompa	Chile	Composite	7.2 ka	36	40	M	Van Wyk de Vries et al. (2001) and Kelfoun and Druitt (2005)	Of which 11 km^3 represents toreva blocks
Meru	Tanzania	Composite	9 ka	20	49	–	Delcamp et al. (2016)	
Planchon-Peteroa	Chile	Composite	11 ka	16	95	Mixed	Tormey (2010)	
Nevado de Colima	Mexico	Composite	18.5 ka	35	30	–	Cortés et al. (2019)	Largest of at least 8 debris avalanches in the past c. 30 kyr
Taranaki	New Zealand	Composite	24.8 ka	7.5	30	M	Zernack et al. (2011) and Roverato et al. (2015)	Largest of 14 debris avalanches in the past c. 130 kyr
Popocatépetl	Mexico	Composite	27.8 ka	10.4	72	M	Siebe et al. (2017)	Youngest of Late Pleistocene debris avalanches
Shiveluch	Kam-chatka, Russia	Composite	10–30 ka	30	32	–	Ponomareva et al. (1998, 2006) and Belousov et al. (1999)	Most likely c. 30 ka; it was the largest of 13 subsequent debris avalanches
Avachinsky	Kam-chatka, Russia	Composite	30 ka	16–20	>10	–	Ponomareva et al. (2006)	
Chimborazo	Ecuador	Composite	60–65 ka	10–12	>35	–	Samaniego et al. (2012)	
El Hierro	Canary Isl., Spain	Shield	40–90 ka	<250	65	–		
Tenerife (Icod)	Canary Isl., Spain	Shield	170 ka	80–150	105	–		
Akagi	Japan	Composite	220 ka	4–8	23.5	–	Yoshida (2016)	
Shasta	USA	Composite	300–400 ka	>45	50	–	Crandell (1989)	Covered an area of>675 km^2

Trigger (type) of collapse:
M: magmatic eruption; Mixed: magmatic and hydrothermal activity and minor eruption; –: unknown

short-lived event which destroys a large portion of the edifice in minutes and produces a volcanic debris avalanche. In the case of repetitive avalanches, the presence of a new cone implies that the volcano has been able to 'heal' following a previous flank failure, as shown for many composite cones in Chap. 5. Since a new cone and its successions tend to bury the preceding erupted products, including the fragmental material of a slope failure, past debris avalanches can only be inferred from exposed deposits, which may not always be the case (Ponomareva et al. 2006; Zernack and Procter 2021).

Recent research has identified hundreds of debris avalanches in historic and prehistoric times (for global compilations, see e.g., Bernard et al. 2008; Dufresne et al. 2008, summarised by Dufresne et al. 2021). Collapse events <0.1 km^3 in volume have occurred on average 5–6 times a century since 1500 AD/CE (Siebert and Roverato 2021; Siebert and Reid 2023). Table 2 contains selected historical debris avalanches ≥ 0.3 km^3 in volume, with their major triggering processes. More than three quarters of the historical flank failures have been associated with magmatic or phreatomagmatic events, and the remainder occurred in the absence of eruptive activity (Siebert and Roverato 2021). It is important to note, however, that going back in time, especially to prehistoric times, there is likely to be much less evidence of magmatic activity.

2.2 Causes of Debris Avalanches

Several factors may be responsible for the instability and failure of growing volcanoes (McGuire et al. 1996; Elsworth and Day 1999; Keating and McGuire 2000; van Wyk de Vries and Davies 2015; Acocella 2021; Roverato et al. 2021; Siebert and Reid 2023). From a mechanical point of view, flank failure occurs when driving forces on the potential sliding surface exceed the resisting forces (Siebert and Reid 2023). The factors and forces can be of both volcanic and non-volcanic origin. Importantly, non-volcanic landslides are comparable to the numerous smaller- to medium-sized flank failures on volcanoes, but they do not reach the large volumes that—much less frequently—may occur on the latter.

Siebert and Reid (2023) plotted failure volumes on volcanoes ≥ 0.01 km^3 from 1500 AD/CE, dividing them according to the trigger mechanism: association with magmatic eruption, phreatic eruption, and no eruption. They compared these volumes with other large historical (from 1900) non-volcanic landslides with volumes ≥ 0.01 km^3 (Fig. 5). Non-volcanic landslides can have many different causes, e.g. earthquakes, steep slopes, unstable rock successions and structural weaknesses. In the case of relatively large landslides, earthquakes were very often the trigger (for the Usoi landslides in Tajikistan and the Diexi landslides in China with $M=7$, as shown in Fig. 5: Siebert and Reid 2023). Other general causes may include large rainstorms, such as typhoons or hurricanes, or snowmelt, but these are not always identifiable from the deposits.

On volcanoes, being sizeable mountains, all of these non-volcanic causes can occur and they can of course be accompanied by volcanocentric triggers of collapse. Overall, the factors and causes of failure can be divided into two broad groups depicted in Fig. 6: (A) instability factors and conditions that weaken the structure and material of the volcanic edifice, usually over a long period of time; (B) short-lived triggering processes that directly cause the debris avalanche.

(A) Instability factors which may be inherently correlated (Reid et al. 2001; Reid 2004; van Wyk de Vries and Davies 2015; Peterson et al. 2021; Roberti et al. 2021; Roverato et al. 2021; Siebert and Reid 2023) include: (1) active (basement) tectonics, e.g. subsidence, uplift, and faulting; (2) edifice loading, in particular accumulation of loose eruptive products; (3) possible gravitational spreading related to tectonic processes or accumulation of structurally weak eruptive products; (4) sloping or weak basement, in part responsible for (5) asymmetric edifice growth and/or locally oversteepened slopes; (6) strength reduction and pore pressure increase by hydrothermal processes, associated with rock alteration, (7) uneven distribution of groundwater or hydrothermal fluid saturation of rocks within the edifice, (8) long-term climate implications such as glaciation, heavy rainstorms, and sea-level changes, partly resulting in (9) asymmetric weakening by erosion or (in the case of coastal setting) wave abrasion of the volcanic edifice.

Table 2 Historical debris avalanches ≥ 0.3 km^3 since 1500 AD/CE in chronologic order, going back in time (modified after Siebert and Roverato 2021). Volumes in parentheses are estimates of collapsed edifice segment volume

Volcano	Location	Type	Year	Volume (km^3)	Area (km^2)	Runout (km)	Trigger (type)	VEI	References
Krakatau	Indonesia	Composite	2018	0.3	–	1[a]	M	3	Priyanto et al. (2021)
Fernandina	Galápagos	Shield	1988	0.9	10	2[a]	M	2	Chadwick et al. (1991)
Mt. St. Helens	Cascades	Composite	1980	2.5	64	24	Mb	5	Voight et al. (1981)
Shiveluch	Kamchatka	Composite	1964	1.5	98	12	M	4	Gorshkov and Dubik (1970)
Bezymianny	Kamchatka	Lava dome complex	1956	0.8	30	18	Mb	5	Gorshkov (1959)
Kharimkotan	Kurile Islands	Lava dome complex	1933	0.5	>20	<6[b]	M	5	Belousova and Belousov (1995)
Ritter Island	Melanesia	Composite	1888	(2.4–4.2)	100	?	M	2?	Johnson (1987)
Bandai	Japan	Composite	1888	0.5	34	11	P	4	Yonechi et al. (1989)
Krakatau	Indonesia	Composite	1883	(3.8)	–	–	M	5	Camus et al. (1992)
Mt. Augustine	Alaska, USA	Composite	1883	0.3	21	7[b]	M	4	Siebert et al. (1995)
Sinarka	Kurile Islands	Composite	1878	0.5?	–	?	M	4	Belousova and Belousov (2011)
Suwanosejima	Japan	Composite	1813	>1.0	–	?	M	4	Shimano et al. (2013)
Tutupaca	Peru	Composite	1802?	0.7?	–	?	M	4?	Samaniego et al. (2015)
Unzen	Japan	Lava dome complex	1792	0.34	15	6	–[c]	–	Ota (1969)
Oshima-Ōshima	Japan	Composite	1741	2.5	69	23[b]	M	4	Satake and Kato (2001)
Callaqui	Chile	Composite	1630?	<0.7	–	14	M	?	Polanco and Naranjo (2008)
Komagatake	Japan	Composite	1640	1.55	126	20	Mb	5	Furukawa et al. (2008)
Ruiz	Colombia	Composite	1595	>0.5	–	?	M	4	Thouret et al. (1990)
Mt. Augustine	Alaska, USA	Lava dome complex	1540 ± 110	0.5	30	<15[a]	Mb	4?	Siebert et al. (1995)

[a] Intracaldera debris avalanche
[b] Volcanic island; debris avalanche flowed into sea
[c] At the Unzen volcano lava dome complex in 1792 there was no eruption at the collapsing Mayuyama dome although there was at neighbouring Fugendake
Triggering mechanism (type of collapse):
M: magmatic eruption; Mb: magmatic eruption with lateral blast; P: phreatic eruption. VEI: Volcanic Explosivity Index

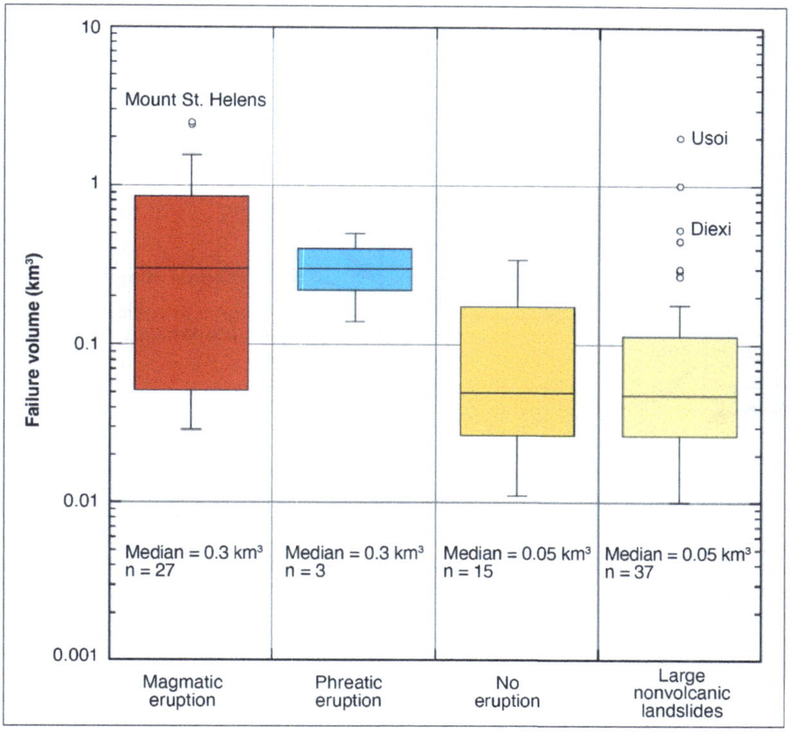

Fig. 5 Box plot of flank failure volumes ≥ 0.01 km^3 since 1500 AD/CE categorised by event type, including failures associated with magmatic eruptions, phreatic eruptions, and no eruption (Siebert and Reid 2023). Volumes of other large (≥ 0.02 km^3), historical (since 1900) non-volcanic landslides are shown for comparison (data from Evans 2006). Potentially more frequent but smaller failures of < 0.01 km^3 are not included. Boxes in each category represent 50% of the data (with median line), horizontal lines represent the upper and lower quartiles, and circles are outliers (with event names)

A growing volcanic edifice incorporates structural weaknesses from the beginning. The accumulation of eruptive products causes topographic loading, which may destabilize the edifice through deformation, increased seismicity, and hydrothermal alteration. Volcanoes that contain large volumes of water, e.g., those in humid climates or accompanied by glaciers, may have increased susceptibility to flank failure (van Wyk de Vries and Davies 2015). Climate effects, e.g., changing frequency of rainfall, extreme weather events, thawing permafrost, can also play a role in (asymmetric) edifice destabilization (Roberti et al. 2021). These factors imply that the failure of a bigger or smaller portion of the flank is not necessarily related to an eruption or magmatic activity in general. Moreover it can also occur during dormant periods, or even on an extinct volcano (Siebert 1992; van Wyk de Vries and Francis 1997).

(B) A number of specific, short-term triggers can be identified that may cause the collapse or failure of a volcano flank. Importantly, these may be independent of the (long-term) causes that lead to edifice instability (Roverato et al. 2021). Triggers may include (1) sudden changes to the stress field, (2) seismic activity, i.e., earthquakes, tremors, (3) magmatic intrusions, e.g., cryptodome or dyke emplacement within the edifice, (4) magmatic eruption, and (5) phreatic eruption.

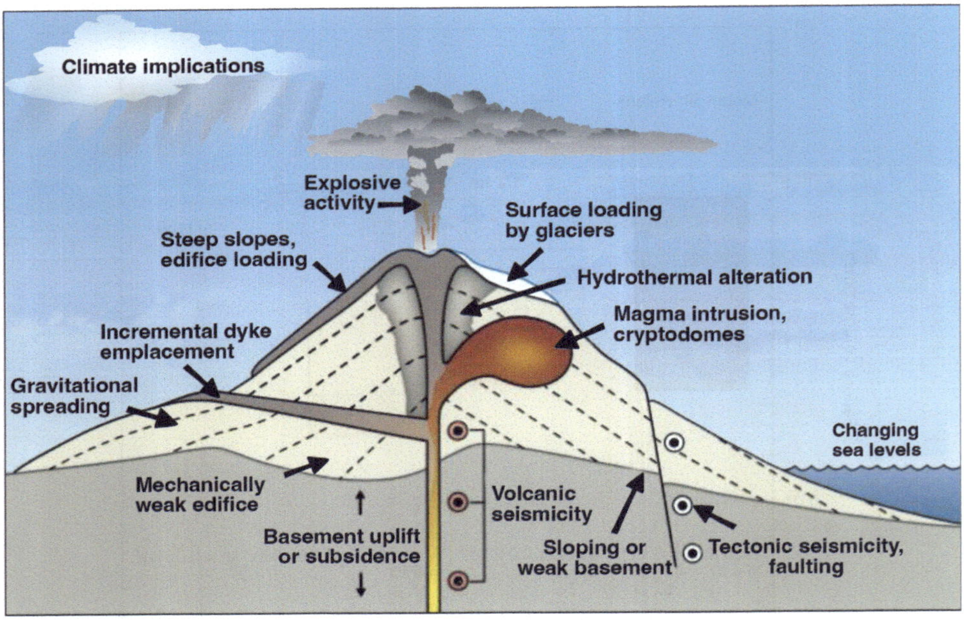

Fig. 6 Cross-section of a volcano illustrating instability factors and triggers. Modified after McGuire et al. (1996), Siebert (2002) and Roverato et al. (2021)

However, identifying a trigger mechanism is not easy, especially since multiple triggers may be present which can be often related. For example, the intrusion of a cryptodome can deform and overload an edifice asymmetrically, leading to brecciation of the deforming rock mass, and subsequently hydrothermal fluids can circulate and reduce rock strength (van Wyk de Vries and Davies 2015; Roverato et al. 2021). Sudden magmatic loading before an eruption, like that during lava dome growth, may lead to smaller-scale flank failure, as occurred at Soufrière Hills, Montserrat in 2009 (Sparks et al. 2002), or Shiveluch, Kamchatka (Russia: Ponomareva et al. 1998, 2006) and Mount Augustine, Alaska (USA: Begét and Kienle 1992; Waitt and Begét 2009). The gravitational lateral collapse and debris avalanche at Bandai-san (Honshū, Japan: Sect. 2.4.2) in 1988 occurred due to decompression in the hydrothermal system, associated with a phreatic blast (Nakamura 1978). This was the latest of a dozen previous flank failure events on the volcano, many of which were connected to magmatic activity (Chiba and Kimura 2001). Krakatau (Sunda Strait, Indonesia) is a good example of a failure scenario caused by magmatic activity. Apart from its 1883 explosive eruption associated with caldera collapse (Somma-type caldera: Chap. 5), the new Anak Krakatau intracaldera cone was truncated laterally in 2018 after a period of intense magmatic activity, triggered by a VEI 3 explosive eruption (Hunt et al. 2021; Priyanto et al. 2021; see Chap. 9). This kind of lateral edifice collapse in a submarine setting often results in tsunami, as was the case with Hokkaido-Komagatake in 1640, Unzen-Mayujama in 1792, Ritter Island (Papua New Guinea) in 1888, and Kharimkotan in 1933 (for a summary, see Paris et al. 2014).

2.3 Frequency of Debris Avalanches: Cyclicity of Edifice Growth and Destruction

Many volcanoes fail or collapse repeatedly. This feature is not necessarily due to edifice overgrowth, as exemplified by the relatively small Mount Augustine composite cone with summit lava domes (Alaska, USA) with a volume of 8 km^3 (Grosse et al. 2012) which shows cyclic lateral collapse. Mount Augustine has been active since 2200 yrs BP, marked by seven major eruptions (Begét and Kienle 1992; Waitt and Begét 2009). The eruptive record shows at least 12 small-scale (0.1–0.5 km^3) debris avalanches with an average recurrence interval of 150–200 yrs. The flank failures typically remove 5–10% of the edifice, and the highly mobile avalanches have travelled up to 18 km including the offshore section (Siebert et al. 1995; see Sect. 4.2).

Overall, the tendency to collapse is the result of a combination of the factors listed above, which may reinforce each other. Failure recurrence times worldwide vary greatly: many active lava domes like Mount Augustine tend to produce frequent but small-volume collapses (see Chap. 2), in contrast to larger composite volcanoes that show lower-frequency, larger-scale collapses that may occur in longer-term cycles (Zernack and Procter 2021). Sooner or later, any large edifice that grows to a certain height will be subjected to flank failure: for instance, 75% of Andean composite volcanoes above 2500 m in height have experienced at least one lateral collapse (Francis and Hawkesworth 1994). Recurrence times on Earth vary from c. 1000 yrs to 200–300 kyrs (Zernack and Procter 2021).

In Japan, volcanic debris avalanches have been identified for half of the active composite volcanoes and lava domes. Age distribution of debris-avalanche deposits suggests that, on average, one per century occurs (Ui et al. 1986; Inokuchi 1988). Within this, mega-collapses with a volume of >0.1 km^3 are estimated to occur every 1000–2000 yrs (Yoshida 2016). Repeatedly collapsing volcanoes include Bandai, (Hokkaido)-Komagatake, Asama, Yatsugatake, and Unzen (see Sect. 2.4).

In Kamchatka, out of the 30 recently active volcanoes, at least eighteen have experienced flank collapses, some of them repeated collapses, such as Shiveluch, Avachinsky, Kamen, Kambalny (Fig. 7). Shiveluch and Avachinsky also produced the largest debris avalanches at 20–30 km^3 in volume (Table 1; Ponomareva et al. 1998, 2006; Belousov et al. 1999). Two well-known volcanoes, Kamen and Bezymianny, are interestingly juxtaposed: the latter, currently active lava dome complex (Sect. 2.4.1) is much lower (2886 m) than its extinct sister (4585 m), and has been built on huge Late Pleistocene megablocks originating from the repeatedly collapsing neighbouring volcano, Kamen. These megablocks are part of what is thought to be a huge debris-avalanche deposit 8–10 km^3 in volume; as it is buried under subsequent eruptive deposits, its dimensions are difficult to calculate precisely (Ponomareva et al 2006). Shiveluch is one of the largest composite edifices (see Chap. 5). Late Pleistocene Old Shiveluch was truncated by a flank failure, the collapse scar of which hosts the overlapping lava domes of Young Shiveluch. The latter is highly active, and has also produced a number of debris avalanches (the last one in 1964: Belousov et al. 1999), with a subsequent, new lava dome in the resultant scar. Multiple collapses of Shiveluch—with a recurrence rate of c. 800 yrs on average—have been explained by a high magma supply rate and related, repetitive lava dome formation (Ponomareva et al. 1998).

Shield volcanoes are characterised by long-term sliding and occasional (mostly submarine) debris avalanches, as presented in Chap. 4. A number of Hawai'ian or Canary Islands shields also display repeated collapses, however these are prolonged events and occur in multiple stages (Moscardelli and Woods 2016; Hunt et al. 2018; León et al. 2019). Although much less is known about the details of these processes, around the Hawai'ian Islands, the volume of the largest debris avalanches can exceed 5000 km^3, and their underwater runout 200 km (Moore et al. 1994; Morgan et al. 2009). The Big Island of Hawai'i has had nine avalanches in the past <700 krs, which equates to a recurrence interval of 70–80 kyrs (Moore et al. 1994). In the Canary Islands, cyclic behaviour of the island shield volcanoes consists of a long-term edifice weakening due to overloading and minor flank eruptions, then retrogressive, multiple flank failures triggered by significant eruptive activity, often associated with caldera collapse (Hunt et al. 2018).

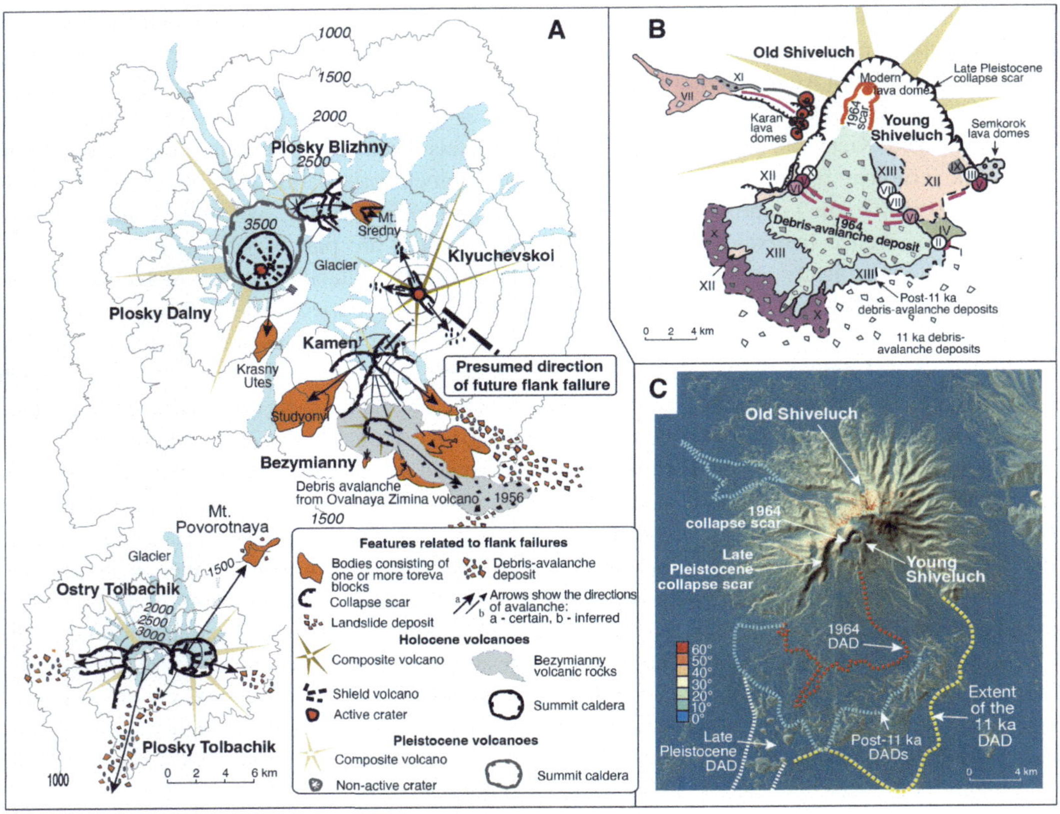

Fig. 7 Volcano-geomorphological sketch of selected volcanoes in Kamchatka (Russia) showing debris-avalanche deposits (DADs) (after Ponomareva et al. 1998, 2006, with updates by Vera Ponomareva, pers. comm.). **A** Klyuchevskoi, Plosky (or Ushkovsky), and Tolbachik volcano; blue markings indicate glaciers. **B** Shiveluch volcano; individual DADs are coloured differently, and marked with either Latin numbers (exposed at the surface) or encircled Latin numbers (where they are identified at the outcrop). Dashed purple lines indicate proposed boundaries of aerial extent of V, VI and X DADs; post-11 ka outline delineates a field within which all the DADs deposited between 11 ka and 1964 were found. **c** Slope map of Shiveluch based on SRTM DEM, with main DAD boundaries also indicated

Overall, edifice growth and destruction may be coupled with magmatic processes in the subvolcanic plumbing system (Watt 2019). This has been proven by post-failure behaviour observed at several terrestrial volcanoes (Zernack and Procter 2021). In many cases, cone re-build yields eruptive products with more mafic compositions, often accompanied by higher eruption rates, before a more evolved upper crustal magma reservoir is established (e.g., Parinacota, Chile: Hora et al. 2007; Pico de Teide, Canary Islands, Spain: Boulesteix et al. 2012; Stromboli, Aeolian Islands, Italy: Vezzoli et al. 2014).

2.4 Main Types of Lateral Flank Failures

Extensive studies following the 1980 eruption of Mount St. Helens highlighted the fact that it differed from the collapse at Bandai-san in 1888, in that there was no magmatic activity at the latter. With regard to the lessons learned from Bandai (Moriya 1980), Siebert et al. (1987) proposed a threefold division which is still in use today. The main flank failure types are *Bezymianny*, *Bandai*, and *Unzen/Ontake*, which, by definition, refer to particular events that took place on these volcanoes. However, based on the large number of triggers (as seen above), it can often be difficult to identify a single initiation mechanism or trigger to edifice collapse on failed volcanoes, thus the threefold grouping seems an oversimplification (Roverato et al. 2021). In addition, submarine flank failures are not included in this division. Nevertheless, it is worth summarising the specific scenarios and characteristics of the three type examples as case studies.

2.4.1 Bezymianny Type

This is associated with volcanic activity, usually a magmatic explosive eruption. Based on the Mount St. Helens (1980) collapse, Belousov and Belousova (1998) reinterpreted the 1956 eruption of Bezymianny (Kamchatka) (first described by Gorshkov 1959) defining it as the 'prototype'. By definition, material from fresh magma is always found in the deposit of Bezymianny-type volcanic debris avalanches, due to their close association with explosive activity, often including pyroclastic density currents (PDCs) and, in particular, blast deposits. Bezymianny collapsed in 1956 during a VEI 5 eruption (Gorshkov 1959). Prior to the failure, a lava dome existed within a smaller-scale older scar. The catastrophic, explosive eruption on 30 March 1956 removed c. 0.8 km^3 of the eastern flank. Subsequent, intense, episodic lava-dome growth, both endo- and exogenous, often associated with VEI 3 explosive activity, has largely filled the horseshoe scar over the past few decades (Shevchenko et al. 2020), the rim of which is now lower than the growing lava dome inside.

The eruption of Mount St. Helens is far better known. The reawakening of this 2950 m-high volcano, after a 123 year-long quiescence, was indicated by earthquakes on 20 March 1980, followed by phreatic eruptions (Lipman and Mullineaux 1981). In early April, the northern side of the mountain began to bulge, then a gradually increasing cryptodome swelled to 120 m horizontally; by mid-May, the bulge reached a height of 150 m above the previous slope, with the Goat Rocks dome at the base of the bulge having moved northward by 106 m (Moore and Albee 1981). At 8:30 in the morning of May 18, a catastrophic (VEI 5) eruption triggered by an earthquake of $M = 5.1$ began with the impressive failure of the northern part of the mountain in three pulses (Fig. 8A–C). Rapid unloading of the hydrothermal-magmatic system in the upper edifice by failure of the first slide block exposed the cryptodome (Siebert and Reid 2023), and resulted in a directed volcanic explosion (*lateral* or *directed blast:* Hoblitt et al. 1981; Belousov et al. 2007) that devastated an area of 600 km^2. The huge megablocks—parts of the collapsed flank—moved at a speed of 50–70 m/s and quickly split into smaller pieces, settling in the upper reaches of Toutle River where they were emplaced as a debris-avalanche deposit. Some of the megablocks representing the original composite volcano sequence can still be seen in the surrounding topography. The total volume of the debris avalanche, based on the missing material from the resulting horseshoe-shaped scar (Fig. 8D and E, Photo 83), is 2.5 km^3. Moments after the flank failure, an open-vent stage began, producing a Plinian eruption column that rose to an altitude of 30 km (Sparks et al. 1986). Collapse of this column resulted in pumiceous PDCs creating a wide apron on the northern front of the volcano, now called Pumice Plain. The first seconds of the flank failure and the generation of the lateral blast were captured by Gary Rosenquist, who camped with his companions 11 km to the north-east. His shots were later turned into a spectacular movie (https://www.youtube.com/watch?v=UNlP9TGZOMI).

The eruption terminated in the evening. Despite the eruption having been forecast, a total of 57 people were killed (Waitt 2015), including volcanologist David A. Johnston; from his position 8 km away his last words were broadcast to his colleagues: "Vancouver! Vancouver! This is it!". Johnston Ridge Observatory was built on that particular site in his memory. A dacite lava dome began to grow in the horseshoe scar of Mount St. Helens in the same year, lasting until 1986, and the same process repeated between 2004 and 2008, affecting

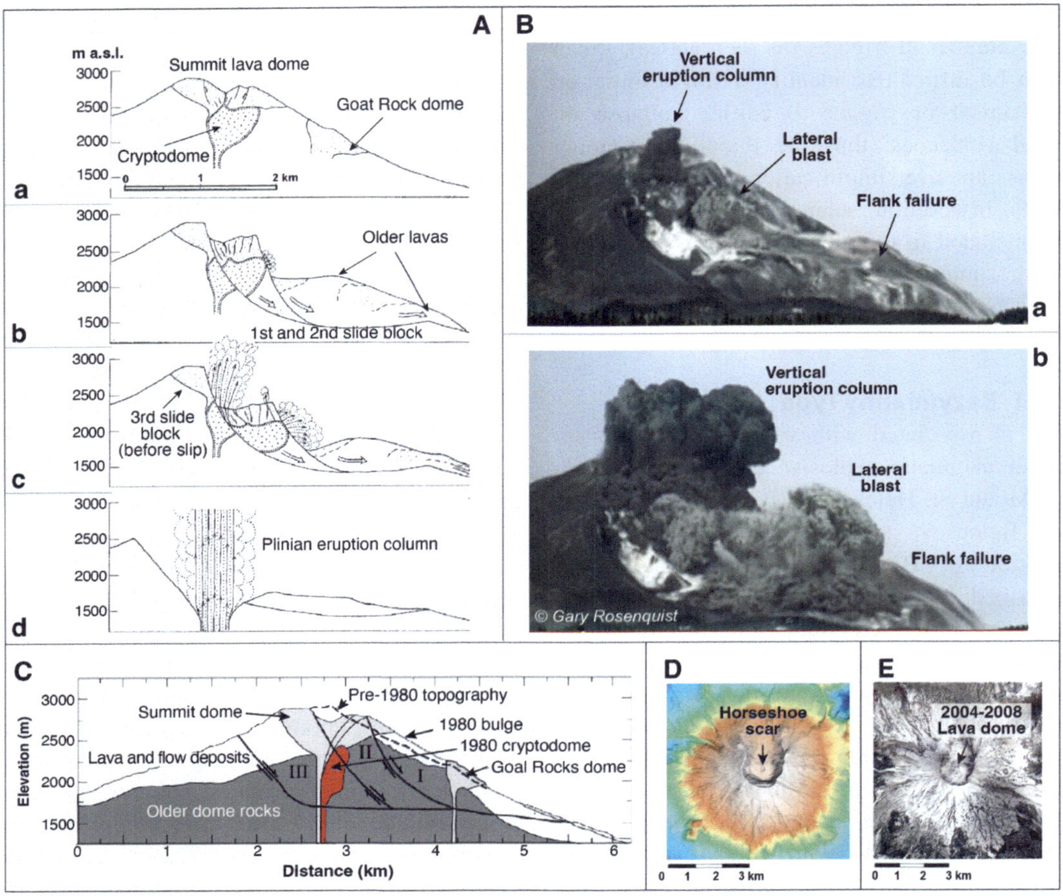

Fig. 8 The 18 May 1980 flank failure of Mount St. Helens (Washington, USA). **A** Changes in profile of the volcano before eruption, 20 and 30 s after landslide, and in the Plinian stage, respectively (after Moore and Albee 1981). **B** Interpretative photos from the Rosenquist photographs (see text) after Walter (2011). **C** Cross section showing north flank bulge, extent of summit dome (light grey) and older domes (grey), inferred extent of 1980 cryptodome (red), and failure surfaces of the three landslide blocks (I, II, and III) (Siebert and Reid 2023). **D** Shaded and coloured image of SRTM DEM of Mount St. Helens's horseshoe scar after the 1980 flank failure (Székely and Karátson 2004); **E** Landsat 8 satellite photo 17 April 2020, credit: NASA Landsat Image Gallery (https://landsat.visibleearth.nasa.gov)

the shape of the internal glacier (Sherrod et al. 2008; Gabrielli et al. 2020; Fig. 8E). However, the dome has remained relatively small compared to Bezymianny.

2.4.2 Bandai Type

The dormant Bandai-san volcano, consisting of overlapping cones, collapsed on the northward flank in 1888 (Fig. 9). Previously, its largest flank failure occurred ca. 40 ka; affecting 30% of the volcano area, creating a horseshoe-shaped scar on the southwestern side and producing the Okinajima debris-avalanche (Yoshida 2016). Within the scar, Ōbandai (Great Bandai) cone was established. After several historical small-scale phreatic eruptions, the type-defining failure event, referred to as Ura-Bandai, began with mild earthquakes in July 1888. Then two strong (M = 5) successive earthquakes shook the area around 7:30 and 7:44 am on 15 July, and 15–20 small phreatic explosions occurred on the north flank of

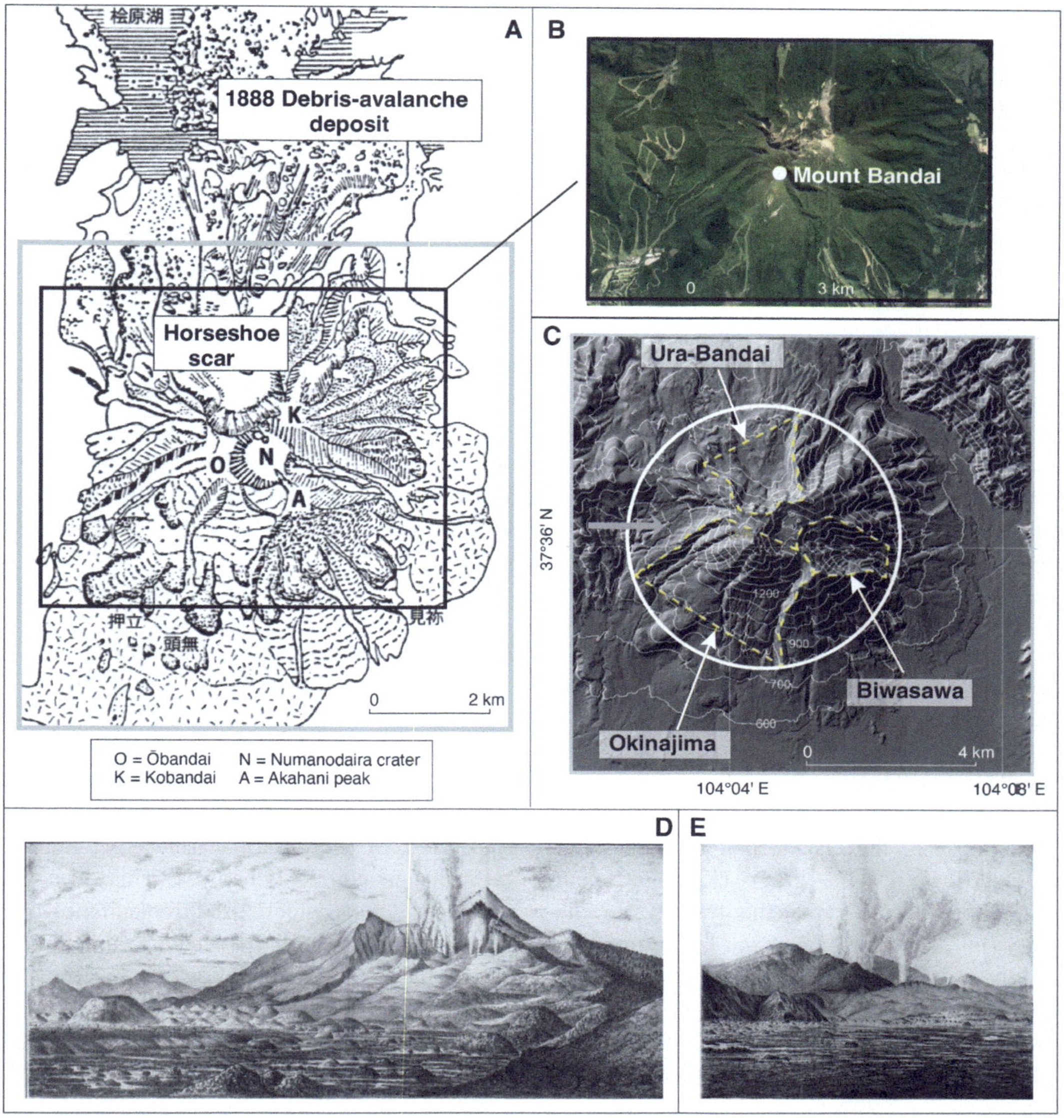

Fig. 9 Geomorphic features of the 1888 Bandai flank failure. **A** Geomorphological map (Moriya 1988). **B** Google Earth image (2023). **C** Scars of three successive sector collapses (marked with dotted yellow lines): Okinajima (40 ka), Biwazawa (2500 BP) and Ura-Bandai (1888 AD/CE), their maximum extents outlined with a white circle, draped over shaded relief image (Yoshida 2016). **D, E** Contemporary drawings from north (left) and northeast (right) of Bandai, depicting the horseshoe scar and post-collapse fumaroles (Sekiya and Kikuchi 1889)

another, smaller cone called Ko-Bandai within one minute, the last one projected almost horizontally northward (Sekiya and Kikuchi 1889). Subsequently, the northern flank and the summit of that cone collapsed in three pulses comprising a total volume of 1.5 km^3, and a debris avalanche flowed downvalley to a distance of 11 km, killing 461 people. Based on the analysis of contemporary photographs and sketches, Yonechi (1987) proposed that the failure events were not instantaneous, but took place over a period of almost one and a half hours. After the first collapse, a 5 km-high eruption column emerged, and phreatic explosions continued into the afternoon (ca. 4 pm; Sekiya and Kikuchi 1889; Yamamoto et al. 1999).

To sum up, the Ura-Bandai collapse event produced only phreatic explosions rather than fresh magma. The failure of Kobandai can be explained by movements and/or intrusion of deep magma (Moriya 1988; Siebert et al. 1987), which might have increased the temperature in the shallow hydrothermal system of the volcano, thus increasing the pore water pressure in the rocks and weakening the volcanic edifice (Reid 2004). In turn, the causal role of phreatic explosions as a direct trigger has already been proposed by Sekiya and Kikuchi (1889), who studied the field a couple of days after the event and published a comprehensive report in English: "it is manifest that the immediate cause of the eruption was the sudden expansion of steam pent up within the mountain". Certainly, this kind of event represents a greater hazard due to the paucity of precursors.

2.4.3 Unzen/Ontake Type

Going back in time another hundred years, the collapse of the coastal part of the Unzen-Mayujama lava dome group (Kyushu, Japan) in 1792 caused one of Japan's most severe disasters (Fig. 10). The Mayujama lava dome rises slightly away from the main dome mass close to the densely populated coast (e.g., Shimabara town). In late 1791, earthquake activity began at the central Fugendake dome and continued in February 1792 with smaller explosive eruptions and lava flows. In April of that year, earthquakes led to rockfalls on Mayujama, a few km to the east, then to the collapse of the eastern flank on the evening of 21 May 1792 (Hoshizumi et al. 1999). Unlike the two other flank failure types, the slope failure removed only a thin layer of rock which, sliding into the sea, formed small islands, comprising a total volume of 0.34 km^3. The debris avalanche generated a tsunami in the sea, causing more than 14,500 deaths in the surrounding coastal area—it was the so-called Shimabara disaster. The direct cause of the landslide was an earthquake and the nearby eruption. The volcano must also have been weakened by movements of groundwater and hydrothermal fluids, as a significant amount of water erupted from the scar left by the summit of the Mayujama lava dome (Hoshizumi et al. 1999).

The other Japanese type example, the active Ontake (Honshū), experienced a small, shallow flank slide in 1984 (Mimura et al. 1988; Voight and Sousa 1994). Ontake is a long-lived compound volcano, which started to grow around 680 ka, with predominantly phreatic explosions in historic times. The 1984 failure event, which was triggered by an earthquake (M = 6.4) without an eruption, removed only 0.034 km^3 from the mid-flank (Yoshida 2016). The failure transformed into a debris avalanche then a lahar that swept down the valleys south and east of the volcano.

2.5 Characteristics of Volcanic Debris-Avalanche Deposits

2.5.1 Geomorphic Characteristics

The most distinctive geomorphic feature resulting from the flank failure is a *horseshoe scar* or *landslide scar* (Fig. 11) that truncates the original volcanic edifice (Chap. 5). This term has become the most widely used, replacing diverse terms like sector collapse caldera, amphitheatre, U-shaped or avalanche caldera, sector collapse scar, etc. (Moriya 1980; Voight et al. 1981; Siebert 1984; Belousov et al. 1999; van Wyk de Vries and Delcamp 2015). The commonly U-shaped, open landform has a semicircular headwall and two sidewalls. The angle between the linear ridges of sidewalls is usually between 0° (if they are parallel) to 70–90° (e.g. Socompa: Photo 84). In general, in the case when the summit is involved in the failure, the angle tends to be greater, and for repeated debris avalanches, irregular or undulating ridges can also be created (e.g., El Hierro and Tenerife, Canary Islands: Carracedo 1999). As a result, the specific, open, horseshoe-shaped depression is different in morphology from collapse calderas at shield or composite volcanoes, which are commonly closed (Siebert 1984), as well as from

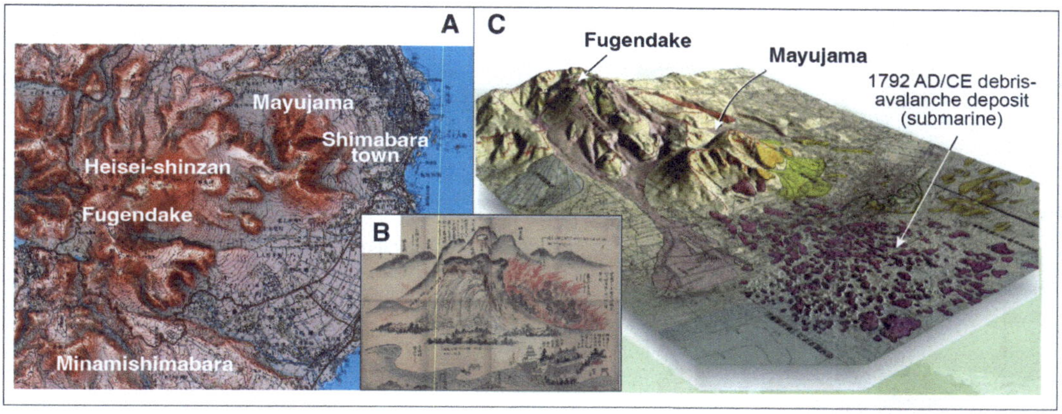

Fig. 10 Geomorphic features of the 1792 Unzen-Mayujama flank failure. **A** Red Relief Image (RR) (Chiba 2018) showing the central lava dome complex (Fugendake and Heisei-Shinzan) and the failed Mayujama dome close to the coast. **B** Contemporary woodcut (credit: http://wwweic.eri.utokyo.ac.jp/dl/meta_pub/G0000002erilib_L000314). **C** Oblique shaded relief image merged with the geomorphological map of Unzen-Mayujama (Inoue 1999)

erosionally modified calderas that are usually breached in one direction (Karátson et al. 1999, see Chap. 8). In the latter case, when a flank failure is followed by long-term fluvial erosion, the originally wide, open mouth may become narrow due to the incision of the breaching valley, and thus the existence of the primary flank failure might become more difficult to detect or identify with certainty (e.g., Mount Taylor, New Mexico, USA: Meyer et al. 2014; Goff et al. 2019). At active volcanoes the horseshoe scar is the site of subsequent eruptions (Photos 85 and 86). Moreover, it is often filled with eruptive products, hence the flank failure event is reconstructed mainly from the deposit. As described above, the post-caldera lava dome growth at Mount St. Helens (2004–2008) filled up about one third of the scar interior; the Bezymiyanny scar has been almost completely filled during the past 30 years of lava dome activity; and the 9 ka DAD of Parinacota (Chile) is only shown by correlative deposits since the former scar has not been preserved (see Chap. 5).

Irregular hummocky terrain occurs below the horsehoe scar over several tens to hundreds of km^2, formed of a *debris-avalanche deposit (DAD)* (Fig. 11). This deposit is an unsorted, heterogenous material composed of preserved parts of the edifice and rip-up clasts derived from the transport path, creating a mixture of diverse lithologies (Bernard et al. 2021). One of the best studied DADs is that of Socompa, which contains c. 60% material from the substrate (van Wyk de Vries et al. 2001; Kelfoun and Druitt 2005). Selected large areas covered by DADs vary in volume by a factor of ten: Nevado de Colima (Mexico) 2200 km^2, Meru (Tanzania) 1250 km^2, Mawenzi (Tanzania) 1150 km^2, Rainier (USA) 5.6 ka, 550 km^2, Socompa (Chile/Argentina) 490 km^2, Shasta (California) 450 km^2, Shiveluch (Kamchatka, Russia) 350 km^2, and Taranaki (New Zealand) 250 km^2 (Siebert 1984; and references in Roverato et al. 2021). Close to the horseshoe scar there are oversized, toppled blocks termed *torevas* (Reiche 1937; Voight et al. 1981); and further away fragmented *hummocks* or *megablocks*, up to tens or hundreds of m in size (Photos 87 and 88), are emplaced by sliding, the latter often having suffered shattering during emplacement (Ui 1983; Mehl and Schmincke 1999; van Wyk de Vries and Davies 2015). In particular, the Socompa DAD has massive toreva blocks up to 3 km across within and at the base of the scar, which is partially filled now by post-collapse eruptive products. These torevas reached a distance of 5–7 km from the scar and blocked the outlet of the 9 km-long, open amphitheatre in

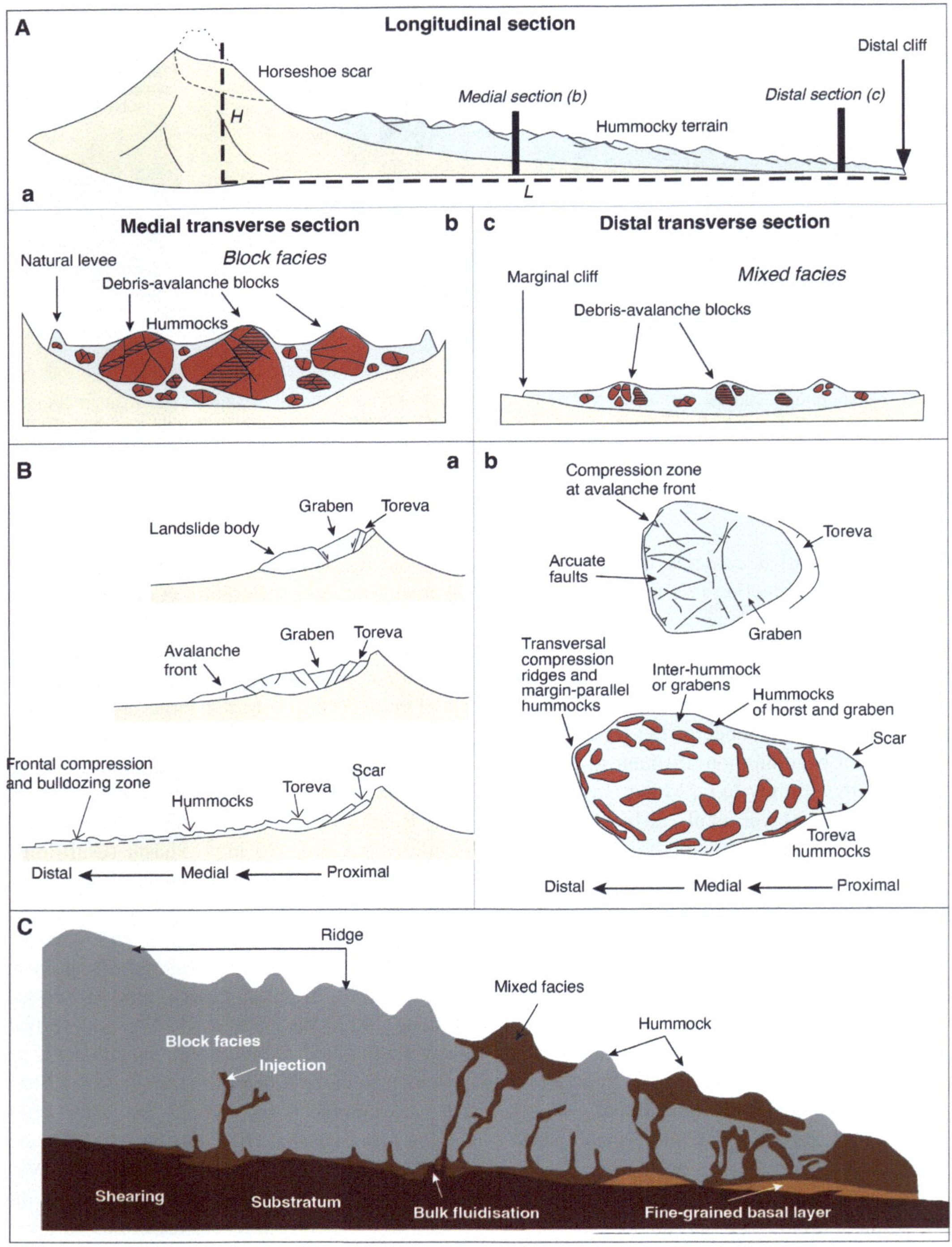

Fig. 11 Geomorphic and sedimentological characteristics of volcanic debris-avalanche deposits. **A** Longitudinal (a) and transversal (b, c) sections showing major geomorphic features, the two main facies types (block and mixed facies), and H (height) and L (length) of avalanche runout (modified from Ui et al. 2000). **B** Longitudinal sections (a) showing the early conditions in the landslide mass, and the development of structural and geomorphic elements during avalanche spreading; and plan view (b) of structural and geomorphic elements (after van Wyk de Vries and Davies 2015). **C** Debris avalanche transport and emplacement showing the injection of the substrate by shearing, fluidisation, and matrix injection, forming the mixed facies (after Bernard et al. 2008)

the last stage of the failure by counter-rotating (Wadge et al. 1995). Torevas and hummocks represent the initial failure blocks that disintegrated due to brittle deformation (Paguican et al. 2021). The longitudinal axis of the hummocks, which generally decreases in length downslope (Yoshida et al. 2012), sometimes points in the direction of transport, constrained by the shear developed in the mass flow (e.g., Shasta: Crandell et al. 1984; see also 2.6).

The surface of some DADs also exhibits longitudinal fissures, i.e., alternating ditches and ridges several metres deep or high and up to 1 km long (e.g., Socompa, Shiveluch). Ridges can also be perpendicular to the spreading (Valderrama et al. 2018). An important geomorphic marker along the margins of the DADs is a natural *levée* (Fig. 11A) formed by the high yield strength of the material. These 'debris dams' at Socompa are substantial walls 40 m high, while Unzen (Mayujama), Usu and Mount St. Helens have smaller levées of around 10 m (Wadge et al. 1995; Ui et al. 2000). Another interesting landform is the *transversal ridges* or *ramparts* (Fig. 11B) formed at the termination of the DAD, described from Shiveluch, Socompa, Parinacota, Jocotitlán etc.; they are formed by the pushing up of loose underlying sediments, e.g., sandstone or fluvial/lacustrine layers ('bulldozer facies'; Belousov et al. 1999; Ui et al. 2000). In other cases, the debris avalanche deposit terminates in steep slopes or scarps (e.g., Socompa: Wadge et al. 1995).

2.5.2 Sedimentological Characteristics

Two facies characterise a DAD based on the predominant grain size and degree of mixing: the block facies (A), and mixed facies (B) (Fig. 11A and C) (Crandell et al. 1984; Glicken 1991, 1996). Dividing the DAD into these two facies allows the failed but mostly intact material of the edifice to be distinguished from the newly formed material (Glicken 1996; Bernard et al. 2021).

(A) The *block facies,* typically forming a hummocky surface after emplacement, is composed of large (or mega)blocks that often preserve the original succession of the volcano and may consist of both lava and volcaniclastics rocks (Photo 89). *Megablocks* can be defined as being over >100 m in size (Mehl and Schmincke 1999), while *megaclast* is another term used for clasts which are >1 m diameter (Ui 1983; Glicken 1996). They are deposited mainly in the upper-intermediate part of the avalanche pathway. In large, ground-scouring or fast-propagating avalanches, non-volcanic material of the bedrock may also be mixed in with this facies (e.g., at Socompa: Francis and Self 1987; Davies et al. 2010 or the late Miocene Cantal DADs: Schneider and Fisher 1998; Nehlig et al. 2001). Therefore, *edifice-block facies* (EBF) was proposed to separate the material coming from the volcanic edifice, from *substratum-block facies* (SBF) containing the material coming from the substrate (Bernard et al. 2008). The blocks break into progressively smaller fragments as they move, but because the DAD is transported with a laminar plug flow mode without turbulence (see 2.6), the material from individual blocks usually remains together. Shattered, fractured blocks are often characterised by a *jigsaw-fit* or *jigsaw-crack* pattern (Ui 1983; Ui et al. 2000). That the blocks tend to stay together during transport is also shown by their very similar paleomagnetic directions (Mimura et al. 1982). Relatively soft blocks, on the other hand, may suffer deformation during transport. This is most common in volcaniclastics, but is also observed for lava blocks (e.g., Mt. Shasta: Crandell et al. 1984; Christiansen et al. 2017).

(B) The *mixed facies,* defined by Glicken (1991), is formed by the disintegration of blocks and the mixing and shearing of this material with the existing or newly forming matrix during movement (Photo 90). As a result, the mixed facies loses its primary structure and sedimentological character. It may also form hummocks. The matrix is highly variable in terms of components and grain size distribution, also influenced by the movement type and the substrate characteristics. The proportion of exotic material increases with distance, and non-volcanic sediments and even plant remains are common. This implies an increasing similarity to debris-flow deposits (Dufresne et al. 2021). Eventually the DAD, filling the valley floors and forming a more evenly distributed surface, becomes somewhat homogenised after travelling tens of kilometres.

Expansion, fragmentation of blocks, the development of matrix facies and the incorporation of exotic material can significantly increase the volume of DADs. Socompa's 25 km^3 DAD is about 30% larger than the volume of material missing from the volcano (Wadge et al. 1995). In the case of Mount St. Helens, the excess volume is about 20% (Voight et al. 1981), and for the clay-rich, cohesive Osceola Mudflow at Rainier, it is close to 50% (Vallance and Scott 1997). The deformation mentioned for the block facies is also typical of the matrix facies, and perhaps occurs more commonly. Incorporation and injection of the softer material and/or the substrate can result in the formation of *clastic dykes*, especially along fault zones (van Wyk de Vries and Davies 2015). Significant deformation can be present especially in the lower part of the DAD, where large blocks are missing. This suggests considerable shear in the basal layer. It has been calculated that half of the total shear in the debris avalanche is concentrated in the lower one-tenth of the deposit (a layer also called pseudo-tachylite or *frictionite:* Legros et al. 2000).

2.6 Movement of Volcanic Debris Avalanches: Transport Mechanisms

The first moment of flank failure involves sliding. This applies to the huge torevas and the largest megablocks that are emplaced over short distances. Then, the actual volcanic debris avalanche starts with a transition from slip to a sort of granular flow, with a progressive disintegration of most of the blocks. Therefore, an overall grainsize reduction occurs with distance, along with an increase of matrix relative to blocks (Perinotto et al. 2015).

The runout of DADs is commonly long or very long (in the order of 100 km) compared to non-volcanic slides. That is, a debris avalanche can travel a considerable distance: while the ratio of vertical drop or height (H, height) to distance travelled (L, length) ranges from 0.4 to 0.6 for smaller non-volcanic landslides, it decreases to 0.1–0.2 for larger landslides and debris avalanches (Fig. 12). Moreover, they can transport much larger volumes (most non-volcanic slides are <0.1 km^3, while the volume carried by debris avalanches is commonly in the order of 1–10 km^3) The increased mobility is due to a rapid loss of resistance at the base of the sliding rock mass: debris avalanches travel much further than would be expected if the motion was constrained by simple granular friction (van Wyk de Vries and Davies 2015).

Such extreme mobility, despite early and recent hypotheses, remains not fully understood. The flow type of the debris avalanche is *laminar plug flow* (Paguican et al. 2021): the material is pushed forward 'en masse'. For example, at Socompa, a typical succession of light dacite overlying dark andesite, which constitutes a large part of the edifice, has been preserved in the DAD from bottom to top. In a laminar flow regime, the spreading of granular material occurs parallel to the flow direction, resulting in the stretching and thinning of units (Makris et al. 2023). The laminar flow behaviour and the extreme mobility might be the result of mechanical fluidisation of the material (Davies 1982).

Fluidisation means that, as the agitation in a granular material increases and the collision of clasts is more frequent compared to frictional contacts, the material dilates and behaves similarly to a fluid (Johnson et al. 2016). Another process (or term) explaining this fluid-like behaviour is dynamic fragmentation during the movement, which reduces friction, particularly in the basal layer. Dynamic fragmentation is caused by the fracturing of intact rock under

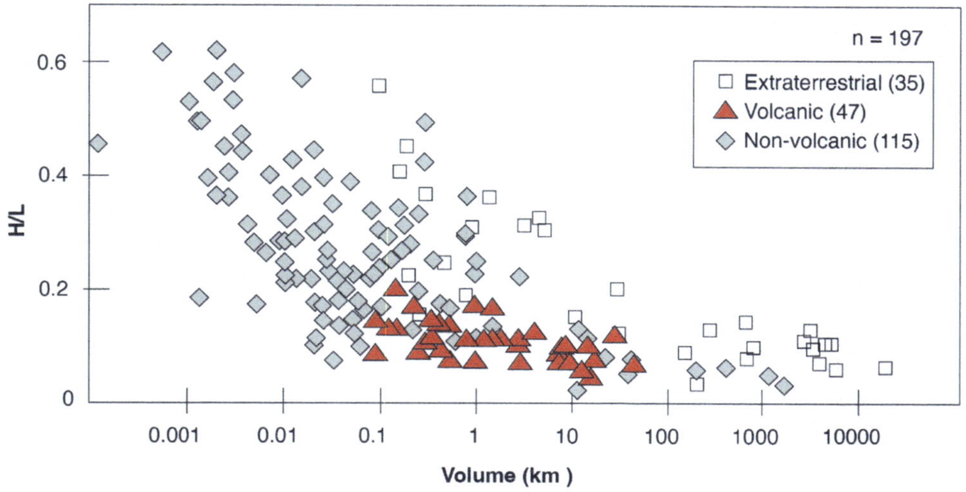

Fig. 12 Vertical drop or height decrease (H)/transport distance (L) of various slope mass movements (van Wyk de Vries and Davies 2015 after Siebert 1984; Francis 1993). The more "mobile" the transport process, the smaller the ratio is: the H/L value of volcanic debris avalanches (as low as 0.1–0.2) is much smaller than that of non-volcanic slides

rapid strain and confining pressure. Numerical models for debris avalanches (Kelfoun and Druitt 2005; Sosio et al. 2012) invoke a retarding stress on the basal layer to achieve long runouts, whereas analogue models use a smooth low-friction base (van Wyk de Vries and Davies 2015). The mechanisms that have been proposed for reducing friction, including various kinds of fluidisation, and lubrication by a number of agents, are summarised in Table 3.

Depending on the prevailing climate and ambient topography, debris avalanches can turn into mudflows or lahars (Scott and Janda 1987; Major et al. 2000) when they flow into fluvial valleys with high discharge (e.g. Osceola Mudflow, Mt. Rainier: Vallance and Scott 1997; Mount St. Helens: Lipman and Mullineaux 1981). Transformation into lahars can occur immediately or post-depositionally, and lahars can also result from the breakout of lakes created by debris avalanches trapping water in a drainage system (for details, see Dufresne et al. 2021). Moreover, debris avalanches (and/or secondary lahars) can generate tsunamis when flowing into the sea or a large lake, as exemplified by the Shimabara catastrophe at Unzen (1792). In submarine settings, i.e. at oceanic islands, much larger tsunamis can be generated, with hundreds of metres of run-up (e.g., Gran Canaria or Mauritius: van Wyk de Vries and Davies 2015).

Following the debris avalanche, an important secondary effect is the decompression within the hydrothermal as well as the magmatic plumbing system of the volcano. Most importantly, this may result in increased eruptive activity (Zernack and Procter 2021) that quickly 'heals' the horseshoe scar, as described for many cases above.

3 Lahars

Lahar flows (the term comes from Indonesian) are a mixture of debris and water that is initiated on the flanks of a volcano and subsequently travels at average velocities of 3–10 m/s, channelled through the drainage system (Vallance and Iverson 2015). Lahar flows can include debris flows (DF), hyperconcentrated flows (HCFs) and other water-rich flows, which may all occur within a single event. Because large volumes of unstable rock, debris and water can spill over from different valleys draining a volcano,

Table 3 Proposed mechanisms explaining the extreme mobility of volcanic debris avalanches*

Fluidisation mechanisms		
Fluidisation with air	Kent (1966)	Fluidisation of the whole avalanche by entrained air
	Shreve (1968)	Trapped, compressed air in the basal layer ("air cushion") could maintain the rocks in suspension
Fluidisation by a molten low-friction basal layer	Erismann (1979)	Sliding is ensured by a basal layer where rocks might melt or become dissociated to form a molten layer called 'frictionite'
Fluidisation with water	Crandell et al. (1984)	Incorporation of fluids (water) during movement
Fluidisation in the rock mass	Rochet (1987)	Fluidisation by reducing the internal friction throughout the transported rock mass
Mechanical fluidisation	Davies (1982)	The avalanche becomes fluidised due to gravity and shear stress in the basal layer
Acoustic fluidisation	Melosh (1979, 2015)	High-frequency vibrations generate sufficient acoustic energy to diminish the effect of frictional contact between clast surfaces and fluidise the debris through local variations in contact forces
Lubrication mechanisms		
Basal self-lubrication	Johnson (1978)	Lubrication of the avalanche with water in a basal layer
	Campbell (1989)	Collisions of numerous, highly agitated particles at low concentration in a thin basal layer maintains the passive avalanche body
Entraining saturated substrate	Hungr and Evans (2004)	Entrainment of saturated soil, clay or debris from substrate increases mobility of rock avalanche
Particle disintegration mechanisms		
Dynamic disintegration	Schneider and Fisher (1998), Davies and McSaveney (2006)	Fragmentation of rock particles maintains the dilatancy of the granular mass and contributes to the extreme mobility
	Liu et al. (2005)	Elastic energy released during particle fragmentation events maintains the dilatancy necessary to keep the granular mass mobile
	Kelfoun and Druitt (2005), Davies et al. (2010), Kelfoun et al. (2010)	Basal fragmentation generates a constant (basal) retarding stress, independent of velocity, which explains the specific rheology
Dynamic disintegration and frictional reduction	Perinotto et al. (2015)	Combined effect of elastic energy release during the dynamic disintegration of the larger clasts, and frictional reduction within the matrix due to interactions between the finer particles

*Threefold division of various hypotheses after Perinotto et al. (2015)

lahar-inundated areas can be larger and wider that those fed by non-volcanic DFs and HCFs (Vallance and Scott 1997; Vallance and Iverson 2015).

Lahars are among the most lethal and destructive volcanic hazards. Since the year 1600 AD/CE, one quarter (56,300) of the total number of casualties related to volcanic activity have been caused by lahars (Brown et al. 2017). This figure places them as the third most deadly volcanic hazard, after pyroclastic density currents (PDCs) and volcanogenic tsunamis (also see Chap. 9). The destructive power of lahars is particularly potent as they can wreak havoc at much greater distances from a volcano than PDCs due to longer run-out distances. The 1985 lahars from Nevado del Ruíz (Colombia) flowed 100 km down valley beyond the Central Cordillera towards the Magdalena River valley (Pierson et al. 1990).

Exceptionally large-volume lahars at Cotopaxi volcano, Ecuador, reached 370 km from their source in 1877 (Mothes et al. 1998). In contrast, channelled PDCs flowed do not exceed 20 km distance (e.g. 16 km at Merapi in 2010, Jousset et al. 2012).

3.1 Types of Lahars

Because lahars are water-saturated, fluid–solid interactions influence their behaviour and distinguish them from debris avalanches (Vallance and Iverson 2015) and jökulhlaups (mass flows related to breakup following a subglacial eruption and reservoir failure within an ice sheet; Gudmundsson 2015; see Chap. 3). Lahars are part of a continuum of water-rich flows in which sediment concentration, particle size distribution and bulk density help distinguish the following two categories (Pierson and Costa 1987; Vallance 2000; Hungr and Jakob 2005; Iverson 2014; Vallance and Iverson 2015):

1. *Debris flows (DFs)* are a mixture of debris and water that moves downslope by gravity in a pulsating pattern. The solid concentration is higher (≥60% volume, >80 wt%) than that of hyperconcentrated flows, with a density of between 1800 and 2400 kg/m^3, twice as much as the muddy water that flows in stream channels during floods (Pierson 1980). The deposits are massive, heterometric and very poorly sorted. The range of particles extends from sand to blocks of several metres, while the proportion of clay and silt remains low in general. A threshold of 3 wt% of silt and clay helps distinguish non-cohesive from cohesive DFs (Scott 1988; Scott et al. 1995).
2. *Hyperconcentrated flows (HCFs)* are two-phase flows, intermediate in sediment concentration between normal stream flows and DFs, with a particle concentration of 20–60% by volume (40–80 wt%) and a density of 1300–1800 kg/m^3. They are dilute, turbulent, and their sandy-gravelly deposits are less heterometric and better sorted than those of DFs (Pierson 2005).

Lahars usually consist of more than one flow regime (Figs. 13 and 14A; Pierson 1986; Vallance 2000; Iverson 2014). An individual lahar pulse can be broken down into three successive segments in time and space: (1) the 'head' or front is characterized by the densest slurry reaching the highest flow height and peak velocity, (2) the 'body' represents the bulk of the lahar and is characterized by packets or pulses driven by variations in incorporation (bulking) and dilution and/or deposition (debulking), and (3) the 'tail' represents the distal part of the slurry, with the lowest sediment concentration due to dilution.

The initiation processes and temporal onset of flows allow lahars to be broken down into three categories:

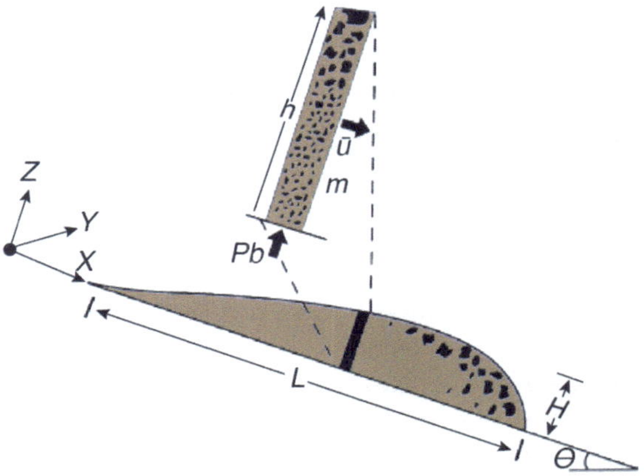

Fig. 13 Sketch diagram after Iverson (2014), after George and Iverson (2011, Fig. 1, p. 416) depicting a depth-averaged model for a debris flow in which the flow velocity $\bar{u}$, depth h, volumetric fraction of the solid component m, and the pore pressure at the base Pb evolve according to location and time. Depth average models can be applied to flows if $H/L \ll 1$, H and L being depth and length of the flow

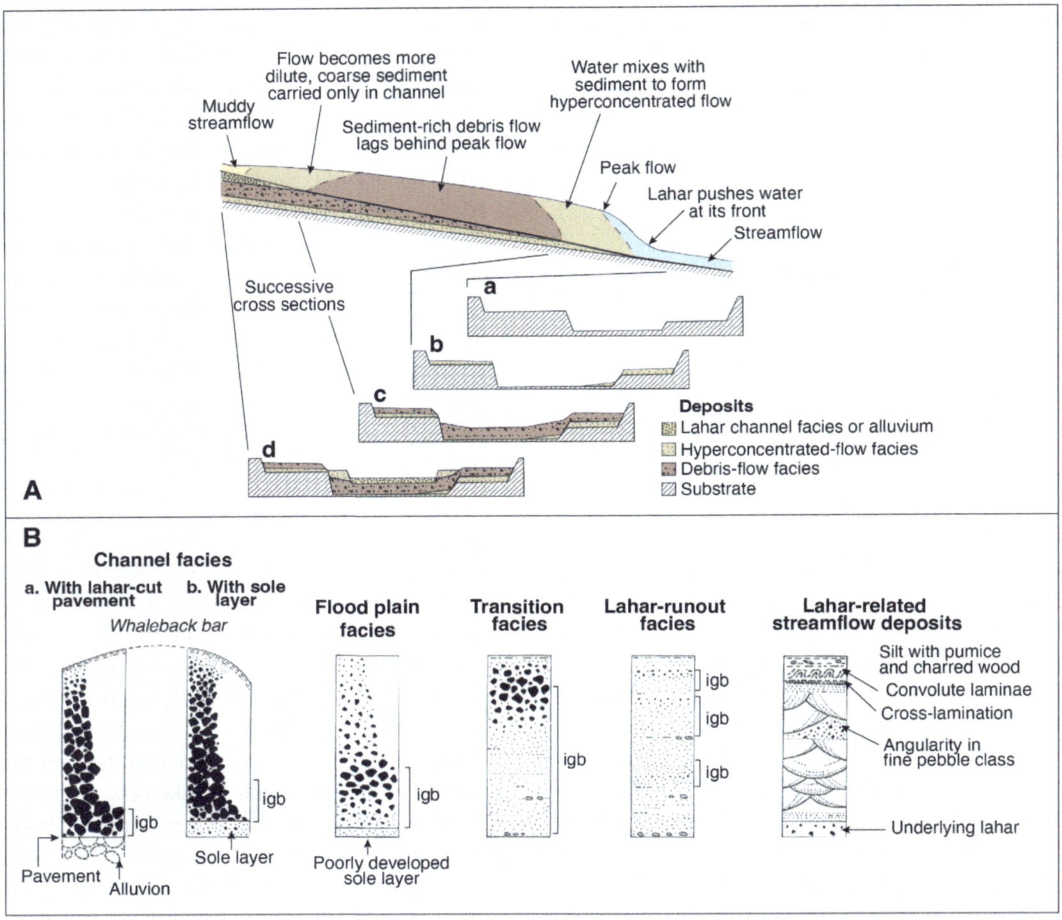

Fig. 14 Lahars and associated flows: deposits, facies. **A** Schematic depiction of spatial and temporal relations between lahar flow and lahar deposits: illustration of a lahar undergoing downstream dilution from debris-flow phase to hyperconcentrated-flow phase and deposit facies. The model shows the expected sequences of hyperconcentrated- and debris-flow deposits in cross sections (a) through (d). From Vallance and Iverson (2015, Fig. 37.4, p. 657). **B** Typical examples of lahar deposits and facies and associated flows, according to the morphology of the valley through which the lahar and the succession of flows from upstream to downstream and over time (Scott 1988, Fig. 10, p. A19). From left to right: channelled facies, flood plain facies, transitional facies, dilute facies in distal channel, and river deposit downstream of lahar facies. igb: inverse graded beds. In a river: a transitional oblong, 'whale-back' shape caused by coarse deposition in a river. Runout, term explained in the text

- *Syn-eruptive or primary lahars* are triggered by the overflow of crater lakes, subglacial eruptions, snowmelt, or the entry of PDCs or landslides into rivers or lakes. For example, large primary lahars (9×10^7 m^3), triggered by melting snow as PDCs passed over the Nevado del Ruiz ice cap, caused the death of 23,000 people in 1985 in Armero, a town located 56 km from the crater (Pierson et al. 1990).
- *Post-eruptive lahars* result from the remobilization of tephra by intense rainfall after large eruptions. They can occur for several years after an eruption. The considerable damage they cause is due to the diversity and repetition of the physical impacts.

- Unrelated to eruptions, *secondary lahars* are triggered on the slopes and deposits of dormant edifices subjected to intense meteorological events (e.g. Hurricane Mitch in 1988 on Casita, Nicaragua: Scott et al. 2005). Some secondary lahars result from remobilization of ash far from its source, for example, about 100 km east of Vesuvius at Sarno in 1998 (Zanchetta et al. 2004), or from the transformation of debris avalanches and landslides on inactive volcanoes such as Casita.

Post-eruptive or secondary lahars are more frequent and protracted than syn-eruptive or primary lahars. Initiation mechanisms include direct transformation of debris avalanches, rapid melting of snow or ice during an eruption, outbreaks of impounded lakes, and rainfall on fresh tephra deposits (Pierson and Major 2014). The most significant factor in triggering these flows is 'critical rainfall' intensity (>10 mm/h or short-term maximum intensity 10 mm/10 min) and duration (<5 h at 10–40 mm/h) depending on the water saturation of the soil (Rodolfo and Arguden 1991). The local, orographic origin of the rainfall triggers short duration and small-volume lahars, whereas extreme hydrometeorological events (e.g. hurricanes) promote long-lasting and voluminous lahars (Capra and Caballero 2021).

In addition, non-eruptive lahars can also occur on dormant volcanoes as the result of transformation of rapid landslides, often termed earthflows (e.g. Iverson and George 2016) or debris avalanches, e.g., at Mt. Ruapehu, New Zealand (Keigler et al. 2011; Tost and Cronin 2015).

3.2 Case Studies of Hazardous Lahars as Agents of Landscape Change

Lessons learned from several case studies have enabled us to build up an inventory of severe and costly damage recorded after flow events (Table 4, all references in: Thouret et al. 2020). Lahars have wreaked havoc in many towns and across diverse environments, including Nevado del Ruiz (Colombia) in 1985, Pinatubo (Philippines) in 1991–1998, Ruapehu (New Zealand) in 1945, 1996 and 2007, Sarno (Italy) in 1998, Santiaguito (Guatemala) between 1987 and 2000, Soufrière Hills (Montserrat) between 2001 and 2004, Semeru (Indonesia) between 2000 and 2003 and in 2020–2021, Mayon (Philippines) in 2006, and Merapi (Indonesia) between 1994 and 1998, again in 2010–2012, and following the dome growth that started in August 2018.

The processes of lahar initiation depend on the local conditions of the watershed. Favourable factors are loose lithology, poorly vegetated soils and a shallow water table. Water input can come from disturbed lakes or rivers, snow or ice melt, but most of the time it derives from long-lasting and intense rainfall, with a critical threshold of 20–30 mm/h, above which lahars can be triggered. Rushing down the volcano slopes, the initial water-debris mixture deforms and liquefies as the pore pressure increases. By eroding flow channels and incorporating sediment, a lahar can increase in volume by 5–10 times, acquiring a maximum velocity of 20–30 m/s and a maximum discharge of 10 times greater than that of floods in a watershed (Costa 1988; Pierson et al. 1990).

3.3 Lahar Emplacement and Associated Deposits

The different sedimentary facies of lahars (Fig. 14B) result from a continuum of flow types, depending on variations in volume, flow rate, sediment load and the distance reached beyond the break in slope between the cone and its foothills, known as the runout distance. Lahars generally have broad, thick fronts (Photo 91), with coarse material exerting high basal friction. They develop downstream of the channel, pushed back by finer debris, less abundant and less slowed by their lower friction on the channel bed. In the case of cohesive lahars (>3% silt and clay), a sediment plug propagates in the channel "en masse" (sudden slowing down of the whole mass of the deposit, with no

Table 4 A selection of lahar events showing hydraulic, physical, and rheological characteristics and related impacts

Volcano location	Event date	Type, occurrence	Surface (km^2)	Volume (10^6 m^3)	Velocity (m/s)	Depth (m)	Peak discharge (m^3/s)	Runout distance (km)	Froude number	Solid concentration (%vol)	Density (kg/m^3)	Fatalities/affected, displaced	Damage to habitat, infrastructure and valuables
Cotopaxi, Ecuador	1877	p LH	440	370[a]				300				1000 / Tens of thousands	Widespread burial
Kelut, Indonesia	1585, 1919 1990, 2014	p LH	131	0.131				40				10,000 5200 / Thousands	105 villages 9000 houses
Mt. Ruapehu, New Zealand	1953	Pe LH	>100	1.9				60				151 />150	Roads and railway Bridges
	1945, 1995–96	p LH	>50	2.9–11 total	2.1–3.6	2–3	42	8–20	<1	35–52	470–910	2 / 25 families	Tourism, electricity facilities, aviation, agriculture, cattle Total ~130 million $NZ (2000)
	2007	s LH		0.44	6.5–19.6		1000–5500	70	<1	56.8–70.5	1900–2200		
Rio Reventado, Irazu, Costa Rica	1963 1965	pe LH, HCF		13–22	2.9–10	8–12	1130–1980		0.5	15–50	1130–1180		
Mount St.Helens, Washington, USA	1980	p, pe LH		50.10^6	10–31.1	3–6	500–2300	100	≥1	35–70	1970–2300	56 / hundreds	Extensive damage: bridges, road network
Nevado del Ruiz, Colombia	1985	p LH	2,100	90	4.9–17	5.2–19.8	15,000–48,000	104	0.7–1.3	30–70	1800–2100	23,030 4420 injured	5092 houses
Mt. Pinatubo, Philippines	1991–1993	pe LH, HCF	200–400	30–40	2.6–3.1	0.5–0.7	10	45	<1	18–31	1110–1500	957/249,370 affected 5000 displaced	112,236 houses
	1997	pe LH, HCF	>400	160[a]	3–6	1–6	20–60	>30		40–57	1840	26/224,610	285 houses
Casita, Nicaragua	1998	s LH	7.5	3.1				30				2513 and 1000 missing	2 buried towns
Mayon, Philippines	2006	pe LH	130	18.8				8–14				1266 / 10,000 displaced	6 towns, 81,000 houses, roads, plantations, dams
Semeru, Indonesia	2000–2004	HCF		0.05–0.5	1.0–5.0	0.6–3.5	50–500	25–35	1–1.7	31–64	1980–2100		Bridges
Chaitén, Chile	2008	pe LH	10–15	1–3				10 + delta				Several/>2000 displaced	Flooded town, roads, bridges, harbour
Merapi, Indonesia	2011–2012	pe LH	>7	>5				20				18 killed or injured / 3000 displaced	860 houses (215 devastated), 70 ha cultivated land buried

Abbreviations: LH (lahars) with HCF hyperconcentrated flow subtype –DF (debris flows), when not indicated–, and occurrence: p primary (syn-eruptive) lahar, pe post-eruptive lahar, s secondary rain-triggered lahar

[a] Reconstructed value. For references of case studies, see Thouret et al. (2020)

stratification) or by progressive aggradation (a rapid succession of more or less stratified deposits). DFs show some cohesive behaviour and relatively little internal mixing. Thus a 'slug' of material moves down the channel and deposition can occur as the result of *en masse* 'freezing' of the flow or pulsed deposition. HCFs have non-Newtonian behaviour, but because their internal cohesion is less than that of DFs, these flows are dominated by friction and interactions between particles (Manville et al. 2009).

3.4 Effects and Impacts of Lahars

The physical impacts of lahars on habitat, infrastructure and lifelines have been analysed in detail from recent case studies (Jenkins et al. 2015; Thouret et al. 2020), following which hazard reduction strategies have been proposed (Pierson et al. 2014). Lahar impact forces are often restricted to dynamic pressure and inundation depth that induce physical actions (Wilson et al. 2014), but other direct and indirect processes must also be considered (Table 5; Photo 92).

The specific characteristics of each lahar determine its geomorphic impact on channels. During a prolonged pulse-marked flow, the transition and succession of HCFs and DFs from upstream to downstream of the channel result in alternating burial and incision phases during each pulse (Manville et al. 2009; Doyle et al. 2010, 2011). The extent of the inundated area depends not only on the mobility of the lahar and its runout distance, but also on the morphology of the channel. Lateral levees and steep channel slopes promote high velocities and long runout distances, while channel widening on the plains reduces these features. Lahars leave the main channel in the case of reduced transport capacity in a channel or when the slope and sinuosity of the riverbed increases (Solikhin et al. 2015). The potentially affected area is thus increased beyond the main valleys by the fact that lahars, having overtopped the interfluvial divides, move through secondary valleys into zones where populations and infrastructure are not usually at risk. This process is termed avulsion, referring to those lahars, or PDCs, which pass through secondary valleys. This phenomenon is common in the foothills and low-lying plains surrounding active volcanoes with a very dense drainage network (e.g. Pinatubo in the Philippines and Merapi in Indonesia).

Lahars contribute to the growth of fans and feed sediment-delivery systems to widespread ring plains, as documented at Mayon following its 1984 eruption (Arguden and Rodolfo 1990), Pinatubo after the large 1991 eruption (Pierson and Janda 1992; Punongbayan and Newhall 1996; Hayes et al. 2002), Santiaguito over the period 1987–2000 (Harris et al. 2006), Popocatéptl where the 1995 and 1997 lahars created and altered gorges (Palacios et al. 2001), Ruapehu after the crater-lake break-out lahar in 2007 (Manville and Cronin 2007; Procter et al. 2010a,b, 2012), Merapi following the 2010 eruption (Fig. 15; De Bélizal et al. 2013; Solikhin et al. 2015; Thouret et al. 2015), Semeru between 1981 and 2011 (Thouret et al. 2014), and Mt. Ruapehu over longer intervals (Cronin and Neall 1997; Keigler et al. 2011; Tost and Cronin 2015).

Unusually high sediment delivery, excess sedimentation rates, and perturbations of the drainage network were the geomorphic responses of the volcanic landscape following the 2008–2009 Chaitén eruption in Chile, as described by Pierson et al. (2013), Pierson and Major (2014) and Major et al. (2016) (see Chap. 9). During the waning phase of explosive activity, modest rainfall triggered an unusual flush of sediment, an HCF lahar, which swiftly aggraded multiple channels by many metres. Ten kilometres from the volcano, the Chaitén River channel aggraded 7 m and the river avulsed through a coastal town. By 26 May 2008, after channel avulsion, about $2.5–3.5 \times 10^6$ m^3 of sediment had been deposited at the mouth of the Chaitén River, forming two deltas; by late 2011 aggradation reached approximately 11×10^6 m^3. However, despite the huge sediment delivery that typically follows large-volume eruptions, disturbed channels usually recover within a few years, irrespective of their location in tropical lowlands, wet and temperate conditions, or in cold climates.

Table 5 Actions, processes, and effects of LH (lahars) and DF (debris flows) on people and living organisms (after Thouret et al. 2020)

Flow processes	Flow properties	Impacts		Consequences
		Direct	Indirect	
Hydrostatic pressure	Water and sediment mixture leading to drowning and trapping of living organisms or fall of elements onto them	Drowning Trauma	Shock	Living organism are drowned, trapped, or buried. Severe trauma (due to building collapse, debris), mutilation and fractures of limbs and skulls
				Injuries due to conveyed heavy objects carried in LHs or DFs that induce broken bones, or from bodies driven against stationary objects and crushed
Boulders as missiles and impact pressure	$E=f(m, u)$ Energy upon impact of block depends on kinetic energy of the missile, hence on its velocity and mass	Trauma	Shock	Injuries, deep lacerations, penetrating wounds, and infections after admission to hospitals
Ablation and undercutting	See undercutting of a structure (Table 5)	Trauma Drowning	Shock	Landslide induced by lateral undercutting and sapping that can lead to people falling from banks into moving flows
Accumulation	Cohesion, density ρ, and ϕ permeability of flow	Induration Cementation	Change in topography	Trapping of living organisms in LH and DF flows that produces traumatic asphyxia or traps people in the valley, alluvial plain or at home surrounded by LH Runoff on indurated crust of deposits may contribute to overrun and avulse towards channels apparently outside main flows
	Leakage of industrial toxic elements and domestic waste in LH flow	Contamination, disease	Contamination and pollution Deterioration of harvest	Contamination of soils, streams and phreatic reservoirs by toxic elements Withering of harvest owing to excess water supply that can lead to food shortage and digestive problems
	Concentration in toxic elements together with their mobility		Remobilisation of material	LH/DF deposits can be remobilised by arrival of additional LH, avulsion, or by heavy rainstorms
Chemical and biological actions Heat and pH	Temperature of hot LHs Contamination of water Low pH of LH/DF traveling across or scouring sulphur-rich fumarolic field or industrial waste	Chemical and temperature burns	Infections Deterioration	Chemical and temperature burns to plants, fauna and people produced by pH and high temperatures. Infections including gas gangrene, tetanus, and necrotizing tissues in victims rescued after being in the mud for days (Baxter and Horwell 2015)

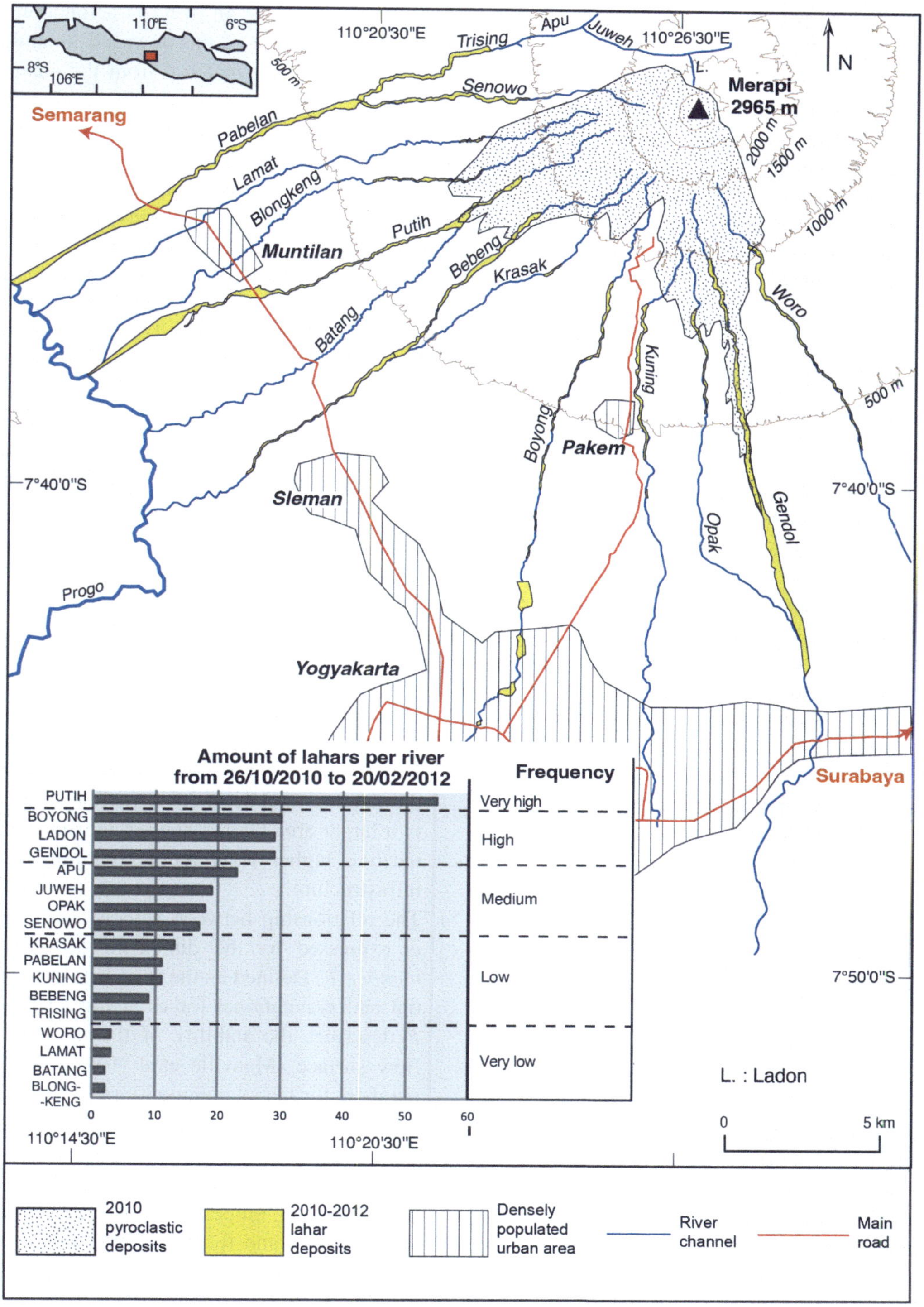

Fig. 15 Map showing the drainage network and pyroclastic (grey) and lahar (yellow) deposits from the 2010 eruption around Merapi (De Bélizal et al. 2013; Thouret et al. 2023). Red quadrangle marks the position of the volcano in Java. The insert shows the number of lahars for each of the 17 rivers for two years after the 2010 eruption. Urban and densely populated areas are dashed

3.5 Characteristics of Lahars and Debris Flows

The characteristics of lahars and debris flows described here have been derived from several case studies (Table 4: Thouret et al. 2020). Hydraulic characteristics of lahars in motion and solid/liquid concentrations have been defined in early classifications (Costa 1988; Coussot and Meunier 1996; Iverson 1997), while rheological parameters have been further measured using small-scale experiments, laboratory tests (Ancey 2007; Hübl et al. 2009; Dumaisnil et al. 2010), and large-scale experiments carried out in channels (Rickenmann 1999; Doyle et al. 2010, 2011; Zhou and Ng 2010) or experimental flumes (Iverson et al. 2010, 2011). Using flume experiment results, Iverson et al. (2011) argued that a key feature of DF behaviour was the development and persistence of dilated, high-friction, coarse-grained flow fronts, pushed from behind by nearly liquid, finer-grained debris. A list of physical and dimensionless parameters of small-scale, experimental DFs and large-scale, natural DFs is provided by de Haas et al. (2015, their Table 2).

3.5.1 Hydraulic Characteristics of Lahars

1. *Velocity* (u, in m/s) of flows can be broken down into three components (Doyle et al. 2007): (i) surface velocity u_{surf}, the instantaneous velocity calculated from video film footage, (ii) travel velocity u_r, corresponding to the total runoff velocity, and (iii) body velocity u_b, the depth-averaged velocity for a given location. The flow-front velocity is often used in equations as the flow head plays a major role in the initial impact intensity on structures. It reaches its maximum when the boulder-rich head is thick and fast-moving (Lavigne and Suwa 2004; Lavigne and Thouret 2003; Thouret et al. 2007). Velocity reaches a peak slightly behind the front because the latter, being thicker and commonly coarser, exerts more friction.
2. *Discharge rate* (Q, in m^3/s) is a function of the mean velocity of the flow and its ability to erode and bulk material from the channel bed and banks. Coupled with high velocities, discharge rate can influence the degree of erosion along the flow path; it is therefore a key parameter for understanding lahar impact (Doyle et al. 2011). Variations in discharge rate with time and space depend on cycles of bulking and debulking. Bulking is the amount of pre-existing material removed from the channel bed and banks and entrained by the flow, whereas debulking is achieved through dilution and deposition of material down valley, together with water infiltration into the channel bed. Bulking is controlled by the erodibility of the channel bed and bank material, the volume of sediment available for erosion, and flow parameters (Pierson 1995; Fagents and Baloga 2006). The capacity for entrainment of material from channel beds and banks can lead to a dramatic growth in flow volume and peak discharge, e.g. fourfold increase over a distance of ~100 km at Nevado del Ruiz (Pierson et al. 1990).
3. *Flow depth* (H, in m), or hydraulic radius during emplacement, can reach 4–5 times the deposit thickness. A deeper flow results in a larger area being exposed and flows that reach a higher level around buildings and infrastructure.
4. The relationship between velocity and depth is expressed by the dimensionless *Froude number, F*. Defined as the ratio between inertial and gravitational forces acting on flows, F describes the stability of the deformable flow surface (Manville et al. 1998). Froude numbers for lahars commonly range between 0.5 and 1.5 (Tiberghien et al. 2007). There are two different regimes characterized by F for cold and hot lahar cases (Fig. 16), with identical slopes but different velocities (Arguden and Rodolfo 1990). For a subcritical flow ($F < 1$), runoff will be deep and slow with a smooth, flat surface, whereas in a supercritical flow ($F > 1$), runoff will be

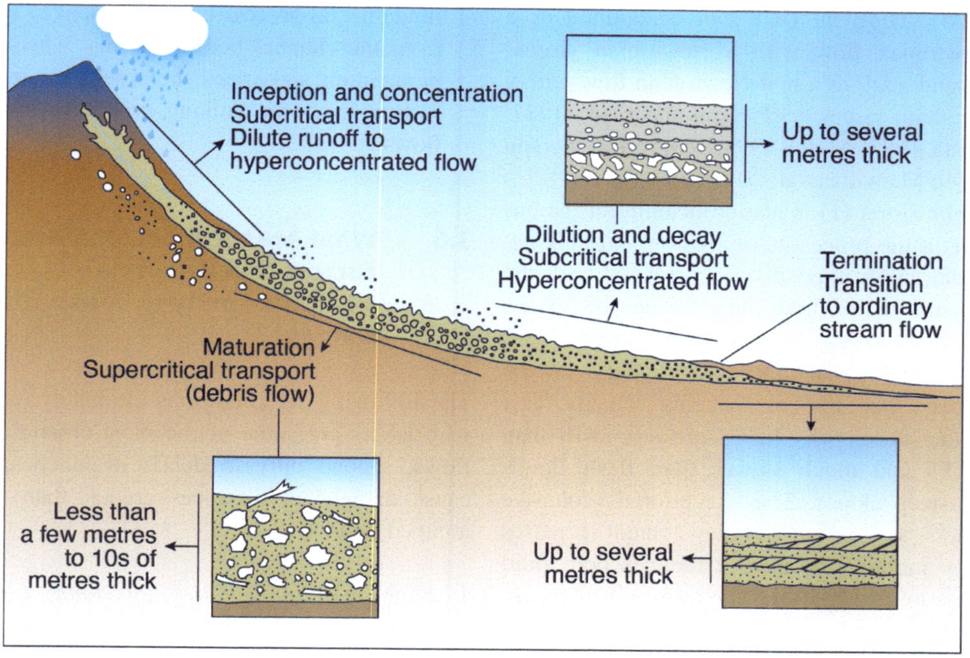

Fig. 16 The relationships of lahars, hyperconcentrated flows, and water floods as a function of sedimentation concentration and criticality regime (according to Froude number). Lockwood and Hazlett (2010)

shallow and fast, and its surface will exhibit waves and disturbances such as hydraulic jumps (Manville et al. 1998).

5. The velocity, discharge rate, flow depth, and the width/depth ratio of the channel influence *flow mobility or momentum* (dimensionless ratio of elevation/travel distance, $\Delta H/L$), which also varies due to the river bed characteristics such as slope angle, basal friction angle and bed roughness. In turn, the flow momentum will determine the flow runout distance, a pivotal parameter for delineating at-risk areas and assets down valley.

3.5.2 Physical and Rheological Parameters and Lahar (DF and HCF) Dynamics

1. *Bulk density* (ρ) of a lahar, i.e. the sum of the fluid density (ρf) and the solid component density (ρs), varies between 1500 and 2500 kg/m^3 on average; the calculated bulk flow density ranges between 1330 and 1800 kg/m^3 for HCFs and increases to 1800–2300 kg/m^3 for DFs, depending on the lithological components and concentration of solids (Iverson 1997; Zanchetta et al. 2004). However, flume experiments have revealed that debris flow porosity and bulk density change constantly in the dilated front and the more stabilized bulk density of the body.
2. *Viscosity* (μ in Pa s, commonly 0.001–0.1) is the property of a fluid that slows the settling of suspending particles and allows it to resist shear deformation, thus controlling the shear and flow rates (Pierson 2005). The latter parameters correspond to the erosion exerted by the flow on the stream bed, which ranges between 10 and 5000 Pa s (Costa 1988; Table 5). Viscosity is a function of the concentration of sediment or solid, particularly the silt and clay contents which play an important role in fluid cohesion and flow mobility. Viscosity depends on inter-particle interactions, and particle-bed and particle–fluid interactions, the combination of which is a complex and continuous process (Pierson

2005). Thus, an HCF can be defined as a Newtonian flow with a Newtonian viscosity and a DF as a non-Newtonian flow with a Bingham viscosity (Pierson and Costa 1987; Costa 1988; Major 1997; Major and Iverson 1999; Manville et al. 2013).

3. *Shear stress* (τ) is also a meaningful parameter in the processes of erosion and bulking. Shear stress is positively correlated with the density of the flow; thus, the denser a flow, the more erosive it is. However, higher density and viscosity increase friction with the channel bed, thereby reducing velocity. This is the reason why lahars are less swift than HCFs and much slower than flash floods on steep slopes. The exception is cohesive lahars, as their silt and clay content increases flow mobility because it reduces pore fluid pressure and friction. The transport of coarse particles, including boulders, is mostly controlled by the cohesion and buoyancy of the flow (Johnson 1970; Hampton 1979), while both are positively correlated with flow density. Hence, denser flows will erode the stream bed and transport particles, which gives rise to an increase in volume (bulking).
4. *Pore-fluid pressure* (pfp) is spatially and temporally heterogeneous while fluctuations buffer or harden flow resistance through the interplay of frictional, collisional or viscous particle–particle contacts. Iverson et al. (2010, 2011) argued that the prominent role of evolving pore fluid pressure is the chief mechanical trait distinguishing DF behaviour from that of granular avalanches, which lack significant pore fluid pressure (pfp). High pfp renders DFs more mobile as it decreases the frictional resistance by buffering grain interactions. Large positive values of pfp have been shown to play a fundamental role in flow dynamics in flume experiments: steep, highly resistant, surge fronts of coarse-grained material without measurable pore-fluid pressure are pushed along by relatively fine-grained and water-rich tails that have a wide range of pore fluid pressures (McCoy et al. 2010; Iverson et al. 2011). Pore fluid pressure variation in flows influences the fundamental process of entrainment of debris from the channel bed and banks when flow momentum and speed increase. Entrainment in turn feeds the bulking process of erosive flows.

3.6 What Are the Most Significant Factors that Determine a Lahar's Ability to Cause Damage?

The following six conditions ultimately explain why lahars are some of the most erosive mass flows, second only to debris avalanches, and cause damage at distances greater than those attained by PDCs.

1. *Flow mobility and/or momentum* (ΔH/L, elevation/travel distance, values between 1 and 20) and flow volume strongly influence the runout distance and inundated area (Rickenmann 1999, 2005; Lube et al. 2009; de Haas et al. 2015). Both runout distance and inundated area are pivotal for predicting the potential extent of damage (Toyos et al. 2007; Bettella et al. 2012; Iverson et al. 2016; Han et al. 2017a, b). Runout distance increases with flow type (e.g. more mobile water-rich flows) and grain-size distribution (clay and silt-rich, cohesive lahars), high slope gradient and wide channels on ring plains, and the existence of lateral levees on steep fans (5–15°), influencing flow width/depth ratio and mobility.
2. Lahars are *unsteady flows* that move in a pulsating way, forming packets or surges each with clear variation of flow properties along the length of the surge, which exert erosion and aggradation that fluctuate in space and time (Doyle et al. 2010, 2011; McCoy et al. 2010). Such mass flows erode their channels by cutting debris away from the banks and the base. Channels can suddenly widen from a few metres to tens of metres and undercut banks, thereby inducing costly damage for homes built along river banks, as witnessed during Mt. Pinatubo's lahar events following

the 1991 eruption. Moreover, excavation of the channel bottom will deepen and increase confinement of lahars, allowing them to travel more rapidly. With the larger debris removed, the flow can continue to move with less loss of gravitational energy due to friction.

3. Unsteady and pulsating lahar behaviour depends on *bulking/debulking*; as the flow volume, flow density and grain size distribution changes, so too does its power of erosion through pressure changes (Manville et al. 2013). Bulking, counterbalanced by infiltration into the channel bed, will increase volume and peak discharge by between a factor of three to ten relative to initial values. For example, bulking rates increased from 160 to 500–680 m^3/m over a distance from 10 to 69 km along two valleys that conveyed Nevado del Ruiz lahars on 13 November 1985 (Pierson et al. 1990). This happens often at the flow head, which exerts the maximum hydrodynamic pressure, in particular when the snout is boulder rich. In contrast, debulking will decrease the pressure, but less concentrated flow pulses or tails will induce incision of the deposit and undercutting of structures such as bridges. Changes in volume and velocity, bulking/debulking and peak discharge tend to decrease downstream, as shown by flatter and longer-lasting lahar discharge hydrographs. However, changes in flow type (from HCF to DF and back again) induce erosion. This leads to burial followed by incision, which can occur within each lahar event, e.g. during a 90-min period at Semeru in 2008, or several lahar pulses over a 16-h period at Ruapehu in 2007 (Manville and Cronin 2007; Cole 2011; Cole et al. 2009).
4. Lahar *composition, grain-size distribution and segregation* (i.e. clay/silt and large particle concentration) derives from the distinctive sources of morphological and sedimentological features, but also fundamentally affects flow dynamics (Iverson et al. 2010). Particle segregation can be affected by size, shape, and density during flow stratification, and can create boulder-rich snouts that push water ahead of them. Lahars can deposit sediment *en masse*, thereby invading a structure, or by step-wise vertical accretion, applying a hydrostatic load to structures, causing bending or breakage with time. Hydrostatic pressure acts together with dynamic pressure on walls and bridge piles. Hydrostatic forces can lead to equalization of lahar depths on the inside and outside of buildings. As equalization usually takes a reasonable amount of time, hydrostatic force is thus often not accounted for in numerical modeling of flow impacts on structures (e.g. Mead et al. 2017). In contrast, HCFs deposit sediment progressively by accretion of bedload and settling from suspension, thereby exerting less head pressure when coming into contact with a structure. In this case, dynamic and hydrostatic pressure exerted by HCFs remains high for long time intervals, instead of the brief, high peak of dynamic pressure exerted by lahar fronts on the outer surface of structures.
5. *Channel morphology* plays a role in runout distance and flooding area. De Haas et al. (2015) explored DF morphology using small-scale experiments which were compared to natural cases. Channel morphology typically consists of high, steep levees and flat, rough beds, while flow fronts create stubby lobes. Longitudinal levees confine flows, thereby promoting high velocity and long runout distance. Steeper channel slopes lead to longer runout and higher flow velocities, induced by the increased gravitational potential energy, thereby increasing dynamic pressure. Runout distance and inundation area increase with increasing channel width, but very wide channels will exhibit a reduced flow runout as the flow becomes thinner. Three geometric parameters of the channel-bed system determine how confined lahars can spill over banks and avulse to cause widespread impacts: reduced channel capacity, change in the longitudinal channel confinement, and change in channel sinuosity (see Chap. 9; Lube et al. 2011; Solikhin et al. 2015). Such geomorphological conditions change flow

dynamics, decoupling the more mobile upper flow from the basal one, leading to overbank flow. Overspill not only creates hazards for areas adjacent to the river channel, but also for secondary channels beyond the main valley, leading to further threats due to flow avulsion, e.g. during and after the 2006 and 2010 Merapi eruptions (Lube et al. 2011; Charbonnier and Gertisser 2011; De Bélizal et al. 2013; Solikhin et al. 2015; Thouret et al. 2015).

6. *Variability in flow-path topography* can strongly influence flow velocity, sediment erosion and deposition, and hence the evolution of flow characteristics and behaviour. Flow-path topography influences the behaviour of the front and rear of the flow during the debris-rich erosional phases. The head is often higher than the rear portion owing to friction with the channel topography and to the fact that the front will first incorporate loose debris. The flow tail does not incorporate excess debris, instead becoming more dilute due to deposition. The more water-rich tail acts like a flood, therefore becoming more rapid, providing efficiency in terms of channel and bank erosion. As a result, flow fronts and tails show contrasting behaviour: the head carries most of the debris, up to a threshold at which the amount can be held or erosion decreases. The water-rich tail continues to erode channels, but a large quantity of debris cannot be removed because the energy is much less than at the front. Thus, the tail velocity decreases, preventing it from removing as much debris, and instead favouring incision of the deposits that were pushed aside and deposited on the channel bed by the head.

3.7 Lessons Gleaned from Surveys, Experiments and Case Studies

1. Because their volumetric sediment concentrations exceed 40%, maximum speeds surpass 10 m/s, and volumes can range up to ~10^9 m^3, the destructive power of lahar ranks second after PDCs in volcanic environments worldwide.
2. Despite remarkable advances, our understanding of lahar behaviour is far from complete. A key feature of flow behaviour is the development and persistence of dilated, high-friction, coarse-grained flow fronts, pushed from behind by nearly liquid, finer-grained debris.
3. Methods applied in the field and in laboratory settings have helped monitor hydraulic characteristics, rheological properties and flow impacts with obstacles. Mechanical testing of flow deposits and construction materials, as well as numerical modelling, helps understand how lahars denude slopes, bury floodplains, cause fatalities and devastate property.
4. Critical conditions enabling lahar events to cause damage at a distance from the source are: flow momentum and composition, grain-size distribution and segregation, which all affect flow dynamics, channel morphology, and variability in flow path/channel topography.
5. Flows not only cause flooding and burial of land, but also scouring and undercutting, as well as chemical and thermal effects. Geomorphic processes including erosion, deposition, overbanking and avulsion, alter river channels and adjacent tributaries, terraces and fans. The duration of inundation, inducing changes in hydrostatic pressure, is the most important factor besides the structural quality of buildings and infrastructure.
6. Lahar damage is caused by three principal forces driving impacts: (i) hydrodynamic pressure, incorporating both the lateral and vertical components, (ii) hydrostatic pressure, and (iii) collisional forces of boulders acting as missiles.
7. Lahar impact dynamics depend on the kinetics of flows, hence the Froude number. Two impact regimes can be distinguished for a flow encountering a rigid structure, governing the hydrodynamic pressure. Flow density, velocity, depth, and the least angle between the flow direction and an obstacle are the

parameters that account for the impact pressure. Instead of a constant value, impact force evolves within the flow segments, as well as vertically.

4 Subaqueous Mass Flows Related to Volcanic Activity

4.1 Terminology and Major Features

Short-term destructive processes on land, like debris avalanches and lahars, can continue their way into lakes or sea as *subaqueous mass-flows*. Subaqueous (generally submarine) flank failures and eruptive processes can also transform into underwater mass-flows. In both cases these flows can develop into *turbidity currents*. Subaqueous flows can move at velocities of up to 50–100 km/h and reach distances of over 1000–2000 km (Stevenson et al. 2018; Clare et al. 2023).

By definition, concentrated, grain-dominated subaqueous mass flows are at the end of the wide spectrum of gravity-driven or simply *gravity flows* (Iverson 1978; Shanmugam 2021a). These flows comprise various mixtures of sediment and water. In this final part of the chapter, the range of subaqueous mass flows of volcanic origin are summarised. Two important features should be highlighted:

(1) their transport mechanisms can change during a single event, e.g., debris flows may evolve into turbidity currents, slides into debris flows, and slumps into debris avalanches (Lamarche et al. 2008);
(2) they play a significant role in the degradation of constructional landforms, shown by the fact that the removed total volume of deposits accumulated offshore can be comparable to that of respective subaerial edifices, particularly around volcanic islands (Le Friant et al. 2009, 2010; Anagnostopoulos and Anastasakis 2020).

4.2 A Wide Variety of Subaqueous Processes

4.2.1 Underwater Debris Avalanches

To get a sense of the wide range of submarine processes, let us start with on-land-generated debris avalanches and debris flows, which may continue under water (e.g., Brunet et al. 2016; Hunt et al. 2018; León et al. 2019). Here, we present a few well-known cases.

As referred to above, Augustine volcano (Alaska) has produced a dozen small- to medium-sized debris avalanches over the past 2000 years. Around the island, submarine hummocky topography extends to a depth of 45 m and as far as 3.8 km off the east coast, and to a depth of 25 m as far as 2.8 km off the north coast (Fig. 17A: Waitt and Begét 2009). West Island, off the western coast of Augustine volcano, associated with a submarine hummocky seafloor, comprises debris-avalanche deposits that are only 350 years old.

Most recently, on 6 October 1883, about 0.3 km^3 of rock debris (Table 2) came down from the volcano at an altitude of c. 1300 m, and travelled a distance of about 6–7 km in two pulses, creating 2 km of new land (named Burr Point for the burr-like appearance of the accumulated mounds: Fig. 17A), and a further distance of 2.4 km underwater. The landscape of the north-northeast shore of Augustine Island changed dramatically (Watts and Begét 2009): randomly distributed hummocks up to 30 m in height were emplaced on land and offshore. Most of the debris-avalanche deposit is found seaward of a pre-eruption sea cliff. For a couple of months after the eruption there may have been a lagoon between the old sea cliff and the new hummocky deposit seawards; eyewitnesses reported that it was as if the mountain 'split in two' with the waterway subsequently filled to above sea level by PDC deposits later in the eruption (Waitt and Begét 2009). On-land hummocks are often overlain by soil which contains a 1–3 cm thick whitish ash layer from the more recent Katmai eruption in 1912.

The debris avalanche entered the sea fast and energetically (with a speed of up to 100 m/s). Contemporary accounts suggest that it generated sea waves of unsubstantiated height. Tsunami deposits on the island indicate that waves were almost 20 m high, while in distal sites they testify to 6–8 m wave heights (Waitt and Begét 2009). Numerical modelling predicts travel times of c. 50–75 min to the nearest populated areas (Begét and Kowalik 2006).

Submarine mass-flow processes and deposits around the volcanic islands of the Lesser Antilles, e.g., Montagne Pelée (Martinique) and Soufrière Hills volcano (Montserrat), have been particularly well explored via bathymetric mapping and ocean-floor drilling (Watt et al. 2012; Trofimovs et al. 2013; Brunet et al. 2016). Offshore of the western coast of Martinique, the last (30–45 ka) flank failure of the Montagne Pelée volcano produced a blocky submarine deposit testifying to a 30 km runout, mostly underwater (Fig. 17B). This deposit covers an even larger submarine fan related to a failure of the seafloor sediments, possibly with c. 70 km runout (Brunet et al. 2016). Around Montserrat, results also revealed submarine flank failures and related deposits, and the comprehensive set of data testifies to how volcanic island growth and collapse has involved large-volume submarine mass flows of variable compositions over a 100 kyr time scale (Trofimovs et al. 2013).

We briefly mention two further examples. Firstly, a giant (>3000 km^3) submarine debris avalanche, referred to as Ruatoria and dated at c. 170 ka, is inferred to have occurred in the Hikurangi Trough, offshore of North Island, New Zealand, causing a mega-tsunami (Collot et al. 2001). Secondly, slope failures on the flanks of the western Canary Islands have also resulted in huge submarine debris-avalanche and debris-flow deposits, exemplified by the Canary Debris Flow extending 700 km from El Hierro (Masson et al. 1998; León et al. 2019).

4.2.2 Pyroclastic Density Currents (PDCs) that Continue Under Water

Explosive volcanic eruptions may also have a subaqueous impact and continuation, as already mentioned in the example of Krakatau 2018. If there is a larger water body in the vicinity, explosive processes can transform into subaqueous mass flows (Fig. 17C).

Several such cases have occurred at Montserrat, where deposits from lava dome collapses and associated PDCs between 1995 and 2005 have been thoroughly studied (Le Friant et al. 2010). Moreover, smaller-scale flank failures and related debris avalanches from the lava domes may also have been associated with pyroclastic flows entering the sea (Voight et al. 2002). In particular, the 2003 lava dome collapses provided an opportunity to study both the subaerial and submarine sequence of events (Trofimovs et al. 2006, 2008). When the PDCs reached the sea, phreatic explosions were observed, and a surge cloud decoupled from the main flow body, which in turn was submerged and rapidly mixed with sea water to form a water-saturated mass flow. The coarsest part of the dense granular flow moving on the seabed was then deposited proximally, while the finer components were elutriated, and evolved into a far-reaching turbidity current (Trofimovs et al. 2008). A similar case has been described recently for the newly discovered 520 ± 10 ka Archaeos Tuff eruption on Santorini (Druitt et al. 2024): the PDCs, in addition to producing minor ignimbrite deposits on some islands, transformed into turbidity currents, forming a volcaniclastic megaturbidite up to 150 m thick in the surrounding marine basins.

During the 1883 eruption of Krakatau, hot (about 500 °C) PDCs travelled at speed into the sea and deposited tuff on the seabed more than 15 km from the source, generating tsunamis in the process (Carey 2000; Freundt 2003).

This explosive activity was also associated with submarine flank failure and caldera collapse (Madden-Nadeau et al. 2021; see Chap. 5). In the 2022 Hunga Tonga–Hunga Ha'apai eruption (Chap. 3), rapid emplacement of large volumes of erupted material onto steep submerged slopes triggered extremely fast (122 km per hour) and long-runout (>100 km) seafloor currents (Clare et al. 2023).

4.2.3 Volcaniclastic Mass-Flows in Crater and Caldera Lakes

Crater and caldera lakes may also contain deposits from various subaqueous mass movements, as exemplified by the volcanic infill of Uzon-Geysernaya caldera, which hosts Kamchatka's largest present-day geothermal field (Lebedeva 2022). In such lakes, talus breccias commonly grade upwards into coarse- to fine-grained turbidites and suspension deposits formed during quiescent periods (Manville et al. 2009). In Alaska, crater and caldera wall failures have led to floods and associated water-saturated mass flows in several cases (Waythomas 2022).

4.3 Submarine Mass-Transport Mechanisms: From Mass Flows to Turbidites

Subaqueous mass flows produce mass-flow deposits commonly referred to as mass-transport deposits (MTDs: e.g., Hampton et al. 1996; Mulder and Cochonat 1996). The term 'mass flow' means that during movement the driving force is the weight of the sediment itself, not flowing water, and the sediment moves *en masse* (the clasts do not move independently and freely). Due to the high concentration, most of these flows are characterized by laminar motion, although turbulent regimes may also occur (Elverhoi et al. 2010). Internal deformation resulting from laminar motion is extensive (Dasgupta 2003; Lamarche et al. 2008).

MTDs develop via various emplacement mechanisms. Generally, those that are related to submarine slope failure can be described as having a deposit divided into three parts: headwall, translational and toe part (or domain), with the dominant stress changing from compressional through shear to extensional, respectively (Fig. 17D; Bull et al. 2009; Cardona et al. 2019). These domains show increasing internal disruption, and characteristic landforms include a headwall scarp, rotational blocks, ramp structures, pop-up blocks (several hundreds of metres high), grooves etc. In the case of debris flows entering a water body, the flow interacts with the overlying water column, hence its upper part will be affected by shearing and dilution. Mud-rich versus clast (or sand)-rich subaqueous debris flows behave differently based on lab experiments (e.g., Sequeiros et al. 2009). The former remain compact, cohesive, and may display a long runout (hundreds of kilometres) due to hydroplaning, that is, acceleration of the head and extension of the flow body, while the latter become less cohesive during movement and segregates into two layers: a denser lower one and a lighter upper one (Elverhoi et al. 2010).

In submarine settings with greater water depth, MTDs transform into turbidite currents, which are not mass flows, but low-concentration, surge-like viscous fluids, entraining ambient seawater as they go due to turbulent mixing (Allen 1985; Boggs 2006). As shown in Fig. 17C, turbidite currents may be the simple continuation of diluted PDCs or other volcaniclastic flows. Turbidity currents can transport a large amount of sediment into deep marine environments, and are thus considered to redistribute more sediment on the Earth's surface, including that of volcanic origin, than any other mass-flow process (Talling et al. 2015).

As a result of recent exploration of deep submarine settings, in particular the Integrated Ocean Drilling Program (IODP), the characteristic processes associated with turbidites have been found to be more complex than initially thought, often showing interbedded stratigraphic successions involving mechanisms that are not fully understood (Mutti et al. 2009; Stow and Smillie 2020; Shanmugam 2021b). There are several turbidite facies, which can be divided into three main types (Fig. 17E: e.g., Stow and Smillie 2020): (1) coarse-grained (gravel-sand)

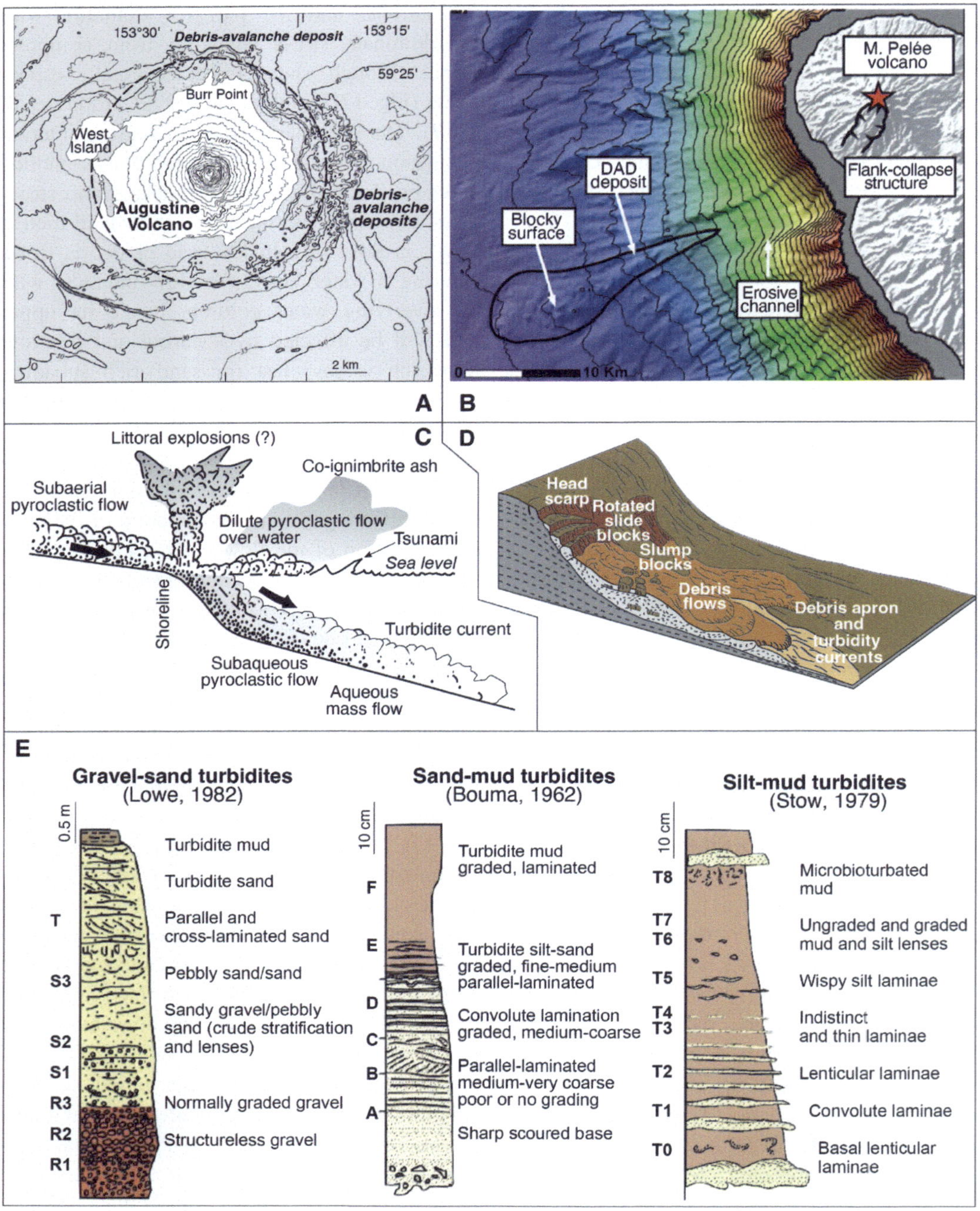

Fig. 17 Subaqueous volcaniclastic processes including mass-flows and turbidity currents. **A** Bathymetry offshore of Augustine Island with the distribution of debris-avalanche deposits, showing a conspicuous hummocky topography (5 m contours of bathymetry traced from USGS 1:250,000 topographic map; Waitt and Begét 2009). **B** Montagne Pelée volcano on the northwest side of Martinique, Lesser Antilles, showing the horseshoe-scar structure of the last flank collapse event (30–45 ka) on-land and the debris-avalanche deposit offshore (Brunet et al. 2016). **C** Pyroclastic flow entering a sea or lake: the denser part continues underwater, whereas the lighter part remains above water; the bottom part may become a turbidity current (based on Schneider 2000; Freundt 2003). **D** Schematic block diagram of submarine volcaniclastic mass-flow deposits (after Morgan et al. 2009; Cardona et al. 2019). **E** Three types of turbidites based on dominant grain size (see text) after Stow and Smillie (2020). Note the scale differences

turbidite family displaying the Lowe facies model (Lowe 1982), with downflow facies separation; (2) medium-grained (sand-mud) turbidite family commonly containing 'classic' Bouma sequences (Bouma 1962) with a pelagite upper layer; and (3) fine-grained silt-mud turbidite family reflecting the Stow model (Stow 1979). These models were developed from siliciclastic sediments, but can also be applied to volcaniclastic turbidites, such as the 2 km-thick Mio/Pliocene Miura Group exposed in southern Honshū, Japan (Stow et al. 1998). There are also well-preserved examples of interbedded associations of explosive silicic shallow-marine to resedimented volcaniclastic facies (including turbidites/debris flows) to subaerial volcaniclastic and fluvial facies in the Miocene Börzsöny Mountains (Hungary) (Karátson and Németh 2001). A more recent example is of deposits formed by turbidity currents offshore of Montserrat, containing well-developed grading, scoured deposit bases, and tractional structures (such as planar and cross-laminae: Trofimovs et al. 2013).

Finally, we note that over the last few years experimental data and theoretical modelling have provided greater understanding of the characteristics of turbidity currents derived from volcanic material (e.g. Clare et al. 2018; Meiburg and Nasr-Azadani 2018; Bougouin et al. 2021).

5 Concluding Remarks

In this chapter, we have reviewed the short-term degradation processes that are typical of especially active volcanoes (i.e. occurring syn-eruptively), although they can also occur on dormant or even extinct ones (i.e. in the inter- or post-eruptive period).

On active volcanoes, much of the surface is either unstable or uncompacted, and due to long-term or rapidly triggered endogenic mechanisms (which may be contributed to by the prevailing climate), material is often transported downhill very suddenly. After eruptive activity, volcanically modified landscapes experience increased erosion, fluvial instability, and sediment yield. From a sedimentological point of view, the processes involved fall under the umbrella term of resedimentation, but this term does not express the multitude of processes that take place within this framework. We have distinguished between terrestrial and underwater settings; we note that in both cases the erosive effects as well as the deposits from sudden mass-flow processes can modify the subaerial and subaqueous topography.

Two main processes with related landforms and deposits have been described in detail. The first is partial flank collapse of volcanic edifices, which might additionally affect their summit regions, resulting in volcanic debris avalanches. This impressive type of mass movement has been explored and studied on all the continents, following the catastrophic eruption of Mount St. Helens in 1980, and we also have a wealth of information on submarine occurrences. The second is the complex processes involving lahars, which comprise water-rich flows of varying material content. While the range of area directly affected by a debris avalanche is mostly limited to the immediate proximity of the volcano (a few tens of km^2), the destructive power of lahars is particularly striking, often reaching great distances (>100 km) from the volcano—greater than most pyroclastic density currents—due to being channelled along valleys.

In the next chapter, we look at the long-term, slower, but ultimately—in terms of degradation—even more effective erosional processes, and describe the resulting secondary (erosional) volcanic landforms.

Photo 82 Aftermath of a six-year-long eruptive period (1990–96) of Unzen volcano (Kyushu, Japan) in the Mizunashi valley. **A** Enlarged river beds, note the damaged fences, poles and bridge walls. **B** Pre-eruptive debris-flow deposits overlain by 1993 block-and-ash flow deposits, in turn covered by remobilized fluvial debris (hammer for scale); **C** Post-eruptive channel development reflecting short-term intense downcutting (*photos* D. Karátson)

Photo 83 Mount St. Helens (Washington, USA), viewed from the north, showing the horseshoe scar resulting from the 18 May 1980 flank failure. Note the truncated, flat summit (reduced by 400 m from an original altitude of 2950 to 2549 m), the resulting central depression (2×1.5 km in diameter), and the increased fluvial erosion in the ring plain (*photo* D. Karátson)

Photo 84 Socompa volcano (6031 m, Chile-Argentina) on oblique satellite image (c. 25 km in diameter at the base). Its striking horseshoe scar, with a 90° opening, faces northwest (to the left). The debris avalanche occurred c. 7200 years ago, but subsequent lava dome activity has partially filled the depression *Credit*: Google Earth

Photo 85 Horseshoe-shaped scar of the 2665 m-high Papandayan volcano, Java (Indonesia). Its first historical eruption occurred in 1772, producing a deadly debris avalanche which destroyed 40 villages and killed 3000 people. The resulting depression shown here hosts the active Kawah Mas ('Golden Crater') with its sulphurous fumarole field (*photo* D. Karátson)

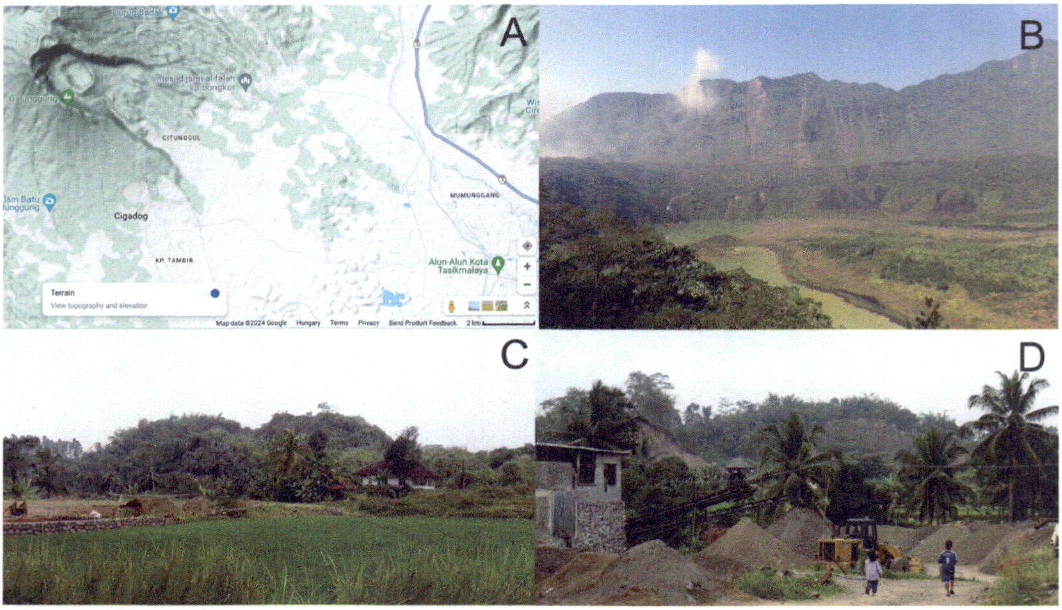

Photo 86 Galunggung volcano, Java (2168 m). **A** Google Map image of the volcano. **B** Lake-filled maar crater from the highly explosive, catastrophic 1982/83 eruption lies inside the striking horseshoe scar (*photo* M. Favalli). **C** Debris-avalanche hummocks from a flank failure about 4200 years ago (*photo* D. Karátson); **D** Material from this debris avalanche deposit is often used in construction (*photo* courtesy of M. Favalli)

Photo 87 Parinacota (Chile-Bolivia) experienced a huge debris avalanche dated at c. 9 ka. The debris-avalanche deposit contains huge, outsized megablocks (inset) called torevas, up to 300 m high and 500–700 m across, with less voluminous hummocks to the left (*photo* D. Karátson)

Photo 88 Mount Shasta in the centre background (4322 m; California, USA), viewed from the northwest. The individual hills in the foreground are huge megablocks 100–150 m high and up to 1–1.5 km across, transported during the eruption c. 300–400 ka (Christiansen et al., 2017) (*photo* D. Karátson)

Photo 89 Characteristics of the Mt. St. Helens 1980 debris-avalanche deposit along the upper reaches of the North Folk Toutle River. The hummocky terrain (**A**) consists of block facies representing coherent parts of the original volcanic succession, e.g. volcaniclastic units (**B**). The hummocky character has not been obliterated despite the intense fluvial downcutting of the past decades (**C**). Composite hummocks (**D**) consist of dark-coloured andesite and lighter gray dacite units (*photos* D. Karátson)

Photo 90 The mixed facies of debris-avalanche deposits (DAD) have characteristically lost their primary structure. Examples here are all Late Miocene, some containing coherent DAD blocks. **A** Is from Călimani Mts, Romania, **B–D** are from the Monts Dore (Massif Central, France). Note the huge volcanoclastic block embedded in matrix facies in **B**. **C** shows a block-rich mixed facies, and **D** a block-poor mixed facies. People give scale (*photos* D. Karátson)

Photo 91 Lahar flow in Curah Lengkong River, southeast flank of Semeru volcano, Java, Indonesia). The boulder-rich lahar front is ~4 m high and ~ 18 m wide, advancing at 4.5 m/s (*photo* courtesy of F. Lavigne, February 2002)

Photo 92 Impact and damage caused by lahars in Kali Putih River, western flank of Merapi volcano, Indonesia, following the 2010 eruption. **A** The red dotted line inside a building indicates the upper flood level from a lahar event, while the red dashed line below marks the inundation from the deposit. **B** A destroyed building shows how reinforced concrete masonry walls failed while deposits remain inside; lahar flowed from right to left. **C** Annotations of lahar damage to the Gejugan dam I upstream of Kali Putih in 2010 (S.D. Hadmoko, *in:* Jenkins et al., 2015). (*photos* courtesy of S. Jenkins et al. 2015, *in:* Thouret et al., 2023)

References

Acocella V (2021) Volcano-tectonic processes. Advances in volcanology series. Springer, 552 pp. ISSN: 2364-3277

Allen JRL (1985) Loose-boundary hydraulics and fluid mechanics: selected advances since 1961. In: Brenchley PJ, Williams PJ (eds) Sedimentology: recent developments and applied aspects. Geol Soc Lond, Blackwell Scientific Publications, Oxford, pp 7–28

Allen SR, Hayward BW, Mathews E (2007) A facies model for a submarine volcaniclastic apron: the Miocene Manukau Subgroup, New Zealand. GSA Bull 119(5–6):725–742. https://doi.org/10.1130/B26066.1

Anagnostopoulos VK, Anastasakis G (2020) Volcanogenic mass flow deposits and seafloor diapirism following the largest insular Quaternary eruption of the eastern Mediterranean at Nisyros island, Aegean Volcanic Arc. Mar Geol 425:106185. https://doi.org/10.1016/j.margeo.2020.106185

Ancey C (2007) Plasticity and geophysical flows: a review. J Non-Newt Fluid Mech 142:4–35. https://doi.org/10.1016/j.jnnfm.2006.05.005

Arguden AT, Rodolfo KS (1990) Sedimentologic and dynamic differences between hot and cold laharic debris flows of Mayon Volcano, Philippines. Geol Soc Amer Bull 102(7):865–876. https://doi.org/10.1130/0016-7606(1990)102<0865:SADDBH>2.3.CO;2

Baxter PJ, Horwell CJ (2015) Chapter 60 – Impacts of Eruptions on Human Health. In: Sigurdsson H (ed) The Encyclopedia of Volcanoes (Second Edition), Academic Press, pp 1035–1047.https://doi.org/10.1016/B978-0-12-385938-9.00060-2

Begét JE, Kienle J (1992) Cyclic formation of debris avalanches at Mount St Augustine volcano. Nature 356(701–704):701–704. https://doi.org/10.1038/356701a0

Begét JE, Kowalik Z (2006) Confirmation and calibration of computer modeling of tsunamis produced by Augustine volcano, Alaska. Sci Tsunami Hazards 24(4):257–266

Belousov A, Belousova M (1998) Bezymyannyi eruption on March 30, 1956 (Kamchatka): sequence of events and debris-avalanche deposits. Volc Seism 20(1):29–47

Belousov A, Belousova M, Voight B (1999) Multiple edifice failures, debris avalanches and associated eruptions in the Holocene history of Shiveluch volcano, Kamchatka, Russia. Bull Volcanol 61:324–342. https://doi.org/10.1007/s004450050300

Belousov A, Voight B, Belousova M (2007) Directed blast and blast-generated pyroclastic density currents: a comparison of the Bezymianny 1956, Mount St Helens 1980, and Soufriére Hills, Montserrat 1997 eruptions and deposits. Bull Volcanol 69:701–740. https://doi.org/10.1007/s00445-006-0109-y

Belousova M, Belousov A (1995) Prehistoric and 1933 debris avalanches and associated eruptions of Harimkotan Volcano (Kurile Islands). Per Mineral 64:99–100

Belousova M, Belousov A (2011) Tsunamigenic volcanic landslides of Kurile-Kamchatka arc. Geophys Res Abs 13, EGU2011-11665-3

Bernard B, van Wyk de Vries B, Barba D, Leyrit H, Robin C, Alcaraz S, Samaniego P (2008) The Chimborazo sector collapse and debris avalanche: deposit characteristics as evidence of emplacement mechanisms. J Volcanol Geother Res 176(1):36–43. https://doi.org/10.1016/j.jvolgeores.2008.03.012

Bernard B, Takarada S, Andrade SD, Dufresne A (2021) Terminology and strategy to describe large volcanic landslides and debris avalanches. In: Roverato M, Dufresne A, Procter J (eds) Volcanic debris avalanches: from collapse to hazard. Advances in volcanology. Springer, pp 51–73. https://doi.org/10.1007/978-3-030-57411-6

Bettella F, Bisantino T, D'Agostino V, Gentile, F (2012) Debris-flow runout distance: laboratory experiments on the role of Bagnold, Savage and Friction numbers. In: de Wrachien D, Brebbia CA, Mambretti S (eds) Monitoring, stimulation, prevention and remediation of dense and debris flows IV. WIT Press, Ashurst, Southampton, UK, pp 27–36

Boggs S (2006) Principles of sedimentology and stratigraphy, 4th edn. Pearson Education Inc., Upper Saddle River, 662 pp

Bougouin A, Roche O, Paris R, Huppert HE (2021) Experimental insights on the propagation of fine-grained geophysical flows entering water. J Geophys Res: Oceans 126:e2020JC016838. https://doi.org/10.1029/2020JC016838

Boulesteix T, Hildenbrand A, Gillot P-Y, Soler V (2012) Eruptive response of oceanic islands to giant landslides: new insights from the geomorphologic evolution of the Teide-Pico Viejo volcanic complex (Tenerife, Canary). Geomorphology 138:61–73. https://doi.org/10.1016/j.jvolgeores.2013.03.008

Bouma AH (1962) Turbidites. Dev Sedimentol 3: 247–256. https://doi.org/10.1016/S0070-4571(08)70967-1

Bronto S (1989) Volcanic geology of Galunggung, West Java, Indonesia. PhD thesis, University of Canterbury, NZ, 490 pp

Brown SK, Jenkins SF, Sparks RSJ, Odbert H, Auker MR (2017) Volcanic fatalities database: analysis of volcanic threat with distance and victim classification. J Appl Volcanol 6(15):1–20. https://doi.org/10.1186/s13617-017-0067-4

Brunet M, Le Friant A, Boudon G, Lafuerza S, Talling P, Hornbach M, Ishizuka O, Lebas E, Guyard H, IODP Expedition 340 Science party (2016) Composition, geometry and emplacement dynamics of a large volcanic island landslide offshore Martinique: from volcano flank-collapse to seafloor sediment failure? Geochem Geophys Geosyst 17(3):699–724. https://doi.org/10.1002/2015GC006034

Bull S, Cartwright J, Huuse M (2009) A review of kinematic indicators from mass-transport complexes using 3D seismic data. Mar Pet Geol 26(7):1132–1151. https://doi.org/10.1016/j.marpetgeo.2008.09.011

Campbell CS (1989) Self-lubrication for long runout landslides. J Geol 97(6):653–665. https://doi.org/10.1086/629350

Camus G, Diament M, Gloaguen M, Provost A, Vincent P (1992) Emplacement of a Debris Avalanche during the 1883 eruption of Krakatau (Sunda Straits, Indonesia). GeoJournal 28:123–128. https://doi.org/10.1007/BF00177224

Capra LP, Caballero LG (2021) Chapter 14 – Rainfall on active volcanoes: morphological response and associated processes. In: Rodrigo-Comino J (ed) Precipitation. Elsevier, pp 327–347.https://doi.org/10.1016/B978-0-12-822699-5.00012-4

Cardona S, Wood LJ, Dugan B, Jobe Z, Strachan LJ (2019) Characterization of the Rapanui mass-transport deposit and the basal shear zone: Mount Messenger Formation, Taranaki Basin, New Zealand. Sedimentology 67:2111–2148. https://doi.org/10.1111/sed.12697

Carey S (2000) Volcaniclastic sedimentation around island arcs. In: Sigurdsson H, Houghton B, McNutt SR, Rymer H, Stix J (eds) Encyclopedia of volcanoes. Academic Press, San Diego, CA, pp 627–642. ISBN: 978-0-12-385938-9

Carey SN, Schneider JL (2011) Volcaniclastic processes and deposits in the Deep-Sea. In: Huneke H, Mulder T (eds) Developments in sedimentology, Deep-Sea sediments, vol 63. Elsevier, pp 457–515

Carracedo JC (1999) Growth, structure, instability and collapse of Canarian volcanoes and comparisons with Hawaiian volcanoes. J Volcanol Geother Res 94(1–4):1–19. https://doi.org/10.1016/S0377-0273(99)00095-5

Cas RAF, Wright JV (1988) Volcanic successions: modern and ancient. Springer, Dordrecht, Netherlands, pp 362–410. ISBN-13: 978-0-412-44640-5

Chadwick WW Jr, de Roy T, Carrasco A (1991) The September 1988 intracaldera avalanche and eruption at Fernandina volcano, Galapagos Islands. Bull Volcanol 53:276–286. https://doi.org/10.1007/BF00414524

Charbonnier SJ, Gertisser R (2011) Deposit architecture and dynamics of the 2006 block-and-ash flows of Merapi Volcano, Java, Indonesia. Sedimentology 58(6):1573–1612. https://doi.org/10.1111/j.1365-3091.2011.01226.x

Chiba T (2018) The topographic expression by Red Relief Image Map contributes to outreach. Japan Geological Union Meeting, session General (Education and Outreach [G04-16]

Chiba S, Kimura J-I (2001) Geology and growth history of Bandai volcano, Tohoku-Honshu arc, Japan—analysis of volcanic activity by tephrochronology. Jpn Mag Mineral Petrol Sci 30:126–156 (in Japanese with English abstract)

Christiansen RL, Calvert AT, Grove TL (2017) Geologic field-trip guide to Mount Shasta volcano, northern California: US Geol Surv Sci Invest Rep 2017-5022-K3, 33 p. https://doi.org/10.3133/sir20175022K3

Clare MA, Le Bas T, Price DM, Hunt JE, Sear D, Cartigny MJB, Vellinga A, Symons W, Firth C, Cronin S (2018) Complex and cascading triggering of submarine landslides and turbidity currents at volcanic islands revealed from integration of high-resolution onshore and offshore surveys. Front Earth Sci 6:223. https://doi.org/10.3389/feart.2018.00223

Clare MA, Yeo IA, Watson S, Wysoczanski R, Seabrook S, Mackay K, Hunt JE, Lane E, Talling PJ, Pope E, Cronin S, Ribó M, Kula T, Tappin D, Henrys S, de Ronde C, Urlaub M, Kutterolf S, Fonua S, Panuve S, Veverka D, Rapp R, Kamalov V, Williams M (2023) Fast and destructive density currents created by ocean-entering volcanic eruptions. Science 381(6662):1085–1092. https://doi.org/10.1126/science.adi3038

Cole SE (2011) Geophysical investigation into the internal dynamics of moving lahars. (Unpubl) PhD dissertation, Massey University, Palmerston North, New Zealand

Cole SE, Cronin SJ, Sherburn S, Manville V (2009) Seismic signals of snow-slurry lahars in motion: 25 September 2007, Mt. Ruapehu, New Zealand. Geophys Res Lett 36(9):L09405. https://doi.org/10.1029/2009GL038030

Collot J-Y, Lewis K, Geoffroy L, Lallemand S (2001) The giant Ruatoria debris avalanche on the northern Hikurangi margin, New Zealand: result of oblique seamount subduction. J Geophys Res 106(B9):19271–19297. https://doi.org/10.1029/2001JB900004

Cortés A, Komorowski JC, Macías JL, Capra L, Layer PW (2019) Late Pleistocene-Holocene debris avalanche deposits from Volcán de Colima. In: Varley N, Connor CB, Komorowski JC (eds) Volcán de Colima. Portrait of a Persistently Hazardous Volcano. Springer, pp 55–79. https://doi.org/10.1007/978-3-642-25911-1

Costa JE (1988) Rheologic, geomorphic, and sedimentologic differentiation of water floods, hyperconcentrated flows, and debris flow. In: Hupp CR, Baker VR, Kochel RC, Patton PC (eds) Flood geomorphology. Wiley, New York, pp 268–317

Coussot P, Meunier M (1996) Recognition, classification and mechanical description of debris flows. Earth-Sci Rev 40(209):182–227. https://doi.org/10.1016/0012-8252(95)00065-8

Crandell DR (1989) Gigantic debris avalanche of Pleistocene age from Ancestral Mount Shasta Volcano, California, and Debris-Avalanche Hazard Zonation. US Geol Surv Bull 1861, 32 p

Crandell DR, Miller CD, Glicken HX, Christiansen RL, Newhall CG (1984) Catastrophic debris avalanche from ancestral Mount Shasta volcano. Calif Geol 12(3):143–146. https://doi.org/10.1130/0091-7613(1984)12<143:CDAFAM>2.0.CO;2

Cronin SJ, Neall VE (1997) A late Quaternary stratigraphic framework for the northeastern Ruapehu and eastern Tongariro ring plains, New Zealand. New Zeal J Geol Geophys 40:185–197. https://doi.org/10.1080/00288306.1997.9514752

Dasgupta P (2003) Sediment gravity flow—the conceptual problem. Earth Sci Rev 62:265–281. https://doi.org/10.1016/S0012-8252(02)00160-5

Davies TRH (1982) Spreading of rock avalanche debris by mechanical fluidization. Rock Mech 15(1):9–24. https://doi.org/10.1007/BF01239474

Davies TRH, McSaveney MJ (2006) Runout of rock avalanches and volcanic debris avalanche. In: Picarelli L (ed) Proceedings of the international conference on fast slope movements: prediction and prevention for risk mitigation, vol 2, pp 113–132. ISBN: 8855528335

Davies TRH, McSaveney MJ, Kelfoun K (2010) Runout of the Socompa volcanic debris avalanche, Chile: a mechanical explanation for low basal shear resistance. Bull Volcanol 72:933–944. https://doi.org/10.1007/s00445-010-0372-9

De Bélizal E, Lavigne F, Hadmoko DS, Degeai J-P, Dipayana GA, Mutaqin BW, Marfai MA, Coquet M, Le Mauff B, Robin AK, Cholik N, Nurnaning A (2013) Rain-triggered lahars following the 2010 eruption of Merapi Volcano, Central Java, Indonesia. J Volcanol Geother Res 261:330–347. https://doi.org/10.1016/j.jvolgeores.2013.01.010

De Haas T, Braat L, Leuven JRFW, Lokhorst IR, Kleinhans MG (2015) Effects of debris flow composition on runout, depositional mechanisms, and deposit morphology in laboratory experiments. J Geophys Res Earth Surf 120:1949–1972. https://doi.org/10.1002/2015JF003525

Delcamp A, Delvaux D, Kwelwa S, Macheyeki A, Kervyn M (2016) Sector collapse events at volcanoes in the North Tanzanian divergence zone and their implications for regional tectonics. GSA Bull 128(1–2):169–186. https://doi.org/10.1130/B31119.1

Doyle EE, Cronin SJ, Cole SE, Thouret J-C (2010) The coalescence and organization of lahars at Semeru, Indonesia. Bull Volcanol 72(8):961–970. https://doi.org/10.1007/s00445-010-0381-8

Doyle EE, Cronin SJ, Thouret J-C (2011) Cycles of bulking and debulking in lahars at Semeru, Indonesia. Geol Soc Amer Bull 123:1234–1246. https://doi.org/10.1130/B30227.1

Doyle EE, Huppert HE, Lube G, Mader HM, Sparks RSJ (2007) Static and flowing regions in granular collapses down channels: insights from a sedimenting shallow water model. Phys Fluids 19(10):106601–106616. https://doi.org/10.1063/1.2773738

Druitt T, Kutterolf S, Ronge TA, Hübscher C, Nomikou P. et al. (2024) Giant offshore pumice deposit records a shallow submarine explosive eruption of ancestral Santorini. Commun Earth Environ 5. http://doi.org.10.1038/s43247-023-01171-z

Dufresne A, Siebert L, Bernard B (2021) Distribution and geometric parameters of volcanic debris avalanche deposits. In: Roverato M, Dufresne A, Procter J (eds) Volcanic debris avalanches: from collapse to hazard. Advances in volcanology. Springer, pp 75–90. https://doi.org/10.1007/978-3-030-57411-6

Dufresne A, Siebert L, Bernard B, Sparks RSJ, Takarada J, Clavero J, Belousov A, Belousova M (2008) Volcanic debris avalanche deposit database—a progress report. Abstract P09, IAVCEI General Assembly, Reykjavik, Iceland

Dumaisnil C, Thouret J-C, Chambon G, Doyle EE, Cronin SJ (2010) Hydraulic, physical and rheological characteristics of rain-triggered lahars at Semeru volcano, Indonesia. Earth Surf Proc Landf 35:573–590. https://doi.org/10.1002/esp.2003

Elverhoi A, Breien H, Blasio FVD, Harbitz CB, Pagliardi M (2010) Submarine landslides and the importance of the initial sediment composition for run-out length and final deposit. Ocean Dynam 60:1027–1046. https://doi.org/10.1007/s10236-010-0317-z

Elsworth D, Day SJ (1999) Flank collapse triggered by intrusion: the Canarian and Cape Verde Archipelagoes. J Volcanol Geother Res 94(1–4): 323–340. https://doi.org/10.1016/S0377-0273(99)00110-9

Erismann TH (1979) Mechanisms of large landslides. Rock Mechanics 12, 15–46. https://doi.org/10.1007/BF01241087

Evans SG (2006) Single-event landslides resulting from massive rock slope failure: characterising their frequency and impact on society. In: Evans SG, Mugnozza GS, Strom A, Hermanns RL (eds) Landslides from massive rock slope failure. NATO Science Series, vol 49. Springer, Dordrecht, pp 53–73. https://doi.org/10.1007/978-1-4020-4037-5_2

Fagents SA, Baloga SM (2006) Toward a model for the bulking and debulking of lahars. J Geophys Res: Solid Earth 111:21. https://doi.org/10.1029/2005JB003986

Fink JH (1992) What goes up could come down. Nature 359:102–103. https://doi.org/10.1038/359102a0

Fisher RV, Schmincke H-U (1984) Pyroclastic rocks. Springer, 472 pp. https://doi.org/10.1007/978-3-642-74864-6

Fisher RV, Smith GA (1991) Sedimentation in volcanic settings. Special Publications of Society for Sedimentary Geology 45, Tulsa, Oklahoma, 257 pp

Francis P (1993) Volcanoes. A planetary perspective. Clarendon Press, Oxford, UK, 452 pp. ISBN: 0-19-854033-7

Francis PW, Hawkesworth CJ (1994) Late Cenozoic rates of magmatic activity in the Central Andes and their relationships to continental crust formation and thickening. J Geol Soc 151(5):845–854. https://doi.org/10.1144/gsjgs.151.5.0845

Francis PW, Self S (1987) Collapsing volcanoes. Sci Am 256(6):90–99. https://www.jstor.org/stable/24979404

Francis PW, Wells GL (1988) Landsat Thematic Mapper observations of debris avalanche deposits in the

Central Andes. Bull Volcanol 50:258–278. https://doi.org/10.1007/BF01047488

Freundt A (2003) Entrance of hot pyroclastic flows into the sea: experimental observations. Bull Volcanol 65:144–164. https://doi.org/10.1007/s00445-002-0250-1

Furukawa R, Nanayama F, Yoshimoto M (2008) Volcanic debris avalanche as a cause of a historic tsunami: the A.D. 1640 eruption of the Hokkaido-Komagatake Volcano, Northern Japan. In: Wallendorf L, Jones C, Jaffe B, Ewing L (eds) Coastal disasters and solutions: tsunamis, pp 235–247. https://doi.org/10.1061/40978(313)

Gabrielli S, Spagnolo M, De Siena ML (2020) Geomorphology and surface geology of Mount St. Helens Volcano. J Maps 16(2):585–594. https://doi.org/10.1080/17445647.2020.1790048

George DL, Iverson RI (2011) A two-phase debris-flow model that includes coupled evolution of volume fractions, granular dilatancy, and pore-fluid pressure. It J Eng Geol Environ 415–424. 5th int conf on debris flows: hazards mitigation, mechanics and prediction assessment. https://doi.org/10.4408/IJEGE.2011-03.B-047

Glicken H (1991) Sedimentary architecture of large volcanic debris avalanches. In: Fisher RV, Smith GA (eds) Sedimentation in volcanic settings. Spec Pub SEPM 45:99–106. https://doi.org/10.2110/pec.91.45.0099

Glicken H (1996) Rockslide-debris avalanche of May 18, 1980, Mount St. Helens volcano, Washington. US Geol Surv Open File Report 96-677, 90 pp. https://doi.org/10.3133/ofr96677

Goff F, Kelley SA, Goff CJ, McCraw DJ, Osburn GR, Lawrence JR, Drakos PG, Skotnicki SJ (2019) Geologic map of Mount Taylor volcanic area, 1:36,000 scale. GM 80, New Mexico Bureau of Geology and Mineral Resources. ISBN: 978-1-883905-44-6. https://doi.org/10.58799/GM-80

Gomez C, Shinohara Y, Tsunetaka H, Hotta N, Bradak B, Sakai Y (2021) Twenty-five years of geomorphological evolution in the Gokurakudani Gully (Unzen Volcano): topography, subsurface geophysics and sediment analysis. Geosciences 11:457. https://doi.org/10.3390/geosciences11110457

Gorshkov GS (1959) Gigantic eruption of the volcano Bezymianny. Bull Volcanol 20:77–109. https://doi.org/10.1007/BF02596572

Gorshkov GS, Dubik YM (1970) Gigantic directed blast at Shiveluch volcano (Kamchatka). Bull Volcanol 34:261–288. https://doi.org/10.1007/BF02597790

Gran KB, Montgomery DR (2005). Spatial and temporal patterns in fluvial recovery following volcanic eruptions: Channel response to basin-wide sediment loading at Mount Pinatubo, Philippines. Bul Geol Soc Amer 117(1–2):195–211. https://doi.org/10.1130/B25528.1

Gran KB, Montgomery DR, Halbur JC (2011) Long-term elevated post-eruption sedimentation at Mount Pinatubo, Philippines. Geology 39:367–370. https://doi.org/10.1130/G31682.1

Grosse P, van Wyk de Vries B, Euillades PA, Kervyn M, Petrinovic IA (2012) Systematic morphometric characterization of volcanic edifices using digital elevation models. Geomorphology 136(1):114–131. https://doi.org/10.1016/j.geomorph.2011.06.001

Gudmundsson M (2015) Hazards from lahars and jökulhlaups. In: Sigurdsson H, Houghton B, McNutt SR, Rymer H, Stix J (eds) The encyclopedia of volcanoes, 2nd edn. Academic Press, Elsevier, pp 971–984. ISBN: 978-0-12-385938-9

Hampton MA (1979) Buoyancy in debris flows. J Sediment Petrol 49(3):753–758. https://doi.org/10.1306/212F7838-2B24-11D7-8648000102C1865D

Hampton MA, Lee HJ, Locat J (1996) Submarine landslides. Rev Geophys 34(1):33–59. https://doi.org/10.1029/95RG03287

Han Z, Chen J, Xu P, Zhan J (2017a) A well-balanced numerical scheme for debris flow run-out prediction in Xiaojia Gully considering different hydrological designs. Landslides 14:2105–2114. https://doi.org/10.1007/s10346-017-0850-7

Han Z, Li L, Huang J, Chen G, Xu L, Tang C, Zhang H, Shang Y (2017b) Numerical simulation for run-out extent of debris flows using an improved cellular automaton model. Bull Eng Geol Environ 76:961–974. https://doi.org/10.1007/s10064-016-0902-6

Harris AJL, Vallance JW, Kimberly PG, Rose WI, Matías O, Flynn LP, Garbeil H (2006) Downstream aggradation owing to lava dome extrusion and rainfall runoff at Volcan Santiaguito, Guatemala. Geol Soc Amer Spec Pap 412:85–104. https://doi.org/10.1130/2006.2412(05)

Hayes SK, Montgomery DR, Newhall CG (2002) Fluvial sediment transport and deposition following the 1991 eruption of Mount Pinatubo. Geomorphology 45(3–4):211–224. https://doi.org/10.1016/S0169-555X(01)00155-6

Hoblitt RP, Miller CD, Vallance JW (1981) Origin and stratigraphy of the deposit produced by the May 18 directed blast. In: Lipman PW, Mullineaux DR (eds) The 1980 eruptions of Mount St. Helens, Washington. US Geol Surv Prof Pap, Washington DC, USA, vol 1250, pp 401–419

Hora JM, Singer BS, Wörner G (2007) Volcano evolution and eruptive flux on the thick crust of the Andean Central Volcanic Zone: $^{40}Ar/^{39}Ar$ constraints from Volcán Parinacota, Chile. GSA Bull 119(3–4):343–362. https://doi.org/10.1130/B25954.1

Hoshizumi H, Uto K, Watanabe K (1999) Geology and eruptive history of Unzen volcano, Shimabara Peninsula, Kyushu, SW Japan. J Volcanol Geother Res 89:81–94

Hsü KJ (1975) Catastrophic debris avalanches (Sturzströms) generated by rockfalls. GSA Bull 86:129–140

Hübl J, Suda J, Proske D, Kaitna R, Scheidl C (2009) Debris flow impact estimation. In: International symposium on water management and hydraulic engineering, Ohrid, Macedonia, 1–5 September 2009 A56, pp 137–148

Hungr O, Evans SG (2004) Entrainment of debris in rock avalanches: an analysis of a long run-out mechanism. GSA Bull 116(9–10):1240–1252. https://doi.org/10.1130/B25362.1

Hungr O, Jakob M (2005) Debris flows hazards and related phenomena. Praxis, Springer, 728 pp. ISBN: 978-3-540-20726-9

Hunt JE, Cassidy M, Talling PJ (2018) Multi-stage volcanic island flank collapses with coeval explosive caldera-forming eruptions. Sci Rep 8:1146. https://doi.org/10.1038/s41598-018-19285-2

Hunt JE, Tappin DR, Watt SFL, Susilohadi S, Novellino A, Ebmeier SK, Cassidy M, Engwell SL, Grilli ST, Hanif M, Priyanto WS, Clare MA, Abdurrachman UU (2021) Submarine landslide megablocks show half of Anak Krakatau island failed on December 22nd, 2018. Nat Commun 12:2827. https://doi.org/10.1038/s41467-021-22610-5

Inbar M, Enriquez AR, Graniel JHG (2001) Morphological changes and erosion processes following the 1982 eruption of El Chichón volcano, Chiapas, Mexico. Géomorphol Relief Proc Environ 3:175–184. https://doi.org/10.3406/morfo.2001.1100

Inokuchi T (1988) Gigantic landslides and debris avalanches on volcanoes in Japan. In: Proceedings of the Kagoshima international conference on volcanoes. National Institute for Research Administration, Japan, pp 456–459

Inokuchi T (2006) Properties of sector-collapse and debris avalanches on Quaternary volcanoes in Japan. J Jpn Landslide Soc 42(5):409–420. https://doi.org/10.3313/jls.42.5_409

Inoue K (1999) Shimabara-Shigatusaku earthquake and topographic changes by Shimabara Catastrophe in 1792. J Jpn Soc Erosion Control Eng 52(4):45–54. https://doi.org/10.11475/sabo1973.52.4_45

Iverson RM (1978) Gravity-driven mass flows. In: Sedimentology. Encyclopedia of earth science. Springer, Berlin, Heidelberg. https://doi.org/10.1007/3-540-31079-7_106

Iverson RM (1997) The physics of debris flows. Rev Geophys 35(3):245–296. https://doi.org/10.1029/97RG00426

Iverson RM (2014) Debris flows: behavior and hazard assessment. Geol Today 30(1):15–20. https://doi.org/10.1111/gto.12037

Iverson RM, George DL (2016) Modelling landslide liquefaction, mobility bifurcation and the dynamics of the 2014 Oso disaster. Geotechnique 66:175–187

Iverson RM, George DL, Logan M (2016) Debris-flow run-up on vertical barriers and adverse slopes. J Geophys Res Earth Surf 121:2333–2357. https://doi.org/10.1002/2016JF003933

Iverson RM, Logan M, LaHusen RG, Berti M (2010) The perfect debris flow? Aggregated results from 28 large-scale experiments. J Geophys Res 115(F3):1–29. https://doi.org/10.1029/2009JF001514

Iverson RM, Reid ME, Logan M, LaHusen RG, Godt JW, Griswold JP (2011) Positive feedback and momentum growth during debris-flow entrainment of wet bed sediment. Nat Geosci 4(2):116–121. https://doi.org/10.1038/ngeo1040

Jenkins SF, Phillips JC, Price R, Feloy K, Baxter PJ, Sri Hadmoko D, de Bélizal E (2015) Developing building-damage scales for lahars: application to Merapi volcano, Indonesia. Bull Volcanol 77(75):1–17. https://doi.org/10.1007/s00445-015-0961-8

John DA, Sisson TW, Breit GN, Rye RO, Vallance JW (2008) Characteristics, extent and origin of hydrothermal alteration at Mount Rainier Volcano, Cascades Arc, USA: implications for debris-flow hazards and mineral deposits. J Volcanol Geother Res 175:289–314. https://doi.org/10.1016/j.jvolgeores.2008.04.004

Johnson AM (1970) Physical process in geology. Freeman and Cooper, San Francisco, 577 pp

Johnson B (1978) Blackhawk landslide, California, U.S.A. Developments in geotechnical engineering, vol 14(A), pp 481–504. https://doi.org/10.1016/B978-0-444-41507-3.50022-2

Johnson B, Campbell CS, Melosh JH (2016) The reduction of friction in long runout landslides as an emergent phenomenon. J Geophys Res Earth Surf 121:881–889. https://doi.org/10.1002/2015jf003751

Johnson RW (1987) Large-scale volcanic cone collapse: the 1888 slope failure of Ritter volcano, and other examples from Papua New Guinea. Bull Volcanol 49:669–679. https://doi.org/10.1007/BF01080358

Jousset P, Pallister J, Boichu M, Buongiorno MF, Budisantoso A, Costa F, Lavigne F (2012) The 2010 explosive eruption of Java's Merapi volcano—a '100-year' event. J Volcanol Geother Res 241:121–135. https://doi.org/10.1016/j.jvolgeores.2012.06.018

Karátson D, Németh K (2001) Lithofacies associations of an emerging volcaniclastic apron in a Miocene volcanic complex: an example from the Börzsöny Mountains, Hungary. Int J Earth Sci (Geol Rundsch) 90:776–794. https://doi.org/10.1007/s005310100193

Karátson D, Thouret J-C, Moriya I, Lomoschitz A (1999) Erosion calderas: origins, processes, structural and climatic control. Bull Volcanol 61:174–193. https://doi.org/10.1007/s004450050270

Kataoka K (2005) Distal fluvio-lacustrine volcaniclastic resedimentation in response to an explosive silicic eruption: the Pliocene Mushono tephra bed, central Japan. GSA Bull 117(1–2):3–17. https://doi.org/10.1130/B25379.1

Kataoka KS, Manville V, Nakajo T, Urabe A (2009) Impacts of explosive volcanism on distal alluvial sedimentation: examples from the Pliocene-Holocene volcaniclastic successions of Japan. Sed Geol 220:306–317. https://doi.org/10.1016/j.sedgeo.2009.04.016

Keating BH, McGuire WJ (2000) Island edifice failures and associated tsunami hazards. Pure Appl Geophys 157:899–955. https://doi.org/10.1007/s000240050011

Keigler R, Thouret J-C, Hodgson K, Neall V, Lecointre J, Procter JN, Cronin SJ (2011) The Whangaehu Formation: debris-avalanche and lahar deposits from ancestral Ruapehu volcano, New Zealand.

Geomorphology 133(1):57–79. https://doi.org/10.1016/j.geomorph.2011.06.019

Kelfoun K, Druitt TH (2005) Numerical modeling of the emplacement of Socompa rock avalanche, Chile. J Geophys Res 110:B12202. https://doi.org/10.1029/2005JB003758

Kelfoun K, Giachetti T, Labazuy P (2010) Landslide-generated tsunamis at Réunion Island. J Geophys Res Earth Surf 115(1–17):F04012. https://doi.org/10.1029/2009JF001381

Kent PE (1966) The transport mechanism in catastrophic rock falls. J Geol 74(1):79–83. https://doi.org/10.1086/627142

Kuenzi WD, Horst OH, McGehee RV (1979) Effect of volcanic activity on fluvial-deltaic sedimentation on a modern arc-trench gap, southwestern Guatemala. GSA Bull 90(9):827–838. https://doi.org/10.1130/0016-7606(1979)90<827:EOVAOF>2.0.CO;2

Lamarche G, Joanne C, Collot J-Y (2008) Successive, large mass-transport deposits in the south Kermadec fore-arc basin, New Zealand: the Matakaoa submarine instability complex. Geochem Geophys Geosyst 9(4):Q04001. https://doi.org/10.1029/2007GC001843

Lavigne F, Suwa H (2004) Contrasts between debris flows, hyperconcentrated flows and stream flows at a channel of Mount Semeru, East Java, Indonesia. Geomorphology 61(1–2):41–58. https://doi.org/10.1016/j.geomorph.2003.11.005

Lavigne F, Thouret J-C (2003) Sediment transportation and deposition by rain-triggered lahars at Merapi volcano, central Java, Indonesia. Geomorphology 49(1–2):45–69. https://doi.org/10.1016/S0169-555X(02)00160-5

Le Friant A, Deplus C, Boudon G, Feuillet N, Trofimous J, Komorowski J-C, Sparks RSJ, Talling P, Loughlin S, Palmer M, Ryan G (2010) Eruption of Soufrière Hills (1995–2009) from an offshore perspective: insights from repeated swath bathymetry surveys. Geophys Res Lett Solid Earth 37(19):L11307. https://doi.org/10.1029/2010GL043580

Le Friant A, Deplus C, Boudon G, Sparks RSJ, Trofimovs J, Talling JP (2009) Submarine deposition of volcaniclastic material from the 1995–2005 eruptions of Soufriere Hills volcano, Montserrat. J Geol Soc 166:171–182. https://doi.org/10.1144/0016-76492008-047

Lebedeva EV (2022) Present sedimentation in the volcanic lakes of the Kurile-Kamchatka region (Russia) as a basis for paleoreconstructions. Limnol Freshw Biol 4:1464–1466. https://doi.org/10.31951/2658-3518-2022-A-4-1464

Legros F, Cantagrel JM, Devouard B (2000) Pseudotachylite (Frictionite) at the base of the Arequipa volcanic landslide (Peru): implications for emplacement mechanisms. J Geol 108:601–611. https://doi.org/10.1086/314421

León R, Palomino D, Vázquez J-T, Medialdea T, Somoza L (2019) A new scenario for the mass transport deposits west Canary volcanic province. Earth Planet Sci Lett 509:27–37. https://doi.org/10.1016/j.epsl.2018.12.020

Lipman PW, Mullineaux (eds) (1981) The 1980 eruptions of Mount St. Helens, Washington. US Geol Surv Prof Pap 1250, Washington, DC, USA, 844 pp. https://doi.org/10.3133/pp1250

Liu HY, Kou SG, Lindqvist P-A (2005) Numerical studies on the inter-particle breakage of a confined particle assembly in rock crushing. Mech Mater 37(9):935–954. https://doi.org/10.1016/j.mechmat.2004.10.002

Lockwood JP, Hazlett RW (2010) Volcanoes, Global perspectives. Wiley-Blackwell, UK, pp 539. ISBN 978-1-4051-6249-4

Lowe NR (1982) Sediment gravity flows: II. Depositional models with special reference to the deposits of high-density turbidity currents. J Sediment Res 52:279–297. https://doi.org/10.1306/212F7F31-2B24-11D7-8648000102C1865D

Lube G, Cronin SJ, Procter JN (2009) Explaining the extreme mobility of volcanic ice-slurry flows, Ruapehu volcano, New Zealand. Geology 37(1):15–18. https://doi.org/10.1130/G25352A.1

Lube G, Cronin SJ, Thouret JC, Surono S (2011) Kinematic characteristics of pyroclastic density currents at Merapi and controls on their avulsion from natural and engineered channels. Geol Soc Amer Bull 123(5):1127–1140. https://doi.org/10.1130/B30244.1

Madden-Nadeau AL, Cassidy M, Pyle DM, Mather TA, Watt SFL, Engwell SL, Abdurrachman M, Nurshal MEM, Tappin DR, Ismail T (2021) The magmatic and eruptive evolution of the 1883 caldera-forming eruption of Krakatau: integrating field- to crystal-scale observations. J Volcanol Geother Res 411. https://doi.org/10.1016/j.jvolgeores.2021.107176

Major JJ (1997) Depositional processes in large-scale debris-flow experiments. J Geol 105:345–366. https://doi.org/10.1086/515930

Major JJ, Bertin D, Pierson TC, Amigo A, Iroumé A, Ulloa H, Castro J (2016) Extraordinary sediment delivery and rapid geomorphic response following the 2008–2009 eruption of Chaíten volcano, Chile. Water Resour Res 52(7):5075–5094. https://doi.org/10.1002/2015WR018250

Major JJ, Iverson RM (1999) Debris-flow deposition: effects of pore-fluid pressure and friction concentrated at flow margins. Geol Soc Amer Bull 111:1424–1434. https://doi.org/10.1130/0016-7606(1999)111<1424:DFDEOP>2.3.CO;2

Major JJ, Pierson TC, Dinehart RL, Costa JE (2000) Sediment yield following severe volcanic disturbance—a two-decade perspective from Mount St. Helens. Geology 28:819–822. https://doi.org/10.1130/0091-7613(2000)28<819:SYFSVD>2.0.CO;2

Major JJ (2022) Subaerial volcaniclastic deposits—influences of initiation mechanisms and transport behaviour on characteristics and distributions. Geol Soc Lond Spec Publ 520:29–100. https://doi.org/10.1144/SP520-2021-142

Makris S, Roverato M, Dávila-Harris P, Cole P, Manzella I (2023) Distributed stress fluidisation: Insights into the propagation mechanisms of the Abona volcanic debris avalanche (Tenerife) through a novel method for indurated deposit sedimentological analysis. Front Earth Sci 11:1177507. https://doi.org/10.3389/feart.2023.1177507

Manville V, Cronin SJ (2007) Break-out lahar from New Zealand's Crater Lake. EOS, Transactions, AGU 88(43):441–442. https://doi.org/10.1029/2007EO430001

Manville V, Hodgson KA, White JDL (1998) Rheological properties of a remobilized tephra lahar associated with the 1995 eruptions of Ruapehu volcano, New Zealand. New Zeal J Geol Geophys 41:157–164. https://doi.org/10.1080/00288306.1998.9514800

Manville V, Major JJ, Fagents SA (2013) Modeling lahar behavior and hazards. Chap. 4. In: Fagents SA, Gregg TKP, Lopes RMC (eds) Modeling volcanic processes. Cambridge University Press, pp 300–330. ISBN: 978-0-521-89543-9

Manville V, Németh K, Kano K (2009) Source to sink: a review of three decades of progress in the understanding of volcaniclastic processes, deposits, and hazards. Sed Geol 220(3–4):136–161. https://doi.org/10.1016/j.sedgeo.2009.04.022

Martí J, Groppelli G, Brum da Silveira A (2018) Volcanic stratigraphy: a review. J Volcanol Geother Res 357:68–91. https://doi.org/10.1016/j.jvolgeores.2018.04.006

Masson DG, Canals M, Lonsoà BA, Urgeles R, Huhnerbach V (1998) The Canary Debris flow: source area morphology and failure mechanisms. Sedimentology 45:411–432. https://doi.org/10.1046/j.1365-3091.1998.0165f.x

McCoy SW, Kean JW, Coe JA, Staley DM, Wasklewicz TA, Tucker GE (2010) Evolution of a natural debris flow: in situ measurements of flow dynamics, video imagery, and terrestrial laser scanning. Geology 38(8):735–738. https://doi.org/10.1130/G30928.1

McGuire WJ, Jones AP, Neuberg J (1996) Volcano instability on the Earth and other planets. Geol Soc London Spec Publ 110, 388 pp. https://doi.org/10.1017/S0016756897216468

McPhie J, Doyle M, Allen R (1993) Volcanic textures: a guide to the interpretation of textures in volcanic rocks. Centre for Ore Deposit and Exploration Studies, University of Tasmania, Hobart, 198 pp

Mead SRM, Magill C, Lemiale V, Thouret J-C, Prakash M (2017) Quantifying lahar damage using numerical modelling. Nat Hazards Earth Sci Syst 17:703–719. https://doi.org/10.5194/nhess-2016-282

Mehl KW, Schmincke H-U (1999) Structure and emplacement of the Pliocene Roque Nublo debris avalanche deposit, Gran Canaria, Spain. J Volcanol Geother Res 94(1–4):105–134. https://doi.org/10.1016/S0377-0273(99)00100-6

Meiburg E, Nasr-Azadani MM (2018) Gravity and turbidity currents: numerical simulations and theoretical models. In: Clercx HJH, Van Heijst GF (eds) Mixing and dispersion in flows dominated by rotation and buoyancy. Springer, Cham, Switzerland, pp 129–180. https://doi.org/10.1007/978-3-319-66887-1_6

Melosh HJ (1979) Acoustic fluidization: a new geologic process? J Geophys Res: Solid Earth 84(B13):7513–7520. https://doi.org/10.1029/JB084iB13p07513

Melosh HJ (2015) Acoustic fluidization: what it is, and is not. Bridging the Gap III: impact cratering in nature, experiments, and modeling, held 21–26 September, 2015 at University of Freiburg, Germany. LPI Contribution 1861:1004

Meyer GA, Watt PM, Wilder M (2014) Was Mount Taylor glaciated in the Late Pleistocene? An analysis based on field evidence and regional equilibrium line altitudes. New Mexico Geol 36(2):32–39. https://doi.org/10.58799/NMG-v36n2.32

Mimura K, Kano K, Nakano S, Hoshizumi H (1988) Ontake debris avalanche in 1984—flow and deposition mechanism inferred from the deposit. Bull Geol Surv Jpn 39:495–523

Mimura K, Kawachi S, Fujimoto U, Taneichi M, Hyuga T, Ichikawa S, Koizumi M (1982) Debris avalanche hills and their natural remnant magnetisation—Nirasaki debris avalanche, central Japan. J Geol Soc Jpn 88:653–663

Miyaji N, Togashi S, Chiba T (2004) A large-scale collapse event at the Eastern Slope of Fuji Volcano about 2900 years ago. Bull Volcanol Soc Jpn 49(5):237–248. https://doi.org/10.18940/kazan.49.5_237

Moore JG, Albee WC (1981) Topographic and structural changes, March-July 1980—photogrammetric data. In: Lipman PW, Mullineaux (eds) The 1980 eruptions of Mount St. Helens, Washington. US Geol Surv Prof Paper 1250, US Government Printing Office, Washington DC, USA, pp 123–134. https://doi.org/10.3133/pp1250

Moore JG, Normark WR, Holcomb RT (1994) Giant Hawaiian landslides. Annu Rev Earth Planet Sci 22:119–144. 1994AREPS..22..119M

Morgan JK, Silver E, Camerlenghi A, Dugan B, Kirby S, Shipp C, Suyehiro K (2009) Addressing geohazards though ocean drilling. Sci Drill 7:15–30. https://doi.org/10.2204/iodp.sd.7.01.2009

Moriya I (1980) "Bandaian eruption" and landforms associated with it. Collection of articles in memory of retirement of Prof. K. Nishimura from Tohoku Univ, pp 214–219 (in Japanese with English abstr)

Moriya I (1988) Geomorphological development of Bandai volcano. J Geogr 97(4):293–300 (in Japanese)

Moscardelli L, Woods L (2016) Morphometry of mass-transport deposits as a predictive tool. Geol Soc Amer Bull 128:47–80. https://doi.org/10.1130/B31221.1

Mothes PA, Hall ML, Janda RJ (1998) The enormous Chillos Valley lahar: an ash-flow-generated debris flow from Cotopaxi Volcano, Ecuador. Bull Volcanol 59:233–244. https://doi.org/10.1007/s004450050188

Mulder T, Cochonat P (1996) Classification of offshore mass movements. SEPM J Sed Res 66:3–57. https://doi.org/10.1306/d42682ac-2b26-11d7-8648000102c1865d

Mutti E, Bernoulli D, Ricci Lucchi F, Tinterri R (2009) Turbidites and turbidity currents from Alpine "Flysch" to the exploration of continental margins. In: McKenzie JA, Bernoulli D, Cita MB (eds) Symposium on major discoveries in sedimentary geology in the Mediterranean realm from a historical perspective to new development. Sedimentology 56(1):267–318. https://doi.org/10.1111/j.1365-3091.2008.01019.x

Nakamura Y (1978) Geology and petrology of Bandai and Nekoma volcanoes. Sci Rep Tohoku Univ Ser 3(14):67–119

Nehlig P, Leyrit H, Dardon A, Freour G, de Goer de Herve A, Huguet D, Thieblemont D (2001) Constructions et destructions du stratovolcan du Cantal. Bulletin de la Société Géologique de France 172(3):295–308. https://doi.org/10.2113/172.3.295

Oehler J-F, Lénat J-F, Labazuy P (2008) Growth and collapse of the Reunion Island volcanoes. Bull Volcanol 70:717–742. https://doi.org/10.1007/s00445-007-0163-0

Orton GJ (1995) Facies models in volcanic terrains: time's arrow versus time's cycle. In: Plint AG (ed) Sedimentary facies analysis. Int Assoc Sed Spec Publ 22:157–193. https://doi.org/10.1002/9781444304091.ch7

Ota K (1969) Study on the collapses in the Mayu-yama. Sci Rpt Shimabara Volcano Observ Fac Sci Kyushu Univ 5:6–35

Paguican EMR, Roverato M, Yoshida H (2021) Volcanic debris avalanche transport and emplacement mechanisms. In: Roverato M, Dufresne A, Procter J (eds) Volcanic debris avalanches: from collapse to hazard. Advances in volcanology. Springer, pp 143–173. https://doi.org/10.1007/978-3-030-57411-6

Palacios D, Zamorano JJ, Gómez A (2001) The impact of present lahars on the geomorphologic evolution of proglacial gorges: Popocatépetl, Mexico. Geomorphology 37(1–2):15–42. https://doi.org/10.1016/S0169-555X(00)00061-1

Paris R, Switzer AD, Belousova M, Belousov A, Ontowirjo B, Whelley PL, Ulvrova M (2014) Volcanic tsunami: a review of source mechanisms, past events and hazards in Southeast Asia Indonesia, Philippines, Papua New Guinea. Nat Hazards 70:447–470. https://doi.org/10.1007/s11069-013-0822-8

Perinotto H, Jean-Luc S, Bachelery P, Le Bourdonnec FX, Famin V, Michon L (2015) The extreme mobility of debris avalanches: a new model of transport mechanism. J Geophys Res Solid Earth 120(8110–8119):2212–2219

Peterson DE, Finn CA, Bedrosian PA (2021) Airborne geophysical imaging of weak zones on Iliamna Volcano, Alaska: implications for slope stability. J Geophys Res Solid Earth 126:e2020JB020807. https://doi.org/10.1029/2020JB020807

Pierson TC (1980) Debris flows: an important process in high country gully erosion. J Tussok Grassland Mt Lands Inst (NZ) 39:3–14

Pierson TC (1986) Flow behavior of channelized debris flows, Mount St. Helens, Washington. In: Abrahams AD (ed) Hillslope processes. Allen and Unwin, pp 269–296. ISBN: 978-1003028840

Pierson TC (1995) Flow characteristics of large eruption-triggered debris flows at snow-clad volcanoes: constraints for debris-flow models. J Volcanol Geother Res 66(1–4):283–294. https://doi.org/10.1016/0377-0273(94)00070-W

Pierson TC (2005) Hyperconcentrated flow—transitional process between water flow and debris flow. Chapter 8. In: Jakob M, Hungr O (eds) Debris-flow hazards and related phenomena. Springer, pp 159–208. ISBN: 978-3-540-20726-9

Pierson TC, Costa JE (1987) A rheologic classification of subaerial sediment-water flows. In: Costa JE, Wieczorek GF (eds) Debris flows/avalanches: process, recognition and mitigation. Geol Soc Amer Rev Eng Geol 7:1–12. ISBN: 0-8137-4107-6

Pierson TC, Janda RJ (1992) Immediate and long-term hazards from lahars and excess sedimentation in rivers draining Mt. Pinatubo, Philippines. Water-Res Investig Rep 92(9):15–18

Pierson TC, Janda RJ, Thouret J-C, Borrero CA (1990) Perturbation and melting of snow and ice by the 13 November 1985 eruption of Nevado del Ruiz, Colombia, and consequent mobilization, flow and deposition of lahars. J Volcanol Geother Res 41:17–66. https://doi.org/10.1016/0377-0273(90)90082-Q

Pierson TC, Major JJ (2014) Hydrogeomorphic effects of explosive volcanic eruptions on drainage basins. Annu Rev Earth Planet Sci 42:469–507. https://doi.org/10.1146/annurev-earth-060313-054913

Pierson TC, Major JJ, Amigo A, Moreno H (2013) Acute sedimentation response to rainfall following the explosive phase of the 2008–2009 eruption of Chaitén volcano, Chile. Bull Volcanol 75:1–17. https://doi.org/10.1007/s00445-013-0723-4

Pierson TC, Wood N, Driedger CL (2014) Reducing risk from lahar hazards: concepts, case studies, and roles for scientists. J Appl Volcanol 3(16):1–25. https://doi.org/10.1186/s13617-014-0016-4

Polanco E, Naranjo JA (2008) Collapse holocene in the Callaqui volcano (37°55′S), Andes del Sur. In: Acta XVII Argentine Geol Cong, pp 1157–1158

Ponomareva VV, Melekestsev IV, Dirksen OV (2006) Sector collapses and large landslides on Late Pleistocene-Holocene volcanoes in Kamchatka, Russia. J Volcanol Geother Res 158(1–2):117–138. https://doi.org/10.1016/j.jvolgeores.2006.04.016

Ponomareva VV, Pevzner MM, Melekestsev IV (1998) Large debris avalanches and associated eruptions in the Holocene eruptive history of Shiveluch volcano, Kamchatka, Russia. Bull Volcanol 59(7):490–505. https://doi.org/10.1007/s004450050206

Priyanto WS, Hunt JE, Hanif M, Tappin DR, Permana H, Susilohadi Cassidy M, Yulianto E (2021) Bathymetry and shallow seismic imaging of the 2018 flank collapse of Anak Krakatau. Front Earth Sci 8:577448. https://doi.org/10.3389/feart.2020.577448

Procter J, Cronin SJ, Fuller IC, Lube G, Manville V (2010a) Quantifying the geomorphic impacts of a lake-breakout lahar, Mount Ruapehu, New Zealand. Geology 38(1):67–70. https://doi.org/10.1130/G30129.1

Procter J, Cronin SJ, Fuller IC, Sheridan M, Neall VE, Keys H (2010b) Lahar hazard assessment using Titan2D for an alluvial fan with rapidly changing geomorphology: Whangaehu River, Mt. Ruapehu. Geomorphology 116:162–174. https://doi.org/10.1016/j.geomorph.2009.10.016

Procter J, Cronin SJ, Sheridan M (2012) Evaluation of Titan2D modelling forecasts for the 2007 Crater Lake break-out lahar, Mt. Ruapehu, New Zealand. Geomorphology 136(1):95–105. https://doi.org/10.1016/j.geomorph.2011.05.001

Punongbayan RS, Newhall CG (1996) Fire and mud. Eruptions and lahars of Mount Pinatubo, Philippines. PHIVOLCS, Quezon City and University of Washington, Seattle, 1126 pp. ISBN: 0-295-97585-7

Reiche P (1937) The Toreva-block—a distinctive landslide type. J Geol 45(5):538–548

Reid ME (2004) Massive collapse of volcano edifices triggered by hydrothermal pressurization. Geology 32(5):373–376. https://doi.org/10.1130/G20300.1

Reid ME, Sisson TW, Brien DL (2001) Volcano collapse promoted by hydrothermal alteration and edifice shape, Mount Rainier, Washington. Geology 29:779–782. https://doi.org/10.1130/0091-7613(2001)029<0779:VCPBHA>2.0.CO;2

Rickenmann D (1999) Empirical relationships for debris flows. Nat Hazards 19:47–77. https://doi.org/10.1023/A:1008064220727

Rickenmann D (2005) Runout prediction methods. In: Jakob M, Hungr O (eds) Debris-flow hazards and related phenomena. Praxis Springer, pp 305–324. ISBN: 978-3-540-20726-9

Roberti G, Roberts NJ, Lit C (2021) Climatic influence on volcanic landslides. In: Roverato M, Dufresne A, Procter J (eds) volcanic debris avalanches: from collapse to hazard. Advances in volcanology. Springer, pp 121–141. https://doi.org/10.1007/978-3-030-57411-6

Rochet L (1987) Application des modéles numériques de propagation à l'étude des éboulements rocheux. Bull Liaison Lab Ponts Chaussées 150:84–95

Rodolfo KS, Arguden AT (1991) Rain-lahar generation and sediment delivery systems at Mayon Volcano, Philippines. In: Fisher RV, Smith GA (eds) Sedimentation in volcanic settings, pp 71–87. SEPM Spec Publ 45. ISBN 0–918985–89–7

Roverato M, Cronin S, Procter J, Capra L (2015) Textural features as indicators of debris avalanche transport and emplacement, Taranaki Volcano. GSA Bull 127(1–2):3–18. https://doi.org/10.1130/B30946.1

Roverato M, Dufresne A, Procter J (eds) (2021) Volcanic debris avalanches: from collapse to hazard. Advances in volcanology. Springer, 358 pp. https://doi.org/10.1007/978-3-030-57411-6

Samaniego P, Barba D, Robin C, Fornari N, Bernard B (2012) Eruptive history of Chimborazo volcano (Ecuador): a large, ice-capped and hazardous compound volcano in the Northern Andes. J Volcanol Geother Res 221–222:33–51. https://doi.org/10.1016/j.jvolgeores.2012.01.014

Samaniego P, Valderrama P, Mariño J, van Wyk de Vries B, Roche O, Manrique N, Chédeville C, Liorzou C, Fidel L, Malnati J (2015) The historical (218 ± 14 aBP) explosive eruption of Tutupaca volcano (Southern Peru). Bull Volcanol 77:51. https://doi.org/10.1007/s00445-015-0937-8

Satake K, Kato Y (2001) The 1741 Oshima-Oshima Eruption: extent and volume of submarine debris avalanche. Geophys Res Lett 28(3):427–430. https://doi.org/10.1029/2000GL012175

Schmincke H-U, Sumita M (1998) Volcanic evolution of Gran Canaria reconstructed from apron sediments: synthesis of VICAP project drilling. In: Weaver PPE, Schmincke H-U, Firth JV, Duffield W (eds) Proceedings of the Ocean Drilling Program, Scientific Results 157:443–469

Schneider J-L (2000) Volcaniclastic sedimentation in submarine settings: products and processes. In: Leyrit H, Montenat C (eds) Volcaniclastic rocks from magmas to sediments. Gordon and Breach Science Publishers, Amsterdam, The Netherlands, pp 175–192

Schneider J-L, Fisher RV (1998) Transport and emplacement mechanisms of large volcanic debris avalanches: evidence from the northwest sector of Cantal volcano (France). J Volcanol Geother Res 83(1–2):141–165. https://doi.org/10.1016/S0377-0273(98)00016-X

Scott KM (1988) Origins, behavior, and sedimentology of lahars and lahar-runout flows in the Toutle-Cowlitz river system. US Geol Surv Prof Paper 1447-A, 74 pp. https://doi.org/10.3133/pp1447A

Scott KM, Janda RJ (1987) Multiple origins of catastrophic lahars at Cascade Range volcanoes—probabilities of precursor events and hazards in downstream areas. Sympos How Volcanoes Work, Hilo, Hawaii, 229 pp

Scott KM, Vallance JW, Kerle N, Macías JL, Strauch W, Devoli G (2005) Catastrophic precipitation-triggered lahar at Casita volcano, Nicaragua: occurrence, bulking and transformation. Earth Surf Proc Land 30(1):59–79. https://doi.org/10.1002/esp.1127

Scott KM, Vallance JW, Pringle PT (1995) Sedimentology, behavior, and hazards of debris flows at Mount Rainier, Washington. US Geol Surv Prof Paper 1547, 56 pp. https://doi.org/10.3133/pp1547

Sekiya S, Kikuchi Y (1889) The eruption of Bandai-san. Tokyo Imperial Univ Coll Sci J 3:91–172

Sequeiros OE, Naruse H, Endo N, Garcia MH, Parker G (2009) Experimental study on self-accelerating turbidity currents. J Geophys Res 114:C05025. https://doi.org/10.1029/2008JC005149

Shanmugam G (2021a) Chapter 3—Gravity flows: debris flows, grain flows, liquefied/fluidized flows, turbidity currents, hyperpycnal flows, and contour currents. In: Shanmugam G (ed) Mass transport, gravity flows, and bottom currents—downslope and alongslope

processes and deposits. Elsevier, pp 89–148. https://doi.org/10.1016/B978-0-12-822576-9.00003-5
Shanmugam G (2021b) The turbidite-contourite-tidalite-baroclinite-hybridite problem: orthodoxy vs. empirical evidence behind the "Bouma Sequence". J Palaeogeogr 10:9. https://doi.org/10.1186/s42501-021-00085-1
Sherrod DR, Scott WE, Stauffer PH (eds) (2008) A volcano rekindled: the renewed eruption of Mount St. Helens, 2004–2006. USGS Geol Surv Prof Pap, Washington, DC, USA, 872 pp. https://doi.org/10.3133/pp1750
Shevchenko AV, Dvigalo VN, Walter TR, Mania R, Maccaferri F, Svirid IY, Belousov AB, Belousova MG (2020) The rebirth and evolution of Bezymianny volcano, Kamchatka after the 1956 sector collapse. Commun Earth Environ 1:15. https://doi.org/10.1038/s43247-020-00014-5
Shimano T, Geshi N, Kobayashi T (2013) Geological map of Suwanosejima volcano. Geol Surv Japan, 1:20,000 map and explanatory text (in Japanese)
Shreve RL (1968) Leakage and fluidization in air-layer lubricated avalanches. GSA Bull 79(5):653–658. https://doi.org/10.1130/0016-7606(1968)79[653:LAFIAL]2.0.CO;2
Siebe C, Salinas S, Arana-Salinas L, Macías JL, Gardner J, Bonasia R (2017) The 23,500 yr 14C BP, White Pumice Plinian eruption and associated debris avalanche and Tochimilco lava flow of Popocatépetl volcano, México. J Volcanol Geother Res 333–334:66–95. https://doi.org/10.1016/j.jvolgeores.2017.01.011
Siebert L (1984) Large volcanic debris avalanches: characteristics of source areas, deposits, and associated eruptions. J Volcanol Geother Res 22(3–4):163–197. https://doi.org/10.1016/0377-0273(84)90002-7
Siebert L (1992) Threats from debris avalanches. Nature 356(6371):658–659. https://doi.org/10.1038/356658a0
Siebert L (2002) Landslides resulting from structural failure of volcanoes. In: Evans SG, DeGraff JV (eds) Catastrophic landslides: effects, occurrence, and mechanisms. Geol Soc Amer Rev Eng Geol 15:209–235. https://doi.org/10.1130/REG15-p209
Siebert L, Begét JE, Glicken H (1995) The 1883 and late-prehistoric eruptions of Augustine volcano, Alaska. J Volcanol Geother Res 66(1–4):367–395. https://doi.org/10.1016/0377-0273(94)00069-S
Siebert L, Glicken H, Ui T (1987) Volcanic hazards from Bezymianny- and Bandai-type eruptions. Bull Volcanol 49:435–459. https://doi.org/10.1007/BF01046635
Siebert L, Reid ME (2023) Lateral edifice collapse and volcanic debris avalanches: a post-1980 Mount St. Helens perspective. Bull Volcanol 85:61. https://doi.org/10.1007/s00445-023-01662-z
Siebert L, Roverato M (2021) A historical perspective on lateral collapse and volcanic debris avalanches. In: Roverato M, Dufresne A, Procter J (eds) Volcanic debris avalanches: from collapse to hazard. Advances in volcanology. Springer, pp 11–50. https://doi.org/10.1007/978-3-030-57411-6
Smith GA (1987) The influence of explosive volcanism on fluvial sedimentation: the Deschutes Formation (Neogene) in Central Oregon. J Sed Res 57(4):613–629. https://doi.org/10.1306/212F8BBB-2B24-11D7-8648000102C1865D
Smith GA (1991) Facies sequences and geometries in continental volcaniclastic sediments. In: Fisher RV, Smith GA (eds) Sedimentation in volcanic settings. Special Publications of Society for Sedimentary Geology 45, Tulsa, Oklahoma, pp 109–121
Smith GA, Lowe DR (1991) Lahars: volcano-hydrologic events and deposition in the debris flow-hyperconcentrated flow continuum. In: Fisher RV, Smith GA (eds) Sedimentation in volcanic settings. Special Publications of Society for Sedimentary Geology 45, Tulsa, Oklahoma, pp 59–70
Solikhin A, Thouret J-C, Liew SC, Gupta A, Sri Sayudi D, Oehler J-F (2015) Deposits and effects from the 2010 pyroclastic density currents and lahars at Merapi Volcano: mapped and analyzed from high-spatial resolution imagery. Bull Volcanol 77:20. https://doi.org/10.1007/s00445-015-0908-0
Sosio R, Crosta GB, Hungr O (2012) Numerical modeling of debris avalanche propagation from collapse of volcanic edifices. Landslides 9:315–334. https://doi.org/10.1007/s10346-011-0302-8
Sparks RSJ, Barclay J, Calder ES, Herd RA, Komorowski J-C, Lucket R, Norton GE, Ritchie LJ, Voight B, Woods AW (2002) Generation of a debris avalanche and violent pyroclastic density current on 26 December (Boxing Day) 1997 at Soufriére Hills Volcano, Montserrat. Geol Soc London Pubs Memoirs 21:409–434. https://doi.org/10.1144/GSL.MEM.2002.021.01.18
Sparks RSJ, Moore JG, Rice CJ (1986) The initial giant umbrella cloud of the May 18th, 1980, explosive eruption of Mount St. Helens. J Volcanol Geother Res 28:257–274. https://doi.org/10.1016/0377-0273(86)90026-0
Staudigel H, Schmincke H-U (1984) The Pliocene seamount series of La Palma/Canary Islands. J Geophys Res: Solid Earth 89(B13):11195–11215. https://doi.org/10.1029/JB089iB13p11195
Stevenson CJ, Feldens P, Georgiopoulou A, Schönke M, Krastel S, Piper DJW, Lindhost K, Mosher D (2018) Reconstructing the sediment concentration of a giant submarine gravity flow. Nat Commun 9:2616. https://doi.org/10.1038/s41467-018-05042-6
Stow DAV (1979) Distinguishing between fine-grained turbidites and contourites on the Nova Scotian deep water margin. Sedimentology 26:371–387. https://doi.org/10.1111/j.1365-3091.1979.tb00915.x
Stow DAV, Smillie Z (2020) Distinguishing between deep-water sediment facies: turbidites, contourites and hemipelagites. Geoscience 10(2):68. https://doi.org/10.3390/geosciences10020068
Stow DAV, Taira A, Ogawa Y, Soh W, Taniguchi H, Pickering KT (1998) Volcaniclastic sediments, process interaction and depositional setting of the

Mio-Pliocene Miura Group, SE Japan. Sediment Geol 115:351–381. https://doi.org/10.1016/S0037-0738(97)00100-0

Székely B, Karátson D (2004) DEM-based morphometry as a tool for reconstructing primary volcanic landforms: examples from the Börzsöny Mountains, Hungary. Geomorphology 63(1–2):25–37. https://doi.org/10.1016/j.geomorph.2004.03.008

Talling PJ, Allin J, Armitage DA, Arnott RWC, Cartigny MJB, Clare MA, Felletti F, Covault JA, Girardclos S, Hansen E, Hill PR, Hiscott RN, Hogg AJ, Clarke JH, Jobe ZR, Malgesini G, Mozzato A, Naruse H, Parkinson S, Peel FJ, Piper DJW, Pope E, Postma G, Rowley P, Sguazzini A, Stevenson CJ, Sumner EJ, Sylvester Z, Watts C, Xu J (2015) Key future directions for research on turbidity currents and their deposits. J Sed Res 85(2):153–169. https://doi.org/10.2110/jsr.2015.03

Thouret J-C, Aisyah N, Jenkins SF, de Bélizal E, Sri Sayudi D, Sulistiyani, Nandaka A, Charbonnier SJ, Mainsant G, Solikhin A (2023) Merapi's lahars: characteristics, behavior, monitoring, impacts, hazard modeling and risk assessment, Chapter 17. In: Gertisser R, Troll VR, Walter TR, Nandaka IGMA, Ratdomopurbo A (eds) Merapi volcano—geology, eruptive activity, and monitoring of a high-risk volcano. IAVCEI Active Volcanoes of the World, Springer, pp 501–552. ISBN: 978-3-031-15039-5

Thouret J-C, Antoine S, Magill C, Ollier C (2020) Lahars and debris flows: characteristics and impacts. Earth-Sci Rev 201:103003. https://doi.org/10.1016/j.earscirev.2019.103003

Thouret J-C, Cantagrel JM, Salinas R, Murcia A (1990) Quaternary eruptive history of Nevado del Ruiz (Colombia). J Volcanol Geoth Res 41(1–4):225–251. https://doi.org/10.1016/0377-0273(90)90090-3

Thouret J-C, Kassouk Z, Gupta A, Liew SC, Solikhin A (2015) Tracing the evolution of 2010 Merapi volcanic deposits (Indonesia) based on object-oriented classification and analysis of multi-temporal, very high resolution images. Rem Sens Environ 70:350–371. https://doi.org/10.1016/j.rse.2015.09.028

Thouret J-C, Lavigne F, Suwa H, Sukatja B, Surono (2007) Volcanic hazards at Mount Semeru, East Java (Indonesia), with emphasis on lahars. Bull Volcanol 70:221–244. https://doi.org/10.1007/s00445-007-0133-6

Thouret J-C, Oehler J-F, Gupta A, Solikhin A, Procter JN (2014) Erosion and aggradation on persistently active volcanoes—a case study from Semeru Volcano, Indonesia. Bull Volcanol 76:857. https://doi.org/10.1007/s00445-014-0857-z

Tiberghien D, Laigle D, Naaim N, Thibert E, Ousset F (2007) Experimental investigations of interaction between mudflow and an obstacle. In: Chen CL, Major JJ (eds) Debris flow hazard mitigation: mechanics, prediction and assessment. Millpress Science Publishers, pp 281–292

Tormey D (2010) Managing the effects of accelerated glacial melting on volcanic collapse and debris flows: Planchon-Peteroa volcano, Southern Andes. Glob Planet Change 74(2):82–90

Tost M, Cronin SJ (2015) Linking distal volcaniclastic sedimentation and stratigraphy with the development of Ruapehu volcano, New Zealand. Bull Volcanol 77:95. https://doi.org/10.1007/s00445-015-0977-0

Toyos G, Dorta DO, Oppenheimer C, Pareschi MT, Sulpizio R, Zanchetta G (2007) GIS-assisted modelling for debris flow hazard assessment based on the events of May 1998 in the area of Sarno, Southern Italy: Part I. Maximum run-out. Earth Surf Proc Landf 32:1491–1502. https://doi.org/10.1002/esp.1472

Trofimovs J, Amy L, Boudon G, Deplus C, Doyle E, Fournier N, Hart MB, Komorowski JC, Le Friant A, Lock EJ, Pudsey C, Ryan G, Sparks RSJ, Talling PJ (2006) Submarine pyroclastic deposits formed at the Soufrière Hills Volcano, Montserrat (1995–2003): what happens when pyroclastic flows enter the ocean? Geology 34(7):549–552. https://doi.org/10.1130/G22424.1

Trofimovs J, Sparks RSJ, Talling P (2008) Anatomy of a submarine pyroclastic flow and associated turbidity current: July 2003 dome collapse, Soufrière Hills volcano, Montserrat, West Indies. Sedimentology 55:617–634. https://doi.org/10.1111/j.1365-3091.2007.00914.x

Trofimovs JP, Talling J, Fisher JK, Sparks RSJ, Watt SFL, Hart MB, Smart CW, Le Friant A, Cassidy M, Moreton SG, Leng MJ (2013) Timing, origin and emplacement dynamics of mass flows offshore of SE Montserrat in the last 110 ka: implications for landslide and tsunami hazards, eruption history, and volcanic island evolution. Geochem Geophys Geosyst 14:385–406. https://doi.org/10.1002/ggge.20052

Ui T (1983) Volcanic dry avalanche deposits—identification and comparison with nonvolcanic debris stream deposits. J Volcanol Geoth Res 18(1–4):135–150. https://doi.org/10.1016/0377-0273(83)90006-9

Ui T, Takarada S, Yoshimoto M (2000) Debris avalanches. In: Sigurdsson H, Houghton B, McNutt SR, Rymer H, Stix J (eds) Encyclopedia of volcanoes. Academic Press, San Diego, CA, pp 617–626. ISBN: 978-0-12-385938-9

Ui T, Yamamoto H, Suzuki-Kamata K (1986) Characterization of debris avalanche deposits in Japan. J Volcanol Geother Res 29(1–4):231–243. https://doi.org/10.1016/0377-0273(86)90046-6

Valderrama P, Roche O, Samaniego P, van Wyk des Vries B, Araujo G (2018) Granular fingering as a mechanism for ridge formation in debris avalanche deposits: laboratory experiments and implications for Tutupaca volcano, Peru. J Volcanol Geother Res 349:409–418. https://doi.org/10.1016/j.jvolgeores.2017.12.004

Vallance JW (2000) Lahars. In: Sigurdsson H, Houghton B, McNutt SR, Rymer H, Stix J (eds) Encyclopedia of volcanoes. Academic Press, San Diego, CA, pp 601–616. ISBN: 978-0-12-385938-9

Vallance JW, Iverson R (2015) Lahars and their deposits. In: Sigurdsson H, Houghton B, McNutt SR, Rymer H, Stix J (eds) The encyclopedia of volcanoes, 2nd edn. Academic Press, Elsevier, pp 649–664. ISBN: 978-0-12-385938-9

Vallance JW, Scott KM (1997) The Osceola Mudflow from Mount Rainier: sedimentology and hazard implications of a huge clay-rich debris flow. Geol Soc Amer Bull 109:143–163

van Wyk de Vries B, Davies T (2015) Landslides, debris avalanches, and volcanic gravitational deformation. In: Sigurdsson H, Houghton B, McNutt SR, Rymer H, Stix J (eds) Encyclopedia of volcanoes. Academic Press, San Diego, CA, pp 665–685. ISBN: 978-0-12-385938-9

van Wyk de Vries B, Delcamp A (2015) Volcanic debris avalanches. In: Shroder JF, Davies T (eds) Landslide hazards, risks, and disasters. A volume in hazards and disasters series. Elsevier, pp 131–157

van Wyk de Vries B, Francis PW (1997) Catastrophic collapse at stratovolcanoes induced by gradual volcano spreading. Nature 387:387–390. https://doi.org/10.1038/387387a0

van Wyk de Vries B, Self S, Francis PW, Keszthelyi L (2001) A gravitational spreading origin for the Socompa debris avalanche. J Volcanol Geother Res 105(3):225–247. https://doi.org/10.1016/S0377-0273(00)00252-3

Vessell RK, Davies DK (1981) Nonmarine sedimentation in an active forearc basin. Soc Econ Paleont Mineral Pub 31:31–45

Vezzoli L, Renzulli A, Menna M (2014) Growth after collapse: the volcanic and magmatic history of the Neostromboli lava cone (island of Stromboli, Italy). Bull Volcanol 76:821. https://doi.org/10.1007/s00445-014-0821-y

Voight B (1978) Rockslides and avalanches 1. Natural Phenomena. Developments in Geotechnical Engineering 14A, Elsevier Scientific Publishing Company, Amsterdam, The Netherlands, 833 pp

Voight B, Glicken H, Janda RJ, Douglass PM (1981) Catastrophic rockslide avalanche of May 18. In: Lipman PW, Mullineaux DR (eds) The 1980 eruptions of Mount St. Helens, Washington: US Geol Surv Prof Pap 1250:347–377

Voight B, Komorowski JC, Norton GE, Belousov AB, Belousova M, Boudon G, Francis PW, Franz W, Heinrich P, Sparks RSJ, Young SR (2002) The 26 December (Boxing Day) 1997 sector collapse and debris avalanche at Soufrère Hills Volcano, Montserrat. In: Druitt TH, Kokelaar BP (eds) The eruption of Soufrière Hills Volcano, Montserrat, from 1995 to 1999. Geol Soc London Mem 21:363–407. https://doi.org/10.1144/GSL.MEM.2002.021.01.17

Voight B, Sousa J (1994) Lessons from Ontake-san: a comparative analysis of debris avalanche dynamics. Eng Geol 38(3–4):261–297. https://doi.org/10.1016/0013-7952(94)90042-6

Wadge G, Francis PW, Ramirez CF (1995) The Socompa collapse and avalanche event. J Volcanol Geother Res 66:309–336. https://doi.org/10.1016/0377-0273(94)00083-S

Waitt RB (2015) In the path of destruction: eyewitness chronicles of Mount St. Helens. Washington State University Press, 416 pp. ISBN: 978-0-87422-323-1

Waitt RB, Begét JE (2009) Volcanic processes and geology of Augustine volcano, Alaska. US Geol Surv Prof Pap 1762:1–78, 2 pl, scale 1:25,000

Walter TR (2011) Structural architecture of the 1980 Mount St. Helens collapse: an analysis of the Rosenquist photo sequence using digital image correlation. Geology 39(8):767–770. https://doi.org/10.1130/G32198.1

Watt S, Talling PJ, Vardy ME, Heller V, Hühnerbach V, Urlaub M, Sarkar S, Masson DG, Henstock TJ, Minshull TA, Paulatto M, Le Friant A, Lebas E, Berndt C, Crutchley GJ, Karstens J, Stinton AJ, Maeno F (2012) Combinations of volcanic-flank and seafloor-sediment failure offshore Montserrat, and their implications for tsunami generation. Earth Planet Sci Lett 319–320:228–240. https://doi.org/10.1016/j.epsl.2011.11.032

Watt SFL (2019) The evolution of volcanic systems following sector collapse. J Volcanol Geother Res 384:280–303.https://doi.org/10.1016/j.jvolgeores.2019.05.012

Waythomas CF (2022) Selected crater and small caldera lakes in Alaska: characteristics and hazards. Front Earth Sci 9:751216. https://doi.org/10.3389/feart.2021.751216

Wilson G, Wilson TM, Deligne NJ, Cole JV (2014) Volcanic hazard impacts to critical infrastructure: a review. J Volcanol Geother Res 286:148–182. https://doi.org/10.1016/j.jvolgeores.2014.08.030

Yamamoto M, Kawakatsu H, Kaneshima S, Mori T, Tsutsui T, Sudo Y, Morita Y (1999) Detection of a crack-like conduit beneath the active crater at Aso Volcano Japan. Geophys Res Lett 26(24):3677–3680. https://doi.org/10.1029/1999GL005395

Yonechi F (1987) A new hypothesis on the collapse of Bandai-san volcano in 1888. The Science Reports of the Tohoku University, 7th series (Geography) 37(2): 159–173

Yonechi F, Chiba N, Ozawa A, Ishimaru S (1989) Large scale slope failure caused by Mt. Bandai's eruption in 1888. Abstracts of the 27th Annual Meeting of the Japan Landslide Society, 20–21 (in Japanese)

Yoshida H (2016) Magnitude-frequency distribution of slope failures in Japan: statistical approach to a true perspective on volcanic mega-collapses. In: Németh K (ed) Updates in volcanology from volcano modelling to volcano geology. InTechOpen, Rijeka, Croatia, pp 191–218. https://doi.org/10.5772/63131

Yoshida H, Sugai T, Ohmori H (2012) Size–distance relationships for hummocks on volcanic rockslide-debris avalanche deposits in Japan. Geomorphology 136(1):76–87. https://doi.org/10.1016/j.geomorph.2011.04.044

Zanchetta G, Sulpizio R, Pareschi MT, Leoni FM, Santacroce R (2004) Characteristics of May 5–6, 1998 volcaniclastic debris flows in the Sarno area (Campania,

southern Italy): relationships to structural damage and hazard zonation. J Volcanol Geoth Res 133:377–393. https://doi.org/10.1016/S0377-0273(03)00409-8
Zernack AV, Cronin SJ, Neall VE, Procter JN (2011) A medial to distal volcaniclastic record of an andesite stratovolcano: detailed stratigraphy of the ring plain succession of south-west Taranaki. Int J Earth Sci 100:1936–1966. https://doi.org/10.1007/s00531-010-0610-6
Zernack AV, Procter JN (2021) Cyclic growth and destruction of volcanoes. In: Roverato M, Dufresne A, Procter J (eds) Volcanic debris avalanches: from collapse to hazard. Advances in volcanology, pp 311–355. https://doi.org/10.1007/978-3-030-57411-6
Zhou GG, Ng CW (2010) Dimensional analysis of natural debris flows. Can Geotech J 47(7):719–729. https://doi.org/10.1139/T09-134

Chapter 8 Long-Term Degradation of Volcanoes

1 Long-Term Erosion Processes Gradually Take Over in Volcanic Terrains

In this chapter, we address the issue of long-term *erosion processes* and resulting *erosional landforms* particularly in terrestrial environments.

As discussed in the previous chapter, the predominance of rapid erosion processes on an active volcano, mainly in the proximal and medial areas, is due, in addition to endogenous factors (e.g. seismicity), to the steep slopes, instability, and erodibility of the accumulated volcanic pile. Lahars and flood flows, which typically move further down the drainage system, commonly affect the distal areas, too. Somewhat in contrast, volcanic deposits accumulated in medial and distal settings (i.e. on the lower flanks and the ring plain) are more subject to long-term erosion processes—i.e. fluvial downcutting, hillslope processes—particularly in periods of quiescence and after the volcano becomes extinct. However, as long as the volcano remains active, the geomorphic consequences of these erosional processes are often overridden by renewed activity (e.g. Cas and Wright 1988; Fisher and Smith 1991; Manville et al. 2009). For instance, deposition of fine tephra reduces the surface infiltration rate (Pierson and Major 2014). Once volcanic activity declines, the compaction/lithification of volcanic material and the appearance of vegetation protect against short-term degradation processes. As pointed out by case studies in Japan and the USA, intense hillside erosion of fresh tephra decreases quickly, after only a few years (Chinen 1986; Collins and Dunne 2019; Major 2022).

Over time, these slow, long-term erosion processes take hold. Their presence and rate in a given area is largely determined by the prevailing climate and altitude, in other words by the pattern of latitudinal and vertical zonation. Other long-term but intense processes may also be present, related to (1) tectonics, in particular seismic activity (sometimes influenced by climate) such as rockfalls and landslides on long-extinct volcanoes (e.g. Crandell and Fahnestock 1965; Norris 1994; Petley 2011; Korup et al. 2019), and (2) glaciation, which—in high latitudes and at high-altitude—can be more efficient than fluvial processes in terms of denudation (e.g. Geirsdóttir et al. 2007; Cunningham et al. 2019; Quesada-Román et al. 2019). However, these processes are less effective than the rapid, large-scale mass movements such as debris-avalanches described in Chap. 7.

Long-term erosion processes that degrade volcanoes are an integral part of the Earth's rock cycle (Skinner et al. 2004; Cas 2005) (Fig. 1).

D. Karátson and J.-C. Thouret, *Volcano Geomorphology*, Advances in Volcanology,
https://doi.org/10.1007/978-3-031-77693-9_9

They involve two independent, but often sequential steps:

(a) transformation of volcanic rocks/deposits by external factors represented by physical and chemical *weathering;*
(b) *removal and redeposition* of weathered material by gravity, via slope processes (slow mass movements), fluvial, aeolian, periglacial and glacial erosion, complemented offshore by slow submarine slope processes.

On our planet, *fluvial erosion* is the leading agent among long-term degradation processes. Fluvial erosion can be relatively significant even under semiarid climate conditions. For example, rates of long-term fluvial incision and projected average denudation (surface lowering) calculated for the eroded Mio-Pliocene volcanoes of Altiplano–Puna amount to no less than 20–30% of the typical values under temperate climates (Karátson et al. 2012; see Sect. 4.3). There are arid-hyperarid areas on Earth, such as the present-day Sahara (Zaki et al. 2022 and references therein), the Namib desert (Bierman et al. 2014) or the Atacama desert (Dunai et al. 2005; Kober et al. 2007) where fluvial erosion is apparently nil due to the extremely dry climate, but these environments are scarce, and may change over thousands of years due to climate fluctuations.

This means that in volcanic landscapes fluvial erosion is the key to degradation of landforms. The accumulated volcanic material, which is removed by weathering, mass movements, and other surface/subsurface processes, is transported from source to sink along the drainage network system (Manville et al. 2009; Major 2022). Since this is a linear process concentrated along defined pathways, it transforms the primary (initial) volcanic landforms according to a peculiar geometric pattern (Cotton 1944, 1952; Ollier 1988; Francis 1993).

This chapter covers three main topics: (1) the break-down of volcanic rocks by physical–chemical weathering; (2) characteristics of fluvial erosion on volcanoes and volcanic landscapes and the resultant erosional volcanic landforms; and (3) the role of other erosive agents such as wind, glacial-periglacial and coastal erosion in modifying the surface of volcanoes being degraded.

2 Weathering of Volcanic Rocks

The lava and volcaniclastic successions of volcanic edifices become compacted/lithified over a shorter or longer period of time (on the order of 10^2–10^3 yrs), depending mainly on rock properties, confining topography, layer or bed geometry, and climate. Their weathering and subsequent long-term erosion follows general rules, but the specific characteristics of igneous rocks have a profound influence on the processes involved. These characteristics include the chemical composition as well as physical, structural, and sedimentological features such as any existing fracture system, permeability, porosity, bedding, stratification, grain size distribution, etc. (Quane et al. 2009; Farquharson et al. 2016).

Weathering of rocks means their 'adaptation' or transformation to the prevailing, whether constant or changing, physico-chemical characteristics in a given environment, determined largely by the climatic as well as hydrogeologic conditions (Fig. 2). In volcanic landscapes,

Fig. 1 **A** The rock cycle: an interplay of internal and external processes on Earth. Rock material in the continental crust can follow any of the arrows from one phase to another. Within the mantle circuit, magma rises from below and forms new igneous rocks in the lithosphere. The old lithosphere descends back down to the mantle, where it is eventually remixed (after Skinner and Porter 2004). **B** Diagram illustrating the connection between subduction activity, magmatism, and sedimentation. Subduction leads to magma generation and associated volcanic activity, and a large influx of volcaniclastic materials causes increased rates of sedimentation. A composite volcano built up of successions of lava flows and tephra-fall, pyroclastic density currents (PDC) and lahar deposits is depicted in the diagram. Transfer of volcaniclastics from the active volcano to depositional basin occurs directly or via a step-wise transfer through flow transformations (adapted from Fisher and Smith 1991) ▶

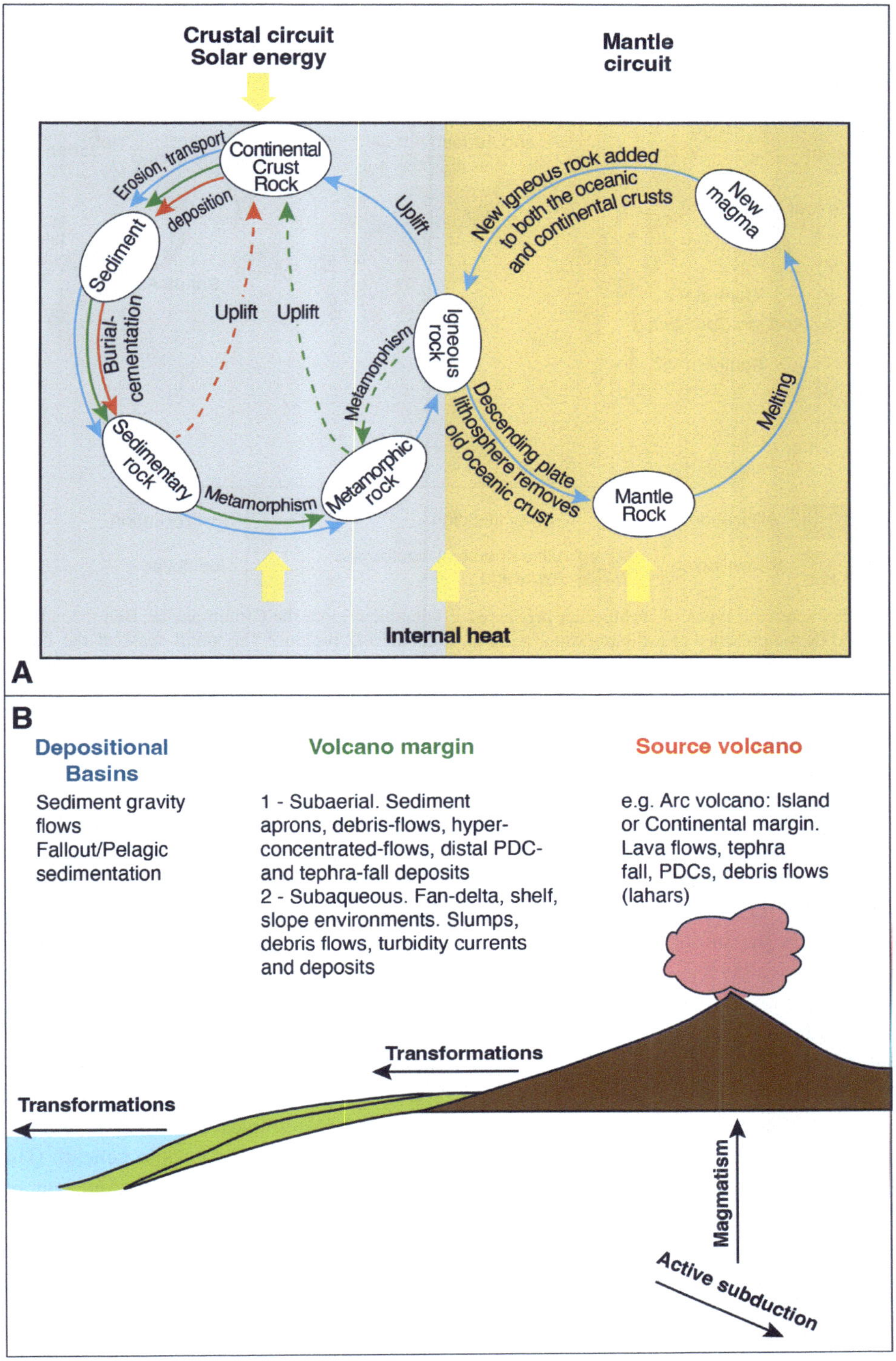

Crustal circuit
Solar energy
Mantle circuit
Continental Crust Rock
Erosion, transport
deposition
Sediment
Burial-cementation
Sedimentary rock
Metamorphism
Metamorphic rock
Uplift
Uplift
Uplift
Metamorphism
Igneous rock
New igneous rock added to both the oceanic and continental crusts
New magma
Melting
Mantle Rock
Descending plate lithosphere removes old oceanic crust
Internal heat
A
B
Depositional Basins
Sediment gravity flows
Fallout/Pelagic sedimentation
Volcano margin
1 - Subaerial. Sediment aprons, debris-flows, hyper-concentrated-flows, distal PDC- and tephra-fall deposits
2 - Subaqueous. Fan-delta, shelf, slope environments. Slumps, debris flows, turbidity currents and deposits
Source volcano
e.g. Arc volcano: Island or Continental margin. Lava flows, tephra fall, PDCs, debris flows (lahars)
Transformations
Transformations
Magmatism
Active subduction

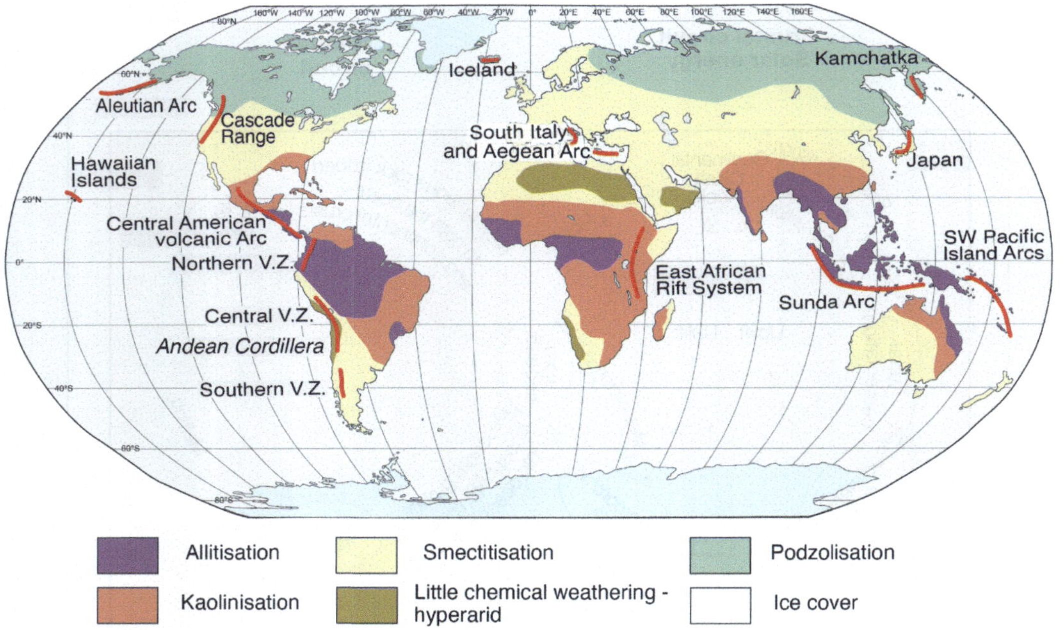

Fig. 2 The characteristic types of weathering processes are determined by the presence of available water and the climate. Some of the most important volcanic areas are marked in red. It should be noted that weathering characteristics apply only to extinct volcanoes and volcanic landscapes. The volcanoes of the largest continuous volcanic region, the Circumpacific Belt extend across all weathering regions. The small scale of the map means that it is not possible to denote that the highest low-latitude volcanoes might be covered by snow and ice caps (e.g. Kilimanjaro, Nevado del Ruíz). (Adapted from Thomas 1974 and Huggett 2011)

weathering is crucial to modification in shape and the final degradation of volcanic edifices, because its processes usually loosen the rocks and prepare them for removal by erosion, which in all cases alters the initial morphology of the volcano. Thus weathering, coupled with erosional processes, not only redistributes the volcanic material but is fundamental in creating the erosional volcanic landforms.

The main agent of weathering is H_2O, the availability and state of which is ensured by the prevailing climate. In nature, weathering includes various processes, which can be grouped for convenience into two, sometimes co-existing, categories (for a recent summary, see Frings and Buss 2019): (1) physical and (2) chemical weathering (Price 1995; Anderson and Anderson 2010).

Physical weathering involves mechanical disaggregation/fragmentation of rocks without compositional changes by exfoliation, pressure changes, thermal expansion (e.g. freeze–thaw), frost or root wedging, other effects of plant growth, and salt crystallization.

Chemical weathering (or chemical alteration) dissolves the fragmented rocks by processes of hydrolysis, oxidation and solution, which induce chemical and mineralogical changes in the rock.

Overall, these processes are self-perpetuating: as soon as weathering has penetrated into the disintegrating rock, it can act on a progressively larger rock surface, and the weathered products are removed by erosion, leaving new intact rock surfaces accessible beneath (Huggett 2011). Less frequently, weathering can protect and preserve the volcanic rock in a specific case of hydrothermal weathering where cementation with silica hardens the rock. Similarly, hardening of tuff surfaces by lichenic acids may also occur, coating them with a weathered crust and thereby preventing further erosion of the rock. For summary and recent literature, see Huggett (2011); Colon (2012) and Trudgill et al. (2022).

2.1 Physical Weathering

The factors involved in physical weathering, i.e. those that disintegrate and fragment volcanic rocks, are as follows, in approximate order of importance in volcanic areas:

(1) the direct effect of various erosion processes (e.g. surface runoff, rill erosion, ice grinding, disintegration due to mass movements);
(2) expansion–shrinkage of rock bodies mostly due to water/moisture uptake and release;
(3) insolation, i.e. the alternation of repeated heating and cooling of rock over daily cycles, possibly enhanced by a low albedo, due to the typically dark colour of many lava rocks;
(4) effect of frost action, i.e. alternate freezing and thawing of moisture in rocks;
(5) decrease in lithostatic pressure (e.g. the exposure of subvolcanic or plutonic rocks);
(6) the effect of hydration (incorporation of water into a mineral, or disintegration of the mineral);
(7) the shattering effect of salt crystals, significant during strong surface evaporation (less frequently, the salt can act as a cement for the rock);
(8) the effect of living organisms, biological activity: e.g. movement of organisms on the surface or in the soil, effects of root acids, which only become significant after the volcano becomes extinct and is colonised by vegetation.

In volcanic rocks, the most spectacular phenomenon of fragmentation in response to a decrease in pressure is *spheroidal weathering* (Ollier 1971). The rock mass solidifying at different depths reveals preserved directions and planes (generally related to cooling and the development of internal shear), which are exploited by weathering once the pressure reduces as the rock becomes exposed, resulting in the formation of a three-dimensional (often cubic) joint system on a dm or m scale. This happens because weathering along joints separates the rock into distinct blocks, often in the form of rough cubes or rectangular prisms, and the rock becomes progressively weakened and altered inwards. Differences in weathering rates between the exposed corners, edges, and faces of the blocks result in the formation of spheroidal layers, which tend to peel off from the outside inwards over time (Photo 93). Occasionally, the original cubic joint system with perpendicular cracks is also observed. This kind of rock disintegration is most common in subvolcanic bodies, and less frequent in thick lava flows. It can thus serve as an important criterion for the field identification of plutonic or subvolcanic rocks, especially in ancient volcanic terrains (Photo 93). Another, related process is *honeycomb weathering* which is typical on sea coasts, most spectacularly in basalts (Goudie 2004). There, pressure from salt crystallization (i.e. salt spray from sea wave pounding) weakens the rock, often along microfractures, and creates surface pits and hollows in a honeycomb pattern.

Amongst the volcanic rocks, basalt tends to have the most developed fracture systems and is overall the most permeable. This is due to the fact that its flow units are commonly thin and well-defined, which creates potential fracture planes. On the other hand, the actual permeability of a basalt succession depends on the the cooling rate, the number of planes that develop within a flow unit, and the total thickness of the flow(s). Generally, andesite has a less developed fracture system and lower permeability, and dacite and rhyolite have the lowest permeability of all. However, significant variations in permeability can occur, with even slight compaction of rock significantly inceasing the permeability due to the improved pore connectivity, i.e. distributed microcracking (Farquharson et al. 2015; Heap and Kennedy 2016).

As a result of physical weathering, basaltic and andesitic lava breaks apart forming rock sheets separated by *platy jointing*. In northern hemisphere volcanic mountains with temperate climates, the most recent glaciations (or, at higher altitude, present-day periglacial climate) have produced a significant amount of weathered rock debris typically over flattish areas, called *block seas* (or *felsenmeers* in German after the eponymous locality in the Odenwald Hills, Central German Uplands). During weathering, in particular frost shattering, polygonal columnar basalts, or more rarely andesites, may develop a set of fragmented horizontal planes within the columns, i.e. perpendicular to the columnar joints, causing them to resemble 'rolls of coins'. These planes are related to flow banding or fractures like striae or chisel marks formed during cooling (see Chap. 2). At the same time, the polygonal pattern of weathered columns becomes rounded and cylindrical in shape: these forms, called *basalt organs* or organ pipes, are a frequent phenomenon in ancient basalt terrains worldwide (Photo 93).

Unwelded pyroclastic deposits are generally porous and thus permeable in their loose, uncompacted state, but this gradually changes as they become hardened and lithified. However, there can be large differences according to sedimentological characteristics (e.g. grain size). Differences in permeability of pyroclastics are also related to the extent of zeolitization (Vinciguerra et al. 2009; Heap et al. 2014). Scoriaceous and pumiceous deposits are relatively permeable, whereas fine ash is much less so. Consequently, as particle size decreases away from the vent, water permeability of pyroclastics generally decreases. Equally, permeability also decreases with increasing grainsize variability. Unlike lava, volcaniclastic rocks with mixed grain size are more resistant to weathering, as expansion and shrinkage of larger clasts (due to the daily insolation cycle) is well tolerated by the fine matrix. Coarse-grained block-and-ash flow breccias and cohesive lahar deposits are particularly resistant to weathering. For example, in Hungary, erosion has carved out spectacular *mushroom rock* towers in block-and-ash flow deposits in the Miocene Börzsöny and Visegrád Mountains (Karátson et al. 2006), located on opposite sides of the Danube Bend that cuts across these mountains in the Visegrád Gorge. These rocky landforms have withstood erosion for over fifteen million years (Photo 94).

2.2 Chemical Weathering

Water entering the fracture system of rocks promotes fragmentation by hydration, which involves the binding of H_2O molecules into the rock-forming minerals. This process is transitional to chemical weathering. Chemical weathering of volcanic rocks is mainly determined by the mineral composition, grain-size distribution, prevailing climate, associated biological activity, and characteristic pH (Gerrard 1988). In terms of mineral composition, the most resistant rock-forming mineral is quartz, but its role is influenced by the size, and size difference, of the minerals, for instance, a porphyritic versus aphanitic texture; silica-poor, quartz-free basalt may also be resistant to erosion because of the typical aphanitic texture. Other common minerals in volcanic rocks, such as feldspars, are also relatively resistant, but amphibole, biotite, olivine (in this order) are more easily altered (Birkenland and Noller 2000; Earle and Panchuk 2019) (Table 1).

Table 1 Typical weathering products of common minerals in volcanic rocks (after Eearle and Panchuk 2019)

Mineral	*Typical weathering products*
Quartz	Quartz as sand grains
Feldspar	Clay minerals plus potassium, sodium, and calcium in solution
Biotite and amphibole	Chlorite plus iron and magnesium in solution
Pyroxene and olivine	Serpentine plus iron and magnesium in solution

The three main processes of long-term chemical weathering are as follows (for a textbook summary, see Earle and Panchuk 2019):

(1) *Hydrolysis*, which is characteristic of all rocks, and leads either to the formation of stable clay minerals, most importantly the fine-grained aggregate of kaolinite via the replacement of feldspar (referred to as *kaolinisation* or *argillisation)*, or, at higher latitudes, to its breakdown into silica (SiO_2) and Al-, Fe-oxyhydroxides *(podzolisation:* see below).
(2) *Oxidation* (e.g. transformation of iron from Fe_2^+ to stable, insoluble Fe_3^+), which is very typical of volcanic rocks. Because the mafic (ferro-magnesian) mineral components of volcanic rocks (olivine, pyroxene, amphibole, biotite) all contain iron, their oxidation results in a specific colour. In ancient volcanic terrains with a temperate climate, oxidised iron-containing minerals are the cause of the very common yellowish, reddish hue of weathered rocks, while in the tropical and subtropical zones the resulting iron oxides (hematite, goethite) give *laterite* and *terra rossa* their characteristic colour.
(3) *Solubilisation* during or after hydrolysis and/or oxidation, which is less important in volcanic rocks.

These processes are influenced by climate, most importantly by the presence of water (in the air and on the ground surface), oxygen, and carbon dioxide. Higher precipitation (larger amounts of available water) and higher temperatures usually lead to faster chemical reactions. As a result, humid, warm climates tend to have more intense weathering than cold, dry climates.

Looking at the most important process of hydrolysis, under temperate climates and neutral or weakly acidic pH, this produces clay minerals, referred to as *siallitisation.* When the chemistry is more alkaline (e.g. in temperate climates with less precipitation), a clay mineral called smectite is formed *(smectitisation).* Under conditions of extremely acidic chemistry and low mean annual temperatures but with sufficient moisture, especially at higher latitudes, clay minerals break down and decompose into silica, aluminium and iron that migrate downward *(podzolisation).* Finally, in the tropics, the downward leaching of minerals into the soil solution results in the complete breakdown of rock-forming silicate minerals. Iron and aluminium minerals accumulate at the surface, while kaolinite forms in the deeper levels—a process which is known as *ferralitisation* or *lateritisation*, including allitic versus ferritic weathering based on the Al/Fe ratio, whose end product is *laterite*. Weathering of young (less than 1000 years old) volcanic rocks, in particular tuffs, produces amorphous Al-silicates (allophanes and imogolite) in the temperate zone, and free Al-oxides and hydroxides in the tropical zone. These play a significant role in the formation of andosols (see below).

Volcanic rocks are generally only weakly soluble. A potential exception to this is the phenomenon of the poorly studied *'volcano-karst'*, originally described in the Călimani Mountains, Romania (Naum et al. 1962). As reviewed by Reffay (2001), only a few examples from Indonesia, the East African Rift System, or Hawaii can be considered as real karst features (termed orto- or parakarsts), implying the formation of surface cavities, depressions, and deep caves, e.g. those developed in the very soluble carbonatites. On the contrary, the majority of such lava landforms are considered as *pseudokarst* (Ford and Williams 2007), since most of these features are formed by collapse, while others are due to the movement of carbon dioxide-rich hydrothermal fluids through the fracture system of massive andesite, and the dissolution of iron-bearing minerals as a result of post-volcanic action (e.g. Szakács 1999).

2.3 From Regolith to Volcanic Soils

Physical and chemical weathering produce a *weathering profile* of varying thickness—up to several metres—on the rock surface, known as *regolith* (e.g. Ehlen 2005), the Greek term for 'blanket rock'. This contains fragmental rock material (either residual or transported) that has been altered to varying degrees, and generally has a loose texture and is suitable for erosional removal. However, precipitates from solutions can render its surface hard and resistant, resulting in *duricrust,* 'hard crust', or *ferricrust,* 'iron crust', in the presence of iron, which is typical of volcanic rocks (Huggett 2011).

The stable clay minerals that form as part of the regolith, together with decomposed material from biological activity, are essential components of soil formation in almost all climates. Soil can only be formed from the disintegration of rock when a certain amount of organic matter has been incorporated, producing a specific dark-coloured, colloidal material: *humus.* This results in the formation of a nutrient-rich system—in volcanic environments, *volcanic soil*—that contains solid debris and is moist, while also being well ventilated.

Volcanic soils are rich in potassium, nitrogen, iron, magnesium and other trace elements, and have been recognised for thousands of years for their special agricultural value (e.g. olive and tomato plantations around Vesuvius, kiwi fruit in New Zealand, tea in Java, pineapples in Hawai'i, etc.). Although these soils cover only c. 1% of the surface of our planet (about 110–120 million hectares), they are home to 10% of the world's population (Neall 2006)—and contain some of the most densely populated areas on Earth: the Naples agglomeration, the island of Java, Kyushu and central Honshū, significant areas of Central America, and some coasts of Hawai'i. However, in terms of fertility, a distinction must be made between the soils formed on lava and pyroclastic rocks, because the former are generally of poorer quality. This is due to the dense structure and the low porosity of lavas compared to that of pyroclasts. Pyroclastic material such as scoriae, pumice fragments and ash layers deposited further away from the volcano retains water after rainfall and releases it slowly to the root system. The weathering capability of vesicular clasts and glassy matrix of pyroclastic deposits is also important: after an eruption, rapid soil formation/regeneration can occur over centuries or millennia (Tilling et al. 1987).

Black, fertile soils of volcanic landscapes, formed in pyroclastic material, are commonly referred to as *andosols.* Indeed, originating from *an* (暗, 'dark') and *do* (土, 'soil'), this term is internationally understood to refer to dark volcanic soils (Takahashi and Shoji 2002), the colour being due to the rich organic matter content. (Note that the terms 'andisol' and 'black dust soil' refer to the same.) The criteria for andosols were initially (FAO/UNESCO Soil Map of the World 1974; Soil Taxonomy of the US Department of Agriculture Handbook, 1999) low density (<0.83 g/cm^3 for the fraction below 2 mm) and at least 60% vitreous volcanic ash. A more complete modern criterion for andosol (World Soil Resources Report 94 2001; Dahlgren et al. 2008; Stoops et al. 2018) is the presence of a diagnostic layer, an 'andic' or 'vitric horizon' at least 30 cm thick in the soil profile at shallow depth, characterized by low density due to the presence of allophanes, adequate clay and volcanic glass content, and its phosphate-binding capacity. (The latter property is one of the few disadvantages of andosols and can be improved by adding lime or through fertilisation.) The geographic distribution of andosols is predominantly around active or dormant volcanoes (Takahashi and Shoji 2002), but they also occur in acidic non-volcanic rocks (e.g. granite, phyllite, amphibolite). Interestingly, they may occur not only in young volcanic regions but also in ancient terrains, such as along the Miocene Inner Carpathians in Europe above 1600 m a.s.l. (Fehér et al. 2006).

3 Fluvial Erosion on Volcano Surfaces and Drainage Network Formation

On the pristine, unvegetated, bare rock surface of volcanoes, weathering can occur simultaneously with the action of running water due to the erosive activity of precipitation. Erosion by running water (sheet wash on bare surfaces) coalesces into a spectacular, linear pattern of rills almost from time zero. That is, although the removal of loose volcanic material by slope processes and rain erosion is surficial, rill erosion eventually causes the removed material to follow linear flow paths following a 'channelisation' process (Horton 1945), which is particularly pronounced and rapid on the steep slopes of active volcanoes (Segerstrom 1950; Ollier 1988; Collins and Dunne 2019). This explains why temporary fluvial erosion is already present during the active lifetime of a volcano, even if eruptions constantly renew the surface and override the drainage pattern (e.g. Thouret et al. 2014).

In recent years, conceptual models of fluvial response to volcanic eruptions have been refined, both through studying the aftermaths of recent explosive eruptions (e.g. Calbuco 1965, 2015: Iroumé et al. 2023; Puyehue—Cordón Caulle 2015: Baumann et al. 2020, both in Chile) and of past effusive eruptions with lava damming events (e.g. Kenya: Veldkamp et al. 2012; Turkey: van Gorp et al. 2016). In particular, focussing on the analysis of three contrasting scenarios in the Tweed caldera catchment area (East Australia), Manley et al. (2020) modelled how fluvial erosion of volcanic edifices and landscape dynamics react to changes in the climate.

3.1 Evolution of Drainage on Volcanoes

The evolution of fluvial drainage is of crucial importance for volcano erosion, as it results in the slow erosional removal and eventual degradation of an edifice, involving an overall lowering of the surface (Sect. 4.1).

Firstly, we concentrate on specific changes in the drainage, which are used to divide it into characteristic 'stages' based on modifications to the channel patterns as well as topography (Fig. 3A). The successive fluvial or geomorphic changes can be considered as stages within the lifetime of a volcano, although the individual steps are not always present, sometimes being obscure or transitional as they are strongly influenced by renewed volcanism, occasional tectonic movements, and climate changes. We avoid viewing the 'stages' described here as belonging to an erosion cycle—this latter being an old concept, developed by William Morris Davis (Davis 1899), whose relevance declined from the 1940s. (For more information, an assessment of the Davisian cyclic view is provided by Inkpen and Collier, 2007, Orme 2007). Nowadays, landforms, including volcanic landforms, may be viewed as responses to the interaction between tectonic activity, surface denudation, and isostatic adjustment (Orme 2007). On the other hand, we note that, depending on fluctuations in climate, weathering and erosion may potentially lead to multiple phases of volcanogenic material redeposition to increasingly distant areas, up to thousands of kilometres away from the eruptive source (Di Capua et al. 2023).

The specific drainage pattern sequence is also important because the drainage pattern persists long after the volcano becomes extinct, even after the positive topography has completely disappeared, thus its identification may be crucial for the reconstruction of the primary volcanic landform. The subsequent stages, qualitatively described by Ollier (1988) following pioneer work by Cotton (1952), have recently been analysed using drainage basin morphometry at the summit and on the flanks of volcanoes (O'Hara et al. 2024). The latter authors modelled the development of drainage basins and related topographic changes using selected composite volcanoes; by establishing common hydrographic metrics, they determined stages of maturation during basin evolution, and derived a new generalized model for edifice degradation.

Hereafter, the different types of symmetrical, sizeable volcanic edifices (including shield volcanoes, composite cones, and large pyroclastic cones) are discussed together for the purposes of fluvial erosion, since it is the degree of impermeability of the material rather than the differences in shape (slope, size) that is important. While fragmented rocks (commonly pyroclastic deposits with large- to medium grain size, e.g. bombs or lapilli) may be permeable prior to contact with water, they may become clogged by weathering, leading to the development of linear erosion at the surface. As a result, very few volcanic edifices, other than some scoria cones, remain permeable over the long term. The mixed or fine-grained material of composite volcanoes, tuff cones and even spatter cones generally favours linear erosion, as does the rock surface of lava domes and shield volcanoes. However, even on the surface of scoria cones, a drainage network can develop if the intensity and duration of heavy rainfall exceeds the infiltration rate (Collins and Dunne 1986). The characteristic drainage that forms on the surface of volcanic cones can be classified according to whether it develops on the outer flanks of the volcano or within its crater or caldera.

3.1.1 The Flanks of the Volcanic Cone: Development of a Radial Drainage

(1) On a volcanic cone with a smooth surface, rainfall splash quickly turns into sheet wash and *rill erosion*. This type of erosion is favoured by minimal surface roughness. Sheet and rill erosion involves only the uppermost few centimetres, although rill erosion can cut down through a tephra layer and incise into the underlying pre-eruption surface (e.g. Baumann et al. 2020). The initial drainage network consists of a system of temporary rills, which can very quickly develop into grooves and ravines, then astonishingly deep gullies (Photo 95). This early stage of gullying is characterised by a high drainage density (1.5 km/km^2 or more: O'Hara et al. 2024).

(2) The initial drainage pattern on a volcanic cone becomes a radially spreading or *centrifugal* network, which is characterised by the alternation of gullies (*barrancos* in Spanish) and intervening sharp ridges, known as *parasol ribbing* (Fig. 3B; Cotton 1952; Ollier 1988: Photo 96). A classic example is the scoria cone of Mount Batok in Bromo-Tengger caldera, Java (Indonesia: Photo 97). Gully development follows the general rules observed at badlands (Schumm 1956). At active volcanoes, loose ash layers resulting from repeated pyroclastic fallout might also be washed away in this radial system, as was shown by the numerous fresh 'ash avalanches' after the 1906 eruption of Vesuvius (Cotton 1952; Hazlett et al. 1991). It is important to stress the temporary nature of these early-formed gullies: the drainage network frequently undergoes disturbance as a result of unstable slopes composed of loose material, the infilling of gullies caused by occasional recurrent volcanic activity, and slope failures on any unvegetated surfaces. Where there is only episodic rainfall, in semiarid to arid climates, the V-shaped grooves of the gullies flatten distally, even filling up towards the lower flanks, and plunging into the surrounding volcaniclastic apron. Gully downcutting can be enhanced by asymmetrically distributed rainfall (i.e. on one preferential slope sector), and the vicinity of the seashore in the case of coastal erosion. In addition, volcanic activity itself may influence or even initiate gully formation: for example, after a small eruption (1959) of the Sinmoedake volcano (Kyushu Island, Japan) a newly formed flank fissure and the deposition of uncompacted ash around it triggered the incision of gullies (Photo 98; Imura and Kobayashi 1991; Karátson et al. 1999).

(3) Steeper gullies with higher energy, collecting their water supply from a relatively large catchment area, soon start to take over their smaller counterparts and capture their gullies. As shown for gully development in non-volcanic areas (e.g. Campo-Bescós et al. 2013; Rafaello and Reis 2016), certain small drainage basins will expand to become dominant, while less-dominant basins become passive and are pushed downflank as nested basins, adjusting to the geometry imposed by their more dominant neighbours (O'Hara et al. 2024). This brings about a reduction in drainage network density. Such gully evolution is also linked to the spreading of vegetation cover.

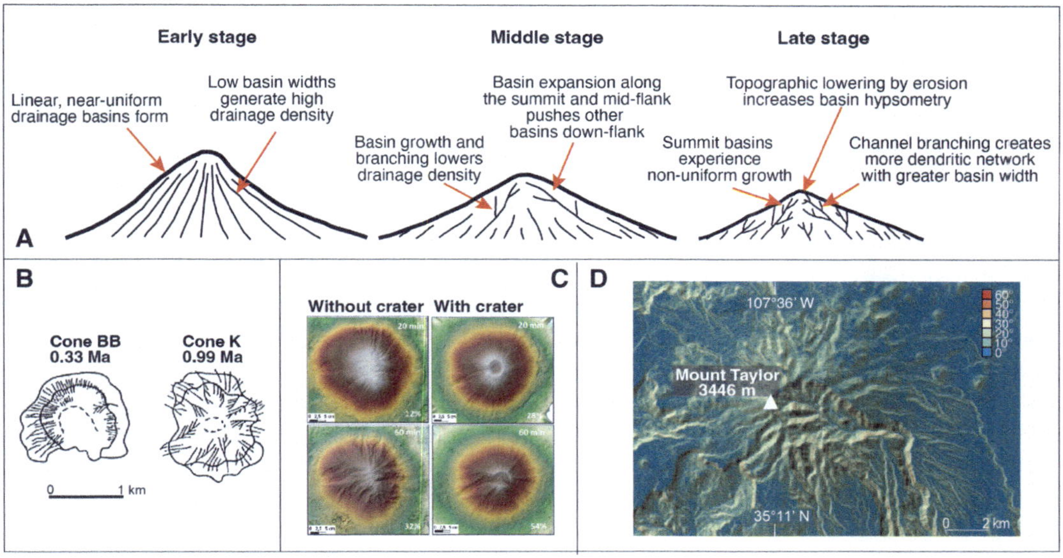

Fig. 3 Evolution of the drainage network from radial to dendritic on the outer slopes of symmetric, circular volcanic edifices. **A** General evolution of drainage pattern based on numerical modelling (O'Hara et al. 2024). See text for description of successive stages. **B** Pleistocene scoria cones of different ages in the Cima Volcanic Area (Mexico), with ages in Ma. The outer double line is the basal apron of the cones, while the dashed line marks the crater (Dohrenwend et al. 1986). **C** Evolution of drainage using analogue modelling (van Wees et al. 2023). Note the effect of a central crater, and the increased average denudation rate (marked in %). **D** Well-established drainage on Pliocene Mount Taylor composite volcano (New Mexico, USA): there is centripetal drainage on the outer flanks and dendritic drainage inside the central depression, which has been significantly reshaped by fluvial erosion and is termed an erosionally modified caldera. Slope map based on SRTM DEM

When precipitation is sufficient (at least 500–600 mm/yr), vegetation absorbs the precipitation and prevents surface runoff, hence erosion slows down over time, leading to the formation of a smaller but permanent number of watercourses (Ohmori 1983). On the volcano flanks, a dendritic drainage network eventually develops, which is fed by tributaries in relatively small drainage basins. Over time, this pattern becomes stable and, once it has occupied the volcano flanks, remains roughly constant for up to some millions of years depending on the climate.

(4) A very important change that accompanies the formation of a dendritic drainage network is that a watercourse with the greatest hydrological potential, if a crater or caldera is present, will breach its rim through retrogressive erosion. Figure 3C shows results from analogue modelling of how summit dissection occurs in the presence or absence of a crater (van Wees et al. 2023). The location of the breaching is determined by the pre-existing topographic unevenness of the rim, the stratigraphic/structural features and any rainfall asymmetry on the flanks. In many cases, the primary breach is the same as that opened by the outpouring lava flow, and is thus merely exploited by fluvial erosion.

3.1.2 Inside the Crater or Caldera: Development of Centripetal Drainage

The process described above also leads to the formation of a simple parallel drainage pattern on the internal walls of a crater or caldera, consisting of adjacent gullies, which, due to the geometry of inner slopes, is convergent or *centripetal* (Photo 99). Feeding the central depression, this drainage network usually creates an internal lake, which can persist for a long time. Some of the many examples (e.g. Branstator 2022; Kozlov and Lebedeva 2024) are: crater lakes: Laacher See (Germany), Nyos (Cameroon), Licancabur Lake (Chile), Lake Yeak Laom (Cambodia), Kelimutu (Flores, Indonesia), Okama (Japan), Lac Pavin (France), and Lake Sf. Ana (Romania) (Photo 100); caldera lakes: Lago di Bracciano, Lago di Vico (Italy), Crater Lake and Katmai (Oregon, USA), Atitlán (Guatemala), Ksudach (Kanchatka, Russia), Kussharo and Shikotsu (Hokkaido, Japan), Segara Anak, Rinjani (Lombok, Indonesia), Toba (Sumatra, Indonesia) and Cuicocha (Ecuador).

Over time, the gullies on the inner slopes become progressively collected into a dendritic drainage pattern, in a similar way to the outer slopes. This process is aided and significantly accelerated when one of the water courses on the outer flanks, mentioned above, breaches the crater or caldera rim because it has a larger catchment area with more erosional power compared to the small gullies in the inner basin. The breaching stream captures and diverts the drainage towards itself and, through its outlet valley, becomes the main artery of the whole drainage system. Adjusting to the outlet, it is not uncommon for both the dendritic drainage pattern and the originally circular crater or caldera depression to gradually elongate over time (Fig. 3D).

3.1.3 Preservation of the Drainage Network

The resulting water network can be preserved for a long time, even after the volcano is completely eroded. Despite the asymmetry of volcanic relief, the irregularity of the exposed inner structure, as well as rocks of different resistance and permeability all affecting the drainage pattern, both the outside centripetal and the inside centrifugal drainage may be preserved.

After the volcano has become extinct, the drainage pattern may be inherited by the underlying non-volcanic rocks. A good example of this is the Early Miocene (19 Ma: Tapu et al. 2023) Ebor volcano (New South Wales) in Australia (Fig. 4), where 90% of the former shield volcano is estimated to have been destroyed (Ashley et al. 1995). The erosion has been marked by the volcano being bisected by the Great Escarpment of the east coast of Australia, which has been subjected to significant erosion by rivers flowing into the Pacific Ocean from its westward-retreating margin over the last tens of Myrs. To the west of the escarpment, the New England highlands—which still have an elevation of 1000–1200 m—have experienced much less erosion, as evidenced by the lopsided preservation of Ebor volcano. While the northwestern part of the volcano has largely remained, the eastern part has disappeared

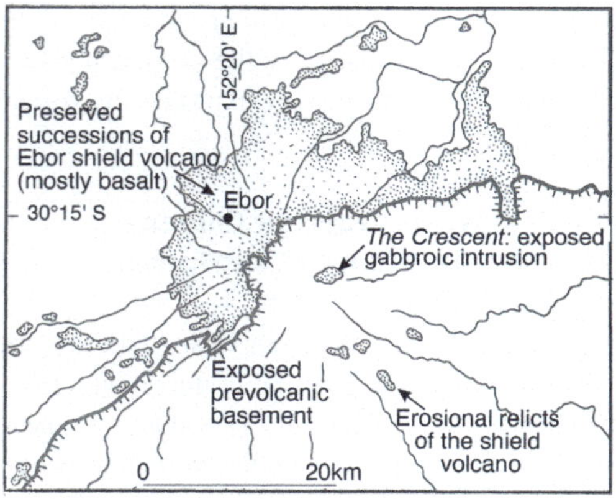

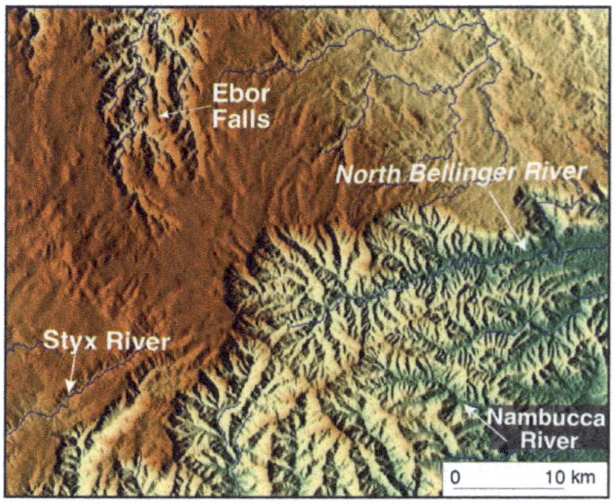

Fig. 4 Drainage of the Early Miocene Ebor shield volcano. **A** Present-day centrifugal drainage shows an inherited pattern down to the prevolcanic bedrock (Ollier 1988; Ashley et al. 1995); **B** Shaded and coloured relief image (*Source* Open Street Map)

completely: beneath the latter, mostly Permian prevolcanic basement rocks together with the Miocene central intrusives are exposed. The pattern of the drainage network has nevertheless been broadly preserved and still shows the former centrifugal pattern inherited by the bedrock.

3.2 Fluvial Erosion of Volcanic Plateaus/Tablelands

The essential topographic features of large lava plains, plateaus and ignimbrite sheets are their generally low-angle slopes and non-circular surfaces. Therefore, the drainage network is randomly generated on them according to minimal differences in slope or other structural factors (e.g. faults or fractures): an *insequent* or *discordant drainage pattern* forms. In particular, many ignimbrites have a markedly parallel drainage pattern due to long, low-angle, uniform slopes on a homogeneous material as well as to fissures and fractures acquired during cooling of the ignimbrite flow unit (Ollier 1969, 1988). However, as this system also evolves with time, the direction of water courses will adjust to the local erosion base level of the surrounding region, and eventually the parallel drainage pattern of large areas becomes a convergent system in one direction on different sides of the plateau. One example of insequent drainage is that which developed on the Mamatu ignimbrite (Ollier 1988; Milner et al. 2003), next to the lake-filled Rotorua caldera (North Island of New Zealand), shown in Fig. 5.

3.3 Modification of the Drainage by Volcanic Activity

The drainage network at and around volcanoes and volcanic plateaus can be overprinted by renewed activity. This may result in the displacement and diversion of water courses and watersheds, or changes in drainage patterns. An area may be blocked and even dammed by

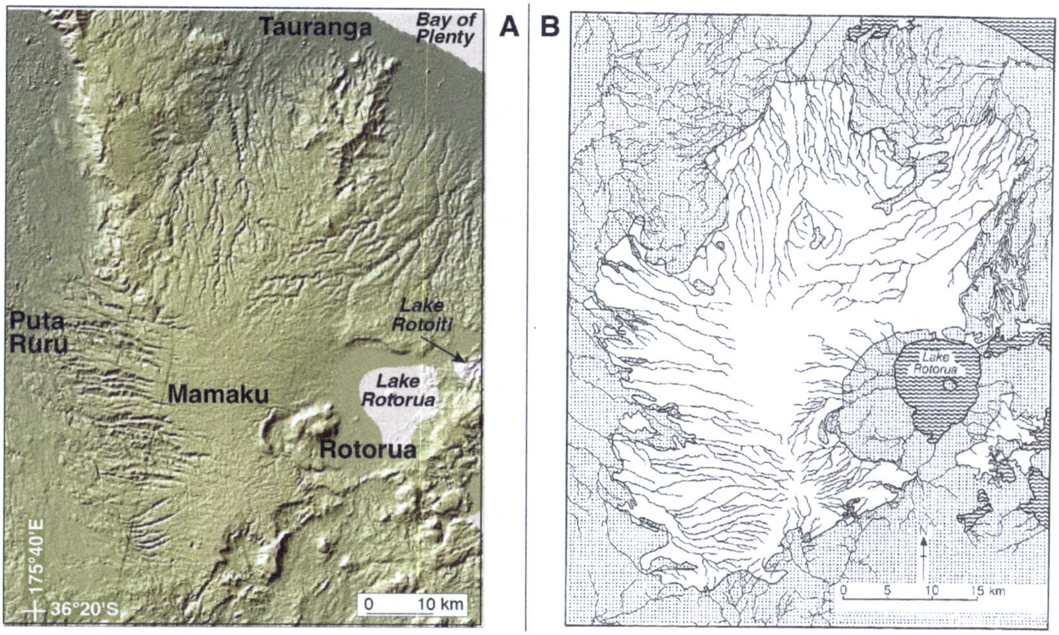

Fig. 5 Drainage network of the Mamaku ignimbrite emplaced by the 220 ka eruption of the Rotorua caldera in New Zealand. **A** Enlarged detail of topography in shaded relief image using GeoMapApp (www.geomapapp.org) **B** Sketch by Ollier (1969): drainage pattern shows an overall radial pattern or slightly dendritic in different directions towards the local erosion bases (Ollier 1969; Milner et al. 2003)

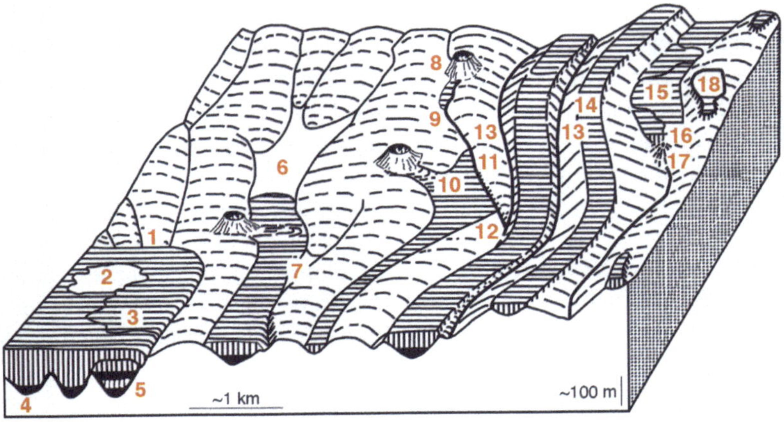

Fig. 6 Diversion/modification of drainage due to volcanic activity (after Ollier 1969, 1988) showing major features of diverted drainage and formation of inverted lava flows. Legend: (1) diversion of drainage by large lava flow or lava field; (2,3) accidental, insequent drainage on lava surface; (4) deep lead; (5) multiple deep lead; (6) valley section with dammed lake due to lava flow; (7) pond dammed by lava flow; (8) small volcanic cone and deep lead beneath lava flow; (9) spring; (10,11) valley-fill lava flow and lava-diverted watercourse; (12) possible gorge formation; (13) watercourse diverted by lava flow to the margin of valley; (14) inverted lava flow; (15) watercourse formed on lava surface; (16) waterfall; (17) alluvial fan; 18) relict inverted surface of older lava flow. Black horizontal layers = aquifer

volcanic material accumulating from eruptions, and existing watercourses may be diverted and captured. Of the large variety of potential hydrological and topographic changes (Fig. 6), two typical scenarios are briefly described: the creation of dammed basins and the diversion of a drainage network.

3.3.1 Lakes/Water Bodies Dammed by Volcanic Activity

Large-scale lava flows, the emplacement of ignimbrite lobes or the build-up of voluminous volcanic edifices can block the existing drainage system, thereby creating a closed basin, which usually accumulates lake water. Lake Managua and the huge Lake Nicaragua (Cocibolca) were formed behind part of the Central American volcanic chain. Lake Van in Turkey was also formed as an undrained continental depression by repeated damming, mainly by the eruptions of Nemrut Dagi volcano, but its extent was also influenced by lava flows from the Süphan volcano. The history of this lake dates back to 600 ka (Stockhecke et al. 2014). The several hundred metre-deep Lake Kivu, whose basin has existed for longer than its present form, was most recently dammed by volcanoes in the Virunga Mountains in the latest Pleistocene, ca. 10–20 ka ago (Snoeks et al. 1997). About 9,000 yrs ago, headward erosion connected it to the Tanganyika tectonic graben. The very shallow Lake Omapare in New Zealand was formed by the damming of the Waitangi River by lava flows some 80 ka (Newnham et al. 2004). The lake has changed its surface extent several times, with its present form being only 600 yrs old.

Not only lava flows but also pyroclastic flows, debris avalanches, and lahars can dam lakes. One of the most recent and best known examples is Spirit Lake at Mount. St. Helens volcano, which was first blocked by a pyroclastic flow 3,000 years ago, and most recently by the huge 1980 debris avalanche described in Chap. 7. Much earlier the 25 ka Cougar debris avalanche dammed the Lewis River to the south (Fig. 7); downcutting of the dam caused flooding into the Columbia River, and filled the lower Lewis River valley with thick volcanic debris (Clynne et al. 2005). The 1980 eruption not only blocked Spirit Lake, but also created Castle Lake and Coldwater Lake, which soon had to be drained using artificial spillways to avoid the risk of sudden lake outbreak (Dahm et al. 2005). The formation of a number of volcanogenic

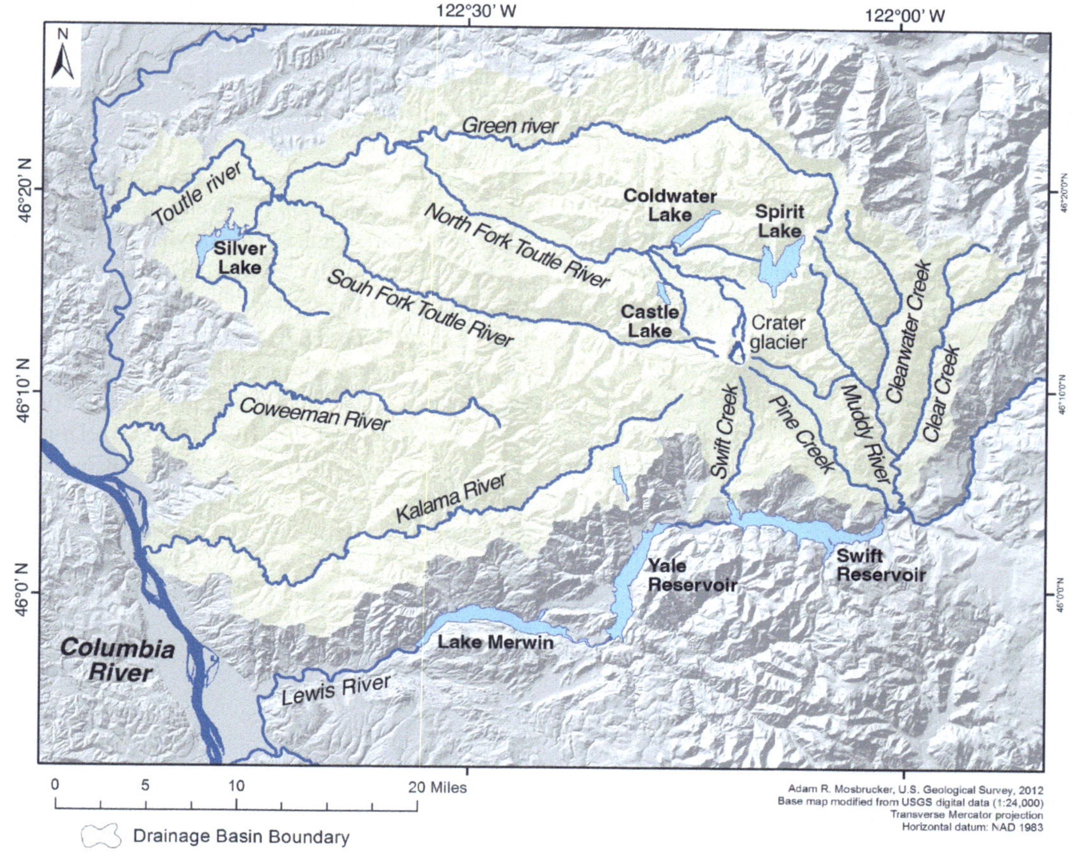

Fig. 7 Drainage basin of the Toutle River—Kalama River system around Mount St. Helens with dammed lakes and river sections mentioned in text (Mosbrucker 2013)

lakes due to lava damming, pyroclastic-flow and lahar deposit accumulation, often associated with break-out floods, have been described from the North Island of New Zealand by Manville et al. (2007).

Of the large number of ancient examples, we cite the Late Pleistocene Ciomadul (Csomád) volcano, East Carpathian Mountains (Romania). This lava dome complex, built over the past 850 ka, repeatedly blocked the upstream Ciuc (Csíki) Basin, which drains the Harghita volcanic chain. In the main dome-building stage, about 100–200 ka, the damming was so extensive that water from the lake swelled and then breached the Ciuc Basin, leading to the release of huge lahars to the south (Karátson et al. 2022).

3.3.2 Diversion of the Drainage Network

The construction of a new volcanic edifice can induce a new drainage network around its base. For example, start of activity at Etna over half a million-years ago has diverted water courses from inland Sicily to the Ionian Sea; some of them now flow northwards collected by the Alcantara river, and others find their way to the south via the Simeto river (Chester and Duncan 1982). In many cases, valley filling of major lava flows has led to the displacement of the existing drainage network. Ollier (1988) provided detailed examples from Victoria, Australia, such as Emu Creek by the lava flows of Mt. Hamilton, Shaw and Eumerella rivers by lava flows of Mt. Rousemu.

Cones that develop next to each other, but with an interval of time, result in the displacement of the pre-existing drainage network. A good example is Gough Island in the South Atlantic. Its history above sea level dates back to ca. 2.5 Ma (Maund et al. 1988), but most of the surface is much younger. The abundant rainfall encourages lush vegetation and a rapidly evolving drainage system (Ollier 1984). The oldest (1–2 Ma) rocks in the southern part of the island constitute the basaltic centre of South Peak. Over the past 300 kyr, its radial drainage network was first destroyed by the trachyte lavas of Green Hill to the southwest (Fig. 8), then by the even younger Richmond Hill centre to the southeast, which split the streams running from Green Hill to the east in two.

4 Degradation of Volcanic Cones and Surfaces, and Formation of Erosional Landforms

Degradation of volcanic edifices and volcanic plateaus or tablelands is predominantly controlled by the patterns of fluvial erosion, as outlined above. Here, we describe the progress of this process under balanced, 'average' conditions of exogenic agents, for example a temperate climate with moderate annual precipitation (about 500–2000 mm) and temperatures (5–20 °C) (Cotton 1952; Kear 1957; MacDonald 1972; Ollier 1988; Francis 1993; Karátson et al. 1999). The general process is common almost everywhere on Earth, except for arid/semi-arid

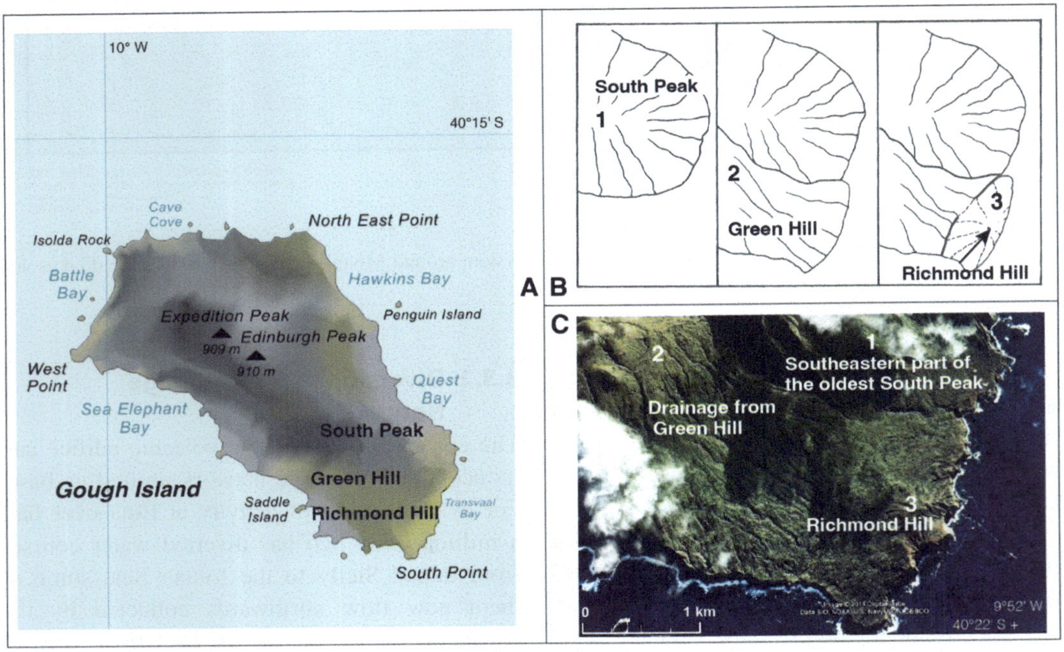

Fig. 8 Drainage evolution of the southern part of Gough Island, south Atlantic Ocean. **A** Sketch map of Gough Island (credit: http://commons.wikimedia.org/wiki/Image:Tristan_Map.png). **B** Subsequent stages of evolving drainage showing adjacent eruptive centres of 1: South Peak, 2: Green Hill, 3: Richmond Hill (Ollier 1984); **c** Google Earth image of the southern part of Gough Island. Note that the surface of young lava flows of Richmond Hill, unlike the older volcanoes, is still only slightly dissected, which constrasts with the extensive coastal erosion on the eastern side

and ice-covered land surfaces. Although the surface evolution is continuous, it can generally be divided into four distinct 'stages', each showing specific geomorphic features (Cotton 1952; Kear 1957; Ollier 1988; Fig. 9A), with climate differences affecting the time required to reach a particular stage.

In terms of geomorphic evolution of volcanic edifices, the observed variations on Earth are controlled by their structure, lithology, the

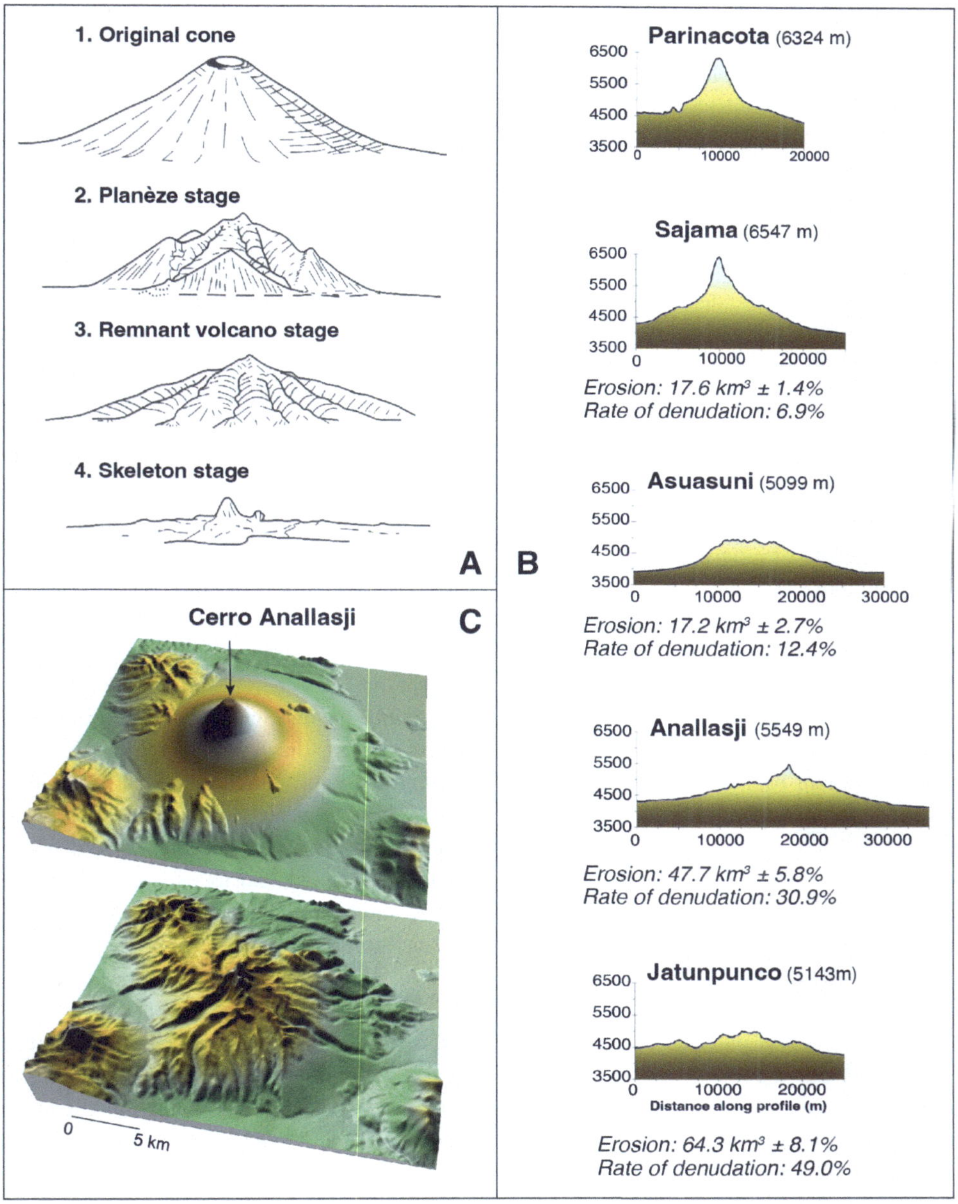

Fig. 9 Erosion stages of volcanoes. **A** Stages of an 'ideal' volcanic cone under average climatic conditions (based on Cotton 1952; MacDonald 1972; Ollier 1988). **B** Real topographic cross sections through composite volcanoes in the Western Cordillera of the Central Andes showing successive degradation (calculated proportions of denudation given for each); **C** Reconstructed original and present-day landform of Cerro Anallajsi volcano (Karátson et al. 2012)

hydrogeological characteristics, and climate. While humid climates are characterised by elevated erosion rates (see Sect. 4.3) resulting in the rapid degradation of the original landforms, erosion is much slower under semiarid (≤400 mm/yr) or arid (≤200 mm/yr) climates. As an example of low-rate erosion, quantitative results from the Western Cordillera of the Central Andes (the Altiplano–Puna Plateau) are shown in Fig. 9B, C (Karátson et al. 2012), depicting representative volcano profiles that reveal successive degradation (not necessarily corresponding to distinct stages).

When we refer to an extinct volcano or an old volcanic formation/deposit, it is of crucial importance whether we are dealing with:

(1) a primary, initial surface (in the case of ancient terrains which undergo significant volcano erosion, this is commonly referred to as a *palaeosurface*)—which, even if erosionally dissected, preserves remnants of the original surface envelope; or
(2) a more degraded topography where erosion has already incised below the original surface, of which there is no longer any trace. In the latter case, as shown for example by Ebor volcano, drainage of the primary landform has been inherited by the underlying (volcanic/subvolcanic) rocks and ultimately by the basement rocks. In many such cases, fossil drainage can be the only clue left by the original landform.

4.1 Erosion of Volcanic Cones Through Four 'Stages'

4.1.1 Original Cone Stage

An *original cone* infers the presence of an intact primary surface over the majority of the volcanic edifice. As described in Sect. 3.1, at first a radial/centrifugal drainage network develops on the surface envelope, evolving into a convergent/dendritic pattern. Dendritic drainage also develops inside the crater or caldera. After the fluvial breaching of the rim of the central depression, the interior of the crater or caldera acts as the largest drainage sub-basin on the scale of the volcano, and can widen considerably through regressive erosion of the dendritic tributaries. In this way, the circular shape of a crater or caldera (while it widens) may be preserved for millions of years after the volcano has become extinct (see Sect. 4.2).

4.1.2 Planèze Stage

Due to the dendritic drainage, not only the pattern but also the density (e.g. channel length/area, km/km^2) of the drainage network changes on the flanks of the volcano. Since the gullies gradually become captured, forming a dendritic pattern, the average drainage density will decrease downstream over the surface of the volcano as a whole. As a consequence, the interfluves between the steady valleys on the lower flanks, which used to host the now beheaded radial system, become little, or un-affected, by linear erosion. These downward widening, triangular surfaces are known as *planèzes* (de Martonne 1909), a name taken from the Cantal volcano in France. Planèzes (Fig. 10) often form in areas with high precipitation, or, in general, pronounced/accelerated fluvial erosion in either a tropical or humid temperate climate, as well as by glacial erosion. The increasing size of some dominant basins with dendritic drainage provides a larger discharge, resulting in significant fluvial downcut, which is in contrast with decelerating erosion on the beheaded radial drainage on the lower outer flanks. Eventually, these quasi-abandoned surfaces become planèzes. Note, however, that planèzes do not always form in less humid climates.

4.1.3 Remnant Volcano Stage

After hundreds of thousands or even millions of years, general denudation results in a significant lowering of the volcano and the planèzes also become eroded. This leads to a stage when only the *remnants* ('ruins') of the original volcanic landform can be observed, and the original surface is no longer visible. However, the volcanic topography may be inferred, either apparent in plan view or reconstructable from the preserved centrifugal drainage network.

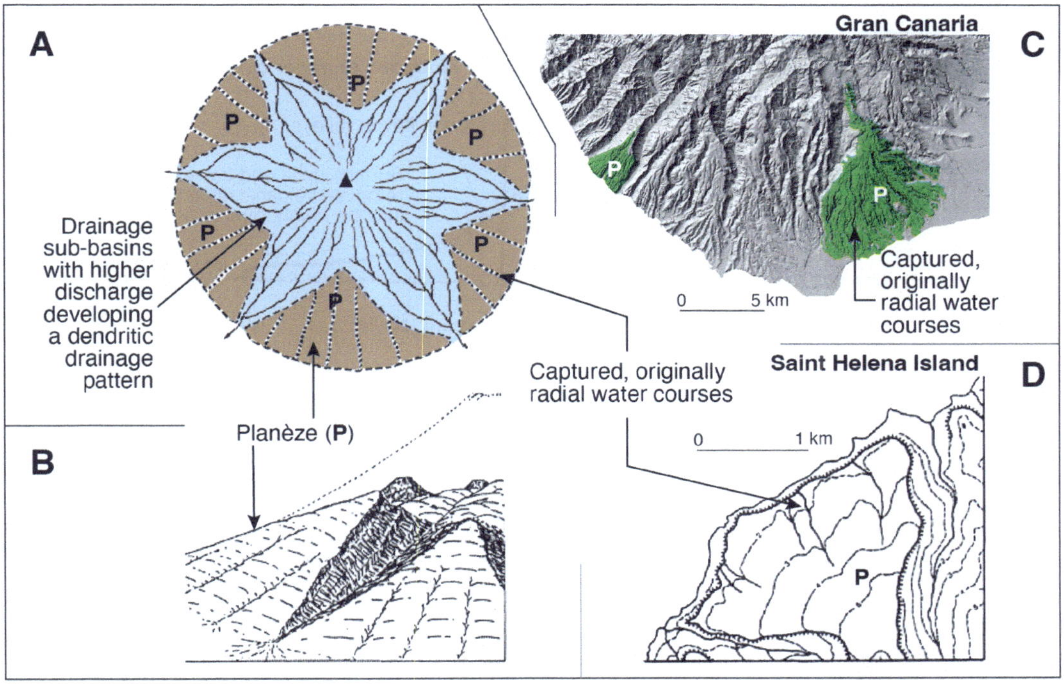

Fig. 10 Planèze formation on volcanic cones. **A** Process of dendritic drainage pattern formation and development of triangular planèzes (Ollier 1988). **B** Ideal section of planèzes formed on the lower flanks of a deeply eroded volcano (Cotton 1952); the surface is no longer the original one because the dip of the strata is intersected by the planèze surface. **C** Planèzes formed on the lower flanks of the c. 11–12 Ma Fataga shield volcano in Gran Canaria (Karátson et al. 2016). **D** Planèze on the western part of the Late Miocene (c. 9 Ma) main shield of Saint Helena island (Ollier 1988; Engbers et al. 2020)

Figure 11A illustrates the generalized morphology in ancient volcanic settings in a temperate climate, showing examples from the Carpathians. Interpretative sketches document two mountain ranges with remnant volcano surfaces: the older (Mid/Upper Miocene) Slánské-Tokaj Mountains and the younger (Upper Miocene/Pleistocene) Călimani-Gurghiu-Harghita Mountains. Figure 11B compares cross sections of the characteristic topography of one of the well-preserved volcano remnants with the active cone of Mount Fuji. It is clear that once early-stage erosion has degraded the upper cone, the surface envelope of the lower flanks still remains close to that of the original. This happens because, subsequent to the planèze stage, the volcano flanks retain the established dendritic drainage system for a long time, with the result that the general lowering of the interfluves is slower than the downcutting of the valleys in between.

Development, relative position and incision of fluvial valleys around a given volcano are primarily determined by the size of the drainage basin. Moreover, the occasional exposure of resistant rocks (e.g. lava breccias, welded ignimbrites, coarse-grained block-and-ash flow deposits) is also important in the context of general surface lowering and eventual valley development, as resistant rocks can be preserved in the long term, and may even become inverted (see Sect. 4.3), thereby contributing to fluvial valleys becoming diverted.

The development of valleys alters when the amount of annual precipitation exceeds a high threshold at around 2000–2500 mm/yr. At this point, *amphitheatre valleys* (also called amphitheatre-headed or Oahu-type valleys: Hinds (1925); Cotton (1952), or amphitheatre-headed canyons: Lamb et al. (2014) develop, which are characterised by a much higher erosive capacity

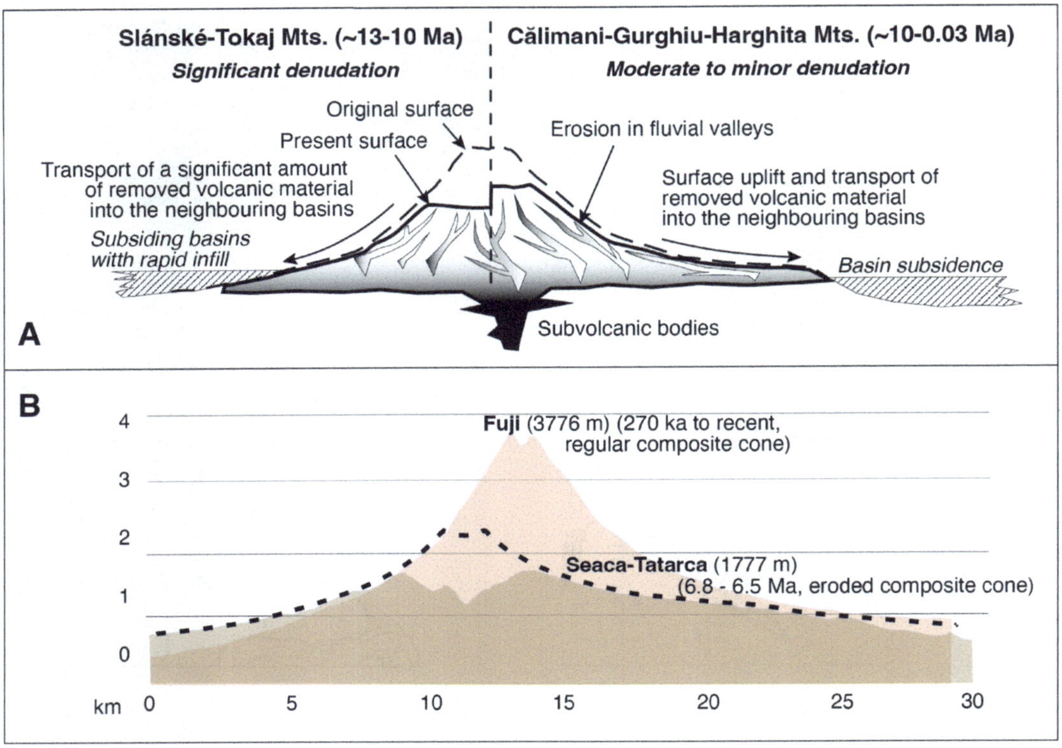

Fig. 11 Geomorphic features of the remnant volcano stage in Miocene to Quaternary segments of the Carpathian Mountains. **A** General comparative sketches of the Slánské-Tokaj (Slovakia/Hungary) and the Călimani-Gurghiu-Harghita Mountains (Romania). **B** Comparison of topographic profile of the well-preserved, Late Miocene (c. 6 Ma) Seăca Tatarca andesitic composite volcano (Gurghiu Mts., Romania) with active Mount Fuji. While much of the upper cone of the Carpathian volcano has been removed by erosion, the present topography of the lower flanks is close to the original (palaeo-) surface (after Karátson and Timár 2005, modified)

than that of common fluvial valleys. On the Big Island of Hawai'i, the average rate of valley headwall retreat can be as high as 6 cm/yr (Lamb et al. 2007). Hence, they degrade the original edifice more quickly.

The term amphitheatre valley was first introduced by Hinds (1925), described later by Cotton (1944, 1952), Ollier (1988), and Karátson et al. (1999), while their origins and development were addressed more recently by Lamb et al. (2007, 2014). These valleys are very deep and wide with a relatively flat bottom, with a cirque- or U-shaped head, while the steep valley slopes are densely dissected into spurs. High erosion rates observed particularly at the amphitheatre valley head are caused by headwall propagation via vertical waterfall erosion based on abrasion by impacting sediment particles in plunge pools (Lamb et al. 2007). The location of amphiteatre valleys can be determined by slumping, e.g. the Pololū Slump at Kohala volcano, Hawai'i (Fig. 12). The onset of rock toppling may also be a tipping point for the formation of amphitheatre-headed canyons, as for a canyon system cut into basalt in Idaho, USA (Lamb et al. 2014). Slumping or toppling creates knickpoints in the valleys, and as dominant streams poured over this headscarp as waterfalls, vertical plunge pool erosion and undercutting causes upstream propagation of the knickpoints, eventually producing amphitheatre-headed valleys (Lamb et al. 2007; Mackey et al. 2014). The rectangular-shaped Haleakala 'Crater' on East Maui, Hawai'i, which is not of volcanic origin, has been created by two coalescing valleys coming from opposite directions (Koolau and Kaupo), possibly initially formed by large submarine slumps (Eakins and Robinson 2006;

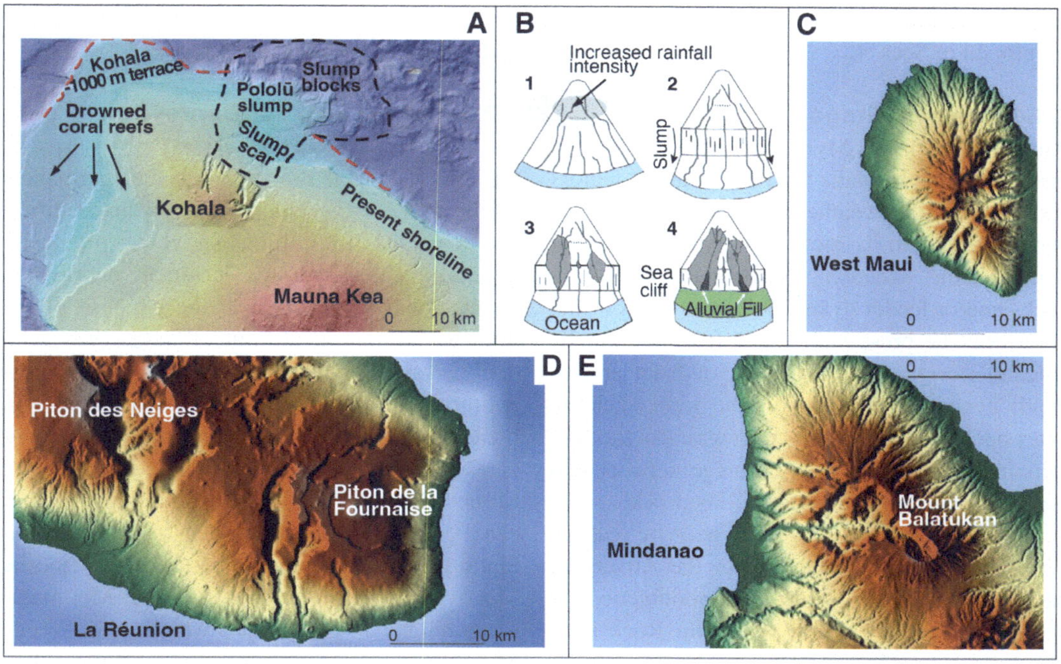

Fig. 12 Remnant volcano surfaces due to amphitheatre-valley evolution. **A** Amphitheatre valley heads incised into the northeast flanks of Kohala volcano, Big Island of Hawai'i, and their connection with the c. 200–250 ka Pololū Slump mapped offshore (Lamb et al. 2007). **B** Amphitheatre valley heads of Kohala volcano might have been initiated or accelerated by the slump. Sketch shows that the slump imposed knickpoints onto the streams, and the knickpoints have propagated upwards in dominant streams by capturing the headwaters of neighbouring valleys (Lamb et al. 2007). Amphitheatre valleys on **C** West Maui (Hawai'i', USA), **D** La Réunion (France), and **E** Mindanao, Philippines. *Source of coloured DEM images:* Open Street Map

Bishop 2021). Filled by recent lavas and scoria cones (of the postshield stage: Photo 46), these valley heads, which have been little deepened by fluvial erosion today, will clearly become amphitheatre valleys in the geologically near future. Moreover, as for common fluvial valleys, the presence of faults may also be crucial: in particular on oceanic islands, e.g. Kauai or West Maui on the Hawai'i Islands, or Piton des Neiges on La Réunion, amphitheatre valley incision can take place along rift zones (Karátson et al. 1999; Michon et al. 2007; Salvany et al. 2012). In addition, the areal coverage of the drainage basin also plays a role. Topographic analysis on Hawai'i suggests a potential drainage area threshold of ~5–8 km^2 for initiating and hosting active amphitheatre valleys (Lamb et al. 2007).

Amphitheatre valleys are found mostly on low-latitude oceanic islands such as Hawai'i, Vanuatu, Tahiti, La Réunion (Photo 101), Mauritius, and the Philippines (Fig. 12). The initial conditions for the formation of amphitheatre valleys are illustrated by the island of Mindañao (Philippines), where these valleys are found in the northern part of the island which receives as much as 2500–3000 mm/yr rainfall, twice as much as in the southern and western part where they are absent (Karátson et al. 1999). However, in addition to rainfall, the location of the groundwater also plays a role in amphitheatre formation. On La Réunion, as well as other oceanic shields, Saint-Ange et al. (2004) and Join et al. (2005) showed that the water table controls the appearance of differently erodible layers, which often results in the formation of 'nested' amphitheatre-heads. In summary, amphitheatre valleys tend to incise retrogressively at a rapid rate, and their efficiency in terms of erosional removal of material has a profound impact on the geomorphic development of a volcanic edifice, especially its central part, as further developed in Sect. 4.2.

4.1.4 Skeleton Stage

After a few tens of Myrs, depending mostly on the prevailing climatic and hydrologic factors and governed by weathering, a volcano that has been degraded to its roots will reveal its internal structure, the *'skeleton'*, occasionally exhuming necks and/or feeder dykes. Different levels of the latter may be preserved, either from within the volcano's original edifice itself, or from greater depth beneath the volcano. The term skeleton dates back to W. M. Davis, who was also the first to point out that the fossil drainage network can often be inherited by the exposed bedrock.

Spectacular neck fills crowned by castles can be found around Le Puy in the Auvergne, but the best known example is undoubtedly the lonely, impressive, ca. ~27 Ma Ship Rock (Tsé Bit' A'í), New Mexico (USA) (Semken 2003). (Photo 102). In this case, the volcanic relief has completely disappeared; three large radial dykes branch out from a central, steep neck, and a drop of ~1000 m relief relative to the original volcano surroundings has been estimated (Semken 2003). Another example is the well-studied, 32–33 Ma Summer Coon in Colorado. It comprises remants of a medium-sized (7 × 10 km) composite volcano (Harp and Valentine 2018). Although the central intrusion is less exposed compared to Ship Rock, the radial dykes are well-preserved. Interestingly, these dykes are not traceable to the intrusion at the present surface; they presumably branch out from it at a deeper level, obliquely upwards (Poland et al. 2008). In addition, outward dipping layers of the original lava and breccia pile of the volcano are preserved in a ring at a distance of 5–10 km from the centre.

4.2 Erosion of Craters and Calderas

The genesis of erosional landforms of composite- and shield volcanoes that host craters or calderas will be determined to a significant extent by the evolution of the central depression, eventually down to the skeletal stage.

4.2.1 Central Volcanic Depressions with One Outlet

Once the dendritic drainage becomes permanent within a crater or caldera of a volcanic edifice, it can persist for millions of years, often significantly enlarging the original depression. Enlargement takes place at the expense of the drainage basins on the outer flanks: their relatively lesser erosive power under temperate and arid climates means that they are unable to breach the crater/caldera from other directions (see e.g. the erosionally enlarged caldera of Mount Taylor in Fig. 3D). Additional breaches can occur only in exceptional cases, such as faulting—a good example is the Mid-Miocene Makovica volcano in the Slánské Mountains (Carpathians), Slovakia, which has been breached by three small, parallel fault-controlled streams (Karátson 1996).

The central depression of these remnant volcanoes, which have been degraded for several Myrs, are considered *erosion(ally) modified volcanic depressions* (Fig. 13). They can be termed either *erosional craters* or *erosional calderas,* depending on the nature of the initial depression (Karátson et al. 1999). Of course, all craters and calderas undergo erosion from their inception, but after significant degradation the dimensions of the initial depression become greatly modified. Thus, erosional craters and calderas are terms used for fluvially breached, enlarged depressions originating, respectively, from primary craters and (collapse) calderas. As a result,

(1) their diameter varies but is considerably larger than the primary depression (a few km for erosion craters, >3–15 km for erosional calderas),
(2) their rounded shape tends to be elongated (as shown by the increased *axis ratio,* the ratio of longer to shorter diameter, being commonly >0.7: for examples, see Karátson et al. 1999),

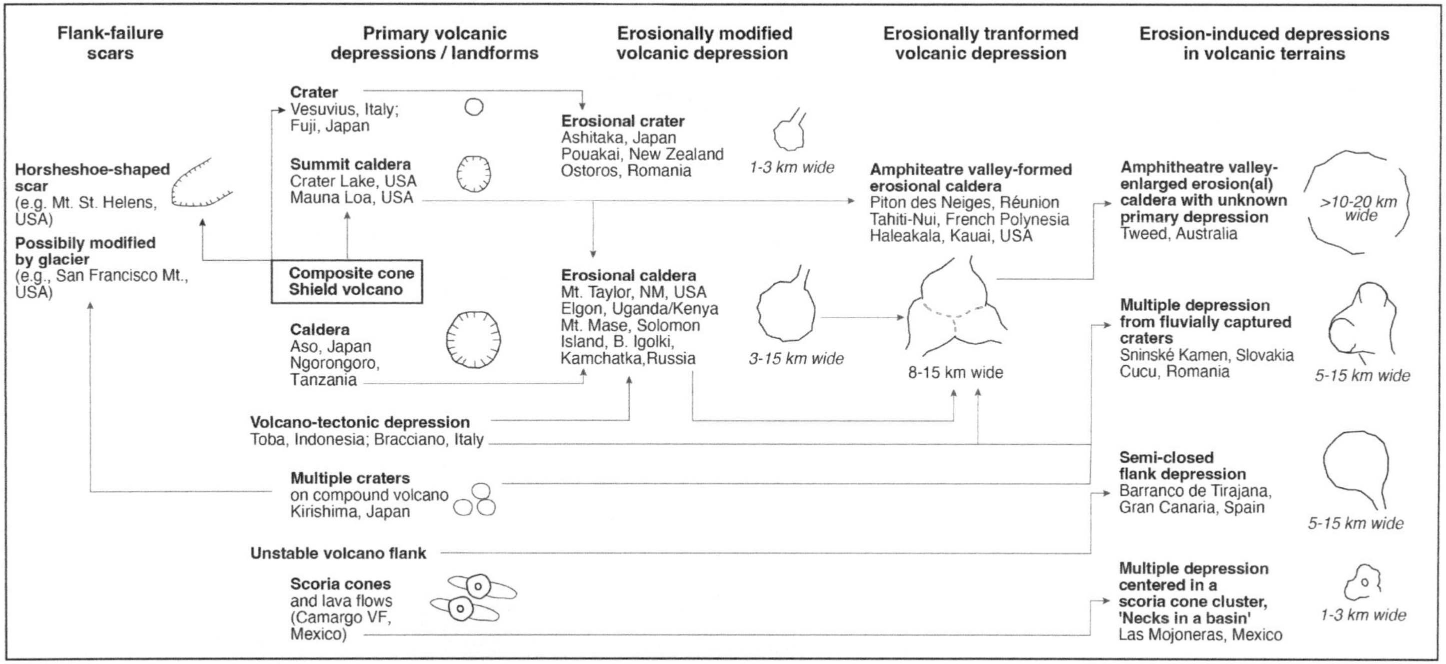

Fig. 13 Flowchart of erosional modification and transformation of primary volcanic landforms including craters and calderas, and erosion-induced depressions in volcanic terrains (modified after Karátson et al. 1999; Aranda-Gómez et al. 2009)

(3) the outlet becomes narrower, reflecting the erosional breach and then incision of a valley of fluvial origin, contrasting with the wide outlet of horseshoe-shaped scars (Siebert 1984; see Chap. 7).

Many authors (e.g. Paris and Carracedo 2001; Lamb et al. 2007; Németh and Cronin 2008; Aranda-Gómez et al. 2009; Manville 2010) accept or have further developed such an erosional landform division. However, some have not: Lockwood and Hazlett (2010) noted that the distinction between erosion(ally modified) calderas and collapse calderas is "difficult as many if not most older… calderas have been extensively modified by erosion and their initial origins may be debatable." This statement is apt, but the term erosional crater or caldera does not indicate the initial origin—it refers only to the fact that erosion plays a major role in the transformation of the original depression, and fluvial processes overprint the primary landform to produce the characteristic shape described above.

Examples of *erosional craters* are found in Japan from the Late Pleistocene: e.g. Mount Ashitaka, south of Fuji (Fig. 3 in Chap. 5), and Mount Azumaya, north of the active Asama volcano. The erosionally enlarged Late Pleistocene Likuruanga crater, New Britain, is also included here (Fig. 14A), although its large (2×3 km) diameter may also correspond to an original summit caldera. The 0.25 Ma Pouakai volcano, northern sister of Mount Taranaki in the North Island of New Zealand (Neall et al. 1986; Gaylord et al. 2014), also has an erosional crater with an outlet to the northeast, with a second valley that has almost breached the crater from the south. Rendova Island (Solomon Islands) has a Plio/Pleistocene basaltic

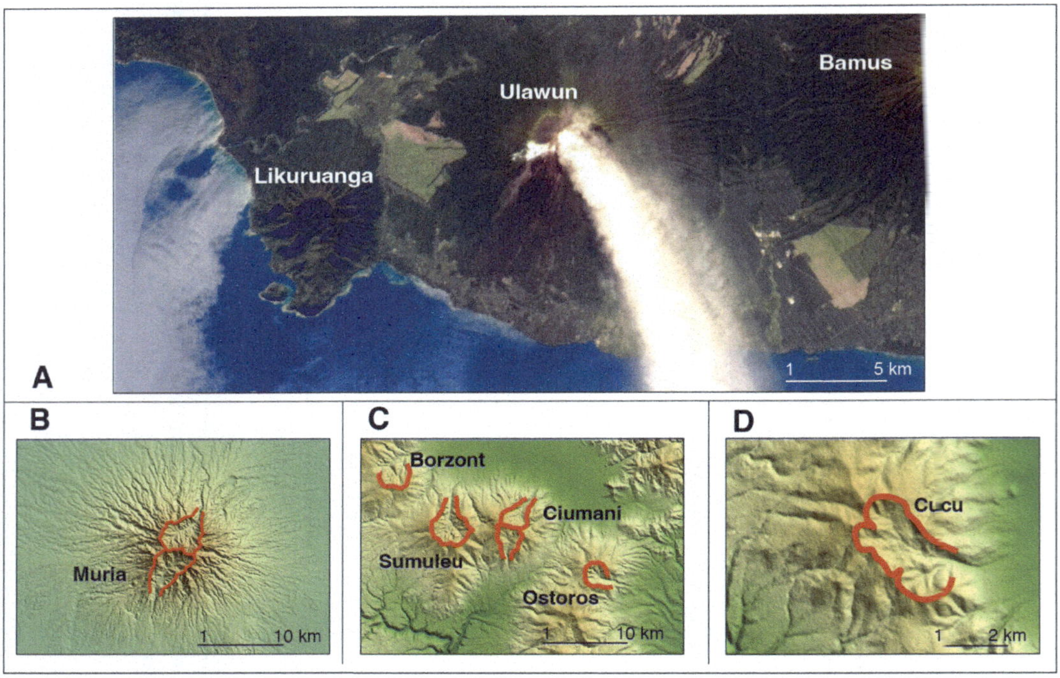

Fig. 14 Examples of erosion craters. **A** Likuruanga (New Britain), in the shadow of the active volcanoes of Ulawun and Bamus (satellite image credit: http://earthobservatory.nasa.gov/NaturalHazards/view.php?id=79903). Note rill erosion even on active volcanoes. Coloured and shaded SRTM DEM images on erosional depressions on composite volcanoes: **B** Muria volcano (Central Java). **C** Examples from the East Carpathians (Romania) showing erosional depressions with a single outlet valley (Karátson 1999). **D** Example of erosionally-induced depression unifying crater remnants: Cucu (Kakukkhegy), East Carpathians (Karátson et al. 1992)

volcano in the north (Global Volcanism Program), and its elongated erosional crater has already been breached by two narrow opposite valleys. The island of Java, with volcanism of different ages, also provides a number of examples, such as the twin erosional craters of Muria volcano, where the two enlarged, elongated depressions are breached in opposite directions (Fig. 14B).

Whereas the above volcanoes in tropical-subtropical climates experience high annual rainfall around or above 2000–3000 mm, the andesitic-dacitic volcanic arc of the Carpathian Mountains, located in a continental moderate climate, shows examples of slower degradation, in accordance with the current rainfall of 1000–2000 mm/yr. Over time, erosion has systematically enlarged the craters of the composite volcanoes, increasing rim diameters to up to ca. 3 km, allowing them to be separated from even larger erosional calderas (Karátson 1996, 1999) (see Sect. 4.3 and Fig. 18A for more detail). In addition to Makovica volcano, referenced above, Late Miocene volcanoes include single erosional craters like Ostoros (Gurghiu Mts), or twin craters like Ciumani (Csomafalvi Délhegy) with the northern and southern craters breached in opposite directions; to the west of them, similar-aged Sumuleu might be an erosionally modified summit caldera (Fig. 14C).

However, since large (primary) craters and small collapse calderas are not uncommon worldwide, further (geophysical, volcanological) information is often required to distinguish between erosional craters and erosional calderas, in addition to the morphometry and the characteristic erosion pattern. Examples of unusual-scale features are the large explosion craters of tuff rings, tuff cones and maars up to 2–2.5 km in diameter in the Kamchatka peninsula, and even, exeptionally, up to 5 km in Alaska (Chap. 3); and on the contrary, relatively small, 2–3 km calderas on shield volcanoes such as Mayor Island (New Zealand) and Longonot (Kenya) (see Chap. 4), or Ohachidaira and Akagi (Honshū, Japan: Moriya 1979).

Examples of clearly identified *erosional calderas* (Photo 103) include Elgon (Uganda/Kenya), Mount Taylor (New Mexico, USA: Fig. 3D), Mount Bosavi (New Guinea), Mount Mase (New Georgia, Solomon Islands: Fig. 15A); for more examples in tropical-subtropical climates, see Karátson et al. (1999); and Seaca-Tatarca (Mezőhavas, Romania) and High Börzsöny (Hungary) in moderate climates (Karátson 1999; Székely and Karátson 2004; Karátson and Timár 2005). Bolshiye Igolki in southern Kamchatka, Russia (Masurenkov 1980), in a subarctic climate, is an erosional caldera formed on a Pleistocene shield volcano (Fig. 15B). Two interesting features in its vicinity can be noted. First, although thousands of glacial cirques occupied a large part of Kamchatka in the Late Quaternary, re-shaping many extinct craters and calderas (Barr and Spagnolo 2015, see Sect. 4.2.3 below), the southern part was subject mainly to fluvial erosion. Second, 40 km southeast of it is the Holocene Ksudach caldera, which has not yet been modified by erosion, but already has a cut in its northern margin due to a late glacial outburst of meltwater (Bindeman et al. 2010), preforming the site of a future outlet valley.

4.2.2 Central Volcanic Depression with a Number of Outlets

The evolution of craters and calderas is different if amphitheatre valleys penetrate into the central depression: their enhanced erosive capacity enables them to breach the rim in one or more direction (Karátson and Székely 2006). This is the case for the Late Pleistocene Kolombangara composite volcano (Feitei and Marau 1982; Fig. 15A). The volcano is still relatively young (Late Pleistocene) so its crater has not yet been breached despite the significant annual rainfall (3000–5000 mm). Breach is anticipated to occur in the southeast, where the rim has already been eroded by about 400 m and one of the outer amphitheatre valleys outside the summit is almost incised. Apart from that valley, there are at least four other amphitheatre valleys cutting headwards from different directions on the flanks (indicated on the slope map of Fig. 15A).

When breaching and draining of the inner depression occurs from more than one direction, the subsequent erosion of the inner basin is not only much more intense, but also significantly transforms and expands its circular shape, to the

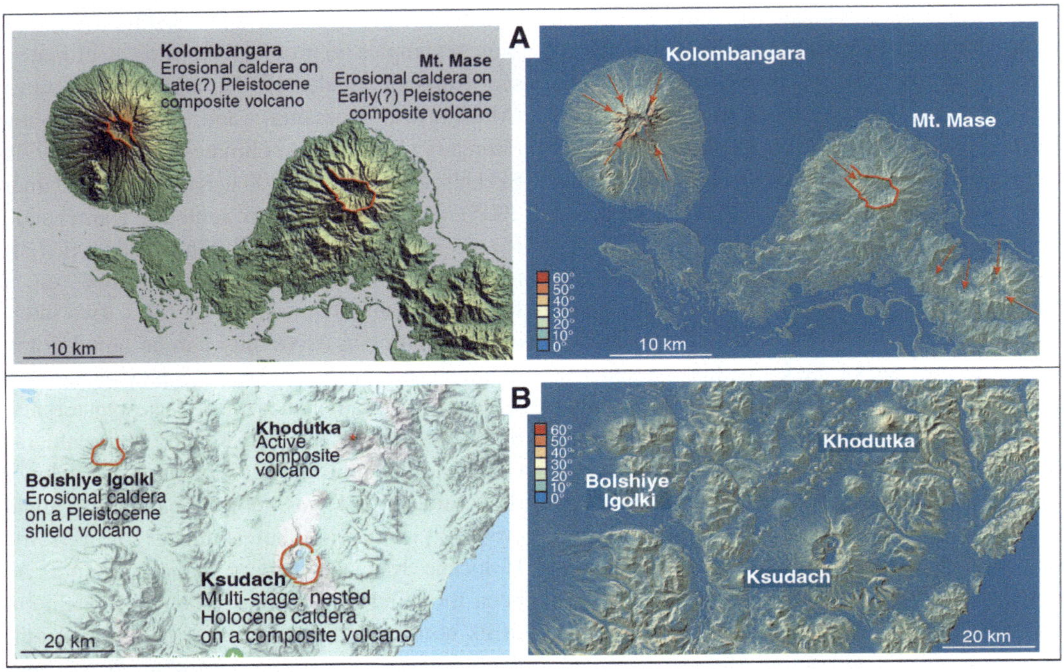

Fig. 15 Examples of erosion calderas. **A** Kolombangara and Mase (New Georgia, Solomon Islands), SRTM DEM image (left) and SRTM DEM-based slope map (right), amphitheatre valleys shown by arrows. **B** Bolshiye Igolki (Kamchatka, Russia); NOTE the pristine landform of Khodutka composite cone and Ksudach caldera (Google Map (left) and SRTM DEM-based slope map (right))

point that the resulting landform might be elongated along the outlet valleys.

Piton des Neiges, the older volcanic edifice at La Réunion, is a clear example of this evolution (Fig. 16A, B). At present, its degradation occurs in three several km-wide and deep central depressions, a type of amphitheatre valley head (so-called 'cirques' in French: the Mafate in the west, the Cilaos in the south and the Salazie on the east side). These enormous depressions, up to 10 km wide, are surrounded by huge 'ramparts', forming scarps that can be up to 1500–2000 m in height. An original caldera on Piton des Neiges is inferred to have existed (Rivals 1950; Kieffer 1989), formed ca. 190 ka (Salvany et al. 2012; Famin et al. 2022), but has been significantly reshaped since then. In the past decades, excavation of the interior of Piton des Neiges has been mainly attributed to giant debris avalanches (Oehler et al. 2008). However, the significant role of extreme rainfall and hurricanes (with as much as 8000–12,000 mm/year rainfall in the eastern half of the island, and around 2000 mm/year in the western side) must also have played a role in destabilizing scarps and producing mass flows. Indeed, more recent research has revealed that the cirques have a complex origin. The efficiency of fluvial and slope erosional processes in their genesis has been highlighted by Salvany et al. (2012), and the occurrence of successive, huge landslides and debris flows enlarging the cirques has been shown by geologic mapping (Lacquement and Nehlig 2014). On the other hand, volcano construction and destruction cannot be separated during the development of Piton des Neiges (Salvany et al. 2012). The amphitheatre valleys had previously carved up the interior of the island, but renewed volcanic activity has always led to infilling by lava and breccia deposits. Although the primary depressions preceding the cirques cannot be inferred with certainty, it is feasible that one of them may correspond to the mentioned caldera (see discussion in Kieffer 1989 and Karátson et al. 1999). The great differences in

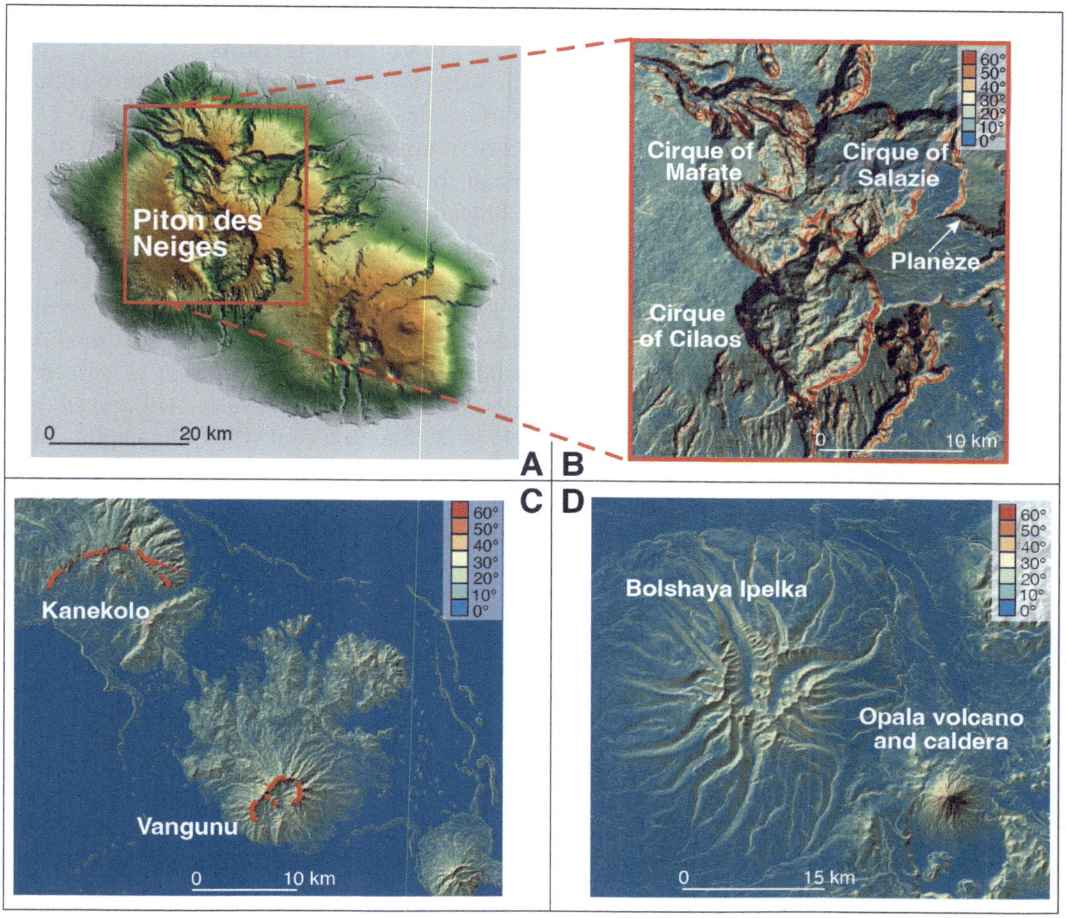

Fig. 16 Erosion-induced depressions. La Réunion: **A** Coloured, shaded relief SRTM DEM image (source: Conseil Régional de la Réunion, MapsLand, https://www.mapsland.com/africa/reunion/large-topographical-map-of-reunion); **B** SRTM DEM-based slope map; **C** Kanekolo and Vangunu, New Georgia (Solomon Islands), related to amphitheatre-valley erosion; **D** Glacial valleys resembling amphitheatre-valley-type erosion on Bolshaya Ipelka, Kamchatka (Russia) (**C, D**: SRTM DEM-based slope map)

elevation favoured amphitheatre valley incision mainly in the central part, but smaller amphitheatre heads can also be found on the lower flanks as well as on the younger edifice of Piton de la Fournaise (see Figs. 12 and 16A). Similar central amphitheatre valleys forming a number of outlets draining the interior of tropical island volcanoes with high annual rainfall include those on the 0.5 Ma-old Tahiti (Hildenbrand et al. 2008), the 1.2 Ma old West Maui, the 1.3 Ma-old East Molokai (Sherrod et al. 2006, 2007), and Vangunu, an older, Plio/Pleistocene edifice with two outlets in New Georgia (Fig. 16C: Feitei and Marau 1982).

On the other hand, huge valleys can also be observed in less humid environments, in particular, when the enhanced erosive capacity is due to glaciation. In Kamchatka, which was heavily glaciated several times, valley widening was mainly caused by the glaciers, and at valley heads there are a large number of U-shaped glacial cirques. In the Late Quaternary, extensive ice-advances were short-lived, but glacier cirques have become a characteristic feature of the region (Barr and Solomina 2014; Barr and Spagnolo 2015). As a fine example of the invasion of more than one glacial valley into a huge volcano, the interior (a possible caldera) of the large, Pleistocene(?) Bolshaja Ipelka shield volcano is shown in Fig. 16D, significantly degraded by three huge glacial valleys with U-shaped heads facing different directions. After the ice melted, the valleys continue to be shaped by fluvial erosion, eventually forming a common drainage basin.

Topographic modelling of amphitheatre valley evolution (Fig. 17A) suggests that, on a conical volcanic edifice, the presence of successive layers with different characteristics is key to valley widening (Karátson and Székely 2006; Karátson et al., in prep.). Thus, wide amphitheatre valley heads can form when paths of regressive erosion cutting into a less erodible upper layer hit a more erodible lower layer with a relatively high critical slope angle. Model results indicate that the intense erosion of amphitheatre valleys on a symmetric edifice not only enlarges a central depression, but eventually results in a regular, ring-like pattern of preserved outer flank fragments through the retreat and enlargement of the rim of the drainage basin (Fig. 17A). Moreover, modelling also shows that such a large landform can be produced without an initial caldera (Karátson and Székely 2006, Karátson et al., in prep.).

4.2.3 Depressions Formed by Erosion in Volcanic Terrains

In ancient volcanic terrains showing significant surface denudation, the site or even the existence of original craters or calderas might be difficult to infer, as long-term erosion completely obliterates the primary endogenic landforms. As shown above, in the case of fragments along a very wide rim showing erosional retreat, it is reasonable to infer a form induced by erosion. Hence, instead of an erosional(ally modified) caldera, the resulting depression is termed an *erosion-induced depression in volcanic terrain* (Fig. 13; Karátson et al. 1999). Examples—with or without an original caldera—may include Mount Kanekolo in the Solomon Islands (Fig. 16C) and the Tweed volcano (or Wollumbin in Aboriginal) in Australia (Fig. 17B).

The original ca. 4000 km^3 Oligo/Miocene (24–21 Ma) shield volcano of Tweed, described in Chap. 4, is one of the oldest preserved edifices on Earth, even though most of it has now been removed by erosion. Despite this, it still covers an area of 7000 km^2. The subtropical climate of the east coast of Australia may have ensured intense, long term fluvial erosion, and as a result, only remnants of the outer flanks have been preserved in a more than 30 km-wide

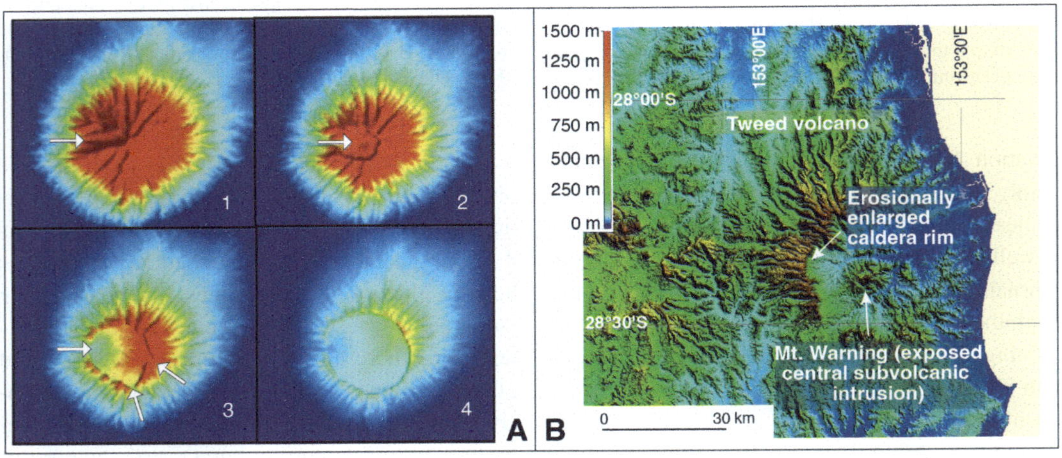

Fig. 17 Amphitheatre valley evolution. **A** Erosional modelling of amphitheatre valleys. Given the high energy of an amphitheatre valley head, when it hits a lower, softer layer, it will excavate an erosional landform at a high rate. Other amphitheatres (see arrows) can also invade the enlarging depression (Karátson and Székely 2006). Example of Mount Tweed, Australia (see text) on coloured and shaded SRTM DEM image (**B**)

circular shape (already termed an 'erosion caldera' by Solomon 1964), in addition to the exposed central subvolcanic intrusion (Mount Warning). The asymmetric preservation of Tweed's flanks may be due to the presence of the Great Australian Escarpment, as for Ebor volcano (see Sect. 3.1.3).

On Earth, there might be other clear examples where an erosional depression has formed either at the summit of the volcano or on the flanks with no primary crater or caldera. It should be emphasized that these are related to fluvial erosion, and lack scars of the large-scale flank failures described in Chap. 7. These possibilities were first explored by Karátson et al. (1999), and then followed up by Aranda-Gómez et al. (2009).

These erosional depressions, which are commonly drained by a single outlet valley, can be formed in several ways (Fig. 13). One is when a group or cluster of craters are amalgamated by a single watercourse via its tributaries capturing the drainage basins of other craters. This process was first described in the Carpathians on the 2 × 5 km-wide, irregular-shaped erosional depression of the 2 Ma Cucu (Kakukkhegy) volcano, Harghita Mountains, Romania (Fig. 14D: Karátson et al. 1992). It was shown, supported by geological mapping, that the poly-lobed shape of the elongated depression is due to the presence of three erosionally modified, enlarged crater remnants grouped together into a joint drainage basin.

Another possibility is where successive landslides cut into the flanks of a (large) volcano, creating a depression, which can be further shaped and drained by fluvial erosion via a relatively narrow exit. The most studied example of this is the Barranco de Tirajana in Gran Canaria (Spain), which Lomoschitz et al. (2002) placed in the Pleistocene, but Schmincke and Sumita (2010) calculated as being earlier, in the Mio/Pliocene.

Finally, there are also erosion-induced depressions that form in monogenetic scoria cone fields. In these cases, one or more scoria cones are surrounded by lava flows, and subsequent erosion, after removing the looser scoriaceous debris aprons, creates a relatively small (<3 km) catchment defined by lava flows, with a central scoria cone remnant or even an exhumed neck rock (termed *neck-in-a-basin*: Aranda-Gómez et al. 2009).

4.3 Erosion Rates in Volcanic Terrains

The efficiency of long-term degradation of volcanoes is testified above all by the prevailing *erosion rates,* or when referring to general surface lowering, *denudation rates*. Denudation is expressed areally, and its rates are usually given in terms of the ratio of either weight mass loss/area per year (e.g. t/km^2/yr), or surface lowering of an area per year (e.g. cm/kyr, m/Myr). Details are beyond the scope of the book, but the results for volcanoes and volcanic landscapes are worth summarising. The main finding is that not only is the short-term erosion of volcanically disturbed landscapes (Chap. 7) quite diverse, but it can also be much higher than that of other landscapes (by up to three or four orders of magnitude: e.g. Pierson and Major 2014). However, climate variability within relatively small areas (for example, an oceanic island) can lead to contrasting erosion rates in the long term.

Based on studies on fluvial erosion, the Earth's average long-term global denudation rate is estimated to be between 43 mm/kyr (or m/Myr: Summerfield and Hulton 1994) and 54 mm/kyr or m/Myr (Wittmann et al. 2020). Based on the latter value, this corresponds to 141 t/(km^2 × yr), which corresponds to a flux of 15.2 ± 2.83 Gt/yr (Wittmann et al. 2020). In the tropics, these values can be two-, five- or even tenfold, depending not only on the climate but also on infiltration (which may be up to 40% of the rainfall: Rad et al. 2007). While in nonvolcanic terrains the preferred method of assessment is geochemical analysis of river sediment load, in volcanic areas, common methods include: sediment volume analysis (e.g. for sediments accumulating from a given catchment into a basin or sea); geochemical analyis such as major and trace element determination or measurement of cosmogenic nuclides from river sediment load (Rad et al. 2007; Ferrier et al. 2013; Litty et al. 2020; Wittmann et al. 2020); dating surfaces of different ages; determining the volume of valleys (i.e. removed volume) cut in surfaces of known age; and tracing morphometric changes in a known geometry over time.

An example of the latter, i.e. geomorphological methods, is the morphometric comparison of initially same-sized geomorphic features (e.g. craters) of progressively older age, such as the Carpathian volcanism which shows mid-Miocene to Late Quaternary along-arc age progression (Pécskay et al. 2006). An early study (Karátson 1996), using topographic maps and focussed on changes in crater diameter and drainage pattern, revealed linear relationships with time and a long-term erosion rate of about 30 m/Myr, which averaged out climate fluctuations over the past 11 Ma (Fig. 18A). In particular, as emphasized by Karátson (1996), the long-term average rates obtained do not factor in the initial, much more intense erosion, which slows down significantly over thousands of years. From approximately the same area of the Carpathians, a recent morphometric analysis (Dibacto et al. 2020) compared reconstructed and current topographies using SRTM DEM, and yielded a similar long-term average erosion rate of 20 m/Myr. The study also revealed the expected fluctuations, ranging between 7 and 38 m/Myr, controlled by the changing climate from continental to occasional semi-arid to humid periods (with interglacial/glacial cycles in the Quaternary).

We also demonstrate the long-term erosion of monogenetic volcanoes through the well-constrained evolution of scoria cones (Fig. 18A). In the semiarid San Francisco Volcanic Field in Arizona, scoria cones up to 3–5 Ma are preserved, providing an excellent opportunity to trace the changes in characteristic morphometric parameters and ratios (Wood 1980). As described in Chap. 3, fresh scoria cones are commonly similar in shape, whereas subsequent morphometric characteristics are mostly controlled by erosional removal. In large areas in the southwest USA, there is little and sporadic rainfall, hence much of the scoriaceous material eroded from the upper flanks is not removed from the base of the cone, a fact well reflected by the low rate of cone height reduction and crater enlargement. Findings from this instructive area have been confirmed by Fornaciai et al. (2012). Recently, Vörös et al. (2022) added other examples of long-term scoria cone degradation, based on four volcanic areas of different ages and different climates (Chaîne des Puys, Kula Volcanic Field, Sierra Chichinautzín, and San Francisco Volcanic Field), and analysed the long-term changes—overall decrease—in average slope pattern by means of statistical tests.

Such results highlight the point that erosion processes are strongly dependent on the prevailing climate (Table 2), and are also constrained by hydrologic and lithologic characteristics. In tropical, mountainous areas with high rainfall, particularly where amphitheatre valleys have formed, erosion rates can rise to several hundred m/Myr. That means that tropical volcanoes undergo the fastest degradation based on the above erosion rates (Fig. 18C). In some areas of New Guinea or Indonesia, Ollier (1969) estimates that it takes only a few tens of thousands of years to reach the residual volcano stage, and even the central vent or intrusive bodies may become eroded. For example, in one of the earliest studies applying radiometric dating in tropical Papua New Guinea, Ruxton and McDougall (1967) calculated erosion rates of more than 700 m/Myr at the higher elevations of the 700 ka Hydrographers Range composite volcano. In the subtropical Sao Antão (Cape Verde Islands), the island-wide differences in rainfall result in contrasting erosion rates: a 20-fold difference in erosion rate has been determined for the wet and dry parts of the island (Litty et al. 2020). In Japan, on the island of Honshū with its subtropical/humid continental climate, even Pleistocene volcanoes have lost not only their planèzes but also their crater remnants (Suzuki 1969), and the former volcanic centres of the Miocene Green Tuff, a distinctive volcanic deposit in Japan, have completely disappeared (Huzioka 1963).

On the contrary, in dry (and occasional cold) areas, erosion rates decrease to as low as a few

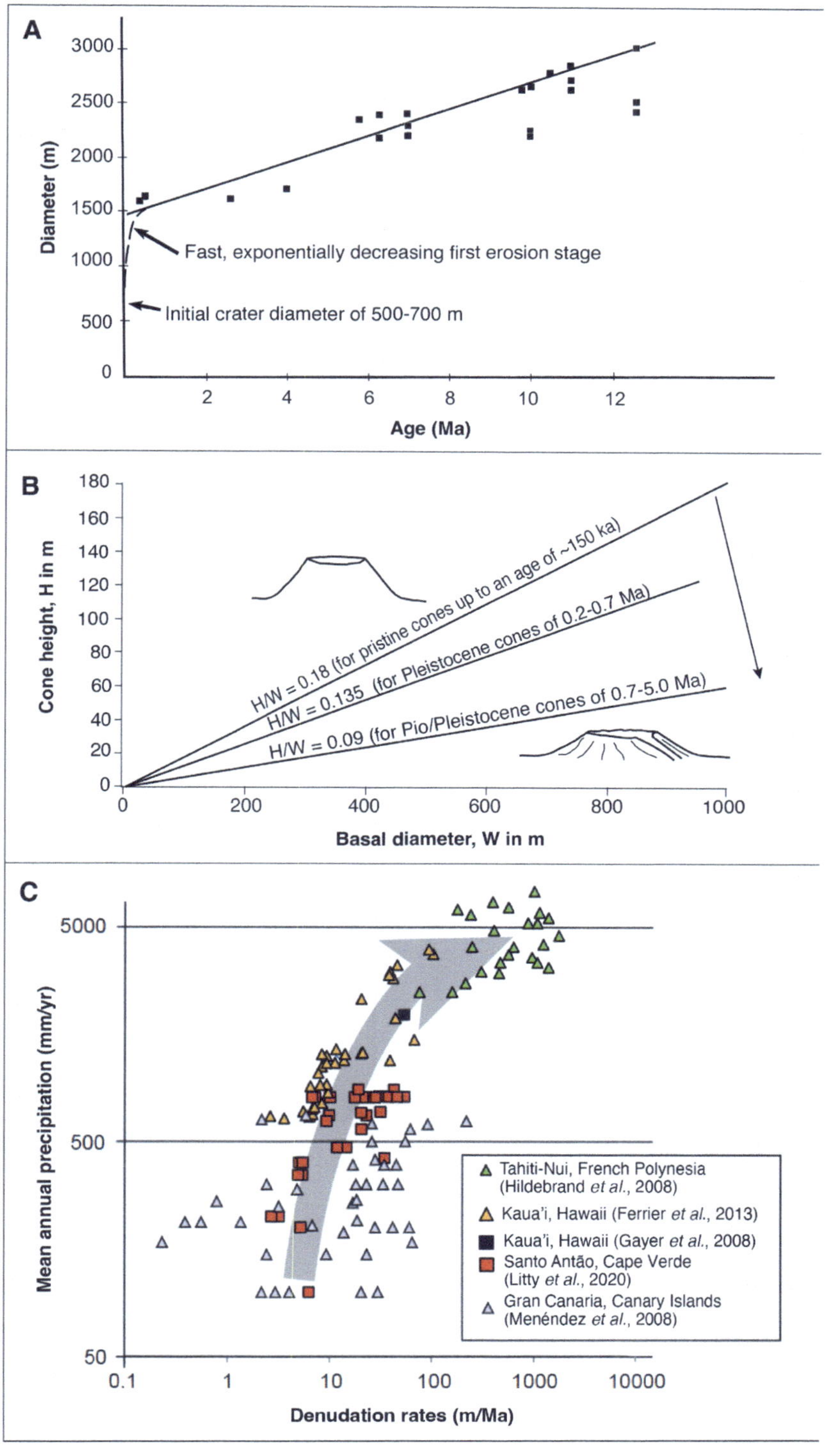

Fig. 18 Quantification of volcano erosion in the light of morphometry and geochemistry. **A** Late Miocene to Pleistocene crater remnants of Carpathian composite volcanoes showing their crater diameter versus age (Karátson 1996). Relationship reveals a long-term linear erosional trend, but the first stage of erosion acting on bare surfaces would have been much more intense (see text). **B** Scoria cone degradation showing a systematic decrease of H/W ratio with time, based mostly on examples from San Francisco Field, Arizona (USA) (Wood 1980, modified). **C** Denudation rates measured on different volcanic islands (Litty et al. 2020, and references therein). Squares: denudation rates based on eroded volume, triangles: denudation rates based on ^{3}He cosmogenic concentration

Table 2 Examples of long-term erosion rates (surface denudation) of selected volcanic areas representing different climate zones

Volcanic area	Erosion rate	Time interval	Prevailing climate	Method	Comment	Reference
Tahiti-Nui, French Polynesia	72 to 2700 m/My	1.3 Myr	Tropical	Morphometry and Gillot-Cassignol K–Ar dating	Depending on rainfall ranging island-wide between 1500–8500 mm/yr	Hildenbrand et al. (2008)
Hydrographers Range, Papua New Guinea	80 to 750 m/My	700 kyr	Tropical	Morphometry and conventional K–Ar dating	Rates depending on altitude (almost 1900 m height difference)	Ruxton and McDougall (1967)
Gran Canaria, Canary Islands	0.1 to 240 m/My	14.5 Myr	Sub-tropical	Drainage basin morphometry	Controlled by climate and also drainage basin size	Menéndez et al. (2008)
Raton-Clayton and Ocate volcanic fields, New Mexico, USA	10 to 114 m/My	9 Myr; highest rates between 6 to 3.5 Ma	Semiarid, mountaineous	Morphometry and Ar–Ar dating	Magmatism and associated uplift played major role relative to climate changes	Nereson et al. (2013)
Kauai, Hawai'i Islands	3 to 112 m/My	5 Myr	Tropical	Cosmogenic ^{3}He from river sediment load	Depending on rainfall ranging island-wide between 500 and 9500 mm/yr	Ferrier et al. (2013)
Santo Antão, Cape Verde Islands	3 to 57 m/My	7.5 Myr	Sub-tropical	Cosmogenic ^{3}He from river sediment load	Depending on rainfall ranging island-wide between 40 and 1100 mm/yr	Litty et al. (2020)
East Carpathian Volcanic Chain, Romania	7 to 38 m/My	10 Myr	Semiarid to humid continental	Morphometry and Gillot-Cassignol K–Ar dating	Rates controlled by long-term climate fluctuations	Karátson (1996), Dibacto et al. (2020)
Altiplano-Puna, CVZ of the Andes	7 to 25 m/My	from 14 to 3 Ma	Arid to hyperarid	Morphometry correlated with K–Ar dating	Lowest rates in hyperarid Puna plateau	Karátson et al. (2012)

m/Myr. There are a number of volcanic landscapes on Earth where the prevailing climate is permanently arid or even hyperarid. For example, along the Western Cordillera of the Central Andes, an extremely dry climate has persisted since Oligo/Miocene times (Dunai et al. 2005), and erosion rates are thought to be as low as <10 m/Myr over large areas (Kober et al. 2007; Karátson et al. 2012).

As a consequence, the duration of the erosional evolution of volcanic edifices, including the time required to reach each of the four 'stages' described in Sect. 4.1, can vary widely. Final degradation, coinciding with the skeleton stage, may take 10–20 Myr in areas with moderate (oceanic or continental) climates; 1–2 Myr in humid (tropical or subtropical) climates; and up to around 30 Myr in arid climates, after which the volcano as a recognisable landform disappears, leaving only the bedrock. To illustrate this with some more regional examples, in the humid temperate climate of the North Island of New Zealand, the planèze stage can be traced back to the Middle Pleistocene, the remnant volcano stage to the Plio/Pleistocene, and the skeleton stage to the Late Miocene (Kear 1957). On the Mio/Pliocene Cantal volcano in central France (Chap. 5), which has a similar climate, well-developed planèzes formed on late-stage basaltic lava flows were incised or inverted during the Pleistocene glaciations and have not yet been obliterated by post-glacial and Holocene fluvial erosion (Nehlig et al. 2001). Thus, what was once the largest volcano in Europe (c. 3500 m), is still in the planèze stage despite a decrease in altitude of nearly 2000 m.

The longest erosional lifetimes of still recognisable volcanic edifices on Earth are found in arid climates, including the Oligocene Summer Coon and Ship Rock in the United States, described in Sect. 4.1.4. The Altiplano–Puna Plateau in the Central Andes, also described above, may contain a number of undated Miocene remnant landforms as well, as suggested by the presence of dozens of preserved mid-Miocene volcanic edifices generally still in the planèze stage (e.g. Cerro Anallajsi, Huicso, Jatunpunco, Asuasuni: Karátson et al. 2012). In these cases (Fig. 9B, C), due to earlier glaciation(s), up to half of their original volume has been removed. In addition, some neighbouring features seem to correspond to the skeletons of former edifices, which are as yet undated.

5 Degradation of Volcanic Plateaus/Tablelands

Large areas of different volcanic plateaus, especially lava plains and ignimbrite sheets, not only have a distinctive drainage pattern, but also show specific degradation features. The most important features controlling erosion are: (1) variations in permeability between successive layers (controlled by cooling history, post-emplacement weathering etc.), which, in addition to the linear erosion pattern, determines the rate of erosion and the location of springs; and (2) the resistance of the rock, which is crucial for the resulting erosional landform.

5.1 Erosion of Lava Plains

In addition to the physical-lithological properties of various lavas, the existence, number and age of successive lava flow units are highly important in determining the location of aquifers, seepage and springs. In flood basalt landscapes, differential erosion of successive layers leads to a phenomenon somewhat similar to that of amphitheatre valley evolution: regressive valley heads develop by undercutting and widening, but on a horizontal, plateau-like surface, rather than a steep volcanic flank slope. Their most spectacular feature with an arcuate form is referred to as an *alcove*, described first in the Columbia and Snake River flood basalt plateaus (Stearns 1936). In the basaltic lava plains of Iceland, valley development may be accelerated in humid climates by frequent landslides, stimulated by meltwater from distant glaciers.

The material composing the lava plains and wider lava flows at the foot of volcanoes, which

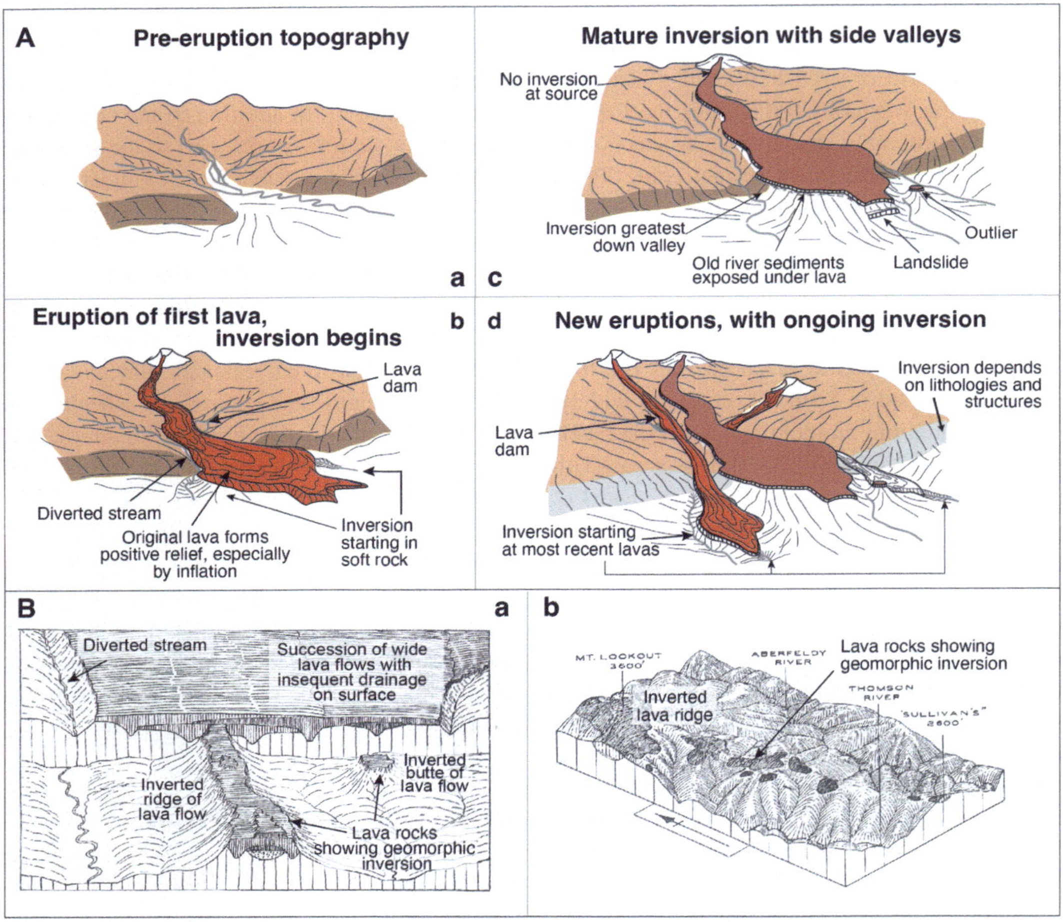

Fig. 19 **A** Sketch explaining the progressive development of inverted relief in lava flows, from original topography through early to mature inversion and finally with a subsequent lava flow (van Wyk de Vries et al. 2023). **B** Historical representation of the formation of inverted lava ridges: **a** general sketch of early (rear) and late (front) stages of destruction of a lava plateau after W. M. Davis; **b** geomorphic inversion of a former valley-filling lava flow in Victoria (Australia) after E. S. Hills (both from Cotton 1952)

is generally harder than the surroundings, eventually becomes exposed and creates inverted relief landforms (Cotton 1952; Ollier 1969; Williams and McBirney 1979). Figure 19 shows a general sketch of the process (van Wyk de Vries et al. 2023) and original drawings by early geomorphologists (W. M. Davis, E. S. Hills) of this well-known phenomenon. In the case of a simple lava flow from a central volcano (compound scoria cones, shield- and composite volcanoes), the valley slopes degrade with time, and over time the resistant volcanic material deposited in a former valley becomes the high ground, while the old high ground becomes the valley forming inverted relief. Such a process also applies to the degradation of lava plains.

Depending on its size, the exhumed area protected by a lava cap is called a butte or a remnant ridge. These are common in arid or semi-arid climates and have names in almost all languages (e.g. *mesa* in Spanish, although this term also describes flat-top sedimentary buttes). One of the most striking examples of inverted relief is the Miocene/Pliocene monogenetic basaltic volcanic field of Hopi Buttes, Arizona, where prevolcanic sedimentary formations are protected by buttes carved into steep, rocky formations (Photo 104). The Pliocene basalt hills of the Bakony-Balaton Hills Uplands display similar landforms. Most of these buttes were formed originally as lava lakes in tuff rings and maars. After the tuff and older, loose Upper Miocene sediments had eroded away, the hard, resistant rock, often composed of columnar basalts at the outer extremes, was carved out as lava caps, protecting the underlying sediments from degradation. For more examples worldwide, see van Wyk de Vries et al. (2023) and references therein.

5.2 Erosion of Ignimbrite Sheets

A permanent drainage pattern develops on ignimbrite sheets when they are covered in vegetation (see Fig. 5), but under arid or semiarid climates, in the absence of vegetation, high-intensity rainfall and hillslope gradient can lead to the formation of temporary gullies. They are found in all continents and have corresponding names in all languages *(cañon, quebrada, gulche, wadi,* etc.). If the rainfall is of sufficient quantity and intensity, the conditions are appropriate for the formation of *badlands,* a system of gullies that densely incise into the surface but are highly variable over time. Between each gully, surface fragments may be unaffected by erosion, resulting in chimney- or needle-like landforms. The larger of these are called *fairy chimneys* in Cappadocia (Photo 105). The smaller occurrences of this kind in the southwest USA are called *tent rocks, wigwam rocks* and *pinnacles.* The Pinnacles, a spectacular badland of narrow needles up to 70 m high, was formed in ignimbrite surrounding Crater Lake, and is explained by the resistance and exhumation of tuff hardened in part by fumarole clusters (Photo 80; Bacon 1983). A particular kind of badland formation is found in pyroclastic successions, where layers of different hardness are exposed. In this case, the hard layer or a larger, harder outsized block protects the lower, less resistant layers to form mushroom rocks (also see Sect. 2.1).

The resistant, welded ignimbrites can be inverted in a similar way to lava, forming individual plateaus or small, individual rock formations protecting the underlying material from degradation. The unwelded, ca. 2 Ma Cerro Galán ignimbrite formation in Argentina (Wright et al. 2011; Chap. 6), which is highly resistant due to vapour-phase recrystallization and devitrification, forms extensive lobes of as much as 10 m-thick valley- or basin-filling deposits (Photo 106). The plateau margins display columnar jointing, and are retreating spectacularly due to erosion. Gully erosion and suffosion of ignimbrite cliffs is also present at the Chuquibamba ignimbrite, southern Peru. Suffosion is due to the presence of early Tertiary sediments beneath the ignimbrite cooling unit, which hosts an aquifer. The weak welding and vapour-phase weathering of the ignimbrite enables water circulation and suffosion within the porous ignimbrite (Thouret et al. 2017).

After emplacement, the ignimbrite splits during cooling, and gas (or steam from groundwater) rising along fractures can cement and harden the ignimbrite material. If this cemented material is subsequently exhumed by erosion, it remains as a rampart-like landform, called *armoured joints* (for examples in Chile, see Francis 1993). If the gas or steam escapes through a chimney rather than along fissures, the tower-like form carved out by erosion is called an *armoured pipe.* The final stage of degradation is when only small, chimney-like towers of the hard ignimbrite remain: these are called *ignimbrite tors,* a term taken from granite geomorphology. Another feature of ignimbrite surface erosion is when the degraded surface is dotted in plan view with thousands of decimetre- to rarer metre-sized depressions, surrounded by rims, due to the action of fumaroles. Where hardening due to hydrothermal mineralization has extended over the whole area of the fumarole vent, small positive landforms called *fumarolic mounds* remain (Sheridan 1970; Jenkins et al. 2021).

6 Wind Erosion, Coastal Abrasion and the Effects of Glaciation on Volcanic Landscapes

6.1 Aeolian Landforms

In arid climates, where fluvial erosion is no longer effective, the prevailing wind erosion creates characteristic positive and negative landforms, instead of the badland surfaces typical of semi-arid areas. These are in many respects identical to the common aeolian landforms (i.e. *cavities, gates, towers,* etc.) that are usually found in harder sedimentary rocks, but in the case of compacted or welded ignimbrites, for example, they are often steeper and form stranger shapes due to the resistance of the rock. Huge *ignimbrite towers* and *pinnacles* have been created in the 4 Ma Atana ignimbrite of La Pacanacaldera (Chile). In other parts of the same area, strong unidirectional wind abrasion forms termed *longitudinal ridges* or *yardangs* form between furrows (Inbar and Risso 2001; see also Chap. 6). Preferrential removal of volcanic ash forms wind-blown *dunes* in many places (Photo 107: e.g. Nakadake, Aso, Japan; Huaynaputina, Peru; Rabaul, New Britain; Hawai'i, Maui, Kauai in Hawai'i Islands, USA; Lassen, USA; the 'seven-coloured sand dunes' of Mauritius; etc.). In the USA, for example, the ash and pumice sand of the Christmas Valley Sand Dunes in Oregon is derived from the ca. 7700 year-old Mazama eruption 100 km or so from Crater Lake caldera. The spectacular sand dunes of the Great Sand Dunes National Park in Colorado are derived from the volcanic tuffs of the San Juan Mountains.

The surface of many ignimbrites is pitted with cavities, holes, and punched 'membranes' that resemble a torn mesh or web covering. This general phenomenon, known as the 'Swiss cheese effect', can occur in two ways. A hard weathering crust forms on the surface of the ignimbrite, which subsequently cracks and peels off, and the softer material underneath is abraded by the wind; or the less resistant secondary minerals and weathered pumices are removed (e.g. Bandelier Tuff, New Mexico; Cerro Galán Ignimbrite, Argentina).

6.2 Coastal Erosional Landforms

On coasts composed of volcanic rocks, wave action exerts abrasion and removal or redeposition of clasts, depending on the physical properties (chiefly hardness) of the rock. In addition to the common abrasional landforms *(gateways, towers, caves, residual rocks, abrasion terraces),* abrasion of columnar basalts located on beaches creates spectacular formations, of which the most famous are the Tertiary lavas of Scotland and Northern Ireland (e.g. Giant's Causeway, Antrim Peninsula; Fingal Cave, Staffa Island). Other famous occurrences include Jeju Island, South Korea; Penghu Nature Reserve, Taiwan; Gành Đá Dĩa, Vietnam; Reynisdrangar and Snaeffels Peninsula, Iceland (Photo 107); Akun Island, South Korea; and Alaska, USA. There were intense coastal abrasion and marine processes following the eruptions of Surtsey in 1963, re-shaping the shoreline and shallow sub-marine levels, but these have slowed down with time (Romagnoli and Jakobsson 2015). Abrasion has reached the skeleton stage at a number of volcanic islands worldwide and exposed the interior, including the vent area, leaving it surrounded by steep cliffs: e.g. Strombolicchio, Lipari (Italy); Ball's Pyramid, Lord Howe Strait, New South Wales (Australia).

On steep volcanic coastlines, hanging valleys with waterfalls can form if degradation by abrasion, in particular the retreat of the coastline, is faster than the incision of river valleys (e.g. Kohala volcano, Hawai'i: MacDonald et al. 1983). Conversely, when the river valleys reach the coast and spread their sediment load, an alluvial fan system forms just off-shore (e.g. Tristan da Cunha).

6.3 Landforms Created by Glacial Erosion

In the case of volcanic surfaces in high latitudes (Alaska, Kamchatka, Southern Andes) and occasionally reaching high altitudes (e.g. Northern Andes, Kilimanjaro), glacial processes are highly efficient means of erosion. On many continents, traces of these processes can still be observed after the ice has retreated following the termination of Late Quaternary glaciation. The wide valley heads of glaciers, like amphitheatre valleys, cut backward much more effectively than river valleys, and the movement of a glacier carrying its transport of sediment both laterally and in the valley bottom grinds and polishes the volcanic bedrock. In just a few tens of thousands of years, distinctive, jagged glacial landforms can be formed: e.g. in Aleutian Islands (USA); northern Cascade Range (Canada–USA); Beerenberg (Jan Mayen Island, Norway); Kilimanjaro (Tanzania), Ross Volcano (Auckland Islands, New Zealand).

The ice sheet and glacial ice that forms in craters and calderas overwrites the former river network by widening the outlet valley. For example, the 7–8 km-wide Veniaminof caldera in the Aleutian Islands, in addition to its wide valley outlet, is probably also characterised by subglacial secondary drainage, and its margins are impacted in several places by glaciers that have heavily retreated (Yount 1990). At the 4392 m-high Mt. Rainier (Washington, USA), retreating glaciers have already degraded the volcano's planèzes into narrow, rocky ridges between 3000 and 4000 m (Lescinsky and Sisson 1998): these are called *cleavers,* suggesting they were splitting up the glaciers. The South Andean Volcanic Zone is still being destroyed by glacial erosion, while in the northern Andes only a few volcanoes have a patchy ice cap. Those of the now ice-free volcanoes of the Central Andes that are higher than 4200–4700 m may previously have been glaciated several times in the Pleistocene (Clayton and Clapperton 1997; Haselton et al. 2002), but these were short-lived processes, and glaciers terminated around 4000 m a.s.l.. Thus, no distinctive alpine glacial landforms have survived: the only traces of former glacial erosion are some specific, large valley heads, once occupied by glaciers and later partially reshaped by river valley erosion. The radial alignment of these short glacial valleys on a deeply eroded volcano is reminiscent of the plant *edelweiss* (Karátson et al. 2012).

7 Concluding Remarks

In this chapter, we have focussed on the conditions, main processes, and resulting erosional landforms involved in the long-term degradation of volcanoes. The main conclusions are as follows:

1. The long-term erosion pattern and eventual degradation of volcanoes and volcanic landscapes corresponds to the high diversity of weathering processes at work in the Earth's climate zones.
2. The most important exogenic agent is fluvial erosion, which is active in almost all areas of the Earth. It is the main process of general surface lowering in most climate zones.
3. In this regard, the specific morphology of most volcanoes as central edifices is of paramount importance, because it determines the pathways of fluvial erosion and the development of long-term drainage patterns.
4. Erosion processes covered in this chapter are ultimately—over millions or tens of millions of years—more effective than short-term degradation processes (see Chap. 7) in view of changes in geomorphology. The time scale for identifiable remnants of volcanic edifices to remain on the Earth's surface is about 30 million years, by which point this is often only possible through the inherited drainage pattern.
5. While volcanic structure, lithological and hydrogeological characteristics play a significant role in short-term processes, long-term erosion processes are overwhelmingly controlled by climate, as shown by the climate-dependent pattern of surface denudation rates worldwide.
6. Long-term erosion can produce specific erosional volcanic landforms that occasionally reflect the surface or subsurface pattern of primary, initial landforms. These may include lower flank remnants of former edifices, exposed subvolcanic bodies, and lava or volcaniclastic landforms that have resisted erosion. Many of the latter can become inverted, providing nice examples of inverted relief.

Photo 93 Features formed by physical weathering. **A** Spheroidal weathering in thick andesite lava flows in the Late Miocene Calimani Mountains (Romania). Watch for scale. **B** Cubic jointing with perpendicular cracks developed in subvolcanic andesite in the Mid-Miocene Börzsöny Mountains (Hungary). **C** Roundish columns termed 'basalt organ columns': Bakony–Balaton Upland Volcanic Field (Hungary). These form when the original polygonal columns become isolated from each other through erosion, and the resulting cylindrical columns often display platy-jointing which is here termed 'rolls of coins' (*photos* D. Karátson)

Photo 94 Mushroom rocks formed by physical weathering of block-and-ash flows, Mid-Miocene Visegrád Mountains (Hungary). **A** Individual rock towers, **B** succession of block-and-ash flow deposits, with **C** explanatory sketch (Karátson et al., 2006). Red rectangle on B and C indicates the largest rock tower on A. During erosion, the outsized block at the top of a given rock tower protects it from further erosion (*photos* D. Karátson)

Photo 95 Temporary gullies are typical features of rill erosion on persistently active volcanoes: the constantly renewed tephra cover prevents a permanent water network forming. The photo shows Sakurajima, one of Japan's most active composite volcanoes, with the inactive Kitadake cone to the right, partly hidden, and the smoking, active twin craters of Minamidake to the left (*photo* D. Karátson)

Photo 96 Examples of initial linear erosion on volcanic cones. **A** Vesuvius cone whose last eruption was in 1944 (Italy; credit: https://www.parconazionaledelvesuvio.it/); **B** Mount Bromo active tuff cone showing centripetal, temporary gullying inside the crater; **C** Mount Batok Holocene scoria cone with well-developed parasol ribbing (both in Java, Indonesia); **D** Merapi's recent lava dome with temporary gully development on outer flanks (*photos* D. Karátson)

Photo 97 Post-caldera tuff and scoria cones developed in the Pleistocene/Holocene Tengger caldera, showing a range of erosional dissection depending on age—ranging from a couple of thousand years to currently active (with the 'fumarolic' Bromo to the left). Batok cone with well-developed parasol ribbing is in the foreground, and Widodaren, the largest cone, is just behind it to the right. Semeru composite volcano, the highest in Java (3676 m), is in the background (*photo* D. Karátson)

Photo 98 Part of the cluster of Kirishima or Kirishima-yama Volcanic Group (Kyushu, Japan), the active Shinmoedake volcano had a phreatomagmatic eruption in 1959. Formation of a fissure, as well as ash deposition on the flanks, triggered intense gullying, clearly visible on the slope facing us in the foreground (*photo* courtesy of R. Imura). Since the photograph was taken (in 1995), the volcano has erupted seven times, the last time being in 2018

Photo 99 Vulcan pumice cone near Rabaul caldera in Papua New Guinea, displaying centripetal drainage inside, and radial drainage outside, the crater. It was created in 1878, then erupted again in 1937. Aerial photograph taken in 1944. In the wet, humid equatorial climate, vegetation appears on the slopes in only a few tens of years (*photo credit* the Illinois Committee on Aerial Photography, Illinois State Geological Survey, IL, USA, stereogram No. 7, USA, 8845245_107-001)

Photo 100 Crater lakes. **A** Kawah-Ijen, East Java, Indonesia (*photo* courtesy of M. Karátson); **B** Sf. Ana (Szent Anna) at Ciomadul (Csomád), Romania (*photo* courtesy of D. Asaftei); **C** Kerid, Iceland (*photo* courtesy of A. Karátson); **D** Kawah Putih, West Java, Indonesia (*photo* D. Karátson). Note the fumarolic activity of acidic lakes within active Javanese craters

Photo 101 Amphiteatre valleys on Réunion: the Cirque de Mafate, looking southeast, towards the peak of Piton des Neiges (*photo* courtesy of D. O'Hara)

Photo 102 Skeleton stage of volcano degradation: Shiprock or, in Navajo, Tsé Bit' A'í (Rock Wing). *Photos* courtesy of P. Chesley (left) and L. Crumpler (right)

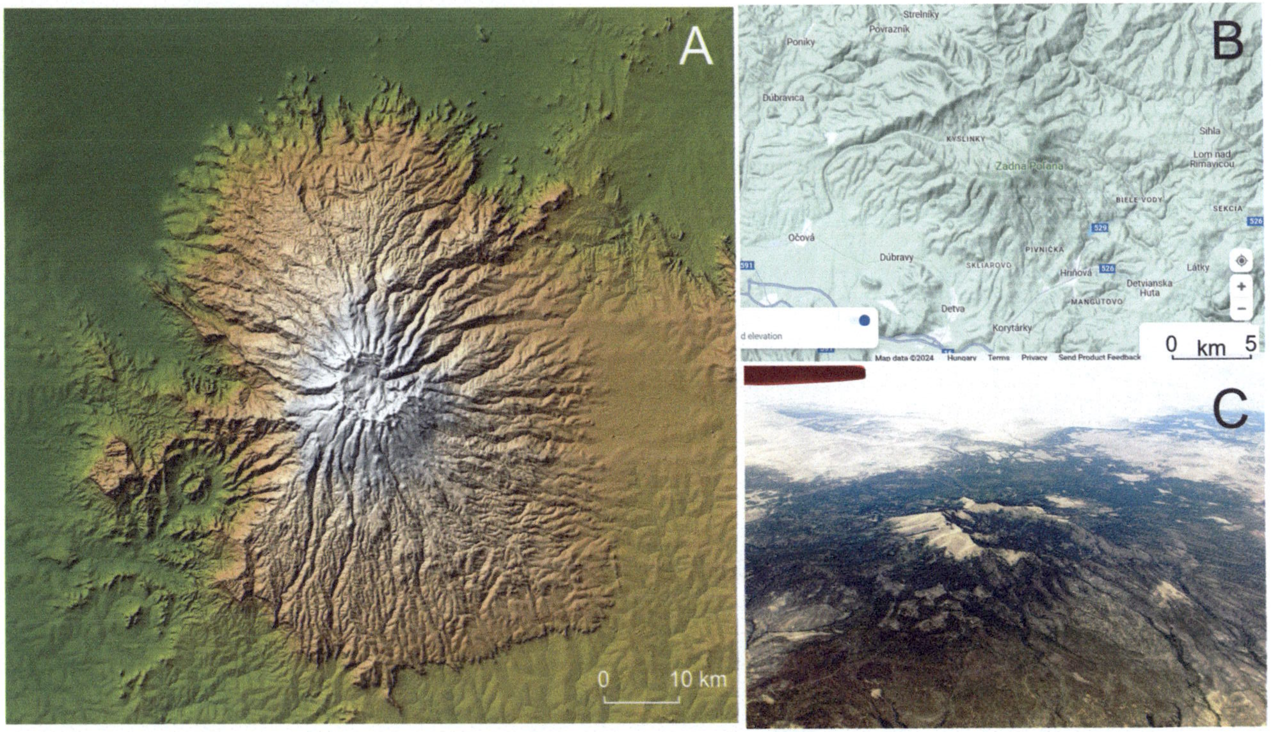

Photo 103 Examples of erosional calderas developed on long-extinct composite volcanoes, showing enlarged rim, centripetal drainage inside the central depression, and radial, partly dendritic drainage on the outer flanks. **A** The Pliocene Mount Elgon, Uganda/Kenya, on SRTM DEM (*photo credit* NASA, Jet Propulson Laboratory, NIMA); **B** The Mid-Miocene Pol'ana, Slovakia (*photo credit* Google Map SRTM DEM); **C** Late Pliocene Mount Taylor, New Mexico, USA, with caldera diameter c. 3 km (aerial *photo* courtesy of L. Crumpler)

Photo 104 Monogenetic basalt volcanoes often form inverted relief as the resistant basalt protects the underlying rocks. Perhaps the best-known example in the world is the Hopi Buttes hills, with protective basalt capping pinkish Mio-Pliocene lake sediments. Photo shows Castle Butte (*photo* D. Karátson)

Photo 105 Typical examples of ignimbrite erosion in Cappadocia, Turkey: loose, non-resistant pyroclastic-flow deposits (ignimbrites) typically form badland surfaces (**A** Kavak Ignimbrite, *photo* D. Karátson); where a more resistant layer is intercalated, it protects the underlying softer layer, and so-called 'fairy chimneys' are often formed (**B** Kavak Ignimbrite is protected by the Zelve plinian fall deposit, *photo* courtesy of S. Lövész)

Photo 106 Long-term post-emplacement erosion can remove a large part of individual ignimbrite sheets, even in arid climates. Here the photo shows the ~2.08 Ma Cerro Galán Ignimbrite, Argentina (*photo* D. Karátson)

Photo 107 Wind erosion and coastal erosion acting on volcanic landscapes. **A** Wind-blown ash dunes on the outer flanks of Nakadake central cone in Aso caldera, Japan (people in red circle give scale); **B** Gattklettur, a series of spectacular rock cliffs carved in basalt lava flows by wave action along the south-western coast of the Snaeffels Peninsula, Iceland (*photos* D. Karátson)

References

Andersen RS, Anderson SP (2010) Geomorphology: The Mechanics and Chemistry of Landscapes. Cambridge Univ Press. Xvi + 637. https://doi.org/10.1017/S0016756810000932

Aranda-Gómez JJ, Housh TB, Luhr JF, Noyola-Medrano C, Rojas-Beltrán MA (2009) Origin and formation of neck in a basin landform: examples from the Camargo volcanic field, Chihuahua (México). J Volcanol Geoth Res 197(1–4):123–132. https://doi.org/10.1016/j.jvolgeores.2009.08.004

Ashley PM, Duncan RA, Eebrey CA (1995) Ebor Volcano and Cresent Complex, northeastern New South Wales: age and geological development. Australian J Earth Sci 42(5):471–480. https://doi.org/10.1080/08120099508728217

Bacon CR (1983) Eruptive history of Mount Mazama and Crater Lake caldera, Cascade Range, USA. J Volcanol Geoth Res 18(1):57–115. https://doi.org/10.1016/0377-0273(83)90004-5

Barr ID, Solomina O (2014) Pleistocene and Holocene glacier fluctuations upon the Kamchatka Peninsula. Global Planetary Change 113:110–120. https://doi.org/10.1016/j.gloplacha.2013.08.005

Barr ID, Spagnolo M (2015) Palaeoglacial and palaeoclimatic conditions in the NW Pacific, as revealed by a morphometric analysis of cirques upon the Kamchatka Peninsula. Geomorphology 192:15–29. https://doi.org/10.1016/j.geomorph.2013.03.011

Baumann V, Bonadonna C, Cuomo S, Moscariello M (2020) Modelling of erosion processes associated with rainfall-triggered lahars following the 2011 Cordon Caulle eruption (Chile). J Volcanol Geoth Res 390:106727. https://doi.org/10.1016/j.jvolgeores.2019.106727

Bierman PR, Coppersmith R, Hanson K, Neveling J, Portenga EW, Rood DH (2014) A cosmogenic view of erosion, relief generation, and the age of faulting in southern Africa. GSA Today 24(9):4–11. https://doi.org/10.1130/GSATG206A.1

Bindeman IN, Leonov VL, Izbekov PE, Ponomareva VV, Watts KE, Shipley NK, Perepelov AB, Bazanova LI, Jicha BR, Singer BS, Schmitt AK, Portnyagin MV, Chen CH (2010) Large-volume silicic volcanism in Kamchatka: Ar-Ar and U-Pb ages, isotopic, and geochemical characteristics of major pre-Holocene caldera-forming eruptions. J Volcanol Geoth Res 189:57–80. https://doi.org/10.1016/j.jvolgeores.2009.10.009

Birkeland PW, Noller JS (2000) Rock and mineral weathering. In: Noller JS, Sowers JM, Lettis WR (eds) Quaternary geochronology: methods and applications. AGU, Reference shelf series 4, Washington DC, pp 293–312. https://doi.org/10.1029/RF004p0293

Bishop KM (2021) Landslide hypothesis for the origin of Haleakala volcano's crater and great valleys. Hawai'i. Geosphere 17(5):1405–1421. https://doi.org/10.1130/GES02215.1

Branstrator DK (2022) Origins of types of lake basins. In: Mehner T, Tockner K (eds) Encyclopedia of Inland Waters. https://doi.org/10.1016/B978-0-12-819166-8.00057-8

Campo-Bescós M, Flores-Cervantes J, Bras R, Casalí J, Giráldez J (2013) Evaluation of a gully headcut retreat model using multitemporal aerial photographs and digital elevation models. J Geophys Res Earth Surf 118:2159–2173. https://doi.org/10.1002/jgrf.20147

Cas RAF (2005) Volcanoes and the geological cycle. In: Marti J, Ernst GJ (eds) Volcanoes and the environment, 1st edn. Cambridge University Press, pp 121–151

Cas RAF, Wright JV (1988) Volcanic successions: modern and ancient. In: A geological approach to processes, products and successions. Allen and Unwin, London, 518 p ISBN-13:978–0–412–44640–5

Chester DK, Duncan AM (1982) The interaction of volcanic activity in quaternary times upon the evolution of the Alcantara and Simeto rivers, Mount Etna Sicily. Catena 9(3):319–342. https://doi.org/10.1016/0341-8162(82)90007-8

Chinen T (1986) Surface erosion associated with tephra deposition on Mt. Usu and other volcanoes. Env Sci Hokkaido Univ 9:137–149. http://hdl.handle.net/2115/37189

Clayton JD, Clapperton CM (1997) Broad synchrony of a Late-glacial glacier advance and the highstand of palaeolake Tauca in the Bolivian Altiplano. J Quat Sci 12(3):169–182. https://doi.org/10.1002/(SICI)1099-1417(199705/06)12:3<169::AID-JQS304>3.0.CO;2-S

Clynne MA, Ramsey DW, Wolfe EW (2005) Pre-1980 eruptive history of Mount St. Helens, Washington. US Geol Surv Fact Sheet 2005–3045. 4. http://pubs.usgs.gOv/fs/2005/3045

Collins BD, Dunne T (1986) Erosion of tephra from the 1980 eruption of Mount St Helens. Bull Geol Soc Am 97:896–905. https://doi.org/10.1130/0016-7606(1986)97<896:EOTFTE>2.0.CO;2

Collins BD, Dunne T (2019) Thirty years of tephra erosion following the 1980 eruption of Mount St Helens. Earth Surf Proc Landf 44:2780–2793. https://doi.org/10.1002/esp.4707

Colon MJ (2012) Weathering: types, processes and effects. In: Earth Sciences in the 21st Century, Nova Science Publishers, pp 215. ISBN 9781613242803

Cotton CA (1944) Volcanoes as landscape forms. Whitcombe and Tombs Ltd, Christchurch, New Zeland, 416 p. ISSN 0028–0836

Cotton CA (1952) Volcanoes as Landscape Forms. John Wiley & Sons Inc, NY, 416 p ISSN 0028–0836

Crandell DR, Fahnestock RK (1965) Rockfalls and avalanches from Little Tahoma Peak on Mount Rainier, Washington. US Geol Surv Bull 1221 – A, US Government Printing Office, Washington DC, pp 33

Cunningham MT, Stark CP, Kaplan MR, Schaefer JM (2019) Glacial limitation of tropical mountain

height. Earth Surf Dynam 7:147–169. https://doi.org/10.5194/esurf-7-147-2019
Dahlgren RA, Macias F, Arbestain MC, Chesworth W, Robarge WP, Bache BW, Emerson WW, Hartmann R, Kostic N et al (2008) Andosols. In: Chesworth W (ed) Encyclopedia of soil science. Encyclopedia of Earth Sciences Series. Springer, Dordrecht, pp 39–46. https://doi.org/10.1007/978-1-4020-3995-9_31
Dahm CN, Larson DW, Petersen RR, Wissmar RC (2005) Response and recovery of lakes. In: Dale VH, Swanson FJ, Crisafulli CM (eds) Ecological responses to the 1980 eruption of Mount St. Helens. Springer. New York, NY, pp 255–274. https://doi.org/10.1007/0-387-28150-9_18
Davis WM (1899) The geographical cycle. Geogr J 14:481–504. https://doi.org/10.2307/1774538
de Martonne E (1909) Traité de Géographie Physique, 1st edn. Armand Colin, Paris, p 910
Di Capua A, De Rosa R, Kereszturi G, Le Pera E, Rosi M, Watt SFL (2023) Volcanic processes in the sedimentary record: when volcanoes meet the environment. Geol Soc Lond Spec Publ 520:1–9. https://doi.org/10.1144/SP520-2022-303
Dibacto S, Lahitte P, Karátson D, Hencz M, Szakács A, Biró T, Kovács I, Veres D (2020) Growth and erosion rates of the East Carpathians volcanoes constrained by numerical models: tectonic and climatic implications. Geomorphology 368:107352. https://doi.org/10.1016/j.geomorph.2020.107352
Dohrenwend JC, Wells SG, Turrin BD (1986) Degradation of Quaternary cinder cones in the Cima volcanic field, Mojave Desert, California. Bull Geol Soc Am 97:421–427. https://doi.org/10.1130/0016-7606(1986)97<421:DOQCCI>2.0.CO;2
Dunai T, González López G, Juez-Larré J (2005) Oligocene-Miocene age of aridity in the Atacama Desert revealed by exposure dating of erosion-sensitive landforms. Geology 33:321–324. https://doi.org/10.1130/G21184.1
Eakins BW, Robinson JE (2006) Submarine geology of Hana Ridge and Haleakala Volcano's northeast flank, Maui. J Volcanol Geoth Res 151: 229–250
Earle S, Panchuk K (2019) Physical geology. 2nd Edition, Kamlops: Bccampus Open Publishing, Victoria, pp 454. https://opentextbc.ca/physicalgeology2ed/
Ehlen J (2005) Above the weathering front: contrasting approaches to the study and classification of weathered mantle. Geomorphology 67(1–2):7–21. https://doi.org/10.1016/j.geomorph.2004.09.026
Engbers YA, Biggin AJ, Bono RK (2020) Elevated paleomagnetic dispersion at Saint Helena suggests long-lived anomalous behavior in the South Atlantic. Proc Nat Acad Sci 117(31):18258–18263. https://doi.org/10.1073/pnas.2001217117
Famin V, Paquez C, Danišík M, Gardiner NJ, Michon L, Kirkland CL, Berthod C, Friedrichs B, Schmitt AK, Monié P (2022) Multitechnique geochronology of intrusive and explosive activity on Piton des Neiges volcano, Réunion Island. Geochem Geophys Geosys 23:e2021GC010214. https://doi.org/10.1029/2021GC010214
Farquharson JIM, Heap MJ, Varley NR, Baud B, Reuschlé T (2015) Permeability and porosity relationships of edifice-forming andesites: a combined field and laboratory study. J Volcanol Geoth Res 297:52–68. https://doi.org/10.1016/j.jvolgeores.2015.03.016
FAO/UNESCO (1974) Soil Map of the World, scale 1:5,000,000. Volume 1, Legend. UNESCO, Paris, 62 p
Farquharson JIM, Heap MJ, Baud P (2016) Strain-induced permeability increase in volcanic rock. Geophys Res Lett 43:11603–11610. https://doi.org/10.1002/2016GL071540
Fehér O, Gy F, Madarász B, Kertész Á (2006) Morphological and diagnostic properties of seven volcanic soil profiles according to the Hungarian soil classification and the world reference base for soil resources (WRB, 1998) (in Hungarian with English summary). Agrokémia És Talajtan (Agrochemistry and Soil Science) 55(2):347–366
Feitei D, Marau PW (eds) (1982) Solomon Islands: New Georgia geological maps, Vella Lavella Island, sheet NG1, 1:100,000, Solomon Islands Survey and Mapping Division
Ferrier KL, Perron JT, Mukhopadhyay S, Rosener M, Stock JD, Huppert KL, Slosberg M (2013) Covariation of climate and long-term erosion rates across a steep rainfall gradient on the Hawai'ian island of Kaua'i. GSA Bull 125(7–8):1146–1163. https://doi.org/10.1130/B30726.1
Fisher RV, Smith GA (eds) (1991) Sedimentation in Volcanic Settings. SEPM Special Publication no. 45. v + 257. Tulsa, SEPM (Society for Sedimentary Geology). ISBN 0 918985 89 7
Ford D, Williams PD (2007) Karst hydrogeology and geomorphology. 2nd ed, Wiley, p 576. https://doi.org/10.1002/9781118684986
Fornaciai A, Favalli M, Karátson D, Tarquini S, Boschi E (2012) Morphometry of scoria cones, and their relation to geodynamic setting: a DEM-based analysis. J Volcanol Geoth Res 157–158:56–72. https://doi.org/10.1016/j.jvolgeores.2011.12.012
Francis PW (1993) Volcanoes: a planetary perspective. Oxford University Press, 443 p. ISBN: 9780199254699
Frings PJ, Buss HL (2019) The central role of weathering in the geosciences. Elements 15(4):229–234. https://doi.org/10.2138/gselements.15.4.229
Gaylord DR, Neall VE, Palmer AS (2014) The middle Pleistocene Maitahi formation, Taranaki, New Zealand: a new formal lithostratigraphic unit. NZJ Geol Geophys 57(4):369–377. https://doi.org/10.1080/00288306.2014.914041
Geirsdóttir Á, Miller GH, Andrews JT (2007) Glaciation, erosion, and landscape evolution of Iceland. J Geodyn 43(1):170–186. https://doi.org/10.1016/j.jog.2006.09.017

Gerrard AJ (1988) Rocks and Landforms. Springer, Dordrecht, p 320. https://doi.org/10.1007/978-94-011-5983-8_5

Goudie AS (2004) Honeycomb weathering. In: Goudie AS (ed) Encyclopedia of geo-morphology, international association of geomorphologists, Routledge, London-New York, pp 530 ISBN 9780415272988

Gran KB, Montgomery DR (2005). Spatial and temporal patterns in fluvial recovery following volcanic eruptions: Channel response to basin-wide sediment loading at Mount Pinatubo, Philippines. Bul Geol Soc Amer 117(1–2):195–211. https://doi.org/10.1130/B25528.1

Harp AG, Valentine G (2018) Emplacement control for the basaltic-andesitic radial dikes of Summer Coon volcano and implications for flank vents at stratovolcanoes. Bull Volcanol 80:16. https://doi.org/10.1007/s00445-018-1194-4

Haselton K, Hilley G, Strecker MR (2002) Average Pleistocene climatic patterns in the southern Central Andes: controls on mountain glaciation and paleoclimate implications. J Geol 110(2):211–226. https://doi.org/10.1086/338414

Hazlett RW, Buesch D, Anderson JL, Elan R, Scandone R (1991) Geology, failure conditions, and implications of seismogenic avalanches of the 1944 eruption at Vesuvius, Italy. J Volcanol Geoth Res 47:249–264. https://doi.org/10.1016/0377-0273(91)90004-J

Heap MJ, Kennedy BM (2016) Exploring the scale-dependent permeability of fractured andesite. Earth Planet Sci Lett 444:139–150. https://doi.org/10.1016/j.epsl.2016.05.004

Heap MJ, Baud P, Meredith PG, Vinciguerra S, Reuschle T (2014) The permeability and elastic moduli of tuff from Campi Flegrei, Italy: implications for ground deformation modelling. Solid Earth 51(5):25–44. https://doi.org/10.5194/se-5-25-2014

Hildenbrand A, Gillot PY, Marlin C (2008) Geomorphological study of long-term erosion on a tropical volcanic ocean island: Tahiti-Nui (French Polynesia). Geomorphology 93:460–481. https://doi.org/10.1016/j.geomorph.2007.03.012

Hinds NEA (1925) Amphitheater valley heads. J Geol 33:816–838

Horton RE (1945) Erosional development of streams and their drainage basins; hydrophysical approach to quantitative morphology. Bull Geol Soc Am 56(3):275–370. https://doi.org/10.1130/0016-7606(1945)56[275:EDOSAT]2.0.CO;2

Huggett RJ (2011) Fundamentals of geomorphology. 3rd ed, Routledge, p 536. https://doi.org/10.4324/9780203860083

Huzioka K (1963) Geology of the Green Tuff Region in Japan. Mining Geology 13(62):358–375 (in Japanese). https://doi.org/10.11456/shigenchishitsu1951.13.62_358

Imura R, Kobayashi T (1991) Eruptions of Shinmoedake volcano, Kirishima volcano group, in the last 300 years. Bull Volcanol Soc Jpn 36:135e148 (in Japanese with English abstract)

Inbar M, Risso C (2001) Holocene yardangs in volcanic terrains in the southern Andes, Argentina. Earth Surf Proc Landf 26:657–666. https://doi.org/10.1002/esp.207

Inkpen R, Collier P (2007) Neo-Lamarckianism and the davisian cycle of erosion. Géomorphologie 13(2):113–124. https://doi.org/10.4000/geomorphologie.902

Iroumé A, Sánchez K, Mazzorana B, Martini L, Picco L (2023) Large wood dynamics in a mountain river disturbed by a volcanic eruption. Geomorphology 422:108551. https://doi.org/10.1016/j.geomorph.2022.108551

Jenkins WT, Klimczak C, Trent PM, Crowe DE (2021) Fumarolic pathways were structurally controlled by a strike-slip fault system beneath the bishop tuff, Bishop, California. Minerals 11:1167. https://doi.org/10.3390/min11111167

Join JL, Folio JL, Robineau B (2005) Aquifers and groundwater within active shield volcanoes. evolution of conceptual models in the Piton de la Fournaise volcano. J Volcanol Geoth Res 147:187–201. https://doi.org/10.1016/j.jvolgeores.2005.03.013

Karátson D (1996) Rates and factors of stratovolcano degradation in a continental climate: a complex morphometric analysis of nineteen Neogene/Pleistocene crater remnants in the Carpathians. J Volcanol Geoth Res 73:65–78

Karátson D (1999) Erosion of primary volcanic depressions in the inner carpathian volcanic chain. Z Geomorphol Suppl-Bd 114:49–62

Karátson D, Thouret JC, Moriya I, Lomoschitz A (1999) Erosion calderas: origins, processes, structural and climatic control. Bull Volcanol 61:174–193. 10.1007/s004450050270

Karátson D, Németh K, Székely B, Zs R-R, Pécskay Z (2006) Incision of a river curvature due to exhumed Miocene volcanic landforms: Danube Bend Humgary. Int J Earth Sci (Geol Rundsch) 95(5):929–944. https://doi.org/10.1007/s00531-006-0075-9

Karátson D, Telbisz T, Wörner G (2012) Erosion rates and erosion patterns of Neogene to quaternary stratovolcanoes in the Western Cordillera of the Central Andes: an SRTM DEM based analysis. Geomorphology 139–140:122–135

Karátson D, Yepes J, Favalli M, Rodríguez-Peces MJ, Fornaciai A (2016) Reconstructing eroded paleovolcanoes on Gran Canaria, Canary Islands, using advanced geomorphometry. Geomorphology 253:123–134. https://doi.org/10.1016/j.geomorph.2015.10.004

Karátson D, Székely B (2006) Amphitheatre valleys on volcanoes: characterization, evolution, Surface Processes Modelling. EGU Vienna, Geophysical Research Abstracts, Vol 8, 09648, SRef-ID: 1607-7962/gra/EGU06-A-09648, Session GMPV16 (Volcano Processes: Observation and Models)

Karátson D, Timár G (2005) Comparative volumetric calculations of two segments of the Carpathian Neogene/Quaternary volcanic chain using SRTM elevation data: implications to erosion and magma output rates. Zeitschrift für Geomorphologie Supplementband 140:19–35. https://www.schweizerbart.de/publications/detail/isbn/9783443211400/Zeitschr_f_Geomorphologie_Suppl_Bd_1

Karátson D, Pécskay Z, Szakács S, Seghedi I (1992) An extinct volcano in Harghita Mountains: Cucu (Kakukkhegy). Tudomány 1:70–79 (Hung issue of Sci Am; in Hungarian language)

Karátson D, Veres D, Gertisser R, Magyari E, Jánosi Cs, Hambach U (2022) Ciomadul (Csomád), the youngest volcano in the Carpathians. Springer, Cham, pp 282. https://doi.org/10.1007/978-3-030-89140-4

Karátson D, O'Hara D, Székely B (in prep.) Can 'erosion calderas' form purely by erosion? Distinguishing endogenic vs exogenic calderas on the basis of surface processes modeling and worldwide examples

Kear D (1957) Erosional stages of volcanic cones as indicators of age. New Zealand J Sci Tech 38:671–682

Kieffer G (1989) La formation des cirques du massif du Piton des Neiges (La Réunion) (The formation of the volcanic cirques in the Reunion island; in French with English abstract). Bull De L'assoc Géogr Français 66(5):361–370. https://doi.org/10.3406/bagf.1989.1502

Kober S, Ivy-Ochs S, Schlunegger F, Baur H, Kubik PW, Wieler R (2007) Denudation rates and a topography-driven rainfall threshold in northern Chile: multiple cosmogenic nuclide data and sediment yield budgets. Geomorphology 83(1–2):97–120. https://doi.org/10.1016/j.geomorph.2006.06.029

Korup O, Seidemann J, Mohr CH (2019) Increased landslide activity on forested hillslopes following two recent volcanic eruptions in Chile. Nat Geosci 12:284–289. https://doi.org/10.1038/s41561-019-0315-9

Kozlov DN, Lebedeva EV (2024) Morphology of crater and caldera lakes in the Far Eastern region of Russia and the features of their development. J Mountain Sci 21(4):246–1258. https://doi.org/10.1007/s11629-023-8253-9

Lacquement F, Nehlig P (2014) Geological map of the cirques of the Piton des Neiges Massif (Île de La Réunion). Geologie De La France 1:67–110

Lamb MP, Mackey BH, Farley KA (2014) Amphitheater-headed canyons formed by megaflooding at Malad Gorge, Idaho. Proc Nat Acad Sci USA 111:57–62. https://doi.org/10.1073/pnas.1312251111

Lamb MP, Howard AD, Dietrich WE, Perron JT (2007) Formation of amphitheater-headed valleys by waterfall erosion after large-scale slumping on Hawai'i. Bull Geol Soc Am 119:805–822

Lescinsky DT, Sisson TW (1998) Ridge-forming, ice-bounded lava flows at Mount Rainier. Washington Geolog 26(4):351–354. 10.1130/0091-7613(1998)026%3c0351:RFIBLF%3e2.3.CO;2

Litty C, Charreau J, Blard PH, Pik R, Nomade S (2020) Spatial variability of Quaternary denudation rates across a volcanic ocean island (Santo Antão, Cape Verde) from cosmogenic 3He. Geomorphology 375:107557. https://doi.org/10.1016/j.geomorph.2020.107557

Lockwood JP, Hazlett RW (2010) Volcanoes, Global perspectives. Wiley-Blackwell, UK, pp 539. ISBN 978–1–4051–6249–4

Lomoschitz A, Meco J, Corominas J (2002) The Barranco de Tirajana basin, Gran Canaria (Spain). a major erosive landform caused by large landslides. Geomorphol 42:117–130

Macdonald GA, Abbott AT, Peterson FL (1983) Volcanoes in the Sea. The Geology of Hawaii, 2nd ed., University of Hawaii Press, Honolulu, HI, pp 517. ISBN 9780824808327

Macdonald GA (1972) Volcanoes. A Discussion of volcanoes, volcanic products, and volcanic phenomena. Prentice-Hall, International, New Jersey, pp xii + 510

Mackey BH, Scheingross JS, Lamb MP, Farley KA (2014) Knickpoint formation, rapid propagation, and landscape response following coastal cliff retreat at the last interglacial sea-level highstand: Kaua'I Hawai'i. Bull Geol Soc Am 126(7–8):925–942. https://doi.org/10.1130/B30930.1

Major JJ (2022) Subaerial volcaniclastic deposits—influences of initiation mechanisms and transport behaviour on characteristics and distributions. Geologic Soc London Spec Public 520:29–100. https://doi.org/10.1144/SP520-2021-142

Manley K, Salles T, Müller RD (2020) Modeling the dynamic landscape evolution of a volcanic coastal environment under future climate trajectories. Front Earth Sci 8:550312. https://doi.org/10.3389/feart.2020.550312

Manville V (2010) An overview of break-out floods from intracaldera lakes. Global and Planet Change 70(1–4):14–23. https://doi.org/10.1016/j.gloplacha.2009.11.004

Manville V, Hodgson KA, Nairn IA (2007) A review of break-out floods from volcanogenic lakes in New Zealand. NZ J Geol Geophys 50:131–150. https://doi.org/10.1080/00288300709509826

Manville V, Németh K, Kano K (2009) Source to sink: a review of three decades of progress in the understanding of volcaniclastic processes, deposits, and hazards. Sed Geol 220(3–4):136–161. https://doi.org/10.1016/j.sedgeo.2009.04.022

Masurenkov YP (1980) Volcanic center: structure, dynamics and products (Karymsky Structure). Moscow, Nauka Pub, pp 299 p (in Russian)

Maund J, Rex DC, Le Roex A, Reid DL (1988) Volcanism on Gough Island: a revised stratigraphy. Geol Mag 125(2):175–181. https://doi.org/10.1017/S0016756800009572

Menéndez I, Silva PG, Martín-Betancor M, Pérez-Torrado FJ, Guillou H, Scaillet S (2008) Fluvial dissection, isostatic uplift, and geomorphological evolution of volcanic islands (Gran Canaria, Canary Islands, Spain). Geomorphology 102(1): 189–203. https://doi.org/10.1016/j.geomorph.2007.06.022

Michon L, Saint-Ange F, Bachèlery B, Villeneuve N, Staudacher T (2007) Role of the structural inheritance of the oceanic lithosphere in the magmato-tectonic evolution of Piton de la Fournaise volcano (La Réunion Island). J Geophys Res 112:B04205. https://doi.org/10.1029/2006JB004598

Milner DM, Cole JW, Wood CP (2003) Mamaku Ignimbrite: a caldera-forming ignimbrite erupted from a compositionally zoned magma chamber in Taupo Volcanic Zone New Zealand. J Volcanol Geoth Res 122(3–4):243–264. https://doi.org/10.1016/S0377-0273(02)00504-8

Moriya I (1979) Geomorphological development and classification of the quaternary volcanoes of Japan. Geogr Rev Jap 52:479–501 (in Japanese)

Mosbrucker A (2013) Lakes and drainages within the watershed of Mount St. Helens. Volcano Hazards Program, US Geol Surv Public Domain. https://www.usgs.gov/media/images/lakes-and-drainages-within-watershed-mount-st-helens

Naum T, Butnaru E, Gierecu M (1962) Vulcanokarstul din masivul Calimanului (Carpatii Orientali). Anulele Univ Bucuresti, Ser Ştiinţele Naturii, Geologie-Geografie 32:143–179

Neall VE, Stewart RB, Smith IEM (1986) History and petrology of the Taranaki volcanoes. In: Smith IEM (ed) Late Cenozoic volcanism in New Zealand. Royal Soc New Zealand Bull 23: 251–263.

Neall VE (2006) Volcanic soils. Encyclopedia of Life Support Systems (UNESCO-EOLSS). In: Verheye WH (ed) Land use, land cover and soil sciences VII: 1–24. http://www.eolss.net/ebooks/sample%20Chapters/C19/E1-05-07-13.pdf

Nehlig P, Leyrit H, Dardon A, Freour G, De Goër de Herve A, Huguet D, Thiéblemont D (2001) Repeated growth and catastrophic destruction of the Cantal stratovolcano (France). Bull Soc Geol France 172:295–308. https://doi.org/10.2113/172.3.295

Németh K, Cronin SJ (2008) Volcanic craters, pit craters and high-level magma-feeding systems of a mafic island-arc volcano: Ambrym, Vanuatu, South Pacific. Geol Soc London 302:87–102. https://doi.org/10.1144/SP302.6

Nereson A, Stroud J, Karlstrom K, Heizler M, McIntosh W (2013) Dynamic topography of the western Great Plains: Geomorphic and ^{40}Ar/ ^{3}9Ar evidence for mantle-driven uplift associated with the Jemez lineament of NE New Mexico and SE Colorado. Geosphere 9:521–545. https://doi.org/10.1130/GES00837.1

Newnham RM, Lowe DJ, Green JD, Turner GM, Harper MA, Froggatt PC (2004) A discontinuous ca. 80 ka record of Late Quaternary environmental change from Lake Omapere, Northland, New Zealand. Palaeogeogr Palaeoecl 207(1–2):165–198. https://doi.org/10.1016/j.palaeo.2004.02.007

Norris RD (1994) Seismicity of rockfalls and avalanches at three Cascade Range volcanoes: implications for seismic detection of hazardous mass movements. Bull Seismol Soc Am 84:1925–1939

O'Hara D, Goren L, van Wees RMJ, Campforts B, Grosse P, Lahitte P, Kereszturi G, Kervyn M (2024) Time-varying drainage basin development and erosion on volcanic edifices. Egusphere 12(2):709–726. https://doi.org/10.5194/esurf-12-709-2024

Oehler JF, Lénat JF, Labazuy P (2008) Growth and collapse of the Reunion Island volcanoes. Bull Volcanol 70:717–742. https://doi.org/10.1007/s00445-007-0163-0

Ohmori H (1983) Erosion rates and their relation to vegetation from the viewpoint of worldwide distribution. Bull Dept Geogr Univ Tokyo 15:7–91

Ollier CD (1971) Causes of spheroidal weathering. Earth Sci Rev 7:127–141. https://doi.org/10.1016/0012-8252(71)90005-5

Ollier CD (1984) Geomorphology of the South Atlantic Volcanic Islands, Part II: Gough Island. Z Geomorphol 28:367–382

Ollier CD (1969) Volcanoes. Australian National University Press, Canberra, pp 177. ISBN: 0708105327

Ollier CD (1988) Volcanoes. Basil Blackweell, Oxford-New York

Orme AR (2007) The Rise and fall of the davisian cycle of erosion: prelude, fugue, coda, and sequel. Physical Geogr 28(6):474–506. https://doi.org/10.2747/0272-3646.28.6.474

Paris R, Carracedo JC (2001) Formation of an erosion caldera and recurring instability on a hotspot-generated island: the caldera de Taburiente, La Palma, Canary Islands (in French with abridged English version). Géomorphologie: Relief Processus Environ 7(2):93–105. https://doi.org/10.3406/morfo.2001.1093

Pécskay Z, Lexa J, Szakács A, Seghedi I, Balogh K, Konečný V, Zelenka T, Kovács M, Póka T, Fülöp A, Márton E, Panaiotu C, Cvetković V (2006) Geochronology of neogene magmatism in the carpathian arc and intra-carpathian area. Geol Carpathica 57(6):511–530. https://www.sav.sk/journals/uploads/05230906GeolCarp_Vol57_No6_511_530.pdf

Petley D (2011) Updated: The video of the week: debris avalanche/lahar on Mount Rainier. AGU, The Landslide Blog, https://blogs.agu.org/landslideblog/2011/06/28/video-of-the-week-debris-avalanche-on-mout-rainier/

Pierson TC, Major JJ (2014) Hydrogeomorphic effects of explosive volcanic eruptions on drainage basins. Ann Rev Earth Planet Sci 42:469–507. https://doi.org/10.1146/annurev-earth-060313-054913

Poland MP, Fink JH, Moats WP (2008) Radial dike emplacement in stratovolcanoes based on observations from the eroded Summer Coon volcano, southern Colorado. Bull Volcanol 70:861–875. https://doi.org/10.1007/s00445-007-0175-9

Price DG (1995) Weathering and weathering processes. Q J Eng Geol Hydrogeol 28:243–252. https://doi.org/10.1144/GSL.QJEGH.1995.028.P3.03

Quane SL, Russell JK, Friedlander EA (2009) Time scales of compaction in volcanic systems. Geology 37(5):471–474. https://doi.org/10.1130/G25625A.1

Quesada-Román A, Ballesteros-Cánovas JA, Stoffel M, Zamorano-Orozco JJ (2019) Glacial geomorphology of the Chirripó National Park Costa Rica. J Maps 15(2):538–545. https://doi.org/10.1080/17445647.2019.1625822

Rad SD, Allègre CJ, Louvat P (2007) Hidden erosion on volcanic islands. Earth Planet Sci Lett 262(1–2):109–124. https://doi.org/10.1016/j.epsl.2007.07.019

Rafaello B, Reis E (2016) Controlling factors of the size and location of large gully systems: a regression based exploration using reconstructed pre-erosion topography. CATENA 147:621–631. https://doi.org/10.1016/j.catena.2016.08.014

Reffay A (2001). Types of karst in volcanic terrains: a review (in French with abridged English version). Géomorphologie: Relief Processus Environ 7(2):121–126. https://doi.org/10.3406/morfo.2001.1095

Rivals P (1950) Histoire géologique de l'Ile de la Réunion. Thèse Univ, Paris (unpublished), p 142

Romagnoli C, Jakobsson PS (2015) Post-eruptive morphological evolution of island volcanoes: surtsey as a modern case study. Geomorphology 250:384–396. https://doi.org/10.1016/j.geomorph.2015.09.016

Ruxton BP, McDougall I (1967) Denudation rates in northeast Papua from potassium-argon dating of lavas. Amer J Sci 265(7):545–561. https://doi.org/10.2475/ajs.265.7.545

Saint-Ange F, Fevre Y, Michon L, Bachelery F, Robineau B (2004) Geomorphology of oceanic volcanic islands from analogue erosion modelling. IAVCEI General Assembly, Pucón, Chile

Salvany T, Lahitte P, Nativel P, Gillot PY (2012) Geomorphic evolution of the Piton des Neiges volcano (Reunion island, Indian Ocean): competition between volcanic construction and erosion since 1.4 Ma. Geomorphology 136:132–147. https://doi.org/10.1016/j.geomorph.2011.06.009

Schmincke HU, Sumita M (2010) Geological evolution of the Canary Islands. Görres Druckerei und Verlag GmbH, A young volcanic archipelago adjacent to the Old African Continent, p 200

Schumm SA (1956) Evolution of drainage systems and slopes in badlands at Perth, Amboy, New Jersey. Bull Geol Soc Am 67:597–646. https://doi.org/10.1130/0016-7606(1956)67[597:EODSAS]2.0.CO;2

Segerstrom K (1950) Erosion studies at paricutin, state of Michoacán, Mexico. US Geol Surv Bull 965:1–163. https://doi.org/10.3133/b965A

Semken S (2003) Black rocks protruding up: the Navajo Volcanic Field. In: Lucas SG, Semken SC, Berglof W, Ulmer-Scholle D (eds) Geology of the Zuni Plateau. New Mexico Geological Society Guidebook, 54th Field Conference, pp 133–138. https://doi.org/10.56577/FFC-54.133

Sheridan MF (1970) Fumarolic mounds and ridges of the Bishop Tuff. California. Bull Geol Soc Am 81(3): 851–868. https://doi.org/10.1130/0016-7606(1970)81[851:FMAROT]2.0.CO;2

Sherrod DR, Hagstrum JT, McGeehin JP, Champion DE, Trusdell FA (2006) Distribution, ^{14}C chronology, and paleomagnetism of latest Pleistocene and Holocene lava flows at Haleakalā volcano, Island of Maui, Hawai'i: a revision of lava flow hazard zones. J Geophys Res 111(B5):B05205. https://doi.org/10.1029/2005JB003876

Sherrod DR, Murai T, Tagami T (2007) New K-Ar ages for calculating end-of-shield extrusion rates at West Maui volcano, Hawaiian island chain. Bull Volcanol 69:627–642

Siebert L (1984) Large volcanic debris avalanches: characteristics of source areas, deposits, and associated eruptions. J Volcanol Geoth Res 22(3–4):163–197. https://doi.org/10.1016/0377-0273(84)90002-7

Skinner BJ, Porter SC, Park J (2004) The dynamic earth: an introduction to physical geology. Wiley, 5th ed, 648 p ISBN-10: 0471152285

Snoeks J, De Vos L, Thys van den Audenaerde DFE (1997) The ichthyogeography of Lake Kivu. South African J Sci 93(11):579–584

Soil taxonomy: A basic system of soil classification for making and interpreting soil surveys, 2nd edition (1999). Ed. Soil Survey Staff, Natural Resources Conservation Service, US Department of Agriculture Handbook 436

Solomon PJ (1964) The mount warning shield volcano: a general geological and geomorphological study of the dissected shield. University of Queensland Papers V 10:1–12

Stearns HT (1936) Origin of large springs and their alcoves along the Snake River in Southern Idaho. J Geol 44:429–450. https://doi.org/10.1086/624441

Stockhecke M, Sturm M, Brunner I, Schmincke HU, Sumita M, Kipfer R, Cukur D, Kwiecien O, Anselmetti FS (2014) Sedimentary evolution and environmental history of Lake Van (Turkey) over the past 600 000 years. Sedimentology 61:1830–1861. https://doi.org/10.1111/sed.12118

Stoops G, Sedov S, Shoba S (2018) Chapter 25—Regoliths and soils on volcanic ash. In: Stoops G, Marcelino V, Mees F (eds) Interpretation of micromorphological features of soils and regoliths (Second Edition). Elsevier, pp 721–751. https://doi.org/10.1016/B978-0-444-63522-8.00025-5

Summerfield MA, Hulton NJ (1994) Natural controls of fluvial denudation rates in major world drainage basins. J Geophys Res 99(B7):13871–13883. https://doi.org/10.1029/94JB00715

Suzuki T (1969) Rate of erosion in strato-volcanoes in Japan. Bull Volcanol Soc Japan 2(14):133–147

Szakács A (1999) Vulcanokarst: a Romanian contribution to speleology. In: Centro Speleologico Etneo: Barone N, Bonaccorso R, Licitra G (eds) Proceedings of the 9th international symposium on vulcanospeleology, Catania, Italy, Sept 11–19, KIP Talks and Conferences, p 228. https://digitalcommons.usf.edu/kip_talks/7

Székely B, Karátson D (2004) DEM-based morphometry as a tool for reconstructingprimary volcanic landforms: examples from the Börzsöny Mountains, Hungary. Geomorphology 63:25–37. https://doi.org/10.1016/j.geomorph.2004.03.008

Takahashi T, Shoji S (2002) Distribution and classification of volcanic ash soils. In: Matsumoto S (ed) Volcanic ashes and their soils. Global Env Res 6:83–97

Tapu AT, Ubide T, Vasconcelos PM (2023) Increasing complexity in magmatic architecture of volcanoes along a waning hotspot. Nat Geosci 16:371–379. https://doi.org/10.1038/s41561-023-01156-9

Thomas MF (1974) Tropical geomorphology: a study of weathering and landform development in warm climates. Macmillan, London, 332 p. ISBN 0470858974, 9780470858974

Thouret JC, Gunnell Y, Jicha BR, Paquette JL, Braucher R (2017) Canyon incision chronology based on ignimbrite stratigraphy and cut-and-fill sediment sequences in SW Peru documents intermittent uplift of the western Central Andes. Geomorphology 298:1–19. https://doi.org/10.1016/j.geomorph.2017.09.013

Thouret JC, Oehler JF, Gupta A, Solikhin A, Procter JN (2014) Erosion and aggradation on persistently active volcanoes—a case study from Semeru Volcano, Indonesia. Bull Volcanol 76:640. https://doi.org/10.1007/s00445-014-0857-z

Tilling RI, Wright TL, Millard HT (1987) Trace-element chemistry of Kilauea and Mauna Loa lava in space and time: a reconnaissance. In: Decker RW, Wright TL, Stauffer PH (eds) Volcanism in Hawai'i, USGS Prof Pap 1350. US Government Printing Office, Washington DC, pp 641–690

Trudgill ST, Goudie AS, Viles HA (2022) Weathering processes and forms. Geol Soc London, Memoirs 58:173–189. https://doi.org/10.1144/M58-2022-2

Van Gorp W, Schoorl JM, Temme AJAM, Reimann T, Wijbrans JR, Maddy D, Demir T, Veldkamp T (2016) Catchment response to lava damming: integrating field observation, geochronology and landscape evolution modelling. Earth Surf Proc Land 41(11):1629–1644. https://doi.org/10.1002/esp.3981

van Wyk de Vries B, Karatson D, Gouard C, Németh K, Vladislav R, Aydar E (2023). Inverted volcanic relief: Its importance in illustrating geological change and its geoheritage potential. Int J Geoheritage Parks 10(1):47–83. https://doi.org/10.1016/j.ijgeop.2022.02.002

Veldkamp A, Schoorl JM, Wijbrans JR, Claessens L (2012) Mount Kenya volcanic activity and the Late Cenozoic landscape reorganisation in the upper Tana fluvial system. Geomorphology 145–146:19–31. https://doi.org/10.1016/j.geomorph.2011.10.026

Vinciguerra S, Del Gaudio P, Mariucci MT, Marra F, Meredith PG, Montone P, Pierdominici S, Scarlato P (2009) Physical properties of tuffs from a scientific borehole at Alban hills volcanic district (central Italy). Tectonophys 471:161–169. https://doi.org/10.1016/j.tecto.2008.08.010

Vörös F, van Wyk de Vries B, Guilbaud M-N, Görüm T, Karátson D, Székely B (2022) DTM-based comparative geomorphometric analysis of four scoria cone areas–suggestions for additional approaches. Rem Sens 14(23):6152. https://doi.org/10.3390/rs14236152

van Wees R, Paguican EMR, O'Hara D, Kereszturi G, Grosse P, Lahitte P, Kervyn M (2023) Insights into volcano degradation from analogue modelling: how volcano shape and size influence runoff erosion and morphology. IAVCEI Scientific Assembly, Rotorua, New Zealand. https://doi.org/10.13140/RG.2.2.29236.83841

Williams H, McBirney AR (1979) Volcanology. Freeman Cooper Co, San Fracisco, CA, p 397

Wittmann H, Oelze M, Gaillardet J, Garzanti E, von Blanckenburg F (2020) A global rate of denudation from cosmogenic nuclides in the Earth's largest rivers. Earth Sci Rev 204:103147. https://doi.org/10.1016/j.earscirev.2020.103147

Wood CA (1980) Morphometric analysis of cinder cone degradation. J Volcanol Geoth Res 8:137–160

Wright HMN, Lesti C, Cas RAF, Porecca M, Viramonte JG, Folkes C, Giordano G (2011) Columnar jointing in vapor phase altered, non-welded Cerro Galan Ignimbrite, Paycuqui, Argentina. In: Cas RAF, Cashman K (eds) The Cerro Galán Ignimbrite and Caldera: characteristics and origins of a very large volume ignimbrite and its magma system. Bull Volcanol 73: 1567–1582. https://doi.org/10.1007/s00445-011-0524-6

Yount ME (1990) Veniaminof/Alaska Peninsula. In: Wood CA, Kienle, J (eds) Volcanoes of North America. Cambridge University Press, New York, 56–58. ISBN: 9780521438117

Zaki AS, Davis JM, Edgett KS, Giegengack R, Roige M, Conway S, Schuster M, Gupta S, Salese F,Sangwan KS, Fairén AG, Hughes CM, Pain CF, Castelltort S (2022) Fluvial depositional systems of the African humid period: an analog for an early, wet Mars in the eastern Sahara. J Geophys Res Planets 127(5):e2021JE007087

Chapter 9 The Role of Geomorphology in Assessing Volcanic Hazards

1 Introduction

There are about 1350 potentially active volcanoes worldwide, besides the mid-ocean ridges. Around 50–70 volcanoes erupt on land every year, and about 500 of the 1350 active volcanoes have erupted in historical time according to the USGS Volcano Hazards Program. As many as 600 million people live within 100 km of an active or potentially active volcano, but most of the volcanoes are not well monitored, if at all (Brown et al. 2015; Freire et al. 2019). Over the past four centuries, volcanoes have caused 280,000 fatalities, including 163,000 related to the five deadliest eruptions (Witham 2005; Auker et al. 2013; Brown et al. 2017): Unzen 1792, Japan; Tambora 1815, Indonesia; Krakatau 1883, Indonesia; Mont Pelée 1902, Martinique; and Nevado del Ruiz 1985, Colombia (Voight 1990; Voight et al. 2013).

The majority of volcanic eruptions are not directly lethal, but severe damage is often caused in densely populated areas due to unanticipated eruptive scenarios, related hazards, and inadequate mitigation strategies. Despite technological and scientific advances in volcanology, recent eruptions (Agung in 2017–2018, Kilauea in May–July 2018, Fuego in June 2018, Grindavik in 2023–2024) remind us that volcanic events are among the most frequent and destructive of the natural hazards. The risk increases if eruptive activity lasts for several months (e.g. Tajogaite-Cumbre Vieja, La Palma Island, September-December 2021: see Case study 1, Chap. 10; Troll and Carracedo) to several years (e.g. Kilauea, East Rift zone from 1983 to 2018; Soufriere Hills, Montserrat from 1995 to 2013). The impacts of volcanic eruptions and subsequent loss of population and assets vary with distance from the active volcano: their flanks can be covered by thick tephra layers, lava flows and pyroclastic density currents (PDCs); the ring plains can be inundated by lahars and floods and affected by erosion and sedimentation processes; distal areas over hundreds to thousands of kilometres from the volcano can be affected by ashfall which can disrupt air traffic and cause respiratory problems.

Despite progress in understanding volcanic phenomena, monitoring eruptive activity and analysing the long-term consequences of eruptions, assessing volcanic and geomorphic hazards still represents a key challenge (e.g. Thouret and Charbonnier 2022):

1. An erupting volcano presents multiple hazards that vary in time and space (e.g. Lirer et al. 2010), each one potentially inducing geomorphic disturbances on surrounding hillslopes, on the drainage network and the ring plains of volcanoes.
2. The duration of eruptions, seven weeks on average (n = 3934, Siebert et al. 2015), is

D. Karátson and J.-C. Thouret, *Volcano Geomorphology*, Advances in Volcanology,
https://doi.org/10.1007/978-3-031-77693-9_10

often longer than commonly expected, and leads to a chain of events affecting the landscape for years.

3. An active volcano can present different eruptive styles and produce a variable range and volume of eruptive products. These products, reworked as volcaniclastic material, are ultimately transported through the drainage network to sedimentary basins.
4. Volcanic activity can generate hazards that are long-lasting, such as lahars that occur decades after the event (Chap. 7), and ones that are far-reaching, such as ashfall that can be deposited hundreds of km from the vent.
5. Knowledge of the eruptive history of active volcanoes is often too sparse to allow a statistical study of the magnitude and recurrence of events. Thus, hazard assessment based on the eruptive record alone remains uncertain, while sequential effects of eruptive events may be unexpected or unprecedented.

1.1 Terminology

In this chapter, we use the international terminology on Disaster Risk Reduction (UN 2016; UNDRR 2017; UNISDR 2017) to which the reader is referred.

'*Hazard* is a potentially damaging physical event, phenomenon or human activity that may cause the loss of life or injury, property damage, social and economic disruption or environmental degradation'. Hazard can be single, sequential or combined, natural or man-made, in its origin and effects. Its components encompass the area of impact, duration, intensity, the magnitude and frequency, and the probability of occurrence. A *volcanic hazard* is produced by eruptive phenomena on and around active volcanoes, among them PDCs and lava flows, as well as widespread tephra fall dispersed from high eruption columns (Jenkins et al. 2012).

A *geomorphic hazard* results from any landform change that 'adversely affects the geomorphic stability of a site and that intersects the human use system with adverse socio-economic impacts' (Slaymaker 1996). Geomorphic hazards are characterized by magnitude, frequency and areal extent. On volcanoes, they are both (1) endogenous, caused by eruptive, seismic or deformation processes, inducing for example flank failure, and (2) exogenous, caused by meteorological and climatic processes leading to weathering and erosion. Geomorphic hazards are therefore manifold and more abundant on volcanoes than on non-volcanic mountains (Thouret 1999, 2004, 2010; Alcántara-Ayala and Goudie 2010): they have been the focus of special issues of Geomorphology journals (Thouret 2001; Thouret and Chester 2005; Thouret et al. 2007a, b; Thouret and Nemeth 2012) and a wealth of articles (e.g. Carrión-Mero et al. 2020).

Geomorphic hazards are more ubiquitous and protracted than volcanic ones. Not only can they take place over a period of decades in affected watersheds, but they can also occur on dormant and inactive volcanoes. Disasters that occurred in 1998 at Casita, Nicaragua and in 2010 at Mt. Meager, British Columbia (Sect. 7) highlight increased risk at the foot of dormant volcanoes. Hazards there result from the combination of unconsolidated, hydrothermally altered volcanic rocks, heavy rainfall, and seismic activity or landslides.

In this chapter on geomorphic hazards, we do not deal with risk, which can be measured as the product of probability × losses, where the probability is a function of hazard and the losses depend on both exposure and vulnerability (Aspinall and Blong 2015). However, we will refer to damage or loss that is defined by the value of the exposed elements, also called valuables or assets. An *impact* of natural hazards is direct or indirect, tangible or intangible, and is usually regarded as a loss to the environment or society (Swiss Re 1998). We use impact as a collective term that includes the direct and indirect actions and consequences that geomorphic hazards effect on vulnerable valuables: buildings, infrastructure or lifelines. We further distinguish the physical effects (e.g. land burial), the processes (e.g. dynamic pressure), and ultimately the damage, i.e. the destruction or removal of valuables, e.g. houses, bridges or vehicles (Table

2). Direct loss deals with the physical effects of the hazard on individuals (death or injury) or infrastructure (reduction in functionality, collapse or removal) (Spence et al. 2005). This chapter will not consider indirect loss.

Prior to carrying out a risk analysis, the hazards, vulnerability and assets at stake must be evaluated, which requires several steps of investigation of which the main one is the modelling of the potentially affected areas. Outlining exposed areas may be useful so that stakeholders can take the necessary preventive measures. However, we emphasize that hazard assessment is uncertain and constantly changing, as its extent and intensity evolve with the eruptive style as the eruption unfolds (Sparks et al. 2011). Impacts also depend on the behaviour, training, and resilience of the exposed community living on active volcanoes.

1.2 Distinguishing Volcanic and Geomorphic Hazards

Hazards linked to active volcanoes have been extensively described in articles and chapters (e.g. Blong 1984, 1996, 2000; Latter 1989; Scott 1989; Aspinall and Blong 2015) and in textbooks: Schmincke (2004), Tilling (2005) *in:* Marti and Ernst (eds, 2005), Lockwood and Hazlett (2010), Loughlin et al. (eds, 2015), Papale et al. (eds, 2015); Sigurdsson et al. (eds, 2015), and Lénat (ed, 2022). We will therefore not delve into direct, eruption-induced volcanic hazards, which include gas emission, lava flows, tephra fall and ballistics, PDCs and blasts, as well as syn-eruptive flank failures and debris avalanches (for the latter see Chap. 7). Volcanic hazard phenomena commonly threaten assets when the eruption magnitude exceeds VEI 2. These hazards impact very different spatial ranges and time spans (Fig. 1A, modified after Chester, 2000 *in:* Thouret 2004; Paris et al. 2022). Indirect volcanic hazards comprise ash in the atmosphere, debris avalanches (Chap. 7), sediment-laden flows (Sect. 3.2), and tsunamis (Sect. 3.4), famine and erosion. Their effects are often more widespread around volcanoes, particularly since they can remobilize primary pyroclastic material during eruptive episodes and continue long after the eruption has stopped. Flank failures, sediment-laden flows and tsunamis can occur in the absence of an eruption, thus they are both indirect and non-eruptive hazards.

Here we analyse volcanic hazards from a geomorphic perspective, in other words the influence of geomorphological characteristics (changes in slope gradient, river channels, substrate strength, erosion power) on a few eruptive, post- and non-eruptive volcanic flows. We also investigate how eruptive dynamics can influence hazardous phenomena through the role of topography or substrate, and geomorphic processes that create transient landforms (e.g. volcaniclastic valley-fill and volcanic debris-dammed rivers). On a geological scale, volcanic landforms are often ephemeral, as erosion resumes as soon as the eruption ceases. Erosion, in the sense of the removal of rocks, debris deposition, and river downcutting, can take place during the eruption, induced by flow processes, and during quiescent intervals between week-long eruptions, for example (Chap. 8). Geomorphic hazards can cause delayed effects at great distances from the vent, such as the destructive debris flows in Sarna in 1998, ~100 km east of Vesuvius and several centuries after its historical eruptions (Zanchetta et al. 2004).

1.3 Objectives and Outline

Understanding geomorphic hazards on volcanoes is complex, as they result from the interplay of a range of volcanic hazards with a combination of post- and non-eruptive processes. Further complexity stems from disturbances that volcanic eruptions cause in drainage basins, and large-scale caldera events which emplace voluminous ignimbrites (>50 to 1000 km^3); all can lead to dramatic changes to the land surface, not only by mantling or obliterating the pre-existing topography but also by creating new landforms (Fig. 1B). In this chapter, we examine six themes related to geomorphic hazards on and around active and dormant volcanoes:

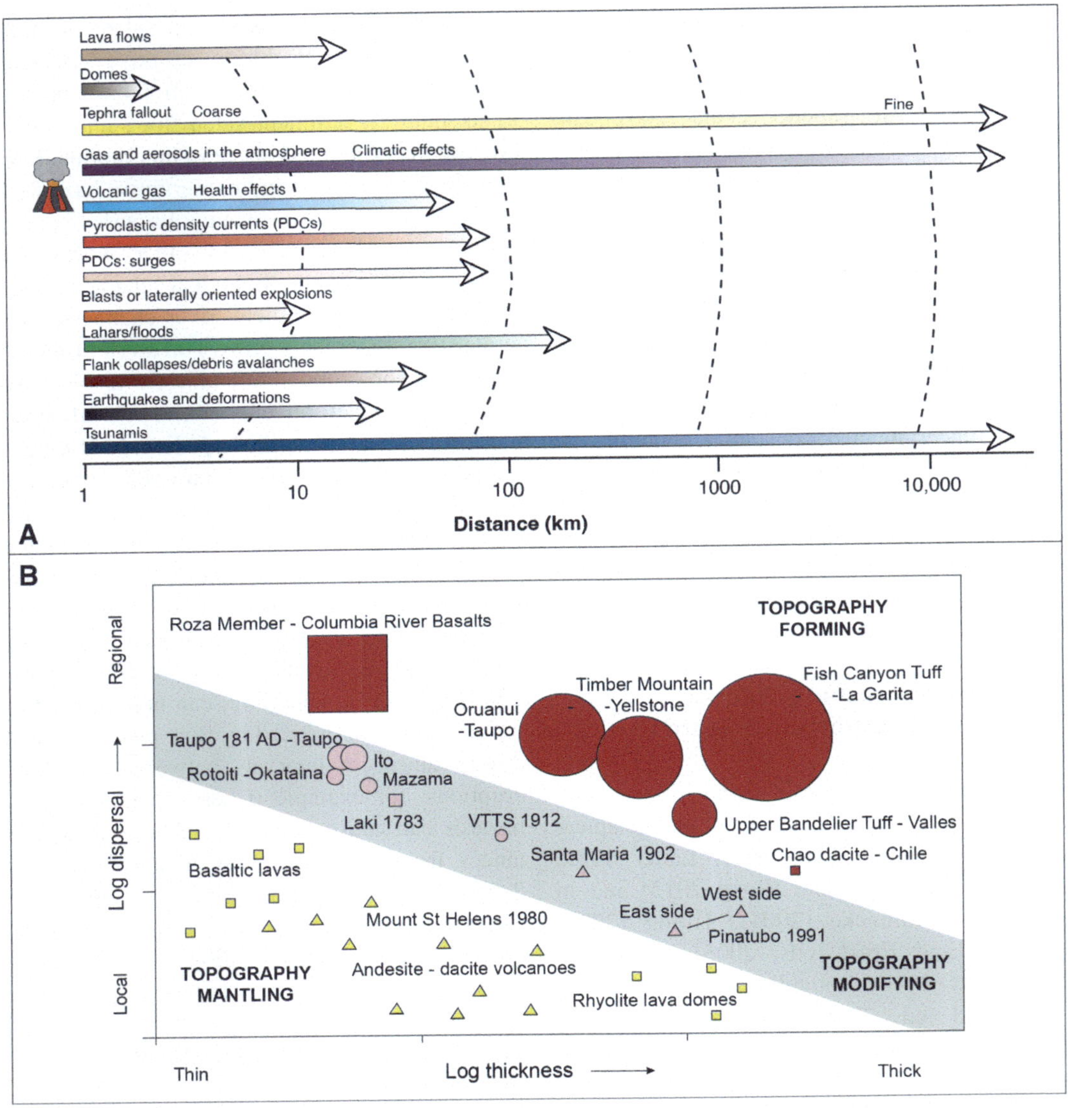

Fig. 1 **A** Range of distances (average: solid lines; exceptional: dashed lines) of the impacts of destructive phenomena directly or indirectly associated with volcanic eruptions. Except for volcanic gas, atmospheric effects and effects on health, all volcanic phenomena may lead to geomorphic hazards. *Source* Reprinted from Global Environmental Change, vol 2, Chester et al. (2000) The increasing exposure of cities to the effects of volcanic eruptions: a global survey, pp 89–103 © with permission from Elsevier. **B** Topography-forming versus topography-mantling eruptions. Size of symbol is proportional to volume of erupted material, axes represent explosivity (dispersal) and mean deposit thickness. Circles represent pyroclastic density current-dominated eruptions, squares represent effusive lava eruptions, and triangles represent dominantly Plinian-style eruptions (with local pyroclastic currents). Intermediate-sized eruptions, where interfluves are temporarily buried, are marked by the shaded field (after Manville et al. 2009a, Fig. 13, p. 151)

1. Direct interplay of geomorphic processes with eruptive activity and volcanic flows, in particular lava flows and PDCs.
2. Geomorphic impact of indirect volcanic hazards: sediment-laden flows, including lahars and jökulhlaups; and volcanic tsunamis.
3. Three categories of post-eruption geomorphic response in watersheds perturbed by small and large eruptions (thresholded at VEI 3), including the sequential geomorphic effects of eruptions of Mount St. Helens (1980), Mount Pinatubo (1991), and Chaitén (2008).
4. Geomorphic impacts from very large eruptions (> VEI 6).
5. Protracted geomorphic hazards in caldera and ignimbrite field contexts.
6. Geomorphic hazards on dormant or inactive volcanoes.

2 Interplay of Geomorphic Processes with Eruptive Activity

At active or potentially active volcanoes, geomorphic hazards are superimposed onto volcanic ones concomitantly and repeatedly during eruptions. Geomorphic processes originally interplay with direct volcanic flow processes, either lava flows, PDCs, or syneruptive lahars. They can also interplay with indirect hazards, such as debris avalanches, volcanic tsunamis and sediment–water flows. The extent and rate of geomorphic hazards that stem from degradation processes are exacerbated on volcanoes due to unconsolidated materials, hydrothermal alteration, weathering, and sesimic activity.

2.1 Effects of Slope, Topography, and Bedrock on Lava Flow Propagation

The morphology and propagation of subaerial lava flows depends on at least three factors: effusion rate, physical and rheological properties, and pre-existing ground slope. We have described the first and second factors in Chap. 2, and alluded to the role of ground slope.

Slope gradient can play a role in the flow width and length: flow width varies inversely with ground slope (Hulme 1974; Gregg and Fink 2000), while the effect of slope on lava length is minor, compared to other factors (Walker 1973; Chap. 2). For angles of 2–10° typical of the slopes of basaltic volcanoes, lengthening of pahoehoe takes about about 1–10 days, with the longer times corresponding to shallower slope angles (Kilburn 2015). Pāhoehoe flows can advance more quickly on low gradient slopes (<3°), but abrupt changes in ground slope, such as breaks-in-slope and cliffs (Glaze et al. 2014), play an important role on the transition from pāhoehoe to 'a'ā flows, as they modify the yield strength (Chap. 2, Sect. 1.2.3). We described the parameters that govern this transition in Chap. 2, Sect. 1.2 (Kilburn 2015, his Fig. 55.5). The threshold advance rate U_{cr} above which pāhoehoe becomes 'a'ā is defined by an equation involving several physical parameters (Chap. 2), in which the angle of slope β acts as a common denominator (Kilburn 2015, his Fig. 55.6).

Changes in lava-flow channel geometry are due to subtle changes in slope angle and topographic constriction or confinement. The high fluidity of Hawaiian lavas makes them susceptible to topographic confinement, which promotes flow lengthening, whereas channel bifurcations caused by topographic obstacles create multiple parallel channels that can limit individual flow-lobe length (Cashman and Mangan 2014). However, slope plays only a secondary role in controlling initial rates of flow advance. Other features encountered in highly fluid pāhoehoe flows can indicate the role of slope such as runup heights on tree moulds and superelevation on channel bends, but they remain minor with respect to effusion rate and volumetric flow rate (Cashman and Mangan 2014).

Subtle changes in topography, with a vertical amplitude of less than the flow thickness, play a role in the susceptibility to lava flow invasion (e.g. Tarquini and Favalli 2010). This prompted

Rümpf et al. (2018) to carry out experiments to investigate the substrate roughness, which had not been hitherto taken into account in most flow emplacement models and hazard estimates. The authors show that substrates with larger grain sizes caused experimental flows to advance more slowly than finer-grained substrates. This confirms field observations that topographic roughness of the order of 15% of flow thickness affects the path of active pāhoehoes. Thus, bed roughness should be included in models because the effective friction parameter may increase accuracy in lava-flow delineation and hazard estimation models.

2.2 Slope, Digital Topography and Lava Flow Hazard Monitoring

Lava-flow eruptions in 2021 at La Palma (Canary Islands) and in 2002 at Nyiragongo (Democratic Republic of Congo, DRC) were deadly and/or destructive events. The compound lava-flow field of Tajogaite-Cumbre Vieja with a surface area in excess of 12 km^2 has rendered the surface sterile for many years to come, while the associated tephra fallout affected both the agriculture and tourist economy of the island (Carracedo et al. 2022; see Case study 1, Chap. 10; Troll and Carracedo). The 2003 flank fissure eruption at Nyiragongo emplaced a complex field of highly fluid lava flows that caused between 70 and 100 deaths, destroyed the homes of 120,000 inhabitants and ultimately devastated 15% of the city of Goma in DRC (Komorowski et al. 2002–2003). Therefore, modelling, monitoring, and mitigation procedures have been implemented for hazardous lava flows around active volcanoes, on the flanks of which there are settlements of various sizes and economic activities.

2.2.1 Digital Topography to Help with Forecasting and Monitoring of Lava-Flow Hazards

Topography and minor slope changes present potential hazards for lava flows (Calvari 2019). At Kilauea volcano, where an effusive eruption lasted for 35 years from 1983, the Hawaiian Volcano Observatory (HVO) scientists in charge of volcano monitoring and hazard assessment produced lava flow maps including potential flow paths based on topographic steepest-descent calculations (Kauahikaua and Tilling 2014; Neal et al. 2018). GIS analysis has incorporated topographically defined drainage channels within which lava will be confined for specific vent positions (Kauahikau 2007; Kauahikau et al. 2003; Harris and Rowland 2015). This is sufficient for forecasting the possible expansion of lava, given that the direction of fast pāhoehoe flows can change rapidly due to minor topographic obstacles, and create an inverted surface topography due to inflation. Thus, high-spatial resolution DEMs (using LiDAR, Terrestrial Laser scanners TLS, or Unmanned Aerial Systems, UAS; Chap. 11) can pinpoint minor ($\leq$ 1 m-scale) changes in topography (breaks-in-slope, and the opposite, gradient flattening) and morphology (channel constrictions and bends, and the opposite, valley widening) that can deflect hazards towards inhabited areas. Hence, it is of utmost importance to acquire timely DEMs with the highest resolution possible before and during effusion phases, including a repeat procedure that helps estimate lava flow extent and volume, lava front advance, and pulses. Such a process was described by Favalli et al. (2009, 2010, 2011) with LiDAR at Etna, and was conducted regularly using UAS at the Kilauea lower East rift zone during the 2018 eruption (Dietterich et al. 2021; Meredith et al. 2022) and TLS at Holuhraun in 2014–2015 (Müller et al. 2017).

Favalli et al. (2010) were the first authors to sequentially map lava flows of a basaltic volcano at Etna with high-spatial resolution LiDAR (<50 cm) at a high temporal frequency of 15 min (Chap. 11). They highlighted the impact of short-time morphostructural modifications between 2005 and 2007 in determining lava flow paths and, consequently, the lava-flow hazards on volcano flanks. However, frequent updating of DEMs is required to account for the effects of multiple overlapping future lava flows as well as of other geomorphic changes to channels, fronts, and inflation that must be taken into account for long-term hazard assessment. This is why an array of techniques was used for hazard assessment during the 2022 eruption at Piton de la Fournaise (Chevrel et al. 2023): aerial structure-for-motion photogrammetry, visible, infrared and radar sensors, and radar multi-viewing angle imagery. These techniques were also used to track the syn-eruptive changes to the Piton Tikal cone morphology from 24 September to 4 October 2022. Where frequent effusive eruptions occur every few years, for example on Etna, satellite measurements of the effusion rate are crucial during the initial phases of a lava-flow eruption (Calvari 2019). These measurements are used to estimate the maximum distance to which a single flow unit can travel and to run lava flow simulations using different models, providing a more accurate and reliable forecast of lava flow expansion on pre-existing topography (Del Negro et al. 2013).

Digital topography forms the basis of all probabilistic models of flow inundation and is required for response to ongoing eruptions (Wadge et al. 1994; Cashman and Mangan 2014; Cordonnier et al. 2016; Dietterich et al. 2017; Neal et al. 2018). Major challenges for developing predictive models include obtaining a timely series of the highest-resolution DEMs required to anticipate the advance of thin flows (on slopes as gentle as 2°) as well as small changes in topography created by earlier lava emplacement. Sequential high-resolution DEMs are key inputs in modelling lava-flow front advances during crises at Kilauea as well as for probabilistic assessments of lava flow inundation (e.g. Chevrel et al. 2021) at Piton de la Fournaise, Réunion Island (Fig. 2), Etna, Italy (Cappello et al. 2013; see Tarquini and de' Michieli Vitturi's contribution, Chap. 11) and Holuhraun in Iceland (Tarquini et al. 2019; see Case study 2, Chap. 10; Pedersen and Gudmundsson).

The most cost-effective method is photogrammetric surveys acquired by low-altitude UAS flights, using low-cost optical sensors, that provide near-real-time digital elevation data on advancing flow fronts (Cashman et al. 2013; Turner et al. 2017; Dietterich et al. 2021). Benefits include: (1) capturing new topographic changes caused by pāhoehoe flow inflation, providing detailed updates for topographic datasets that were not previously possible; (2) supplementing existing satellite and manned aircraft-based remote sensing and field based monitoring of active flows, and; (3) better estimating future flow path behaviour and reducing uncertainty in hazard forecasting, particularly in forested and urbanized areas in which small obstacles may deflect lava flows (Turner et al. 2017).

Mitigation procedures can be implemented to avoid damage, such as the diversion of lava flows both in the field and in the laboratory (Dietterich et al. 2015). Using experimental viscous flows with V-shaped and oblique obstacles with varying orientations, the authors described methods for diverting lava flows: forced branching of flows to redirect lava supply and barriers to confine the flow. They demonstrate that it is possible to slow a flow by constructing a long orthogonal barrier or by dividing it into branches. Flow deceleration will be proportional to the relative flux in each branch. V-shaped flow-splitting barriers should have small internal angles to reduce the potential for overtopping, with long arms to ensure that the two branches remain far enough apart to maintain separate channels.

2.3 Direct Interplay of Geomorphic Parameters with Hazards from PDCs

Pyroclastic density currents, including pyroclastic flows and surges, are laterally moving, buoyantly expanding mixtures of hot gases and fragmented particles, from ash to blocks (Druitt 1998; Branney and Kokelaar 2002; Cole et al. 2015; Dufek et al. 2015; Valentine 2019). These extremely hazardous natural phenomena, often associated with explosive volcanism, are difficult to face, apart from by proactive evacuation of inhabitants, due to their high temperatures and velocities, long runout distances, and their ability to generate buoyant plumes that inject ash into the atmosphere (Andrews and Manga 2011; Cole et al. 2015; Dufek 2016; Lube et al.

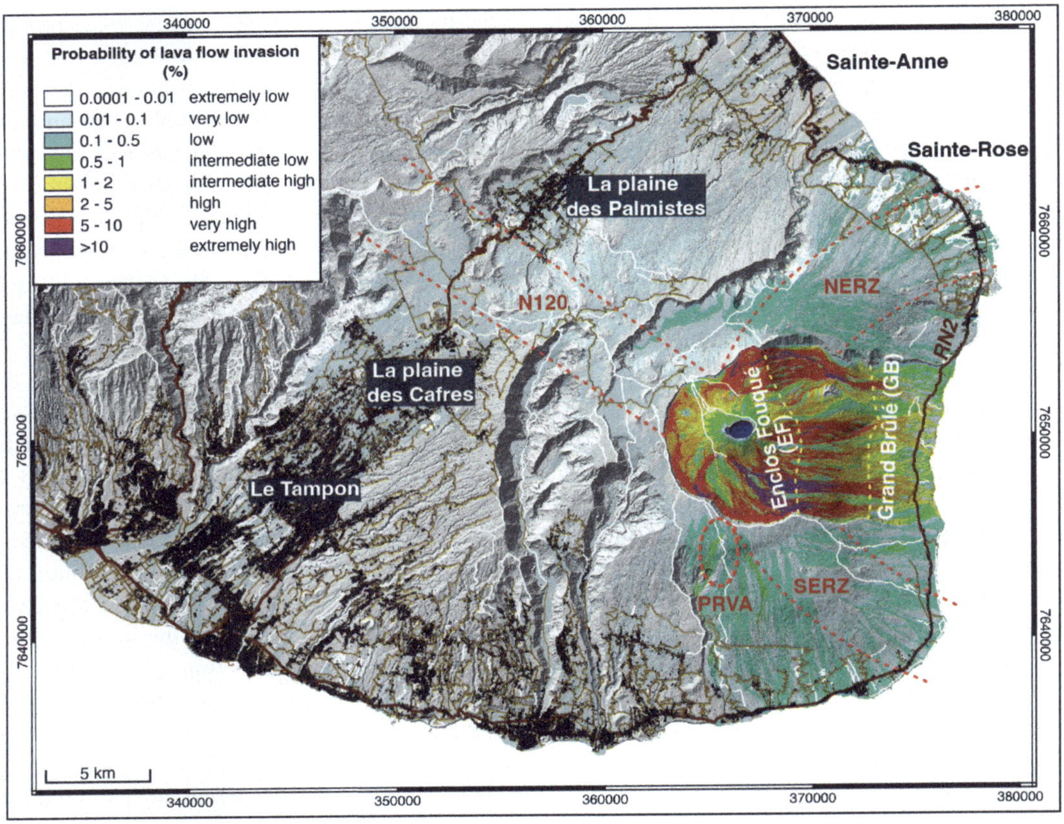

Fig. 2 Lava flow hazard map of Piton de la Fournaise (Courtesy of Chevrel et al. 2021). Probability of lava flow invasion is given as percentages and is colour-coded from extremely low (<0.01%) to extremely high (>10%). Buildings are represented as black polygons, the main roads are in brown (the RN2 national road is marked by a thicker line), and tourist trails are shown as thin black lines (data from BD TOPO® IGN). The dashed red lines outline the rift zones (N120E), to the northeast (NERZ), and to the southeast (SERZ) as well as the Puy Raymond volcanic alignment (PRVA). The Enclos depression is separated into three regions by the dashed yellow lines: the upper Enclos Fouqué (EF), the steep slopes "Grandes Pentes", and the lower part "Grand Brûlé" (GB). The background is the relief shading of the LiDAR-based DEM from IGN—released in 2010

2020; Giordano et al. 2024). Understanding the dynamics and consequent destructive potential of these currents is essential in order to protect populations living around active volcanoes (Tadini et al. 2022).

The same PDC may exhibit properties of both end-members (concentrated flow and dilute surge) during the course of its runout (Burgisser and Bergantz 2002; Sulpizio et al. 2014; Dufek 2016; Lube et al. 2020) (Fig. 3A, B). Lube et al. (2020) summarize the principal findings on multiphase PDC regimes, transport behaviour, and characteristics that affect hazards. The authors highlight concentrations that separate dilute, intermediate and concentrated regimes of transport, and how the intermediate regime controls the evolving transport behaviour of both end-members. Moreover, the identification of pore-pressure variations in experimental PDCs provides insights into the origin of the high mobility and very low effective friction of these flows, both of which factors favour hazardous behaviour. Small- to medium-volume PDCs (less than ~1 km^3) are frequent occurrences and combine a high mobility (runout in excess of 15 km from the vent) with an enormously

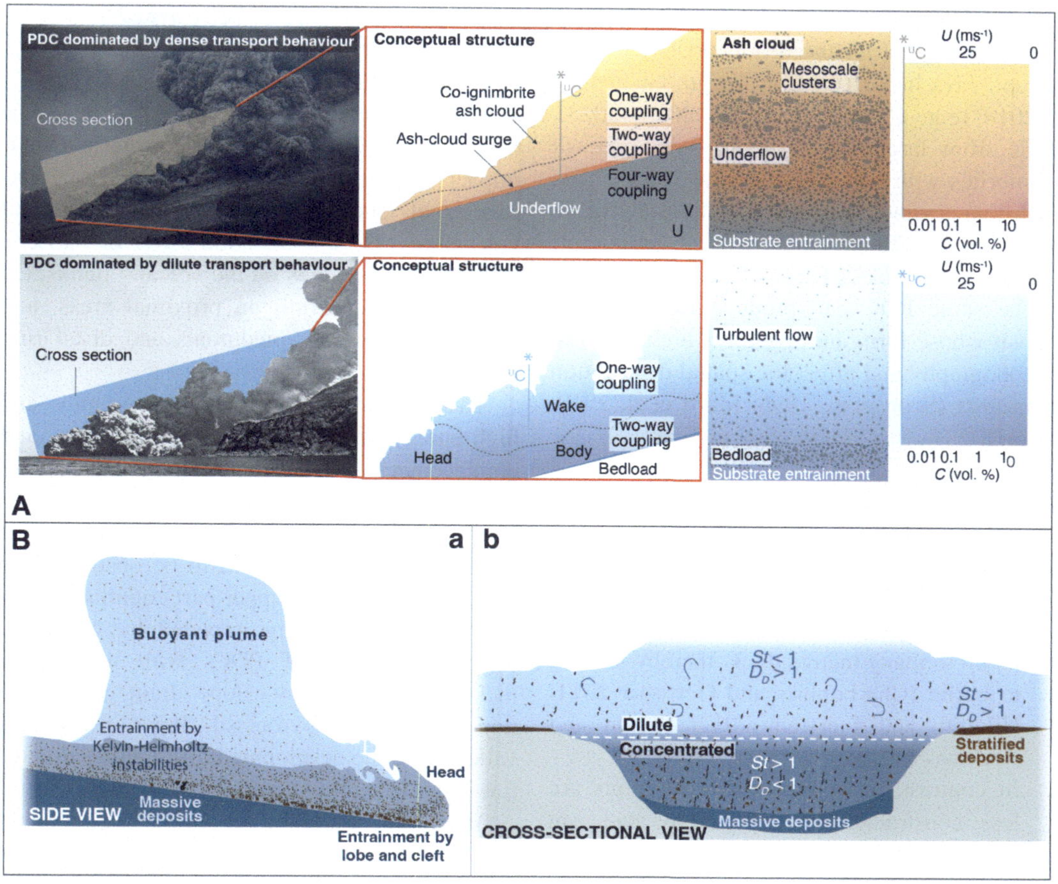

Fig. 3 **A** Contrasting dynamics and deposits of concentrated and dilute PDCs: pyroclastic density current (PDC) from Sinabung volcano in 2014, as an example of a PDC that is dominated by concentrated gas–particle transport (top), and a pyroclastic surge from Stromboli volcano in 2019, exemplifying PDCs with dominantly dilute gas–particle transport (bottom) (Lube et al. 2020, Fig. 1a, p. 350). The left-hand panels show photos of the two PDCs, with coloured rectangles defining cross-sections used to show the conceptual structure of concentrated and dilute PDCs (middle-left panels). They show that gas particle coupling mechanisms vary with height and down the flow differently for the two regimes. Close-ups of the lower regions of the PDCs (middle-right panels) highlight the presence of a granular fluid-based underflow that is modified by gas pore pressure in concentrated PDCs, and a bedload region in dilute PDCs. The right-hand panels display generalized vertical profiles of time-averaged velocity (U) and volumetric particle concentration (C) for predominantly concentrated and dilute PDCs (Lube et al. 2020). The vertical line for the profile is shown in the second panel. **B** Sketches illustrating the anatomy of a generalized PDC (Dufek 2016, from Ann. Rev. Earth Planet. Sci., vol. 42@2014 by Annual Reviews, http://www.annualreviewes.org). **a** Side view depicting the particle concentration gradient resulting in a suspended and bed-load region. The current entrains air at the head of the current owing to lobe and cleft instabilities and entrains air at the top of the current through Kelvin–Helmholtz instabilities (which occur when there is velocity shear in a single continuous fluid) in which shear overcomes stable density stratification. **b** Cross section of a channelized pyroclastic density current. Here the dense-dilute parameter (D_D, Timescale of collisions relative to particle drag response time) is less than 1 in the channel, resulting in deposition from a concentrated current, and the overbank deposits have a collisional Stokes number St ~1 and D_D>1, resulting in stratified deposits from a dilute current (Dufek 2016, modified from Burgisser and Bergantz 2002, Fig. 3, p. 408)

destructive power (e.g. Valentine 1998; Nakada 2000). Examples are the deadly PDC events at Merapi, Java in 2010 and at Fuego, Guatemala in 2018 (e.g. Charbonnier et al. 2013, 2023). PDCs from large-scale eruptions that deposit ignimbrites with volumes of 10^1 to 10^3 km^3 can reach longer runout distances in the order of 60 to 120 km (e.g. Taupo 186 CE, Peach Spring tuff 18.8 Ma; *in:* Dufek 2016; Roche et al. 2016).

The same PDC can produce two types of deposit whose form and thickness are related to the topography and affect the spatial distribution of hazards (Scott et al. 1996; Marti et al. 2019; Major 2023).

(1) High concentration, dense flows commonly channel down topographic lows during eruptions as soon as they leave the proximal vent area; then they tend to follow valleys, where they can leave non-stratified deposits up to many metres thick, thereby leading to burial (Charbonnier and Gertisser 2011; Hildreth and Fierstein 2012; Brown and Andrews 2015).
(2) In contrast, low-concentration currents are less constrained by topography and can flow well beyond channel boundaries and other topographic lows down slope. Dilute surges can expand with devastating effects, allowing them to travel over rugged terrain and surmount multiple ridges and valleys. The most energetic surges can leave deposits that are so thin relative to the volume and coverage area of the initial flows that it is hard to identify them for hazard diagnostic purpose (Hoblitt et al. 1981; Wilson 1986; Fisher 1990, 1995).

The interplay of geomorphic processes with hazardous PDCs will be broken down into three parts: (1) direct effects of high-energy relief on transport and deposition; (2) direct interaction with substrate, as flows exert erosion and entrain particles; (3) channel characteristics that control the runout and behaviour of PDCs, leading to decoupling and *avulsion* (the re-direction into a tributary valley) toward areas hitherto unexposed to hazards.

2.3.1 Direct Effects of Topography on Transport and Deposition of PDCs

Topography exerts a strong influence on PDCs, hence on the distribution of deposits and their impact. The term topography covers four categories of landform and slope on and around volcanoes: (1) high-elevation relief or barriers, (2) valleys and ridges on the volcano flanks, (3) steep slopes, generally in proximal areas, and (4) gentle slopes of piedmonts and distal ring plains.

The influence exerted by topography and high-energy relief on PDC transport and emplacement has long been recognized. Fisher (1990, 1995) showed how topographic obstacles accentuated the decoupling of flows, causing their lower, denser part to be obstructed or deflected, while their upper part continues along its path without being impeded by obstacles. Such interactions of PDCs with topography include blocking, downslope drainage, formation of secondary pyroclastic flows in tributary valleys, and the development of dividing streamlines and decoupling (Fisher 1995).

Voluminous ignimbrites from large Plinian and caldera-forming eruptions (>10 km^3) that are deposited on rough ground illustrate the role of topography on emplacement and sedimentation (Valentine et al. 1992). Ignimbrite sheets typically show massive and poorly sorted valley-fill deposits 100 m thick in depressions, stratified deposits pinching out against valley slopes, and veneer deposits centimetres to decimetres-thick, more stratified, better sorted and depleted in fines, on high ground and ridges. These relationships with topography have been described for the 1912 ignimbrite of the Valley of Ten Thousand Smokes, Alsaka (Hildreth and Fierstein 2012), the 1.8 ka BP Taupo ignimbrite, New Zealand (Wilson and Walker 1985), and the 273 ka Poris eruption, Tenerife (Brown and Branney 2013).

Because the dilute parts of currents are highly inflated and have a thickness in excess of hundreds of metres, PDCs are very mobile and can pass over substantial topographic barriers (Woods et al. 1998), from a couple of hundred metres, such as at Mt. St. Helens (1980) or Unzen (1991–1994), up to 1000 m for large-volume ignimbrites such as the Campanian one (with 150 km^3 DRE; Fedele et al. 2008). Valuables situated leeside and apparently sheltered behind such barriers are exposed to the devastating effects of the most dilute parts of pyroclastic currents (Sect. 2.3.5 below). Using laboratory experiments that mimic natural PDC conditions, Andrews and Manga (2011) show that the critical height of barriers for preventing transit on ridges is ~1.5 times the flow thickness of currents. This threshold provides a means of estimating minimum flow thickness based upon maximum topographic relief between volcanic vents and distal deposits.

2.3.2 PDC Erosion of Substrate, Contributing to Bulking and Thus to Impacts

Two of the most important gaps in understanding of PDC relations with topography are the mechanism of substrate erosion through ablation or scouring, and the influence of entrainment of particles from the substrate on flow dynamics (Pollock et al. 2016). Determining both erosional processes has pivotal implications in the field of flow dynamics and the assessment of PDC impacts, not only around volcanoes, but far away from the source in the case of ignimbrites.

Erosion occurs mostly during waxing phases of PDCs, as demonstrated by case studies at three volcanoes: Mt. St. Helens, USA (Kieffer and Sturtevant 1988; Brand et al. 2014; Pollock et al. 2016), Lascar, Chile (Sparks et al. 1997; Calder et al. 2000), and Tungurahua, Ecuador (Douillet et al. 2013; Bernard et al. 2014; Benage et al. 2016). Erosion contributes to bulking, hence to increased current density, mobility and runout length; in turn, these characteristics help interpret the spatial distribution of impacts and damage.

PDCs can leave erosional landforms such as furrows (flute casts), which formed in the loose sediments under deposits of the lateral blast of 18 May 1980 eruption of Mt. St. Helens (Kieffer and Sturtevant 1988; Kieffer and Simonds 1995). Having identified furrows ~100 m-long from 3.5 to 9 km from the crater, these authors estimated the erosion rate of this blast to be 20.6 $kg/m^2/s$, corresponding to an erosion depth of ~1 m in a time of 30 s. The erosional features are attributed to scouring by longitudinal vortices resulting from flow instabilities induced by complex topography. Another example stems from spectacular erosional features produced by PDCs generated by the 19–20 April 1993 eruption of Lascar Volcano, Chile (Sparks et al. 1997; Calder et al. 2000). The flows eroded and incorporated unconsolidated surface talus and colluvium and then formed striae, percussion marks and furrowed surfaces in the bedrock. Such furrows developed with wavelengths of 0.5–2 m and depths of 0.1–0.3 m. The flows were strongly erosive on steep slopes, and they accelerated at channel constrictions or where local slope increased markedly. The Lascar flows are thought to have segregated into lithic-rich and pumice-rich parts upon propagation (Sparks et al. 1997). Lithic-rich basal portions of flows were less mobile but highly erosive on slopes up to 14°, whereas the pumice-rich parts moved beyond the lithic-rich flows to the 1993 fan margins on slopes of less than 4°, but with little erosive capacity. The development of two deposit types with contrasting erosive power is attributed to internal segregation of the two components within the moving flows based on density.

Recent large eruptions at shoaling volcanoes have provided researchers with new opportunities for investigating the role of PDC erosion underwater. Clare et al. (2023) and Seabrook et al. (2023) found that the rapid emplacement of a large amount of material onto steep submerged slopes triggered extremely fast (122 km per hour) and long-runout (>100 km) seafloor density currents during the major eruption of the Hunga Tonga-Hunga Ha'apai caldera in January 2022. Almost 10 km^3 of seafloor material was removed during the eruption, most of which was redeposited within 20 km of the caldera by long run-out seafloor density currents, destroying telecommunication cables, reshaping the seafloor and killing seafloor life. The deep scours excavated by these currents are similar to those around other submerged volcanoes, providing evidence of large underwater eruptions worldwide.

Bulking (the flow action in which particles from the substrate are ploughed up and incorporated into the flow body) leads to increased PDC mobility on the steep slopes of arc volcanoes, as the augmented flow mass enhances both flow velocity and runout distance, expanding the geographic scope and the scale of devastating impacts. Bernard et al. (2014) investigated the bulking effect of the August 2006 eruption of Tungurahua volcano, Ecuador by means of a twofold approach.

Firstly, bulking and its effect on flow mobility were estimated through the amount of the non-juvenile material incorporated into the scoria flows emplaced in one valley. About 40 vol.% of non-juvenile material, corresponding to a 1-m-thick erodible bed, occurred principally on the steep (>25°) upper slope of Tungurahua. The runout distance of small-volume scoria-flow deposits ($<10^6$ m^3) increased by up to 4–20%, depending on slope angle towards the base of the volcano (0–10°). Therefore, the presence of large amounts of non-juvenile material prior to PDC emplacement helped propagate its destructive effect (Bernard et al. 2014). Secondly, the authors used the VolcFlow numerical code to reproduce PDC behaviour during the 2006 Tungurahua eruption, using two simple depth-averaged mass-flow models to compare bulking mechanisms. The erosion pattern can be simulated by fluctuations of the basal shear stress of the flow during emplacement. Such variations are controlled by waxing-waning processes within the flow, notably by the formation of a thick non-erosive body that pushes a thin frictional erosive front during PDC emplacement.

Relationships between bulking and destructive effects, as analyzed on Tungurahua, Lascar and around other shoaling volcanoes, should be very much taken into account for hazard assessment at other arc volcanoes worldwide where similar erosive PDC events are common.

2.3.3 Implications for Substrate Erosion and Entrainment on Flow Dynamics and Hazards

PDCs can transfer particles from the bed into the current (Branney and Kokelaar 2002). This process—particle entrainment, removal and incorporation of bed material—can alter flow dynamics, in particular the runout. The fundamental force driving PDCs down slope is the current's high density relative to the ambient air medium; thus, changing the bulk density and mass of the flow via substrate entrainment is likely to affect basal flow properties and runout distance, the latter being critical for delineating potentially hazardous areas.

Substrate erosion and entrainment of particles involve a series of physical processes acting inside the flow head and tail that have been inferred from recent laboratory experiments and detailed field analyses on PDC deposits (Roche et al. 2013, 2016; Chédeville and Roche 2015; Roche 2015; Fauria et al. 2016; Pollock et al. 2016; Lube et al. 2020). The main factors affecting the ability of PDCs to pluck and incorporate lithics from the substrate are: internal pore pressure which changes flow mobility, self-channelization which influences the carrying capacity of the currents, and surface roughness which impacts substrate erosion.

As PDCs interact with uneven topography, studies on the evolution of basal shear stress and pore pressure provide a more complete understanding of the mechanisms of substrate erosion, contributing to flow dynamics and hazardous behaviour. A range of experiments has examined entrainment mechanisms as diverse as force chain dynamics (Estep and Dufek 2012), the ejection of particles due to impact by a projectile, which increases current density (Fauria et al. 2016), and erosion of the bed by low-pressure parts of the pyroclastic current due to fluidized transport (Mangeney et al. 2010; Roche et al. 2013). Such processes interact with the composition of the lower boundary of a PDC, and play a significant role in entrainment and runout distance (Fauria et al. 2016), although experiments highlight that pore pressure variations at the flow base are the most significant process.

Basal underpressure at the head of granular flows, which varies with the square of the flow velocity, may be responsible for creating uplift for particles set in motion, as suggested by analogue laboratory experiments (Roche et al. 2013, 2015, 2016). This mechanism explains the apparent dichotomy that granular flows are able to entrain decimeter-sized blocks while often leaving finely grained substrates relatively undisturbed. Experiments also provide a method for inverting data from PDC deposits by constraining flow velocities based on the size and density of entrained blocks (e.g. Pittari et al. 2007). From experiments with the particle uplift model for dense flows, Roche et al. (2015, 2016) have inferred front velocities of up to ~25–30 m/s for the parent PDCs of some ignimbrites. Velocities of parent currents of these ignimbrites are lower than expected, showing that other parameters play a major role in mobility and long runout. Results on bulking via basal underpressure may explain why PDC volume and density increase, contributing to entrainment at the current base, hence to longer runout of some ignimbrites, often modifying or creating new topography in the process (Fig. 1B).

2.3.4 Role of Geomorphic Parameters on PDC Dynamics Along Drainage Networks

Geomorphic parameters may govern PDC dynamics along drainage networks in two ways: (1) Self-channelization increases flow capacity and runout distance; (2) detachment and decoupling of density-stratified PDCs produces unconfined flows, potentially creating surges that may be lethal for populations living beyond the valley edges.

Relationships between valley channelization and morphology

Drainage network may control the propagation of dense PDCs once they are valley confined. Aravena and Roche (2022) evaluated the main topographic characteristics and the drainage network of a set of 40 composite cones worldwide. Results indicate that the channelization efficiency of a PDC is strongly controlled by the relative scale between flow size and topographic features of the volcanoes. The authors recognized four statistical groups of composite cones according to the mode of interaction between their topography and dense PDCs: (1) intense channelization (a) through different valleys down to distal domains (e.g. Colima, Mexico and Planchón-Peteroa, Chile), (b) through a single, dominant valley down to distal domains (e.g. Reventador, Ecuador and Mt. St. Helens, USA), and (c) near the source; (2) moderate distal channelization, frequently involving intertwined drainage networks (e.g. Tungurahua, Ecuador and El Misti, Peru); (3) potentially intense channelization only near the source, typically involving flat distal topographies (e.g. Sinabung, Sumatra and Mayon, Philippines), and; (4) weak channelization in proximal domains, resulting in efficient early energy dissipation and thus reduced flow runout distance (e.g. Kelut, Java and Akagi, Japan).

This classification, combined with morphometric analyses, is relevant in enabling the main processes that control PDC channelization in specific drainage networks to be identified, in particular at poorly documented volcanoes.

Effects of detachment or decoupling of density stratified PDCs

Most PDCs develop density stratification, with a dense, high particle concentration, gravity-driven lower part (valley-confined flow) and a dilute, low particle concentration, upper part (unconfined surge), as shown in Fig. 3. Such differential stratification, which can lead to decoupling with the two parts moving independently and in different directions, has long been recognized. Examples of small-volume flows propagating along narrow, sinuous, medial and distal valley channels include Merapi (Bourdier and Abdurachman 2001), Tungurahua (Stinton and Sheridan 2008) and Unzen (Nakada et al. 1999; Nakada 2000). In regions of rugged topography, the height of barrier ridges, breaks-in-slope on the flanks of composite cones, and the gradient of the ground surface greatly influence the effectiveness of decoupling processes and flow directions. As a result, destructive effects can extend onto the interfluves.

Important examples of PDC detachment that caused many fatalities include the eruptions of Mt. Pelée in 1902 (Fisher and Heiken 1982; Boudon and Lajoie 1989), Unzen in 1991–1994 (Nakada et al. 1999), Soufrière Hills in 1997 (Druitt et al. 2002), Merapi in 1994, 2006 and 2010 (Jenkins et al. 2013; Komorowski et al. 2013), and Fuego in 2018 (Charbonnier et al. 2023). The uncommon behaviour of dilute PDCs became fully recognized on 25 June 1997 at Soufrière Hills (Montserrat) when a detached, dilute current decelerated and rapidly sedimented material that drained from the steep topography to form a dense, highly mobile PDC, which took on a new course and travelled a further 3 km into a populated area (Druitt et al. 2002).

2.3.5 Categories and Effects of Unconfined Flows

Unconfined pyroclastic currents exhibit a range of concentration, velocity and temperature that are important characteristics in determining their physical impact. Lerner et al. (2022) carried out a broad comparative study of the different types of unconfined flows, their deposits, dynamics and impacts.

Taking the well-studied 2010 eruption of Merapi, Indonesia (Gertisser et al. 2023; see Case study 3, Chap. 10; Merapi 2010 eruption and PDCs, Indonesia; Gertisser), Lerner et al. (2022) focused on the Gendol River valley where confined and unconfined PDCs propagated in November 2010 (Fig. 4). Satellite imagery combined with fieldwork was used to determine the presence of both overbank flows and detached surges, which had resulted in the loss of hundreds of lives and the destruction of hundreds of houses in areas that were devastated up to 9 km from the summit. On 5 November 2010, a PDC flowed 15.5 km along the Gendol valley and a dilute surge detached from it at a bend before a 'Sabo dam' (used to regulate sediments in the water flow) and flowed over the bank into Bronggang village, leading to 54 deaths among villagers who had been attempting to evacuate, unaware of the immediacy of the danger (Fig. 4) (Jenkins et al. 2013).

The full spectrum of unconfined PDCs consists of four end-member types (Lerner et al. 2022; see Charbonnier and Gueugneau contribution, Fig. 7 in Chap. 11): (1) Fast overspill flows; (2) Slow overspill flows of dense, unconfined PDC; (3) High-energy, dilute surges, typically unconfined, initiating from a dome explosion; and (4) Low-energy detached surges (LEDs) when the dilute, upper portion of a PDC has decoupled from its denser basal parent flow, allowing it to travel and invade areas not reached by the lower, dense portion. Broad conclusions can be drawn about the key characteristics of, and damage caused by, unconfined currents (Lerner et al. 2022; see Charbonnier and Gueugneau's contribution, Chap. 11). In contrast to the complete destruction of buildings, vegetation and infrastructure in the main path of fast overspill flows and high-energy surges, LEDs are unique amongst the four unconfined flow types in causing relatively little damage and leaving little geological evidence. However, LEDs are still often fatal due to their high heat flux, and should therefore be afforded the same consideration during hazard planning as the

Fig. 4 **A** Areas of the Gendol River valley affected by the November 2010 confined and unconfined PDCs generated by the large eruption of Merapi (Lerner et al. 2022, Fig. 2, p. 7, modified after Jenkins et al. 2013). PDC types described in the key are channelled (confined flows) in dark brown, overspills in light brown, and low-energy detached surges (LEDS) in beige. Types of damage are also shown along the valley and particularly in two villages, Bakalan and Bronggang, on the western margin. Damage caused by PDCs to buildings and infrastructure was induced by dynamic pressure (from 0.5 to 15 kPa); see Jenkins et al. (2013) for a map showing estimated dynamic pressures from PDCs along the Gendol River valley. **B** Longitudinal profile of the Gendol River valley showing reaches 1–5 with average slope values and depicting the principal areas of PDC and lahar deposits. Note that 'Reach 3', shown in Fig. 4B, is mentioned in Sect. 3.1.2 (after Solikhin et al. 2015, Fig. 7b, p. 12) ▶

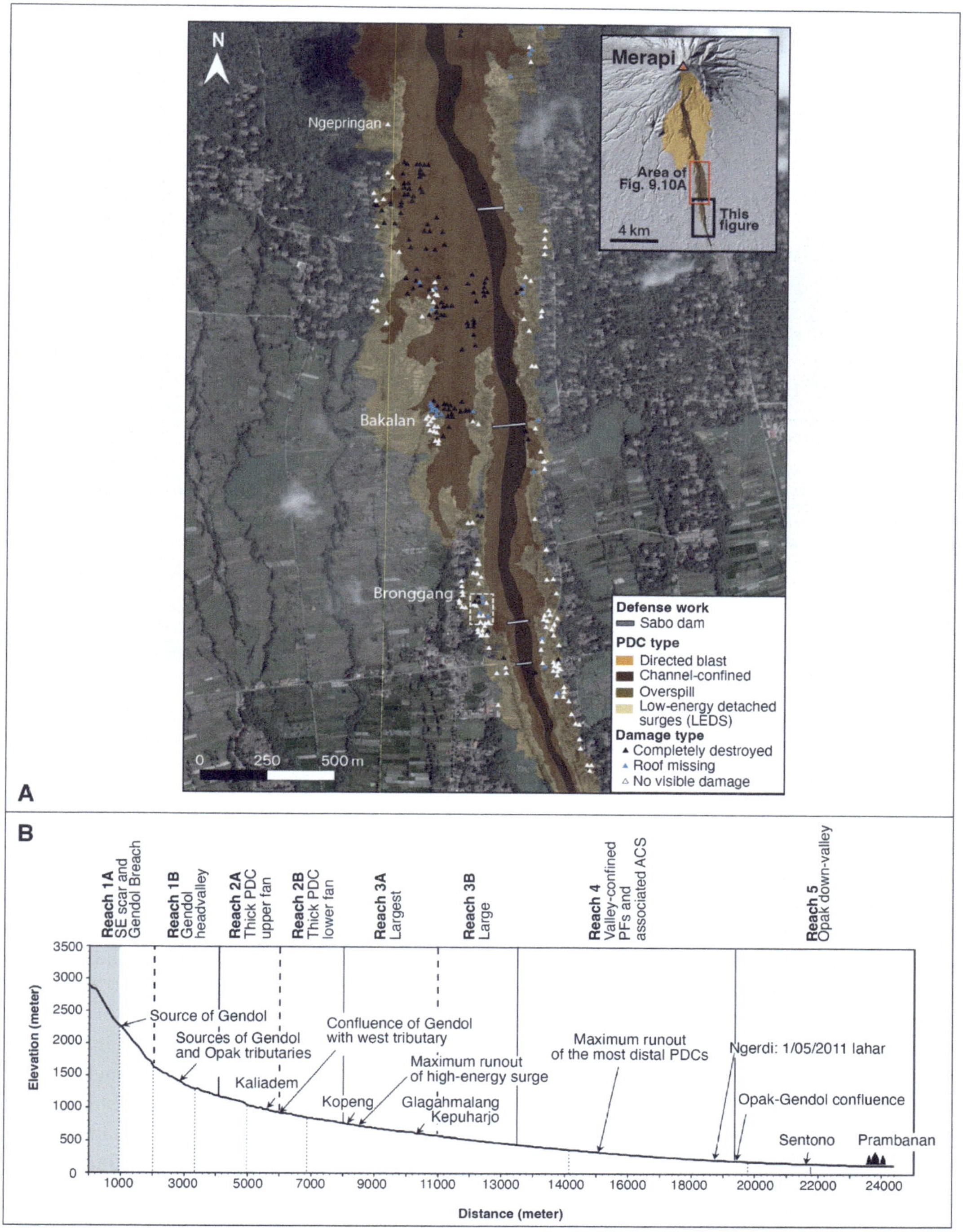
N
Ngepringan
Bakalan
Bronggang
Merapi
Area of Fig. 9.10A
This figure
4 km
0 250 500 m
Defense work
Sabo dam
PDC type
Directed blast
Channel-confined
Overspill
Low-energy detached surges (LEDS)
Damage type
Completely destroyed
Roof missing
No visible damage
A
B
Reach 1A SE scar and Gendol Breach
Reach 1B Gendol headvalley
Reach 2A Thick PDC upper fan
Reach 2B Thick PDC lower fan
Reach 3A Largest
Reach 3B Large
Reach 4 Valley-confined PFs and associated ACS
Reach 5 Opak down-valley
Elevation (meter)
Distance (meter)
Source of Gendol
Sources of Gendol and Opak tributaries
Kaliadem
Confluence of Gendol with west tributary
Kopeng
Glagahmalang
Kepuharjo
Maximum runout of high-energy surge
Maximum runout of the most distal PDCs
Ngerdi: 1/05/2011 lahar
Opak-Gendol confluence
Sentono
Prambanan
3500
3000
2500
2000
1500
1000
500
0
1000
3000
5000
7000
9000
11000
13000
14000
16000
18000
20000
22000
24000

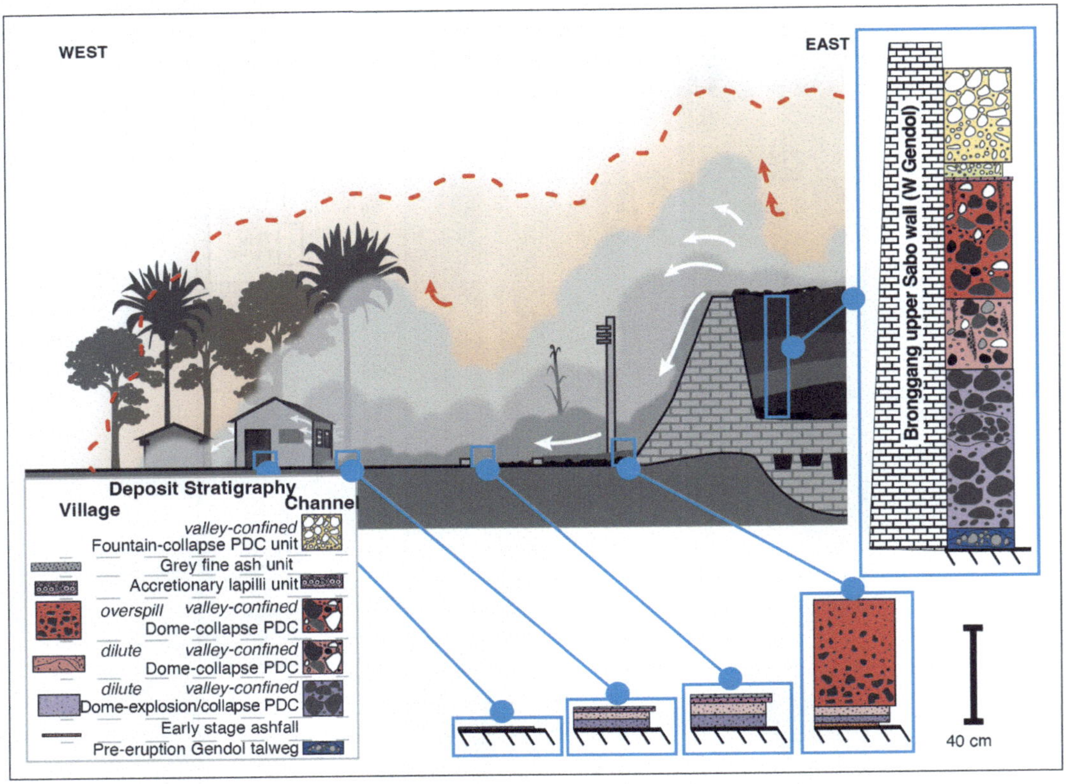

Fig. 5 Sketch demonstrating the dynamics of an LEDS as it leaves the channel and inundates Bronggang (see Fig. 4), showing a stratigraphy of channel-confined, overspill and LEDS deposits in and adjacent to Bronggang ~13.5 km from the volcano summit (Lerner et al. 2022, Fig. 6, p. 11)

unconfined dense flows and high-energy surges that leave more geological evidence (Fig. 5).

2.3.6 Geomorphic Parameters Governing Overbank and Avulsion of PDCs

A factor of particular concern for hazard mitigation around PDC–forming volcanoes is the overbank of valley-confined flows onto interfluvial terraces and ridges and avulsion to tributaries, with subsequent impacts on densely populated areas (Widiwijayanti et al. 2009). Such redirected PDCs have long been reported from observed historical eruptions, e.g. Mount Pelée (Lacroix 1904), Mt. Lamington (Taylor 1983), and Mt. St. Helens (Fisher 1995). Overspills are more likely when a channel is nearing capacity through previous infilling from confined flows, which reduces the height difference between the valley base and top (Lube et al. 2011; Charbonnier et al. 2013). This can be exacerbated by the unsteady, pulsating behaviour of PDCs, and by their velocity and volume (Sulpizio et al. 2008; Charbonnier and Gertisser 2011; Charbonnier et al. 2020, 2023; Kubo-Hutchison and Duffek 2021).

Lube et al. (2011) explained how interactions of pyroclastic currents with the channel topography contribute to the hazardous avulsion mechanism. At Merapi, multiple dome collapses on 14 June 2006 led to the generation of four individual block-and-ash flows (BAFs 1 to 4: Fig. 6A) in the Gendol River valley on the south flank. All four BAFs locally overflowed the channel margins to devastate cultivated terraces along each side of the box-shaped canyon and propagated as far as 2.2 km along adjacent tributaries (Thouret et al. 2010). The greatest destruction

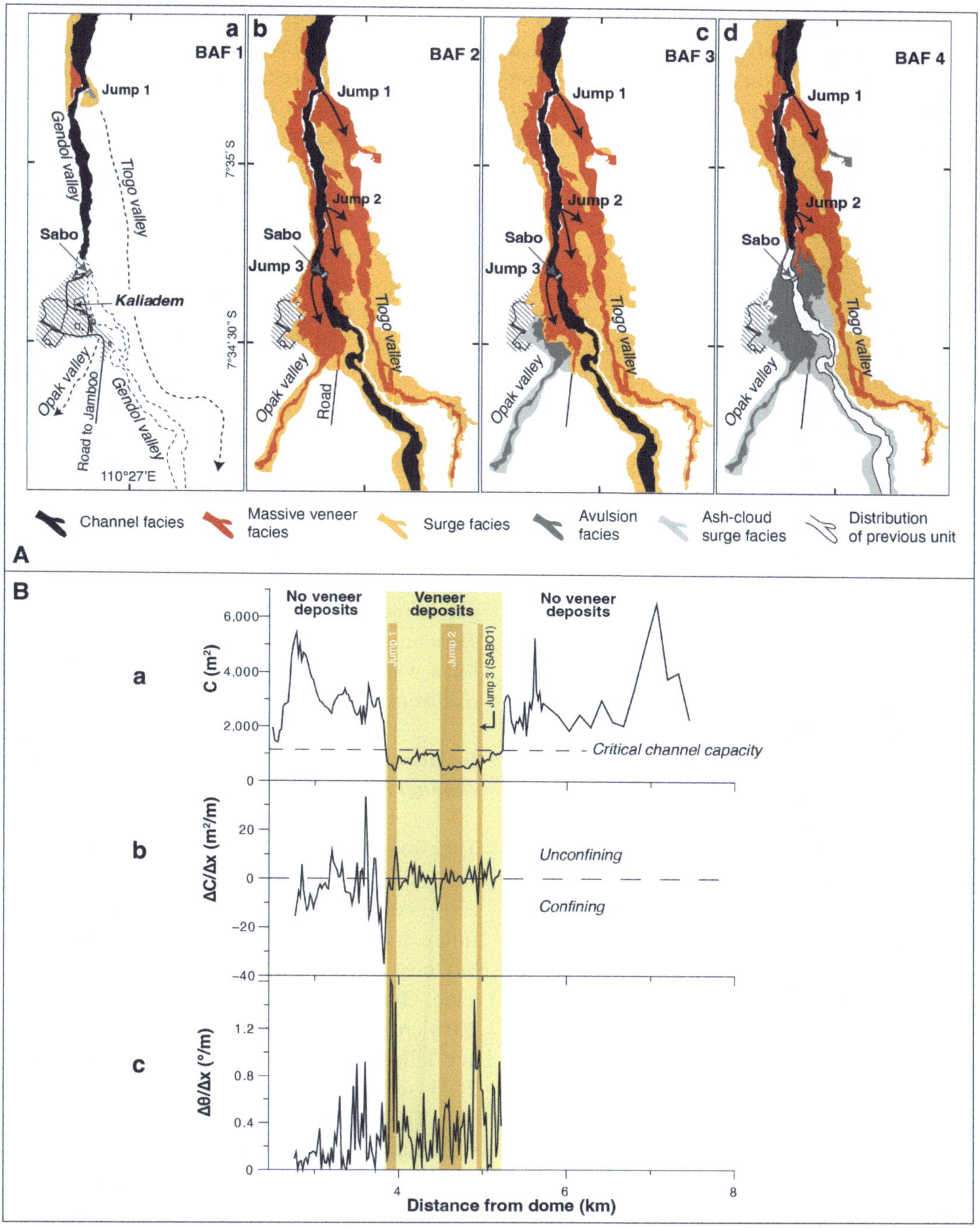

Fig. 6 Extent of the 2006 PDCs in the upper Gendol River watershed (Merapi south flank), and computation of geomorphic parameters along the valley channel. **A** Distribution of three sedimentary facies (channel-confined, massive veneer, and surge facies) of BAF1 (**A**), BAF2 (**B**), BAF3 (**C**), and BAF4 (**D**) in the stretch from 3.7 to 5.8 km from source. Black arrows mark the locations of flow avulsion jumps 1, 2, and 3. Note how each coincides with veneer facies (see Fig. 5B for explanation) (Lube et al. 2011, Fig. 2, p. 1130). **B** Geometric parameters of Gendol valley prior to the emplacement of BAF2, as a function of distance (Lube et al. 2011, Fig. 8, p. 1136): **a** the channel cross-sectional area C (m^2), **b** the longitudinal rate of channel confinement $\Delta C/\Delta x$ (m^2/m), and **c** the change in channel sinuosity $\Delta\theta/\Delta x$ (°/m). The yellow band indicates the mapped extent of massive veneer deposits on the interfluves bordering Gendol valley. The orange bars indicate the locations of focused flow avulsion

was caused by the second and third BAFs: both partially spilled out of the Gendol valley at a travel distance of 4.9 km, bypassing a Sabo dam just upstream of the village of Kaliadem, almost completely destroying it, and leaving it buried under several metres of overbank deposits.

The main mechanism of flow avulsion on 14 June 2006 was the overflow of part of the dense, basal part of the BAF onto the interfluves (Lube et al. 2011), which is different from the decoupling process of the overriding, less dense and tubulent ash (Sect. 2.3.5). The proportions of valley-escaped material increased dramatically thoughout the four successive flows. The critical values of the main geometric parameters controlling avulsion were quantified through high-resolution GPS surveys of the pre-event valley for each of the 14 June flows. Using this dataset, Lube et al. (2011) computed separate channel cross sections as a function of flow distance using six geomorphic parameters: cross-sectional area C, channel confinement $\Delta C/\Delta x$, base slope δ and sinuosity θ, valley width B and depth h (Fig. 6B). The key parameters controlling passive overflow are C, $\Delta C/\Delta x$, and θ, which allow the sites of flow avulsion to be determined.

Figure 6B illustrates the pre-event valley cross-sectional area against distance from source for BAF2 (Lube et al. 2011), in particular the longitudinal change in channel capacity $\Delta C/\Delta x$, where x is travel distance. The progressive infill of the Gendol valley by the subsequent flows BAF3 and BAF4 further decreased the channel's capacity. The mapped extent of the veneer facies (in yellow, Fig. 6A) and the critical area of ~1100 m^2 are in good agreement, thereby defining a threshold value for avulsion of flows with comparable volume fluxes. The data of $\Delta C/\Delta x$ against flow distance show three clear negative minima with capacity $C < 1100$ m^2. These occur immediately upstream of jumps 1, 2 and 3, coinciding with the only avulsion locations between 3.8 and 5.4 km. Channel sinuosity, hence potential for flow acceleration in outer bends, has also two maxima at jumps 1 and 3 (Fig. 6B). Hence, critical changes in channel capacity and sinuosity in valley cross sections can serve as conclusive geometric parameters for identifying potential sites of flow avulsion.

The strong correlation between the channel cross-sectional area and the occurrence as well as the magnitude of flow avulsion supports the mechanism attributed to the passive overflow of a strongly valley-confined current. Arguably, engineering techniques such as Sabo dams can be counterproductive in that they can increase the potential for flow avulsion, as witnessed by the case of Kaliadem at Merapi in 2006. The construction of PDC-proof bunkers is generally not recommended either, though they have occasionally been used as a last resort (e.g. at Kaliadem on the south flank of Merapi in 2006: Fig. 6A; Lube et al. 2011; Cole et al. 2015). The unexpected behaviour of PDCs during many explosive eruptions highlights the dangers of building special structures in an attempt to deflect or shelter people from them. Such mitigation measures may also have unforeseen consequences and show that there is no substitute for a wide and timely evacuation from areas at risk from volcanic flows.

Influence of high mass flux, infilling and channel morphology on overspill

The findings above have encouraged detailed studies of the channel geometry of volcanically active valleys in order to constrain potential future locations for major flow overspill. This has been recently demonstrated at Fuego volcano, Guatemala, where the catastrophic 3 June 2018 eruption flattened villages on the SE flank and killed about 430 people (Charbonnier et al. 2023; see contribution on PDC modelling, Chap. 11). This was an unusually paroxysmal event with much longer-runout PDCs than during previous activity, and its toll on the population and infrastructure was the greatest since the 1932 and 1974 eruptions (Charbonnier et al. 2023).

A map of the small-volume BAF deposits of the 3 June 2018 eruption, which totalled 50 million m^3 (Charbonnier et al. 2023), demonstrates that the flows and their overspill were largely controlled by the topography. The DEM-based map comprises (Fig. 7A): (1) valley-confined deposits in the ravines; (2) overbank deposits onto interfluves at several sites (black arrows) that were subsequently re-channelled into the adjacent tributaries; and (3) thin ash-cloud surge deposits that blanketed large areas on the medial to proximal flanks. Two main eruptive events were responsible for the high death toll from long runout PDCs that devastated the village of San Miguel de Los Lotes (Fig. 7B): (1) the generation of flows with high mass-fluxes from rapid retrogressive collapses of a large portion of the perched pyroclastic material accumulated in the old collapse structure of the SE flank; and (2) the rapid (~10 min) infilling of the distal Las Lajas channel section by the first two PDCs in the afternoon that led to major overspills at 8.2 km from the summit; the channel constriction and 90° bend both caused the subsequent flow pulses to deviate towards Los Lotes.

The recognition of impacts from overspill and unconfined flows should improve hazard assessment of PDCs around active composite cones and domes, in particular along the populated areas down valley. Hazard assessment requires modelling of the extent and effects of confined and unconfined flows around active volcanoes (e.g. Valentine and Wohletz 1989; Wadge et al. 1998; Sheridan et al. 2001; Baxter et al. 2005; Zuccaro et al. 2008; Gueugneau et al. 2020). Modelling PDC hazards should take the following into account: (a) as the succession of sustained PDCs significantly modifies the valley topography during emplacement and alters flow paths, a challenging task is to fast acquire repeated DEMs; (b) frequencies and scales of previous flows determined from pyroclastic successions are likely to be under-estimated because these currents commonly bypass volcano slopes, thus leaving no deposit record.

3 Geomorphic Impact of Indirect Volcanic Hazards

Volcanic hazards are termed indirect when they affect people, housing, transport lines and infrastructure as a consequence of interactions between an eruption and other elements such as the sea, lakes, snow and ice, rainfall, earthquakes, and crust deformation. Indirect hazardous processes affect a wider area than the usual perimeter of the direct effects of eruptive phenomena (Fig. 1A). Most of them do not cause casualties but their aftermath can induce severe impacts on human or animal health, and can significantly alter the climate or the ground surface. Ranked on the basis of impacted surface area and geomorphic effects, indirect hazards include debris avalanches (dealt with in Chap. 7), volcanic sediment-laden flows and other mass flows, volcanic tsunamis, jökulhlaups, and deformation of volcanoes.

3.1 Volcanic Sediment-Laden Flows

Water-rich mass flows (Chap. 7, Sect. 3) are quasi-permanent threats for populations living along the radial valleys as far as 100 km from active volcanoes (e.g. Nevado del Ruiz; Photo 108). After PDCs, these flows are the most destructive hazards associated with volcanic eruptions because they can occur with little or no warning, they travel longer distances downvalley than PDCs and spill far beyond river-channel confines (Pierson and Major 2018). As a result, people in distal areas generally do not expect these events or anticipate their destructive power (e.g. Mt. Rainier; Vallance 2000, 2005). Hydrological disasters have been reported around volcanoes that are ice-clad (Delgado et al. 2021) and also in areas of high rainfall where annual rainfall exceeds 2 m such as the equatorial and monsoon belt. Requiring only the sudden mixture of large amounts of water with abundant, loose and easily eroded debris on a volcano slope, water-rich flows can be formed in a variety of ways (Chap. 7, Sect. 3.1).

In addition to the two principal lahar categories of debris flows and hyperconcentrated flows (see Chap. 7), a combination of eruptive, glacial, and geomorphic processes can lead to the generation of primary sediment-laden flows around snow-capped volcanoes (see Fig. 17 in Chap. 3)

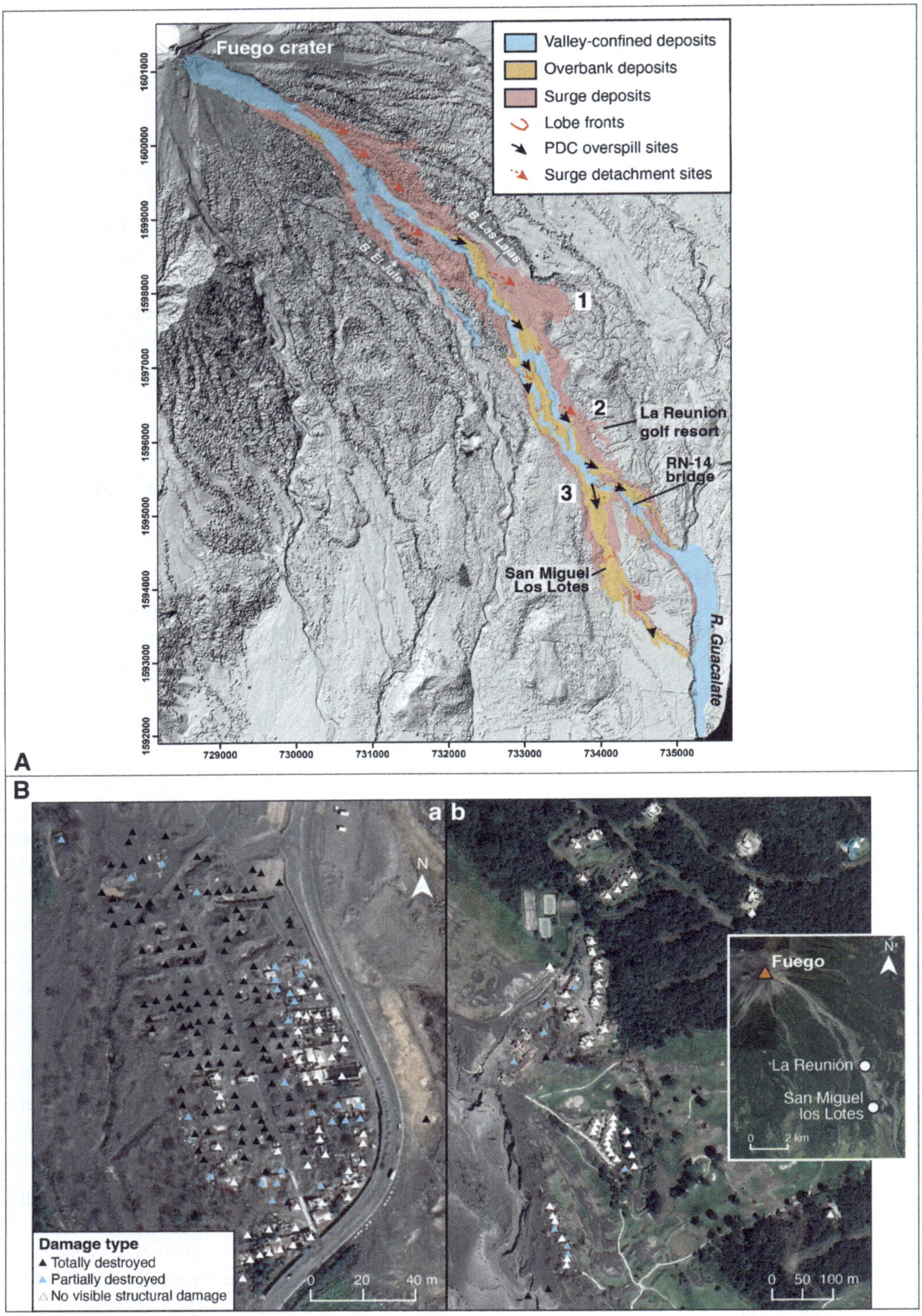
Fuego crater
Valley-confined deposits
Overbank deposits
Surge deposits
Lobe fronts
PDC overspill sites
Surge detachment sites
B. Las Lajas
B. El Jute
1
2
La Reunion golf resort
RN-14 bridge
3
San Miguel Los Lotes
R. Guacalate
1601000
1600000
1599000
1598000
1597000
1596000
1595000
1594000
1593000
1592000
729000
730000
731000
732000
733000
734000
735000
A
B
a
b
N
Damage type
Totally destroyed
Partially destroyed
No visible structural damage
0 20 40 m
Fuego
La Reunión
San Miguel los Lotes
0 2 km
0 50 100 m

◀ **Fig. 7** **A** Geological map of 3 June 2018 PDC deposits on the southeastern flank of Fuego volcano, Guatemala, superimposed onto a digital surface model. The main locations of interest and primary channels are also included. Numbers 1, 2 and 3 indicate the three main PDC overspill sites discussed in the text. The catastrophic PDC overbank event occurred in San Miguel de Los Lotes (Charbonnier et al. 2013, Fig. 3, p. 5). **B** Damage to buildings caused by PDCs in the area of San Miguel de Los Lotes (Lerner et al. 2022, Fig. 8, p. 14). Remotely mapped building damage for the two main flow overspill locations of **a** San Miguel Los Lotes, and **b** La Reunión golf resort, following the 2018 eruption. Google Earth satellite image from November 2018. Insert map shows locations on Fuego volcano

(Major and Newhall 1989; Pierson et al. 1990; Thouret 1990; Pierson 1995; Manville et al. 2000). These hybrid wet flows lie between conventional pyroclastic flows and lahars, and comprise: (1) mixed avalanches that are predominantly granular flows of snow and admixed volcanic debris (Pierson and Janda 1994; Thouret et al. 1995, 2007c; Photo 32, Chap. 3); (2) Other transient, mixed avalanches transformed to initial snow slurry lahars, such as at Ruapehu, New Zealand, in 1995 (Cronin et al. 1996, 1997). (3) Unusual ice diamicts comprising clasts of glacier ice and rock debris in a matrix of ice, snow and coarse ash, as reported after the 1989 and 1992 eruptions of Mt. Redoubt (Waitt et al. 1994) and Mt. Spurr in Alaska (Waitt 1995); These hybrid flows are hazardous because the reduced internal friction helps propagate them down valley beyond the reach of dry PDCs. Although geomorphically distinct, their deposits soon become extensively reworked and hard to recognize in the geological record.

While unsteady, pulsating lahars incorporate unconsolidated materials, increasing the total flow and peak flow in a process termed bulking, and enabling such flows to inundate inhabited valleys farther than the volcano base, the opposite can also happen. This is termed debulking, and is favoured by infiltration into the channel bed and changes in flow regime, which induces a decrease in concentration. This can lead to the development of more dilute flow pulses or tails which start downcutting the channel bed and can undercut structures such as bridges.

3.1.1 Geomorphic Actions and Damage Induced by Lahars

Lahars are more deadly and devastating than PDCs for several reasons: (1) They occur more frequently and over longer periods of time. (2) The rock fragments carried by lahars make them especially destructive; (3) Low friction and water content allow them to flow over gentle slopes (<2°) and inundate densely populated floodplains farther down valley. Despite this, lahar impacts have not been covered in detail in the literature except for the review of Pierson et al. (2014), which primarily dealt with strategies for their hazard reduction, and the inventory of Wilson et al. (2014) that looked at the effects of all volcanic phenomena on infrastructure and lifelines. The impacts on lifelines and infrastructure have not been fully studied, except for lahars around Merapi volcano (Jenkins et al. 2015; Hadmoko et al. 2018). Thus, we focus here on the geomorphic actions of lahars that induce physical impacts and damage.

Table 1 shows the principal geomorphic processes induced by lahars in drainage networks (after Thouret et al. 2020). All bury land and alter the river bed, bank morphology, and eventually valleys. Geomorphic changes that increase hazards within channels are: gradient steepening, headward wall retreat and incision in the steep initiation area; channel progradation and widening due to land burial, or entrenchment due to scouring, and enhanced channel curvature and sinuosity (Table 1) Geomorphic changes that induce hazards outside the channels are: land burial due to channel overspill and flow avulsion towards non-active tributaries, which can extend hazards to hitherto un-prepared people and un-exposed valuables.

A range of damage has been identified in populated areas along river channels draining volcano slopes and ring plains (e.g. Solikhin et al. 2015 and Jenkins et al. 2015 for lahars around Semeru and Merapi; Zanchetta et al. 2004 for non-eruptive lahars in Sarno, Italy). Essential for understanding the mechanisms

Table 1 Geomorphic impacts and changes due to lahars and excess sedimentation

Domain	Geomorphic unit	Acting process	Impacts		Geomorphic changes increasing hazards	
			Direct	Indirect	Mid-term	Long-term
River channel and valley	Bed	Burial scouring	Deposition, sedimentation	Tail cuts down into bed and deposit	Bed raising	Overspill made possible
	Channel and bends	Infilling readjustment	Favours overrun cut through	Debris entrainment	Downcutting	Gradient steepening
	Overtop banks and terraces	Bank steepening and collapse Terrace overtopping Damming	Pressure sterilizes land and habitat of natural life	Volume growth down valley destroy forest	Increasing width	Avulsion made possible
	Profile: transversal longitudinal	Enhanced alteration and steeper banks	Favours overspill accelerates flows	Landslides on adjacent slopes	Valley narrows or enlarged	Headward retreat Entranchment Progradation
Outside and beyond river valley	Terraces and area beyond edges	Overbank sediment supply damming river	Limited deposition Bank collapse	Excess sediment contamination Buries forest	Enlargement	Overspill and avulsion
	Tributary, independent drainage	Avulsion, sudden supply to river	Extensive aggradation	Unexpected burial	Re-activation	Coalescence of multiple valleys
	Volcaniclastic plains and fans	Overland flow Anastomosis Migration	Large valleys and extensive fans Solid charge	Creates and/or erodes	Extension of ring plains	Growth and/or degradation
Volcano foothills and ring plain	Volcano flanks	Erosion and deposition	Topographic changes	Destabilization	Erosion sedimentation	Degradation and aggradation
	Ring plain fans	Volcaniclastic sedimentation	Aggradation Progradation	Hydrological perturbations	Overbank and avulsion	Channel shifting

See Manville et al. (2009a, 2013) and Pierson and Major (2014) for a review of explosive eruption impacts on drainage basins, and Major (2023) for a review of volcaniclastic processes, deposits and hazards in the context of sedimentary systems (Di Capua et al. 2023)

that lead to damage, inventories of impacts have been carried out at the scale of a city, a neighbourhood or a valley using remote sensing techniques and field surveys following disasters (Kerle and Oppenheimer 2002; Ettinger et al. 2015). High-spatial resolution pre- and post-disaster images and survey data are used to detect post-disaster changes with coded damage levels that are automatically recognized through cognitive system engineering embedded in GIS (Kerle and Hoffman 2013). Remote sensing techniques for damage quantification have recently been improved by means of high-spatial resolution Sentinel satellites dedicated to monitoring changes to the Earth's land and sea. The European Copernicus programme helps make rapid and efficient post-catastrophe damage assessment as well as other disaster risk management procedures (Miyazaki et al. 2015).

The principal impacts from lahars that lead to damage are as follows (Thouret et al. 2020, 2022, 2023; see Table 7.5 in Chap. 7): (1) Hydrodynamic pressure increases logarithmically with velocity during an impact (Zanchetta et al. 2004). Pressure exerted on structures induces deformation and failure of walls, and roofs can collapse. (2) Hydrostatic pressure, exerted by flow depth and water content, and acting either normal or lateral to the obstacle and vertical upon burial, persists longer than hydrodynamic pressure (Mead et al. 2017). (3) Burial and subsequent induration last for longer with high-rate or recurrent deposition, and suppress soil infiltration. Accumulation of flow material may block doors and prevent evacuation of residents. (4) The residence time of the flow within the invaded structure eventually soaks and weakens material, in particular wooden structures. (5) Downcutting undercuts or scours buildings and bridge foundations. Removal of soil under edifices leads to deformation by edifice load, while flows can remove protruding house elements. (6) Buoyancy of the viscous and dense fluid/solid mixture can lift up edifices, which is usually followed by capillary rise. Subsequent dampness can soak and deform construction materials of low density. (7) Boulders act as missiles, inducing dynamic and erosion processes. Boulder impacts can damage walls, pillars and roofs, particularly after a series of shocks. (8) The temperature of hot, primary lahars weakens wood and fragile elements, scalds people, burns inflammable material and corrodes structures. The higher dynamic pressure of hot, viscous lahars likely leads to a greater amount of damage (Scott 1988; Vallance and Scott 1997; Bardou et al. 2007). (9) Non-physical processes, such as chemical or biological reactions from hot lahars, low pH water, spillover from industrial plants, all contribute to the weakening of structures and cause delayed deformation.

3.1.2 Geomorphic Characteristics of the Drainage System Can Explain Lahar Overspill

The complex phenomenon of hazardous lahar overspills involves many parameters, including flow depth, volume and solid concentration, as well as the topography and geometry of the flow-confining valley (Lavigne et al. 2000; Lube et al. 2011; Andrews and Manga 2011; Charbonnier et al. 2013; de Bélizal et al. 2013). Rapid changes in river gradient and channel geometry are critical in terms of hazard assessment for people living close to the river banks downstream.

Solikhin et al. (2015) used high-spatial resolution (HSR) imagery from GeoEye, Pléiades, IKONOS, and SPOT5 satellites to assess the extent and effects of the post-eruption lahars at Merapi. The study demonstrates the potential of HSR satellite imagery to delineate lahar deposits, and thus has improved hazard assessment for Merapi, a volcano whose summit and drainage network have changed more following the 2010 eruption than at any other time since 1931 (See Case study 3, Chap. 10). Due to monsoon rainfall that remobilised abundant pyroclastic material on the south and west flanks in 2011 and 2012, overbank lahars affected two valleys in the populated ring plain, devastating villages along the Putih River on the southeast flank and the Gendol River on the south flank (See Photo 92, Chap. 7). Long reaches of narrow channels were inundated by sediment between 16.5 and

19.5 km along the Gendol River course (profile shown in Fig. 4B). Post-eruption lahar disasters can be attributed to the low topographic valley gradient (0.04 m/m), high sinuosity accelerating flows across the <2° ring plain, and the limited capacity (200–300 m^2) of river channels.

Solikhin et al. (2015) investigated the relationships between the lahar overbank process and the morphology of the Gendol River channel (Figs. 4 and 8). They look at similar geomorphic channel characteristics to those used for PDCs: the channel cross-sectional capacity, the longitudinal rate of channel confinement, and sinuosity. Changes in these parameters were analysed along ten transects (Kh0–Kh9; Fig. 8A) at distances of 8 to 11 km from the summit, where November 2010 PDC deposits formed a large apron of overbank and veneer deposits in reach 3A (Fig. 4B). A clear correlation exists between the reduction in channel capacity and the width of areas characterized by overbank lahar deposits (Fig. 8A). The Kh3 cross-section exhibits the most negative value for channel confinement together with a relatively high value for channel sinuosity.

Computation of channel geomorphic parameters based on the HSR images and DEMs is also useful for anticipating the sites of future lahar overspills (Fig. 8B). Overbank is the most hazardous process acting along the lowermost river courses of Kali Putih toward the city of Magelang and along the Gendol-Opak River, which transports sediment toward the area of the iconic Prambanan temple farther down the Opak River. By applying morphometric indices analyzed along the Gendol River, Solikhin et al. (2015) found three spots near the Prambanan temples that are potential sites for future lahar jumps and overbank processes, where channel sinuosity exceeds 1 deg/m while its capacity is below 210 m^2 if the flow volume and velocity are equal to or greater than the 2011–2012 lahar events (Fig. 8B).

3.2 Assessment and Mitigation of Sediment-Laden Volcanic Flows

Assessment and prevention of sediment-laden flow hazards encompasses a number of undertakings such as mapping and modelling hazard zones, protection works, and mitigation procedures.

3.2.1 Delineation and Modelling

Maps of potentially impacted areas convey information to the exposed population, such as those published before the Nevado del Ruiz catastrophe in November 1985, represent good examples (Voight 1990; Thouret and Charbonnier 2022), while maps can also be used as tools for mitigation and land planning, such as those delineating potential lahars around Mt. Rainier (USA) (Iverson et al. 1998; Wood and Soulard 2009). A modern approach to hazard-zone delineation has emerged since the 2000s, following eruptions that had disastrous consequences, such as Soufrière Hills (1995–2010), Sinabung from 2010, Calbuco in 2015, Agung in 2017 and Fuego in 2018. It consists of pre- and post-eruption damage inventories, similar to the method developed for post-earthquake damage analysis (Jenkins et al. 2014, 2015). It requires the identification and statistical analysis of damage, followed by the simulation of potential footprints of lahars, fallout and PDCs. The resulting hazard-zone maps present a

Fig. 8 **A, a** 15 November 2010, GeoEye image of the upper middle course of the Gendol Valley near the village of Kepuharjo showing a series of cross-sectional profiles of Kh0–Kh9 (After Solikhin et al. 2015, Fig. 10, p. 18) **b** Example of pre-event channel cross sections at Kh1, Kh3, and Kh8 used to determine three morphometric indices: channel capacity (C *in* m^2; w, width of affected area, in m, longitudinal change in channel confinement ($\Delta C/\Delta x$ *in* m^2/m), and channel sinuosity ($\Delta\theta/\Delta x$ *in* deg/m). See Lube et al. (2011) for calculation technique. **B** Potential site of lahar overbank (after Solikhin et al. 2015, Fig. 12, p. 20). **a** Low-altitude photograph of the Opak-Gendol Valley near the Prambanan temple (25 km from Merapi summit) with the outlined profile of **b** the three morphometric indices: channel capacity (C *in* m^2), longitudinal change in channel confinement ($\Delta C/\Delta X$ *in* m^2/m), and channel sinuosity ($\Delta\theta/\Delta x$ *in* deg/m) ▶

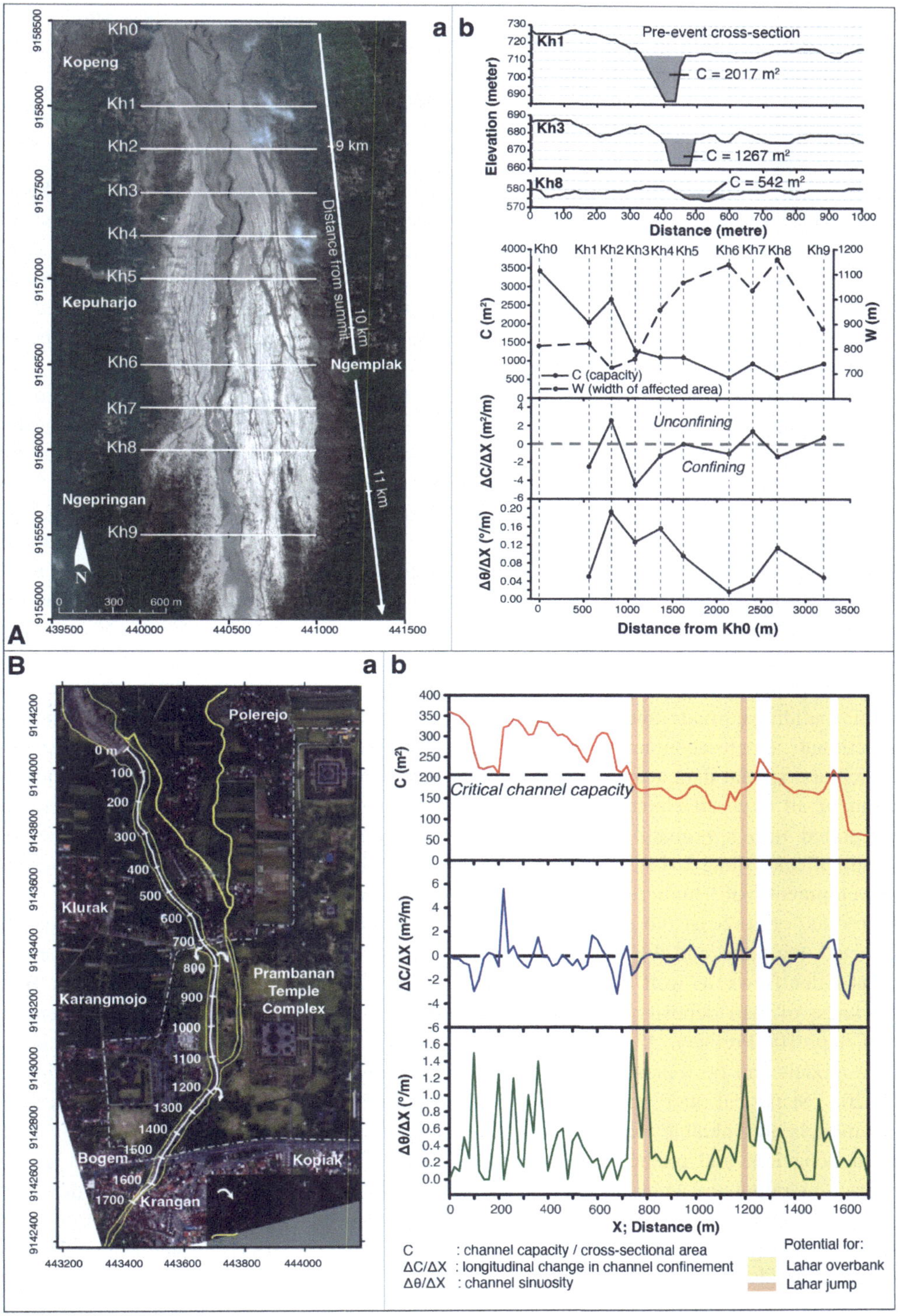
A
a
b
Kh0
Kh1
Kh2
Kh3
Kh4
Kh5
Kh6
Kh7
Kh8
Kh9
Kopeng
Kepuharjo
Ngemplak
Ngepringan
9 km
10 km
11 km
Distance from summit
0 300 600 m
N
Pre-event cross-section
C = 2017 m²
C = 1267 m²
C = 542 m²
Elevation (meter)
Distance (metre)
C (m²)
W (m)
C (capacity)
W (width of affected area)
ΔC/ΔX (m²/m)
Unconfining
Confining
Δθ/ΔX (°/m)
Distance from Kh0 (m)
B
a
b
Polerejo
Klurak
Karangmojo
Prambanan Temple Complex
Bogem
Kopiak
Krangan
Critical channel capacity
X; Distance (m)
C : channel capacity / cross-sectional area
ΔC/ΔX : longitudinal change in channel confinement
Δθ/ΔX : channel sinuosity
Potential for:
Lahar overbank
Lahar jump

distribution of the exposed elements based on a probable impact scenario, leading to the probabilistic analysis of potential damage (Thouret and Charbonnier 2022).

Delineation methods now exploit an array of tools and complementary information: DEMs, ground and space-based remote sensing, numerical codes, GIS and database of valuables, and modelling of loss in the event of a disaster. These methods have brought about the following advances:

(1) Flow simulation models that are very sensitive to the resolution and accuracy of DEMs. Suitable DEMs can be computed using stereo-photogrammetry of sub-metre resolution images from satellites, UAS and LiDAR surveys, which must be frequently updated in the same way as the DEMs used to monitor lava flow invasion (Sect. 2.2).
(2) Digital management of the varied information by means of a GIS or a geomatic platform. In order to exploit the hazard-zone maps in terms of exposure and risk, structural (buildings, roads and networks) and functional (socio-economic vulnerability indicators) spatial data are used. The evaluation of all exposed elements can then be integrated into a probabilistic risk matrix (Aspinall and Blong 2015).
(3) Development of numerical models to reproduce or forecast the area likely to be affected, the paths and the velocity of sediment-laden flows, as well as the extent and thickness of their deposits (See Case Study 4, Chap. 10; Procter). A review of available models can be found in Thouret et al. (2020). Statistical and physical 2D- and 3D-models are valuable tools to investigate lahar behaviour and predict the extent and impact in populated areas (e.g. Takahashi 2014; Turnbull et al. 2015). The most sophisticated models combine digital map-based methods of hazard-zone delineation with spatially distributed hydrologic and slope-stability programmes (e.g. Griswold and Iverson 2008). However, all models are complicated by debris flow rheology (Iverson 2014).

3.2.2 Protection Works and Challenging Mitigation Strategy

The construction of protection features has long been used for hazard mitigation. Sabo dams against lahars, developped in Japan (e.g. at Unzen, Photo 109), have now been applied to many other volcanoes around the world. They are designed to channel and break up the flows in the valleys by reducing their load.

Crater lakes on active volcanoes need monitoring as they favour the formation of lahars (Manville and Cronin 2007; Procter et al. 2010): two emblematic cases are Ruapehu in New Zealand and Kelud in Indonesia. At Ruapehu, the lake and its moraine dam are monitored by automatic cameras and instruments measuring the chemical and thermal properties of the lake, while the downstream infrastructure is protected. At Kelud, successive tunnels have been dug since 1919 to drain the lake (Suryo and Clarke 1985). This system has proven to be very effective, but it was necessary to clear or dig new tunnels regularly until 2014 as they were systematically blocked or damaged after each eruption. Since the 2014 eruption, the recent dome almost completely fills the crater formerly occupied by the lake.

Strategies to mitigate damage or loss from lahars, aptly summarized by Pierson et al. (2014), include four categories of tasks that can also be implemented for other sediment-laden flows: (1) avoidance of hazards through land-use planning, especially for high population concentrations, (2) modification of hazards through engineered protection structures; (3) establishment of warning systems to enable evacuations, and; (4) effective response to the flows when they do occur. Pierson et al. (2014) argued that risk-reduction efforts cannot be effectively accomplished without the active involvement of scientists working in partnership with local people, including elected officials and emergency managers.

3.3 Jökulhlaups

Jökulhlaups, or glacial outbursts in Icelandic, are impressive but much less frequent than lahars because interaction between ice caps and volcanic activity is limited to Iceland, Antarctica, and arctic North America. In these areas, many volcanoes lie beneath ice caps and, due to prolonged geothermal activity or to sudden explosive activity, large quantities of water escape from subglacial reservoirs. These glacial outbursts have been responsible for creating the extensive *sandur*, an outwash plain composed of glacial meltwater deposits, around the south Icelandic coastline (Fig. 9A). Jökulhlaups include water floods as well as hyperconcentrated flows which carry more sediment than normal floods. Their immense size was illustrated by the 1918 jökulhlaup generated by the Katla subglacial volcano and by the November 1996 and 2004 events resulting from Gjálp and Grimsvötn, two volcanoes beneath the Vatnajökull ice cap (Photo 110). Given the intense scientific activity before the Gjálp event, no casualties were incurred, but that was not the case in historical subglacial eruptions (Björnsson 2003).

3.3.1 Types of Jökulhlaups and Meltwater Scenarios for Hazard Assessment

Volcanogenic jökulhlaups are principally of two types (Gudmundsson 2015):

1. Small type 1 jökulhlaups come from subglacial lakes formed over geothermal areas by sustained melting at the base of the ice. They are minor events with a duration of hours to days, and a peak discharge of a few hundreds of m^3/s. Some long-lived reservoirs store 0.1–3 km^3 of water in active calderas beneath ice caps such as Grismsvötn, and drain periodically every few years, resulting in jökulhlaups with peak discharge of the order of 1000–10,000 m^3/s, but they usually do not cause any major damage (Björnsson 2003).
2. Large type 2 jökulhlaups result from direct melting of ice during volcanic eruptions, such as the Grimsvötn caldera release in 1996 and the Eyjafjallajökul composite cone in 2010 (Gudmundsson 2015). These floods usually have higher discharge and sediment load than the geothermal ones, as volcaniclastic debris is carried by the meltwater. Flows on the surface of glaciers can entrain large amounts of snow and firn, resulting in slurries of water, slush, tephra, and ice blocks. The total amount of water involved in jökulhlaups associated with eruptions is often much greater than for lahars. This is due to the very large energy transfer that occurs when a large fraction of the thermal energy of a subglacial eruption is dissipated extremely fast by ice melting (see Fig. 16B, Chap. 3; Gudmundsson 2015; Edwards et al. 2015).

The melting scenario is different for eruptions on tall composite cones with ice-clad summit. PDCs produced by a collapsing eruption column and by partial collapse of a lava dome can flow over glaciers, melting their surface rapidly through entrainment and mechanical melting (Chap. 3, Sect. 5.2.2).

The eruption of Eyjafjallajökull, Iceland, in 2010 is known for the unprecedented disruption to air traffic in Europe, but locally the principal concern in the first few days was the jökulhlaup hazard. Alluvial plains in its lowlands are formed by sediments from glacial rivers, and to a considerable extent by jökulhlaups since the end of the last glaciation (Gudmundsson 2015). The eruption of 14 April-22 May 2010 occurred in the 2.5 km-wide summit caldera covered by a ~200-m-thick ice cap (Chap. 3, Sect. 5.2), melting ice cauldrons in the glacier and leading to meltwater drainage mostly towards the north, down the steep 4-km-long Gigjökull outlet glacier (Magnússon et al. 2012; Oddsson et al. 2016). This was released in repeated jökulhlaups that inundated the 2- to 3-km wide and uninhabited valley of the Markarfljót River (Photo 111). The discharge was 2000–3000 m^3/s 20 km downstream from the outlet glacier. Large morphological changes occurred where the jökulhlaups reached the lowlands and impounded a glacial lagoon, which was filled by volcanic debris in the first 24 h. The mixture of

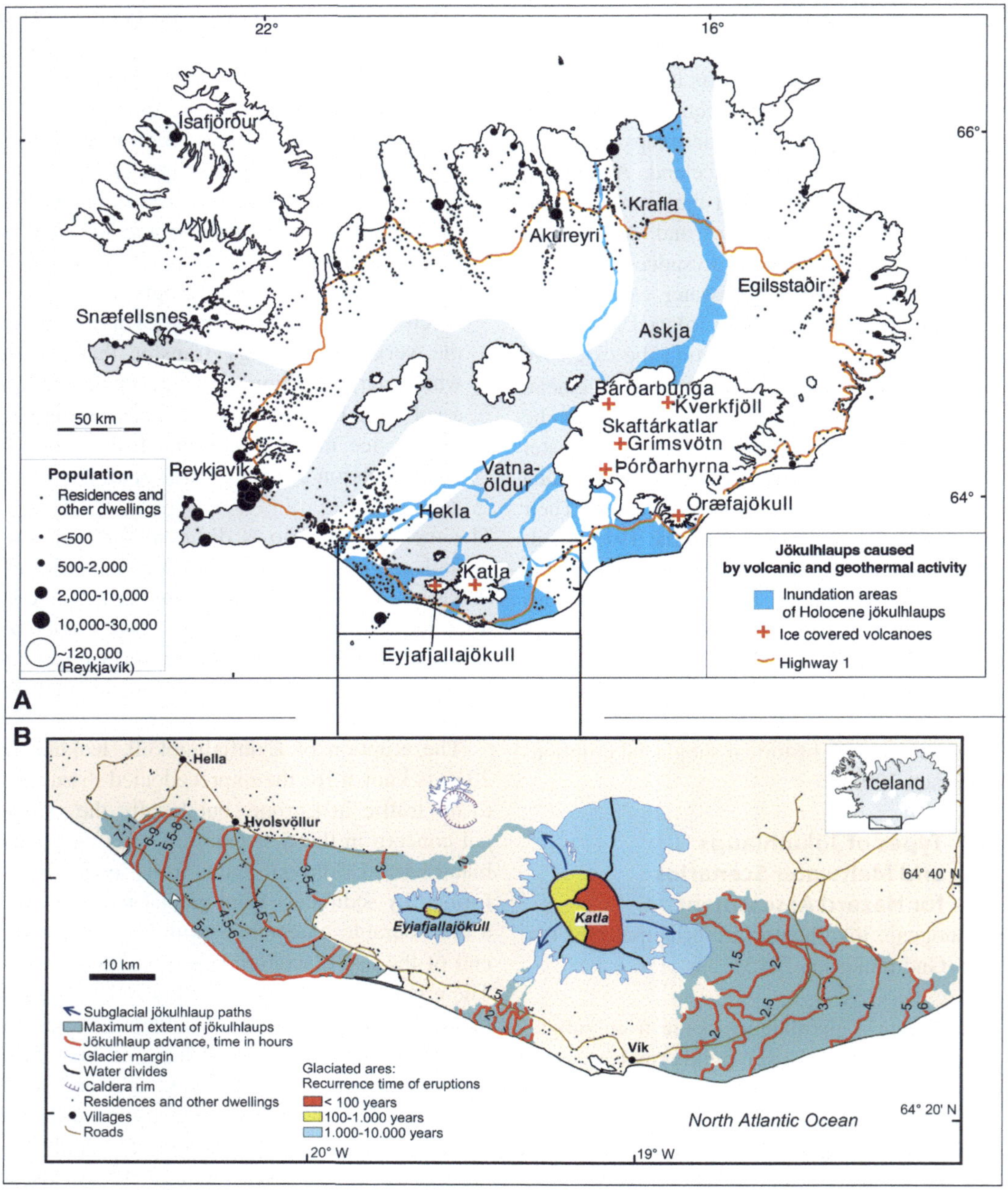

Fig. 9 **A** Areas affected by jökulhlaups (glacial outbursts) attributed to volcanic activity in Iceland during the Holocene. Over the last 1130 years the most severely and frequently impacted areas have been Myrdalssandur east of Katla and Skeidararsandur southeast of Vatnajökull. The box indicates the area shown in B. The volcanic zones are shaded grey (Courtesy of Gudmundsson et al. 2008). **B** The inundated areas and predicted propagation times for jökulhlaups triggered by eruptions within the ice-covered Katla caldera and on the Eyjafjallajökull composite cone on the southeast coast of Iceland (after Gudmundsson 2015, Fig. 56.8, p. 980). The colour of the glacier-covered areas (red, yellow and blue) indicates the recurrence times of eruptions. The expected maximum extent of jökulhlaups is based on flood modelling using the magnitude of the 1918 Katla eruption as a worst-case scenario (with peak discharge of 250,000–300,000 m^3/s). The extent of floods along all three expected pathways out of the Katla caldera are shown. The jökulhlaup advance time refers to the time in hours since the onset of a subglacial eruption (see also Larsen et al. 2021)

water, tephra, and ice blocks had the characteristics of a hyperconcentrated flow for the first few kilometres after it issued from the glacier, and debulking down valley led to a transition to water flow behaviour within 5 km from the former lagoon. The hazards were not dissipated at the end of the eruption. Large volumes of tephra, which had accumulated on the glacier on the southern slopes of the volcano, were remobilized in lahars by heavy rainfall in May. By 2012, two years after eruption, the situation had stabilized.

3.3.2 Mitigation of Hazards from Jökulhlaups in Iceland

The response to the Eyjafjallajökull eruption was considered to have been a success at a local level, due to previous hazard assessment including mapping, modelling of the hydraulic routing of jökulhlaup paths using reasonable worst-case scenarios, extensive response planning and trust built up between local people, officials, and scientists in the years prior to the eruption. Due to danger from a jökulhlaup, a large area with several hundred inhabitants was temporarily evacuated three times during the eruption, but luckily only very minor damage occurred to farmland or dwellings.

More effort has been put recently into volcanic hazard mitigation on Katla volcano in Iceland, a caldera and ice cap about 20 km east of Eyjafjallajökull on the south coast of Iceland (Fig. 9B). The potential for this volcano to generate major jökulhlaups is testified by its recent eruption history (Gudmundsson 2015; Larsen et al. 2021). Eruptions at Katla, on average about once every 50 years over the last millenium, have occurred within the caldera, which is 100 km^2 in area and filled with 400 to 700 m-thick ice (Fig. 9B, Gudmundsson et al. 2008). The jökulhlaup event following the 1 October 1918 eruption is considered to be probably the largest volcanogenic jökulhlaup on Earth: it had a peak discharge of about 300,000 m^3/s, while water carrying tephra and ice inundated an area of ~500 km^2 over the course of three hours (Tómasson 1996). The volcanic material was deposited mostly on the Mýrdalssandur plain and at the coast to the east and southeast of Katla volcano, but part of it was probably carried out by gravity currents into the sea. A large number of livestock perished, although there were no human fatalities. Nowadays, tourists increasingly not only cross but also camp in the sandur plain, unaware of the swiftness and short warning times of Katla eruptions and jökulhlaups (Gudmundsson 2015).

3.4 Effects, Hazards and Source Mechanisms of Volcanogenic Tsunamis

Tsunami (a Japanese word meaning 'great harbour wave') are deadly hazards mainly caused by submarine earthquakes unrelated to volcanic activity, but they may also be generated by large submarine eruptions or massive volcanic debris avalanches or PDCs that enter the ocean. Tsunami from volcanoes, also called 'volcanogenic tsunami' by Freundt et al. (2007) or 'volcanism-induced tsunami' by Nishimura (2020), are giant sea waves, so large that their movements extend downward to the ocean floor. The depth of the sea controls their speed: individual wave sets may travel through the open ocean as fast as 800 km/h. All of this enormous energy is concentrated near the shore, causing wavelengths to shorten and wave crests to rise. In inlets and bays additional wave funneling creates abnormally high waves that may run kilometres up coastal valleys as they dispense their energy before receding (Mastin and Witter 2000; Lockwood and Hazlett 2010; Borrero et al. 2020). The worst tsunami effects on coastal areas are due to flow depth and runup height and distance (Fig. 10A), which erode not only beaches but up to 0.8 km inland and up to 83 m above the sea level (Muhari et al. 2019; Borrero et al. 2020; Photo 112 A and B).

The lethal effect of volcanic tsunami is the unexpected transfer of energy from isolated volcanoes to sea waves, which can travel rapidly to densely populated, far-flung shorelines. Volcanic tsunami represent ~5% of all tsunami listed over the past four centuries (about

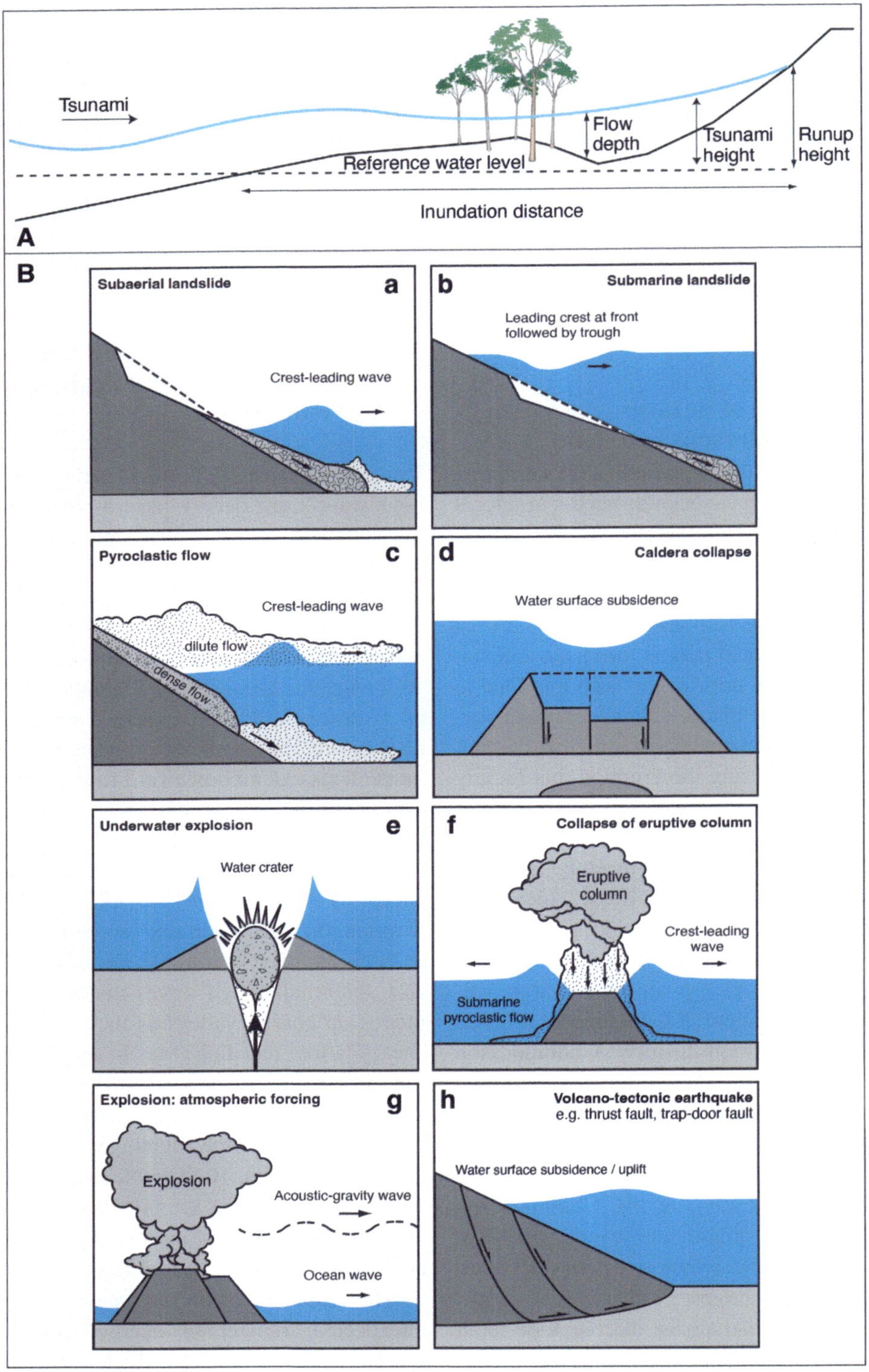

Fig. 10 **A** Profile of a coast with two features of tsunami effects: wave height and runup (from Borrero et al. 2020, Fig. 2, p. 2460). **B** Different source mechanisms for tsunami at volcanoes (Adapted from Paris et al. 2014a, b, 2022): underwater explosions, air shock wave generated by blast, pyroclastic flow, caldera collapse, flank failures. Other mechanisms such as lahars, collapse of lava benches or volcanic earthquakes are not shown (see Table 2 for a complete list of source mechanisms)

Table 2 Inferred source mechanisms of volcanic tsunamis (Paris et al. 2014a; Paris 2015)

Source mechanism	% of events	Source volume (km^3)	Volume flux (m^3/s)	Wave height (m) at the shoreline	Travel distance (km)
Underwater explosion	25	1	$<10^9$	<10	<200
Pyroclastic current	20	1–200	10^5–10^8	<30	<300
Earthquake	<20			<15	<500
Flank failure; dome collapse	15	1–500	10^5–10^6	<100?	<6000
Caldera subsidence	10	1–100	10^6–10^8	<20	<200
Shock wave	5			<3	>1000
Lahar	<5	<1	$<10^5$	<3	<10
Collapse of lava bench	<1	<0.01	$<10^6$	<2	<10

20 events per century), but they have exerted a high human toll. They have accounted for 20% of volcanic casualties, representing 55,300 deaths, caused by five major events (Auker et al. 2013): Krakatau in 1883 (Indonesia, 95% of the 36,500 victims were killed by tsunami), Unzen in 1792 (Japan, 75% of the 15,030 victims died due to tsunami), Ritter Island in 1888 (Papua New Guinea, between 500 and 3000 victims), Oshima-Oshima in 1741 (Japan, 2000 victims) and Iliwerung in 1979 (Indonesia, 580 victims). Yet, volcanic tsunamis are understudied compared to other volcanic hazards and are not sufficiently taken into consideration in volcanic risk assessment (Paris et al. 2022).

3.4.1 Multiple Sources and Decoupled Effects of Historical Tsunamis

The most calamitous volcanogenic tsunami in recorded history occurred during the eruption of Krakatau on 26–27 August 1883. Two large waves swept along the Sunda Strait shorelines, followed by a gigantic wave >15 m high that inundated the entire coastal strip on the northern coast of Java and southern coast of Sumatra. About 36,000 lives were lost at or close to the coast. Its debris washed up on shorelines as far as 4000 km from Krakatau for months afterwards. The tsunami probably originated by three mechanisms (Latter 1981; Schmincke 2004): (1) a large-scale collapse of the northern part of Krakatau Island as part of a caldera-forming process, (2) the discharge of about 12 km^3 of subaerially generated PDCs which were violently emplaced into the sea (Paris et al. 2014a), and (3) a major debris avalanche into the sea north of Krakatau, as shown by a hummocky submarine morphology.

Yokoyama (1981) argued the pyroclastic activity was one of the main causes and distinguished two processes. Because of density differences, the dense basal ash flow components sank and travelled along the seabed, displacing large amounts of water like a submarine landslide, while the lighter ash-cloud surge raced out across the surface, also pushing water ahead of it and generating large amounts of steam. A pyroclastic surge travelled at least 10 km laterally from the vent, but at Katimbang on the SE cape of Sumatra 40 km from Krakatau, a Dutch official and his family, who fled inland and took refuge in a mountain hut, were seriously burned by scalding ash gushing through openings in the floor of the shelter.

Such decoupled processes have been experimentally revealed by Bougouin et al. (2021): gas-fluidized granular flows entering a water body promote the generation of a granular jet over the water surface, creating a large leading wave, and a particle-driven gravity current underwater.

3.4.2 Tsunami Generated by a Variety of Mechanisms

Begét (2000) emphasizes that a tsunami is 'a high wave or surge of water produced by a variety of eruptive and non-eruptive processes at volcanoes'. In the historical record, 92 examples of volcanic tsunami can be attributed to eight causal mechanisms as shown by Table 2 and Fig. 10B (Lockwood and Hazlett 2010; Paris et al. 2014a, b, 2022; Paris 2015; Bougouin et al. 2021; Schindelé et al. 2024): earthquakes accompanying eruptions, pyroclastic flows, and submarine explosions each account for 20% of all volcanogenic tsunami. Caldera collapse is responsible for about 10%, avalanches of cold and hot materials about 14%, with lahars, air shock waves and lava avalanching being more minor.

During major explosive eruptions, it is sometimes difficult to identify the source of a tsunami, especially in the case of a Plinian eruption leading to the formation of a submarine caldera: see the numerous publications on the Krakatau eruption in 1883 or the Minoan eruption of Santorini (Paris et al. 2014b; Minoura et al. 2000; Karstens et al. 2023). Another difficulty can also stem from the recurrence of tsunamis during a single eruption: at least ten were observed on the west coast of Java during the Krakatau eruption between 26 and 27 August 1883 (Verbeek 1888).

An active volcano is considered to be potentially tsunamigenic and hazardous if it satisfies one or more of the following criteria (Paris et al. 2014a):

1. It is a steep-flanked composite cone with an elevated height H/distance D ratio of between 0.2 and 0.4. In addition to the H/D ratio, which is a major simplification, edifice volume, volume flux and density of material emplaced in previous eruptions should also be accounted for when defining the hazard at composite cones near populated coasts.
2. It belongs to a coastal or partly submerged complex of eruptive centres near populated coasts. A distinction can be made between caldera lakes (e.g. Taal), calderas open to the sea (e.g. Rabaul) and submerged calderas with emerged eruptive centres (e.g. Anak Krakatau).
3. It is a submarine edifice, especially if located in shallow waters and characterised by explosive activity. Many dormant volcanoes lie underwater, and their hazard is unknown.

Stromboli Island is one such tsunamigenic volcano, as several rockfalls and landslides, both submarine and subaerial, occurred in the Sciara del Foco in 2002 and again in 2019–2020 (Maramai et al. 2005; Di Traglia et al. 2020, 2022; see Case study 5, Chap. 10; Calvari et al.). Studying the tsunami impacts on the broad scale of the Tyrrhenian Sea, Rosi et al. (2019) bring to light that the present community of Stromboli, together with the communities now settling the southern coasts of the Tyrrhenian Sea, are actually exposed to a much higher tsunami hazard than previously thought. Repeated tsunami and ground-shaking led to the abandonment of the island during the Middle Ages, showing how repeated adverse natural phenomena can overcome the resilience of the small human community, forcing it to leave.

3.4.3 Prevention of Tsunami Hazards

Considering the diversity of source mechanisms, prevention of volcanic tsunamis through monitoring is challenging and must be coupled with policies of population preparedness (Paris 2015). Prevention comprises the following objectives: (1) tsunami hazard should be better accounted for in the monitoring system of active volcanoes; (2) numerical models should be improved, in particular for a complex trigger mechanism such as PDCs; (3) probabilistic methods of tsunami hazards in populated volcanic coastal areas should be developed; (4) people living around a potentially tsunamigenic volcano should be made more aware of the risk, because they often consider this hazard as being marginal.

Modelling volcanic tsunami scenarios to better assess and anticipate this potentially major hazard is a real challenge, as shown by the benchmark study by Esposti Ongaro et al. (2021) to test the relative sensitivity of numerical predictions to the 2002 landslide-generated tsunami event and wave models at Stromboli (see Case Study 5, Chap. 10). Multi-fluid models based on the Navier–Stokes equations can simulate complex tsunami sources and reproduce the interactions with coastal structures and inundation on land. Another type of modelling using VolcFlow was carried out by Giachetti et al. (2012) for a potential tsunami induced by an eruption of Anak Krakatau across the Sunda Straits, Indonesia (Fig. 11). The model results are quite similar to the wave heights and runup on shorelines reported after the 2018 eruption and flank collapse of this volcanic island, which triggered a tsunami. The erosional effects on coasts and forests (Photo 112A and B) have been mapped in detail by Muhari et al. (2019) and Borrero et al. (2020), with a maximum runup of 15 to 83 m and inundation distance of 330 to 800 m inland.

4 Post-eruption Geomorphic Responses and Hazards

The fact that volcanically-induced geomorphic processes continue eroding landforms and remobilising deposits directly after the eruption renders volcanoes very distinct. No other hazardous geological process than volcanic eruptions include such long-lasting impacts after the onset of an event. The effects of explosive eruptions on landscapes near composite volcanoes have been recognized from at least the early 1900s (e.g. Anderson and Flett 1903), but the magnitude of their geological and socio-economic significance has become better understood over the past few decades as a result of well-studied eruptions: Irazú, Costa Rica, 1963–1965; Usu, Japan, 1977–1978; Mt. St. Helens, USA, 1980; Nevado del Ruiz, Colombia, 1985; Mt. Pinatubo, Philippines, 1991; Unzen, Japan, 1991–1995; Mt. Merapi, Indonesia, 2006 and 2010; Soufriere Hills, Montserrat, West Indies,

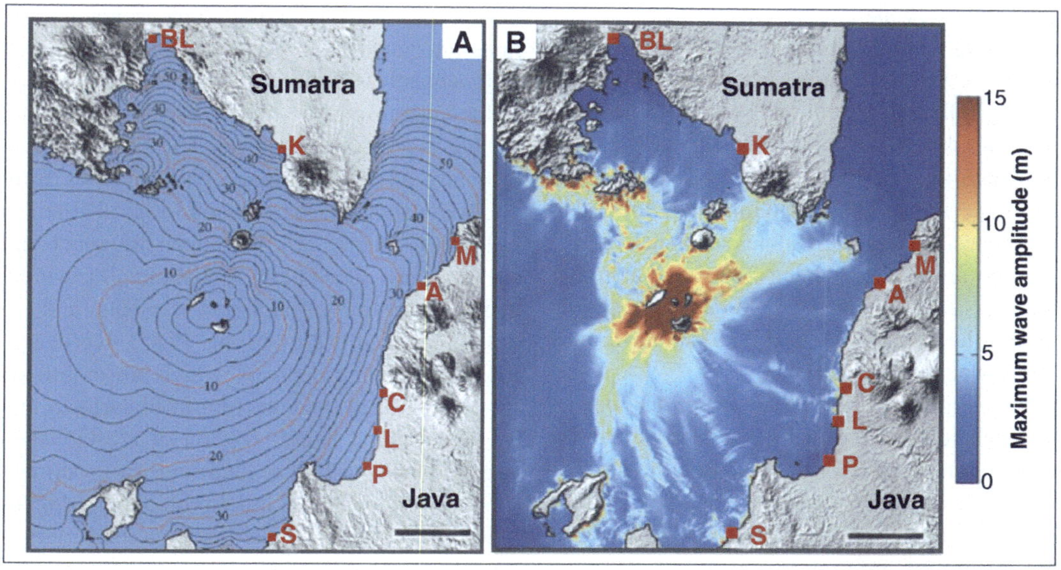

Fig. 11 Modelling of a potential tsunami triggered by an eruption of Anak Krakatau (Giachetti et al. 2012, *in:* Paris et al. 2014a, b, Courtesy of R. Paris). Numerical modelling of a hypothetical 0.28-km^3 failure of the Anak Krakatau cone. **A** Wave travel time in Sunda Strait and main coastal cities (BL Bandar Lampung, K Kalianda, M Merak, A Anyer, C Carita, L Labuhan, P Panimbang, S Sumur). **B** Maximum wave amplitude (m) recorded over 100 min of simulation, using a constant retarding stress of 10 kPa to simulate the landslide propagation (modified from Giachetti et al. 2012). Photos after Muhari et al. (2019) and Borrero et al. (2020) help compare the model results to the actual flow depth and runup induced by the 2018 tsunami (e.g. Photo 112)

1995–2013; Chaitén, Chile, 2008–2009; and Puyehue Cordón-Caulle, Chile, 2011–2012.

Explosive eruptions can affect individual or collective watersheds or drainage basins downwind and downstream of volcanoes, primarily by destroying vegetation and depositing large volumes of erodible debris over broad areas (Fig. 12). As a result, fluxes of water and sediment in affected drainage networks can increase dramatically. Geomorphic responses to larger fluxes generate significant volcano-hydrologic hazards long after eruption and commonly include large, frequent and widespread seasonal floods and high sediment yields (the rate at which sediment leaves a drainage basin; Pierson and Major 2014; Chap. 7). The complex geomorphic response has been modelled conceptually as a fluvial-system response with two phases of variable duration (Smith 1991; Figs. 13 and 14; Pierson and Major 2014): (1) an initial step of accelerated erosion (several hundred times greater than pre-eruption values) coinciding with the short syn-eruptive reaction to destabilization, followed by (2) an exponential decrease over a few years, which represents relaxation during a long inter-eruption period of adjustment and return to equilibrium. In general, the more severely a fluvial system has been disturbed, the longer it takes to reestablish metastable geomorphic conditions: several decades following minor disturbances and centuries to millennia following major ones. However, in some cases such as Mt. Pinatubo and Mt. St. Helens, elevated sediment yield persisted for decades afterwards, suggesting that the fluvial system did not stabilize until after this period (Fig. 14B).

Geomorphic responses to eruptions generate volcaniclastic sediment that can extend far downstream, persist for decades, and be hazardous for downstream communities. The vulnerability of populations and valuables arises from two hazardous effects of disturbed drainage basins (Pierson and Major 2014): (1) blockage of tributary streams by newly emplaced deposits and the subsequent impoundment of lakes by unstable natural dams, and (2) infilling of river valleys with excess sediment as channels aggrade, increasing the frequency and magnitude of floods.

4.1 Hydro-morphological and Sedimentological Concepts

Geomorphic and hydrologic concepts such as excess sedimentation, response of watershed and rivers, sediment yield and basin drainage recovery should be considered before we examine the range of geomorphic responses to volcanic eruptions. There are overviews on sedimentary impacts of eruptions on landscapes by Pierson and Major (2014) and on the impacts of volcaniclastic sediments by Manville et al. (2009a, b) and Major (2023; see Chap. 7).

4.1.1 Excess Sedimentation and Hydrologic Effects in Basins Affected by Volcanism

The amount of sediment exceeding that of a river in an equilibrium condition can be thought of as excess sedimentation. If a river system is loaded with more sediment than it can carry, its transport capacity is overcome, leading to storage of sediment along the channel and valley floor. Erosion of fresh and stored volcaniclastic material can then generate yield per unit-area sediment releases that rival those of the greatest sediment-transporting rivers worldwide (Chap. 7). Excess sediment delivered to rivers in volcanically affected drainage basins comes from two basic sources (Pierson and Major 2014): (1) hillslopes, where it is mobilized by sheet and rill erosion and shallow landsliding of widely dispersed tephra-fall and flow deposits; and (2) more localized valley-filling PDC, debris-avalanche, and lahar deposits, which are mobilized by sheet and rill erosion from infill surfaces and by incision and lateral erosion of reestablished stream channels. This second source of sediment produces a greater and more prolonged sedimentation response than the first, but both subsequently increase lahar frequency.

The hydrologic effects of eruption result from: (a) large-scale destruction of vegetation, which decreases evapotranspiration and interception of precipitation, and (b) the widespread deposition of fine volcanic ash, which reduces hillslope infiltration rates and increases the proportion of surface runoff (Fig. 14A). Increased

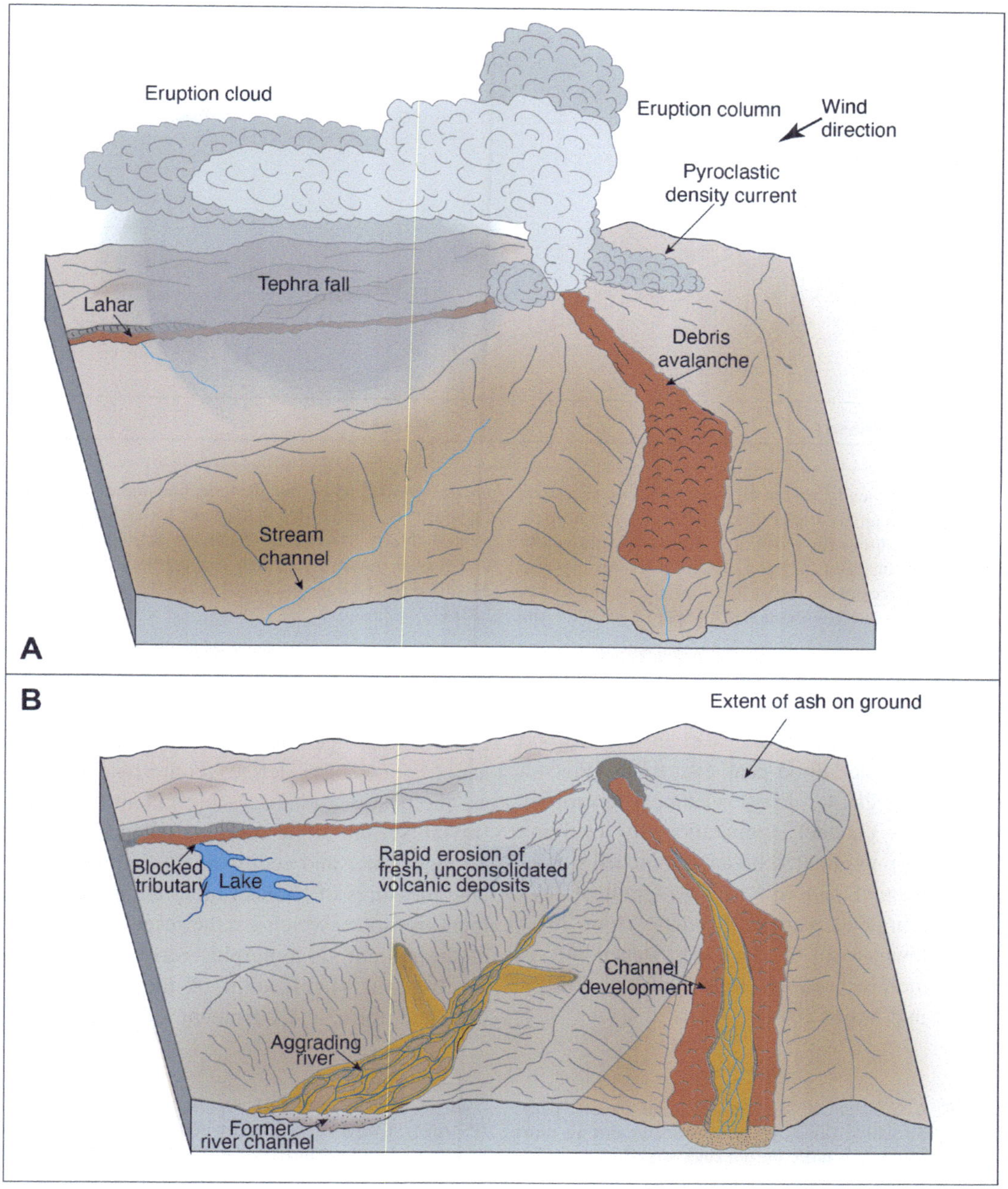

Fig. 12 Diagram illustrating explosive eruption processes and subsequent landscape responses (Pierson and Major 2014, Fig. 1 p. 471, Ann. Rev. Earth Planet. Sci., vol. 42 @2014 by Annual Reviews, httpp://www.annualreviews.org). **A** Volcanic processes from or related to an explosive eruption, including tephra fall, pyroclastic density currents, debris avalanches, and lahars. **B** Landscape changes and responses to disturbances caused by the eruption processes in A include rapid erosion of fresh tephra by rilling and landsliding, erosion of channel fill and development of new channel networks, channel aggradation and formation of braided channel patterns, and impoundment of channel-margin lakes

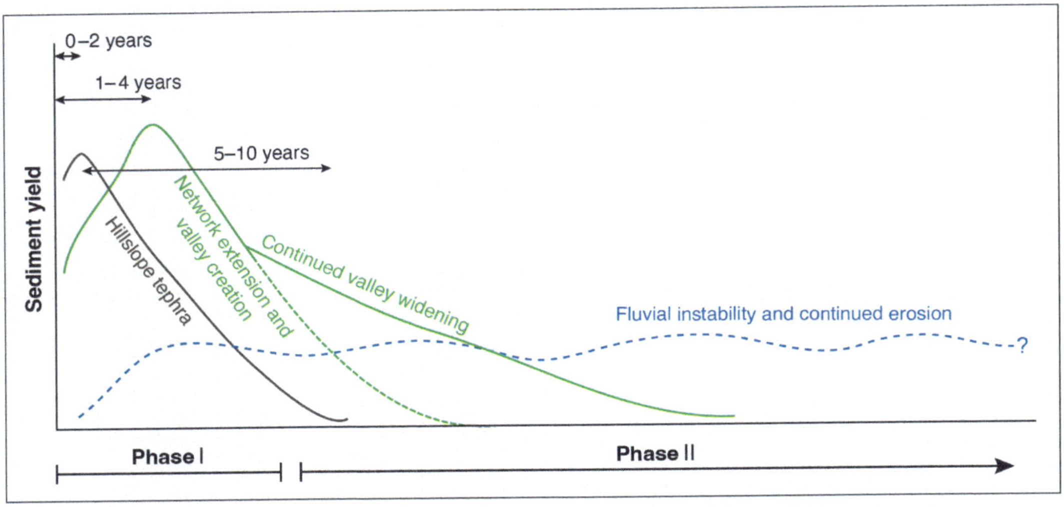

Fig. 13 Volcanically disturbed basin and its recovery. Conceptual diagram illustrating phases of geomorphic change and sources for sediment yield from volcanically disturbed basins, as a function of time. Phase I response is caused by erosion of tephra from hillslopes and development of channel networks. This phase releases the greatest peak yields, but they decline rapidly. Phase II response is caused by persistent lower-level erosion related chiefly to ongoing channel widening and continued fluvial instability, which can continue for decades. Modified after Gran et al. (2011) in: Pierson and Major (2014)

production rates and volumes of runoff on hillslopes, together with hydraulically smoother channels on valley floors, lead to larger flood peaks and faster flood-peak rise times (Pierson and Major 2014). River channels invariably respond to channel aggradation by evolving from a steep gradient, with an entrenched single-thread channel pattern to a low gradient, less confined braided pattern anywhere from 20 to 60 km downstream of volcanoes, depending on the geomorphic setting (Manville et al. 2009a; Pierson and Major 2014).

4.1.2 Drainage Basin and Geomorphic Recovery

From a hydro-geomorphic perspective, Pierson and Major (2014) define geomorphic recovery as 'the post-disturbance achievement by a basin fluvial system of a state of balance between average water and sediment fluxes, and adjustable measures of channel morphology and sediment transport, which are the response variables of the system'. Geomorphic recovery generally involves the stabilization of the channel pattern and form, establishment of an armoured

Fig. 14 **A** Conceptual diagrams of chief modes, pathways, and relative fluxes of water transfer during rainfall or snowmelt runoff, before and after volcanic disturbance. Width of arrows depicts relative magnitude of water transfer. In the post-eruption diagram, note the loss of canopy interception, the greater amount of precipitation reaching the ground surface (throughfall), reduced evapotranspiration, reduced ground surface infiltration, and prominent overland flow (after Pierson and Major 2014, Fig. 5 p. 482, reproduced with permission from the Ann. Rev. Earth Planet. Sci., vol. 42 @2014 by Annual Reviews, httpp://www.annual reviews.org). **B** Comparison of sediment yields from severely disturbed basins at Mt. St. Helens (USA) and Mt. Pinatubo (Philippines) as a function of time since eruption. The point in time labelled SRS represents the time that the US Army Corps of Engineers completed a large sediment retention structure on the North Fork Toutle River, Washington, which trapped both bed load and suspended load. Approximately 10 years after SRS completion, sediment began passing over the structure spillway. Mt. St. Helens sediment yields are based on suspended-sediment data only, whereas Mt. Pinatubo data come mainly from measurements of accumulated deposits and thus represent both suspended-load and bed-load sediment. Note the rapid decline of sediment yield within the first 5 to 10 years after each eruption. Modified from Major et al. (2000) and Gran et al. (2011), *in:* Pierson and Major (2014, Fig. 2, p. 368)

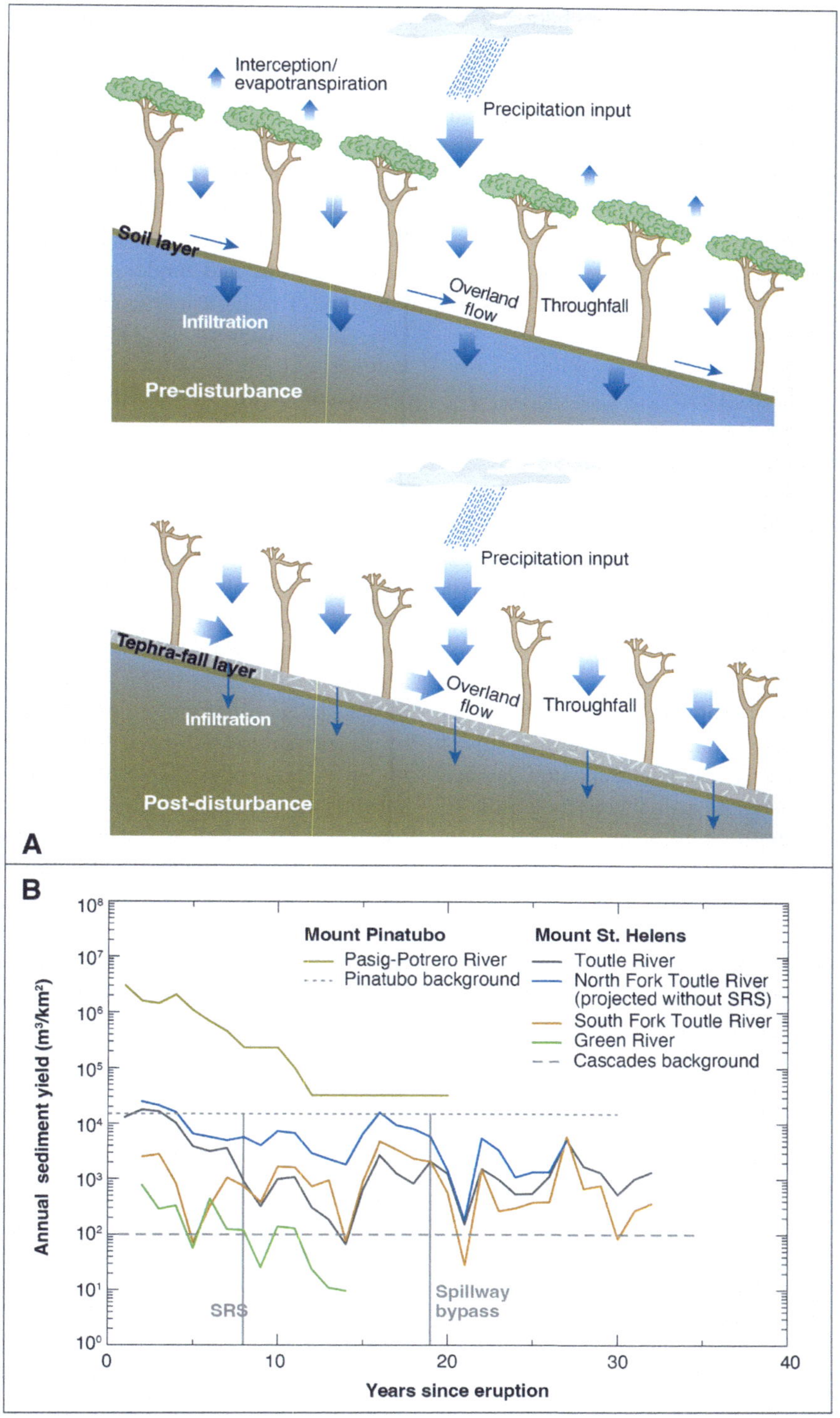
Interception/
evapotranspiration
Precipitation input
Soil layer
Overland flow
Throughfall
Infiltration
Pre-disturbance
Precipitation input
Tephra-fall layer
Overland flow
Throughfall
Infiltration
Post-disturbance
A
B
Mount Pinatubo
Pasig-Potrero River
Pinatubo background
Mount St. Helens
Toutle River
North Fork Toutle River
(projected without SRS)
South Fork Toutle River
Green River
Cascades background
Annual sediment yield (m³/km²)
SRS
Spillway
bypass
Years since eruption
0
10
20
30
40

river bed, and development of a characteristic hydraulic geometry. The added sediment to rivers causes channel conditions and transport processes to promote drainage adjustments in the channel pattern, form, position, and gradient. As a result, channel beds become smoother, wider, steeper, and straighter than their buried predecessors.

We distinguish three categories of geomorphic responses to large/small, infrequent or long-lasting eruptions.

4.2 Three Categories of Geomorphic Response to Volcanically-Disturbed Watersheds

Because 56% of eruptions last more than one month (Siebert et al. 2015), eruptive phases can trigger a remarkable chain of processes, which induce severe impacts several tens of km around an active volcano. The geomorphic effects of sequential eruptive events produce a diversity of cascading impacts on the landscape and damage to populated areas and valuables along valleys. Over the twentieth and twenty-first centuries alone, a number of iconic eruptions on different scales have taken place, which have caused dramatic changes to landforms, infrastructure and society: e.g. the 1980 eruption of Mt. St. Helens (USA), the 1991 eruption of Mt. Pinatubo (Philippines), and the 2008 Chaitén eruption (Chile). Case study 5, Chap. 10 highlights the rapid geomorphic evolution of the unstable Sciara del Fuoco flank between the 2002 and 2020–2022 eruptions of Stromboli.

Geomorphic response to volcanically-disturbed catchments can be broken down into three categories: (1) small sediment input from discrete events, tephra fall and small PDCs generated by moderate explosive eruptions (≤ VEI 3); (2) voluminous input from tephra fall, PDCs, lahars, and occasional debris avalanches caused by large (> VEI 3) but infrequent explosive eruptions; (3) moderate but steady input from tephra supplied by near-continuous eruptive activity.

4.2.1 Category 1: Short-Lasting Geomorphic Response to Moderate Eruptions

Moderate eruptions (≤ VEI 3) induce short-term responses and moderate hazards, as illustrated by five volcanoes: Sakurajima, Unzen, Usu, Mt. Yakedake in Japan and Merapi in Indonesia (Table 3; Suwa and Sumaryono 1996; Mizuyama et al. 1996; Suwa and Yamakoshi 1999; Okano et al. 2012), where erosion, lahar activity, and fluvial transport have been reported during their average moderate eruptive activity. The specific sediment discharge measured by these authors is within the range of published sediment yields of 10^4–10^6 t/km^2/year for catchment areas of 10^1 to 10^3 km^2 on similar volcanoes (Lavigne 2004; Thouret 2004; Gran and Montgomery 2005; Table 3; Fig. 15). In contrast to large eruptions, moderate ones give rise to a fast increase followed by a rapid decrease in lahar frequency. The erosion rate increased abruptly to 10^1 and 10^2 mm/year after the eruptions and decreased exponentially to 10^1 mm/year in less than 10 years at Merapi and to less than 10 mm/year at Yakedake in 32 years (Suwa et al. 2011). The rapid increase may be due to the accelerated runoff on the indurated top surface of pyroclastic deposits and bare slopes. The rapid decrease may be promoted by declining runoff due to fragmentation of soil crust, soil infiltration, vegetation recovery, and depletion of the reservoir of early remobilized material.

Geomorphic hazards following small to moderate eruptions are locally severe but limited in time, usually to around one year. Many studies of recent tephra erosion, such as Paricutín 1943–52 in Mexico (Segerstrom 1950; Inbar et al. 1994), Irazú 1963–65 in Costa Rica (Waldron 1967), and Usu 1977–82 in Japan (Chinen and Kadomura 1986), suggest that the erosion rate peaked soon after the tephra was emplaced and then declined

Table 3 Comparison of sediment discharge and hydro-geomorphological characteristics on three volcanoes with moderate eruptions

Geomorphic and hydrologic parameters	Yakedake Kamikamihori Creek	Merapi Bebeng River	Unzen Mizunashi River
Summit elevation (m asl)	2445	2911	1473
Relative height between summit and observation site (m)	865	2100	1392
Full length of the stream (m)	2500	14,000	7500
Mean stream gradient (°)	20.7	9.9	10.9
Stream gradient at the observation site (°)	7	2.8	2.7
Catchment area for the observation site (km^2)	0.83	5	12
Mean annual precipitation (mm)	2600	4500	3100
Time after last effective eruption (eruption year)	33 (1962)	11 (1984)	0 (1991–1994)
Permeability at the surface (cm/s)	0.7	5	23
Frequency of debris flow per year	0.5	20	210
Total bulk volume of sediment transported ($10^4 m^3$/yr)	0.6	4	17.5
Specific sediment yield ($\times 10^4$ m^3/km^2/year)	10^2–10^3	10^2–10^3	10^3–10^5
Annual depth of sediment yield (mm/year)	6	40	175

Note Mt Yakedake (Japan) is dormant, while Mt Merapi (Java) and Mt Unzen (Japan) are active volcanoes (after Suwa and Sumaryono, 1996 *in:* Thouret 2004)

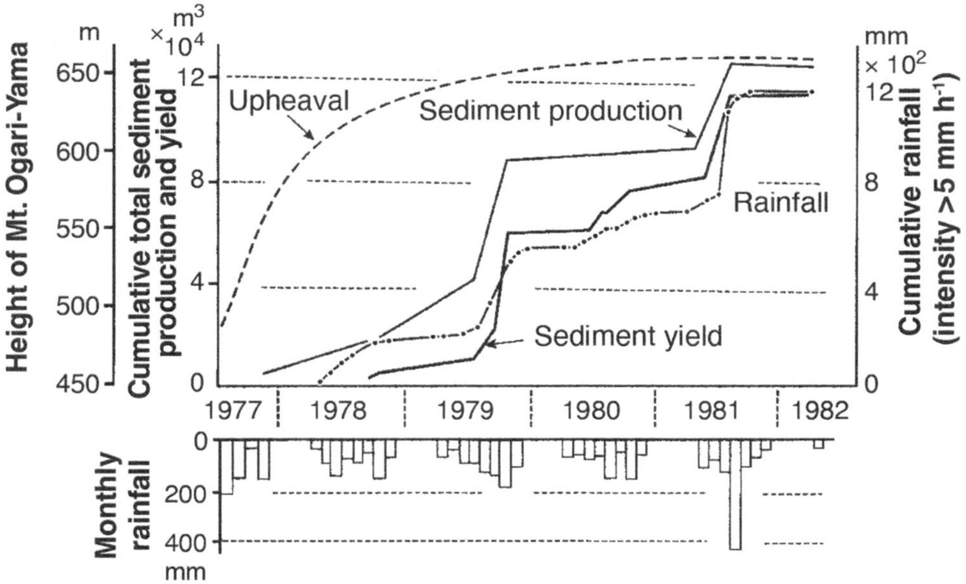

Fig. 15 Sediment production, sediment yield, rainfall, and upheaval of Mt Ogari-yama, August 1977–May 1982, Usu volcano, Hokkaido, Japan. Rainfall data from Nishiyama-gawa watershed gauge, at a distance of 1 km from the summit atrio. *Source* Chinen and Kadomura (1986), reproduced with the permission of Gebrüder Bornatraeger Verlagsbuchhandlung, Stuttgart, *in:* Thouret and Chester (2005)

rapidly, from 25–100 mm/year in the first two years to 1–5 mm/year within five years of the eruption. Sediment yields calculated at several composite cones following moderate eruptions range between 1.1 and 2.7×10^5 m^3/km^2 (e.g. Galunggung in 1982–83, Merapi in 1994–95 and Semeru in 2000; Unzen in 1991–93). The range of sediment yields compares well with data compiled by Major et al. (2000) for catchments affected by eruptions on composite volcanoes in a wide range of humid climatic conditions.

After deposition of tephra, geomorphic hazards tend to concentrate in valley channels whereas they decline rapidly on hillslopes and divides (Collins and Dunne 1986). Yields from basins affected solely by hillslope disturbance typically subside within several months or up to one year, whereas yields from basins that experience dominantly channel disturbance will likely remain elevated for as much as a few decades (Major et al. 2000). The decline on hillslopes is not caused by a recovery of vegetation, but by increased infiltration and decreased erodibility of the tephra layer, exposure of more permeable and less erodible substrates, and by the development of a stable rill network.

Brief geomorphic response to the Chaitén 2008 ashfall and rainfall

The immediate aftermath of the Chaitén eruption in the Southern Volcanic Zone of the Andes emphasizes that hydrologic hazards indirectly associated with eruptions can have more severe consequences than direct impacts (Pierson et al. 2013; Major et al. 2016). The brief eruption did not produce a large volume of deposits (~1.1 km^3; Pallister et al. 2013), but its effects were extensive in Chile and Argentina (Chap. 2, Sect. 3.5). Ashfall on 1 May 2008 was dispersed over >200,000 km^2, shutting down regional airports and resulting in the cancellation of several flights in both countries. Floods and lahars inundated the already evacuated town of Chaitén on 12 May; the economic loss was estimated to reach US$12 million for insured public buildings alone. Floods also disrupted the backbone road of continental Chile and the active aquiculture industry. More than US$36 million were allocated to provide social support for evacuated people (Lara 2010; Major and Lara 2013).

The 2008–2009 eruption of Chaitén involved a variety of hydro-geomorphic processes. Pierson et al. (2013) and Major et al. (2016) reported an acute sedimentation response to rainfall following the 10-day explosive phase at the start of the eruption, which blanketed the steep, rain-forest cloaked, 77-km^2 Chaitén River watershed with 3 to >100 cm of fine rhyolitic ash. Rain falling on this ash blanket generated a hyperconcentrated flow, followed closely by a long, complex muddy flood. Sediment mobilized in this lahar-flood event filled the Chaitén River channel, buried the town of Chaitén 10 km downstream of the volcano in up to 3 m of sediment, and caused the lower 3 km of the channel to avulse through the town (Photo 113).

The sedimentary response was not unprecedented, but its nature and rate were unusual in several ways (Pierson et al. 2013): (1) Nearly 70 percent of the aggradation in the Chaitén River channel was caused by a lahar, triggered by 20 mm of rainfall over a span of about 24 h. (2) Direct damage to the town was caused by the sediment-laden flood phase of the event, not the lahar phase. (3) The volume of sediment eroded from hillslopes and delivered to the Chaitén River channel was at least $3–8 \times 10^6$ m^3, i.e. roughly 15–40% of the minimum tephra volume that mantled the Chaitén River catchment. (4) The emplacement of thick and highly impermeable, very fine ash on the steep watershed slopes appears to have been responsible for the highly efficient rainfall runoff and the acute sedimentation response.

The eruption also damaged nearby forests (Swanson et al. 2013) because it involved a combination of: coarse and fine tephra fall over 50 km^2 and 480 km^2, respectively; a laterally directed blast which damaged forest over an area of 4 km^2; fluvial deposition of >100 cm of remobilized tephra which buried floodplain forest along two rivers; a variety of mass-movement processes, and PDCs. Although some effects of this eruption are very similar to those from the 1980 Mt. St. Helens one (Swanson et al. 1983), interactions between biota and volcano-geomorphic processes at Chaitén produced some unique forest damage: burial by tephra and sediment, canopy loading, heating, abrasion, and impact force leading to uprooting.

After only a few weeks, sediment production from the basin waned and incision of the channel fill began, probably owing to rapid stabilization of the ash mantle in response to continued heavy rainfall. The Chaitén eruption demonstrated that the time lag between basin disturbance due to ashfall and maximum level of aggradation by lahars and subsequent fluvial sedimentation can be very brief. In contrast, the tephra in basaltic regions is commonly coarse, and the post-eruptive sedimentary response is less marked, shorter term and less widespread.

Studies by Pierson et al. (2013) and Major et al. (2016) support previous findings that tephra fall can radically alter a catchment's hydrologic regime and sediment delivery, at least transiently. They emphasize two concepts that were probably not widely known by emergency managers: (1) hazardous mobilization of sediment can be triggered by relatively minor amounts of rainfall, and (2) deleterious impacts from rain-triggered lahars and syneruptive fluvial sedimentation can occur very soon after rain starts falling on fresh hydrophobic tephra, possibly faster than a community can be evacuated.

4.2.2 Category 2: Severe and Protracted Geomorphic Hazards in Catchments Disturbed by Large Eruptions

Prolonged excessive sediment transport after an eruption can cause environmental and socio-economic harm exceeding that caused directly by the eruption. The long-term instability of erupted debris means that effective mitigation measures must remain in action for decades. The largest sediment yields worldwide of from 10^3 to 10^6 t/km^2/yr^1, exceeding pre-eruption ones by several orders of magnitude, are transported in volcano catchments affected by debris flows in the aftermath of explosive eruptions of large magnitude (several km^3 of erupted products).

Geomorphic hazards from cascading effects of the 1980 Mount St. Helens eruption

The 1980 eruption of Mt. St. Helens represented a landmark case study of cascading effects of the 18 May climactic events and the post-18 May evolution (Rosenfeld 1996). Many effects of the eruption were anticipated, but others were not, hence illustrating some advantages and limitations of volcanic hazard assessments based on the past eruptive record alone.

The 18 May 1980 eruption included an unprecedented chain of sequential events (see Chap. 7): swarms of earthquakes, steam-blast eruptions, a massive rockslide avalanche that triggered a low-angle directed blast that devastated up to 600 km^2 on the north flank (Photo 114), a vertical eruption column showered thick tephra east-northeastward as far as 1500 km away from the vent, destructive syn-eruptive lahars and floods, pumice-rich pyroclastic flows, and dome building within the horsheshoe-shaped scar (Christiansen and Peterson 1981). The effects of the 18 May eruption were cataclysmic in some places near the volcano, especially in the North Fork Toutle River. About 60 people were killed or reported missing, mostly in the area devastated by the blast (Waitt 2014).

The multiple lessons gained from this eruption include: (1) Syn-eruptive lahars, fed by glacier ice meltwater (70% of the glacier volume was removed) and the remobilised debris avalanche, were one of the cascading effects of the climactic 18 May events. They radiated out from the composite cone into all the major drainages, filling stream channels with sediment (Janda et al. 1981). Farther downvalley, lahars and floods spread over valley floors, raised channel beds, and destroyed or damaged countless civil works, and eventually obstructed the deep-water navigation in the Columbia River (Tilling et al. 1990). (2) Rain-generated runoff, lahars and floods represent serious hazards at any time during the rainy seasons following an eruption, and they have lingering effects on basins. After 20 years, the average annual suspended sediment yield from the 1980 Mt. St. Helens debris-avalanche deposit remained 100 times (10^4 t/km^2) above typical background levels (10^2 t km^2), while average erosion rates, based on annual yields from valleys coated by lahar deposits, declined but remained about ten times (10^3 t/km^2) above background values for 20 years (Fig. 14B) (Major et al. 2000; Major 2004; Major and Mark 2006).

The chain of eruptive events was not limited to 1980. In September 2004, lava-dome building resumed in the horseshoe-shaped scar of the Mt. St. Helens cone and rapidly interacted with glaciers up to 2008. In addition to the emplacement of several types of dacitic extrusions and the subsequent shrinking of glaciers, detailed geological and geophysical observations contributed to the assessment of the geomorphic evolution of the former crater of the volcano. Abundant data have been gathered in the USGS Professional Paper by Sherrod et al. (2008) and subsequent papers (e.g. Walder et al. 2007; Dzurisin et al. 2015), while Gabrielli et al. (2020) carried out LiDAR-based geomorphological mapping of the crater, domes and glaciers between 2003 and 2009.

Widespread and protracted geomorphic impacts of the 1991 Pinatubo eruption

The geomorphic response to large eruptions (VEI 6) had not been documented prior to the study of the Mt. Pinatubo lahars. Studies suggest that its impacts on landscape exceeded by at least one order of magnitude those of Mt. St. Helens (Wolfe and Hoblitt 1996). One NASA image depicts the extraordinary extent of PDC and lahar deposits around Pinatubo in the lowlands of Luzon Island (Photo 115).

The volcanic unrest and eruption evolved through eight successive phases over eighteen months. Steam explosions announced the initial unrest at Mt. Pinatubo on 2 April 1991. This culminated 10 weeks later in the world's second largest eruption of the twentieth century (Wolfe and Hoblitt 1996). The 15 June climactic eruption over approximately 9 h included the

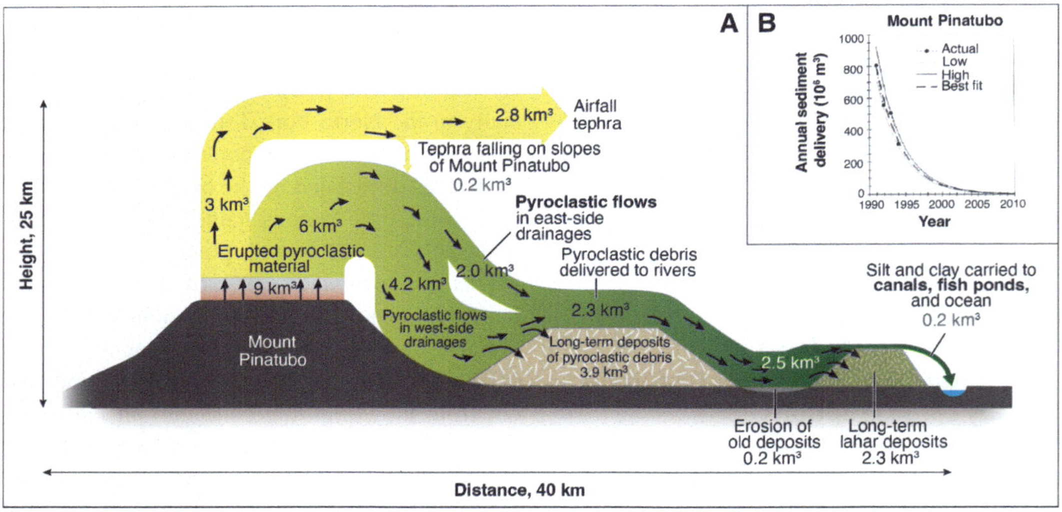

Fig. 16 Sediment-routing flow diagram at Mount Pinatubo. **A** A sediment-routing flow diagram for drainage basins on and near Mount Pinatubo (Philippines). Most of the sediment went into storage in these basins in the first year following the 1991 eruption. Lahars were the dominant sediment-transporting process immediately after the eruption (After Pierson et al. 1992, Fig. 6, p. 23, U.S. Geol. Survey Water Res. Investig. Rep. 92-4039; see also; Pierson and Major 2014). **B** Diagram of lahar frequency and volume 1991–2004: actual volume, observed and predicted (After PHIVOLCS-DOST-IAVCEI 1995, Primer and field guide, 35 pp. Angeles City, Philippines, October 17–21, 1995)

collapse of the volcano's summit, producing a 2.5 km-wide caldera. The July–October 1992 final phase led to the extrusion of a 4×10^6 m^3 dome in the summit caldera. Voluminous PDCs, emplaced during the climactic eruption simultaneously with the Plinian fall, extended as far as 12 to 16 km from the vent down all sectors of the mountain and impacted an area of approximately 400 km^2, dramatically modifying the pre-existing topography (Photo 116). Massive pumice-rich flows filled pre-existing valleys to depths as much as 200 m locally. The climactic eruption also injected approximately 17 Mt of SO_2 into the atmosphere, generating atmospheric and climatic effects that persisted globally for several years (Wolfe and Hoblitt 1996). A total bulk volume of 8.4 to 10.4 km^3 of erupted products (3.7 to 5.3 km^3 of DRE magma) includes 5 to 6 km^3 of PDC and 3.4 to 4.4 km^3 of tephra-fall deposits. This is an order of magnitude greater than the volume of magma erupted in 1980 from Mt. St. Helens, but is smaller than the 13 ± 3 km^3 (DRE) of deposits of the 1912 eruption of Novarupta, Alaska (Hildreth and Fiernstein 2012).

Scientists from the Philippine Institute of Volcanology and Seismology and the US Geological Survey successfully forecasted the eruptive events and their effects, enabling Philippine civil leaders to organize massive evacuations that saved thousand of lives and the Clark Air Base of the US Air Force. Nevertheless, the climactic eruption, coinciding with typhoon Yunya on 15 June, caused almost 300 deaths and extensive property damage, owing to an extraordinarily broad distribution of heavy, wet tephra-fall deposits. A rain-saturated, blanket of tephra fell over about 7500 km^2 of central and western Luzon, and towards the South China Sea (Wolfe and Hoblitt 1996). The weight of the wet tephra caused numerous roofs to collapse in the communities around the volcano.

Rain produced by typhoon Yunya triggered lahars on 15 June in all major drainages around the volcano. The most significant damage was caused by lahars along the Abacan River, with buildings and all bridges in Angeles City being undercut and destroyed by lahar bank erosion (Wolfe and Hoblitt 1996). Lahars left more than 50,000 people homeless, affected more than 1,350,000 people, and caused deaths as well as enormous property losses and social disruption (PHIVOLCS-IAVCEI-DOST 1995; Janda et al. 1996). In 1991, as much as 5 to 6 km^3 of loose PDC deposits in the heads of valleys draining Mt. Pinatubo and ~0.2 km^3 of tephra on its flanks became the primary source of sediment for 2.5 km^3 lahars. Pierson et al. (1992, 1996) detailed the sedimentary budget all the way from the erupted products to lahars and floods that eventually entered the sea (Fig. 16a). After the 1991 eruption, sediment yields set world records during the first three post-eruption years in rivers due to basin-wide sediment loading: 10^6 $m^3/km^2/yr$ in 1991, nearly an order of magnitude greater than the maximum sediment yield computed after the 18 May 1980 eruption of Mt. St. Helens (Pierson et al. 1992).

The prodigious sediment yield from Pinatubo had geomorphic after-effects, such as watershed disruption, channel piracy and avulsion, and blockage of tributaries which then impounded temporary lakes. Floods triggered by breaching of these lakes provided an additional hazard. Unlike rain-induced lahars, breakout floods can occur in the absence of rainfall, and limit the possibility of warning threatened areas. Although the areas affected by lahars had expanded, the frequency of events had decreased and the number of impacted river systems had dwindled to four in 1995, as source materials were depleted. Sediment yields in some basins remained orders of magnitude above the pre-eruption levels even after the major lahars ended in 2000 (Scott et al. 1996; Montgomery et al. 1999; Hayes et al. 2002; Gran and Montgomery 2005).

The Pinatubo case provides at least three significant lessons in terms of hazards (Major et al. 1996): (1) heavy rainfall alone was not responsible for generating the lahars in 1991; the additional necessary condition was altered hydrology from fresh tephra mantling the various catchments; (2) geomorphic processes due to secondary explosions in PDC deposits and channel piracy affecting catchments and river channels play a significant role in the erosion and redistribution of sediments. These fostered more lahars in the ensuing years and more flooding beyond the alluvial fans onto the densely populated plains; and (3) after the geomorphic impact of the devastating 1991 lahars, the subsequent flows triggered by the seasonal monsoon rains and other geomorphic processes had far greater social and economic impacts. Mt. Pinatubo lahars have caused severe socio-economic and institutional disruption (Rodolfo 1995; Leone and Gaillard 1999). The dominant problem was increasing (1991–1994) and slowly decreasing channel aggradation until at least 2000 (Fig. 16b; PHIVOLCS-IAVCEI-DOST 1995; Umbal 1997), which promoted lateral migration of the flows across the low-gradient alluvial fans surrounding the volcano. One NASA image portrays the evolution of the Pasig-Potrero river system and lahars from 1991 to 1996 east of Mt. Pinatubo (Photo 117). Despite the resulting progressive burial of villages and valuable agricultural land, loss of life was minimized by early warning systems.

4.2.3 Category 3: Geomorphic Response to Persistent Explosive Activity

The type and chronology of geomorphic response to near-continuous eruptive activity has been poorly evaluated, which motivates our study of persistently active volcanoes. Around these volcanoes, constant variations in sediment flux supply relatively small but steady amounts of sediment to rivers, so that their cumulative effects may be greater than from episodic large eruptions. This study of the geomorphic response to near continuous volcanic activity was driven by two observations:

(1) At persistently explosive volcanoes such as Semeru, Sakurajima and Santiaguito, studies on production of sediment flux have focused only on short-term variation patterns in sedimentation (Lavigne 2004; Lavigne and Thouret 2002). Lavigne (2004) and Lavigne and Suwa (2004) used two different methods to estimate short-term sediment yield dominated by rain-triggered lahars at Merapi and Semeru, fed by PDC deposits, dome rockfalls, and material removed from the riverbed and banks. Lavigne (2004) calculated a sediment yield in the range of $1{\cdot}5 \times 10^5$ m^3/km^2/year for one catchment at both active volcanoes.

(2) The impact of near-continuous tephra supply and runoff on lahar activity and channel response is not well understood. The declining sediment flux and fluvial adjustment were examined after the 1902 catastrophic eruption of Santa María Volcano in Guatemala (Kuenzi et al. 1979; Vessel and Davies 1981), but further analysis of the fluvial response to steady activity is required. Correlations between supplies of debris from constant dome extrusion, rainfall, and lahar activities and the form and behaviour of fluvial systems have only been established for Santa Maria-Santiaguito (Harris et al. 2006). The authors used a time series of 21 Landsat Thematic Mapper images from 1987 to 2000 to map, measure, and track temporal and spatial variations in the area of lahar impact and river aggradation. Two zones of the fluvial system have been tracked over 13 years: in the proximal zone, a positive correlation has been found between extrusion rate at domes, aggradation area 12 months after the eruption, and rainfall during the intervening 12 months. In the medial zone, the correlation between extrusion rate and/or area of aggradation is negative due to removal of material in valleys. From this study, the authors distinguished two types of volcanic fluvial regimes based on transient and persistent variations in sediment flux.

Geomorphic aggradation and degradation at persistent volcanoes: Semeru volcano

To study the geomorphic response of watersheds to near-continuous supply of sediment, we chose Semeru volcano, Java, which has experienced daily eruptive activity since at least 1967 (Figs. 17 and 18; Solikhin et al. 2012; Thouret et al.

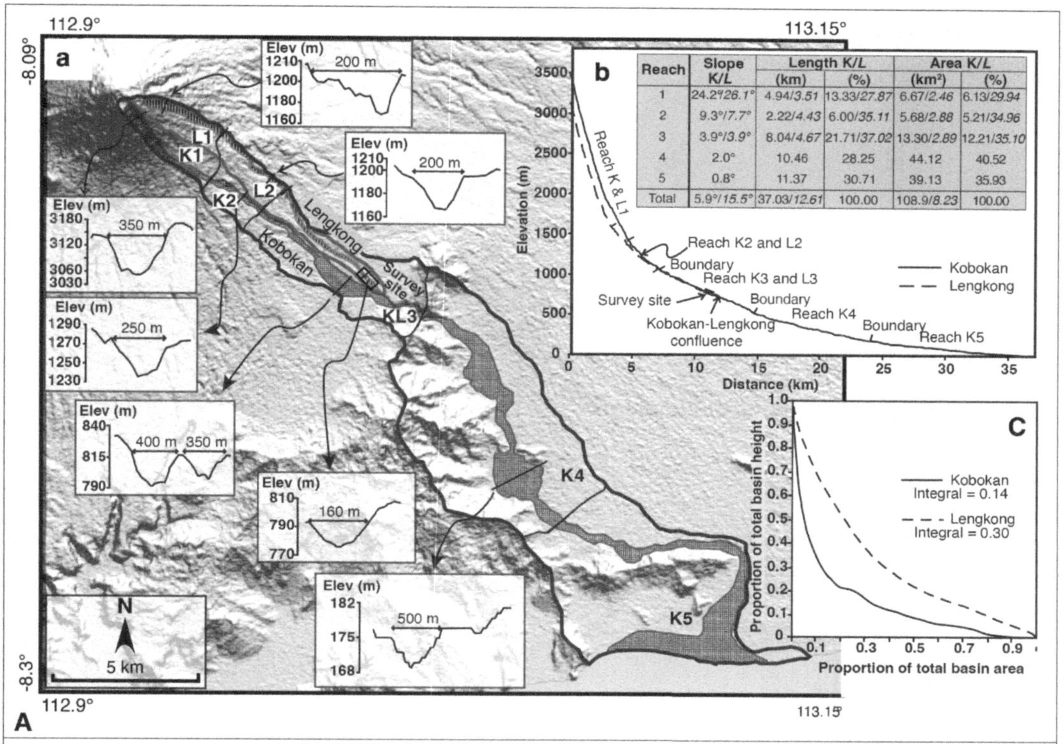

B

Morphometric parameter	Kobokan	Lengkong
Area (A)	108.9 km^2	8.23 km^2
Height of basin (H bv)	3635 m	3610 – 725 = 2885 m
Height of river (h rv)	3472 m	3142 – 725 = 2417 m
Basin length (L b)	33653.6 m	11421.40 m
Perimeter (P)	72.91 km	24.09 km
Relief ratio (R*h*= H/Ln)	0.108	0.253
Relative relief (R*hp*= H/p)	0.05	0.12
Relative basin height (v= h/H)	0.955	0.838
Relative basin area (x= a/A)	*1/3 of E and SE catchments*	0.076
Length area (L=1.4 A$^{0.6}$)	23.35 km	4.96 km
Basin shape (R*f*= A/Lb^2)	0.1	0.06
Basin area ratio (R*a*= An/An$_{-1}$)	13.23	
ΣL	~ 290 km	37.02 km
Drainage density (D= ΣL/A)	2.66	4.49
N	~120	18
Stream frequency (F*s*= N/A)	1.10	2.18
Average stream length avL=avL$_1$R$_L$	37032.42 m	12612.43 m
Length ratio (R$_L$= avL$_n$/L$_{n+1}$)		0.341

Fig. 17 **A** Two selected watersheds, Koboan and Curah Lengkong on the southwest flank of Semeru volcano. Reaches, geomorphic and hydrologic characteristics of the river channels (Thouret et al. 2014). See Google Earth map of the watersheds on the southeast flank (Photo 118). **B** Hydro-geomorphic characteristics of the two selected catchments, Koboan River and Curah Lengkong River, southeast flank of Semeru volcano (Thouret et al. 2014, Fig. 4, p. 7 and Table 2, p. 8)

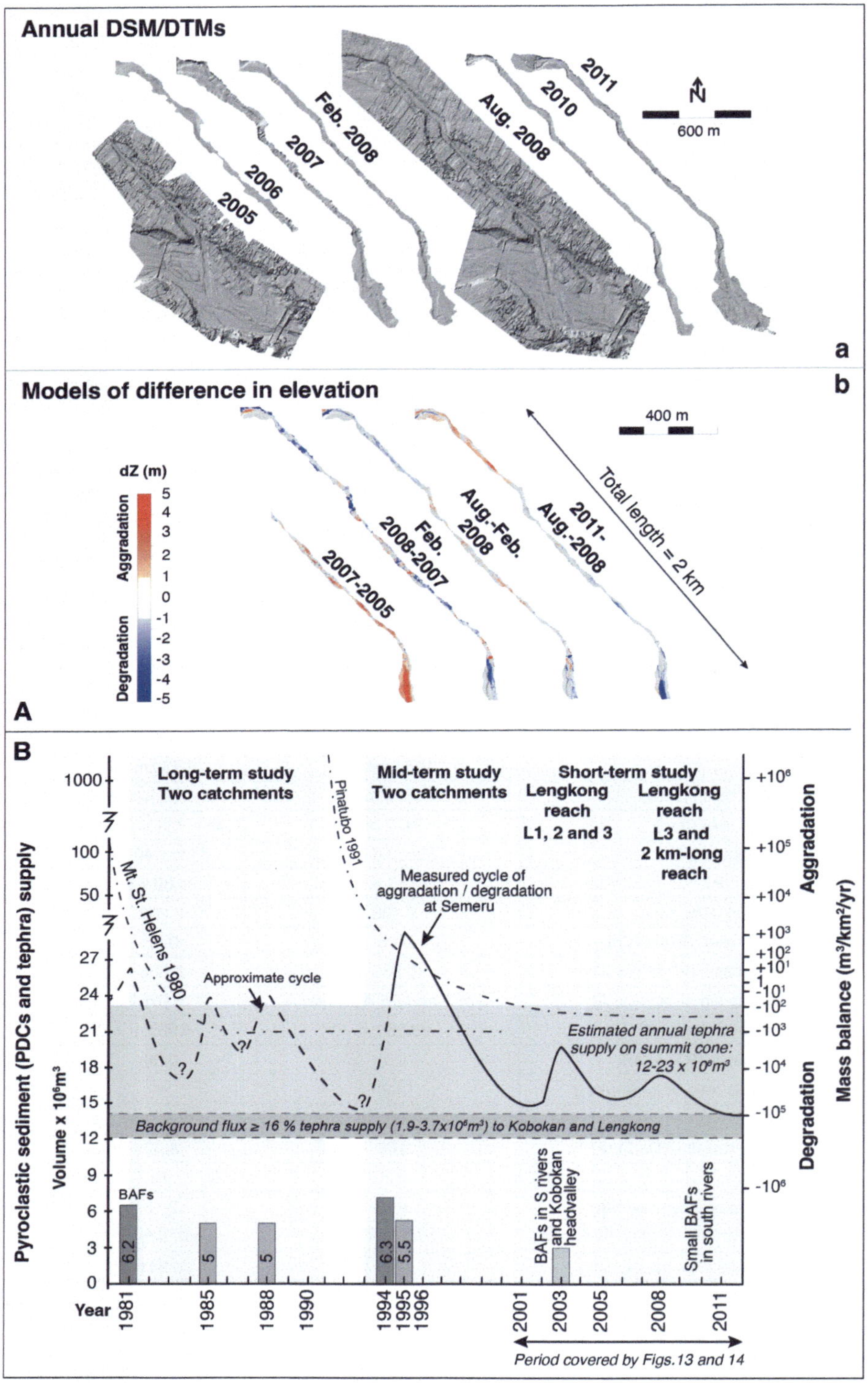
Annual DSM/DTMs
2005
2006
2007
Feb. 2008
Aug. 2008
2010
2011
N
600 m
a
Models of difference in elevation
b
400 m
dZ (m)
Aggradation
Degradation
5
4
3
2
1
0
-1
-2
-3
-4
-5
2007-2005
Feb. 2008-2007
Aug.-Feb. 2008
2011-Aug.-2008
Total length = 2 km
A
B
Long-term study
Two catchments
Mid-term study
Two catchments
Short-term study
Lengkong reach L1, 2 and 3
Lengkong reach L3 and 2 km-long reach
Pinatubo 1991
Mt. St. Helens 1980
Measured cycle of aggradation / degradation at Semeru
Approximate cycle
Estimated annual tephra supply on summit cone: 12-23 x 10⁶m³
Background flux ≥ 16 % tephra supply (1.9-3.7x10⁶m³) to Kobokan and Lengkong
BAFs
BAFs in S rivers and Kobokan headvalley
Small BAFs in south rivers
Pyroclastic sediment (PDCs and tephra) supply
Volume x 10⁶m³
Aggradation
Degradation
Mass balance (m³/km²/yr)
Year
Period covered by Figs.13 and 14

◀ **Fig. 18** Geomorphic response to near continuous supply of erupted products. **A** Aggradation and degradation processes at two selected watersheds on Semeru. **a** Seven annual (shaded relief) DEMs/DSMs from 2005 to 2011 in the 2-km-long Lengkong middle reach L3, most of them including the confluence area. The DEMs were calculated from D-GNSS (2006, 2007, and February 2008, grid step 3 m) and LIDAR TLS data (2010 and 2011, grid steps 3 and 0.5 m, respectively). The DSMs (August 2005 and August 2008, grid step 50 cm) were acquired from low-altitude air photographs using photogrammetric techniques. DEMs and DSMs incorporate the temporal and spatial aggradation and degradation patterns over seven rainy seasons. Note the geomorphic changes in the uppermost check dam area and the fan-like confluence zone. **b** The colour-coded bar shows δz values (in m) with a minimum limit of detection (minLoD) of 1 m. Aggradation is coloured in red and degradation in blue (after Thouret et al. 2014, Fig. 12, p. 18). **B** 'Long'- (30 years), 'mid'- (10 years), and 'short-term' (2–3 years) cycles of aggradation and degradation promoted by two types of pyroclastic sediment input from persistent eruptive activity, exemplified by Semeru over 30 years. The fluctuating curve indicates the approximate, estimated cycle before 2001 and the measured cycle after 2001. Aggradation and degradation rates are summarized at the scale of the Lengkong sub-catchment from Fig. 15, p. 23 in Thouret et al. (2014). In contrast, two generalized curves (dot-dash lines) suggest the steady, exponential decline of sediment mass balance for large but infrequent eruptions of Mt. St Helens (1980) and Pinatubo (1991). The grey area indicates the estimated annual tephra supply. BAF volumes taken from Thouret et al. (2007a, b) and references therein

2007a, b, 2014). High-spatial resolution satellite imagery, aerial photographs and DEMs along with field-based geological mapping over the period 1981–2012 were drawn on to assess sediment budgets in two watersheds of the near-continuously active Semeru volcano (Photo 118).

We place this case study in the hydrogeomorphic context of aggradation and degradation processes at work in the drainage network. Net accumulation of sediment in streams causes channels to become partially infilled with bed-material sediment, which raises bed elevation; this process is termed channel aggradation (Vessel and Davies 1981). Waning of excess sediment supply readjusts the balance between water flow and sediment transport, allowing streams to lower bed elevation by downcutting through previously accumulated sediment; this process is termed channel degradation. Channel aggradation generally occurs relatively quickly, over weeks to a couple of years, whereas degradation is more protracted, typically over years to decades (Pierson and Major 2014).

Geomorphic patterns of aggradation and degradation of Semeru's catchments differ from those of other active volcanoes because (1) both episodic PDCs and continuous supply of tephra fall generate pulses of sediment, (2) sediment is transferred via cycles of aggradation and degradation that continue for >15 years in river channels after each PDC-producing pulse, and (3) rain-triggered lahars remove more material than fluvial transport as a result of long intense monsoon rainfall. This study provides a number of results on geomorphic response to near-continuous volcaniclastic supply:

1. Average sediment yields (10^3–10^5 t/km^2/year) at persistently active volcanoes are two to three orders of magnitude lower than sediment yields following large and infrequent eruptions, but the continuous and steady sediment transfer in rivers removes more sediment on a mid-term (~10 years) to long-term (>30 years) basis. Decay of sediment yields is not constant and river channels do not fully recover at steadily active volcanoes, as episodic input from PDC eruptions, superimposed on the background remobilization of daily tephra, have a greater cumulative effect.
2. The geomorphic response of two of Semeru's rivers to volcanic sediment migration (Fig. 18A) indicates that (1) each river experiences alternating aggradation and degradation cycles following PDC-producing eruptions and (2) spatial patterns of sediment transfer are governed by geomorphic characteristics of the river reaches. Usually high degradation in the steep source reach is followed by a long bypassing middle reach. Aggradation prevails in the depositional reaches further down valley on the ring plain.
3. In addition to constant tephra supply, frequent PDC-producing eruptions at persistently active volcanoes favour high aggradation rates due to repeated waves of

migrating sediment. At Semeru, two categories of river responses are induced by sediment waves (Fig. 18B; Thouret et al. 2014): (a) a slow long-term (>15 years) response to voluminous sediment input tends to choke multi-thread, meandering, low-gradient channels of relatively large catchments, inducing a slow dispersion wave; (b) a quick (<10 year) response to limited sediment input by short, narrow, and single-thread channels of steeper sub-catchments erodes and creates terraces in infilling sediment.

4. Average sediment yields at persistently active volcanoes (based on the Semeru case), remain one to two orders of magnitude above the background level between eruptions, in contrast to the exponential decay of sediment yields computed after large eruptions. Near decadal sediment yields are two to three orders of magnitude less than annual, immediate post-eruption sediment yields after large but infrequent events. During long-term periods of 10 to 30 years, more sediment will be derived from persistently active volcanoes than those with large but infrequent eruptions.

It should be noted that the 2020 and 2021 PDC events at Semeru substantially changed the evolution of the drainage system. Further study is required to determine whether average sediment yields still remain one to two orders of magnitude above the background level between eruptions at near-continuously active volcanoes.

5 Geomorphic Impacts of Very Large Eruptions (VEI > 6)

Very large eruptions yielding more than 450 km^3 of magma have been called super-eruptions, but their record is incomplete, and the most recent known example occurred 26,000 years ago (Self 2006). Depending on where the volcano is located, the effects are likely to be felt globally or at least throughout a hemisphere. Large areas will be devastated by PDCs, and more widely dispersed tephra falls will be laid down over areas equivalent to that of continents. The most widespread effects will be derived from volcanic gases, sulphur gases being particularly important. This gas is converted into sulphuric acid aerosols in the stratosphere and aerosol layers can penetrate the whole global atmosphere within a few weeks to months. These remain for several years and affect atmospheric circulation, causing surface temperatures to fall in many regions (e.g. Shindell et al. 2004). Major disruption of services that society depends upon can be expected for periods of months, and possibly years, after the next very large explosive eruption and the cost to global financial markets will be high and sustained.

The geomorphic impacts of such very large eruptions are unknown since they lie outside of modern experience, thus some aspects of the understanding and prediction of super-eruptions are problematic (Self 2006). Judging from the large eruptions (VEI 6–7) such as Samalas (1257 AD/CE; Lavigne et al. 2013), Huaynaputina (1600 AD/CE; Thouret et al. 2002), Tambora (1815 AD/CE; Oppenheimer 2003), Mt. Pinatubo 1991 (Newhall and Punongbayan 1996), Novarupta 1912 (Hildreth and Fierstein 2012), and from the effects of large caldera-forming eruptions (e.g. Taupo ~1.8 ka; Manville et al. 2009b), the impacts on landscapes and landforms would be of massive scale. For example, the work of Malawani et al. (2023) reconstructs the paleo-topography of western Lombok before the Samalas eruption (VEI 7) and analyses the sedimentation processes that led to its landscape evolution over the last 700 years. Modelling results from a paleo-DEM demonstrate that the relief has been significantly modified in lowland areas, which is the sediment accumulation zone. River channels have experienced changes such as the location of the river mouth and the degree of meandering. Significant changes have occurred at the shoreline which has prograded westwards by approximately 1.6 km over the last 700 years.

Such eruptions are unstoppable, and a very large one would potentially involve several weeks of magma output from the volcano. Moreover, their effects are both immediate (widespread ash fall, ignimbrites) and protracted, creating health problems for years afterwards due to atmospheric aerosols, and re-sedimentation of deposits by fluvial activity. As the next large explosive eruptions will bring about the most widespread health impacts, it is pivotal to carry out further global atmospheric circulation models to evaluate their potential effects on climate and weather, and other potential feedbacks in global environmental cycles. In these ways, the hazards posed by very large eruptions differ from those due to all other natural disasters. New strategies are needed to minimize and cope with the impacts, whose economic costs of recovery will be a major burden on society.

The recurrence of very large eruptions is relatively low at a human scale: less than once a millenium for VEI 7 (e.g. Tambora 1815) and less than once every 10,000 years for VEI 8 (e.g. Taupo, 1.8 ka) (Siebert et al. 2015). However, such statistical frequencies of eruptions should be treated cautiously, as the record based on the Holocene volcanoes is incomplete (Siebert et al. 2015). Yet, as stated by Newhall et al. (2018), 'preparation for such low-probability but high-consequence events is difficult to imagine, but some modest early measures can be considered' in anticipating monitoring of candidate volcanoes for VEI 7 eruptions, at least in the most densely populated countries.

6 Delayed Geomorphic Hazards in Caldera and Ignimbrite Contexts

Temporary storage of water, which is later released in catastrophic floods, commonly takes place in lakes of calderas and dome complexes in temperate to humid environments. The contrast between hydrographic and sedimentary responses to intracaldera eruptions with respect to composite cones is exemplified by the Tarawera fissure eruption in 1886 AD/CE and Taupo caldera 1.8 ka (Manville et al. 2007, 2009b), both in New Zealand (Fig. 19). Post-eruption failures of the topographic rim of

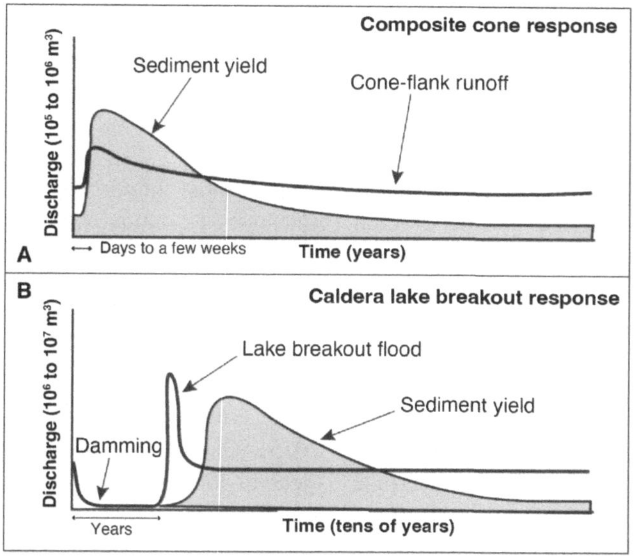

Fig. 19 Contrast between hydraulic and sedimentary response to composite cone eruptions versus intracaldera eruptions as exemplified by Tarawera, New Zealand (White et al. 1997, *in:* Thouret 2004). **A** On a volcanic cone, the hydrograph curve shows an initial increase in runoff owing to soil and vegetation damage followed by a gradual return to background levels. **B** Following a caldera eruption, a caldera lake is initially dammed for years to decades, a breakout flood then follows, succeeded by a return to pre-eruption flow (from White et al. 1997, reproduced with the permission of the Geological Society of America)

the Taupo caldera have produced catastrophic releases of up to 60 km^3 of water, and intra-caldera lakes hosted within the Okataina and Rotorua volcanic centres have also been the source of repeated floods. Crater or caldera lakes, and lakes dammed by volcanic debris can break out many years after the eruptions. Studies of breakout floods from volcanic areas have generally focused on those from intracaldera lakes and those caused by failure of pyroclastic and/or debris-avalanche-dammed riverine lakes (e.g. in Alaska: Waythomas 2015). A common feature of volcanoes in Alaska is their extensive cover of glacier ice, seasonal snow, or both. As a result, the generation of meltwater and a variety of mass flows and sediment-laden floods, are typical outcomes of most types of eruptive activity. Occasionally, such flows can be quite large, with volumes in the range of 10^7–10^9 m^3.

The delayed channel disturbance following the 1886 Tarawera event (White et al. 1997) may characterise the response to Plinian eruptions in humid-environment caldera complexes, which have modest relief, weakly integrated drainage systems and numerous lakes. Sediment yield is minimal for years prior to a breakout flood, but then increases to a maximum only after the event modifies the main stream valley. Sediment yield rises rapidly as gullies sap headwards from the flood-scoured valley, and the increase is reflected in rapid downstream aggradation of stream deposits. Sediment yield decreases gradually as streams restabilise but this gradual decline is distinct at calderas in having an initial period of stability (Fig. 19).

Delayed and indirect volcanogenic hazards are significant in caldera areas with large ignimbrites (e.g. Buesch 1991) that have temporarily dammed rivers around them, even where direct primary consequences of volcanic eruptions are minor. Kataoka et al. (2008) described the breakout flood from an ignimbrite-dammed valley after the 5 ka Numazawako eruption in northeastern Japan. The 4 km^3 of valley-confined ignimbrite dammed the Tadami River to a depth of >100 m, temporarily impounding 1.6 km^3 of water. The catastrophic release of the dammed lake and the ensuing flood redeposited pyroclastic material tens of metres thick as far as the coastal Niigata Plain >150 km downstream of the volcano. Palaeohydraulic estimation techniques indicate a peak discharge of 30,000–50,000 $m^3\,s^{-1}$ at the breach point (Kataoka et al. 2008). Large-scale breakout floods with very high peak discharges and volumes of water released can occur as a result of eruptive activity at small- to moderate-sized volcanoes, even if the eruption itself is moderate and primary eruption impacts cover only a small area. Many large cities in Japan and in comparable volcanic arcs are concentrated on alluvial plains away from volcanoes, reducing the perceived volcanic hazards. This fact points to the need for emergency and land-use planners to reconsider volcano-hydrologic hazards in their hazard mitigation procedures.

7 Geomorphic Hazards on Dormant Volcanoes

Volcanoes that have erupted within the past 10,000 years are commonly regarded as active (Siebert et al. 2010), but there are many more dormant volcanoes. Dormant volcanoes should not be underestimated as sources of hazards and risk, not only because they can reawaken, but they may also be the site of non-eruption mass movements.

The most important criteria for regarding volcanoes as extinct (i.e. not expected to erupt again) or dormant (i.e. could become active again), are the total lifetime of a volcanic complex and the overall frequency of individual eruptions. This requires very precise analysis to reach reasonable estimates of the probability of future eruptions (Schmincke 2004; Siebert et al. 2015). The significance of this problem became clear in the context of the proposed long-term storage of US-generated nuclear waste at Yucca Mountain, in a huge cave system dug into Tertiary ignimbrites in Nevada. The youngest scoria cone, Lathrop Wells, 15 km south of the storage area, is about 0.07–0.1 Ma, which has to be compared to a period of 10^4 to 10^5 years until the radioactivity of high-level waste has decayed. The question of whether or not volcanic eruptions might occur in the area of the deposit is therefore crucial to the final decision to house waste at Yucca Mountain (Connor et al. 2000; Hinze et al. 2008).

The most hazardous volcanoes are commonly those in which the eruptive phases are separated by hundreds or thousands of years of quiescence (e.g. Pinatubo had been dormant for 650 years prior to the climactic 1991 eruption). A major cause of vulnerability is also the attitude of people who live in the vicinity of a long-term dormant volcano (Schmincke 2004). It is thus often difficult to evacuate people from potentially affected areas. When volcanoes believed to be extinct erupt, catastrophes often follow, such as at Mt. Lamington (Papua New Guinea, 1951) or El Chichón (Mexico, 1982). Volcanic fields can be active for a few million years but short periods of volcanic activity alternate with long periods of quiescence, rendering probability assessment challenging.

Geomorphic processes acting on dormant and extinct volcanoes are related to mass movements and debris flows, common on all steep mountains, but exacerbated by unconsolidated and weathered pyroclastic material. These hazards are unrelated to eruptions and stem from the remobilisation of old tephra. Secondary lahars occur during volcanic quiescence and can be generated by crater lake breakout, heavy rainfall from extreme events such as hurricanes, and earthquake-induced avalanches. Characteristically, the volume transported changes during movement due to entrainment and deposition along the flow path, and the sequence of propagation commonly shows stop-start behaviour as a result of landslide dam formation and breaching.

On dormant volcanoes, catastrophic erosion processes are uncommon but can reach dramatic proportions. Examples of landslide disasters on dissected volcanoes include Mt. Ontake (Japan), triggered by an earthquake in 1984 (Voight and Sousa 1994), and Casita volcano (Nicaragua), induced by the heavy rains of Hurricane Mitch in 1998 (Kerle 2002; Scott et al. 2005). The 1998 Casita disaster, Nicaragua, further emphasizes the hazards of highly mobile debris flows that begin as landslides, then can transform and amplify during flow. A single wave of debris flow, triggered by hurricane precipitation, killed more than 2000 residents of two towns near Casita volcano on the 30 October 1998 (Kerle et al. 2003; Scott et al. 2005). The flow wave began as a small flank failure near the summit of the inactive edifice. Peak precipitation from Hurricane Mitch was the flow trigger: almost one metre of rain had fallen between 26 and 30 October. Downstream, the flow volume increased by a factor of nine, transforming into a catastrophic debris-flow deposit 3–6 m thick and over 1 km wide. Only land-use planning based on debris-flow hazard assessment could have prevented the Casita disaster. The towns, established in the 1980s on apparently safe terrain 4 km from the volcano, were found afterwards to have been located in the pathway of prehistoric debris flows.

Unlike other volcanic hazards, debris avalanches and lahars do not require an eruption, as they can be triggered by rainfall or by edifice failure long after an eruption. At least eight high-velocity rock avalanches have occurred in the dormant Garibaldi volcanic belt of southwest British Columbia since 1855. The 1984 event of Mt. Cayley volcano in southwest British Columbia illustrates the movement mechanisms typical of rapid landslides on dissected volcanoes, a significant entrainment of debris following initial failure, high velocity, and a reach extended by a distal debris flow (Evans et al. 2001). A disintegrating rock mass of Quaternary pyroclastic rock slid from the western flank, entrained more material and then formed a rock avalanche that travelled at 42 m/s a horizontal distance of 3.45 km from its source over a vertical elevation difference of 1.2 km. The rock avalanche was partially transformed into a distal debris flow that travelled a further 2.6 km down a narrow path to the main river, temporarily blocking it. Another example is the 12.7 km-long rock-slide avalanche from the flank of Mt. Meager volcanic complex that occurred on 6 August 2010, with a volume estimated at 48.5×10^6 m^3 (Guthrie et al. 2012). The predicted outburst flood of a resulting impounded lake led to the evacuation of a town 65 km downstream.

8 Concluding Remarks

In this chapter we have seen how geomorphology can play a fundamental role in assessing volcanic hazards, in particular the interplay of primary volcanic flows and indirect sediment-laden flows. Field surveys, HSR mapping, geophysical techniques, and laboratory analysis and experiments contribute to delineating hazard zones, and assessing and mitigating risk using new modelling approaches (e.g. Thouret and Charbonnier 2022).

1. The geomorphic hazards of volcanoes are specific to cascading eruptive events in addition to those which are common to all mountains. Geomorphic hazards often recur many years after an eruption, and they can be severe around dormant or even extinct volcanic edifices.
2. Rates of erosion are very rapid at active volcanoes (Chaps. 7 and 8), especially at frequently erupting composite cones and around calderas, thus they increase non-eruptive hazards on hillslopes, along channels and across ring plains.
3. The sequential chain of events during long-lasting or large explosive eruptions trigger geomorphic hazards during volcano development. Sediment yields from volcanically disturbed basins can initially be several hundred times greater than typical pre-eruption yields, but they usually diminish rapidly within a few years of disturbance.
4. Volcaniclastic sediments have an unusually high geomorphic and sedimentologic impact on drainage network, across ring plains and in depositional basins. Volcaniclastic deposits can mantle, modify, or create new topography, and can adversely affect communities over great distances downstream and downwind of volcanoes.
5. Post-eruption erosion and sediment remobilization that endure for decades after large eruptions can cause more social and economic harm than their direct impacts.
6. Defining volcanic hazards based on past eruption behaviour is subject to great uncertainty, as is the analysis of hazard recurrence based on VEIs. However, the calculation of quiescent intervals preceding events suggests that measures taken before frequent small eruptions (< VEI 3) are critical in the most populated areas, whereas challenging mitigation for very large events (≥ VEI 6) requires basic monitoring to be in place to provide warning. Unfortunately, the cost of monitoring all active volcanoes in populated areas is prohibitive.
7. Recent disasters indicate that, despite some success in forecasting events and in mitigating hazardous processes at individual volcanoes, the fatalities and loss still remain too high.

Photo 108 Lahars flowing down from the Nevado del Ruiz ice cap (Colombia) on 18 December 1985, three days after the eruption (*photo* J.-C. Thouret, looking SSE across Río Gualí headvalley)

Photo 109 Sabo dams and engineering techniques (retention and filtration) constructed on the eastern flank of Unzen, Kyushu, Japan, following the 1990–1995 eruption (*photo* courtesy of R. Paris)

Photo 110 **A** Ice cauldron (subsidence) at the site of the onset of the 1996 Gjálp eruption on the Vatnajökull ice cap, Iceland (*photo* courtesy of MT Gudmundsson, 1 October 1996). **B** Vent and degassing at the site of the 1996 Gjálp eruption through the fractured Vatnajökull ice cap mantled by tephra, Iceland (*photo* courtesy of M.T. Gudmundsson, 3 October 1996)

Photo 111 Jökulhlaup at the base of the Gigjökull glacier and along the Markarfljót River, following the Eyjafjallajökull eruption on 17 April 2010 in southern Iceland (*photo* courtesy of M.T. Gudmundsson)

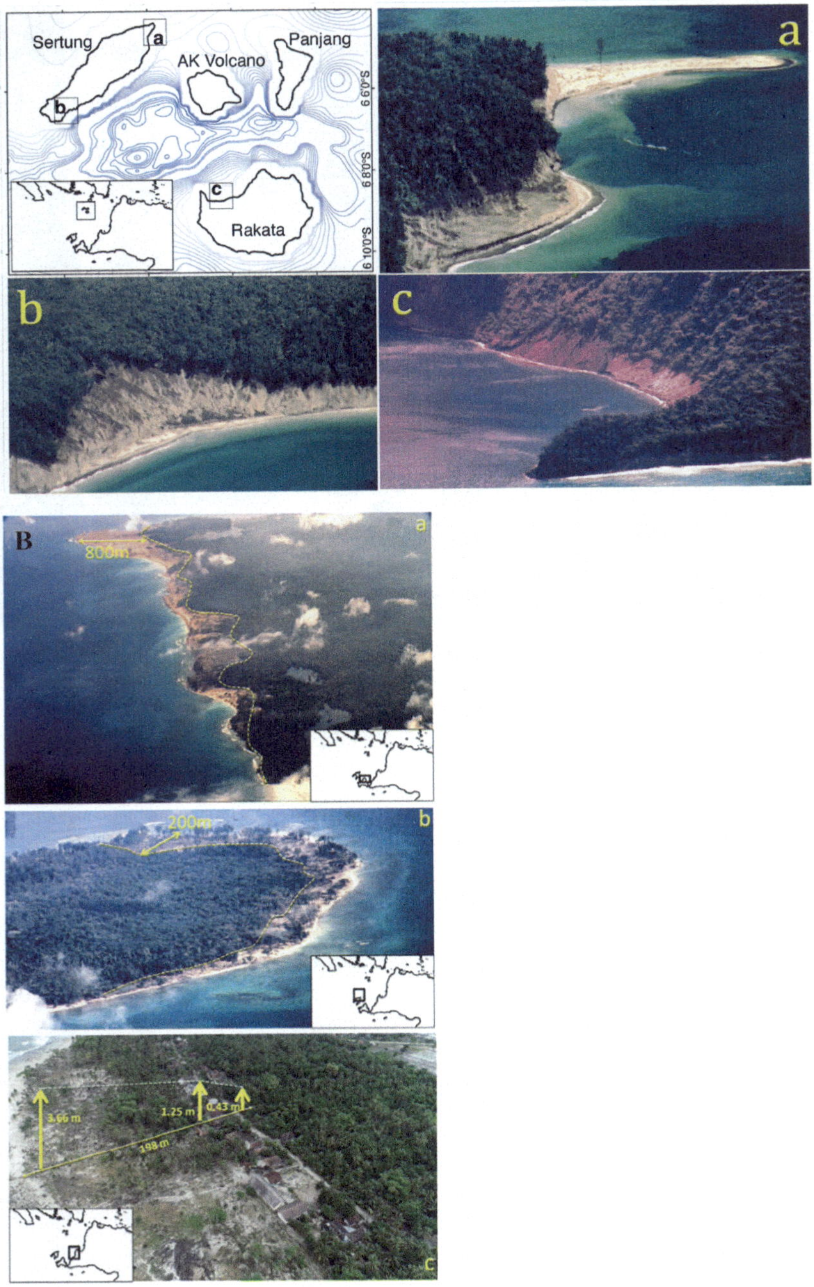

Photo 112 Effects of the 2018 tsunami from Anak Krakatau on coastal areas of the Straits, Indonesia. **A** Tsunami impacts on islands surrounding Anak Krakatau Volcano (inset in the map a). **a, b** Destruction at the north part of Sertung Island, **c** Outcrop at the south part of Sertung Island, and **d** Damage to vegetation at Rakata Island (*photo credit* Muhari et al., 2019, Fig. 5). **B**. **a** Observed damage at Ujung Kulon Peninsula where tsunami penetrated and destroyed dense vegetation up to 800 m inland at the tip of the peninsula. **b** Damage on Panaitan Island located at the north of Ujung Kulon Peninsula. **c** Effect of coastal forest in reducing tsunami impact (for tsunami of less than 5 m in height) to residential areas during the December 2018 Anak Krakatau tsunami (*photo credit* Muhari et al., 2019, Figs. 5, 6)

Photo 113 Flood sedimentation along the ChaiténRiver, central Chile, following the Chaitén dome eruption. **a** Aggradational channel fill along the Chaitén River1–2 km upstream of Chaitén town. Lava dome visible at the head of valley (*photo credit* SERNAGEOMIN, J. Muñoz, 24 May 2008). **b** Top surface of aggradational channel-fill section deposited in May 2008, about 1 km upstream of Chaitén town (*photo credit* USGS, T.C. Pierson, 6 March 2011). **c** Deposition in Chaitén town, looking upstream. Note filled and abandoned channel on right (*black arrow*) and newly avulsed channel following streets on left (*white arrow*) (*photo credit* SERNAGEOMIN, H. Moreno, 26 May 2008). **d** New channel cutting through town and widening, following avulsion and abandonment of the lower 3 km of channel. Flow direction away from camera toward the bay (*photo credit* SERNAGEOMIN, H. Moreno, 23 May 2008). *Source* Pierson et al. (2013) Bull Volcanol 75, 723. https://doi.org/10.1007/s00445-013-0723-4

Photo 114 Effects of the 18 May 1980 blast (lateral surge of a complex mixture) on the conifer forest surrounding the Mt. St. Helens volcano (Washington, USA), which affected forests up to 20–25 km northwards Trees were uprooted and knocked down along the main direction of the flow (photo Getty images, *credit*: J.T. Barr)

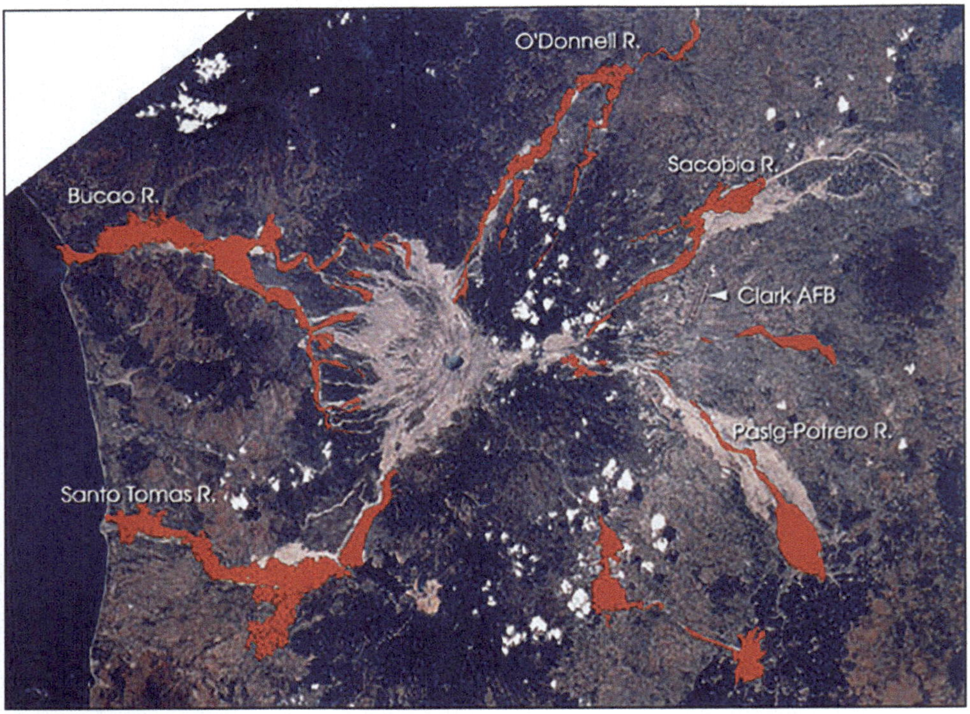

Photo 115 Lahars remobilising PDC deposits around Mount Pinatubo, Philippines. The areas covered by lahars in early August 1992 (shown in red, and mapped from STS046-75-79A) are overlaid on a 14 March 1995 image of Pinatubo (STS067-721a-31). By 1995, more than 400 km^2 of land had been buried by lahars. Image ~80 × 60 km across from NASA Earth Observatory

A

B

Photo 116 Rapid geomorphic changes around Mount Pinatubo (Luzon, Philippines), 1991–1994 (Punongbayan, Newhall and Hoblitt, 1996, Photographic record of rapid geomorphic changes around Mount Pinatubo, 1991–1994, Photos 16A and D). **A** View to the southeast, down the north fork of the Pasig-Potrero River, to its confluence with the south fork. The lightest-coloured material on the valley floor is a several-metre-thick pyroclastic-flow deposit from 15 June 1991. Slightly darker coloured material is from early lahars across the 1991 pyroclastic-flow surface. Hills with columnar joints exposed along the valley walls are from the older (Inararo) pyroclastic-flow deposit. Terraces in the background are underlain by the post-Inararo lahar deposits (*photo credit* V. Gempis, July 1991). **B** Same view as in A on 30 September 1994. Rapid lahar deposition in August and September 1994, as well as deposition from a secondary pyroclastic flow, refilled the Timbo valley and reburied the sabo dam. Much of this rapid deposition was accomplished by continuous lahars with discharges of only 10 m^3/s or less, related to the reappearance of springs along a several-kilometre-long upstream reach. The level of fill was about 15 m higher than in 1991; the L-shaped extension visible in A disappeared, as did two lower terraces in the background (*photo credit* C.G. Newhall, 1994)

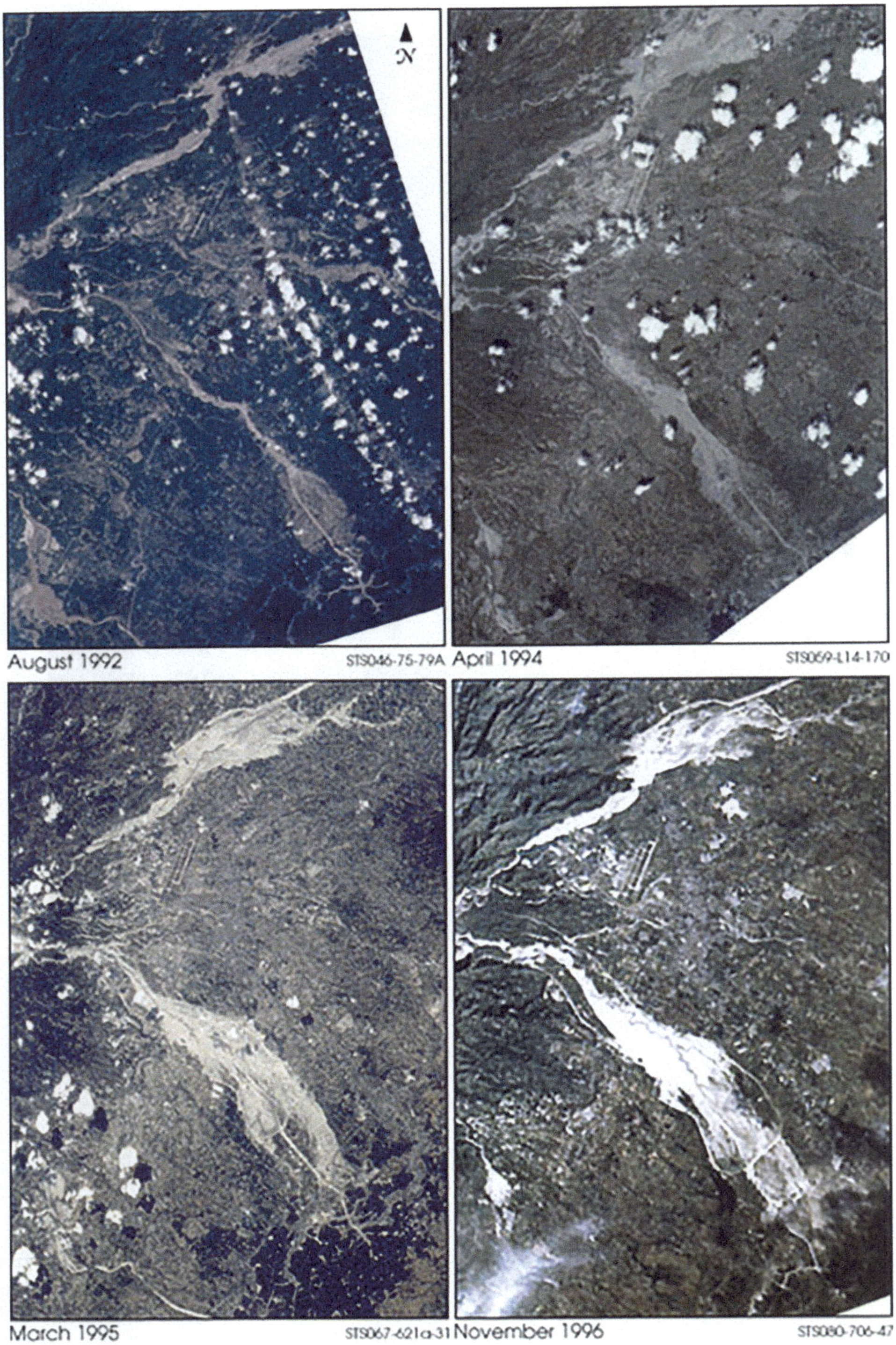

Photo 117 Floods and lahars in the Pasig-Potrero River, eastern slopes of Mt. Pinatubo, Philippines. The time series shown here focuses on the river and the progressively flooded regions surrounding the Pasig-Potrero from August 1992 to November 1996 (*photo credit* NASA Earth Observatory, Images August 1992 STS046-75-79A; April 1994 STS059-L14-170; March 1995 STS067-621a-31; November 1996 STS080-706–47)

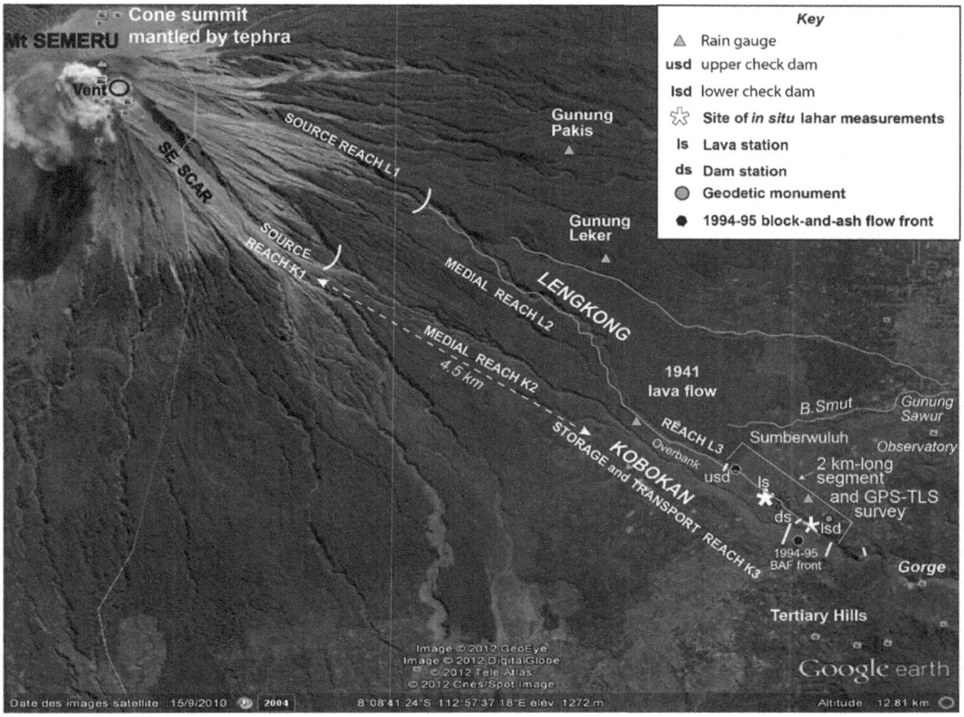

Photo 118 Eastern flank of Semeru volcano, Java, Indonesia. Drainage network of the Kobokan River and its tributary, the Lengkong River. Sites of lahar measurements and devices deployed for an ongoing study are marked (*credit*: Google Earth map from Thouret et al. 2014)

References

Alcántara-Ayala I, Goudie A (eds) (2010) Geomorphological hazards and disaster prevention. Cambridge University Press, pp 282. https://doi.org/10.1017/CBO9780511807527

Anderson T, Flett JS (1903) Report on the eruptions of Soufrière, in St. Vincent, in 1902, and on a visit to Montagne Pelée in Martinique—Part 1. Philos Trans R Soc A 200:353–553

Andrews BJ, Manga M (2011) Effects of topography on pyroclastic density current runout and formation of coignimbrites. Geology 39:1099–1102. https://doi.org/10.1130/G32226.1

Aravena A, Roche O (2022) Influence of the topography of stratovolcanoes on the propagation and channelization of dense pyroclastic density currents analyzed through numerical simulations. Bull Volcanol 84. https://doi.org/10.1007/s00445-022-01576-2

Aspinall W, Blong R (2015) Volcanic Risk Assessment. In: Sigurdsson H, Houghton BF, McNutt SR, Rymer H, Stix J (eds) The Encyclopedia of Volcanoes, Chapter 70, pp 1215–1231, 2nd edn, Academic Press Elsevier. https://doi.org/10.1016/b978-0-12-385938-9.00070-5

Auker MR, Sparks RSJ, Siebert L, Crosweller HS, Ewert J (2013) A statistical analysis of the global historical volcanic fatalities record. J Appl Volcanol 2(1):2. https://doi.org/10.1186/2191-5040-2-2

Bardou E, Bowen P, Boivin P, Banfil P (2007) Impact of small amounts of swelling clays on the physical properties of debris-flow-like granular materials. Implications for the study of alpine debris flows. Earth Surf Proc Landf 32(5):698–710. https://doi.org/10.1002/esp.1412

Baxter PJ, Boyle R, Cole P, Neri A, Spence R, Zuccaro G (2005) The impacts of pyroclastic surges on buildings at the eruption of the Soufrière Hills volcano. Montserrat. Bull. Volcanol 67(4):292–313. https://doi.org/10.1007/s00445-004-0365-7

Begét JE (2000) Volcanic tsunamis. In: Sigurdsson H, Houghton B, Mc Nutt SR, Rymer H, Stix J (eds) The Encyclopedia of Volcanoes, Part VII, pp 1005–1010, 1st edn, Academic Press, New York. ISBN 978–008054–798–5

Benage MC, Dufek J, Mothes PA (2016) Quantifying entrainment in pyroclastic density currents from the

Tungurahua eruption, Ecuador: integrating field proxies with numerical simulations. Geophys Res Lett 43(13):6932–6941. 10.1002/ 2016GL069527

Bernard J, Kelfoun K, Le Pennec JL, Vallejo Vargas S (2014) Pyroclastic flow erosion and bulking processes: comparing field-based versus modeling results at Tungurahua volcano, Ecuador. Bull Volcanol 76:1–16. https://doi.org/10.1007/S00445-014-0858-Y

Björnsson H (2003) Subglacial lakes and jökulhlaups in Iceland. Global Planet Change 35(3–4):255–271. https://doi.org/10.1016/S0921-8181(02)00130-3

Blong RJ (1984) Volcanic hazards. A sourcebook on the effects of eruptions. Academic Press Australia, pp 423. ISBN 0–12–107180–4

Blong RJ (1996) Volcanic Hazards risk assessment. In: Scarpa R, Tilling RI (eds) Monitoring and mitigation of volcanic hazards, pp 675–698, Springer. ISBN 978–3–642–80089–4

Blong RJ (2000) Volcanic hazards and risk management. Section eruption response and mitigation. In: Sigurdsson H, Houghton BF, McNutt SR, Rymer H, Stix J (eds) The Encyclopedia of Volcanoes, 1st edn, Academic Press, San Diego, pp 1215–1227. ISBN 978–008054–798–5

Borrero JC, Solihuddin T, Fritz HM et al (2020) Field survey and numerical modelling of the December 22, 2018 Anak Krakatau Tsunami. Pure Appl Geophys 177:2457–2475. https://doi.org/10.1007/s00024-020-02515-y

Boudon G, Lajoie J (1989) The 1902 Peléean deposits in the Fort Cemetery of St. Pierre, Martinique: a model for the accumulation of turbulent nuées ardentes. J Volcanol Geoth Res 38(1–2):113–29. https://doi.org/10.1016/0377-0273(89)90033-4

Bougouin A, Roche O, Paris R, Huppert HE (2021) Experimental insights on the propagation of fine-grained geophysical flows entering water. J Geophys Res: Oceans 126:e2020JC016 838. https://doi.org/10.1029/2020JC016838

Bourdier JL, Abdurachman E (2001) Decoupling of small-volume pyroclastic flows and related hazards at Merapi volcano, Indonesia. Bull Volcanol 63:309–325. https://doi.org/10.1007/s004450100133

Brand BD, Mackaman-Lofland C, Pollock NM, Bendaña S, Dawson B, Wichgers P (2014) Dynamics of pyroclastic density currents: conditions that promote substrate erosion and self-channelization–Mount St. Helens, Washington (USA). J Volcanol Geoth Res 276:189–214. https://doi.org/10.1016/j.jvolgeores.2014.01.007

Branney MJ, Kokelaar P (2002) Pyroclastic density currents and the sedimentation of ignimbrites. Geol Soc London Memoir 27:152. ISBN 1–86239–097–5

Brown RJ, Andrews GDM (2015) Deposits of pyroclastic density currents. In: Sigurdsson H, Houghton B, McNutt SR, Rymer H, Stix J (eds) The Encyclopedia of Volcanoes, Chapter 36, pp 631–648, 2nd edn, Academic Press Elsevier. ISBN 978–0–12–385938–9

Brown RJ, Branney MJ (2013) Internal flow variations and diachronous sedimentation within extensive, sustained, density stratified pyroclastic density currents flowing down gentle slopes, as revealed by the internal architecture of ignimbrites on Tenerife. Bull Volcanol 75:727. https://doi.org/10.1007/s00445-013-0727-0

Brown SK, Jenkins SF, Sparks RSJ, Odbert H, Auker MR (2017) Volcanic fatalities database: analysis of volcanic threat with distance and victim classification. J Appl Volcanol 6(1):1–20. https://doi.org/10.1186/s13617-017-0067-4

Brown SK, Loughlin SC, Sparks RSJ, Vye-Brown C, Barclay J, Calder E, Cottrell E, Jolly G, Komorowski JC, Mandeville CW, Newhall C, Palma J, Potter S, Valentine G (2015) Global volcanic hazards and risk. In: Loughlin SC et al (eds) Global volcanic hazards and risk, Chapter 2, pp 81–172, Cambridge University Press. https://doi.org/10.1017/CBO9781316276273.004

Buesch DC (1991) Changes in depositional environments resulting from emplacement of a large-volume ignimbrite. In: Fisher RV, Smith GA (eds) Sedimentation in volcanic settings. SEPM Spec Publ 45, pp 59–70. ISBN 0–918985–89–7

Burgisser A, Bergantz GW (2002) Reconciling pyroclastic flow and surge: the multiphase physics of pyroclastic density currents. Earth Plan Sci Lett 202:405–418. https://doi.org/10.1016/S0012-821X(02)00789-6

Calder ES, Sparks RSJ, Gardeweg MC (2000) Erosion, transport and segregation of pumice and lithic clasts in pyroclastic flows inferred from ignimbrite at Lascar Volcano Chile. J Volcanol Geoth Res 104(1–4):201–235. https://doi.org/10.1016/S0377-0273(00)00207-9

Calvari S (2019) Understanding basaltic lava flow morphologies and structures for hazard assessment. Ann Geophys 62(2):VO217: 1–17. https://doi.org/10.4401/ag-8048

Cappello A, Bilotta G, Neri M, Del Negro C (2013) Probabilistic modelling of future volcanic eruptions at Mount Etna. J Geophys Res, Solid Earth 118: 1925–1935

Carracedo JC, Troll VR, Day JM, Geiger H, Aulinas M, Soler V, Deegan FM, Perez-Torrado FJ, Gisbert G, Gazel E, Rodriguez-Gonzalez A (2022) The 2021 eruption of the Cumbre Vieja volcanic ridge on La Palma, Canary Islands. Geology Today 38(3):94–107. https://doi.org/10.1111/gto.12388

Carrión-Mero P, Montalván-Burbano N, Paz-Salas N, Morante-Carballo F (2020) Volcanic geomorphology: a review of worldwide research. Geosci 10(9):347. https://doi.org/10.3390/geosciences10090347

Cashman KV, Soule SA, Mackey BH, Deligne NI, Deardorff ND, Dietterich HR (2013) How lava flows: new insights from applications of Lidar technologies to lava flow studies. Geosphere 9(6):1664–1680. https://doi.org/10.1130/GES00706.1

Cashman KV, Mangan MT (2014) A Century of studying effusive eruptions in Hawai'i. In: Poland MP, Takahashi TJ, Landowski CM (eds) Characteristics of Hawaiian volcanoes, Chapter 9, pp 357–394. US Geol Survey Prof Paper 1801, 455 pp. ISSN 1044–9612

Charbonnier SJ, Gertisser R (2011) Deposit architecture and dynamics of the 2006 block-and-ash flows of Merapi Volcano, Java, Indonesia. Sediment 58(6):1573–15612. https://doi.org/10.1111/j.1365-3091.2011.01226.x

Charbonnier SJ, Germa A, Connor CB, Gertisser R, Preece K, Komorowski JC et al (2013) Evaluation of the impact of the 2010 pyroclastic density currents at Merapi volcano from high-resolution satellite imagery, field investigations and numerical simulations. J Volcanol Geoth Res 261:295–315. https://doi.org/10.1016/j.jvolgeores.2012.12.021

Charbonnier SJ, Thouret JC, Gueugnon V, Constantinescu R (2020) New insights into the c.2070 cal yr BP pyroclastic currents at El Misti volcano (Peru) from field investigations, satellite imagery and probabilistic modeling. Front Earth Sci 8. https://doi.org/10.3389/feart.2020.557788

Charbonnier SJ, Garin F, Rodríguez LA, Ayala K, Cancel S, Escobar-Wolf R, Chigna G, Chun-Quinillo C, González D, Chigna W, Mérida R, Juarez F, Calder ES (2023) Unravelling the dynamics and hazards of the June 3rd, 2018, pyroclastic density currents at Fuego volcano (Guatemala). J Volcanol Geoth Res 436:107991. https://doi.org/10.1016/j.jvolgeores.2023.107791

Chédeville C, Roche O (2015) Influence of slope angle on pore pressure generation and kinematics of pyroclastic flows: insights from laboratory experiments. Bull Volcanol 77:96. https://doi.org/10.1007/s00445-015-0981-4

Chester DK, Degg M, Duncan AM, Guest JE (2000) The increasing exposure of cities to the effects of volcanic eruptions: a global survey. Environ Haz 2(3):89–103. https://doi.org/10.3763/ehaz.2000.0214

Chevrel MO, Favalli M, Villeneuve N, Harris AJL, Fornaciai A, Richter N, Derrien A, Boissier P, Di Muro A, Peltier A (2021) Lava flow hazard map of Piton de la Fournaise volcano. Nat Haz Earth Syst Sci 21:2355–2377.https://doi.org/10.5194/nhess-21-2355-2021

Chevrel MO, Villeneuve N, Grandin R, Froger JL, Coppola D, Massimetti F, Campus A, Hrysiewics A, Peltier A (2023) Report on the lava flow daily monitoring of the 19 September–05 October 2022 eruption at Piton de la Fournaise. Volcanica 6(2):391–404. https://doi.org/10.30909/vol.06.02.39140

Chinen T, Kadomura H (1986) Post-eruption sediment budget of a small catchment on Mt. Usu, Hokkaido. Zeit Für Geomorph, Erosion Budgets and Their Hydrological Basis. Suppl-Band 60:217–232

Christiansen RL, Peterson DW (1981) Chronology of the 1980 eruptive activity. In: Lipman PW, Mullineaux DR (eds) The 1980 eruptions of Mount St. Helens, Washington, pp 17–31. Geol Survey Prof Paper 1250, 844 pp. Library of Congress Catalog no 81–600142

Clare MA, Yeo IA, Watson S, Wysoczanski R, Seabrook S, Mackay K, Hunt J, Lane E, Talling PJ, Williams M et al (2023) Fast and destructive density currents created by ocean-entering volcanic eruptions. Science 381(6662):1085–1092

Cole PD, Neri A, Baxter PJ (2015) Hazards from pyroclastic density currents. In: Sigurdsson H, Houghton B, McNutt SR, Rymer H, Stix J (eds) The Encyclopedia of Volcanoes, Chapter 54, pp 943–956, 2nd edn, Academic Press Elsevier

Collins BD, Dunne T (1986) Erosion of tephra from the 1980 eruption of Mount St Helens. Geol Soc Amer Bull 97(7):896–905. https://doi.org/10.1130/0016-7606(1986)97<896:EOTFTE>2.0.CO;2

Connor CB, Stamatakos JA, Ferrill DA, Hill BE, Ofoegbu GI, Conway FM, ...Trapp J (2000) Geologic factors controlling patterns of small-volume basaltic volcanism: Application to a volcanic hazards assessment at Yucca Mountain, Nevada. J Geophys Res, Solid Earth 105(B1):417–432. https://doi.org/10.1029/1999JB900353

Cordonnier B, Lev E, Garel F (2016) Benchmarking lava-flow models. Geol Soc London, Spec Publ 426(1):425–445. https://doi.org/10.1144/SP426.7

Cronin SJ, Neall VE, Lecointre JA, Palmer AS (1996) Unusual 'snow-slurry' lahars from Ruapehu volcano, New Zealand, September 1995. Geology 24(12): 1107–1110

Cronin SJ, Neall VE, Lecointre JA, Palmer AS (1997) Changes in Whangaehu river lahar characteristics during the 1995 eruption sequence, Ruapehu volcano, New Zealand. J Volcanol Geoth Res 76:47–61. https://doi.org/10.1016/S0377-0273(96)00064-9

de Bélizal E, Lavigne F, Hadmoko DS, Degeai JP, Dipayana GA, Mutaqin BW et al (2013) Rain-triggered lahars following the 2010 eruption of Merapi volcano, Indonesia: a major risk. J Volcanol Geoth Res 261:330–347. https://doi.org/10.1016/j.volgeores.2013.01.010

Delgado Granados H, Julio Miranda, P., Carrasco Nuñez G, Pulgarín Alzate B, Mothes P, Moreno Roa R, Cáceres Correa BE, Cortés Ramos J (2021) Hazards at ice-clad volcanoes: phenomena, processes, and examples from Mexico, Colombia, Ecuador, and Chile. In: Haeberli W, Whiteman C (eds) Snow and ice-related hazards, risks, and disasters, Chapter 17, pp 607–646. Hazards and disaster series, Elsevier. https://doi.org/10.1016/B978-0-12-394849-6.00017-2

Del Negro C, Cappello A, Neri M, Bilotta G, Hérault A, Ganci G (2013) Lava flow hazards at Mount Etna: constraints imposed by eruptive history and numerical simulations. Scient Rep 3:3493. https://doi.org/10.1038/srep03493

Di Capua A, De Rosa R, Kereszturi G, Le Pera E, Rosi M, Watt SFL (2023) From volcanoes to sedimentary systems. In: Di Capua A, De Rosa R, Kereszturi G, Le Pera E, Rosi M, Watt SFL (eds) Volcanic processes in the sedimentary record: when volcanoes meet the environment, pp 11–27. Geol Soc London, Spec Publ 520. https://doi.org/10.1144/SP520-2022-303

Dietterich HR, Cashman KV, Rust AC, Lev E (2015) Diverting lava flows in the lab. Nat Geosci 8:494–496. https://doi.org/10.1038/ngeo2470

Dietterich HR, Lev E, Chen J, Richardson JA, Cashman KV (2017) Benchmarking computational fluid dynamics models of lava flow simulation for hazard assessment, forecasting, and risk management. J Appl Volcanol 6:9. https://doi.org/10.1186/s13617-017-0061-x

Dietterich HR, Diefenbach AK, Soule SA et al (2021) Lava effusion rate evolution and erupted volume during the 2018 Kīlauea lower East Rift Zone eruption. Bull Volcanol 83:25. https://doi.org/10.1007/s00445-021-01443-6

Di Traglia F, Fornaciai A, Favalli M, Nolesini T, Casagli N (2020) Catching geomorphological response to volcanic activity on steep slope volcanoes using multi-platform remote sensing. Rem Sens 12:438. https://doi.org/10.3390/rs12030438

Di Traglia F, Fornaciai A, Casalbore D, Favalli M, Manzella I, Romagnoli C, Chiocci FL, Cole P, Nolesini T, Casagli N (2022) Subaerial-submarine morphological changes at Stromboli volcano (Italy) induced by the 2019–2020 eruptive activity. Geomorph 400:108093. https://doi.org/10.1016/j.geomorph.2021.108093

Douillet GA, Tsang-Hin-Sun E, Kueppers U, and 11 Authors (2013) Sedimentology and geomorphology of the deposits from the August 2006 pyroclastic currents at Tungurahua volcano, Ecuador. Bull Volcanol 75:1–21. https://doi.org/10.1007/S00445-013-0765-7

Druitt TH (1998) Pyroclastic density currents. In: Gilbert JS, Sparks RSJ (eds) The physics of explosive volcanic eruptions, pp 145–182. Geol Soc London, Spec Publ 145, 186 pp. ISBN 1–86239–020–7

Druitt TH, Calder ES, Cole PD, Hoblitt RP, Loughlin SC, Norton GE, Ritchie LJ, Sparks RSJ, Voight B (2002) Small-volume, highly mobile pyroclastic flows formed by rapid sedimentation from pyroclastic surges at Soufriere Hills Volcano, Montserrat: an important volcanic hazard. In: Druitt TH, Kokelaar BP (eds) The eruption of Soufriere Hills volcano, Montserrat, from 1995 to 1999, pp 263–280. Geol Soc London Memoir 21, pp 645. ISBN 1–86239–098–3

Dufek J (2016) The fluid mechanics of pyroclastic density currents. Ann Rev Fluid Mech 48:459–485. https://doi.org/10.1146/annurev-fluid-122414-034252

Dufek J, Esposti Ongaro T, Roche O (2015) Pyroclastic density currents: Processes and models. In: Sigurdsson H, Houghton B, McNutt SR, Rymer H, Stix J (eds) The Encyclopedia of Volcanoes, Chapter 35, pp 617–629, 2nd edn, Academic Press Elsevier

Dzurisin D, Moran SC, Lisowski M, Schilling SP, Anderson KR, Werner C (2015) The 2004–2008 dome-building eruption at Mount St. Helens, Washington: epilogue. Bull Volcanol 77:89

Edwards BR, Gudmundsson MT, Russel JK (2015) Glaciovolcanism. In: Sigurdsson H, Houghton B, McNutt SR, Rymer H, Stix J (eds) The Encyclopedia of Volcanoes, Chapter 20, pp 377–394, 2nd edn, Academic Press Elsevier. ISBN 978–0–12–385938–9

Esposti Ongaro T, de'Michieli Vitturi M, Cerminara M, Fornaciai A, Nannipieri L, Favalli M, Calusi B, Macías J, Castro MJ, Ortega S, González-Vida JM (2021) Modeling tsunamis generated by submarine landslides at Stromboli Volcano (Aeolian Islands, Italy): a numerical benchmark study. Front Earth Sci 9:628652. https://doi.org/10.3389/feart.2021.628652

Estep J, Dufek J (2012) Substrate effects from force chain dynamics in dense granular flows. J Geophys Res, Earth Surf 117(F1):F01028. https://doi.org/10.1029/2011JF002125

Ettinger S, Mounaud L, Magill C, Yao-Lafourcade A-F, Thouret J-C, Manville V, Negulescu C, Zuccaro G, De Gregorio D, Nardone S, Uchuchoque JAL, Arguedas A, Macedo L, Manrique Llerena N (2015) Building vulnerability to hydro-geomorphic hazards: Estimating damage probability from qualitative vulnerability assessment using logistic regression. J Hydrol 541A:563–581. https://doi.org/10.1016/j.jhydrol.2015.04.017

Evans SG, Hungr O, Clague JJ (2001) Dynamics of the 1984 rock avalanche and associated distal debris flow on Mount Cayley, British Columbia, Canada: implications for landslide hazard assessment on dissected volcanoes. Eng Geol 61:29–51. https://doi.org/10.1016/S0013-7952(00)00118-6

Fauria KE, Manga M, Chamberlain M (2016) Effect of particle entrainment on the runout of pyroclastic density currents. J Geophys Res, Solid Earth 121: 6445–6461

Favalli M, Tarquini S, Fornaciai A (2011) DOWNFLOW code and LIDAR technology for lava flow analysis and hazard assessment at Mount Etna. Ann Geophys 54:5. https://doi.org/10.4401/ag-5339

Favalli M, Tarquini S, Fomaciai A, Boschi E (2009) A new approach to risk assessment of lava flow at Mount Etna. Geology 37(12):1111–1114. https://doi.org/10.1130/G30187A.1

Favalli M, Fornaciai A, Mazzarini F, Harris A, Neri M, Behncke B, Pareschi MT, Tarquini S, Boschi E (2010) Evolution of an active lava flow field using a multitemporal LIDAR acquisition. J Geophys Res, Solid Earth 115:B11203. https://doi.org/10.1029/2010JB007463

Fedele L, Lanphere M, Melluso L, Morra V, Perrotta A, Scarpati C, Ricci G (2008) The Breccia Museo formation, Campi Flegrei, southern Italy: Chemostratigraphical and geochronological characterization and its relationship with the Campanian Ignimbrite eruption. Bull Volcanol 70:1189–1219. https://doi.org/10.1007/s00445-008-0197-y

Fisher RV (1990) Transport and deposition of a pyroclastic surge across an area of high relief: the 18 May 1980 eruption of Mount St. Helens Washington. Geol Soc Am Bull 102:1038–1054. https://doi.org/10.1130/0016-7606(1990)102<1038:TADOAP>2.3.CO;2

Fisher RV (1995) Decoupling of pyroclastic currents: hazards assessments. In: Ida Y, Voight B (eds) Models of magmatic processes and volcanic eruptions, pp 257–263, J Volcanol Geoth Res 66:1–4. https://doi.org/10.1016/0377-0273(94)00075-R

Fisher RV, Heiken G (1982) Mt. Pelée, Martinique: May 8 and 20, 1902, pyroclastic flows and surges. J Volcanol Geoth Res 13:339–371. https://doi.org/10.1016/0377-0273(82)90056-7

Freire S, Florczyk AJ, Pesaresi M, Sliuzas R (2019) An improved global analysis of population distribution in proximity to active volcanoes, 1975–2015. ISPRS Inter J Geo-Infor (MDPI) 8, 341. https://doi.org/10.3390/ijgi8080341

Freundt A, Strauch W, Kutterolf S, Schmincke HU (2007) Volcanogenic tsunamis in lakes: examples from Nicaragua and general implications. Pure Appl Geophys 164:527–545. https://doi.org/10.1007/s00024-006-0178-z

Gabrielli S, Spagnolo M, De Siena L (2020) Geomorphology and surface geology of Mount St Helens Volcano. J Maps 16(2):585–594. https://doi.org/10.1080/17445647.2020.1790048

Gertisser R, Troll VR, Walter TR, Nandaka IGMA, Ratdomopurbo A (eds) (2023). Merapi volcano: geology, eruptive activity, and monitoring of a high-risk volcano. Springer Nature, Collection Active volcanoes of the World, 553 p. ISBN 978–3–031–15039–5

Giachetti T, Paris R, Kelfoun K, Ontowirjo B (2012) Tsunami hazard related to a flank collapse of Anak Krakatau volcano, Sunda Strait, Indonesia. In: Terry JP, Goff J (eds) Natural hazards in the Asia–pacific region: recent advances and emerging Concepts, pp 79–90. Geol Soc London Spec Publ 361, 1. https://doi.org/10.1144/SP361

Giordano G, Cas R, Wright JV (2024). Subaerial pyroclastic density currents (Pyroclastic Flows and Surges). In: Cas R, Giordano, Wright JV, Volcanology: processes, deposits, geology and resources. Textbooks in earth sciences, geography and environment. Springer, Cham. https://doi.org/10.1007/978-3-319-66613-6_10

Glaze LS, Baloga SM, Fagents SA, Wright R (2014) The influence of slope breaks on lava flow surface disruption. J Geophys Res Solid Earth 119:1837–1850. https://doi.org/10.1002/2013JB010696

Gran KB, Montgomery DR (2005) Spatial and temporal patterns in fluvial recovery following volcanic eruptions: channel response to basin-wide sediment loading at Mount Pinatubo, Philippines. Geol Soc Amer Bull 117(1–2):195–211. https://doi.org/10.1130/B25528.1

Gran KB, Montgomery DR, Halbur JC (2011) Long-term elevated post-eruption sedimentation at Mount Pinatubo, Philippines. Geology 39:367–370. https://doi.org/10.1130/G31682.1

Gregg TKP, Fink JH (2000) A laboratory investigation into the effects of slope on lava flow morphology. J Volcanol Geoth Res 96:145–159. https://doi.org/10.1016/S0377-0273(99)00148-1

Griswold JP, Iverson RM (2008) Mobility statistics and automated hazard mapping for debris flows and rock avalanches. US Geol Survey Scient Invest Rep 2007–5276, 59

Gudmundsson MT (2015) Hazards from lahars and jökulhlaups. In: Sigurdsson H, Houghton BF, McNutt SR, Rymer H, Stix J (eds) The Encyclopedia of Volcanoes, Chapter 56, pp 971–984, 2nd edn, Academic Press Elsevier. ISBN 978–0–12–385938–9

Gueugneau V, Kelfoun K, Charbonnier SJ, Germa A, Carazzo G (2020) Dynamic and impacts of the May 8th, 1902 pyroclastic current at Mount Pelée (Martinique): new insights from numerical modeling. Front Earth Sci 8:279. https://doi.org/10.3389/feart.2020.00279

Gudmundsson MT, Larsen G, Höskuldsson A, Gylfason AG (2008) Volcanic hazards in Iceland. Jökull 58:251–268

Guthrie RH, Friele P, Allstadt K, Roberts N, Evans SG, Delaney KB, ... Jakob M (2012) The 6 August 2010 Mount Meager rockslide-debris flow, Coast Mountains, British Columbia: characteristics, dynamics, and implications for hazard and risk assessment. Nat Haz Earth Sys Sci 12(5):1277–1294. https://doi.org/10.5194/nhess-12-1277-2012

Hadmoko DS, De Bélizal E, Mutaqin BW, Dipayana GA, Marfai MA, Lavigne F et al (2018) Post-eruptive lahars at Kali Putih following the 2010 eruption of Merapi volcano, Indonesia: occurrences and impacts. Nat Haz 94:419–444. https://doi.org/10.1007/s11069-018-3396-7

Harris AJL, Rowland SK (2015) Flowgo 2012: An updated framework for thermo-rheological simulations of channel-contained lava. In: Carey R, Cayol V, Poland M, Weis D (eds) Geophys Monogr Series 208, pp 457–481

Harris AJL, Vallance JW, Kimberly P, Rose WI, Matıas O, Flynn, L.P., Garbeil H (2006) Downstream aggradation owing to lava dome extrusion and rainfall runoff at Volcan Santiaguito, Guatemala. In: Rose WI, Bluth GJS, Carr MJ, Ewert JW, Patino LC, Vallance JW (eds) Volcanic Hazards in Central America, Geol Soc Amer, Spec Paper 412: 85–104. ISBN 978–0–8137–2412–6

Hayes SK, Montgomery DR, Newhall CG (2002) Fluvial sediment transport and deposition following the 1991 eruption of Mount Pinatubo. Geomorph 45:211–224. https://doi.org/10.1016/S0169-555X(01)00155-6

Hildreth W, Fierstein J (2012) The Novarupta-Katmai eruption of 1912: largest eruption of the twentieth century: centennial perspectives. US Geol Survey Prof Paper 1791, pp 259. https://pubs.usgs.gov/pp/1791/

Hinze WJ, Marsh BD, Weiner R, Coleman NM (2008) Evaluating Igneous activity at Yucca Mountain: technical basis for decision making. US Nuclear Regulatory Commission, Advisory Committee on Nuclear Waste and Materials. HERO Id 5385837

Hoblitt RP, Miller CD, Vallance JW (1981) Origin and stratigraphy of the deposit produced by the May 18 directed blast. In: Lipman PW, Mullineaux DR (eds) The 1980 Eruptions of Mount St. Helens, Washington, pp 401–419. US Geol Survey Prof Paper 1250, 844 pp. Library of Congress Catalog no. 81–600142

Hulme G (1974) The interpretation of lava flow morphology. Geophys J R Astron Soc 39(2):361–383. https://doi.org/10.1111/j.1365-246X.1974.tb05460.x

Inbar M, Hubp JL, Ruiz LV (1994) The geomorphological evolution of the Paricutín cone and lava flows, Mexico, 1943–1990. Geomorph 9(1):57–76. https://doi.org/10.1016/0169-555X(94)90031-0

Iverson RM (2014) Debris flows: behavior and hazard assessment. Geol Today 30(1):15–20

Iverson RM, Schilling SP, Vallance JW (1998) Objective delineation of lahar-inundation hazard zones. Geol Soc Amer Bull 110(8):972–984. https://doi.org/10.1130/0016-7606(1998)110<0972:ODOLIH>2.3.CO;2

Janda RJ, Scott KM, Nolan KM, Mattison HA (1981) Lahar movement, effects, and deposits. In: Lipman PW, Mullineaux DR (eds) The 1980 eruptions of Mount St. Helens, Washington, pp 461–478. US Geol Survey Prof Paper 1250, 844 pp. Library of Congress Catalog no 81–600142.

Janda RJ, Daag AS, Delos Reyes PJ, Newhall CG, Pierson TC, Punongbayan RS, Rodolfo KS, Solidum RU, Umbal JV (1996) Assessment and response to lahar hazard around Mount Pinatubo, 1991 to 1993. In: Newhall CG, Punongbayan RS (eds) Fire and Mud: eruptions and lahars of Mount Pinatubo, Philippines, pp 107–139. Phivolcs, Quezon City and University of Washington Press, Seattle, 1126 pp. ISBN 0–295–97585–7

Jenkins S, Magill C, McAneney J, Blong R (2012) Regional ash fall hazard I: a probabilistic assessment methodology. Bull Volcanol 74(7):1699–1712. https://doi.org/10.1007/s00445-012-0627-8

Jenkins SF, Komorowski JC, Baxter PJ, Spence R, Picquout A, Lavigne F, Surono S (2013) The Merapi 2010 eruption: an interdisciplinary impact assessment methodology for studying pyroclastic density current dynamics. J Volcanol Geoth Res 261:316–329. https://doi.org/10.1016/j.jvolgeores.2013.02.012

Jenkins SF, Spence RJS, Fonseca JFBD, Solidum RU, Wilson TM (2014) Volcanic risk assessment: quantifying physical vulnerability in the built environment. J Volcanol Geoth Res 276:105–120. https://doi.org/10.1016/j.jvolgeores.2014.03.002

Jenkins SF, Phillips JC, Price R, Feloy K, Baxter PJ, Hadmoko DS, de Bélizal E (2015) Developing building-damage scales for lahars: application to Merapi volcano, Indonesia. Bull Volcanol 77(9):1–17. https://doi.org/10.1007/s00445-015-0961-8

Karstens J, Preine J, Carey S, Bell KL, Nomikou P, Hübscher C, Lampridou D, Urlaub M (2023) Formation of undulating seafloor bedforms during the Minoan eruption and their implications for eruption dynamics and slope stability at Santorini. Earth Planet Sci Lett 616:118215. https://doi.org/10.1016/j.epsl.2023.118215

Kataoka KS, Urabe A, Manville V, Kajiyama A (2008) Large-scale breakout flood from an ignimbrite-dammed river: aftermath of the Numazawako eruption (3400 BC), northeast Japan. Geol Soc Amer Bull 120(9–10):1233–1247. https://doi.org/10.1130/B26159.1

Kauahikaua J (2007) Lava flow hazard assessment, as of August 2007, for Kilauea east rift zone eruptions, Hawai'i Island. US Geol Survey Open-File Report 2007–1264, p 9. ISSN 2331–1258

Kauahikaua J, Tilling RI (2014) Natural Hazards and Risk Reduction in Hawai'i. In: Poland MP, Takahashi TJ, Landowski CM (eds) US Geol Survey Prof Paper 1801: 397–427. ISSN 1044–9612

Kauahikaua J, Sherrod DR, Cashman KV, Heliker CC, Hon K, Marttox TN, Johnson JA (2003) Hawaiian lava-flow dynamics during the Pu'u 'O'o-Kupaianaha eruption of Kilauea volcano, Hawai'i; the first 20 years. In: Heliker C, Swanson DA, Takahashi TJ (eds) The Pu'u 'O'o-Kupaianaha eruption of Kilauea volcano, Hawai'i: The first 20 years, pp 63–87. US Geol Survey Prof Paper 1676, 203 pp. QE.523KSP89 2003

Kerle N (2002) Volume estimation of the 1998 flank collapse at Casita volcano, Nicaragua: a comparison of photogrammetric and conventional techniques. Earth Surf Proc Landf 27(7):759–772. https://doi.org/10.1002/esp.351

Kerle N, Oppenheimer C (2002) Satellite remote sensing as a tool in lahar disaster management. Disasters 26(2): 140–160. https://doi.org/10.1111/1467-7717.00197

Kerle N, Hoffman RR (2013) Collaborative damage mapping for emergency response: the role of cognitive systems engineering. Nat Haz Earth Syst Sci 13: 97–113. https://doi.org/10.5194/nhess-13-97-2013

Kerle N, Froger JL, Oppenheimer C, Van Wyk de Vries B (2003) Remote sensing of the 1998 mudflow at Casita volcano, Nicaragua. Int J Rem Sens 1 – 26. https://doi.org/10.1080/01431160310000068187

Kieffer SW, Sturtevant B (1988) Erosional furrows formed during the lateral blast at Mount St. Helens, May 18, 1980. J Geophys Res 93(B12):14793–14816. https://doi.org/10.1029/JB093iB12p14793

Kieffer SW, Simonds CH (1995) Constraints on the rate of erosion of high-speed volcanic flows. In: Volcanoes in Town (Roma, September 1995), Period Mineral 64:209–211

Kilburn CRJ (2015) Lava flow hazard and modeling. In: Sigurdsson H, Houghton B, McNutt SR, Rymer H, Stix J (eds) The Encyclopedia of Volcanoes, Chapter 55, pp 957–969, 2nd edn, Academic Press Elsevier. ISBN ISBN 978–0–12–385938–9

Komorowski JC, Tedesco D, Kasereka M et al. (2002/2003) The January 2002 flank eruption of Nyiragongo Volcano (Democratic Republic of Congo): Chronology, evidence for a tectonic rift trigger, and impact of lava flows on the city of Goma. Acta Vulcanol 14(1–2): 1000–1035 and 15, 1–2: 27–62 (2003). https://doi.org/10.1400/19077

Komorowski J-C, Jenkins SF, Baxter PJ, Picquout A, Lavigne F, Charbonnier S et al (2013) Paroxysmal dome explosion during the Merapi 2010 eruption: Processes and facies relationships of associated high-energy pyroclastic density currents. J Volcanol Geoth Res 261:260–294. https://doi.org/10.1016/j.jvolgeores.2013.01.007

Kubo Hutchison A, Duffek J (2021) Generation of overspill pyroclastic density currents in sinuous channels. J Geophys Res, Solid Earth 126. https://doi.org/10.1029/2021JB022442

Kuenzi WD, Horst OH, McGehee RV (1979) Effect of volcanic activity on fluvial–deltaic sedimentation in a modern arc-trench gap, southwestern Guatemala. Geol Soc Amer Bull 90:827–838. https://doi.org/10.1130/0016-7606(1979)90<827:EOVAOF>2.0.CO;2

Lacroix A (1904) La Montagne Pelée et ses éruptions, vol 1. Masson, Paris

Lara E (2010) The 2008 eruption of the Chaitén Volcano, Chile: a preliminary report. Andean Geol 36(1): 125–129

Larsen G, Janebo M, Gudmundsson MT (2021) The explosive basaltic Katla eruption in 1918, south Iceland I: course of events, tephra fall and flood routes. Jökull 71:1–20. https://doi.org/10.33799/jokull2021.71.001

Latter JN (1981) Tsunamis of volcanic origin: summary of causes with particular references to Krakatoa, 1883. Bull Volcanol 44(3):467–490. https://doi.org/10.1007/BF02600578

Latter JK (ed) (1989) Volcanic Hazards; Assessment and Monitoring. IAVCEI Proceedings in Volcanology 1, pp 625, Springer Verlag. ISBN 3–540–19337–5

Lavigne F (2004) Rate of sediment yield following small-scale volcanic eruptions: a quantitative assessment at the Merapi and Semeru stratovolcanoes, Java, Indonesia. Earth Surf Proc Landf 29:1045–1058. https://doi.org/10.1002/esp.1092

Lavigne F, Suwa H (2004) Contrasts between debris flows, hyperconcentrated flows and stream flows at a channel of Mount Semeru, East Java, Indonesia. Geomorph 61:41–58. https://doi.org/10.1016/j.geomorph.2003.11.005

Lavigne F, Thouret J-C (2002) Sediment transportation and deposition by rain-triggered lahars at Merapi Volcano, Central Java, Indonesia. Geomorph 49(1–2): 45–69

Lavigne F, Thouret JC, Voight B, Suwa H, Sumaryono A (2000) Lahars at Merapi volcano, Central Java: an overview. J Volcanol Geoth Res 100(1–4):423–456. https://doi.org/10.1016/S0377-0273(00)00150-5

Lavigne F, Degeai J-P, Komorowski J-C, Guillet S, Robert V, Lahitte P, Oppenheimer C, Stoffel M, Vidal CM, Surono, Pratomo I, Wassmer P, Hajdas I, Hadmoko DS, de Belizal E (2013) Source of the great A.D. 1257 mystery eruption unveiled, Samalas volcano, Rinjani Volcanic Complex, Indonesia. Proc Nat Acad Sci 110(42):16742–16747

Lénat J-F (ed) (2022) Hazards and monitoring of volcanic activity 1. Geological and historic approaches. Geosciences, Sciences, ISTE Wiley, 232 pp. Library of Congress Control no. 2022932275

Leone F, Gaillard C (1999) Analysis of the institutional and social responses to the eruption and the lahars of Mount Pinatubo volcano from 1991 to 1998 (Central Luzon, Philippines). GeoJournal 49:223–238. https://doi.org/10.1023/A:1007076704752

Lerner GA, Jenkins SF, Charbonnier SJ, Komorowski JC, Baxter PJ (2022) The hazards of unconfined pyroclastic density currents: a new synthesis and classification according to their deposits, dynamics, and thermal and impact characteristics. J Volcanol Geoth Res 421:107429. https://doi.org/10.1016/j.jvolgeores.2021.1074290377-0273

Lirer L, Petrosino P, Alberico I (2010) Hazard and risk assessment in a complex multi-source volcanic area: the example of the Campania Region, Italy. Bull Volcanol 72(4):411–429. https://doi.org/10.1007/s00445-009-0334-2

Lockwood JP, Hazlett RW (2010) Volcanoes, Global perspectives. Wiley-Blackwell, UK, pp 539. ISBN 978–1–4051–6249–4

Loughlin SC, Sparks S, Brown SK, Jenkins SF, Vye-Brown C (eds) (2015) Global Volcanic Hazards and Risk, Cambridge University Press, pp 391. ISBN 978–1–107–11175–2

Lube G, Cronin SJ, Thouret JC, Surono S (2011) Kinematic characteristics of pyroclastic density currents at Merapi and controls on their avulsion from natural and engineered channels. Bull Geol Soc Amer 123(5):1127–1140

Lube G, Breard ECP, Esposti-Ongaro T, Dufek J, Brand B (2020) Multiphase flow behaviour and hazard prediction of pyroclastic density currents. Nat Rev Earth and Envir 1(7):348–436. https://doi.org/10.1038/s43017-020-0064-8

Magnússon E, Gudmundsson MT, Sigurdsson G, Roberts M, Höskuldsson F, Oddsson B (2012) Ice-volcano interactions during the 2010 Eyjafjallajökull eruption, as revealed by airborne radar. J Geophys Res 117:B07405. https://doi.org/10.1029/2012JB009250

Major JJ (2004) Posteruption suspended sediment transport at Mount St. Helens: decadal-scale relationships with landscape adjustments and river discharges. J Geophys Res 109:F01002

Major JJ (2023) Subaerial volcaniclastic deposits–influences of initiation mechanisms and transport behaviour on characteristics and distributions. In: Di Capua A, De Rosa R, Kereszturi G, Le Pera E, Rosi M, Watt SFL (eds) Volcanic Processes in the Sedimentary Record: When Volcanoes Meet the Environment, pp 29–100. Geol Soc London, Spec Publ 520. https://doi.org/10.1144/SP520-2021-142

Major JJ, Newhall CG (1989) Snow and ice perturbation during historical volcanic eruptions and the formation of lahars and floods: a global review. Bull Volcanol 52:1–27. https://doi.org/10.1007/BF00641384

Major JJ, Mark LE (2006) Peak flow responses to landscape disturbances caused by the cataclysmic 1980 eruption of Mount St. Helens, Washington. Geol Soc Amer Bull 118:938–958. https://doi.org/10.1130/B25914.1

Major JJ, Lara LE (2013) Overview of Chaitén Volcano, Chile, and its 2008–2009 eruption. Andean Geol 40(2):196–215

Major JJ, Janda RJ, Daag AS (1996) Watershed disturbance and lahars on the east side of Mount Pinatubo during the mid-June 1991 eruptions. In: Newhall CG, Punongbayan RS (eds) Fire and Mud: eruptions and lahars of Mt Pinatubo, Philippines, pp 895–918. Phivolcs, Quezon City and University of Washington Press, Seattle, 1126 pp. ISBN 0–295–97585–7

Major JJ, Pierson TC, Dinehart RL, Costa JE (2000) Sediment yield following severe volcanic disturbance: a two-decade perspective from Mount St Helens. Geology 28:819–822. https://doi.org/10.1130/0091-7613(2000)28<819:SYFSVD>2.0.CO;2

Major JJ, Bertin D, Pierson TC, Amigo A, Iroumé A, Ulloa H, Castro J (2016) Extraordinary sediment delivery and rapid geomorphic response following the 2008–2009 eruption of Chaíten volcano, Chile. Wat Res Res 52(7):5075–5094. https://doi.org/10.1002/2015WR018250

Malawani MN, Lavigne F, Hadmoko DS, Syamsuddin S, Handayani L, Sudrajat Y, Virmoux C, Saulnier S, Copard KK (2023) Coastal sedimentation and topographic changes in the Mataram Plain, Lombok (Indonesia) following the 1257 CE eruption of Samalas volcano. Earth Surf Proc Landf 48:2100–2116. https://doi.org/10.1002/esp.5592

Mangeney A, Roche O, Hungr O, Mangold N, Faccanoni G, Lucas A (2010) Erosion and mobility in granular collapse over sloping beds. J Geophys Res, Earth Surf 115:F03040. https://doi.org/10.1029/2009JF001462

Manville V, Cronin SJ (2007) Break-out lahar from New Zealand's Crater Lake. EOS, Transactions, AGU 88(43): 441–442. https://doi.org/10.1029/2007EO430001

Manville V, Hodgson KA, Nairn IA (2007) A review of break-out floods from volcanogenic lakes in New Zealand. NZ J Geol Geophys 50:131–50. https://doi.org/10.1080/00288300709509826

Manville V, Németh K, Kano K (2009a) Source to sink: A review of three decades of progress in the understanding of volcaniclastic processes, deposits, and hazards. Sedim Geol 220:136–161. https://doi.org/10.1016/j.sedgeo.2009.04.022

Manville V, Major JJ, Fagents SA (2013) Modeling lahar behavior and hazards. In: Fagents SA, Gregg TKP, Lopes RMC (eds) Modeling Volcanic Processes, Chapter 4, pp 300–330. Cambridge University Press. ISBN 978–0–521–89543–9

Manville V, Hodgson KA, Houghton BF, Keys JR, White JDL (2000) Tephra, snow and water: complex sedimentary responses at a cone volcano, Ruapehu, New Zealand. Bull Volcanol 62:278–293. https://doi.org/10.1007/s004450000096

Manville V, Segschneider B, Newton EH, White JLD, Houghton BF, Wilson CJN (2009b) Environmental impact of the 1.8 ka Taupo eruption: landscape responses to a large-scale explosive rhyolitic eruption. Sedim Geol 220:318–336. https://doi.org/10.1016/j.sedgeo.2009.04.017

Maramai A, Graziani L, Alessio G, Burrato P, Colini L, Nappi R et al (2005) Near- and farfield survey report of the 30 December 2002 Stromboli (Southern Italy) tsunami. Marine Geol 215:93–106. https://doi.org/10.1016/j.margeo.2004.11.009

Marti J, Ernst GGJ (eds) (2005) Volcanoes and the Environment. Cambridge University Press, pp 471. ISBN 13–978–0–521–59254–3

Marti J, Doronzo DM, Pedrazzi D, Colombo F (2019) Topographic controls on small-volume pyroclastic flows. Sedim 66:2297–2317. https://doi.org/10.1111/SED.12600

Mastin LG, Witter JB (2000) The hazards of eruptions through lakes and seawater. J Volcanol Geoth Res 97(1–4):195–214. https://doi.org/10.1016/S0377-0273(99)00174-2

Mead SRM, Magill C, Lemiale V, Thouret J-C, Prakash M (2017) Quantifying lahar damage using numerical modelling. Nat Haz Earth Sci Sys 17:703–719. https://doi.org/10.5194/nhess-2016-282

Meredith ES, Jenkins SF, Hayes JL et al (2022) Damage assessment for the 2018 lower East Rift Zone lava flows of Kīlauea volcano, Hawai'i. Bull Volcanol 84:65. https://doi.org/10.1007/s00445-022-01568-2

Minoura K, Imamura F, Kuran U, Nakamura T, Papadopoulos G A, Takahashi T, Yalciner A C (2000) Discovery of Minoan tsunami deposits. Geology 28(1):59–62

Miyazaki H, Nagai M, Shibasaki R (2015) Reviews of geospatial information technology and collaborative data delivery for disaster risk management. ISPRS Int J Geo-Inf 4:1936–1964. https://doi.org/10.3390/ijgi4041936

Mizuyama T, Kobashi S, Walling DE, Webb BW (1996) Sediment yield and topographic change after major volcanic activity. In: Erosion and sediment yield: Global and regional perspectives (Proceedings of the Exeter Symposium), Intern Assoc Hydrol Sci (IAHS) Series Proceed Rep Publication 236:295–302

Montgomery DR, Panfil MS, Hayes SK (1999) Channel-bed mobility response to extreme sediment loading at Mount Pinatubo. Geology 27:271–274. https://doi.org/10.1130/0091-7613(1999)027<0271:CBMRTE>2.3.CO;2

Muhari A, Heidarzadeh M, Susmoro H, Nugroho HD, Kriswati E, Supartoyo WAB, Imamura F, Arikawa T (2019) The December 2018 Anak Krakatau volcano tsunami as inferred from post-tsunami field surveys and spectral analysis. Pure Appl Geophys 176:5219–5233. https://doi.org/10.1007/s00024-019-02358-2

Müller D, Walter TR, Schöpa A, Witt T, Steinke B, Gudmundsson MT, Dürig T (2017) High-resolution digital elevation modeling from TLS and UAV campaign reveals structural complexity at the 2014/2015 Holuhraun eruption site, Iceland. Front Earth Sci 5:59. https://doi.org/10.3389/feart.2017.00059

Nakada S (2000) Hazards from pyroclastic flows and surges. In: Sigurdsson H, Houghton B, Rymer H, McNutt SR, Stix J (eds) The Encyclopedia of Volcanoes, pp 945–955, 1st edn, Academic Press, San Diego. ISBN 978–0–12–385938–9

Nakada S, Shimizu H, Ohta K (1999) Overview of the 1990–1995 eruption at Unzen Volcano. J Volcanol Geoth Res 89(1–4):1–22. https://doi.org/10.1016/S0377-0273(98)00118-8

Neal CA, Brantley SR, Antolik L et al (2018) The 2018 rift eruption and summit collapse of Kilauea Volcano. Science 363(6425):367–374. https://doi.org/10.1126/science.aav7046

Newhall CG, Punongbayan RS (eds) (1996) Fire and Mud. Eruptions and lahars of Mount Pinatubo, The Philippines. Phivolcs, Quezon City and University of Washington Press, Seattle, 1126 pp. ISBN 0–295–97585–7

Newhall C, Self S, Robock A (2018) Anticipating future Volcanic Explosivity Index (VEI) 7 eruptions and

their chilling impacts. Geosphere 14(2):572–603. https://doi.org/10.1130/GES01513.1
Nishimura Y (2020) Volcanism-induced tsunamis and tsunamiites. In: Shiki T, Tsuji Y, Yamazaki T, Nanayama F (eds) Tsunamiites−features and implications, Chapter 15, pp 239–260, 2nd edn, Elsevier, Amsterdam, pp 453. ISBN 978–0–12–823939–1
Oddsson B, Gudmundsson MT, Edwards BR, Thordarson T, Magnússon E, Sigurðsson G (2016) Subglacial lava propagation, ice melting and heat transfer during emplacement of an intermediate lava flow in the 2010 Eyjafjallajökull eruption. Bull Volcanol. 78:48. https://doi.org/10.1007/s00445-016-1041-4
Okano K, Suwa H, Kanno T (2012) Characterization of debris flows by rainstorm condition at a torrent on the Mount Yakedake Volcano, Japan. Geomorph 136:88–94. https://doi.org/10.1016/j.geomorph.2011.04.006
Oppenheimer C (2003) Climatic, environmental and human consequences of the largest known historic eruption: Tambora volcano (Indonesia) 1815. Prog Phys Geogr 27(2):230–259. https://doi.org/10.1191/0309133303pp379r
Pallister JS, Diefenbach AK, Burton WC, Muñoz J, Griswold JP, Lara LE, Lowenstern JB, Valenzuela CE (2013) The Chaitén rhyolite lava dome: Eruption sequence, lava dome volumes, rapid effusion rates and source of the rhyolite magma. Andean Geol 40(2): 277–294
Papale P, Eichelberger JC, Nakada S, Loughlin S, Yepes H (2015) Volcanic hazards, risks, and disasters. Hazards and Disasters Studies, Elsevier, pp 505. ISBN 978–0–12–396453–3
Paris R (2015) Source mechanisms of volcanic tsunamis. Philos Trans Royal Soc, A Math Phys Eng Sci 373:20140380. https://doi.org/10.1098/rsta.2014.038
Paris R, Switzer AD, Belousova M, Belousov A, Ontowirjo B, Whelley PL, Ulvrova M (2014a) Volcanic tsunami: A review of source mechanisms, past events and hazards in Southeast Asia (Indonesia, Philippines, Papua New Guinea). Nat Haz 70:447–470. https://doi.org/10.1007/s11069-013-0822-8
Paris R, Wassmer P, Lavigne F, Belousov A, Belousova M, Iskandarsyah Y, Mhammed Benbakkar M, Ontowirjo B, Mazzoni N (2014b) Coupling eruption and tsunami records: the Krakatau 1883 case study, Indonesia. Bull Volcanol 76:1–23. https://doi.org/10.1007/s00445-014-0814-x
Paris R, Bani P, Chevrel O, Donnadieu F, Eychenne J, Gauthier PJ, Gouhier M, Jessop D, Kelfoun K, Moune S, Roche O, Thouret J-C (2022) Volcanic hazards. In: Lénat J-F (ed) Hazards and monitoring of volcanic activity 1. Geological and historic approaches, Chapter 2, pp 75–140. ISTE Wiley. Library of Congress control number 2022932275
PHIVOLCS-DOST-IAVCEI (1995) International field workshop on Pinatubo lahars and watershed processes. Primer and field guide, pp 35. Angeles City, Philippines, October 17–21
Pierson TC (1995) Flow characteristics of large eruption-triggered debris flows at snow-clad volcanoes; constraints for debris-flow models. In: Ida Y, Voight B (eds) Models of magmatic processes and volcanic eruptions. J Volcanol Geoth Res 66(1–4):283–294. https://doi.org/10.1016/0377-0273(94)00070-W
Pierson TC, Janda JJ (1994) Volcanic mixed avalanches: a disaster eruption-triggered mass-flow process at snow-clad volcanoes. Geol Soc Amer Bull 106:1351–1358
Pierson TC, Major JJ (2014) Hydrogeomorphic effects of explosive volcanic eruptions on drainage basins. Ann Rev Earth Planet Sci 42(1):469–507
Pierson TC, Major JJ (2018) Lahar; River of volcanic mud and debris. USGS Reducing the risk from volcano hazards. Fact Sheet 2018–3024, April 2018, p 5
Pierson TC, Wood N, Driedger CL (2014) Reducing risk from lahar hazards: concepts, case studies, and roles for scientists. J Appl Volcanol 3(16):1–25. https://doi.org/10.1186/s13617-014-0016-4
Pierson TC, Janda RJ, Thouret JC, Borrero CA (1990) Perturbation and melting of snow and ice by the 13 November 1985 eruption of Nevado del Ruiz, Colombia, and consequent mobilization, flow and deposition of lahars. J Volcanol Geoth Res 41(1–4):17–66. https://doi.org/10.1016/0377-0273(90)90082-Q
Pierson TC, Janda RJ, Umbal JV, Daag AS (1992) Immediate and long-term hazards from lahars and excess sedimentation in rivers draining Mount Pinatubo, Philippines. US Geol Survey Water-Res Invest Report 92–4039, pp 37. PHIVOLCS and USAID
Pierson TC, Major JJ, Amigo A, Moreno H (2013) Acute sedimentation response to rainfall following the explosive phase of the 2008–2009 eruption of Chaitén volcano, Chile. Bull Volcanol 75:1–7. https://doi.org/10.1007/s00445-013-0723-4
Pierson TC, Daag AS, Delos Reyes PJ, Regalado MT, Solidum RU, Tubianosa BS (1996) Flow and deposition of posteruption hot lahars on the East side of Mount Pinatubo, July-October 1991. In: Newhall CG, Punongbayan RS (eds) Fire and Mud: eruptions and lahars of Mt Pinatubo, Philippines, pp 921–950. Phivolcs, Quezon City and University of Washington Press, Seattle, pp 1126. ISBN 0–295–97585–7
Pittari A, Cas RAF, Monaghan JJ, Martí J (2007) Instantaneous dynamic pressure effects on the behaviour of lithic boulders in pyroclastic flows: The Abrigo Ignimbrite, Tenerife, Canary Islands. Bull Volcanol 69:265–279. https://doi.org/10.1007/s00445-006-0072-7
Pollock NM, Brand BD, Roche O (2016) The controls and consequences of substrate entrainment by pyroclastic density currents at Mount St Helens, Washington (USA). J Volcanol Geoth Res 325:135–147. https://doi.org/10.1016/j.jvolgeores.2016.06.012
Procter JN, Cronin SJ, Fuller IC, Lube G, Manville V (2010) Quantifiying the geomorphic impact of a lake breakout lahar, Mount Ruapehu, New Zealand. Geology 38:67–70
Roche O (2015) Nature and velocity of pyroclastic density currents inferred from models of entrainment of substrate lithic clasts. Earth Planet Sci Lett 418:115–125. https://doi.org/10.1016/j.epsl.2015.03.0010012-821X

Roche O, Buesch DC, Valentine GA (2016) Slow-moving and far-travelled dense pyroclastic flows during the Peach Spring super-eruption. Nature Com 7, 10890: 1–8

Roche O, Nino Y, Mangeney A, Brand B, Pollock N, Valentine G (2013) Dynamic pore-pressure variations induce substrate erosion by pyroclastic flows. Geology 41:1107–1110. https://doi.org/10.1130/G34668.1

Rodolfo KS (1995) Pinatubo and the Politics of Lahar. Eruption and aftermath, 1991. University of the Philippines Press, Quezon City, pp 341

Rosenfeld Ch-L (1996) Monitoring of geomorphic effects of the 1980 eruptions of Mount St. Helens, Washington, USA. Geojournal 38(3):321–328. https://doi.org/10.1007/BF00204724

Rosi M, Levi ST, Pistolesi M, Bertagnini A, Brunelli D, Cannavò V, Di Renzoni A, Ferranti F, Renzulli A, Yoon D (2019) Geoarchaeological evidence of middle-age tsunamis at Stromboli and consequences for the tsunami Hazard in the southern Tyrrhenian Sea. Scient Rep 9(1):677. https://doi.org/10.1038/s41598-018-37050-3

Rümpf ME, Lev E, Wysicki R (2018) The influence of topographic roughness on lava flow emplacement. Bull Volcanol 80:63. https://doi.org/10.1007/s00445-018-1238-9

Schindelé F, Kong L, Lane E, Paris R, Ripepe M, Titov V, Bailey R (2024) A review of tsunamis generated by volcanoes (TGV): source mechanism, modelling, monitoring and warning systems. Pure App Geophys 181:1745–1792. https://doi.org/10.1007/s00024-024-03515-y

Schmincke H-U (2004) Volcanism. Springer-Verlag, Berlin, pp 324. ISBN 3–540–43650–2.

Scott KM (1988) Origins, behavior, and sedimentology of lahars and lahar-runout flows in the Toutle-Cowlitz River system. US Geol Surv Prof Paper 1447-A, pp 7

Scott KM, Janda RJ, de la Cruz EG, Gabinete E, Eto I, Isada M, Sexon M, Hadley KC (1996) Channel and sedimentation responses to large volumes of 1991 volcanic deposits on the east flank of Mount Pinatubo. In: Newhall CG, Punongbayan RS (eds) Fire and Mud: eruption and lahars of Mount Pinatubo, Philippines, pp 971–988. PHIVOLCS and Washington Univ Press, Seattle. ISBN 0–295–97585–7

Scott KM, Vallance JW, Kerle N, Macías JL, Strauch W, Devoli G (2005) Catastrophic precipitation-triggered lahar at Casita volcano, Nicaragua: occurrence, bulking and transformation. Earth Surf Proc Landf 30(1): 59–79. https://doi.org/10.1002/esp.1127

Scott WE (1989) Volcanic hazard zonation and long-term forecasts. In: Tilling RI (ed) Volcanic hazards, short courses in geology V1, Chapter 3, pp 25–49, AGU. https://doi.org/10.1029/SC001p0025

Seabrook S, Mackay K, Watson SJ, Clare MA, Hunt JE, Yeo IA, ... Williams MJ (2023) Volcaniclastic density currents explain widespread and diverse seafloor impacts of the 2022 Hunga Volcano eruption. Nature Comm 14(1):7881. https://doi.org/10.1038/s41467-023-43607-2

Segerstrom K (1950) Erosion studies at Paricutín, state of Michoacán, Mexico. US Geol Surv Bull 965-A, 164 pp. https://doi.org/10.3133/b965A

Self S (2006) The effects and consequences of very large explosive volcanic eruptions. Phil Trans Royal Soc, A: Math Phys Eng Sci 364(1845):2073–2097. https://doi.org/10.1098/rsta.2006.1814

Sheridan MF, Hubbard B, Bursik MI, Abrams M, Siebe C, Macías JL, Delgado H (2001) Gauging short-term volcanic hazards at Popocatepetl. Eos 82(16):185–189. https://doi.org/10.1029/01EO00097

Sherrod DR, Scott WE, Stauffer PH, eds (2008) A volcano rekindled: the renewed eruption of Mount St. Helens, 2004–2006. US Geol Surv Prof Paper 1750, pp 856. https://doi.org/10.3133/pp1750

Shindell DT, Schmidt GA, Mann ME, Faluvegi G (2004) Dynamic winter climate response to large tropical volcanic eruptions since 1600. J Geophys Res, Atmospheres 109(D5). https://doi.org/10.1029/2003JD004151

Siebert L, Simkin T, Kimberly P (2010) Volcanoes of the World. 3rd edn, Smithsonian Institution, University of California Press, 551 p. ISBN 978–0–520–26877–7

Siebert L, Cottrell E, Venzke E, Andrews B (2015) Earth's volcanoes and their eruptions: an overview. In: Sigurdsson H, Houghton B, McNutt SR, Rymer H, Stix J (eds) The Encyclopedia of Volcanoes, Chapter 12, pp 239–255, 2nd edn, Academic Press Elsevier. ISBN 978–0–12–385938–9

Sigurdsson H, Houghton B, McNutt SR, Rymer H, Stix J (eds) (2015) The Encyclopedia of Volcanoes. 2nd edn, Academic Press, Elsevier, pp 1421. ISBN 978–012–395839–9

Slaymaker O (1996) Introduction. In: O. Slaymaker (ed) Geomorphic Hazards, pp 1–7. Wiley. ISBN 978–0–521–87812–8

Solikhin A, Thouret J-C, Harris A, Liew SC, Gupta A (2012) Geology, tectonics, and the 2002–2003 eruption of Semeru volcano, Indonesia: interpreted from high-spatial resolution satellite imagery. Geomorph 138:364–372. https://doi.org/10.1016/j.geomorph.2011.10.001

Solikhin A, Thouret J-C, Liew SC, Gupta A, Sayudi DS, Oehler J-F, Kassouk Z (2015) High-spatial-resolution imagery helps map deposits of the large (VEI 4) 2010 Merapi Volcano eruption and their impact. Bull Volcanol 77(3):20. https://doi.org/10.1007/s00445-015-0908-0

Smith GA (1991) Facies sequences and geometries in continental volcaniclastic sedimentation. In: Fisher RV, Smith GA (eds) Sedimentation in volcanic settings, pp 109–121. SEPM Spec Publ 45. ISBN 0–918985–89–7

Sparks RS, Aspinall WP, Crosweller HS, Hincks TK (2011) Risk and uncertainty assessment of volcanic hazards. In: Rougier J, Sparks RJS, Hill L (eds) Risk and uncertainty assessment for natural hazards, pp 364–397. Cambridge University Press, Cambridge. https://doi.org/10.1017/CBO9781139047562.012

Sparks RSJ, Gardeweg MC, Calder ES, Matthews SJ (1997) Erosion by pyroclastic flows on Lascar Volcano, Chile. Bull Volcanol 58:557–565. https://doi.org/10.1007/s004450050162

Spence R, Kelman I, Calogero E, Toyos G, Baxter PJ, Komorowski JC (2005) Modelling expected physical impacts and human casualties from explosive volcanic eruptions. Nat Haz Earth Syst Sci 5:1003–1015. https://doi.org/10.5194/nhess-5-1003-2005

Stinton AJ, Sheridan MF (2008) Implications of long-term changes in valley geomorphology on the behavior of small-volume pyroclastic flows. J Volc Geoth Res 176:134–140. https://doi.org/10.1016/j.jvolgeores.2008.01.010

Sulpizio R, De Rosa R, Donato P (2008) The influence of variable topography on the depositional behaviour of pyroclastic density currents: the examples of the Upper Pollara eruption (Salina Island, southern Italy). J Volcanol Geoth Res 175:367–385. https://doi.org/10.1016/j.jvolgeores.2008.03.018

Sulpizio R, Dellino P, Doronzo DM, Sarocchi D (2014) Pyroclastic density currents: state of the art and perspectives. J Volcanol Geoth Res 283:36–65. https://doi.org/10.1016/j.jvolgeores.2014.06.014

Suryo I, Clarke MCG (1985) The occurrence and mitigation of volcanic hazards in Indonesia as exemplified at the Mount Merapi, Mount Kelut and Mount Galunggung volcanoes. Quart J Eng Geol Hydrogeol 18(1):79–98. https://doi.org/10.1144/GSL.QJEG.1985.018.01.09

Suwa H, Sumaryono A (1996) Sediment discharge by storm runoff from a creek on Merapi volcano. J Japan Soc Erosion Control Eng 48 (Special vol):117–128. 10.11475/ sabo1973.48.Special_117

Suwa H, Yamakoshi T (1999) Sediment discharge by storm runoff at volcanic torrents affected by eruption. Zeit Für Geomorph Suppl Bd 114:63–88

Suwa H, Okano K, Kanno T (2011) Forty years of debris flow monitoring at Kamikamihorizawa Creek, Mt Yakedake, Japan. In: Fifth Conference on Debris flows: disasters, risk, forecast, protection, pp 605–613. Ital J Eng Geol and Envir, Universita La Sapienza. https://doi.org/10.4408/IJEGE.2011-03.B-066

Swanson FJ, Collins B, Dunne T, Wicherski BP (1983) Erosion of tephra from hillslopes near Mount St. Helens and other volcanoes. In: Symposium on erosion control in volcanic areas, pp 183–221, Seattle, July 1982. Ibaraki, Japan, Public Works Research Institute

Swanson FJ, Jones JA, Crisafulli CM, Lara A (2013) Effects of volcanic and hydrologic processes on forest vegetation: Chaitén Volcano, Chile. Andean Geol 40(2):359–391. https://doi.org/10.5027/andgeoV40n2-a10

Swiss Re (1998) Floods—an insurable risk? Swiss Reinsurance Company, Zurich, pp 48

Tadini A, Gurioli L, Charbonnier SJ, Thivet S, Thouret J-C (2022) Monitoring of eruptive products: pyroclastic density currents and their deposits. In: Lénat J-F (ed) Hazards and Monitoring of Volcanic Activity 3. Gravimetric, electric and magnetic fluids, products and methods, Chapter 3, pp 183–233. ISTE Wiley. https://doi.org/10.1002/9781394173730.ch3

Takahashi T (2014) Debris Flow, Mechanics, Prediction and Countermeasures. CRC Press Taylor and Francis Group, 2nd edn, Balkema, pp 448. https://doi.org/10.1201/9780203946282

Tarquini S, Favalli M (2010) Changes of the susceptibility to lava flow invasion induced by morphological modifications of an active volcano: the case of Mount Etna, Italy. Nat Haz 54:537–546. https://doi.org/10.1007/s11069-009-9484-y

Tarquini S, de'Michieli Vitturi M, Jensen EH, Pedersen GB, Barsotti S, Coppola D, Pfeffer MA (2019) Modeling lava flow propagation over a flat landscape by using MrLavaLoba: the case of the 2014–2015 eruption at Holuhraun, Iceland. Ann Geophys 62(2), 228:1–18

Taylor GA (1983) The 1951 Eruption of Mount Lamington, Papua. Australian Gov Pub Service, vol 38

Thouret J-C (1990) Effects of the 13 November 1985 eruption on the ice cap and snow pack of Nevado del Ruiz, Colombia. J Volcanol Geoth Res 41(1–4):177–201. https://doi.org/10.1016/0377-0273(90)90088-W

Thouret J-C (1999) Volcanic Geomorphology. Earth-Sci Rev 47:95–131. https://doi:10.1016/S0012-8252(99)00014-8

Thouret J-C (2004) Geomorphic processes and hazards on volcanic mountains. In: Slaymaker O, Owens Ph (eds) Mountain geomorphology, Chapter 11, pp 242–273. Arnold Publishers, London. https://doi.org/10.4324/9780203764824

Thouret J-C (2010) Volcanic hazards and risks: a geomorphological perspective. In: Alcantara-Ayala I, Goudie A (eds) Geomorphological hazards and disaster prevention, Chapter 3, pp 13–32. Cambridge University Press, Cambridge, pp 282. ISBN 978–0–521–76925–9

Thouret J-C (ed) (2001) Formes de relief, processus et risques volcaniques (Volcano landforms, processes, and related hazards and risks). Géomorph: relief, proc, envir 7(2): 75–78 and 7, 3: 155–158 www.persee.fr/doc/morfo_1266-5304_2001_num_7_2_1091

Thouret J-C, Chester DK (eds) (2005) Volcanic landforms, processes and hazards. Zeit für Geomorph NF, Suppl band 140, 231 pp. Gebrüder Borntraeger, Berlin

Thouret J-C, Nemeth K (eds) (2012) Editorial. Special Issue: Volcanic Geomorphology: landforms, processes and hazards. Geomorph 136(1):1–5. https://doi.org/10.1016/j.geomorph.2011.08.011

Thouret J-C, Charbonnier SJ (2022) Assessment, delineation of hazard zones and modeling of volcanic hazards. In: Lénat J-F (ed) Hazards and monitoring of volcanic activity 1. Geological and historic approaches, Chapter 3, pp 151–184. ISTE Wiley. Library of Congress control no. 2022932275

Thouret J-C, Chester DK, Duncan A, Ollier C (eds) (2007a). Volcanic landforms, processes and hazards on volcanoes. Géomorph: relief, proc, envir 13(3):211–215. https://doi.org/10.4000/geomorphologie.3682

Thouret J-C, Lavigne F, Suwa H, Bambang RE, Surono (2007b) Assessment of volcanic hazards at Semeru, Java, with emphasis on lahars. Bull Volcanol 70(2):221–244. https://doi:10.1007/ s00445-007-0133-6

Thouret J-C, Gupta A, Lube G, Liew SC, Cronin SJ, Surono (2010) Analysis of the 2006 eruption deposits of Merapi Volcano, Java, Indonesia, using high-resolution IKONOS images and complementary ground-based observations. Rem Sens Envir 114:1949–1967. https://doi.org/10.1016/j.rse.2010.03.016

Thouret J-C, Vandemeulebrouck J, Komorowski JC, Valla F (1995) Volcano-glacier interactions: field survey, remote sensing and modelling – a case study (Nevado del Ruiz, Colombia). In: Slaymaker O (ed) Steepland Geomorphology, Chapter 5, pp 63–88. J Wiley & Sons, 283 pp. ISBN 0–471–95752–6

Thouret J-C, Ramirez J, Naranjo JL, Gibert-Malengreau B, Vandemeulebrouck J, Funk M, Valla F (2007c) The Nevado del Ruiz ice cap, Colombia, 21 years after: volcano-glacier interactions, meltwater generation, and lahar hazards. Ann Glaciol 45:115–127. https://doi.org/10.3189/172756407782282589

Thouret J-C, Dávila J, Juvigné E, Gourgaud A, Boivin P (2002) Reconstruction of the AD 1600 explosive eruption at Huaynaputina volcano, Peru, based on geologic evidence and Spanish chronicles. J Volcanol Geoth Res 115(3–4):529–570. https://doi.org/10.1016/S0377-0273(01)00323-7

Thouret J-C, Oehler J-F, Solikhin A, Gupta A, Procter J (2014) Aggradation and erosion at persistently active volcanoes: the Semeru case study, Indonesia. Bull Volcanol 76, 857, pp 26. https://doi.org/10.1007/s00445-014-0857-z

Thouret J-C, Antoine S, Magill C, Ollier C (2020) Lahars and debris flows: characteristics and impacts. Earth-Sci Rev 201:479–502. https://doi.org/10.1016/j.earscirev.2019.103003

Thouret J-C, Arapa E, Charbonnier S, Guerrero A, Kelfoun K, Cordoba G, Rodriguez D, Santoni O (2022) Modeling tephra fall and sediment-water flows to assess their impacts on a vulnerable building stock in the city of Arequipa, Peru. Front Earth Sci 10:865989. https://doi.org/10.3389/feart.2022.865989

Thouret J-C, Aisyah N, Jenkins SF, de Belizal E, Sulistiyani, Charbonnier SJ, Sri Sayudi, Nandaka A, Mainsant G, Solikhin A (2023) Merapi's lahars: characteristics, behaviour, monitoring, impacts, hazard modeling and risk assessment. In: Gertisser R, Troll VR, Walter TR, Made Agung IG, Nandaka, Ratdomopurbo A (eds) Merapi Volcano: Geology, eruptive history and monitoring of a high-risk volcano, Chapter 16, pp 501–552. Springer, Series Active Volcanoes of the World, 571 pp. ISBN 978–3–031–15039–5

Tilling RI (2005) Volcanic hazards. In: Marti J, Ernst GJ (eds) Volcanoes and the Environment, pp 55–89. Cambridge University Press, Cambridge. ISBN 0–521–59254–2

Tilling RI, Topinka LJ, Swanson DA (1990) Eruptions of Mount St. Helens: Past, Present and Future. US Dept of the Interior, General Interest Publications of the US Geological Survey

Tomassón H (1996) The jökulhlaup from Katla in 1918. Ann Glaciol 22:249–254. https://doi.org/10.3189/1996AoG22-1-249-254

Turnbull B, Bowman ET, McElwaine JN (2015) Debris flows: Experiments and modelling. CR Physics 16:86–96. https://doi.org/10.1016/j.crhy.2014.11.006

Turner NR, Perroy RL, Hon K (2017) Lava flow hazard prediction and monitoring with UAS: a case study form the 2014–2015 Pāhoa lava flow crisis, Hawai'i. J Appl Volc 6:17. https://doi.org/10.1186/s13617-017-0068-3

Umbal JV (1997) Five years of lahars at Pinatubo volcano: declining but still potentially lethal hazards. J Geol Soc Philippines 52(1):1–19

UN United Nations (2016) Report of the open-ended intergovernmental expert working group on indicators and terminology relating to disaster risk reduction. United Nations: Geneva, Switzerland, pp 41

UNDRR United Nations Office for Disaster Risk Reduction (2017) Terminology. we/inform/ terminology

UNISDR (2 February 2017) Terminology on Disaster Risk Reduction. Basic definitions on disaster risk reduction to promote a common understanding on the subject for use by the public, authorities and practitioners. https://www.preventionweb.net/files/50

Valentine GA, Wohletz KH, Kieffer SW (1992) Effects of topography on facies and compositional zonation in caldera-related ignimbrites. Geol Soc Am Bull 104(2):154–165. https://doi.org/10.1130/0016-7606(1992)104<0154:EOTOFA>2.3.CO;2

Valentine GA (1998) Damage to structures by pyroclastic flows and surges, inferred from nuclear weapons effects. J Volcanol Geoth Res 87:117–140. https://doi.org/10.1016/S0377-0273(98)00094-8

Valentine GA (2019) Preface to the topical collection – pyroclastic current models: benchmarking and validation. Bull Volcanol 81:69. https://doi.org/10.1007/s00445-019-1328-3

Valentine GA, Wohletz KH (1989) Environmental hazards of pyroclastic flows determined by numerical models. Geology 17(7):641–644. https://doi.org/10.1130/0091-7613(1989)017<0641:EHOPFD>2.3.CO;2

Vallance JW, Scott KM (1997) The Osceola Mudflow from Mount Rainier: Sedimentology and hazard implications of a huge clay-rich debris flow. Geol Soc Amer Bull 109(2):143–163. https://doi.org/10.1130/0016-7606(1997)109<0143:TOMFMR>2.3.CO;2

Vallance JW (2000) Lahars. In: Sigurdsson H, Houghton B, McNutt SR, Rymer H, Stix J (eds) Encyclopedia of Volcanoes, Part IV, pp 601–616, 1st edn, Academic Press, San Diego. ISBN 978–008054–798–5

Vallance JW (2005) Volcanic debris flows. In: Jakob M, Hungr O (eds) Debris-flows hazards and related phenomena, Chapter 10, pp 247–274. Springer Praxis Books, Springer, Berlin. https://doi.org/10.1007/3-540-27129-5_10

Verbeek RDM (1888) Krakatau. Uitgeven op last van zijne excellentie den gouverneur-generaal van Nederlandsch-Indië. Batavia, Landsdrukkerij, pp 546

Vessel RK, Davies DK (1981) Nonmarine sedimentation in an active fore arc basin. In: Ethridge FG, Flores RM (eds) Recent and ancient nonmarine depositional environments: models for exploration, pp 31–45. Soc Econ Paleont Mineral, Spec Publ 31. https://doi.org/10.2110/pec.81.31

Voight B (1990) The 1985 Nevado del Ruiz Volcano catastrophe: anatomy and retrospection. J Volcanol Geoth Res 42:151–188. https://doi.org/10.1016/0377-0273(90)90027-D

Voight B, Sousa J (1994) Lessons from Ontake-san: a comparative analysis of debris avalanche dynamics. Eng Geol 38(3–4):261–297. https://doi.org/10.1016/0013-7952(94)90042-6

Voight B, Calvache ML, Hall ML, Monsalve ML (2013) Nevado del Ruiz Volcano, Colombia 1985. In: Bobrowsky PT (ed) Encyclopedia of Natural Hazards, Encyclopedia of Earth Sciences Series, Springer, Dordrecht. https://doi.org/10.1007/978-1-4020-4399-4_253

Wadge G, Young P, McKendrick I (1994) Mapping lava flow hazard using computer simulation. J Geophys Res 99(B1):489–504. https://doi.org/10.1029/93JB01561

Wadge G, Jackson P, Bower SM, Woods AW, Calder E (1998) Computer simulations of pyroclastic flows from dome collapse. Geophys Res Lett 25(19):3677–3680. https://doi.org/10.1029/98GL00710

Waitt RB (1995) Hybrid wet flows formed by hot pyroclasts interacting with snow during the 1992 eruptions of Crater Peak, Mount Spurr volcao, Alaska. In: Keith TEC (ed) The 1992 eruptions of Crater Peak vent, Mount Spurr Volcano, Alaska, pp 107–118. US Geol Survey Bull 2139, pp 223. Library of Congress Catalog QE75.B9 no.2139

Waitt RV (2014) In the path of destruction. Eyewitness chronicles of Mount St. Helens. WSU Press, Washington State University, p 413. ISBN 978-0-87422-23-1

Waitt RB, Gardner CA, Pierson TC, Major JJ, Neal CA (1994) Unusual ice diamicts emplaced during the December 15, 1989 eruption of Redoubt volcano, Alaska. J Volcanol Geoth Res 62:409–428. https://doi.org/10.1016/0377-0273(94)90045-0

Walder JS, LaHusen RG, Vallance JW, Schilling SP (2007) Emplacement of a silicic lava dome through a crater glacier: Mount St Helens, 2004–2006. Ann Glaciol 45:14–20. https://doi.org/ https://doi.org/10.3189/172756407782282426

Waldron HH (1967) Debris flow and erosion control problems caused by the ash eruptions of Irazu Volcano, Costa Rica. Geol Survey Bull 1241(I):137. https://doi.org/10.3133/b1241I

Walker GPL (1973) Mount Etna and the 1971 eruption – Lengths of lava flows. Phil Trans R Soc London 274:107–118. https://doi.org/10.1098/rsta.1973.0030

Waythomas C (2015) Geomorphic consequences of volcanic eruptions in Alaska: a review. Geomorph 246:123–145. https://doi.org/10.1016/j.geomorph.2015.06.004

White JDL, Houghton BF, Hodgson KA, Wilson CJN (1997) Delayed sedimentary response to the AD 1886 eruption of Tarawera, New Zealand. Geology 25: 459–462

Widiwijayanti C, Voight B, Hidayat D, Schilling SP (2009) Objective rapid delineation of areas at risk from block-and-ash pyroclastic flows and surges. Bull Volcanol 71(6):687–703. https://doi.org/10.1007/s00445-008-0254-6

Wilson CJN (1986) Pyroclastic flows and ignimbrites. Sci Progr 70(2):171–202. www.jstor.org/stable/43420636

Wilson CJN, Walker GP (1985) The Taupo eruption, New Zealand I. General aspects. Phil Trans Royal Soc London, Series A, Math Phys Eng Sci 314(1529):199–228. https://doi.org/10.1098/rsta.1985.0019

Wilson G, Wilson TM, Deligne NJ, Cole JV (2014) Volcanic hazard impacts to critical infrastructure: a review. J Volcanol Geoth Res 286:148–182. https://doi.org/10.1016/j.jvolgeores.2014.08.030

Witham CS (2005) Volcanic disasters and incidents: a new database. J Volcanol Geoth Res 148(3–4):191–233. https://doi.org/10.1016/j.jvolgeores.2005.04.017

Wolfe EW, Hoblitt RP (1996) Overview of the eruptions. In: Newhall CG, Punongbayan RS (eds) Fire and Mud – Eruptions and lahars of Mount Pinatubo, Philippines, pp 3–20. Phivolcs, Quezon City and University of Washington Press, Seattle, pp 1126. ISBN 0–295–97585–7

Wood N, Soulard C (2009) Variations in population exposure and sensitivity to lahar hazards from Mount Rainier, Washington. J Volcanol Geoth Res 188(4):367–378. https://doi.org/10.1016/j.jvolgeores.2009.09.019

Woods AW, Bursik MI, Kurbatov AV (1998) The interaction of ash flows with ridges. Bull Volcanol 60(1):38–51. https://doi.org/10.1007/s004450050215

Yokoyama I (1981) A geophysical interpretation of the 1883 Krakatau eruption. J Volcanol Geoth Res 9:359–378. https://doi.org/10.1016/0377-0273(81)90044-5

Zanchetta G, Sulpizio R, Pareschi MT, Leoni FM, Santacroce R (2004) Characteristics of May 5–6, 1998 volcaniclastic debris flows in the Sarno area (Campania, southern Italy): relationships to structural damage and hazard zonation. J Volcanol Geoth Res 133:377–393. https://doi.org/10.1016/S0377-0273(03)00409-8

Zuccaro G, Cacace F, Spence RJS, Baxter PJ (2008) Impact of explosive eruption scenarios at Vesuvius. J Volcanol Geoth Res 178(3):416–453. https://doi.org/10.1016/j.jvolgeores.2008.01.005

Chapter 10 Case Studies Illustrating the Role of Geomorphology in Assessing Volcanic Hazards

J. -C. Thouret, D. Karátson
J. C. Carracedo, V. R. Troll
G. B. M. Pedersen, M. T. Gudmundsson
R. Gertisser
J. N. Procter
S. Calvari, D. Casalbore, F. Di Traglia, A. Costa

This chapter, in relation to Chap. 9, contains five case studies from contributors who depict a range of geomorphic hazards and responses to eruptive and non-eruptive events at active volcanoes. Related to Sect. 2 of Chap. 9, the contributions are: the 2021 cone building and lava-flow eruption of Tajogaite-Cumbre Vieja, La Palma Island; the 2014–2015 Holuhraun and Bárðarbunga eruption and lava-flow field, Iceland; and the 2010–2018 PDCs and dome building at Merapi, Indonesia. Related to Sect. 3 of Chap. 9, changes induced by the 2012 Te Maari eruption, New Zealand are described, while related to Sect. 4, emphasis is on the rapid evolution of the Sciara del Fuoco at Stromboli, Italy, due to eruptions and a tsunamigenic landslide between 2012 and 2021.

1 Case Study 1

Volcanic and Geomorphic Controls on the 2021 Eruption of Tajogaite Volcano, La Palma

Juan Carlos Carracedo[1] and Valentin R. Troll[1,2]

[1]Instituto de Estudios Ambientales y Recursos Naturales (i-UNAT), University of Las Palmas de Gran Canaria (ULPGC), Las Palmas de Gran Canaria, Spain

[2]Department of Earth Sciences, Natural Resources and Sustainable Development (NRHU), Uppsala University, Uppsala, Sweden

D. Karátson and J.-C. Thouret, *Volcano Geomorphology*, Advances in Volcanology,
https://doi.org/10.1007/978-3-031-77693-9_11

The Island of La Palma, located at the western end of the Canary Island chain, is the second youngest island in the archipelago and the one that has experienced the most historical eruptions (in the last 500 years). The most recent eruption (19 September to 13 December 2021) was located, like all the recorded historical eruptions, in the south of La Palma, on the Cumbre Vieja Ridge, and was the longest-lasting and most damaging of them all. Both the persistence of volcanism on La Palma, specifically along the Cumbre Vieja Ridge, and the considerably greater destruction caused in the most recent eruptive event, can be explained by the volcanic growth pattern of the island and by its morphology. The southbound directed volcanic activity on the Cumbre Vieja makes future volcanic activity there highly likely.

The Structure of La Palma Island

La Palma developed in three stages, corresponding to an initial cone-shaped shield volcano, the Taburiente Volcano, showing an uplifted submarine basement, and two successive southward extensions, the Cumbre Nueva and the Cumbre Vieja volcanoes. Activity has thus migrated progressively from north to south with the subaerial Taburiente shield active from about 2.1 to 0.5 Ma, initiating a southward extension from about 0.8 Ma to develop the Cumbre Nueva Ridge. This resulted in a giant westbound lateral collapse some 0.56 Ma ago, but the collapse embayment was later partially filled by the nested Bejenado Volcano. Volcanic activity then migrated further southward to form the Cumbre Vieja Ridge from about 0.45 Ma up to the present day.

Regarding the morphology, the island of La Palma thus displays successive volcanic edifices in a progressive and continuous extension towards the south, a process which persists today. This may be the result of a propagating fracture which, once initiated, continues to advance in the same direction, a fundamental process in the configuration of the island of La Palma and in the distribution of volcanism. The recent continuation of eruptive activity implies that volcanism along the Cumbre Vieja Ridge will inevitably reccur in the coming years, especially as the ridge continues on the ocean floor as an elevated raised narrow mountain range for >20 km. The eruptive centres are clustered along the ridge crest and are fed via a dense swarm of dykes, inside the ridge. These feeder dykes act as the conduits that supply successive eruptions (Fig. 1), thus increasing the anisotropy of the volcano and promoting the forceful intrusion of new dykes (Carracedo and Troll 2016, 2021).

The 2021 Tajogaite Eruption

The 2021 Tajogaite volcanic crisis started in mid September, and began with eruptions from a series of vents along a ca. 500 m long fissure on 19 September. During the three months of the eruption a major scoria cone of over 300 m in height was built up and an extensive lapilli-fall blanket and lava field was constructed (e.g. Carracedo et al. 2022). Thick lapilli deposits mantled roads and houses in the Valle de Aridane area. The almost constant lapilli airfall strongly affected the western flank of the Cumbre Vieja ridge, but also reached the airport and the island's capital Santa Cruz on the

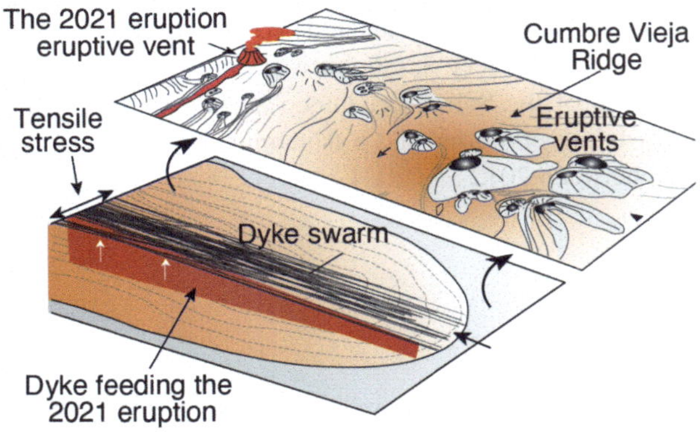

Fig. 1 The Cumbre Vieja volcanic ridge has been largely constructed within the last 125,000 years and is composed of a group of densely packed, N-S aligned cinder cones and associated lava flows. The internal structure beneath the cones is composed of a dense swarm of feeder dykes, many of which represent the conduits of past eruptions. Image modified from Carracedo and Troll (2016)

eastern side during periods when winds were blowing from the west. The lava flows had relatively high eruption temperatures (max. 1140 °C) and were consequently very fluid at times (Fig. 2), with velocities of some channelized lava flows exceeding 30 km/h (Gisbert et al.

Fig. 2 Lava flows were one of the most destructive features during the 2021 La Palma eruption. **A** Rivers of lava were emitted from the central cone system and flowed downhill at 30 km/h during several episodes of the eruption, as the lava temperature commonly exceeded 1100°C near the vent (max. 1140°C). **B** These channellised (thermally insulated) lava flows allowed rapid and efficient lava transport all the way to the sea, where coastal lava platforms developed. **C, D** Frontal lava tongues and overbank facies, in turn, cooled rapidly and became blocky lava flows that advanced in caterpillar mode, slowly progressing towards buildings and destroying all man-made structures in their way, like this house in the Las Manchas area. **E** If the supply of lava at the vent ceases, lava fronts and tongues stop and freeze irrespective of topography, like here in La Laguna village, where the edge of a lava tongue entered the town. This shift to 'a'ā style emplacement, acting in a similar way to a huge bulldozer at the flow front, can sweep away everything in the path and demolish man-made structures such as houses with great ease, and contrasts with the mode of thinner lava emplacement at higher temperatures that could flow around houses, **F** As a consequence, islands of relict infrastructure (kipukas) within the main lava field are not uncommon. Photos from Carracedo et al. (2022) and Troll et al. (2024)

2023). Lava flows were one of the most destructive features during the 2021 eruption. Frontal lava tongues and overbank facies, in turn, cooled rapidly and became blocky lava flows that advanced in caterpillar mode, slowly progressing towards buildings and destroying all man-made structures in their way.

A major determining factor for the destructive nature of the 2021 eruption was the morphology of the Cumbre Vieja ridge prior to the volcanic event (Fig. 3). The ridge is triangular in shape, 1944 m high, 17 km wide in the north, and 19 km wide above sea level. In the north, the Cumbre Vieja rests on the southern edge of the Cumbre Nueva, which has collapsed and created a wide, gently sloping valley (the Aridane valley), which is densely populated and occupied by extensive plantations (e.g. Troll et al. 2024). Another densely populated area in a similar low-slope area is found on the eastern side and includes the capital, Santa Cruz de La Palma, the airport and the main harbour. The rest of Cumbre Vieja, with steeply sloping sides, is sparsely populated as there is little land suitable for cultivation. These differences in the configuration of the ridge explain why historical eruptions have had a more limited impact on infrastructure and human settlements (Fig. 3). The 2021 eruption of the Tajogaite volcano changed this pattern which may be a key

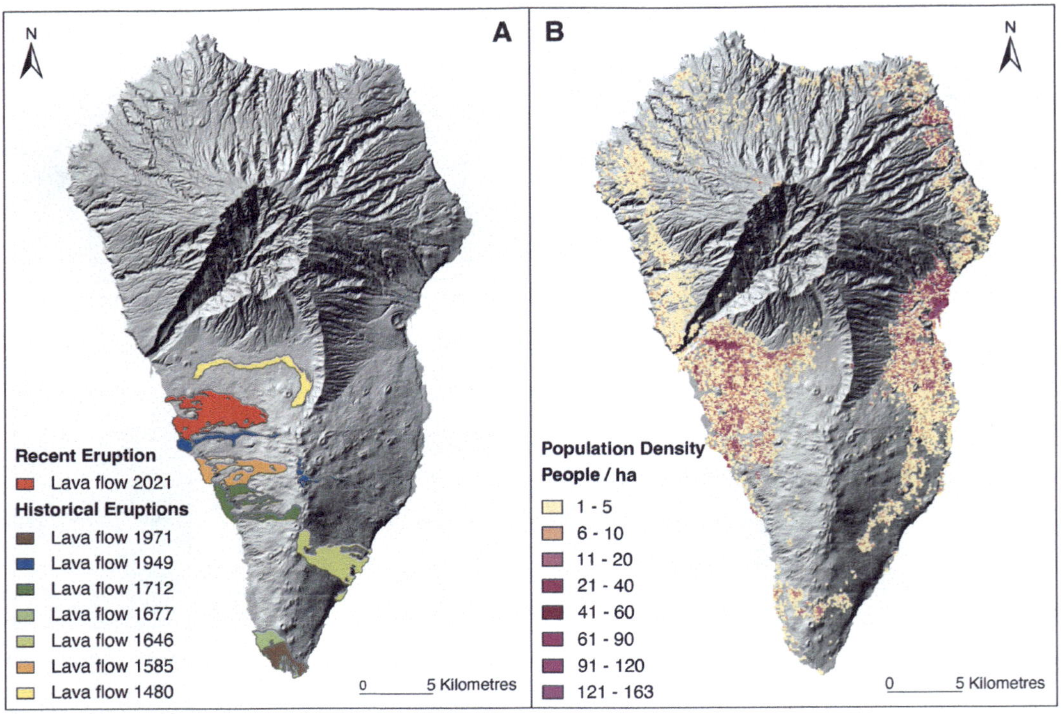

Fig. 3 **A** Slope map of the Cumbre Vieja ridge with historical eruptions outlined in colour. Past eruptions located in the southern part of the Cumbre Vieja caused no casualties and little damage to settlements and infrastructure. **B** In contrast, the eruption of the Tajogaite Volcano in 2021 was considerably more destructive, as it was located in the much flatter upper third of Cumbre Vieja, which is more densely populated and contains extensive banana plantations. From Troll et al. (2024). The eruption destroyed over 2900 buildings, devastated some 12 km^2 of farmland and settlements through lava flows and thick lapilli cover, destroyed over 90 km of vital roads, and led to the evacuation of almost 8000 people (e.g. Carracedo et al. 2022), with a total estimated cost of damage between 900 million and 1 billion euros. Since future eruptions along the Cumbre Vieja are almost inevitable, it is imperative that action is taken to ensure the safety of the island's growing population from similar future events

reason for many people having underestimated the potentially devastating effects of eruptions on the island.

For these reasons, it is likely that future eruptions will occur again on La Palma, most probably on the Cumbre Vieja ridge, with the expected damage depending primarily on the location of the eruptive centres and the consequent lava flow routes. The devastation caused by the eruption in 2021 was not a result of eruption magnitude, which was only marginally larger than other historical events, but because of the increasing vulnerability due to population growth and increasing rural land use on the lower slopes of the volcanically active Cumbre Vieja Ridge (Fig. 3).

2 Case Study 2

Holuhraun Eruption 2014–15: Lessons Learned from the Development of a Large Lava Flow Field in Iceland

Gro Birkefeldt Moller Pedersen[1] and Magnús Tumi Gudmundsson[1]

[1]Nordic Volcanological Center, Institute of Earth Sciences, University of Iceland, Sturlugata 7, 102 Reykjavík, Iceland

The 2014–2015 eruption at Holuhraun in the Bárðarbunga-Veiðivötn volcanic system was the largest eruption in Iceland since the 1783–1784 AD/CE Laki eruption. The lava field has an area of 84 km^2, an estimated bulk volume of 1.44 $km^3 \pm 0.07$ and the mean output rate over the six months was ~90 m^3/s (Gudmundsson et al. 2016; Pedersen et al. 2017). Bárðarbunga central volcano, located in the northwest corner of the large Vatnajökull ice cap, is covered by up to 800 m of ice within its 11 by 8 km wide caldera (Björnsson and Einarsson 1990). The associated fissure swarm is mostly subaerial and lies on the boundary between the North American and Eurasian plates, extending 80 km to the southwest and 80–125 km to the north (Hjartardóttir et al. 2016).

An intense seismic swarm began at Bárðarbunga on 16 August 2014. Over the following 11 days the formation of a 48 km-long segmented dyke could be tracked by observations of seismicity and deformation (Sigmundsson et al. 2015). A sequence of large earthquakes (M4-M5.8) started on 20 August and marked the start of a gradual caldera collapse (Gudmundsson et al. 2016). A few ice cauldrons developed in the glacier above the dyke path suggesting minor subglacial eruptions (Sigmundsson et al. 2015). The dyke terminated at a shallowly-dipping floodplain, 8 km north of Vatnajökull. On 29 August a minor, 4-h long, eruption took place along a 600 m fissure and on 31 August the volcanic fissure was re-activated, reaching 1.8 km in length, marking the onset of the 6-month-long effusive eruption, which ended on 27 February 2015. Minor effusive eruption was observed on 5–6 September 2 km south of the main fissure (Pedersen et al. 2017).

The first phase of the eruption (31 August to mid-October) was typified by channelized lava flows, 1–2 km in width and up to 17 km in length (Pedersen et al. 2017). In the first few days slabby pāhoehoe sheet flows (flow velocities of 800–1000 m/h) changed into rubbly pāhoehoe and then into a'ā downstream. Subsequently short-lived morphological changes from a'ā lava to rubbly pāhoehoe to a'ā were observed multiple times at the active flow front (flow velocities up to 65 m/h). During this phase the 1.8 km fissure partly closed down and activity became concentrated into an ~500 m-long elongated crater with a lava lake, which fed the rest of the eruption. A 1 km^2 lava pond ~1 km east of the vent formed in the second phase of the eruption (mid-October to end of November) that controlled the emplacement of subsequent flows, which slowed down and shortened in length over time (to ~5 km). By the end of November inflation plateaus developed in the lava channels that were active in September 2014 (Pedersen et al. 2017). These were surface manifestations of a growing lava tube system, which became the dominant lava transport system in the third phase (December to end-February 2015). These tubes fed breakouts of primarily spiny pāhoehoe (flow velocities of 1–4 m/h) that resurfaced >19 km^2 of lava and extended the entire length of the flow field. Thus the lava field changed from horizontal expansion, to growth, primarily by inflation, and vertical stacking of lobes (Pedersen et al. 2017).

The 2014–15 Holuhraun lava is classified as olivine tholeiite and the composition was remarkably uniform throughout the eruption (Halldorsson et al. 2018), consistent with an eruption fed from a homogenized magma chamber beneath the Bárðarbunga central volcano. There, a downsag-type caldera subsidence was observed reaching a maximum of 65 m in the centre (Gudmundsson et al. 2016). The uniform composition and the shallow dipping substrate suggest that morphological changes in the lava were caused by variations in flow velocity and local flux influencing crust formation (Pedersen et al. 2017).

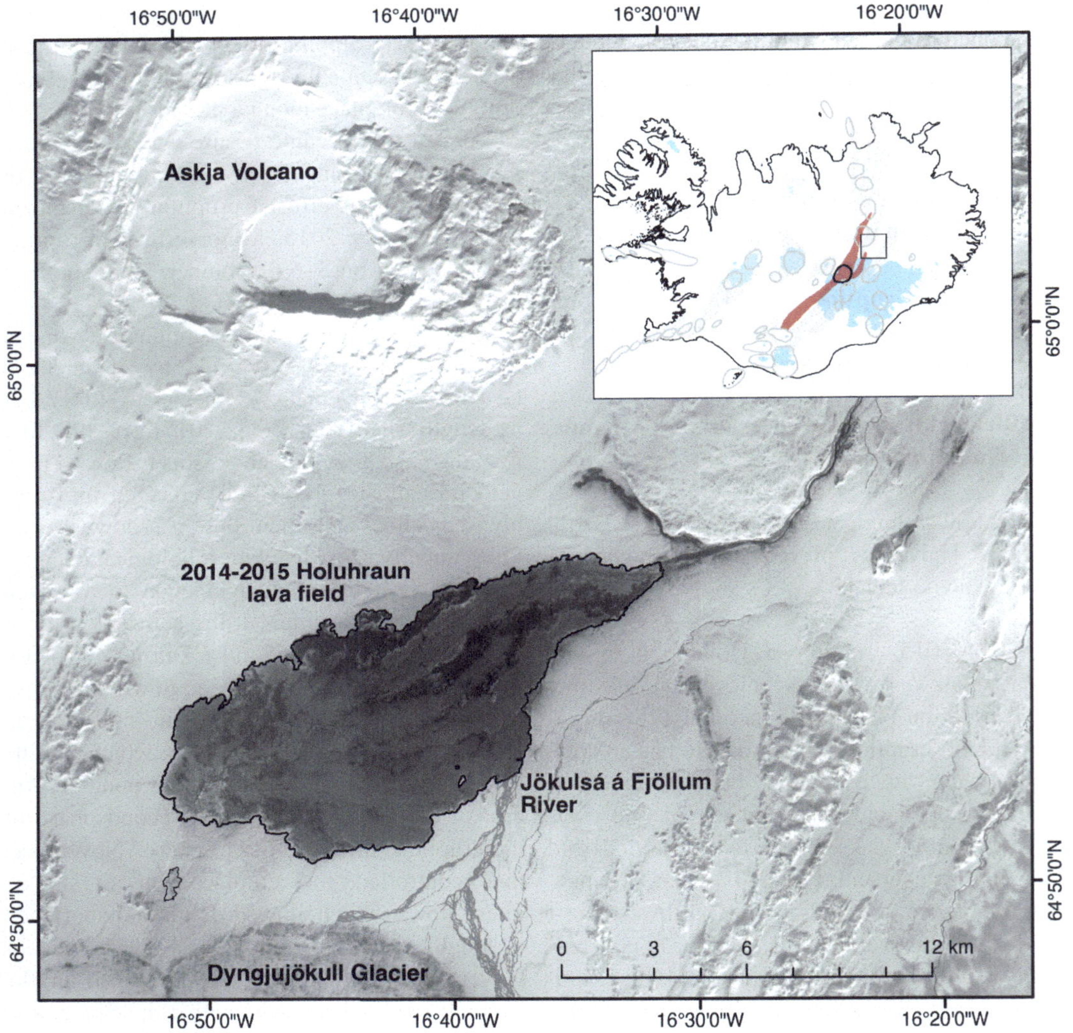

Fig. 1 The new Holuhraun lava field imaged by a LANDSAT on 2 April 2015, 33 days after the eruption stopped. The location of Holuhraun is marked by the black box on the inset map of Iceland in the upper right corner. The glaciers are shown in blue and the Icelandic volcanic systems are marked in grey. The fissure swarm of the Bárðarbunga-Veiðivötn volcanic system is highlighted in red and its central volcano is marked with a black circle

Fig. 2 First day of the Holuhraun eruption, with a near-continous fire curtain along the 1.8 km-long fissure (31 August 2015; Photo by Gro B. M. Pedersen)

At the beginning of this volcanic event the main concern was the potential danger of the eruption migrating southwards, under the Vatnajökull ice cap, causing ice melting and subsequent flooding coupled with tephra fallout. However, this did not happen. Instead, the main hazard turned out to be the release of volcanic gases with a total SO_2 emission of ca. 9.6 Mt. (Barsotti et al. 2022). Gas impacted air quality, with SO_2 concentration exceeding 1000 μg/m^3 in several instances, causing a need for increase in health care visits and treatment for asthma in Iceland. In Europe, the hourly air quality standard of 350 μg/m^3 was also exceeded in a few places in September 2014 (Barsotti et al. 2022) (Figs. 1 and 2).

3 Case Study 3

The Large-Magnitude 2010 Eruption and Post-2010 Activity of Merapi Volcano, Indonesia

Ralf Gertisser[1]

[1]School of Geography, Geology and the Environment, Keele University, Keele, ST5 5BG, UK

Rising almost 3000 m above the city of Yogyakarta in Central Java, Indonesia, Merapi volcano is renowned for its almost continuous eruptive activity and frequent occurrence of hazardous pyroclastic density currents (PDCs).

Most eruptions in the 20th and early twenty-first century followed Merapi's usual pattern of prolonged lava dome growth and subsequent gravitational dome failure to produce PDCs that mostly affected the W, SW and S slopes of the volcano, up to 10 km from the summit. The most significant recent eruption of this type occurred in 2006, when PDCs flowed down the SW and S flanks, causing destruction and fatalities in the village of Kaliadem on the bank of the Gendol River (Kali Gendol) valley (Gertisser et al. 2023).

The 2010 Eruption of Merapi and Its Environmental Effects

When Merapi erupted again in 2010, it produced the volcano's largest eruption since 1872, characterised by a Volcano Explosivity Index of 4, and the deadliest eruptive event since 1930. Moreover, the style of the 2010 eruption differed markedly from previous eruptive events (Gertisser et al. 2023). Most notably, the eruption began with partially laterally directed explosions on 26 October, producing PDCs channelled into Kali Gendol and toward the village of Kinahrejo (Fig. 1a). A lava dome formed only a few days after the initial explosions, with unprecedented growth rates of 25 m^3 s^{-1} and frequent explosive and gravitational dome failures that produced PDCs channelled mainly into Kali Gendol. During the climactic eruption phase on 5 November, a series of explosions rapidly destroyed this dome and produced a large summit crater and its downslope extension—the Gendol breach—both open to the SE (Fig. 1b). Dilute, high-energy (blast-like) PDCs were generated at the peak of the eruption and concomitant dense, valley-confined PDCs, of which at least one travelled about 16 km along Kali Gendol, caused widespread devastation on Merapi's south flank and about 400 casualties (Fig. 1c). In a later phase on 5 November, PDCs rich in pumice and scoria formed, followed by renewed dome extrusion at a rate exceeding the already high rates of the pre-climax dome extrusion phase. The eruption intensity declined on 8 November, but lahars, which had already started during the eruption, continued for several years, sweeping down all major valleys, and causing considerable environmental impact. Lahar runouts exceeded 15 km and even reached Yogyakarta on some occasions, destroying Sabo dams and bridges, affecting the road network, and destroying more than 800 houses. As a result of the eruption, many areas that were affected by PDCs have remained uninhabitable, and nearly 400,000 people were displaced from their homes and accommodated in temporary or permanent residences.

Post-2010 Activity of Merapi

After the 2010 events, volcanic activity at Merapi continued (Gertisser et al. 2023; Global Volcanism Program 2023). Following a gradual return to background levels, the background activity was interrupted by a period of phreatic explosions from 2012 to 2014, and again in 2018. A common feature of these short-lived eruptions was the lack of obvious early warning signs, making such activity difficult to forecast. A Vulcanian explosion on 1 June 2018 signalled the beginning of a new magmatic episode, with a new lava dome appearing at the summit on 11 August 2018, marking a return to typical 'Merapi-type' activity following the 2010 eruption. The new dome piled up on top of the 2010 lava dome and grew intermittently and at a relatively low rate. The activity increased

Fig. 1 **A** ASTER false-colour satellite image from 15 November 2010, showing pyroclastic density current (PDC) and minor lahar deposits of the 2010 eruption of Merapi filling the valleys of Kali Gendol, Kali Opak and Kali Woro on Merapi's S flank (light grey). The dark grey area north of the Merapi Golf course was affected by dilute PDCs, causing almost total devastation. — Source: NASA Earth Observatory (Digital Global, November 2010). (**B, C**) Impacts of the 2010 eruption: **B** Summit crater, open to the SE, formed during the 2010 eruption, with its downslope extension, the Gendol breach. **C**. House in Kinahrejo, destroyed during the climactic eruption phase on 5 November 2010. **D** UAV image from 16 September 2021, showing two simultaneously growing, active lava domes at the summit of Merapi: the SW dome and the central dome (after Gertisser et al. 2023) ▶

A
Merapi
Ash plume
Cloud
5 km
Kali Bebeng
Destroyed vegetation
Kinahrejo
Kali Woro
Kali Boyong
Kaliadem
10 km
Kali Opak
Merapi golf course
PDC (and minor lahar) deposits
Kali Gendol
15 km
B
2010 crater
Gendol breach
C
D
N
200 m
The Central Dome
The Southwest Dome
Bebeng
Krasak
Bedog
Boyong
Gendol
Woro

in September 2019, causing several explosive eruptions in the months to follow. During November and December 2020, the deformation rate continued to accelerate, and numerous rock avalanches were reported. Incandescence at the summit on 31 December 2020 hinted at the arrival of new magma at the surface, and on 7 January 2021, the appearance of a new dome in the SW part the summit just below the SW rim (SW lava dome) was reported. Several PDCs, accompanied by a significant explosion with an ash column of 12 km altitude and ash fall in the areas surrounding Merapi, occurred on 27 January 2021. On 17 February 2021, two domes were noted in the summit area, the SW lava dome, and a new dome in the centre of the 2010 summit crater (central lava dome). This has led to the remarkable situation of two simultaneously growing, spatially separate active lava domes at the summit of Merapi (Fig. 1d).

Both domes continued to grow in the following months, accompanied by incandescent rock avalanches and short-runout PDCs up to 2–3 km from the summit in the S and SW sector, particularly from the SW lava dome, but occasionally also down Kali Gendol due to failure of the central lava dome. By early 2022, both the SW lava dome and the central lava dome reached their maximum volume of 1.67 million m^3 and 3.23 million m^3, respectively. Since then, magma extrusion, dome growth and partial dome destruction by rock falls, rock avalanches, small PDCs, and explosions, particularly from the SW lava dome, have continued at high levels. Rock avalanches and small PDCs derived from the SW lava dome typically travel no more than 2 km in the SW and S sector, reaching the upper parts of the Bebeng and Boyong drainages. Geomorphic changes at Merapi's summit have typically been minor during this period due to continuing lava effusion and dome growth, balanced by concurrent partial dome collapse. More significant geomorphic changes have occurred only during either a period of dome growth or a period of dome failure. In December 2023, due to the continuing high-level of activity, the volcanic alert level at Merapi was still at level 3 (on a scale of 1–4), and the population was advised, depending on their location, to stay clear of areas between 3–7 km from the summit (Global Volcanism Program 2023).

4 Case Study 4

Mt. Tongariro, New Zealand: Geomorphic Changes Associated with the Te Maari Eruption of 6 August 2012

Jonathan N. Procter [1]

[1]Volcanic Risk Solutions, School of Agriculture and Environment, Massey University, Palmerston North, New Zealand

Mt. Tongariro, New Zealand is situated within the Taupo Volcanic Zone. The edifice has been constructed over the last 500 ka through the formation of different volcanic centres that are nested within a complex graben structure. Each centre has a history of discrete explosive and effusive periods related to the formation and construction of volcanic craters or cones (Hobden 1997). The Tongariro complex is deeply eroded in places, with numerous glacio-geomorphic features. Glacial retreat contributed to the generation of lahars by freeing up available water and exposing loose unconsolidated material in valleys and the massif slopes. The depositional sequences of laharic activity described in distal areas are sedimentologically and texturally different to debris flow and volcanic induced landslides preserved on more active proximal areas of the volcano. Lahars (including hyperconcentrated flows and debris flows) and debris avalanches are not commonly associated with this volcanic complex. Where they do occur in the stratigraphic record, they are a result of secondary processes of remobilisation of material or weathering/erosion of a stable edifice/cone rather than associated with explosive activity or vent formation.

The Te Maari Craters were most likely formed from activity that started at around 14 ka following the last glacial maximum. The area consists of an upper crater and an older lower crater, with the upper crater being more active historically. The 6 August 2012 eruption of the upper Te Maari Crater and the formation of new vents were similar to past eruptive episodes (1892, 1896; Hobden 1997). Small phreatic eruptions like the one that occurred on 6 August 2012 are relatively less hazardous than other phenomena in the Tongariro Volcanic Centre due to hazards from hydrothermal/phreatic eruptions being more localised (with deposits being <1 km radius). The 6 August Upper Te Maari eruption was a short-lived phreatic eruption that produced an eruption column between 6 and 10 km in height. There were distinctive events associated with the 6 August eruption; a small flank collapse that resulted in a debris flow, excavation and formation of vents and blasts/surges, ejection of ballistics and the phreatic eruption which distributed ash. Hill (1893) highlighted related eruption products such as surges (or pyroclastic flows) and debris flows (or lahars) associated with past eruptions.

Siebert et al. (1987) classifies the production of mass flows caused by collapse into two categories: those that are magmatic, which have syneruptive directed blast deposits (Bezymianny type), and those that are solely phreatic eruptions, commonly with debris avalanches associated with the onset of the eruptive activity (Bandai-type). Regarding the Bandai eruption of 1888, Yamamoto et al. (1999) suggested that the precursory seismic activity and the collapse of the volcanic edifice released the hydrothermal system, producing a debris avalanche and pyroclastic flow. The 6 August sequence of events could be classified as a Bandai-type.

The 6 August 2012 eruption produced measurable geomorphic changes (Fig. 1). LiDAR and aerial photos were collected after the event and compared to pre-existing aerial and elevation data (10 m, contour derived digital elevation models from topographic maps). Digital elevation model (DEM) analysis of the change (Procter et al. 2010) of the Upper Te Maari Crater (280,000 m^3) identified a new vent (200,000 m^3) and fissures (148,000 m^3) as well as a significant wedge of material (683,000 m^3) that slipped from the western flank of the crater that can be attributed to the source of a debris flow that propagated 2 km from source. The debris-flow deposit, from its morphology, can be subdivided into two sections: an upper (700 m long) segment and a lower segment, with the lower segment being further divided into three sections representing individual flow pulses that show a pattern of erosion on steep channels and deposition on flatter areas. The individual pulses of the flow (~1 min for each pulse) entrained material (up to 50% and 100% of their original volume) to runout up to 2 km from the source. The dynamics of the flow involved a change in elevation of 202 m and a runout (planar) of 769 m, indicating an H/L ratio of 0.26, consistent with the mean mobility of other cold rock, debris flows/avalanches. The area/volume ratio 0.125 indicates a flow similar to other debris flows (at the lower end of the lahar spectrum) where its runout is controlled by frictional regimes.

Velocities calculated from super-elevations indicate a pattern of pulses of material accelerating (down steep slopes) up to 11 ms^{-1} away from depositional areas and decelerating (9–10 ms^{-1}) on depositional, flatter areas. The initial stages of the debris flow or collapsing mass indicate higher velocities of 15 ms^{-1}. The average accelerations calculated would indicate a value of 11.5 $m.s^{-1}$ and an emplacement time of 215 s (3.5 min). Jolly et al. (2014a) highlight in seismic records that a lower frequency, low amplitude signal was recorded for approximately 4 min before the actual eruption at 11.52 UTC. If ~11:49 UTC marks the failure of the upper scarp and onset of the debris avalanche, and 11:51:50 marks the time that the debris flow entered the lower channel, then the debris avalanche would have propagated at approximately 18 m/s. This propagation time is consistent with the average calculated velocity of 11.5 ms^{-1} and an emplacement time of 215 s (3.5 min).

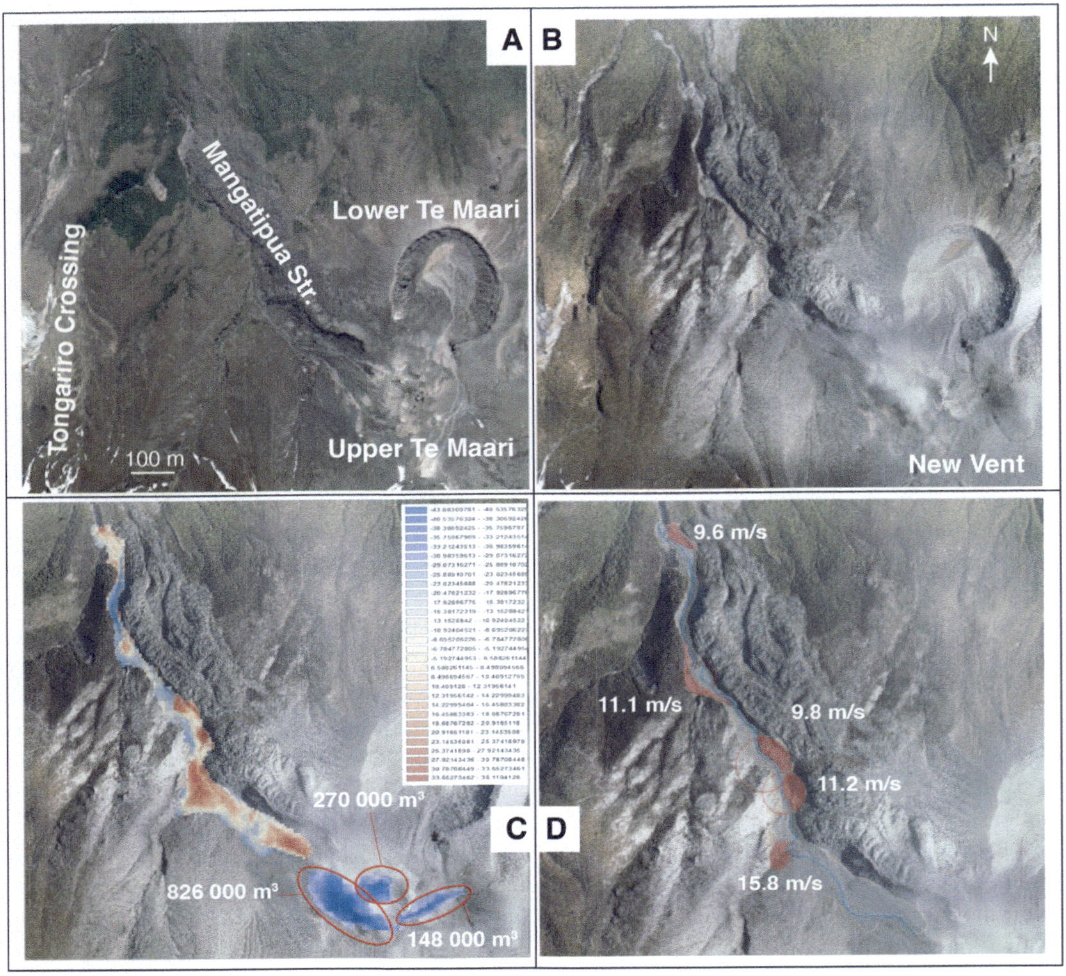

Fig. 1 Te Maari Craters and the Mangatipua stream. **A** 2004 orthophoto, pre-event. **B** 2012 orthophoto, post-event. **C** DEM difference map showing areas of volumetric change and volumes associated with the creation of new vents. **D** Velocities associated with superelevation of pulses within the flow

Using the volume displaced from the source area and the density of the material from hydraulic conductivity measurements, combined with the area of the displaced mass, we can calculate that a lithostatic pressure applied by that mass would be equivalent to 0.5 Mpa. Questions remain as to whether the release of this pressure and mass capping the area, retarding the heating and expanding the hydrothermal system could be enough to trigger the eruption. This process of collapse along weakened structures of the edifice of a cone is a commonly recognised trigger for debris avalanches, and while this process modifies the geomorphology of the edifice, it can also alter volcanic processes inducing the rise of magma and eruptions (e.g. Zemeny et al. 2023).

Jolly et al. (2014b) highlights the importance of undertaking geomorphological assessments prior to and after the eruption. Quantifying the geomorphic impact is an important contributor for assessing hazard impact, and can also be incorporated into models and mitigation strategies for future risk management.

5 Case Study 5

Rapid Geomorphic Evolution of the Sciara del Fuoco, Stromboli

Sonia Calvari[1], Daniele Casalbore[2], Federico Di Traglia[3], Antonio Costa[4]

[1]Istituto Nazionale di Geofisica e Vulcanologia, Osservatorio Etneo—Sezione di Catania

[2]Dipartimento di Scienze della Terra, Università degli Studi di Roma "La Sapienza"

[3]Istituto Nazionale di Geofisica e Vulcanologia, Osservatorio Vesuviano—Sezione di Napoli

[4]Istituto Nazionale di Geofisica e Vulcanologia, Sezione di Bologna

Stromboli volcano, famous worldwide for its mild and persistent Strombolian explosive eruptions, is an active volcanic island in the Mediterranean Sea. It also has occasional effusive eruptions (Calvari and Nunnari 2023), as well as rapid landscape build up and erosion (Di Traglia et al. 2022), summit crater failures (Di Traglia et al. 2018), flank collapses (Di Traglia et al. 2023), and tsunamigenic landslides (Chiocci et al. 2008). Effusive eruptions span in volume from a few tens to a few millions of m^3 (Calvari et al. 2005; Di Traglia et al. 2018), and can just accumulate within the crater depression (Calvari and Nunnari 2023), or spread along the barren upper Sciara del Fuoco slope. More rarely, long-lasting flank eruptions involve the opening of eruptive fissures along the flanks of the volcano (Calvari et al. 2005), leading to the creation of new land and coastal modifications and its gradual erosion (Casalbore et al. 2021).

To assess the rapid geomorphic phenomena taking place within the Sciara del Fuoco, the steep and unstable northwestern submarine-subaerial slope of the volcano, we looked at multi-platform and multi-temporal topo-bathymetric surveys, along with remote sensing data, from the past decade. This dataset encompasses a variety of topographic data sources, including multi-temporal digital elevation models (DEMs) generated through Light Detection and Ranging (LiDAR) technology and photogrammetry, utilizing images acquired from PLÉIADES satellites and Unmanned Aerial Vehicles (UAVs) (Di Traglia et al. 2022). In addition, high-spatial-resolution (HSR) optical imagery from QUICKBIRD and PLÉIADES satellites, as well as COSMO-SkyMed Synthetic Aperture Radar backscattering images, have been used to detect topographic changes (Casalbore et al. 2021). Offshore data was collected using repeated multibeam bathymetric surveys (MBES) and Side Scan Sonar (SSS) backscatter mosaics of the submarine flank (Di Traglia et al. 2022). The measurements were validated through direct observations of eruptive activity and geomorphic processes using monitoring webcams and repeated field inspections conducted over the time period (Calvari et al. 2005).

Based on the available geological data, the Sciara del Fuoco was formed through a lateral collapse event approximately 5.5 ka. Subsequently, significant landslides, accompanied by tsunami events, occurred during the Middle Ages (around 1 ka), and on 30 December 2002 (Casalbore et al. 2020 and reference therein). This latter event was triggered by the intrusion of magma into the Sciara del Fuoco, which coincided with the onset of the 2002–2003 flank eruption (Di Traglia et al. 2023). The landslide that caused the tsunami waves that struck the northern coast of the island involved a complex movement, characterised by the displacement of some 100–300 million m^3 of material, which varied between different stages of the landslide. Of this volume, only 21.1 million m^3 of material collapsed, of which approximately 9.5 million m^3 were mobilised in the submarine zone and 11.6 million m^3 were re-deposited between the coastal and submarine zones (Casalbore et al. 2020; Di Traglia et al. 2023).

Following the 30 December 2002 landslide, Stromboli experienced phases of significant material accumulation in the Sciara del Fuoco, primarily linked to flank eruptions (2007, 2014, 2019), as well as periods of increased eruptive activity characterized by frequent explosions, lava overflows from the crater, and collapses of the crater rim (e.g. 2012–13, 2021, 2022).

The integrated analysis of this large dataset revealed that fluctuations in 'typical' volcanic activity led to the gradual accumulation of erupted material on the Sciara del Fuoco slope during periods of increased explosion frequency. Conversely, during periods of reduced explosion frequency, the dominant processes observed were subaerial weathering and marine-driven reworking. 'Anomalously intense' eruptive phases, characterized by the emission of lava flows and/or high-intensity explosive activity resulting in pyroclastic density currents (PDCs), rapidly led to significant accumulation and erosion of the Sciara del Fuoco slope, mainly down to a depth of 300 m underwater. This dynamic has two main implications: (i) the alternation of brecciated lava and heterogeneous volcaniclastic deposits, often lacking lateral continuity, may influence the stability of the Sciara del Fuoco slope (Di Traglia et al. 2023); (ii) lava flows or PDCs entering the sea possess the potential for seafloor erosion and/or failure (Casalbore et al. 2020), which has implications for modeling both PDCs and tsunami generation (Figs. 1, 2 and 3).

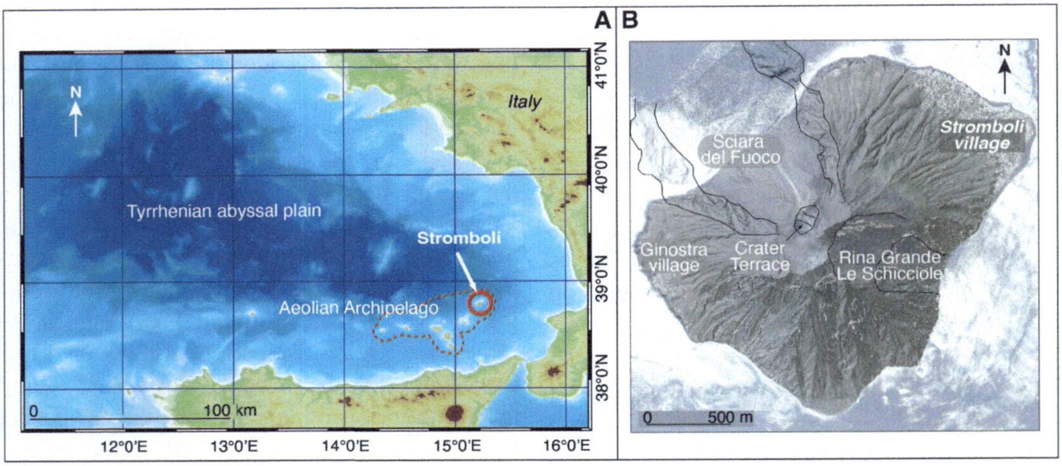

Fig. 1 A Location of Stromboli in the South Tyrrhenian Sea. **B** Orthophoto (PLEIADES-1) of the island of Stromboli, superimposed on a topo-bathymetric model, highlighting the primary geological-volcanological features and the two inhabited areas (modified from Di Traglia et al. 2023)

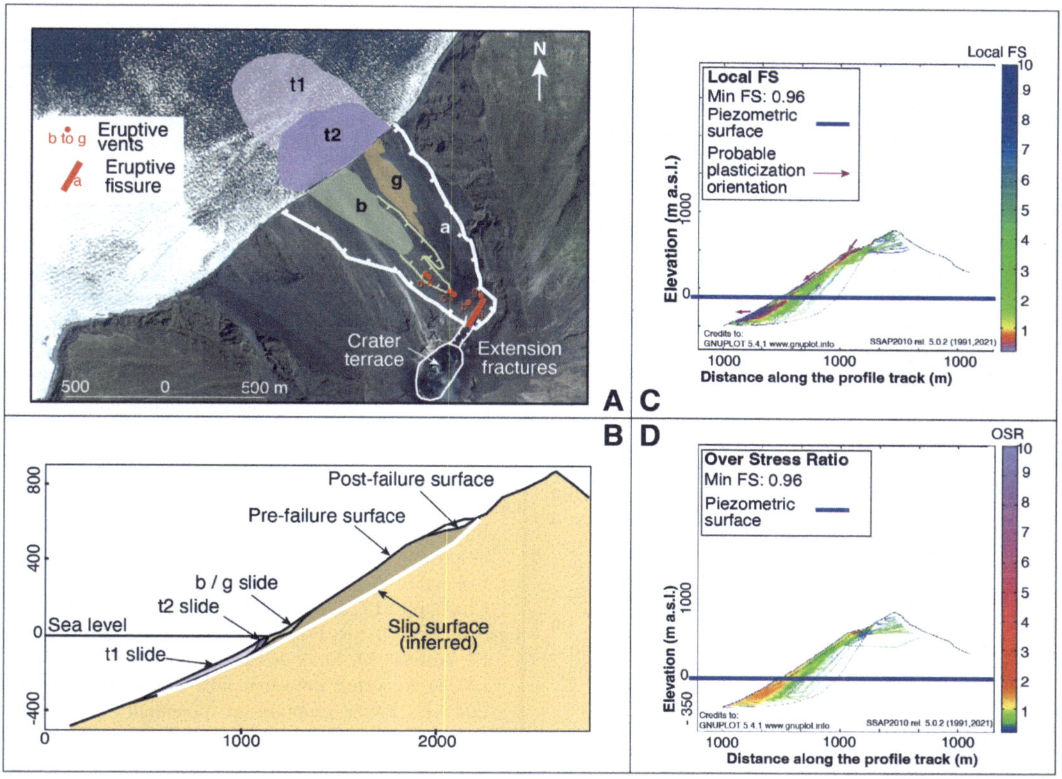

Fig. 2 **A** Map and **B** Schematic profile reconstructing the 2002 landslide sequence. The results of the back-analysis related to the 2002 landslide sequence are reported in **C** as a Factor of Safety profile and **D** Over Stress Ratio profile. Both profiles show the 2D stability results of the Sciara del Fuoco, accounting for the heterogeneity of the slope deposits, characterized by the irregular alternation of layers of volcaniclastic deposits and underlying lava levels (from Di Traglia et al. 2023)

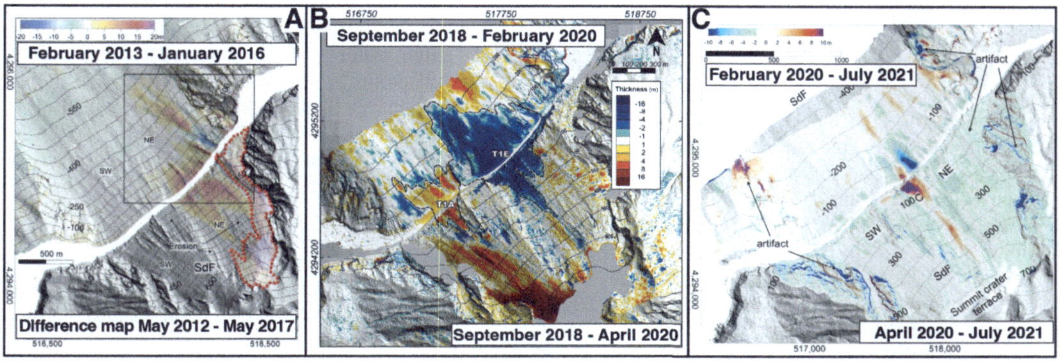

Fig. 3 Difference maps of the Sciara del Fuoco collapse scar (Stromboli), showing the main morphological changes induced by: **A** the 2014 effusive eruption (Casalbore et al. 2021), **B** the 2019 pyroclastic density currents and subsequent lava overflows (Di Traglia et al. 2022), and **C** the May 2021 pyroclastic density current due to crater rim failure and subsequent lava overflows (Casalbore et al. 2022). Dates in the upper and lower parts refer to the pre- and post-eruption topographic and bathymetric surveys, respectively

References

Barsotti S, Oddsson B, Gudmundsson MT, Pfeffer MA, Parks MM, Ófeigsson BG, Sigmundsson F, Reynisson,V, Jónsdóttir K, Roberts MJ, Heiðarsson EP, Jónasdóttir EB, Einarsson P, Jóhannsson T, Gylfason ÁG, Vogfjörd K (2022) Operational response and hazards assessment during the 2014–2015 volcanic crisis at Bárðarbunga volcano and associated eruption at Holuhraun, Iceland. J Volcanol Geoth Res 390:106753. https://doi.org/10.1016/j.jvolgeores.2019.106753

Björnsson H, Einarsson P (1990) Volcanoes beneath Vatnajökull, Iceland: evidence from radio echo-sounding, earthquakes and jökulhlaups. Jokull 40:147–168

Calvari S, Spampinato L, Lodato L, Harris AJL, Patrick MR, Dehn J, Burton MR, Andronico D (2005) Chronology and complex volcanic processes during the 2002–2003 flank eruption at Stromboli volcano (Italy) reconstructed from direct observations and surveys with a handheld thermal camera. J Geophys Res 110:B02201. https://doi.org/10.1029/2004JB003129

Calvari S, Nunnari G (2023) Statistical insights on the eruptive activity at Stromboli Volcano (Italy) recorded from 1879 to 2023. Rem Sens 15:4822. https://doi.org/10.3390/rs15194822

Carracedo JC, Troll VR (2016) The geology of the Canary Islands. Elsevier, ISBN 9780128096635

Carracedo JC, Troll VR, Day JMD, Geiger H, Aulinas M, Soler V, Deegan FM, Perez-Torrado FJ, Gisbert G, Gazel E, Rodriguez-Gonzalez A, Albert H (2022) The 2021 eruption of the Cumbre Vieja volcanic ridge on La Palma, Canary Islands. Geol Today 38:94–107. https://doi.org/10.1111/gto.12388

Carracedo JC, Troll VR (2021) North-East Atlantic Islands: The macaronesian archipelagos. In: Encyclopedia of Geology, vol 1–6, 2nd edn, vol 4: 674–699. Elsevier. https://doi.org/10.1016/B978-0-08-102908-4.00027-8

Casalbore D, Passeri F, Tommasi P, Verrucci L, Bosman A, Romagnoli C, Chiocci FL (2020) Small-scale slope instability on the submarine flanks of insular volcanoes: The case-study of the Sciara del Fuoco slope (Stromboli). Int J Earth Sci 109:2643–2658. https://doi.org/10.1007/s00531-020-01853-5

Casalbore D, Di Traglia F, Bosman A, Romagnoli C, Casagli N, Chiocci FL (2021) Submarine and subaerial morphological changes associated with the 2014 eruption at Stromboli Island. Rem Sens 13:2043. https://doi.org/10.3390/rs13112043

Casalbore D, Di Traglia F, Romagnoli C, Favalli M, Gracchi T, Tacconi Stefanelli C, Chiocci FL (2022) Integration of remote sensing and offshore geophysical data for monitoring the short-term morphological evolution of an active volcanic flank: a case study from Stromboli Island. Rem Sens 14(18):4605. https://doi.org/10.3390/rs14184605

Chiocci FL, Romagnoli C, Tommasi P, Bosman A (2008) The Stromboli 2002 tsunamigenic submarine slide: Characteristics and possible failure mechanisms. J Geophys Res, Solid Earth 113, B10102. htpps://doi.org/10.1029/2007JB005172

Di Traglia F et al (2018) Lava delta deformation as a proxy for submarine slope instability. Earth Planet Sci Lett 488:46–58. https://doi.org/10.1016/j.epsl.2018.01.038

Di Traglia F et al (2022) Subaerial-submarine morphological changes at Stromboli volcano (Italy) induced by the 2019–2020 eruptive activity. Geomorph 400:108093. https://doi.org/10.1016/j.geomorph.2021.108093

Di Traglia F et al (2023) Assessing flank instability of Stromboli volcano (Italy) by the reappraising of the 30 December 2002 tsunamigenic landslides. Landslides 20:1363–1380. https://doi.org/10.1007/s10346-023-02043-5

Gertisser R, Troll VR, Walter T, Nandaka IGMA, Ratdomopurbo A (2023). Merapi volcano: geology, eruptive activity, and monitoring of a high-risk volcano. Collection Active Volcanoes of the World. Springer Cham, pp XIX, 572. ISBN 978-3-031-15039-5.

Gisbert G, Troll VR, Day JMD, Geiger H, Perez-Torrado FJ, Aulinas M, Deegan FM, Albert H, Carracedo JC (2023) Reported ultra-low lava viscosities from the 2021 La Palma eruption are potentially biased. Nat Commun 14:6453. https://doi.org/10.1038/s41467-023-42022-x

Global Volcanism Program (2023). Merapi (263250). In: [Database] Volcanoes of the World (v. 5.1.4; 9 Nov 2023). Distributed by Smithsonian Institution, compiled by E. Venzke. https://doi.org/10.5479/si.GVP.VOTW5-2023.5.1

Gudmundsson MT, Jónsdóttir K, Hooper A, Holohan EP, Halldórsson SA, Ófeigsson BG, Cesca S, Vogfjörd KS, Sigmundsson F, Högnadóttir T, Einarsson P, Sigmarsson O, Jarosch AH, Jónasson K, Magnússon E, Hreinsdóttir S, Bagnardi M, Parks MM, Hjörleifsdóttir V, Pálsson F, Walter TR, Schöpfer MPJ, Heimann S, Reynolds HI, Dumont S, Bali E, Gudfinnsson GH, Dahm T, Roberts MJ, Hensch M, Belart JMC, Spaans K, Jakobsson S, Gudmundsson GB, Fridriksdóttir HM, Drouin V, Dürig T, Aðalgeirsdóttir G, Riishuus MS, Pedersen GBM, Boeckel T van, Oddsson B, Pfeffer MA, Barsotti S, Bergsson B, Donovan A, Burton MR, Aiuppa A (2016) Gradual caldera collapse at Bárdarbunga volcano, Iceland, regulated by lateral magma outflow. Science 353. https://doi.org/10.1126/science.aaf8988

Halldórsson SA, Bali E, Hartley ME, Neave DA, Peate DW, Guðfinnsson GH, Bindeman I, Whitehouse MJ, Riishuus MS, Pedersen GBM, Jakobsson S, Askew R, Gallagher CR, Guðmundsdóttir ER, Gudnason J, Moreland WM, Óskarsson BV, Nikkola P, Reynolds HI, Schmith J, Thordarson T (2018) Petrology and geochemistry of the 2014–2015 Holuhraun eruption, central Iceland: compositional and mineralogical characteristics, temporal variability and magma storage. Contrib Mineral Petrol 173:64. https://doi.org/10.1007/s00410-018-1487-9

Hill H (1893) The volcanic outburst at Te Mari, Tongariro, in November, 1892. Trans NZ Inst 26:388–392

Hjartardóttir ÁR, Einarsson P, Gudmundsson MT, Högnadóttir T (2016) Fracture movements and graben subsidence during the 2014 Bárðarbunga dike intrusion in Iceland. J Volcanol Geoth Res 310:242–252. https://doi.org/10.1016/j.jvolgeores.2015.12.002

Hobden BJ (1997) Modelling magmatic trends in time and space: eruptive and magmatic history of Tongariro Volcanic complex, New Zealand. Unpublished PhD thesis, University of Canterbury

Jolly GE, Keys HJR, Procter JN, Deligne NI (2014b) Overview of the co-ordinated risk-based approach to science and management response and recovery for the 2012 eruptions of Tongariro volcano, New Zealand. J Volcanol Geoth Res 286:184–207. https://doi.org/10.1016/j.jvolgeores.2014.08.028

Jolly AD, Jousset P, Lyons JJ, Carniel R, Fournier N, Fry B, Miller C (2014a) Seismo-acoustic evidence for an avalanche driven phreatic eruption through a beheaded hydrothermal system: an example from the 2012 Tongariro eruption. J Volcanol Geotherm Res 286:331–347. https://doi.org/10.1016/j.jvolgeores.2014.04.007

Pedersen GBM, Höskuldsson A, Dürig T, Thordarson T, Jónsdóttir I, Riishuus MS, Óskarsson BV, Dumont S, Magnusson E, Gudmundsson MT, Sigmundsson F, Drouin VJPB, Gallagher C, Askew R, Gudnason J, Moreland WM, Nikkola P, Reynolds HI, Schmith J (2017) Lava field evolution and emplacement dynamics of the 2014–2015 basaltic fissure eruption at Holuhraun, Iceland. J Volcanol Geoth Res 340: 155–169. https://doi.org/10.1016/j.jvolgeores.2017.02.027

Procter J, Cronin SJ, Fuller IC, Lube G, Manville V (2010) Quantifying the geomorphic impacts of a lake-breakout lahar, Mount Ruapehu, New Zealand. Geology 38(1):67–70. https://doi.org/10.1130/G30129.1

Siebert L, Glicken H, Ui T (1987) Volcanic hazards from Bezymianny and Bandai-type eruptions. Bull Volcanol 49:435–459. https://doi.org/10.1007/BF01046635

Sigmundsson F, Hooper A, Hreinsdóttir S, Vogfjörd KS, Ófeigsson BG, Heimisson ER, Dumont S, Parks M, Spaans K, Gudmundsson GB, Drouin V, Árnadóttir T, Jónsdóttir K, Gudmundsson MT, Högnadóttir T, Fridriksdóttir HM, Hensch M, Einarsson P, Magnússon E, Samsonov S, Brandsdóttir B, White RS, Ágústsdóttir T, Greenfield T, Green RG, Hjartardóttir ÁR, Pedersen R, Bennett RA, Geirsson H, La Femina PC, Björnsson H, Pálsson F, Sturkell E, Bean CJ, Möllhoff M, Braiden AK, Eibl EPS (2015) Segmented lateral dyke growth in a rifting event at Bárðarbunga volcanic system, Iceland. Nature 517:191–195. https://doi.org/10.1038/nature14111

Troll VR, Aulinas M, Carracedo JC, Geiger H, Perez-Torrado FJ, Soler V, Deegan FM, Bloszies C, Weis F, Albert H, Gisbert G, Day JMD, Rodríguez-González A, Gazel E, Dayton K (2024) The 2021 La Palma eruption; social dilemmas resulting from living close to an active volcano. Geol Today 40(3):96–111. https://doi.org/10.1111/gto.12472

Yamamoto T, Nakamura Y, Glicken H (1999) Pyroclastic density current from the 1888 phreatic eruption of Bandai volcano, NE Japan. J Volcanol Geoth Res 90(3–4):191–207. https://doi.org/10.1016/S0377-0273(99)00025-6

Zemeny A, Zellmer GF, Ubide T, Smith IEM, Procter J, Tapu AT, Zernack AV (2023) Investigation of the Mid-Age (65–34 ka) Period of Taranaki Volcano, New Zealand: indications for the effect of volcano growth. J Petrol 64(5). https://doi.org/10.1093/petrology/egad027

Chapter 11 Methods and Techniques for Studying Volcanic Landforms

D. Karátson, J.-C. Thouret
P. Grosse, M. Kervyn, T. Telbisz, P. Lahitte, D. O'Hara
S. Tarquini, M. de' Michieli Vitturi
S. J. Charbonnier, V. Gueugneau
B. Van Wyk de Vries
Q. Dumont, V. Cayol, J.-L. Froger
P. Lahitte, X. Quidelleur
D. Miallier
I. Schimmelpfennig, G. Bromley
J.-P. Dumoulin, J.-C. Lefèvre

1 Introduction

During the twentieth and twenty-first centuries, we have seen a dramatic development and continuous improvement in the methodologies used in Earth sciences, including volcanology. Most of the methods used are now essential for high-quality research. In this chapter the most important methods currently in use in volcano geomorphology are reviewed with the aim of covering the multiple fields of science that facilitate the study of volcanoes as landforms.

Section 2 is devoted to aspects of digital elevation model (DEM)-based volcano morphometry. Its Part 1, Morphometric analysis of volcanic landforms, is a summary by Pablo Grosse and Matthieu Kervyn, who present the currently available DEMs, whose resolution is improving continuously, of how the parameters of volcanic shapes can be obtained, and how the analysis helps constrain the evolution and degradation of volcanic edifices. Part 2, written by Tamás Telbisz, is devoted to the volumetry of volcanic landforms, showing how spatial evolution (in terms of volume growth) and then degradation (in terms of volume loss) can be quantified, and how the errors in the calculations can be assessed. Part 3 by Pierre Lahitte shows how the original shape of an extinct volcano can be reconstituted, and the constructed and eroded volumes calculated, using surface fitting. Finally, Part 4, by Daniel O'Hara, covers volcano erosion, primarily through soil creep and drainage evolution (channel incision), as well as landscape evolution modelling.

Section 3, written by Simone Tarquini and Mattia de' Michieli Vitturi, shows how lava flow propagation can be modelled by numerical simulations. Recent examples of Holuhraun (Iceland) and Etna (Italy) are used to demonstrate the application of deterministic and probabilistic codes.

Section 4 by Sylvain Charbonnier and Valentin Gueugneau is devoted to a highly

D. Karátson and J.-C. Thouret, *Volcano Geomorphology*, Advances in Volcanology,
https://doi.org/10.1007/978-3-031-77693-9_12

important volcanic process: how to model the poorly predictable hazard of pyroclastic density current (PDC) overspill using field and experimental studies and numerical modelling.

Section 5 is devoted to unravelling volcano structures. Part 1 is a summary of structural geology and geomorphology methods by Benjamin van Wyk de Vries; and Part 2 covers monitoring volcano deformation and magmatic processes using modern InSAR methods by Quentin Dumont, Valérie Cayol and Jean-Luc Froger.

Section 6 reviews the basic geochronological methods used to date the evolution of volcanoes over time. Part 1 introduces perhaps the most widely-used K–Ar and Ar–Ar method techniques by Pierre Lahitte and Xavier Quidelleur. Part 2 deals with two techniques of luminescence dating: thermoluminescence (TL) and optically stimulated luminescence (OSL), written by Didier Miallier. Part 3 covers cosmogenic nuclide dating methods, in particular surface-exposure dating, by Irene Schimmelpfennig and Gordon Bromley. Finally, Part 4 lays out the fundamentals and applicability of radiocarbon (^{14}C) dating by Jean-Pascal Dumoulin and Jean-Claude Lefèvre.

2 DEM-Based Morphometry

2.1 Morphometric Analysis of Volcanic Landforms

Pablo Grosse[1,2], **Matthieu Kervyn**[3]

[1]Consejo Nacional de Investigaciones Científicas y Técnicas (CONICET), Argentina

[2]Fundación Miguel Lillo, Miguel Lillo 251, (4000) Tucumán, Argentina

[3]Department of Geography, Vrije Universiteit Brussel, Pleinlaan 2, 1050 Elsene, Belgium

2.1.1 Volcano Morphometry and Digital Elevation Models

Volcano morphometry deals with the quantitative description and analysis of the shape and size of volcanic landforms (e.g. scoria cones, craters, calderas, collapse scars) and eruptive products (e.g. lava flows, debris-avalanche deposits). The purpose of volcano morphometry is to contribute to the understanding of the volcanic processes that define the growth, evolution and degradation of volcanoes. This can be done through repeat measurements of rapidly evolving volcanic features (e.g. active lava flows, lava domes, craters) or systematic quantitative analyses of volcanoes of various morphologies, evolutionary stages or different ages.

Morphometric data has been used for a wide variety of geomorphological studies of many types of volcanic landforms. Main applications include: (1) morphometric classification of volcano types and interpretation of related growth or constructional processes and rates in particular for monogenetic cones, shield volcanoes, strato- or composite volcanoes, subglacial volcanoes and extra-terrestrial volcanoes; (2) identification of systematic morphometric evolution related to volcano age or erosion stage; (3) control of tectonism and magmatism on the spatial and temporal distribution of volcano volume and shape; and (4) quantitative constraints on the short-term geomorphologic changes and growth rates of volcanoes for hazard assessment.

The main source of data for volcano morphometric studies is digital elevation models (DEM). DEMs are digital representations of the Earth's topography, typically consisting of regularly spaced elevation values in a raster. DEMs of the Earth surface can be obtained from ground surveys, existing topographic maps and airborne or satellite-based remote sensing (Table 1). Methods to generate DEMs include digitizing topographic maps, stereo-photogrammetry or multi-image photogrammetry (i.e. structure-from-motion), synthetic aperture radar (SAR) interferometry, and laser scanning. DEMs can also be obtained underwater (bathymetry) through sonar systems. Each acquisition technique has its own spatial resolution, defined as the size of the pixels or spacing between points for which a unique elevation value is known, and vertical and horizontal accuracies. DEM accuracy is the difference between the actual X–Y location (horizontal

Table 1 Summary of digital elevation model (DEM) sources and characteristics useful for studies in volcano morphometry; updated from Nelson et al. (2009) and Grosse et al. (2012)

Source	Method	Spatial resolution	Scene size	Coverage	Cost
Topographic map	Digitizing	Variable	–	–	Variable
Airborne photography	Photogrammetry	<1 m	Relatively small	–	Expensive
Spaceborne optical sensors					
ASTER GDEM	Photogrammetry	30 m	1° × 1°	Near-global (83°N-83°S)	Free
ALOS World 3D	Photogrammetry	30 m	1° × 1°	Near-global (82°N-82°S)	Free
High-resolution optical sensors	Photogrammetry	0.5–2 m	Relatively small	–	Expensive
Airborne radar	Single-pass interferometry	Variable	–	–	Expensive
Spaceborne synthetic aperture radar					
Shuttle Radar Topography Mission	Single-pass interferometry	30, 90 m	1° × 1°, 5° × 5°	Near-global (60°N-56°S)	Free
TerraSAR-X / TanDEM-X (WorldDEM)	Single-pass interferometry	12, 30 m	1° × 1°	Near-global (87°N-87°S)	Free / limited cost
Several radar sensors	Repeat-pass interferometry	~10 to 30 m	~100 × 100 km²	–	Variable
Airborne LiDAR	Laser scanning	~1 m	Relatively small	–	Expensive
Unmanned Aerial Vehicle; Unoccupied Aircraft System	Photogrammetry	<1 m	Very small	–	Expensive
National Datasets	Several	~10 to 30 m	Variable	Country-wide	Variable

accuracy) and elevation (vertical accuracy) of points measured on the ground and their representation in the DEM; it can be affected by the calibration methods used to georeference the DEM.

DEMs that are widely used for volcano morphometry are the free and near-global coverage Shuttle Radar Topography Mission (SRTM), Advanced Spaceborne Thermal Emission and Reflection Radiometer (ASTER GDEM) and Advanced Land Observing Satellite (ALOS) World 3D datasets. Their 30 m spatial resolution has proved suitable for morphometric studies of polygenetic volcanoes (i.e. composite volcanoes, shield volcanoes), but is of limited use for characterizing smaller volcanic features. The study of relatively small, sub-kilometre-sized volcanic landforms (e.g. scoria cones, lava domes, craters, maars and lava flows) requires higher spatial resolution DEMs. The reason is that coarse spatial resolution may lead to artificial smoothing of the topography, causing inaccurate local slope estimates and inexact identification or delineation of the small-scale volcanic features. The TerraSAR-X / TanDEM-X 12 m DEM (also known as WorldDEM) is adequate for these types of landforms given its higher resolution and near-global coverage. Other options are national datasets of variable sources and resolutions, such as the 10 m resolution National Elevation Dataset for the USA. Metre- to sub-metre-resolution DEMs can be obtained through photogrammetry of high-resolution optical satellite images (e.g. Pleiades, WorldView, Ikonos), airborne Light Detection and Ranging (LiDAR) or Unmanned Aerial Vehicles (UAVs), although often at high cost and/or workload. The latter are increasingly used for detailed morphometric studies on small features, and in particular for the analysis of morphometric changes with time (i.e. growth stage of a lava dome or emplacement of a newly erupted lava flow).

2.1.2 Volcano Landform Delineation

The spatial extent of each volcanic landform of interest must be defined to quantify the morphometry. This is a key issue because the resulting morphometric parameter values can vary considerably depending on the landform outline (e.g. Grosse et al. 2012). Landform delineation is typically carried out manually on a GIS platform. There are two main approaches for defining volcanic landforms: geology-based and topography-based. Delineation based on geology uses existing geological maps and/or aerial or satellite optical or radar images that can discriminate between lithologies, together with field knowledge. Delineation based on topography uses one or a combination of DEM-derived maps, such as shaded relief, slope, aspect or curvature, to draw outlines following breaks-in-slope along the landform base (in the case of volcano edifices or lava flows) or rim (in the case of craters, calderas or maars) (Smith and Clark 2005; Grosse et al. 2012). A few semi-automatic methods of DEM-based volcano edifice delineation have been developed, such as MBOA (Bohnenstiehl et al. 2012), NETVOLC (Euillades et al. 2013), and object-based image analysis (Pedersen 2016). These methods are less subjective, but suffer limitations when applied to complex topographic settings. Delineation of volcanic landforms can be used as a tool for geological mapping and as a basis for the identification of potentially hazardous areas.

2.1.3 Volcano Morphometric Parameters Derived from DEMs

A wide range of morphometric parameters can be obtained for elementary landforms (lava flows), volcanic edifices (scoria cones, lava domes, composite volcanoes), and volcanic depressions (craters, maars, calderas) (Fig. 1). Most parameters are computed from the DEM elevation dataset, whereas others are obtained from DEM-derived rasters, such as slope, and from derived elevation contours. Morphometric parameters can be obtained for the entire landform within the delineated area or for different elevation ranges. Morphometric data are generally computed using GIS or specifically designed programs such as MorVolc (Grosse et al. 2009, 2012), ShapeVolc (Lahitte et al. 2012; see 10.1.3) and DrainageVolc (O'Hara et al. 2024; see 10.1.4).

Size parameters, with a metric dimension, quantify the extent or magnitude of the landform. We describe edifices by the basal area, basal width or diameter, height and volume. We characterize depressions using the outline area and width, depth and volume. Flows are characterized by their length, width, thickness, areal coverage and volume. Estimates of volume and height of edifices, depth of depressions, and thickness of flows require the computation of a basal surface. Instead of using a horizontal plane at a chosen elevation for the basal surface, it is more accurate to construct a 3D basal surface by interpolation techniques, which consider the landform outline and the surrounding topography. Different interpolation methods can be used such as Triangular Irregular Network (TIN), Inverse Distance Weighting (IDW) or kriging (see 11.1.2). This is because derived metrics (height, depth, volume) are highly sensitive to such basal surfaces, especially when the landform has a sloping or irregular base.

DEM-derived slope maps can be used to compute different slope statistics, including the mean or median slope of the whole landform, of parts of the landform, and at successive elevation intervals (e.g. Rowland and Garbeil 2000; Grosse et al. 2012, 2014). Other DEM-derived rasters can be used to obtain additional parameters such as roughness or openness, highly useful for volcano geomorphological mapping (e.g. Favalli and Fornaciai 2017).

Non-dimensional ratios between size parameters quantify the overall landform profile shape. These include ratios between (1) edifice height and basal width, which is connected to flank steepness; (2) crater (or summit) width and basal width, which estimates edifice truncation or flat-toppedness; (3) crater, maar or caldera depth and width; and (4) lava thickness and length or area, which is controlled by effusion rate and

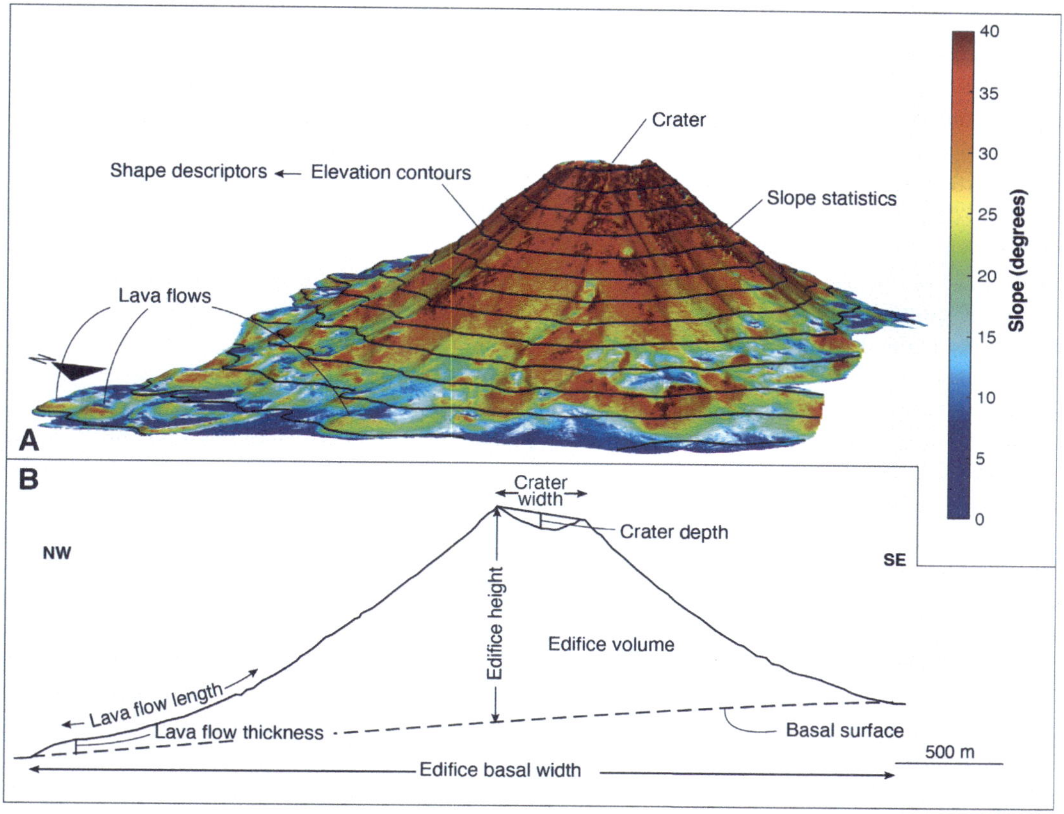

Fig. 1 3D and profile views of Chiliques volcano (Central Andes, Northern Chile) showing some DEM-derived morphometric parameters of the edifice, the crater and a lava flow

rheological parameters (Grosse et al. 2009, 2012; Cashman et al. 2013; Bemis and Ferencz 2017).

Edifice plan shape can be characterized by the shape of the landform outline and the shape of the closed elevation contours contained within the landform. The shape of complex contours, such as elevation contours, can be quantified by combinations of simple geometric measures (e.g. axial ratios, perimeter, and area) into dimensionless indexes known as shape descriptors. Descriptors that have been applied to volcano landforms include those that characterize contour elongation, such as ellipticity or circularity indexes (Grosse et al. 2009, 2012; Karátson et al. 2010; Graettinger 2018), and those that characterize contour complexity, such as dissection or irregularity indexes (Wright et al. 2006; Grosse et al. 2009, 2012). More complex indexes applying Fourier descriptors/transforms have also been proposed (Zarazúa-Carbajal and De la Cruz-Reyna 2020; Vörös et al. 2022). For outlines and elevation contours that are elongated, the azimuth of the major axis indicates the orientation of elongation, which is related to tectonic structures or stress conditions (Corazzatto and Tibaldi 2006).

2.1.4 Applications of DEM-Derived Morphometric Datasets in Volcano Geomorphology

Several studies have attempted to link volcano morphometry with constructional processes and evolutions. Grosse et al. (2009) identified evolutionary trends for composite arc volcanoes and inferred that they are controlled by two balancing factors: magma pressure versus lithostatic pressure, and conduit resistance versus edifice resistance. Karátson et al. (2010) recognized two types of symmetrical composite cones and linked their different morphometries to different eruption styles. Rowland and Garbeil (2000) analyzed the slopes of large shield volcanoes and identified the influence of several constructional and destructive processes, whereas Grosse and Kervyn (2018) classified shield volcanoes into 12 morphometric types and suggested a variety of controlling factors. A summary of the results obtained by these studies is given in Chaps. 4 and 5.

Other studies have focused on the degradation or erosion of volcanic landforms. As a first approach, morphometric analyses of scoria cones have highlighted the systematic decrease of the cone height-to-width ratio and flank slopes with increasing age (see Chap. 3 for details). A similar concept was used for regular-shaped composite cones to quantify long-term degradation (Chap. 8). For scoria cones, in addition, plan view irregularity indicators have also been shown to systematically increase through incision of the cone by gullies (Zarazúa-Carbajal and De la Cruz-Reyna 2020). However, different climatic and eruption conditions may lead to complex spatial relationships (Kervyn et al. 2012; Kereszturi et al. 2013).

Morphometric parameters have also been used to infer eruption styles (Whelley et al. 2015) and to assess analogies between volcanoes (Tierz et al. 2019). Other applications include analyses of links between morphometry and tectonic structures (Corazzatto and Tibaldi 2006) and of volcano volumes for arc-scale quantifications of magmatic additions to the crust (Völker et al. 2011; O'Hara et al. 2020).

High-resolution DEMs of active lava flows and domes provide constraints on eruption and growth rates and related processes with an unprecedented level of detail, making them fundamental for hazard mapping and assessment. Examples include the recent eruptions of Fagradalsfjall in Iceland (Pedersen et al. 2022) and Cumbre Vieja in the Canary Islands (Civico et al. 2022), the successive lava flows in the 2000s at El Reventador (Ecuador; Naranjo et al. 2016) and the recent lava dome growths of Merapi (Indonesia; Darmawan et al. 2018) and Santiaguito (Guatemala; Zorn et al. 2020).

2.2 Volcano Volumetry

Tamás Telbisz[1]

[1]Department of Physical Geography, Eötvös University, Pázmány s. 1/ C, H-1117 Budapest, Hungary

2.2.1 Introduction

Although height and base area are more easily imaginable parameters than volume, the real 3D dimension of a volcano can be characterised by its volume. Therefore, volume should be one of the most important parameters when comparing the size of volcanoes (e.g. Pike and Clow 1981; Carr 1984; Grosse et al. 2009; Karátson et al. 2010). The growth process of volcanoes can be described by the volume of magma brought to the surface and magma supply variations over time. The erosion of volcanoes can also be expressed in terms of volume and volume rates. In the following, data inputs, methods and possible errors related to volume calculations are briefly described.

The data types that can be used for volume calculation were presented in Sect. 2.1. The simplest approach is to model the volcanic edifices with simple 3D geometric bodies, i.e. a cone or a truncated cone. In this case, the base area (A_b), relative height (h), and the area (A_c) of the crater (or caldera, if present) are required. The base area (or crater area) can be converted to a radius using the formula $A = r^2\pi$ (R_b is the radius of the base; R_c is the radius of the crater/caldera). Then, the volume is

(1) in case of a cone: $V = A_b \cdot h/3 = R^2\pi \cdot h/3$

(2) in case of a truncated cone:

$$V = \left(A_b + \sqrt{A_b A_c} + A_c\right) \cdot \tfrac{h}{3} = \left(R_b^2 + R_b R_c + R_c^2\right) \cdot \tfrac{h\pi}{3}$$

2.2.2 The Concept of Volume Calculation

The volume calculations are nowadays mostly based on DEMs (see 2.1.1 for DEM data sources). The volcano volume can be given as a double integral between two surfaces (Karátson et al. 2012; Telbisz et al. 2012). The upper surface is the actual topography, the lower surface is the basal surface on which the given volcanic feature (the whole volcano or a smaller unit such as a scoria cone, or a lava flow) has been built. In the case of currently active volcanoes, if a DEM has been created before the most recent eruption, the pre-eruption DEM can be used as a perfect lower surface if the volume of this eruption is to be calculated. In the case of older volcanoes, the lower surface can be estimated by interpolation from the volcano boundary line and/or from geological data (e.g. boreholes, geophysical profiles). For numerical integration, three approximation methods are often used: the rectangle method, the trapezoidal rule, and Simpson's rule (which has a slightly modified version, Simpson's 3/8 rule). The rectangular rule in 3D can be thought of as distributing the DEM into vertical prisms and adding their volumes, as follows:

$$V = \sum_{px \in A_b} A_{px} \cdot h_{px}$$

where A_{px} is the pixel area, and h_{px} is the relative height of a given pixel. In the case of fixed size pixels A_{px} can be extracted. n is the number of pixels in the base area. Then, rearranging the above formula, we obtain:

$$V = A_{px} \cdot \sum_{px \in A_b} h_{px} = A_{px} \cdot n \cdot \overline{h} = A_b \cdot \overline{h}$$

where $\overline{h}$ is the mean height.

This is the easiest method to apply based on DEMs, and corresponds to calculating the difference between the upper and lower surfaces for the total area of the feature, and then multiplying the mean value ($\overline{h}$) by the base area of the feature. The mean value can also be obtained by first calculating the means of both surfaces separately and then subtracting the mean of the lower surface from that of the upper surface. The trapezoidal rule assumes a sloping flat-topped prism, while Simpson's rule fits a second- or third-order polynomial surface to the top of the prisms, so in theory they are capable of providing a more accurate volume calculation.

2.2.3 Error Assessment

In order to quantify the error magnitudes attributed to the above factors (methods and input data), we carried out test volume calculations on some volcanic landforms including small (lava domes) and large (composite volcanoes) edifices (Fig. 2). We incrementally changed the

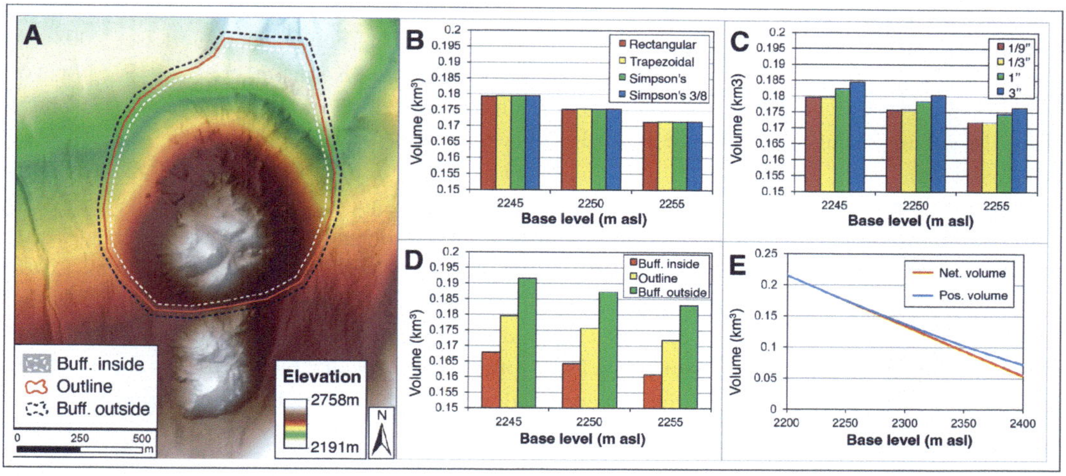

Fig. 2 Volume calculation tests on a small edifice (lava dome on Mt. Shasta, N41.4541, W122.1945). Horizontal axis: elevation asl (m), vertical axis: volume (km^3). **A** Topography; **B** Different integration methods; **C** Different DEM resolutions; **D** Different outlines; **E** different base levels (note that the vertical scale is different in this diagram)

resolution, the boundary outline, and the base level of the target landforms, and used different methods of volume calculation to assess the effect of spatial resolution, the integration method, the horizontal delimitation and the base level, on the final result (see Table 2).

Our comparison based on 1/9″ (~3m), 1/3″ (~10 m) LiDAR and 1″ (~30 m) and 3″ (~90 m) SRTM data shows that in most cases the volumes can be appropriately calculated using SRTM 1″ and even SRTM 3″: the average volume difference between SRTM 1″ and the best available spatial resolution dataset is only 2.8%. The integration method of volume calculation (rectangular versus trapezoidal; Simpson's rule vs trapezoidal) yields negligible differences, which are generally less than 1‰.

On the other hand, delimitation of the horizontal boundaries has a more significant effect on the volume; we found that 1% variation in the average edifice radius introduces an average error of 0.8% in terms of volume. In the case of field-based studies (i.e. when reliable geological maps are available), the error of the average edifice radius may be in the 1–5% range, thus the resultant error on volume is less than 4%. Finally, the most critical factor is the base level. For instance, using a theoretical approach, error (*dV*) of volume estimation (*V*) of a regular cone-shaped volcano (with a slope of α) due to error (*dh*) in height measurement (*h*) yields a relative error in volume:

$$dV/V = \left(\pi \cdot ctg^2(\alpha) \cdot h^2 \cdot dh\right)/\left(\pi \cdot ctg^2(\alpha) \cdot h^2/3\right) = 3dh/h$$

where the cotangent function is the reciprocal of the tangent (tg).

This represents, for instance, a relative error of 10% for a difference of 10 m at the base level of a 300-m-high volcano. In our test calculations of worldwide volcanic edifices, if the base level error is assumed to be 1% of the height, then the inferred volumetric error is on average 6.1%. Since the base level uncertainty might exceed 1% if the basement is not constrained by field evidence, field constraints on this measurement are considered a crucial issue in order to obtain precise values for the volume.

In order to estimate the volume loss of a volcano due to erosion, the first step is to reconstruct the 'original maximum size' of the volcano (see 10.1.3. about volcano landform reconstruction). Then the eroded volume can be calculated as the difference between the reconstructed original volume and the actual volume (Karátson et al. 2012). The degree of denudation is the proportion of the eroded volume with respect to the reconstructed volume. Surface lowering is the eroded volume divided by the area, therefore it characterizes the mean surface lowering by erosion over the area of the entire volcano and not that of a single topographic point.

The volume of volcanic landforms has been investigated in a number of studies using the simple or DEM-based methods described above for different climatic zones and different landform size ranges. Lava dome volumes have been calculated for examples in Chile (de Silva et al. 1994), USA (Christiansen et al. 2007; Scott et al. 2008), Japan (Nakada et al. 1999; Tatsumi et al. 2018), Mexico (Capra et al. 2015), Guatemala (Harris et al. 2003), Papua New Guinea (Taylor 1958), Lesser Antilles (Wadge et al. 2010, Boudon et al. 2015, Indonesia (Charbonnier and Gertisser 2008a, b) and Romania (Karátson et al. 2019). For volume calculations of larger edifices, see Chaps. 4 and 5; here, we highlight only a few: Grosse et al. (2009), Karátson et al. (2012) for composite volcanoes; Grosse and Kervyn (2018) for shield volcanoes; Broz et al. (2015), Grosse et al. (2020) and Becerra-Ramírez et al. (2022) for scoria cones. Moreover, volume calculations are important not only for entire edifices but for other types of volcanic landforms (such as lava flows) or products (debris-flow deposits).

Table 2 Error factors in volume calculations

Volcano name	V_{ref} (km³)	Area (km²)	Height (m)	Best resolution	Volume differences (dV, %) relative to V_{ref} due to							
					Resolution difference		Integration method		Delimitation difference (1% of R)		Base level difference (1% of height)	
					dV_{1s}	dV_{3s}	dV_{rect}	dV_{Simp}	dR (m)	dV_{del} (%)	dBL (m)	dV_{BL} (%)
Unnamed dome (Shasta North; USA)	0.07	0.8	464	1/9"	1.45%	2.31%	0.13%	0.00%	5	1.2%	5	3.3%
Puu Waawaa (USA)	0.18	2.0	407	1/3"	4.61%	2.24%	0.05%	0.01%	8	0.9%	4	5.0%
Murgul Mare (ROM)	0.27	3.0	326	1"		0.44%	0.16%	0.02%	10	0.4%	3	3.1%
Black Butte (Shasta; USA)	1.12	5.5	764	1/3"	0.95%	1.01%	0.04%	0.00%	13	0.3%	8	3.3%
Putauaki (New Zealand)	1.62	9.6	773	1/3"	7.12%	6.64%	0.00%	0.00%	17	0.4%	8	4.7%
Lookout (Or, USA)	3.36	27.3	184	1/3"	6.01%	6.10%	0.00%	0.00%	29	0.4%	2	1.7%
Ciomadul (ROM), whole	7.80	72.9	721	1"		0.02%	0.14%	0.00%	48	1.6%	7	5.4%
Crater Mt (Ca, USA)	12.03	87.4	578	1/3"	5.23%	5.19%	0.00%	0.00%	53	0.2%	6	4.4%
Sutter Buttes (Ca, USA)	16.43	182.5	634	1/3"	0.44%	0.45%	0.00%	0.00%	76	0.1%	6	7.8%
Burney (USA)	18.81	117.4	1472	1/3"	2.28%	2.40%	0.01%	0.00%	61	1.1%	15	9.8%
Tongariro (New Zealand)	101.65	496.4	1816	1/3"	1.24%	1.04%	0.00%	0.00%	126	1.1%	18	10.1%
Ruapehu (New Zealand)	244.13	1176.9	2383	1/3"	1.83%	1.91%	0.00%	0.00%	194	1.0%	24	12.5%
Adams (USA)	310.22	756.0	3182	1/3"	1.18%	1.23%	0.00%	0.00%	155	1.0%	32	8.3%
Shasta (USA), whole	327.84	767.8	3510	1/3"	1.07%	1.08%	0.01%	0.00%	156	1.0%	35	5.8%

V_{ref}: reference volume calculated after the best resolution DTM, the most reliable delimitation and base level, using trapezoidal rule. dV_{1s}, dV_{3s}: volume difference between V_{ref} and the volume obtained by using the 1″ (3″) SRTM; dV_{rect}: volume difference between V_{ref} and the volume obtained by using the rectangular integration method; dV_{Simp}: volume difference between V_{ref} and the volume obtained by using Simpson's rule integration method; dV_{del}: volume difference between V_{ref} and the volume obtained by changing the volcano outline by 1% of the mean radius (dR); dV_{BL}: volume difference between V_{ref} and the volume obtained by changing the volcano base level by 1% of the height (dBL)

2.3 Landform Reconstruction, Surface Fitting

Pierre Lahitte[1]

[1]Université Paris-Saclay, CNRS, GEOPS, 91,405 Orsay, France

2.3.1 Introduction—Main Purposes of Volcano Reconstruction

Since the 1990s, the availability of high-spatial resolution digital elevation models (DEMs), presented in 2.1.1, have provided a sound foundation for quantitative geomorphological approaches, including those that deal with volcanic surfaces. Such data have significantly improved our ability to study the surface morphology of volcanoes, enabling accurate modelling and analysis. Beyond other morphometric approaches, volcano reconstruction aims to provide a model of the edifice shape as it was at the end of its volcanic activity (= pre-erosional volcanic surface), i.e. prior to being degraded by erosion.

The reconstruction of volcanic surfaces provides insight into and constraints on various geological processes. The comparison between the pre-erosional volcanic surface with the pre-volcanic surface (i.e. the one that existed prior to the volcanic activity) documents the volume of volcanic construction. This bulk volume, which can be converted to dry rock equivalent volume, is a good proxy for magmatic activity, which in turn is conditioned by the geotectonic context (White et al. 2006; Dibacto et al. 2020; Bablon et al. 2020). On the other hand, comparison between the pre-erosional volcanic surface and the current one allows the 'missing' material, i.e. the volume that has been lost through erosion, to be quantified. Erosion combines all kinds of geological processes that remove and transport material, inducing a local decrease in elevation. If the removed volume is due to an instantaneous event, such comparison allows the magnitude of the given event, such as a caldera collapse (Lavigne et al. 2013) or a giant landslide (Germa et al. 2015; Grosse et al. 2022), to be calculated. In the case of long-term degradation through continuous erosion processes (Fig. 3A), such comparison constrains the volume of removed material or the degree of denudation, which is the ratio between the eroded and constructed volume. Once divided by the time since the end of volcanic activity, the obtained ratio defines the long-term erosion and degradation rate, which is often correlated to the prevailing local climate or its evolution through time (see Chap. 8; Karátson et al. 2012; Dibacto et al. 2020). In addition, pre-erosional volcanic surfaces can be used as initial surfaces for numerical modelling of erosion (see 2.4). In the case of complex and polyphased volcanoes (a volcanic massif for example, Fig. 3B), successive pre-erosional volcanic surfaces can be computed in order to document the volcanic evolution through its construction and degradation stages (Lahitte et al. 2012; Germa et al. 2015).

2.3.2 Main Methods for Volcano Reconstruction

Volcano reconstruction is typically carried out using digital elevation models (DEM). The reconstruction of volcanic surfaces is a complex and multi-faceted challenge. At its core, the process involves identifying the characteristics of the construction phase of a volcano and then applying mathematical modelling to create a theoretical representation of the volcano's initial surface.

Method 1: convex hull

The simplest method consists of computing a set of elevation contour lines, and then transforming them using the convex hull polygon. This assumes that the concave parts of the contours correspond to degraded sectors given that the global shape of a volcano is convex. This method is not very accurate as it replaces these portions by straight lines, instead of a convex shape, and does not consider any possible missing volume above the present edifice summit.

Method 2: fitting of elliptical contours

Another relatively simple method aims to model the volcano shape by computing a series of elliptical contours that best fit each of the

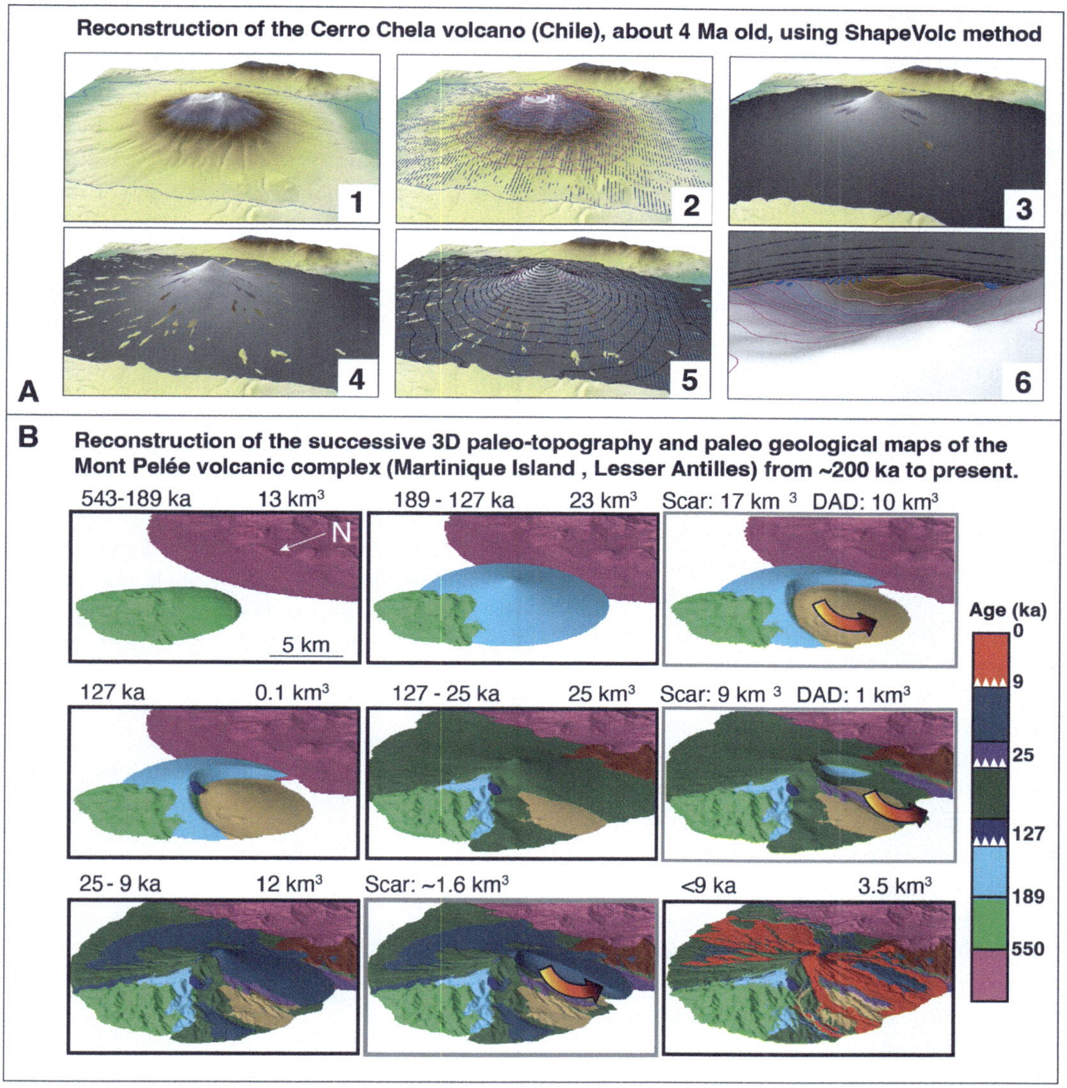

Fig. 3 Illustration of reconstructing a simple composite volcano and a more complex massif. **A** Main steps of volcanic reconstruction using the ShapeVolc algorithm. (1) current shaded and coloured DEM and volcano boundary (blue line); (2) Extraction of constraining points (in blue); the pink contours highlight the degree of erosion; (3) First-order model of reconstruction (defined by a single set of parameters); (4) Second-order model of reconstruction (a different set of parameters for each cell of the model); (5) Comparison between the second-order model, the CPs, the current contours (in pink), and the contours computed from the model (in black). Areas where pink contours are not visible are the more eroded ones, areas with both pink and black contours are those where erosion is minimal; (6) View from the summit toward the main southern valley. The space between the grey and brown/white surfaces highlights the loss of elevation since the end of volcanic activity by degradation processes, the integration of which documents the total volume of erosion. **B** Example of reconstruction of a complex polyphase massif, drawn from the model computed in Germa et al. (2015). The models bounded by black boxes show the main construction phases (with period of activity and constructed volume), those bounded by grey boxes illustrate large flank collapse events, with age and volume of the collapse (DAD = debris-avalanche deposit)

current contours of the volcano. This can be obtained for each contour by computing the ellipse centre (from the centre of mass for instance) and then the major and minor axes as well as the major axis azimuth.

Method 3: surface of revolution from preserved sectors

A more sophisticated method giving more realistic results (Fig. 3) consists of computing a surface of revolution around the volcano's centre (Lahitte et al. 2012; Favalli et al. 2014) from a characteristic elevation profile (e.g. Karátson et al. 2010).

Defining preserved sectors: The first step of this method is to identify the sectors of the current volcanic surface (Fig. 3A1) that have undergone ideally no, or minimal, degradation since the end of the construction phase. Extraction of these so-called constraining points (CPs) is a critical step in the process of reconstructing the volcano's initial surface, and as such it requires careful analysis of the current surface. Once these sectors have been identified, the next step is to model the numerical parameters to be used to create a mathematical representation of the initial surface. This is often done using computer software that can apply a model of interpolation based on these preserved sectors. The final step involves applying these numerical parameters to calculate the initial surface of the entire volcanic area.

2.3.3 Extraction of Constraining Points

The constraining points (CPs) are located either on quasi-planar surfaces or planèzes (Chap. 8), i.e. areas relatively protected from long-term erosion, or along the highest crests that define sectors where erosion was minimal (Fig. 3A2) (Lahitte et al. 2012; Karátson et al. 2016a, b). The selection of CPs is carried out considering one or a combination of DEM-derived morphometric parameters (Favalli and Fornaciai 2017), and manual checking to remove outliers where necessary. The two main morphometric parameters used for extraction of CPs are:

(1) *Openness parameter:*

Openness characterizes the degree to which a point on the terrain is surrounded by higher or lower terrain (Yokoyama et al. 2002), by measuring the amount of unobstructed space around a point. A point having a clear view of the surrounding terrain has a high openness value; one surrounded by higher terrain and obstructed or hidden from view has a low openness value; finally, one with an almost null value belongs to a planar surface (Chiba et al. 2008). Areas with low openness values should be systematically excluded from reconstructions as they are considered to have suffered intense erosion (i.e. valleys or gullies). Large areas with near-zero values are characteristic of the most well-preserved sectors, i.e. formed by lava flows and ash-fall deposits. Typically, such flat or gently sloping areas are bordered by steep cliffs or scarps that isolate and consequently protect these planèzes from the most intense erosional processes. In the context of more intensely eroded volcanoes where no planèzes remain, an alternative consists of the extraction of the interfluve boundaries, which are defined by the highest openness values. Indeed, as interfluves are the highest topography between adjacent watersheds, they define the less eroded areas locally.

(2) *Relative Topographic Position Index (RTPI):*

Introduced by Gallant and Wilson (2000), the RTPI index measures the relative vertical position of a point in relation to the surrounding terrain. The magnitude of the RTPI value indicates the steepness of the terrain around the central point. It is computed at each cell by dividing the difference between its elevation and the average elevation of its surrounding area (within a certain radius) by the standard deviation of the elevations within this area. An RTPI value >1 occurs where the cell is significantly higher than the surrounding area, i.e. along crests, ridges, or at isolated local peaks. On the other hand, an RTPI value <-1 identifies valleys or gullies.

2.3.4 Modelling of the Preserved Surface

Interpolation techniques are used to calculate the mathematical surface that best fits a set of points. Specific methods must be used to generate a surface that preserves the specific characteristics of volcanoes, namely a central symmetry around the summit and decreasing altitudes, to the first order, from the latter. Classical interpolation methods (i.e. kriging, local polynomial, or inverse distance weighting) would only be used if the degraded sectors are relatively minor and localized on the intermediate flanks of the volcano, as they are often unable to correctly model a dismantled summit area.

Theoretically, as eruptions mostly occur from a single and central vent, volcanic products are considered to spread relatively uniformly around the vent to develop a symmetric, conical edifice. Assuming such an edifice geometry, modelling of the preserved surface consists of the following main steps:

- *Defining the vent location:* there are several possible scenarios for identifying the location of a volcano's eruptive centre:
  - If the summit is preserved, the eruptive centre corresponds to the highest point, or the centre of the crater if such a landform occupies the top of the volcano.
  - If the central sector has been eroded, the most likely paleo-summit must be determined. This can be done by searching for the convergence point of the main valleys or ridges, which are usually radially oriented around the summit (Hampton and Cole 2009; Bossis et al. 2022). Alternatively, the palaeo-summit can be defined as the weighted average of the centres of the best fit circles or ellipses of the elevation contours, taking into account only those portions of the contours located in non-eroded areas (Karátson et al. 2010).
- *Defining one or several average profiles:* either a unique profile for the whole edifice or one for each main direction is needed to model the pre-erosional topography. The average profile of a volcano is determined by calculating the mathematical function that best fits all the CPs, once they have been projected onto an altitude-distance diagram from the eruptive centre. The nature of the volcanic edifice determines the optimal profile shape, which can be described by an affine function (in the case of a perfect cone), a polynomial function (such as a parabola), an inverse exponential function, or a Gaussian curve (Karátson et al. 2010; Grosse and Kervyn 2018). The volcano profile can be fitted using classical methods, such as least squares fitting, where the fitted profile lies as closely as possible to the set of points, i.e. passes through them (Karátson et al. 2010). Alternatively, more sophisticated methods can be used to define the best profile passing above maximum points (Lavigne et al. 2013; Favalli et al. 2014). This latter approach assumes that the volcano has undergone some erosion and hence the initial pre-erosional surface must have been higher than the set of points currently defining its shape.
- *Interpolating the whole surface:* Once the best profile has been computed (or the best ones, in the case of a sectorial approach), the overall shape is interpolated through the definition of a revolution surface (i.e. Lahitte et al. 2012; Karátson et al. 2016a, b). Such a surface is a three-dimensional one created by rotating a two-dimensional curve around an axis in space (Fig. 3A3). The resulting shape has a radial symmetry around the summit of the volcano. A more sophisticated approach consists of rotating the profile not simply in a circle but along an ellipse (Lahitte et al. 2012; Favalli et al. 2014).

Some methods have been specifically developed to compute all parameters defining this pristine volcano shape, i.e. summit location, profile equation, circular-elliptical shape parameters. The most advanced of them is ShapeVolc (Fig. 3A; Lahitte et al. 2012; Germa et al. 2015; Dibacto et al. 2020). Its main approach consists of defining the set of parameters that minimize the height differences between each CP and the elevations on the fitted surface computed at these points (Fig. 3A1-3). It results either in a volcano topography that passes through the CPs or one that drapes a majority of them, assuming that the pristine volcanic topography should be above the CPs. A more sophisticated approach (Fig. 3A4) consists of not only defining a unique set of parameters that optimizes the cost function for the whole edifice, but locally adjusting a specific set of parameters for each cell of the reconstructed surface (Dibacto et al. 2020; Grosse et al. 2022). This second-order model takes into account the complexity or irregularity of the volcanic landform caused by local elevation variations and details such as the presence of flank cones or domes, or particularly thick lava flows (Fig. 3A5).

Finally, from these reconstructed surfaces, studies of erosion processes (Fig. 3A) as well as definition of a complete a palaeo-topography and palaeo-geology can be carried out (Fig. 3B).

2.4 Numerical Modelling of Volcanic Edifice Erosion

Daniel O'Hara[1]

[1]Department of Geography, Vrije Universiteit Brussel, Pleinlaan 2, 1050 Elsene, Belgium

2.4.1 Processes of Volcano Degradation

Volcanoes weather and erode through both chemical and physical processes (see Chap. 8). Chemical and physical weathering degrades volcanic material, altering the composition, forming a soil profile, and making it more susceptible to erosion. This process, for example, decomposes ash deposits into clay or amorphous material (e.g. Wood 1980) and alters initially-permeable lava flows and scoria cones making it possible for them to support overland water runoff and drainage development (Lohse and Dietrich 2005; Jefferson et al. 2010).

However, in a broad sense, erosion (i.e. the removal of material) on volcanic edifices is characterised through three families of drivers—slope stability, eruptions, and climate. Mass-wasting occurs on steep terrain when the underlying structure of the subsurface is unable to support the shear stress induced by gravity on topography (often modulated by pore water pressure), producing both local-scale (e.g. rock falls, landslides) and edifice-scale (e.g. flank collapses and debris avalanches) events (see Chap. 7). Eruption-related erosion occurs through scouring by pyroclastic density currents, thermal erosion of the surface by hot material, or evacuation of material during explosive eruptions and caldera formation (e.g. Bacon and Lanphere 2006; Gase et al. 2017). Mass-wasting and volcanic eruptions are often related (such as during the 1980 Mount St. Helens eruption), but can also occur independently.

Climate-based erosion occurs on a spectrum of scale, precipitation, and temperature. Common drivers include wind, glaciation, soil creep, and overland water flow. Wind-driven sediment transport, scouring, and deposition are most common in arid environments where aeolian processes dominate over fluvial ones (e.g. the arid Central Andes). Glacial sculpting is one of the fastest and most efficient methods to remove material from edifices and leads to distinct morphologies (e.g. U-shaped valleys, arêtes or ridges, horns). However, this process is limited to colder climates associated with high latitudes, high elevations, or cooler periods of the Earth's history. Soil creep occurs within sedimentary regions in association with the alteration of fresh rock into regolith through a chemical weathering front, and is often viewed as a diffusive process with down-hill movement and accumulation related to bioturbation, rain splash, diurnal expansion and contraction, as well as water-based shrinking and swelling (e.g. Culling 1963). Finally, overland water flow occurs when precipitation overcomes the infiltration capacity of the soil surface. In regions of the edifice where topography converges, runoff can concentrate to generate rills that evolve into gullies and eventually drainage channels (Horton 1945) that incise into the edifice bedrock (i.e. channel incision) and transport material away from the volcano. Other transport processes (e.g. debris flows: see Chap. 7) can occur near channel heads where soil creep and flow convergence overlap.

Of the processes that erode landscapes, many of these either do not have well-constrained physical laws, have not been investigated in relation to volcanic edifices, or have not been analysed within the context of erosion (an exception to this is erosion of the substrate by PDCs, which has been examined in the field, by experiments, and theoretically; see Sect. 4.1). On the contrary, two common processes that have been modelled for volcanoes in depth include soil creep and channel incision. Their formulation and application are the focus of the rest of this section.

2.4.2 Soil Creep

Soil creep is typically modelled as a diffusion equation, with erosion dependent on the change in slope (i.e. curvature) of topography (Gilbert 1909). In its simplest form (and in one spatial dimension, x), sediment flux (Q) is modelled as being proportional to topographic slope (S) through a coefficient that is the soil's diffusivity (D). Erosion (E) is then calculated as the divergence in sediment flux (Culling 1963):

$$E = -\frac{dQ}{dx} = -D\frac{dS}{dx} \tag{1}$$

This equation predicts that erosion occurs when the slope gradient is negative (such as at the top of a ridge), and deposition occurs when the change in slope is positive (such as in a valley). Equation (1) also suggests that sediment flux and slope are linearly related. Although this assumption is valid in low-relief areas, it often breaks down in steep terrain where slopes are limited to an angle of stability (Roering et al. 1999). This inconsistency in steep terrain, as well as other issues related to topographic form, have led researchers to consider nonlinear diffusivity models. Various nonlinear relationships between slope and sediment flux have been proposed (e.g. Hooper and Sheridan 1998); a commonly-used model assumes a limiting slope factor as (Roering et al. 1999):

$$Q = \frac{DS}{1 - (S/S_c)^2} \tag{2}$$

where S_c is a critical slope related to the material's internal friction angle. This model suggests that sediment flux follows the linear diffusion model when slopes are low, but takes on a nonlinear relationship as slopes increase.

Figure 4A demonstrates the topographic evolution of a scoria cone using both the linear and nonlinear diffusion models based on the setup of Pelletier and Cline (2007) and diffusivity value of 1 m^2/yr. From an initial cone height of 280 m, radius of 450 m, and crater depth and radius of 120 m and 140 m, respectively, both models predict that the crater fills and disappears after ~5 kyr, with the cone decreasing to a height of ~180 m after 300 kyr. Although topographies are similar between the two models, comparing curvature values through time shows the models differ when slopes are steeper (Fig. 4B, time steps 1 and 5 kyr).

2.4.3 Channel Incision

When overland flow converges to form drainage channels, we assume that erosion occurs through shear stress induced on the channel bed by the overlying water column and local slope. Expanding on this assumption by incorporating geometric observations that channel discharge, depth, width, gradient, and upstream drainage area (A) are all related by power-law relationships (e.g. Leopold and Maddock 1953; Hack 1957; Flint 1974), a common formulation of river channel erosion is the 'stream power law', given as follows (Howard and Kerby 1983; Whipple and Tucker 1999):

$$E = KA^m|S|^n \tag{3}$$

in which K, m, and n are empirical constants related to lithology, erosional process, basin hydrology, and channel geometries. K is commonly referred to as 'rock erodibility', a measure of the rock's resistance to erosion, and the ratio between m and n is commonly used as a measure for the channel's longitudinal concavity.

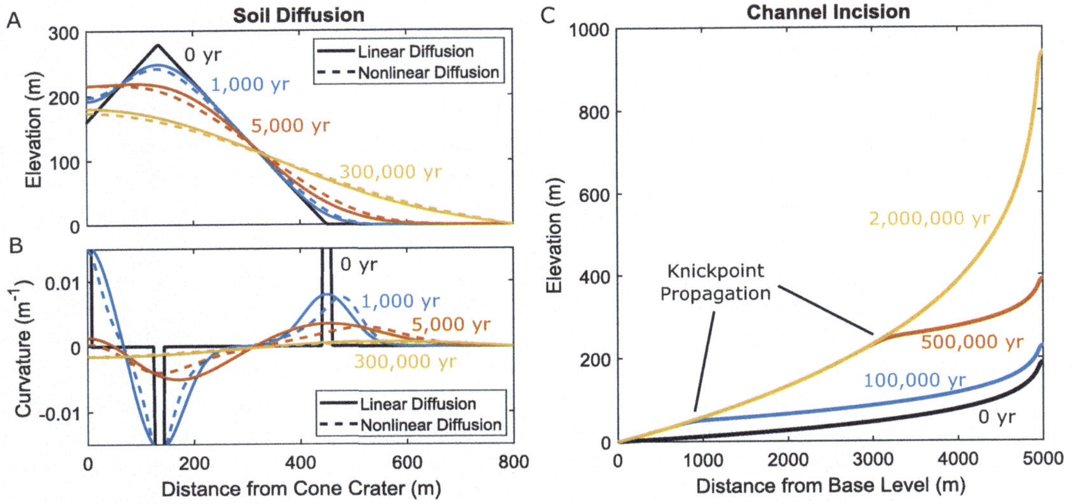

Fig. 4 Example of numerical models. **A, B** Evolution of (**A**) topography and (**B**) topographic curvature values of an eroding scoria cone with diffusion modelling. Solid and dashed lines are the linear and nonlinear diffusion models, respectively. **C** Evolution of a channel profile using the stream power law. In all panels, colours correspond to specific time steps. Model parameters provided in the text

Equation (3) is an advection equation, and predicts that a perturbation that alters the state of the channel (e.g. changes in channel concavity referred to as 'knickpoints') will migrate upstream to the channel head, adapting the landscape to a new equilibrium. Despite its simplicity, the stream power law is often used within orogenic settings to analyse the relationships between climate, tectonics, and topography (e.g. Kirby and Whipple 2001; Willett et al. 2014). However, an underlying assumption of this model is that sediment flux is lower than the channel's capacity for transport, such that deposition never occurs (i.e. a 'detachment-limited' setting). Other models have been developed that do not contain this assumption and simulate river erosion with sediment deposition (i.e. 'transport-limited' settings) (e.g. Shobe et al. 2017).

Figure 4C shows the evolution of a channel profile with $K = 10^{-5}\ m^{1-2m}/yr$, m = 0.5, and n = 1. Starting from an initial steady state (where surface uplift and erosion are equal), surface uplift is increased by a factor of five. This uplift increases local slope at the base level, inducing higher erosion rates. As the lower reaches of the channel adjust to the new uplift, the slope directly upstream becomes steeper, thus creating a knickpoint and incision wave that propagates upstream to the channel's head. During each time step, channel slopes are in equilibrium with the increased uplift rate everywhere downstream of the knickpoint, while slopes are still equilibrated to the previous uplift rate upstream of the knickpoint; in this manner, the channel retains some 'memory' of its previous conditions. After 2 Myr, the knickpoint has propagated through the entire system such that erosion and uplift are again in steady-state everywhere and all signs of the channel's previous state are gone.

2.4.4 Landscape Evolution Modelling Applied to Volcanoes

Landscape evolution modelling is a useful method for understanding how landscape topography changes through time (t). The general landscape evolution equation takes the form

$$\frac{dz}{dt} = U - E, \tag{4}$$

where U represents surface growth or uplift (if applicable), and E here is broadly defined to represent some erosion model—the stream power law or hillslope diffusion model, a combination of the two, or a different model not discussed here.

Volcanic edifices are often modelled for their conical initial shapes, datable surfaces, and rapid growth to analyze geomorphic principles. For example, diffusion-based models have been tested on scoria cones: Pelletier and Cline (2007) and de' Michieli Vitturi and Arrowsmith (2013) analysed cone morphologies to develop nonlinear diffusion models and derive their evolution, while McGuire et al. (2016) and Ben-Asher et al. (2017) analysed the effects of slope orientation, vegetation, and precipitation on cone morphology across cone fields. Similarly, the Hawaiian Islands have been used to test bedrock erosion models: Seidl et al. (1994) and Ferrier et al. (2013) analysed river channels with the stream power law to investigate longitudinal river profile evolution and the positive relationships between bedrock incision and precipitation, respectively. Huppert et al. (2020) utilised gradients in tidal force and a wave power model to analyse sea-cliff erosion and retreat.

Other studies have employed landscape evolution models as tools to explore edifice morphologies through time, and to understand how volcano growth and erosion affect regional landscape evolution. Hooper and Sheridan (1998) fit linear and nonlinear diffusion models to scoria cones from two fields in Arizona to analyze the evolution in crater dimensions and cone slopes (see Chap. 3). Comparatively, O'Hara et al. (2019) explored the effects of rapid intrusion emplacement (a proxy for volcano growth) on the evolution of surrounding drainage basins, while Manley et al. (2020) tested climate and sea level effects on volcanic sediment erosion and deposition along the Australian coast.

3 Numerical Modelling of Lava Flows

Simone Tarquini[1], **Mattia de' Michieli Vitturi**[1]

[1]Istituto Nazionale di Geofisica e Vulcanologia, Sezione di Pisa, 53 Via Cesare Battisti, 56126 Pisa, Italy

Topography in general and local slopes exert a straightforward effect on the path followed by all kinds of gravity-driven flows, including those occurring on the flanks of volcanoes, such as lava flows. Lava flows are undoubtedly one of the most frequent and spectacular phenomena at active volcanoes. Globally, millions of people live with the threat posed by future lava flows. For this reason, ever since the emergence of digital representations of volcano topography (i.e. DEMs), volcanologists have used numerical models to forecast the areas potentially at risk of inundation by future or ongoing lava flows (Macedonio et al. 1990; Wadge et al. 1994). Since then, considerable progress has been made, and nowadays the most advanced volcano observatories perform numerical modelling of lava flows as part of their standard tools for hazard and risk assessment purposes (e.g. Harris et al. 2019).

How do lava flow numerical simulations work? Several scientific codes to run simulations can be downloaded from the web for free (e.g. Cordonnier et al. 2016; Mossoux et al. 2016; de' Michieli Vitturi and Tarquini 2018); additional options are to ask the authors for permission to use their codes, or to use online facilities such as Vhub (https://theghub.org/resources/tools) and Victor (https://victor.ldeo.columbia.edu/) to carry out simulations through the web. And there are several free downloadable resources for DEMs of volcanoes and nearby areas, which have been available from the beginning of the twenty-first century (e.g. Kervyn et al. 2008).

Existing lava flow numerical codes can be divided into 'deterministic' and 'probabilistic' codes (Costa and Macedonio 2005). Deterministic codes account for the physics that rules the evolution of the cooling lava, while probabilistic codes approximate the whole process of the lava flow emplacement incorporating interplay between the gravitational pull driving the flow along the steepest descent path and the formation of the lava deposit which modifies the existing topography (Favalli et al. 2005; de' Michieli Vitturi and Tarquini 2018). Some deterministic codes, in principle, can mimic a step by step evolution of the lava flow system with time, but the complexity of the involved dynamics requires some approximation of the actual processes driving lava flows. The deterministic codes can be divided into full 3D models (e.g. Hidaka et al. 2005) and 2D 'shallow-water' models, based on a depth-averaging of the governing equations to reduce the computational cost of the numerical solution (e.g. Kelfoun and Vargas 2016). In contrast, probabilistic codes have a limited connection with the physics of the cooling lava, but they also illustrate how the topography drives their emplacement, with the advantage of a lower computational cost. Indeed, lava flows can be viewed as the building blocks that construct the volcano edifice over time, and such blocks constitute a substantial portion of many composite volcanoes and most shield edifices. Lava flows mimic running water, initially find their way downhill by following the existing valleys (Fig. 5). But unlike water, lava cools down to a solid state, progressively forming thick lava deposits along their pathways. Thus, after a given amount of time, what was initially a valley becomes firstly levelled topography, then eventually a ridge (termed 'topographic inversion', see inset in Fig. 5B).

From August 2014 to February 2015, the Bardarbunga volcanic system (central Iceland) discharged about 1.5 km^3 of lava into the bare floodplain of Holuhraun (Pedersen et al. 2017). This scenario was simulated using the MrLavaLoba code (de' Michieli Vitturi and Tarquini 2018). Despite the very gentle slope (0.2°), the initial progression of lavas was strongly controlled by the pre-eruption topography. On the other hand, the emplacement of late-stage lavas was influenced by the build-up of previously emplaced lavas, which provided additional topographic confinement and forced the flow field to spread out. An additional complication was the evidence that, despite the surficial morphology, a significant portion of the lava supply was transported further away from the eruptive vent through subsurface lava tubes which supplied distal ephemeral vents. As a result, the best-fit simulation of the final flow field was obtained by including computational vents in given places to account for the actual lava distribution system (Fig. 5A).

Lava flow simulation codes, besides being used to assess the progression of ongoing eruptions, are widely used to derive hazard and risk maps to help with environmental and land use planning. Similar uses are nowadays common practice at Mount Etna, one of the most active and densely populated volcanoes of the world. Creating these maps involves the processing of a large number of simulations obtained from an array of possible future vents. Each possible vent can be assigned with a given probability of activation and/or given flow characteristics, and the results convey the probability for any given area to be inundated by a future lava flow. Figure 6 (modified after Tarquini and Favalli 2010) illustrates an example of this probabilistic assessment. In this case, the focus is on the effect of geomorphic modifications in the areas prone to future lava flow invasion. An array of 523 computational vents is defined around the summit area of Mount Etna, and the DOWNFLOW code (Favalli et al. 2005) is used to derive a simulation from each vent. The whole set of simulations (i.e. from the same vents) is repeated using, as computational domains, four different DEMs of the volcano, the first created in 1986, and subsequent ones from 1998, 2005 and 2007. For each of these DEMs, a map of the susceptibility of areas exposed to future lava flow inundation from summit vents is then obtained (see top right panel in Fig. 6 for the 2005 DEM). Then, DEM-difference maps (left column in Fig. 6), and susceptibility-difference maps are created (right column in Fig. 6). The obtained results illustrate how, in this case, the build-up of lava flows correlates with a decrease in the probability of the area being inundated by future lava flows.

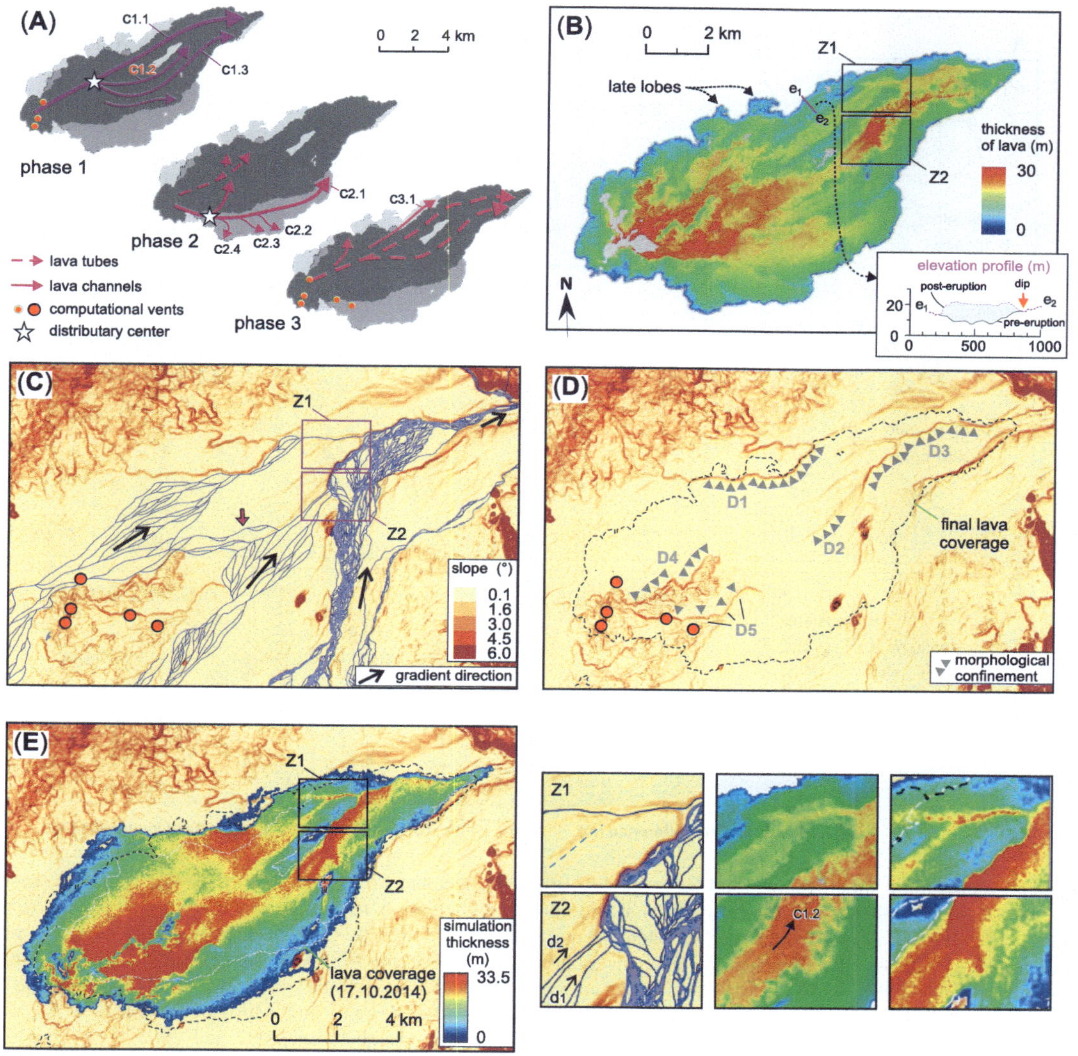

Fig. 5 Cross-comparison between DEM, real lava flow field of Holuhraun, and a MrLavaLoba simulation (modified after Tarquini et al. 2019). **A** Main direction of lava transport during the three phases of the eruption (younger lavas shown in lighter tones); **B** Real lava flow thickness mapped at the end of the eruption; **C** Hydrological drainage network overlain on slope map from pre-eruption DEM; **D.** Pre-emplacement morphological confinements; **E** Simulation of 1.5 km^3 of lava, vents as in (**A**), phase 3. Z1 and Z2 zoom in on selected details for comparison

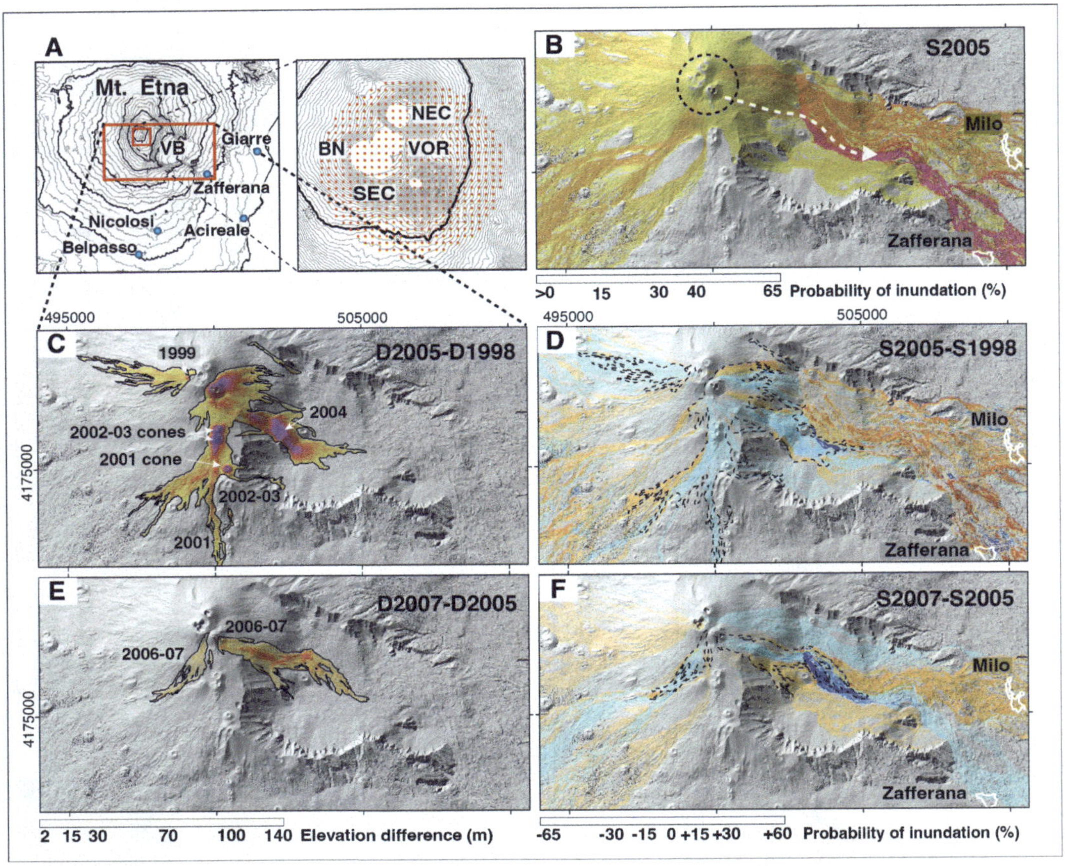

Fig. 6 Geomorphic modifications and susceptibility of areas to being inundated by future lava flows at Mount Etna (1986–2007). **A** The study area at Mount Etna, VB = Valle del Bove, main towns are labelled; small inset shows the array of 523 computation vents; crater rims in orange, BN = Bocca Nuova, VOR = Voragine, NEC = North East Crater, SEC = South East Crater. **B** Example of susceptibility map obtained for the 2005 DEM; white dashed arrow is the main direction of the emplacement for the 2006–2007 flows; note that these flows follow a maximum susceptibility direction from the vent downhill. **C** DEM-difference map between the years 2005 and 1998 with dates of main lava deposits marked. **D** Susceptibility-difference map (S2005-S1998); towns of Zafferana and Milo are labelled. **E** DEM-difference map between 2007 and 2005 with dates of main lava deposits marked. **F.** Susceptibility-difference map (S2007-S2005)

4 The Overspill Hazard of Small-volume PDCs

Charbonnier, S. J.[1], **Gueugneau, V.**[2]

[1]School of Geosciences, University of South Florida, Tampa, FL, USA

[2]Sol Solution, ZA des Portes de Riom Nord, France

4.1 Introduction

Even though small-volume (usually $<10^7$ m^3) PDCs are channelised in valleys of volcanic edifices, most fatalities are due to the unpredicted overspill of the flows from the channel confines and a rapid inundation of the interfluves, as unconfined PDCs can travel up to a few kilometres from their overflow point. Table 3 summarises notable past PDC overspills, the

Table 3 Summary of notable PDCs and morphometry of the corresponding channels they flowed into. For each case, the topographic feature identified as being responsible for an overspill is described

Eruption	Topographic feature type	Distance of source to obstacle (km)	Flow runout distance (km)	Height/ width of obstacle	Ratio of distance of obstacle to flow runout	H/L flow	References
Colima 2015	Bend	3.5	10.4	0.33	0.33	0.28	Macorps et al. (2018)
Soufrière Hills 1997	Bend	1.8	5.3	0.33	0.34	0.14	Loughlin et al. (2002); Ogburn et al. (2014)
Unzen 1991	Bend	1.6	7.2	0.22	0.22	0.31	Yamamoto et al. (1993); Nakada and Fujii (1993)
Merapi 1994	Break in slope	1.8	6.5	0.15	0.28	0.29	Bourdier and Abdurachman (2001); Kelfoun et al. (2000)
Merapi 2006	Obstacle + constriction	4.5	7	0.39	0.64	0.28	Charbonnier and Gertisser (2008a, b, 2011); Lube et al. (2011)
Merapi 2010	Bend + constriction	1.8	16.2	0.15	0.11	0.16	Komorowski et al. (2013); Jenkins et al. (2016); Gertisser et al. (2012)
Fuego 2018	Bend	7.8	11.4	0.29	0.68	0.26	Albino et al. (2020)
Ngauruhoe 1975	Break in slope	1.2	2.24	0.14	0.53	0.45	Nairn et Self (1978); Lube et al. (2007)
Lascar 1993	Constriction	1	6	0.16	0.17	0.29	Sparks et al. (1997); Jessop et al. (2012)

H/L refers to the ratio between the initial height (H) of the material over the flow runout length (L). Modified from Gueugneau et al. (2022), all references therein

morphometry of the corresponding channels they flowed into, as well as the topographic feature identified as being responsible for the overspill.

As detailed in Gueugneau et al. (2021), Kubo Hutchison and Dufek (2021), and Lerner et al. (2021), unconfined PDCs exhibit various depositional and dynamic characteristics, and we distinguish here two main types based on their solid concentration (Fig. 7):

(i) *PDC overspill* when the concentrated basal layer of a PDC escapes the valley at a specific location, usually, but not always, accompanied by its upper ash-cloud surge. This overspill continues to flow along valley banks, volcaniclastic terraces, and interfluves, and is usually named *overbank flow* to distinguish it from its parental *valley-confined flow*, as at Merapi during the 2006 and 2010 eruptions (Charbonnier and Gertisser 2008a, b; Lube et al. 2011; Gertisser et al. 2012; Charbonnier et al. 2013. Solikhin et al. 2015), at Volcán de Colima (Mexico) in 2015 (Macorps et al. 2018), and at Fuego in 2018 (Charbonnier et al. 2023).

(ii) *Ash-cloud surge detachment* when only the dilute ash-cloud surge decouples from the concentrated basal layer and escapes the confining valley to propagate on its own, as at Montserrat in 1997 (Druitt et al. 2002; Loughlin et al. 2002; Ogburn et al. 2014), Unzen in 1991 (Nakada and Fujii 1993), Merapi in 2010 (Komorowski et al. 2013), and Montagne Pelée in 1902 (Gueugneau et al. 2020).

Fig. 7 Examples of unconfined PDCs with schematics showing the interaction between PDC and topography, some impacts and an indication of the remaining associated deposits. Black diamonds indicate the different types of unconfined PDCs, with one historical example given for each type. The red diamond represents the characteristics of a high-energy channel-confined PDC for comparison. *Note* Many of the eruptions provided as examples for each particular PDC type contained other PDC types as well. Citations for each eruption are as follows: **B** Gertisser et al. (2012), **D** Anderson and Flett (1902), **E** Nakada and Fujii (1993), **F** Rosi et al. (1993). Modified from Lerner et al. (2022)

4.2 Recent Findings and Perspectives

The physics behind PDC overspills remains poorly known despite their danger. Most of our knowledge is based on field studies of PDC deposits (Charbonnier and Gertisser 2008a, b; Lube et al. 2011; Gertisser et al. 2012; Charbonnier et al. 2013; Ogburn et al. 2014; Charbonnier et al. 2023). Such studies have pointed out that a sudden change in the channel geometry (from natural causes or as a result of human intervention) can potentially reduce the channel capacity, causing the PDC to overspill. Recently, Kubo Hutchison and Dufek (2021) and Gueugneau et al. (2021) numerically studied the overspill mechanism. Kubo Hutchison and Dufek (2021) have demonstrated that a sinuous valley can cause a significant increase in local flow thickness around the outside of a bend, called superelevation, leading to the overspilling of the channelized flow. Such sudden superelevation was also inferred from field studies at Merapi in 2006 (Charbonnier and Gertisser 2008a, b; Lube et al. 2011).

Using a combination of synthetic topographies, field data compiled from three recent eruptions, and numerical models, Gueugneau et al. (2021) highlighted the link between the increase in local PDC volume flux (calculated along the cross-sectional area of the channel) and the occurrence of an overspill (Fig. 8). Unsteady flow conditions and increases in flow velocity, thickness, and volume flux can also promote overspilling (Charbonnier et al. 2013; Kubo Hutchison and Dufek 2021). Moreover, Gueugneau et al. (2021) showed that overspill processes occur either after a drop in the channel capacity (constriction) or a peak in channel sinuosity (bend). The same observations have been made for several other natural block-and-ash flows (BAFs): the 14 June 2006 BAFs (Charbonnier and Gertisser 2008a, b, 2011; Lube et al. 2011) and the 5 November 2010 BAFs at Merapi (Charbonnier et al. 2013; Cronin et al. 2013; Solikhin et al. 2015) and the 11 July 2015 BAFs at Colima (Macorps et al. 2018). Gueugneau et al. (2022) showed that the formation and dynamics of overbank flows are linked to the dynamics of their parent confined flows, and a correlation exists between their respective flow runouts. Overbank flow is caused by both the formation of a granular jump (sudden thickness variations of a granular

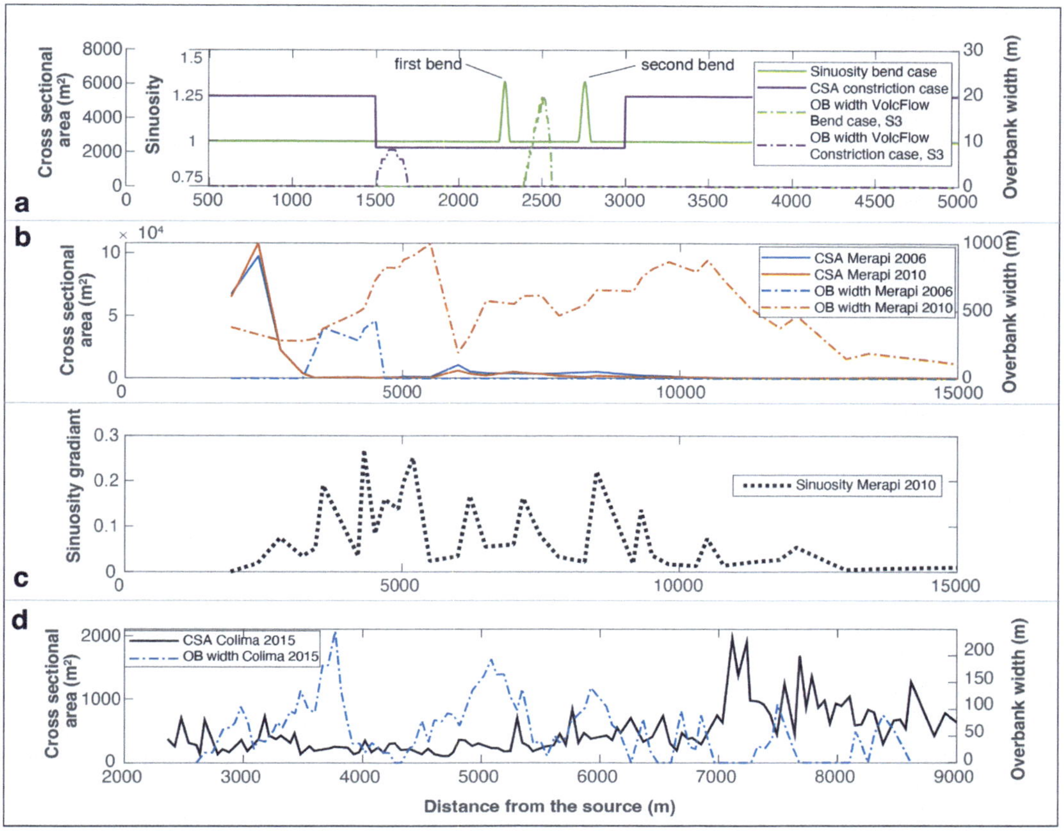

Fig. 8 Morphometric data extracted from the benchmark results of Gueugneau et al. (2021) compared to a set of field data compiled from three recent eruptions (Merapi 2006 and 2010, Colima 2015) associated with block-and-ash flows (BAFs): **A** Channel sinuosity and cross-sectional area measured along the synthetic channel in the 'bend' and the 'constriction' benchmark cases, respectively, as well as the corresponding overbank width extracted from two VolcFlow simulations. **B** Cross-sectional area measured along the Gendol River channel at Merapi, and the corresponding overbank width measured before and after the emplacement of the 2006 and 2010 BAFs, respectively. **C** Sinuosity gradient measured along the Gendol River at Merapi in 2010 before the eruption. **D.** Cross-sectional area measured along the Montegrande channel at Colima in 2005, and the corresponding overbank width measured after the emplacement of the 2015 BAFs (from Macorps et al. 2018). Modified from Gueugneau et al. (2021)

flow impacting an obstacle) and a peak in discharge rate due to flow unsteadiness, which contribute to an increase in granular material thickness along the impacted side of the bend and cause an overspill. These recent studies suggest that the PDC properties (volume, flow rate) and the geometry of the valley channel are key parameters controlling overspills. A deeper understanding of the interaction between the valley-confined PDCs and the topography, as well as the evolution of the internal flow conditions during emplacement, is needed to better characterise the mechanisms of PDC overspills and assess related hazards.

Future studies on PDC overspills should focus on determining the conditions that promote the formation of a superelevation wedge in the vicinity of a sudden topographic change. Using up-to-date, high-resolution DEMs (such as LiDAR-derived maps), an automatic recognition of hazardous overspill zones along valleys of PDC-generating volcanoes from numerical simulations could lead to a new generation of enhanced and dynamic hazard and risk maps.

The fact that a granular jump is one of the factors contributing to the formation of overbank flows confirms the ability of shallow-water numerical models to simulate PDC features, since such models are able to reproduce hydraulic jumps (as seen for example in Gueugneau et al. 2021). The possible presence of a basal stagnation zone, however, would require the use of other modelling approaches, such as two-phase continuum (Aravena et al. 2021), multiphase discrete element (Lube et al. 2019; Breard et al. 2022), or smooth-particle hydrodynamics (Zhu et al. 2021) simulations.

Further experimental studies should address some key parameters in more detail. First, the channel geometry: as shown by earlier studies, the angle and the sinuosity of the bend (Peruzzetto et al. 2021; Kubo Hutchinson and Dufek 2021) as well as the shape of the topographic feature (obstacle, break in slope, constriction; Gueugneau et al. 2021) control the generation and type of overspill. Comparison with the results of other existing experimental facilities such as that of Sulpizio et al. (2016), for which flows are only confined along half of their path, could better constrain the effect of flow confinement on the flow runout. It would also be important to run experiments with polydisperse granular materials in order to investigate the dynamics of two-layer flows, and in particular the detachment of the upper dilute part on impact with different topographic features.

5 Approaches for Unravelling Volcano Structures

5.1 Volcano Structure—Methods of Analysis in Geomorphology

Benjamin van Wyk de Vries[1]

[1]Laboratoire Magmas et Volcans, Observatoire du Physique du Globe de Clermont, Université Clermont, Auvergne, IRD, UMR6524-CNRS, 63,178 Aubiere, France

Volcanoes are built up by eruption and intrusion, in contrast to other mountains, which are formed through erosion of uplifted terrain. The built nature of volcanoes generates a number of consequences that are seen in the structure and have impacts on the methods to study their landforms. As built-up mountains, volcanoes have a range of geomorphic features, such as lava flows, lobes of pyroclastic density currents and accumulated fragments that are studied through their detailed topography, as these features are mostly of fractured material (lavas and intrusions) or loose material (the ejected pyroclastic or remobilized epiclastic fragments). Such landforms can be studied by morphometry, using digital elevation models (DEMs); and global datasets have become available today to chart a full range of volcano morphologies (see Chap. 11, Sect. 1). This method also allows structural analysis, as volcano shape is not only due to construction, but also erosion and deformation (Fig. 9; Grosse et al. 2012).

A consequence of the loose nature of rock means that structural deformation, e.g. by faults, will spread out, and only in terrain with lava flows are faults seen as well-localised features. Even in such hard lavas, the surface faulting generally separates already-formed cooling fractures, thus influencing the path of faults and their small-scale morphological expression. This means that structural analysis cannot follow normal methods as used in more compact rock, and a greater accent on geomorphological methods is necessary.

The condition of having loose rock has the consequence that volcanoes can collapse and slide easily. Thus, most volcanoes have landslide scars, and are often associated with debris avalanches, which form highly distinctive hummocky terrain at their base. The landside scars and extensive debris-avalanche deposits deeply and rapidly modify the volcano landform and its surroundings. The structural analysis of landslides has been very important in the understanding of the kinematics of these geomorphologically significant events. Most especially, faulting studies can be used to analyse how the distinctive hummocky topography is produced (Paguican et al. 2012; Yoshida et al. 2012).

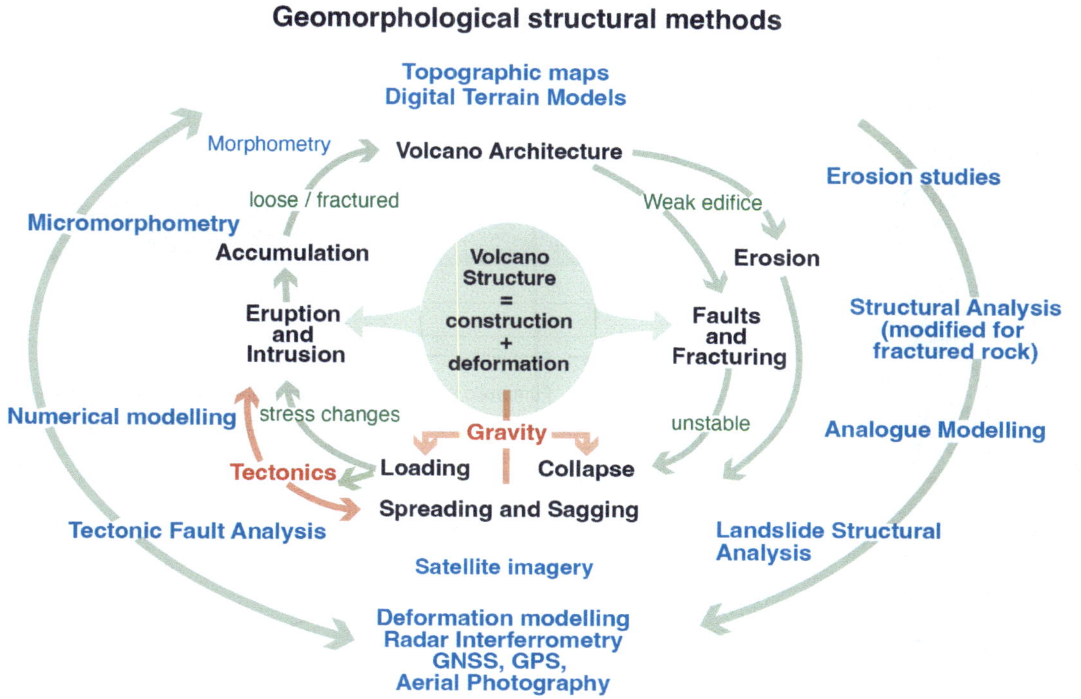

Fig. 9 System diagram showing the essential elements of geomorphology related to structural features and processes on volcanoes in the centre, and surrounding these, the methods used to study them. While specific methods are adapted to specific phenomena, a combination of methods is generally best for studying structural geomorphology

While avalanches are obvious geomorphic and sedimentological features, detecting their onset is largely a geomorphological problem, where changes in slope and the appearance of structures needs to be tracked. Mapping and geomorphic features can help (Tortini et al. 2015), but dynamic monitoring of deformation is needed to detect landslide movement. Photogrammetry has been used successfully to locate sites of developing landslides (e.g. Roberti et al. 2021), or large-scale movements on volcanoes such as Hawai'i, Etna, La Réunion, through GPS, GNS, and interferometry (Carnemolla et al. 2023). However, as yet most volcanoes are not monitored enough to cover the eventuality of such events, and detection of unstable areas can still be successfully made with standard geomorphological methods (Fig. 10: van Wyk de Vries et al. 2000). A global automated interferometry programme can eventually provide an early warning for such events, and a global topographic mapping programme can identify many latent ones.

A consequence of volcano construction and growth is that the new mountain loads its base. This causes the base to flex, and, if the substrate is ductile enough, to flow outwards, squeezed under the volcano's weight. This process is especially visible in the geomorphology and morphometry of sizeable volcanoes, where the mass has most effect on edifices built on very weak rock. If only flexure occurs, the volcano flanks are squeezed together by the bending, and may form terraces of thrust faults—these are seen at the base of the Galápagos volcanoes, and extensively on Martian volcanoes, both of which flex on thick crust. Such features are very hard to detect in the field, and are mainly studied by DEMs and bathymetry. Analogue modelling has been a particularly potent method for studying the dynamics of these processes (Galland and Scheibert 2013).

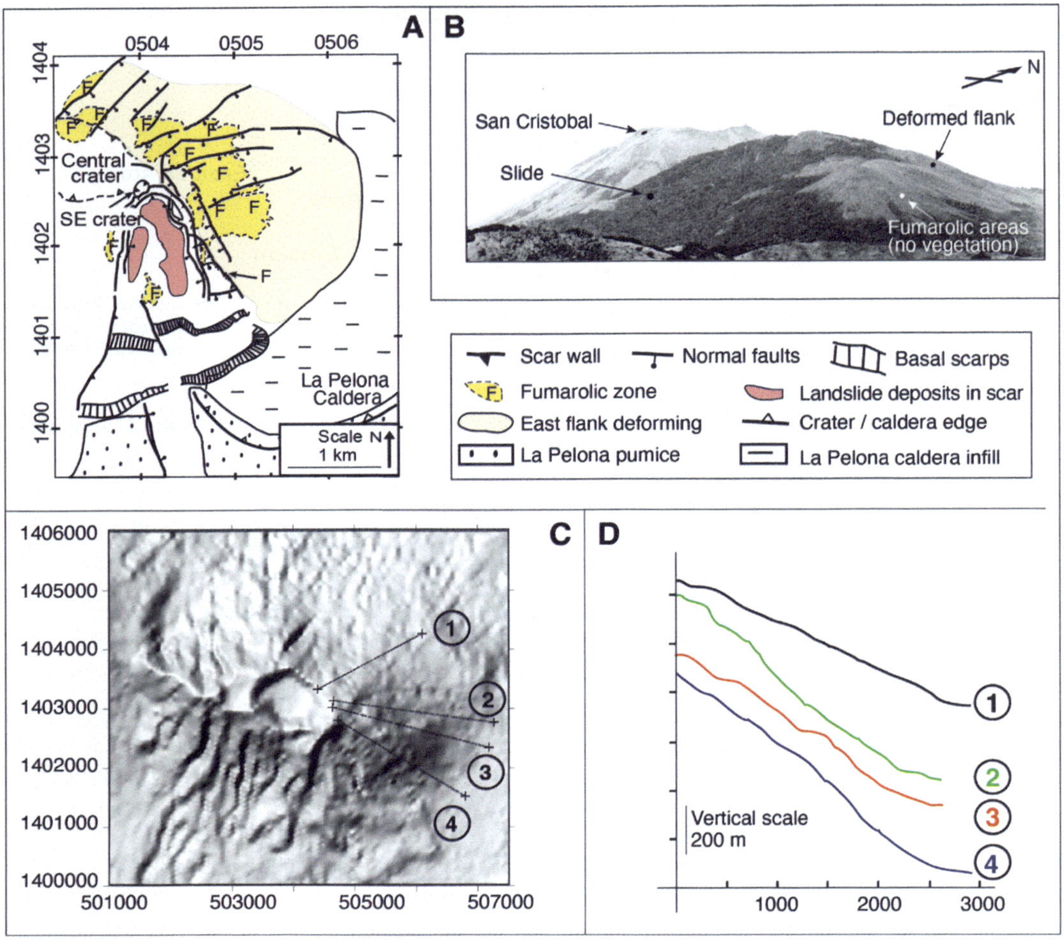

Fig. 10 Example of volcano-structural geomorphological studies from the classical case of Casita, Nicaragua (van Wyk de Vries et al. 2000). **A** Map of deformed Casita flanks (with caption), showing location of basal scarps, cut-in-half summit crater, and gravitational slump structure. Fumarolic areas and landslide deposits within scar are also shown. Complex pattern of basal scarps may have been caused by irregularity of La Pelona caldera and its pumice layer; **B** Photograph of Casita massif looking northwest, with major features; **C.** DEM shaded relief image with the localities of different slope profiles shown in **D**

If rock is weak enough to spread outwards, then the flanks extend, and normal faults can produce leaf-shaped radial rifts ('leaf grabens'), while anticlinal hills can form at the base where material pushes out. In this scenario, structures on the edifice are much more obvious, and while DEM analysis is useful, direct field-work can also be done. However, such structures are readily eroded where the rock is weak, and so the structures become obscured by deep ravines that follow the structural weaknesses. Back erosion means that structures are not located at their escarpments, and the diffuse nature of faulting complicates structural analysis. Another return to modelling and DEM analysis is necessary, combined with erosion studies of valley formation.

The built-up volcano has an important effect on tectonic faulting, as the load acts to create an addition stress field. Faults approaching volcanoes will be deflected by the stress field of the volcano load, and will bend towards or away from the volcano, depending on the type of substrate present. Notably, strike-slip faults that cross volcanoes form elegant sigmoid shapes that can control the direction of intrusions, of landslides, and thus condition the morphology of a volcano. Here, large-scale tectonic mapping can be applied, but will always be backed up by geomorphological and morphometric approaches, as the shape of the volcano will change with tectonic deformation, unless it is resurfaced by frequent eruptions (Norini and Lagmay 2005).

Likewise, intrusions of magma rising towards the cone may be affected by the volcano load and substrate strength. When a volcano is flexing, this can halt magma rise, and causes magma to spread laterally, forming lateral vents if it reaches the surface. In the spreading case, magma is drawn in and up, and rift zones form where magma pushes out and the edifice spreads at the same time. Analogue modelling has been particularly useful in the study of the geomorphic effects of such processes (Delcamp et al. 2011).

There is a particular type of volcano, formed principally by intrusion, where magma mostly collects without eruption, and bulges the surface upwards. This forms a very specific morphology with steep sides and a faulted summit graben. First identified by von Buch (1820) and opposed by Lyell (1830), the models of 'craters of elevation' and 'volcanoes of eruption', respectively, are now unified into a general model of volcano growth, shape change and geomorphic expression (Van Wyk de Vries et al. 2014). All such features can be studied through DEMs and surface deformation methods; e.g. the rise of Alu volcano, Ethiopia, was observed by interferometry (Pagli et al. 2012) and confirmed by satellite mapping. Importantly, seismic studies of buried topography have proved very useful in determining the connection between surface processes and internal structure (Magee et al. 2013).

In conclusion, the structural geomorphology of volcanoes is based on the fact that they are rapidly built-up mountains (rather than eroded), which not only provides them with many specific landforms, but makes them weak, prone to structural deformation, collapse and landslides, all of which are studied by geomorphological, morphometric analysis and ground deformation monitoring. At the same time, such work has been backed up by analogue and numerical modelling, the former especially useful to see geomorphic changes, the latter for the internal stresses that cause them. Standard structural geology techniques are complicated by the loose and fractured nature of most volcanic rocks, but seismic reflection studies have been especially useful for seeing connections between internal and external processes (Magee et al. 2013).

5.2 Monitoring Volcano Deformation and Internal Magmatic Processes Using InSAR

Quentin Dumont[1], **Valérie Cayol**[2], **Jean-Luc Froger**[3]

[1]Department of Geophysics, Graduate School of Science, Tohoku University, Sendai 980–8578, Japan

[2]Université Clermont Auvergne, CNRS, IRD, OPGC, Laboratoire Magmas et Volcans, F-63000 Clermont-Ferrand, France

[3]Université Jean Monnet—Faculté des Sciences et Techniques, Laboratoire de géologie de Lyon: Terre, Planètes, Environnement—UMR CNRS 5276 LGL-TPE, 42,023 Saint-Etienne, France

5.2.1 Introduction

Ground deformation is one of the most frequently recorded elements at volcanoes, thus its measurement is one of the major monitoring techniques. Typically, volcano deformation indicates the accumulation and/or emptying of magma in the storage system, and its transfer to the surface through tabular intrusions or cylindrical conduits. Deformation can also be related to hydrothermal activity, fault or flank movement, or subsidence of recent lava flows, etc. (Dzurisin 2006; Cayol et al. 2022). As volcanoes form in part by endogenous growth and may undergo mass wasting events, deformation monitoring is used to track the evolution of volcano morphology. This monitoring, combined with mechanical models, is also a powerful tool for gaining insight into the internal structure of volcanoes, along with the processes that lead to eruptions and flank collapse.

5.2.2 Methods: InSAR and Modelling

Ground deformation measurements can be carried out by permanent networks of continuous GNSS or tiltmeters, or by occasional measurements over large areas, such as those provided by Interferometric Synthetic Aperture Radar (InSAR) and photogrammetry (Dzurisin 2006; Cayol et al., 2022). InSAR has been widely used since the early 1990s, and provides a spatial resolution ranging from 1–100 m, covering a surface area of 10–10^4 km^2, with centimetric to sub-centimetric accuracy, together with a reiteration frequency in the order of ten days. As it is a remote sensing technique, InSAR can be used to survey remote and dangerous volcanoes. As of 2020, this monitoring technique had been used to detect more than 300 deformation events at 160 volcanoes (Biggs and Wright 2020).

InSAR is an active remote sensing method which uses radar waves, and is generally performed by a satellite, or more rarely by plane. The method is based on the acquisition of two radar images covering the same area at different dates (Hanssen 2001; Merryman-Boncori 2019). Each pixel records both the intensity and the phase of the wave emitted by the radar to the ground and subsequently backscattered by the ground back to the satellite. The recorded phase depends on the distance travelled by the wave, and on the interaction between the wave and the ground, which is highly variable from pixel to pixel, but is expected to vary little in time. Subtracting one radar image from the other eliminates the variable pixel component, and generates a third image, the interferogram (Fig. 11a), which represents the travel time difference between the Earth and the satellite. Any displacement of the ground between the two acquisition times will produce a travel time difference, usually recognizable as concentric interferometric fringes on the interferogram. Although recognized as being a powerful method for measuring ground surface displacement, InSAR has some limitations mainly related to: (1) variations in the humidity content of the atmosphere, which modify the wave propagation speed, (2) temporal changes of the ground properties within the pixels (e.g. vegetation, snowfall/snowmelt, etc.), and (3) large displacement gradients when more than one fringe occurs in an individual pixel. Another shortcoming is that InSAR can only measure changes in distance along the Earth-Satellite direction. However, the use of multiple interferograms acquired along different directions makes it possible to build up the three-dimensional components of displacements, albeit with much larger errors on the northward component (Wright et al. 2004).

Models are sometimes used for the characterization and interpretation of the deformation sources. They enable the prediction, in a given mechanical framework, of the surface deformation caused by one or several sources, the characteristics of which are given as parameters (location, shape, pressure variation of the source, dislocation amplitude, etc.). This approach is referred to as a forward approach. However, in Earth sciences, we seek to infer the parameters of the source that best explains the surface observations. This is referred to as an inverse approach (or inversion). There are numerous deformation modelling methods (Segall 2010; Cayol et al. 2022) and inversion methods (Tarantola 2005) but these will not be discussed further here.

5.2.3 Case Study of Piton de la Fournaise

The relevance of measuring and modelling deformation at volcanoes is particularly well illustrated at Piton de la Fournaise (Réunion Island, France). Piton de la Fournaise is a basaltic shield volcano, with eruptions typically associated with fissures opening along an arcuate NE-SE main rift-zone. There has been continuous InSAR monitoring at Piton de la Fournaise since 2003 (Richter and Froger 2020), with additional RADARSAT images acquired from 1998 to 2002 (Fukushima et al. 2010; Dumont et al.

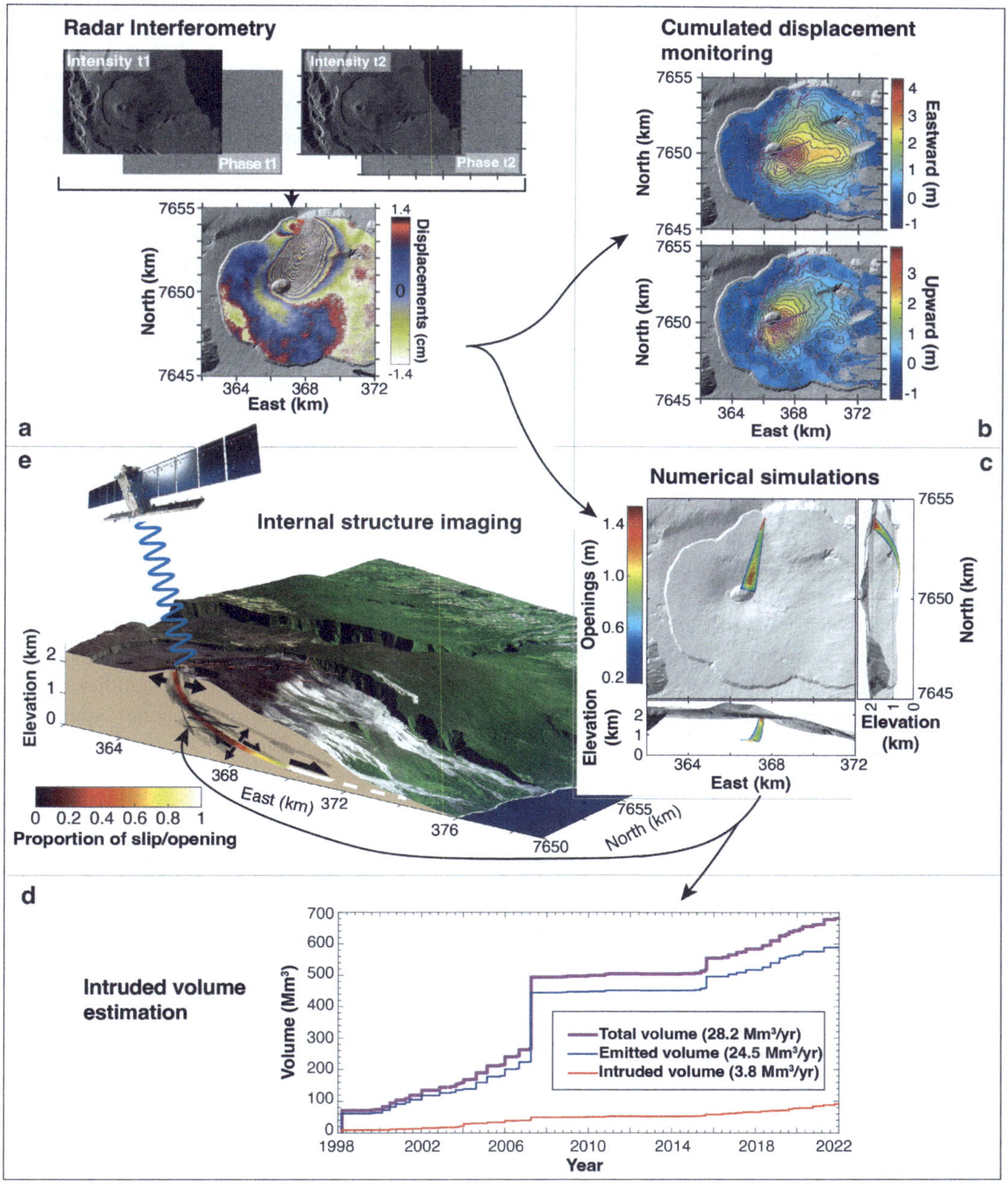

Fig. 11 Examples of applications of deformation measurements by InSAR at Piton de la Fournaise volcano. **a** Generation of interferogram from two radar images acquired at t1 and t2. The eruption of 3 April 2018 is shown as an example. Each fringe represents one half-wavelength of displacement in the Earth-Satellite direction. **b** Cumulated eastward and upward co-eruptive displacement related to 50 eruptions between 1998 and 2022. Contour lines are drawn at 0.25 m intervals. Pink segments represent the eruptive fissures. **c** Geometry of the 3 April 2018 intrusion as calculated using inverse modelling. **d** Estimation of the total volume produced by the eruptive activity. **e** Image of the inner spoon-shaped intrusion zone determined from the geometry of 22 intrusions (Dumont et al. 2022)

2022). This long-term monitoring, together with the intense activity of the volcano (2.3 eruptions/year on average), has revealed displacements relating to 59 eruptions that took place between 1998 and 2022. For fifty of them, a number of InSAR images were acquired, enabling the eastward and upward components of the displacement to be calculated (Fig. 11b). The cumulative co-eruptive displacements over the 1998–2022 period clearly show that the eastern side of the main rift-zone is subject to more intense displacement than the western side: there is up to 4 m cumulative eastward displacement east of the rift-zone, while there is only 50 cm westward displacement west of the rift-zone. This intense and asymmetric displacement, together with other lines of evidence based on the bathymetry of around fifty mass wasting events (Oehler et al. 2008), raises questions concerning the long-term stability of the eastern flank of the volcano.

Further analysis of the co-eruptive displacement of each intrusive event by inverse modelling (Fig. 11c) led to the 3D imaging of preferential intrusion zones with unprecedented accuracy (Dumont et al. 2022). The models revealed the continuation at depth of the NE-SE rift zone as a spoon-shaped structure open towards the sea (Fig. 11e). This structure extends under the eastern flank as a normal fault plane, intruded by sills, similar to that observed at the neighbouring, extinct volcano, Piton des Neiges (Famin and Michon 2010; Chaput et al. 2017). On a geological timescale, repeated injections of magma into this structure could lead to major flank collapse, which may be associated with earthquakes.

The models also allow the volume of magma intruded into the edifice to be estimated. In combination with mapping of the lava flow thicknesses (by InSAR or photogrammetry, e.g. Richter and Froger 2020), the evolution of the total volume, both emitted and intruded, associated with the shallow magmatic activity has been calculated (Fig. 11d). Over the 24 year-long period, the volcano has gained 0.7 km^3 at a fairly constant rate, with 10% of this volume occurring as magmatic intrusions and 90% as lava flows.

These results provide a better understanding of (1) the hazards associated with repeated intrusions, and (2) the long-term evolution of the morphology of the edifice. InSAR acquisitions at Piton de la Fournaise have also been successfully used for other applications (Richter and Froger 2020) such as: (1) mapping lava flow locations, along with their thicknesses and post-emplacement deformation, (2) measurement of the slow slip of the eastern flank of the volcano (Chen et al. 2017), and (3) identification of post-eruptive displacements after intrusion events. The increase in the number of monitored volcanoes, and the development in modelling capacity to create models closer to the physical reality and able to integrate different types of data, such as continuous GNSS data (Smittarello et al. 2019), continuous gravity data (Carbone et al. 2015), and geochemical and petrological data (Anderson et Segall 2013; Kilbride et al. 2016; Wong et al. 2017), contributes to our increased understanding of magmatic plumbing systems, and of the parameters that control the flank stability of volcanoes. This understanding will better equip us to assess hazards related to volcanic activity.

6 Geochronological Approaches for Studying Volcano Evolution

6.1 K–Ar and ^{40}Ar/^{39}Ar Dating of Volcanic Rocks

Pierre Lahitte[1], **Xavier Quidelleur**[1]

[1]Université Paris-Saclay, CNRS, GEOPS, Orsay, 91405, France

6.1.1 Introduction

The dating of volcanic materials is a prerequisite for studying the geomorphology of volcanoes. Volcanic activity can create a wide variety of processes and surfaces that, once dated, provide chronological markers for interpreting the dynamics of construction and dismantling of volcanic systems. For instance, the ages of basal and summit flows of a sequence constrain the duration of construction of a given volcanic phase, while ages obtained on valley flows are valuable time constraints for a quiescence period, and thus for erosion phases. In addition, dating a tephra deposit can provide a topographical marker over a wide area of a volcanic region.

Sampling of the freshest sample in the field is required to obtain a precise and meaningful radio-isotopic age. There must be no alteration, and the mineral phases selected for dating must be validated by thin section examination. For differentiated rocks, K-feldspars can be used, but for basic K-poor rocks, carefully selected groundmass is preferable. This is beause the groundmass crystallized last during the cooling of erupted lavas, and as K is an incompatible element, it is concentrated in the groundmass. Whole-rock dating should be avoided as the early-crystalizing minerals, such as olivine and pyroxene, can carry a significant amount of inherited radiogenic argon that will bias the ages towards older values (e.g. Lahitte et al. 2019; Quidelleur et al. 2022).

6.1.2 The Potassium-Argon Chronometer

The decay of potassium-40 into argon-40 was demonstrated by Aldrich and Nier (1948), and since then, the potassium-argon (K–Ar) chronometer has become a widely used dating method for volcanic rocks. Potassium has three naturally occurring isotopes (^{39}K: 93.26%, ^{40}K: 0.0012% and ^{41}K: 6.73%), of which only ^{40}K is radioactive. About 11.20% of ^{40}K decays to stable ^{40}Ar by electron capture (11% to an excited state and 0.2% directly to the ground state), while 88.8% produces stable ^{40}Ca by β- emission. Because ^{40}Ca is the most abundant isotope of Ca, the K-Ca method cannot be used for dating volcanic rocks. On the other hand, since argon is a gas, the ^{40}Ar produced by ^{40}K decay continuously diffuses out of the crystal lattice until the mineral reaches its closure temperature. In volcanic systems, such cooling occurs during the eruption. Once the material is solidified, which is considered as instantaneous on a geological timescale, the $^{40}Ar^*$ can no longer escape and remains trapped in the minerals, where it accumulates over time. Therefore, by measuring the amount of K and of radiogenic ^{40}Ar in the mineral, and, knowing the decay constant of ^{40}K ($\lambda = 5.543 \times 10^{-10}$ a^{-1}; Steiger and Jaëger 1977), one can determine the age of any volcanic rock using the following formula: $t = \frac{1}{\lambda} \cdot \ln\left(1 + \frac{\lambda}{\lambda_{Ar}} \cdot \frac{^{40}Ar^+}{^{40}K}\right)$, with λ_{Ar} the sum of both decay constants of ^{40}K to ^{40}Ar.

Due to the fact that potassium is a relatively abundant element in the Earth's crust and that the half-life of ^{40}K is quite long ($t_{1/2} = 1.25 \times 10^9$ years), the K–Ar method can be applied to date rocks as young as the Holocene, as well as rocks as old as the Precambrian. Based on the potassium-argon chronometer, both the K–Ar and $^{40}Ar/^{39}Ar$ techniques are used for dating volcanic successions. For the former, ^{40}K and ^{40}Ar contents are determined independently for two different aliquots of the sample, while for the latter, the ratio of daughter and parent isotopes is determined simultaneously for the same aliquot. Each technique has its own advantages and drawbacks when applied to volcanic rocks.

6.1.3 The K–Ar Dating Technique

The calculation of ^{40}K is made using the isotopic abundances of K content commonly obtained by flame spectroscopy in atomic absorption mode. It is a relatively straightforward measurement with a relative uncertainty (σ_K) of about 1%. On the other hand, the determination of the amount of radiogenic argon ($^{40}Ar^*$) by mass spectrometry is more challenging, especially for young rocks with less than a few percent $^{40}Ar^*$. Following melting of the sample above 1400 °C, gases released are purified in several steps using Ti foam heated to about 800 °C and Al-Zr getters present in each section of the ultra-high vacuum line.

The Conventional K–Ar Technique

The conventional K–Ar technique is based on the isotopic dilution approach using an ^{38}Ar spike. A known number of atoms of ^{38}Ar ('spike') is mixed with the argon released from the sample to be dated. Then the number of atoms of ^{40}Ar of the sample can be determined by measuring isotopic ratios ($^{40}Ar/^{38}Ar$ and $^{38}Ar/^{36}Ar$) of both the unknown and the spike. This technique was widely used when K–Ar dating started, but it has been almost abandoned over the last few decades because it does not allow young (<0.1 Ma) rocks with low (<5%) radiogenic content to be dated.

The Cassignol-Gillot Technique

The Cassignol-Gillot ('unspiked') K–Ar technique was specially developed for dating volcanic rocks in the ^{14}C range and up to the Holocene (Cassignol and Gillot 1982; Gillot and Cornette 1986; Gillot et al. 2006). This technique relies on the detection of a very small difference between the isotopic $^{40}Ar/^{36}Ar$ ratio extracted from the sample and the atmospheric ratio. A specially designed 180° sector mass spectrometer is used to reach very high stability during measurements in order to measure quantities of $^{40}Ar^*$ as low as 0.1% (Quidelleur et al. 2001). Its main features are as follows: a small radius of 60 mm, a volume of less than 200 ml, a Nier-type source with a low emission filament current, a high voltage of only 530 V, and simultaneous collection of both ^{40}Ar and ^{36}Ar isotopes using Faraday cups equipped with sub-femto ampmeters. Such stability renders the use of the ^{38}Ar spike unnecessary and allows the ^{40}Ar signal to be converted into a number of atoms by measuring a ^{40}Ar aliquot calibrated using standard minerals with a known age. The uncertainty associated with this calibration ($\sigma^{40}Ar_{cal}$), mainly due to the uncertainty on the age of the standard as well as the decay constants, is only about 1%. Finally, the radiogenic content ($\%^{40}Ar^*$) of the sample is obtained from the precise comparison of its $^{40}Ar/^{36}Ar$ ratio with that of an air aliquot of atmospheric argon, measured under the same argon pressure as the sample. Note that such direct comparison between the sample and the atmosphere makes it unnecessary to know the exact $^{40}Ar/^{36}Ar$ ratio of the atmosphere. Similarly, as only two argon isotopes (^{40}Ar and ^{36}Ar) are measured for (1) the signal calibration, (2) the sample, and (3) the atmospheric correction, no mass discrimination correction is needed with the Cassignol-Gillot technique. The relative uncertainty for each age determination (σ_t) includes the three sources of error mentioned above and is calculated using:

$$\sigma_t = \sqrt{\sigma_K^2 + \sigma_{^{40}Ar_{cal}}^2 + \sigma_{^{40}Ar*}^2}, \text{ with}$$

$$\sigma_{^{40}Ar*} = \frac{0.10}{\%^{40}Ar*}$$

The total relative age uncertainty for old samples, hence with high $\%^{40}Ar^*$, tends to be about 1.4% ($=\sqrt{2}$), while for very young samples with $\%^{40}Ar^*$ as low as 0.1 it becomes close to 100%. Details of the analytical procedure, standards, and calculations, can be found in Germa et al. (2011) and Bablon et al. (2018).

6.1.4 The $^{40}Ar/^{39}Ar$ Dating Technique

The $^{40}Ar/^{39}Ar$ technique, first proposed by Merrihue and Turner (1966), is based on the production of ^{39}Ar by irradiation of K-bearing minerals with fast neutrons in a nuclear reactor. The ^{39}Ar produced from ^{39}K is a function of the amount of K in the sample, the duration of irradiation, and the neutron flux of the reactor. It is quantified using the *J*-factor, which in turn is determined using standards (or flux monitors) of known age irradiated along with the sample to be dated. As both the parent ^{40}K and daughter $^{40}Ar^*$ isotopes can be determined from isotopic measurements of argon, an apparent age can be obtained by step-heating from about 500 °C to complete fusion of the sample. At each temperature step, the age is calculated using:

$$t = \frac{1}{\lambda} \cdot \ln\left(1 + J \cdot \frac{^{40}Ar^*}{^{39}Ar_K}\right)$$

If, for a significant part of the total ^{39}Ar fraction released (>50%), the ages from at least three successive steps are similar within uncertainty, then a plateau age can be calculated. Such plateaus attests to the system having remained undisturbed since t_0. In addition, using the data from the step-heating measurements, the inverse isochron plot showing $^{36}Ar/^{40}Ar$ as a function of $^{39}Ar/^{40}Ar$ can be used to obtain the pure trapped and pure radiogenic components in the sample.

When compared with the atmospheric ratio, the y-axis intercept of the inverse isochron provides a check that no excess argon is present and the trapped component results from atmospheric contamination alone. More details can be found in McDougall and Harrison (1999) and Schaen et al. (2020).

Using high-sensitivity mass spectrometers coupled with laser-fusion, it is posible to carry out single crystal dating of K-rich minerals, even for relatively young rocks of only a few 100 ka. This is especially interesting for explosive volcanic deposits, which often consist of a mixture of juvenile crystals and inherited crystals from the pre-eruption edifice.

6.1.5 Dating Young Volcanic Rocks

The K–Ar and $^{40}Ar/^{39}Ar$ techniques both have advantages and drawbacks, and are complementary. When the presence of extraneous argon or disturbance of the system since initial cooling is suspected, which is rare for Quaternary volcanic rocks, the $^{40}Ar/^{39}Ar$ technique must be used. This technique must also be used when

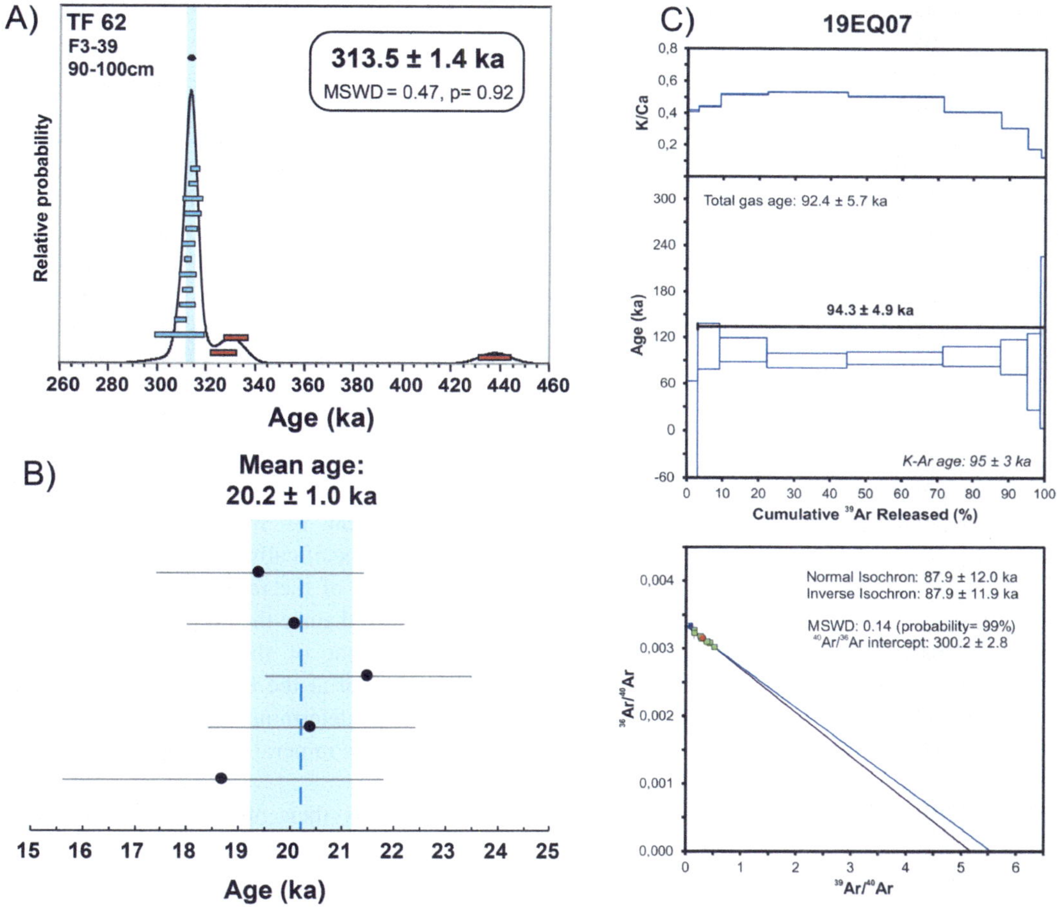

Fig. 12 **A** Relative probability distribution of $^{40}Ar/^{39}Ar$ ages of single sanidine crystals. Each age is shown with a $\pm 1\ \sigma$ wide box. Blue boxes represent crystals from the juvenile magma, whereas red boxes show ages from xenocrysts removed from the weighted mean age (Leicher et al. 2023). **B** K–Ar ages obtained for 5 aliquots of glassy rocks and the resulting weighted mean age (Karátson et al. 2018). **C** Results of step-heating $^{40}Ar/^{39}Ar$ analyses (Santamaria et al. 2023). K/Ca ratio and apparent age spectra (in ka) are shown as a function of the cumulative ^{39}Ar content (in %); inverse isochrons are shown in the lower graph

dating K-feldspars from tephra deposits because the single-grain approach makes it posible to detect inherited crystals (Fig. 12A). But when applied to Ca-rich minerals, such as plagioclase or groundmass, the $^{40}Ar/^{39}Ar$ technique has strong limitations. This is due to the production of ^{36}Ar from ^{40}Ca during irradiation, which renders the atmospheric contamination correction delicate, especially for very young rocks for which it represents a large part of the total ^{36}Ar budget. Another irradiation-induced drawback is the ^{39}Ar recoil out of the minerals, which can limit the use of the $^{40}Ar/^{39}Ar$ technique for small-size and/or poorly retentive minerals and/or glassy samples. On the other hand, the K–Ar Cassignol-Gillot technique is better for dating effusive rocks with basaltic or andesitic compositions, which represent more than 90% of the volcanic rocks on the Earth's surface. This is because this technique does not need any nuclear irradiation, since the total ^{36}Ar measured originates from the atmosphere and can therefore be used for the atmospheric correction. It allows detection of ^{40}Ar radiogenic contents as low as 0.1%, i.e. a few ka for K-poor minerals (Quidelleur and Famin 2024), making this technique relevant in the 0–50 ka range where ^{14}C dating is classically used (Fig. 12B). Naturally, where material is fresh and does not suffer from extraneous argon, both $^{40}Ar/^{39}Ar$ and K–Ar techniques yield comparable results (Fig. 12C).

6.2 Dating Volcanoes and Volcanic Deposits by Means of Luminescence Techniques

Didier Miallier[1]

[1]Laboratoire de Physique, Université Clermont-Auvergne, UMR6533 CNRS, 63,178 Aubiere, France

6.2.1 Introduction

Luminescence-dating techniques: thermoluminescence (TL) and optically stimulated luminescence (OSL) determine the time elapsed since a mineral was last exposed to heat or light (Aitken 1985, 1998; Preusser et al. 2008). The physical process involved in these techniques is based on the ability of many minerals to record the amount of ionizing radiation they have experienced since a resetting event consisting of exposure to heat or light, and to then reveal this radiation dose in the form of a luminescence signal when stimulated by heat (TL) or light (OSL). The luminescence yield is an increasing function of the dose. The relationship between time, i.e. the 'age', and radiation dose is linear as long as the natural radioactivity of the environment of the mineral has been stable over time.

6.2.2 Method

According to the above principles, TL dating can apply to minerals once heated by a volcanic event, e.g. quartz grains embedded in a soil baked by a lava flow, or juvenile minerals crystallized within lava, such as feldspar phenocrysts. OSL is better for indirect dating of tephra layers by evaluating the time elapsed since the underlying sediment was last exposed to sunlight (Fig. 13).

In practice, the procedure comprises two distinct series of manipulations and measurements:

(1) Measurement of the TL or OSL of a selected, specifically prepared mineral, for assessment of the radiation dose Q (Gy) it has received since the eruption.
(2) Measurement of the radioactivity of the environment at the location of the mineral, in order to determine the radiation dosage D ($Gy.y^{-1}$) the mineral has received annually.

The age, A, is then obtained by the simple equation:

$$A(y) = Q/D$$

Actually, Q and D are both combinations of many intermediate parameters, the evaluation of which has led to the development of a large variety of procedures and experimental protocols by

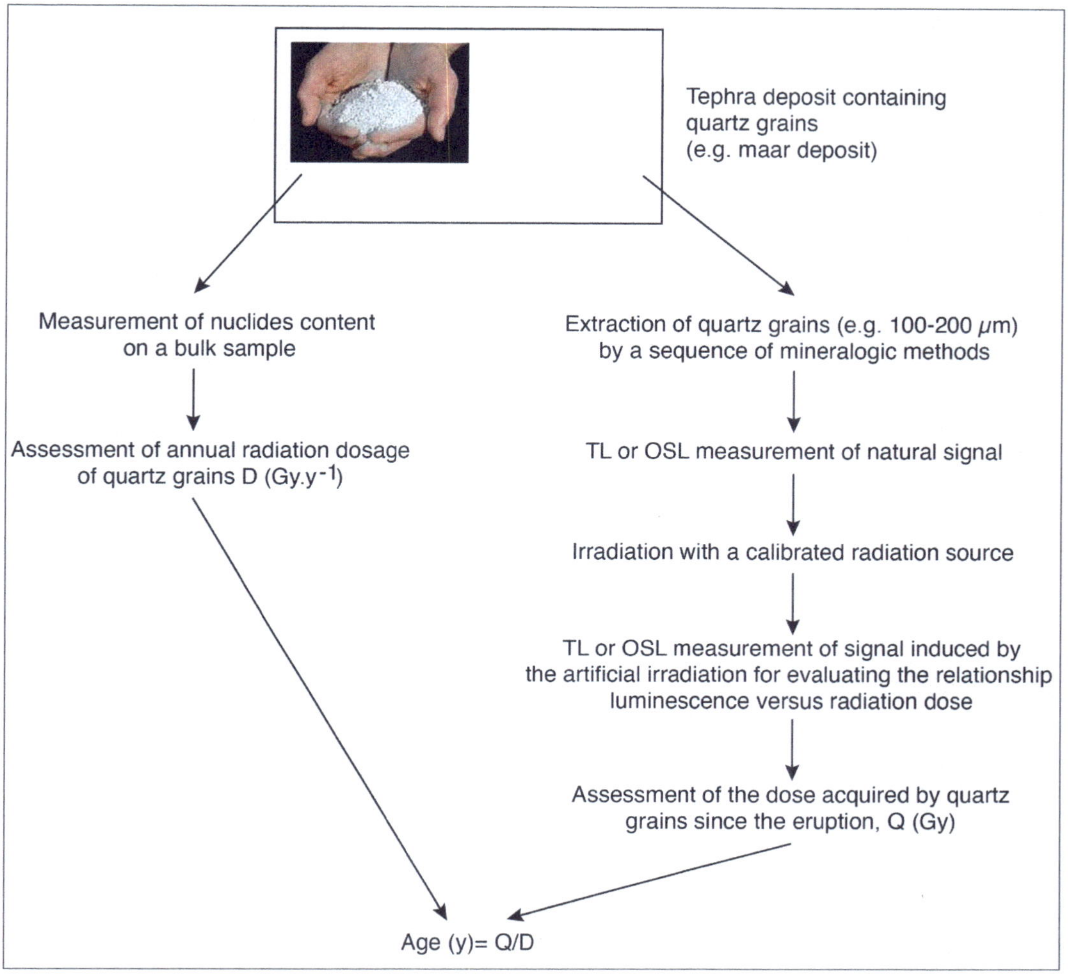

Fig. 13 Simplified flowchart of the luminescence dating process

different laboratories over the world (Bösken and Schmidt 2020).

Due to their complexity, luminescence-dating techniques have variable accuracy, from good to mediocre, typically in the range of 7% to 15%, but this modest achievement is counterbalanced by a wide spectrum of potential applications in an extended chronological interval, from a few centuries up to several hundred thousand years. Luminescence-dating techniques are potentially operational whenever the usual ones, such as radiocarbon, K–Ar or the ^{238}U-^{230}Th disequilibrium techniques (U/Th), are not available.

6.2.3 TL Dating in Action

The utility of TL dating can be illustrated by the Quaternary volcanic field of the Chaîne des Puys (French Massif central), which comprises about 80 volcanoes dated at between ~8.6 ka and ~90 ka (Boivin et al. 2017). Many edifices of this monogenic volcanic field cannot be dated by ^{14}C, either because they are too old (> ~45 ka) or because appropriate material, typically charred wood, is lacking from their products or associated material (Table 4). On the other hand, many volcanoes and deposits cannot be dated by K/Ar techniques, either because they are too young (< ~40 ka) — which is the case with

Table 4 Dating results of the Chaîne des Puys volcanic field (modified from Boivin et al. 2017)

Site/volcano	Age (ka)	±	Methods	Nature
Chanat	90	13	**TL**	TB
Jumes (?)	72	12	**TL**	TB
Argnat	71.6	2.9	K	TB
Mas d'Argnat	70	13	**TL**	TB
Aydat 2	67.1	2.8	K	TB
Montchié (southern crater)	63.7	9.0	**TL**	B
Volvic (Le Fraisse)	63.4	9.0	**TL**	TB
Gravenoire	60	6	**TL**, ESR	TB
Chanonat	61	8	**TL**	B
Beaunit	54	7	**TL**	TB
Manzat	51	15	**TL**	B
Grosmanaux (?)	46	8	**TL**	B
La Toupe	45.5	7.0	**TL**	B
Aydat 1	44.6	3.3	K	TB
Chuquet Genestoux	44.0	12.0	**TL**	B
Randanne (Montgy)	44.0	7.0	**TL**	TB
Puy Gonnard	44	10	**TL**	TB
Salomon	43	7	**TL**	TB
Bravant	42.5	1.0	K	TB
Volvic (Les Riaumes)	42.3	1.8	K	TB
Louchadière	41.4	1.7	K	TB
Tiretaine (Royat lava flow)	41.3	1.3	**TL**	TB
Blanzat (southern lava flow)	41.2	1.2	K	B
Peschadoires (South)	41.0	4.0	U/Th	TB
Peschadoires (North)	41.0	4.0	**TL**,U/Th	TB
Olby	40.4	1.1	**TL**, K. U/Th	TB
Montmeyre	40.3	1.9	K–Ar	TB
Laschamp	40. 2	1.3	**TL**, K, U/Th	TA
Combegrasse	38.9	7.0	**TL**	B
Blanzat (northern lava flow)	37.8		**TL**	B
Nébouzat	36.7	5.8	**TL**	TA
St Pierre le Chastel	36.0	4.7	U/Th	TB
Ceyssat	35.4	1.7	K	TB
Theix carrefour	35.3	1.7	K	TB
Tiretaine (La Font de l'Arbre lava flow)	33.8	1.7	K	TB
Petit Sarcoui	31.5	5.0	**TL**	TB
Paugnat	30.5	7.0	**TL**	TB
Lantégy 1	30	4	**TL**	TB
Puy des Gouttes	~30		Teph	B

(continued)

Table 4 (continued)

Site/volcano	Age (ka)	±	Methods	Nature
Lantégy 2	29.6	6.0	**TL**	TA
Charmont	29.9	5.2	**TL**	TB
Theix	28.2	2.2	K	B
La Rodde	27.1	6.2	**TL**	B
Les quatre routes de Randanne	24.3	1.9	K	B
Moulet-Marcenat	18.5	3.1	K	TB
Blanzat (southern lava flow)	18.4	5.4	K	B
Fontfreide (Mercoeur ?)	18.1	4.5	**TL**	TB
Cheire de Mazayes (Filhu or Grand Suchet ?)	17.3	0.4	**TL**	TB
Clierzou	~15		Teph	T
Aumône	~15		Teph	T
Côme	13.7	0.9	**TL**, ^{14}C	TA
Montchié (NE crater)	13.7	1.4	**TL**	TA
La Nugère (CF1 tephra)	13.3	0.3	^{14}C, Teph	TA
Bois d'Epinoux (Filhu or Grand Suchet ?)	13.2	2.6	**TL**	TB
Grand Sarcoui (dome)	12.6	1.0	**TL**, U/Th	T
CF2	12–13		Teph	TA
CF3	12.1	0.3	Teph	TB
CF4	~11.4		Teph	TA
Puy de Dôme (dome)	11.0	0.2	**TL**,U/Th, ^{14}C	T
Puy de Dôme (final summit eruption)	10.7	0.1	^{14}C	T
Grand Sarcoui (final summit eruption)	10.4	0.1	^{14}C	T
Nid de la Poule	~10		Teph	TB
Pariou	~10		Teph, ^{14}C	TA
La Mey	9.7	2.0	**TL**	TA
Chopine	9.54	0.15	^{14}C	T
Kilian	9.4	0.2	^{14}C	T
Vasset	9.2	0.4	^{14}C	T
Téphra des Chaves	9.2	1.1	**TL**	B ?
Lassolas et La Vache	8.6	0.2	^{14}C	TB

Methods: **TL** = Thermoluminescence; Teph = tephrostratigraphy; K = K/Ar or $^{39}Ar/^{40}Ar$; U/Th = disequilibrium in U and Th series. Composition of lava: B = basalt, TA = trachyandesite; TB = trachybasalt; T = trachyte. The chronology of the eruptions illustrates the temporal evolution of the composition of the lavas due to a differentiation process, from the least evolved magma to the most evolved one, according to the following sequence: basalt, trachybasalt, trachyandesite, and trachyte. The chronology owes a lot to thermoluminescence, especially in the older part of the series probably more than half of the volcanoes listed in Table 4—, or because their products are contaminated by a significant xenolith content, which is quite common. Therefore, about 50 dates of the more than 70 dates obtained for volcanic products in the Chaîne des Puys derive from TL dating (Boivin et al. 2017). Among them, 35 TL dates are based on juvenile feldspar phenocrysts (Guérin and Valladas 1980; Guérin 1983), the other ones being mainly based on xenolithic quartz grains, which were extracted from baked soils or alternatively from tephra including country rocks and minerals.

The Grand Sarcouy lava dome (Fig. 14) provides a good example. The eruption of this short-lived monogenetic, trachytic volcano was estimated at 12.5 ± 0.5 ka using the TL dating method on quartz grains embedded within tephra, emitted during the explosive opening of the vent through granitic bedrock (Miallier et al. 2017). This result is consistent with an age derived from U/Th (12.8 ± 0.7 ky; Miallier et al. 2003). Despite close investigation, no plant fragments have been found to date the eruption by means of radiocarbon. This absence can be explained by the fact that the growth of the dome did not generate a large enough amount of tephra to cover and burn vegetation, which would anyway have been scarce at the time of the eruption (the cold Dryas III chronozone).

The potential of luminescence methods for dating Quaternary volcanoes can also be seen for several other volcanic countries around the world, like Japan (Nitta et al. 2020), Cameroon (Schmidt et al. 2017), New Zealand (Lindsay and Leonard 2009), Romania (Karátson et al. 2016a, b) and Germany (Paulick et al. 2008).

Fig. 14 Grand Sarcoui lava dome looking east (Chaîne des Puys, French Massif central)

6.3 Cosmogenic Nuclide Dating

Irene Schimmelpfennig[1], **Gordon Bromley**[2]

[1]Aix-Marseille Université, CNRS, IRD, INRAE, CEREGE, Aix en Provence, France

[2]Geography, University of Galway, Galway H91 TK33, Ireland

6.3.1 Cosmogenic Nuclides: Production, Isolation, and Measurement

Cosmogenic nuclides are formed by interactions between cosmic-ray particles and atoms within common minerals forming the Earth's surface, and provide a relatively straightforward means to resolve the ages of terrestrial surfaces, such as relict lavas and volcanic landforms. Owing to the detectability of even tiny concentrations of these nuclides, cosmogenic nuclide dating is particularly useful for cases in which volcanic rocks are too young to be dated effectively with the common radiometric methods (e.g. $^{40}Ar/^{39}Ar$, K–Ar, U–Pb) and where organic material for bracketing ^{14}C ages is unavailable. In this contribution, we outline the principles of cosmogenic nuclide geochronology, specifically surface-exposure dating, describe the methodology, and demonstrate the nature and scope of data generated via a case study from the Central Andean Volcanic Zone in Peru.

Earth is constantly being bombarded by high-energy particles originating from outside our Solar System. Dominated by protons and travelling close to the speed of light, this influx of particles is collectively termed primary cosmic radiation, and the interaction of cosmic radiation with elements at the Earth's surface is the basis for cosmogenic nuclide geochronology (Gosse and Phillips 2001; Dunai 2010). When primary cosmic-ray particles encounter Earth's atmosphere, they collide with the nuclei of common gases such as oxygen and nitrogen to produce high-energy spallation reactions. During spallation reactions, target nuclei (e.g. O) are transformed into different elements known as atmospheric 'cosmogenic nuclides'. Spallation reactions also generate secondary high-energy particles that continue to collide with other nuclei, resulting in a cascade of secondary cosmic radiation. At the Earth's surface, the interaction between secondary cosmic-ray particles and terrestrial nuclei, also called 'target elements', produces in situ cosmogenic nuclides that, in many cases, would not otherwise occur in abundance within minerals at the Earth's surface. Terrestrial cosmogenic nuclides include beryllium-10 (^{10}Be), aluminium-26 (^{26}Al), chlorine-36 (^{36}Cl), helium-3 (^{3}He), neon-21 (^{21}Ne), and carbon-14 (^{14}C). While the majority of these nuclides are produced via spallation reactions induced by high-energy neutrons, smaller—yet sometimes important—contributions are also made by muons and thermal neutron capture. Whereas ^{10}Be is generally the most common nuclide for dating surfaces in quartz-rich lithologies, here, we focus on ^{36}Cl and ^{3}He, which are typically the most suitable nuclides applied in volcanic environments.

The premise of cosmogenic surface-exposure dating is that, upon exposure to cosmic radiation, a surface begins to accumulate *in situ* cosmogenic nuclides in those rocks and minerals containing target elements, and at a roughly constant rate (known as the production rate). Where the concentration (C) of a specific cosmogenic nuclide in these minerals has been measured and the local production rate at the sampling site (PR) is known, the time of exposure (T) (and thus lava flow emplacement in our specific case) can be calculated with the basic equation $T = C / PR$. Whereas ^{3}He is a stable isotope, for applications of ^{36}Cl, its radioactive decay (half-life = 301 kyr) needs to be accounted for.

The choice of the cosmogenic nuclide depends primarily on the lithology and mineralogy of the rock surface to be dated, in addition to the analytical facilities available for the study. Taking the example of lava flows, cosmogenic ^{3}He can be analysed in the lattices of the mafic minerals, pyroxene and olivine, while ^{36}Cl can be analysed in various minerals and silicate rocks containing at least one of the major target elements for ^{36}Cl production (i.e. Ca and K), in particular Ca- and K-feldspars, Ca-pyroxene and whole-rock basalts. If quartz is sufficiently abundant (e.g. as xenocrysts), ^{10}Be can also be used to exposure-date the surface (Alcalá-Reygosa et al. 2018).

Sample collection for cosmogenic nuclide dating consists of detaching the upper few centimetres of the rock surface, because the flux of cosmic-ray particles attenuates rapidly in rock. Practitioners must be confident that the sampled features have not moved, been covered by considerable amounts of sediment, or significantly eroded since deposition/formation, as these processes would all impact the sample's exposure to the cosmic-ray flux, thereby interrupting or modifying the accumulation of cosmogenic nuclides and resulting in erroneously young surface-exposure ages (Dunai 2010). In most volcanic settings, satisfying these criteria is relatively straightforward and relies on established geologic/geomorphologic principles. Because sample location, elevation, and the degree to which surrounding topography obstructs the cosmic ray flux all influence the production rate at a specific site, these parameters are measured with a GPS and clinometer. Sample preparation in the laboratory starts with crushing, sieving and, if needed, mineral separation using magnetic and density-based methods and acid leaching (Dunai 2010; Bromley et al. 2014). The nuclide of interest is then isolated from the mineral or rock material with a specific method, and its concentration determined after counting (via mass spectrometry) its abundance relative to a naturally abundant isotope of the same element. In the case of ^{3}He, isolation from the mafic minerals is achieved with a dedicated gas extraction and purification system and analysis of the $^{3}He/^{4}He$ ratios by noble gas mass spectrometry (Kurz 1986). Following dissolution of rock material and several chemical purification and precipitation steps, ^{36}Cl is analysed in the final silver chloride precipitate via measurement of $^{36}Cl/^{35}Cl$ and $^{35}Cl/^{37}Cl$ ratios by isotope dilution accelerator mass spectrometry (AMS) (Finkel and Suter 1993; Desilets et al. 2006). The advantages of the AMS technique, which is also used for measuring ^{14}C, ^{10}Be, and ^{26}Al, are (1) that minute amounts of these cosmogenic isotopes can be detected—concentrations are typically 10^{11} to 10^{15} times lower than the abundant isotope of the same element—and (2) that the cosmogenic isotopes can be efficiently isolated from interfering isotopes of the same mass (sulphur-36 in the case of ^{36}Cl). The concentrations deduced from the isotopic ratios, together with the site- and sample-specific information needed to determine the local PR, then allow calculation of the exposure age, usually via one of several freely available calculators (Balco et al. 2008; Martin et al. 2017; Schimmelpfennig et al. 2009).

Chlorine-36 is the cosmogenic nuclide with the most complex production systematics, being produced from interactions of various cosmic-ray particles with several target elements, while both ^{36}Cl and ^{3}He are also produced via non-cosmogenic (nucleogenic) reactions within the rock (Schimmelpfennig et al. 2009; Blard 2021). Furthermore, the associated abundant isotopes (^{35}Cl and ^{4}He, respectively) are naturally present in rocks and minerals in highly varying concentrations. Therefore, the composition of lava samples also needs to be quantified in order to calculate surface-exposure ages effectively. The available calculators dispose of appropriate production calculation models and parameters and permit uniform methods to be used by different researchers.

The production rates of cosmogenic nuclides are calibrated from their concentrations measured in surface samples for which exposure histories have been robustly constrained via other dating methods (e.g. ^{14}C; K–Ar). Volcanic rocks are most often used for the production rate calibration of ^{36}Cl and ^{3}He (e.g. Schimmelpfennig et al. 2011; Delunel et al. 2016). Due to the spatial and temporal variability of geomagnetic field strength and deceleration of cosmic-ray particles passing through the atmosphere, production rates vary considerably with latitude and elevation, as well as over time. Therefore, reference values of calibrated production rates are reported relative to sea level and high latitude, whereupon they can be extrapolated to other study sites by means of specific scaling models (Stone 2000; Lifton et al. 2014). While the production rates of some cosmogenic nuclides, including ^{3}He, are relatively well known (Blard 2021), others are still characterised by significant discrepancies between the calibration results, in particular those of ^{36}Cl (Schimmelpfennig et al. 2011; Marrero et al. 2016). Therefore, a high priority for the cosmogenic nuclide community is the ongoing refinement of global production rate datasets (Borchers et al. 2016; Balco 2020).

6.3.2 A Volcanic Application of Cosmogenic Nuclide Surface-Exposure Dating

Having introduced the principles and basic laboratory methods, we now give an example of the application of cosmogenic surface-exposure dating from our own work in the Central Volcanic Zone of the Andes in South America (Bromley et al. 2019), highlighting as we go some of the standard practices of this method. In this case study, we sought to establish the eruption chronologies of five volcanoes in the Peruvian Andes, including Nevado Coropuna, Nevado Casiri and the Purupurini lava domes (Fig. 15). The dominance of pyroxene-bearing andesite at Coropuna makes ^{3}He surface-exposure dating the most appropriate tool for that site; at Casiri and Purupuni, where the local andesite is relatively pyroxene poor, we applied the ^{36}Cl method to the abundant Ca-rich feldspars. At both sites, we used a hammer and chisel to remove 0.5–1 kg of material from relict lava surfaces, and, after crushing and sieving, separated out pyroxenes and feldspars via mechanical, densimetric and chemical means. All ^{3}He/^{4}He ratios were measured by melting aliquots of purified pyroxene in a noble gas mass spectrometer at Woods Hole Oceanographic Institute, whereas ^{36}Cl/^{35}Cl ratios were measured using the CEREGE accelerator mass spectrometer ASTER.

The resulting surface-exposure ages (Fig. 15) reveal how long each lava surface has been exposed to cosmic radiation and provide robust

Fig. 15 Surface-exposure ages calculated using two different cosmogenic nuclides for an andesitic lava on Nevado Coropuna, Peru. Samples COR-07-05, COR-15-39, COR-15-40, and COR-16-70 are ^{3}He ages based on pyroxenes (Bromley et al. 2019), while sample NCNE03 is a whole-rock ^{36}Cl age (Ubeda et al. 2018). Samples CAS-14-02 and 03 are ^{36}Cl ages based on feldspars. Map panel adapted from Bromley et al. (2019)

estimates for the age of each landform. Most of the emplacements of the dated lava flows occurred during the Holocene, between ~6 ka and 1.5 ka; two flows at the Coropuna lava dome are older, returning ages of c. 12.5 ka and ~120 ka. These results indicate widespread Late Pleistocene and Holocene occurrence of large effusive eruptions of the studied Peruvian volcanoes. At the Sabancaya complex, two stratigraphically consistent ^{3}He ages are in agreement with a previously published maximum radiocarbon age from a peat buried by lava in the late Holocene. This shows that cosmogenic nuclide surface-exposure dating provides a powerful means for reconstructing eruptive histories on timescales of centuries to millions of years. Thus, this method is particularly invaluable where target landforms are too young to permit ^{40}Ar/^{39}Ar, K–Ar, or U–Pb dating or insufficient organic material exists for bracketing ^{14}C ages. Finally, we note that it also provides a novel opportunity to cross-check ages provided by other techniques.

6.4 Radiocarbon Dating

Jean-Pascal Dumoulin[1], **Jean-Claude Lefèvre**[2]

[1]LMC14, LSCE/IPSL, CEA-CNRS-UVSQ, Université Paris Saclay

[2]UMR8148 GEOPS Université Paris Saclay

6.4.1 Introduction

The development of radiocarbon dating by W. F. Libby in the early 1950s followed the identification of the radioactive isotope ^{14}C by M.D. Kamen and S. Ruben in 1940. The first measurements in natural organic matter were made in 1947 by E. Anderson and W. F. Libby, establishing the principle of radiocarbon dating.

From the beginning of the 1960s, following the creation of numerous laboratories, the use of radiocarbon extended to the dating of volcanic events. For example, the first ^{14}C dating list from the Saclay Laboratory (Delibrias et al. 1964) includes eight dates in a volcanic study. The Puy de la Vache in the Chaîne des Puys, Auvergne (no. Sa-90) was dated by charcoal found beneath the lava flow at Saint-Saturnin. This dating result, with that of the Puy Pariou (no. Sa-94), determined the end of the volcanic activity of the Chaîne des Puy at around 7650 ± 350 and 8580 ± 350 BP, respectively (−9398 −7828), (−10,483 −8631) cal BP. Since the beginning of the 1980s, ^{14}C dating using Accelerator Mass Spectrometer (AMS) has made it possible to expand the dating method by reducing the mass of the sample to be dated from a few grams to a few milligrams, or even a few tenths of a microgram for AMS equipped with a gas source. It has enabled a larger range of samples to be studied. For example, in addition to direct dating of small charcoal fragments from volcanic flows or pyroclastic fallout deposits, ^{14}C dating of foraminifera can provide an indirect chronology for tephra deposits in marine drilling cores.

6.4.2 Principle

The cosmic particles, penetrating the Earth's magnetic field, collide with nitrogen and oxygen atoms, producing neutrons. In the high atmosphere, once neutrons have slowed down by successive shocks, they interact with nitrogen atoms and produce ^{14}C atoms, according to the nuclear reaction:

$$^{14}N_7 + ^1 n_0 \rightarrow ^{14} C_6 + ^1 P_1$$

The ^{14}C isotope, radioactive with a period (T) of 5730 ± 40 years (Godwins 1962), rapidly oxidizes into carbon dioxide (CO_2). It mixes with atmospheric CO_2 and enters the carbon cycle by assimilation into the biosphere through photosynthesis and then the food chain. The production and disintegration tend towards an equilibrium referenced in 1950 at $A_0 = 13.56$ disintegrations per minute and gram of carbon. This equilibrium is broken at the time of the death of the organisms by ending carbon exchange, and the decay of ^{14}C starts according to the following equation:

$$A(t) = A_0 \exp(-\lambda t) \text{ (}\lambda\text{ radioactive decay constant)}$$

$$\text{with } \lambda = \ln(2)/T$$

The radiocarbon age (t) is expressed as follows:

$$t(\text{years}) = -8033 \times \ln(A_{ech}/A_0)$$

$$(A_{ech} = \text{residual activity in a sample; } A_0 \text{ activity of the atmosphere})$$

The ^{14}C age is expressed in years BP (Before Present) with the year 0 BP referring to 1950 AD or CE (Stuiver and Polach 1977). The temporal range of ^{14}C dating extends from the present to 50 ka.

The ^{14}C production in the upper atmosphere and the CO_2 exchanges with the different carbon reservoirs (ocean, atmosphere and biosphere) have changed in the past. This has led the scientific community to quantify the offset between the ^{14}C ages and the calendar ages, namely the calibration procedure. The latest calibration curve published in 2020 (INTCAL20) covers the time interval 0 to 50 ka BP (Reimer et al. 2020). Once calibrated, the date is expressed in cal BP, AD/BC or CE/BCE. Several programs can be used to calibrate a ^{14}C age including Calib, Chronomodel, Oxcal etc.

6.4.3 Sample Preparation

Wood, charcoal or sediments are commonly used to date volcanic eruptions. It is recommended that users contact the laboratory to receive guidance on appropriate sampling strategies and sample storage. Clean tools and receptacles (aluminium foil or glass bottles) should be used during sample collection to avoid adding exogenous carbon to the sample. For ^{14}C dating, charcoal or wood are preferred over bulk soil which can be heterogeneous, with different sources of carbon. It is better to analyse dry samples whenever possible and if wet samples must be sent, they should be stored in a fridge at 4°C to avoid contamination through the formation of mould, fungi or bacteria.

Depending on the material to be dated, chemical treatment is needed to clean the sample and remove contaminants that may alter the age of the sample. A microscopic inspection is sufficient to remove macro-contaminants such as rootlets, hairs or fibres that may be present in the sample. To remove molecular-scale contaminants such as carbonates or humic acids, chemical pre-treatment is necessary.

Most often the classic ABA (acid, base, acid) treatment is used:

(1) The geological carbonates present in the soil that contain dead carbon (namely carbon without ^{14}C) are very old and affect the sample age. As a consequence, all organic material is carefully decarbonated with acid (0.5 M HCl, for 1 h at 80 °C) and then water-washed to raise the pH to a neutral value.
(2) Humic acids are highly mobile organic compounds in soils and can be deposited in the porous structures of samples such as burnt wood or charcoal. They are soluble in an alkaline solution so a basic treatment (NaOH, 0.1 M at 80 °C for 1 h) is recommended to remove this type of contaminant.
(3) Finally, a second acid treatment (HCl 0.5 M, for 1 h at 80 °C) is applied to remove atmospheric CO_2 which could have been adsorbed by the sample during the basic treatment.

Submarine eruptions or terrestrial eruptions that have projected tephra into the sea provide chronostratigraphic markers in deep-sea sediment cores. In this case, marine organisms like shells, corals or foraminifera can be used to date the event (Paterne et al. 1988; Siani et al. 2001). These carbonate samples must also be carefully cleaned before dating. For corals, any recrystallization of calcite on the primary aragonite must be removed with sandblasting and all the carbonates undergo chemical treatment with nitric acid, whose concentration depends on the size of the sample (HNO_3 from 0.01N to 0.3N, cold).

6.4.4 Application to the Case Study of Solfatara of Pozzuoli (Campi Flegrei, Italy)

We present a case study using ^{14}C to monitor active volcanic areas. Usually, this work is done by measuring variations in the gas emissions. After water vapour, carbon dioxide (CO_2) is the most abundant gas emitted, due to its weak solubility in magmas. Among the different techniques used to measure CO_2 emissions from soil, one of the most common procedures is the accumulation chamber method (ACM). An alternative approach for long-term CO_2 flux measurement is based on the lack of ^{14}C concentration in plants growing in active volcanic zones due to their assimilation of ^{14}C-free CO_2 from the gas emanating from the soil (Pasquier-Cardin et al. 1999; Ewans et al. 2010; Lewicki et al. 2014). It is thus possible to estimate the local CO_2 emission by comparing the activity of the emitted CO_2 in the chamber with that of the atmosphere.

Here we present the results of a series of radiocarbon measurements on plants sampled within and near to the Solfatara of Pozzuoli (Campi Flegrei, Italy), an active volcano affected by intense degassing (Lefevre et al. 2018) (Fig. 16). The results are compared with the data obtained, on the same sites, by repeated ACM CO_2 flux measurements. Table 5 shows the comparative data of outgassing values measured by ACM and ^{14}C activities in plants at the same location for three samples (data were taken in May 2013, May 2014 and May 2015). Figure 17 shows the ACM measurement curve of degassed CO_2 versus plant ^{14}C activity. These results show that the ^{14}C activity decreases relative to an increase in CO_2 outgassing. This ^{14}C-free CO_2 from the soil can account for up to 50% of the CO_2 fixed by plant photosynthesis, giving an apparent age of around 5000 years. The measurement of ^{14}C in plants shows a lower dispersion of measurements from ACM results. In fact, determination of the ^{14}C deficiency in plants provides an estimation of the CO_2 rate from the soil, within a few square metres, integrating CO_2 soil degassing variations and meteorological data over a few months. It can therefore be an efficient

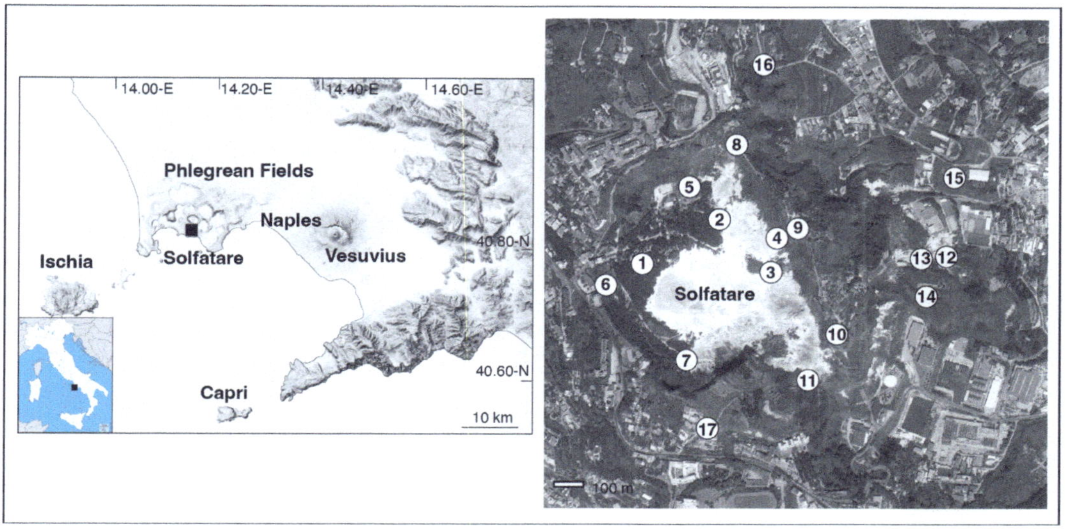

Fig. 16 Location of the Phlegrean Field and map of sampling sites around Solfatara

Table 5 Measurement of CO_2 fluxes by ACM and ^{14}C activities of plants sampled at the same location over three successive years

Samples	S1	S2	S3	S4	S5	S7	S8	S9	S10	S11	S12	S13	S14	S15	S17
ACM flux May 2013 (g/m^2.j)	193	594	213	974	364	222	35	20	137	495	103	159	243	54	64
^{14}C activity (%)	83.98 ± 0.31	65.61 ± 0.31	77.21 ± 0.29	52.72135	8,671,103	84.72	87.43 ± 03	92.43 ± 0.32	82.15 ± 0.28	66.64 ± 0.24	81.11 ± 0.28	69.1 ± 0.25	8732 ± 0.32	96.34 ± 0.32	97.39 ± 0.33
^{14}C age (BP)	1400 ± 30	3387 ± 35	2080 ± 30	5145	1147 ± 30	1,330,135	1080 ± 30	630 ± 30	1580 ± 25	3260 ± 30	1685 ± 25	2970 ± 30	1090 ± 30	300 ± 25	210 ± 25
ACM flux May 2014 (g/m^2.j)	81.4	304.1	74.8	580.3	289.6	–	57.9	29.6	156.0	643.8	126.6	300.9	252.6	33.9	48.6
^{14}C activity (%)	84.94 ± 0.3	73.24 ± 0.26	76.36 ± 0.26	55.74 ± 0.21	84.58 ± 0.29	–	87.64 ± 03	90.03 ± 0.3	81.55 ± 029	64.7 ± 0.24	82.35 ± 0.29	71.27 ± 0.25	85.12 ± 0.32	98.07 ± 0.34	95.61 ± 032
^{14}C age (BP)	1311 ± 30	2502 ± 30	2167 ± 30	4695 ± 30	1345 ± 30	–	1060 ± 30	844 ± 26	1658 ± 30	3490 ± 30	1560 ± 30	2720 ± 30	1291 ± 30	156 ± 30	361 ± 28
ACM flux May 2015 (g/m^2.j)	171.7	426.7	124.9	523.2	272.2	227.6	41.9	19.1	99.5	485.7	428.0	227.1	117.0	247.3	385
^{14}C activity (%)	83.21 ± 0.29	71.03 ± 0.25	75.74 ± 0.29	51.0310.2	84.5310.29	82.93 ± 0.29	86.87 ± 029	89.77 ± 0.31	8087 ± 0.28	63.65 ± 0.23	72.39 ± 0.26	67.13	87.4 ± 15	94.98 ± 0.34	95.04 ± 0.32
^{14}C age (BP)	1476 ± 28	2748 ± 28	2232 ± 31	5,404,332	1350 ± 28	1504 ± 28	1130 ± 27	867 ± 28	1706 ± 28	2117 ± 29	2596 ± 29	3201 ± 28	1082 ± 30	414 ± 30	409 ± 27

Reference for modern activity (1950): 100% or 13.56 dpm
Modern 'normal' activity for plants is 104% or 14.1dpm for 2013. 103% or 14.0 dpm for 2014

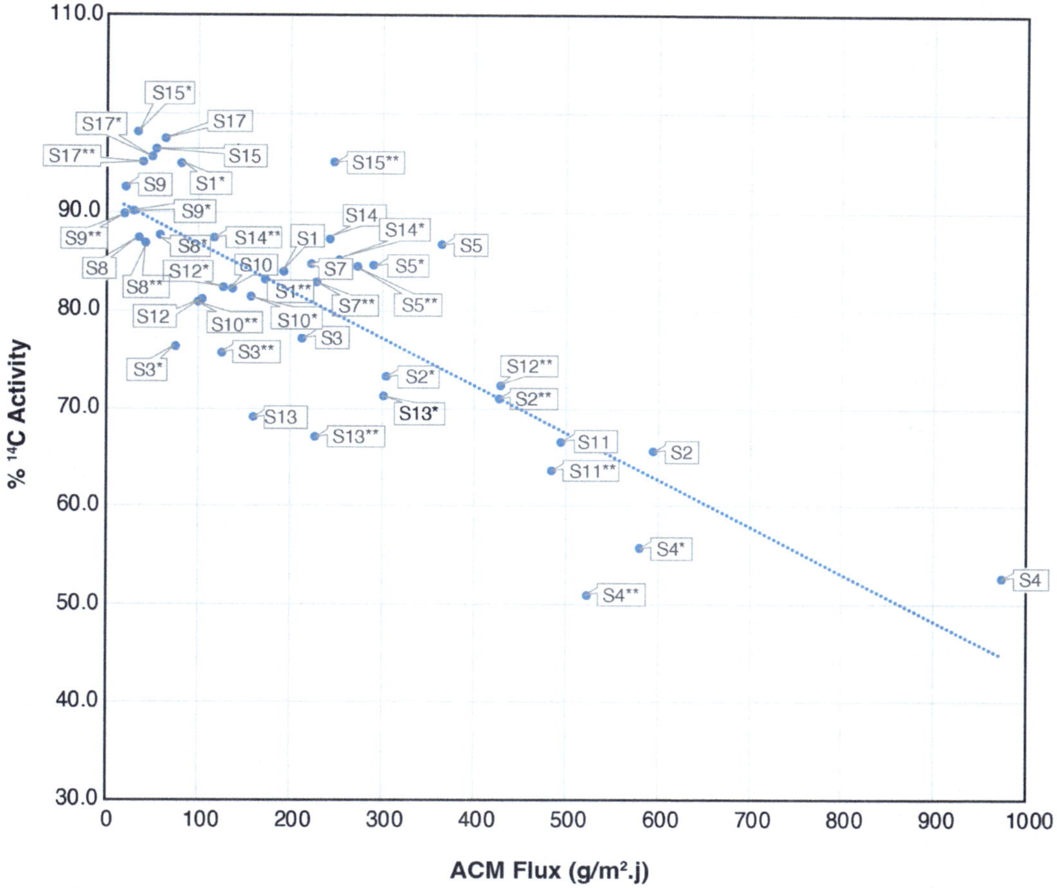

Fig. 17 Graph of the ^{14}C activity measured in the plants versus the CO_2 fluxes measured by ACM in each site

bio-sensor and can be used as a proxy to map the soil CO_2 emissions and to determine its variations through time.

References

Aitken M (1985) Thermoluminescence Dating. In: Göksu HY, Oberhofer M, Regulla D (eds) Scientific dating methods, advanced scientific techniques, vol 1, Kluwer Academic Publishers, pp 121–140. ISBN 0-7923-1461-1

Aitken MJ (1998) An Introduction to Optical Dating: the Dating of Quaternary Sediments by the Use of Photon-simulated luminescence. Clarendon Press

Albino F, Biggs J, Escobar-Wolf R, Naismith A, Watson M, Philips JC, Chigna Marroquin GA (2020) Using TanDEM-X to measure pyroclastic flow source location, thickness, and volume: application to the 3 June 2018 eruption of Fuego volcano Guatemla. J Volcanol Geoth Res406:107063.https://doi.org/10.1016/j.jvolgeores.2020.107063

Alcalá-Reygosa J, Arce JL, Schimmelpfennig I, Salinas EM, Rodríguez MC, Léanni L, Aumaitre G, Bourles D, Keddadouche K and Aster Team (2018) Revisiting the age of the Jumento volcano, Chichinautzín Volcanic Field (Central Mexico), using in situ-produced cosmogenic. ^{10}Be. J Volcanol Geoth Res 366:112–119

Aldrich LT, Nier AO (1948) Argon 40 in potassium minerals. Phys Rev 74:876–877. https://doi.org/10.1103/PhysRev.74.876

Anderson T, Flett JS (1902) Preliminary report on the recent eruption of the Soufrière in St. Vincent, and of a visit to Mont Pelée, in Martinique. Proc R Soc Lond 70:423–445. https://doi.org/10.1098/rspl.1902.0042

Anderson K, Segall P (2013) Bayesian inversion of data from effusive volcanic eruptions using physics-based models: Application to Mount St. Helens 2004–2008. J Geophys Res Solid Earth 118(5):2017–2037. https://doi.org/10.1002/jgrb.50169

Aravena A, Chupin L, Dubois T, Roche O (2021) The influence of gas pore pressure in dense granular flows: numerical simulations versus experiments and implications for pyroclastic density currents. Bull Volcanol 83:77. https://doi.org/10.1007/s00445-021-01507-7

Bablon M, Quidelleur X, Samaniego P, Le Pennec JL, Lahitte P, Liorzou C, Bustillos JE, Hidalgo S (2018) Eruptive chronology of Tungurahua volcano (Ecuador) revisited based on new K Ar ages and geomorphological reconstructions. J Volcanol Geoth Res 357:378–398. https://doi.org/10.1016/j.jvolgeores.2018.05.007

Bablon M, Quidelleur X, Samaniego P, Le Pennec JL, Santamaría S, Liorzou C et al (2020) Volcanic history reconstruction in northern Ecuador: Insights for eruptive and erosion rates on the whole Ecuadorian arc. Bull Volcanol 82(1):11. https://doi.org/10.1007/s00445-019-1346-1

Bacon CR, Lanphere MA (2006) Eruptive history and geochronology of Mount Mazama and the Crater Lake region, Oregon. Bull Geol Soc Am 118:1331–1359. https://doi.org/10.1130/B25906.1

Balco G (2020) A prototype transparent-middle-layer data management and analysis infrastructure for cosmogenic-nuclide exposure dating. Geochron 2(2):169–175. https://doi.org/10.5194/gchron-2-169-2020

Balco G, Stone JO, Lifton NA, Dunai TJ (2008) A complete and easily accessible means of calculating surface exposure ages or erosion rates from ^{10}Be and ^{26}Al measurements. Quat Geochron 3:174–195. https://doi.org/10.1016/j.quageo.2007.12.001

Becerra-Ramírez R, Dóniz-Páez J, González E (2022) Morphometric analysis of scoria cones to define the 'volcano-type' of the campo de calatrava volcanic Region (Central Spain). Land 11:917. https://doi.org/10.3390/land11060917

Bemis KG, Ferencz M (2017) Morphometric analysis of scoria cones: the potential for inferring process from shape. In: Németh K, Carrasco-Núñez G, Aranda-Gómez JJ, Smith IEM (eds) Monogenetic volcanism. Geol Soc London Spec Publ 446, pp 61–100. https://doi.org/10.1144/SP446.9

Ben-Asher M, Haviv I, Roering JJ, Crouvi O (2017) The influence of climate and microclimate (aspect) on soil creep efficiency: cinder cone morphology and evolution along the eastern Mediterranean Golan Heights. Earth Surf Process Landf 42:2649–2662. https://doi.org/10.1002/esp.4214

Biggs J, Wright TJ (2020) How satellite InSAR has grown from opportunistic science to routine monitoring over the last decade. Nat Comm 11(1):10–13. https://doi.org/10.1038/s41467-020-17587-6

Blard PH (2021) Cosmogenic ^{3}He in terrestrial rocks: a review. Chem Geol 586:120543. ISSN 0009–2541. https://doi.org/10.1016/j.chemgeo.2021.120543

Bohnenstiehl DR, Howell JK, White SM, Hey RN (2012) A modified basal outlining algorithm for identifying topographic highs from gridded elevation data, Part 1: Motivation and methods. Comput Geosci 49:308–314. https://doi.org/10.1016/j.cageo.2012.04.023

Boivin P, Besson JC, Briot D, Deniel C, Gourgaud A, Langlois E, Labazuy P, de Larouzière FD, Livet M, Merciecca C, Mergoil J, Miallier D, Morel JM, Thouret JC, Vernet G (2017) Volcanology of the Chaîne des Puys. Parc Naturel Régional de la Chaîne des Puys, pp 196

Borchers B, Marrero S, Balco G, Caffee M, Goehring B, Lifton N, Nishiizumi K, Phillips F, Schaefer J, Stone J (2016) Geological calibration of spallation production rates in the CRONUS-Earth project. Quat Geochron 31:188–198. https://doi.org/10.1016/j.quageo.2015.01.009

Bösken JJ, Schmidt C (2020) Direct and indirect luminescence dating of tephra: a review. J Quat Sci 35:1–2. Special issue: tephrochronology as a global geoscientific research tool, 39–53. https://doi.org/10.1002/jqs.3160

Bossis R, Regard V, Carretier S (2022) Initial shape reconstruction of a volcanic island as a tool for quantifying long-term coastal erosion: the case of Corvo Island (Azores). Earth Surf Dynam Discuss 2022:1–32. https://doi.org/10.5194/esurf-2022-18

Boudon G, Balcone-Boissard H, Villemant B, Morgan DJ (2015) What factors control superficial lava dome explosivity? Sci Rep 5(1):14551. https://doi.org/10.1038/srep14551

Bourdier JL, Abdurachman E (2001) Decoupling of small-volume pyroclastic flows and related hazards at Merapi volcano, Indonesia. Bull Volcanol 63:309–325. https://doi.org/10.1007/s004450100133

Breard ECP, Fullard L, Dufek J, Tennenbaum M, Fernandez Nieves A, Dietiker JF (2022) Investigating the rheology of fluidized and non-fluidized gas-particle beds: implications for the dynamics of geophysical flows and substrate entrainment. Granul Matter 24:34. https://doi.org/10.1007/s10035-021-01192-5

Bromley GR, Winckler G, Schaefer JM, Kaplan MR, Licht KJ, Hall BL (2014) Pyroxene separation by HF leaching and its impact on helium surface-exposure dating. Quat Geochron 23:1–8. https://doi.org/10.1016/j.quageo.2014.04.003

Bromley GR, Thouret JC, Schimmelpfennig I, Mariño J, Valdivia D, Rademaker K, del Pilar Vivanco Lopez S, Team A.S.T.E.R., Aumaître G, Bourlès D, Keddadouche K (2019) In situ cosmogenic ^{3}He and ^{36}Cl and radiocarbon dating of volcanic deposits refine the Pleistocene and Holocene eruption chronology of SW Peru. Bull Volcanol 81:1–16. https://doi.org/10.1007/s00445-019-1325-6

Brož P, Čadek O, Hauber E, Rossi AP (2015) Scoria cones on Mars: Detailed investigation of morphometry based on high-resolution digital elevation models. J Geophys Res Planets 120:1512–1512–1527. https://doi.org/10.1002/2015JE004873

Capra L, Roverato M, Groppelli G, Caballero L, Sulpizio R, Norini G (2015) Glacier melting during lava dome growth at Nevado de Toluca volcano (Mexico): evidences of a major threat before main eruptive phases at ice-caped volcanoes. J Volcanol Geoth Res 294:1–10. https://doi.org/10.1016/j.jvolgeores.2015.02.005

Carbone D, Zuccarello L, Messina A, Scollo S, Rymer H (2015) Balancing bulk gas accumulation and gas output before and during lava fountaining episodes at Mt Etna. Sci Rep 5:1–11. https://doi.org/10.1038/srep18049

Carnemolla F, De Guidi G, Bonforte A, Brighenti F, Briole P (2023) The ground deformation of the south-eastern flank of Mount Etna monitored by GNSS and SAR interferometry from 2016 to 2019. Geophys J Intern 234:664–682. https://doi.org/10.1093/gji/ggad088

Carr MJ (1984) Symmetrical and segmented variation of physical and geochemical characteristics of the Central American Volcanic Front. J Volcanol Geoth Res 20:231–252. https://doi.org/10.1016/0377-0273(84)90041-6

Cashman KV, Soule SA, Mackey BH, Deligne NI, Deardorff ND, Dietterich HR (2013) How lava flows: new insights from applications of lidar technologies to lava flow studies. Geosphere 9(6):1664–1680. https://doi.org/10.1130/GES00706.1

Cassignol C, Gillot PY (1982) Range and effectiveness of unspiked potassium-argon dating: experimental groundwork and applications, Chapter 9, pp 159–179. In: Odin GS (ed) Numerical dating in Stratigraphy. John Wiley & Sons. ISBN-10.0471100854

Cayol V, Peltier A, Froger J, Beauducel F (2022) Monitoring volcano deformation. Chap. 2, pp 95–165. In Lénat JF (ed) Hazards and monitoring of volcanic activity 2: seismology, deformation and remote sensing. Wiley

Chaput M, Famin V, Michon L (2017) Sheet intrusions and deformation of Piton des Neiges, nd their implication for the volcano-tectonics of La Réunion. Tectonophys 717. https://doi.org/10.1016/j.tecto.2017.08.039

Charbonnier SJ, Gertisser R (2008a) Field observations and surface characteristics of pristine block-and-ash flow deposits from the 2006 eruption of Merapi Volcano, Java, Indonesia. J Volcanol Geoth Res 177:971–982. https://doi.org/10.1016/j.jvolgeores.2008.07.008

Charbonnier SJ, Gertisser R (2008b) Field observations and surface characteristics of pristine block-and-ash flow deposits from the 2006 eruption of Merapi Volcano, Java Indonesia. J Volcanol Geoth Res 177(4):971–982. https://doi.org/10.1016/j.jvolgeores.2008.07.008

Charbonnier SJ, Gertisser R (2011) Deposit architecture and dynamics of the 2006 block-and-ash flows of Merapi Volcano, Java, Indonesia. Sedimentology 58:1573–1612. https://doi.org/10.1111/j.1365-3091.2011.01226.x

Charbonnier SJ, Germa A, Connor CB, Gertisser R, Komorowski JC, Preece K, Lavigne F, Dixon T, Connor L (2013) Evaluation of the impact of the 2010 pyroclastic density currents at Merapi volcano from high-resolution satellite imagery, field investigations and numerical simulations. J Volcanol Geoth Res 261:295–315. https://doi.org/10.1016/j.jvolgeores.2012.12.021

Charbonnier SJ, Garin F, Rodríguez LA, Ayala K, Cancel S, Escobar-Wolf R, Chigna G, Chun-Quinillo C, González D, Chigna W, Chun-Quinillo K, Mérida R, Juarez F, Calder ES (2023) Unravelling the dynamics and hazards of the June 3rd, 2018, pyroclastic density currents at Fuego volcano (Guatemala). J Volcanol Geoth Res 436:107791. https://doi.org/10.1016/j.jvolgeores.2023.107791

Chen Y, Remy D, Froger JL, Peltier A, Villeneuve N, Darrozes J, Perfettini H, Bonvalot S (2017) Long-term ground displacement observations using InSAR and GNSS at Piton de la Fournaise volcano between 2009 and 2014. Rem Sens Envir 194:230–247. https://doi.org/10.1016/j.rse.2017.03.038

Chiba T, Kaneta SI, Suzuki Y (2008) Red relief image map: new visualization method for three dimensional data. Int Arch Photogramm Remote Sens Spat Inf Sci 37:1071–1076

Christiansen RL, Lowenstern JB, Smith RB, Heasler H, Morgan LA, Nathenson M, Mastin LG, Muffler Ljp, Robinson JE (2007) Preliminary assessment of volcanic and hydrothermal hazards in Yellowstone National Park and vicinity. Reston, Virginia, US Geol Surv Open-file Report 2007–1071, pp 94. https://doi.org/10.3133/ofr20071071

Civico R, Ricci T, Scarlato P, Taddeucci J, Andronico D, Del Bello E, D'Auria L, Hernández, PA, Pérez NM (2022) High-resolution digital surface model of the 2021 eruption deposit of Cumbre Vieja volcano, La Palma, Spain. Scientific Data 9(1):435. https://doi.org/10.1038/s41597-022-01551-8

Corazzato C, Tibaldi A (2006) Fracture control on type, morphology and distribution of parasitic volcanic cones: an example from Mt. Etna, Italy. J Volcanol Geoth Res 158(1):177–194. https://doi.org/10.1016/j.jvolgeores.2006.04.018

Cordonnier B, Lev E, Garel F (2016) Benchmarking lava-flow models. Geol Soc London, Spec Publ 426(1):425–445. https://doi.org/10.1144/SP426.7

Costa A, Macedonio G (2005) Computational modeling of lava flows: a review. Geol Soc Am 396:209. https://doi.org/10.1130/0-8137-2396-5.209

Cronin SJ, Lube G, Dayudi DS, Sumarti, S, Subrandiyo SS (2013) Insights into the October–November 2010 Gunung Merapi eruption (Central Java, Indonesia) from the stratigraphy, volume and characteristics of its pyroclastic deposits. In: Jousset P, Pallister J, Surono (eds) Merapi eruption. J Volcanol Geoth Res 261:244–259. https://doi.org/10.1016/j.jvolgeores.2013.01.005

Culling WEH (1963) Soil creep and the development of hillside slopes. J Geol 71:127–161. https://doi.org/10.1086/626891

Darmawan H, Walter TR, Brotopuspito KS, Subandriyo NIGMA (2018) Morphological and structural changes at the Merapi lava dome monitored in 2012–15 using unmanned aerial vehicles (UAVs). J Volcanol Geoth Res 349:256–267. https://doi.org/10.1016/j.jvolgeores.2017.11.006

Delcamp A, van Wyk de Vries B, James M, Gailler L, Lebas E (2011) Relationships between volcano gravitational spreading and magma intrusion. Bull Volcanol 74:743–765. https://doi.org/10.1007/s00445-011-0558-9.

Delibrias G, Guillier M, Labeyrie J (1964) Saclay natural radiocarbon measurements I. Radiocarbon 6:233–250. https://doi.org/10.1017/S0033822200010717

Delunel R, Blard PH, Martin LC, Nomade S, Schlunegger F (2016) Long term low latitude and high elevation cosmogenic 3He production rate inferred from a 107 ka-old lava flow in northern Chile; 22 S-3400 m asl. Geochim et Cosmochim Acta, 184:71–87. https://doi.org/10.1016/j.gca.2016.04.023

Desilets D, Zreda M, Almasi PF, Elmore D (2006) Determination of cosmogenic ^{36}Cl in rocks by isotope dilution: innovations, validation and error propagation. Chem Geol 233:185–195. https://doi.org/10.1016/j.chemgeo.2006.03.001

de' Michieli Vitturi M, Arrowsmith JR (2013). Two-dimensional nonlinear diffusive numerical simulation of geomorphic modifications to cinder cones. Earth Surf Proc Landf 38:1432–1443. https://doi.org/10.1002/esp.3423

de' Michieli Vitturi M, Tarquini S (2018) MrLavaLoba: A new probabilistic model for the simulation of lava flows as a settling process. J Volcanol Geoth Res 349:323–334. https://doi.org/10.1016/j.jvolgeores.2017.11.016

de Silva SL, Self S, Francis PW, Drake RE, Carlos RR (1994) Effusive silicic volcanism in the Central Andes: The Chao dacite and other young lavas of the Altiplano-Puna Volcanic Complex. J Geophys Res Solid Earth, 99 (B9):17805–17825. https://doi.org/10.1029/94JB00652

Dibacto S, Lahitte P, Karátson D, Hencz M, Szakács A, Biró T, Kovács I, Veres D (2020) Growth and erosion rates of the East Carpathians volcanoes constrained by numerical models: tectonic and climatic implications. Geomorphology 368:107352. https://doi.org/10.1016/j.geomorph.2020.107352

Druitt TH, Calder ES, Cole PD, Hoblitt RP, Loughlin SC, Norton GE, Ritchie LJ, Sparks RSJ, Voight B (2002) Small-volume, highly mobile pyroclastic flows formed by rapid sedimentation from pyroclastic surges at Soufriere Hills Volcano, Montserrat: an important volcanic hazard. Geol Soc Lond Mem 21:263–279. https://doi.org/10.1144/gsl.mem.2002.021.01.12

Dumont Q, Cayol V, Froger J, Peltier A (2022) 22 years of satellite imagery reveal a major destabilization structure at Piton de la Fournaise. Nat Comm 13:1–11. https://doi.org/10.1038/s41467-022-30109-w

Dunai TJ (2010) Cosmogenic nuclides: principles, concepts and applications in the earth surface sciences. Cambridge University Press, Cambridge. ISBN 978-0-521-87380-2

Dzurisin D (2006) Volcano deformation: new geodetic monitoring techniques. Springer Berlin, Heidelberg. https://doi.org/10.1007/0-306-47633-9

Euillades LD, Grosse P, Euillades PA (2013) NETVOLC: an algorithm for automatic delimitation of volcano edifice boundaries using DEMs. Comp Geosci 56:151–160. https://doi.org/10.1016/j.cageo.2013.03.011

Evans WC, Bergfeld D, McGeehan JP, King J, Heasler H (2010) Tree-ring ^{14}C links seismic warm to CO_2 spike at Yellowstone, USA. Geology 38:1075–1078. https://doi.org/10.1017/10.1130/G31345.1

Famin V, Michon L (2010) Volcano destabilization by magma injections in a detachment. Geology 38(3):219–222. https://doi.org/10.1130/G30717.1

Favalli M, Pareschi MT, Neri A et al (2005) Forecasting lava flow paths by a stochastic approach. Geophys Res Lett 32(3). https://doi.org/10.1029/2004GL021718.

Favalli M, Karátson D, Yepes J, Nannipieri L (2014) Surface fitting in geomorphology-examples for regular-shaped volcanic landforms. Geomorphology 221:139–149. https://doi.org/10.1016/j.geomorph.2014.06.009

Favalli M, Fornaciai A (2017) Visualization and comparison of DEM-derived parameters. Application to volcanic areas. Geomorphology 290:69–84. https://doi.org/10.1016/j.geomorph.2017.02.029

Ferrier KL, Huppert KL, Perron JT (2013) Climatic control of bedrock river incision. Nature 496:206–209. https://doi.org/10.1038/nature11982

Finkel RC, Suter M (1993) AMS in earth sciences: technique and applications. Adv Anal Geochemistr 1:1–114. ISBN 9781559383325.

Flint JJ (1974) Stream gradient as a function of order, magnitude, and discharge. Water Resour Res 10:969–973. https://doi.org/10.1029/WR010i005p00969

Fukushima Y, Cayol V, Durand P, Massonnet D (2010) Evolution of magma conduits during the 1998–2000 eruptions of Piton de la Fournaise volcano, Réunion Island. J Geophys Res Solid Earth 115:10. https://doi.org/10.1029/2009JB007023

Galland O, Scheibert J (2013) Analytical model of surface uplift above axisymmetric flat-lying magma intrusions: Implications for sill emplacement and geodesy. J Volcanol Geoth Res 253:114–130. https://doi.org/10.1016/j.jvolgeores.2012.12.006

Gallant JC, Wilson JP (2000) Primary topographic attributes. In: Wilson JP, Gallant JC (eds) Terrain analysis: principles and applications. Wiley, New York, pp 51–85

Gase AC, Brand BD, Bradford JH (2017) Evidence of erosional self-channelization of pyroclastic density currents revealed by ground-penetrating radar imaging at Mount St. Helens, Washington (USA). Geophys Res Lett 44:2220–2228. https://doi.org/10.1002/2016GL072178

Germa A, Lahitte P, Quidelleur X (2015) Construction and destruction of Mont Pelée volcano: volumes and rates constrained from a geomorphologicalmodel of evolution. J Geophys Res, Earth Surface 120:1206–1226. https://doi.org/10.1002/2014JF003355

Germa A, Quidelleur X, Labanieh S, Chauvel C, Lahitte P (2011) The volcanic evolution of Martinique Island: Insights from K-Ar dating into the Lesser Antilles arc migration since the Oligocene. J Volcanol Geoth Res 208:122–135. https://doi.org/10.1016/j.jvolgeores.2011.09.007

Gertisser R, Cassidy NJ, Charbonnier SJ, Nuzzo L, Preece K (2012) Overbank block-and-ash flow deposits and the impact of valley-derived, unconfined flows on populated areas at Merapi volcano, Java, Indonesia. Nat Hazards 60:623–648. https://doi.org/10.1007/s11069-011-0044-x

Gillot PY, Cornette Y (1986) The Cassignol technique for potassium-argon dating, precision and accuracy: examples from the late Pleistocene to recent volcanism from southern Italy. Chem Geol (isotope geoscience section) 59:205–222. https://doi.org/10.1016/0168-9622(86)90072-2

Gillot PY, Hildenbrand A, Lefèvre JC, Albore-Livadie C (2006) The K/Ar dating method: principle, analytical techniques, and application to Holocene volcanic eruptions in southern Italy. Acta Vulcanologica 18(2):55–66

Gilbert GK (1909). The Convexity of Hilltops. J Geol 17(4):344–350. https://www.jstor.org/stable/30057970

Godwin H (1962) Half-life of radiocarbon. Nature 195:984. https://doi.org/10.1038/195984a0

Gosse JC, Phillips FM (2001) Terrestrial in situ cosmogenic nuclides: theory and application. Quat Sci Rev 20:1475–1560. https://doi.org/10.1016/S0277-3791(00)00171

Graettinger AH (2018) Trends in maar crater size and shape using the global Maar Volcano Location and Shape (MaarVLS) database. J Volcanol Geoth Res 357:1–13. https://doi.org/10.1016/j.jvolgeores.2018.04.002

Grosse P, Kervyn M (2018) Morphometry of terrestrial shield volcanoes. Geomorphology 304:1–14. https://doi.org/10.1016/j.geomorph.2017.12.017

Grosse, P, van Wyk de Vries B, Petrinovic IA, Euillades PA, Alvarado GE (2009) Morphometry and evolution of arc volcanoes. Geology 37:651–654. https://doi.org/10.1130/G25734A.1

Grosse P, van Wyk de Vries B, Euillades PA, Kervyn M, Petrinovic IA (2012) Systematic morphometric characterization of volcanic edifices using digital elevation models. Geomorphology 136:114–131. https://doi.org/10.1016/j.geomorph.2011.06.001

Grosse P, Euillades PA, Euillades LD, van Wyk de Vries B (2014) A global database of composite volcano morphometry. Bull Volcanol 76:784. https://doi.org/10.1007/s00445-013-0784-4

Grosse P, Ochi Ramacciotti ML, Escalante Fochi F, Guzmán S, Orihashi Y, Sumino H (2020) Geomorphology, morphometry, spatial distribution and ages of mafic monogenetic volcanoes of the Peinado and Incahuasi fields, southernmost Central Volcanic Zone of the Andes. J Volcanol Geoth Res 401:106966. https://doi.org/10.1016/j.jvolgeores.2020.1069660377-0273

Grosse P, Danišík M, Apaza FD, Guzmán SR, Lahitte P, Quidelleur X, Self S, Siebe C, van Wyk de Vries B, Ureta G, Guillong M, De Rosa R, Le Roux P, Wotzlaw JF, Bachmann O (2022) Holocene collapse of Socompa volcano and pre- and post-collapse growth rates constrained by multi-system geochronology. Bull Volcanol 84:85. https://doi.org/10.1007/s00445-022-01594-0

Guérin G (1983) La thermoluminescence des plagioclases, méthode de datation du volcanisme. Applications au domaine volcanique français: Chaîne des Puys, Mont-Dore et Cézallier, Bas Vivarrais. PhD thesis. Université Pierre et Marie Curie, Paris VI, Paris

Guérin GG, Valladas G (1980) Thermoluminescence dating of volcanic plagioclases. Nature 286:697–699. https://doi.org/10.1038/286697a0

Gueugneau V, Kelfoun K, Charbonnier SJ, Germa A, Carazzo G (2020) Dynamic and impacts of the May 8th, 1902 pyroclastic current at Mount Pelée (Martinique): new insights from numerical modeling. Front Earth Sci 8:279. https://doi.org/10.3389/feart.2020.00279

Gueugneau V, Charbonnier S, Esposti Ongaro T, de' Michieli Vitturi M, Peruzzetto M, Mangeney A, Bouchut F, Patra A, Kelfoun, K (2021) Synthetic benchmarking of concentrated pyroclastic current models. Bull Volcanol 83:1–32. https://doi.org/10.1007/s00445-021-01491-y

Gueugneau V, Charbonnier S, Roche O (2022) PyroCLAST: a new experimental framework to investigate overspilling of channelized, concentrated pyroclastic currents. Bull Volcanol 85:5. https://doi.org/10.1007/s00445-022-01623-y

Hack JT (1957) Studies of longitudinal stream profiles in Virginia and Maryland. US Geol Surv Prof Pap 249:97. https://doi.org/10.3133/pp294B

Hampton SJ, Cole JW (2009) Lyttelton Volcano, Banks Peninsula, New Zealand: primary volcanic landforms and eruptive centre identification. Geomorphology 104(3–4):284–298. https://doi.org/10.1016/j.geomorph.2008.09.005

Hanssen RF (2001) Radar Interferometry—data interpretation and error analysis. In: J Physics A: Mathematical and Theoretical, 2: 8. Kluwer Academic. https://doi.org/10.1007/0-306-47633-9

Harris AJ, Rose WI, Flynn LP (2003) Temporal trends in lava dome extrusion at Santiaguito 1922–2000. Bull Volcanol 65(2):77–89. https://doi.org/10.1007/s00445-002-0243-0

Harris A, Chevrel M, Coppola D et al (2019) Validation of an integrated satellite-data-driven response to an effusive crisis: the April–May 2018 eruption of Piton de la Fournaise. Ann Geophys-Italy 61:1–18. https://doi.org/10.4401/ag-7972

Hidaka M, Goto A, Umino S et al (2005) VTFS project: Development of the lava flow simulation code LavaSIM with a model for three-dimensional convection, spreading, and solidification. Geochem Geophys Geosys 6(7). https://doi.org/10.1029/2004GC000869

Hooper DM, Sheridan MF (1998) Computer-simulation models of scoria cone degradation. J Volcanol Geoth Res 83:241–267. https://doi.org/10.1016/S0377-0273(98)00031-6

Horton RE (1945) Erosional development of streams and their drainage basins; hydrological approach to quantative morphology. Bull Geol Soc Am 56:275–370. https://doi.org/10.1130/0016-7606(1945)56[275:EDOSAT]2.0.CO;2

Howard AD, Kerby G (1983) Channel changes in badlands. Bull Geol Soc Am 94:739–752. https://doi.org/10.1130/0016-7606(1983)94<739:CCIB>2.0.CO;2

Huppert KL, Perron JT, Ashton AD (2020) The influence of wave power on bedrock sea-cliff erosion in the Hawaiian islands. Geology 48:499–503. https://doi.org/10.1130/G47113.1

Jefferson A, Grant GE, Lewis SL, Lancaster ST (2010) Coevolution of hydrology and topography on a basalt landscape in the Oregon Cascade Range, USA. Earth Surf Proc Landf 35:803–816. https://doi.org/10.1002/esp.1976

Jenkins SF, Komorowski JC, Baxter P, Charbonnier SJ, Cholik N, Surono (2016) The devastating impact of the 2010 eruption of Merapi volcano, Indonesia. In: Duarte JC, Schellart WP (eds) Plate boundaries and natural hazards, AGU Book Chapters, Geophys Monogr 219, 1st Edition, pp 259–269. https://doi.org/10.1002/9781119054146.ch12

Jessop DE, Kelfoun K, Labazuy P, Mangeney A, Roche O, Tillier JL, Trouillet M, Thibault G (2012) LiDAR derived morphology of the 1993 Lascar pyroclastic flow deposits, and implication for flow dynamics and rheology. J Volcanol Geoth Res 245–246:81–97. https://doi.org/10.1016/j.jvolgeores.2012.06.030

Karátson D, Favalli M, Tarquini S, Fornaciai A, Wörner G (2010) The regular shape of stratovolcanoes: a DEM-based morphometrical approach. J Volcanol Geoth Res 193:171–181. https://doi.org/10.1016/j.jvolgeores.2010.03.012

Karátson D, Telbisz T, Wörner G (2012) Erosion rates and erosion patterns of neogene to quaternary stratovolcanoes in the Western Cordillera of the Central Andes: an SRTM DEM based analysis. Geomorphology 139(140):122–135. https://doi.org/10.1016/j.geomorph.2011.10.010

Karátson D, Yepes J, Favalli M, Rodríguez-Peces MJ, Fornaciai A (2016a) Reconstructing eroded paleovolcanoes on Gran Canaria, Canary Islands, using advanced geomorphometry. Geomorphology 253:123–134. https://doi.org/10.1016/j.geomorph.2015.10.004

Karátson D, Wulf S, Veres D, Magyari EK, Gertisser R, Timar-Gabor A, Novothny Á, Telbisz T, Szalai Z, Anechitei-Deacu V, Appelt O, Bormann M, Jánosi C, Hubay K, Schäbitz F (2016b) The latest explosive eruptions of Ciomadul (Csomad) volcano, East Carpathians—a tephrostratigraphic approach for the 51–29 ka BP time interval. J Volcanol Geoth Res 319:29–51. https://doi.org/10.1016/j.jvolgeores.2016.03.005

Karátson D, Gertisser R, Telbisz T, Vereb V, Quidelleur X, Druitt T, Nomikou P, Kósik S (2018) Towards reconstruction of the lost Late Bronze Age intra-caldera island of Santorini Greece. Sci Rep 8:7026. https://doi.org/10.1038/s41598-018-25301-2

Karátson D, Telbisz T, Dibacto S, Lahitte P, Szakács A, Veres D, Gertisser R, Jánosi Cs, Timár G (2019) Eruptive history of the Late Quaternary Ciomadul (Csomád) volcano, East Carpathians, part II: magma output rates. Bull Volcanol 81(4):28, 20 p. https://doi.org/10.1007/s00445-019-1287-8

Kelfoun K, Legros F, Gourgaud A (2000) A statistical study of trees damaged by the 22 November 1994 eruption of Merapi volcano (Java, Indonesia): relationships between ash-cloud surges and block-and-ash flows. J Volcanol Geoth Res 100: 379–393. https://doi.org/10.1016/s0377-0273(00)00147-5

Kelfoun K, Vargas SV (2016) VolcFlow capabilities and potential development for the simulation of lava flows. Geol Soc London Spec Publ 426(1):337–343. https://doi.org/10.1144/SP426.8

Kereszturi G, Geyer A, Martí J, Németh K, Dóniz-Páez JF (2013) Evaluation of morphometry-based dating of monogenetic volcanoes–a case study from Bandas del Sur, Tenerife (Canary Islands). Bull Volcanol 75(7):734. https://doi.org/10.1007/s00445-013-0734-1

Kervyn M, Ernst GGJ, Goossens R et al (2008) Mapping volcano topography with remote sensing: ASTER vs SRTM. Int J Remote Sens 29:6515–6538. https://doi.org/10.1080/01431160802167949

Kervyn M, Ernst GGJ, Carracedo JC, Jacobs P (2012) Geomorphometric variability of "monogenetic" volcanic cones: Evidence from Mauna Kea, Lanzarote and experimental cones. Geomorphology 136:59–75. https://doi.org/10.1016/j.geomorph.2011.04.009

Kilbride BMC, Edmonds M, Biggs J (2016) Observing eruptions of gas-rich compressible magmas from space. Nat Comm 7:1–8. https://doi.org/10.1038/ncomms13744

Kirby E, Whipple K (2001) Quantifying differential rock-uplift rates via stream profile analysis. Geology 29:415–418. https://doi.org/10.1130/0091-7613(2001)029<0415:QDRURV>2.0.CO;2

Komorowski JC, Jenkins S, Baxter P, Picquout A, Lavigne F, Charbonnier SJ, Gertisser R, Preece K, Cholik N, Budi-Santoso A, Surono, A (2013) Paroxysmal dome explosion during the Merapi 2010 eruption: processes and facies relationships of associated high-energy pyroclastic density currents. J Volcanol Geoth Res 261:260–294. https://doi.org/10.1016/j.jvolgeores.2013.01.007

Kubo Hutchison A, Dufek J (2021) Generation of overspill pyroclastic density currents in sinuous channels. J Geophys Res Solid Earth 126: e2021JB022442

Kurz MD (1986) Cosmogenic helium in a terrestrial igneous rock. Nature 320:435–439. https://doi.org/10.1038/320435a0

Lahitte P, Samper A, Quidelleur X (2012) DEM-based reconstruction of southern Basse-Terre volcanoes (Guadeloupe archipelago, FWI): Contribution to the Lesser Antilles Arc construction rates and magma production. Geomorphology 136:148–164. https://doi.org/10.1016/j.geomorph.2011.04.008

Lahitte P, Dibacto S, Karátson D, Gertisser R, Veres D (2019) Eruptive history of the Late Quaternary Ciomadul (Csomád) volcano, East Carpathians, part I: timing of lava dome activity. Bull Volcanol 81:27. https://doi.org/10.1007/s00445-019-1286-9

Lavigne F Degeai JP, Komorowski JC Guillet S, Robert V, Lahitte P, Oppenheimer C, Stoffel M, Vidal CM, Surono, Pratomo I, Wassmer P, Hajdas I, Hadmoko DS, de Belizal E (2013) Source of the great A.D. 1257 mystery eruption unveiled, Samalas volcano, Rinjani Volcanic Complex, Indonesia. Proc Natl Acad Sci USA 110, 16742. https://doi.org/10.1073/pnas.1307520110

Lefèvre JC, Gillot PY, Cardellini C, Gresse M, Lesage L, Chiodini G, Oberlin C (2018) Use of the radiocarbon activity deficit in vegetation as a sensor of CO_2 soil degassing: example from la Solfatara (Naples, Southern Italy). Radiocarbon 60(2):549–560. https://doi.org/10.1017/RDC.2017.76

Leicher N, Giaccio B, Pereira A, Nomade S et al (2023) Central Mediterranean tephrochronology between 313 and 366 ka: New insights from the Fucino palaeolake sediment succession. Boreas 52:240–271. https://doi.org/10.1111/bor.12610

Leopold LB, Maddock TJ (1953) The Hydraulic Geometry of Stream Channels and Some Physiographic Implications. US Geol Surv Prof Pap 252:57

Lerner GA, Jenkins SF, Charbonnier SJ, Komorowski JC, Baxter PJ (2022) The hazards of unconfined pyroclastic density currents: a new synthesis and classification according to their deposits, dynamics, and thermal and impact characteristics. J Volcanol Geoth Res 421:107429. https://doi.org/10.1016/j.jvolgeores.2021.107429

Lewicki JL, Hilley GE, Shelly DR, King JC, McGeehin JP, Mangan M, Evans WC (2014) Crustal migration of CO_2-rich magmatic fluids recorded by tree-ring radiocarbon and seismicity at Mammoth Mountain, CA, USA. Earth Planet Sci Lett 390:52–8. https://doi.org/10.1016/j.epsl.2013.12.035

Lifton N, Sato T, Dunai TJ (2014) Scaling in situ cosmogenic nuclide production rates using analytical approximations to atmospheric cosmic-ray fluxes. Earth Planet Sci Lett 386:149–160. https://doi.org/10.1016/j.epsl.2013.10.052

Lindsay J, Leonard G (2009) Age of the Auckland volcanic field. IESE Report 1–2009.02 June 2009

Lohse KA, Dietrich WE (2005) Contrasting effects of soil development on hydrological properties and flow paths. Water Resour Res 41:1–17. https://doi.org/10.1029/2004WR003403

Loughlin SC, Calder ES, Clarke A et al (2002) Pyroclastic flows and surges generated by the 25 June 1997 dome collapse, Soufriere Hills Volcano, Montserrat. Geol Soc Lond Mem 21:191–209. https://doi.org/10.1144/GSL.MEM.2002.021.01.09

Lube G, Cronin SJ, Platz T, Freundt A, Procter JN, Henderson C, Sheridan MF (2007) Flow and deposition of pyroclastic granular flows: a type example from the 1975 Ngauruhoe eruption, New Zealand. J Volcanol Geoth Res 161:165–186. https://doi.org/10.1016/j.jvolgeores.2006.12.003

Lube G, Cronin SJ, Thouret J-C, Surono, (2011) Kinematic characteristics of pyroclastic density currents at Merapi and controls on their avulsion from natural and engineered channels. Geol Soc Am Bull 123:1127–1140. https://doi.org/10.1130/b30244.1

Lube G, Breard ECP, Jones J, Fullard L, Dufek J, Cronin SJ, Wang T (2019) Generation of air lubrication within pyroclastic density currents. Nat Geosci 12:381–386. https://doi.org/10.1038/s41561-019-0338-2

Lyell J (1830) Principles of Geology. 1st edition, 1st vol. Jan. 1830, John Murray, London. eBook ISBN 9781315859095

Macedonio G, Pareschi M, Santacroce R (1990) A simple model for lava hazard assessment: Mount Etna, in Proceedings: Mainz, Germany, International Association of Volcanology and Chemistry of the Earth's Interior (IAVCEI)

Macorps E, Charbonnier SJ, Varley NR, Capra L, Atlas Z, Cabre J (2018) Stratigraphy, sedimentology and inferred flow dynamics from the July 2015 block-and-ash flow deposits at Volcan de Colima, Mexico. J Volcanol Geoth Res 349:99–116. https://doi.org/10.1016/j.jvolgeores.2017.09.025

Magee C, Hunt-Stewart E, Jackson CAL (2013) Volcano growth mechanisms and the role of sub-volcanic intrusions: Insights from 2D seismic reflection data. Earth Planet Sci Lett 373:41–53. https://doi.org/10.1016/j.epsl.2013.04.041

Manley K, Salles T, Müller RD (2020) Modeling the dynamic landscape evolution of a volcanic coastal environment under future climate trajectories. Front Earth Sci 8:1–21. https://doi.org/10.3389/feart.2020.550312

Marrero SM, Phillips FM, Borchers B, Lifton N, Aumer R, Balco G (2016) Cosmogenic nuclide systematics and the CRONUScalc program. Quat Geochron 3:160–187. https://doi.org/10.1016/j.quageo.2015.09.005

Martin LCP, Blard PH, Balco G, Lavé J, Delunel R, Lifton N, Laurent V (2017) The CREp program and the ICE-D production rate calibration database: a fully parameterizable and updated online tool to compute cosmic-ray exposure ages. Quat Geochron 38:25–49. https://doi.org/10.1016/j.quageo.2016.11.006

McDougall I, Harrison TM (1999) Geochronology and Thermochronology by the $^{40}Ar/^{39}Ar$ Method. Oxford University Press, 269, ISBN 0195109201

McGuire LA, Pelletier JD, Roering JJ (2014) Development of topographic asymmetry: Insights from dated cinder cones in the western United States.

J Geophys Res Earth Surf 119:1725–1750. https://doi.org/10.1002/2014JF003081

McGuire JL, Lawler JJ, McRae BH, Nuñez TA, Theobald DM (2016) Achieving climate connectivity in a fragmented landscape. P Natl Acad Sci USA 113: 7195–200. https://doi.org/10.1073/pnas.1602817113

Merrihue C, Turner G (1966) Potassium-argon dating by activation with fast neutrons. J Geophys Res 71(11): 2852–2857. https://doi.org/10.1029/JZ071i011p02852

Merryman-Boncori JP (2019) Measuring coseismic deformation with spaceborne synthetic aperture radar: a review. Front Earth Sci 7:1–20. https://doi.org/10.3389/feart.2019.00016

Miallier D, Condomines M, Pilleyre T, Sanzelle S, Guittet J (2003) Concordant thermoluminescence and ^{238}U–^{230}Th ages for a trachytic dome (Grand Sarcoui) from the Chaîne des Puys (French Massif Central). Quat Sci Rev 23:709–715. https://doi.org/10.1016/j.quascirev.2003.06.002

Miallier D, Pilleyre T, Boivin P, Labazuy P, Gailler L, Rico J (2017) Grand Sarcoui volcano (Chaîne des Puys, Massif Central, France), a case study for monogenetic trachytic lava domes. J Volcanol Geoth Res 345:125–141. https://doi.org/10.1016/j.jvolgeores.2017.07.015

Mossoux S, Saey M, Bartolini S et al (2016) Q-LAVHA: a flexible GIS plugin to simulate lava flows. Comput Geosci 97:98–109. https://doi.org/10.1016/j.cageo.2016.09.003

Nairn IA, Self S (1978) Explosive eruptions and pyroclastic avalanches from Ngauruhoe in February 1975. J Volcanol Geoth Res 3:39–60. https://doi.org/10.1016/0377-273(78)90003-3

Nakada S, Fujii T (1993) Preliminary report on the activity at Unzen Volcano (Japan), November 1990-November 1991: Dacite lava domes and pyroclastic flows. J Volcanol Geoth Res 54:319–333. https://doi.org/10.1016/0377-0273(93)90070-8

Nakada S, Shimizu H, Ohta K (1999) Overview of the 1990–1995 eruption at Unzen Volcano. J Volcanol Geoth Res 89(1):1–22. https://doi.org/10.1016/S0377-0273(98)00118-8

Naranjo MF, Ebmeier SK, Vallejo S, Ramón P, Mothes P, Biggs J, Herrera F (2016) Mapping and measuring lava volumes from 2002 to 2009 at El Reventador Volcano, Ecuador, from field measurements and satellite remote sensing. J Applied Volc 5(1):8. https://doi.org/10.1186/s13617-016-0048-z

Nelson A, Reuter HI, Gessler P (2009). DEM production methods and sources. In Hengl T, Reuter HI (eds) Geomorphometry: concepts, software, applications. Developments in Soil Science. Elsevier, Amsterdam, pp 65–85. https://doi.org/10.1016/S0166-2481(08)00003-2

Nitta H, Saito T, Shitaoka Y (2020) Recent eruption history inferred from eruption ages of the two latest lava flows using multi-dating at Yokodake Volcano Japan. Earth Planets Space 72:103. https://doi.org/10.1186/s40623-020-01220-3

Norini G, Lagmay AMF (2005) Deformed symmetrical volcanoes. Geology 33(7):605–608. https://doi.org/10.1130/G21565.1

Oehler JF, Lénat JF, Labazuy P (2008) Growth and collapse of the Reunion Island volcanoes. Bull Volcanol 70(6):717–742. https://doi.org/10.1007/s00445-007-0163-0

Ogburn SE, Calder ES, Cole PD, Stinton AJ (2014) The effect of topography on ash-cloud surge generation and propagation. Geol Soc London, Mem. 39, Chapter 10: 179–194. https://doi.org/10.1144/m39.10

O'Hara D, Karlstrom L, Roering JJ (2019) Distributed landscape response to localized uplift and the fragility of steady states. Earth Planet Sci Lett 506:243–254. https://doi.org/10.1016/j.epsl.2018.11.006

O'Hara D, Karlstrom L, Ramsey DW (2020) Time-evolving surface and subsurface signatures of Quaternary volcanism in the Cascades arc. Geology 48(11):1088–1093. https://doi.org/10.1130/G47706.1

O'Hara D, Goren L, van Wees RMJ, Campforts B, Grosse P, Lahitte P, Kereszturi G, Kervyn M (2024). Time-varying drainage basin development and erosion on volcanic edifices. ESurf 12:709–726. https://esurf.copernicus.org/articles/12/709/2024.

Pagli C, Wright TJ, Ebinger CJ, Yun SH, Cann JR, Barnie T, Ayele A (2012) Shallow axial magma chamber at the slow-spreading Erta Ale Ridge. Nat Geosci 5, 4: 284–288. https://doi.org/10.1038/ngeo1414

Paguican ME, van Wyk de Vries B, Lagmay AFM (2012) Hummocks: how they form and how they evolve in rockslide-debris avalanches. Landslides 11:67–80. https://doi.org/10.1007/s10346-012-0368-y

Pasquier-Cardin A, Allard P, Ferreira T, Hatté C, Coutinho R, Fontugne M, Jaudon M (1999) Magma derived CO_2 emissions recorded in ^{14}C and ^{13}C content of plants growing in Furnas caldera Azores. J Volcanol Geoth Res 92(1–2):195–207. https://doi.org/10.1016/S0377-0273(99)00076-1

Paterne M, Guichard F, Labeyrie J (1988) Explosive activity of the South Italian volcanoes during the past 80,000 years as determined by marine tephrochonology. J Volcanol Geoth Res 34:153–172. https://doi.org/10.1016/0377-0273(88)90030-3

Paulick H, Ewen C, Blanchard H, Zöller L (2008) The Middle-Pleistocene (~300 ka) Rodderberg maar-scoria cone volcanic complex (Bonn, Germany): Eruptive history, geochemistry and thermoluminescence dating. Int J Earth Sci 8:1879–1899.https://doi.org/10.1007/s00531-008-0341-0.

Pedersen GBM, Höskuldsson A, Dürig T et al (2017) Lava field evolution and emplacement dynamics of the 2014–2015 basaltic fissure eruption at Holuhraun, Iceland. J Volcanol Geoth Res 340:155–169. https://doi.org/10.1016/j.jvolgeores.2017.02.027

Pedersen GBM (2016) Semi-automatic classification of glaciovolcanic landforms: an object-based mapping approach based on geomorphometry. J Volcanol Geoth Res 311:29–40. https://doi.org/10.1016/j.jvolgeores.2015.12.015

Pedersen GBM, Belart JM, Óskarsson BV, Gudmundsson MT, Gies N, Högnadóttir T, Hjartardóttir ÁR, Pinel V, Berthier E, Dürig T, Reynolds HI, Hamilton CW, Valsson G, Einarsson P, Ben-Yesohua D, Gunnarson A, Odsson B (2022) Volume, effusion rate, and lava transport during the 2021 Fagradalsfjall eruption: Results from near real-time photogrammetric monitoring. Geophys Res Lett 49(13):e2021GL097125. https://doi.org/10.1029/2021GL097125

Pelletier JD, Cline ML (2007) Nonlinear slope-dependent sediment transport in cinder cone evolution. Geology 35:1067–1070. https://doi.org/10.1130/G23992A.1

Peruzzetto M, Mangeney A, Bouchut F, Grandjean G, Levy C, Thiery Y, Lucas A (2021) Topography curvature effects in thin-layer models for gravity driven flows without bed erosion. J Geophys Res Solid Earth 126:e2020JF005657. https://doi.org/10.1029/2020JF005657

Pike RJ, Clow GD (1981) Revised classification of terrestrial volcanoes and catalog of topographic dimensions, with new results on edifice volume. US Geol Survey Open-File Report 81–1038, 40 p

Preusser F, Degering D, Fuchs M, Hilgers A, Kadereit A, Klasen N, Krbetschek M, Richter D, Spencer JQG (2008) Luminescence dating: basics, methods and applications. Eiszeitalter Gegenwart Quat Sci J 57:95–149

Quidelleur X, Gillot PY, Soler V, Lefèvre JC (2001) K/Ar dating extended into the last millennium: application to the youngest effusive episode of the Teide Volcano (Spain). Geophys Res Lett 28(16):3067–3070. https://doi.org/10.1029/2000GL012821

Quidelleur X, Michon L, Famin V, Geffray MC, Danišík M, Gardiner N, Rusquet A, Mohamed Gou Z (2022) Holocene volcanic activity in Anjouan Island (Comoros archipelago) revealed by new Cassignol-Gillot groundmass K-Ar and ^{14}C ages. Quat Geochron 67:101236. https://doi.org/10.1016/j.quageo.2021.101236

Quidelleur X, Famin V (2024) Last 150 kyr volcanic activity on Mauritius Island (Indian Ocean) revealed by new Cassignol-Gillot unspiked K-Ar ages. Quat Geochron 82. https://doi.org/10.1016/j.quageo.2024.101534

Reimer PJ et al (41 co-authors) (2020) The IntCal20 Northern Hemisphere radiocarbon age calibration curve (0–55 cal kBP). Radiocarbon 62(4):725–757. https://doi.org/10.1017/RDC.2020.41

Richter N, Froger JL (2020) The role of Interferometric synthetic aperture radar in detecting, mapping, monitoring, and modelling the volcanic activity of piton de la Fournaise La Reunion: a review. Rem Sens 12:6. https://doi.org/10.3390/rs12061019

Roberti G, van Wyk de Vries B et al (2021) Structure from Motion used to revive archived aerial photographs for geomorphological analysis: an example from Mount Meager volcano, British Columbia, Canada. Canadian J Earth Sci 58:1253–1267. https://doi.org/10.1139/cjes-2020-0140.

Roering JJ, Kirchner JW, Dietrich WE (1999) Evidence for nonlinear, diffusive sediment transport on hillslopes and implications for landscape morphology. Water Resour Res 35:853–870. https://doi.org/10.1029/1998WR900090

Rosi M, Principe C, Vecci R (1993) The 1631 Vesuvius eruption. A reconstruction based on historical and stratigraphical data. In: De Vivo B, Scandone R, Trigila R (eds) Mount Vesuvius, J Volcanol Geoth Res 58:1–4: 151–182. https://doi.org/10.1016/0377-0273(93)90106-2.

Rowland SK, Garbeil H (2000) Slopes of oceanic basalt volcanoes. In: Mouginis-Mark PJ, Crisp JA, Fink JH (Eds) Remote Sensing of Active Volcanism. AGU Geophys Monograph Ser 116:223–247. https://doi.org/10.1029/GM116p0223

Santamaria S, Quidelleur X, Samaniego P, Audin L, Le Pennec JL, Hidalgo S, Liorzou C, Guillou H (2023) Timing of Pleistocene volcanism and its relationship with tectonics in the central segment of the Ecuadorian Andes. J Volcanol Geoth Res 442:107895. https://doi.org/10.1016/j.jvolgeores.2023.107895

Schaen AJ et al. (42 authors) (2020) Interpreting and reporting ^{40}Ar/^{39}Ar geochronologic data. GSA Bull 133:461–487. https://doi.org/10.1130/B35560.1

Schimmelpfennig I, Benedetti L, Finkel R, Pik R, Blard PH, Bourlès D, Burnard P, Williams A (2009) Sources of in-situ ^{36}Cl in basaltic rocks. implications for calibration of production rates. Quat Geochron 4:441–461. https://doi.org/10.1016/j.quageo.2009.06.003

Schimmelpfennig I, Benedetti L, Garreta V, Pik R, Blard PH, Burnard P, Bourlès D, Finkel R, Ammon K, Dunai T (2011) Calibration of cosmogenic ^{36}Cl production rates from Ca and K spallation in lava flows from Mt. Etna (38 N, Italy) and Payun Matru (36 S, Argentina). Geochim Cosmoch Acta 75:2611–2632. https://doi.org/10.1016/j.gca.2011.02.013

Schmidt C, Tchouankoue JP, Nkouamen Nemzoue PN, Ayaba F, Nformidah-Ndah SS, Nformi Chifu E (2017) New thermoluminescence age estimates for the Nyos maar eruption (Cameroon volcanic line). PLoS ONE 12(5):e0178545. https://doi.org/10.1371/journal.pone.0178545

Scott WE, Sherrod DR, Gardner CA (2008) Overview of the 2004 to 2006, and continuing, eruption of Mount St. Helens, Washington. In: Sherrod DR, Scott WE, Stauffer PH (eds) A volcano rekindled: The renewed eruption of Mount St. Helens, 2004–2006, US Geol Survey Prof Pap 1750, pp. 71–95, Vancouver

Segall P (2010) Earthquake and volcano deformation. Princeton University Press. https://doi.org/10.1515/9781400833856

Seidl MA, Dietrich WE, Kirchner JW (1994) Longitudinal profile development into bedrock: an analysis of Hawaiian channels. J Geol 102:457–474. https://doi.org/10.1086/629686

Shobe CM, Tucker GE, Barnhart KR (2017) The SPACE 1.0 model: a Landlab component for 2-D calculation of sediment transport, bedrock erosion, and landscape evolution. Geosci Model Dev 10:4577–4604. https://doi.org/10.5194/gmd-10-4577-2017

Siani G, Paterne M, Michel E, Sulpizio R, Sbrana A, Arnold M, Haddad G (2001) Mediterranean Sea surface radiocarbonreservoir age changes since the last glacial maximum. Science 294:1917–1920. https://doi.org/10.1126/science.1063649

Smith MJ, Clark CD (2005) Methods for the visualization of digital elevation models for landform mapping. Earth Surf Proc Landf 30:885–900. https://doi.org/10.1002/esp.1210

Smittarello D, Cayol V, Pinel V, Froger JL, Peltier A, Dumont Q (2019) Combining InSAR and GNSS to track magma transport at basaltic volcanoes. Rem Sens 11(19):1–26. https://doi.org/10.3390/rs11192236

Solikhin A, Pinel V, Vandemeulebrouck J, Thouret JC, Hendrasto M (2015) Mapping the 2010 Merapi pyroclastic deposits using dual-polarization Synthetic Aperture Radar (SAR) data. Rem Sens Environ 158:180–192. https://doi.org/10.1016/j.rse.2014.11.002

Sparks RSJ, Gardeweg MC, Calder ES, Matthews SJ (1997) Erosion by pyroclastic flows on Lascar Volcano, Chile. Bull Volcanol 58:557–565. https://doi.org/10.1007/s004450050162

Steiger RH, Jäger E (1977) Subcommission on geochronology: convention on the use of decay constants in geo and cosmochronology. Earth Planet Sci Lett 36(3):359–362. https://doi.org/10.1016/0012-821X(77)90060-7

Stone JO (2000) Air pressure and cosmogenic isotope production. J Geophys Res Solid Earth 105:23753–23759. https://doi.org/10.1029/2000JB900181

Stuiver M, Polach HA (1977) Reporting of ^{14}C. Radiocarbon 19:355–363. https://doi.org/10.1017/S0033822200003672

Sulpizio R, Castioni D, Rodriguez-Sedano LA, Sarocchi D, Lucchi F (2016) The influence of slope-angle ratio on the dynamics of granular flows: insights from laboratory experiments. Bull Volcanol 78:77. https://doi.org/10.1007/s00445-016-1069-5

Tarantola A (2005) Inverse problem theory. In: Society for industrial and applied mathematics

Tarquini S, Favalli M (2010) Changes of the susceptibility to lava flow invasion induced by morphological modifications of an active volcano: the case of Mount Etna, Italy. Nat Haz 54:537–546. https://doi.org/10.1007/s11069-009-9484-y

Tarquini S, de'Michieli Vitturi M, Jensen EH, Pedersen G, Barsotti S, Coppola D, Pfeffer MA (2019) Modeling lava flow propagation over a flat landscape by using MrLavaLoba: the case of the 2014–2015 eruption at Holuhraun, Iceland. Ann Geophys-Italy 61: 1–18. https://doi.org/10.4401/ag-7812

Tatsumi Y, Suzuki-Kamata K, Matsuno T, Ichihara H, Seama N, Kiyosugi K and 11 authors (2018) Giant rhyolite lava dome formation after 7.3 ka supereruption at Kikai caldera, SW Japan. Sci Rep 8(1):2753. https://doi.org/10.1038/s41598-018-21066-w

Taylor GA (1958) The 1951 eruption of Mt. Lamington, Papua. Comm Australia Bur Min Res Geol Geophys Bull 38:1–117

Telbisz T, Kovács G, Székely B, Karátson D (2012) The method of swath analysis based on digital terrain models (A sávszelvényelemzés módszere digitális terepmodell (DTM) alapján; in Hungarian with English abstract). Földtani Közlöny (Bull Hung Geol Soc) 142(2):193–200

Tierz P, Loughlin SC, Calder ES (2019) VOLCANS: an objective, structured and reproducible method for identifying sets of analogue volcanoes. Bull Volcanol 81:1–22. https://doi.org/10.1007/s00445-019-1336-3

Tortini R, van Wyk de Vries B, Carn S (2015) Seeing the faults from the hummocks: tectonic or landslide fault discrimination with LiDAR at Mt Shasta, California. Front Earth Sci 3:48. https://doi.org/10.3389/feart.2015.00048

Úbeda J, Bonshoms M, Iparraguirre J, Sáez L, De la Fuente R, Janssen L, Concha R, Vásquez P, Masías P (2018) Prospecting glacial ages and paleoclimatic reconstructions northeastward of Nevado Coropuna (16 S, 73 W, 6377 m). Arid Tropical Andes Geosci 8:307. https://doi.org/10.3390/geosciences8080307

van Wyk de Vries B, Márquez A, Herrera R, Bruña JG, Llanes P, Delcamp A (2014) Craters of elevation revisited: forced-folds, bulging and uplift of volcanoes. Bull Volcanol 76(11):1–20. https://doi.org/10.1007/s00445-014-0875-x

van Wyk de Vries B, Kerle N, Petley D (2000) A sector-collapse forming at Casita volcano, Nicaragua. Geology 28:167–170. https://doi.org/10.1130/0091-7613

von Buch L (1820) Ueber die Zusammenensetzung der basaltischen Inseln und über Ehrebungs-Cratere, Prussian Academy of Science.

Völker D, Kutterolf S, Wehrmann H (2011) Comparative mass balance of volcanic edifices at the southern volcanic zone of the Andes between 33°S and 46°S. J Volcanol Geoth Res 205:114–129. https://doi.org/10.1016/j.jvolgeores.2011.03.011

Vörös F, van Wyk de Vries B, Guilbaud MN, Görüm T, Karátson D, Székely B (2022) DTM-Based comparative geomorphometric analysis of four scoria cone areas—suggestions for additional approaches. Remote Sens 14:6152

Wadge G, Young P, McKendrick I (1994) Mapping lava flow hazard using computer simulation. J Geophys Res 99(B1):489–504. https://doi.org/10.1029/93JB01561

Wadge G, Herd R, Ryan, G, Calder ES, Komorowski JC (2010) Lava production at Soufrière Hills Volcano, Montserrat: 1995–2009. Geophys Res Lett 37(19):L00E03. https://doi.org/10.1029/2009GL041466

Whelley PL, Newhall CG, Bradley KE (2015) The frequency of explosive volcanic eruptions in Southeast Asia. Bull Volcanol 77:1–11. https://doi.org/10.1007/s00445-014-0893-8

Whipple KX, Tucker GE (1999) Dynamics of the stream-power river incision model: Implications for height limits of mountain ranges, landscape response time-scales, and research needs. Earth Surf Proc Landf 38:1432–1443. https://doi.org/10.1002/esp.3423

White SM, Crisp JA, Spera FJ (2006) Long-term volumetric eruption rates and magma budgets. Geochem Geophys Geosyst 7:Q03010. https://doi.org/10.1029/2005GC001002

Willett SD, McCoy SW, Perron TJ, Goren L, Chen CY (2014) Dynamic reorganization of River Basins. Science 343. https://doi.org/10.1126/science.1248765

Wong YQ, Segall P, Bradley A, Anderson K (2017) Constraining the magmatic system at Mount St. Helens (2004–2008) using Bayesian inversion with physics-based models including gas escape and crystallization. J Geophys Res Solid Earth 122, 10: 7789–7812. https://doi.org/10.1002/2017JB014343

Wood CA (1980) Morphometric analysis of cinder cone degradation. J Volcanol Geoth Res 8:137–160. https://doi.org/10.1016/0377-0273(80)90101-8

Wright TJ, Parsons BE, Zhong L (2004) Toward mapping surface deformation in three dimensions using InSAR. Geophys Res Lett 31(1):L01607

Wright R, Garbeil H, Baloga SM, Mouginis-Mark PJ (2006) An assessment of shuttle radar topography mission digital elevation data for studies of volcano morphology. Rem Sens Envir 105:41–53. https://doi.org/10.1016/j.rse.2006.06.002

Yamamoto T, Takarada S, Suto S (1993) Pyroclastic flows from the1991 eruption of Unzen volcano, Japan. Bull Volcanol 55:166–175. https://doi.org/10.1007/bf00301514

Yokoyama R, Shirasawa M, Pike RJ (2002) Visualizing topography by openness: a new application of image processing to digital elevation models. Photogramm Eng Remote Sens 68:251–266. https://api.semanticscholar.org/CorpusID:9906987

Yoshida H, Sugai T, Ohmori H (2012) Size-distance relationships for hummocks on volcanic rockslide-debris avalanche deposits in Japan. Geomorphology 136(1):76–87

Zarazúa-Carbajal MC, De la Cruz-Reyna S (2020) Morpho-chronology of monogenetic scoria cones from their level contour curves. Applications to the Chichinautzin monogenetic field, Central Mexico. J Volcanol Geoth Res 407:107093. https://doi.org/10.1016/j.jvolgeores.2020.107093

Zhu C, Chen Z, Huang Y (2021) Coupled moving particle simulation-finite-element method analysis of fluid-structure interaction in geodisasters. Int J Geomech 21:04021081. https://doi.org/10.1061/(asce)gm.1943-5622.0002041

Zorn EU, Walter TR, Johnson JB, Mania R (2020) UAS-based tracking of the Santiaguito Lava Dome Guatemala. Sci Rep 10(1):1–13. https://doi.org/10.1038/s41598-020-65386-2

Appendix

Fig. A

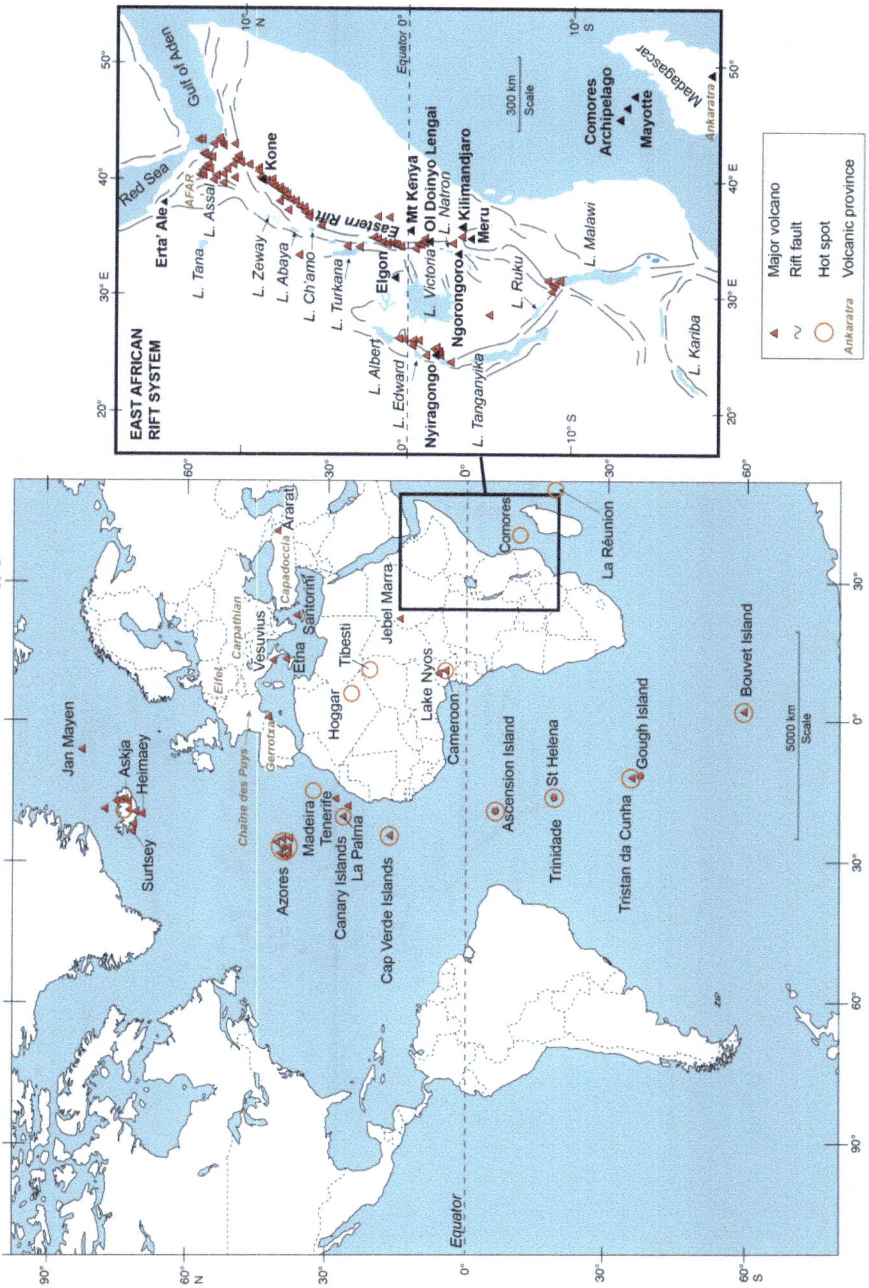

D. Karátson and J.-C. Thouret, *Volcano Geomorphology*, Advances in Volcanology,
https://doi.org/10.1007/978-3-031-77693-9

Fig. B

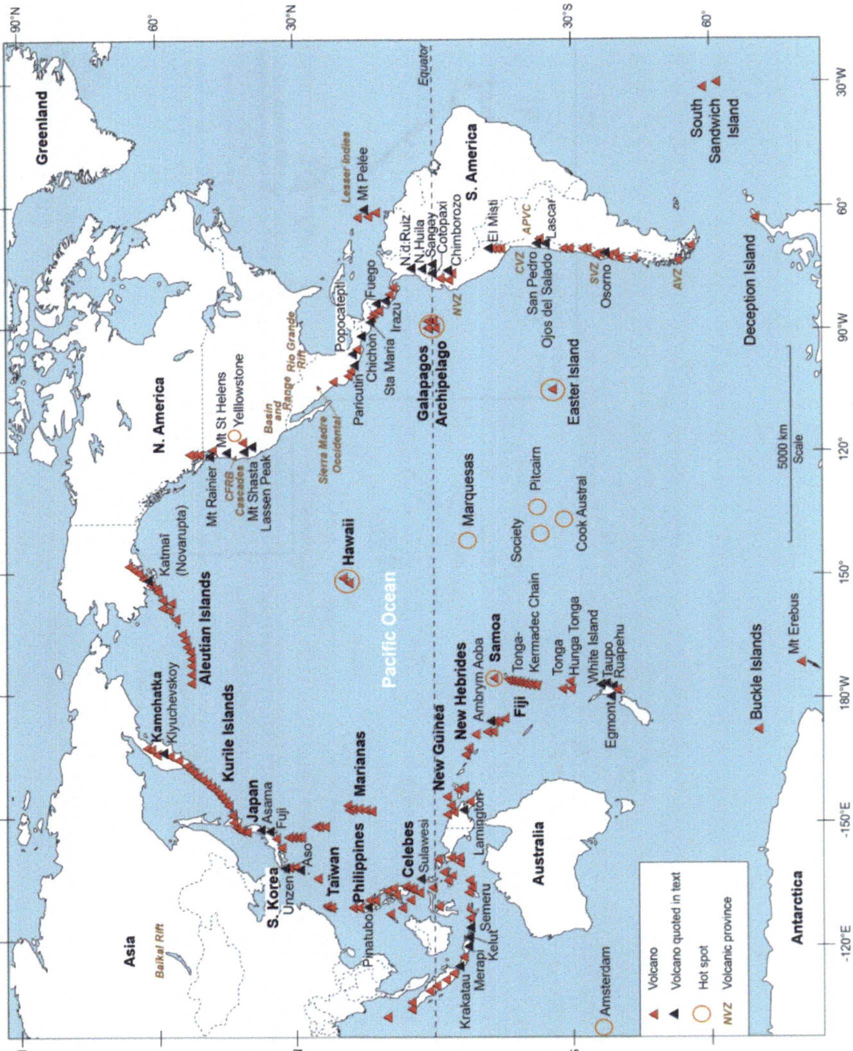

Greenland
Asia
Baikal Rift
N. America
S. America
Australia
Antarctica
Pacific Ocean
Equator
Kamchatka
Klyuchevskoy
Kurile Islands
Aleutian Islands
Katmai (Novarupta)
Japan
Asama
Fuji
Aso
Unzen
S. Korea
Taiwan
Philippines
Pinatubo
Marianas
Celebes
Sulawesi
New Guinea
Lamington
Krakatau
Merapi
Kelut
Semeru
New Hebrides
Ambrym Aoba
Samoa
Fiji
Tonga-Kermadec Chain
Tonga
Hunga Tonga
White Island
Taupo
Ruapehu
Egmont
Mt Erebus
Buckle Islands
Amsterdam
Hawaii
Marquesas
Society
Pitcairn
Cook Austral
Easter Island
Galapagos Archipelago
Mt Rainier
Mt St Helens
Yelllowstone
CFRB
Cascades
Mt Shasta
Lassen Peak
Basin and Range
Rio Grande Rift
Sierra Madre Occidental
Paricutin
Popocatepti
Chichon
Sta Maria
Fuego
Irazu
Lesser Indies
Mt Pelée
N.d.Ruiz
N.Huila
Sangay
Cotopaxi
Chimborozo
NVZ
El Misti
CVZ
APVC
San Pedro
Ojos del Salado
Lascar
SVZ
Osorno
AVZ
Deception Island
South Sandwich Island
5000 km
Scale
Volcano
Volcano quoted in text
Hot spot
NVZ Volcanic province
90°N
60°
30°N
30°S
60°
-120°E
-150°E
180°W
150°
120°
90°W
60°
30°W

Index

A

'A'ā (slabby, scoriaceous or spiny, clinkery, blocky subtypes), 45, 77, 513
Abyssal hills, 175, 176
Acatenango composite volcano, Guatemala, 271
Accretionary lapilli, 152
Accretionary lava ball, 78, 88
Active, inactive lava lake, 102
Adventitious landform (cone, dome), 4
Aeolian Islands, dacite lavas, 84
Aeolian landforms, 488
Afar, Ethiopia, 32, 333
Aggradation, degradation processes in channels, drainage basins, 542
Agua composite volcano, Guatemala, 271
Aguas Calientes caldera, Argentina, 365
Agung volcano, Bali, Indonesia, 509, 532
Aira caldera, Japan, 365
Ajoya composite volcano, erosional caldera, Chile, 274
Akagi volcano, Japan, 521
Alcedo shield volcano, caldera, Isabela Island, Galapagos, 226, 229, 354
Alcove, 485
Aleutians island arc, 35, 236
Alkali basalts, 60, 61
Altiplano–Puna Miocene/Pliocene volcanoes, 391, 485
  Cerro Anallajsi, Huicso, Jatunpunco, Asuasuni, 485
Altiplano–Puna Volcanic Complex (APVC), 37, 366, 470
Alu volcano, Ethiopia, 627
Amatitlán caldera, Guatemala, 272
Ambae (or Aoba) shield volcano, caldera, Vanuatu Islands, 35, 158, 237, 343
  hydrothermal caldera, 343
Ambrym, Marum crater, lava lake, Vanuatu, 102
Ambrym, shield volcano, Vanuatu Islands, 35, 51, 210, 237
Ampato–Sabancaya volcanic cluster, Peru, *see* Nevado Sabancaya
Amphitheatre valley, 471–473, 477, 478, 480, 482, 485, 489
Anahim Volcanic Belt, shield volcanos, Turkana subtype, Ilgachuz Range, Itczha Range, Rainbow Range, 235
Anak Krakatau, 2018 debris avalanche, tsunami, effects, modelling, Indonesia, 402, 541
Andean Volcanic Zones, North, Central, South, Austral, 26, 37
Andesitic-dacitic caldera, 346
Andosol, 460
Aniakchak, composite volcano, caldera, Alaska, USA, 269, 292
Antisana, Ecuador, 272
Apoyo caldera, Nicaragua, 329
Ararat compound volcano, Turkey, 34, 279, 294
  Arc volcano, 35, 36
Arc shield (Newberry-subtype), 237
Ardnamurchan Peninsula, ring complexes, Scotland, 332
Arenal composite volcano, Costa Rica, 146, 272
Armoured joints, pipes, 487
Asama volcano, Japan, 35
Ascension volcano, 30
Ash-cloud surge detachment, 527, 621
Ash cone, *see* tuff cone, 139
Ash-flow or ignimbrite caldera (rhyolitic composition), 16, 345, 352
Ash in the atmosphere, volcanic hazard, 511
Ashitaka composite volcano, Honshu, Japan, 276, 476
Askja volcano, caldera, Iceland, 29, 76, 333
Aso volcano, caldera, Japan, 35, 140, 327, 329, 354
Aspect ratio of lava flow, 89, 357, 360
  Aspect ratio of ignimbrite, 357, 360, 362
Aspect ratio (of resurgent calderas), relationship magma chamber width / roof thickness, 357, 360
Assessment (delineation, modelling) of sediment-laden volcanic flows, 532, 534
Atacama desert, 454
Atana ignimbrite, La Pacana caldera, Altiplano-Puna Volcanic Complex, 328, 366
  ignimbrite erosion, 488, 500
Atitlán, caldera (lake), Guatemala, 271
Atitlán–Tolimán compound volcano, Guatemala, 271
Atolls, Hawai'i–Emperor ridge, 226
Auchanquilcha volcanic cluster, Chile, 297
Auckland basaltic Volcanic Field, New Zealand, 53, 163
Aumône dome, Chaîne des Puys, France, 110

D. Karátson and J.-C. Thouret, *Volcano Geomorphology*, Advances in Volcanology,
https://doi.org/10.1007/978-3-031-77693-9

Auto-breccia, lava flow, autobrecciation, 78, 83, 178
Avachinsky, Kamchatka, debris avalanche, Russia, 403
Avulsion (of PDC, lahar), 524, 526
Axial and Brown Bear seamounts, Juan de Fuca Ridge, Pacific Ocean, 57
Azores or Archipelago, volcanoes, hot spot, plateau, 39, 52
Azores volcanism, volcanoes, 155, 280

B
Back-arc (seeting, volcanism), 36
Badland, 487
Baikal Rift, Russia, 31
Baja California, volcanic fields, Mexico, 163
Bakony-Balaton Uplands volcanic field, Hungary, 140, 151, 152, 487
Ballistics, volcanic hazard, 511
Bamus composite volcano, New Britain, Papua New Guinea, 277
Bandai-san composite volcano, Honshu, Japan, 276, 403
Bandai-type flank failure, 406
Bandelier Tuff, ignimbrite, Valles caldera, New Mexico, USA, 355, 488
  erosion, 488
Bárðarbunga, eruption, caldera, Vatnajökull ice cap, Iceland, 29, 58, 333, 365
Bárðarbunga-Holuhraun, 2014–2015 eruption, Iceland, 170
  modelling (MrLavaLoba code), 618
Barranco de Tirajana, Canary Islands, Spain, 481
Basalt composition, 61, 64, 73
Basaltic volcanic field, 162, 163
Basalt organs, 458
Basin and Range, western USA, 31, 34, 38, 51, 333
Batok scoria cone, Bromo-Tengger caldera, Java, 462
Ba volcano, Viti Luevu, Fiji, 183
Bazman and Taftan composite volcanoes, Makran arc, Iran, 279
Beagle, Tagus volcanoes, tuff cones, Galápagos, 157
Beerenberg composite volcano, Jan Mayen Island, Norway, 281
Belknap volcano (and Little Belknap), Oregon, USA, 232
Bezymianny composite volcano, Kamchatka, Russia, 275, 403
Bezymianny-type flank failure, 405
Big Ben, Heard Island, Indian Ocean, Australia, 282
Big Obsidian Flow, Newberry caldera, Oregon, 110, 236
Bingham viscosity (of lava), 89
Bishop Tuff, ignimbrite, Long Valley caldera, California, 335, 355
Blast (type of PDC), volcanic hazard, Mt. St. Helens, 1980 eruption, 519, 548–550
Block-and-ash flow, deposit, overbank, overspill, 524, 622
Block lava, 74, 83
Block seas (or felsenmeers), 458
Bocca, lava flow opening, 76
Bohemia volcanoes, Czech Republic, 34
Bolshaja Ipelka, Kamchatka, Russia, 480
Bolshije Igolki erosional caldera, Kamchatka, Russia, 477
Börzsöny Mountains, Hungary, 431, 458, 477
Bosavi composite volcano, Papua New Guinea, 280
Boset-Bericha-Gudda volcanic complex, Ethiopia, 53, 297
Breccias (in tuya), 165, 167, 329
Bulking (by PDC), 520, 521
Bulking/debulking (by lahar), 425, 542
Bulk modulus, deformability/compressibility of liquid water, 175

C
Calbuco composite volcano, Chile, 274, 461
  2015 eruption, hazard delineation, 532
Calc-alkaline magmas, 60, 62
Caldeira do Cabeço Grande, Faial Island, Azores, Portugal, 333
Caldeira Grande, Pico Island, Azores, Portugal, 333
Caldera, 2, 32, 34–36, 321–325, 365
Caldera classification, types, morphology, 373
Caldera collapse (process, mechanism, multi-phase), 340, 346, 347, 359, 370
Caldera complex, compound or polygenetic caldera, 3, 323, 365
Caldera-forming eruption sequence, 328
Caldera geometry (plate or piston, piecemeal, downsag, trapdoor, funnel), 323, 363
Caldera lake, breakout, hazards, Alaska, 558
Caldera margins, 329, 333
  Caldera, relationships with tectonics, regional structures, 333
Caldera resurgence, 354
Caldera types (basaltic shield, summit of composite cones, ash-flow or ignimbrite), 321
Călimani–Gurghiu–Harghita Mountains, Romania, 471
Călimani Mountains, Romania, 459
Cameroon Volcanic Line, Cameroon, 53, 152
Campanian ignimbrite, Phlegrean Fields, Italy, 338
  interactions with topography, 519
Campi Flegrei caldera, *see* Phlegrean Fields, 643
Canada, shield volcanoes, 210
Canary debris flow (submarine debris avalanche), offshore El Hierro, 428
Canary Islands, Canary shield and composite volcanoes, 39, 214, 216, 220, 221, 238, 514, 584
Cantal, Auvergne, France, 411, 474
Capelinhos volcano, tuff ring, Faial, Azores, Portugal, 155
Cape Verde Islands, Atlantic Ocean, 30, 280
Cappadocia volcanic field, ignimbrites, Anatolia, Turkey, 34, 140, 345
  ignimbrite erosion, 487
Carlsberg Ridge or Indian Ocean Ridge, 27
Carribean-Columbian plateau, LIP, 42

Cascade (arc, volcanic range or belt), USA, 36, 37, 51, 55, 208, 236, 270
Cascading or sequential effect of eruption, 546
Casita volcano, 1998 disaster, Nicaragua, 417, 559
structural geomorphological study, 626
Cauldron, 19, 323
Cauldron (and caldera) subsidence, 332, 374
Cauldron roots, 330
Cayambe composite volcano, Ecuador, 272
Central America, volcanoes, 208
Central Atlantic Magmatic Province (Marocco), 43
Central caldera, shield volcano, Hawai'i, Galápagos, 209
Central Volcanic Zone of the Andes (CAVZ or CVZ), 37, 640
application of cosmogenic nuclide dating, 640
Centrifugal drainage, 462
Centripetal drainage, 463
Cerro Azul, shield volcano, caldera, Galápagos, 226, 228, 229
Cerro Galán caldera, caldera complex, volcanic system, Argentina, 329, 355, 365, 368
ignimbrite erosion, 329
Cerro Negro, compound cone, Nicaragua, 138, 143, 144, 146
Cerro Panizos, ignimbrite shield-like volcano, Bolivia, 345
Cerros de La Pacana, resurgent block, La Pacana caldera, APVC, 366
Chaîne des Puys, Auvergne (France), 26, 73, 138, 641
Chaitén River, Chaitén town, 2008-2009 eruption, flood disaster, 548
Chaiten volcano, lava dome, Chile, 106, 110, 115, 274, 419
hazards, post-2008-2009 eruption response, ashfall, 104, 542, 548
Changbaisan (or Paektusan) composite volcano and Lake Tianchi caldera, Korea, 283
Channel aggradation, 555
Channel degradation, 555
Channelization (of PDCs), 521
Chao dacite flow, coulee, Chile, 83, 110
Chemical weathering (of volcanic rocks), 510
Chichinautzín volcanic field, Mexico, 26, 73, 163
Chicxulub asteroid impact crater, Mexico, 48
Chiliques composite volcano, Chile, 605
Chimborazo composite volcano, Ecuador, 272
Chisel mark (in lava column), 98
Christiana–Santorini–Kolumbo volcanic field, Aegean Sea, Greece, 347
Chukogu volcanic field, southwest Honshu, Japan, 163
Cima volcanic field, California, USA, 140
Cinder cone
scoria cone, 74, 140
Ciomadul (Csomád) lava dome complex, Romania, 280, 467
Circumferential or ring fissures (of shield), 228
Citlaltépetl or Pico de Orizaba composite volcano, Mexico, 270
Cladistic classification (of volcanoes), 18
Classification of volcanic landforms, 16, 18
Cliersou dome, Chaîne des Puys, France, 109
Clinker, *see* 'a'ā, 78
Coastal erosional landforms, 488
Coatepeque caldera, El Salvador, 272
Cocos plate, Cocos Ridge, Carnegie Ridge, 226
Coexistence of shield and composite volcanoes, 239
Colima composite volcano, Mexico, 271, 397, 521, 621
2015 eruption, Montegrande River, 623
Collapse pit, 323
Colli Albani volcanoes, Latium, Italy, 57
Colonnade (lower, upper), 95
Columbia River Basalt Plateau or Province (CRBP) or Group, 40, 73
Columbia River Flood Basalts (CRFB), 26
Columbia River, North Fork Toutle River, Mt. St. Helens drainage system, debris avalanche, lahars, floods, 549
Columnar jointing, 95
Columns of lava flow, 95
Comoros Archipelago, Islands, volcanoes, North Mozambique, 176
Composite volcano classification (simple, truncated, somma, compound cones), 16, 260
Composite volcano, cone, 16, 30, 257
Composite volcanoes at active subduction zones, 269
at atypical subduction zones: section, 268
at collision zones, 279
at intraplate and rift zone settings, 280
Compound cones (of composite volcanoes), 293
Compound lava flow, 74
Compound shield volcanoes, 231
Concentration (soild, particles) in volcanic flows (e.g. high or low concentration PDCs), 516
Cone sheet dykes, 332
Conical landform (glaciovolcanic type), 164
Constructive, constructional processes, 7
Continental arc, 35, 36
Continental flood basalt, 42
Continental Flood Basalt Plateau, Province (CFBP), 43
Continental intraplate and arc shields (medium sized, subtypes), 210, 233
Continental rift, 26, 31
Cooling rate (of lava), 89, 178
Coral reef (seamount, elmergent island), 185
Corvo Island composite volcano, Azores, Portugal, 53, 281
Cosmogenic nuclide dating 3He, 36Cl, 10Be, techniques, application, 638
Costa Giardini diatreme, Iblei Mountains, Sicily, Italy, 158
Cotahuasi-Ocoña-Marán volcanic province, ignimbrites, 345
Cotopaxi composite volcano, Ecuador, 272
Coulee (high-silica lava flow), 74, 83
Crater, 323
Crater Lake caldera, Oregon, USA, 321, 335, 347, 365
Crater lake monitoring, Ruapehu, New Zeland, Kelut, Indonesia, 534

'Crater of elevation', 627
Crater row, 4
Crater wall facies, 142
Creede caldera, San Juan Mountains, Colorado, USA, 328, 329
Cristallinity, crystal content (of lava), 90
Cryptodome, 105, 181
Crystal-rich dome, 113, 115
Cucu (Kakukkhegy), Carpathians, Romania, 481
Cumulo-dome, 111

D
Dalsheidi type sheet-like sequence (subglacial fissure eruption), 170
Damage, impact, loss, 510, 529
Damávand composite volcano, Alborz Mts., Iran, 263, 279, 289
Dammed lakes, 466
Darwin shield volcano, Galápagos, 157, 228–230
Debris avalanches, 397, 511, 513, 624
Debris flow, hyperconcentrated flow, 415, 422, 527
Debris flow, Sarno (1998), Italy, 511, 527
Debulking (by lahar), 425, 542
Debundscha maar, Cameroon, 152
Deccan traps, plateau, LIP, flood basalt province, 42, 73
Deception Island, shield volcano, Antarctica, 229
Degradation and erosion of volcanoes (terminology), 391
DEM-based morphometry, 602
DEM Digital Elevation Model, DTM Digital Terrestrial Model, DSM Digital Surface Model, 514
Destructional, destructive or excavational processes, 7
Detachment or decoupling (of PDCs), effects, 521
Devil's Postpile, California, USA, 95
Devil's Tower, Wyoming, USA, 95
Diamictite (e.g. tillite), 170
Diamond Head volcano, tuff ring, Oahu, Hawai'i, 154
Diatreme, 153
Digital Elevation Model (DEM), 144, 261, 514, 534, 601, 603
Digital topography in lava-flow hazard assessment, 514
Divergent plate boundaries, 24
Diversion of drainage, 465
Dome coulee, 109
Dome (endogenous, exogenous), 105
Dome (simple, piton or spine, bristling, nested, cracked and dispalced, asymmetrical), 106, 181
Domes in calderas, 329
Dormant volcanoes, indirect volcanic hazards, 511
Drainage basin, disturbed by eruptions, 544
Drainage-basin recovery, 544
Drainage caldera (lateral drainage of magma in mafic caldera), 218, 340
Drainage (on volcanoes), 461
Duricrust, 460
Dyke (sill), 55, 170, 329
Dyngjujökull glacier, Vatnajökull, Iceland, 170

E
East African Rift System (EARS), 30, 31, 83, 208, 210
East and West Eiffel basaltic volcanic fields, Germany, 163
Eastern Australia, shield volcanoes of Turkana subtype, 210, 233
  Buckland, Hillsborough, Nebo, Springsure shield volcanoes, Queensland; Canobalas, Ebor, Nandewar, Tweed, Warrumbungles shield volcanoes, New South Wales, 234
East Pacific Rise, 26, 58, 82, 333
Ebor shield volcano, New South Wales, Australia, 464
Ecuador shield volcano, Galápagos, 228
Effusion rate (instantaneous effusion rate, time-averaged discharge rate, mean output rate), 74
Effusive eruption, volcano, 9
Eifel (volcanic field), Germany, 23
El Abrigo caldera, Las Cañadas caldera, Canary Islands, 365
Elbrus composite volcano, Russia, 279
El Chichón volcano, Mexico, 35, 271, 292, 559
Eldgjá eruption, Iceland, 45
Elementary (monogenetic) volcanic landforms, 18, 140
Elevated rhyolite dome (Tardso Abéki, Tibesti), 344
Elgon volcano, Tanzania, 31, 477
El Hierro volcano, Canary Islands, 214, 217, 223
El Hoyo or Las Pilas volcanic complex or massif, 144
Elliptical caldera, 333
El Misti composite volcano, Peru, 273, 521
Emi Koussi shield-like volcano, Tibesti, Chad, 344
Emuruangogolak shield volcano, EARS, 233
Enclos Fouqué caldera, Piton de la Fournaise, La Réunion Island, 220, 343
Endogenous dome, 105
Englacial processes, drainage, 174
Entablature, 95
Entrainment of particles (by PDCs), 519
Epiclastic deposits, processes, 391
Ercyies Dağı composite volcano, Turkey, 279
Erebus composite volcano and lava lake, Ross Island, Antarctica, 98, 101, 102
Erosion
  long term, 392, 453, 489
  short term, 481
Erosional craters and calderas, 474
Erosional landforms (in general), 7, 453, 474
Erosion exerted by PDCs on substrate, 519
Erosion (of volcanic cones), 487
Erosion rates, 481, 482
Erosion (removal of rocks, debris deposition, river down-cutting), volcanic hazards, 511
Erta Ale volcano, lava lake, Afar, Ethiopia, 29, 31, 99, 102, 104, 333
Eruption frequency (of composite cones), 295
Eruptive fissure, 2
Eruptive, post-eruptive, non-eruptive processes, volcanic and geomorphic hazards, 511

Eruptive stages, preshield, shield, postshield, rejuvenated, Hawaiian shield volcano, 221
Espenberg volcanic field, maars (Devil Mountain Lakes), Alaska, USA, 152
Ethiopian Rift (part of EARS), Main Ethiopian Rift (MER), 31, 333
Etna volcano, Sicily, Italy, 27, 83, 88, 89, 98, 140, 279, 467, 514, 601
  Probabilistic Downflow code, 618
European Copernicus Programme, damage assessment, 531
Evolution of composite cones, 293
Evolution of Galápagos shield, 230
Evolution of Hawaiian shield, 221
Excess sedimentation, 542
Exogenous dome, 105
Explosive caldera, 17, 323
Explosive eruption, 18
Extracaldera region of a caldera, 328
Extrusive eruption, 9
Eyjafjallajökull volcano, Iceland, 168, 281, 535
  eruption, jökulhlaup, 170, 535, 536

F

Fagradalsfjall, Reykjanes peninsula, Iceland, 29, 231
Faial (or Fayal) Island, volcano, Azores, Portugal, 53, 281
Fairy chimneys (tent rocks, pinnacles), 487
Fani Maoré, seamount, east of Mayotte, 176
Faroe Islands, shield volcanoes, North Atlantic Ocean, 209, 231
Fault, faulting, stress field in volcanic terranes, 51, 55, 57, 626
Felsic caldera, felsic caldera complex, 345, 370
Felsic lava, 73
Fernandina Island, shield volcano, Galápagos, 226, 228, 229
Ferricrust, 460
Fieale volcano, Asal rift, Djibouti, 57
Fire fountain, 76
Fish Canyon Tuff eruption, ignimbrite, La Garita caldera, Colorado, USA, 323, 336
Fisher composite volcano, caldera, Aleutians, Alaska, USA, 269, 292
Flank failure, sector collapse, volcanic hazard, 511
'Flare up' in magma production, 47
Flood basalt, 16, 73
Flow foliation (rhyolite coulee, ramping, flow structures), 84
Flow toe or front, 86
Flow unit, 73, 74
Fluvial erosion, 454, 461
Fogo volcano, Cape Verde Islands, 225, 282
Fountain (subaqueous volcanism), 179
Fuego composite volcano, Guatemala, 271, 296, 621
  2018 eruption, 514
  detachment, overspill of PDCs, 518, 526
  hazard delineation, 532
Fuerteventura Island, shield volcano, Canary Islands, 216, 222, 225
Fuji composite volcano, Honshu, Japan, 3, 27, 140, 263, 276, 287
Furrow (flute cast), Mt. St. Helens 1980 eruption, Lascar 1993 eruption, 519

G

Galápagos caldera type, 321, 341
Galápagos Islands, 19, 23
  Eastern, western islands and shields:hot spot, Galápagos Spreading Centre, Galápagos Platform, 226, 230
  morphometry, 228
  evolution, 230
Galápagos shield, 2, 98, 214, 221, 226, 228–230, 233, 238, 625
Galunggung volcano, Java, Indonesia, 548
Garibaldi volcano, Bristih Columbia, 167
Garrotxa volcanic field, Catalonia, Spain, 34
Gas emission, volcanic hazards, 511
Geisha seamounts, Nortwest Pacific, 177
Geitlandsjökull (shield), Iceland, 232
Gendol-Opak River, Bronggang, Kaliadem villages, south flank of Merapi volcano, Indonesia, 522, 623
Geochronology, dating methods and techniques (K/Ar, 40Ar/39Ar; luminescence, thermoluminescence TL and optically stimulated luminescence OSL, tephrostratigraphy, disequilibrium in U and Th series U/Th, cosmogenic nuclide dating, Radiocarbon dating 14C), 638
Geodynamic setting, *see* tectonic setting, 18, 57
Geomorphic aggradation, degradation, 552
Geomorphic hazards, 510, 511, 513, 546, 560
Geomorphic hazards (delayed) in caldera and ignimbrite contexts, 557
Geomorphic hazards on dormant volcanoes, 558
Geomorphic impacts (cascading, sequential) and responses, 513, 550
Geomorphic impacts of very large eruptions, 556
Geomorphic parameters acting on PDCs along drainage networks, 521
Geomorphic parameters of river channels, PDC behaviour, overbank and avulsion, 524, 532
Geomorphic processes, hazards, interplay with eruptive activity, volcanic flows, 154
Geomorphic recovery, 544
Geomorphic response (to volcanic eruptions, to persistent volcanic activity), 513, 546, 550, 552
Geyser, 3
Giant's Causeway, Northern Ireland (columnar jointing), 95, 488
Gígjökull glacier, Eyjafjallajökull 2010 eruption, 171, 535
GIS Geographical Information System, geomorphic, volcanic hazards, 514, 531, 534

Gjálp 1996 eruption, Grimsvötn 1996, 2004 eruptions, Vatnajökull ice cap, Iceland, 168, 535
Glacial erosional landforms, 489
Glaciovolcanic (landforms, processes), 139
Glass Mountain, obsidian dome, flow, California, USA, 236, 335
Glencoe, cauldron, ring complex, Scotland, 322, 332
Goma (city), RDC, 2002-2003 eruption, 98, 514
Goosenest shield volcano, California, USA, 232
Gough Island, South Atlantic, British Overseas Territory, 468
Graben caldera, 333
Gran Canaria, Canary Islands, 152, 214, 225, 238, 343
  caldera, 183
Great Basin SLIP, western USA, 47
Great Sumatra fault, Indonesia, 327
Grindavik 2023-204 eruption, Iceland, 509
Grísmvötn eruption, caldera, jökulhlaup, Iceland, 168, 535
Grosmanaux-Vasset volcanoes, shallow intrusion, Chaîne des Puys, France, 58, 109
Ground deformation of volcanoes, monitoring, modeling, 627
Growth rate of shield volcanoes, 217, 236
Growth rates of composite volcanoes, 294, 297, 300
Guacha caldera complex, Bolivia, 366
Gully (quebrada, wadi, gulche) evolution, 462, 487
Guyot, 176, 177

H

Hackle (in lava column), 98
Hakone composite volcano, caldera, Japan, 276, 293, 322
Hakuto caldera, Kagoshima Bay, Japan, 370
Haleakalā volcano, Island of Maui, Hawai'i, 215, 224, 472
Halemaumau crater, caldera, lava lake, Kilauea, Hawai'i, 98, 102
Halla shield volcano, Jeju Island, South Korea, 152
Haruj volcanic field, shield volcanoes, Lybia, 231
Hasan Dağı composite volcano, Turkey, 279
Havre volcano, submarine, Kermadec Arc, New Zealand, 178
Hawaiian (eruptive style), 9, 139, 241
Hawaiian shield, 16, 212, 214, 217, 239
Hawaiian Volcano Observatory (HVO), 514
Hawai'i caldera type, 321, 343
Hawai'i–Emperor chain, ridge (volcanoes, atolls, seamounts), 39, 214
Hawai'i Islands, Hawai'i Big Island, 5, 27, 73, 214, 215, 218, 220, 222, 223, 403, 472
  debris avalanches, 488
Hawaiite (lava), 88
Hazard (natural), 510
Hazard-zone map, 532, 534
H/D ratio (for tsunamigenic volcano assessment), 540
Heat transfer through ice, magma fragmentation, 170
Hekla composite volcano, Iceland, 281
Helgafell ridge, tindar, Iceland, 167
Hellenic island arc, 26, 51
High Lava Plain, Oregon (USA), 42, 43
Hoggar mountains, volcanoes, 3, 23, 39
Hokkaido-Komagatake, Hokkaido, Japan, 402
Holuhraun (Bárðarbunga) lava field, eruption, 29, 587, 588
  case study, 514
Honeycomb weathering, 457
Hoodo, pillow lava delta, Bristih Columbia, 168
Hopi Buttes, New Mexico, USA, 98, 104, 486
Hornito, 88
Horseshoe scar, 408
Hot spot, 26, 27
Hot-spot volcano, 38
Hot spring, 3
Hualālai volcano, Hawai'i, 39, 216, 222, 224
Huallatiri lava dome complex, Chile, 274
Huaynaputina volcano, 1600 AD/CE eruption, Peru, 274, 556
Hummock, hummocky topography, 409, 624
Hunga Tonga-Hunga Ha'Apai volcano, Tonga-Kermadec Archipelago, 35, 183
  erosion, 519
Hyaloclastite, hyaloclastite tuff, 74, 82, 156
Hybrid setting, 38
Hybrid wet flows (mixed avalanches, snow slurry lahars, ice diamicts), 529
Hydroclastic cone, 2, 138
Hydrodynamic pressure of lahars, 531
Hydrographers Range, Papua New Guinea, 484
Hydrological effects of eruptions, 542
Hydrolysis (of volcanic rocks), 459
Hydro-morphological and sedimentalogical concepts, 542
Hydrostatic pressure (water column), 175
  of lahars, 531
Hydrothermal activity (geyser, hot spring, etc.), 19
Hydrothermal activity (in caldera), 335
Hydrothermal caldera, 220, 341
Hydrothermal caldera (mafic shield), 341
Hydrothermal system (in caldera), 329
Hydrovolcanic, hydroclastic eruption, 152
Hyperconcentrated flows, 415

I

Icelandic shield, 19, 230–232, 238
Iceland volcanism, volcanoes, 39, 232, 280
Ice-rich mixed diamicts (e.g. mixed avalanche), magma and ice interactions, 172
Ignimbrite, 42
Ignimbrite dammed valleys, Tadami River, lake breakout, Numazawako eruption, Japan, 558
Ignimbrite feeder dykes, Kilehegé volcano, Tibesti, 344
Ignimbrite in cauldron, 332
Ignimbrite plateau, 19, 345
Ignimbrite sheet, 17
Ignimbrite shield-like volcano, ignimbrite shield, 16, 344
Ignimbrite tor, 487

Ignimbrite, volcanic hazard, 511
Ilchulbong (Sunrise Peak), tuff cone, Jeju Island, South Korea, 156
Impact (of natural hazard), 510
Impacts of lahars, 529
Inactive volcanoes, indirect volcanic hazards, 513
Increased fluvial erosion, 393, 542, 546
Indirect volcanic hazards (debris avalanche, sediment-laden flows, tsunamis, jökulhlaups), geomorphic impacts, 527
Inflation mechanism (lava flow), 93
Inflation pit, 75, 88
InSAR, Interferometric Synthetic Aperture Radar, method for volcano deformation analysis and monitroing, 628
Interactions, magma and water (passive, active), 174
Interferometry (technique for deformation detection, InSAR and photogrammetry), 602
Intermediate-silicic multivent centre, 370
Intra-caldera structure, deposit, 329
Intraglacial (landforms, processes), 163
Intraplate Large Ignous Province (LIP), 41
Intraplate rift and hot spot shield (Turkana-subtype), 233
Intrusion (shallow, effect on volcano), 3, 58, 630
Inverse volcano (volcano-tectonic depression), 19, 370
Inverted relief landforms (e.g. volcanic mesa, planèze), 19
Inyo domes, California, USA, 110
Irazú composite volcano, Costa Rica, 272
  1963-1965 eruption, 541, 546
Ischia Island, volcano, resurgent caldera, Italy, 333, 354, 359, 360
Island arc, 35, 36
Isle of Mull (ring complex), Scotland, 330
Ithnayn volcanic field, Saudi Arabia, 231
Iwo-Jima resurgent caldera, Japan, 354, 360
Izu-Oshima Island, composite volcano, caldera, Japan, 293

J
Jebel Marra volcano, shield-like edifice, Sudan, 39
Jemez resurgent caldera, 359
Joint or crack in columnar lava flow, 95, 97
Jökulhlaup, glacial outburst, mitigation, 170, 535
  (Gjálp, Grimsvötn, Katla, Eyjafjallajökull), 535, 536
Joya Honda maar, Mexico, 153
Juan de Fuca Ridge (Pacific Ocean), 27
Juan Fernandez Ridge (Pacific Ocean), 37

K
Kagoshima Bay, volcano-tectonic graben, Kyushu Island, Japan, 370
Kaho'olawe, West Maui, East Moloka'i, Ko'olau, Wai'anae calderas, Hawai'i, 220
Kaimondake volcano, Kagoshima Bay, Japan, 370
Kaiserstuhl, Upper Rhine Graben, Germany, 235
Kambalny, Kamchatka, debris avalanche, Russia, 403
Kamchatka peninsula, Kamchatka-Kurile arc, volcanoes, 35, 154
  Shield volcanoes, Snegovoy and Maly Palyalpan, Tsirk, 238
Kameni Island (lava flows and domes), Santorini, Greece, 83, 108
Kameni Vouno, dome, Santorini, Cyclades Islands, Greece, 106
Kamen, Kamchatka, debris avalanche, Russia, 403
Kapenga caldera, Taupo Volcanic Zone, New Zealand, 371
Karkar shield volcano, Papua New Guinea, 228
Katla volcano, ice cap, jökulhlaup, 537
Katmai composite volcano, caldera, Alaska, USA, 269, 292, 322, 332, 365
  eruption, 332, 366
Kauai, Wai'anae, volcanoes, Island of Oahu, Hawai'i, 215
Keanakakoï Ash Member eruption, Kilauea, Hawai'i, 224
Kelut (or Kelud), Java, Indonesia, 98, 521
Kenya volcano, Kenya, 31
Kerguelen plateau, LIP, 42
Kerinci composite volcano, Sumatra, Indonesia, 276, 521, 539
Kermadec arc (submarine volcanoes), Pacific Ocean, 182
Kikai-Akahoya submarine rhyolite lava dome, 182
Kikai caldera, Japan, 365
Kīlauea East Rift Zone, West Rift Zone, 102, 217
Kīlauea, Mauna Iki, Mauna Alae, Mauna Ulu, Makaopuhi craters, Ailaau eruption, Hawai'i, 101
Kīlauea volcano, caldera, Hawai'i, 39, 216, 509, 514
Kilbourne Hole, maar, New Mexico, USA, 154
Kilimanjaro volcano, Tanzania, 31, 489
Kimberley diatremes, South Africa, 153
Kiplombe shield volcano, EARS, 233, 234
Kirishima compound volcano, Kagoshima Bay, Japan, 370
Klyuchevskoy composite volcano, Kamchatka, Russia, 276, 287, 294
Klyuchevskoy–Tolbachik volcanoes, Kamchatka, Russia, 37
Kobayashi caldera, Kagoshima Bay, Japan, 370
Kohala volcano, Hawai'i, 39, 224
Koko crater, tuff cone, Oahu Island, Hawai'i, 156
Kolombangara composite volcano, Solomon Islands, 477
Kolumbo volcano, Christiana–Santorini-Kolumbo volcanic field, Greece, 347
Kone volcano, caldera, Garoboldi, Ethiopia (MER), 32, 343
Koolau volcano, Nuuanu landslide, Oahu Island, Hawai'i, 220
Krafla volcano, caldera, Iceland, 29, 333
Krakatau composite volcano and caldera, Sunda Strait, Indonesia, 276, 293, 322, 335, 346, 428, 539, 541
  caldera, tsunami, 1883 and 1918 eruptions, 539
Kronotsky composite volcano, Kamchatka, Russia, 289

Ksudach caldera, Kamchatka, Russia, 329, 477
Kumano caldera, Japan, 359
Kupaianaha lava pond, Kilauea, 82

L
Laacher, *see* volcano, tuff ring, Eifel, Germany, 156
Laboratory (analogue) experiments, on lava flows, PDCs, lahars, 515, 519, 520
La Garita, caldera complex, San Juan Mountains, Colorado, USA, 323, 327, 329, 336
La Gomera shield volcano, Canary Island, 224, 238
Laguna Colorada, ignimbrite shield-like volcano, Bolivia, 345
Laguna de Bay volcano-tectonic depression, Philippines, 370
Lahar (flow, event, hazards), 171, 413
  hot, primary lahar, 416
  syn-eruptive, post-eruptive, non-eruptive hot, primary lahar, 513
Lahar damage, geomorphic actions in river channels, 529
Lake breakout (glacier-dammed lake, maar lake), 172
Lake City caldera, Colorado, USA, 359
Lake Myvatn, Iceland, 89
Lake Nyos, crater, maar, Cameroon, 39, 152
Lake Pavin maar, 151
Laki, Lakagigar eruption, Iceland, 46
Lamington volcano, Papua New Guinea, 35
Lanai, Molokai, inactive Islands, Hawai'i, 216
Lanai volcano, Hawai'i, 217, 226
Landform reconstruction, methods, modeling, 608
Landscape evolution modeling, 616
Landslides of composite cones (including non-volcanic ones), 397, 614
Landslides of Hawaiian shields, debris avalanches, 220
Lanzarote volcanoes, Gran Canaria volcanoes, Canary Islands, 152, 225
La Pacana resurgent caldera, Altiplano-Puna Volcanic Complex, Chile, 354, 365
La Palma Island, Canary Islands, 39, 214, 222, 223, 238, 584
La Primavera caldera, volcano-tectonic depression, Mexico, 3, 322, 325, 370, 371
La Réunion Island, shield volcanoes, 18, 39, 73, 208, 214, 217, 220, 223, 473
Large Igneous Province LIP, 3, 26
Large, long-lived Volcanic Clusters (LVCs) or volcano complexes, 19, 296
La Roza member flow, Columbia River Basalt Province, 43
La Sauvetat, maar, scoria cone, Devès, France, 153
Las Cañadas caldera, Tenerife, Canary Islands, 220, 332, 333, 343
Lascar compound volcano, Chile, 274, 293, 519, 608, 621
Lascar volcano, Chile, 519
Lassen Peak, dome, California, USA, 109
Las Sierras shield volcano, Nicaragua, 341
Late Bronze Age, ~1600 BC/BCE Minoan eruption, Santorini caldera, Greece, 347
Lateiki, Metis Shoal Island, shoaling volcano, Tofua Arc, 183
Lateritisation, 459
Latium, volcanic complexes, Italy, 56
Lau-Havre Trough, Pacific Ocean, 51
Lava blister, 79, 88
Lava channels (flow toe or front, zone of dispersed flow, transitional channel zone, stable channel zone), 75, 86
La Vache and Lassolas twin cones, Chaîne des Puys, France, 143
Lava cone, 207
Lava delta, 89
Lava dome (characteristics, structure, textures), 111
Lava-fed delta, Antarctica (volcano-ice interactions), 170
Lava field (simple, compound), 16, 73
Lava flow front, advance rate, 90
Lava flow hazard, monitoring, 514
Lava flow inflation, 75
Lava flow, lava flow field, 73
Lava flow length, 73
Lava flow modeling, deterministic, probabilistic codes, 601
Lava flow propagation, slope, topography, bedrock, 513
Lava lake, 98
Lava levees (initial, rubble, overflow, and accretionary), 86
Lava mound, 16
Lava rise terrace, 75
Lava shield, 207
Lava tree, 89
Lava tube, tunnel, 86, 87
Lemptégy volcano, Chaîne des Puys, France, 143
Lesser Antilles, 26, 35
Levee (initial, rubble, oveflow), 86
Licancábur composite volcano, Chile/Bolivia, 274, 287
LiDAR Light Detection and Ranging (for survey and DEM), 514, 534
Likuruanga erosional crater, New Britain, 476
Limu o Pele or Pele's seaweed, 178
Linear ridge (glaciovolcanic landform), 178
Lithofacies, 266, 393
Little Glass Mountain dome coulee, Medicine Lake, California, USA, 110
Lobate lava flow, 82, 177
Loch Ba ring dyke, Isle of Mull, Scotland, 330
Loihi or Kama'ehuakanaloa, volcano, seamount, Hawai'i, USA, 98, 176, 214
Longonot shield volcano, EARS, 233
Long Valley resurgent caldera, California, USA, 329, 335, 355, 358
Look Seamount, Western Pacific, 185
Louisville Islands, 39
Low lava dome (pancake), 110
Lunar Crater basaltic volcanic field, USA, 163

M

Maar, 3, 138
Macauley volcano, Kermadec Arc, Pacific Ocean, 182
Madagascar Rift, 176
Madeira archipelago, Mid-Atlantic Ocean, 183
Maderas volcano, Nicaragua, 57
Mafic caldera (on shield volcano), 9, 339
Mafic lava, 73
Magma chamber, 1, 46
Magma generation, transport and storage, 60
Magma-ice interaction, 164
Magma pressurization (mresurgence model), 359
Magmatic rebound and/or decompression (resurgence model), 359
Magmatism, magma composition, 5, 47
Magma-water interaction, 148, 151
Magma-water ratio, 151
Main Ethiopian Rift (MER), 32
Makovica erosional crater, Carpathians, Slovakia, 474
Malha maar, Sudan, 153
Mammoth Mountains, California, USA, 98
Mangakino compound caldera, Taupo Volcanic Zone, New Zealand, 371
Marianas Trough, Island arc, 26, 51
Markarfljót River, Eyjafjallajökull 2010 eruption, 535
Marquesas Islands, shield volcano, 222
Marsabit shield volcano, Kenya, 209
Marshall-Samoan Islands, 39
Marun, Benbow volcanoes, tuff cones, Ambrym, Vanuatu, New Hebrides, 158
Masao Mimatsu, diagram of Showa Sin-Zan dome growth, Hokkaido, Japan, 109
Masaya pit craters (Masaya, Nindiri, Santiago and San Pedro) in caldera, Nicaragua, 341
Masaya Triple Layer, Masaya caldera (eruption), Nicaragua, 343
Masaya Tuff/Ticuantepe Lapilli, Masaya caldera (eruption), Nicaragua, 341
Masaya-type shield, 211, 344
Masaya volcano, caldera, lava lake, Nicaragua, 102, 322, 326, 343
Mascarenes Islands, plateau, 39
Massif (of composite cones), 261, 300
Mass-transport deposit, 429
Maui shield volcano (East, West), Island, Hawa'i, 217, 224
Mauna Kea shield volcano, Hawai'i, scoria cones, 39, 140, 208, 213, 215, 216, 222–224, 238
Mauna Loa shield volcano, Hawai'i, 39, 86, 89, 207, 208, 215–217, 219, 221–225, 238
Mauritius Island, shield volcano, 214
  erosion, 488
Mayon composite volcano, Philippines, 277, 289, 541
Mayor Island, shield volcano, New Zealand, 238
Medicine Lake shield volcano, caldera, California, USA, 210, 232, 236
Meltwater (ice, magma and ice or snow interactions), 171
Menengai shield volcano, caldera, Menengai Tuff, Kenya Rift, EARS, 233
Merapi case study 3, 621
  2006 and 2010 PDCs overspill modelling, 620
  decoupling of PDCs, 2010 eruption, overbank and avulsion, 518, 524
  lahars, 529
  post-2006, 2010 eruption response, 546, 552
Merapi lava dome complex, Java, Indonesia, 104, 106, 110, 116, 269, 417
  volcanic hazards, 522
Mesa (volcanic table), 105
Methods (field, laboratory, numerical) and techniques for volcano study, 19, 640
Michoacán-Guanajuato volcanic field, Mexico, 151, 163, 207, 209, 212, 231
Mid-Atlantic Ridge (MAR), 52
Middle Atlas volcanic field, Morocco, 140
Mid-Ocean Ridge (MOR), 16, 27, 29
Mid-Ocean Ridge Basalt (MORB), 60, 61
Milos Island, Greece, 181
Minoan or Late Bronze Age eruption at Santorini ~1600 years BC/BCE, tsunami, 35, 540
Mitigation of sediment-laden volcanic flows, 532
Miura Group, Honshu, Japan, 431
Miyakejima caldera, Island, 2000 eruption, Izu-Bonin volcanic chain, Japan, 330, 345
Miyakeyima volcano, Izu Island, Japan, 155
Móberg, 164
Móberg (Palagonite) Formation, Iceland, 164
Mokuaweoweo caldera, Mauna Loa, Hawai'i, 89, 216, 219
Molokai shield volcano (East, West), Island, Hawai'i, 216, 217
Momotombo composite volcano, Nicaragua, 272
Monogenetic (cones, volcano, volcanic field), 4, 16, 138
Monoun crater, maar, Cameroon, 152
Montaña Rajada, dome, Tenerife, Canary Islands, 106
Monte Nuovo 1538 eruption, Phlegrean Fields, Italy, 339
Monte Pilato, Lipari, Aeolian Islands, Italy, pumice hills, 148
Mont Gerbier de Jonc, volcano, Velay, France, 105
Monts Dore, Auvergne, France, 283
Montserrat Island, *see* Soufrière Hills volcano
MOR Mid-Ocean Ridge Basalt, 27, 177
Morphometry analysis of volcanic landforms, volcano morphometry, 601
Morphometry of composite cones, 287, 290, 293
Morphometry of maars, 153, 207, 209, 214, 233
Morphometry of scoria cones, 143
Mt. Ararat composite volcano, Turkey/Armenia, 34
Mt. Augustine lava dome complex, Alaska, USA, 269, 402, 403, 427
Mt. Baker composite volcano, Washington, USA, 36, 295, 297
Mt. Bosavi composite volcano, Papua New Guinea, 280
Mt. Cameroon shield volcano, Cameroon, 39, 140, 209, 214, 217

Mt. Carlisle composite volcano, Alaska, USA, 269
Cayley volcano, 1984 landslide, debris flow, Garibaldi volcanic belt, British Columbia, Canada, 559
Mt. Cleveland composite volcano, Alaska, USA, 269
Mt. Dixon, Heard Island, Indian Ocean, Australia, 282
Mt. Edziza composite volcano, Brisith Columbia, Canada, 283
Mt. Egmont, *see* Tanaraki
Mt. Hamilton, Australia, 467
Mt. Herbert composite volcano, Alaska, USA, 269
Mt. Kanaga composite volcano, Alaska, USA, 269
Mt. Kanekolo, Solomon Islands, 480
Mt. Karthala shield volcano, Comores Islands, 214, 217
Mt. Lamington composite volcano, Papua New Guinea, 109, 110, 280, 524
Mt. Mase erosional caldera, Solomon Islands, 477
Mt. Mazama composite volcano (prior to Crater Lake caldera) and volcanic field, 292, 301, 335, 346
Mt. McLoughlin composite volcano, Oregon, USA, 287
Mt. Meager composite volcano, 2010 rocklside avalanche, British Columbia, Canada, 510, 559
Mt. Melbourne composite volcano, Ross Island, Antarctica, 284
Mt. (Montagne) Pelée volcano / lava dome complex, Martinique, FWI, 35, 106, 108, 109, 115, 258, 272, 428, 509, 621
1902 eruption, PDC decoupling, overbank, avulsion, 509
Mt. (Mount) Adams composite volcano, Washington, USA, 272, 294–297
Mt. Ontake, flank failure, landslide disaster, Honshu, Japan, 408, 559
Mt. Pavlof and Pavlof Sister, composite volcanoes, Alaska, USA, 269
Mt. Pelée volcanic complex, reconstruction, modelling, 621
Mt. Pinafore volcano, sheet-like sequence, Antarctic, 170
Mt. Pinatubo composite volcano, The Philippines, 35, 329, 424
post-1991 eruption response, 541, 546
Mt. Pordon volcano, Newer Volcanic Province, Victoria, Australia, 153
Mt. Rainier composite volcano, debris avalanche, Washington, USA, 269, 412, 413, 489, 527, 532
glacial erosion, 488
Mt. Rechesnoi composite volcano, Alaska, USA, 269
Mt. Redoubt composite volcano, Alaska, USA, 115, 168, 172, 529
Mt. Ruapehu composite volcano, New Zealand, 35, 98, 172, 417, 529
Mt. Saint (St.) Helens composite volcano, debris avalanche, lava dome, lake damming, Washington, USA, 35, 104, 108, 109, 115, 172, 269, 397, 405, 466
Mt. Shasta, composite volcano, California, USA, 270, 290, 607
Mt. Shishaldin composite volcano, Alaska, USA, 269
Mt. Spurr composite volcano, Alaska, USA, 172, 529
Mt. St. Helens, barriers, interaction of PDC with topography, erosion, PDC decoupling, overbank, avulsion, 511, 518, 519, 551
post-1980 eruption response, 550
Mt. Taylor composite volcano, New Mexico, USA, 409
Mt. Tongariro composite volcano, New Zealand, 592
Mt. Vesuvius composite volcano, Italy, 258
Mt. Vsevidof composite volcano, Alaska, USA, 269
Mt. Yakedake composite volcano, geomorphic response to moderate eruptions, Japan, 546
Mud volcano, 3
Mull volcano, cauldron, Scotland, 58
Multi-crustal intrusion sills and laccoliths, *see* Santorini models, 351
Multi-vent centres of intermediate to silicic composition, 371
Muria volcano, erosional crater, Java, Indonesia, 477
Mushroom rocks, 487
Mýrdalssandur plain, coast, Katla volcano, 1918 jökulhlaup, 537

N

Nabeyama tuff cone, Sakurajima volcano, Kyushu Island, Japan, 148
Namib desert, 454
Narse d'Espinasse, maar and Puy de l'Enfer, scoria cone, Chaîne des Puys, France, 153
Nazca plate, 29, 226
Nea and Palea Kameni Islands, Santorini caldera, Greece, 329
Neapolitan Yellow Tuff, ignimbirte, Phlegrean Fields, Italy, 338
Nejapa and Ticomo maars, Nicaragua, 153
Nemrut Dagi volcano, Turkey, 34
Neptunian (subaqueous volcanism), 179
Nevado Ampato–Sabancaya volcanic cluster, Peru, 297, 641
Nevado Casiri volcano, Peru, 640
Nevado Chachani, volcanic cluster, southern Peru, 106
Nevado Coropuna composite volcano, Peru, 35, 273, 640
Nevado del Ruiz composite volcano, 1985 eruption, lahar, Colombia, 35, 168, 272, 414, 509, 527, 541
Nevados de Payachata, Chile, 274, 293
Nevado Solimana composite volcano, Peru, 273
Newberry shield volcano, caldera (Central pumice cone), Oregon, USA, 148, 236, 322
Newberry-subtype shield, arc shield volcanoes, 19, 211, 229, 233, 235, 239
Newer basaltic volcanic province, Victoria, southeast Australia, 6, 38, 73, 152, 153
shield volcanoes, 231
New Hebrides, Islands, volcanoes, 208, 210, 236, 237, 239
Ngauruhoe composite volcano, New Zealand, 35, 621
Niahue eruption, maar, Chile, 152
Ni'ihau shield volcano, Island, Hawai'i, 225

Niuafo'ou atoll, shield volcano, Tonga islands, 229
Northeast Honshu volcanoes, 37
Novarupta plug dome, 1912 eruption, Katmai volcano, Alsaka, 35, 105
Nuku-Hiva (hydrothermal caldera), French Polynesia, Pacific Ocean, 343
Numerical models and modeling, 360, 363, 534, 540, 616, 617
Nyamuragira (Nyamulagira) shield volcano, DRC, 31, 239
Nyiragongo composite volcano, lava lake, DRC, 31, 98, 102, 239
  2002-2003 eruption, 514
Nyos crater, maar, Cameroon, 152

O

Oahu Island, shield volcano, Hawai'i, USA, 208, 217
Obsidian dome, 115
Obsidian flow, coulee, 74, 84
Oceanic and continental shields (Icelandic type), 212
Oceanic hot spot shield volcano (large, medium-sized), 73, 209, 212
Oceanic-Island Basalt (OIB), 26, 39
Odenwald Hills, Germany, 458
Ogive (on coulee), 84
Ojos del Salado compound volcano, Chile, 35, 274, 293
Okataina caldera, Taupo Volcanic Zone, New Zealand, 333, 365, 371, 558
Okmok caldera, Aleutians, USA, 269, 321, 354
Okmok, shield volcano, Aleutioans, Alaska, USA, 210, 237
Ol Doinyo Lengai composite volcano, Tanzania, 31
Ontake composite volcano, Honshu, Japan, 276
Ontong Java plateau, LIP, 5, 42
Öræfajökull composite volcano, Iceland, 280
Original cone (stage), 470
Osorno composite volcano, Chile, 295
Ossippee Mountains, cauldron, New Hampshire, USA, 330
Ostoros and Ciumani erosional craters, Carpathians, 477
Overbank (lahar, PDC), fast, slow overspill (PDC), 621
  Overspill of small-volume PDCs, 524
Overpressure caldera, 365
Oxidation (of volcanic rocks), 459

P

Pacaya composite volcano and Mackenney cone, Guatemala, 272
Pāhoehoe (S- and P-types and subtypes\sheet, shelly, slabby, rough or rubbly, finger or entrail, ropey), 74, 75
  Pāhoehoe toe, 88
Pāhoehoe sheet lobe, 93, 513
Paka caldera, shield volcano, EARS, 333, 335
Palaeosurface (of volcano), 470
Pali Aike basaltic volcanic field, Argentina – Chile border, 153
Pali, cliff, fault, Hilina pali, Kilauea, Hawai'i, 220
Panarea volcano, Aeolian Islands, Italy, 57
Pantelleria resurgent caldera, Italy, 322, 333, 343, 354, 357, 359
Paraná–Etendeka flood basalt (Brazil and Namibia), 41
Parasol ribbing, 462
Paricutín volcano, scoria cone, lava flow, Mexico, 138
  geomorphic response to moderate eruptions, 546
Parinacota composite volcano, Nevados de Payachata, Chile, 274, 287, 294
Pasig-Potrero River, Pinatubo volcano, drainage evolution, response to 1991 eruption, lahars, floods, 552
Pastos Grandes caldera complex, Bolivia, 366
Paulina Peak, dome, Newberry volcano, USA, 106
Payún Matrú shield volcano, Argentina, 236
Peach Spring Tuff, ignimbrite, California, Arizona, Nevada, USA, 518
Þeistareykjarbunga shield volcano, Iceland, 231
Pelean dome or composite dome, 109
Pelean extrusive style, 9
Peperite, 167
Peralkaline caldera, 343
Phlegrean Fields (Campi Flegrei), caldera complex, Italy, 322, 329, 335, 336, 338, 354, 360, 643
  Solfatare Pozzuoli (14C dating and monitoring method), 643
Photogrammetric surveys, UAS, real-time digital elevation data, 515
Phreatomagmatic (eruption, eruption style), 140, 148, 149, 152–154, 224
Phreatoplinian, shallow water eruption style, 179
Physical weathering (of volcanic rocks), 457
Pico Island, volcano, Azores, Portugal, 53, 281
Pico Teide volcano, Pico Viejo volcano, Tenerife, Canary Islands, 225
Pillow lava, 29, 74, 82, 164, 177, 178
Pillow lava, breccia, 176
Pillow lava cone, 177
Pillow mound (volcano-ice interactions), 170
Pillow volcano, 177
Pinacate and Camargo basaltic volcanic field, Mexico, 163
Pinzón, Rábida, Santiago shield volcanoes, eastern Galápagos, 226, 229, 230
Piton de la Fournaise, La Réunion Island, 39, 76, 214, 223, 628
  deformation monitoring and modeling, lava flow mapping, long-term evolution, 610
  (hydrothermal caldera), 343, 515
Piton des Neiges, La Réunion Island, 39, 223, 478
Piton Tikal cone, Piton de la Fournaise, 2022 eruption, 515
Planchón-Peteroa volcano, Chile, 521
Planèze stage, 470, 471, 485
Platoro Caldera complex, San Juan Mountains, Colorado, USA, 354
Platy jointing, 457

Plinian eruptive style, 9, 156, 179, 180, 265, 279, 284, 292, 296, 328, 335, 336, 343, 345, 346, 348, 354, 357, 364
Plug (upheaved dome), plug dome, eroded plug, 19, 105
Plumbing system, 4
Plumose (in lava columns), 98
Pluton (beneath caldera), 322, 323
Poas volcano, Costa Rica, 98
Podzolisation, 459
Poladpur pulse, Deccan flood basalt, K/Pg boundary, 50
Polygenetic (volcano), 16, 257, 258, 297
Pomerape composite volcano, Nevados de Payachata, Chile, 274
Popocatépetl composite volcano, Mexico, 35, 110, 270
Poris eruption 273 ka, Tenerife, Canary Islands, 518
Porto Santo, Madera archipelago, 183
Post-eruption response (short-lasting, moderate, severe and protracted), hazards, 541, 546, 551, 556
Post-shield stage (of Hawaiian shield), 221
Pouakai erosional crater, North Island, New Zealand, 476
Pozzuoli (town), Phelgrean Fieldfs, Italy, 339
Prambanan temple, Merapi volcano, potential lahar assessment, 532
Pre-shield stage (of Hawaiian shields), 216
Pressure Compensation Level (PCL), 183
Pressure ridge, 88
Probabilistic method (for assessment of PDCs, lava flows, tsunamis, lahars), 540
Protection works: bunkers for PDCs, Sabo dam (check dam), 534
Pseudocrater, rootless vent, 89
Pumice cone, 138, 148
Pumice cone within Newberry Caldera, Oregon, USA, 148
Purico, ignimbrite shield-like volcano, Chile, 345
Purupurini dome cluster, torta (dome type), southern Peru, 105, 110
Putih River, Merapi volcano, 2011-2012 lahar disaster, 531
Puu Waawaa pumice cone, Hualalai shield, Hawai'i, USA, 148
Pu'u' Ō'ō (Kilauea, Hawai'i) spatter cone, vent, lava, 86, 87, 102
Puy Chopine, upheaved plug, piton, spine, Chaîne des Puys, France, 106, 109
Puy de Dôme, cumulo-dome, Petit Puy de Dôme volcano, Chaîne des Puys, France, 58, 106, 111
Puy de la Vache, scoria cone and lava flow, 641
Puyehue –Cordón Caulle volcano, Chile, 274, 461
  Chile post-2011-2012 eruption response, 541
Puy Pariou, compound cone, Chaîne des Puys, France, 26, 143, 641
Pyroclastic cone(s), volcano, 16, 17, 138
Pyroclastic Density Current (PDC) multiphase regimes, hazards, geomorphic hazards, 509, 515
Pyroclastic Density Current (PDC), pyroclastic flow, pyroclastic surge, 106, 171, 172
Pyroclastic flow, 515
Pyroclastic surge, 515
  Low energy detached surges, high-energy dilute surges, 522
Pyroduct (lava tube), 86

Q

Quench fragmentation, quenched lava, 174, 178, 179, 181

R

Rabaul caldera, tsunami, Papua New Guinea, 146, 540
Radial vents (of Hawaiian shield), 216, 229
  (of Galápagos shield), 229
Rangitoto Island shield-like volcano, New Zealand, 232
Recurrence of large eruptions, 557, 560
Redondo Dome, resurgent dome, Valles caldera, New Mexico, USA, 355
Red Sea Rift, 31
Regional detumescence (resurgence model), 359
Regolith, 459
Rejuvenated stage (of Hawaiian shield), 225
Remnant volcano ('stage'), 471
Rendova Island erosional crater, Solomon Islands, 474
Reporoa caldera, New Zealand, 326
Resedimentation, 391
Resurgence, 335, 354, 355, 357, 374
Resurgence, tumescence, doming, modes, mechanism, models, 357, 359, 360
Resurgent caldera or cauldron, 19, 354
Resurgent dome, block, 329, 358
Reventador composite volcano, Ecuador, 292, 521
Reykjanes Peninsula, Iceland, 207
Rheological parameters of lava flow, 89
Rheology (of magma), 175
Rhine Rift, Germany, 31
Rift zones (of Hawaiian and Canarian shields), 220
Rill erosion, 462
Ring dyke, 209, 228, 330
Ring fault, *see* structural boundary, caldera, 327
'Ring of Fire', Pacific Ocean, 35
Ring structure, 323
Rio Grande Rift, Basin and Range, USA, 31, 355
Ritter Island, debris avalanche, Papua New Guinea, 402
Roche Rosse obsidian lava flow, 148
Roof aspect ratio for caldera, thickness/width t/w, 512
Roque Nublo volcano, Gran Canaria Island, Canary Islands, 225
Rota-1 volcano, submaerine, Marianas Arc, 178
Rotohana eruption, maar, Tarawera volcano, New Zealand, 152
Rotorua caldera, Taupo volcanic zone, New Zealand, 333, 336, 465, 558
Ruapehu compound volcano, New Zealand, 279, 293
Ruatoria submarine debris avalanche, offshore North Island, New Zealand, 428
Runout distance for PDCs, 515

S

Sabaloka cauldron, ring dykes, Sudan, 332
Sabatini volcanoes, Latium, Italy, 57
Sabo dam, check dam, 522
Sæfell tuff cone, Iceland, 158
Sahara desert, 454
Sajama composite volcano, Bolivia, 274
Sakurajima volcano, Kyushu, Japan, geomorphic response to moderate eruptions and persistent explosive activity, 546, 552
Samalas volcano, 1257 AD/CE eruption, Lombok, Indonesia, 556
Samoa Islands, Savaii, Upolu, Manua volcanoes, 217
Samosir Island, Uluan peninsula, resurgent domes, Toba caldera, 357
San Antonio Tephra (eruption), Masaya caldera, Nicaragua, 343
San Carlos shield volcano, Bioko Island, Equatorial Guinea, 228
San Cristóbal composite volcano, Nicaragua, 271
San Cristóbal shield volcano, eastern Galápagos, 230
Sandur, outwash plain, e.g. Iceland, 535
San Francisco Peaks (or Mountain) composite volcano, Arizona, USA, 283
San Francisco volcanic field, Arizona, USA, 163, 482
Sangay composite volcano, Ecuador, 36, 272, 295
San Juan Mountains, caldera complexes, Colorado, USA, 48, 323, 330, 336, 345, 365, 373
San Miguel composite volcano, El Salvador, 271
San Miguel de los Lotes, Fuego volcano, Guatemala, 2018 eruption, disaster, 527
San Pedro composite volcano, Guatemala, 271
San Pedro–San Pablo composite volcanoes, Chile, 274
San Pedro–Tatara or San Pedro–Pellado compound volcano, Chile, 274, 294
Santa Ana composite volcano and Izalco cone, El Salvador, 272
Santa Cruz shield volcano, eastern Galápagos, 229, 238
Santa Maria – Santiaguito composite volcano, Guatemala, 552
Santiaguito lava dome complex, 83, 105, 110, 116
  response to persistent volcanic activity, 552
Santorini caldera, Cyclades Islands, Greece, 35, 321, 329, 332, 347
  Therasia Island, and Nea and Palea Kameni Islands, 347
Santorini multi-phase caldera formation (Archaeos Tuff, Lower Pumice 2, Upper Scoria 1, Cape Riva, and Late Bronze Age or Minoan eruptions), 346, 365, 428
Sao Jorge and Sao Miguel Islands volcanoes, Azores, Portugal, 53
Saray, Sahand and Sabalan, Zagros Mountains, Iran, 279
Sarcoui (or Grand Sarcoui), lava dome, Chaîne des Puys, France, 106, 637
  TL and OSL Luminescence dating techniques, 635
Satellite imagery, high-spatial resolution (Landsat TM, GeoEye, Pleaides, Ikonos, SPOT5 satellites), 531, 555
Scafell caldera, England, 363
Sciara del Fuoco, Stromboli, eruptions, tusnami, Italy, 546, 595
Scoria cone (cinder cone), 16, 138
Scotia island arc, 35, 51
Scutulum (-type) shield, 209
Seaca-Tatarca (Mezőhavas) erosional caldera, Carpathians, Romania, 228
Seamount, 74, 176
Sediment yield, 393, 542, 546, 548–552
Selvedge (quenched glassy margins), 82
Semeru composite volcano, Java, Indonesia, 36, 295, 417, 529, 548, 555
Sete Cidades volcano, caldera, Azores, Portugal, 281
Shatter ring, 88
Shear strain rate (of lava), 79
Shear stress (in lava), 89
Sheet lava flow, 29, 82, 177
Sheet (lava) volcano, 177
Shield stage (of Hawaiian shield), 223
Shield volcano, 16, 231
  morphology, types, 208
Shimabara catastrophe, *see* Unzen/Ontake-type flank failure
Shinjima Island (submarine caldera), Kagoshima Bay, Japan, 182
Shinmoedake volcano (Kyushu Island, Japan), 462
Ship Rock (Tsé Bit' A'í), New Mexico (USA), 474
Shiribetsu volcanic complex, Hokkaido, Japan, 345
Shiveluch composite volcano, Kamchatka, Russia, 275, 397, 398, 402, 403
Shoaling (island, phase), emergent volcanic island, 177, 183
Showa Sin-Zan, dome, Hokkaido, Japan, 109
Siberian traps LIP, 42
Sideromelane (in glass-rich basalt), 167
Sierra Madre Occidental SLIP, 41
Sierra Negra shield volcano, caldera, 212, 229
Silali shield volcano, caldera, EARS, 229, 333, 335
Silent Canyon caldera, Nevada, USA, 326
Silicic Large Ignimbrite Province (SLIP), 41
Silicic lava flow, 83
Silicic multi-vent complex, caldera complex, 17, 370
Sill, 176, 329
Simple, irregular cones (of composite volcanoes), 289
Simple lava flow, 73
Simple, regular cones (of composite volcanoes), 289
Simple, terrestrial volcanic landforms, 137
Sinabung composite volcano, Sumatra, Indonesia, 276, 521, 532
Siwi caldera, Vanuatu Islands, 354
Skjaldbreiður shield volcano, Icelandic type, 231
Slánské–Tokaj Mountains, 471
Slope (ground, gradient), effects on lava flow, geomorphic hazard, 513
Snaeffels Peninsula, Iceland, 488
Snake River Plain lava field, Idaho, USA, 40
  long-term erosion, 485
  shield volcanoes, 210, 231

Snow avalanche, snowmelt, snow-slab avalanche, slush-flow, slurry, 172
Society Islands, shield volcanoes, 222
Socompa composite cone, debris avalanche, Chile, 292, 408, 412
Solubilisation (of volcanic rocks), 459
Somma volcanoes, 292
Soufrière de Guadeloupe, dome, FWI, 108
Soufrière de Saint-Vincent volcano, dome, Saint Vincent Island, Carribean Sea, 108
Soufrière Hills volcano, lava dome complex, debris avalanche, Montserrat Island, Carribean Sea, 104, 106, 109, 110, 272, 402, 428, 431, 509, 614
  decoupling, 1997 eruption, 522
  hazard delineation, 532
  post-1995-2013 eruption response, 541
South Andean Volcanic Zone, 37
Spatter cone, 4, 17, 76, 88, 140
Spatter rampart, 76
Spheroidal weathering, 457
Spine, type of dome, 105
Spinose protrusion, *see* clinker of 'a'ā flow, 78
Springerville basaltic volcanic field, Arizona, USA, 140, 151
Squeeze-up (on lava flow), 88
Stable channel zone, 86
Steam-driven, deep subaqueous eruption, 179
St Helena, South Atlantic Ocean, 30
Stratovolcano versus composite volcano, 257
Strombolian (eruptive style, cone), 9, 140–142, 222, 239, 264, 282, 284, 296
Strombolian-like (subaqueous volcanism), 179
Stromboli composite volcano, Island, Aeolian archipelago, Italy, 57, 279, 546, 595
Stromboli, Tyrrhenian Sea, tsunami prevention, 540, 541
Structural boundary (of a caldera), 327
Structural geomorphology of volcanoes, 627
Subaerial volcanism, 7, 174
Subaqueous mass flow, 429
Subaqueous (submarine and lacustrine) volcanism, 18, 174
Subcaldera intrusions, plutons, 330, 354
Subcritical, supercritical state (gas, water), 175
Subduction zone, 24, 26
Subglacial (landform, process, eruption), 9, 163
Subglacial volcanic landforms, 168
Submarine caldera, 182
Submarine eruption, deposits, 177
Submarine explosive eruption (silicic magma), 182
Submarine fire fountain spatter deposits, 181
Submarine volcanism, 174
Subplinian eruptive style, 140, 143, 146
Subsidence mechanisms, 360
Subsidence (of shields), 215, 216
Subsidence versus diameter of caldera, s/d ratio, 360
Substrate erosion and entrainment (by PDC), 518–520
Substrate erosion by PDCs, i.e. force chain dynamics, impact by projectiles, low-pressure parts of PDCs, 520
Sumatra volcanoes, 37
Summer Coon, Colorado (USA), 474
Summit caldera (of shields), 217, 229, 233, 257, 264, 292, 293
  (of composite cones), 257, 263, 288
Sunset Crater, San Francisco Volcanic Field, Arizona, USA, 146
Supraglacial (eruption, process), 168, 171
Surge (wet, dry), 172
Surtseyan (shallow water eruption style, tuff cone), 156, 158, 160
Surtsey Island, volcano, tuff ring, eruption, Iceland, 155, 162
Suswa shield volcano, caldera, Kenya Rift, EARS, 333, 335, 343
Suwolbong, Songaksan tuff ring, South Korea, 155
Swabian Alb, diatremes, Germany, 153

T

Taalian eruption (phreatomagmatic), tuff cone, 160
Taal volcano-tectonic depression, tsunami, Luzon Island, Philippines, 370, 540
Taapacá lava dome complex, Chile, 274
Table and Badger Mountains, shield volcanoes, California, USA, 232
Tahake, pillow lava delta, Antarctica, 168
Tahiti-Tuamotu-Society, Pacific Ocean, 39, 479
Tajogaite - Cumbre Vieja volcano, La Palma Island, Canary Islands, 39, 74, 89, 143, 146, 217, 509, 513, 584
Talus (debris-slope) facies, 142
Tambora composite volcano, Sumbawa Island, Indonesia, 35, 276, 292, 509, 556
  1815 eruption, 509, 557
Taranaki (or Mt. Egmont) composite volcano, New Zealand, 35, 279, 398, 409
Tarawera volcano, 1886 fissure eruption, 557
Taupo 186 AD/CE eruption, 1.8 ka BP Taupo ignimbrite, 518, 556, 557
Taupo caldera, Taupo Volcanic Zone (TVZ), New Zealand, 3, 37, 325, 326, 333, 365, 371
Tavurvur volcano, scoria cone, Papua New Guinea, 146
Taxonomy, classification of volcanoes, 17
Taylor, Kelvin-Helmholtz instabilities (in fluids), 178
Tectonics (and relationships with volcanoes), 50
Tectonic setting (of volcanoes), 7, 23, 24, 27, 38, 41, 55, 60, 61
Teide volcano, Canary Islands, 39
Telica composite volcano, Nicaragua, 272
Te Maari 2012 eruption, Tongariro massif, New Zealand, 593
Tenerife Island, Canary Islands, 39, 208, 214, 217, 220, 225, 238
Tengger caldera, Java, Indonesia, 322
Tephra-fall, deposits (interactions with snow, ice and water), 172
Tephra fall hazard, 511, 546, 548, 551
Terceira Island, Azores, 53

Terrestrial Laser Scanner (TLS), 514
The Philippines, Island arc, 35
Thermal energy (of magma), 175
Thermal structure (lava flow), 93
Tholoid (dome type), 110
Thórisjökull (shield), 232
Tibesti calderas (Tarso Voon, Emi Koussi, Yirrigué), ignimbrites, Tarso Toussidé volcanic complex, Chad, 344
Tibesti Mountains, volcanic field, 23
Tibesti Volcanic Field, ignimbrite shield-like volcanoes, 344
Tihany and Szigliget, maar and tuff ring, Bakony–Balaton Uplands, Hungary, 153
Timescale for calderas, 347
Tindar (elongate), 164
Toba caldera forming eruptions (1.2 Ma Haranggaol Dacite Tuff, 0.840 Ma Oldest Toba Tuff, 0.501 Ma Middle Toba Tuff, and 0.074 Ma Youngest Toba Tuff), Sumatra, Indonesia, 357
Toba coumpound caldera, Sumatra, Indonesia, 57, 321, 327, 329, 354, 355
Toconquis Group ignimbrite, Cerro Galán caldera complex, 368
Tohoku region, volcanoes, northeast Japan, 18
Tokachi-Dake volcano, 1926 eruption, Hokkaido, Japan, 172
Tok Island, shoaling volcano, South Korea, 183
Tolbachik eruption, Kamchatka, Russia, 37
Tolimán composite volcano, Guatemala, 271
Toluca composite volcano, Mexico, 270
Tondano caldera, Sulawesi, Indonesia, 57
Tonga-Kermadec, Island arc, 35
Tongariro composite volcano, New Zealand, 35, 279
Toothpaste lava, 74, 79
Topographic boundary (of a caldera), 327
Topography effects on lava flows, 513
Topography effects on PDCs, 518
Topography on volcanoes, i.e. high elevation barriers, valleys and ridges, steep flank slopes, gentle piedmont or ring plain slopes, 518
Torta (flat-topped dome), 105
Transitional channel zone, 86
Transitional pahoehoe, 75
Trans-Mexican Volcanic Belt, volcanic field, Mexico (south of Mexico city), 163
Transpression, 38
Transtension, 38, 52, 53
Traps (Deccan, Siberia, Yemen, Ethiopia, Marocco), 17, 42
Tree moulds, 89
Tristan da Cunha composite or shield volcano, South Atlantic Ocean, British Overseas Territory, 29, 280
Trölladyngja shield volcano, Iceland, 231
Truncated cones (of composite volcanoes), 290
Tsirk shield volcano, Kamchatka, Russia, 210
Tsunami (effect, hazard, prevention), volcanic, volcanogenic, 511, 513, 537, 540
Tsunamigenic volcano, 540
Tsunami mechanisms (earthquake, underwater explosion, caldera collapse, PDCs, flank failure, dome collapse, schock wave, lahar), 540
Tuff cone, 139, 146, 148, 149, 154, 156–158, 160, 162, 182, 183
Tuff ring, 16, 138, 154
Tumulus, tumuli, 3, 75, 88
Tungurahua volcano, Ecuador, 520, 521
Tupungato–Tupungatito compound cone, Chile/Argentina, 293
Turbidite, 429
Turkana shield volcanoes, Kenya Rift, EARS, 233
Turkana-subtype shield, 19, 210, 211, 233, 239
Turrialba composite volcano, Costa Rica, 295
Tuya (table mountain), flat-topped, conical, 164
Tuya Mountain, British Columbia, Canada, 165
Tweed (or Wollumbin) shield volcano and Mt. Warning, erosion-induced caldera, South Wales, Australia, 480
Typhoon Yunya, 15 June eruption of Pinatubo, lahars, Luzon, Philippines, 551

U

UAS Unmanned Aerial system (for DEM), 514, 534
Ubehebe craters, maar, California, USA, 153
Ubinas volcano, Peru, 273
Udo tuff cone, south Korea, 156
Ukinrek eruption, maar, Alska, USA, 152
Ulawun composite volcano, New Britain, Papua New Guinea, 277
Unconfined pyroclastic flows, 521
Undara lava flood plateau, Queensland, eastern Australia, 42, 88
Underpressure caldera, 364
Underwater (subaqueous) volcanic landforms, environment, 176
Unmanned Aerial Systems (UAS), 514
Unrest at active calderas (e.g. Phlegrean Fields), 338
Unzen/Ontake-type flank failure, 408, 411, 413
Unzen (Unzendake) lava dome complex, Kyushu, Japan, 35, 104, 106, 276
Unzen volcano, 509, 621
  Barriers, interactions of PDCs with topography, decoupling, post-1991-1995 eruption response to moderate eruptions, 518, 521, 522, 541, 546
Upheaved plug (spine, piton), 109
Uplift (of shields), 215
Upper Rhine Graben (European Cenozoic Rift System), shield volcanoes, Vogelsberg, 235
Ushkovsky compound volcano, Kamchatka, Russia, 293
Usu composite volcano and Toya caldera, Hokkaido, Japan, 109, 293, 411, 541, 546
  1977-1978 eruption, 109, 409, 546
Uzon-Geysernaya caldera, Kamchatka, Russia, 429

V

Valle del Bravo volcanic field, domes, Mexico, 106
Valles-Toledo resurgent calderas, New Mexico, USA, 321, 328, 329, 354, 358, 359
Valley channelization of PDCs, lahars, 521
Valley of Ten Thousand Smokes, ignimbrites, 1912 eruption, Katmai, Alaska, 518
Vangunu, Solomon Islands, 479
Velay, volcanic field, France, 156
Vent (summit, flank), crater, 2
Vent or crater facies, 141
Vesicle in lava, 74
Vesicular tuff, lapilli, 152
Vestmannaeyjar Island, volcanoes, Iceland, 156
Vesuvius volcano, somma (caldera), Italy, 89, 292, 296, 345, 462
Vico volcano, Italy, 322
Victoria scoria mound, SE Asutralia, 143
Villarrica volcano, Chile, 98, 102, 274, 289
  lava lake, 98, 101
Violent Strombolian (eruptive style), 140
Virunga Volcanic Province, volcanoes, DRC, 32, 466
Viscosity (of magma, lava), 74, 175, 265
Viscosity contrast and vesiculation (resurgence model), 359
Volatile (gas) content, 9, 175
Volcanically-disturbed basin, 557
Volcanic arc, 25, 35, 36
Volcanic Cluster (LVC), 17
Volcanic construct, 6
Volcanic debris avalanche, 172, 397
Volcanic edifice erosion, numerical modelling, 615
Volcanic flows, volcanic hazards, 511
Volcanic hazards, direct, indirect, 509, 511
Volcanic landform, 1
Volcaniclastic rocks, sediments, 176, 542
Volcanic-sedimentary deposits, 391
Volcanic soil, 460
Volcano (definition), volcanoes, 1
Volcano degradation processes (soil creep, channel incision), 614
Volcano evolution, geochronology methods, 630
Volcano ground deformation (monitoring, modeling), 627
Volcano-ice interactions, processes, 168, 170, 542
Volcano-magma system, 6
Volcano morphometric parameters, 604
Volcano structures, analysis, 624
Volcano-tectonic depression, inverse volcano, 7, 325, 365, 370
Volcano volumetry, 606
VolcFlow numerical code (applied to PDCs), 520
Volume calculations and errors, 606
Voui lake, tuff cone, Aoba (Ambae), Vanuatu, New Hebrides, 160
Vulcanian eruptive style, 9, 140
Vulcan Island, Taal caldera, Philippines, 155
Vulcano volcano, Aeolian Islands, Italy, 98, 321
Vulcan tuff cone, New Britain, Papua New Guinea, 146
Vulsini volcanoes, Latium, Italy, 56

W

Wai 'anae volcano, O 'ahu Island, Hawai'i, 224
Wapi Lava Field, Hell's Half Acre, shield volcanoes, Iceland, 232
Water-magma ratio, 160
Water thermal conductivity, heat capacity, 175
Weathering (of volcanic rocks), 454
West Antarctic Rift, 23
Westdahl caldera, Alaska, USA, 269
West, East Maui shield volcanoes, Hawai'i, 217
West Mata Volcano, submarine, Lau Basin, Tofua arc, 178
Whaleback shield volcano, California, 232
Wizard Island, cone, Crater Lake, Oregon, USA, 329
Wolff shield volcano, caldera, Galápagos, 229, 230
Wudalianchi Geopark in Northern China, lava landforms, 88
Wuhlandada volcanic field, NE China, 140

X

Xico volcano, tuff ring, Mexico, 154

Y

Yellowstone and Snake River Plateau (YSRP) or Plain system, 41, 357
Yellowstone volcanic centre, resurgent caldera, plateau, 3, 40, 61, 329, 336, 354, 357, 359, 371
Yield strength of domes, 108
Yield strength (of lava), 89
Yucca Mountain, ignimbrites, Lathrop Wells scoria cone, Nevada, USA, 558
Yunaska caldera, Alaska, USA, 269

The manufacturer's authorised representative in the EU is Springer Nature Customer Service Centre GmbH, Europaplatz 3, 69115 Heidelberg, Germany. If you have any concerns regarding our products, please contact ProductSafety@springernature.com

Printed and bound by CPI Group (UK) Ltd, Croydon, CR0 4YY
01/10/2026
02225971-0020